MW00562431

ENERGY CONVERSION

Mechanical Engineering Series
Frank Kreith & Roop Mahajan - Series Editors

Published Titles

Distributed Generation: The Power Paradigm for the New Millennium
Anne-Marie Borbely & Jan F. Kreider
Elastoplasticity Theory
Vlado A. Lubarda
Energy Audit of Building Systems: An Engineering Approach
Moncef Krarti
Energy Conversion
D. Yogi Goswami and Frank Kreith
Finite Element Method Using MATLAB, 2nd Edition
Young W. Kwon & Hyochoong Bang
Fluid Power Circuits and Controls: Fundamentals and Applications
John S. Cundiff
Fundamentals of Environmental Discharge Modeling
Lorin R. Davis
Handbook of Energy Efficiency and Renewable Energy
Frank Kreith and D. Yogi Goswami
Heat Transfer in Single and Multiphase Systems
Greg F. Naterer
Introductory Finite Element Method
Chandrakant S. Desai & Tribikram Kundu
Intelligent Transportation Systems: New Principles and Architectures
Sumit Ghosh & Tony Lee
Mathematical & Physical Modeling of Materials Processing Operations
Olusegun Johnson Ilegbusi, Manabu Iguchi & Walter E. Wahnsiedler
Mechanics of Composite Materials
Autar K. Kaw
Mechanics of Fatigue
Vladimir V. Bolotin
Mechanism Design: Enumeration of Kinematic Structures According to Function
Lung-Wen Tsai
MEMS: Applications
Mohamed Gad-el-Hak
MEMS: Design and Fabrication
Mohamed Gad-el-Hak
The MEMS Handbook, Second Edition
Mohamed Gad-el-Hak
MEMS: Introduction and Fundamentals
Mohamed Gad-el-Hak
Multiphase Flow Handbook
Clayton T. Crowe
Optomechatronics: Fusion of Optical and Mechatronic Engineering
Hyungsuck Cho
Practical Inverse Analysis in Engineering
David M. Trujillo & Henry R. Busby
Pressure Vessels: Design and Practice
Somnath Chattopadhyay
Principles of Solid Mechanics
Rowland Richards, Jr.

Thermodynamics for Engineers
Kau-Fui Wong
Vibration and Shock Handbook
Clarence W. de Silva
Viscoelastic Solids
Roderic S. Lakes

ENERGY CONVERSION

Edited by

D. Yogi Goswami

Frank Kreith

CRC Press is an imprint of the
Taylor & Francis Group, an **informa** business

CRC Press
Taylor & Francis Group
6000 Broken Sound Parkway NW, Suite 300
Boca Raton, FL 33487-2742

Printed in the United States of America on acid-free paper
10 9 8 7 6 5 4 3 2 1

International Standard Book Number-13: 978-1-4200-4431-7 (Hardcover)

Library of Congress Cataloging-in-Publication Data

Energy conversion / Eds, D. Yogi Goswami and Frank Kreith.
p. cm.
Includes bibliographical references and index.
ISBN-13: 978-1-4200-4431-7 (alk. paper)
ISBN-10: 1-4200-4431-1 (alk. paper)
1. Energy conversion I. GoswaMi, Yogi. II. Kreith, Frank. III. Title.

TK2896.E485 2007
621.042--dc22 2007017629

Visit the Taylor & Francis Web site at
http://www.taylorandfrancis.com

and the CRC Press Web site at
http://www.crcpress.com

Preface

Energy is universally acknowledged to be the mainstay of an industrial society. Without an adequate supply of energy, the stability of the social and economic order, as well as the political structure of a society, is in jeopardy. As the world supply of fossil energy sources decreases, the need for energy conservation, efficient energy conversion, and developing renewable energy technologies becomes ever more critical. This book deals with energy conversion from traditional fossil fuels and nuclear as well as renewable energy resources.

Recently, the issue of energy efficiency has emerged as a serious engineering challenge because it is now generally accepted that human activities, mainly burning of fossil fuels, are the main contributors to global climate change. Global warming is largely the result of the emission of radiation-trapping gases, such as carbon dioxide and methane, into the atmosphere. It is the consensus of the scientific community that human activities are largely responsible for the increase in the average global temperature. Thus, improving the efficiency of energy conversion from fossil fuels and the development of renewable energy technologies is becoming ever more important for the engineering community.

This book is divided into two parts: energy resources and energy conversion. The first seven chapters deal with available energy resources including fossil fuels, nuclear, and renewable. Chapters 8–11 cover conventional energy conversion technologies of steam power plants, gas turbines, internal combustion engines, and hydraulic turbines. Advanced conversion technologies such as advanced coal power plants, combined cycle power plants, Stirling engines, and advanced nuclear power are covered in Chapters 12–17. Chapter 15 covers various storage technologies.

Renewable energy technologies including solar thermal power, photovoltaics, wind energy conversion, biomass and biofuels, geothermal energy conversion, as well as waste-to-energy combustion are covered in Chapters 18–24. Chapter 26 presents fundamentals as well as technology assessment of fuel cells. Unconventional energy conversion systems still under development, including nuclear fusion, ocean energy, and direct energy conversion by thermionic, thermoelectric, and magneto-hydrodynamic methods, are covered in Chapters 17, 25, and 27.

The material presented in this handbook has been extracted from the two previous handbooks edited by us. The first is the *Handbook of Mechanical Engineering* and the other the *Handbook of Energy Efficiency and Renewable Energy* published in 2007. It is hoped that bringing all the information for energy resources and conversion under one roof will be useful to engineers in designing and building energy generation systems from traditional and renewable resources. The editors would like to express their appreciation to the authors for their forbearance and diligence in preparing their work for publication.

In a work of this type, scope errors and omissions are unavoidable. The editors would therefore appreciate feedback from the readers to rectify any errors and improve the coverage of future editions.

D. Yogi Goswami
Frank Kreith

Editors-in-Chief

D. Yogi Goswami, Clean Energy Research Center, University of South Florida, Tampa, Florida

Frank Kreith, Department of Mechanical Engineering, University of Colorado, Boulder, Colorado

Contributors

Elsayed M. Afify
North Carolina State University
Raleigh, North Carolina

Anthony F. Armor
Electric Power
Research Institute
Palo Alto, California

Roger E.A. Arndt
University of Minnesota
Minneapolis, Minnesota

Richard Bajura
West Virginia University
Morgantown, West Virginia

Riccardo Battisti
University of Rome
"La Sapienza"
Rome, Italy

Dale E. Berg
Sandia National Laboratories
Albuquerque, New Mexico

Desikan Bharathan
National Renewable Energy
Laboratory
Golden, Colorado

Robert C. Brown
Center for Sustainable
Environmental Technologies
Iowa State University
Ames, Iowa

Philip C. Crouse
Philip C. Crouse and Associates Inc.
Dallas, Texas

Steven I. Freedman
Gas Research Institute
Deerfield, Illinois

Nitin Goel
Intel Technology India Pvt. Ltd.
Bangalore, India

D. Yogi Goswami
Clean Energy Research Center
University of South Florida
Tampa, Florida
and
University of Florida
Gainesville, Florida

Leonard M. Grillo
Grillo Engineering Company
Hollis, New Hampshire

Roel Hammerschlag
Institute for Lifecycle
Environmental Assessment
Seattle, Washington

Edwin A. Harvego
Idaho National Laboratory
Idaho Falls, Idaho

Massoud Kayhanian
Center for Environmental and
Water Resources Engineering
Department of Civil and
Environmental Engineering
University of California at Davis
Davis, California

John Kern
Siemens Power Generation
Milwaukee, Wisconsin

Kevin Kitz
U.S. Geothermal Inc.
Boise, Idaho

David E. Klett
North Carolina A&T State
University
Greensboro, North Carolina

Kenneth D. Kok
WSMS Mid-America
Oak Ridge, Tennessee

Alex Lezuo
Siemans Power Generation
Erlangen, Germany

Xianguo Li
University of Waterloo
Waterloo, Ontario, Canada

Robert McConnell
National Center for Photovoltaics
National Renewable
Energy Laboratory
Golden, Colorado

Roger Messenger
Department of
Electrical Engineering
Florida Atlantic University
Boca Raton, Florida

Jeffrey H. Morehouse
Department of Mechanical
Engineering
University of South Carolina
Columbia, South Carolina

Ralph P. Overend
National Renewable Energy
Laboratory
Golden, Colorado

Takhir M. Razykov
Physical Technical Institute
Uzbek Academy of Sciences
Tashkent, Uzbekistan

T. Agami Reddy
Department of Civil, Architectural and Environmental Engineering
Drexel University
Philadelphia, Pennsylvania

Marshall J. Reed
U.S. Department of Energy
Washington, D.C.

Joel L. Renner
Idaho National Engineering Laboratory
Idaho Falls, Idaho

Robert Reuther
U.S. Department of Energy
Morgantown, West Virginia

Manuel Romero-Alvarez
Plataforma Solar de Almeria-CIEMAT
Madrid, Spain

Christopher P. Schaber
Institute for Lifecycle Environmental Assessment
Seattle, Washington

Hans Schweiger
AIGUASOL Engineering
Active Solar Systems Group
Barcelona, Spain

Thomas E. Shannon
University of Tennessee
Knoxville, Tennessee

William B. Stine
California State Polytechnic University
Pasadena, California

George Tchobanoglous
Department of Civil and Environmental Engineering
University of California at Davis
Davis, California

Ayodhya N. Tiwari
Centre for Renewable Energy Systems Technology (CREST)
Department of Electronic and Electrical Engineering
Loughborough University
Loughborough, Leicestershire, United Kingdom

James S. Tulenko
University of Florida
Gainesville, Florida

Hari M. Upadhyaya
Centre for Renewable Energy Systems Technology (CREST)
Department of Electronic and Electrical Engineering
Loughborough University
Loughborough, Leicestershire, United Kingdom

Charles O. Velzy
Private Consultant
White Haven, Pennsylvania

Sanjay Vijayaraghavan
Intel Technology India Pvt. Ltd.
Bangalore, India

Werner Weiss
AEE INTEC
Feldgasse, Austria

Roland Winston
University of California
Merced, California

Lynn L. Wright
Oak Ridge National Laboratory
Oak Ridge, Tennessee

Federica Zangrando
National Renewable Energy Laboratory
Golden, Colorado

Eduardo Zarza
Plataforma Solar de Almeria-CIEMAT
Madrid, Spain

Contents

1 Introduction *D. Yogi Goswami* **1-1**
1.1 Energy Use by Sectors **1-3**
1.2 Electrical Capacity Additions to 2030 .... **1-4**
1.3 Present Status and Potential of Renewable Energy **1-5**
1.4 Role of Energy Conservation **1-7**
1.5 Energy Conversion Technologies **1-10**

SECTION I Energy Resources

2 Fossil Fuels **2-1**
2.1 Coal *Robert Reuther* **2-1**
2.2 Environmental Aspects *Richard Bajura* **2-14**
2.3 Oil *Philip C. Crouse* **2-16**
2.4 Natural Gas *Philip C. Crouse* **2-21**

3 Biomass Energy *Ralph P. Overend and Lynn L. Wright* **3-1**
3.1 Biomass Feedstock Technologies **3-1**
3.2 Biomass Conversion Technologies **3-4**

4 Nuclear Resources *James S. Tulenko* **4-1**
4.1 The Nuclear Fuel Cycle **4-1**
4.2 Processing of Nuclear Fuel **4-2**

5 Solar Energy Resources *D. Yogi Goswami* **5-1**
5.1 Solar Energy Availability **5-1**
5.2 Earth–Sun Relationships **5-2**
5.3 Solar Time **5-4**
5.4 Solar Radiation on a Surface **5-4**
5.5 Solar Radiation on a Horizontal Surface **5-5**
5.6 Solar Radiation on a Tilted Surface **5-5**
5.7 Solar Radiation Measurements **5-6**
5.8 Solar Radiation Data **5-6**

6 Wind Energy Resources *Dale E. Berg* **6-1**
6.1 Wind Origins **6-1**

6.2 Wind Power **6-1**
6.3 Wind Shear **6-2**
6.4 Wind Energy Resource **6-2**
6.5 Wind Characterization **6-6**
6.6 Wind Energy Potential **6-6**

7 Geothermal Energy *Joel L. Renner and Marshall J. Reed* **7-1**
7.1 Heat Flow **7-1**
7.2 Types of Geothermal Systems **7-2**
7.3 Geothermal Energy Potential **7-2**
7.4 Geothermal Applications **7-4**
7.5 Environmental Constraints **7-4**
7.6 Operating Conditions **7-6**

SECTION II Energy Conversion

8 Steam Power Plant *John Kern* **8-1**
8.1 Introduction **8-1**
8.2 Rankine Cycle Analysis **8-2**
8.3 Topping and Bottoming Cycles **8-5**
8.4 Steam Boilers **8-5**
8.5 Steam Turbines **8-7**
8.6 Heat Exchangers, Pumps, and Other Cycle Components **8-11**
8.7 Generators **8-14**

9 Gas Turbines *Steven I. Freedman* **9-1**
9.1 Overview **9-1**
9.2 History **9-1**
9.3 Fuels and Firing **9-2**
9.4 Efficiency **9-2**
9.5 Gas Turbine Cycles **9-3**
9.6 Cycle Configurations **9-4**
9.7 Components Used in Complex Cycles **9-6**
9.8 Upper Temperature Limit **9-9**
9.9 Materials **9-10**
9.10 Combustion **9-10**
9.11 Mechanical Product Features **9-11**

10 Internal Combustion Engines *David E. Klett and Elsayed M. Afify* **10-1**
10.1 Introduction **10-1**
10.2 Engine Types and Basic Operation **10-2**
10.3 Air Standard Power Cycles **10-7**
10.4 Actual Cycles **10-10**
10.5 Combustion in IC Engines **10-12**
10.6 Exhaust Emissions **10-15**
10.7 Fuels for SI and CI Engines **10-17**
10.8 Intake Pressurization—Supercharging and Turbocharging **10-20**

11 Hydraulic Turbines *Roger E.A. Arndt* **11**-1
11.1 General Description **11**-1
11.2 Principles of Operation **11**-5
11.3 Factors Involved in Selecting a Turbine **11**-8
11.4 Performance Evaluation **11**-12
11.5 Numerical Simulation **11**-14
11.6 Field Tests **11**-17

12 Stirling Engines *William B. Stine* **12**-1
12.1 Introduction **12**-1
12.2 Thermodynamic Implementation of the Stirling Cycle **12**-2
12.3 Mechanical Implementation of the Stirling Cycle **12**-4
12.4 Future of the Stirling Engine **12**-9

13 Advanced Fossil Fuel Power Systems *Anthony F. Armor* **13**-1
13.1 Introduction **13**-1
13.2 Fuels for Electric Power Generation in the U.S. **13**-2
13.3 Coal as a Fuel for Electric Power (World Coal Institute 2000) **13**-3
13.4 Clean Coal Technology Development **13**-4
13.5 Pulverized-Coal Plants **13**-5
13.6 Emissions Controls for Pulverized Coal Plants **13**-9
13.7 Fluidized Bed Plants **13**-13
13.8 Gasification Plants **13**-16
13.9 Combustion Turbine Plants **13**-20
13.10 Central Station Options for New Generation **13**-24
13.11 Summary **13**-26

14 Combined-Cycle Power Plants *Alex Lezuo* **14**-1
14.1 Combined-Cycle Concepts **14**-1
14.2 Combined-Cycle Thermodynamics **14**-2
14.3 Combined-Cycle Arrangements **14**-4
14.4 Combined Heat and Power from Combined-Cycle Plants **14**-7
14.5 Environmental Aspects **14**-8

15 Energy Storage Technologies *Roel Hammerschlag and Christopher P. Schaber* **15**-1
15.1 Overview of Storage Technologies **15**-1
15.2 Principal Forms of Stored Energy **15**-3
15.3 Applications of Energy Storage **15**-3
15.4 Specifying Energy Storage Devices **15**-4
15.5 Specifying Fuels **15**-6
15.6 Direct Electric Storage **15**-7
15.7 Electrochemical Energy Storage **15**-8
15.8 Mechanical Energy Storage **15**-13
15.9 Direct Thermal Storage **15**-15
15.10 Thermochemical Energy Storage **15**-18

16 Nuclear Power Technologies *Edwin A. Harvego and Kenneth D. Kok* **16**-1
16.1 Introduction **16**-1
16.2 Development of Current Power-Reactor Technologies **16**-2
16.3 Next-Generation Technologies **16**-8

16.4 Generation-IV Technologies **16**-11
16.5 Fuel Cycle **16**-20
16.6 Nuclear Waste **16**-26
16.7 Nuclear Power Economics **16**-29
16.8 Conclusions **16**-29

17 Nuclear Fusion *Thomas E. Shannon* **17**-1
17.1 Introduction **17**-1
17.2 Fusion Fuel **17**-1
17.3 Confinement Concepts **17**-2
17.4 Tokamak Reactor Development **17**-2
17.5 Fusion Energy Conversion and Transport **17**-4

18 Solar Thermal Energy Conversion **18**-1
18.1 Active Solar Heating Systems *T. Agami Reddy* **18**-1
18.2 Solar Heat for Industrial Processes *Riccardo Battisti, Hans Schweiger, and Werner Weiss* **18**-49
18.3 Passive Solar Heating, Cooling, and Daylighting *Jeffrey H. Morehouse* **18**-59
18.4 Solar Cooling *D. Yogi Goswami and Sanjay Vijayaraghavan* **18**-121

19 Concentrating Solar Thermal Power *Manuel Romero-Alvarez and Eduardo Zarza* **19**-1
19.1 Introduction and Context **19**-2
19.2 Solar Concentration and CSP Systems **19**-6
19.3 Solar Concentrator Beam Quality **19**-9
19.4 Solar Concentration Ratio: Principles and Limitations of CSP Systems **19**-13
19.5 Solar Thermal Power Plant Technologies **19**-15
19.6 Parabolic Trough Solar Thermal Power Plants **19**-18
19.7 Central Receiver Solar Thermal Power Plants **19**-50
19.8 Volumetric Atmospheric Receivers: PHOEBUS and Solair **19**-80
19.9 Solar Air Preheating Systems for Combustion Turbines: The SOLGATE Project **19**-82
19.10 Dish/Stirling Systems **19**-85
19.11 Market Opportunities **19**-91
19.12 Conclusions **19**-92

20 Photovoltaics Fundamentals, Technology and Application **20**-1
20.1 Photovoltaics *Roger Messenger and D. Yogi Goswami* **20**-1
20.2 Thin-Film PV Technology *Hari M. Upadhyaya, Takhir M. Razykov, and Ayodhya N. Tiwari* **20**-28
20.3 Concentrating PV Technologies *Roland Winston, Robert McConnell, and D. Yogi Goswami* **20**-54

21 Wind Energy Conversion *Dale E. Berg* **21**-1
21.1 Introduction **21**-1
21.2 Wind Turbine Aerodynamics **21**-4
21.3 Wind Turbine Loads **21**-16
21.4 Wind Turbine Structural Dynamic Considerations **21**-16
21.5 Peak Power Limitation **21**-18
21.6 Turbine Subsystems **21**-20
21.7 Other Wind-Energy Conversion Considerations **21**-23

22 Biomass Conversion Processes For Energy Recovery **22-1**
22.1 Energy Recovery by Anaerobic Digestion *Massoud Kayhanian and George Tchobanoglous* **22-2**
22.2 Power Generation *Robert C. Brown* **22-37**
22.3 Biofuels *Robert C. Brown* **22-51**

23 Geothermal Power Generation *Kevin Kitz* **23-1**
23.1 Introduction **23-2**
23.2 Definition and Use of Geothermal Energy **23-2**
23.3 Requirements for Commercial Geothermal Power Production **23-3**
23.4 Exploration and Assessment of Geothermal Resources **23-15**
23.5 Management of the Geothermal Resource for Power Production **23-18**
23.6 Geothermal Steam Supply (from Wellhead to Turbine) **23-25**
23.7 Geothermal Power Production—Steam Turbine Technologies **23-32**
23.8 Geothermal Power Production—Binary Power Plant Technologies **23-38**
23.9 Environmental Impact **23-43**
23.10 Additional Information on Geothermal Energy **23-46**

24 Waste-to-Energy Combustion *Charles O. Velzy and Leonard M. Grillo* **24-1**
24.1 Introduction **24-1**
24.2 Waste Quantities and Characteristics **24-2**
24.3 Design of WTE Facilities **24-6**
24.4 Air Pollution Control Facilities **24-24**
24.5 Performance **24-32**
24.6 Costs **24-34**
24.7 Status of Other Technologies **24-36**
24.8 Future Issues and Trends **24-38**

25 Ocean Energy Technology *Desikan Bharathan and Federica Zangrando* **25-1**
25.1 Ocean Thermal Energy Conversion **25-1**
25.2 Tidal Power **25-2**
25.3 Wave Power **25-2**
25.4 Concluding Remarks **25-3**

26 Fuel Cells *Xianguo Li* **26-1**
26.1 Introduction **26-1**
26.2 Principle of Operation for Fuel Cells **26-2**
26.3 Typical Fuel Cell Systems **26-3**
26.4 Performance of Fuel Cells **26-4**
26.5 Fuel Cell Electrode Processes **26-25**
26.6 Cell Connection and Stack Design Considerations **26-27**
26.7 Six Major Types of Fuel Cells **26-29**
26.8 Summary **26-44**

27 Direct Energy Conversion **27-1**
27.1 Thermionic Energy Conversion *Mysore L. Ramalingam* **27-1**
27.2 Thermoelectric Power Conversion *Jean-Pierre Fleurial* **27-7**
27.3 Magnetohydrodynamic Power Generation *William D. Jackson* **27-15**

Appendices *Nitin Goel*
Appendix 1 The International System of Units, Fundamental Constants, and Conversion Factors ... **A1**-1
Appendix 2 Solar Radiation Data ... **A2**-1
Appendix 3 Properties of Gases, Vapors, Liquids and Solids ... **A3**-1
Appendix 4 Ultimate Analysis of Biomass Fuels ... **A4**-1

Index ... **I**-1

I

Energy Resources

2 **Fossil Fuels** *Robert Reuther, Richard Bajura, Philip C. Crouse* **2-1**
Coal • Environmental Aspects • Oil • Natural Gas

3 **Biomass Energy** *Ralph P. Overend, Lynn L. Wright* **3-1**
Biomass Feedstock Technologies • Biomass Conversion Technologies

4 **Nuclear Resources** *James S. Tulenko* **4-1**
The Nuclear Fuel Cycle • Processing of Nuclear Fuel

5 **Solar Energy Resources** *D. Yogi Goswami* **5-1**
Solar Energy Availability • Earth–Sun Relationships • Solar Time • Solar Radiation on a Surface • Solar Radiation on a Horizontal Surface • Solar Radiation on a Tilted Surface • Solar Radiation Measurements • Solar Radiation Data

6 **Wind Energy Resources** *Dale E. Berg* **6-1**
Wind Origins • Wind Power • Wind Shear • Wind Energy Resource • Wind Characterization • Wind Energy Potential

7 **Geothermal Energy** *Joel L. Renner, Marshall J. Reed* **7-1**
Heat Flow • Types of Geothermal Systems • Geothermal Energy Potential • Geothermal Applications • Environmental Constraints • Operating Conditions

1

Introduction

1.1 Energy Use by Sectors 1-3
1.2 Electrical Capacity Additions to 2030 1-4
Transportation
1.3 Present Status and Potential of Renewable Energy 1-5
1.4 Role of Energy Conservation 1-7
Forecast of Future Energy Mix
1.5 Energy Conversion Technologies 1-10
Defining Terms 1-10
References 1-10
For Further Information 1-11

D. Yogi Goswami
University of South Florida

Global energy consumption in the last half-century has increased very rapidly and is expected to continue to grow over the next 50 years. However, we expect to see significant differences between the last 50 years and the next. The past increase was stimulated by relatively "cheap" fossil fuels and increased rates of industrialization in North America, Europe, and Japan, yet while energy consumption in these countries continues to increase, additional factors have entered the equation making the picture for the next 50 years more complex. These additional complicating factors include the very rapid increase in energy intensity of China and India (countries representing about a third of the world's population); the expected depletion of oil resources in the not-too-distant future; and, the global climate change. On the positive side, the renewable energy (RE) technologies of wind, biofuels, solar thermal, and photovoltaics (PV) are finally showing maturity and the ultimate promise of cost competitiveness.

Statistics from the International Energy Agency (IEA) World Energy Outlook 2004 show that the total primary energy demand in the world increased from 5536 MTOE in 1971 to 10,345 MTOE in 2002, representing an average annual increase of 2% (see Figure 1.1 and Table 1.1).

Of the total primary energy demand in 2002, the fossil fuels accounted for about 80% with oil, coal and natural gas being 35.5, 23, and 21.2%, respectively. Biomass accounted for 11% of all the primary energy in the world, almost all of it being traditional biomass in the developing countries which is used very inefficiently.

The last 10 years of data for energy consumption from British Petroleum (BP) Corp. also shows that the average increase per year is 2%. However, it is important to note (from Table 1.2) that the average worldwide growth from 2001 to 2004 was 3.7% with the increase from 2003 to 2004 being 4.3%. The rate of growth is rising mainly due to the very rapid growth in Asia Pacific which recorded an average increase from 2001 to 2004 of 8.6%.

More specifically, China increased its primary energy consumption by 15% from 2003 to 2004. Unconfirmed data show similar increases continuing in China, followed by increases in India. Fueled by high increases in China and India, worldwide energy consumption may continue to increase at rates between 3 and 5% for at least a few more years. However, such high rates of increase cannot continue for too long. Various sources estimate that the worldwide average annual increase in energy consumption for

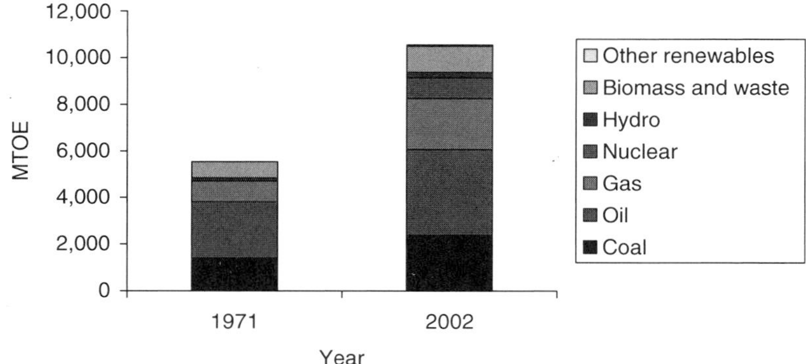

FIGURE 1.1 World primary energy demand (MTOE). (Data from IEA, *World Energy Outlook*, International Energy Agency, Paris, France, 2004.)

the next 25 years will be 1.6%–2.5% (IEA 2004; IAEA 2005). The Energy Information Agency (EIA), U.S. Department of Energy projects an annual increase of 2% from now until 2030 (Figure 1.2).

Based on a 2% increase per year (average of the estimates from many sources), the primary energy demand of 10,345 MTOE in 2002 will double by 2037 and triple by 2057. With such high-energy demand expected 50 years from now, it is important to look at the available resources to fulfill the future demand, especially for electricity and transportation.

TABLE 1.1 World Total Energy Demand (MTOE)

Energy Source/Type	1971	2002	Change 1971–2002 (%)
Coal	1,407	2,389	1.7
Oil	2,413	3,676	1.4
Gas	892	2,190	2.9
Nuclear	29	892	11.6
Hydro	104	224	2.5
Biomass and waste	687	1,119	1.6
Other renewables	4	55	8.8
Total	5,536	10,345	2.0

Source: Data from IEA, *World Energy Outlook*, International Energy Agency, Paris, France, 2004.

TABLE 1.2 Primary Energy Consumption (MTOE)

Region	2001	2002	2003	2004	Average Increase/ Year (%)	2004 Change Over 2003 (%)
North America including U.S.A.	2,681.5	2,721.1	2,741.3	2,784.4	1.3	1.6
U.S.A.	2,256.3	2,289.1	2,298.7	2,331.6	1.1	1.4
South and Central America	452	454.4	460.2	483.1	2.2	5
Europe and Euro-Asia	2,855.5	2,851.5	2,908	2,964	1.3	1.9
Middle East	413.2	438.7	454.2	481.9	5.3	6.1
Africa	280	287.2	300.1	312.1	3.7	4
Asia Pacific	2,497	2,734.9	2,937	3,198.8	8.6	8.9
World	9,179.3	9,487.9	9,800.8	10,224.4	3.7	4.3

This data does not include traditional biomass which was 2229 MTOE in 2002, according to IEA data.
Source: Data from BP Statistical Review of World Energy, 2006.

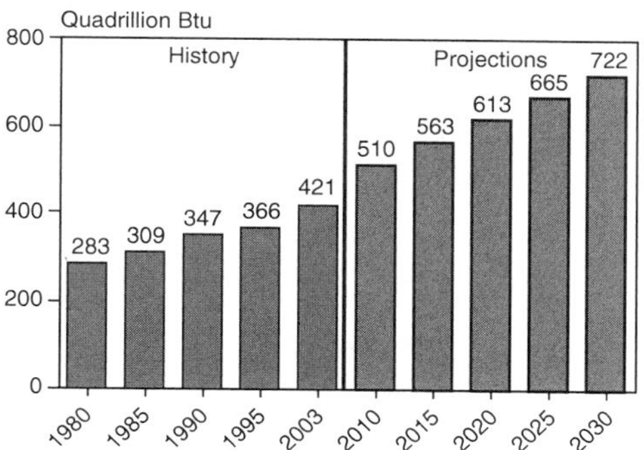

FIGURE 1.2 Historical and projected energy consumption in the world. (From EIA, *Energy Information Outlook 2006*, Energy Information Agency, U.S. Department of Energy, Washington, DC, 2006.)

1.1 Energy Use by Sectors

The major sectors using primary energy sources include electrical power, transportation, heating, industrial and others, such as cooking. The IEA data show that the electricity demand almost tripled from 1971 to 2002. This is not unexpected as electricity is a very convenient form of energy to transport and use. Although primary energy use in all sectors has increased, their relative shares except for transportation and electricity have decreased (Figure 1.3). Figure 1.3 shows that the relative share of primary energy for electricity production in the world increased from about 20% in 1971 to about 30% in 2002. This is because electricity is becoming the preferred form of energy for all applications.

Figure 1.4 shows that coal is presently the largest source of electricity in the world. Consequently, the power sector accounted for 40% of all emissions in 2002. Emissions could be reduced by increased use of RE sources. All RE sources combined accounted for only 17.6% share of electricity production in the world, with hydroelectric power providing almost 90% of it. All other RE sources provided only 1.7% of

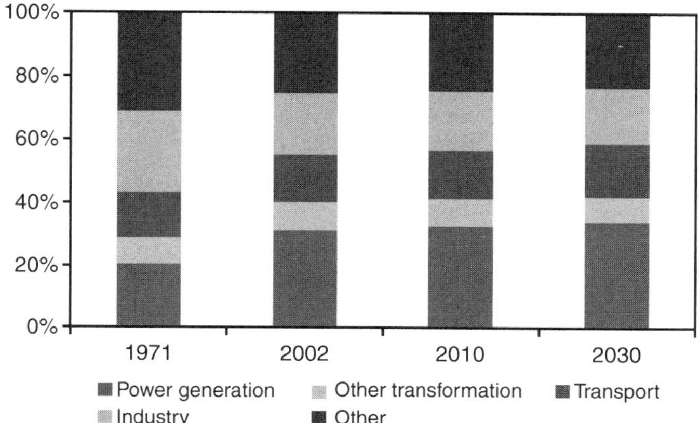

FIGURE 1.3 Sectoral shares in world primary energy demand. (From IEA, *World Energy Outlook*, International Energy Agency, Paris, France, 2004.)

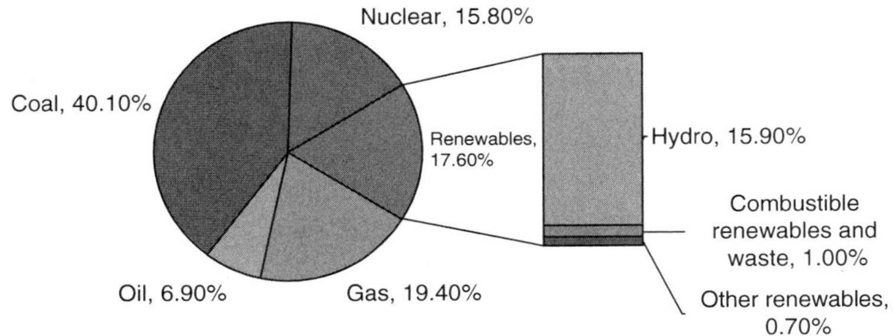

FIGURE 1.4 World electricity production by fuel in 2003. (From IEA, *Renewables Information 2005*, International Energy Agency, Paris, France, 2005.)

electricity in the world. However, the RE technologies of wind power and solar energy have vastly improved in the last two decades and are becoming more cost effective. As these technologies mature and become even more cost competitive in the future they may be in a position to replace major fractions of fossil fuels for electricity generation. Therefore, substituting fossil fuels with RE for electricity generation must be an important part of any strategy of reducing CO_2 emissions into the atmosphere and combating global climate change.

1.2 Electrical Capacity Additions to 2030

Figure 1.5 shows the additional electrical capacity forecast by IEA for different regions in the world. The overall increase in the electrical capacity is in general agreement with the estimates from International Atomic Energy Agency (IAEA 2005) which project an average annual growth of about 2%–2.5% up to 2030. It is clear that of all countries, China will add the largest capacity with its projected electrical needs accounting for about 30% of the world energy forecast. China and India combined will add about 40% of all the new capacity of the rest of the world. Therefore, what happens in these two countries will have important consequences on the worldwide energy and environmental situation. If coal provides as much as 70% of China's electricity in 2030, as forecasted by IEA (IEA 2004), it will certainly increase worldwide CO_2 emissions which will further affect global climate.

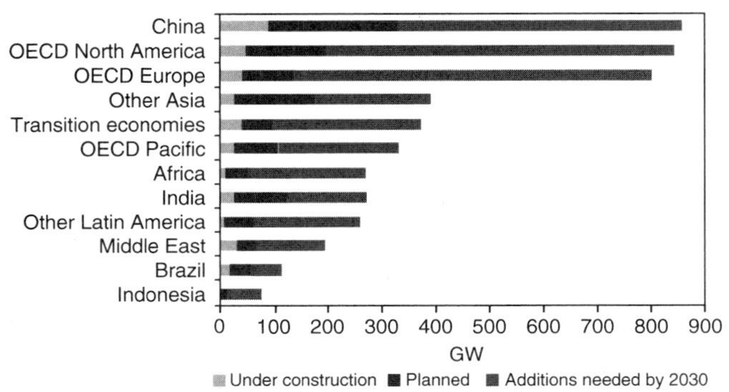

FIGURE 1.5 Electrical capacity requirements by region. (From IEA, *World Energy Outlook*, International Energy Agency, Paris, France, 2004.)

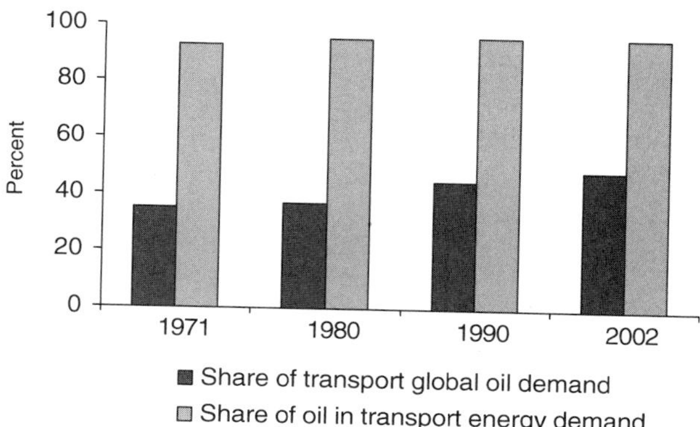

FIGURE 1.6 Share of transport in global oil demand and share of oil in transport energy demand. (Data and Forecast from IEA, *World Energy Outlook*, International Energy Agency, Paris, France, 2004.)

1.2.1 Transportation

Transportation is another sector that has increased its relative share of primary energy. This sector has serious concerns as it is a significant source of CO_2 emissions and other airborne pollutants—and it is almost totally based on oil as its energy source (Figure 1.6). In 2002, the transportation sector accounted for 21% of all CO_2 emissions worldwide. An important aspect of future changes in transportation depends on what happens to the available oil resources, production, and prices. At present 95% of all energy for transportation originates from oil. Since oil production is expected to peak in the near future, there is an urgent need for careful planning for an orderly transition away from oil as the primary transportation fuel. An obvious replacement for oil would be biofuels, such as ethanol, methanol, biodiesel, and biogases. Hydrogen is another alternative which has been claimed by some to be the ultimate answer as they propose a "hydrogen-based economy" to replace the present "carbon-based economy" (Veziroglu and Barbir 1994). However, other analysts (Kreith and West 2004; Hammerschlag and Mazza 2005; West and Kreith 2006) dispute this based on the infrastructure requirements of hydrogen, and the lower efficiency of hydrogen vehicles as compared to hybrid or fully electric vehicles. Electric transportation presents another viable alternative to the oil-based transportation (West and Kreith 2006). Already hybrid-electric automobiles are becoming popular around the world as petroleum becomes more expensive. Complete electric transportation will require development of long-range and long-life batteries. However, any large-scale shift to electric transportation will require large amounts of additional electrical generation capacity.

1.3 Present Status and Potential of Renewable Energy

According to the data in Table 1.3, 13.3% of the world's total primary energy supply came from RE in 2003. However, almost 80% of the RE supply was from biomass (Figure 1.7), and in developing countries it is mostly converted by traditional open combustion which is very inefficient. Because of its inefficient use, biomass resources presently supply only about 20% of what they could if converted by modern, more efficient, and available technologies. As it stands, biomass provides only 11% of the world total primary energy which is much less than it's real potential. The total technologically sustainable biomass energy potential for the world is 3–4 TW_e (UNDP 2004), which is more than the entire present global generating capacity of about 3 TW_e.

In 2003, shares of biomass and hydro in the total primary energy of the world were about 11 and 2%, respectively. All of the other renewables including solar thermal, solar PV, wind, geothermal and ocean combined, provided only about 0.5% of the total primary energy. During the same year, biomass

TABLE 1.3 2003 Fuel Shares in World Total Primary Energy Supply

Source	Share (%)
Oil	34.4
Natural Gas	21.2
Coal	24.4
Nuclear	6.5
Renewables	13.3

Source: Data from IEA, *World Energy Outlook*, International Energy Agency, Paris, France, 2004.

combined with hydroelectric resources provided more than 50% of all the primary energy in Africa, 29.2% in Latin America, and 32.7% in Asia (Table 1.4). However, biomass is used very inefficiently for cooking in these countries. Such use has also resulted in significant health problems, especially for women.

The total share of all renewables for electricity production in 2002 was about 17%, a vast majority (89%) of it being from hydroelectric power (Table 1.5).

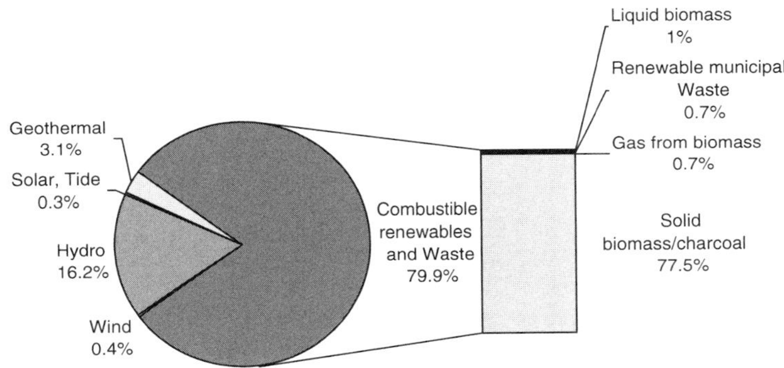

FIGURE 1.7 2003 Resource shares in world renewable energy supply. (Data from IEA, *World Energy Outlook*, International Energy Agency, Paris, France, 2004.)

TABLE 1.4 Share of Renewable Energy in 2003 Total Primary Energy Supply (TPES) on a Regional Basis

	MTOE		
Region	TPES	Renewables	(%)
Africa	558.9	279.9	50.1
Latin America	463.9	135.5	29.2
Asia	1,224.4	400	32.7
India	553.4	218	39.4
China	1,425.9	243.4	17.1
Non-OECD[a] Europe	103.5	9.7	9.4
Former USSR	961.7	27.5	2.9
Middle East	445.7	3.2	0.7
OECD	5,394.7	304.7	5.6
U.S.A.	2,280.8	95.3	4.2
World	10,578.7	1,403.7	13.3

[a]Organization for Economic Cooperation and Development.
Source: IEA, *Renewables Information 2005*, International Energy Agency, Paris, France, 2005.

TABLE 1.5 Electricity from Renewable Energy in 2002

Energy Source	2002 TWh	%
Hydropower	2610	89
Biomass	207	7
Wind	52	2
Geothermal	57	2
Solar	1	0
Tide/wave	1	2
Total	2927	100

Source: Data from IEA, *World Energy Outlook*, International Energy Agency, Paris, France, 2004.

Table 1.6 summarizes the resource potential and the present costs and the potential future costs for each renewable resource.

1.4 Role of Energy Conservation

Energy conservation can and must play an important role in future energy use and the consequent impact on the environment. Figure 1.8 and Figure 1.9 give us an idea of the potential of the possible energy efficiency improvements. Figure 1.9 shows that per capita energy consumption varies by as much as a factor

TABLE 1.6 Potential and Status of Renewable Energy Technologies

Technology	Annual Potential	Operating Capacity 2005	Investment Costs US$ per kW	Current Energy Cost	Potential Future Energy cost
Biomass Energy					
Electricity	276–446 EJ	~44 GW_e	500–6000/kW_e	3–12 ¢/kWh	3–10 ¢/kWh
Heat	Total or 8–13 TW	~225 GWth	170–1000/kWth	1–6 ¢/kWh	1–5 ¢/kWh
Ethanol	MSW ~6 EJ	~36 bln lit.	170–350/kWth	25–75 ¢/lit(ge)[a]	6–10 $/GJ
Bo-Diesel		~3.5 bln lit.	500–1000/kWth	25–85 ¢/lit.(de)[b]	10–15 $/GJ
Wind Power	55 TW Theo. 2 TW Practical	59 GW	850–1700	4–8 ¢/kWh	3–8 ¢/kWh
Solar Energy	>100 TW				
Photovoltaics		5.6 GW	5000–10000	25–160 ¢/kWh	5–25 ¢/kWh
Thermal Power		0.4 GW	2500–6000	12–34 ¢/kWh	4–20 ¢/kWh
Heat			300–1700	2–25 ¢/kWh	2–10 ¢/kWh
Geothermal					
Electricity	600,000 EJ useful resource base	9 GW	800–3000	2–10 ¢/kWh	1–8 ¢/kWh
Heat	5,000 EJ economical in 40–50 years	11 GWth	200–2000	0.5–5 ¢/kWh	0.5–5 ¢/kWh
Ocean Energy					
Tidal	2.5 TW	0.3 GW	1700–2500	8–15 ¢/kWh	8–15 ¢/kWh
Wave	2.0 TW		2000–5000	10–30 ¢/kWh	5–10 ¢/kWh
OTEC	228 TW		8000–20000	15–40 ¢/kWh	7–20 ¢/kWh
Hydroelectric					
Large	1.63 TW Theo.	690 GW	1000–3500	2–10 ¢/kWh	2–10 ¢/kWh
Small	0.92 TW Econ.	25 GW	700–8000	2–12 ¢/kWh	2–10 ¢/kWh

[a] ge, gasoline equivalent liter.
[b] de, diesel equivalent liter.

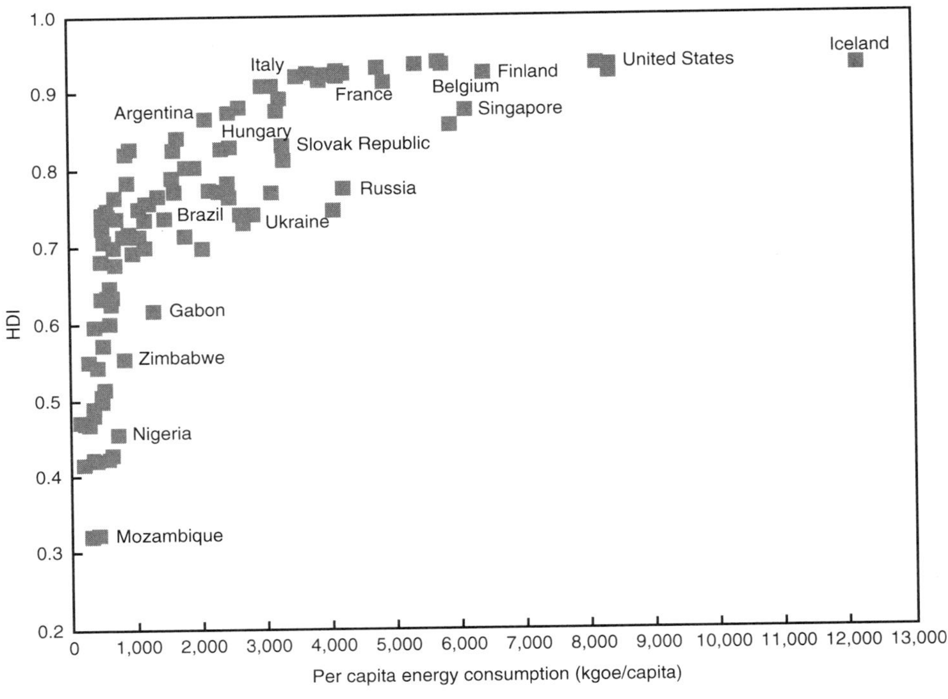

FIGURE 1.8 Relationship between human development index and per capita energy use, 1999–2000. (From UNDP, *World Energy Assessment: Energy and the Challenge of Sustainability*, 2004.)

of three between the U.S.A. and some European countries with almost the same level of human development index. Even taking just the Organization for Economic Cooperation and Development (OECD) European countries combined, the per capita energy consumption in the U.S.A. is twice as much. It is fair to assume that the per capita energy of the U.S.A could be reduced to the level of OECD Europe of 4.2 kW by a combination of energy efficiency improvements and changes in the transportation infrastructure. This is significant because the U.S.A. uses about 25% of the energy of the whole world.

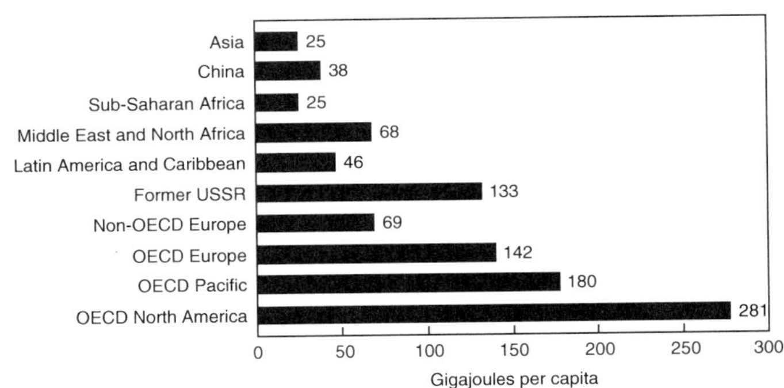

Notes: Asia excludes Middle East, China, and OECD countries; Middle East and North Africa comprises Algeria, Bahrain, Egypt, Iran, Iraq, Israel,Jordan,Kuwait,Lebanon, Libya, Morocco, Oman,Qatar,Saudi Arabia, Syria, Tunisia, United Arab Emirates adn Yemen; Latin America and Caribbean excludes Mexico; OECD Pacific comprises Australia, Japan Korea, and New Zealand; Former USSR comprises Armenia, Azerbaijan, Belarus, Estonia, Georgia, Kazakhstan, Kyrgyzstan, Latvia, Lithuania, Moldova, Russia, Tajikistan, Turkmenistan, Ukraine, and Albania, Bosnia and Herzegovina, Bulgaria, Croatia, Cyprus, Gibraltar, Macedonia, Malta, Romania, and Slovenia; OECD North America includes Mexico.

FIGURE 1.9 Per capita energy use by region (commercial and non-commercial) 2000. (From UNDP, *World Energy Assessment: Energy and the Challenge of Sustainability*, 2004.)

The present per capita energy consumption in the U.S.A is 284 GJ which is equivalent to about 9 kW per person while the average for the whole world is 2 kW. The Board of Swiss Federal Institutes of Technology has developed a vision of a 2 kW per capita society by the middle of the century (UNDP 2004). The vision is technically feasible. However, to achieve this vision will require a combination of increased R&D on energy efficiency and policies that encourage conservation and use of high-efficiency systems. It will also require some structural changes in the transportation systems. According to the 2004, World Energy Assessment by UNDP, a reduction of 25%–35% in primary energy in the industrialized countries is achievable cost effectively in the next 20 years, without sacrificing the level of energy services. The report also concluded that similar reductions of up to 40% are cost effectively achievable in the transitional economies and more than 45% in developing economies. As a combined result of efficiency improvements and structural changes, such as increased recycling, substitution of energy intensive materials, etc., energy intensity could decline at a rate of 2.5% per year over the next 20 years (UNDP 2004).

1.4.1 Forecast of Future Energy Mix

Since oil comprises the largest share of world energy consumption and may remain so for a while, its depletion will cause a major disruption unless other resources can fill the gap. Natural gas and coal production may be increased to fill the gap, with the natural gas supply increasing more rapidly than coal. However, that will hasten the time when natural gas production peaks. Additionally, any increase in coal consumption will worsen the global climate change situation. Although research is going on in CO_2 sequestration, it is doubtful that there will be any large-scale application of this technology anytime in the next 20–30 years.

Presently, there is a resurgence of interest in nuclear power; however, it is doubtful that it alone will be able to fill the gap. Forecasts from IAEA show that nuclear power around the world will grow at a rate of 0.5%–2.2% over the next 25 years (IAEA 2005). This estimate is in the same range as that of IEA.

Based on this information, it seems logical that the RE technologies of solar, wind, and biomass will not only be essential but also hopefully be able to provide the additional resources to fill the gap and provide a clean and sustainable energy future. Wind and Photovoltaic Power are growing at rates of over 30%–35% per year for the last few years, keeping in mind that this growth rate is based on very small existing capacities for these sources. There are many differing views on the future energy mix. The IEA estimates (Figure 1.10) that the present mix will continue until 2030 (IEA 2004).

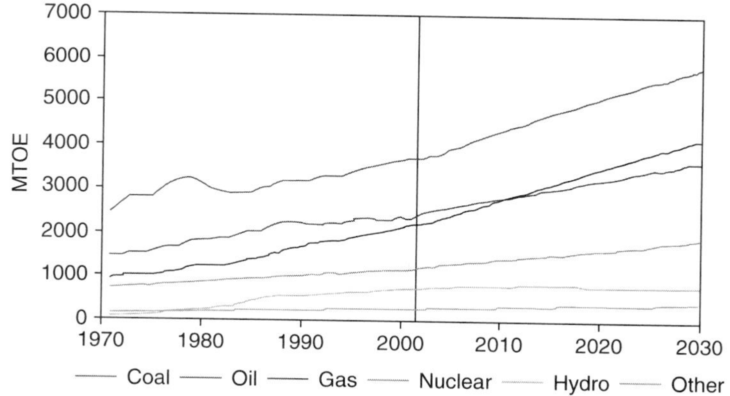

FIGURE 1.10 **(See color insert following page 19-52.)** World primary energy demand by fuel types. (According to IEA, *World Energy Outlook*, International Energy Agency, Paris, France, 2004.)

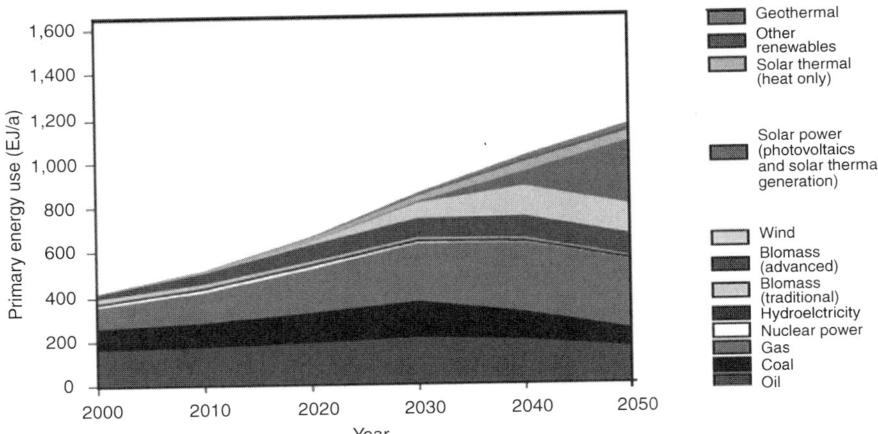

FIGURE 1.11 The global energy mix for year 2050. (According to WBGU, *World in Transition—Towards Sustainable Energy Systems*, German Advisory Council on Global Change, Berlin, 2003.)

On the other hand, the German Advisory Council on global change (WBGU) estimates that as much as 50% of the world's primary energy in 2050 will come from RE, (Figure 1.11). However to achieve that level of RE use by 2050 and beyond will require worldwide effort on the scale of a global Apollo Project.

1.5 Energy Conversion Technologies

It is clear that in order to meet the ever growing energy needs of the world, we will need to use all of the available resources including fossil fuels, nuclear and RE sources for the next 20–40 years. However, we will need to convert these energy resources more efficiently. It is also clear that renewable resources will have to continue to increase their share of the total energy consumption. There are many new developments in the conversion technologies for solar, wind, biomass, and other RE resources. In addition, there are newer and improved technologies for the conversion of fossil fuel and nuclear resources. This Handbook provides a wealth of information about the latest technologies for the direct and indirect conversion of energy resources into other forms such as thermal, mechanical, and electrical energy.

Defining Terms

MTOE: Mega tons of oil equivalent; 1 MTOE $= 4.1868 \times 10^4$ TJ (Terra Joules)
$= 3.968 \times 10^{13}$ BTU

GTOE: Giga tons of oil equivalent and 1 GTOE $=$ 1000 MTOE

Quadrilion Btu: 10^{15} British thermal units or Btu, also known as **Quad**; 1 Btu $=$ 1055 J

References

BP Statistical Review of World Energy 2006.

EIA 2006. *International Energy Outlook 2006*. Energy Information Agency, U.S. Department of Energy, Washington, DC. Website: http://eia.doe.gov

Hammererschlag, R. and Mazza, P. 2005. Questioning hydrogen. *Energy Policy*, 33, 2039–2043.

IAEA. 2005. *Energy, Electricity and Nuclear Power Estimates to 2030*. Reference data Series No. 1, July 2005.

IEA 2004. *World Energy Outlook*. International Energy Agency, Paris, France.

IEA 2005. *Renewables Information 2005*. International Energy Agency, Paris, France.

Kreith, F. and West, R. E. 2004. Fallacies of a hydrogen economy. *JERT*, 126, 249–257.

UNDP. 2004. *World Energy Assessment: Energy and the Challenge of Sustainability*. United Nations Development Program, New York, NY.

Veziroglu, T. N. and Barbir, F. 1992. Hydrogen: The wonder fuel. *International Journal of Hydrogen Energy*, 17(6), 391–404.

WBGU 2003. *World in Transition—Towards Sustainable Energy Systems*. German Advisory Council on Global Change, Berlin. Report available at http://www.wbgu.de

West, R. E. and Kreith, F. 2006. *A Vision for a Secure Transportation System without Hydrogen. ASME Journal of Energy Resources Technologies*, 128, 236–243.

For Further Information

Historical energy consumption data are published annually by the Energy Information Agency (EIA), U.S. Department of Energy, Washington, DC, International Energy Agency (IEA), Paris, and BP Corp. EIA and IEA also publish forecasts of future energy consumption and energy resources. However, their projections for individual fuels differ from other projections available in the literature.

2

Fossil Fuels

2.1 Coal 2-1
Coal Composition and Classification • Coal Analysis and Properties • Coal Reserves • Important Terminology: Resources, Reserves, and the Demonstrated Reserve Base • Transportation

2.2 Environmental Aspects 2-14

Defining Terms 2-14

References 2-15

For Further Information 2-15

2.3 Oil 2-16
Overview • Crude Oil Classification and World Reserves • Standard Fuels

2.4 Natural Gas 2-21
Overview • Reserves and Resources • Natural Gas Production Measurement • World Production of Dry Natural Gas • Compressed Natural Gas • Liquefied Natural Gas (LNG) • Physical Properties of Hydrocarbons

Defining Terms 2-25

For Further Information 2-25

Robert Reuther
U.S. Department of Energy

Richard Bajura
West Virginia University

Philip C. Crouse
Philip C. Crouse and Associates, Inc.

2.1 Coal

Robert Reuther

2.1.1 Coal Composition and Classification

Coal is a sedimentary rock formed by the accumulation and decay of organic substances, derived from plant tissues and exudates, which have been buried over periods of geological time, along with various mineral inclusions. Coal is classified by **type** and **rank**. Coal type classifies coal by the plant sources from which it was derived. Coal rank classifies coal by its degree of metamorphosis from the original plant sources and is therefore a measure of the age of the coal. The process of metamorphosis or aging is termed **coalification**.

The study of coal by type is known as coal petrography. Coal type is determined from the examination of polished sections of a coal sample using a reflected-light microscope. The degree of reflectance and the color of a sample are identified with specific residues of the original plant tissues. These various residues are referred to as **macerals**. Macerals are collected into three main groups: vitrinite, inertinite, and exinite (sometimes referred to as liptinite). The maceral groups and their associated macerals are listed in Table 2.1, along with a description of the plant tissue from which each distinct maceral type is derived.

TABLE 2.1 Coal Maceral Groups and Macerals

Maceral Group	Maceral	Derivation
Vitrinite	Collinite	Humic gels
	Telinite	Wood, bark, and cortical tissue
	Pseudovitrinite	? (Some observers place in the inertinite group)
Exinite	Sporinite	Fungal and other spores
	Cutinite	Leaf cuticles
	Alginite	Algal remains
Inertinite	Micrinite	Unspecified detrital matter, <0 μ
	Macrinite	Unspecified detrital matter, 10–100 μ
	Semifusinite	"Burned" woody tissue, low reflectance
	Fusinite	"Burned" woody tissue, high reflectance
	Sclerotinite	Fungal sclerotia and mycelia

Source: Modified from Berkowitz, N., *An Introduction to Coal Technology*, Academic Press, New York, 1979. With permission.

Coal rank is the most important property of coal because rank initiates the classification of coal for use. Coalification describes the process that the buried organic matter undergoes to become coal. When first buried, the organic matter has a certain elemental composition and organic structure. However, as the material becomes subjected to heat and pressure, the composition and structure slowly change. Certain structures are broken down, and others are formed. Some elements are lost through volatilization, while others are concentrated through a number of processes, including exposure to underground flows, which carry away some elements and deposit others. Coalification changes the values of various properties of coal. Thus, coal can be classified by rank through the measurement of one or more of these changing properties.

In the United States and Canada, the rank classification scheme defined by the American Society of Testing and Materials (ASTM) has become the standard. In this scheme, the properties of **gross calorific value** and **fixed carbon** or **volatile matter** content are used to classify a coal by rank. Gross calorific value is a measure of the energy content of the coal and is usually expressed in units of energy per unit mass. Calorific value increases as the coal proceeds through coalification. Fixed carbon content is a measure of the mass remaining after heating a dry coal sample under conditions specified by the ASTM.

Fixed carbon content also increases with coalification. The conditions specified for the measurement of fixed carbon content result in being able, alternatively, to use the volatile matter content of the coal, measured under dry, ash-free conditions, as a rank parameter. The rank of a coal proceeds from lignite, the "youngest" coal, through sub-bituminous, bituminous, and semibituminous, to anthracite, the "oldest" coal. The subdivisions within these rank categories are defined in Table 2.2. (Some rank schemes include meta-anthracite as a rank above, or "older" than, anthracite. Others prefer to classify such deposits as graphite—a minimal resource valuable primarily for uses other than as a fuel.)

According to the ASTM scheme, coals are ranked by calorific value up to the high-volatile A bituminous rank, which includes coals with calorific values (measured on a moist, mineral matter-free basis) greater than 14,000 Btu/lb (32,564 kJ/kg). At this point, fixed carbon content (measured on a dry, mineral matter-free basis) takes over as the rank parameter. Thus, a high-volatile A bituminous coal is defined as having a calorific value greater than 14,000 Btu/lb, but a fixed carbon content less than 69 wt%. The requirement for having two different properties with which to define rank arises because calorific value increases significantly through the lower-rank coals, but very little (in a relative sense) in the higher ranks; fixed carbon content has a wider range in higher rank coals, but little (relative) change in the lower ranks. The most widely used classification scheme outside North America is that developed under the jurisdiction of the International Standards Organization, Technical Committee 27, Solid Mineral Fuels.

2.1.2 Coal Analysis and Properties

The composition of a coal is typically reported in terms of its **proximate analysis** and its **ultimate analysis.** The proximate analysis of a coal is made up of four constituents: volatile matter content; fixed

TABLE 2.2 Classification of Coals by Rank

Class	Group	Fixed Carbon Limits, % (dmmf)		Volatile Matter Limits, % (dmmf)		Gross Calorific Value Limits, Btu/lb (moist, mmf)		Agglomerating Character
		Equal to or Greater Than	Less Than	Greater Than	Equal to or Less Than	Equal to or Greater Than	Less Than	
Anthracitic	Meta-anthracite	98	—	—	2	—	—	Nonagglomerating
	Anthracite	92	98	2	8	—	—	Nonagglomerating
	Semianthracite	86	92	8	14	—	—	Nonagglomerating
Bituminous	Low-volatile bituminous	78	86	14	22	—	—	Commonly agglomerating
	Medium-volatile bituminous	69	78	22	31	—	—	Commonly agglomerating
	High-volatile A bituminous	—	69	31	—	14,000	—	Commonly agglomerating
	High-volatile B bituminous	—	—	—	—	13,000	14,000	Commonly agglomerating
	High-volatile C bituminous	—	—	—	—	11,500	13,000	Commonly agglomerating
	High-volatile C bituminous	—	—	—	—	10,500	11,500	Agglomerating
Subbituminous	Subbituminous A	—	—	—	—	10,500	11,500	Nonagglomerating
	Subbituminous B	—	—	—	—	9,500	10,500	Nonagglomerating
	Subbituminous C	—	—	—	—	8,300	9,500	Nonagglomerating
Lignitic	Lignite A	—	—	—	—	6,300	8,300	Nonagglomerating
	Lignite B	—	—	—	—	—	6,300	Nonagglomerating

Source: From the American Society for Testing and Materials' Annual Book of ASTM Standards. With permission.

carbon content; moisture content; and ash content, all of which are reported on a weight percent basis. The measurement of these four properties of a coal must be carried out according to strict specifications codified by the ASTM. Note that the four constituents of proximate analysis do not exist, per se, in the coal, but are measured as analytical results upon treating the coal sample to various conditions.

ASTM volatile matter released from coal includes carbon dioxide, inorganic sulfur- and nitrogen-containing species, and organic compounds. The percentages of these various compounds or species released from the coal varies with rank. Volatile matter content can typically be reported on a number of bases, such as moist; dry, mineral matter-free (dmmf); moist, mineral matter-free; moist, ash-free; and dry, ash-free (daf), depending on the condition of the coal on which the measurements were made.

Mineral matter and ash are two distinct entities. Coal does not contain ash, even though the ash content of a coal is reported as part of its proximate analysis. Instead, coal contains mineral matter, which can be present as distinct mineral entities or inclusions and as material intimately bound with the organic matrix of the coal. Ash, on the other hand, refers to the solid inorganic material remaining *after combusting* a coal sample. Proximate ash content is the ash remaining after the coal has been exposed to air under specific conditions codified in ASTM Standard Test Method D 3174. It is reported as the mass percent remaining upon combustion of the original sample on a dry or moist basis.

Moisture content refers to the mass of water released from the solid coal sample when it is heated under specific conditions of temperature and residence time as codified in ASTM Standard Test Method D 3173.

The fixed carbon content refers to the mass of organic matter remaining in the sample after the moisture and volatile matter are released. It is primarily made up of carbon. However, hydrogen, sulfur, and nitrogen also are typically present. It is reported by difference from the total of the volatile matter, ash, and moisture contents on a mass percent of the original coal sample basis. Alternatively, it can be reported on a dry basis; a dmmf basis; or a moist, mineral matter-free basis.

The values associated with a proximate analysis vary with rank. In general, volatile matter content decreases with increasing rank, while fixed carbon content correspondingly increases. Moisture and ash also decrease, in general, with rank. Typical values for proximate analyses as a function of the rank of a coal are provided in Table 2.3.

The ultimate analysis provides the composition of the organic fraction of coal on an elemental basis. Like the proximate analysis, the ultimate analysis can be reported on a moist or dry basis and on an ash-containing or ash-free basis. The moisture and ash reported in the ultimate analysis are found from the corresponding proximate analysis. Nearly every element on Earth can be found in coal. However, the important elements that occur in the organic fraction are limited to only a few. The most important of these include carbon; hydrogen; oxygen; sulfur; nitrogen; and, sometimes, chlorine. The scope, definition of the ultimate analysis, designation of applicable standards, and calculations for reporting results on different moisture bases can be found in ASTM Standard Test Method D 3176M. Typical values for the ultimate analysis for various ranks of coal found in the U.S. are provided in Table 2.4. Other important properties of coal include swelling, caking, and coking behavior; ash fusibility; reactivity; and calorific value.

Calorific value measures the energy available in a unit mass of coal sample. It is measured by ASTM Standard Test Method D 2015M, Gross Calorific Value of Solid Fuel by the Adiabatic Bomb Calorimeter, or by ASTM Standard Test Method D 3286, Gross Calorific Value of Solid Fuel by the Isothermal-Jacket Bomb Calorimeter. In the absence of a directly measured value, the gross calorific value, Q, of a coal (in Btu/lb) can be estimated using the Dulong formula (Elliott and Yohe 1981):

$$Q = 14,544C + 62,028[H-(O/8)] + 4,050S$$

where C, H, O, and S are the mass fractions of carbon, hydrogen, oxygen, and sulfur, respectively, obtained from the ultimate analysis.

Swelling, caking, and coking all refer to the property of certain bituminous coals to change in size, composition, and, notably, strength, when slowly heated in an inert atmosphere to between 450 and 550

TABLE 2.3 Calorific Values and Proximate Analyses of Ash-Free Coals of Different Rank

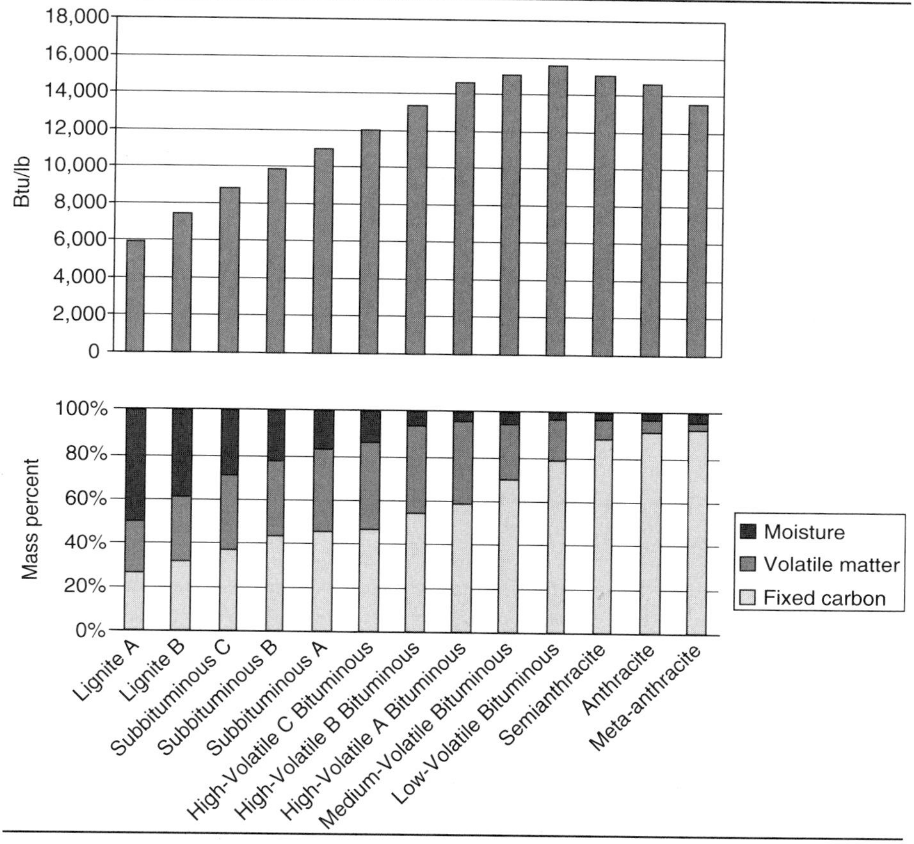

Btu/lb

18,000
16,000
14,000
12,000
10,000
8,000
6,000
4,000
2,000
0

Mass percent

100%
80%
60%
40%
20%
0%

Moisture
Volatile matter
Fixed carbon

Lignite A
Lignite B
Subbituminous C
Subbituminous B
Subbituminous A
High-Volatile C Bituminous
High-Volatile B Bituminous
High-Volatile A Bituminous
Medium-Volatile Bituminous
Low-Volatile Bituminous
Semianthracite
Anthracite
Meta-anthracite

Source: From Averitt, P., Coal Resources of the United States, January 1, 1974. U.S. Geological Survey Bulletin 1412, Government Printing Office, Washington, DC, 1975.

or 600°F. Under such conditions, the coal sample initially becomes soft and partially devolatilizes. With further heating, the sample takes on a fluid characteristic. During this fluid phase, further devolatilization causes the sample to swell. Still further heating results in the formation of a stable, porous, solid material with high strength. Several tests have been developed, based on this property, to measure the degree and

TABLE 2.4 Ultimate Analysis in Mass Percent of Representative Coals of the U.S.

Component	Fort Union Lignite	Powder River Subbituminous	Four Corners Subbituminous	Illinois C Bituminous	Appalachia Bituminous
Moisture	36.2	30.4	12.4	16.1	2.3
Carbon	39.9	45.8	47.5	60.1	73.6
Hydrogen	2.8	3.4	3.6	4.1	4.9
Nitrogen	0.6	0.6	0.9	1.1	1.4
Sulfur	0.9	0.7	0.7	2.9	2.8
Oxygen	11.0	11.3	9.3	8.3	5.3
Ash	8.6	7.8	25.6	7.4	9.7
Gross calorific value, Btu/lb	6,700	7,900	8,400	10,700	13,400

Source: Modified from Probstein, R. and Hicks, R., *Synthetic Fuels*, McGraw-Hill, New York, 1982. With permission.

suitability of a coal for various processes. Some of the more popular tests are the free swelling index (ASTM Test Method D 720); the Gray–King assay test (initially developed and extensively used in Great Britain); and the Gieseler plastometer test (ASTM Test Method D 2639), as well as a host of dilatometric methods (Habermehl et al. 1981).

The results of these tests are often correlated with the ability of a coal to form a coke suitable for iron making. In the iron-making process, the high carbon content and high surface area of the coke are used in reducing iron oxide to elemental iron. The solid coke must also be strong enough to provide the structural matrix upon which the reactions take place. Bituminous coals that have good coking properties are often referred to as metallurgical coals. (Bituminous coals without this property are, alternatively, referred to as steam coals because of their historically important use in raising steam for conversion to mechanical energy or electricity generation.)

Ash fusibility is another important property of coals. This is a measure of the temperature range over which the mineral matter in the coal begins to soften, eventually to melt into a slag, and to fuse together. This phenomenon is important in combustion processes; it determines if and at what point the resultant ash becomes soft enough to stick to heat exchanger tubes and other boiler surfaces or at what temperature it becomes molten so that it flows (as slag), making removal as a liquid from the bottom of a combustor possible.

Reactivity of a coal is a very important property fundamental to all coal conversion processes (such as combustion, gasification, and liquefaction). In general, lower rank coals are more reactive than higher rank coals. This is due to several different characteristics of coals, which vary with rank as well as with type. The most important characteristics are the surface area of the coal, its chemical composition, and the presence of certain minerals that can act as catalysts in the conversion reactions. The larger surface area present in lower rank coals translates into a greater degree of penetration of gaseous reactant molecules into the interior of a coal particle. Lower rank coals have a less aromatic structure than higher ranks. This corresponds to the presence of a higher proportion of lower energy, more reactive chemical bonds. Lower rank coals also tend to have higher proximate ash contents, and the associated mineral matter is more distributed, even down to the atomic level. Any catalytically active mineral matter is thus more highly dispersed.

However, the reactivity of a coal also varies depending upon what conversion is attempted. That is, the reactivity of a coal toward combustion (or oxidation) is not the same as its reactivity toward liquefaction, and the order of reactivity established in a series of coals for one conversion process will not necessarily be the same as that for another process.

2.1.3 Coal Reserves

Coal is found throughout the U.S. and the world. It is the most abundant fossil energy resource in the U.S. and the world, comprising 95% of U.S. fossil energy resources and 70% of world fossil energy resources on an energy content basis. All coal ranks can be found in the U.S. The largest resources in the U.S. are made up of lignite and sub-bituminous coals, which are found primarily in the western part of the country, including Alaska. Bituminous coals are found principally in the Midwest states, northern Alaska, and the Appalachian region. Principal deposits of anthracite coal are found in northeastern Pennsylvania.

The Alaskan coals have not been extensively mined because of their remoteness and the harsh climate. Of the other indigenous coals, the anthracite coals have been heavily mined to the point that little economic resource remains. The bituminous coals continue to be heavily mined in the lower 48 states, especially those with sulfur contents less than 2.5 wt%. The lignite and subbituminous coals in the western U.S. have been historically less heavily mined because of their distance from large population centers and because of their low calorific values and high moisture and ash contents. However, with the enactment of the 1990 Clean Air Act Amendments, these coals are now displacing high sulfur-containing coals for use in the eastern U.S. A map showing the general distribution of coal in the U.S. is included as Figure 2.1.

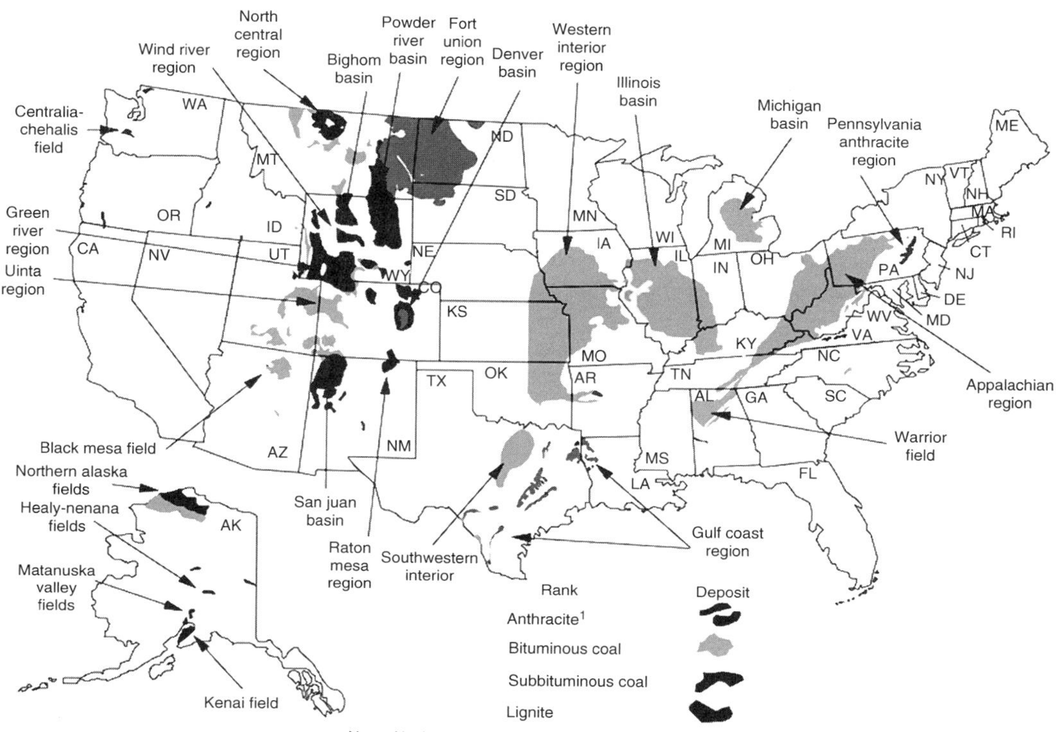

FIGURE 2.1 U.S. coal deposits.

The amount of coal that exists is not known exactly and is continually changing as old deposits are mined out and new deposits are discovered or reclassified. Estimates are published by many different groups throughout the world. In the U.S., the Energy Information Administration (EIA), an office within the U.S. Department of Energy, gathers and publishes estimates from various sources. The most commonly used definitions for classifying the estimates are provided below.

2.1.4 Important Terminology: Resources, Reserves, and the Demonstrated Reserve Base[1]

Resources are naturally occurring concentrations or deposits of coal in the Earth's crust, in such forms and amounts that economic extraction is currently or potentially feasible.

Measured resources refers to coal for which estimates of the rank and quantity have been computed to a high degree of geologic assurance, from sample analyses and measurements from closely spaced and geologically well-known sample sites. Under the U.S. Geological Survey (USGS) criteria, the points of

[1]For a full discussion of coal resources and reserve terminology as used by EIA, USGS, and the Bureau of Mines, see U.S. Coal Reserves, 1996, Appendix A, "Specialized Resource and Reserve Terminology."*Sources*: U.S. Department of the Interior, Coal Resource Classification System of the U.S. Bureau of Mines and the U.S. Geological Survey, Geological Survey Bulletin 1450-B (1976). U.S. Department of the Interior, Coal Resource Classification System of the U.S. Geological Survey, Geological Survey Circular 891 (1983) U.S. Department of the Interior, A Dictionary of Mining, Mineral, and Related Terms, Bureau of Mines (1968).

observation are no greater than ½ mile apart. Measured coal is projected to extend as a ¼-mile-wide belt from the outcrop or points of observation or measurement.

Indicated resources refers to coal for which estimates of the rank, quality, and quantity have been computed to a moderate degree of geologic assurance, partly from sample analyses and measurements and partly from reasonable geologic projections. Under the USGS criteria, the points of observation are from ½ to 1½ miles apart. Indicated coal is projected to extend as a ½-mile-wide belt that lies more than ¼ mile from the outcrop or points of observation or measurement.

Demonstrated resources are the sum of measured resources and indicated resources.

Demonstrated reserve base (DRB; or simply "reserve base" in USGS usage) is, in its broadest sense, defined as those parts of identified resources that meet specified minimum physical and chemical criteria related to current mining and production practices, including those for quality, depth, thickness, rank, and distance from points of measurement. The "reserve base" is the in-place demonstrated resource from which reserves are estimated. The reserve base may encompass those parts of a resource that have a reasonable potential for becoming economically recoverable within planning horizons that extend beyond those that assume proven technology and current economics.

Inferred resources refers to coal of a low degree of geologic assurance in unexplored extensions of demonstrated resources for which estimates of the quality and size are based on geologic evidence and projection. Quantitative estimates are based on broad knowledge of the geologic character of the bed or region from which few measurements or sampling points are available and on assumed continuation from demonstrated coal for which geologic evidence exists. The points of measurement are from 1½ to 6 miles apart. Inferred coal is projected to extend as a 2¼-mile-wide belt that lies more than ¾ mile from the outcrop or points of observation or measurement. Inferred resources are not part of the DRB.

Recoverable refers to coal that is, or can be, extracted from a coalbed during mining.

Reserves relates to that portion of demonstrated resources that can be recovered economically with the application of extraction technology available currently or in the foreseeable future. Reserves include only recoverable coal; thus, terms such as "minable reserves," "recoverable reserves," and "economic reserves" are redundant. Even though "recoverable reserves" is redundant, implying recoverability in both words, EIA prefers this term specifically to distinguish recoverable coal from in-ground resources, such as the demonstrated reserve base, that are only partially recoverable.

Minable refers to coal that can be mined using present-day mining technology under current restrictions, rules, and regulations.

The demonstrated reserve base for coals in the U.S. as of January 1, 2001, is approximately 501.1 billion (short) tons. It is broken out by rank, state, and mining method (surface or underground) in Table 2.5. As of December 31, 1999 (December 31, 2000, for the U.S.), the world recoverable reserves are estimated to be 1083 billion (short) tons. A breakdown by region and country is provided in Table 2.6. The recoverability factor for all coals can vary from approximately 40 to over 90%, depending on the individual deposit. The recoverable reserves in the U.S. represent approximately 54% of the demonstrated reserve base as of January 1, 2001. Thus, the U.S. contains approximately 25% of the recoverable reserves of coal in the world.

2.1.5 Transportation

Most of the coal mined and used domestically in the U.S. is transported by rail from the mine mouth to its final destination. In 1998, 1119 million short tons of coal were distributed domestically. Rail constituted 58.3% of the tonnage, followed by water at 21.4%; truck at 11.0%; and tramway, conveyor, or slurry pipeline at 9.2%. The remaining 0.1% is listed as "unknown method." Water's share includes transportation on the Great Lakes, all navigable rivers, and on tidewaters (EIA 1999).

In general, barge transportation is cheaper than rail transportation. However, this advantage is reduced for distances over 300 miles (Villagran 1989). For distances less than 100 miles, rail is very inefficient, and trucks are used primarily, unless water is available as a mode of transport.

TABLE 2.5 U.S. Coal Demonstrated Reserve Base, January 1, 2001

		Bituminous Coal		Subbituminous Coal		Lignite	Total		
Region and State	Anthracite	Underground	Surface	Underground	Surface	Surface[a]	Underground	Surface	Total
Appalachian	7.3	72.9	23.7	0.0	0.0	1.1	76.9	28.1	105.0
Appalachian	7.3	7.40	24.0	0.0	0.0	1.1	78.0	28.5	106.5
Alabama	0.0	1.2	2.1	0.0	0.0	1.1	1.2	3.2	4.4
Kentucky, eastern	0.0	1.7	9.6	0.0	0.0	0.0	1.7	9.6	11.3
Ohio	0.0	17.7	5.8	0.0	0.0	0.0	17.7	5.8	23.5
Pennsylvania	7.2	19.9	1.0	0.0	0.0	0.0	23.8	4.3	28.1
Virginia	0.1	1.2	0.6	0.0	0.0	0.0	1.3	0.6	2.0
West Virginia	0.0	30.1	4.1	0.0	0.0	0.0	30.1	4.1	34.2
Other[b]	0.0	1.1	0.4	0.0	0.0	0.0	1.1	0.4	1.5
Interior	0.1	117.8	27.5	0.0	0.0	13.1	117.9	40.7	158.6
Illinois	0.0	88.2	16.6	0.0	0.0	0.0	88.2	16.6	104.8
Indiana	0.0	8.8	0.9	0.0	0.0	0.0	8.8	0.9	9.7
Iowa	0.0	1.7	0.5	0.0	0.0	0.0	1.7	0.5	2.2
Kentucky, western	0.0	16.1	3.7	0.0	0.0	0.0	16.1	3.7	19.7
Missouri	0.0	1.5	4.5	0.0	0.0	0.0	1.5	4.5	6.0
Oklahoma	0.0	1.2	0.3	0.0	0.0	0.0	1.2	0.3	1.6
Texas	0.0	0.0	0.0	0.0	0.0	12.7	0.0	12.7	12.7
Other[c]	0.1	0.3	1.1	0.0	0.0	0.5	0.4	1.6	2.0
Western	(s)	22.3	2.3	121.3	61.8	29.6	143.7	93.7	237.4
Alaska	0.0	0.6	0.1	4.8	0.6	(s)	5.4	0.7	6.1
Colorado	(s)	8.0	0.6	3.8	0.0	4.2	11.8	4.8	16.6

(continued)

TABLE 2.5 ***(Continued)***

		Bituminous Coal		Subbituminous Coal		Lignite	Total		
Region and State	Anthracite	Underground	Surface	Underground	Surface	Surface[a]	Underground	Surface	Total
Montana	0.0	1.4	0.0	69.6	32.8	15.8	71.0	48.5	119.5
New Mexico	(s)	2.7	0.9	3.5	5.2	0.0	6.2	6.1	12.3
North Dakota	0.0	0.0	0.0	0.0	0.0	9.2	0.0	9.2	9.2
Utah	0.0	5.4	0.3	0.0	0.0	0.0	5.4	0.3	5.6
Washington	0.0	0.3	0.0	1.0	(s)	(s)	1.3	0.0	1.4
Wyoming	0.0	3.8	0.5	38.7	23.2	0.0	42.5	23.7	66.2
Other[d]	0.0	0.1	0.0	(s)	(s)	0.4	0.1	0.4	0.5
U.S. total	7.5	213.1	53.5	121.3	61.8	43.8	338.5	162.5	501.1
States east of the Mississippi River	7.3	186.1	44.8	0.0	0.0	1.1	190.1	49.3	239.4
States west of the Mississippi River	0.1	27.0	8.7	121.3	61.8	42.7	148.4	113.3	261.7

Notes: (s) = Less than 0.05 billion short tons. Data represent known measured and indicated coal resources meeting minimum seam and depth criteria, in the ground as of January 1, 2001. These coal resources are not totally recoverable. Net recoverability ranges from 0% to more than 90%. Fifty-four percent of the demonstrated reserve base of coal in the United States is estimated to be recoverable. Totals may not equal sum of components due to independent rounding.

Source: Energy Information Administration, Coal Reserves Data Base.

[a] Lignite resources are not mined underground in the U.S.

[b] Georgia, Maryland, North Carolina, and Tennessee.

[c] Arkansas, Kansas, Louisiana, and Michigan.

[d] Arizona, Idaho, Oregon, and South Dakota.

Source: Energy Information Administration, Coal Reserves Data Base.

TABLE 2.6 World Recoverable Reserves of Coal

Region/Country	Recoverable Anthracite and Bituminous	Recoverable Lignite and Subbituminous	Total Recoverable Coal
North America			
Canada	3,826	3,425	7,251
Greenland	0	202	202
Mexico	948	387	1,335
U.S.	126,804	146,852	273,656
Total	131,579	150,866	282,444
Central and South America			
Argentina	0	474	474
Bolivia	1	0	1
Brazil	0	13,149	13,149
Chile	34	1,268	1,302
Colombia	6,908	420	7,328
Ecuador	0	26	26
Peru	1,058	110	1,168
Venezuela	528	0	528
Total	8,530	15,448	23,977
Western Europe			
Austria	0	28	28
Croatia	7	36	43
France	24	15	40
Germany	25,353	47,399	72,753
Greece	0	3,168	3,168
Ireland	15	0	15
Italy	0	37	37
Netherlands	548	0	548
Norway	0	1	1
Portugal	3	36	40
Slovenia	0	303	303
Spain	220	507	728
Sweden	0	1	1
Turkey	306	3,760	4,066
United Kingdom	1,102	551	1,653
Yugoslavia	71	17,849	17,919
Total	27,650	73,693	101,343
Eastern Europe and former U.S.S.R.			
Bulgaria	14	2,974	2,988
Czech Republic	2,330	3,929	6,259
Hungary	0	1,209	1,209
Kazakhstan	34,172	3,307	37,479
Kyrgyzstan	0	895	895
Poland	22,377	2,050	24,427
Romania	1	1,605	1,606
Russia	54,110	118,964	173,074
Slovakia	0	190	190
Ukraine	17,939	19,708	37,647
Uzbekistan	1,102	3,307	4,409
Total	132,046	158,138	290,183
Middle East			
Iran	1,885	0	1,885
Total	1,885	0	1,885
Africa			

(continued)

TABLE 2.6 ***(Continued)***

Region/Country	Recoverable Anthracite and Bituminous	Recoverable Lignite and Subbituminous	Total Recoverable Coal
Algeria	44	0	44
Botswana	4,740	0	4,740
Central African Republic	0	3	3
Congo (Kinshasa)	97	0	97
Egypt	0	24	24
Malawi	0	2	2
Mozambique	234	0	234
Niger	77	0	77
Nigeria	23	186	209
South Africa	54,586	0	54,586
Swaziland	229	0	229
Tanzania	220	0	220
Zambia	11	0	11
Zimbabwe	553	0	553
Total	60,816	216	61,032
Far East and Oceania			
Afghanistan	73	0	73
Australia	46,903	43,585	90,489
Burma	2	0	2
China	68,564	57,651	126,215
India	90,826	2,205	93,031
Indonesia	871	5,049	5,919
Japan	852	0	852
Korea, North	331	331	661
Korea, South	86	0	86
Malaysia	4	0	4
Nepal	2	0	2
New Caledonia	2	0	2
New Zealand	36	594	631
Pakistan	0	2,497	2,497
Philippines	0	366	366
Taiwan	1	0	1
Thailand	0	1,398	1,398
Vietnam	165	0	165
Total	208,719	113,675	322,394
World total	571,224	512,035	1,083,259

Notes: The estimates in this table are dependent on the judgment of each reporting country to interpret local economic conditions and its own mineral assessment criteria in terms of specified standards of the World Energy Council. Consequently, the data may not all meet the same standards of reliability, and some data may not represent reserves of coal known to be recoverable under current economic conditions and regulations. Some data represent estimated recovery rates for highly reliable estimates of coal quantities in the ground that have physical characteristics like those of coals currently being profitably mined. U.S. coal rank approximations are based partly on Btu content and may not precisely match borderline geologic ranks. Data for the U.S. represent recoverable coal estimates as of December 31, 2000. Data for other countries are as of December 31, 1999.

Millions of tons.

Sources: World Energy Council, Survey of Energy Resources 2001, October 2001. U.S. Energy Information Administration. Unpublished file data of the Coal Reserves Data Base (February 2002).

Prior to the signing of the 1990 Clean Air Act Amendments, most coal was transported to the closest power plant or other end-use facility to reduce transportation costs. Because most coal-fired plants are east of the Mississippi River, most of the coal was transported from eastern coal mines. However, once the Amendments, which required sulfur emissions to be more strictly controlled, began to be enforced, the potential economic advantage of transporting and using low-sulfur western coals compared to installing

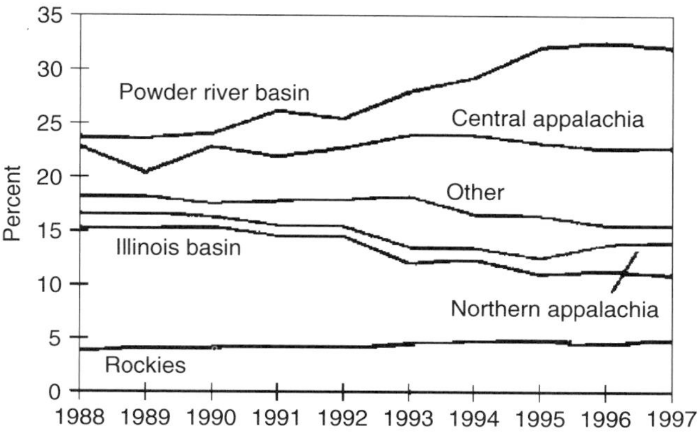

FIGURE 2.2 Supply region shares of domestic coal distribution. (From Energy Information Administration, EIA-6, "Coal Distribution Report.")

expensive cleanup facilities in order to continue to use high-sulfur eastern coals began to be considered. This resulted in increasing the average distance coal was shipped from 640 miles in 1988 to 793 miles in 1997.

In comparing shipments from coal-producing regions, the trend of Figure 2.2 shows that an increasing share of coal was shipped from the low-sulfur coal producing Powder River Basin between 1988 and 1997 and that less coal was shipped from the high-sulfur coal producing Central Appalachian Basin. Overall, coal use continued to increase at about 2.2% per year over this timeframe.

The cost of transporting coal decreased between 1988 and 1997, due to the increased competition from the low-sulfur western coals following passage of the Clean Air Act Amendments in 1990. This decrease held for all sulfur levels, except for a slight increase in medium sulfur B coals over the last couple of years, as shown in Figure 2.3.

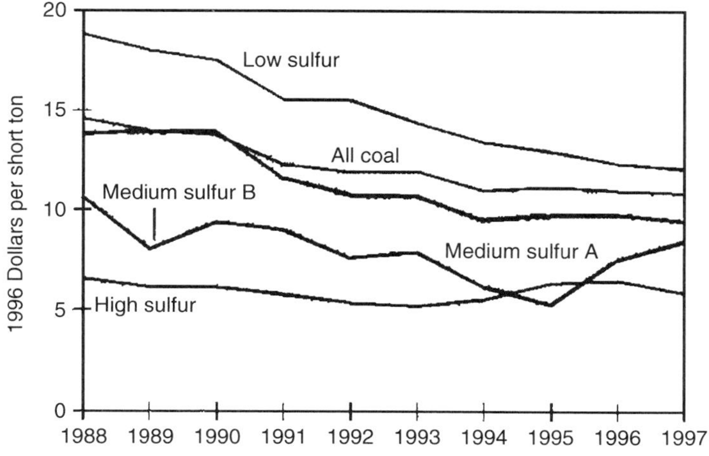

FIGURE 2.3 Average rate per ton for contract coal shipments by rail, by sulfur category, 1988–1997. Notes: low sulfur = less than or equal to 0.6 lb of sulfur per million Btu; medium sulfur A = 0.61–1.25 lb per million Btu; medium sulfur B = 1.26–1.67 lb per million Btu; high sulfur = greater than 1.67 lb per million Btu. 1997. (From Energy Information Administration, Coal Transportation Rate Database.)

2.2 Environmental Aspects

Richard Bajura

Along with coal production and use comes a myriad of potential environmental problems, most of which can be ameliorated or effectively addressed during recovery, processing, conversion, or reclamation. Underground coal reserves are recovered using the two principal methods of room-and-pillar mining (60%) and longwall mining (40%). In room-and-pillar mining, coal is removed from the seam in a checkerboard pattern (the "room") as viewed from above, leaving pillars of coal in an alternate pattern to support the roof of the mine. When using this technology, generally half of the reserves are left underground. Depending upon the depth of the seam and characteristics of the overburden, subsidence due to the removal of the coal may affect the surface many years after the mining operation is completed. Because of the danger of collapse and movement of the surface, undermined lands are not used as building sites for large, heavy structures.

Longwall mining techniques employ the near-continuous removal of coal in rectangular blocks with a vertical cross section equal to the height of the seam multiplied by the horizontal extent (width) of the panel being mined. As the longwall cutting heads advance into the coal seam, the equipment is automatically moved forward. The roof of the mine collapses behind the shields, and most of the effects of subsidence are observed on the surface within several days of mining. If the longwall mining operation proceeds in a continuous fashion, subsidence may occur smoothly so that little damage occurs to surface structures. Once subsidence has occurred, the surface remains stable into the future. Longwall mining operations may influence water supplies as a result of fracturing of water-bearing strata far removed from the panel being mined.

When coal occurs in layers containing quartz dispersed in the seam or in the overburden, miners are at risk of exposure to airborne silica dust, which is inhaled into their lungs. Coal workers' pneumonoconiosis, commonly called black lung disease, reduces the ability of a miner to breathe because of the effects of fibrosis in the lungs.

Surface mining of coal seams requires the removal of large amounts of overburden, which must eventually be replaced into the excavated pit after the coal resource is extracted. When the overburden contains large amounts of pyrite, exposure to air and water produces a discharge known as acid mine drainage, which can contaminate streams and waterways. Iron compounds formed as a result of the chemical reactions precipitate in the streams and leave a yellow- or orange-colored coating on rocks and gravel in the streambeds. The acid caused by the sulfur in the pyrite has been responsible for significant destruction of aquatic plants and animals. New technologies have been and continue to be developed to neutralize acid mine drainage through amendments applied to the soil during the reclamation phases of the mining operation. Occasionally, closed underground mines fill with water and sufficient pressure is created to cause "blowouts" where the seams reach the surface. Such discharges have also been responsible for massive fish kills in receiving streams.

The potential for acid rain deposition from sulfur and nitrogen oxides released to the atmosphere during combustion is a significant concern. About 95% of the sulfur oxide compounds can be removed through efficient stack gas cleaning processes such as wet and dry scrubbing. Also, techniques are available for removing much of the sulfur from the coal prior to combustion. Combustion strategies are also being developed that reduce the formation and subsequent release of nitrogen oxides.

The potential for greenhouse warming due to emissions of carbon dioxide during combustion (as well as methane during mining and mine reclamation) has also been raised as a significant concern. Because coal is largely composed of carbon with relatively little hydrogen, its combustion leads to a higher level of carbon dioxide emissions per unit of energy released than for petroleum-based fuels or natural gas.

Defining Terms

Coalification: The physicochemical transformation that coal undergoes after being buried and subjected to elevated temperature and pressure. The classification of a particular coal by rank is a measure of the extent of its coalification. Thus, coalification is a measure of the "age" of a particular coal.

Fixed carbon content: One of the constituents that make up the proximate analysis of a coal. It is normally measured by difference. That is, one measures the volatile matter content and the moisture and ash contents, if the fixed carbon content is reported on a basis containing one or both of those constituents, and subtracts the result(s) from 100% to find the fixed carbon content. One should not confuse the fixed carbon content of a coal with its (elemental) carbon content found in the ultimate analysis. Although carbon is certainly in the material making up the fixed carbon content, it is not all of the carbon present in the original coal, and other elements are also present.

Gross calorific value: Calorific value is a measure of the energy content of a material—in this case, a coal sample. Calorific value is measured by ASTM Standard Test Method D 2015M, Gross Calorific Value of Solid Fuel by the Adiabatic Bomb Calorimeter, or by ASTM Standard Test Method D 3286, Gross Calorific Value of Solid Fuel by the Isothermal-Jacket Bomb Calorimeter. The *gross* calorific value takes into account the additional heat gained by condensing any water present in the products of combustion, in contrast to the *net* calorific value, which assumes that all water remains in the vapor state.

Maceral: An organic substance or optically homogeneous aggregate of organic substance in a coal sample that possesses distinctive physical and chemical properties.

Proximate analysis: A method to measure the content of four separately identifiable constituents in a coal: volatile matter content; fixed carbon content; moisture content; and ash content, all of which are reported on a weight percent basis. The standard method for obtaining the proximate analysis of coal or coke is defined by the ASTM in Standard Test Method D 3172.

Rank: A classification scheme for coals that describes the extent of coalification that a particular coal has undergone. The structure, chemical composition, and many other properties of coals vary systematically with rank. The standard method for determining the rank of a coal sample is defined by the ASTM in Standard Test Method D 388.

Type: A classification scheme for coals that references the original plant material from which the coal was derived.

Ultimate analysis: A method to measure the elemental composition of a coal sample. Typical ultimate analyses include carbon, hydrogen, oxygen, sulfur, and nitrogen contents, but other elements can also be reported. These other elements are usually not present to any appreciable extent. However, if they are reported, the sum of all the elements reported (including moisture and ash content) should equal 100%. The standard method for the ultimate analysis of coal or coke is defined by the ASTM in Standard Test Method D 3176.

Volatile matter content: The mass of material released upon heating the coal sample under specific conditions, defined by the ASTM Standard Test Method D 3175.

References

Elliott, M. A. and Yohe, G. R. 1981. The coal industry and coal research and development in perspective. In *Chemistry of Coal Utilization. Second Supplementary Volume*, M. A. Elliott, ed., pp. 26–328. Wiley, New York.

Habermehl, D., Orywal, F., and Beyer, H.-D. 1981. Plastic properties of coal. In *Chemistry of Coal Utilization. Second Supplementary Volume*, M. A. Elliott, ed., pp. 319–328. Wiley, New York.

Villagran, R. A. 1989. *Acid Rain Legislation: Implications for the Coal Industry*, pp. 37–39. Shearson, Lehman, Button, New York.

For Further Information

An excellent resource for understanding coal, its sources, uses, limitations, and potential problems is the book by Elliott referenced under Elliott and Yohe (1981) and Habermehl et al. (1981). A reader wishing an understanding of coal topics could find no better resource. Another comprehensive book, which includes more-recent information but is not quite as weighty as Elliott's (664 pages vs. 2374 pages), is *The Chemistry and Technology of Coal*, edited (second edition, revised and expanded) by James G. Speight. For information specific to the environmental problems associated with the use of coal, the reader is referred to Norbert Berkowitz's chapter entitled "Environmental Aspects of Coal Utilization" in *An Introduction to Coal Technology*. For information on the standards for coal analyses and descriptions of the associated procedures, the reader is referred to any recent edition of the ASTM's *Annual Book of ASTM Standards*. Section 5 covers petroleum products, lubricants, and fossil fuels, including coal and coke.

2.3 Oil

Philip C. Crouse

2.3.1 Overview

The U.S. Department of Energy's Energy Information Administration (EIA) annually provides a wealth of information concerning most energy forms including fossil fuels. The oil and natural gas sections are extracted summaries for the most germane information concerning oil and natural gas. Fossil fuel energy continues to account for over 85% of all world energy in 2000. The EIA estimates that in 2025, fossil fuels will still dominate energy resources with natural gas having the most growth. The base case of the EIA predicts that world energy consumption will grow by 60% over the next two decades. Figure 2.4 shows steady growth in global energy consumption. The projections show that in 2025 the world will consume three times the energy it consumed in 1970.

In the United States, wood served as the preeminent form of energy for about half of the nation's history. Around the 1880s, coal became the primary source of energy. Despite its tremendous and rapid expansion, coal was overtaken by petroleum in the middle of the 1900s. Natural gas, too, experienced rapid development into the second half of the 20th century, and coal began to expand again. Late in the 1900s, nuclear electric power was developed and made significant contributions.

Although the world's energy history is one of large-scale change as new forms of energy have been developed, the outlook for the next couple of decades is for continued growth and reliance on the three major fossil fuels of petroleum, natural gas, and coal. Only modest expansion will take place in renewable resources and relatively flat generation from nuclear electric power, unless major breakthroughs occur in

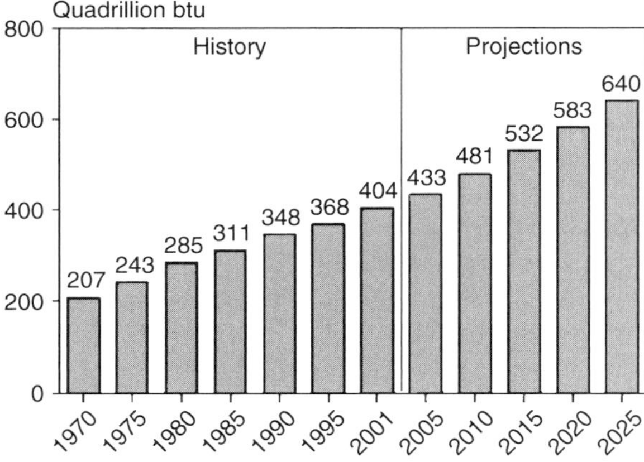

FIGURE 2.4 World energy consumption, 1970–2025. (History from EIA, International Energy Annual 2001, DOE/EIA-0219(2001), Washington, DC, Feb. 2003, www.eia.doe.gov/iea/. Projections from EIA, System for the analysis of Global Energy Markets (2003).)

TABLE 2.7 World Total Energy Consumption by Region and Fuel, Reference Case, 1990–2025

	History			Projections					
Region/Country	1990	2000	2001	2005	2010	2015	2020	2025	Average Annual Percent Change, 2001–2025
					Industrialized Countries				
North America									
Oil	40.4	46.3	45.9	48.3	54.2	59.7	64.3	69.3	1.7
Natural Gas	23.1	28.8	27.6	30.6	34.0	37.9	42.0	46.9	2.2
Coal	20.7	24.5	23.9	24.9	27.3	28.7	30.0	31.8	1.2
Nuclear	6.9	8.7	8.9	9.4	9.6	9.7	9.7	9.5	0.3
Other	9.5	10.6	9.4	11.3	12.0	12.7	13.4	13.9	1.7
Total	**100.6**	**118.7**	**115.6**	**124.6**	**137.2**	**148.7**	**159.4**	**171.4**	**1.7**
Western Europe									
Oil	25.8	28.5	28.9	29.2	29.7	30.3	30.6	31.6	0.4
Natural gas	9.7	14.9	15.1	15.9	17.5	20.1	23.4	26.4	2.4
Coal	12.4	8.4	8.6	8.3	8.2	7.5	6.8	6.7	−1.0
Nuclear	7.4	8.8	9.1	8.9	9.1	8.8	8.1	6.9	−1.1
Other	4.5	6.0	6.1	6.8	7.5	8.0	8.4	8.8	1.5
Total	**59.9**	**66.8**	**68.2**	**69.1**	**72.1**	**74.7**	**77.3**	**80.5**	**0.7**
Industrialized Asia									
Oil	12.1	13.2	13.0	13.5	14.3	15.1	15.8	16.7	1.1
Natural gas	2.5	4.0	4.1	4.4	4.6	5.0	5.3	5.9	1.5
Coal	4.2	5.7	5.9	5.8	6.3	6.7	7.0	7.4	0.9
Nuclear	2.0	3.0	3.2	3.2	3.6	3.9	4.0	3.9	0.9
Other	1.6	1.6	1.6	1.9	2.0	2.1	2.3	2.4	1.7
Total	**22.3**	**27.5**	**27.7**	**28.8**	**30.8**	**32.8**	**34.4**	**36.4**	**1.1**
Total industrialized									
Oil	78.2	88.1	87.8	90.9	98.2	105.1	110.7	117.6	1.2
Natural Gas	35.4	47.7	46.8	50.9	56.1	63.0	70.7	79.2	2.2
Coal	37.3	38.6	38.5	39.1	41.9	42.9	43.7	45.9	0.7
Nuclear	16.3	20.5	21.2	21.5	22.3	22.3	21.8	20.4	−0.2
Other	15.6	18.2	17.1	20.0	21.6	22.8	24.0	25.2	1.6
Total	**182.8**	**213.0**	**211.5**	**222.5**	**240.1**	**256.2**	**271.1**	**288.3**	**1.3**
EE/FSU									
Oil	21.0	10.9	11.0	12.6	14.2	15.0	16.5	18.3	2.1

(continued)

TABLE 2.7 ***(Continued)***

Region/Country	History			Projections					Average Annual Percent Change, 2001–2025
	1990	2000	2001	2005	2010	2015	2020	2025	
Natural gas	28.8	23.3	23.8	27.9	31.9	36.9	42.0	47.0	2.9
Coal	20.8	12.2	12.4	13.7	12.7	12.5	11.2	10.2	−0.8
Nuclear	2.9	3.0	3.1	3.3	3.3	3.3	3.0	2.6	−0.7
Other	2.8	3.0	3.2	3.6	3.7	3.9	4.0	4.1	1.1
Total	**76.3**	**52.2**	**53.3**	**61.1**	**65.9**	**71.6**	**76.7**	**82.3**	**1.8**
					Developing Countries				
Developing Asia									
Oil	16.1	30.2	30.7	33.5	38.9	45.8	53.8	61.9	3.0
Natural gas	3.2	6.9	7.9	9.0	10.9	15.1	18.6	22.7	4.5
Coal	29.1	37.1	39.4	41.3	49.4	56.6	65.0	74.0	2.7
Nuclear	0.9	1.7	1.8	2.6	3.1	4.1	4.5	5.0	4.3
Other	3.2	4.5	5.1	6.1	7.8	8.9	10.0	11.0	3.2
Total	**52.5**	**80.5**	**85.0**	**92.5**	**110.1**	**130.5**	**151.9**	**174.6**	**3.0**

Quadrillion Btu.
Source: International Energy Outlook-2003, U.S. Dept. of Energy, Energy Information Administration.

energy technologies. Table 2.7 shows EIA's estimate of growth of selected energy types with oil needs dominating the picture over the next 20 years.

2.3.2 Crude Oil Classification and World Reserves

Obtaining accurate estimates of world petroleum and natural gas resources and reserves is difficult and uncertain, despite excellent scientific analysis made over the years. Terminology standards used by industry to classify resources and reserves has progressed over the last 10 years with the Society of Petroleum Evaluation Engineers leading an effort to establish a set of standard definitions that would be used by all countries in reporting reserves. Classifications of reserves, however, continue to be a source of controversy in the international oil and gas community. This subsection uses information provided by the Department of Energy classification system. The next chart shows the relationship of resources to reserves. **Recoverable reserves** include discovered and undiscovered resources. **Discovered resources** are those resources that can be economically recovered. Figure 2.5 shows the relationship of petroleum resource and reserves terms.

Discovered resources include all production already out of the ground and reserves. Reserves are further broken down into proved reserves and other reserves. Again, many different groups classify reserves in different ways, such as measured, indicated, internal, probable, and possible. Most groups break reserves into producing and nonproducing categories. Each of the definitions is quite voluminous and the techniques for qualifying reserves vary globally. Table 2.8 shows estimates made by the EIA for total world oil resources.

2.3.3 Standard Fuels

Petroleum is refined into petroleum products that are used to meet individual product demands. The general classifications of products are:

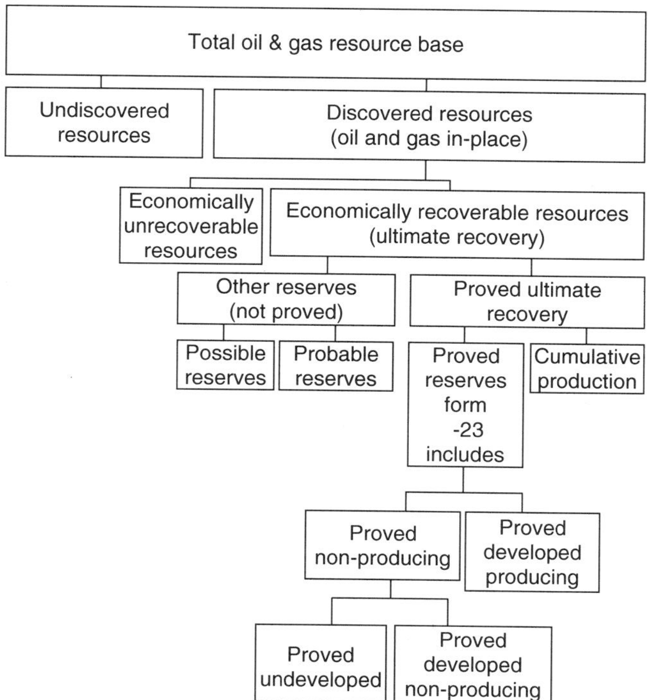

FIGURE 2.5 Components of the oil and gas resource base. (From EIA, Office of Gas and Oil.)

TABLE 2.8 Estimated World Oil Resources, 2000–2025

Region and Country	Proved Reserves	Reserve Growth	Undiscovered
Industrialized			
U.S.	22.45	76.03	83.03
Canada	180.02	12.48	32.59
Mexico	12.62	25.63	45.77
Japan	0.06	0.09	0.31
Australia/New Zealand	3.52	2.65	5.93
Western Europe	18.10	19.32	34.58
Eurasia			
Former Soviet Union	77.83	137.70	170.79
Eastern Europe	1.53	1.46	1.38
China	18.25	19.59	14.62
Developing countries			
Central and South America	98.55	90.75	125.31
India	5.37	3.81	6.78
Other developing Asia	11.35	14.57	23.90
Africa	77.43	73.46	124.72
Middle East	685.64	252.51	269.19
Total	1,212.88	730.05	938.90
OPEC	819.01	395.57	400.51
Non-OPEC	393.87	334.48	538.39

Note: Resources include crude oil (including lease condensates) and natural gas plant liquids.
Billion barrels.
Source: U.S. Geological Survey, *World Petroleum Assessment 2000*, web site http://greenwood.cr.usgs.gov/energy/WorldEnergy/DDS-60.

1. *Natural gas liquids and liquefied refinery gases.* This category includes ethane (C_2H_6); ethylene (C_2H_4); propane (C_3H_8); propylene (C_3H_6); butane and isobutane (C_4H_{10}); and butylene and isobutylene (C_4H_8).
2. *Finished petroleum products.* This category includes motor gasoline; aviation gasoline; jet fuel; kerosene; distillate; fuel oil; residual fuel oil; petrochemical feed stock; naphthas; lubricants; waxes; petroleum coke; asphalt and road oil; and still gas.
   - *Motor gasoline* includes reformulated gasoline for vehicles and oxygenated gasoline such as gasohol (a mixture of gasoline and alcohol).
   - *Jet fuel* is classified by use such as industrial or military and naphtha and kerosene type. Naphtha fuels are used in turbo jet and turbo prop aircraft engines and exclude ram-jet and petroleum rocket fuel.
   - *Kerosene* is used for space heaters, cook stoves, wick lamps, and water heaters.
   - *Distillate fuel oil* is broken into subcategories: No. 1 distillate, No. 2 distillate, and No. 4 fuel oil, which is used for commercial burners.
   - *Petrochemical feedstock* is used in the manufacture of chemicals, synthetic rubber, and plastics.
   - *Naphthas* are petroleums with an approximate boiling range of 122°F–400°F.
   - *Lubricants* are substances used to reduce friction between bearing surfaces, as process materials, and as carriers of other materials. They are produced from distillates or residues. Lubricants are paraffinic or naphthenic and separated by viscosity measurement.
   - *Waxes* are solid or semisolid material derived from petroleum distillates or residues. They are typically a slightly greasy, light colored or translucent, crystallizing mass.
   - *Asphalt* is a cement-like material containing bitumens. *Road oil* is any heavy petroleum oil used as a dust pallatine and road surface treatment.

TABLE 2.9 World Crude Oil Refining Capacity, January 1, 2002

Region/Country	Number of Refineries	Thousand Barrels per Day			
		Crude Oil Distillation	Catalytic Cracking	Thermal Cracking	Reforming
North America	180	20,254	6,619	2,450	4,140
Central and South America	70	6,547	1,252	435	447
Western Europe	112	15,019	2,212	1,603	2,214
Eastern Europe and Former U.S.S.R.	87	10,165	778	516	1,353
Middle East	46	6,073	312	406	570
Africa	46	3,202	195	88	387
Asia and Oceania	203	20,184	2,673	421	2,008
World Total	744	81,444	14,040	5,918	11,119

Source: Last updated on 3/14/2003 by DOE/EIA.

- *Still gas* is any refinery by-product gas. It consists of light gases of methane; ethane; ethylene; butane; propane; and the other associated gases. Still gas is typically used as a refinery fuel.

Table 2.9 shows *world refining capacity* as of January 1, 2002. The number of oil refineries continues to grow as demands for petroleum products have continued to grow.

2.4 Natural Gas

Philip C. Crouse

2.4.1 Overview

Natural gas has been called the environmentally friendly fossil fuel because it releases fewer harmful contaminants. World production of dry natural gas was 73.7 trillion ft^3 and accounted for over 20% of world energy production. In 1990 Russia accounted for about one third of world natural gas. With about one quarter of the world's 1990 natural gas production, the second largest producer was the U.S.

According to the U.S. Department of Energy, natural gas is forecast to be the fastest growing primary energy. Consumption of natural gas is projected to nearly double between 2001 and 2025, with the most robust growth in demand expected among the developing nations. The natural gas share of total energy consumption is projected to increase from 23% in 2001 to 28% in 2025.

Natural gas traded across international borders has increased from 19% of the world's consumption in 1995 to 23% in 2001. The EIA notes that pipeline exports grew by 39% and liquefied natural gas (LNG) trade grew by 55% between 1995 and 2001. LNG has become increasingly competitive, suggesting the possibility for strong worldwide LNG growth over the next two decades. Figure 2.6 shows projections of natural gas consumption in 2025 to be five times the consumption level in 1970.

2.4.2 Reserves and Resources

Since the mid-1970s, world natural gas reserves have generally trended upward each year As of January 1, 2003, proved world natural gas reserves, as reported by *Oil & Gas Journal*, were estimated at 5501 trillion ft^3. Over 70% of the world's natural gas reserves are located in the Middle East and the EE/FSU, with Russia and Iran together accounting for about 45% of the reserves. Reserves in the rest of the world are fairly evenly distributed on a regional basis.

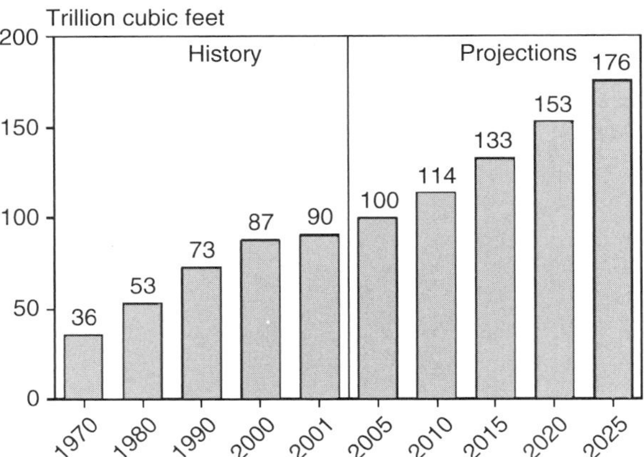

FIGURE 2.6 World natural gas consumption, 1970–2025. (History from EIA, International Energy Annual 2001, DOE/EIA-0219(2001), Washington, DC, Feb. 2003, www.eia.doe.gov/iea/. Projections from EIA, System for the analysis of Global Energy Markets (2003).)

The U.S. Geological Survey (USGS) regularly assesses the long-term production potential of worldwide petroleum resources (oil, natural gas, and natural gas liquids). According to the most recent USGS estimates, released in the *World Petroleum Assessment 2000*, the mean estimate for worldwide undiscovered gas is 4839 trillion ft^3. Outside the U.S. and Canada, the rest of the world reserves have been largely unexploited. Outside the U.S., the world has produced less than 10% of its total estimated natural gas endowment and carries more than 30% as remaining reserves. Figure 2.7 shows world natural gas reserves by region from 1975 to 2003. Table 2.10 shows natural gas reserves of the top 20 countries compared to world reserves. Russia, Iran, and Qatar account for over half of estimated world gas reserves.

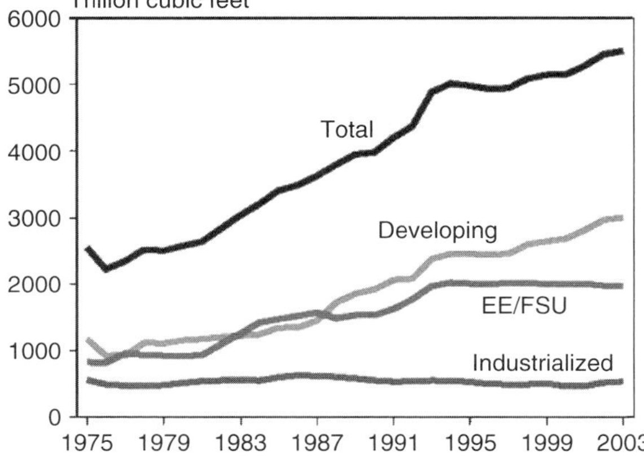

FIGURE 2.7 World natural gas reserves by region, 1975–2003. (Data for 1975–1993 from Worldwide oil and gas at a glance, *International Petroleum Encyclopedia*, Tulsa, OK: PennWell Publishing, various issues. Data for 1994–2003 from *Oil & Gas Journal*, various issues.)

TABLE 2.10 World Natural Gas Reserves by Country as of January 1, 2003

Country	Reserves (trillion ft^3)	Percent of World Total
World	5501	100.0
Top 20 countries	4879	88.7
Russia	1680	30.5
Iran	812	14.8
Qatar	509	9.2
Saudi Arabia	224	4.1
United Arab Emirates	212	3.9
U.S.	183	3.3
Algeria	160	2.9
Venezuela	148	2.7
Nigeria	124	2.3
Iraq	110	2.0
Indonesia	93	1.7
Australia	90	1.6
Norway	77	1.4
Malaysia	75	1.4
Turkmenistan	71	1.3
Uzbekistan	66	1.2
Kazakhstan	65	1.2
Netherlands	62	1.1
Canada	60	1.1
Egypt	59	1.1
Rest of World	622	11.3

Source: *Oil Gas J.*, 100 (December 23, 2002), 114–115.

2.4.3 Natural Gas Production Measurement

Natural gas production is generally measured as "dry" natural gas production. It is determined as the volume of natural gas withdrawn from a reservoir less (1) the volume returned for cycling and repressuring reservoirs; (2) the shrinkage resulting from the removal of lease condensate and plant liquids; and (3) the nonhydrocarbon gases. The parameters for measurement are 60°F and 14.73 lb standard per square inch absolute.

2.4.4 World Production of Dry Natural Gas

From 1983 to 1992, dry natural gas production grew from 54.4 to 75 trillion ft^3. The breakdown by region of world is shown in Table 2.11.

TABLE 2.11 World Dry Natural Gas Production

Country/Region	1983	1992	2000
North, Central, and South America	21.20	25.30	30.20
Western Europe	6.20	7.85	10.19
Eastern Europe and former U.S.S.R.	21.09	28.60	26.22
Middle East and Africa	2.95	6.87	12.01
Far East and Oceania	2.96	6.38	9.48
World total	54.40	75.00	88.10

Trillion ft^3.

Source: From EIA, Annual Energy Review 1993, EIA, Washington, DC, July 1994, 305, and International Energy Outlook-2003.

TABLE 2.12 Relation of API Gravity, Specific Gravity, and Weight per Gallon of Gasoline

Degree API	Specific Gravity	Weight of Gallon (lb)
8	1.014	8.448
9	1.007	8.388
10	1.000	8.328
15	0.966	8.044
20	0.934	7.778
25	0.904	7.529
30	0.876	7.296
35	0.850	7.076
40	0.825	6.870
45	0.802	6.675
50	0.780	6.490
55	0.759	6.316
58	0.747	6.216

Note: The specific gravity of crude oils ranges from about 0.75 to 1.01.

2.4.5 Compressed Natural Gas

Environmental issues have countries examining and supporting legislation to subsidize the development of cleaner vehicles that use compressed natural gas (CNG). Even with a push toward the use of CNG-burning vehicles, the numbers are quite small when compared with gasoline vehicles. Recent efforts toward car power have been focused on hybrid electric-gasoline cars and fuel cell vehicles.

2.4.6 Liquefied Natural Gas (LNG)

Natural gas can be liquefied by lowering temperature until a liquid state is achieved. It can be transported by refrigerated ships. The process of using ships and providing special-handling facilities adds significantly to the final LNG cost. LNG projects planned by a number of countries may become significant over the next 20 years, with shipments of LNG exports ultimately accounting for up to 25% of all gas exports.

2.4.7 Physical Properties of Hydrocarbons

The most important physical properties from a crude oil classification standpoint are density or specific gravity and the viscosity of liquid petroleum. Crude oil is generally lighter than water. A Baume-type scale is predominantly used by the petroleum industry and is called the API (American Petroleum Institute) gravity scale (see Table 2.12). It is related directly to specific gravity by the formula:

$$\varphi = \frac{(141.5)}{(131.5 + {}^{\circ}\mathrm{API})}$$

where $\phi =$ specific gravity. Temperature and pressure are standardized at 60°F and 1 atm pressure.

Other key physical properties involve the molecular weight of the hydrocarbon compound and the boiling point and liquid density. Table 2.13 shows a summation of these properties.

TABLE 2.13 Other Key Physical Properties of Hydrocarbons

Compound	Molecular Weight	Boiling Point at 14.7 psia in °F	Liquid Density at 14.7 psia and 60°F-lb/gal
Methane	16.04	−258.7	2.90
Ethane	30.07	−125.7	4.04
Propane	44.09	−43.7	4.233
Isobutane	58.12	10.9	4.695
n-Butane	58.12	31.1	4.872
Isopentane	72.15	82.1	5.209
n-Pentane	72.15	96.9	5.262
n-Hexane	86.17	155.7	5.536
n-Heptane	100.2	209.2	5.738
n-Octane	114.2	258.2	5.892
n-Nonane	128.3	303.4	6.017
n-Decane	142.3	345.4	6.121

Defining Terms

API gravity: A scale used by the petroleum industry for specific gravity.
Discovered resources: Include all production already out of the ground and reserves.
Proved resources: Resources that geological and engineering data demonstrate with reasonable certainty to be recoverable in future years from known reservoirs under existing economic and operating conditions.
Recoverable resources: Include discovered resources.

For Further Information

The Energy Information Agency of the U.S. Department of Energy, Washington, DC, publishes *International Energy Outlook* and other significant publications periodically.

3

Biomass Energy

Ralph P. Overend
National Renewable Energy Laboratory

Lynn L. Wright
Oak Ridge National Laboratory

3.1 Biomass Feedstock Technologies 3-1
Photosynthesis • Biomass Residue Resources • Potential Forestry Biomass Resources Worldwide • Potential Energy Crop Production • Terrestrial and Social Limitations • Biomass Facility Supply Considerations

3.2 Biomass Conversion Technologies 3-4
Physical Characteristics • Chemical Characteristics • Combustion Applications • Electric Power Generation From Biomass • Thermal Gasification Technologies • Biological Process Technologies • Liquid Fuels and Bioproducts From Lignocellulosics

References 3-16

For Further Information 3-17

Biomass energy encompasses a wide variety of energy technologies that use renewable plant matter, or phytomass, derived from photosynthesis as a feedstock to produce solid, liquid, and gaseous biofuels or used directly as an energy source producing heat and electricity. Biorefinery concepts are being developed that could result in the production of multiple energy carriers, as well as bioproducts from the same biomass resource.

3.1 Biomass Feedstock Technologies

3.1.1 Photosynthesis

Biomass fuels are derived from green plants, which capture solar energy and store it as chemical energy through the **photosynthetic** reduction of atmospheric carbon dioxide. Plant leaves are biological solar collectors while the stems, branches, and roots are the equivalent of batteries storing energy-rich complex carbon compounds. Elemental analysis shows that wood and grasses are about 50% carbon. The average photosynthetic efficiency of converting solar energy into energy stored in organic carbon compounds on an annual basis varies from less than 0.5% in temperate and tropical grasslands to about 1.5% in moist tropical forests (Cralle and Vietor 1989). Although seemingly quite low, the worldwide annual storage of photosynthetic energy in terrestrial biomass is huge, representing approximately 10 times world annual use of energy (Hall et al. 1993). This annual energy storage reflects the diversity and adaptability of terrestrial plants in many different climate zones, from the polar regions to the tropics.

3.1.2 Biomass Residue Resources

The majority of biomass energy used today is derived from residues associated with the production of timber and food crops in the field and the forest, as well as in their processing into final products. On the average, for each tonne of grain or timber harvested, there is an equivalent amount of stalk, straw, or forest residues such as branch materials. Residues from agricultural crops, such as cereal straws, are already used for bioenergy in many parts of the world and represent a large, immediately accessible resource for bioenergy in the United States. Agriculture residue recovery in the U.S. will have the most potential in high-yield cropland areas, especially those in which no-till management systems are used. Under such conditions, some portion of the crop residue may be sustainably and economically recoverable for bioenergy.

Under conventional management practices, corn stover, wheat straw, and other crop residues often have greater economic value because they are left on the land to restore nutrients, reduce erosion, and stabilize soil structure. Sustainably recoverable agricultural crop residues in the U.S. are estimated to be less than 136 Mt (Walsh et al. 2000; Gallagher et al. 2003) (all biomass amounts are on a dry basis). Crop residues worldwide are estimated to have an energy value of 12.5 EJ (Hall et al. 1993).

Recoverable wood residues from all sources in the U.S. are estimated to be approximately 94.3 Mt (McKeever 2003). This includes 67.6 Mt of logging residues; 16.4 Mt of construction and demolition wastes; and 8.7 Mt of recoverable solid wood from the postconsumer urban residue stream (or municipal solid waste stream). Primary timber mills produce about 73.7 Mt of residues; however, about 97% of this is already utilized.

A potentially large wood residue resource consists of the small-diameter growing trees and woody debris that need to be removed from U.S. forests to reduce hazardous fuel loading in forest fire mitigation. Small-diameter wood is already harvested for fuelwood in many parts of the world. The estimated forest residues, worldwide have an energy value of 13.5 EJ (Hall et al. 1993). Forest industry processing residues (e.g., sawdust or black liquor from pulping processes) are already used for energy in most cases.

3.1.3 Potential Forestry Biomass Resources Worldwide

The amount of harvestable woody biomass produced by natural forests on an annual basis ranges from about 2 to 6 t ha^{-1} y^{-1} (with the higher yields usually in tropical regions); this could be increased to 4–12 t ha^{-1} y^{-1} if brought under active management. Such management would include optimizing harvesting strategies for production, fertilization, and replanting natural stands with faster-growing species. As of 1990, 10% of world forests, or 355 Mha, were actively managed. If managed forests were increased to 20%, and if 20% of the harvested material were used for energy, the annual worldwide resource of wood for energy from currently forested land would amount to between 284 and 852 Mt of available wood, or about 5.6–17 EJ of primary energy based on potential yield ranges of managed forests. The most optimistic estimates of biomass potential would provide a total primary energy resource of about 30 EJ from managed forests (Sampson et al. 1993).

3.1.4 Potential Energy Crop Production

Significantly increasing the world's biomass energy resources will require the production of high-yield crops dedicated to conversion to bioenergy. The most environmentally beneficial dedicated crop production systems will be the production of perennial plants, using genetically superior materials, established on previously cropped land, and managed as agricultural crops. Perennials such as annually harvested grasses or short rotation trees harvested on a cycle of 3–10 years minimize soil disturbance, increase the build-up of soil carbon, provide wildlife habitat, and generally require fewer inputs of chemicals and water for a given level of production.

Energy crop yields will be highest in locations where genetically superior material is planted on land with plenty of access to sunshine and water. In the moist tropics and subtropics, as well as in irrigated desert regions, yields in the range of 20–30 t ha^{-1} y^{-1} are achievable with grasses and trees (e.g., eucalypts and tropical grasses, and hybrid poplars in the irrigated Pacific Northwest of the U.S.). Without irrigation, temperate woody crops would be expected to yield 9–13 t ha^{-1} y^{-1} of harvestable biomass, though several experimental trials have demonstrated yields of improved genotypes of poplars in the range of 12–20 t ha^{-1} y^{-1} and best yields up to 27 t ha^{-1} y^{-1} (Wright 1994). Unimproved temperate perennial grasses with no irrigation typically yield about 7–11 t ha^{-1} y^{-1}; selected grasses in experimental trials have commonly achieved yields of 11–22 t ha^{-1} y^{-1} with occasional yields of 16–27 t ha^{-1} y^{-1} in some years. Although grass yields vary from year to year, the Alamo variety of switchgrass has averaged around 23 t ha^{-1} y^{-1} (10 dry ton/acre/year) over 12 years in the southeastern U.S., where a long growing season favors warm-season grasses (McLaughlin and Kszos personal communication).

Temperate shelterbelts and tropical agroforestry plantings could also contribute greatly to biomass energy resources. Assuming a range of possible yields, the primary energy resource potential of converting 10%–15% of cropland to energy crops has been estimated to range from a low of 18 EJ to a high of 49 EJ. A further 25–100 EJ has been estimated to be available from the conversion of grasslands and degraded areas to the production of biomass energy resources worldwide (Sampson et al. 1993).

3.1.5 Terrestrial and Social Limitations

M.J.R. Cannell provides estimates of theoretical, realistic, and conservative/achievable capacity of biofuel plantations of trees or annual crops to produce primary energy and offset global carbon emissions between 2050 and 2100 (Cannell 2003). The theoretical estimate of biomass energy potential assumes 600–800 Mha, or a maximum of 55% of current cropland area (although including large areas of previous crop land that are now degraded) and average yields increasing from 10 to 25 t ha^{-1} y^{-1} providing 150–300 EJ. The realistic estimate of biomass energy potential assumes 200–400 Mha (14%–28% of current cropland area) and average yields of 10 t ha^{-1} y^{-1} (37–74 EJ); the conservative/achievable estimate assumes 50–200 Mha of current cropland and average yields of 10 t ha^{-1} y^{-1} (9–37 EJ).

Cannell's conservative estimate recognizes that developing countries continue to face food and water shortages and increasing populations, and that substantial economic and policy obstacles must be overcome. Additionally, the life cycle environmental and social impacts are not always strongly positive, depending on the crop management systems used and the conversion process employed.

3.1.6 Biomass Facility Supply Considerations

The location of large biorefinery facilities will be limited by local biomass supply availability and price considerations. Siting opportunities for facilities in the U.S. that require 500 t d^{-1} or less of biomass delivered at prices of 40$ t^{-1} or greater will be abundant, especially if dedicated crops are included in the supply mix. U.S. locations with abundant supplies of biomass delivered at 20$ t^{-1} or less will be quite limited. As facility size increases to 2000 t d^{-1} (the smallest size being considered for most biorefineries), up to 10,000 dt/day, suitable locations will become increasingly limited and/or prices will be considerably higher (Graham et al. 1995).

Facilities with the capability of processing a wide variety of feedstock types may gain a price benefit from seasonal feedstock switching (such as using corn stover in fall, woody crops in winter, and grasses in summer) and from using opportunity feedstock (such as trees blown down from storms). Facilities with a large demand for feedstocks with specific characteristics will depend upon crops genetically selected and tailored for the specific process. The higher feedstock prices would presumably be offset by the economies of scale leading to reductions in processing costs (Jenkins 1997).

Acceptance of biomass facility siting will depend in part on the land area changes required to supply a facility. For economic analysis, it is normally assumed that biomass supplies will be available within an 80 km radius (although in many situations worldwide supply is coming from longer distances, with international trading emerging in some areas). An approximate idea of the hectares and percent land area within an 80 km radius required as a function of the biomass yield that can be sustainably harvested/collected from the land is provided in Table 3.1. The lower yield is representative of levels of agricultural residues (such as corn stover and wheat straw) that could be collected in some areas.

TABLE 3.1 Impact of Facility Size and Energy Crop Productivity on Area Required

Yield t ha^{-1} y^{-1}	Plant Throughput 500 t d^{-1}		Plant Throughput 2000 t d^{-1}		Plant Throughput 10,000 t d^{-1}	
	Hectares	% Area[a]	Hectares	% Area[a]	Hectares	% Area[a]
3.3	44,535	2.2	178,141	8.8	890,705	44.3
9.0	16,330	0.8	65,318	3.3	326,592	16.3
18.0	8,165	0.3	32,659	1.6	163,296	8.13

[a] Percent area required within 80-km radius of demand center.

Dedicated energy crops will vary between 9 and 18 t ha^{-1} y^{-1} in the near future and could rise higher. Facility demands would be expected to decrease in the future as conversion technology efficiencies increase, requiring less land per unit of final product output. Table 3.2 shows the expected yield of typical secondary energy products from the same daily biomass inputs. The efficiencies are described as *present*, representing the state of the art today, and *future*, in which there are foreseeable changes in the technology.

3.2 Biomass Conversion Technologies

The energy services that can be satisfied by biomass are the supply of heat and combined heat and power, as well as the conversion of biomass into other energy forms such as electricity, liquid, and gaseous fuels identified (see Table 3.2). The transformation of the energy in the raw biomass to other forms of energy and fuels requires an acknowledgment of the physical and chemical differences between biomass resources in order to effect the most efficient transformation at the desired scale of operation. Each conversion process must take into account the differences in biomass physical characteristics on receipt at the processing facility and its proximate composition, which can vary widely in terms of moisture and ash content. Many processes require knowledge of the elemental composition and may even need information at the ultra-structural and polymeric level.

3.2.1 Physical Characteristics

Physical properties are important in the design and handling of biomass fuels and feedstocks at processing plants. In the raw state, woody biomass has a relatively low bulk density compared with fossil fuels. Bituminous coal or crude oil, for example, has a volume of 30 dm^3 GJ^{-1}, while solid wood has around 90 dm^3 GJ^{-1}. In chip form, the volume increases to 250 dm^3 GJ^{-1} for hardwood species and 350 dm^3 GJ^{-1} for coniferous species. Straw has even less energy density, ranging from 450 dm^3 GJ^{-1} for large round bales to 1–2 m^3 GJ^{-1} for chopped straw, similar to bagasse. The cell walls of biomass fibers

TABLE 3.2 Impact of Facility Size and Process Efficiency on Expected Outputs of Fuels and Electricity

Scale td^{-1}	Electricity[a]		Ethanol		Fischer–Tropsch Liquids		Hydrogen[b]	
	Capacity MW		Production kL/d		Production Bbl/d		Production t/d	
	Present	Future	Present	Future	Present	Future	Present	Future
500	30.4	45.6	165.9	186.6	453.5	538.5	6.5	7.7
2,000	121.5	182.3	663.5	746.4	1,813.8	2,153.9	25.9	30.8
10,000	607.6	911.5	3,317.5	3,732.2	9,069.0	10,769.5	129.7	154.1

[a] Efficiency: present = 30%; future = 45%.
[b] Efficiency: present = 50%; future = 60%.

have densities of about 1.5 g cm^{-3}; however, the large volume of the vascular elements that make up the fibers results in the low energy densities described previously. The fibrous nature also precludes close packing, which is the reason that chips and chopped straw have densities that are so low.

Size reduction of biomass resources is often more difficult than with minerals because the materials are naturally strong fibers, and the production of uniform particle size feedstocks is correspondingly difficult. The size reduction challenge is often mitigated in combustion and gasification on account of the high chemical reactivity of the biomass materials. However, when it is necessary to have penetration of chemicals into the structure of the biomass or to create access for biological agents as in acid and enzymatic hydrolysis of biomass, a higher degree of size reduction will be required. The energy requirements for size reduction are proportional to the surface exposure and rise in inverse proportion to the cube of the average length, so although moderate amounts of energy are required at particle sizes of several millimeters, the energy penalty at the 100-μm size is considerable.

3.2.2 Chemical Characteristics

3.2.2.1 Proximate Analysis

The value of the proximate analysis is that it identifies the fuel value of the as-received biomass material; provides an estimate of the ash handling and water removal requirements; and describes something of the characteristics in burning (Table 3.3). Generally, highly volatile fuels, such as biomass, need to have specialized combustor designs to cope with the rapid evolution of gas. Coals with very high fixed carbon need to be burnt on a grate because they take a long time to burn out if they are not pulverized to a very small size.

The energy content of biomass is always reported for dry material; however, most woody crops are harvested in a green condition with as much as 50% of their mass water. Two different heating value reporting conventions are in use. The term *higher heating value*, or HHV, refers to the energy released in combustion when the water vapor resulting from the combustion is condensed, thus releasing the latent heat of evaporation. Much of the data from North America are reported in this way. The *lower heating value*, or LHV, reports the energy released when the water vapor remains in a gaseous state. The HHV and LHV are the same for pure carbon, which only produces carbon dioxide when burned. For hydrogen, one molecule of water is produced for each molecule of hydrogen, and the HHV is 18.3% greater than the LHV. The HHV is 11.1% greater than the LHV for the combustion of one molecule of methane (CH_4).

3.2.2.2 Ultimate Analysis

The elemental composition of the fuel biomass is usually reported on a totally dry basis and includes the ash. Table 3.4 gives the ultimate analysis for several different fuels.

3.2.2.3 Polymeric Composition

Biomass conversion technologies that depend on the action of microorganisms are very sensitive to the ultrastructure and polymeric composition of biomass feedstocks. In ethanol production and in anaerobic digestion, a complex sequence of biochemical events take place in which the polymers are broken down into smaller units and hydrolysed to give simple sugars, alcohols, and acids that can be processed into fuel

TABLE 3.3 Proximate Analysis of Solid Fuels

Fuel	Ash Content (%)	Moisture (%)	Volatiles (%)	Heating Value (HHV GJ Mg^{-1})[a]
Anthracite coal	7.83	2.80	1.3	30.90
Bituminous coal	2.72	2.18	33.4	34.50
Sub-bituminous	3.71	18.41	44.3	21.24
Softwood	1.00	20.00	85.0	18.60

[a] On a moisture and ash free basis (maf).

TABLE 3.4 Ultimate Analysis Data for Biomass and Selected Solid Fuels

Material	C	H	N	S	O	Ash	HHV (GJ Mg^{-1})
Bituminous coal	75.5	5.0	1.20	3.10	4.90	10.3	31.67
Sub bituminous coal	77.9	6.0	1.50	0.60	9.90	4.1	32.87
Charcoal	80.3	3.1	0.20	0.00	11.30	3.4	31.02
Douglas fir	52.3	6.3	0.10	0.00	40.50	0.8	21.00
Douglas fir bark	56.2	5.9	0.00	0.00	36.70	1.2	22.00
Eucalyptus grandis	48.3	5.9	0.15	0.01	45.13	0.4	19.35
Beech	51.6	6.3	0.00	0.00	41.50	0.6	20.30
Sugar cane bagasse	44.8	5.4	0.40	0.01	39.60	9.8	17.33
Wheat straw	43.2	5.0	0.60	0.10	39.40	11.4	17.51
Poplar	51.6	6.3	0.00	0.00	41.50	0.6	20.70
Rice hulls	38.5	5.7	0.50	0.00	39.80	15.5	15.30
Rice straw	39.2	5.1	0.60	0.10	35.80	19.2	15.80

Dry basis, weight percent.

molecules. Wood, grasses, and straws are collectively called lignocellulosics and consist mainly of fibers composed of lignin, cellulose, and hemicellulose. Cellulose, hemicellulose, and lignin are different carbon–hydrogen–oxygen polymers with differing energy contents and chemical reactivities.

Lignocellulosics typically contain cellulose (40%–50%) and hemicellulose (25%–30%) along with lignin (20%–30%), some extractives, and inorganic materials. Cellulose is depolymerized and hydrolyzed to glucose, a six-carbon sugar (C-6), while hemicellulose is a complex mixture of mainly five-carbon sugar (C-5) precursors with xylose, a major product. Lignin is a complex polymer based on phenyl propane monomeric units. Each C_9 monomer is usually substituted with methoxy groups on the aromatic ring, occurring mainly in syringyl (dimethoxy) and guaiacyl (monomethoxy) forms for which the ratio is a signature characteristic of grasses, hardwoods, and softwoods.

Other major plant components containing polymers of carbon, hydrogen, and oxygen that are also used for energy include:

- Starches that are the major part of the cereal grains, as well as the starch from tubers such as mannioc and potatoes
- Lipids produced by oil seed-bearing plants such as soya, rape, or palms, which are possible diesel fuel substitutes when they are esterified with simple alcohols such as methanol and ethanol
- Simple sugars produced by sugar beet in temperate climates or sugar cane in the tropics, which can be directly fermented to alcohols.

In addition to the carbon, hydrogen, and oxygen polymers, there are more complex polymers such as proteins (which can contain sulfur in addition to nitrogen), extractives, and inorganic materials. The inorganic materials range from anions such as chlorine, sulphate, and nitrates, and cations such as potassium, sodium, calcium, and magnesium as major constituents. Many trace elements, including manganese and iron, are the metallic elements in key enzyme pathways involved in cell wall construction.

3.2.3 Combustion Applications

More than 95% of the worldwide use of biomass is through its combustion. The majority is used in developing countries for use in *daily living*, to provide heat for cooking and space heating. In the industrialized countries, a parallel in the cold regions of the northern hemisphere is the *community applications* of district heating. *Industrial applications*—especially of CHP (combined heat and power)—are important in industrialized countries as well as in the agricultural processing sector of developing countries. In several countries, biomass is burned directly to produce electricity without the coproduction of heat; however, like coal and oil-fired steam electricity generation, this is relatively inefficient.

3.2.3.1 Daily Living

Half of all solid biomass combustion worldwide goes into cooking and space heating at the household level, serving the primary fuel needs of over 2 billion people. The typical final energy demand for cooking is around 12 MJ a day for a household in Southeast Asia. This demand can be met with around 20 MJ d^{-1} of primary energy input using LPG (propane or butane) or kerosene with a typical conversion efficiency of 60%–75%. Biomass efficiencies are much lower, ranging between 10 and 20%; thus, the required primary energy input is between 60 and 120 MJ d^{-1}. The three-stone stove has been extensively studied and has a peak power rating of 5 kW in the startup and a simmer rate of around 2 kW. The measured efficiency with a skilled operator is 10%–15%. Daily household biomass consumption is therefore in the region of 2–4 kg d^{-1} or about 0.5–1.0 t y^{-1}.

The low efficiency of use is mainly due to poor combustion efficiency and inadequate heat transfer to the cooking vessels. The poor combustion results in the loss of fuel as solid particulate (soot), unburnt hydrocarbons and carbon monoxide; these are often referred to as the products of incomplete combustion (PIC). The heat transfer problem is related to the poor contact between the hot gases from the burner and the pot, and its fouling with PIC (Prasad 1985). The PICs are not only a loss of fuel value, but are also toxic chemicals to which human exposure should be minimized.

Carbon monoxide is a chemical asphyxiant, and the aldehydes, phenols, and other partial combustion products are skin, eye, and lung irritants; some, such as benzopyrenes, are even carcinogens. Because many stoves are used in enclosed spaces and do not have chimneys, the household is exposed to these hazards; eye and lung damage is experienced especially by women and children, who work closely with the stoves while cooking. The epidemiological data on this are extensive (Ezzati and Kammen 2001), and large efforts to improve the efficiency and minimize the emissions in cookstoves are underway. The health impacts are of such a magnitude that subsidized alternative fuels such as LPG could be a significant social benefit while reducing the impacts of biomass harvesting.

3.2.3.2 Space Heating

Space heating with wood in the form of logs is very common in forested areas of the northern hemisphere and Latin America with wood stoves supplying a household heat requirement of 30–100 kW. The thermal efficiencies are about 60%–70%. Log burning is controlled by air starvation to reduce power output and may result in severe emissions of PIC, causing poor air quality in regions with high densities of wood-fueled space heating. Residential wood stoves using advanced combustion system designs or catalysts reduce PIC. Legislation has mandated emissions testing of new appliances before sale to ensure that they meet new source performance standards (NSPS) in the U.S. The use of dry prepared fuels such as pellets (manufactured by drying, grinding, and compression into a pellet about 1–2 cm in diameter) allows combustion to be efficiently managed so that PIC are extremely low with 75%–80% thermal efficiency.

3.2.3.3 Community Systems

Biomass use in community district heating (DH) systems results in improved thermal efficiency with reduced emissions because of the scale of the equipment. Europe has seen large-scale adoption of biomass-fired district heating. There are over 300 round wood-fired DH schemes in Austria with an average heating effect of 1 MW or less, serving villages of 500–3000 inhabitants with minigrids using hot water distribution. Using straw, 54 Danish systems cover larger population centers and range in size from 600 kW to 9 MW. Larger scale DH systems can be CHP units producing process steam for industries, heating for households, and electricity for the grid with high-quality emissions control systems. In Finland, even larger units are serving cities, e.g., the Forssa wood-fired CHP–DH plant is rated at 17.2 MW_e and 48 MW_{th} district heating, generating steam at 6.6 MPa and 510°C, with a bubbling fluidized bed boiler. The seasonal efficiency, which includes a condensing steam cycle in the summer, is 78%.

3.2.3.4 Industrial Applications—Combined Heat and Power (CHP) or Cogeneration

Many industrial processes, such as pulp production or cane sugar manufacture, have large heat requirements that are satisfied by generating low pressure steam at 1.7 MPa and 250° using their own process residues such as hog fuel (a mix of wood and bark); black liquor (the lignin by-product of pulp manufacture in kraft pulp mills); or bagasse (the fibrous residue after the pressing of sugar cane). A proportion of the steam is used in direct drives for the process plant's needs. Often, because the primary goal has been to eliminate residue disposal, the systems have not operated at maximum efficiency.

In industries such a pulp and paper (P&P), the trend is already toward requiring higher power–heat (P/H) ratios because steam demand is trending down due to energy conservation practices and electricity demand is rising to supply the increasing environmental and process needs. The thermal demand reduction is a consequence of improvements in drying and evaporation technologies, which have decreased from 8.6 to only 4 GJ adt^{-1} (adt=air dry tonne of pulp). The average pulp mill energy consumption has gone from 17 GJ adt^{-1} in the 1980s to 7–8 GJ adt^{-1} in the 1990s.

The application of advanced power generation cycles in the pulp and paper industry using black liquor and wood residues as feedstocks would give large increases in electricity export from that industry. Similar energy conservation trends in the sugar industry can reduce the requirement of steam to process cane from 58.5% (tonne steam/tonne of cane processed) to less than 35%. If investments in energy conservation and adoption of integrated gasification combined cycle (IGCC) technology were applied worldwide, the annual amount of power that could be produced from the large sugar cane resource would be >650 TWh based on the current >1.0 Gt of cane. Current cane generation is estimated to be 30–50 TWh y^{-1} because only sufficient electricity is generated in most instances to power the mill in season.

3.2.4 Electric Power Generation From Biomass

3.2.4.1 Prime Mover Systems and Fuels

Power generation takes place in *prime movers*, a technical term to describe engines of all kinds attached to alternators to generate electricity. Prime movers include the:

- Steam engine
- Steam turbine engine (the Rankine cycle)
- Internal combustion engine (ICE), which comes in two types: the Otto or gasoline-based spark ignition (SI) engine, and the diesel or compression ignition (CI) engine
- Gas turbine engine (Brayton cycle)
- Stirling engine.

Each of these requires that the raw biomass be processed to some level and then used in the prime mover. Eventually fuel cells will replace the prime mover and alternator requirement by generating electricity directly from biomass-derived hydrogen fuels.

The steam cycle (already discussed) uses combustor and boiler combinations to generate steam, requiring that the fuel be reduced in size (perhaps dried to some level) and have physical contaminants removed. The high-pressure steam is then expanded through a steam engine at small scales or through a turbine at larger scales. The efficiency of conversion technologies for combustion steam boiler systems is very scale dependent—a small steam engine or turbine would not exceed 10%. However, typically sufficient biomass is available over a transportation radius of 10–80 km to operate a unit in the range of 10–50 MW electricity output. Current Rankine cycles, at that scale, operate in the range of 25%–30% efficiency and, as a consequence, require approximately 0.75–1 t of dry biomass to produce 1 MWh of electricity.

Industrial and power generation boilers range in size from 100 to 300 MW thermal output. The major types of boilers are: pile burners; grate boilers; suspension fired boilers; fluidized beds; and circulating fluid beds. Recent trends in power generation are to use circulating and bubbling fluidized bed combustors, although the majority of units in current service are stoker-fired moving grate units.

Biomass resources can be used in ICEs and gas turbine systems only if they are converted into clean liquid or gaseous fuels. Ethanol, biodiesel from lipids, or Fischer–Tropsch liquids can be used with little alteration to SI or CI engines. Gaseous fuels include the mixture of methane and carbon dioxide (biogas) produced by the action of microorganisms on biomass in anaerobic digestion (AD). AD is conducted at an industrial scale, using sewage or effluent streams containing high levels of soluble sugars, alcohols, and acids, as well as in landfills.

The installed landfill power generation capacity in the U.S. is now in excess of 1 GW (Goldstein 2002). The energy content of biogas is typically 20–25 MJ Nm^{-3} or between 50 and 60% that of natural gas. Fuel gases can also be produced by thermal gasification; when this is carried out at small scales, the gasifying agent is usually air. The product gas, which has a low calorific value (LCV) with a heating value of 12%–15% that of natural gas, is often called a producer gas. LCV gas at about 5–6 MJ Nm^{-3} has carbon monoxide (CO), hydrogen (H_2), and methane (CH_4) as the main fuel components, diluted with a lot of nitrogen and carbon dioxide. Larger scale processes can utilize pure oxygen, enriched air, or an indirect heating method to produce medium calorific value (MCV) gas in the range of 15–25 MJ Nm^{-3} heating value with essentially the same fuel gas components but with much less inert diluent. Clean MCV gases can be burnt without much modification of ICEs or gas turbines.

3.2.4.2 Cofiring Biomass With Coal

In a biomass/coal cofiring operation, a biomass fuel is used to replace a portion of the coal fed to an existing coal-fired boiler. This has been practiced, tested, or evaluated for a variety of boiler technologies. There are four types of coal boilers: pulverized coal (PC); cyclone; stoker; and fluidized bed. PC boilers are the most common type in the U. S., representing about 92% of the U.S. coal-generating capacity, with cyclone boilers as the next most common, with about 7% representation. Demonstrations have been undertaken at coal plants ranging in size from about 30 MW_e through to 700 MW_e. Wood residues, hybrid poplar, and switchgrass have all been tested as supplemental fuels, and several utilities have biomass cofiring plants in commercial operation.

Solid biomass can be fed to coal boilers by blending biomass on the coal pile or separately injecting biomass into the boiler. Introducing blended feeds to PC boilers can cause operational problems with the pulverizer, so the biomass proportion is limited to no more than 2%–3% by heat input (4%–5% by mass). Separate injection allows for the introduction of higher biomass percentages to the PC boiler—typically up to 15% on a heat input basis (about 30% by mass).

However, separate injection requires additional fuel handling equipment and increased fuel preparation. Capital costs for the blended feed approach are typically \$50/100 kW^{-1}. For the separate feed approach, capital costs are typically higher, in the range of \$175/200 kW^{-1}. Cofiring can reduce boiler efficiency to some degree. For example, cofiring in a PC boiler with 10% of the heat input from wood may decrease the boiler efficiency by 0.5%–1.5%; after "tuning" the boiler's combustion output and adjusting for any efficiency losses, the combustion efficiency to electricity would be approximately 33%.

Because coal plants comprise more than half of U.S. power plant capacity currently in operation, cofiring technology has the advantage of improving environmental performance at existing power plants, while providing fuel source flexibility and using proven and familiar equipment. It reduces air pollution emissions, GHG emissions, and the amount of waste ash generated as a by-product of the combustion. In addition, it requires relatively low up-front capital expenses compared to other renewable energy options; this makes it a straightforward and inexpensive way to diversify the fuel supply and divert biomass from landfill disposal.

3.2.5 Thermal Gasification Technologies

The conversion of biomass into a gaseous fuel opens up modern applications in electric power generation, the manufacture of liquid fuels, and the production of chemicals from biomass. The chemistry of gasification of biomass is best viewed as an extension of the pyrolysis process. Pyrolysis is simply defined as the chemical changes occurring in the solid biomass when heat is applied to a material in the absence of oxygen. The products of biomass pyrolysis include water, charcoal (or more correctly *char*, a carbonaceous solid), oils or tars, and permanent gases including methane, hydrogen, carbon monoxide, and carbon dioxide.

The majority of gasifiers are partial oxidation reactors, in which just sufficient air or oxygen is introduced to burn part of the input biomass to provide the heat for pyrolysis and gasification. If the oxidant is air, the product gas is diluted by the nitrogen present. Although air is 79% nitrogen, the stoichiometry of partial oxidation is such that the final LCV product, gas, has about 50% nitrogen as a diluent. The energy content of the typical gases produced in biomass gasification is shown in Table 3.5. The use of pure oxygen as the gasification agent eliminates the nitrogen diluent and can produce medium calorific value (MCV) gases in the range of 10–20 MJ Nm^{-3}.

An alternative strategy is to carry out the gasification process by means of indirect heat the product stream is even higher in calorific value, because neither nitrogen nor the carbon dioxide produced from the combustion in situ of the partial oxidation processes is present in the product gas stream. The challenges to achieve a clean and useable fuel gas have been addressed through gasifier design and postgasification processing to remove tar and particulate contaminants from the gas stream.

3.2.5.1 Gasifier Systems

The main challenge in gasification is enabling the pyrolysis and gas-reforming reactions to take place, using the minimum amount of energy, in reactors that are economical to construct. During a history dating back to the late 18th century, an extraordinary number of different designs and process configurations have been proposed. Prior to the development of fluidized bed technologies in the 1920s, the majority of the gasifiers were so-called fixed bed units. The flow of gasifying agents, usually air and steam, could be cocurrent with the biomass feed, or countercurrent; these are often described, respectively, as downdraft and updraft gasifiers.

Downdraft gasification was widely used during the Second World War as an on-board fuel gas generator to offset the lack of gasoline. Millions of units were constructed and then abandoned as soon as petroleum supplies resumed. Units derived from the automotive application are marketed today as stationary generating sets equipped with ICEs, with SI or CI for power production in remote locations and in developing countries without grid systems.

The simplest and oldest gasifier is the counterflow moving bed, which consists of an insulated shaft into which the feedstock (typically pieces larger than 3 cm on a side) are fed. The shaft is filled to a depth of 0.5–2 times the vessel diameter, and the mass of material is supported on a grate. The oxidant (air or enriched air/oxygen) and sometimes steam are introduced below the grate. The grate material is ignited and the hot gases flow up through the bed, exchanging heat with the down-flowing biomass material and, at the same time, pyrolyzing and drying it. At steady state, the bed level is maintained by continuous additions of the feed.

TABLE 3.5 Heating Values of Fuel Gases

	HHV[a] Btu ft^{-3} [c]	LHV[b] Btu ft^{-3} [c]	HHV[a] MJ Nm^{-3} [d]	LHV[b] MJ Nm^{-3} [d]
Hydrogen	325	275	12.75	10.79
Carbon monoxide	322	322	12.63	12.63
Methane	1013	913	39.74	35.81
Ethane	1792	1641	69.63	63.74
Propane	2590	2385	99.02	91.16
Butane	3370	3113	128.39	118.56

Note: Conversion factors for 1 MJ Nm^{-3} at 273.15 K and 101.325 kPa. 25.45 Btu ft^{-3} at 60°F and 14.73 psia. Inverse 1 Btu ft^{-3} at 60°F and 30 in. Hg. 0.0393 MJ Nm^{-3}.

[a] Higher heating value.

[b] Lower heating value.

[c] Standard temperatures and pressure of dry gas are 60°F and 14.73 psia (NIST 2004).

[d] S.I. units.

The product gases include the desired fuel gases (methane, carbon monoxide, hydrogen, and C_2 hydrocarbons), nitrogen, carbon dioxide, and water vapor, which exit at relatively low temperatures (<250°C). The water arises from the moisture content of the feed and the products of combustion. The gas also contains several percent of condensable tars. Because the temperatures in the combustion zone can be very high, problems with the formation of slags (molten mixtures that can impede the oxidant flow) may occur. Some of the systems are constructed deliberately to form a molten ash or slag, and this is handled by dropping the slag into a receiving tank, where it solidifies and shatters to an easily handled aggregate.

The throughput of air-based systems is relatively low, handling about 150 kg m^{-2} h^{-1} of dry biomass. This operational limitation is due to the need to reduce the rate of heat generation at the grate to avoid slagging in dry systems. The lower throughput rate also helps to manage the problems of maintaining a uniform flow of the descending material to avoid channeling and even greater reductions in throughput. The largest units are typically 3–4 m in diameter and handle between 50 and 100 t d^{-1} of material with a rating of 10–20 MW_{th} input. The hot gas efficiency can be as high as 80% conversion of the input energy into a fuel gas and sensible heat. The cold gas efficiency will be about 70%.

If the gas is maintained at temperature and then fired in a kiln or boiler, the presence of tars and particulate are of little consequence because the tars are combusted in the excess air and the particulate is removed in the kiln exhaust or boiler flue gas system. Such close-coupled systems are quite common applications of biomass gasification in several industries and the basis of district heating systems used in Finland. The gas typically has a heating value of 6 MJ Nm^{-3} on a dry basis and is richer in methane, ethene, and ethyne than most other gasifier systems using air. For engine applications the gas must be cleaned up and cooled to ambient temperatures. Typically, this is done with some sort of aqueous quench and scrubbing system (e.g., a wet electrostatic precipitator) to remove tar and particulate matter, and the tar/water removed must be separated and then treated for release to the environment.

Fluidized beds can remove the grate limitation of the moving bed gasifier because, due to their high mixing rates and outstanding heat transfer rates, they can achieve through-puts of over 1500 kg m^{-2} h^{-1} of dry biomass. The two modes of fluidized bed operation involve bubbling (BFB) and circulating (CFB) solids. In a BFB, the inert fluid bed medium (silica sand, dolomite, alumina, and olivine) is retained in the body of the bed, which is expanded and in motion (literally bubbling) due to the flow of the fluidizing gas medium into the bed. The terminal velocity of a particle in the flow is still higher than the linear flow rate of the gas and thus remains in the body of the unit. In the CFB, the gas flow is increased to the point that the terminal velocity of the particle is exceeded and it is transferred out of the expanded bed and captured in a cyclone for return to cycle.

The effective rating of a 3-m diameter unit is over 50 MW_{th} input of biomass. Because the combustion process and the pyrolysis process are now mixed, the exit gas temperatures are typically 700°C–800°C. Due to cracking processes, the tar production is lower and mainly composed of secondary and tertiary tar components; however, the tolerance of high moisture content feedstocks is much more limited than the countercurrent moving bed unit. The system design is also more complex, requiring blowers to inject the oxidant at the base of the fluidized bed. In addition to the expansion of the gas due to temperature, the gas volume increases due to the formation of the fuel gas. This necessitates careful design of the CFB and the freeboard in a BFB; in a CFB, this is accommodated by the fact that the entire bed is in circulation. The oxidant requirement is typically 0.3 kg O_2 kg^{-1} of dry feedstock.

The highest reactivity gasification systems are entrained flow units in which the feedstock is finely divided and burnt in a substoichometric flame at high temperatures. The postcombustion gases are then allowed to reach chemical equilibrium before being quenched. Such units produce very few higher hydrocarbons and tar materials. The challenge for biomass is to obtain a sufficient rate of fuel injection to the flame. One configuration is the Koppers–Totzek gasifier. For fossil fuels, the Texaco gasifier uses a dense slurry of coal or petroleum coke in water as the feed to a pressurized oxygen flame, producing a hydrogen and carbon monoxide mixture for use as a syngas or as a fuel in gas turbines.

Cocurrent moving bed gasifiers (also known as downdraft and cross-draft units) were widespread during World War II (various estimates put the number at over 1 million). Like the other gasifiers described earlier, the design is a partial oxidation system. The physical arrangement is a shaft reactor into which the biomass is introduced at the top; the material then flows down to an oxidation zone, where the air is admitted. This zone feeds heat back to the incoming biomass to pyrolyze the material into gases, tars, and charcoal. The charcoal is partially combusted in the oxidation zone, and the products of pyrolysis flow through the incandescent combustion zone to be cracked into, primarily, hydrogen and carbon monoxide as the fuel gases, with few or no tars passing through. The combusted gases flow on through the hot charcoal, reacting with it, until the rate of reaction effectively goes to zero because the reactions are very endothermic.

The net result is a gas at about 500°C that is very low in condensible tars; other than carbon dioxide, water vapor, and nitrogen, the fuel gases are mainly hydrogen and carbon monoxide with only a small amount of methane and higher hydrocarbons. The low tar production is the reason this was the favored system for on-road vehicles in the 1940s. Like all moving bed reactors, the operation depends on the presence of an adequate void space between particles. The wood feedstock must produce a char that has sufficient strength to support the column of feedstock without crushing and blocking the flow.

The overall efficiency is lower than the countercurrent system for two reasons. Chemically, it is not possible to convert the charcoal fully; the sensible heat from the combustion zone is not transferred to the incoming feed material, but rather must be heat exchanged with air oxidant prior to being quenched. Because of the need for uniformity of mass and heat transfer in the high-temperature reaction zone, the diameter of such units is generally limited to less than 1.5 m and, with typical throughputs of 300 kg m^{-2} h^{-1} of dry biomass, the size of units ranges from 100 kW_{th} to 2 MW_{th} input. The chemistry of this gasifier has a lot in common with iron blast furnaces and is significantly different from the other gasifiers, which have very high burdens of liquids such as tars in the product gas.

Medium calorific value (MCV) gas can be produced without using oxygen or air in gasification. This requires an independent source of heat to be transferred to the biomass to pyrolize the feedstock under conditions of high severity, i.e., long residence times at high temperatures. To this end, proposals have been made to use concentrated solar energy or high-temperature nuclear reactors as the heat source in the gasification of a wide range of feedstocks. In the case of biomass, the pyrolysis gasification process will provide a char stream with slow or fast pyrolysis to temperatures of 600°C–750°C, in amounts representing between 12 and 25% of the input biomass. This contains sufficient energy to drive the pyrolysis process.

A number of developments in which biomass is pyrolized using a solid heat carrier in an inert atmosphere or steam or product MCV gas to generate the MCV fuel gas and char are already in the demonstration phase. The char and the cooled heat carrier, which can be silica sand, alumina, or a mineral such as olivine, are then transferred to a separate air combustion unit in which the char is combusted to heat up the heat carrier, which is then returned to the pyrolysis gasification unit. Mechanical separation and isolation devices in the heat carrier circulation loops prevent the passage of fuel gas from the gasifier to the oxidation conditions of the combustor. The transportation mechanism is via fluidization of the heat carrier.

To this end, a group in Austria constructed a version with a CFB combustion unit and BFB pyrolysis gasifier; in the U.S., units have been constructed with a CFB pyrolysis reactor and a BFB combustor as well as a dual CFB configuration scaled up to 60 MW_{th} biomass input. Another variant, developed by Manufacturing and Technology Conversion International, Inc. (MTCI), a U.S. company, was a BFB pyrolysis and steam char reaction system with the heat supply from immersed heat exchanger bundles in the bed. The heat exchanger bundles are fueled with part of the product gas, and the char from pyrolysis is consumed, as in the cocurrent gasifier.

3.2.5.2 Commercialization

Use of gasifiers as precombustors in which the hot gases are passed through a high-temperature duct into a boiler or industrial process kiln has been widespread; cofiring is an example. Foster Wheeler Energia Oy CFB has been supplying a LCV gas for several years to an existing large-scale utility boiler at the Kymijärvi 167 MW_e and 240 MW_{th} fossil-fired plant close to Lahti, Finland. A combination of forestry residues, industrial wood residues, and recycled fuel (a clean fraction of the municipal waste stream composed mainly of wood, paper, and cardboard separated at the source) is the fuel source. This project builds on many years of successful operation of biomass CFBs in thermal applications and substitutes for about 15% of the total fuel used in the boiler.

Applications in which clean gases (free of particulate and tar) are utilized occur mainly in the development and demonstration stages, with the exception of deployment of cocurrent gasifier–ICE generator sets. In 2000, Sydkraft A.B. successfully completed a high-pressure LCV gas biomass integrated gasification and combined cycle (IGCC) operation, which was part of the Värnamo (Sweden) district heating system. This first complete biomass-fueled pressurized CFB IGCC was fueled with about 18 MW_t equivalent of wood residues and produced about 6 MW_e (4 from the gas turbine, 2 from the steam cycle) and 9 MW_{th}. The LCV gas is cooled to 350°C–400°C before it is cleaned in metal filters and then passed to the Typhoon gas turbine, which is manufactured by Alstom. The project demonstrated more than 1500 h of IGCC operation on the product gas prior to its shutdown at the end of the demonstration phase (Stahl et al. 2000).

Small-scale gasifiers for the production of fuel gas for cooking, as well as the demonstration of electricity in ICEs, are already commercial in India and China. Their introduction has been accelerated by the use of moderate government subsidies, especially for communities not served by the electricity grid. In development circles, such systems are often referred to as village power systems. Considerable discussion among the OECD nations concerns the use of distributed energy resources, which would provide local power and heat (e.g., CHP units) while remaining connected to the grid. Many of these units are based on natural gas; however, when the economics and availability of biomass are appropriate, gasification is considered; demonstrations of gasfier–ICEs with CHP applications in district heating have taken place in Denmark.

3.2.6 Biological Process Technologies

The polymers that make up biomass are produced through photosynthetically driven biological reactions at the cellular level in the plant. The resulting phytomass is then fuel and food (or feedstock) for a large number of the fauna on Earth. From simple bacteria to large mammals, the processes to digest and metabolize plant polymers have evolved over billions of years. In a number of instances, the biological degradation processes have been harnessed in industrial conversion processes to produce biofuels and bioproducts to serve humanity. Of the many processes, only two have attained large-scale use: anaerobic digestion and ethanol production.

3.2.6.1 Anaerobic Digestion

Anaerobic microorganisms convert the biodegradable fraction of organic materials to carbon dioxide and methane in the absence of oxygen. Commercial-scale anaerobic digestion has been used in many parts of the world for many years and is only beginning to be commercial in the U.S. Several utilities are currently producing "green energy" with biogas from landfills. Biogas from anaerobic digesters on dairy and hog farms and cattle feed lots is beginning to contribute to the energy mix in the U.S.

3.2.6.2 Ethanol Production

Ethanol is currently the most important liquid fuel produced from biomass. It is produced by fermentation of sugars. The fermentable sugars can be obtained directly from sugar cane and sugar beet, or they can be produced by means of the hydrolysis of starches or cellulose and hemicelluloses. Starch crops include cereals, and manioc in tropical countries, while cellulose and hemicellulose are the major polymers of wood, grasses, straws, and stalks. The production from sugar cane in Brazil has been in the range of 11–13 GL y^{-1} for the last few years; in the U.S., the annual production has grown rapidly from 5.42 GL in 2000 to 8.08 GL in 2002. This growth is primarily due to legislation banning the use of methyl tertiary-butyl ether as an octane enhancer in several state markets.

Over 90% of the U.S. production is from corn, which is processed in wet or dry mills. Corn kernels have about 70%–75% starch content; approximately 10% protein (gluten); 4%–5% germ (the source of corn oil); and 3%–4% fiber content. Dry milling consists of grinding the whole kernel into a flour and then slurrying this with enzymes to break down the starch to fermentable sugars. The stillage separated from the beer is composed of coarse grain and a wide range of soluble materials. The solubles are concentrated by evaporation into a syrup (condensed distillers solubles), which is then added to the coarse grain to produce dried distillers grain with solubles or DDGS. This is marketed as a feed for animals. Dry milling theoretically produces about 460 L of ethanol per tonne of corn.

Wet milling is an example of a biomass refinery. The corn kernels are steeped in dilute sulfurous acid for 48 h in a mild prehydroylsis process prior to a grinding step to separate the corn germ. Corn oil is recovered from the germ as high-value cooking oil. The starch and the gluten protein are then separated. The gluten is dried and sold as a 60% protein feed to mainly poultry markets. The starch fraction is the source of dried or modified cornstarches and a syrup converted enzymatically into high-fructose corn syrup or passed to a fermentation step producing ethanol and other fermentation-derived chemicals.

Yields of ethanol are somewhat lower than those from dry milling at around 440 L t^{-1} of corn. According to the National Corn Growers Association (www.ncga.com), corn utilization rates of the 246.25 Mt, 2002 harvest (9.695 billion bushel) were: 58% as feed; 19% as export grain; 6% as high-fructose corn syrup; 9% as fuel ethanol; and 8% from all other applications (including breakfast cereal, cornstarches, and seed corn).

Since the inception of the corn-to-ethanol process technology in the 1970s, continuous improvement has occurred in terms of yield, process efficiency, reduced capital investment, and nonfeed operating costs. The industry average yield has gone from 360 L t^{-1} (2.4 USgal/bu) to 410 L t^{-1} (2.75 USgal/bu), while the capital cost has declined from 220 to about 115\$ L^{-1} d^{-1} capacity (2.5 down to 1.3 USD per annual gallon of capacity), according to the industry.

3.2.7 Liquid Fuels and Bioproducts From Lignocellulosics

Lignocellulosics (such as wood, straw, and grasses) contain cellulose (40%–50%) and hemicellulose (25%–30%). There is considerable ethanol potential through fermentation of the sugars derived from these polymers, producing similar theoretical yields per tonne as corn, i.e., 450 L t^{-1} and a price structure, which is more stable relative to the fuel markets than foodstuffs. Alternative processes would include thermochemical routes through the production of synthesis gas and its catalytic synthesis to mixed alcohols, Fischer–Tropsch liquids, or hydrogen—a future transportation fuel (Table 3.2).

Current RD&D in the U.S. is focused on the development of two product platforms: a sugar platform that inherits much of the prior research into the production of ethanol from lignocellulosic biomass, and a syngas platform evolved from the development of biomass gasifiers primarily for power production coupled with the extensive downstream catalysis developments of the petrochemical industry. The syngas platform is also the basis for the combustion units for the utilities component of the platforms, which are shown diagramatically in Figure 3.1.

The sugar platform inherits over two decades of work on the ethanol conversion process from lignocellulosics (Wyman 1994). The polymers involved are much more recalcitrant than starches and, because of the complex nature of lignocellulosics, they require extensive effort to break down the lignin, cellulose, and hemicellulose structure so that the individual polymers become available for hydrolysis. Cellulose is hydrolyzed to glucose, a six-carbon sugar (C-6), while hemicellulose is a complex mixture of mainly five-carbon sugar (C-5) precursors, with xylose as a major product. Although the C-6 sugars are relatively easy to ferment with yeasts, such as *Saccharomyces* spp., the C-5 sugars have not been as easy to ferment to ethanol (Zhang et al. 1995). The pretreatment stages can include steam, acid, and alkali treatments, while the hydrolysis steps can be carried out with acids or enzymes. Because of inhibition of the enzymatic hydrolysis by the sugars produced, the National Renewable Energy Laboratory (NREL) developed a simultaneous saccharification and fermentation process to remove the sugars as they are

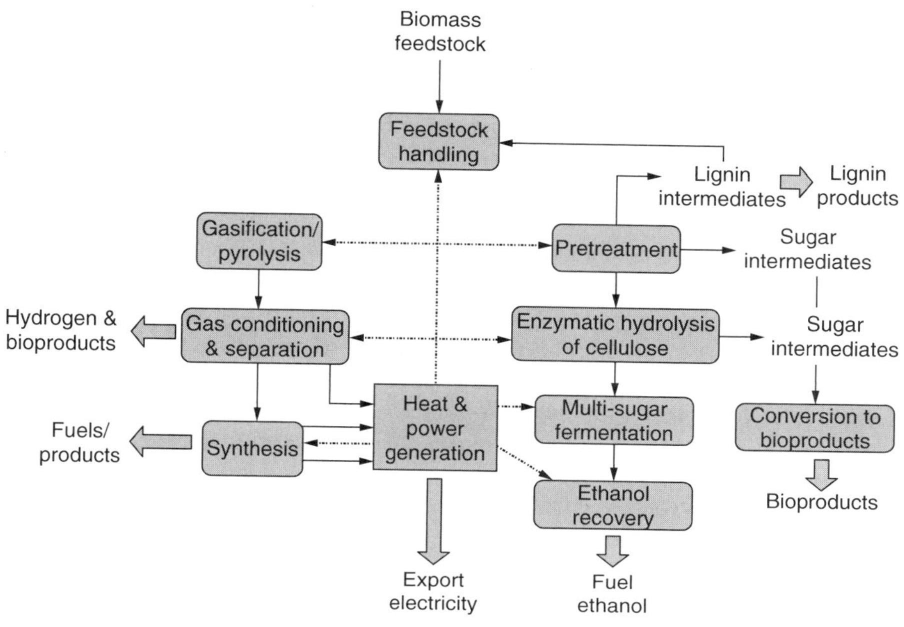

FIGURE 3.1 The U.S. RD&D approach to the sugar and syngas platforms.

formed by producing ethanol in the same reactor. Such process integration will be the key to producing low-cost sugars and thus ethanol or other bioproducts in the future (Lynd et al. 1999).

The syngas platform using biomass would enable the production of renewable liquid fuels and chemicals such as ammonia and hydrogen by well-established and proven syngas technologies. Today these processes utilize the syngas produced by the steam reforming of natural gas or use syngas from coal gasification. The production of liquid fuels from syngas has a long history dating from the pioneering work of Fischer and Tropsch to synthesize hydrocarbon fuels in Germany in the 1920s. The process was implemented in Germany in the 1930s where nine plants operated during World War II on syngas from coal gasification; after the war, until the mid 1950s, they operated on syngas from the gasification of heavy oil. In the 1950s, South Africa constructed the first of the South African Synthetic Oil Limited (SASOL) plants, an activity that has continued its development to the present day using syngas produced from coal and natural gas. Both of these developments were initiated during times of petroleum shortages, to fuel the transportation systems of Germany and South Africa.

Continuing research and development is based on the environmental advantages of producing zero sulfur fuels, with combustion properties tailored to specific engine designs to minimize emissions from petroleum-derived fuels. At the time of writing, there are at least six developers of what are known as Fischer–Tropsch liquids (FTL) plants, which are scaled to use remote and difficult-to-get-to-market sources of natural gas as the feedstock for the syngas. FTL processes use catalysts based mainly on iron, cobalt, ruthenium, and potassium and have been extensively characterized. They operate at high pressures between 2.5 and 4.5 MPa, and temperatures between 220 and 450°C. The reactors range from fixed catalyst beds to bubbling as well as circulating fluidized beds. The product distribution depends on the polymerization on the surface of the catalyst. The hydrocarbon chains are built up one carbon atom at a time—a process known as homologation. Polymer growth of this type is now understood to take place via what is known as the Anderson–Shulz–Flory mechanism, and it is not possible to produce a single compound with a fixed number of carbon atoms or even a narrowly defined distribution of molecular weights.

Extensive product recycling and recirculation of unreacted syngas can be used to reduce the number of small hydrocarbon molecules (e.g., C_1 to C_5); however, production of wax-like molecules with as many as 30–50 carbon atoms also exists. The conversion per pass of the syngas over the catalysts is often limited, due in part to the fact that the polymerization reaction is exothermic and the equilibrium mixtures obtained are very temperature sensitive, affecting the yield and proportions of the desired products. The recycle processes are very energy intensive and reduce the throughput of the catalyst system.

Recent progress in enhancing the yields of gasoline and diesel products has been made partly by optimization of the catalysts and operating conditions, and also by companies such as Shell with its Shell Middle Distillate Synthesis (SMDS) process in which the higher molecular weight waxes are hydrotreated over selective catalysts to increase the yield of the desired range of hydrocarbons C_8 through C_{20}. The smaller carbon number products, C_1 to C_5, are also reformed to increase the hydrogen to the overall process. Other innovations in the FTL arena include slurry reactors, which use very finely divided catalysts suspended in an oil medium. Such reactors often demonstrate much increased conversion in a single pass and offer economies with respect to recycle energy and investment. However, from the chemical engineering perspective; such three-phase reaction systems have required a lot of R&D and development.

An alternative production process would use a high conversion single-pass catalyst such as a slurry reactor and then utilize the tail gas after the catalyst as fuel for a gas turbine combined cycle. Such a process would then coproduce electricity, heat, and a liquid fuel and offer lower investment costs and higher overall efficiency.

Higher alcohols are produced from syngas over typical Fischer–Tropsch catalysts under conditions of greater severity than those used predominantly in the production of hydrocarbons. Pressures are between 5 and 15 MPa and are generally at temperatures higher than FTL synthesis. The Anderson–Shulz–Flory mechanism does not appear to determine the product distribution, although homologation through the addition of one carbon atom at a time is followed, so that product methanol is recycled over the catalyst to grow into higher alcohols.

Changes in catalyst composition can determine the isomer composition of the higher ($>C_3$) alcohols. Also, mixed thermochemical syngas and biotechnology routes to ethanol are under development. Ethanol can be synthesized from syngas using an anaerobic bacterium, *Clostridium ljungdahlii*. The growth conditions of the bacterium are managed so as to maximize the ethanol yield over the production of acetate (as acetic acid) from the key biological pathway intermediate acetyl-CoA, which is derived from the syngas. Although not yet commercialized, many of the development challenges of producing the bacterium and making effective gas–liquid contactors and bioreactors have been overcome. Yields are similar to those from the inorganic high temperature Fischer–Tropsch catalysts.

References

Anonymous. 1994. Sweden's largest biofuel-fired cfb up and running, Tampere, Finland, 16–17.

Cannell, M. G. R. 2003. Carbon sequestration and biomass energy offset: theoretical, potential and achievable capacities globally, in Europe and the U.K. *Biomass Bioenergy*, 24, 97–116.

Cralle, H. T. and Vietor, D. M. 1989. Productivity: solar energy and biomass. In *Biomass Handbook*, O. Kitani and C. W. Hal, eds., pp. 11–20. Gordon and Breach Science Publishers, New York.

Ezzati, M. and Kammen, D. M. 2001. Quantifying the effects of exposure to indoor air pollution from biomass combustion on acute respiratory infections in developing countries. *Environ. Health Perspect.*, 109 (5), 481–488.

Gallagher, P., Dikeman, M., Fritz, J., Wailes, E., Gauther, W., and Shapouri, H. 2003. Biomass from crop residues: cost and supply estimates. U.S. Department of Agriculture, Office of the Chief Economist, Office of energy Policy and New Uses. Agricultural Economic Report No. 819.

Goldstein, J. 2002. Electric utilities hook up to biogas. *Biocycle*, March, 36–37.

Graham, R. L. et al. 1995. The Effect of location and facility demand on the marginal cost of delivered wood chips from energy crops: a case study of the state of Tennessee. In *Proc. 2nd Biomass Conf. Am.: Energy, Environ., Agric., Ind.*, 1324–1333.

Hall, D. O., Rosillo-Calle, F., Williams, R. H., and Woods, J. 1993. Biomass for energy: supply prospects. In *Renewable Energy: Sources for Fuels and Electricity*, T. B. Johansson, H. Kelly, A. K. N. Reddy, and R. H. Williams, eds., pp. 593–651. Island Press, Washington, DC.

Jenkins, B. M. 1997. A comment on the optimal sizing of a biomass utilization facility under constant and variable cost scaling. *Biomass Bioenergy*, 13 (1/2), 1–9.

Lynd, L. R., Wyman, C. E., and Gerngross, T. U. 1999. Biocommodity engineering. *Biotechnol. Prog.*, 15 (5), 777–793.

McKeever, D. 2003. Taking inventory of woody residuals. *Biocycle*, July, 31–35.

McLaughlin, S. and Kszos, L. personal communication 2003. (These managers of switchgrass research at Oak Ridge National Laboratory have summarized the past 15 years of switchgrass research in a Biomass and Bioenergy 2005 paper in press).

NIST 2004. Handbook 44, Specifications, Tolerances, and Other Technical Requirements for Weighing and Measuring Devices, T. Butcher, L. Crown, R. Suiter, and J. Williams, eds.

Prasad, K. 1985. Stove design for improved dissemination. In *Wood-Stove Dissemination*, R. Clarke, ed., pp. 59–74. Intermediate Technology Publications, London.

Sampson, R. N. et al. 1993. Biomass management and energy. *Water, Air, Soil Pollut.*, 70, 139–159.

Stahl, K., Neergard, M., and Nieminen, J. 2000. Final report: Varnamo demonstration programme. In *Progress in Thermochemical Biomass Conversion*, A. V. Bridgwater, ed., pp. 549–563. Blackwell Sciences Ltd, Oxford, U.K.

Walsh, M. E., Perlack, R. L., Turhollow, A., de la Torre Ugarte, D. G., Becker, D. A., Graham, R. L., Slinsky, S. E., and Ray, D. E. 2000. Biomass feedstock availability in the United States: 1999 state level analysis. Report prepared for the U.S. Department of Energy found at: http://bioenergy.ornl.gov/pubs/resource_data.html

Wright, L. L. 1994. Production technology status of woody and herbaceous crops. *Biomass Bioenergy*, 6 (3), 191–209.

Wyman, C. E. 1994. Ethanol from lignocellulosic biomass: technology, economics, and opportunities. *BioResource Technol.*, 50 (1), 3–16.

Zhang, M., Eddy, C., Deanda, K., Finkelstein, M., and Picataggio, S. 1995. Metabolic engineering of a pentose metabolism pathway in ethanologenic *Zymomonas mobilis*. *Science*, 267 (5195), 240–243. 13 January.

For Further Information

Combustion is widely described in the literature. The International Energy Agency recently produced an extremely useful reference book on the topic of biomass combustion: S. van Loo and J. Koppejan, eds. 2002. *Handbook of Biomass Combustion and Cofiring*. Enschede, Netherlands, Twente University Press (A multiauthor IEA Bioenergy collaboration Task 32 publication).

Power generation is described and analyzed in: EPRI 1997. *Renewable Energy Technology Characterizations*. Electric Power Research Institute, Washington, DC.

Gasification in general and synthetic liquid fuels are well described in: Probstein, R. F. and Hicks, R. E. 1982. *Synthetic Fuels*. McGraw-Hill Inc., New York.

Anaerobic digestion is the subject of a wonderful handbook that is only available in German at present: Schulz, H. and Eder, B. 2001. *Bioga Praxis: grundlagen, planung, anlagenbau, beispiele*. Freiburg, Germany, Ökobuch Verlag, Staufen bei Freiburg.

Ethanol from lignocellulosics as well as a useful section on starch ethanol can be obtained in: C.E. Wyman, ed. 1996. *Handbook on Bioethanol: Production and Utilization*. Applied Energy Technology Series. Taylor & Francis, Washington, DC.

Additional information on ethanol can be obtained from the Web sites of the National Corn Growers Association (http://www.ncga.com) and the Renewable Fuels Association (http://www.ethanolrfa.org/).

The last three decades of biomass activity in the United States are described in: Chum, H. L. and Overend, R. P. 2003. *Biomass and Bioenergy in the United States*. In *Advances in Solar Energy: an Annual Review of Research and Development*. Y. Goswami, ed. American Solar Energy Society, Boulder, CO. U.S.A. 83–148.

Additional information on energy efficiency and renewable energy technologies can be obtained from the U.S. Energy Efficiency and Renewable Energy Websites http://www.eere.energy.gov/biomass.html, and http://www.eere.energy.gov/RE/bioenergy.html. Also the Website http://bioenergy.ornl.gov/ provides useful resource information and many links to other bioenergy Websites.

4

Nuclear Resources

James S. Tulenko
University of Florida

4.1 The Nuclear Fuel Cycle 4-1
Sources of Nuclear Fuels and World Reserves
4.2 Processing of Nuclear Fuel 4-2

4.1 The Nuclear Fuel Cycle

4.1.1 Sources of Nuclear Fuels and World Reserves

Nuclear power can use two naturally occurring elements, uranium, and thorium, as the sources of its fissioning energy. Uranium can be a fissionable source (fuel) as mined (Candu Reactors in Canada), while thorium must be converted in a nuclear reactor into a fissionable fuel. Uranium and thorium are relatively plentiful elements ranking about 60th out of 80 naturally occurring elements. All isotopes of uranium and thorium are radioactive. Today, natural uranium contains, in atomic abundance, 99.2175% uranium-238 (U^{238}); 0.72% uranium-235 (U^{235}); and 0.0055% uranium-234 (U^{234}). Uranium has atomic number 92, meaning all uranium atoms contain 92 protons, with the rest of the mass number being composed of neutrons. Uranium-238 has a half-life of 4.5×10^9 years (4.5 billion years), U-235 has a half-life of 7.1×10^8 years (710 million years), and U-234 has a half-life of 2.5×10^5 years (250 thousand years). Since the age of the earth is estimated at 3 billion years, roughly half of the U-238 present at creation has decayed away, while the U-235 has changed by a factor of sixteen. Thus, when the earth was created, the uranium-235 enrichment was on the order of 8%, enough to sustain a natural reactor of (there is evidence of such an occurrence in Africa). The U-234 originally created has long disappeared, and the U-234 currently present occurs as a product of the decay of U-238.

Uranium was isolated and identified in 1789 by a German scientist, Martin Heinrich Klaproth, who was working with pitchblend ores. No one could identify this new material he isolated, so in honor of the planet Uranus which had just been discovered, he called his new material uranium. It wasn't until 1896, when the French scientist Henri Becquerel accidentally placed some uranium salts near some paper-wrapped photographic plates, that radioactivity was discovered.

Until 1938, when the German scientists Otto Hahn and Fritz Shassroen succeeded in uranium fission by exposure to neutrons, uranium had no economic significance except in coloring ceramics, where it proved valuable in creating various shades of orange, yellow, brown, and dark green. When a uranium atom is fissioned it releases 200 million electron volts of energy; the burning of a carbon (core) atom releases 4 eV. This difference of 50 million times in energy release shows the tremendous difference in magnitude between chemical and nuclear energy.

Uranium is present in the earth's crust to the extent of four parts per million. This concentration makes uranium about as plentiful as beryllium, hafnium, and arsenic; and greater in abundance than tungsten, molybdenum, and tantalum. Uranium is an order of magnitude more plentiful than silver and a hundred times more plentiful than gold. It has been estimated that the amount of uranium in the earth's crust to a depth of 12 miles is of the order of 100 trillion tons.

Thorium, which is composed of only one isotope, thorium-232, has a half-life of 14 billion years (1.4×10^{10} years), is more than three times more abundant than uranium, and is in the range of lead and gallium in abundance. Thorium was discovered by Berjelius in 1828 and named after Thor, the Scandinavian god of war. For reference, copper is approximately five times more abundant than thorium and twenty times more abundant than uranium.

Uranium is chemically a reactive element; therefore, while it is relatively abundant, it is found chemically combined as an oxide (U_3O_8 or UO_2) and never as a pure metal. Uranium is obtained in three ways, either by underground mining, open pit mining, or in situ leaching. An economic average ore grade is normally viewed as 0.2% (4 pounds per short ton), though recently ore grades as low as 0.1% have been exploited. A large quantity of uranium exists in sea-water which has an average concentration of 3×10^{-3} ppm, yielding an estimated uranium quantity available in sea-water of 4000 million tons. A pilot operation was successfully developed by Japan to recover uranium from sea-water, but the cost was about $900/lb, and the effort was shut down as uneconomical.

The major countries with reserves of uranium in order of importance are Australia, United States, Russia, Canada, South Africa, and Nigeria. The countries with major thorium deposits are India, Brazil, and the United States. It is estimated that for a recovery value of $130/kg ($60/lb), the total uranium reserves in these countries are approximately 1.5 million tonnes of uranium in the U.S., 1 million tonnes of uranium in Australia, 0.7 million tonnes of uranium in Canada, and 1.3 million tonnes of uranium in the former Soviet Union. As mentioned earlier, thorium reserves are approximately four times greater. With the utilization of breeder reactors, there is enough uranium and thorium to provide electrical power for the next thousand years at current rates of usage.

4.2 Processing of Nuclear Fuel

Once the uranium ore is mined it is sent to a concentrator (mill) where it is ground, treated, and purified. Since the ore is of a grade of 0.1%–0.2% uranium, a ton of ore contains only between 1 and 2 kg of uranium per 1000 kg of ore. Thus, thousands to tonnes of ore have to be extracted and sent to a mill to produce a relatively small quantity of uranium. In the concentration process approximately 95% of the ore is recovered as U_3O_8 (yellowcake) to a purity grade of about 80%. Thus, assuming 0.15% uranium ore, the milling and processing of a metric ton (1000 kg) of ore yields a concentrate of 1.781 kg (1.425 kg of uranium and 0.356 kg of impurities). For this reason the mills must be located relatively close to the mine site. The ore tailings (waste) amounts to 998.219 kg and contains quantities of radon and other uranium decay products and must be disposed of as a radioactive waste.

The U_3O_8 concentrate is then taken to a conversion plant where the concentrate is further purified (the 20% impurities are removed) and the uranium yellowcake is converted to uranium hexafluoried UF_6). The uranium hexafluoride is a gas at fairly low temperature and is an ideal material for the U-235 isotope enriching processes of either gaseous diffusion or gaseous centrifuge. The UF_6 is shipped in steel cylinders in a solid state, and UF_6 is vaporized by putting the cylinder in a steam bath.

If the uranium is to be enriched to 4% U^{235}, then 1 kg of 4% U^{235} product will require 7.4 kg of natural uranium feed and will produce 6.4 kg of waste uranium (tails or depleted uranium) with a U^{235} isotope content of 0.2%. This material is treated as a radioactive waste. Large quantities of tails (depleted uranium) exist as UF_6 in their original shipping containers at the enriching plants. Depleted uranium (a dense material) has been used as shields for radioactive sources, armor piercing shells, balancing of helicopter rotor tips, yacht hold ballast, and balancing of passenger aircraft.

The enriched UF_6 is then sent to a fabrication plant where it is converted to a uranium dioxide (UO_2) powder. The powder is pressed and sintered into cylindrical pellets which are placed in zircaloy tubes (an alloy of zirconium), pressurized with helium, and sealed. The rods are collected in an array ($\sim 17 \times 17$) bound together by spacer grids, with top and bottom end fittings connected by tie rods or guide tubes.

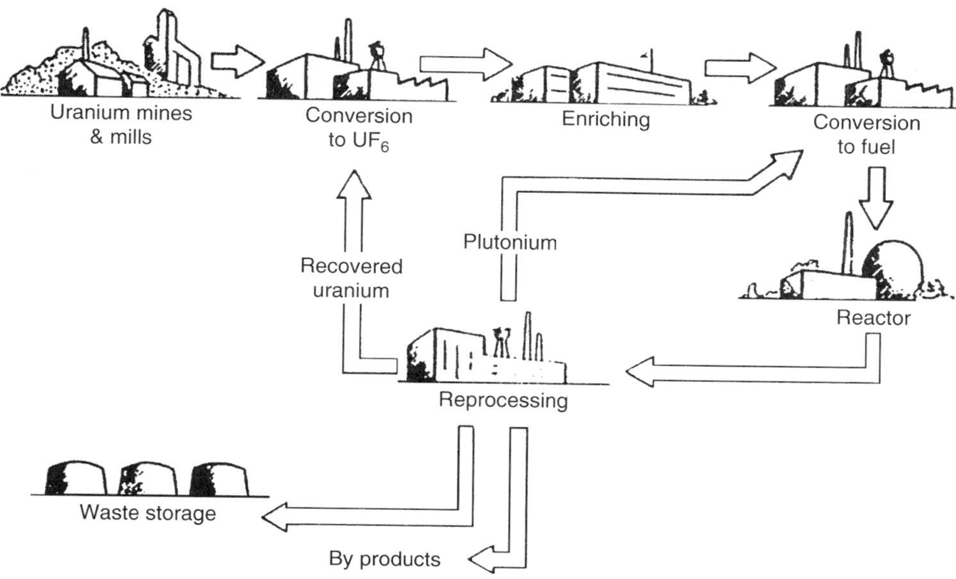

FIGURE 4.1 The nuclear fuel cycle.

Pressurized water reactor fuel assemblies, each containing approximately 500 kg of uranium, are placed in a reactor for 3–4 years. A single fuel assembly produces 160,000,000 kilowatt hours of electricity and gives 8000 people their yearly electric needs for its three years of operation. When the fuel assembly is removed from the reactor it must be placed in a storage pond to allow for removal of the decay heat. After approximately five years of wet storage, the fuel assembly can be removed to dry storage in concrete or steel containers. In the United States the current plan is to permanently store the nuclear fuel, with the Department of Energy assuming responsibility for the "spent" fuel. The money for the government to handle the storage comes from a fee of 1 mill per kilowatt hour paid by consumers of nuclear-generated electricity. A mill is a thousandth of a dollar or a tenth of a penny. Thus, the fuel assembly described above would have collected $160,000 in the waste fund for the Department of Energy to permanently store the fuel. In Europe, when the fuel is taken out of wet storage it is sent to a reprocessing plant where the metal components are collected for waste disposal; and the fuel is chemically recovered as 96% uranium, which is converted to uranium dioxide for recycling to the enrichment plant, 1% plutonium, which is converted to fuel or placed in storage, and 3% fission products which are encased in glass and permanently stored.

The important thing to remember about the fuel cycle is the small quantity of radioactive fission products (1.5 kg) which are created as radioactive waste in producing power which can serve the yearly electricity needs of 8000 people for the three years that it operates. The schematic of the entire fuel cycle showing both the United States system (once-through) and the European (recycle) system is given in Figure 4.1.

5

Solar Energy Resources

5.1 Solar Energy Availability 5-1
5.2 Earth–Sun Relationships 5-2
5.3 Solar Time 5-4
5.4 Solar Radiation on a Surface 5-4
5.5 Solar Radiation on a Horizontal Surface 5-5
5.6 Solar Radiation on a Tilted Surface 5-5
5.7 Solar Radiation Measurements 5-6
5.8 Solar Radiation Data 5-6
Defining Terms 5-8
References 5-9
For Further Information 5-9

D. Yogi Goswami
University of Florida

The sun is a vast nuclear power plant of the fusion variety which generates power in the form of radiant energy at a rate of 3.8×10^{23} kW. An extremely small fraction of this is intercepted by Earth, but even this small fraction amounts to the huge quantity of 1.8×10^{14} kW. On the average, about 60% of this energy incident at the outer edge of the atmosphere, reaches the surface. To compare these numbers with our energy needs, consider the present electrical-generating capacity in the United States, which is approximately of 7×10^8 kW. This is equivalent to an average solar radiation falling on only 1000 square miles in a cloudless desert area. It must, however, be remembered that solar energy is distributed over the entire surface of Earth facing the sun, and it seldom exceeds 1.0 kW/m^2. Compared to other sources, such as fossil fuels or nuclear power plants, solar energy has a very low energy density. However, solar radiation can be concentrated to achieve very high energy densities. Indeed, temperatures as high as 3000 K have been achieved in solar furnaces.

Solar energy technology has been developed to a point where it can replace most of the fossil fuels or fossil fuel-derived energy. In many applications it is already economical, and it is a matter of time before it becomes economical for other applications as well.

This section deals in the availability of solar radiation, including methods of measurement, calculation, and available data.

5.1 Solar Energy Availability

Detailed information about solar radiation availability at any location is essential for the design and economic evaluation of a solar energy system. Long-term measured data of solar radiation are available for a large number of locations in the United States and other parts of the world. Where long-term measured data are not available, various models based on available climatic data can be used to estimate the solar energy availability. The solar energy is in the form of electromagnetic radiation with the wavelengths ranging from about 0.3 μm (10^{-6} m) to over 3 μm, which correspond to ultraviolet (less

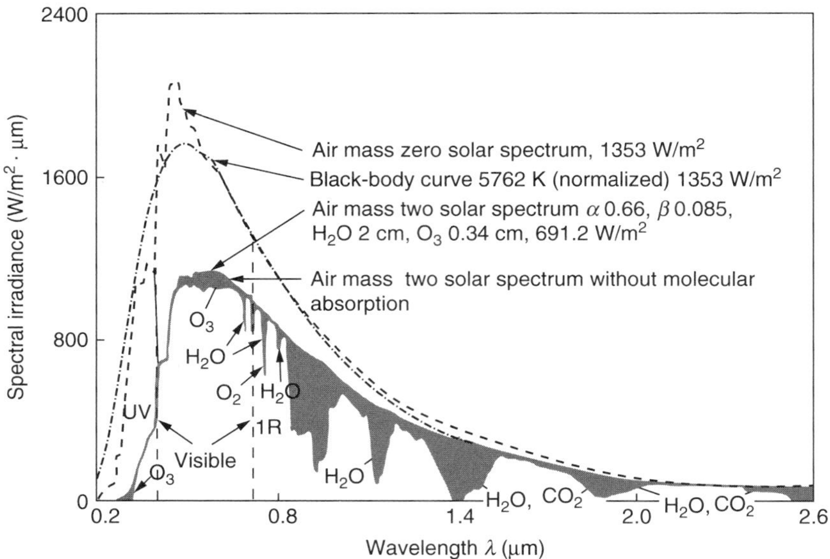

FIGURE 5.1 Spectral distribution of solar energy at sea level. (Reprinted From Goswami, D. Y., Kreith, F., and Kreider, J., *Principles of Solar Engineering*, Taylor & Francis, Philadelphia, PA, 2000. with permission)

than 0.4 μm), visible (0.4 and 0.7 μm), and infrared (over 0.7 μm). Most of this energy is concentrated in the visible and the near-infrared wavelength range (see Figure 5.1). The incident solar radiation, sometimes called **insolation**, is measured as irradiance, or the energy per unit time per unit area (or power per unit area). The units most often used are watts per square meter (W/m^2), British thermal units per hour per square foot (Btu/h-ft^2), and langleys (calories per square centimeter per minute, cal/cm^2-min).

The amount of solar radiation falling on a surface normal to the rays of the sun outside the atmosphere of the earth (extraterrestrial) at mean Earth–sun distance (D) is called the **solar constant**, I_0. Measurements by NASA indicated the value of solar constant to be 1353 W/m^2 ($\pm$1.6%). This value was revised upward and the present accepted value of the solar constant is 1377 W/m^2 (Quinlan 1979) or 437.1 Btu/h-ft^2 or 1.974 langleys. The variation in seasonal solar radiation availability at the surface of Earth can be understood from the geometry of the relative movement of Earth around the sun.

5.2 Earth–Sun Relationships

Figure 5.2 shows the annual motion of Earth around the sun. The **extraterrestrial solar radiation** varies throughout the year because of the variation in the Earth–sun distance (D) as:

$$I = I_0(D/D_0)^2 \tag{5.1}$$

which may be approximated as (Spencer 1971)

$$(D/D_0)^2 = 1.00011 + 0.034221\cos(x) + 0.00128\sin(x) + 0.000719\cos(2x) + 0.000077\sin(2x) \tag{5.2}$$

where

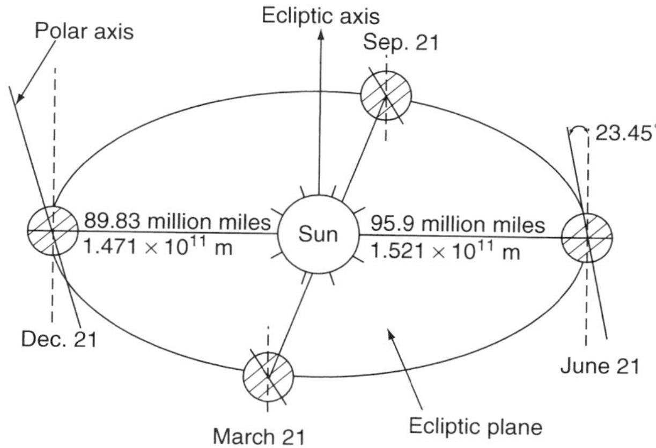

FIGURE 5.2 Annual motion of the Earth around the sun. (Adapted from Goswami, D. Y., Kreith, F., and Kreider, J., *Principles of Solar Engineering*, Taylor & Francis, Philadelphia, PA, 2000.)

$$x = 360(N - 1)/365° \tag{5.3}$$

and N = Day number (starting from January 1 as 1). The axis of the Earth is tilted at an angle of 23.45° to the plane of its elliptic path around the sun. This tilt is the major cause of the seasonal variation of solar radiation available at any location on Earth. The angle between the Earth–sun line and a plane through the equator is called **solar declination**, δ. The declination varies between −23.45° to +23.45° in 1 year. It may be estimated by the relation:

$$\delta = 23.45° \sin[360(284 + N)/365°] \tag{5.4}$$

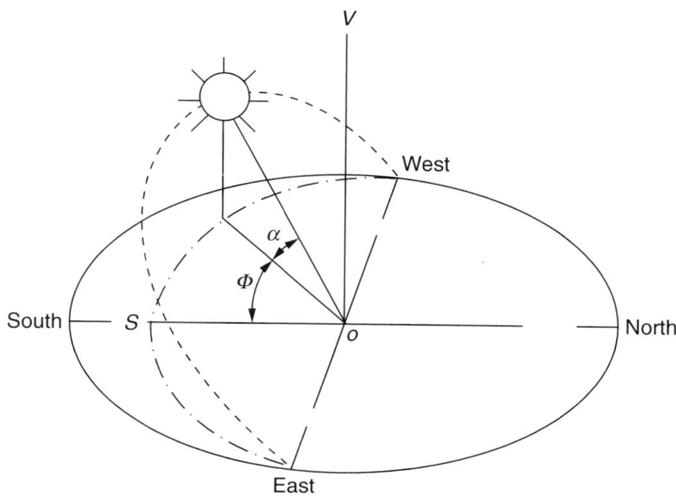

FIGURE 5.3 Apparent daily path of the sun across the sky from sunrise to sunset, showing the solar altitude and azimuth angles.

The apparent motion of the sun around the earth is shown in Figure 5.3. The **solar altitude angle,** β, and the **solar azimuth angle,** Φ, describe the position of the sun at any time.

5.3 Solar Time

The sun angles are found from the knowledge of solar time, which differs from the local time. The relationship between solar time and local standard time (LST) is given by

$$\text{Solar Time} = \text{LST} + \text{ET} + 4(L_{st} - L_{loc}) \tag{5.5}$$

where ET is the **equation of time,** which is a correction factor in minutes that accounts for the irregularity of the motion of the Earth around the sun. L_{st} is the standard time meridian and L_{loc} is the local longitude. ET can be calculated from the following empirical equation:

$$\text{ET (in minutes)} = 9.87 \sin 2B - 7.53 \cos B - 1.5 \sin B \tag{5.6}$$

where $B = 360(N-81)/365°$.

The sun angles α (altitude) and Φ (azimuth) can be found from the equations:

$$\sin \alpha = \cos \ell \cos \delta \cos H + \sin \ell \sin \delta \tag{5.7}$$

where ℓ = latitude angle,

$$\sin \Phi = \cos \delta \sin H / \cos \alpha \tag{5.8}$$

and

$$H = \text{Hour angle} = \frac{\text{Number of minutes from local solar noon}}{4 \text{ min/degree}} \tag{5.9}$$

(At solar noon, $H=0$, so $\alpha = 90 - |\ell - \delta|$ and $\Phi = 0$.)

5.4 Solar Radiation on a Surface

As solar radiation, I, passes through the atmosphere, some of it is absorbed by air and water vapor, while some gets scattered by molecules of air, water vapor, aerosols, and dust particles. The part of solar radiation that reaches the surface of the Earth with essentially no change in direction is called **direct or beam normal radiation,** I_{bN}. The scattered radiation reaching the surface from the atmosphere is called **diffuse radiation,** I_d.

I_{bN} can be calculated from the extraterrestrial solar irradiance, I, and the atmospheric optical depth τ as (Goswami et al. 1981; ASHRAE 1995)

$$I_{bN} = I e^{-\tau \sec \theta_z} \tag{5.10}$$

where θ_z is the solar zenith angle (angle between the sun rays and the vertical). The atmospheric optical depth determines the attenuation of the solar radiation as it passes through the atmosphere. Threlkeld and Jordan (1958) calculated values of τ for average atmospheric conditions at sea level with a moderately dusty atmosphere and amounts of precipitable water vapor equal to the average value for the United States for each month. These values are given in Table 5.1. To account for the differences in local conditions from the average sea level conditions Equation 5.10 is modified by a parameter called Clearness Number, Cn, introduced by Threlkeld and Jordan (1958):

TABLE 5.1 Average Values of Atmospheric Optical Depth (τ) and Sky Diffuse Factor (C) for 21st Day of Each Month

Month	1	2	3	4	5	6	7	8	9	10	11	12
τ	0.142	0.144	0.156	0.180	0.196	0.205	0.207	0.201	0.177	0.160	0.149	0.142
C	0.058	0.060	0.071	0.097	0.121	0.134	0.136	0.122	0.092	0.073	0.063	0.057

Source: From Threlkeld, J. L. and Jordan, R. C., *ASHRAE Trans.*, 64, 45, 1958.

$$I_{\mathrm{bN}} = \mathrm{Cn}Ie^{-\tau \sec \theta_z} \quad (5.11)$$

values of Cn vary between 0.85 and 1.15.

5.5 Solar Radiation on a Horizontal Surface

Total incident solar radiation on a horizontal surface is given by

$$I_{\mathrm{t,\ Horizontal}} = I_{\mathrm{bN}} \cos \theta_z + C I_{\mathrm{bN}} \quad (5.12)$$

$$= I_{\mathrm{bN}} \sin \beta + C I_{\mathrm{bN}} \quad (5.13)$$

where θ_z is called the solar zenith angle and C is called the sky diffuse factor, as given in Table 5.1.

5.6 Solar Radiation on a Tilted Surface

For a surface of any orientation and tilt as shown in Figure 5.4, the angle of incidence, θ, of the direct solar radiation is given by

$$\cos \theta = \cos \alpha \cos \gamma \sin \beta + \sin \alpha \cos \beta \quad (5.14)$$

where γ is the angle between horizontal projections of the rays of the sun and the normal to the surface. β is the tilt angle of the surface from the horizontal.

For a tilted surface with angle of incidence θ, the total incident solar radiation is given by

$$I_{\mathrm{b}} = I_{\mathrm{bN}} \cos \theta + I_{\mathrm{diffuse}} + I_{\mathrm{reflected}} \quad (5.15)$$

where

$$I_{\mathrm{diffuse}} = C I_{\mathrm{bN}}(1 + \cos \beta)/2 \quad (5.16)$$

and

$$I_{\mathrm{reflected}} = \rho I_{\mathrm{bN}}(C + \sin \alpha)(1 - \cos \beta)/2 \quad (5.17)$$

where ρ is the reflectivity of the surroundings. For ordinary ground or grass, ρ is approximately 0.2 while for ground covered with snow it is approximately 0.8.

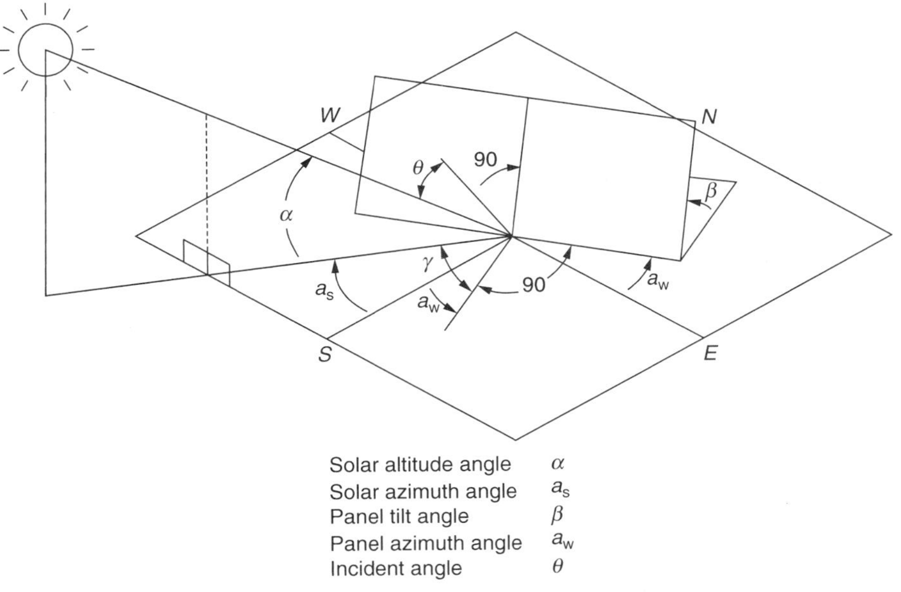

FIGURE 5.4 Definitions of solar angles for a tilted surface.

5.7 Solar Radiation Measurements

Two basic types of instruments are used in measurements of solar radiation. These are (see Figure 5.5):

1. *Pyranometer*: An instrument used to measure global (direct and diffuse) solar radiation on a surface. This instrument can also be used to measure the diffuse radiation by blocking out the direct radiation with a shadow band.
2. *Pyrheliometer*: This instrument is used to measure only the direct solar radiation on a surface normal to the incident beam. It is generally used with a tracking mount to keep it aligned with the sun.

More-detailed discussions about these and other solar radiation measuring instruments can be found in Zerlaut (1989).

5.8 Solar Radiation Data

Measured values of solar radiation data for locations in the United States are available from the National Climatic Center in Asheville, NC. A number of states have further presented solar radiation data for locations in those states in readily usable form. Weather services and energy offices in almost all the countries have available some form of solar radiation data or climatic data that can be used to derive solar radiation data for locations in those countries. Table 5.2 through Table 5.4 give solar radiation data for clear days for south-facing surfaces in the Northern Hemisphere (and northern-facing surfaces in the Southern Hemisphere) tilted at 0°, 15°, 30°, 45°, 60°, 75°, and vertical, for latitudes 0°, 30°, and 60°. The actual average solar radiation data at a location is less than the values given in these tables because of the cloudy and partly cloudy days in addition to the clear days. The actual data can be obtained either from long-term measurements or from modeling based on some climatic parameters, such as percent sunshine.

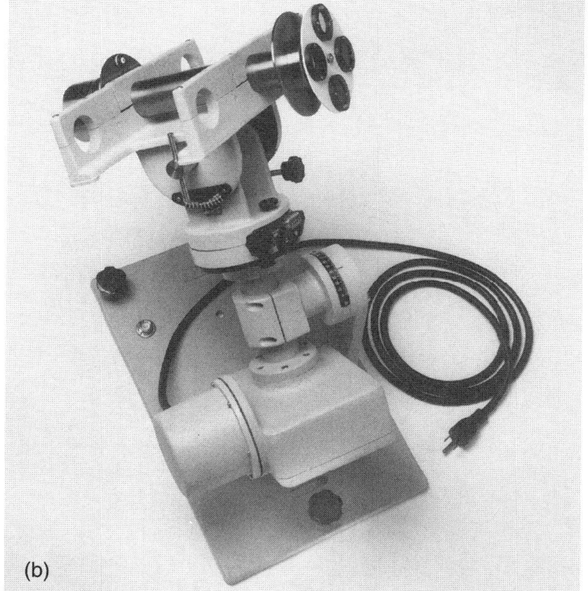

FIGURE 5.5 Two basic instruments for solar radiation: (a) pyranometer; (b) pyrheliometer.

TABLE 5.2 Average Daily Total Solar Radiation on South-Facing Surfaces in Northern Hemisphere; Latitude = 0°N

Month	Horiz.	15°	30°	45°	60°	75°	90°
1	31.11	34.13	35.13	34.02	30.90	25.96	19.55
2	32.34	33.90	33.45	31.03	26.80	21.05	14.18
3	32.75	32.21	29.79	25.67	20.12	13.53	6.77
4	31.69	29.13	24.93	19.39	12.97	6.59	4.97
5	29.97	26.08	20.81	14.64	8.34	4.92	5.14
6	28.82	24.43	18.81	12.54	6.66	5.07	5.21
7	29.22	25.08	19.66	13.48	7.45	5.17	5.31
8	30.59	27.48	22.87	17.13	10.82	5.58	5.32
9	31.96	30.51	27.34	22.65	16.78	10.18	5.33
10	32.18	32.82	31.54	28.44	23.73	17.72	10.84
11	31.33	33.80	34.28	32.72	29.24	24.08	17.58
12	30.51	33.90	35.27	34.53	31.73	27.05	20.83

TABLE 5.3 Average Daily Total Solar Radiation on South-Facing Surfaces in Northern Hemisphere; Latitude = 30°N

Month	Horiz.	15°	30°	45°	60°	75°	90°
1	17.19	22.44	26.34	28.63	29.15	27.86	24.85
2	21.47	26.14	29.25	30.59	30.06	27.70	23.68
3	26.81	30.04	31.50	31.09	28.84	24.90	19.54
4	31.48	32.71	32.06	29.57	25.44	19.96	13.60
5	34.49	33.96	31.56	27.49	22.08	15.82	9.49
6	35.61	34.24	31.03	26.28	20.40	13.97	8.02
7	35.07	34.06	31.21	26.76	21.11	14.77	8.68
8	32.60	33.00	31.54	28.35	23.68	17.89	11.57
9	28.60	30.87	31.35	30.02	26.97	22.42	16.67
10	23.41	27.38	29.74	30.33	29.10	26.14	21.66
11	18.50	23.48	27.05	28.98	29.14	27.51	24.20
12	15.90	21.19	25.21	27.68	28.44	27.43	24.71

TABLE 5.4 Average Daily Total Solar Radiation on South-Facing Surfaces in Northern Hemisphere; Latitude = 60°N

Month	Horiz.	15°	30°	45°	60°	75°	90°
1	1.60	3.54	5.26	6.65	7.61	8.08	8.03
2	5.49	9.38	12.71	15.25	16.82	17.32	16.72
3	12.82	17.74	21.60	24.16	25.22	24.73	22.71
4	21.96	26.22	28.97	30.05	29.38	27.00	23.09
5	30.00	32.79	33.86	33.17	30.73	26.72	21.45
6	33.99	35.82	35.93	34.29	31.00	26.26	20.46
7	32.26	34.47	34.97	33.71	30.78	26.36	20.80
8	25.37	28.87	30.80	31.02	29.53	26.42	21.94
9	16.49	21.02	24.34	26.22	26.54	25.27	22.51
10	8.15	12.39	15.90	18.45	19.85	20.01	18.92
11	2.70	5.27	7.53	9.31	10.51	11.03	10.84
12	0.82	2.06	3.16	4.07	4.71	5.05	5.07

Note: Values are in megajoules per square meter. Clearness number = 1.0; ground reflection = 0.2.

Worldwide solar radiation data is available from the World Radiation Data Center (WRDC). WRDC has been archiving data from over 500 stations and operates a website in collaboration with NREL (wrdc-mgo.nrel.gov).

Defining Terms

Diffuse radiation: Scattered solar radiation coming from the sky.

Direct or beam normal radiation: Part of solar radiation coming from the direction of the sun on a surface normal to the sun's rays.

Equation of time: Correction factor in minutes, to account for the irregularity of the Earth's motion around the sun.

Extraterrestrial solar radiation: Solar radiation outside Earth's atmosphere.

Insolation: Incident solar radiation measured as W/m^2 or $Btu/h\text{-}ft^2$.

Solar altitude angle: Angle between the solar rays and the horizontal plane.

Solar azimuth angle: Angle between the true south horizontal line and the horizontal projection of the sun's rays.

Solar constant: Extraterrestrial solar radiation at the mean Earth–sun distance.

Solar declination: Angle between the Earth–sun line and a plane through the equator.

References

ASHRAE 1995. *1995 HVAC Applications.* ASHRAE, Atlanta, GA.

Goswami, D. Y. 1986. *Alternative Energy in Agriculture., Vol. 1*, CRC Press, Boca Raton, FL.

Goswami, D. Y., Klett, D. E., Stefanakos, E. K., and Goswami, T. K. 1981. Seasonal variation of atmospheric clearness numbers for use in solar radiation modelling. *AIAA J. Energ.*, 5 (3), 185.

Goswami, D. Y., Kreith, F., and Kreider, J. 2000. *Principles of Solar Engineering.* Taylor & Francis, Philadelphia, PA.

Kreith, F. and Kreider, J. F. 1978. *Principles of Solar Engineering.* Hemisphere Publishing, Washington, DC.

Quinlan, F. T. ed. 1979. *SOLMET Volume 2: Hourly Solar Radiation—Surface Meteorological Observations*, National Oceanic and Atmospheric Administration, Asheville, NC.

Spencer, J. W. 1971. Fourier series representation of the position of the sun. *Search*, 2, 172.

Threlkeld, J. L. and Jordan, R. C. 1958. Direct radiation available on clear days. *ASHRAE Trans.*, 64, 45.

Zerlaut, G. 1989. Solar radiation instrumentation. In *Solar Resources*, R. L. Hulstrom, ed., MIT Press, Cambridge, MA, chap. 5.

For Further Information

Hulstrom, R. H. ed. 1989. *Solar Resources*, MIT Press, Cambridge, MA.

World Radiation Data Center (WRDC), St. Petersburg, Russia: WRDC, operating under the auspices of World Meteorological Organization (WMO), has been archiving data over 500 stations and operates a website in collaboration with NREL (wrdc-mgo.nrel.gov).

6

Wind Energy Resources*

6.1 Wind Origins 6-1
6.2 Wind Power 6-1
6.3 Wind Shear 6-2
6.4 Wind Energy Resource 6-2
6.5 Wind Characterization 6-6
6.6 Wind Energy Potential 6-6
Defining Terms 6-6
References 6-6
For Further Information 6-7

Dale E. Berg
Sandia National Laboratories

6.1 Wind Origins

The primary causes of atmospheric air motion, or wind, are uneven heating of the Earth by solar radiation and the Earth's rotation. Differences in solar radiation absorption at the surface of the Earth and transference back to the atmosphere create differences in atmospheric temperature, density, and pressure, which in turn create forces that move air from one place to another. For example, land and water along a coastline absorb radiation differently, and this is the dominant cause of the light winds or breezes normally found along a coast. The Earth's rotation gives rise to semipermanent global wind patterns such as trade winds, westerlies, easterlies, and subtropical and polar jets.

6.2 Wind Power

The available power in the wind with air density ρ, passing through an area A, perpendicular to the wind, at a velocity U, is given by

$$\text{Power} = \frac{1}{2}\rho A U^3 \tag{6.1}$$

Air density decreases with increasing temperature and increasing altitude above sea level. The effect of temperature on density is relatively weak and is normally ignored because these variations tend to average out over the period of a year. The density difference due to altitude, however, is significant; it does not average out and cannot be ignored. For example, the air density at Denver, Colorado (elevation 1600 m, or 5300 ft., above sea level), is approximately 14% lower than at sea level, so wind at Denver contains 14% less power than wind of the same velocity at sea level.

*This work was supported by the United States Department of Energy under Contract DE-AC04-94AL85000.

From Equation 6.1, it is obvious that the most important factor in the available wind power is the velocity of the wind—an increase in wind velocity of only 20%, e.g., from 5 to 6 m/s (11.2–13.4 mph), yields a 73% increase in available wind power.

6.3 Wind Shear

Wind moving across the Earth's surface is slowed by trees, buildings, grass, rocks, and other obstructions in its path. The result is a wind velocity that varies with height above the Earth's surface—a phenomena known as **wind shear**. For most situations, wind shear is positive (wind speed increases with height), but situations in which the wind shear is negative or inverse are not unusual. In the absence of actual data for a specific site, a commonly used approximation for wind shear in an open area is:

$$\frac{U}{U_0} = \left(\frac{h}{h_0}\right)^{\alpha} \tag{6.2}$$

where

U = the velocity at a height h
U_0 = the measured velocity at height h_0
α = the wind shear exponent.

The wind shear exponent, α, varies with terrain characteristics, but usually falls between 0.10 and 0.25. Wind over a body of open water is normally well modeled by a value of α of about 0.10; wind over a smooth, level, grass-covered terrain such as the U.S. Great Plains by an α of about 0.14; wind over row crops or low bushes with a few scattered trees by an α of 0.20; and wind over a heavy stand of trees, several buildings, or hilly or mountainous terrain by an α of about 0.25. Short-term shear factors as large as 1.25 have been documented in rare, isolated cases.

The available wind power at a site can vary dramatically with height due to wind shear. For example, for $\alpha = 0.20$, Equation 6.1 and Equation 6.2 reveal that the available wind power at a height of 50 m is approximately $\{(50/10)^{0.2}\}^3 = 2.63$ times the available wind power at a height of 10 m.

6.4 Wind Energy Resource

The amount of energy available in the wind (the **wind energy resource**) is the average amount of power available in the wind over a specified period of time—commonly 1 year. If the wind speed is 20 m/s, the available power is very large at that instant, but if it only blows at that speed for 10 h per year and the rest of the time the wind speed is near zero, the resource for the year is small. Therefore, the site **wind speed distribution**, or the relative frequency of occurrence for each wind speed, is very important in determining the resource. This distribution is often presented as a probability density function, such as the one shown in Figure 6.1. The probability that the wind occurs in any given wind speed range is given by the area under the density function for that wind speed range. If the actual wind speed probability density distribution is not available, it is commonly approximated with the Rayleigh distribution, given by:

$$f(U) = \frac{\pi}{4}\frac{U}{\overline{U}}\exp\left[-\frac{\pi}{4}\frac{U^2}{\overline{U^2}}\right] \tag{6.3}$$

where

$f(U)$ = the frequency of occurrence of wind speed U
$\overline{U}$ = the yearly average wind speed.

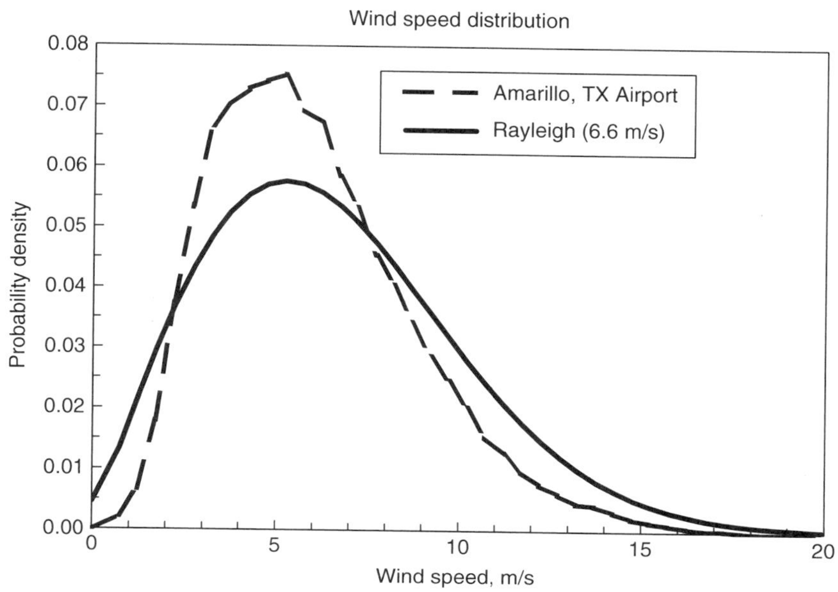

FIGURE 6.1 Rayleigh and measured wind speed distributions.

The measured wind speed distribution at the Amarillo, Texas, airport (yearly average wind speed of 6.6 m/s) is plotted in Figure 6.1, together with the Rayleigh distribution for that wind speed. It is obvious that the Rayleigh distribution is not a good representation for the Amarillo airport.

How large is the wind energy resource? Even though wind energy is very diffuse, the total resource is very, very large. In the U.S. and many other countries around the world, the resource is large enough to supply the entire current energy consumption of the country, potentially. In 1987, scientists at Batelle Pacific Northwest Laboratory (PNL) in the U.S. carefully analyzed and interpreted the available long-term wind data for the U.S. and summarized their estimate of the wind energy resources in the *Wind Energy Resource Atlas of the United States* (Elliott et al. 1987). Their summary for the entire U.S. is reproduced in Figure 6.2. The results are presented in terms of wind power classes based on the annual average power available per square meter of intercepted area (see the legend on Figure 6.2).

Scientists at Denmark's Risø National Laboratory have produced a European wind atlas (Troen and Petersen 1989) that estimates the wind resources of the European Community countries and summarizes the resource available at a 50 m height for five different topographic conditions. A summary of these results is reproduced in Figure 6.3. The estimates presented in Figure 6.2 and Figure 6.3 are quite crude and have been superceded in recent years by much higher resolution maps, made possible by improvements in wind resource computer modeling programs and increases in computer speed.

Many countries around the world have recently embarked on high-resolution mapping efforts to quantify their wind resources and identify those areas of highest resource accurately. The resultant resource maps are frequently available to the public, but in some cases a payment is required to obtain them. High-resolution wind resource maps of the individual states in the U.S. may be found on the Web at www.eere.energy.gov/windpoweringamerica/wind_resources.html. Similar maps for some other countries may be found at www.rsvp.nrel.gov/wind_resources.html, and information on where to find maps and/or data for other countries may be found at www.windatlas.dk/index.htm.

Remember that even the highest resolution resource estimates are just that—estimates. The actual wind resources in any specific area can vary dramatically from those estimates and should be determined with long-term, site-specific measurements.

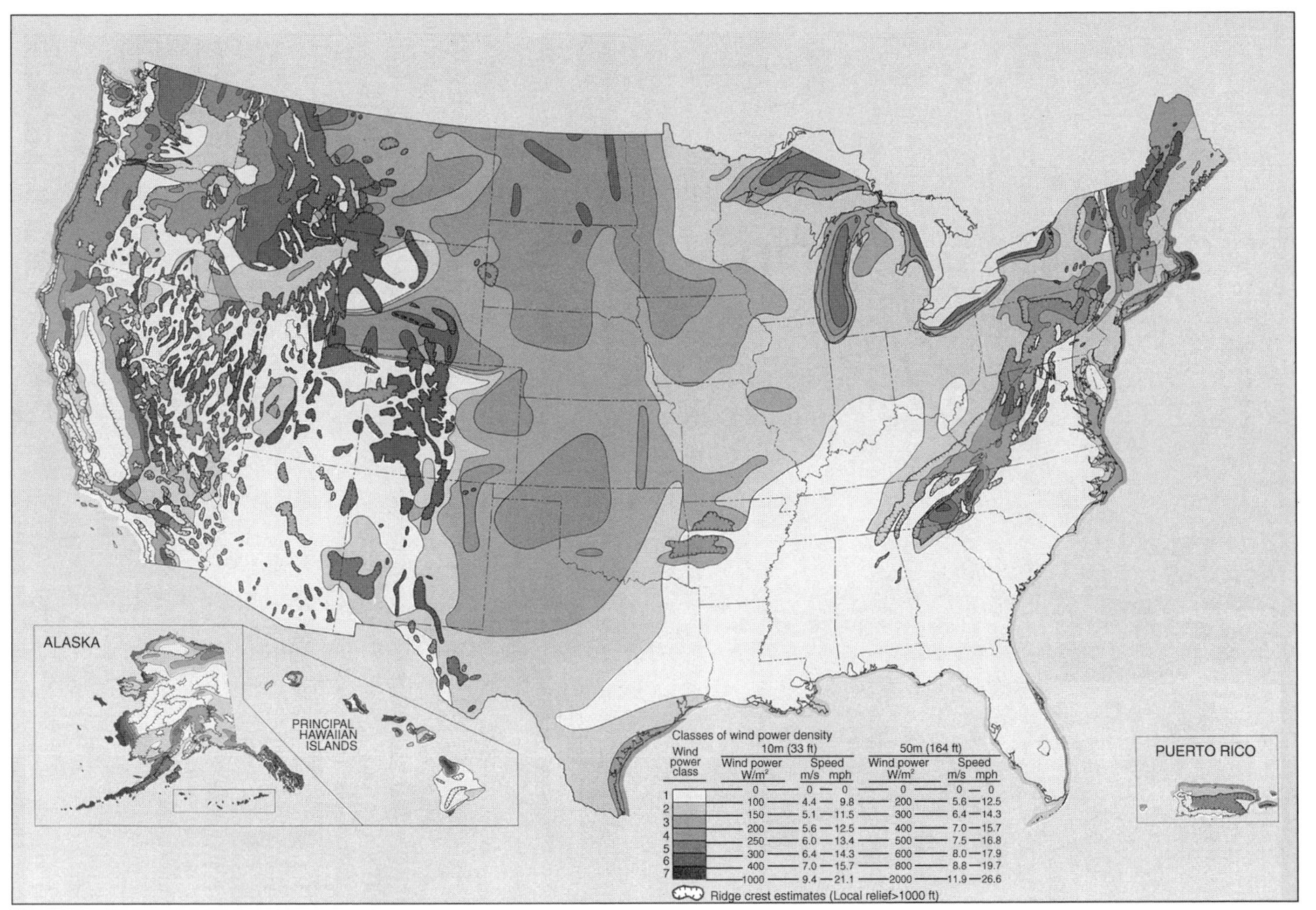

FIGURE 6.2 Map of U.S. wind energy resources. Reproduced from Elliott et al. *Wind Energy Resource Atlas of the United States.* (Courtesy of National Renewable Energy Laboratory, Golden, Colorado.)

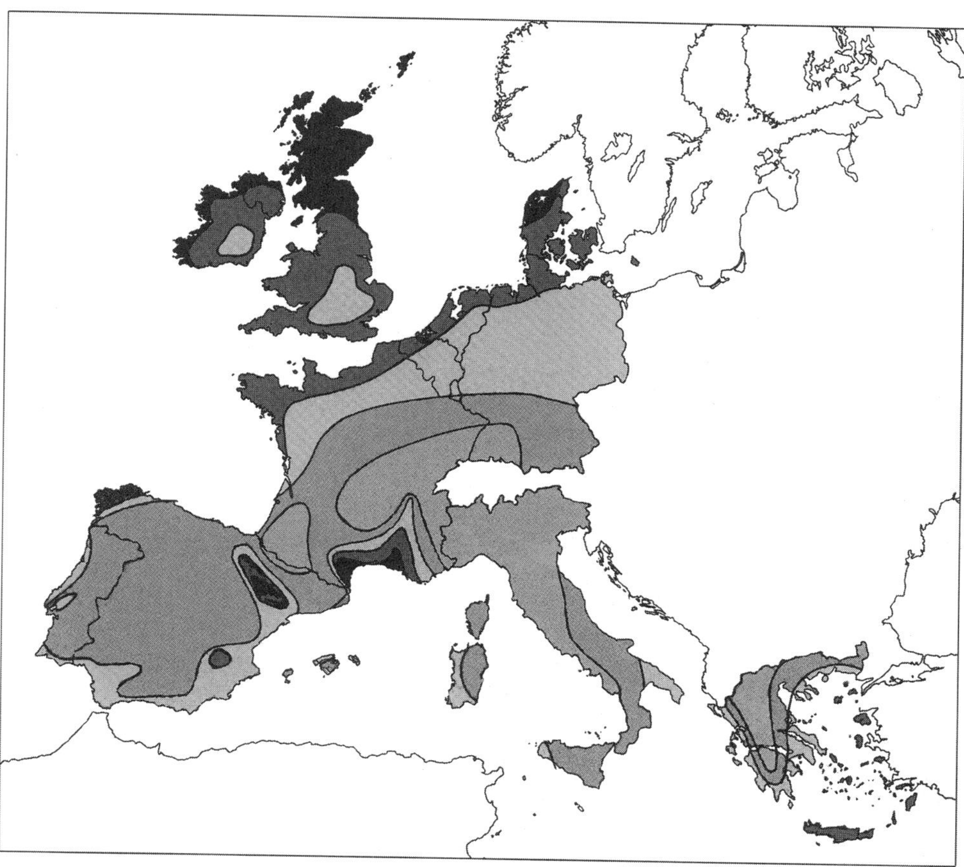

Wind resources[1] at 50 metres above ground level for five different topographic conditions										
	Sheltered terrain[2]		Open plain[3]		At a sea coast[4]		Open sea[5]		Hills and ridges[6]	
	$m\,s^{-1}$	Wm^{-2}	$m\,s^{-1}$	Wm^{-2}	$m\,s^{-1}$	Wm^{-2}	$m\,s^{-1}$	Wm^{-2}	$m\,s^{-1}$	Wm^{-2}
	> 6.0	> 250	> 7.5	> 500	> 8.5	> 700	> 9.0	> 800	> 11.5	> 1800
	5.0–6.0	150–250	6.5–7.5	300–500	7.0–8.5	400–700	8.0–9.0	600–800	10.0–11.5	1200–1800
	4.5–5.0	100–150	5.5–6.5	200–300	6.0–7.0	250–400	7.0–8.0	400–600	8.5–10.0	700–1200
	3.5 – 4.5	50 – 100	4.5–5.5	100–200	5.0–6.0	150–250	5.5–7.0	200–400	7.0 – 8.5	400 – 700
	< 3.5	< 50	< 4.5	< 100	< 5.0	< 150	< 5.5	< 200	< 7.0	< 400

1. The resources refer to the power present in the wind. A wind turbine can utilize between 20 and 30% of the available resource. The resources are calculated for an air density of 1.23 kg m^{-3}, corresponding to standard sea level pressure and a temperature of 15°C. Air density decreases with height but up to 1000 m a.s.l. the resulting reduction of the power densities is less than 10%.
2. Urban districts, forest and farm land with many windbreaks (roughness class 3).
3. Open landscapes with few windbreaks (roughness class 1). In general, the most favourable inland sites on level land are found here.
4. The classes pertain to a straight coastline, a uniform wind rose and a land surface with few windbreaks (roughness class 1). Resources will be higher, and closer to open sea values, if winds from the sea occur more frequently, i.e. the wind rose is not uniform and/or the land protrudes into the sea. Conversely, resources will generally be smaller, and closer to land values, if winds from land occur more frequently.
5. More than 10 km offshore (roughness class 0).
6. The classes correspond to 50% overspeeding and were calculated for a site on the summit of a single axisymmetric hill with a height of 400 metres and a base diameter of 4 km. The overspeeding depends on the height, length and specific setting of the hill.

FIGURE 6.3 Map of European wind energy resources. Reproduced from Troen and Petersen, 1989. *European Wind Atlas.* (Courtesy of Risø National Laboratory, Roskilde, Denmark.)

6.5 Wind Characterization

Wind speed, direction, distribution, and shear can vary significantly over fairly short distances in the horizontal or vertical directions, so in order to get the best possible estimate of the wind energy resource at a particular location, it is important to measure the wind resource at the specific site and height of interest. However, a comprehensive site characterization normally requires measuring the wind for at least 12 months, according to meteorologists at PNL (Wegley et al. 1980). This is a very time-consuming and potentially expensive effort. Long-term data from the nearest airport or weather recording station can help determine whether the data obtained at a site are representative of normal winds for the site or of higher or lower than average winds. Wegley et al. (1980) and Gipe (1993) give suggestions on methods of using available data from nearby sites to estimate site wind speed with minimal on-site data.

Sites of wind power class 4 or above (at least 200 W/m^2 at 10 m height or 400 W/m^2 at 50 m height) are often considered economic for utility-scale wind power development with available wind technology. Sites of wind power class 3 (150–200 W/m^2 at 10 m height or 300–400 W/m^2 at 50 m height) are not considered economic for utility development today but are likely to become economic with near-term wind technology advances. Sites of wind power class 2 or lower (less than 150 W/m^2 at 10 m height or 300 W/m^2 at 50 m height) are usually considered economic only for remote or hybrid wind power systems.

6.6 Wind Energy Potential

With a wind speed distribution and a turbine power curve (the electrical power generated by the turbine at each wind speed) properly adjusted for the local air density, the **wind energy potential**, or gross annual wind energy production, for a specific site can be estimated as:

$$\text{Energy} = 0.85\left[8760\sum_{i=1}^{n} f(U_i)\,\Delta\,U_i P(U_i)\right] \tag{6.4}$$

where

8760 = the number of hours in a year
n = the number of wind speeds considered
$f(U_i)\Delta U_i$ = the probability of a wind speed occurring in the wind-speed range ΔU_i
$P(U_i)$ = the electrical power produced by the turbine at wind speed U_i, the center of the range ΔU_i.

The leading 0.85 factor assumes 15% in losses (10% due to power transfer to the grid, control system losses, and decreased performance due to dirty blades; 5% due to operation within an array of wind turbines). If the turbine is not inside an array, replace 0.85 with 0.90. Wind energy potential is typically 20%–35% of the wind energy resource.

Defining Terms

Wind energy potential: Total amount of energy that can actually be extracted from the wind, taking into account the efficiency of the wind turbine.
Wind energy resource: Total amount of energy present in the wind.
Wind shear: Change in wind velocity with increasing height above the ground.
Wind speed distribution: Probability density of occurrence of each wind speed over the course of a year for the site in question.

References

Elliott, D. L., Holladay, C. G., Barchet, W. R., Foote, H. P., and Sandusky, W. F. 1987. *Wind Energy Resource Atlas of the United States*, DOE/CH10094-4. Solar Energy Research Institute, Golden, Colorado.

Gipe, P. 1993. *Wind Power for Home & Business—Renewable Energy for the 1990s and Beyond*. Chelsea Green Publishing Company, Post Mills, VT.

Troen, I. and Petersen, E. L. 1989. *European Wind Atlas*. Risø National Laboratory, Roskilde, Denmark.

Wegley, H. L., Ramsdell, J. V., Orgill, M. M., and Drake, R. L. 1980. *A Siting Handbook for Small Wind Energy Conversion Systems*, PNL-2521. Pacific Northwest Laboratory, Richland, WA.

For Further Information

Wind Characteristics—An Analysis for the Generation of Wind Power, Rohatgi, J. S. and Nelson, V., Alternative Energy Institute, West Texas A&M University, is an excellent source for additional information on the wind resource.

Wind Turbine Technology, Fundamental Concepts of Wind Turbine Engineering, Spera, D., ed., ASME Press, New York, contains a wealth of information on wind energy resources, history, and technology, together with extensive reference lists.

Extensive information on wind energy resources and technology may also be found on the World Wide Web. Excellent sites to start with include those of the U.S. National Renewable Energy Laboratory Wind Energy Technology Center at www.nwtc.nrel.gov; the Danish Risø National Laboratory at www.risoe.dk/vea/index.htm; the American Wind Energy Association at www.awea.org; the British Wind Energy Association at www.britishwindenergy.co.uk; and the European Wind Energy Association at www.ewea.org.

7

Geothermal Energy

7.1 Heat Flow ... 7-1
7.2 Types of Geothermal Systems ... 7-2
7.3 Geothermal Energy Potential ... 7-2
7.4 Geothermal Applications ... 7-4
7.5 Environmental Constraints ... 7-4
7.6 Operating Conditions ... 7-6
Defining Terms ... 7-8
Acknowledgments ... 7-8
References ... 7-9
For Further Information ... 7-10

Joel L. Renner
Idaho National Engineering Laboratory

Marshall J. Reed
U.S. Department of Energy

The word *Geothermal* comes from the combination of the Greek words *gê*, meaning Earth, and *thérm*, meaning heat. Quite literally, geothermal energy is the heat of the Earth. Geothermal resources are concentrations of the Earth's heat, or geothermal energy, that can be extracted and used economically now or in the reasonable future. Currently, only concentrations of heat associated with water in permeable rocks can be exploited. Heat, fluid, and permeability are the three necessary components of all exploited geothermal fields. This section of Energy Resources will discuss the mechanisms for concentrating heat near the surface, the types of geothermal systems, and the environmental aspects of geothermal production.

7.1 Heat Flow

Temperature within the Earth increases with depth at an average of about 25°C/km. Spatial variations of the thermal energy within the deep crust and mantle of the Earth give rise to concentrations of thermal energy near the surface of the Earth that can be used as an energy resource. Heat is transferred from the deeper portions of the Earth by conduction of heat through rocks, by the movement of hot, deep rock toward the surface, and by deep circulation of water. Most high-temperature geothermal resources are associated with concentrations of heat caused by the movement of magma (melted rock) to near-surface positions where the heat is stored.

In older areas of continents, such as much of North America east of the Rocky Mountains, heat flow is generally 40–60 mW/m^2 (milliwatts per square meter). This heat flow coupled with the thermal conductivity of rock in the upper 4 km of the crust yields subsurface temperatures of 90°C–110°C at 4 km depth in the Eastern United States. Heat flow within the Basin and Range (west of the Rocky Mountains) is generally 70–90 mW/m^2, and temperatures are generally greater than 110°C at 4 km. There are large variations in the Western United States, with areas of heat flow greater than 100 mW/m^2 and areas which have generally lower heat flow such as the Cascade and Sierra Nevada Mountains and the West Coast. A more detailed discussion of heat flow in the United States is available in Blackwell et al. (1991).

7.2 Types of Geothermal Systems

Geothermal resources are hydrothermal systems containing water in pores and fractures. Most hydrothermal resources contain liquid water, but higher temperatures or lower pressures can create conditions where steam and water or only steam are the continuous phases (White et al. 1971; Truesdell and White 1973). All commercial geothermal production is expected to be restricted to hydrothermal systems for many years because of the cost of artificial addition of water. Successful, sustainable geothermal energy usage depends on reinjection of the maximum quantity of produced fluid to augment natural recharge of hydrothermal systems.

Other geothermal systems that have been investigated for energy production are (1) geopressured-geothermal systems containing water with somewhat elevated temperatures (above normal gradient) and with pressures well above hydrostatic for their depth; (2) magmatic systems, with temperature from 600°C–1400°C; and (3) hot dry rock geothermal systems, with temperatures from 200°C–350°C, that are subsurface zones with low initial permeability and little water. These types of geothermal systems cannot be used for economic production of energy at this time.

7.3 Geothermal Energy Potential

The most recent report (Huttrer 1995) shows that 6800 MW_e (megawatts electric) of geothermal electric generating capacity is on-line in 21 countries (Table 7.1). The expected capacity in the year 2000 is 9960 MW_e. Table 7.2 lists the electrical capacity of U.S. geothermal fields. Additional details of the U.S. generating capacity are available in McClarty and Reed (1992) and DiPippo (1995). Geothermal resources also provide energy for agricultural uses, heating, industrial uses, and bathing. Freeston (1995) reports that 27 countries had a total of 8228 MW_t (megawatts thermal) of direct use capacity.

TABLE 7.1 Installed and Projected Geothermal Power Generation Capacity

Country	1995	2000
Argentina	0.67	n/a[a]
Australia	0.17	n/a
China	28.78	81
Costa Rica	55	170
El Salvador	105	165
France	4.2	n/a
Greece[b]	0	n/a
Iceland	49.4	n/a
Indonesia	309.75	1080
Italy	631.7	856
Japan	413.705	600
Kenya	45	n/a
Mexico	753	960
New Zealand	286	440
Nicaragua	35	n/a
Philippines	1227	1978
Portugal (Azores)	5	n/a
Russia	11	110
Thailand	0.3	n/a
Turkey	20.6	125
U.S.	2816.775	3395
Total	6797.975	9960

[a] n/a, information not available.
[b] Greece has closed its 2.0 MW_e Milos pilot plant.
Source: From Huttrer, G. W., in *Proceedings of the World Geothermal Congress, 1995*, International Geothermal Association, Auckland, New Zealand, 1995, 3–14. With permission.

TABLE 7.2 U.S. Installed Geothermal Electrical Generating Capacity in MW_e

Rated State/Field	Plant Capacity	Type
California		
Casa Diablo	27	B
Coso	240	2F
East Mesa	37	2F
East Mesa	68.4	B
Honey Lake Valley	2.3	B
Salton Sea	440	2F
The Geysers	1797	S
Hawaii		
Puna	25	H
Nevada		
Beowawe	16	2F
Brady Hot Springs	21	2F
Desert Peak	8.7	2F
Dixie Valley	66	2F
Empire	3.6	B
Soda Lake	16.6	B
Steamboat	35.1	B
Steamboat	14.4	1F
Stillwater	13	B
Wabuska	1.2	B
Utah		
Roosevelt	20	1F
Cove Fort	2	B
Cove Fort	9	S

Note: S, natural dry steam; 1F, single flash; 2F, double flash; B, binary; H, hybrid flash and binary.

The total energy used is estimated to be 105,710 TJ/year (terajoules per year). The thermal energy used by the ten countries using the most geothermal resource for direct use is listed in Table 7.3.

The U.S. Geological Survey has prepared assessments of the geothermal resources of the U.S. Muffler (1979) estimated that the identified hydrothermal resource, that part of the **identified accessible base** that could be extracted and utilized at some reasonable future time, is 23,000 MW_e for 30 years. This resource would operate power plants with an aggregate capacity of 23,000 MW_e for 30 years. The undiscovered U.S. resource (inferred from knowledge of Earth science) is estimated to be 95,000–150,000 MW_e for 30 years.

TABLE 7.3 Geothermal Energy for Direct Use by the Ten Largest Users Worldwide

Country	Flow Rate (kg/s)	Installed Power (MW_t)	Energy Used (TJ/year)
China	8,628	1,915	16,981
France	2,889	599	7,350
Georgia	1,363	245	7,685
Hungary	1,714	340	5,861
Iceland	5,794	1,443	21,158
Italy	1,612	307	3,629
Japan	1,670	319	6,942
New Zealand	353	264	6,614
Russia	1,240	210	2,422
U.S.	3.905	1,874	13.890
Total	37,050	8,664	112,441

Source: From Freeston, D. H., in *Proceedings of the World Geothermal Congress, 1995*, International Geothermal Association, Auckland, New Zealand, 1995, 15–26. With permission.

7.4 Geothermal Applications

In 1991, geothermal electrical production in the United States was 15,738 GWh (gigawatt hours), and the largest in the world (McLarty and Reed 1992).

Most geothermal fields are water dominated, where liquid water at high temperature, but also under high (hydrostatic) pressure, is the pressure-controlling medium filling the fractured and porous rocks of the reservoir. In water-dominated geothermal systems used for electricity, water comes into the wells from the reservoir, and the pressure decreases as the water moves toward the surface, allowing part of the water to boil. Since the wells produce a mixture of flashed steam and water, a separator is installed between the wells and the power plant to separate the two phases. The flashed steam goes into the turbine to drive the generator, and the water is injected back into the reservoir.

Many water-dominated reservoirs below 175°C used for electricity are pumped to prevent the water from boiling as it is circulated through heat exchangers to heat a secondary liquid that then drives a turbine to produce electricity. **Binary geothermal plants** have no emissions because the entire amount of produced geothermal water is injected back into the underground reservoir. The identified reserves of lower-temperature geothermal fluids are many times greater than the reserves of high-temperature fluids, providing an economic incentive to develop more-efficient power plants.

Warm water, at temperatures above 20°C, can be used directly for a host of processes requiring thermal energy. Thermal energy for swimming pools, space heating, and domestic hot water are the most widespread uses, but industrial processes and agricultural drying are growing applications of geothermal use. In 1995, the United States was using over 500 TJ/year of energy from geothermal sources for direct use (Lienau et al. 1995). The cities of Boise, ID; Elko, NV; Klamath Falls, OR; and San Bernardino and Susanville, CA have geothermal district-heating systems where a number of commercial and residential buildings are connected to distribution pipelines circulating water at 54°C–93°C from the production wells (Rafferty 1992).

The use of geothermal energy through ground-coupled heat pump technology has almost no impact on the environment and has a beneficial effect in reducing the demand for electricity. Geothermal heat pumps use the reservoir of constant temperature, shallow groundwater and moist soil as the heat source during winter heating and as the heat sink during summer cooling. The energy efficiency of geothermal heat pumps is about 30% better than that of air-coupled heat pumps and 50% better than electric-resistance heating. Depending on climate, advanced geothermal heat pump use in the United States reduces energy consumption and, correspondingly, power-plant emissions by 23%–44% compared to advanced air-coupled heat pumps, and by 63%–72% compared with electric-resistance heating and standard air conditioners (L'Ecuyer et al. 1993).

7.5 Environmental Constraints

Geothermal energy is one of the cleaner forms of energy now available in commercial quantities. Geothermal energy use avoids the problems of acid rain, and it greatly reduces greenhouse gas emissions and other forms of air pollution. Potentially hazardous elements produced in geothermal brines are removed from the fluid and injected back into the producing reservoir. Land use for geothermal wells, pipelines, and power plants is small compared with land use for other extractive energy sources such as oil, gas, coal, and nuclear. Geothermal development projects often coexist with agricultural land uses, including crop production or grazing. The average geothermal plant occupies only 400 m^2 for the production of each gigawatt hour over 30 years (Flavin and Lenssen 1991). The low life-cycle land use of geothermal energy is many times less than the energy sources based on mining, such as coal and nuclear, which require enormous areas for the ore and processing before fuel reaches the power plant. Low-temperature applications usually are no more intrusive than a normal water well. Geothermal development will serve the growing need for energy sources with low atmospheric emissions and proven environmental safety.

All known geothermal systems contain aqueous carbon dioxide species in solution, and when a steam phase separates from boiling water, CO_2 is the dominant (over 90% by weight) **noncondensible gas.** In most geothermal systems, noncondensible gases make up less than 5% by weight of the steam phase. Thus, for each megawatt-hour of electricity produced in 1991, the average emission of carbon dioxide by plant type in the United States was 990 kg from coal, 839 kg from petroleum, 540 kg from natural gas, and 0.48 kg from geothermal flashed-steam (Colligan 1993). Hydrogen sulfide can reach moderate concentrations of up to 2% by weight in the separated steam phase from some geothermal fields.

At The Geysers geothermal field in California, either the Stretford process or the incineration and injection process is used in geothermal power plants to keep H_2S emissions below 1 ppb (part per billion). Use of the Stretford process in many of the power plants at The Geysers results in the production and disposal of about 13,600 kg of sulfur per megawatt of electrical generation per year. Figure 7.1, shows a typical system used in the Stretford process at The Geysers (Henderson and Dorighi 1989).

The incineration process burns the gas removed from the steam to convert H_2S to SO_2, the gases are absorbed in water to form SO_3^{-2} and SO_4^{-2} in solution, and iron chelate is used to form $S_2O_3^{-2}$ (Bedell and Hammond 1987). Figure 7.2 shows an incineration abatement system (Bedell and Hammond 1987). The major product from the incineration process is a soluble thiosulfate which is injected into the reservoir with the condensed water used for the reservoir pressure-maintenance program. Sulfur emissions for each megawatt-hour of electricity produced in 1991, as SO_2 by plant type in the United States was 9.23 kg from coal, 4.95 kg from petroleum, and 0.03 kg from geothermal flashed-steam (Colligan 1993). Geothermal power plants have none of the nitrogen oxide emissions that are common from fossil fuel plants.

The waters in geothermal reservoirs range in composition from 0.1 to over 25 wt% dissolved solutes. The geochemistry of several representative geothermal fields is listed in Table 7.4. Temperatures up to 380°C have been recorded in geothermal reservoirs in the United States, and many chemical species have a significant solubility at high temperature. For example, all of the geothermal waters are saturated in silica with respect to quartz. As the water is produced, silica becomes supersaturated, and, if steam is flashed, the silica becomes highly supersaturated. Upon cooling, amorphous silica precipitates from the

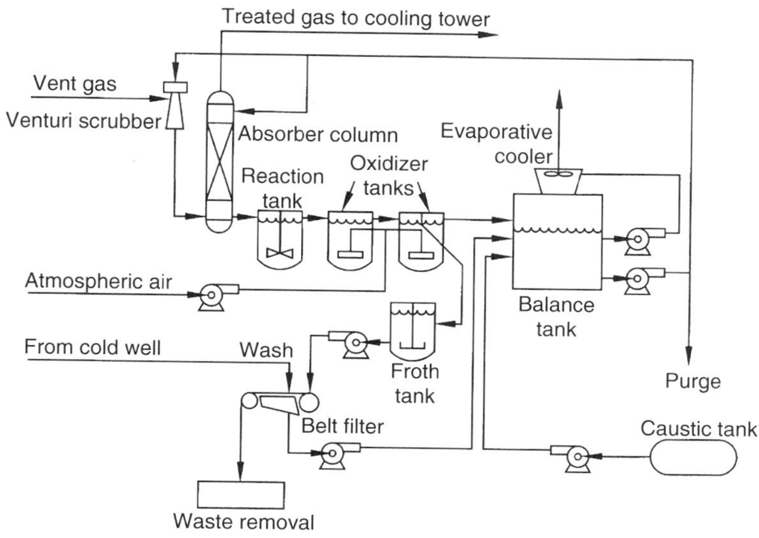

FIGURE 7.1 Typical equipment used in the Stretford process for hydrogen sulfide abatement at The Geysers geothermal field. (Based on the diagram of Henderson, J. M. and Dorighi, G. P., *Geotherm. Resour. Counc. Trans.*, 13, 593–595, 1989.)

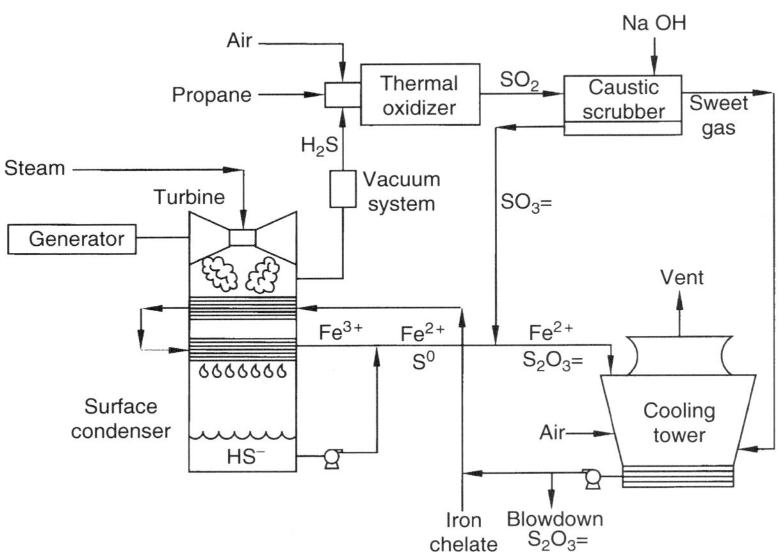

FIGURE 7.2 Equipment used in the incineration process for hydrogen sulfide abatement at The Geysers geothermal field. (Based on the diagram of Bedell, S. A. and Hammond, C. A., *Geotherm. Resour. Counc. Bull.*, 16 (8), 3–6, 1987.)

supersaturated solution. The high flow rates of steam and water from geothermal wells usually prevent silica from precipitating in the wells, but careful control of fluid conditions and residence time is needed to prevent precipitation in surface equipment. Silica precipitation is delayed in the flow stream until the water reaches a crystallizer or settling pond. There the silica is allowed to settle from the water, and the water is then pumped to an injection well.

7.6 Operating Conditions

For electrical generation, typical geothermal wells in the United States have production casing pipe in the reservoir with an inside diameter of 29.5 cm, and flow rates usually range between 150,000 and 350,000 kg/h of total fluid (Mefferd 1991). The geothermal fields contain water, or water and steam, in the reservoir, and production rates depend on the amount of boiling in the reservoir and the well on the way to the surface. The Geysers geothermal field in California has only steam filling fractures in the reservoir, and, in 1987 (approximately 30 years after production began), the average well flow had decreased to 33,000 kg/h of dry steam (Mefferd 1991) supplying the maximum field output of 2000 MW_2. Continued pressure decline has decreased the production.

In the Coso geothermal field near Ridgecrest, CA initial reservoir conditions formed a steam cap at 400–500 m depth, a two-phase (steam and water) zone at intermediate depth, and a liquid water zone at greater depth. Enthalpy of the fluid produced from individual wells ranges from 840 to 2760 kJ/kg (Hirtz et al. 1993), reservoir temperatures range from 200 to 340°C, and the fluid composition flowing from the reservoir into the different wells ranges from 100% liquid to almost 100% steam. Production wells have a wide range of flow rates, but the average production flow rate is 135,000 kg/h (Mefferd 1991). Much of the produced fluid is evaporated to the atmosphere in the cooling towers of the power plant, and only about 65% of the produced mass is available for injection into the reservoir at an average rate of 321,000 kg/h (Mefferd 1991).

The Salton Sea geothermal system in the Imperial Valley of southern California has presented some of the most difficult problems in brine handling. Water is produced from the reservoir at temperatures

TABLE 7.4 Major Element Chemistry of Representative Geothermal Wells

Field	T (°C)	Na	K	Li	Ca	Mg	Cl	F	Br	SO_4	Total[a] CO_2	Total[a] SiO_2	Total[a] B	Total[a] H_2S
Reykyavik, Iceland	100	95	1.5	<1	0.5	—	31	—	—	16	58	155	0.03	—
Hveragerdi, Iceland	216	212	27	0.3	1.5	0.0	197	1.9	0.45	61	55	480	0.6	7.3
Broadlands, New Zealand	260	1,050	210	1.7	2.2	0.1	1,743	7.3	5.7	8	128	805	48.2	<1
Wairekai, New Zealand	250	1,250	210	13.2	12	0.04	2,210	8.4	5.5	28	17	670	28.8	1
Cerro Prieto, Mexico	340	5,820	1,570	19	280	8	10,420		14.1	0	1,653	740	12.4	700
Salton Sea, California	340	50,400	17,500	215	28,000	54	155,000	15	120	5	7,100	400	390	16
Roosevelt, Utah[b]	<250	2,320	461	25.3	8	<2	3,860	6.8	—	72	232	563	—	—

[a] Total CO_2, Sio_2, etc., is the total $CO_2 + HCO_2^- + CO_3^{2-}$ expressed as CO_2, silica + silicate as SIO_2, etc.
[b] From Wright (1991); remainder of data from Ellis and Mahon (1977).

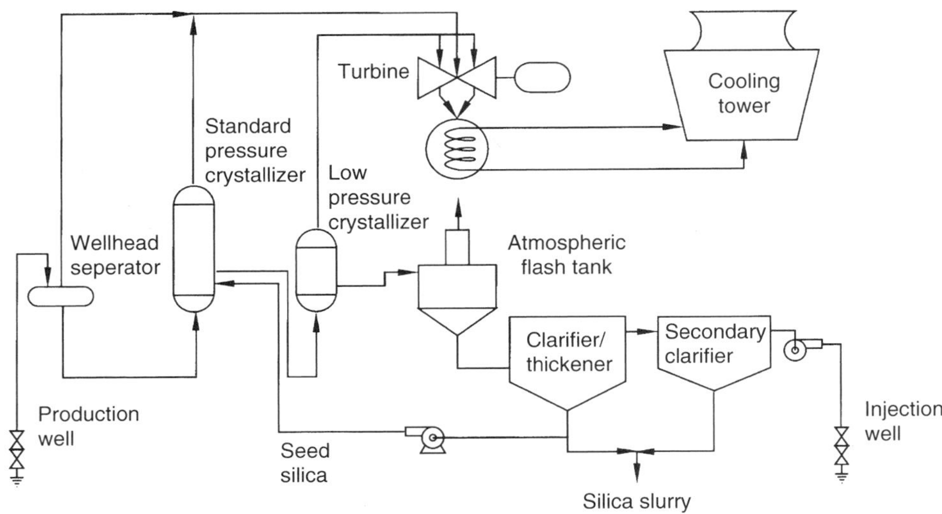

FIGURE 7.3 The flow stream for removal of solids from the vapor and brine in typical power plants in the Salton Sea geothermal field. (Modified from the diagram of Signorotti, V. and Hunter, C. C., *Geotherm. Resour. Counc. Bull.*, 21 (9), 277–288, 1992.)

between 300 and 350°C and total dissolved solid concentrations between 20 and 25% by weight at an average rate of 270,000 kg/h (Mefferd 1991). When up to 20% of the mass of brine boils during production, the salts are concentrated in the brine causing supersaturation with respect to several solid phases. Crystallizers and clarifier and thickener tanks are needed to remove solids from the injection water. Figure 7.3 shows the flow stream for removal of solids from the vapor and brine (Signorotti and Hunter 1992). Other power plants use the addition of acid to lower the pH and keep the solutes in solution (Signorotti and Hunter 1992). The output from the crystallizers and clarifiers is a slurry of brine and amorphous silica. The methods used to dewater the salt and silica slurry from operations in the Salton Sea geothermal system are described by Benesi (1992). Approximately 80% of the produced water is injected into the reservoir at an average rate of 310,000 kg/h.

Defining Terms

Binary geothermal plant: A geothermal electric generating plant that uses the geothermal fluid to heat a secondary fluid that is then expanded through a turbine.

Identified accessible base: That part of the thermal energy of the Earth that is shallow enough to be reached by production drilling in the foreseeable future. *Identifed* refers to concentrations of heat that have been characterized by drilling or Earth science evidence. Additional discussion of this and other resource terms can be found in Muffler (1979).

Noncondensible gases: Gases exhausted from the turbine into the condenser that do not condense into the liquid phase.

Acknowledgments

This study was supported in part by the U.S. Department of Energy, Assistant Secretary for Energy Efficiency and Renewable Energy, Geothermal Division, under DOE Idaho Operations Office Contract DE-AC07-941D13223.

References

Bedell, S. A. and Hammond, C. A. 1987. Chelation chemistry in geothermal H_2S abatement. *Geotherm. Resour. Counc. Bull.*, 16 (8), 3–6.

Benesi, S. C. 1992. Dewatering of slurry from geothermal process streams. *Geotherm. Resourc. Counc. Trans.*, 16, 577–581.

Blackwell, D. B., Steele, J. L., and Carter, L. S. 1991. Heat-flow patterns of the North American continent; a discussion of the geothermal map of North America. In *Neotectonics of North America*, D. B. Slemmons, E. R. Engdahl, M. D. Zoback, and D. B. Blackwell, eds., Vol. 1, pp. 423–436. Geological Society of America, Boulder, CO.

Colligan, J. G. 1993. U.S. electric utility environmental statistics. In *Electric Power Annual 1991*, U.S Department of Energy, Energy Information Administration, DOE/EIA-0348(91), Washington, DC.

DiPippo, R. 1995. Geothermal electric power production in the United States: a survey and update for 1990–1994. In *Proceedings of the World Geothermal Congress, 1995*, pp. 353–362. International Geothermal Association, Auckland, New Zealand.

Ellis, A. J. and Mahon, W. A. J. 1977. *Chemistry and Geothermal Systems*. Academic Press, New York.

Flavin, C. and Lenssen, N. 1991. Designing a sustainable energy system. In *State of the World, 1991, A Worldwatch Institute Report on Progress Toward a Sustainable Society*, pp. 1–595. W. W. Norton and Company, New York.

Freeston, D. H. 1995. Direct uses of geothermal energy 1995—preliminary review. In *Proceedings of the World Geothermal Congress, 1995*, pp. 15–26. International Geothermal Association, Auckland, New Zealand.

Henderson, J. M. and Dorighi, G. P. 1989. Operating experience of converting a Stretford to a Lo-Cat(R) H_2S abatement system at Pacific Gas and Electric Company's Geysers unit 15. *Geotherm. Resour. Counc. Trans.*, 13, 593–595.

Hirtz, P., Lovekin, J., Copp, J., Buck, C., and Adams, M. 1993. Enthalpy and mass flowrate measurements for two-phase geothermal production by tracer dilution techniques. In *Proceedings of 18th Workshop on Geothermal Reservoir Engineering*, pp. 17–27. Stanford University, Palo Alto, CA, SGPTR-145.

Huttrer, G. W. 1995. The status of world geothermal power production 1990–1994. In *Proceedings of the World Geothermal Congress, 1995*, pp. 3–14. International Geothermal Association, Auckland, New Zealand.

L'Ecuyer, M., Zoi, C., and Hoffman, J. S. 1993. *Space Conditioning—The Next Frontier*. U.S. Environmental Protection Agency, Washington, DC, EPA430-R-93-004.

Lienau, P. J., Lund, J. W., and Culver, G. G. 1995. Geothermal direct use in the United States, update 1990–1995. In *Proceedings of the World Geothermal Congress, 1995*, pp. 363–372. International Geothermal Association, Auckland, New Zealand.

McLarty, L. and Reed, M. J. 1992. The U.S. geothermal industry: three decades of growth. *Energ. Sources*, 14, 443–455.

Mefferd, M. G. 1991. *76th Annual Report of the State Oil & Gas Supervisor: 1990*, California Division of Oil & Gas, Pub. 06, Sacramento, CA.

Muffler, L. J. P., ed. 1979. Assessment of geothermal resources of the United States—1978, U.S. Geological Survey Circular 790, Washington, DC.

Rafferty, K. 1992. A century of service: the Boise Warm Springs water district system. *Geotherm. Resour. Counc. Bull.*, 21 (10), 339–344.

Signorotti, V. and Hunter, C. C. 1992. Imperial Valley's geothermal resource comes of age. *Geotherm. Resour. Counc. Bull.*, 21 (9), 277–288.

Truesdell, A. H. and White, D. E. 1973. Production of superheated steam from vapor-dominated reservoirs. *Geothermics*, 2, 145–164.

White, D. E., Muffler, L. T. P., and Truesdell, A. H. 1971. Vapor-dominated hydrothermal systems compared with hot-water systems. *Econ. Geol.*, 66 (1), 75–97.

Wright, P. M. 1991. Geochemistry. *Geo-Heat Cent. Bull.*, 13 (1), 8–12.

For Further Information

Geothermal education materials are available from the Geothermal Education Office, 664 Hilary Drive, Tiburon, CA 94920.

General coverage of geothermal resources can be found in the proceedings of the Geothermal Resources Council's annual technical conference, *Geothermal Resources Council Transactions*, and in the Council's *Geothermal Resources Council Bulletin*, both of which are available from the Geothermal Resources Council, P.O. Box 1350, Davis, CA 95617-1350.

Current information concerning direct use of geothermal resources is available from the Geo-Heat Center, Oregon Institute of Technology, Klamath Falls, OR 97601.

A significant amount of geothermal information is also available on a number of geothermal home pages that can be found by searching on "geothermal" through the Internet.

II

Energy Conversion

8 **Steam Power Plant** *John Kern* 8-1
Introduction • Rankine Cycle Analysis • Topping and Bottoming Cycles • Steam Boilers • Steam Turbines • Heat Exchangers, Pumps, and Other Cycle Components • Generators

9 **Gas Turbines** *Steven I. Freedman* 9-1
Overview • History • Fuels and Firing • Efficiency • Gas Turbine Cycles • Cycle Configurations • Components Used in Complex Cycles • Upper Temperature Limit • Materials • Combustion • Mechanical Product Features

10 **Internal Combustion Engines** *David E. Klett, Elsayed M. Afify* 10-1
Introduction • Engine Types and Basic Operation • Air Standard Power Cycles • Actual Cycles • Combustion in IC Engines • Exhaust Emissions • Fuels for SI and CI Engines • Intake Pressurization—Supercharging and Turbocharging

11 **Hydraulic Turbines** *Roger E.A. Arndt* 11-1
General Description • Principles of Operation • Factors Involved in Selecting a Turbine • Performance Evaluation • Numerical Simulation • Field Tests

12 **Stirling Engines** *William B. Stine* 12-1
Introduction • Thermodynamic Implementation of the Stirling Cycle • Mechanical Implementation of the Stirling Cycle • Future of the Stirling Engine

13 **Advanced Fossil Fuel Power Systems** *Anthony F. Armor* 13-1
Introduction • Fuels for Electric Power Generation in the U.S. • Coal as a Fuel for Electric Power (World Coal Institute 2000) • Clean Coal Technology Development • Pulverized-Coal Plants • Emissions Controls for Pulverized Coal Plants • Fluidized Bed Plants • Gasification Plants • Combustion Turbine Plants • Central Station Options for New Generation • Summary

14 **Combined-Cycle Power Plants** *Alex Lezuo* 14-1
Combined-Cycle Concepts • Combined-Cycle Thermodynamics • Combined-Cycle Arrangements • Combined Heat and Power from Combined-Cycle Plants • Environmental Aspects

15 **Energy Storage Technologies** *Roel Hammerschlag, Christopher P. Schaber* 15-1
Overview of Storage Technologies • Principal Forms of Stored Energy • Applications of Energy Storage • Specifying Energy Storage Devices • Specifying Fuels • Direct Electric Storage • Electrochemical Energy Storage • Mechanical Energy Storage • Direct Thermal Storage • Thermochemical Energy Storage

16 **Nuclear Power Technologies** *Edwin A. Harvego and Kenneth D. Kok* **16-1**
Introduction • Development of Current Power-Reactor Technologies • Next-Generation Technologies • Generation-IV Technologies • Fuel Cycle • Nuclear Waste • Nuclear Power Economics • Conclusions

17 **Nuclear Fusion** *Thomas E. Shannon* **17-1**
Introduction • Fusion Fuel • Confinement Concepts • Tokamak Reactor Development • Fusion Energy Conversion and Transport

18 **Solar Thermal Energy Conversion** *T. Agami Reddy, Riccardo Battisti, Hans Schweiger, Werner Weiss, Jeffrey H. Morehouse, Sanjay Vijayaraghavan D. Yogi Goswami* **18-1**
Active Solar Heating Systems • Solar Heat for Industrial Processes • Passive Solar Heating, Cooling, and Daylighting • Solar Cooling

19 **Concentrating Solar Thermal Power** *Manuel Romero-Alvarez, Eduardo Zarza* **19-1**
Introduction and Context • Solar Concentration and CSP Systems • Solar Concentrator Beam Quality • Solar Concentration Ratio: Principles and Limitations of CSP Systems • Solar Thermal Power Plant Technologies • Parabolic Trough Solar Thermal Power Plants • Central Receiver Solar Thermal Power Plants • Volumetric Atmospheric Receivers: PHOEBUS and Solair • Solar Air Preheating Systems for Combustion Turbines: The SOLGATE Project • Dish/Stirling Systems • Market Opportunities • Conclusions

20 **Photovoltaics Fundamentals, Technology and Application** *Roger Messenger, D. Yogi Goswami, Hari M. Upadhyaya, Takhir M. Razykov, Ayodhya N. Tiwari, Roland Winston, Robert McConnell* **20-1**
Photovoltaics • Thin-Film PV Technology • Concentrating PV Technologies

21 **Wind Energy Conversion** *Dale E. Berg* **21-1**
Introduction • Wind Turbine Aerodynamics • Wind Turbine Loads • Wind Turbine Structural Dynamic Considerations • Peak Power Limitation • Turbine Subsystems • Other Wind-Energy Conversion Considerations

22 **Biomass Conversion Processes For Energy Recovery** *Massoud Kayhanian, George Tchobanoglous, Robert C. Brown* **22-1**
Energy Recovery by Anaerobic Digestion • Power Generation • Biofuels

23 **Geothermal Power Generation** *Kevin Kitz* **23-1**
Introduction • Definition and Use of Geothermal Energy • Requirements for Commercial Geothermal Power Production • Exploration and Assessment of Geothermal Resources • Management of the Geothermal Resource for Power Production • Geothermal Steam Supply (from Wellhead to Turbine) • Geothermal Power Production—Steam Turbine Technologies • Geothermal Power Production—Binary Power Plant Technologies • Environmental Impact • Additional Information on Geothermal Energy

24 **Waste-to-Energy Combustion** *Charles O. Velzy, Leonard M. Grillo* **24-1**
Introduction • Waste Quantities and Characteristics • Design of WTE Facilities • Air Pollution Control Facilities • Performance • Costs • Status of Other Technologies • Future Issues and Trends

25 **Ocean Energy Technology** *Desikan Bharathan and Federica Zangrando* **25-1**
Ocean Thermal Energy Conversion • Tidal Power • Wave Power • Concluding Remarks

26 **Fuel Cells** *Xianguo Li* **26-1**
Introduction • Principle of Operation for Fuel Cells • Typical Fuel Cell Systems • Performance of Fuel Cells • Fuel Cell Electrode Processes • Cell Connection and Stack Design Considerations • Six Major Types of Fuel Cells • Summary

27 **Direct Energy Conversion** *Mysore L. Ramalingam, Jean-Pierre Fleurial William D. Jackson* **27-1**
Thermionic Energy Conversion • Thermoelectric Power Conversion • Magnetohydrodynamic Power Generation

8

Steam Power Plant

8.1 Introduction **8-1**
8.2 Rankine Cycle Analysis **8-2**
References **8-5**
8.3 Topping and Bottoming Cycles **8-5**
For Further Information **8-5**
8.4 Steam Boilers **8-5**
Drum-Type Boilers • Once-Through Boilers • Major Boiler Components
8.5 Steam Turbines **8-7**
General • Blading • Rotors • Choosing the Turbine Arrangement • Materials • Cylinders and Bolting • Valves
For Further Information **8-11**
8.6 Heat Exchangers, Pumps, and Other Cycle Components **8-11**
Heat Exchangers • Pumps
For Further Information **8-13**
8.7 Generators **8-14**
Generator Ventilation • Generator Auxiliaries • Excitation
For Further Information **8-16**

John Kern
Siemens Power Generation

8.1 Introduction

This section provides an overview of the steam power cycle. There are noteworthy omissions in the section: site selection; fuel handling; activities related to civil engineering (such as foundations); controls; and nuclear power. Thermal power cycles take many forms, but the majority are fossil steam, nuclear, simple-cycle gas turbine, and combined cycle. Of those listed, conventional coal-fired steam is the predominant power producer—especially in developing countries that have indigenous coal or can import coal inexpensively. A typical steam power plant is shown in Figure 8.1.

Because the Rankine cycle is the overwhelmingly preferred process for steam power generation, it is discussed first. Topping and bottoming cycles, with one exception, are rare and mentioned only for completeness. The exception is the combined cycle, in which the steam turbine cycle is a bottoming cycle. Developed countries have been moving to the combined cycle because of relatively low capital costs when compared with coal-fired plants; its high thermal efficiency, which approaches 60%, and low emissions.

The core components of a steam power plant are boiler; turbine; condenser and feedwater pump; and generator. These are covered in successive subsections. The final subsection is an example of the layout and contents of a modern steam power plant.

As a frame of reference, the following efficiencies are typical for modern, subcritical, fossil fuel steam power plants. The specific example chosen has steam conditions of 2400 psia; 1000°F main steam

FIGURE 8.1 Modern steam power plant.

temperature; and 1000°F reheat steam temperature: boiler thermal 92; turbine/generator thermal 44; turbine isentropic 89; generator 98.5; boiler feedwater pump and turbine combined isentropic 82; condenser 85; plant overall 34 (Carnot 64). Supercritical steam plants operate with main steam above the "critical" pressure for water where water and steam have the same density and no longer exist as separate phase states. They are generally used when higher efficiency is desired. Modern supercritical coal plants with main steam conditions of 3600 psia at 1050 and 1050°F for reheat steam can exceed 40% in overall net plant efficiency.

Nuclear power stations are so unique that they are worthy of a few closing comments. Modern stations are all large, varying from 600 to 1500 MW. The steam is low temperature and low pressure (~600°F and ~1000 psia), compared with fossil applications, and hovers around saturation conditions. Therefore, the boilers, superheater equivalent (actually a combined moisture separator and reheater), and turbines are unique to this cycle. The turbine generator thermal efficiency is around 36%.

8.2 Rankine Cycle Analysis

Modern steam power generation is based on the Rankine cycle and thermodynamics govern the ultimate performance of the cycle whether used in a coal-fired steam plant or the bottoming cycle of a combined-cycle plant. The basic, ideal Rankine cycle is shown in Figure 8.2. The ideal cycle comprises the processes from state 1:

- 1–2: Saturated liquid from the condenser at state 1 is pumped isentropically (i.e., $S_1 = S_2$) to state 2 and into the boiler.
- 2–3: Liquid is heated at constant pressure in the boiler to state 3 (saturated steam).
- 3–4: Steam expands isentropically (i.e., $S_3 = S_4$) through the turbine to state 4, where it enters the condenser as a wet vapor.
- 4–1: Constant-pressure transfer of heat in the condenser takes place to return the steam to state 1 (saturated liquid).

If changes in kinetic and potential energy are neglected, the total heat added to the Rankine cycle can be represented by the shaded area on the *T–S* diagram in Figure 8.2; the work done by this cycle can be

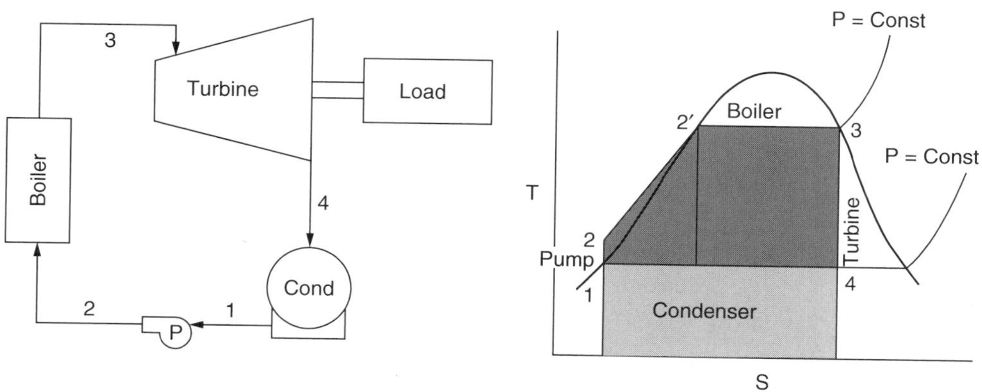

FIGURE 8.2 Basic Rankine cycle.

represented by the crosshatching within the shaded area. The thermal efficiency of the cycle (η) is defined as the work (W_{NET}) divided by the heat input to the cycle (Q_{H}).

The Rankine cycle is preferred over the Carnot cycle for the following reasons:

- The heat transfer process in the boiler must be at constant temperature for the Carnot cycle, whereas in the Rankine cycle it is superheated at constant pressure. Superheating the steam can be achieved in the Carnot cycle during heat addition, but the pressure must drop to maintain constant temperature. This means the steam is expanding in the boiler while heat is being added, which is not a practical method.
- The Carnot cycle requires that the working fluid be compressed at constant entropy to boiler pressure. This would require taking wet steam from point $1'$ in Figure 8.2 and compressing it to saturated liquid condition at $2'$. A pump required to compress a mixture of liquid and vapor isentropically is difficult to design and operate. In comparison, the Rankine cycle takes the saturated liquid and compresses it to boiler pressure. This is more practical and requires much less work.

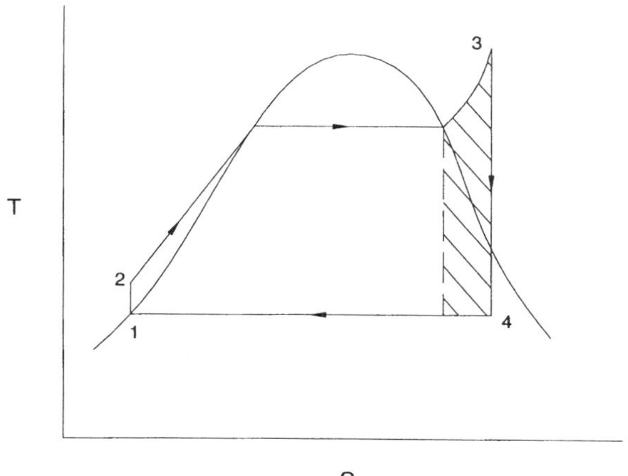

FIGURE 8.3 Rankine cycle with superheat.

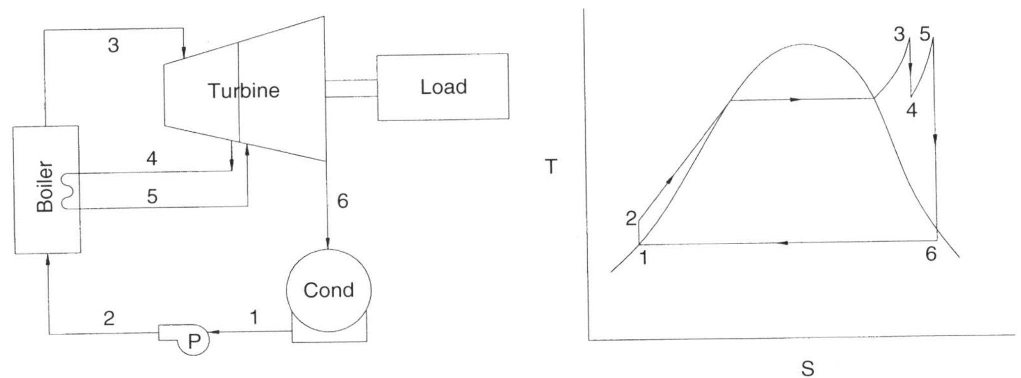

FIGURE 8.4 Rankine cycle with reheat.

The efficiency of the Rankine cycle can be increased by utilizing a number of variations to the basic cycle. One such variation is superheating the steam in the boiler. The additional work done by the cycle is shown in the crosshatched area in Figure 8.3.

The efficiency of the Rankine cycle can also be increased by increasing the pressure in the boiler. However, increasing the steam generator pressure at a constant temperature will result in the excess moisture content of the steam exiting the turbine. To take advantage of higher steam generator pressures and keep turbine exhaust moistures at acceptably low values, the steam is expanded to some intermediate pressure in the turbine and then reheated in the boiler. Following reheat, the steam is expanded to the cycle exhaust pressure. The reheat cycle is shown in Figure 8.4.

Another variation of the Rankine cycle is the regenerative cycle, which involves the use of feedwater heaters. The regenerative cycle regains some of the irreversible heat lost when condensed liquid is pumped directly into the boiler by extracting steam from various points in the turbine and heating the condensed liquid with this steam in feedwater heaters. Figure 8.5 shows the Rankine cycle with regeneration.

The actual Rankine cycle is far from ideal because losses are associated with the cycle. They include piping losses due to friction and heat transfer; turbine losses associated with steam flow; pump losses due to friction; and condenser losses when condensate is subcooled. The losses in the compression (pump) and expansion process (turbine) result in an increase in entropy. Also, energy is lost in heat addition (boiler) and rejection (condenser) processes as they occur over a finite temperature difference.

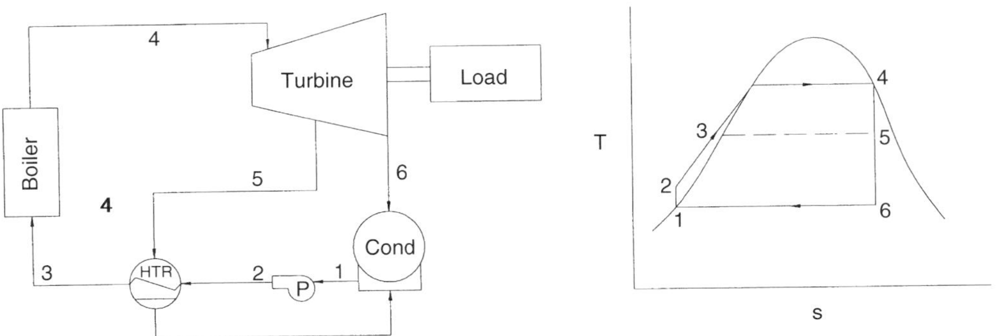

FIGURE 8.5 Rankine cycle with regeneration.

Most modern power plants employ some variation of the basic Rankine cycle in order to improve thermal efficiency. For larger power plants, economies of scale will dictate the use of one or all of these variations to improve thermal efficiency. In most cases, power plants in excess of 200,000 kW will have 300°F superheated steam leaving the boiler reheat and seven to eight stages of feedwater heating.

References

Salisbury, J. K. 1950. *Steam Turbines and Their Cycles*, reprint 1974. Robert K. Krieger Publishing, Malabar, FL.

Van Wylen, G. J. and Sonntag, R. E. 1986. *Fundamentals of Classical Thermodynamics, 3rd Ed.* Wiley, New York.

8.3 Topping and Bottoming Cycles

Steam Rankine cycles can be combined with topping and/or bottoming cycles to form binary thermodynamic cycles. These topping and bottoming cycles use working fluids other than water. Topping cycles change the basic steam Rankine cycle into a binary cycle that better resembles the Carnot cycle and improves efficiency. For conventional steam cycles, state-of-the-art materials allow peak working fluid temperatures higher than the supercritical temperature for water. Much of the energy delivered into the cycle goes into superheating the steam, which is not a constant-temperature process. Therefore, a significant portion of the heat supply to the steam cycle occurs substantially below the peak cycle temperature.

Adding a cycle that uses a working fluid with a boiling point higher than water allows more of the heat supply to the thermodynamic cycle to be near the peak cycle temperature, thus improving efficiency. Heat rejected from the topping cycle is channeled into the lower-temperature steam cycle. Thermal energy not converted to work by the binary cycle is rejected to the ambient-temperature reservoir. Metallic substances are the working fluids for topping cycles. For example, mercury has been used as the topping cycle fluid in a plant that operated for a period of time but has since been dismantled. Significant research and testing has also been performed over the years toward the eventual goal of using other substances, such as potassium, sodium, or cesium, as a topping-cycle fluid, but none has proven to be commercially successful.

Steam power plants in a cold, dry environment cannot take full advantage of the low heat rejection temperature available. The very low pressure to which the steam would be expanded to take advantage of the low heat sink temperature would increase the size of the low-pressure (LP) turbine to such an extent that it is impractical or at least inefficient. A bottoming cycle that uses a working fluid with a vapor pressure higher than water at ambient temperatures (such as ammonia or an organic fluid) would enable smaller LP turbines to function efficiently. Thus, a steam cycle combined with a bottoming cycle may yield better performance and be more cost effective than a stand-alone Rankine steam cycle. However, again, these techniques are not at present commercially viable and are not being broadly pursued.

For Further Information

Fraas, A. P. 1982. *Engineering Evaluation of Energy Systems*. McGraw-Hill, New York.

Horlock, J. H. 1992. *Combined Power Plants, Including Combined Cycle Gas Turbine (CCGT) Plants*. Pergamon Press, Oxford.

Lezuo, A. and Taud, R. 2001. Comparative evaluation of power plants with regard to technical, ecological and economical aspects. In *Proceedings of ASME Turbo Expo 2001*, New Orleans.

8.4 Steam Boilers

A boiler, also referred to as a steam generator, is a major component in the plant cycle. It is a closed vessel that efficiently uses heat produced from the combustion of fuel to convert water to steam. Efficiency is the

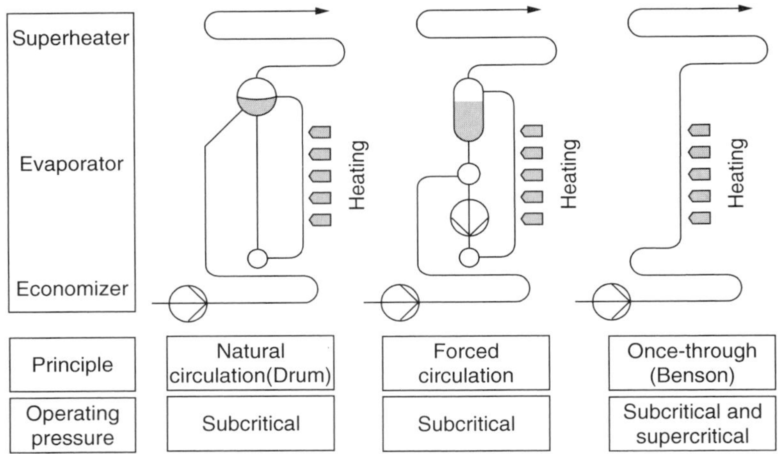

FIGURE 8.6 Boiler principles.

most important characteristic of a boiler because it has a direct bearing on electricity production. Boilers are classified as drum-type or once-through (Figure 8.6). Major components of boilers include an economizer, superheaters, reheaters, and spray attemperators.

8.4.1 Drum-Type Boilers

Drum-type boilers depend on constant recirculation of water through some of the components of the steam–water circuit to generate steam and keep the components from overheating. These boilers circulate water by natural or controlled circulation.

Natural Circulation Boilers. Natural circulation boilers use the density differential between water in the downcomers and steam in the waterwall tubes for circulation.

Controlled Circulation Boilers. Controlled circulation boilers use boiler-water-circulating pumps to circulate water through the steam–water circuit.

8.4.2 Once-Through Boilers

Once-through boilers convert water to steam in one pass through the system rather than re-circulating through the drum. Current designs for once-through boilers use a spiral-wound furnace to assure even heat distribution across the tubes.

8.4.3 Major Boiler Components

- *Economizer.* The economizer is the section of the boiler tubes in which feedwater is first introduced into the boiler and flue gas is used to raise the temperature of the water.
- *Steam drum (drum units only).* The steam drum separates steam from the steam–water mixture and keeps the separated steam dry.
- *Superheaters.* Superheaters are bundles of boiler tubing located in the flow path of the hot gases created by the combustion of fuel in the boiler furnace. Heat is transferred from the combustion gases to the steam in the superheater tubes. Superheaters are classified as primary and secondary. Steam passes first through the primary superheater (located in a relatively cool section of the boiler) after leaving the steam drum. There the steam receives a fraction of its final superheat and then passes through the secondary superheater for the remainder.

- *Reheaters.* Reheaters are bundles of boiler tubes that are exposed to the combustion gases in the same manner as superheaters.
- *Spray attemperators.* Attemperators, also known as desuperheaters, are spray nozzles in the boiler tubes between the two superheaters. These spray nozzles supply a fine mist of pure water into the flow path of the steam to prevent tube damage from overheating. Attemperators are provided for the superheater and the reheater.

Worldwide, the current trend is to use higher temperatures and pressures to improve plant efficiency, which in turn reduces emissions. Improvements in high-temperature materials such as T-91 tubing provide high-temperature strength and improved corrosion resistance permitting reliable operation in advanced steam cycles. In addition, the development of reliable once-through Benson type boilers has resolved most of the operational problems experienced with first- and second-generation supercritical plants.

Steam plant boilers burning coal require advanced exhaust gas clean-up systems to meet today's strict environmental emissions limits. A typical plant burning high-sulfur eastern coal will have an SCR (selective catalytic reduction) for NO_x control, a precipitator for particulate control, and a wet limestone scrubber to reduce SO_x. A typical plant burning low-sulfur western coal might include an SCR, a baghouse filter for particulate control, and a dry scrubber for SO_x reduction.

8.5 Steam Turbines

8.5.1 General

Each turbine manufacturer has unique features in its designs that affect efficiency, reliability, and cost. However, the designs appear similar to a non-steam-turbine engineer. Figure 8.7 shows a modern steam turbine generator as used in a coal-fired steam power plant. Steam turbines for power plants differ from most prime movers in at least three ways:

- All are extremely high powered, varying from about 70,000 to 2 million hp, and require a correspondingly large capital investment, which puts a premium on reliability.
- Turbine life is normally between 30 and 40 years with minimal maintenance.

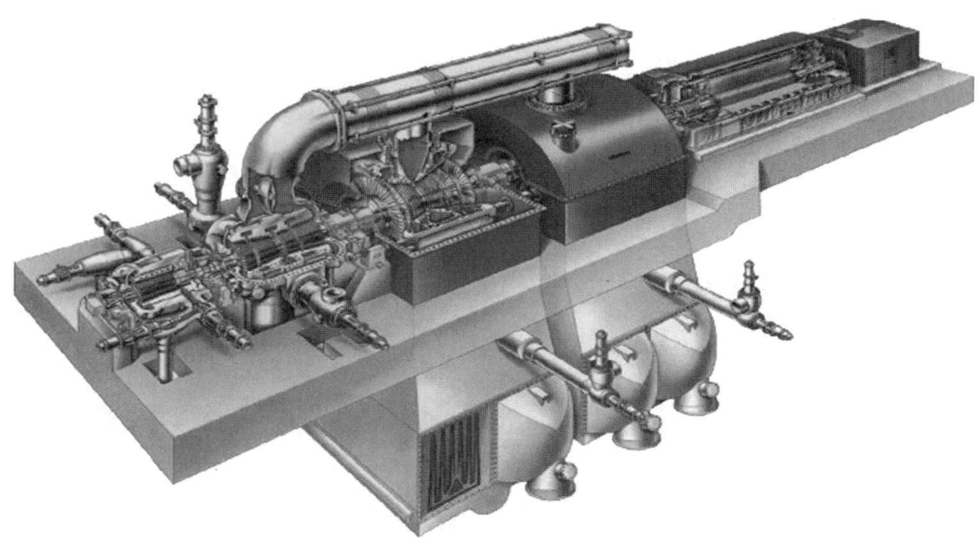

FIGURE 8.7 Modern steam turbine generator for a coal-fired steam plant.

- Turbines spend the bulk of their lives at constant speed, normally 3600 or 1800 rpm for 60-Hz operation.

These three points dominate the design of the entire power station, particularly of the steam turbine arrangement and materials. Figure 8.8 shows the dramatic increase of steam turbine power output for one manufacturer over the past 50 years. This is reasonably typical of the industry.

In an earlier subsection it was shown that high steam-supply temperatures make for more efficient turbines. In Europe and Japan, the trend is to use increasingly higher steam-supply temperatures to reduce fuel cost and emissions.

8.5.2 Blading

The most highly stressed component in steam turbines is the blades. Blades are loaded by centrifugal and steam-bending forces and also harmonic excitation (from nonuniform circumferential disturbances in the blade path). All blades are loaded by centrifugal and steam-bending loads, and smaller blades are designed to run when the harmonic excitation is resonant with the natural modes of the blade. If harmonic excitation is permitted on very long blades, however, the blades become impractically large. Fortunately, because the turbine runs at constant speed, the blade modes can be tuned away from resonant conditions so that the harmonic loads are significantly reduced. This forms a split in blade design, commonly referred to as tuned and untuned blading.

Blades guide steam throughout the turbine in as smooth and collision-free a path as possible. Collisions with blades (incidence) and sudden expansions reduce the energy available for doing work. Until recently, designers would match flow conditions with radially straight blades (called parallel-sided blades). Turbine physics does not recognize this convenience for several reasons. The most visually obvious is the difference in tangential velocity between blade hub and tip. The latest blades address the full three-dimensional nature of the flow by curving in three dimensions (bowed blades). Three dimensional design techniques allow for better matching of the flow (and area) conditions and now, with the use of numerical control machine tools to make it more cost competitive, three-dimensional blading is used extensively in many modern turbines. Examples of three-dimensional and parallel-sided blades are shown in Figure 8.9.

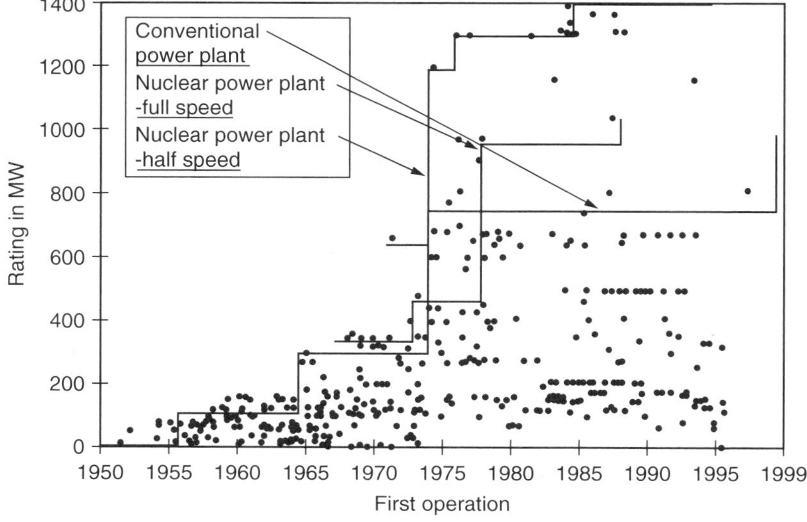

FIGURE 8.8 Increase of steam turbine power.

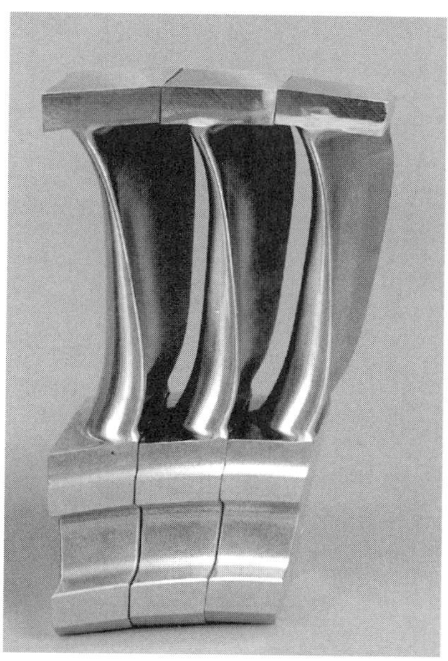
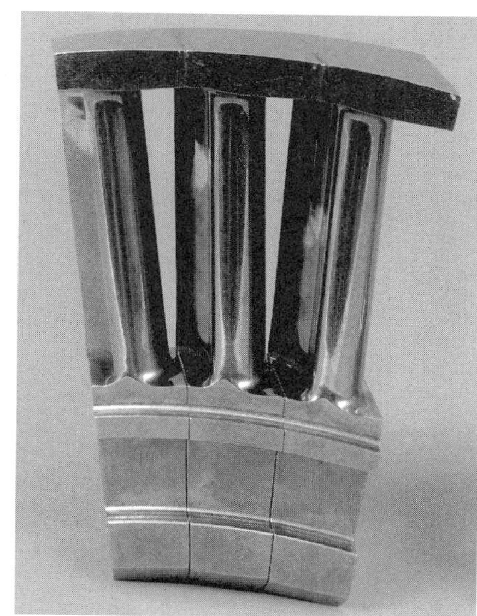

FIGURE 8.9 Typical steam turbine blades.

8.5.3 Rotors

After blades, steam turbine rotors are the second most critical component in the machine. Rotor design must take into account

- The large high-strength alloy steel rotor forging that must have uniform chemistry and material properties
- Centrifugal force from the rotor body and the increased centrifugal pull from the attached blades
- The need to have high resistance to brittle fracture, which could occur when the machine is at high speed and the material is still not up to operating temperature
- Creep deformation of the high-pressure (HP) and intermediate-pressure (IP) rotors under steady load while at high temperature.

The life cycle is further complicated by transient fatigue loads that occur during power changes and start-up. Two further events are considered in rotor design: torsional and lateral vibrations caused by harmonic steam and electrical loads. As with tuned blades, this is normally addressed by tuning the primary modes away from resonance at full running speed.

8.5.4 Choosing the Turbine Arrangement

Because the turbine shaft would be too long and flexible if it were built in one piece with all the blades in sequence, the rotor is separated into supportable sections. The "cuts" in the shaft result in HP (high pressure), IP (intermediate pressure), and LP (low pressure) cylinders. Manufacturers address the grouping of cylinders in many different ways, depending upon steam conditions. It is common practice to combine HPs and IPs into one cylinder for subcritical units in the power range of about 250 to 600 MW. One manufacturer's grouping, shown in Figure 8.10, is fairly representative of the industry.

So far, the text has discussed the steam flow as though it expanded monotonically through the turbine. This is usually not the case for two reasons. First, the most common steam conditions would cause steam exiting the last row of blades to be very wet, creating excessive erosion. Second, thermal efficiency can be raised by removing the steam from the turbine, reheating, and then returning it to the blade path; this

Type	Basic Configuration	Power
HP-IP-LP (with reheat)	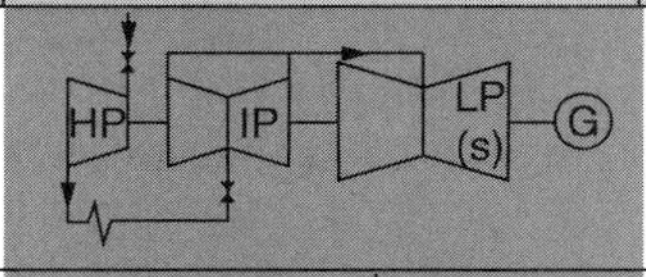	200Ð1200 MW
HP/IP-LP (with reheat)	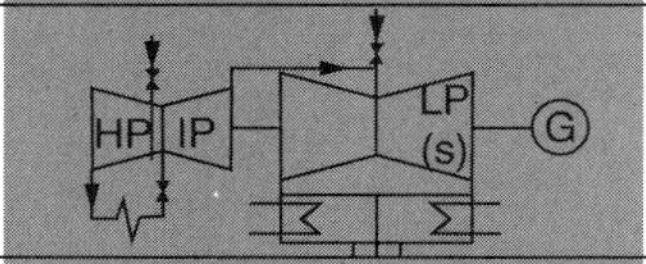	160Ð700 MW
HP-IP/LP (with reheat)	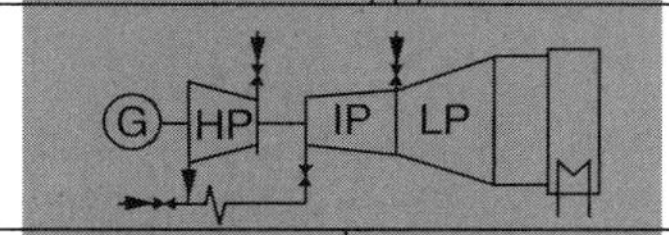	120Ð260 MW
Single case (without reheat)	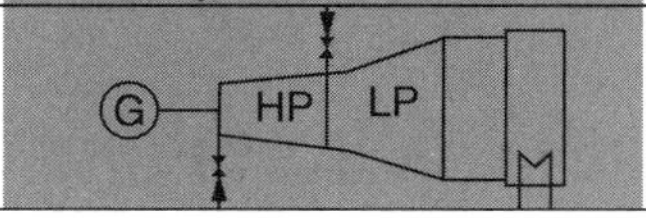	50Ð300 MW

FIGURE 8.10 Steam turbine product combinations.

increases the "average" heat supply temperature and reduces moisture levels in the turbine exhaust. The turbine position for reheat is normally between the HP and IP turbines.

There is one further geometric arrangement. Cylinders need not be all on one shaft with a single generator at the end. A cross-compound arrangement exists in which the steam path is split into two separate parallel paths with a generator on each path. Commonly, the split will be with paths of HP–LP generator and IP–LP generator. Torsional and lateral vibrations are more easily analyzed with shorter trains, which make the foundation more compact. The primary shortcoming is the need for two generators, two control systems, and a larger power house—all of which increase overall plant cost.

Historically, steam turbines have been split into two classes, reaction and impulse, as explained in Basic Power Cycles. This difference in design makes an observable difference between machines. Impulse turbines have fewer, wider stages than reaction machines. As designs have been refined, the efficiencies and lengths of the machines are now about the same. For a variety of reasons, the longer blades in the LP ends are normally reaction designs. Because each stage may now be designed and fabricated separately, the line between impulse and reaction turbines is diminishing with most manufacturers supplying blading that has characteristics of both technologies. Turbine blading is broadly split between machines as shown in the following table.

	Cylinder			
			LP	
	HP	IP	Short Blades	End Blade(s)
Reaction turbines	Reaction	Reaction	Reaction	Reaction
Impulse turbines	Impulse	Impulse	Impulse	Reaction

8.5.5 Materials

Materials are among the most variable of all turbine parts, with each manufacturer striving to improve performance by using alloying and heat-treatment techniques. It follows that accurate generalizations are difficult. Even so, the following table is reasonably representative for steam turbines with 1000°F–1050°F inlet temperatures:

Item	Common Material Description								
High-temperature HP and IP blades	Moderate- and cold-temperature stator blades	Moderate-temperature rotating blades	Cold LP rotating blades	High-temperature rotors	Low-temperature rotors	Hot	LP	High-temperature bolting	Cold bolting
Mod'd SS403	SS304	SS403	SS403 or 17/4 PH	1CrMoV, occasionally 10Cr	3.5 NiCrMoV	1.25Cr or 2.25Cr	Carbon, steel	SS422	B16

8.5.6 Cylinders and Bolting

These items are relatively straightforward, except for the very large sizes and precision required for the castings and fabrications. In a large HP–IP cylinder, the temperature and pressure loads split between an inner and outer cylinder. In this case, finding space and requisite strength for the bolting presents a challenge for the designer.

8.5.7 Valves

The turbine requires many valves for speed control, emergency control, drains, hydraulics, bypasses, and other functions. Of these, four valves are distinguished by their size and duty: throttle or stop; governor or control; reheat stop; and reheat interceptor. The throttle, reheat stop, and reheat interceptor valves normally operate fully open, except in some control and emergency conditions. Their numbers and design are selected for the appropriate combination of redundancy and rapidity of action. The continuous control of the turbine is accomplished by throttling the steam through the governor valve. This irreversible process detracts from cycle efficiency. In more modern units, the efficiency loss is reduced by reducing the boiler pressure (normally called sliding pressure) rather than throttling across the valves when reducing output.

For Further Information

Japikse, D. and Nicholas, C. B. 1994. *Introduction to Turbomachinery.* Concepts ETI, Norwich, VT.

Kutz, M. 1986. *Mechanical Engineers' Handbook.* Wiley, New York.

Stodola, A. and Loewenstein, L. C. 1927. *Steam and Gas Turbines*, reprint of 6th Ed., 1945. Peter Smith, New York.

8.6 Heat Exchangers, Pumps, and Other Cycle Components

8.6.1 Heat Exchangers

Heaters. The two classifications of condensate and feedwater heaters are the open or direct contact heater and the closed or shell-and-tube heater.

Open Heaters. In an open heater, the extraction or heating steam comes in direct contact with the water to be heated. Although open heaters are more efficient than closed heaters, each requires a pump to feed the outlet water ahead in the cycle. This adds cost and maintenance and increases the risk of water induction to the turbine, making the closed heater the preferred heater for power plant applications.

Closed Heaters. These heaters employ tubes within a shell to separate the water from the heating steam (see Figure 8.11). They can have three separate sections in which the heating of the feedwater occurs.

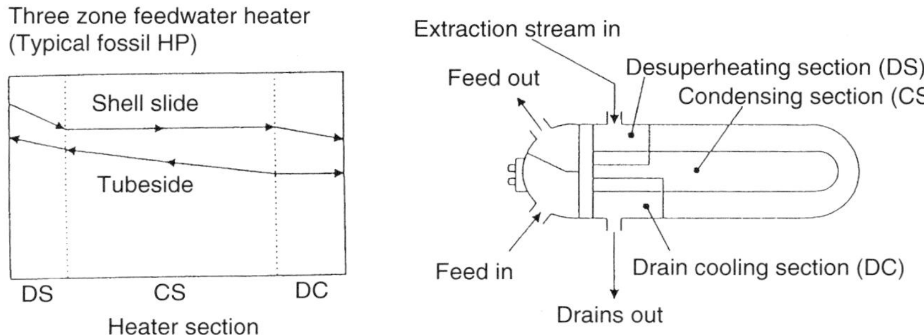

FIGURE 8.11 Shell-and-tube feedwater heater.

First is the drain cooler section where the feedwater is heated by the condensed heating steam before cascading back to the next-lower-pressure heater. The effectiveness of the drain cooler is expressed as the drain cooler approach (DCA), which is the difference between the temperature of the water entering the heater and the temperature of the condensed heating steam draining from the heater shell. In the second section (condensing section), the temperature of the water is increased by the heating steam condensing around the tubes. In the third section (desuperheating section), the feedwater reaches its final exit temperature by desuperheating the extraction steam. Performance of the condensing and superheating sections of a heater is expressed as the terminal temperature difference (TTD). This is the difference between the saturation temperature of the extraction steam and the temperature of the feedwater exiting the heater. Desuperheating and drain cooler sections are optional depending on the location of the heater in the cycle (for example, desuperheating is not necessary in wet extraction zones) and economic considerations.

The one exception is the deaerator (DA), which is an open heater used to remove oxygen and other gases that are insoluble in boiling water. The DA is physically located in the turbine building above all other heaters, and the gravity drain from the DA provides the prime for the boiler feed pump (BFP).

Two other critical factors considered in heater design and selection are (1) venting the heater shell to remove any noncondensable gases; and (2) the protection of the turbine caused by malfunction of the heater system. Venting the shell is required to avoid air-binding a heater, which reduces its performance. Emergency drains to the condenser open when high water levels are present within the shell to prevent back-flow of water to the turbine, which can cause serious damage. Check valves on the heating steam line are also used with a water-detection monitor to alert operators to take prompt action when water is present.

Condenser. Steam turbines generally employ surface-type condensers comprising large shell-and-tube heat exchangers operating under vacuum. The condenser (1) reduces the exhaust pressure at the last-stage blade exit to extract more work from the turbine; and (2) collects the condensed steam and returns it to the feedwater-heating system. Cooling water circulates from the cooling source to the condenser tubes by large motor-driven pumps. Multiple pumps, each rated less than 100% of required pumping power, operate more efficiently at part load and are often used to allow for operation if one or more pumps are out of service. Cooling water is supplied from a large heat sink water source, such as a river, or from cooling towers. The cooling in the cooling tower is assisted by evaporation of 3%–6% of the cooling water. Airflow is natural draft (hyperbolic towers) or forced draft. Noncondensable gases are removed from the condenser with a motor-driven vacuum pump or, more frequently, steam jet air ejectors, which have no moving parts.

When adequate cooling water is not available, a dry condenser can be used. This device uses large motor-driven fans to move air across a large radiator-like heat exchanger to condense the steam at

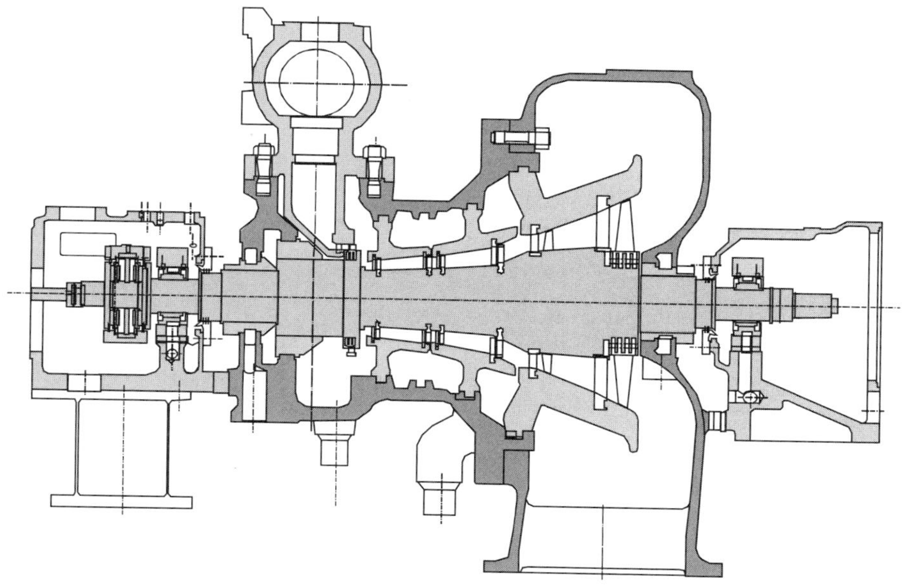

FIGURE 8.12 Boiler feed pump turbine.

ambient temperature. Air condensers are significantly more expensive than wet condensers and generally decrease overall plant efficiency, so they are used only when necessary.

8.6.2 Pumps

Condensate Pump. Condensate is removed from the hot well of the condenser and passed through the LP heater string via the condensate pump. Typically, two or more vertical (larger units) or horizontal (medium and small units) motor-driven centrifugal pumps are located near the condenser hot well outlet. Depending on the size of the cycle, condensate booster pumps may be used to increase the pressure of the condensate on its way to the DA.

Feedwater Booster Pump. The DA outlet supplies the feedwater booster pump, which is typically a motor-driven centrifugal pump. This pump supplies the required suction head for the BFP (boiler feed pump).

Boiler Feed Pump. These pumps are multiple-stage centrifugal pumps that, depending on the cycle, can be turbine or motor driven. BFP turbines (BFPT; Figure 8.12), are single-case units that draw extraction steam from the main turbine cycle and exhaust to the main condenser. Typical feed pump turbines require 0.5% of the main unit power at full-load operation. Multiple pumps rated at 50%–100% each are typically used to allow the plant to operate with one pump out of service.

With the increasing reliability of large electric motors, many plant designers are now using motors to drive the feed pumps for plants up to about 800 MW. Although the cycle is not quite as efficient as using a turbine drive, the overall plant capital cost is significantly less when motor BFP drives are used.

For Further Information

British Electricity International 1992. *Modern Power Station Practice 3rd Ed.* Pergammon Press, Oxford.
Lammer, H. B. and Woodruff 1967. *Steam Plant Operation, 3rd Ed.* McGraw-Hill, New York.
Powell, C. 1955. *Principles of Electric Utility Operation.* Wiley, New York.

8.7 Generators

The electric generator converts rotating shaft mechanical power of the steam turbine to three-phase electrical power at voltages between 11.5 and 27 kV, depending upon the power rating. The generator comprises a system of ventilation, auxiliaries, and an exciter. Figure 8.13 shows an installed hydrogen-cooled generator and brushless exciter of about 400 MW. Large generators greater than 25 MW usually have a solid, high-strength steel rotor with a DC field winding embedded in radial slots machined into the rotor. The rotor assembly then becomes a rotating electromagnet that induces voltage in stationary conductors embedded in slots in a laminated steel stator core surrounding the rotor (see Figure 8.14).

The stator conductors are connected to form a three-phase AC armature winding. The winding is connected to the power system, usually through a step-up transformer. Most steam turbines driven by fossil-fired steam use a two-pole generator and rotate at 3600 rpm in 60-Hz countries and 3000 rpm in 50-Hz countries. Most large steam turbines driven by nuclear steam supplies use a four-pole generator and rotate at 1800 or 1500 rpm for 60 and 50 Hz, respectively.

8.7.1 Generator Ventilation

Cooling the active parts of the generator is of such importance that generators are usually classified by the type of ventilation they use. Air-cooled generators are used commonly up to 300 MW. Some use ambient air, drawing air through filters, and others recirculate air through air-to-water heat exchangers. Above 250 MW, most manufacturers offer hydrogen for overall cooling. Hydrogen has 14 times the specific heat of air and is 14 times less dense. This contributes to much better cooling and much lower windage and blower loss. The frame must be designed to withstand the remote circumstance of a hydrogen explosion

FIGURE 8.13 Generator and exciter.

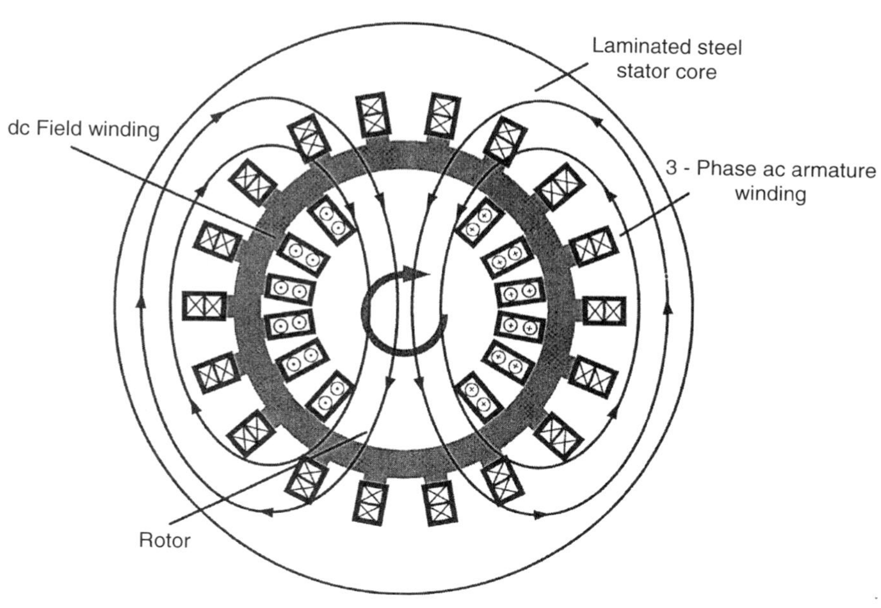

FIGURE 8.14 Generator magnetic paths.

and requires shaft seals. Hydrogen is noncombustible with purities greater than 70%. Generator purities are usually maintained well above 90%. Depending upon the manufacturer, generators with ratings above 500 MW generally have water-cooled stator winding; the remaining components are cooled with hydrogen.

8.7.2 Generator Auxiliaries

Large generators must have a lubrication oil system for the shaft journal bearings. Major components of this system are pumps, coolers, and a reservoir. In most cases, the turbine and generator use a combined system. For hydrogen-cooled generators, a shaft seal system and hydrogen supply system are needed. The shaft seal system usually uses oil pumped to a journal seal ring at a pressure somewhat higher than the hydrogen pressure. Its major components are pumps, coolers, and reservoir, similar to the lubrication system. The hydrogen supply system consists of a gas supply and regulators. A CO_2 supply is used to purge the generator when going from air to hydrogen or vice versa to avoid a combustible hydrogen–air mixture. The stator winding water supply again uses pumps, coolers, and a reservoir. It requires demineralizers to keep the water nonconducting because the water flow provides a path between the high-voltage conductors and ground. Depending upon the design approach, it may also include chemistry or oxygen content control to avoid corrosion in the winding cooling passages.

8.7.3 Excitation

The rotor field winding must have a DC source. Many generators use rotating "collector" rings with stationary carbon brushes to transfer DC current from a stationary source, such as a thyristor-controlled "static" excitation system, to the rotor winding. A rotating exciter, known as a brushless exciter, is used for many applications. It is essentially a small generator with a rotating rectifier and transfers DC current through the coupling into the rotor winding without the need for collectors and brushes.

For Further Information

Fitzgerald, A. E., Kingsley, C. F., and Umans, S. D. 2002. *Electric Machinery, 6th Ed.* McGraw-Hill, New York.

IEEE C50.13-2003 Standard for Cylindrical-Rotor 50 and 60 Hz Synchronous Generators Rated 10 MVA and Above.

IEEE 67-1990 (R1995) Guide for Operation and Maintenance of Turbine Generators.

IEEE 37.102-1995 Guide for AC Generator Protection.

Nelson, R. J., et al. 2000. Matching the capabilities of modern large combustion turbines with air- and hydrogen-cooled generators. In *Proceedings of the American Power Conference*, Chicago, April 2000.

9

Gas Turbines

9.1 Overview 9-1
9.2 History 9-1
9.3 Fuels and Firing 9-2
9.4 Efficiency 9-2
9.5 Gas Turbine Cycles 9-3
9.6 Cycle Configurations 9-4
9.7 Components Used in Complex Cycles 9-6
9.8 Upper Temperature Limit 9-9
9.9 Materials 9-10
9.10 Combustion 9-10
9.11 Mechanical Product Features 9-11
Defining Terms 9-11
For Further Information 9-12
Appendix 9-12

Steven I. Freedman
Gas Research Institute

9.1 Overview

Gas turbines are steady-flow power machines in which a gas (usually air) is compressed, heated, and expanded for the purpose of generating power. The term *turbine* is the component which delivers power from the gas as it expands; it is also called an expander. The term *gas turbine* refers to a complete power machine. The term gas turbine is often shortened to simply turbine, which can lead to confusion with the term for an expander.

The basic thermodynamic cycle on which the gas turbine is based is known as the Brayton cycle. Gas turbines may deliver their power in the form of torque or one of several manifestations of pneumatic power, such as the thrust produced by the high-velocity jet of an aircraft propulsion gas turbine engine.

Gas turbines vary in size from large, 250,000-hp utility machines, to small 5 hp, automobile and motorcycle turbochargers. Microturbines, 25–250 kW, recuperated gas turbines are now being sold.

Gas turbines are used in electric power generation, propulsion, and compressor and pump drives. The most efficient power generation systems in commercial service are gas turbine combined cycle plants with power-to-fuel energy efficiencies of more than 54% (higher heating value basis) or 59% (lower heating value basis). Systems five points higher in efficiency are now under development and are being offered commercially, and systems of even higher efficiency are considered feasible.

9.2 History

The fourth quarter of the 19th century was one of great innovation in power machinery. Along with the spark-ignited gasoline engine, the compression-ignited diesel engine, and the steam turbine, engineers applied their skills to several hot-air engines. Charles Curtis received the first U.S. patent for a complete gas turbine on June 24, 1895. Aegidius Elling built the first gas turbine in 1903, which produced 11 hp.

The first commercial stationary gas turbine engineered for power generation was a 4000-kW machine built by the Brown Boverei Company in Switzerland in 1939.

Aviation provided the impetus for gas turbine development in the 1930s. In Germany, Hans von Ohain's first engine ran in March 1937. Frank Whittle's first engine ran in England in April 1937. The first airplane flight powered by a gas turbine jet engine was in Germany on August 27, 1939. The first British airplane powered by a gas turbine flew on May 15, 1941.

A Swiss railway locomotive using a gas turbine was first run in 1941. The first automobile powered by a gas turbine was a British Rover, which ran in 1950. And, in 1956, a gas turbine-powered Plymouth car drove over 3000 miles on a coast-to-coast exhibition trip in the United States.

9.3 Fuels and Firing

The first heat engines were external combustion steam engines. The combustion products never came in contact with the working fluid, so ash, corrosive impurities, and contaminants in the fuel or exhaust did not affect the internal operation of the engine. Later, internal combustion (piston) engines were developed. In these engines, a mixture of air and fuel burned in the space enclosed by the piston and cylinder walls, thereby heating the air. The air and combustion products formed the working fluid, and contacted internal engine parts.

Most gas turbines in use today are internal combustion engines and consequently require clean fuels to avoid corrosion and erosion of critical turbine components. Efforts were made to develop gas turbines rugged enough to burn residual or crude oil. However, due to the higher efficiencies obtainable by burning extremely clean fuel at higher temperatures, there is little current interest in using fuel other than clean gas and distillate oil in gas turbines. Interest in the use of coal and residual oil is now centered on gasifying and cleaning these fuels prior to use.

A few external combustion gas turbines have been built for use with heavy oil, coal, nuclear reactor, radioisotope, and solar heat sources. However, none of these has become commercial. The added cost and pressure drop in the fired heater make externally fired gas turbines expensive. Because the working fluid temperature cannot be greater than that of the walls of the fired heater, externally fired gas turbines are substantially less efficient than modern internal combustion gas turbines with internally cooled blades.

The only internal combustion coal-fired gas turbine of current interest is the pressurized fluidized bed (PFB) combustion system. In the PFB, air discharged from the compressor of the turbine is used to fluidize a bed of limestone or dolomite in which coal is burned. The bed is maintained at modest temperature so that the ash in the coal does not form sticky agglomerates. Fortuitously, this temperature range also minimizes NO_x formation and allows capture of sulfur dioxide (SO_2) in the bed. Bed temperature is maintained in the desired range by immersed boiler tubes. Carryover fly ash is separated from gaseous combustion products by several stages of cyclone inertial separators and, in some cases, ceramic filters. The power turbine is modified to accommodate the combustion products, which after mechanical cleanup may still contain particles as large as 3–5 μm.

The most common gas turbine fuels today are natural gas and distillate oil. To avoid hot corrosion by alkali metal sulfates, the total sodium and potassium content of the fuel is typically limited to less than 5 ppm. Liquid fuels may also contain vanadium, which also causes corrosion. Fuels must be ash-free because particles larger than 3–5 μm rapidly erode blades and vanes.

9.4 Efficiency

The term *efficiency* is applied not only to complete power generation machines but also to the individual compression, expansion, and combustion processes that make up the gas turbine operating cycle.

Different definitions of efficiency apply in each case. In an **expansion process**, the **turbine efficiency** is the ratio of the actual power obtained to the maximum power that could have been obtained by expanding the gas reversibly and adiabatically between the same initial and final pressures.

Gas turbines typically involve high-speed gas flows, so appreciable differences exist between the static pressure and temperature and the total (or stagnation) pressure and temperature. Care must be taken in interpreting data to be sure that the pressure condition—static or stagnation—at each component interface is properly used.

Irreversible losses in one stage of an expansion process show up as heat (increased temperature) in later stages and add to the power delivered by such stages. Hence, a distinction exists between the polytropic efficiency (used to describe the efficiency of a process of differential pressure change) and the adiabatic (complete pressure change) efficiency. The efficiency of compressors and turbines based on their inlet and outlet pressures is called the isentropic or adiabatic efficiency. Unfortunately, both terms are reported in the literature, and confusion can exist regarding the meaning of the term *efficiency*.

Combustion efficiency in well-engineered and well-built internal combustion gas turbines is almost always close to 100%. The combustion losses appear as carbon monoxide, unburned hydrocarbons, and soot, which are typically below 50 ppm, with clean fuels.

The **gas turbine or engine efficiency** is the ratio of the net power produced to the energy in the fuel consumed. The principal gas turbine fuels are liquid and gaseous hydrocarbons (distillate oil and natural gas) which have high hydrogen content. Consequently, the term *engine efficiency* needs to be qualified as to whether it is based on the higher or the lower heat content of the fuel (the difference between the two being the latent heat of condensation of the water vapor in the products of combustion). Utility fuel transactions are traditionally based on higher heating values, and most engine publications presume the lower heating value of the fuel as the efficiency basis. In the case of natural gas fuel, the higher heating value (HHV) efficiency is greater than the lower heating value (LHV) efficiency by 10% of the value of the HHV efficiency.

Engineers analyze gas turbine machines to evaluate improvements in component performance, in higher temperature and pressure ratio designs, and in innovative cycles. Ideal case cycle calculations generally assume the following:

- Air (with either constant or temperature-dependent specific heats) is the working fluid in both turbine and compressor (with equal mass flows);
- Air is the working fluid in both turbine and compressor but with the turbine mass flow greater by the amount of fuel used.

Components are modeled with or without frictional pressure drops, and heat transfer effectiveness may be ideal (unity) or actual, depending on the purpose of the analysis. Use of compressor air for cooling of high-temperature structure, nozzles, and blades are modeled in varying degrees of complexity. Two-dimensional temperature profiles or pattern factors exist. Component inlet and exit total pressure losses should be included in cycle analyses.

9.5 Gas Turbine Cycles

Gas turbine cycles are usually plotted on temperature–entropy (T–S) coordinates. Readers unfamiliar with entropy are referred to the chapter on thermodynamics. The T–S plot is useful in depicting cycles because in an adiabatic process—as is the case for turbines and compressors—the power produced or consumed is the product of the mass flow and the enthalpy change through the process. Thus, temperature difference, which is found on a T–S plot, is proportional to the power involved. Additionally, the heat exchange in a process involving zero power—such as a combustor or heat exchanger—is the product of the absolute temperature and the entropy change. On a T–S chart, the area under a process line for a combustor or heat exchanger is the heat exchanged.

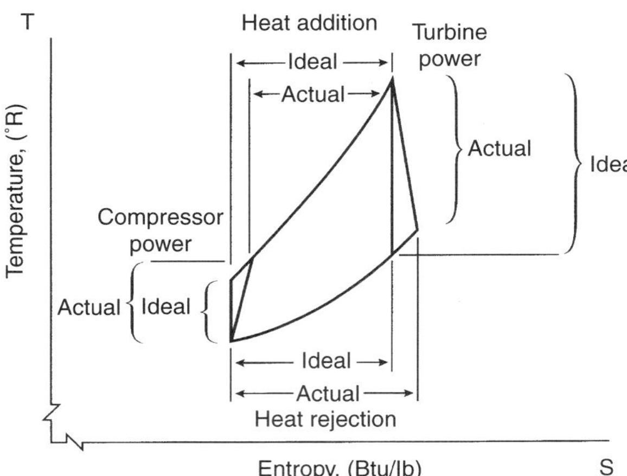

FIGURE 9.1 T–S diagram for a simple cycle illustrating the differences in compressor and turbine power for ideal (100% efficient) and actual components.

The slope of a constant-pressure line on a T–S diagram is proportional to the absolute temperature. Consequently, lines of constant pressure become steeper, and diverge as the temperature increases. This illustrates that more work is obtained expanding a gas between fixed pressures at higher temperatures than at lower temperatures. Figure 9.1 shows a comparison of the process of an ideal and an actual simple cycle gas turbine on a T–S diagram. The increased compressor power consumption and the decreased turbine power generation in the actual cycle are shown to provide an understanding of the differences that component efficiencies make on machine performance.

The incremental amount of power produced per differential pressure change in the gas is given by

$$d\left(\frac{\text{Power}}{\text{mass flow}}\right) = -RT\frac{dp}{p}$$

Two phenomena are illustrated by this equation. First, power is proportional to the absolute temperature of the gas. Second, power is proportional to the percent change in pressure. This latter point is important in understanding the effect of pressure losses in cycle components. In heat exchangers, the proper measure of power lost is the percent pressure drop.

9.6 Cycle Configurations

The basic Brayton cycle consists of a compressor, a combustor or burner, and an expander. This configuration is known as the simple cycle. In idealizing the actual cycle, combustion is replaced by constant-pressure heat addition, and the cycle is completed by the assumption that the exhaust to ambient pressure could be followed by a zero-pressure-loss cooling to inlet conditions.

A T–S diagram of the simple cycle gas turbine with an upper temperature limit set by metallurgical conditions is illustrated in Figure 9.2 for cycles of low, medium, and high pressure ratios. The heat addition is only by fuel combustion, simplified here to be without mass addition or change in specific heat of the working fluid.

It is seen that the low-pressure-ratio cycle requires a large heat addition, which leads to a low efficiency, and the high-pressure-ratio cycle has turbine power output barely greater than the compressor power requirement, thereby leading to low net output and low efficiency. At intermediate pressure ratios, the turbine power output is substantially higher than the compressor power requirement, and the heat addition is modest in comparison with the difference between the turbine and compressor powers.

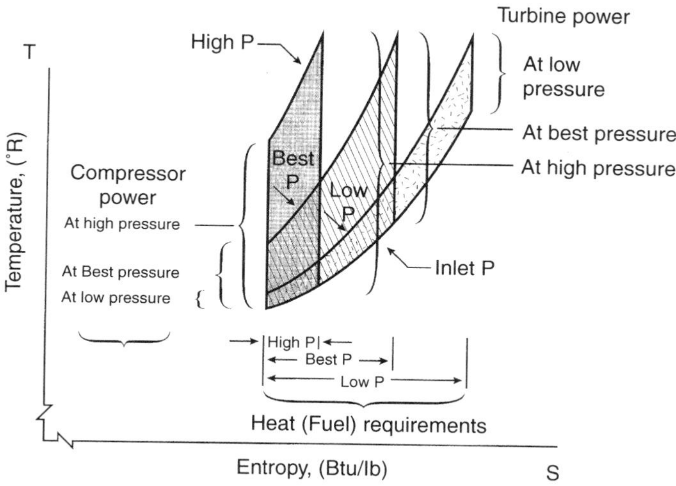

FIGURE 9.2 T–S diagram illustrating the power and heat (fuel) requirements at low, best, and high cycle pressures.

There is an optimum pressure ratio for maximum efficiency, which is a function mainly of the maximum gas temperature in the machine, and to a lesser extent, by the component efficiencies, internal pressure losses, and the isentropic exponent. There is another optimum pressure ratio for maximum specific power (power per unit mass flow).

As the achievable turbine inlet temperature increases, the optimum pressure ratios (for both maximum efficiency and maximum specific power) also increase. The optimum pressure ratio for maximum specific power is at a lower pressure level than that for maximum efficiency for all cycles not employing a recuperator. For cycles with a recuperator, the reverse is true: maximum efficiency occurs at a lower pressure ratio than maximum specific power. Heavy-duty utility and industrial gas turbines are typically designed to operate near the point of maximum specific power, which approximates lowest equipment cost, while aeroderivative gas turbines are designed to operate near the point of maximum efficiency, approximating highest thrust. Figure 9.3 shows a performance map (efficiency as a function of

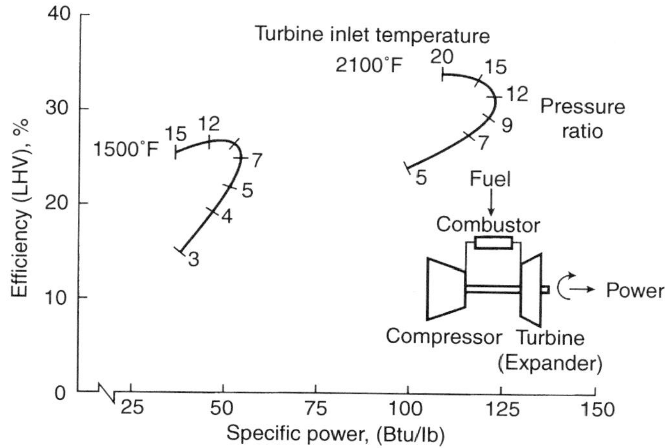

FIGURE 9.3 Performance map of a simple cycle gas turbine.

power per unit of air flow) for a simple cycle gas turbine for two turbine inlet temperatures. It is seen that at higher temperature, both the efficiency and specific power increase, as well as the optimum pressure ratios for both the maximum efficiency and maximum specific power conditions.

Aircraft gas turbines operate at temperatures above the limit of turbine materials by using blades and vanes with complex internal cooling passages. The added cost is economically justified because these machines can command high prices in the aircraft propulsion marketplace. Aeroderivative engines have higher pressure ratios, higher efficiencies, and lower exhaust temperatures than heavy-duty machines. The stationary power markets served by aeroderlvative gas turbines are principally pipeline compressor stations and oil/gas production wells. Aeroderivative gas turbines also are economically advantageous for intermediate-duty electric power generation applications.

9.7 Components Used in Complex Cycles

Recuperators and **regenerators** recover heat from the turbine exhaust and use it to preheat the air from the compressor before it enters the combustor, thereby saving fuel. This heat transfer is shown in Figure 9.4. While recuperators and regenerators are quite similar thermodynamically, they are totally different in design. Recuperators are conventional heat exchangers in which hot and cold gases flow steadily on opposite sides of a solid (usually metal) wall.

Regenerators are periodic-flow devices. Fluid streams flow in opposite directions through passages in a wheel with heat storage walls. The wheel rotates, transferring heat from one stream to the other. Regenerators usually use a nest of very small parallel passages oriented axially on a wheel which rotates between hot and cold gas manifolds. Such regenerators are sometimes used in industrial processes for furnace heat recovery, where they are referred to as heat wheels. Because regenerators are usually more compact than recuperators, they were used in experimental automotive gas turbines. The difficulty in using regenerators on gas turbines intended for long life is that the two gas streams are at very different pressures. Consequently, the seals between the manifolds and the wheel must not leak excessively over the maintenance overhaul interval of the engine. If they do, the power loss due to seal leakage will compromise engine power and efficiency. Figure 9.5 shows a performance map for the regenerative

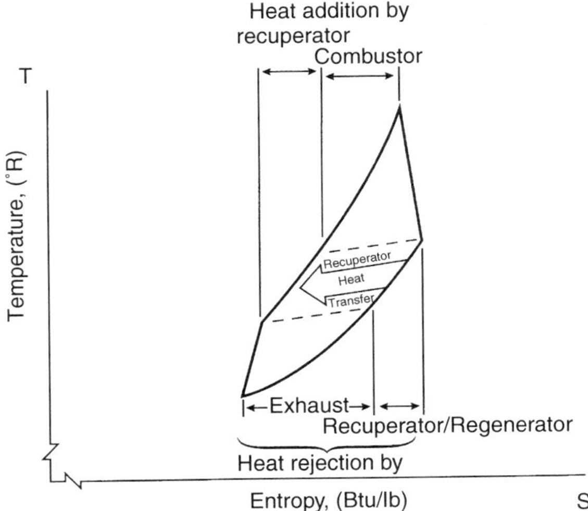

FIGURE 9.4 T–S diagram illustrating the heat transfer from the turbine exhaust to the compressor discharge accomplished by a recuperator/regenerator.

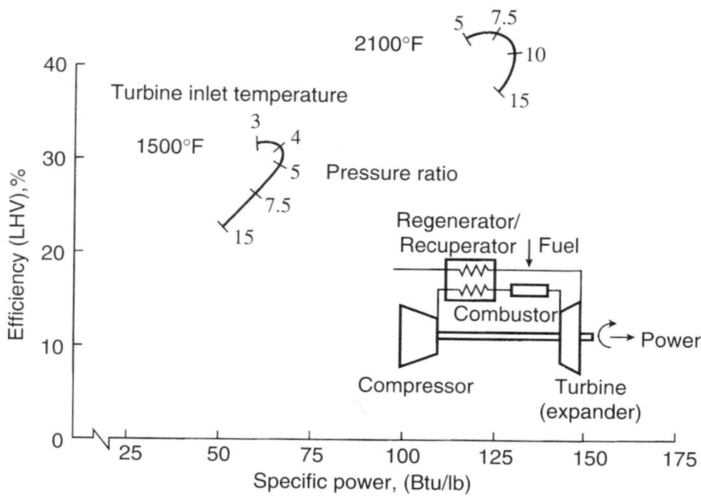

FIGURE 9.5 Performance map of a regenerative cycle gas turbine.

gas turbine cycle for two temperatures. It is seen that as the temperature increases, the efficiency, specific power, and optimum pressure ratio all increase.

Current research on the recovery of gas turbine exhaust heat includes examination of thermochemical recuperation, where exhaust heat is used to effect a chemical reaction (reforming) of the fuel with steam, thereby increasing the heating value of the fuel. Although this process is feasible, research is underway to determine if it is practical and economic.

Industrial process compressors frequently use **intercoolers** to reduce compressor power when the compressor has a high pressure ratio and operates for a large number of hours per year. When analyzing cycles with intercoolers, the added pressure drops in the compressor interstage entrance and exit diffuser and scroll and the pressure drop in the intercooler itself should be included.

In a similar manner, turbine reheat can be used to increase the power output of a large-pressure-ratio turbine. This is the thermodynamic principle in turbojet afterburner firing. Turbine reheat increases power, but decreases efficiency unless the turbine exhaust heat is used for additional power generation, as is the case with a combined cycle, or is used with a recuperator to preheat combustor inlet air.

Intercoolers and reheat burners increase the temperature difference between the compressor and turbine discharges, thereby increasing the opportunity to use a recuperator to preheat the burner air with exhaust heat. An intercooled recuperated (ICR) machine is at present in development. The efficiency decrease at part load of an ICR gas turbine is much less than of conventional simple cycle machines.

Small gas turbines have uncooled turbine blades as a result of the difficulty in manufacturing extremely small cooling passages in small blades. This results in low efficiencies, making it difficult for such turbines to compete with high-volume production (low-cost) reciprocating (piston) engines. The low-pressure-ratio recuperated cycle has greater efficiency, although at higher cost. The recuperated cycle is finding favor in programs for small (under 300-kW) gas turbines used for stationary power.

Because of their compact size, low emissions, and light weight, gas turbines are also being considered for hybrid engine-battery vehicles. Proponents are pursuing the low-pressure-ratio recuperated gas turbine as the way to obtain high efficiency and low emissions in a compact power plant.

An ingenious gas turbine cycle is the closed cycle in which the working fluid is sealed in the system. Heat is added to the fluid with an externally fired heater and extracted from the fluid through heat exchangers. The working fluid may be any gas, and the density of the gas may be varied—to vary the power delivered by the machine—by using a gas storage cylinder connected to the compressor discharge

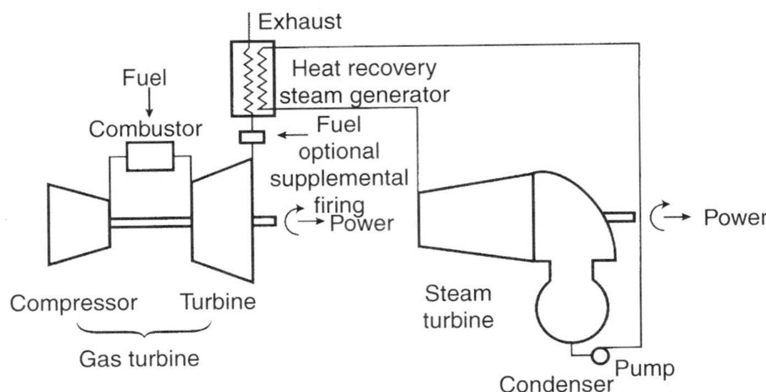

FIGURE 9.6 Combined (Brayton–Rankine) cycle.

and inlet. The gas storage system is at an intermediate pressure so that it can discharge gas into the lowest pressure point in the cycle and receive gas from the highest pressure point in the cycle. About ten such units were built between 1938 and 1968; however, in spite of its sophistication, the added cost and low efficiency inherent in external combustion systems prevented this system from becoming economic.

The exhaust from a gas turbine is quite hot and can be used to raise steam, which can then be used to generate additional power with a steam turbine. Such a compound gas turbine-steam turbine system is referred to as a **combined cycle**. Figure 9.6 shows a schematic diagram of the equipment in a combined cycle. Because the exhaust of heavy-duty machines is hotter than that of aeroderivative machines, the gain in combined cycle system efficiency through use of the steam bottoming cycle described above is greater for heavy-duty machines than for aeroderivatives. Indeed, heavy-duty machines are designed with two criteria in mind: achieving lowest cost for peaking (based on the simple cycle configuration) and achieving highest efficiency in combined cycle configuration for baseload use. The optimum pressure ratios for these two system configurations are very close. Steam bottoming cycles used in combined cycles usually use steam at multiple pressure levels to increase efficiency.

Another system in which the power and efficiency of a gas turbine is increased through the use of steam is the **steam-injected gas turbine**. Figure 9.7 shows a schematic diagram of a steam-injected gas turbine cycle. Here the turbine exhaust flows into a heat recovery steam generator (HRSG) operating at a pressure somewhat higher than the compressor discharge pressure. The steam is introduced into the gas turbine at the combustor. The steam-air mixture then passes into the turbine, where the augmented mass flow increases the power produced by the turbine. Additional fuel is required by the combustor because the steam must be heated from the HRSG delivery temperature to the combustor discharge temperature.

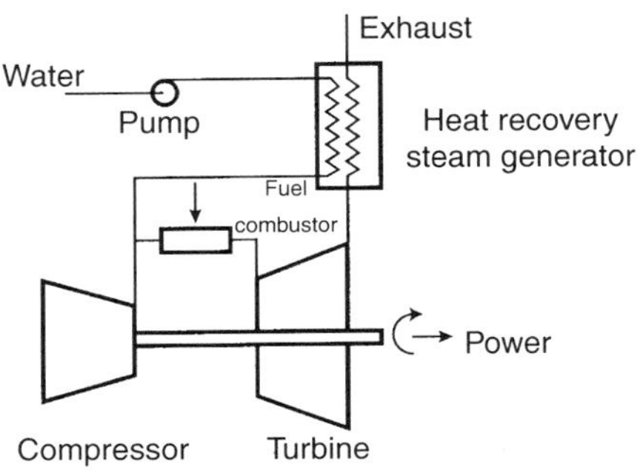

FIGURE 9.7 Steam-injected gas turbine.

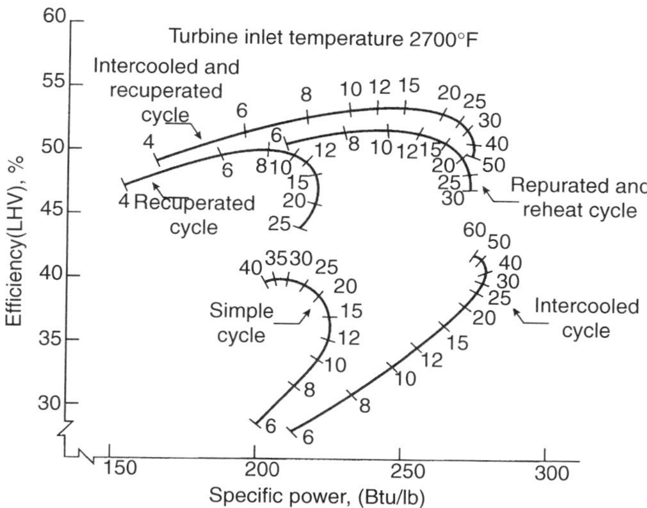

FIGURE 9.8 Specific power (Btu/lb).

Typical turbines can accommodate only a limited additional mass flow—from 5 to 15%, depending on the design of the original gas turbine. Steam-injected gas turbines enable the host to use the steam for industrial purposes, space heating, or for the generation of additional power.

A group of cycles under consideration for development involve the use of **adiabatic saturators** to provide steam at compressor discharge pressure to augment greatly the mass flow through the turbine, and consequently increase cycle power and efficiency. In the adiabatic saturator, water flows in a countercurrent path to the compressor discharge air in a mass transfer tower. Such equipment is often used in the chemical processing industries. The saturated air is preheated in a turbine exhaust heat recuperator. This cycle is called the **humid air turbine**, or HAT, cycle. The HAT cycle is particularly useful in using the low-temperature heat generated in coal-gasification-fueled gas turbine power plants. As the mass flow through the turbine is significantly augmented, engineers can no longer use the expansion turbine which was matched to the compressor in a conventional simple cycle gas turbine.

Figure 9.8 shows performance maps for the gas turbine cycles of major interest for a turbine inlet temperature typical of new products. Intercooling increases the specific power appreciably when compared with a simple cycle; however, such improvement requires an increase in pressure ratio. Recuperated cycles have considerably higher efficiency than similar cycles without recuperation. The effect of pressure ratio on the performance of recuperated cycles is opposite to that of similar cycles without recuperation. For recuperated cycles, the pressure ratio for maximum efficiency is considerably lower than for maximum specific power. Performance maps such as these are used in screening cycle alternatives for improved performance. Individual curves are generated for specific component performance values for use as a guide in developing new or improved machines.

9.8 Upper Temperature Limit

Classically, gas turbine engineers often spoke of a metallurgical limit in reference to maximum turbine inlet temperature. Later, turbine vane and blade cooling became standard on large machines. This situation creates a temperature difference between the combustion products flowing through the turbine and the turbine blade wall. Thus, because heat can be removed from the blades, the turbine can be operated with a combustion gas temperature higher than the metallurgical limit of the blade material. As a rule, the blades and vanes in new large gas turbines contain complex internal passages, through which up to 20% of compressor discharge air is directed. The cooling air first flows through internal convective cooling passages, then through impingement passages, where the air is directed at the blade and vane walls, and finally through small holes in the blade, where it is used to provide a low-temperature film over the blade surface. This film cooling of the surface reduces heat transfer to the blade.

The design of blade and vane cooling passages is an extremely competitive endeavor because greater cooling enables use of higher combustion temperatures without exceeding the metallurgical limit of the blade material. However, a balance between air flow for cooling and air flow for power must be achieved; the cooling air flowing within a blade drops in pressure without producing any power within that stage (although it is available for power in later stages). In the newest gas turbines, blade cooling, the difference between turbine inlet gas temperature and blade metal temperature, is around 1000°F.

Some of the latest large gas turbines being introduced to the market in the 2005 period are being offered for combined cycle application with closed-circuit **steam cooling** of selected hot section parts. Steam cooling reduces the need for air cooling, so that more of the compressor discharge air can be used for NO_x reduction in the combustor and for power generation. The heat transferred to the steam increases the efficiency of the bottoming cycle. The additional combustion products which flow through the high-pressure portions of the turbine generate substantially more power, thereby increasing both the power output and the efficiency of the machine. With more air for combustion, the fuel can be burned as a leaner mixture, with either less NO_x produced, or, as is preferred, with higher-temperature gases going to the turbine and the same NO_x (or a combination of these benefits).

9.9 Materials

The high-technology parts of a gas turbine are its hot section parts: blades, vanes, combustors and transition pieces. Gas turbine power, efficiency, and economics increase with the temperature of the gas flowing through the turbine blade passages. It is in the fabrication of these hot section parts that manufacturers are most competitive. Materials are selected to survive in serviceable condition for over 50,000 h and associated numbers of thermal cycles. Ceramic coatings protect materials from oxidation and corrosion and provide thermal insulation, permitting higher gas temperatures.

Gas turbine alloys are frequently referred to as superalloys because of their extremely high strength at high temperatures. These superalloys are nickel based (such as IN 738), cobalt based (such as FSX-414), or with a nickel-iron base such as Inconel 718. Nickel resists oxidation and is creep resistant, but is subject to corrosive sulfidation. Alloy and manufacturing advancements have been led by the needs of military aircraft engines. Coating developments for corrosion resistance have been led by the needs of stationary power for overhaul intervals as large as 50,000 h. The developmental needs of automotive gas turbines have led to significant advances in strength and reliability of high-temperature ceramic components, including radial inflow turbines. Ceramic materials, principally silicon nitride, are of interest to the developer of service soon in small gas turbines.

9.10 Combustion

Gas turbine combustors appear to be simple in design, yet they solve several difficult engineering challenges. Until relatively recently, gas turbine combustors employed a (turbulent) diffusion flame design approach, which created the most compact flame. European heavy-duty gas turbine manufacturers—with substantial interest in burning heavy fuel oils—preferred large, off-engine combustors, often called silo combustors because of their appearance, in order to obtain lower flame velocities and longer residence times. American heavy-duty gas turbine manufacturers use compact on-engine combustors and design for gaseous and clean (distillate) liquid fuels. Aeropropulsion gas turbines require the smallest possible frontal area and use only clean liquid fuels; hence, they use on-engine combustors.

Recently, stationary engines have been required to reduce NO_x emissions to the greatest extent possible, and combustors on stationary gas turbines first modified their diffusion flame combustors and employed water and steam injection to quench flame hot spots. Most recently, designs changed to the lean-premixed process. With the improved blade cooling, materials, and coatings now in use, the material limits on turbine inlet temperature and the NO_x emission limits on combustor temperature appear to be converging on a combustion-temperature asymptote around 2700°F (1482°C). This may be increased if catalytic combustors prove practical.

9.11 Mechanical Product Features

In view of the need to achieve all the performance features described above, one must keep in mind that a gas turbine is a high-speed dynamic machine with numerous machine design, materials, and fabrication features to consider. Major issues include the following: critical shaft speed considerations, bearing rotational stability, rotor balancing, thrust bearing design, bearing power loss, oil lubrication system, oil selection, air filter design and minimization of inlet and exhaust diffuser pressure drops, instrumentation, controls, diagnostic systems, scheduled service and inspection, overhaul, and repair. All of these topics must be addressed to produce a cost-effective, reliable, long-lived, practical gas turbine product that will satisfy users while also returning to investors sufficient profit for them to continue to offer better power generation products of still higher performance.

Defining Terms

Adiabatic saturator: A combined heat-and-mass-exchanger whereby a hot gas and a volatile liquid pass through a series of passages such that the liquid is heated and evaporates into the gas stream.

Combined cycle: An arrangement of a gas turbine and a stream turbine whereby the heat in the exhaust from the gas turbine is used to generate steam in a heat recovery boiler which then flows through a steam turbine, thereby generating additional power from the gas turbine fuel.

Combustion efficiency: Ratio of rate of heat delivered in a device which burns fuel to the rate of energy supplied in the fuel.

Expansion process: Process of power generation whereby a gas passes through a machine while going from a condition of high pressure to one of low pressure, usually the machine produces power.

Gas turbine or engine efficiency: The ratio of the net power delivered (turboexpander power minus compressor and auxiliary power) to the rate of energy supplied to the gas turbine or engine in the form of fuel, or, in certain cases such as solar power, heat.

Humid air turbine: A gas turbine in which the flow through the expander is augmented by large amounts of steam generated by use of an adiabatic saturator.

Intercooler: A heat exchanger used to cool the flow between sections of a compressor such that the high pressure section acts on a stream of reduced volumetric flow rate, thereby requiring less overall power to compress the stream to the final pressure.

Recuperator: A heat exchanger in which the hot and cold streams pass on opposite sides of a wall through which heat is conducted.

Regenerator: A heat exchanger in which the hot and cold streams alternately heat and cool a wall whose temperature rises and falls, thereby transferring heat between the streams.

Steam cooling: A process in which steam is used as the heat transfer fluid to cool a hot component.

Steam-injected gas turbine: A system in which the gas turbine flow is augmented by steam, thereby generating additional power.

Turbine efficiency: Ratio of the power delivered in an expansion process employing a turbine as the expander to the maximum power which could be produced by expanding the gas in a reversible adiabatic (isentropic) process from its initial pressure and temperature to its final pressure to the actual power.

For Further Information

Wilson, D. G. 1984. *The Design of High-Efficiency Turbomachinery and Gas Turbines*. MIT Press, Cambridge, MA.

Kerrebrock, J. 1992. *Aircraft Engines and Gas Turbines*. MIT Press, Cambridge, MA.

Boyce, M. P. 1982. *Gas Turbine Engineering Handbook*. Gulf Publishing, Houston, TX.

Sawyer's Gas Turbine Engineering Handbook, Vol. 1: *Theory and Design*, Vol. 2: *Section and Application*, Vol. 3: *Accessories and Support*, Turbomachinery International Publications, Norwalk, CT, 1985.

Appendix

Equations for gas turine calculations based on the use of a perfect gas as the working fluid.

Perfect gas law	$pv = RT$
Gas constant	$R = \check{R}$/molecular weight
For air (molecule weight of 28.97)	$R = 286.96$ J/kg K
	$= 0.06855$ Btu/lb$_m$ °R
	$= 53.32$ ft lb$_f$/lb$_m$ °R
Universal gas constant	$\check{R} = 8313$ J/kg mol K
	$= 1.986$ Btu/lb mol °R
	$= 1545$ ft lb$_f$/lb mol °R
Relationships of properties	$c_p = c_v + R$
Isentropic exponet	$\gamma = c_p/c_v$ (air, $\gamma = 1.4$)
	$(\gamma - 1)/\gamma = R/c_p$
Isentropic process	$pv^\gamma = \text{constant}$
	$P_2/P_1 = (T_2/T_1)^{\gamma/(\gamma-1)}$
Prolytropic process	$pv^n = \text{constant}$
	$P_2/P_1 = (T_2/T_1)^{n/(n-1)}$
Pressure ratio	$r = P_2/P_1$
Ratio of stagnation T° and p° to static T and p	$\frac{T^\circ}{T} = 1 + \frac{\gamma - 1}{2} M^2$
	$\frac{p^\circ}{p} = \left(1 + \frac{\gamma - 1}{2} M^2\right)^{\gamma/(\gamma-1)}$
Match number	$M = V/\sqrt{g_c \gamma R T}$
Gravitational constant	$g_c = \text{ma/F}$
Subscripts	t = turbine
	c = compressor
	f = fuel
	i = inlet
	e = exit

Cycle efficiency:

$$\eta = \frac{\dot{m}_t \Delta h_t - \dot{m}_c \Delta h_c}{\dot{m}_f \text{HV}}$$

where HV = heating value of fuel.

For specific heat independent of temperature and small mass flow of fuel in comparison to air:

$$\eta = \frac{\Delta T_t - \Delta T_c}{\Delta T_b}$$

Isentropic efficiency (finite pressure ratio):

$$\eta_t = \Delta T \text{ actual}/\Delta T \text{ isentropic}$$

$$\eta_t = \frac{1 - T_e/T_i}{1 - r^{(\gamma-1)/\gamma}}$$

or

$$\eta_t = \frac{1 - r^{(n-1)/n}}{1 - r^{(\gamma-1)/\gamma}}$$

and

$$\eta_c = \Delta T \text{ isentropic}/\Delta T \text{ actual}$$

$$\eta_c = \frac{r^{(\gamma-1)/\gamma} - 1}{T_e/T_i - 1}$$

or

$$\eta_c = \frac{r^{(\gamma-1)/\gamma} - 1}{r^{(\eta-1)/\eta} - 1}$$

Polytropic efficiency (differential pressure ratio):

$$\eta_t = \frac{(n-1)/n}{(\gamma-1)/\gamma}$$

and

$$\eta_c = \frac{(\gamma-1)/\gamma}{(n-1)/n}$$

Relationships between isentropic and polytropic efficiencies:

$$\eta_{s,c} = \frac{r^{(\gamma-1)/\gamma} - 1}{r^{(\gamma-1)/\gamma \eta_{pc}} - 1}$$

$$\eta_{s,t} = \frac{1 - r^{(\gamma-1)/\gamma \eta_{p,t}}}{1 - r^{(\gamma-1)/\gamma}}$$

$$\eta_{p,c} = \frac{\ln r^{(\gamma-1)/\gamma}}{\ln\left[\frac{r^{(\gamma-1)/\gamma} - 1}{\eta_{s,c}} + 1\right]}$$

$$\eta_{p,t} = \frac{\ln[1 - \eta_{s,t}(1 - r^{(\gamma-1)/\gamma})]}{\ln r^{(\gamma-1)/\gamma}}$$

10

Internal Combustion Engines

David E. Klett
North Carolina A&T State University

Elsayed M. Afify
North Carolina State University

10.1 Introduction **10-1**
10.2 Engine Types and Basic Operation **10-2**
Four-Stroke SI Engine • Two-Stroke SI Engine • Compression Ignition Engine
10.3 Air Standard Power Cycles **10-7**
Constant-Volume Heat Addition—Ideal Otto Cycle • Constant-Pressure Heat Addition—Ideal Diesel Cycle
10.4 Actual Cycles **10-10**
10.5 Combustion in IC Engines **10-12**
Combustion in Spark Ignition Engines • Combustion in Compression Ignition Engines
10.6 Exhaust Emissions **10-15**
Harmful Constituents • Control of Emissions from IC Engines
10.7 Fuels for SI and CI Engines **10-17**
Background • Gasoline • Diesel Fuels
10.8 Intake Pressurization—Supercharging and Turbocharging **10-20**
Background • Supercharging • Turbocharging
Defining Terms **10-23**
References **10-25**
For Further Information **10-25**

10.1 Introduction

This section discusses the two most common reciprocating internal combustion (IC) engine types in current use: the **spark ignition** (SI) and the **compression ignition** (CI or diesel) **engines**. The Stirling engine (technically, an external combustion engine) and the gas turbine engine are covered in other sections of this chapter. Space limitations do not permit detailed coverage of the very broad field of IC engines. For a more detailed treatment of SI and CI engines and for information on variations, such as the Wankel rotary engine and the Miller cycle engine (a variation on the reciprocating four-stroke SI engine introduced in production by Mazda in 1993), several excellent textbooks on the subject, technical papers, and other sources are included in the list of references and the section on further information.

Basic SI and CI engines have not fundamentally changed since the early 1900s with the possible exception of the introduction of the Wankel rotary SI engine in the 1960s (Norbye 1971). However, major advances in the areas of materials, manufacturing processes, electronic controls, and computer-aided design have led to significant improvements in dependability, longevity, thermal efficiency, and emissions during the past decade. Electronic controls, in particular, have played a major role in efficiency gains in SI automotive engines through improved control of the fuel injection and ignition systems that control the combustion process. Electronic control of diesel fuel injection systems is also now quite common and is producing improvements in diesel emissions and fuel economy.

This section presents the fundamental theoretical background of IC engine function and performance, including **four-stroke** and **two-stroke** SI and CI **engines**. Sections on combustion, emissions, fuels, and intake pressurization (**turbocharging** and **supercharging**) are also included.

10.2 Engine Types and Basic Operation

IC engines may be classified by a wide variety of characteristics; the primary ones are SI vs. CI; four stroke vs. two stroke; and reciprocating vs. rotary. Other possible categories of classification include intake type (naturally aspirated vs. turbocharged or supercharged); number of cylinders; cylinder arrangement (in-line, vee, opposed); cooling method (air vs. water); fueling system (injected vs. carbureted); valve gear arrangement (overhead cam vs. pushrod); type of **scavenging** for two-stroke engines (cross, loop, or uniflow); and type of injection for diesel engines (direct vs. indirect).

10.2.1 Four-Stroke SI Engine

Figure 10.1 is a cross-section schematic of a four-stroke SI engine. The SI engine relies on a spark plug to ignite a volatile air–fuel mixture as the piston approaches **top dead center (TDC)** on the compression stroke. This mixture may be supplied from a carburetor, a single throttle-body fuel injector, or by individual fuel injectors mounted above the intake port of each cylinder. One combustion cycle involves two revolutions of the crankshaft and thus four strokes of the piston, referred to as the intake, compression, power, and exhaust strokes. Intake and exhaust valves control the flow of mixture and exhaust gases into and out of the cylinder, and an ignition system supplies a spark-inducing high voltage to the spark plug at the proper time in the cycle to initiate combustion.

On the intake stroke, the intake valve opens and the descending piston draws a fresh combustible charge into the cylinder. During the compression stroke, the intake valve closes and the fuel–air mixture

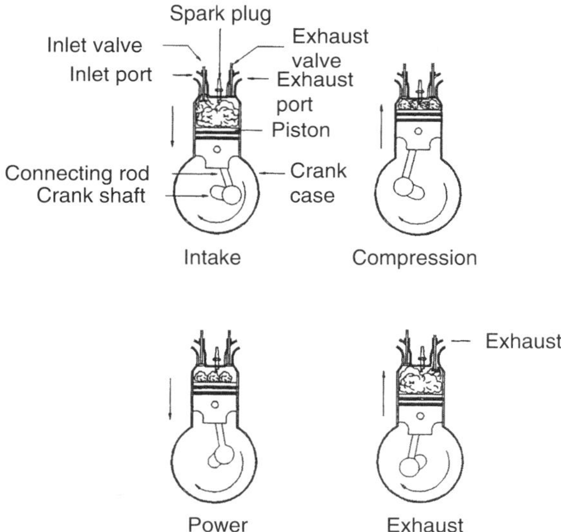

FIGURE 10.1 Schematic diagram of four-stroke SI engine.

is compressed by the upward piston movement. The mixture is ignited by the spark plug, typically before TDC. The rapid, **premixed, homogeneous combustion** process causes a sharp increase in cylinder temperature and pressure that forces the piston down for the power stroke. Near **bottom dead center** (BDC), the exhaust valve opens and the cylinder pressure drops rapidly to near atmospheric. The piston then returns to TDC, expelling the exhaust products. At TDC, the exhaust valve closes and the intake valve opens to repeat the cycle again. Figure 10.2 is a cutaway drawing of a modern high-performance automotive SI engine. This is a fuel-injected normally aspirated aluminum alloy V-8 engine of 3.9 L displacement with dual overhead cams for each cylinder bank and four valves per cylinder. Peak power output is 188 kW at 6100 rpm and peak torque is 354 N-m at 4300 rpm.

10.2.2 Two-Stroke SI Engine

The two-stroke SI engine completes a combustion cycle for every revolution of the crankshaft by essentially overlapping the power and exhaust functions in one downward stroke and the intake and compression processes in one upward stroke. A single-cylinder, crankcase-scavenged, two-stroke SI engine is illustrated schematically in Figure 10.3. The operation is as follows.

On the upward stroke, the piston first covers the transfer port and then the exhaust port. Beyond this point, the fresh charge is compressed and ignited near TDC. During the upward stroke, the negative pressure created in the crankcase below the piston draws in a fresh charge of fuel–air mixture through a one-way valve. On the downward power stroke, the mixture in the crankcase is pressurized. The piston uncovers the exhaust port and the high-pressure exhaust gases exit. Near BDC, the transfer port is uncovered

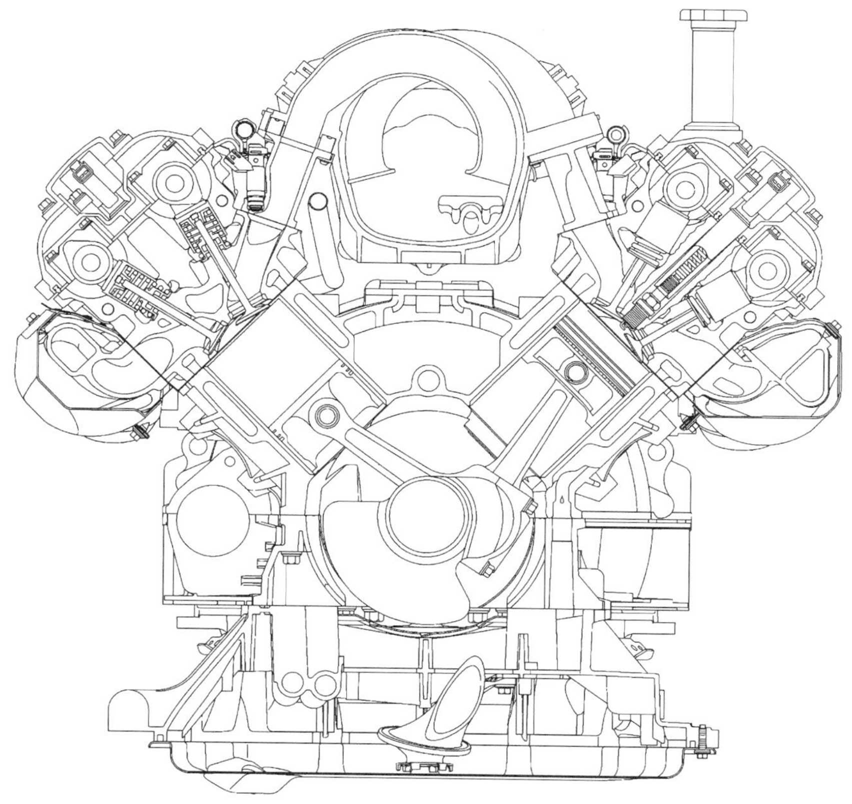

FIGURE 10.2 Ford 4.6-L aluminum V-8 SI engine. (Courtesy of Ford Motor Company.)

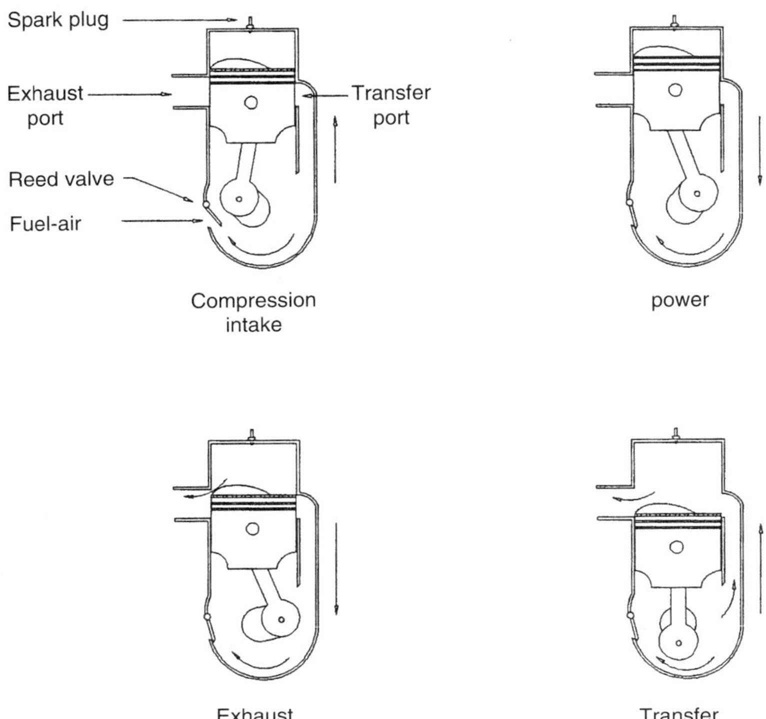

FIGURE 10.3 Schematic drawing of two-stroke SI engine.

and the pressurized mixture flows from the crankcase into the cylinder and the cycle repeats. Because the crankcase is part of the induction system, it does not contain oil, and lubrication is accomplished by mixing oil with the fuel. With the cross-flow scavenging configuration illustrated in Figure 10.3, there will be a certain degree of mixing of the fresh charge with the combustion products remaining in the cylinder and some loss of fresh charge out the exhaust port.

Because two-stroke engines produce twice the power impulses of four-stroke engines for the same rpm, a two-stroke engine has a higher **power density** and is thus smaller and lighter than a four-stroke engine of equal output. The disadvantages of the two-stroke engine have historically been lower fuel efficiency and higher exhaust emissions because of overlapping intake and exhaust processes and the loss of some fresh intake mixture with the exhaust products. For this reason, two-stroke SI engines have largely been confined to small-displacement applications, such as small motorcycles, outboard marine engines, and small equipment. Several manufacturers have addressed these shortcomings in recent years and have achieved significant improvements in two-stroke engine fuel economy and emissions (Blair 1988).

The orbital combustion process (OCP), as illustrated in Figure 10.4, applies air-assisted direct injection of the fuel into the cylinder of a two-stroke engine and, in conjunction with a high turbulence combustion chamber design, achieves very favorable fuel economy and significantly reduced levels of hydrocarbon emissions. This system, in use today on single-cylinder scooters and on 2-, 3-, and 6-cylinder marine two-stroke engine applications, is also applicable to four-stroke engines.

10.2.3 Compression Ignition Engine

The basic valve and piston motions are the same for the CI, or diesel, engine as discussed earlier for the SI engine. The CI engine relies on the high temperature and pressure of the cylinder air resulting from

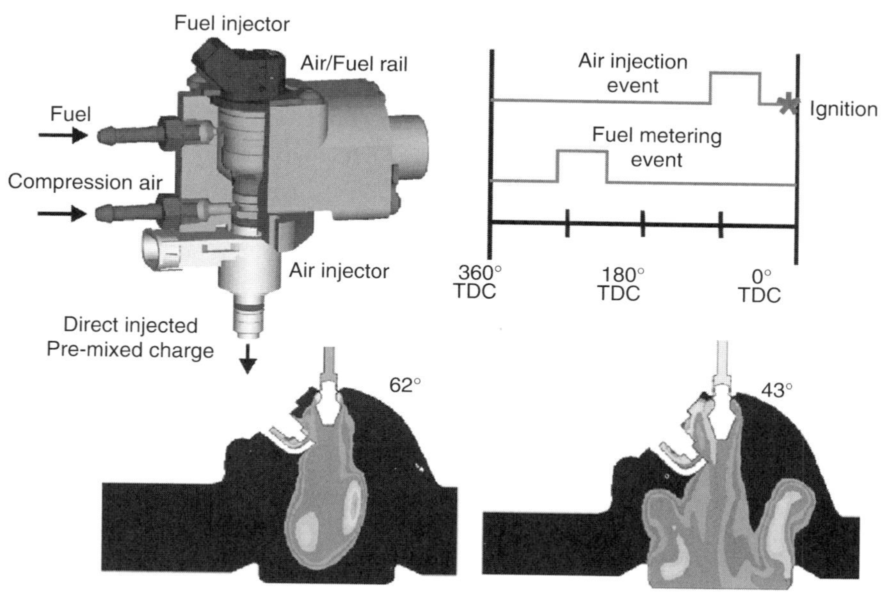

FIGURE 10.4 Orbital OCP combustion system. (Courtesy of Orbital Engine Company.)

the compression process to cause **autoignition** of the fuel, which is injected directly into the combustion chamber of **direct injection** (**DI**) engines or into the prechamber of **indirect injection** (**IDI**) engines, when the piston approaches TDC on the compression stroke. Compression ratios are typically much higher for CI than for SI engines to achieve the high air temperatures required for autoignition, and the fuels used must have favorable autoignition qualities.

The time period between the start of fuel injection and the occurrence of autoignition is called the **ignition delay period**. Long ignition delay periods allow more time for fuel vaporization and fuel–air mixing and result in objectionable diesel knock when this larger premixed charge autoignites. Combustion chambers and fuel injection systems must be designed to avoid extended ignition delay periods. Diesel engines may be classified as DI or IDI. In DI engines, the combustion chamber consists of a bowl formed in the top of the piston; the fuel is injected into this volume. The injector tip generally has from four to eight holes to form multiple spray cones.

Two variations are illustrated in Figure 10.5. The quiescent chamber engine utilizes a large-diameter shallow bowl shape that produces low **swirl** and low turbulence of the air during compression. Fuel is injected at high pressure through a multihole nozzle; mixing of the fuel and air relies primarily on the energy of the injected fuel to cause air entrainment in the spray cone and diffusion of vaporized fuel into the air. This system is suited to large slow-speed engines that are operated with significant excess air.

The toroidal bowl combustion chamber is used in conjunction with intake ports and/or valve shrouds designed to produce air swirl to enhance fuel–air mixing. The **swirl ratio** is defined by swirl ratio = swirl speed (rpm)/engine speed (rpm). The swirl velocity component is normal to the fuel spray direction and tends to promote mixing in the regions between the individual spray cones. This system makes better use of the available air and is utilized extensively in moderate-speed engines such as over-the-road truck engines. DI does not lend well to high-speed operation because less time is available for proper mixing and combustion. Diesel engines for passenger car applications are generally designed for higher speed operation to produce higher specific output. They typically utilize IDI combustion systems, two of which are illustrated in Figure 10.6.

IDI systems make use of small prechambers incorporated in the cylinder head to promote rapid mixing of fuel and air and shorten the ignition delay period. Swirl chambers are designed to produce

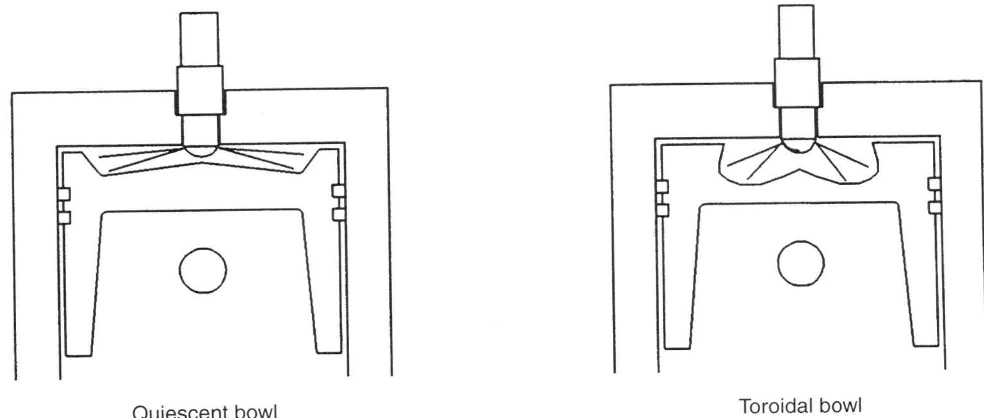

FIGURE 10.5 Examples of DI diesel combustion chamber design.

a strong vortex in the prechamber during compression. The fuel is sprayed into the chamber through a single-hole nozzle and the high vorticity promotes rapid mixing and short ignition delay periods. Precombustion chambers do not attempt to generate an orderly vortex motion within the chamber; instead, to promote mixing, they rely on a high level of turbulence created by the rush of air into the chamber during compression. Both types of prechambers generally include a lining of low-conductivity material (ceramic) to increase the surface temperature to promote fuel evaporation. Prechambers can be used in small-displacement diesel engines to achieve operating speeds up to 5000 rpm.

Disadvantages of the IDI system include poor cold-start characteristics due to high heat-transfer rates from the compressed air to the chamber wall that result from the high velocities and turbulence levels in the chamber. **Glow plugs** are often installed in each prechamber to heat the air to improve cold starting. Higher compression ratios are also used for IDI engines to improve cold starting. The compression ratios, typically 18–24, are higher than the optimum for fuel efficiency (due to decreased mechanical efficiency resulting from higher friction forces), and IDI engines are typically less efficient than larger, slower, DI engines. The use of IDI is generally restricted to high-speed automotive engines, with displacements in

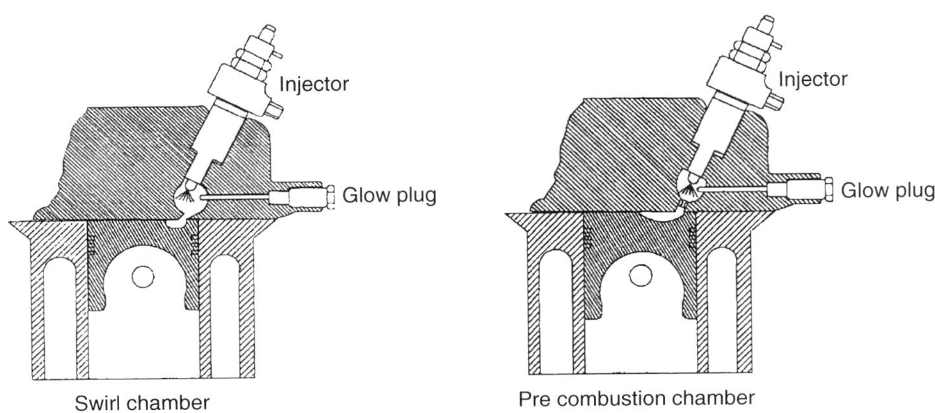

FIGURE 10.6 Two examples of IDI combustion chambers.

the range of 0.3–0.8 L per cylinder, and some degree of fuel economy is sacrificed in the interest of improved driveability.

CI engines are produced in two-stroke and four-stroke versions. Because the fuel is injected directly into the combustion chamber of CI engines just prior to TDC, two-stroke CI engines do not suffer the same emission and efficiency shortcomings as do older crankcase-scavenged two-stroke SI engines. Thus, they are available in much larger displacements for high-power-requirement applications such as locomotive and ship propulsion and electric power generation systems. Two-stroke CI engines are generally of the DI type because the use of IDI in a two-stroke engine would lead to aggravated cold-start problems due to lower compression ratios.

10.3 Air Standard Power Cycles

The actual operation of IC engines is idealized at a very basic level by the air standard power cycles (ideal thermodynamic models for converting heat into work on a continuous basis). The following simplifying assumptions are common to the air standard cycles: (1) the working substance is air, (2) the air is assumed to behave as an ideal gas with constant specific heats, (3) heat is added to the cycle from an external source, and (4) expansion and compression processes not involving heat transfer occur isentropically. The air standard cycles, while grossly oversimplified in terms of the complex processes occurring within actual engines, are nevertheless useful in understanding some fundamental principles of SI and CI engines. The simplified models also lend insight into important design parameters, e.g., **compression ratio**, that govern theoretical maximum cycle thermal efficiencies.

10.3.1 Constant-Volume Heat Addition—Ideal Otto Cycle

The theory of operation of the SI engine is idealized by the Otto cycle, which assumes that heat is added to the system at constant volume. Constant-volume heat addition is approximated in the SI engine by virtue of the combustion process taking place rapidly when the piston is near TDC. A *P–V* diagram for the Otto cycle is illustrated in Figure 10.7. The cycle consists of the following processes: $1 \rightarrow 2$ isentropic compression; $2 \rightarrow 3$ constant-volume heat addition; $3 \rightarrow 4$ isentropic expansion; and $4 \rightarrow 1$

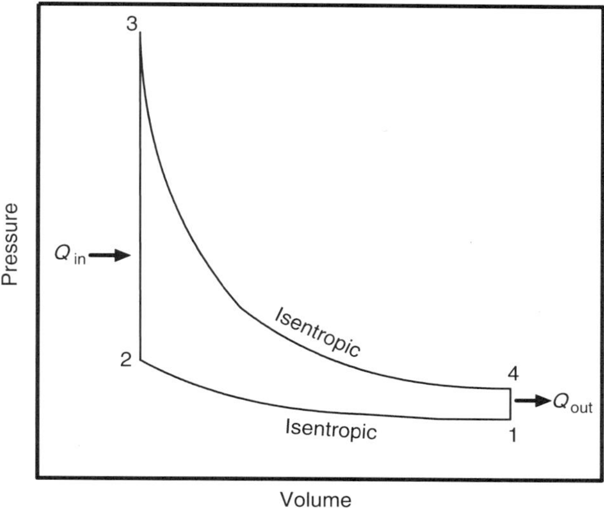

FIGURE 10.7 Schematic pressure–volume diagram for the ideal Otto cycle.

constant-volume heat rejection. The constant-volume heat rejection process is approximated in SI engines by the exhaust valve opening near BDC and the rapid blow-down of exhaust gases.

Thermal efficiency for a power cycle is defined as the ratio of work output to heat input per cycle,

$$\eta = \frac{W_{\text{out}}}{Q_{\text{in}}} \tag{10.1}$$

For the Otto cycle, the basic efficiency expression can be manipulated into the form

$$\eta = 1 - \frac{1}{r^{\gamma-1}} \tag{10.2}$$

where γ is the ratio of specific heats ($\gamma = C_p/C_v$) and r is the compression ratio, or ratio of the maximum to minimum cycle volumes ($r = V_1/V_2$).

In actual IC engines, the minimum cycle volume is referred to as the **clearance volume** and the maximum cycle volume is the **cylinder volume**. The ideal Otto cycle efficiency for air, with $\gamma = 1.4$, is shown plotted in Figure 10.8. The theoretical efficiency of the constant volume heat addition cycle increases rapidly with compression ratio, up to about $r = 8$. Further increases in compression ratio bring moderate gains in efficiency. Compression ratios in practical SI engines are limited because of autoignition (knock) and high NO_x emission problems that accompany high compression ratios. Production SI automotive engines typically have compression ratios in the range of 8–10, whereas high-performance normally aspirated racing engines may have compression ratios as high as 14, but require the use of special fuels to avoid autoignition.

10.3.2 Constant-Pressure Heat Addition—Ideal Diesel Cycle

The air standard diesel cycle is the idealized cycle underlying the operation of CI or diesel engines. The diesel cycle, illustrated by the *P–V* diagram in Figure 10.9, consists of the following processes: $1 \rightarrow 2$ isentropic compression from the maximum to the minimum cycle volume; $2 \rightarrow 3$ constant-pressure heat addition during an accompanying increase in volume to V_3; $3 \rightarrow 4$ isentropic expansion to the maximum cycle volume; and $4 \rightarrow 1$ constant-volume heat rejection.

Actual diesel engines approximate constant-volume heat addition by injecting fuel for a finite duration that continues to burn and release heat at a rate that tends to maintain the pressure in the cylinder over a period of time during the expansion stroke. The efficiency of the ideal diesel cycle is given by

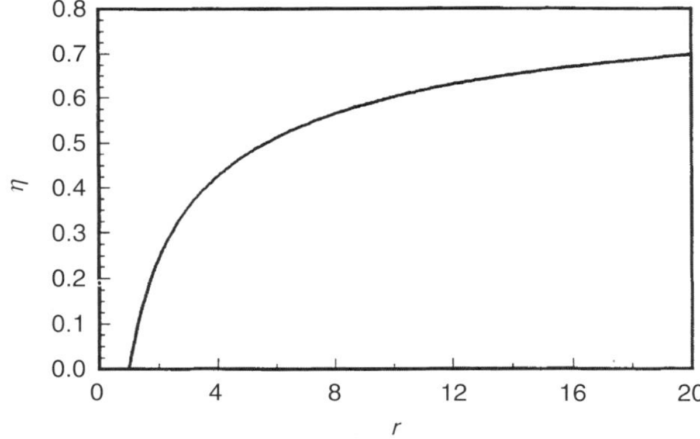

FIGURE 10.8 Efficiency of the ideal Otto cycle.

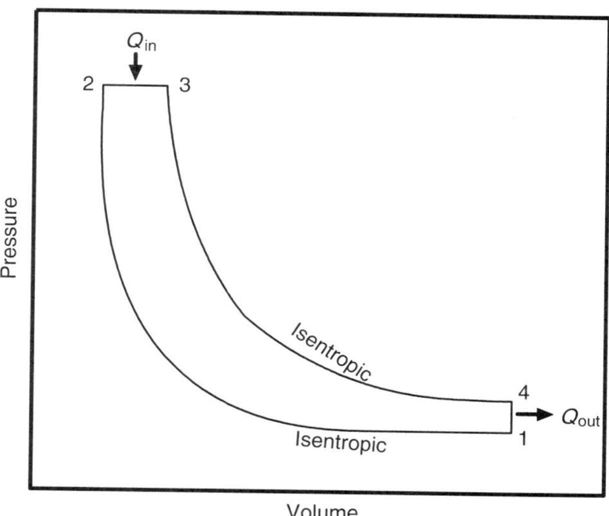

FIGURE 10.9 Schematic pressure–volume diagram of ideal diesel cycle.

$$\eta = 1 - \frac{1}{r^{\gamma-1}}\left[\frac{r_c^{\gamma}-1}{\gamma(r_c-1)}\right] \tag{10.3}$$

The efficiency of the ideal diesel cycle depends not only on the compression ratio, r, but also on the **cut-off ratio**, $r_c = V_3/V_2$, the ratio of the volume when heat addition ends to the volume when it begins. Equation 10.3 is shown plotted in Figure 10.10 for several values of r_c and for $\gamma = 1.4$. An r_c value of 1 is equivalent to constant-volume heat addition—i.e., the Otto cycle. The efficiency of the ideal Diesel cycle is less than the efficiency of the ideal Otto cycle for any given compression ratio and any value of the cut-off ratio greater than 1. The fact that CI engines, by design, operate at much higher compression ratios than SI engines (generally between 12 and 24) accounts for their typically higher operating efficiencies relative to SI engines.

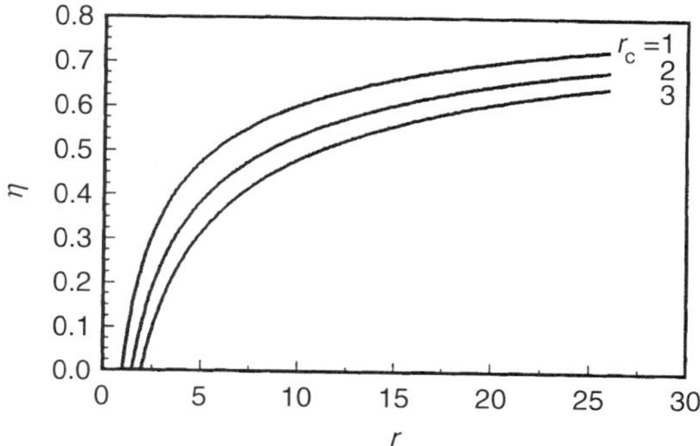

FIGURE 10.10 Efficiency of the ideal diesel cycle.

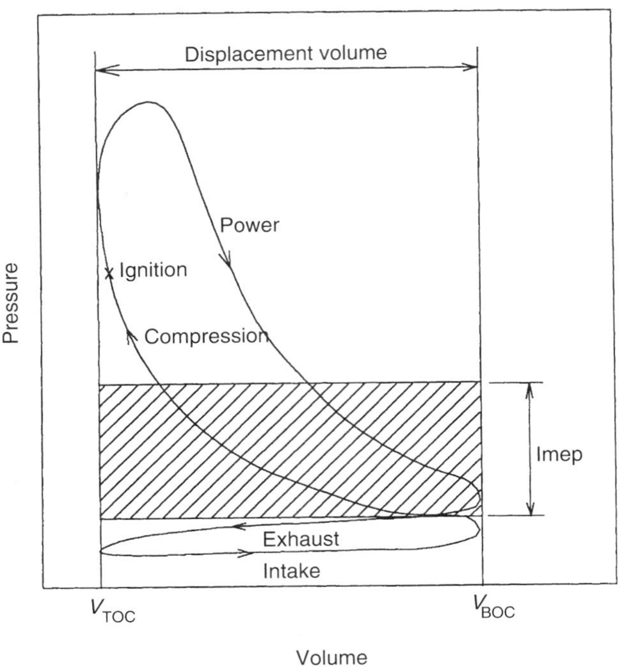

FIGURE 10.11 Schematic indicator diagram.

10.4 Actual Cycles

IC engines do not operate on closed thermodynamic cycles, such as the air standard power cycles, but rather on open mechanical cycles; heat addition occurs neither at constant volume nor at constant pressure. Figure 10.11 is a schematic representation of an **indicator diagram** (pressure–volume history) of a four-stroke IC engine; it could be SI or CI. The pressure changes during the intake and exhaust strokes are exaggerated in the diagram. The **indicated work** performed per cycle can be calculated by taking the integral of *PdV* for the complete cycle. The **indicated mean effective pressure, imep,** is defined as the ratio of the net indicated work output to the **displacement volume**:

$$\text{imep} = \frac{\text{indicated work output per cycle}}{\text{displacement volume}} \tag{10.4}$$

The shaded area in Figure 10.11 thus represents the net indicated work output per cycle. During the intake and exhaust processes of a normally aspirated engine, the negative work performed represents pumping losses and acts to decrease the net work output of the engine. The magnitude of the pumping losses depends on the flow characteristics of the intake and exhaust systems, including the valves, ports, manifolds, piping, mufflers, etc. The more restrictive these passages are, the higher the pumping losses will be.

SI engines control power output by throttling the intake air. Thus, under partial-load conditions, the pressure drop resulting from the air throttling represents a significant increase in pumping loss with a corresponding decrease in operating efficiency. SI engines are therefore less efficient at partial-load operation than at full load. The power level of CI engines, on the other hand, is controlled by varying the amount of fuel injected, as opposed to throttling the intake air, making them significantly more efficient than SI engines under partial-load conditions.

Brake work (or power) is the actual work (or power) produced at the output shaft of an engine, as measured by a dynamometer. The brake work will be less than the indicated work due to friction losses and any parasitic power requirements for oil pumps, water pumps, etc. The **brake mean effective pressure, bmep**, is defined as

$$\text{bmep} = \frac{\text{brake work output per cycle}}{\text{displacement volume}} \tag{10.5}$$

The mechanical efficiency can then be defined as

$$\eta_{\text{m}} = \frac{\text{brake work (power)}}{\text{indicated work (power)}} = \frac{\text{bmep}}{\text{imep}} \tag{10.6}$$

Engine thermal efficiency can be determined from the ratio of power output to rate of fuel energy input, or

$$\eta_{\text{t}} = \frac{\text{Power}}{m_{\text{f}} Q_{\text{c}}} \tag{10.7}$$

where m_{f} is the rate of fuel consumption per unit time and Q_{c} is the heat of combustion per unit mass of fuel. The thermal efficiency in Equation 10.7 could be indicated or brake depending on the nature of the power used in the calculation. Uncertainty associated with variations of energy content of fuels may present a practical difficulty with determining engine thermal efficiency. In lieu of thermal efficiency, **brake-specific fuel consumption** (**bsfc**) is often used as an efficiency index.

$$\text{bsfc} = \frac{\text{fuel consumption rate (kg/h)}}{\text{brake power (kW)}} \tag{10.8}$$

The efficiency of engines operating on the same fuel may be directly compared by their bsfc.

Volumetric efficiency, η_{v}, is an important performance parameter for four-stroke engines defined as

$$\eta_{\text{v}} = \frac{m_{\text{actual}}}{m_{\text{d}}} \tag{10.9}$$

where m_{actual} is the mass of intake mixture per cycle and m_{d} is the mass of mixture contained in the displacement volume at inlet conditions (pressure and temperature near the inlet port).

For SI engines, the mixture mass includes air and fuel; for CI engines only air is present during intake. With the intake mixture density determined at inlet conditions, η_{v} accounts for pressure losses and charge heating associated with the intake ports, valves, and cylinder. Sometimes, for convenience, the mixture density is taken at ambient conditions. In this case, η_{v} is called the overall volumetric efficiency and includes the flow performance of the entire intake system.

Because a certain minimum amount of air is required for complete combustion of a given amount of fuel, it follows that the maximum power output of an engine is directly proportional to its air-flow capacity. Therefore, although not affecting the thermal efficiency of the engine in any way, the volumetric efficiency directly affects the maximum power output for a given displacement and thus can affect the efficiency of the overall system in which the engine is installed because of the effect on system size and weight. Volumetric efficiency is affected primarily by intake and exhaust valve geometry; valve lift and timing; intake port and manifold design; mixing of intake charge with residual exhaust gases; engine speed; ratio of inlet pressure to exhaust back pressure; and heat transfer to the intake mixture from warmer flow passages and combustion chamber surfaces. For further information on the fundamentals of IC engine design and operation, see Taylor (1985), Heywood (1988), Stone (1993), and Ferguson and Kirkpatrick (2001).

10.5 Combustion in IC Engines

10.5.1 Combustion in Spark Ignition Engines

Background. In SI engines, combustion of the fuel–air mixture is initiated by a spark generated between the electrodes of a spark plug. The intake and compression strokes are designed to prepare the mixture for combustion by completely vaporizing the fuel and heating the mixture to just below its autoignition temperature. This is one reason, in addition to controlling emissions, for the current practice of limiting the maximum compression ratio of nonracing SI engines to about 10:1. Near the end of compression, the mixture is well conditioned for combustion and the spark is discharged to initiate the combustion process. For best fuel economy, the combustion process must be completed as close as possible to TDC. This requires that the spark timing be controlled for varying operating speed and load conditions of the engine. Fuel metering and control, according to the engine load requirements and with minimum variation from cylinder to cylinder and cycle to cycle, are essential for good fuel economy, power output, and emission control of the engine.

Carburetors and fuel injection systems are used for fuel-metering control. Because of the superior control capabilities of fuel injection systems, they are nearly universally used today in production automotive applications. Carburetors are used for applications with less-stringent emission requirements, e.g., small engines for lawn and garden equipment.

Figure 10.12 illustrates the effect of **fuel–air** ratio on the indicated performance of an SI engine. The **equivalence ratio** (ϕ) is defined by the ratio fuel–air$_{actual}$/fuel–air$_{stoichiometric}$. Rich mixtures have fuel–air ratios greater than stoichiometric ($\phi > 1$) and lean mixtures have fuel–air ratios less than stoichiometric ($\phi < 1$). Optimum fuel economy, coinciding with maximum thermal efficiency, is obtained at part throttle with a lean mixture as a result of the fact that the heat release from lean mixtures suffers minimal losses from dissociation and variation of specific heat effects when compared with stoichiometric and rich fuel–air ratios.

Maximum power is obtained at full throttle with a slightly rich mixture—an indication of the full utilization of the air within the cylinders. Idling with a nearly closed throttle requires a rich mixture due to the high percentage of residual exhaust gas that remains in the cylinders. The fuel–air mixture requirement under transient operation, such as acceleration, requires a rich mixture to compensate for

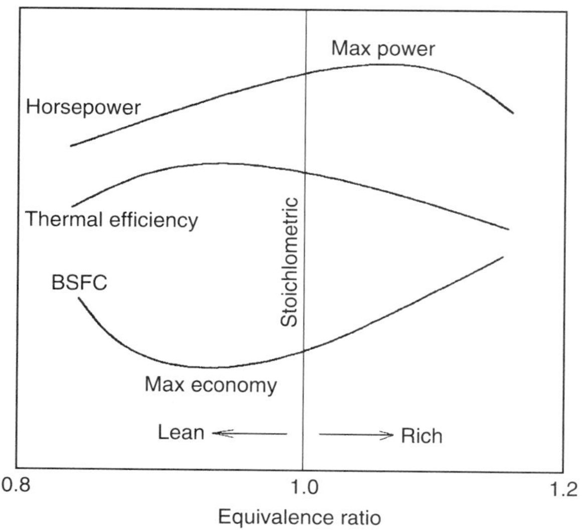

FIGURE 10.12 Effect of fuel–air mixture on indicated performance of an SI engine.

the reduced evaporation caused by the sudden opening of the throttle. Cold starting also requires a rich mixture to ensure the vaporization of sufficient amounts of the highly volatile components in the fuel to achieve proper ignition.

Normal Combustion Process. The combustion processes in SI engines can be divided into two categories: normal and abnormal. The normal combustion process occurs in three stages: initiation of combustion; flame propagation; and termination of combustion. Combustion normally starts within the spark plug gap when the spark is discharged. The fuel molecules in and around the spark discharge zone are ignited and a small amount of energy is released. The important criterion for the initial reaction to be self-sustaining is that the rate of heat release from the initial combustion be larger than the rate of heat transfer to the surroundings. The factors that play an important role in making the initial reaction self-sustaining, and thereby establishing a flame kernel, are the ignition energy level; the spark plug gap; the fuel–air ratio; the initial turbulence; and the condition of the spark plug electrodes.

After a flame kernel is established, a thin spherical flame front advances from the spark plug region progressively into the unburned mixture zone. Flame propagation is supported and accelerated by two processes. First, the combined effect of the heat transfer from the high-temperature flame region and the migration of active radicals from the flame front into the adjacent unburned zone raise the temperature and accelerate the reactivity of the unburned mixture region directly ahead of the flame front. This helps to condition and prepare this zone for combustion.

Second, the increase in the temperature and pressure of the burned gases behind the flame front will cause it to expand and progressively create thermal compression of the remaining unburned mixture ahead of the flame front. The flame speed will be slow at the start of combustion, then reach a maximum at about half the flame travel, and finally decrease near the end of combustion. Overall, the flame speed is strongly influenced by the level of turbulence in the combustion chamber; the shape of the combustion chamber; the mixture strength; the type of fuel; and the engine speed.

When the flame front approaches the walls of the combustion chamber, the high rate of heat transfer to the walls slows down the flame propagation and, finally, the combustion process terminates close to the walls because of surface quenching. This leaves a thin layer of unburned fuel close to the combustion chamber walls that shows up in the exhaust as unburned hydrocarbons.

Abnormal Combustion. Abnormal combustion may occur in SI engines associated with two combustion phenomena: **knock** and **surface ignition**. Knock occurs near the end of the combustion process if the end portion of the unburned mixture, which is progressively subjected to thermal compression and seeding by active radicals, autoignites prematurely before the flame front reaches it. As a result of the sudden energy release, a violent pressure wave propagates back and forth across the combustion chamber, causing the walls or other parts of the engine to vibrate, producing a sharp metallic noise called knock. If knock persists for a period of time, the high rate of heat transfer caused by the traveling high pressure and temperature wave may overheat the spark plug electrode or ignite carbon deposits that may be present in the combustion chamber, causing uncontrolled combustion and surface ignition. As a result, loss of power and serious engine damage may occur.

Knock is sensitive to factors that increase the temperature and pressure of the end portion of the unburned mixture, as well as to fuel composition and other time factors. Factors that increase the probability of knock include: (1) increasing the temperature of the mixture by increasing the charge intake temperature, increasing the compression ratio, or turbo/supercharging; (2) increasing the density of the mixture by turbo/supercharging or increasing the load; (3) advancing the spark timing; (4) increasing the time of exposure of the end portion of the unburned mixture to autoignition conditions by increasing the length of flame travel or decreasing the engine speed and turbulence; and (5) using low-octane fuel and/or maximum power fuel–air ratios.

Other engine design factors that affect knock in SI engines include the shape of the combustion chamber and the location of the spark plug and inlet and exhaust valves relative to the location of the end portion of the unburned mixture. Modern computerized engine management systems that incorporate a knock sensor can automatically retard the ignition timing at the onset of knock, greatly reducing the possibility of engine damage due to knock.

Surface ignition is the ignition of the unburned mixture by any source in the combustion chamber other than the normal spark. Such sources could include overheated exhaust valves or spark plug electrodes, glowing carbon deposits, or other hot spots. Surface ignition will create secondary flame fronts, which cause high rates of pressure rise resulting in a low-pitched, thudding noise accompanied by engine roughness. Severe surface ignition, especially when it occurs before spark ignition, may cause serious structural and/or component damage to the engine.

10.5.2 Combustion in Compression Ignition Engines

Unlike the SI engine, in which the charge is prepared for combustion as a homogeneous mixture during the intake and compression strokes, fuel preparation for combustion in CI engines occurs in a very short period of time called the ignition delay period, which lasts from the beginning of fuel injection until the moment of autoignition. During this period, the fuel injected into the high-temperature air near the end of the compression stroke undergoes two phases of transformation. A physical delay period, during which the fuel is vaporized, mixed with the air, and raised in temperature, is followed by a chemical delay period during which fuel cracking and decomposition occur; this leads to autoignition and combustion of the fuel.

The combustion process is **heterogeneous** and involves two modes, usually identified as premixed combustion and diffusion combustion. Premixed combustion occurs early in the process when the fuel that has evaporated and mixed with air during the ignition delay period autoignites. This mode is characterized by uncontrolled combustion and is the source of combustion noise because it is accompanied by a high rate of heat release, which produces a high rate of pressure rise. When the premixed fuel–air mixture is depleted, diffusion combustion takes over, characterized by a lower rate of heat release and producing controlled combustion during the remainder of the process. Figure 10.13 depicts the different stages of the combustion process in CI engines.

The ignition delay period plays a key role in controlling the time duration of the two modes of combustion. Prolonging the ignition delay, through engine design factors or variations in operating conditions, will generate a larger portion of premixed fuel–air mixture and thus tend to increase the premixed combustion mode duration and decrease the diffusion mode duration. This may lead to higher peak cylinder pressure and temperature; this may improve thermal efficiency and reduce CO and **unburned hydrocarbon** (**UHC**) emissions at the expense of increased emissions of oxides of nitrogen (**NO_x**).

Large increases in the ignition delay period will cause high rates of pressure rise during premixed combustion and may lead to objectionable diesel knock. Reducing the ignition delay period causes

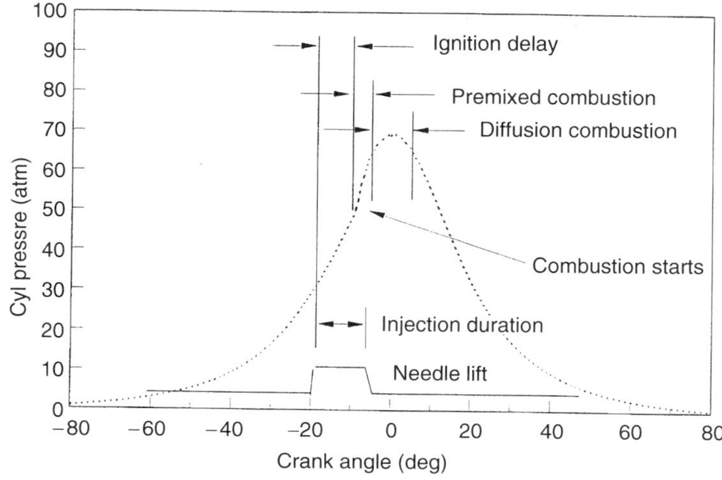

FIGURE 10.13 Combustion process in a CI engine.

the premixed combustion duration to decrease while increasing the diffusion combustion duration. A large reduction in ignition delay may lead to loss of power, decrease in thermal efficiency, and possible deterioration of exhaust emissions. Several factors related to the fuel–air mixture temperature and density, engine speed, combustion chamber turbulence, injection pressure, rate of injection, and fuel composition influence the duration of the ignition delay period.

Knock in CI Engines. Because the combustion process in CI engines is triggered by autoignition of the fuel injected during the ignition delay period, factors that prolong the ignition delay period will increase the premixed combustion duration, causing very high rates of energy release and thus high rates of pressure rise. As a result, diesel knock may occur. The phenomenon is similar to knock in SI engines except that it occurs at the beginning of the combustion process rather than near the end, as observed in SI combustion. Factors that reduce the ignition delay period will reduce the possibility of knock in diesel engines. Among them are increasing the compression ratio; supercharging; increasing combustion chamber turbulence; increasing injection pressure; and using high-**cetane-number** (**CN**) fuel. For a more detailed discussion of the combustion process in IC engines, see Henein (1972), Lenz (1992), and Keating (1993).

10.6 Exhaust Emissions

10.6.1 Harmful Constituents

The products of combustion from IC engines contain several constituents that are considered hazardous to human health, including CO, UHCs NO_x, and **particulates** (from diesel engines). These emission products are discussed briefly next, followed by a description of the principal schemes for their reduction.

Carbon Monoxide. CO is a colorless, odorless, and tasteless gas that is highly toxic to humans. Breathing air with a small volumetric concentration (0.3%) of CO in an enclosed space can cause death in a short period of time. CO results from the incomplete combustion of hydrocarbon fuels. One of the main sources of CO production in SI engines is the incomplete combustion of the rich fuel mixture that is present during idling and maximum power steady-state conditions and during such transient conditions as cold starting, warm-up, and acceleration. Fuel maldistribution, poor condition of the ignition system, and slow CO reaction kinetics also contribute to increased CO production in SI engines. CO production is not as significant in CI engines because these engines are always operated with significant excess air.

Unburned Hydrocarbons. When UHCs combine with NO_x (see following) in the presence of sunlight, ozone and photochemical oxidants form that can adversely affect human health. Certain UHCs are also considered to be carcinogenic. The principal cause of UHC in SI engines is incomplete combustion of the fuel–air charge, resulting in part from flame quenching of the combustion process at the combustion chamber walls and engine misfiring. Additional sources in four-stroke engines may include fuel mixture trapped in crevices of the top ring land of the piston and outgassed fuel during the expansion (power) stroke that was absorbed into the lubricating oil film during intake. In two-stroke SI engines, the scavenging process often results in a portion of the fresh mixture exiting the exhaust port before it closes, resulting in high UHC emissions.

The presence of UHC in CI engines is related to the heterogeneous nature of the fuel–air mixture. Under certain conditions, fuel–air mixtures that lie outside the flammability limits at the lean and rich extremes can exist in portions of the combustion chamber and escape combustion, thus contributing significantly to UHC in the exhaust. Fuel injected near the end of the combustion process and fuel remaining in the nozzle **sac volume** at the end of injection contribute to UHC emission in CI engines. Engine variables that affect UHC emissions include the fuel–air ratio; intake air temperature; and cooling water temperature.

Oxides of Nitrogen. Nitric oxide (NO) is formed from the combination of nitrogen and oxygen present in the intake air under the high-temperature conditions that result from the combustion process. As the gas temperature drops during the expansion stroke, the reaction is frozen, and levels of NO persist in the exhaust products far in excess of the equilibrium level at the exhaust temperature. In the presence of additional oxygen in the air, some NO transforms to nitrogen dioxide (NO_2), a toxic gas.

The combined NO and NO_2 are referred to as oxides of nitrogen or NO_x. The production of NO_x is in general aggravated by conditions that increase the peak combustion temperature. In SI engines, the most important variables that affect NO_x production are the air/fuel ratio; spark timing; intake air temperature; and amount of residual combustion products remaining in the cylinder after exhaust. In CI engines, ignition delay, which affects the degree of premixed combustion, plays a key role in NO_x formation. A larger premixed combustion fraction will produce higher combustion temperatures and higher levels of NO_x.

Particulates. Particulates are a troublesome constituent in the exhaust from CI engines. They are defined by the U.S. Environmental Protection Agency (EPA) as any exhaust substance (other than water) that can be trapped on a filter at temperatures of 325 K or below. Particulates trapped on a filter may be classified as soot plus an organic fraction of hydrocarbons and their partial oxidation products. Soot consists of agglomerates of solid uncombusted carbon particles. Particulates are of concern because their small size permits inhalation and entrapment in the lung walls, making them potential lung carcinogens.

Soot is formed in CI engines under conditions of heavy load when the gas temperature is high and the concentration of oxygen is low. Smoke production is affected by such parameters as fuel CN; rate of fuel injection; inlet air temperature; and the presence of secondary injection.

10.6.2 Control of Emissions from IC Engines

Figure 10.14 depicts the relative concentrations of CO, NO_x, and UHC in the exhaust products of an SI engine as a function of the fuel–air mixture. Lean mixture combustion, which promotes good thermal efficiency, also results in low UHC and CO production but causes high levels of NO_x emission. Increasing the fuel/air ratio to reduce NO_x results in increased CO and UHC emission. Approaches to reduce total emissions fall under two categories: (1) engine design and fuel modifications; and (2) treatment of exhaust gases after they leave the engine.

In SI engines, the first approach focuses on addressing engine variables and design modifications, which improve in-cylinder mixing and combustion in an effort to reduce CO and UHC emissions. To

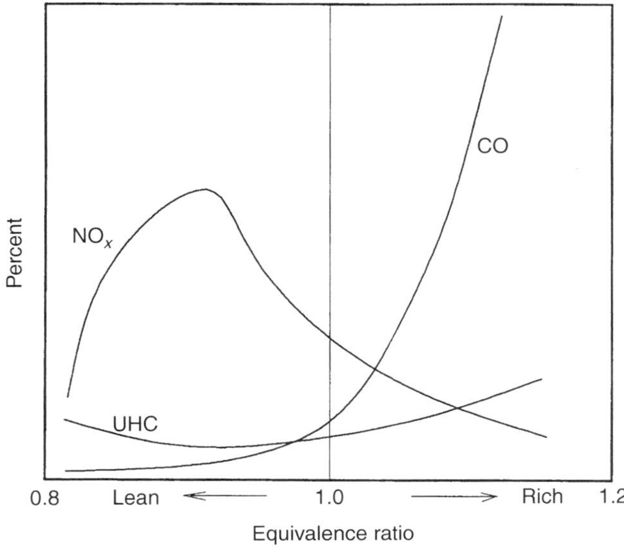

FIGURE 10.14 Emission levels from an SI engine vs. fuel–air mixture.

reduce NO_x, attention is focused on factors that reduce peak combustion temperature and reduce the oxygen available in the flame front. Design and operating parameters implemented or modified for decreased emissions include compression ratio reduction, increased coolant temperature, modification of the combustion chamber shape to minimize surface-to-volume ratio and increase turbulence, improvement of intake manifold design for better charge distribution, use of fuel injection instead of carburetors for better mixture control, use of exhaust gas recirculation to reduce NO_x by lowering combustion temperatures, positive crankcase ventilation to reduce UHC, and increased aromatic content in gasoline.

Engine modifications that have been implemented to reduce emissions from CI engines include modifications to the combustion chamber shape to match the air swirl pattern and fuel spray pattern for better mixing and complete combustion; use of exhaust gas recirculation to limit NO_x production; use of higher injection pressure for better atomization to reduce soot and UHC; and the use of precise injection timing with electronic control.

In the second approach, several devices have been developed for after treatment of exhaust products. A thermal reactor may be used to oxidize UHC and CO. These typically consist of a well-insulated volume placed close to the exhaust manifold, with internal baffles to increase the gas residence time and an air pump to supply fresh oxygen for the oxidation reactions. Thermal reactors are ineffective for NO_x reduction and thus have limited application.

Catalytic converters utilize a catalyst, typically a noble metal such as platinum, rhodium, or palladium, deposited on a ceramic substrate to promote reactions at lower temperatures. Two types are in use: oxidation converters and reduction converters. Oxidation catalytic converters use the excess air available in lean mixtures (or supplied from an external air pump) to oxidize CO and UHC emissions. Reduction catalytic converters operate with low levels of oxygen to cause reduction of NO_x. Sometimes, dual catalytic converters are employed to treat all three pollutants with a reducing converter, to reduce NO_x, placed upstream of an oxidation converter for treating CO and UHC. This arrangement requires that the engine be operated with a rich mixture, which decreases fuel economy.

Three-way catalytic converters are a recent development that permits treatment of NO_x, CO, and UHC in a single device, thus reducing size and weight of the exhaust system. Proper operation of a three-way catalyst requires very nearly stoichiometric combustion. If the combustion is too lean, NO_x is not adequately reduced, and if it is too rich, UHC and CO are not adequately oxidized. Within a narrow band for equivalence ratio (from about 0.999 to 1.007), conversion efficiency is 80% or better for all three pollutants (Kummer 1980). Maintaining engine operation within this narrow mixture band requires a closed-loop fuel-metering system that utilizes an oxygen sensor placed in the exhaust system to monitor excess oxygen and control the fuel injection to maintain near stoichiometric combustion.

Reduction catalytic converters cannot be used with CI engines to reduce NO_x because they normally run lean with significant amounts of excess oxygen in the exhaust. Thus, engine design factors must be relied on to keep NO_x as low as possible. Soot emission may be reduced by after treatment using a device called a trap oxidizer. A trap oxidizer filters particulate matter from the exhaust stream and oxidizes it, usually with the aid of a catalyst for reducing the oxidation temperature. These have been used on small, high-speed automotive diesel engines, but their application to larger, slower speed engines is limited because of the higher level of particulate production and the lower exhaust temperature. For additional information on emissions, see Henein (1972), Obert (1973), and *SAE Surface Vehicle Emissions Standards Manual* (1993).

10.7 Fuels for SI and CI Engines

10.7.1 Background

The primary distinguishing factor between SI and CI engines is the fundamental difference in the combustion process. SI engines rely on homogeneous, spark-ignited, premixed combustion, while CI engines are designed for heterogeneous combustion with an autoignited premixed combustion period followed by a diffusion combustion period. The differences in the combustion process call for quite different qualities in the fuels to achieve optimum performance.

By far the most common fuel for SI engines is gasoline, although other fuels can be used in special circumstances, including alcohol, natural gas, and propane. Even such low-grade fuels as wood gas and coal gas have been used to fuel SI engines during wartime when conventional fuels were in short supply. Diesel fuel is the predominant fuel for CI engines, but they too can be designed to operate on a variety of other fuels, such as natural gas, bio-gas, and even coal slurries. This discussion is confined to gasoline and diesel fuel, both of which are distilled from crude oil.

Crude oil is composed of several thousand different hydrocarbon compounds that, upon heating, are vaporized at different temperatures. In the distillation process, different "fractions" of the original crude are separated according to the temperatures at which they vaporize. The more volatile fraction, naphtha, is followed in order of increasing temperature of vaporization by fractions called distillate, gas oil, reduced crude, and residual oil. These fractions may be further subdivided into light, middle, and heavy classifications. Light virgin naphtha can be used directly as gasoline, although it has relatively poor antiknock quality. The heavier fractions can be chemically processed through coking and catalytic cracking to produce additional gasoline. Diesel fuel is derived from the light to heavy virgin gas oil fraction and from further chemical processing of reduced crude.

10.7.2 Gasoline

Gasoline fuels are mixtures of hydrocarbon compounds with boiling points in the range of 32°C–215°C. The two most important properties of gasoline for SI engine performance are volatility and octane rating. Adequate volatility is required to ensure complete vaporization, as required for homogeneous combustion, and to avoid cold-start problems. If the volatility is too high, however, vapor locking in the fuel delivery system may become a problem. Volatility may be specified by the distillation curve (the distillation temperatures at which various percentages of the original sample have evaporated). Higher volatility fuels will be characterized by lower temperatures for given fixed percentages of evaporated sample or, conversely, by higher percentages evaporated at or below a given temperature. Producers generally vary the volatility of gasoline to suit the season, increasing the volatility in winter to improve cold-start characteristics and decreasing it in summer to reduce vapor locking.

The octane rating of a fuel is a measure of its resistance to autoignition or knocking; higher octane fuels are less prone to autoignition. The octane rating system assigns the value of 100 to iso-octane (C_8H_{18}, a fuel that is highly resistant to knock) and the value 0 to *n*-heptane (C_7H_{16}, a fuel that is prone to knock). Two standardized methods are employed to determine the octane rating of fuel test samples: the research method and the motor method; see ASTM Standards Part 47—Test Methods for Rating Motor, Diesel and Aviation Fuels (ASTM 1995).

Both methods involve testing the fuel in a special variable compression-ratio engine (cooperative fuels research or CFR engine). The test engine is operated on the fuel sample and the compression ratio is gradually increased to obtain a standard knock intensity reading from a knock meter. The octane rating is obtained from the volumetric percentage of iso-octane in a blend of iso-octane and *n*-heptane that produces the same knock intensity at the same compression ratio.

The principal differences between the research method and the motor method are the higher operating speed, higher mixture temperature, and greater spark advance employed in the motor method. Ratings obtained by the research method are referred to as the **research octane number** (RON); those obtained with the motor method are called the motor octane number (MON). MON ratings are lower than RON ratings because of the more stringent conditions, i.e., higher thermal loading of the fuel. The octane rating commonly advertised on gasoline pumps is the **antiknock index**, $(R+M)/2$, which is the average of the values obtained by the two methods. The typical range of antiknock index for automotive gasolines currently available at the pump is 87–93. In general, higher compression SI engines require higher octane fuels to avoid autoignition and to realize full engine performance potential from engines equipped with electronic control systems incorporating a knock sensor.

Straight-run gasoline (naphtha) has a poor octane rating on the order of 40–50 RON. Higher octane fuels are created at the refinery by blending with higher octane components produced through alkylation wherein light olefin gases are reacted with isobutane in the presence of a catalyst. Iso-octane, for example, is formed by reacting isobutane with butene. Aromatics with double carbon bonds shared between more than one ring, such as naphthalene and anthracene, serve to increase octane rating because the molecules are particularly difficult to break.

Additives are also used to increase octane ratings. In the past, a common octane booster added to automotive fuels was lead alkyls—tetraethyl or tetramethyl lead. For environmental reasons, lead has been removed from automotive fuels in most countries. It is, however, still used in aviation fuel. Low-lead fuel has a concentration of about 0.5 g/L, which boosts octane rating by about five points. The use of leaded fuel in an engine equipped with a catalytic converter to reduce exhaust emissions will rapidly deactivate the catalyst (typically a noble metal such as platinum or rhodium), quickly destroying the utility of the catalytic converter. Octane-boosting additives in current use include the oxygenators methanol, ethanol, and methyl tertiary butyl ether (MTBE).

RON values of special-purpose, high-octane fuels for racing and aviation purposes can exceed 100 and are arrived at through an extrapolation procedure based on the knock-limited indicated mean effective pressure (klimep). The klimep is determined by increasing the engine intake pressure until knock occurs. The ratio of the klimep of the test fuel to that for iso-octane is used to extrapolate the octane rating above 100.

10.7.3 Diesel Fuels

Diesel fuels are blends of hydrocarbon compounds with boiling points in the range of 180°C–360°C. Properties of primary importance for CI fuels include the density, viscosity, cloud point, and ignition quality (CN). Diesel fuel exhibits a much wider range of variation in properties than does gasoline. The density of diesel fuels tends to vary according to the percentages of various fractions used in the blend. Fractions with higher distillation temperatures tend to increase the density. Variations in density result in variations in volumetric energy content and thus fuel economy, because fuel is sold by volume measure. Higher density fuel will also result in increased soot emission.

Viscosity is important to proper fuel pump lubrication. Low-viscosity fuel will tend to cause premature wear in injection pumps. Too high viscosity, on the other hand, may create flow problems in the fuel delivery system. Cloud point is the temperature at which a cloud of wax crystals begins to form in the fuel. This property is critical for cold-temperature operation because wax crystals will clog the filtration system. ASTM does not specify maximum cloud point temperatures, but rather recommends that cloud points be no more than 6°C above the 10th percentile minimum ambient temperature for the region for which the fuel is intended; see ASTM D 975 (ASTM 1995).

CN provides a measure of the autoignition quality of the fuel and is the most important property for CI engine fuels. The CN of a fuel sample is obtained through the use of a CI CFR engine in a manner analogous to the determination of octane rating. The test method for CN determination is specified in standard ASTM D 613. *n*-Cetane (same as hexadecane, $C_{16}H_{34}$) has good autoignition characteristics and is assigned the cetane value of 100. The bottom of the cetane scale was originally defined in terms of α-methyl naphthalene ($C_{11}H_{10}$), which has poor autoignition characteristics and was assigned the value 0. In 1962, for reasons of availability and storability, the poor ignition quality standard fuel used to establish the low end of the cetane scale was changed to heptamethylnonane (HMN), with an assigned CN of 15. The CN of a fuel sample is determined from the relative volumetric percentages of cetane and HMN in a mixture that exhibits the same ignition delay characteristics as the test sample using the relation

$$\text{CN} = \%n\text{-cetane} + 0.15\,(\%\text{HMN}) \tag{10.10}$$

ASTM standard D 976 (ASTM 1995) provides the following empirical correlation for calculating the **cetane index** of straight petroleum distillate fuels (no additives) as an approximation to the measured CN:

$$\text{Cetane index} = 454.74 - 1641.416D + 774.74D^2 - 0.554B + 97.803(\log B)^2 \quad (10.11)$$

where D=density at 15°C (g/mL) and B=mid-boiling temperature (°C).

ASTM standard D 975 (ASTM 1995) establishes three classification grades for diesel fuels (No. 1-D, No. 2-D, and No. 4-D) and specifies minimum property standards for these grades. No. 1-D is a volatile distillate fuel for engines that must operate with frequent changes in speed and load. No. 2-D is a lower volatility distillate fuel for industrial and heavy mobile service engines. No. 4-D is a heavy fuel oil for low- and medium-speed engines. Nos. 1-D and 2-D are principally transportation fuels, while No. 4-D is for stationary applications. The ASTM minimum CN for No. 1-D and No. 2-D is 40, and for No. 4-D the minimum is 30. Typical CNs for transportation fuels lie in the range of 40–55. Use of a low-cetane fuel aggravates diesel knock because of the longer ignition delay period, which creates a higher fraction of premixed combustion.

Antiknock quality (octane number) and ignition quality (CN) are opposing properties of distillate fuels. The CN increases with decreasing octane rating of various fuels. Gasoline, with good antiknock quality, has a CN of approximately 10, while a diesel fuel with a CN of 50 will have an octane number of about 20. Thus, gasoline is not a suitable fuel for CI engines because of its poor autoignition quality, and diesel fuel is inappropriate for use in SI engines as a result of its poor antiknock quality. For additional information on fuels for IC engines see the *SAE Fuels and Lubricants Standards Manual* (1993) and Owen and Coley (1995).

10.8 Intake Pressurization—Supercharging and Turbocharging

10.8.1 Background

Pressurizing the intake air (or mixture) by means of a compressor may be used to boost the specific power output of SI and CI engines. Supercharging generally refers to the use of compressors that are mechanically driven from the engine crankshaft, while turbocharging refers to compressors powered by a turbine, which extracts energy from the exhaust stream. Increasing the intake pressure increases the density and thus the mass flow rate of the intake mixture; this allows an increase in the fueling rate, thereby producing additional power.

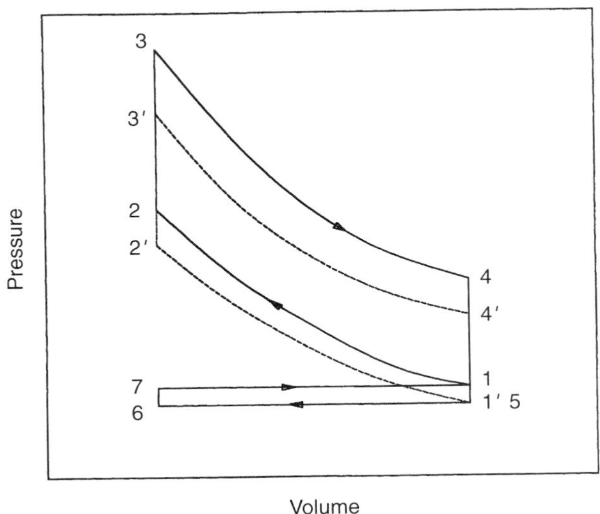

FIGURE 10.15 Comparison of supercharged and naturally aspirated Otto cycle.

The mere process of increasing the cylinder pressure results in increased work output per cycle, as illustrated in the $P–V$ diagram in Figure 10.15, which compares supercharged and naturally aspirated, air standard Otto cycles having the same compression ratio. The work done for the compressed intake cycle (Area 1, 2, 3, 4, 1 and Area 5, 6, 7, 1, 5) is greater than that for the naturally aspirated cycle (Area $1'$, $2'$, $3'$, $4'$, $1'$) due to the boost of the intake pressure. Positive-displacement superchargers are capable of producing higher boost pressures than turbochargers, which are nearly always centrifugal-type fans. From a practical standpoint, the maximum useful boost pressure from either system is limited by the onset of autoignition in SI engines and by the permissible mechanical and thermal stresses in CI engines.

10.8.2 Supercharging

The principal applications of supercharging SI engines are in high-output drag-racing engines and in large aircraft piston engines to provide high specific output at takeoff and to improve power output at high altitudes. A few high-performance production automobiles also use a supercharger in lieu of the more common turbocharger to achieve their increased performance. For diesel applications, supercharging is used mainly in marine and land-transportation applications. It is common to use supercharging or turbocharging to improve the scavenging process in two-stroke diesel engines. Figure 10.16 is a schematic of an engine with a mechanically driven supercharger. Superchargers may be belt, chain, or gear driven from the engine crankshaft.

Two types of superchargers are in use: the positive displacement type (Roots blower) and the centrifugal type. Roots blowers may be classified as: (1) straight double lobe; (2) straight triple lobe; and (3) helix triple lobe (twisted 60%). The helix triple-lobe type runs more quietly than the others and is generally recommended, especially for diesel engines operating under high torque at various speed conditions. Because of its high capacity and small weight and size, the centrifugal type is best suited for applications in which power and volumetric efficiency improvement are required at high engine speed, e.g., with aircraft engines. A centrifugal blower will also survive a backfire more readily than a Roots blower in SI

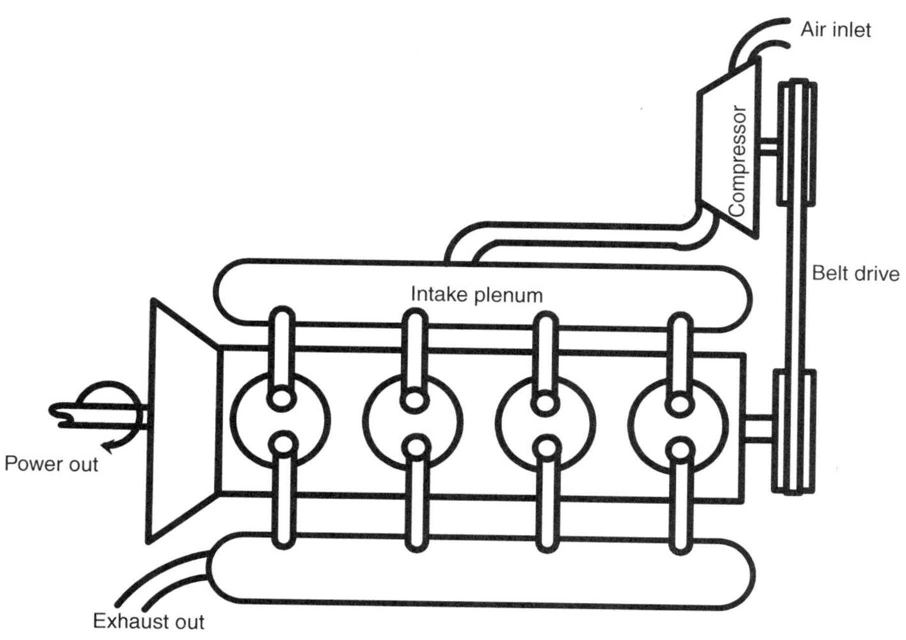

FIGURE 10.16 Schematic diagram of supercharged engine.

applications. Because superchargers are directly driven from the engine output shaft, no inherent lag in the rate of pressure increase with engine speed is present, as is typically the case with turbochargers.

10.8.3 Turbocharging

Turbochargers utilize a centrifugal compressor directly connected to a turbine that extracts energy from the exhaust gases of the engine and converts it to the shaft work necessary to drive the compressor. Turbocharging is widely used to increase power output in automotive and truck applications of four-stroke SI and CI engines and to improve scavenging of two-stroke CI engines.

There are three methods of turbocharging: the constant pressure; the pulse; and the pulse converter. In the constant-pressure method, as illustrated in Figure 10.17, the exhaust pressure is maintained at a nearly constant level above atmospheric. To accomplish this, the exhaust manifold must be large enough to damp out the pressure fluctuations caused by the unsteady flow characteristic of the engine exhaust process. In this method, the turbine operates efficiently under steady-flow conditions; however, some engine power is lost because of the increased backpressure in the exhaust manifold.

The pulse turbocharger, as illustrated in Figure 10.18, utilizes the kinetic energy generated by the exhaust blow-down process in each cylinder. This is accomplished by using small exhaust lines grouped together in a common manifold to receive the exhaust from the cylinders, which are blowing down sequentially. In this method, the pressure at the turbine inlet tends to fluctuate; this is not conducive to good turbine efficiency. This is offset to a large degree, however, by improved engine performance as a result of the lower exhaust backpressure relative to the constant-pressure method. The pulse converter method represents a compromise between the previous two techniques. In principle, this is accomplished by converting the kinetic energy in the blow-down process into a pressure rise at the turbine by utilizing one or more diffusers. Details of the different methods of turbocharging may be found in Watson and Janota (1982).

Recent advances in turbocharging technology have focused mainly on (1) improving turbine transient response (turbo-lag); (2) improving torque-speed characteristics of the engine; and (3) increasing the power output by increasing the boost pressure and using charge cooling (intercooling). The use of ceramic materials in fabricating turbine rotors improves the turbine transient response because they are lighter in weight and have less rotational inertia. Ceramic rotors also have greater thermal operating range because of their lower thermal expansion. The use of variable-geometry turbochargers can improve

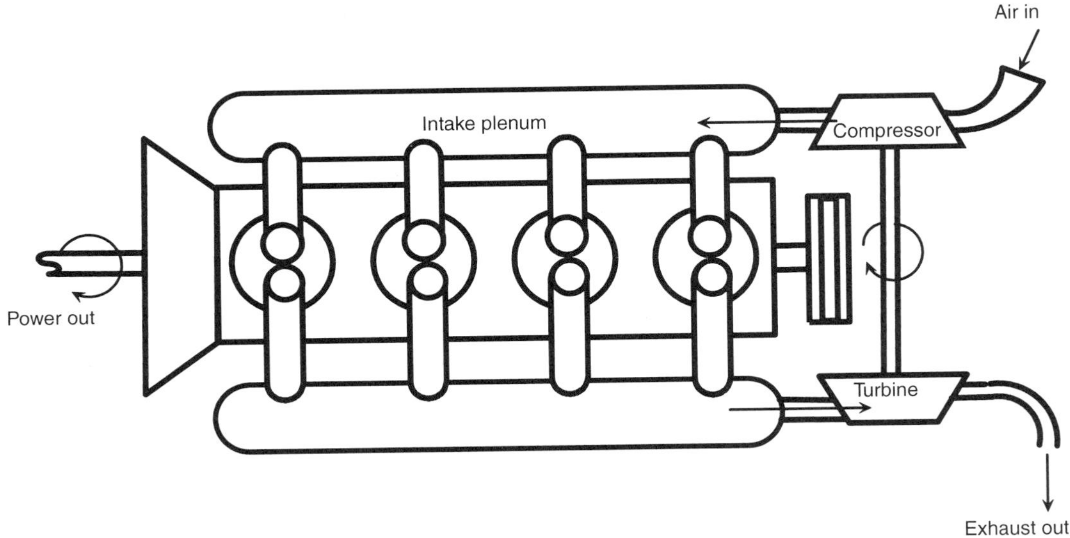

FIGURE 10.17 Schematic diagram of a constant-pressure turbocharger.

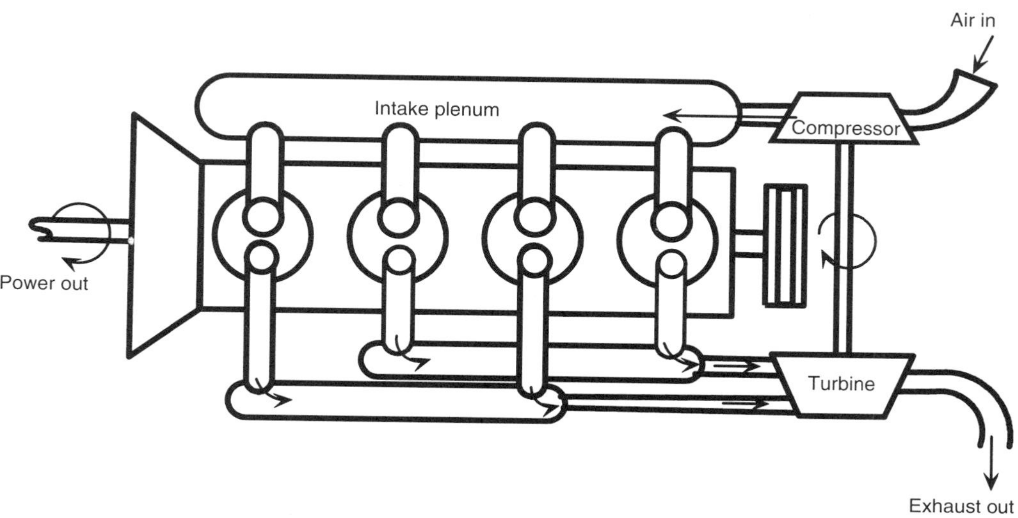

FIGURE 10.18 Schematic diagram of a pulse turbocharger.

the low-speed torque characteristics of the engine and help reduce the transient response time. This is due to the ability of the variable-geometry turbocharger to change its internal geometry to accommodate low flow rates at low engine speeds and higher volume flow rates at high engine speeds.

However, because the geometry of the turbine rotor remains unchanged while the internal geometry varies, the turbine efficiency will be reduced for all internal geometries other than the optimum design geometry. In response to increased demand for diesel engines with high boost pressure and with size constraints, advances in the aerothermodynamics of axial/radial flow and of two-stage turbochargers, as well as in the design of compressor and turbine blades, have allowed high boost pressure at improved overall turbocharger efficiency.

Charge cooling by means of a heat exchanger (intercooler) between the compressor and the intake ports is effective in reducing NO_x emissions and improving the power output of turbocharged diesel engines and in reducing the probability of knock in SI engines. Two types of charge cooling are in use: air–air and air–water. Air-to-water cooling is used in marine applications, in which a source of cool water is available; air-to-air intercoolers are used for automotive and truck applications.

Defining Terms

Antiknock index: The average of the two octane numbers obtained by the research method and the motor method.

Autoignition: The ability of a fuel–air mixture to ignite spontaneously under conditions of high temperature and pressure.

Bottom dead center (BDC): Piston located at its lowest position in the cylinder. Cylinder volume is maximum at BDC.

Brake mean effective pressure (bmep): Ratio of brake work output per cycle to the displacement volume.

Brake specific fuel consumption (bsfc): The ratio of fuel consumption rate in kilograms per hour to the engine output in kilowatts.

Brake work: Work produced at the output shaft of an IC engine as measured by a dynamometer.

Cetane index: An approximation to the measured cetane number determined from an empirical relationship specified in ASTM D 976.

Cetane number: A measure of the autoignition quality of a fuel important for proper performance of CI engines determined experimentally through use of a CI CFR test engine.

Clearance volume: Combustion chamber volume remaining above the piston at TDC.

Compression ignition (CI) engine: Air alone is compressed in the cylinder and fuel is injected near TDC. Combustion results from autoignition of the fuel–air mixture due to the high temperature of the air.

Compression ratio: The ratio of the cylinder volume at BDC to the volume at TDC.

Cut-off ratio: Ratio of cylinder volume at the end of heat addition to the volume at the start of heat addition in the ideal diesel cycle.

Cylinder volume: Volume above piston at BDC; equals displacement volume plus clearance volume.

Direct injection (DI): Method of fuel injection in low- and medium-speed CI engines wherein fuel is injected into the main combustion chamber formed by a bowl in the top of the piston.

Displacement volume: Difference in cylinder volume between TDC and BDC.

Equivalence ratio: Actual fuel–air ratio divided by stoichiometric fuel–air ratio.

Four-stroke engine: Entire cycle completed in two revolutions of the crankshaft and four strokes of the piston.

Fuel–air ratio: Ratio of mass of fuel to mass of air in the cylinder prior to combustion.

Glow plug: Electric heater installed in prechamber of an IDI diesel engine to aid cold starting.

Heterogeneous combustion: Refers to the mixture of liquid fuel droplets and evaporated fuel vapor and air mixture present in CI engine combustion chambers prior to ignition.

Ignition delay period: Period between start of injection and onset of autoignition in a CI engine.

Indicated mean effective pressure (imep): Ratio of net indicated work output of an IC engine to the displacement volume.

Indicated work: Work output of an IC engine cycle determined by an area calculation from an indicator diagram.

Indicator diagram: Pressure–volume trace for an IC engine cycle; area enclosed by diagram represents work.

Indirect injection (IDI): Method of fuel injection used in high-speed CI engines wherein the fuel is injected into a precombustion chamber to promote fuel–air mixing and reduce ignition delay.

Knock: In SI engines: the noise that accompanies autoignition of the end portion of the uncombusted mixture prior to the arrival of the flame front. In CI engines: The noise that accompanies autoignition of large premixed fractions generated during prolonged ignition delay periods. Knock is detrimental to either type of engine.

NO_x: Harmful oxides of nitrogen (NO and NO_2) appearing in the exhaust products of IC engines.

Octane number: Antiknock rating for fuels important for prevention of autoignition in SI engines.

Particulates: Any exhaust substance, other than water, that can be collected on a filter. Harmful exhaust product from CI engines.

Power density: Power produced per unit of engine mass.

Premixed homogeneous combustion: Fuel and air are mixed in an appropriate combustible ratio prior to ignition process. This is the combustion mode for SI engines and for the initial combustion phase in CI engines.

Sac volume: Volume of nozzles below the needle of a diesel fuel injector that provides a source of UHC emissions in CI engines.

Scavenging: The process of expelling exhaust gases and filling the cylinder with fresh charge in two-stroke engines. This is often accomplished in SI engines by pressurizing the fresh mixture in the crankcase volume beneath the piston and in CI engines by using a supercharger or turbocharger.

Spark ignition (SI) engine: Homogeneous charge of air–fuel mixture is compressed and ignited by a spark.

Stroke: Length of piston movement from TDC to BDC; equal to twice the crankshaft throw.

Supercharging: Pressurizing the intake of an IC engine using a compressor that is mechanically driven from the crankshaft.

Surface ignition: A source of autoignition in SI engines caused by surface hot spots.

Swirl: Circular in-cylinder air motion designed into CI engines to promote fuel–air mixing.
Swirl ratio: Ratio of rotational speed of in-cylinder air (rpm) to engine speed (rpm).
Top dead center (TDC): Piston located at its uppermost position in the cylinder. Cylinder volume (above the piston) is minimum at TDC.
Turbocharging: Pressurizing the intake of an IC engine with a compressor driven by a turbine that extracts energy from the exhaust gas stream.
Two-stroke engine: Entire cycle completed in one revolution of the crankshaft and two strokes of the piston.
Unburned hydrocarbons (UHC): Harmful emission product from IC engines consisting of hydrocarbon compounds that remain uncombusted.
Volumetric efficiency: Ratio of the actual mass of air intake per cycle to the displacement volume mass determined at inlet temperature and pressure.

References

ASTM. 1995. *Annual Book of ASTM Standards*. American Society for Testing and Materials, Philadelphia.
Blair, G. P. ed. 1988. Advances in Two-Stroke Cycle Engine Technology, Society of Automotive Engineers, Inc., Warrendale, PA.
Ferguson, C. R. and Kirkpatrick, A. T. 2001. *Internal Combustion Engines, Applied Thermosciences*. Wiley, New York.
Henein, N. A. 1972. *Emissions from Combustion Engines and Their Control*. Ann Arbor Science Publishers, Ann Arbor, MI.
Heywood, J. B. 1988. *Internal Combustion Engine Fundamentals*. McGraw-Hill, New York.
Keating, E. L. 1993. *Applied Combustion*. Marcel Dekker, New York.
Kummer, J. T. 1980. Catalysts for automobile emission control. *Prog. Energy Combust. Sci.*, 6, 177–199.
Lenz, H. P. 1992. *Mixture Formation in Spark-Ignition Engines*. Springer, New York.
Norbye, J. P. 1971. *The Wankel Engine*. Chilton Press, Philadelphia.
Obert, E. F. 1973. *Internal Combustion Engines and Air Pollution. 3rd Ed.* Harper & Row, New York.
Owen, K. and Coley, T. 1995. *Automotive Fuels Reference Book. 2nd Ed.* Society of Automotive Engineers, Inc., Warrendale, PA.
SAE Fuels and Lubricants Standards Manual. 1993. Society of Automotive Engineers, Inc., Warrendale, PA.
SAE Surface Vehicle Emissions Standards Manual. 1993. Society of Automotive Engineers, Inc., Warrendale, PA.
Stone, R. 1993. *Introduction to Internal Combustion Engines. 2nd Ed.* Society of Automotive Engineers, Inc., Warrendale, PA.
Taylor, C. F. 1985. *2nd Ed. The Internal Combustion Engine in Theory and Practice., Vols. I and II*, MIT Press, Cambridge, MA.
Watson, N. and Janota, M. S. 1982. *Turbocharging the Internal Combustion Engine*. Wiley, New York.

For Further Information

The textbooks on IC engines by Obert (1973), Taylor (1985), Heywood (1988), Stone (1993), and Ferguson and Kirkpatrick (2001) listed under the references provide excellent treatments of this subject. In particular, Stone's book is up to date and informative. The *Handbook of Engineering* (1966) published by CRC Press, Boca Raton, Florida, contains a chapter on IC engines by A. Kornhauser. The Society of Automotive Engineers (SAE) publishes transactions, proceedings, and books related to all aspects of automotive engineering, including IC engines. Two very comprehensive handbooks distributed by SAE are the *Bosch Automotive Handbook* and the *SAE Automotive Handbook*. For more information contact: SAE Publications, 400 Commonwealth Drive, Warrendale, PA, 15096-0001; (412)776–4970.

11

Hydraulic Turbines

11.1 General Description 11-1
Typical Hydropower Installation • Turbine Classification
11.2 Principles of Operation 11-5
Power Available, Efficiency • Similitude and Scaling Formulae
11.3 Factors Involved in Selecting a Turbine 11-8
Performance Characteristics • Speed Regulation • Cavitation and Turbine Setting
11.4 Performance Evaluation 11-12
Model Tests
11.5 Numerical Simulation 11-14
11.6 Field Tests 11-17
Defining Terms 11-18
References 11-18
For Further Information 11-19

Roger E.A. Arndt
University of Minnesota

A hydraulic turbine is a mechanical device that converts the potential energy associated with a difference in water elevation (**head**) into useful work. Modern hydraulic turbines are the result of many years of gradual development. Economic incentives have resulted in the development of very large units (exceeding 800 MW in capacity) with efficiencies that are sometimes in excess of 95%.

The emphasis on the design and manufacture of very large turbines is shifting to the production of smaller units, especially in developed nations, where much of the potential for developing large base-load plants has been realized. At the same time, the escalation in the cost of energy has made many smaller sites economically feasible and has greatly expanded the market for smaller turbines. The increased value of energy also justifies the cost of refurbishment and increasing the capacity of older facilities. Thus, a new market area is developing for updating older turbines with modern replacement runners that have higher efficiency and greater capacity. The introduction of modern computational tools in the last decade has had considerable influence on turbine design.

11.1 General Description

11.1.1 Typical Hydropower Installation

As shown schematically in Figure 11.1, the hydraulic components of a hydropower installation consist of an intake, penstock, guide vanes or distributor, turbine, and **draft tube**. Trash racks are commonly provided to prevent ingestion of debris into the turbine. Intakes usually require some type of shape transition to match the passageway to the turbine and also incorporate a gate or some other means of stopping the flow in case of an emergency or turbine maintenance. Some types of turbines are set in an open flume; others are attached to a closed-conduit penstock.

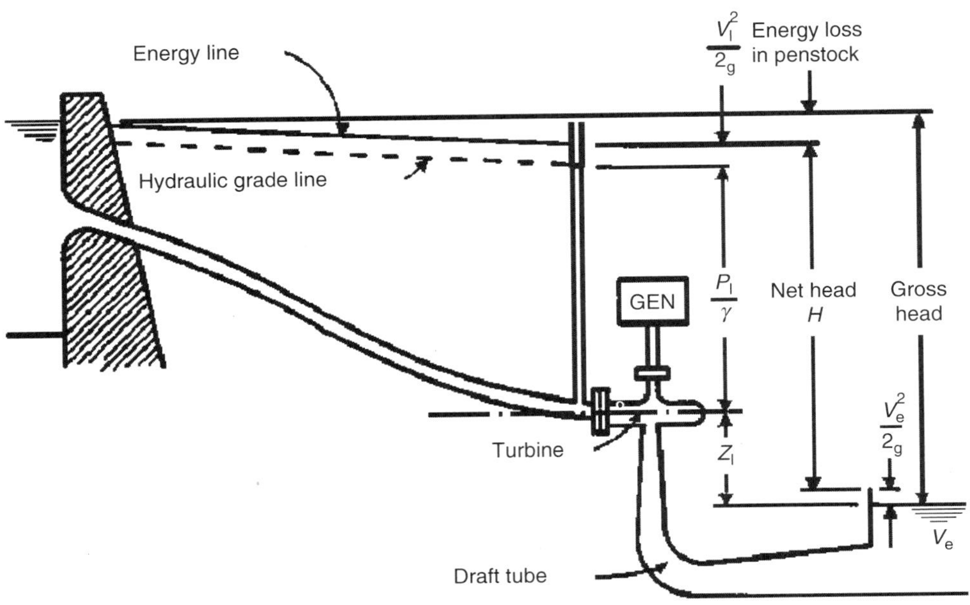

FIGURE 11.1 Schematic of a hydropower installation.

11.1.2 Turbine Classification

The two types of turbines are denoted as impulse and reaction. In an *impulse turbine*, the available head is converted to kinetic energy before entering the **runner**; the power available is extracted from the flow at approximately atmospheric pressure. In a *reaction turbine*, the runner is completely submerged and the pressure and the velocity decrease from inlet to outlet. The velocity head in the inlet to the turbine runner is typically less than 50% of the total head available.

Impulse Turbines. Modern impulse units are generally of the Pelton type and are restricted to relatively high head applications (Figure 11.2). One or more jets of water impinge on a wheel containing many curved buckets. The jet stream is directed inwardly, sideways, and outwardly, thereby producing a force on the bucket, which in turn results in a torque on the shaft. All kinetic energy leaving the runner is "lost." A draft tube is generally not used because the runner operates under approximately atmospheric pressure and the head represented by the elevation of the unit above tailwater cannot be utilized. (In principle, a draft tube could be used; this requires the runner to operate in air under reduced pressure. Attempts at operating an impulse turbine with a draft tube have not met with much success.) Because this is a high-head device, this loss in available head is relatively unimportant. As will be shown later, the Pelton wheel is a **low specific speed** device. Specific speed can be increased by the addition of extra nozzles—the specific speed increasing by the square root of the number of nozzles. Specific speed can also be increased by a change in the manner of inflow and outflow. Special designs such as the Turgo or crossflow turbines are examples of relatively high specific speed impulse units (Arndt 1991).

Most Pelton wheels are mounted on a horizontal axis, although newer vertical-axis units have been developed. Because of physical constraints on orderly outflow from the unit, the number of nozzles is generally limited to six or less. Whereas **wicket gates** control the power of a reaction turbine, the power of the Pelton wheel is controlled by varying the nozzle discharge by means of an automatically adjusted needle, as illustrated in Figure 11.2. Jet deflectors or auxiliary nozzles are provided for emergency unloading of the wheel. Additional power can be obtained by connecting two wheels to a single generator or by using multiple nozzles. Because the needle valve can throttle the flow while maintaining essentially

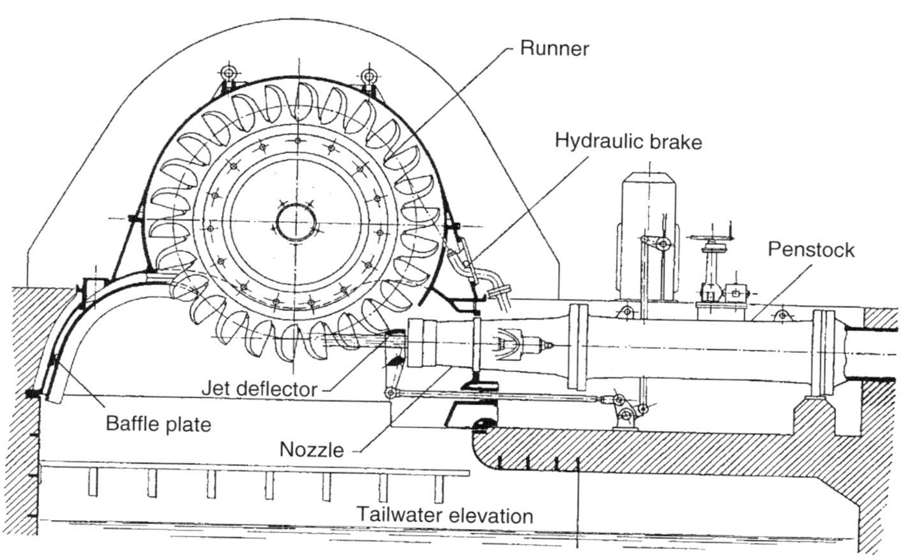

FIGURE 11.2 Cross section of a single-wheel, single-jet Pelton turbine. This is the third highest head Pelton turbine in the world, $H = 1447$ m; $n = 500$ rpm; $P = 35.2$ MW; $N_S \sim 0.038$. (Courtesy of Vevey Charmilles Engineering Works. Adapted from Raabe, J. 1985. *Hydro Power: The Design, Use, and Function of Hydromechanical, Hydraulic, and Electrical Equipment.* VDI Verlag, Dusseldorf, Germany.)

constant jet velocity, the relative velocities at entrance and exit remain unchanged, producing nearly constant efficiency over a wide range of power output.

Reaction Turbines. Reaction turbines are classified according to the variation in flow direction through the runner. In radial- and mixed-flow runners, the flow exits at a radius different from the radius at the

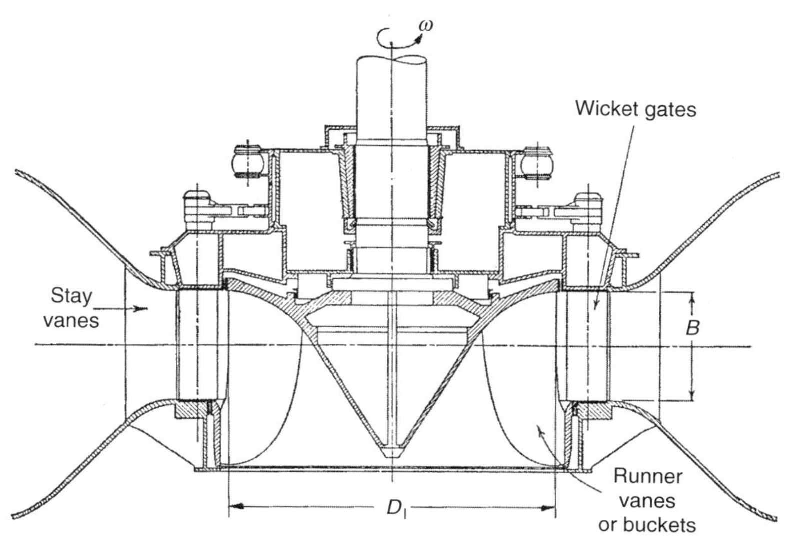

FIGURE 11.3 Francis turbine, $N_S \sim 0.66$. (Adapted from Daily, J. W. 1950. In *Engineering Hydraulics,* H. Rouse, ed. Wiley, New York. Reprinted with permission.)

inlet. If the flow enters the runner with only radial and tangential components, it is a radial-flow machine. The flow enters a mixed-flow runner with radial as well as axial components. Francis turbines are of the radial- and mixed-flow type, depending on the design specific speed. A Francis turbine is illustrated in Figure 11.3.

Axial-flow propeller turbines are generally of the fixed-blade or Kaplan (adjustable-blade) variety. The "classical" propeller turbine, illustrated in Figure 11.4, is a vertical-axis machine with a scroll case and a radial wicket gate configuration that is very similar to the flow inlet for a Francis turbine. The flow enters radially inward and makes a right-angle turn before entering the runner in an axial direction. The Kaplan turbine has adjustable runner blades and adjustable wicket gates. The control system is designed so that the variation in blade angle is coupled with the wicket gate setting in a manner that achieves best overall efficiency over a wide range of flow rates.

Some modern designs take full advantage of the axial-flow runner; these include the tube, bulb, and Straflo types illustrated in Figure 11.5. The flow enters and exits the turbine with minor changes in direction. A wide variation in civil works design is also permissible. The tubular type can be fixed-propeller, semi-Kaplan, or fully adjustable. An externally mounted generator is driven by a shaft that extends through

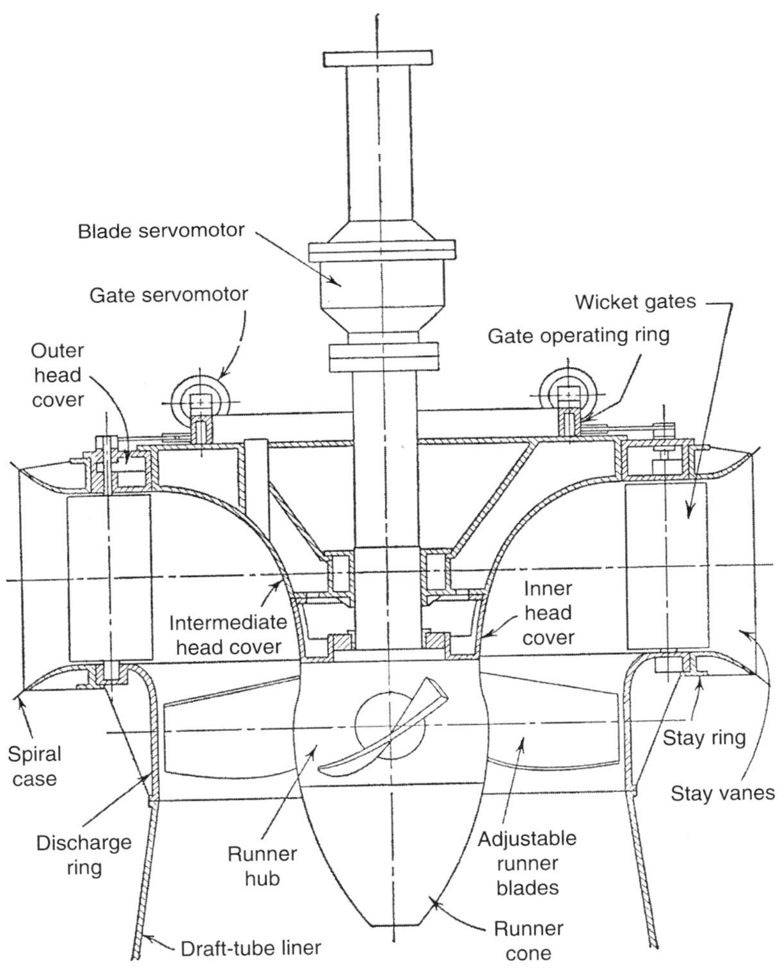

FIGURE 11.4 Smith–Kaplan axial-flow turbine with adjustable-pitch runner blades, $N_S \sim 2.0$. (From Daily, J. W. 1950. In *Engineering Hydraulics*, H. Rouse, ed. Wiley, New York. Reprinted with permission.)

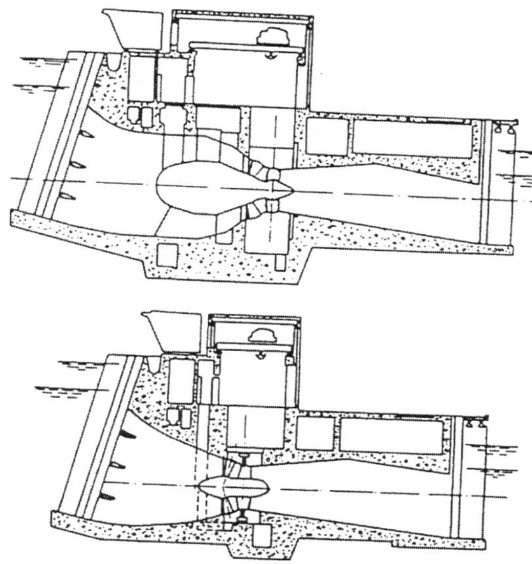

FIGURE 11.5 Comparison between bulb (upper) and Straflo (lower) turbines. (Courtesy U.S. Department of Energy.)

the flow passage upstream or downstream of the runner. The bulb turbine was originally designed as a high-output, low-head unit. In large units, the generator is housed within the bulb and is driven by a variable-pitch propeller at the trailing end of the bulb. Pit turbines are similar in principle to bulb turbines, except that the generator is not enclosed in a fully submerged compartment (the bulb). Instead, the generator is in a compartment that extends above water level. This improves access to the generator for maintenance.

11.2 Principles of Operation

11.2.1 Power Available, Efficiency

The power that can be developed by a turbine is a function of the head and flow available:

$$P = \eta \rho g Q H \tag{11.1}$$

where η is the turbine efficiency; ρ is the density of water (kg/m^3); g is the acceleration due to gravity (m/s^2); Q is the flow rate (m^3/s); and H is the net head in meters. *Net head* is defined as the difference between the *total head* at the inlet and the tailrace, as illustrated in Figure 11.1. Various definitions of net head are used in practice; these depend on the value of the exit velocity head, $V_e^2/2g$, that is used in the calculation. The International Electrotechnical Test Code uses the velocity head at the draft tube exit.

The efficiency depends on the actual head and flow utilized by the turbine runner; flow losses in the draft tube; and the frictional resistance of mechanical components.

11.2.2 Similitude and Scaling Formulae

Under a given head, a turbine can operate at various combinations of speed and flow depending on the inlet settings. For reaction turbines the flow into the turbine is controlled by the wicket gate angle, α.

The nozzle opening in impulse units typically controls the flow. Turbine performance can be described in terms of nondimensional variables,

$$\Psi = \frac{2gH}{\omega^2 D^2} \tag{11.2}$$

$$\phi = \frac{Q}{\sqrt{2gHD^2}} \tag{11.3}$$

where ω is the rotational speed of the turbine in radians per second and D is the diameter of the turbine.

The hydraulic efficiency of the runner alone is given by

$$\eta_h = \frac{\phi}{\sqrt{\Psi}}(C_1 \cos \alpha_1 - C_2 \cos \alpha_2) \tag{11.4}$$

where C_1 and C_2 are constants that depend on the specific turbine configuration, and α_1 and α_2 are the inlet and outlet angles that the absolute velocity vectors make with the tangential direction. The value of $\cos \alpha_2$ is approximately zero at peak efficiency. The terms ϕ, ψ_2, α_1, and α_2 are interrelated. Using model test data, isocontours of efficiency can be mapped in the $\phi\psi$ plane. This is typically referred to as a *hill diagram*, as shown in Figure 11.6.

The **specific speed** is defined as

$$N_s = \frac{\omega\sqrt{Q}}{(2gH)^{3/4}} = \sqrt{\frac{\phi}{\Psi}} \tag{11.5}$$

A given specific speed describes a specific combination of operating conditions that ensures similar flow patterns and the same efficiency in geometrically similar machines regardless of the size and rotational speed of the machine. It is customary to define the design specific speed in terms of the value at the design head and flow where peak efficiency occurs. The value of specific speed so defined permits a classification of different turbine types.

The specific speed defined here is dimensionless. Many other forms of specific speed exist are dimensional and have distinct numerical values depending on the system of units used (Arndt 1991). (The literature also contains two other minor variations of the dimensionless form. One differs by a factor of $1/\pi^{1/2}$ and the other by $2^{3/4}$.) The similarity arguments used to arrive at the concept of specific speed indicate that a given machine of diameter D operating under a head H will discharge a flow Q and produce a torque T and power P at a rotational speed ω given by

$$Q = \phi D^2 \sqrt{2gH} \tag{11.6}$$

$$T = T_{11}\rho D^3 2gH \tag{11.7}$$

$$P = P_{11}\rho D^2 (2gH)^{3/2} \tag{11.8}$$

$$\omega = \frac{2u_1}{D} = \omega_{11}\frac{\sqrt{2gH}}{D}, \quad \left[\omega_{11} = \frac{1}{\sqrt{\Psi}}\right] \tag{11.9}$$

with

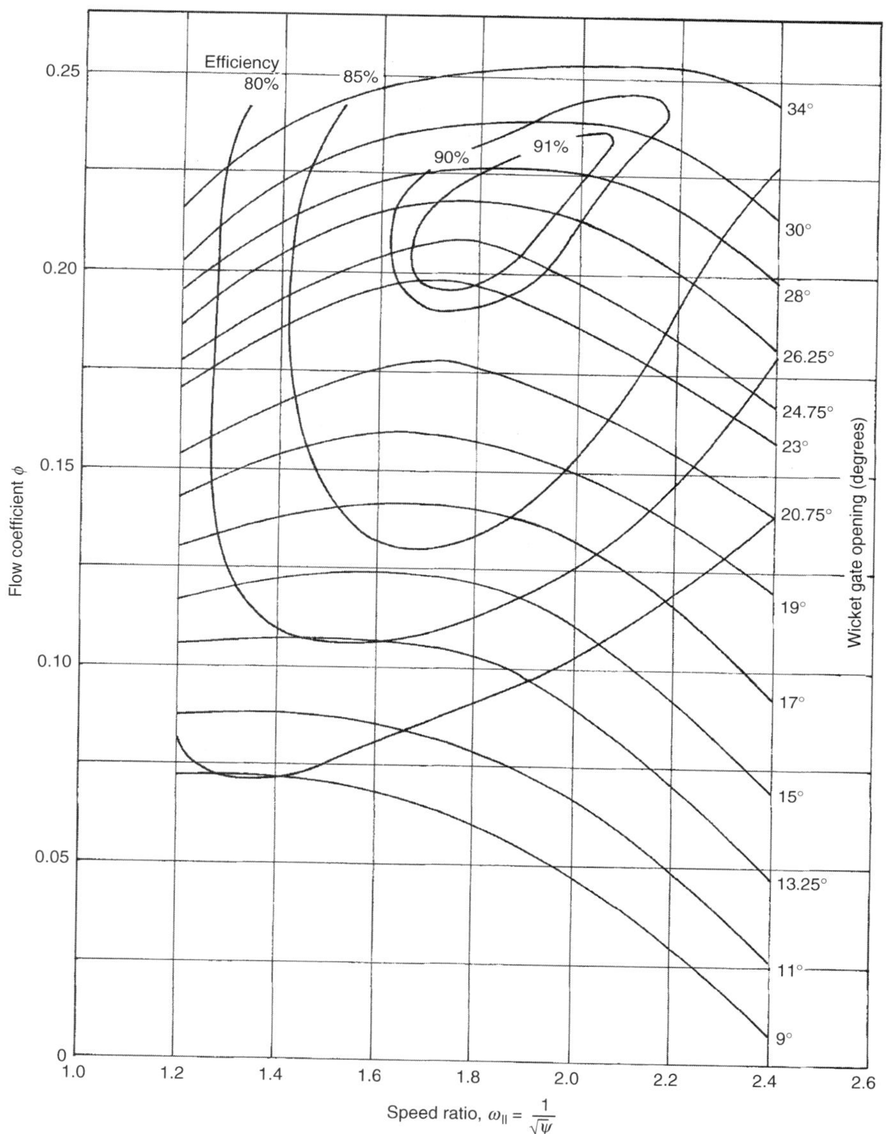

FIGURE 11.6 Typical hill diagram. Information of this type is obtained in turbine test stand (see Figure 11.9). (Adapted from Wahl, T. L. 1994. Draft tube surging times two: the twin vortex problem. *Hydro Rev.* 13, 60–69, 1994. With permission.)

$$P_{11} = T_{11}\omega_{11} \tag{11.10}$$

where T_{11}, P_{11}, and ω_{11} are also nondimensional. (The reader is cautioned that many texts, especially in the American literature, contain dimensional forms of T_{11}, P_{11}, and ω_{11}.) In theory, these coefficients are fixed for a machine operating at a fixed value of specific speed, independent of the size of the machine. Equation 11.6 through Equation 11.10 can be used to predict the performance of a large machine using the measured characteristics of a smaller machine or model.

11.3 Factors Involved in Selecting a Turbine

11.3.1 Performance Characteristics

Impulse and reaction turbines are the two basic types of turbines. They tend to operate at peak efficiency over different ranges of specific speed, due to geometric and operational differences.

Impulse Turbines. Of the head available at the nozzle inlet, a small portion is lost to friction in the nozzle and to friction on the buckets. The rest is available to drive the wheel. The actual utilization of this head depends on the velocity head of the flow leaving the turbine and the setting above tailwater. Optimum conditions, corresponding to maximum utilization of the head available, dictate that the flow leaves at essentially zero velocity. Under ideal conditions, this occurs when the peripheral speed of the wheel is one half the jet velocity. In practice, optimum power occurs at a speed coefficient, ω_{11}, somewhat less than 1.0 (illustrated in Figure 11.7). Because the maximum efficiency occurs at fixed speed for fixed H, V_j must remain constant under varying flow conditions. Thus, the flow rate Q is regulated with an adjustable nozzle. However, maximum efficiency occurs at slightly lower values of ω_{11} under partial power settings. Present nozzle technology is such that the discharge can be regulated over a wide range at high efficiency.

A given head and penstock configuration establishes the optimum jet velocity and diameter. The size of the wheel determines the speed of the machine. The design specific speed is approximately

$$N_s = 0.77 \frac{d_j}{D} \text{ (Pelton turbines)} \tag{11.11}$$

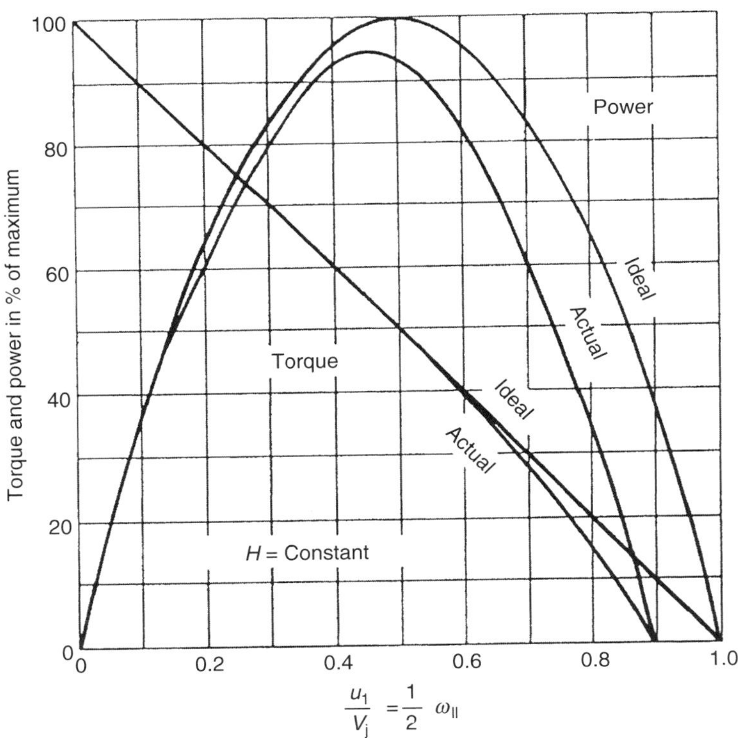

FIGURE 11.7 Ideal and actual variable-speed performance for an impulse turbine. (Adapted from Daily, J. W. 1950. In *Engineering Hydraulics*, H. Rouse, ed. Wiley, New York. With permission.)

Practical values of d_j/D for Pelton wheels to ensure good efficiency are in the range 0.04–0.1, corresponding to N_S values in the range 0.03–0.08. Higher specific speeds are possible with multiple nozzle designs. The increase is proportional to the square root of the number of nozzles. In considering an impulse unit, one must remember that efficiency is based on net head; the net head for an impulse unit is generally less than the net head for a reaction turbine at the same gross head because of the lack of a draft tube.

Reaction Turbines. The main difference between impulse units and reaction turbines is that a pressure drop takes place in the rotating passages of the reaction turbine. This implies that the entire flow passage from the turbine inlet to the discharge at the tailwater must be completely filled. A major factor in the overall design of modern reaction turbines is the draft tube. It is usually desirable to reduce the overall equipment and civil construction costs by using high specific speed runners. Under these circumstances, the draft tube is extremely critical for flow stability and efficiency. (This should be kept in mind when retrofitting on older, low specific speed turbine with a new runner of higher capacity.) At higher specific speed, a substantial percentage of the available total energy is in the form of kinetic energy leaving the runner. To recover this efficiently, considerable emphasis should be placed on the draft tube design.

The practical specific speed range for reaction turbines is much broader than for impulse wheels. This is due to the wider range of variables that control the basic operation of the turbine. The pivoted guide vanes allow for control of the magnitude and direction of the inlet flow. Because of a fixed relationship among blade angle, inlet velocity, and peripheral speed for shock-free entry, this requirement cannot be completely satisfied at partial flow without the ability to vary blade angle. This is the distinction between the efficiency of fixed-propeller and Francis types at partial loads and the fully adjustable Kaplan design.

In Equation 11.4, optimum hydraulic efficiency of the runner would occur when α_2 is equal to 90°. However, the overall efficiency of the turbine is dependent on the optimum performance of the draft tube as well, which occurs with a little swirl in the flow. Thus, the best overall efficiency occurs with $\alpha_2 \approx 75°$ for high specific speed turbines.

The determination of optimum specific speed in a reaction turbine is more complicated than for an impulse unit because there are more variables. For a radial-flow machine, an approximate expression is

$$N_s = 1.64\left[C_v \sin\alpha_1 \frac{B}{D_1}\right]^{1/2} \omega_{11} \text{ (Francis turbines)} \tag{11.12}$$

where C_v is the fraction of net head that is in the form of inlet velocity head and B is the height of the inlet flow passage (see Figure 11.3). N_S for Francis units is normally found to be in the range 0.3–2.5.

Standardized axial-flow machines are available in the smaller size range. These units are made up of standard components, such as shafts and blades. For such cases,

$$N_s \sim \frac{\sqrt{\tan\beta}}{n_B^{3/4}} \text{ (Propeller turbines)} \tag{11.13}$$

where β is the blade pitch angle and n_B is the number of blades. The advantage of controllable pitch is also obvious from this formula; the best specific speed is simply a function of pitch angle.

It should be further noted that ω_{11} is approximately constant for Francis units and N_S is proportional to $(B/D_1)^{1/2}$. It can also be shown that velocity component based on the peripheral speed at the throat, ω_{11e}, is proportional to N_S. In the case of axial-flow machinery, ω_{11} is also proportional to N_S. For minimum cost, peripheral speed should be as high as possible—consistent with cavitation-free performance. Under these circumstances N_S would vary inversely with the square root of head (H is given in meters):

$$N_s = \frac{C}{\sqrt{H}} \tag{11.14}$$

where the range of C is 8–11 for fixed-propeller units and Kaplan units and 6–9 for Francis units.

Performance Comparison. The physical characteristics of various runner configurations are summarized in Figure 11.8. It is obvious that the configuration changes with speed and head. Impulse turbines are efficient over a relatively narrow range of specific speed, whereas Francis and propeller turbines have a wider useful range. An important consideration is whether a turbine is required to operate over a wide range of load. Pelton wheels tend to operate efficiently over a wide range of power loading because of their nozzle design. In the case of reaction machines that have fixed geometry, such as Francis and propeller turbines, efficiency can vary widely with load. However, Kaplan and Deriaz (an adjustable-blade mixed-flow turbine; see Arndt 1991) turbines can maintain high efficiency over a wide range of operating conditions. The decision of whether to select a simple configuration with a relatively "peaky" efficiency curve or incur the added expense of installing a more complex machine with a broad efficiency curve will depend on the expected operation of the plant and other economic factors.

Note that in Figure 11.8 an overlap in the range of application of various types of equipment is shown. This means that either type of unit can be designed for good efficiency in this range, but other factors, such as generator speed and cavitation, may dictate the final selection.

11.3.2 Speed Regulation

The speed regulation of a turbine is an important and complicated problem. The magnitude of the problem varies with size; type of machine and installation; type of electrical load; and whether the plant is tied into an electrical grid. It should also be kept in mind that runaway or no-load speed can be higher than the design speed by factors as high as 2.6. This is an important design consideration for all rotating parts, including the generator.

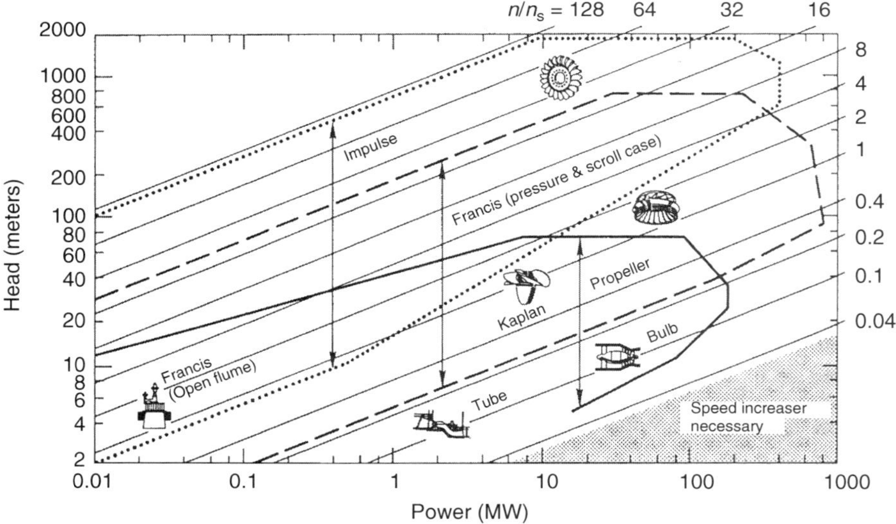

FIGURE 11.8 Application chart for various turbine types (n/n_S is the ratio of turbine speed in rpm, n, to specific speed defined in the metric system, $n_S = nP^{1/2}/H^{5/4}$, with P in kilowatts). (From Arndt, R. E. A. 1991. In *Hydropower Engineering Handbook*, J. S. Gulliver and R. E. A. Arndt, eds., pp. 4.1–4.67. McGraw-Hill, New York. With permission.)

The speed of a turbine must be controlled to a value that matches the generator characteristics and the grid frequency:

$$n = \frac{120f}{N_p} \tag{11.15}$$

where n is turbine speed in rpm; f is the required grid frequency in Hertz; and N_p is the number of poles in the generator. Typically, N_p is in multiples of 4. There is a tendency to select higher speed generators to minimize weight and cost. However, consideration must be given to speed regulation.

It is beyond the scope of this section to discuss the question of speed regulation in detail. Regulation of speed is normally accomplished through flow control. Adequate control requires sufficient rotational inertia of the rotating parts. When load is rejected, power is absorbed, accelerating the flywheel; when load is applied, some additional power is available from deceleration of the flywheel. Response time of the governor must be carefully selected, because rapid closing time can lead to excessive pressures in the penstock.

Opening and closing the wicket gates, which vary the flow of water according to the load, control a Francis turbine. The actuator components of a governor are required to overcome the hydraulic and frictional forces and to maintain the wicket gates in fixed position under steady load. For this reason, most governors have hydraulic actuators. On the other hand, impulse turbines are more easily controlled because the jet can be deflected or an auxiliary jet can bypass flow from the power-producing jet without changing the flow rate in the penstock. This permits long delay times for adjusting the flow rate to the new power conditions. The spear on needle valve controlling the flow rate can close quite slowly, e.g., in 30–60 s, thereby minimizing any pressure rise in the penstock.

Several types of governors are available that vary with the work capacity desired and the degree of sophistication of control. These vary from pure mechanical to mechanical-hydraulic and electrohydraulic. Electrohydraulic units are sophisticated pieces of equipment and would not be suitable for remote regions. The precision of governing necessary will depend on whether the electrical generator is synchronous or asynchronous (induction type). The induction type of generator has its advantages. It is less complex and therefore less expensive, but typically has slightly lower efficiency. Its frequency is controlled by the frequency of the grid into which it feeds, thereby eliminating the need for an expensive conventional governor. It cannot operate independently but can only feed into a network and does so with lagging power factor, which may be a disadvantage, depending on the nature of the load. Long transmission lines, for example, have a high capacitance and, in this case, the lagging power factor may be an advantage.

Speed regulation is a function of the flywheel effect of the rotating components and the inertia of the water column of the system. The start-up time of the rotating system is given by

$$t_s = \frac{I\omega^2}{P} \tag{11.16}$$

where I = moment of inertia of the generator and turbine, kg m^2 (Bureau of Reclamation 1966).

The start-up time of the water column is given by

$$t_p = \frac{\sum LV}{gH} \tag{11.17}$$

where L = the length of water column and V = the velocity in each component of the water column.

For good speed regulation, it is desirable to keep $t_s/t_p > 4$. Lower values can also be used, although special precautions are necessary in the control equipment. It can readily be seen that higher ratios of t_s/t_p can be obtained by increasing I or decreasing t_p. Increasing I implies a larger generator, which also results in higher costs. The start-up time of the water column can be reduced by reducing the length of the flow system, using lower velocities, or by adding surge tanks, which essentially reduce the effective length of the conduit. A detailed analysis should be made for each installation because, for a given length, head, and discharge, the flow area must be increased to reduce t_p, which leads to associated higher construction costs.

11.3.3 Cavitation and Turbine Setting

Another factor that must be considered prior to equipment selection is the evaluation of the turbine with respect to tailwater elevations. Hydraulic turbines are subject to pitting, loss in efficiency, and unstable operation due to cavitation (Arndt 1981, 1991; Arndt et al. 2000). For a given head, a smaller, lower cost, high-speed runner must be set lower (i.e., closer to tailwater or even below tailwater) than a larger, higher cost, low-speed turbine runner. Also, atmospheric pressure or plant elevation above sea level is a factor, as are tailwater elevation variations and operating requirements. This is a complex subject that can only be accurately resolved by model tests. Every runner design will have different cavitation characteristics. Therefore, the anticipated turbine location or setting with respect to tailwater elevations is an important consideration in turbine selection.

Cavitation is not normally a problem with impulse wheels. However, by the very nature of their operation, cavitation is an important factor in reaction turbine installations. The susceptibility for cavitation to occur is a function of the installation and the turbine design. This can be expressed conveniently in terms of Thoma's sigma, defined as

$$\sigma_T = \frac{H_a - H_v - z}{H} \tag{11.18}$$

where H_a is the atmospheric pressure head; H is the vapor pressure head (generally negligible); and z is the elevation of a turbine reference plane above the tailwater (see Figure 11.1). Draft tube losses and the exit velocity head have been neglected.

The term σ_T must be above a certain value to avoid cavitation problems. The critical value of σ_T is a function of specific speed (Arndt 1991). The Bureau of Reclamation (1966) suggests that cavitation problems can be avoided when

$$\sigma_T > 0.26N_s^{1.64} \tag{11.19}$$

Equation 11.19 does not guarantee total elimination of cavitation, only that cavitation is within acceptable limits. Cavitation can be totally avoided only if the value of σ_T at an installation is much greater than the limiting value given in Equation 11.19. The value of σ_T for a given installation is known as the plant sigma, σ_P. Equation 11.19 should only be considered a guide in selecting σ_P, which is normally determined by a model test in the manufacturer's laboratory. For a turbine operating under a given head, the only variable controlling σ_P is the turbine setting z. The required value of σ_P then controls the allowable setting above tailwater:

$$z_{allow} = H_a - H_v - \sigma_P H \tag{11.20}$$

It must be borne in mind that H_a varies with elevation. As a rule of thumb, H_a decreases from the sea-level value of 10.3 m by 1.1 m for every 1000 m above sea level.

11.4 Performance Evaluation

11.4.1 Model Tests

Model testing is an important element in the design and development phases of turbine manufacture. Manufacturers own most laboratories equipped with model turbine test stands. Major hydro projects have traditionally had proof-of-performance tests in model scale (at an independent laboratory or the manufacturer's laboratory) as part of the contract. In addition, it has been shown that competitive model testing at an independent laboratory can lead to large savings at a major project because of improved efficiency.

Recently, turbine design procedures have been dramatically improved through the use of sophisticated numerical analysis of the flow characteristics. These analysis techniques, linked with design programs, provide the turbine designer with powerful tools for achieving highly efficient turbine designs. In spite of this progress, computational methods require fine-tuning with model tests. In addition, model testing is necessary for determining performance over a range of operating conditions and for determining quasitransitory characteristics. Model testing can also be used to eliminate or mitigate problems associated with vibration, cavitation, hydraulic thrust, and pressure pulsation (Fisher and Beyer 1985).

A typical turbine test loop is shown in Figure 11.9. All test loops perform basically the same function. A model turbine is driven by high-pressure water from a head tank and discharges into a tail tank, as shown. The flow is recirculated by a pump, usually positioned well below the elevation of the model to ensure cavitation-free performance of the pump while performing cavitation testing with the turbine model. One important advantage of a recirculating turbine test loop is that cavitation testing can be done over a wide range of cavitation indices at constant head and flow.

The extrapolation of model test data to prototype values has been a subject of considerable debate for many years. Equation 11.6 through Equation 11.10 can be used to predict prototype values of flow, speed, power, etc., from model tests. Unfortunately, many factors lead to scale effects, i.e., the prototype efficiency and model efficiency are not identical at a fixed value of specific speed. The cited scale-up

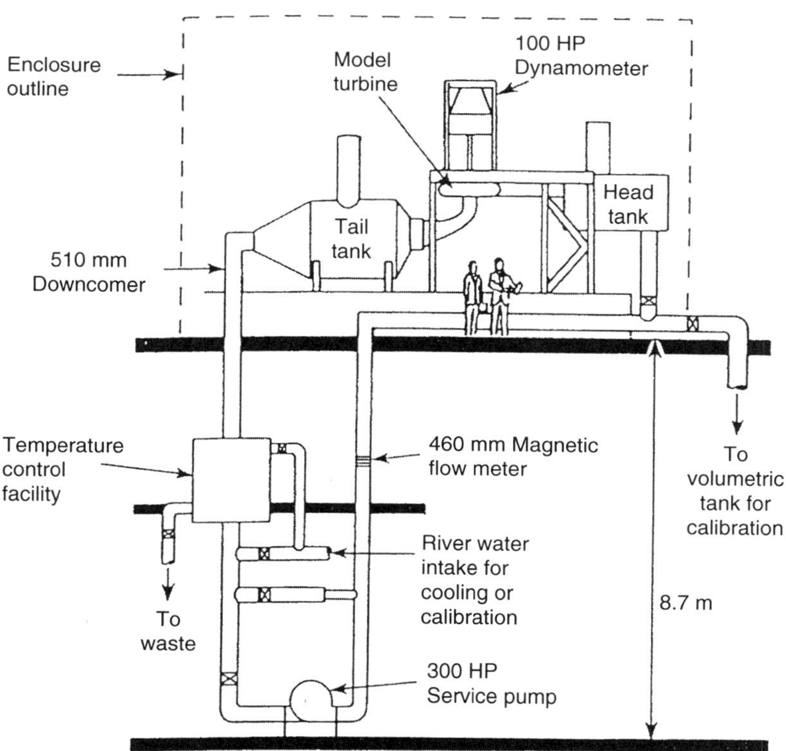

FIGURE 11.9 Schematic of the SAFL independent turbine test facility. (Courtesy of the St. Anthony Falls Laboratory, University of Minnesota.)

formulae are based on inviscid flow. Several sources of energy loss lead to an efficiency that is less than ideal. All of these losses follow different scaling laws and, in principle, perfect similitude can only be achieved by testing the prototype. Several attempts at rationalizing the process of scaling up model test data have been made. The International Electrotechnical Test Code and various ASME publications outline in detail the differences in efficiency between model and prototype. It should also be pointed out that other losses, such as those in the draft tube and "shock" losses at the runner inlet, might not be independent of Reynolds number.

11.5 Numerical Simulation[1]

Until very recently, the analysis of turbines and other components of hydropower facilities was largely dependent on approximate models such as the Euler equation and Reynolds averaged Navier–Stokes (RANS) models because the complete Navier–Stokes equations were considered to be too difficult to be solved for hydropower components (see Chapter 4). The Euler equation model has been applied with reasonable success for turbine runner simulation, but energy losses and the efficiency could not be calculated. RANS models have been applied for the spiral case and the draft tube simulations with limited success. Goede et al. (1991) contains a good summary of experiences with the application of these computational methods to various hydropower components. Very recently, reliable commercial codes have become available and are used with increasing frequency. Many of the current commercial codes rely on variants of RANS models although rapid progress is being made in adapting LES.

The large eddy simulation method (LES) is a step forward in the application of CFD Song et al. (1995). This technique is able to capture the effects of turbulent flow in a turbine more accurately than previous techniques. At the present time, it requires a supercomputer to achieve sufficient resolution and good accuracy for final design purposes. Parallel processing with desktop computers shows promise and can presently be used for relatively simple geometry or for preliminary evaluation purposes. However, progress in its application is very rapid and it is anticipated that an entire computation may be carried out on a high-end desktop computer in the near future.

The components that require simulation include the spiral case, wicket gates, the runner, and the draft tube. Often the spiral case, including stay vanes and wicket gates, is modeled as a unit. This is necessary because the stay vanes and wicket gates are so close to each other that their mutual interactions cannot be ignored; each stay vane may be of slightly different shape and orientation and cannot be modeled separately. A typical spiral case contains more than 20 stay vanes and an equal number of wicket gates, requiring extensive computational resources for a complete simulation. A sample calculation is shown in Figure 11.10. In this example, the calculated energy loss through this device is 2.62% of the net available energy for this particular case. This is significant and justifies additional computational effort to minimize the losses.

The runner is the most extensively studied component of a turbine. Because all the blades in a runner are of the same geometrical shape, only one- or two-flow passage models are commonly used for runner simulation. A complete model is required if vibration or cavitation due to nonsymmetrical modes of interactions between blades and vortices are to be studied. An important application of computer simulation is in the design of runners for units used for pumped storage.

Pumped-storage schemes are becoming a very popular for smoothing out the difference between energy demand and supply. Special care is required in the design because a runner must be designed to act efficiently as a turbine and a pump. Because of viscous effects, a runner optimized under the turbine mode may have poor efficiency in the pump mode. Flow in the pumping mode can be unstable and more difficult to calculate. An LES-based analysis greatly facilitates the optimum design of this type of runner. Figure 11.11 is an example of the calculated flow in the pump mode. A small flow separation near the

[1]This subsection was prepared by Professor Charles Song of the Saint Anthony Falls Laboratory, University of Minnesota.

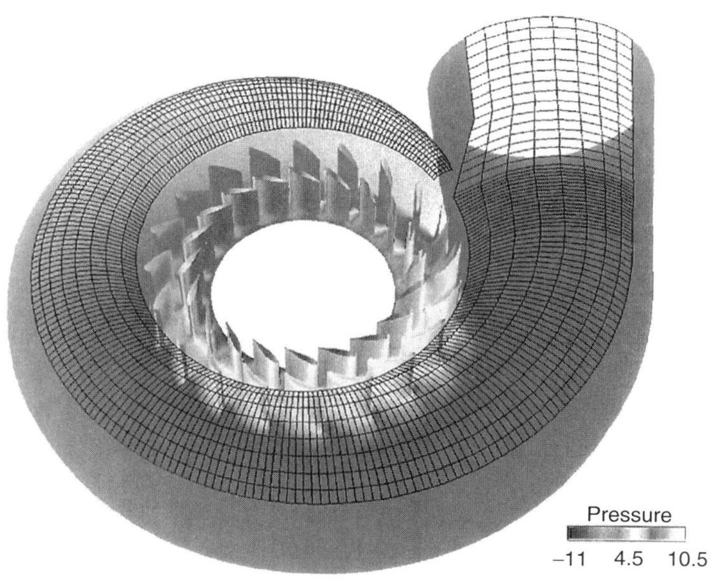

FIGURE 11.10 Simulated velocity field in a spiral case showing the pressure distribution on the boundaries of the spiral case, the stay vanes, and wicket gates. (Adapted from Song et al. *Hydro Rev.* XIV, 104–111. With permission.)

entrance can be observed. This kind of information is very useful to determine how the blade geometry can be modified to improve performance.

Draft tube design is an important factor in the efficiency and stability of operation of a turbine. Although a typical draft tube geometry is somewhat simpler than that of a runner, it takes much more advanced computational techniques to simulate its performance accurately. This is because the diffuser-like flow produces secondary currents, three-dimensional vortex shedding, and horseshoe vortices that

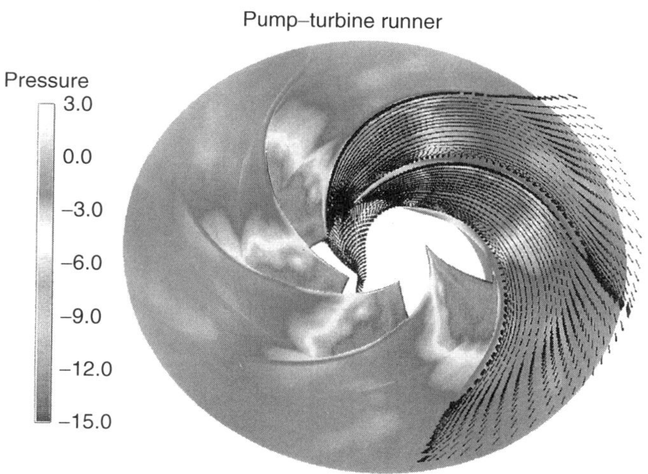

FIGURE 11.11 Simulation of the velocity and pressure distribution in a pump turbine runner operating in the pumping mode. (Adapted from Song et al. *Hydro Rev.* XIV, 104–111. With permission.)

are very important contributors to energy loss. Current RANS models are ineffective for this flow; at the present time only the LES model can fulfill the requirements for draft tube simulation.

Figure 11.12 illustrates the complexity of the problem. The instantaneous pressure distribution on the walls of an elbow-type draft tube with a divider wall is compared with the same draft tube with the divider wall removed. A dramatic change in the flow pattern and pressure distribution occurs when the divider wall is removed. By removing the wall, the draft tube becomes a diffuser of large angle with very unstable flow. Clearly, the divider wall stabilizes the flow and reduces the energy loss due to vortex shedding. These types of simulations are invaluable in evaluating draft tube performance. This is underscored by the fact that many projects involve refurbishing existing units. Typically, only the runner is replaced, usually with increased design flow. On many occasions, the existing draft tube is unable to operate efficiently at higher flow rates, thus canceling out any improvements that a new runner can provide.

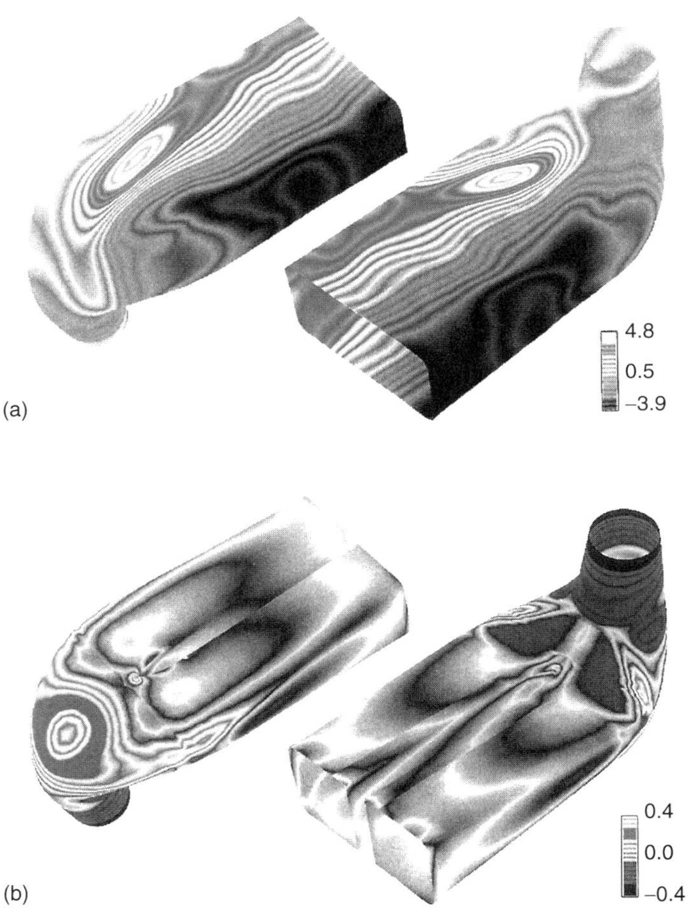

FIGURE 11.12 Simulation showing a comparison of the flow in a draft tube with and without a dividing wall. (a) The pressure patterns on the walls indicate a very asymmetrical flow pattern without a divider wall. (b) With a wall in place, a very uniform flow pattern is evident. (Adapted from Song et al. *Hydro Rev.* XIV, 104–111. With permission.)

11.6 Field Tests

Model tests and numerical simulations are only valid when geometric similitude is adhered to, i.e., the prototype machine is not guaranteed to be an accurate reproduction of the design. In addition, approach flow conditions, intake head losses, the effect of operating other adjacent units, etc., are not simulated in model tests. For these reasons, field performance tests are often performed. The several different types of field tests serve different purposes. The *absolute* efficiency is measured for acceptance or performance tests. *Relative* efficiency is often measured when operating information or fine-tuning of turbine performance is desired. Field tests are also carried out for commissioning a site and for various problem-solving activities. Basic procedures are covered by codes of the American Society of Mechanical Engineering (1992) and the International Electrotechnical Commission (1991).

The major difference between an "absolute" and a relative or index test is in the measurement of flow rate. Net head is evaluated in the same manner for each procedure. A variety of methods for measuring flow are code accepted. These include the pressure-time technique; tracer methods (salt velocity, dye dilution); area velocity (Pitot tubes or current meters); volumetric (usually on captive pumped storage sites); Venturi meters; and weirs. The thermodynamic method is actually a direct measure of efficiency. Flow is not measured. In addition to the code-accepted methods, it has been demonstrated that acoustic meters can measure flow in the field with comparable accuracy.

The *pressure-time technique* relies on measuring the change in pressure necessary to decelerate a given mass of fluid in a closed conduit. The method requires the measurement of the piezometric head at two cross sections spaced a distance L apart. A downstream valve or gate is necessary for this procedure. This technique requires load rejection for each test point and the need to estimate or measure any leakage. An adequate length of conduit is required and the conduit geometry must be accurately measured (Hecker and Nystrom 1991).

The *salt velocity method* is based on measuring the transit time, between two sensors, of an injected cloud of concentrated salt solution. Given the volume of the conduit between sensors, the flow rate may be calculated from the average transit time. Electrodes that measure the change in conductivity of the liquid detect the passage of the salt cloud at a given location.

The *dye-dilution method* is based on conservation of a tracer continuously injected into the flow. A sufficient length for complete mixing is necessary for accurate results. The data required are the initial concentration and injection flow rate of the tracer and the measured concentration of the fully mixed tracer at a downstream location. The method is quite simple, but care is necessary to achieve precise results.

In principle, *area-velocity measurements* are also quite simple. Pitot tubes or propeller-type current meters are used to measure point velocities that are integrated over the flow cross section. The method is applicable to closed conduits or open channels. A relatively uniform velocity distribution is necessary for accurate results. A single unit can be traversed across the conduit or a fixed or movable array of instruments can be used to reduce the time for data collection.

The *thermodynamic method* is a direct indication of turbine efficiency. Flow rate is not measured. In its simplest form, the method assumes adiabatic conditions, i.e., no heat transfer from the flow to its surroundings. Under these conditions, the portion of available energy not utilized in the machine to produce useful work results in increased internal energy of the fluid, which is sensed as an increase in temperature.

Acoustic flow meters have been developed that produce results with a precision equal to or greater than the code-accepted methods. Flow velocity is determined by comparing acoustic travel times for paths diagonally upstream and downstream between pairs of transducers. The speed of sound is assumed constant. The difference in travel time is related to the component of flow velocity along the acoustic path (increased travel time upstream, decreased travel time downstream). An extensive evaluation and comparison of this method has been reported (Sullivan 1983).

Index tests circumvent the problem of accurate flow measurement by measuring relative flow detected by the differential pressure between two points in the water passages leading to the runner. Often the differential pressure is measured with Winter–Kennedy taps positioned at the inner and outer radii of the spiral case of a turbine. Calibration of properly placed Winter–Kennedy taps shows that flow rate is very closely proportional to the square root of the pressure difference. Index testing is useful for calibration of relative power output vs. gate opening and for optimizing the various combinations of gate opening and blade setting in Kaplan units. The use of index testing to optimize cam settings in Kaplan turbines has resulted in substantial increases in weighted efficiency (i.e., a flatter efficiency curve over the full range of operation).

Defining Terms

Draft tube: the outlet conduit from a turbine that normally acts as a diffuser. This is normally considered an integral part of the unit.

Forebay: the hydraulic structure used to withdraw water from a reservoir or river. This can be positioned a considerable distance upstream from the turbine inlet.

Head: the specific energy per unit weight of water. *Gross head* is the difference in water surface elevation between the forebay and tailrace. *Net head* is the difference between *total head* (the sum of velocity head, $V^2/2g$, pressure head, $p/\rho g$, and elevation head, z), at the inlet and outlet of a turbine. Some European texts use specific energy per unit mass, for example, specific kinetic energy is $V^2/2$.

Pumped storage: a scheme in which water is pumped to an upper reservoir during off-peak hours and used to generate electricity during peak hours.

Runner: the rotating component of a turbine in which energy conversion takes place.

Specific speed: a universal number for a given machine design.

Spiral case: the inlet to a reaction turbine.

Surge tank: a hydraulic structure used to diminish overpressures in high-head facilities due to water hammer resulting from the sudden stoppage of a turbine.

Wicket gates: pivoted, streamlined guide vanes that control the flow of water to the turbine.

References

American Society of Mechanical Engineers. 1992. Power Test Code 18.

Arndt, R. E. A. 1981. Cavitation in fluid machinery and hydraulic structures. *Ann. Rev. Fluid Mech.*, 13, 273–328.

Arndt, R. E. A. 1991. Hydraulic turbines. In *Hydropower Engineering Handbook*, J. S. Gulliver and R. E. A. Arndt, eds., pp. 4.1–4.67. McGraw-Hill, New York.

Arndt, R. E. A., Keller, A., and Kjeldsen, M. 2000. Unsteady operation due to cavitation. *Proc. 20th IAHR Symp. Hydraulic Machinery Syst.*, Charlotte, NC, August.

Bureau of Reclamation. 1966. *Selecting Hydraulic Reaction Turbines*. Engineering Monograph No. 20.

Daily, J. W. 1950. Hydraulic machinery. In *Engineering Hydraulics*, H. Rouse, ed., Wiley, New York.

Fisher, R. K. and Beyer, J. R. 1985. The value of model testing for hydraulic turbines. *Proc. Am. Power Conf.*, ASME 47, 1122–1128.

Goede, E., Cuenod, R., Grunder, R., and Pestalozzi, J. 1991. A new computer method to optimize turbine design and runner replacement. *Hydro Rev.*, X (1), 76–88. February.

Hecker, G. E. and Nystrom, J. B. 1991. Which flow measurement technique is best? *Hydro Rev.*, 6 (3), June.

International Electrotechnical Commission. 1991. International Code for the Field Acceptance Tests of Hydraulic Turbines. Publication 41. International Electrotechnical Commission.

Raabe, J. 1985. *Hydro Power: The Design, Use, and Function of Hydromechanical, Hydraulic, and Electrical Equipment*. VDI Verlag, Dusseldorf, Germany.
Song, C. C. S., Chen, X., He, J., Chen, C., and Zhou, F. 1995. Using computational tools for hydraulic design of hydropower plants. *Hydro Rev.*, XIV (4), 104–111, July.
Sullivan, C. W. 1983. Acoustic flow measurement systems: economical, accurate, but not code accepted. *Hydro Rev.*, 6 (4), August.
Wahl, T. L. 1994. Draft tube surging times two: the twin vortex problem. *Hydro Rev.*, 13 (1), 60–69.

For Further Information

J. Fluids Eng. Published quarterly by the ASME.
ASME Symposia Proc. on Fluid Machinery and Cavitation. Published by the Fluids Eng. Div.
Hydro Rev., Published eight times per year by HCI Publications, Kansas City, MO.
Moody, L. F. and Zowski, T. 1992. Hydraulic machinery. In *Handbook of Applied Hydraulics*, C. V. Davis and K. E. Sorenson, eds., McGraw-Hill, New York.
Waterpower and Dam Construction. Published monthly by Reed Business Publishing, Surrey, U.K.

12

Stirling Engines

12.1	Introduction	12-1
12.2	Thermodynamic Implementation of the Stirling Cycle Working Gases • Heat Exchange • Power Control	12-2
12.3	Mechanical Implementation of the Stirling Cycle Piston/Displacer Configurations • Piston/Displacer Drives • Seals and Bearings • Materials	12-4
12.4	Future of the Stirling Engine	12-9
Defining Terms		12-9
References		12-10
For Further Information		12-11

William B. Stine
California State Polytechnic University

12.1 Introduction

The Stirling engine was patented in 1816 by Rev. Robert Stirling, a Scottish minister (Figure 12.1). Early Stirling engines were coal-burning, low-pressure air engines built to compete with saturated steam engines for providing auxiliary power for manufacturing and mining. In 1987, John Ericsson built an enormous marine Stirling engine with four 4.2-m-diameter pistons. Beginning in the 1930s, the Stirling engine was brought to a high state of technology development by Philips Research Laboratory in Eindhoven, The Netherlands, with the goal of producing small, quiet electrical generator sets to be used with high-power-consuming vacuum tube electronic devices. Recently, interest in Stirling engines has resurfaced, with solar electric power generation (Stine and Diver 1994) and hybrid automotive applications in the forefront.

In theory, the **Stirling cycle** engine can be the most efficient device for converting heat into mechanical work with high efficiencies requiring high-temperatures. In fact, with regeneration, the efficiency of the Stirling cycle equals that of the Carnot cycle, the most efficient of all ideal thermodynamic cycles. [See West (1986) for further discussion of the thermodynamics of Stirling cycle machines.]

Since their invention, prototype Stirling engines have been developed for automotive purposes; they have also been designed and tested for service in trucks, buses, and boats (Walker 1973). The Stirling engine has been proposed as a propulsion engine in yachts, passenger ships, and road vehicles such as city buses (Meijer 1992). The Stirling engine has also been developed as an underwater power unit for submarines, and the feasibility of using the Stirling for high-power space-borne systems has been explored by NASA (West 1986). The Stirling engine is considered ideal for solar heating, and the first solar application of record was by John Ericsson, the famous British–American inventor, in 1872 (Stine and Diver 1994).

Stirling engines are generally externally heated engines. Therefore, most sources of heat can be used to drive them, including combustion of just about anything, radioisotopes, solar energy, and exothermic chemical reactions. High-performance Stirling engines operate at the thermal limits of the materials used

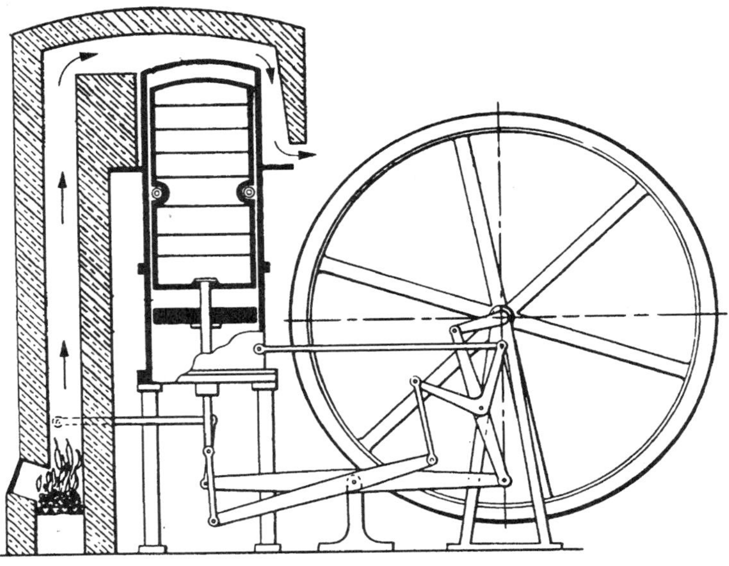

FIGURE 12.1 The original patent Stirling engine of Rev. Robert Stirling.

for their construction. Typical temperatures range from 650 to 800°C (1200°F–1470°F), resulting in engine conversion efficiencies of around 30%–40%. Engine speeds of 2000–4000 rpm are common.

12.2 Thermodynamic Implementation of the Stirling Cycle

In the ideal Stirling cycle, a **working gas** is alternately heated and cooled as it is compressed and expanded. Gases such as helium and hydrogen, which permit rapid heat transfer and do not change phase, are typically used in the high-performance Stirling engines. The ideal Stirling cycle combines four processes, two constant-temperature heat-exchange processes and two constant-volume heat-exchange processes. Because more work is done by expanding high-temperature, high-pressure gas than is required to compress low-temperature, low-pressure gas, the Stirling cycle produces net work, which can drive an electric alternator or other mechanical devices.

12.2.1 Working Gases

In the Stirling cycle, the working gas is alternately heated and cooled in constant-temperature and constant-volume processes. The traditional gas for Stirling engines has been air at atmospheric pressure. At this pressure, air has a reasonably high density and therefore can be used directly in the cycle with loss of working gas through seals a minor problem. However, internal component temperatures are limited because of the oxygen in air which can degrade materials rapidly.

Because of their high heat-transfer capabilities, hydrogen and helium are used for high-speed, high-performance Stirling engines. To compensate for the low density of these gases, the mean pressure of the working gas is raised by charging the gas spaces in the engine to high pressures. Compression and expansion vary above and below this **charge pressure**. Hydrogen, thermodynamically a better choice, generally results in more-efficient engines than does helium (Walker 1980). Helium, on the other hand, has fewer material-compatibility problems and is safer to work with. To maximize power, high-performance engines typically operate at high pressure, in the range of 5–20 MPa (725–2900 psi).

Operation at these high gas pressures makes sealing difficult, and seals between the high-pressure region of the engine and those parts at ambient pressure have been problematic in some engines. New designs to reduce or eliminate this problem are currently being developed.

12.2.2 Heat Exchange

The working gas is heated and cooled by heat exchangers that add heat from an external source, or reject heat to the surroundings. Further, in most engines, an internal heat storage unit stores and rejects heat during each cycle.

The **heater** of a Stirling engine is usually made of many small-bore tubes that are heated externally with the working gas passing through the inside. External heat transfer by direct impingement of combustion products or direct adsorption of solar irradiation is common. A trade-off between high heat-transfer rate using many small-bore tubes with resulting pumping losses, and fewer large-bore tubes and lower pumping losses drives the design. A third criterion is that the volume of gas trapped within these heat exchangers should be minimal to enhance engine performance. In an attempt to provide more uniform and constant-temperature heat transfer to the heater tubes, **reflux** heaters are being developed (Stine and Diver 1994). Typically, by using sodium as the heat-transfer medium, liquid is evaporated at the heat source and is then condensed on the outside surfaces of the engine heater tubes.

The **cooler** is usually a tube-and-shell heat exchanger. Working gas is passed through the tubes, and cooling water is circulated around the outside. The cooling water is then cooled in an external heat exchanger. Because all of the heat rejected from the power cycle comes from the cooler, the Stirling engine is considered ideal for cogeneration applications.

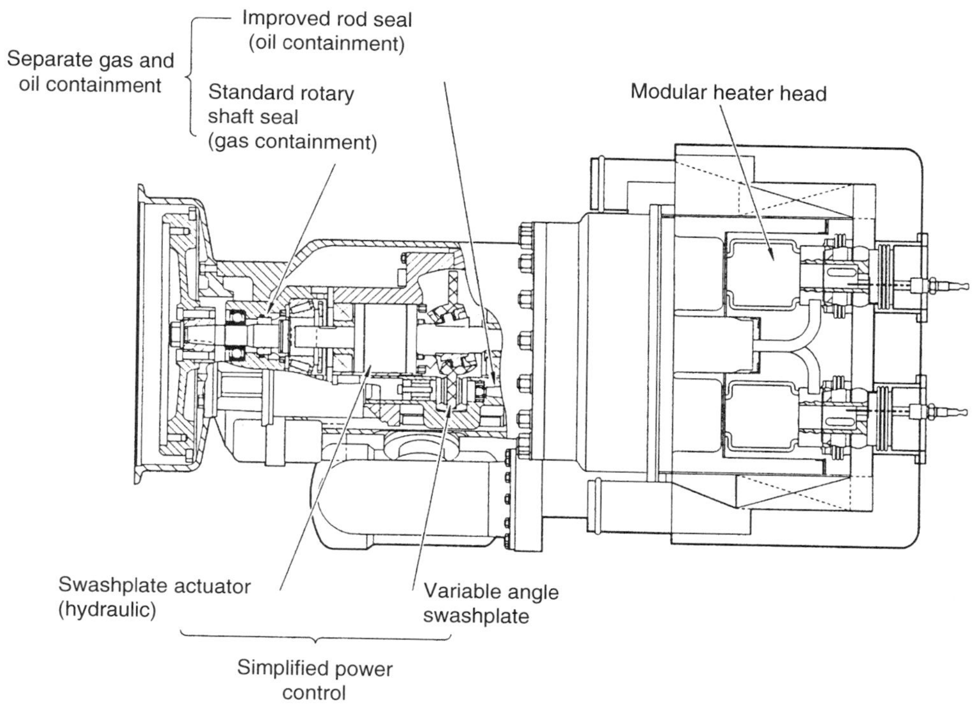

FIGURE 12.2 Stirling Thermal Motors 4-120 variable swash plate Rinia configuration engine. (Courtesy Stirling Thermal Motors, Ann Arbor, Michigan.)

Most Stirling engines incorporate an efficiency-enhancing **regenerator** that captures heat from the working gas during constant-volume cooling and replaces it when the gas is heated at constant volume. Heating and cooling of the regenerator occurs at over 60 times a second during high-speed engine operation. In the ideal cycle, all of the heat-transferred during the constant volume heating and cooling processes occurs in the regenerator, permitting the external heat addition and rejection to be efficient constant-temperature heat-transfer processes. Regenerators are typically chambers packed with fine-mesh screen wire or porous metal structures. There is enough thermal mass in the packing material to store all of the heat necessary to raise the temperature of the working gas from its low to its high temperature. The amount of heat stored by the regenerator is generally many times greater than the amount added by the heater.

12.2.3 Power Control

Rapid control of the output power of a Stirling engine is highly desirable for some applications such as automotive and solar electric applications. In most Stirling engine designs, rapid power control is implemented by varying the density (i.e., the mean pressure) of the working gas by bleeding gas from the cycle when less power is desired. To return to a higher power level, high-pressure gas must be reintroduced into the cycle. To accomplish this quickly and without loss of working gas, a complex system of valves, a temporary storage tank, and a compressor are used.

A novel method of controlling the power output is to change the length of stroke of the power piston. This can be accomplished using a variable-angle swash plate drive as described below. Stirling Thermal Motors, Inc., uses this method on their STM 4-120 Stirling engine (Figure 12.2).

12.3 Mechanical Implementation of the Stirling Cycle

12.3.1 Piston/Displacer Configurations

To implement the Stirling cycle, different combinations of machine components have been designed to provide for the constant-volume movement of the working gas between the high- and low-temperature regions of the engine, and compression and expansion during the constant-temperature heating and cooling. The compression and expansion part of the cycle generally take place in a cylinder with a piston. Movement of the working gas back and forth through the heater, regenerator, and cooler at constant volume is often implemented by a **displacer**. A displacer in this sense is a hollow plug that, when moved to the cold region, displaces the working gas from the cold region causing it to flow to the hot region and vice versa. Only a small pressure difference exists between either end of the displacer, and therefore, sealing requirements and the force required to move it are minimal.

Three different design configurations are generally used (Figure 12.3). Called the alpha-, beta-, and gamma-configurations. Each has its distinct mechanical design characteristics, but the thermodynamic cycle is the same. The **alpha-configuration** uses two pistons on either side of the heater, regenerator, and the cooler. These pistons first move uniformly in the same direction to provide constant-volume processes to heat or cool the gas. When all of the gas has been moved into one cylinder, one piston remains fixed and the other moves to compress or expand the gas. Compression work is done by the **cold piston** and expansion work done on the **hot piston**. The alpha-configuration does not use a displacer. The Stirling Power Systems V-160 engine (Figure 12.4) is an example of this configuration.

A variation on using two separate pistons to implement the alpha-configuration is to use the front and back side of a single piston called a **double-acting piston**. The volume at the front side of one piston is connected, through the heater, regenerator, and cooler, to the volume at the back side of another piston. With four such double-acting pistons, each 90° out of phase with the next, the result is a four-cylinder alpha-configuration engine. This design is called the *Rinia* or *Siemens configuration* and the United Stirling 4-95 (Figure 12.5) and the Stirling Thermal Motors STM 4-120 (Figure 12.2) are current examples.

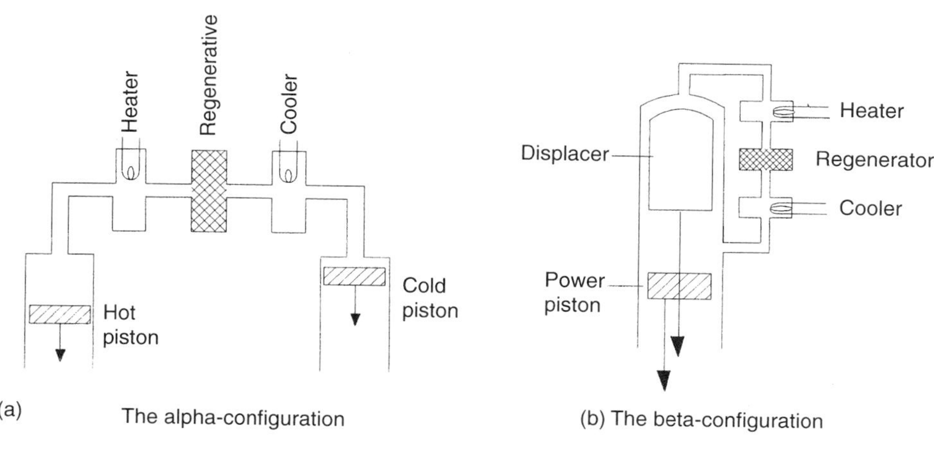

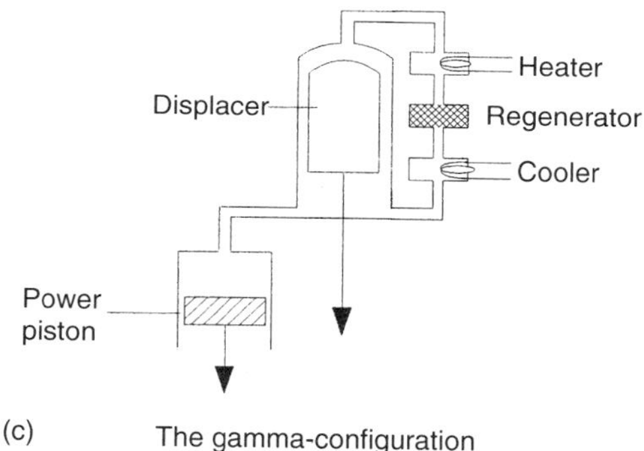

FIGURE 12.3 Three fundamental mechanical configurations for Stirling engines.

The **beta-configuration** is a design incorporating a displacer and a power piston in the same cylinder. The displacer shuttles gas between the hot end and the cold end of the cylinder through the heater, regenerator, and cooler. The power piston, usually located at the cool end of the cylinder, compresses the working gas when the gas is in the cool end and expands the working gas when the gas has been moved to the hot end. The original patent engine by Robert Stirling and most free-piston Stirling engines discussed below are of the beta-configuration.

The third configuration, using separate cylinders for the displacer and the power piston, is called the **gamma-configuration**. Here, the displacer shuttles gas between the hot end and the cold end of a cylinder through the heater, regenerator, and cooler, just as with the beta-configuration. However, the power piston is in a separate cylinder, pneumatically connected to the displacer cylinder.

12.3.2 Piston/Displacer Drives

Most Stirling engine designs dynamically approximate the Stirling cycle by moving the piston and displacer with **simple harmonic motion**, either through a crankshaft, or bouncing as a spring/mass second-order mechanical system. For both, a performance penalty comes from the inability of simple harmonic motion to perfectly follow the desired thermodynamic processes. A more desirable dynamic from the cycle point of view, called overdriven or **bang–bang motion**, has been implemented in some designs, most notably the **Ringbom configuration** and engines designed by Ivo Kolin (West 1986).

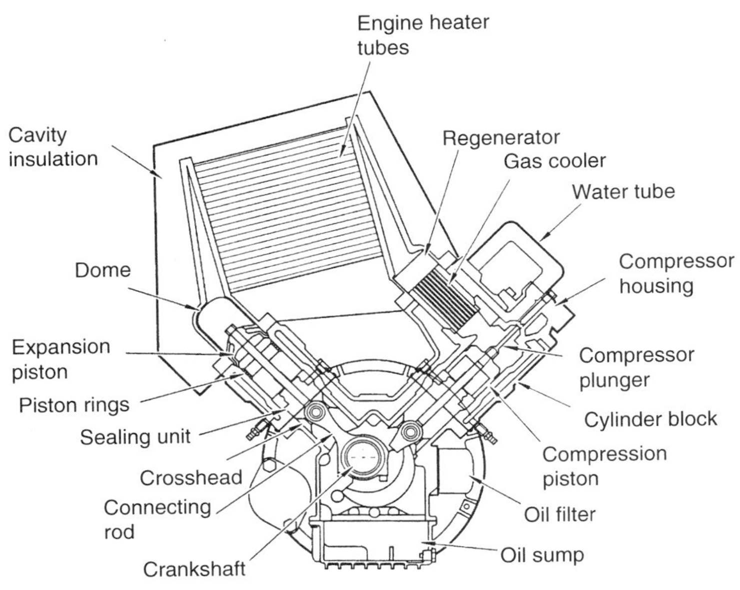

FIGURE 12.4 Stirling Power Systems/Solo Kleinmotoren V-160 alpha-configuration Stirling engine.

Kinematic Engines. Stirling engine designs are usually categorized as either kinematic or free-piston. The power piston of a **kinematic Stirling engine** is mechanically connected to a rotating output shaft. In typical configurations, the power piston is connected to the crankshaft with a connecting rod. In order to eliminate side forces against the cylinder wall, a **cross-head** is often incorporated, where the connecting rod connects to the cross-head, which is laterally restrained so that it can only move linearly in the same direction as the piston. The power piston is connected to the cross-head and therefore experiences no lateral forces. The critical sealing between the high-pressure and low-pressure regions of the engine can now be created using a simple **linear seal** on the shaft between the cross-head and the piston. This design also keeps lubricated bearing surfaces in the low-pressure region of the engine, reducing the possibility of fouling heat-exchange surfaces in the high-pressure region of the engine. If there is a separate displacer piston as in the beta- and gamma-configurations, it is also mechanically connected to the output shaft.

A variation on crankshaft/cross-head drives is the **swash plate** or **wobble-plate drive**, used with success in some Stirling engine designs. Here, a drive surface affixed to the drive shaft at an angle, pushes fixed piston **push rods** up and down as the slanted surface rotates beneath. The length of stroke for the piston depends on the angle of the plate relative to the axis of rotation. The STM 4-120 engine (Figure 12.2) currently being commercialized by Stirling Thermal Motors incorporates a **variable-angle swash plate drive** that permits variation in the length of stroke of the pistons.

Free-Piston Engine/Converters. An innovative way of accomplishing the Stirling cycle is employed in the free-piston engine. In this configuration, the power piston is not mechanically connected to an output shaft. It bounces alternately between the space containing the working gas and a spring (usually a gas spring). In many designs, the displacer is also free to bounce on *gas springs* or mechanical springs (Figure 12.6). This configuration is called the Beale free-piston Stirling engine after its inventor, William Beale. Piston stroke, frequency, and the timing between the two pistons are established by the dynamics of the spring/mass system coupled with the variations in cycle pressure. To extract power, a magnet can be attached to the power piston and electric power generated as it moves past stationary coils. These Stirling engine/alternator units are called **free-piston Stirling converters**. Other schemes for extracting power from free-piston engines, such as driving a hydraulic pump, have also been considered.

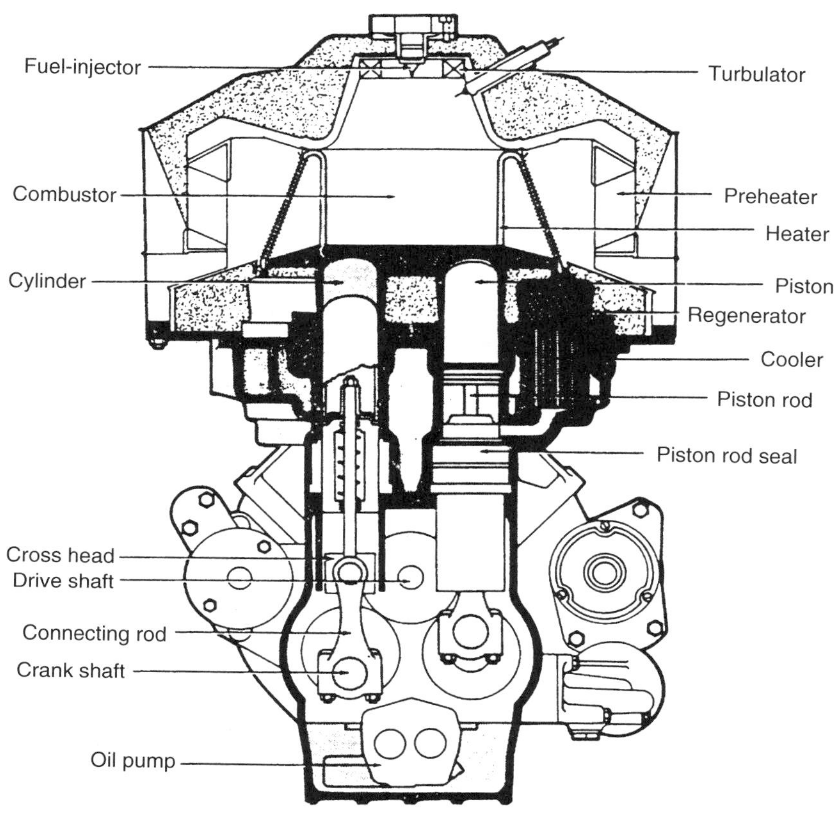

FIGURE 12.5 The 4-95 high-performance Siemens configuration Rinia engine by United Stirling (Malmo, Sweden).

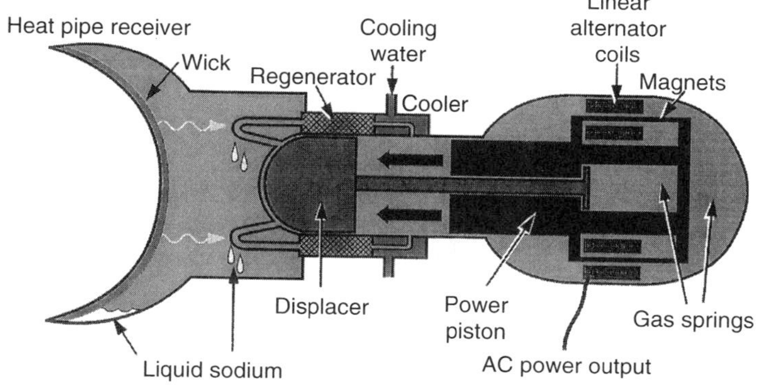

FIGURE 12.6 Basic components of a Beale free-piston Stirling converter incorporating a sodium heat pipe receiver for heating with concentrated solar energy.

Free-piston Stirling engines have only two moving parts, and therefore the potential advantages of simplicity, low cost, and ultra-reliability. Because electricity is generated internally, there are no dynamic seals between the high-pressure region of the engine and ambient, and no oil lubrication is required.

This design promises long lifetimes with minimal maintenance. A number of companies are currently developing free-piston engines including Sunpower, Inc. (Figure 12.7), and Stirling Technology Company.

12.3.3 Seals and Bearings

Many proposed applications for Stirling engine systems require long-life designs. To make systems economical, a system lifetime of at least 20 years with minimum maintenance is generally required. Desired engine lifetimes for electric power production are 40,000–60,000 h—approximately ten times longer than that of a typical automotive internal combustion engine. Major overhaul of engines, including replacement of seals and bearings, may be necessary within the 40,000- to 60,000-h lifetime, which adds to the operating cost. A major challenge, therefore, in the design of Stirling engines is to reduce the potential for wear in critical components or to create novel ways for them to perform their tasks.

Piston seals differ from those used in internal combustion engines in a number of ways. Sealing of the power piston is critical since **blow-by loss** of the hydrogen or helium working gas must be captured and recompressed, or replaced from a high-pressure cylinder. Displacer sealing is less critical and only necessary to force most of the working gas through the heater, regenerator, and cooler. Oil for friction reduction or sealing cannot be used because of the danger of it getting into the working gas and fouling the heat-exchange surfaces. This leads to two choices for sealing of pistons, using **polymer sealing rings** or **gas bearings** (simply close tolerance fitting between piston and wall).

Free-piston engines with gas springs have special internal sealing problems. Small leakage can change the force-position characteristics of the "spring" and rapidly upset the phase and displacement dynamics designed into the engine. In order to prevent this, *centering ports* are used to ensure that the piston stays at a certain location; however, these represent a loss of potential work from the engine.

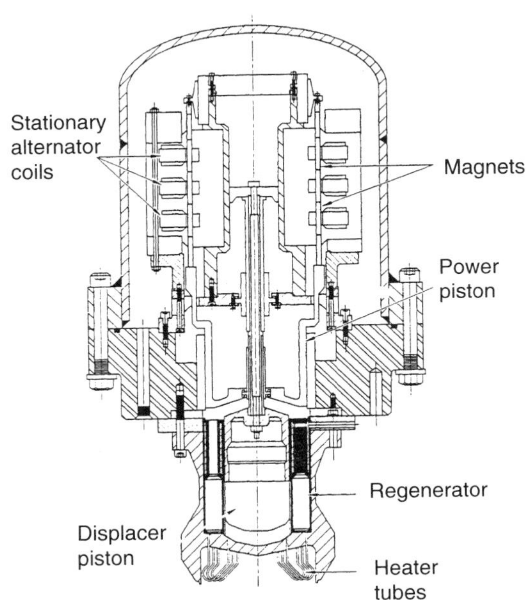

FIGURE 12.7 The Sunpower 9-kWe free-piston beta-configuration Stirling engine.

12.3.4 Materials

Materials used in Stirling engines are generally normal steels with a few exceptions. Materials that can withstand continuous operation at the cycle high temperature are required for the heater, regenerator, and the hot side of the displacement volume. Because most engines operate at high pressure, thick walls are often required. In the hot regions of the engine, this can lead to *thermal creep* due to successive heating and cooling. In the cool regions, large spaces for mechanical linkages can require heavy, thick walls to contain the gas pressure. Use of composite structure technology or reducing the size of the pressurized space can eliminate these problems.

12.4 Future of the Stirling Engine

The principal advantages of the Stirling engine, external heating and high efficiency, make this the engine of the future, replacing many applications currently using internal combustion engines and providing access to the sun as an inexpensive source of energy (Figure 12.6). For hybrid-electric automotive applications, the Stirling engine is not only almost twice as efficient as modern spark-ignition engines, but because of the continuous combustion process, it burns fuel more cleanly and is not sensitive to the quality or type of fuel. Because of the simplicity of its design, the Stirling engine can be manufactured as an inexpensive power source for electricity generation using biomass and other fuels available in developing nations.

Most importantly, the Stirling engine will provide access to inexpensive solar energy. Because it can receive its heat from a resource 93 million miles away using concentrating solar collectors, and because its manufacture is quite similar to the gasoline or diesel engine, and because economies of scale are certain when producing thousands of units per year, the Stirling engine is considered to be the least expensive alternative for solar energy electric generation applications in the range from 1 kWe to 100 MWe.

Defining Terms

Alpha-configuration: Design of a Stirling engine where two pistons moving out of phase, and cause the working gas between them to go through the four processes of the Stirling cycle.

Beale free-piston Stirling engine: Stirling engine configuration where the power piston and displacer in a single cylinder are free to bounce back and forth along a single axis, causing the enclosed working gas to go through the four processes of the Stirling cycle. Restoration forces are provided by the varying pressure of the working gas, springs (gas or mechanical), and the external load which can be a linear alternator or a fluid pump.

Beta-configuration: Design of a Stirling engine where the displacer and power piston are located in the same cylinder and cause the enclosed working gas to go through the four processes of the Stirling cycle.

Blow-by: The gas that leaks past a seal, especially a piston-to-cylinder seal.

Charge pressure: Initial pressure of the working gas.

Cooler: Heat exchanger that removes heat from the working fluid and transfers it out of the cycle.

Cross-head: A linear sliding bearing surface connected to a crankshaft by a connecting rod. It is purpose is to provide linear reciprocating motion along a single line of action.

Displacer: Closed volume 'plug' that forces the working fluid to move from one region of the engine to another by displacing it.

Double-acting piston: A piston in an enclosed cylinder where pressure can be varied on both sides of the piston, resulting a total amount of work being the sum of that done on or by both sides.

Dynamic seals: Seals that permit transfer of motion without permitting gas or oil leakage. These can be either *linear seals* permitting a shaft to move between two regions (i.e., the piston rod seals in a Stirling engine), or *rotating seals* that permit rotating motion to be transmitted from one region to another (i.e., the output shaft of a Stirling engine).

Free-piston Stirling converters: A name given to a hermetically sealed free-piston Stirling engine incorporating an internal alternator or pump.

Gamma-configuration: A design of a Stirling engine where the displacer and power piston are located in separate, connected cylinders and cause the enclosed working gas to go through the four processes of the Stirling cycle.

Gas bearing: A method of implementing the sliding seal between a piston and cylinder as opposed to using piston rings. Uses a precision-fitting piston that depens on the small clearance and long path for sealing and on the viscosity of the gas for lubrication.

Gas spring: A piston that compresses gas in a closed cylinder where the restoration force is linearly proportional to the piston displacement. This is a concept used in the design of free-piston Stirling engines.

Heater: A heat exchanger which adds heat to the working fluid from an external source.

Kinematic Stirling engine: Stirling engine design that employ physical connections between the power piston, displacer, and a mechanical output shaft.

Linear seal: see **dynamic seals**.

Overdriven (bang–bang) motion: Linear motion varying with time as a square-wave function. An alternative to simple harmonic motion and considered a better motion for the displacer of a Stirling engine but difficult to implement.

Phase angle: The angle difference between displacer and power piston harmonic motion with a complete cycle representing 360°. In most Stirling engines, the harmonic motion of the power piston follows (lags) the motion of the displacer by approximately 90°.

Push rod: A thin rod connected to the back of the piston that transfers linear motion through a dynamic linear seal, between the low- and high-pressure regions of an engine.

Reflux: A constant-temperature heat-exchange process where a liquid is evaporated by heat addition and then condensed as a result of cooling.

Regenerator: A heat-transfer device that stores heat from the working gas during part of a thermodynamic cycle and returns it during another part of the cycle. In the Stirling cycle the regenerator stores heat from one constant-volume process and returns it in the other constant-volume process.

Ringbom configuration: A Stirling engine configuration where the power piston is kinematically connected to a power shaft, and the displacer is a free piston that is powered by the difference in pressure between the internal gas and atmospheric pressure.

Simple harmonic motion: Linear motion varying with time as a sine function. Approximated by a piston connected to a crankshaft or a bouncing of a spring mass system.

Stirling cycle: A thermodynamic power cycle with two constant-volume heat addition and rejection processes and two constant-temperature heat-addition and rejection processes.

Swash plate drive: A disk on a shaft, where the plane of the disk is tilted away from the axis of rotation of the shaft. Piston push rods that move parallel to the axis of rotation but are displaced from the axis of rotation, slide on the surface of the rotating swash plate and therefore move up and down.

Variable-angle swash plate drive: A swash plate drive where the tilt angle between the drive shaft and the plate can be varied, resulting in a change in the displacement of the push rods.

Wobble plate drive: Another name for a swash plate drive.

Working gas: Gas within the engine that exhibits pressure and temperature change as it is heated or cooled and/or compressed or expanded.

References

Meijer, R. F. 1992. Stirling engine. In *McGraw-Hill Encyclopedia of Science and Technology, 7th Ed.*, pp. 440–445. McGraw-Hill, New York.

Stine, W. B. and Diver, R. E. 1994. *A Compendium of Solar Dish Stirling Technology*, Report SAND94-7026, Sandia National Laboratories, Albuquerque, NM 87185.

Walker, G. 1973. The Stirling engine. *Sci. Am.*, 229 (2), 80–87.
Walker, G. 1980. *Stirling Engines.* Clarendon Press, Oxford.
West, C. D. 1986. *Principles and Applications of Stirling Engines.* Van Nostrand Reinhold, New York.

For Further Information

Books

Hargraves, C. M. 1991. *The Philips Stirling Engine.* Elsevier Press, London.
Organ, A. J. 1992. *Thermodynamics and Gas Dynamics of the Stirling Cycle Machine.* Cambridge University Press, Cambridge.
Senft, J. R. 1993. *Ringbom Stirling Engines.* Oxford University Press, Oxford.
Stine, W. B. and Diver, R. E. 1994. *A Compendium of Solar Dish/Stirling Technology. SAND93-7026.* Sandia National Laboratory, Albuquerque.
Urieli, I. and Berchowitz, D. M. 1984. *Stirling Cycle Engine Analysis.* Adam Hilger, Bristol.
Walker, G. 1980. *Stirling Engines.* Clarendon Press, Oxford.
Walker, G. and Senft, J. R. 1985. *Free-Piston Stirling Engines.* Springer, New York.
Walker, G., Reader, G., Fauvel, O. R., and Bingham, E. R. 1994. *The Stirling Alternative.* Bordon & Breach, New York.
West, C. D. 1986. *Principles and Applications of Stirling Engines.* Van Nostrand Reinhold, New York.

Periodicals

Proceedings of the Intersociety Energy Conversion Engineering Conference (IECEC), published annually.
Stirling Machine World, a quarterly newsletter devoted to advancements in Stirling engines, edited by Brad Ross, 1823 Hummingbird Court, West Richland, WA 99353-9542.

Stirling Engine Developers

Stirling Technology Company, 4208B W. Clearwater Ave., Kennewick, WA 99336.
Stirling Thermal Motors, 275 Metty Drive, Ann Arbor, MI 48103.
Sunpower Incorporated, 6 Byard Street, Athens, OH 45701.
Clever Fellows Innovation Consortium, 302 Tenth St., Troy, NY 12180.
Mechanical Technologies Inc., 968 Albany-Shaker Rd., Latham, NY 12110.
Solo Kleinmotoren GmbH, Postfach 60 0152; D-71050 Sindelfingen; Germany.
Aisin-Seiki Ltd., 1, Asahi-Mach: 2-chome; Kariya City Aich: Pref 448; Japan.

13

Advanced Fossil Fuel Power Systems

Anthony F. Armor
Electric Power Research Institute

13.1 Introduction 13-1
13.2 Fuels for Electric Power Generation in the U.S. 13-2
13.3 Coal as a Fuel for Electric Power (World Coal Institute 2000) 13-3
13.4 Clean Coal Technology Development 13-4
Coal Cleaning
13.5 Pulverized-Coal Plants 13-5
Materials Advances • Cycle Selection • Supercriticals
13.6 Emissions Controls for Pulverized Coal Plants 13-9
Conventional Lime/Limestone Wet Scrubber • Spray Drying • Control of Nitrogen Oxides • Control of Mercury
13.7 Fluidized Bed Plants 13-13
Atmospheric Fluidized Bed Combustion (AFBC) • Pressurized Fluidized Bed Combustion (PFBC)
13.8 Gasification Plants 13-16
Polk County IGCC • Buggenum IGCC
13.9 Combustion Turbine Plants 13-20
Advanced Combustion Turbines • Humidified Air Power Plants • Other Combustion Turbine Cycle Enhancements
13.10 Central Station Options for New Generation 13-24
Overall Plant Performance and Cost Estimates
13.11 Summary 13-26
Defining Terms 13-26
References 13-28
For Further Information 13-29

13.1 Introduction

The generation of electric power from fossil fuels has seen continuing, and in some cases dramatic, technical advances over the last 20–30 years. Technology improvements in fossil fuel combustion have been driven largely by the need to reduce emissions and conserve fossil fuel resources, as well as by the economics of the competitive marketplace. The importance of fossil fuel-fired electric generation to the world is undeniable—more than 70% of all power in the U.S. is fossil fuel-based; worldwide the percentage is higher, and growing. Today, although most large power plants worldwide burn coal, generating companies increasingly are turning to natural gas, particularly when the cost of gas-fired generation and the long-term supply of gas appear favorable. This section reviews the current status and likely future deployment of competing generation technologies based on fossil fuels.

It is likely, particularly in the developed world, that gas turbine-based plants will continue to dominate the new generation market in the immediate future. The most advanced combustion turbines now achieve more than 40% **lower heating value** (LHV) efficiency in simple cycle mode and greater than 50% efficiency in **combined cycle** mode. In addition, combustion turbine/combined cycle (CT/CC) plants offer siting flexibility, swift construction schedules, and capital costs between $400 and $800/kW. These advantages, coupled with adequate natural gas supplies (though new wells and pipelines will be needed in the U.S.) and the assurance, in the longer term, of **coal gasification** back-up, make this technology currently the prime choice for green field and repowered plants in the U.S. and Europe.

However, the developing world, particularly China and India, have good reasons why the coal-fired power plant may still be the primary choice for many generation companies. Fuel is plentiful and inexpensive, and sulfur dioxide scrubbers have proved to be more reliable and effective than early plants indicated. In fact, up to 99% SO_2 removal efficiency is now possible. Removal of nitrogen oxides is also well advanced with over 95% removal possible using selective catalytic reduction (SCR). Ways to remove mercury are currently under study, and the issue of carbon dioxide control and sequestration from fossil plants is receiving renewed attention as ways to control global warming are pursued.

Combustion of coal can occur in three basic forms: direct combustion of pulverized coal; combustion of coal in a suspended bed of coal and inert matter; and coal gasification. The pulverized coal (PC) plant, the most common form of coal combustion, has the capability for much improved efficiency even with full **flue gas desulfurization** because ferritic materials technology has now advanced to the point at which higher steam pressures and temperatures are possible. Advanced supercritical PC plants are moving ahead commercially, particularly in Japan and Europe. Even higher steam conditions for PC plants, perhaps using nickel-based superalloys, are under study.

Worldwide, the application of atmospheric fluidized bed combustion (AFBC) plants has increased; these plants offer reductions in SO_2 and NO_x while permitting the efficient combustion of vast deposits of low-rank fuels such as lignites. Since the early 1990s, AFBC boiler technology power—with its advantage of in-furnace SO_2 capture with limestone—has become established worldwide as a mature, reliable technology for the generation of steam and electricity. In fact, the major impetus in the widespread deployment of this relatively new boiler technology since the mid-1980s has been its resemblance to a conventional boiler with the added capability for in-situ SO_2 capture, which could eliminate or reduce the need for flue gas desulfurization.

Coal gasification power plants are operating at the 250- to 300-MW level. Much of the impetus came from the U.S. DOE clean coal program where two gasification projects are in successful commercial service. Large gasification plants for power are also in operation in Europe. Gasification with combined cycle operation not only leads to minimum atmospheric (SO_2 and NO_x) and solid emissions, but also provides an opportunity to take advantage of new gas turbine advances. With the rapid advances now being introduced in combustion turbine technology, the coal gasification option is seen as a leading candidate for new plant construction within the first half of the 21st century.

13.2 Fuels for Electric Power Generation in the U.S.

The Energy Information Administration listed 498 GW of fossil-steam generating facilities in the U.S. in 1999. This included 407 GW at utilities and 91 GW at nonutilities. Coal-fired units dominated with 1393 units capable of generating 305 GW. All told, fossil-steam plants generate more than 70% of all electric energy in the country (Figure 13.1); these aging units (on average more than 30 years old) will remain the foundation of the power industry for the immediate future.

The U.S. electric power industry burns about $30 billion worth of fossil fuels each year, accounting for 70%–80% of the operating costs of fossil-fired plants. Coal dominates and recent changes to the fuel mixes include:

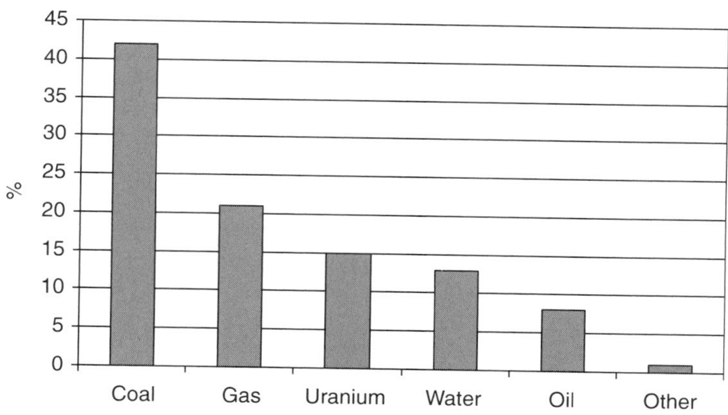

FIGURE 13.1 U.S. installed capacity by fuel percentage.

- A mix of eastern high-sulfur coal with low-sulfur, low-cost western coals, often from Powder River Basin (PRB) deposits in Montana and Wyoming. Compared with eastern bituminous coals, PRB coals have lower heating value, sulfur and ash, but higher moisture content and finer size.
- A mix of 10%–20% gas with coal in a boiler designed for coal firing.
- Orimulsion, a bitumen-in-water emulsion produced only from the Orinoco Basin in Venezuela. This fuel is relatively high in sulfur and vanadium. Power plants that use it need to add scrubbers.
- A mix of coal with petroleum coke, a by-product of refining, whose cost is currently low but whose sulfur content is high.

13.3 Coal as a Fuel for Electric Power (World Coal Institute 2000)

Coal is the altered remains of prehistoric vegetation that originally accumulated as plant material in swamps and peat bogs. The accumulation of silt and other sediments, together with movements in the Earth's crust (tectonic movements), buried these swamps and peat bogs, often to great depth.

With burial, the plant material was subjected to elevated temperatures and pressures, which caused physical and chemical changes in the vegetation, transforming it into coal. Initially, peat, the precursor of coal, was converted into **lignite** or **brown coal**—coal-types with low organic maturity. Over time, the continuing effects of temperature and pressure produced additional changes in the lignite, progressively increasing its maturity and transforming it into what is known as **sub-bituminous coals**. As this process continued, further chemical and physical changes occurred until these coals became harder and more mature; at this point they are classified as **bituminous** coals. Under the right conditions, the progressive increase in the organic maturity continued, ultimately forming **anthracite**.

The degree of metamorphism or coalification undergone by a coal as it matures from peat to anthracite has an important bearing on its physical and chemical properties, and is referred to as the "rank" of the coal. Low-rank coals, such as lignite and sub-bituminous coals, are typically softer, friable materials with a dull, earthy appearance; they are characterized by high moisture levels and low carbon content, and thus a low energy content. Higher rank coals are typically harder and stronger and often have a black vitreous luster. Increasing rank is accompanied by a rise in the carbon and energy contents and a decrease in the moisture content of the coal. Anthracite is at the top of the rank scale and has a correspondingly higher carbon and energy content and a lower level of moisture.

Large coal deposits only formed after the evolution of land plants in the Devonian period, some 400 million years ago. Significant accumulations of coal occurred during the Carboniferous period (280–350 million years ago) in the Northern Hemisphere; the Carboniferous/Permian period (225–350 million

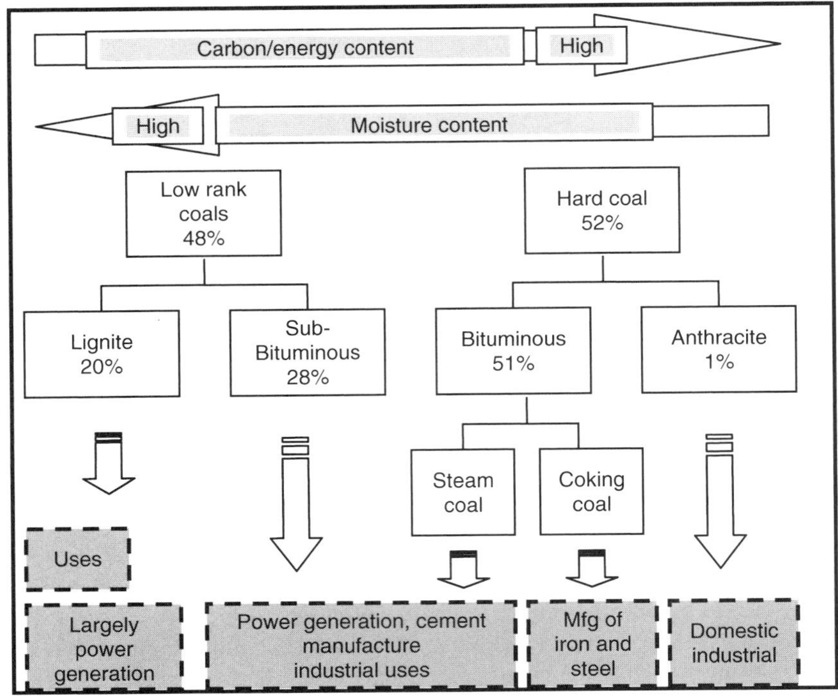

FIGURE 13.2 Coals cover a spectrum from lignites (high-moisture, lower-carbon content) to bituminous and anthracite coals (lower moisture, higher carbon content). All the coals can be burned to generate electricity.

years ago) in the Southern Hemisphere; and, more recently, the late Cretaceous period to early Tertiary era (approximately 15–100 million years ago) in areas as diverse as the U.S., South America, Indonesia, and New Zealand. Of all the fossil fuels, coal is the most plentiful in the world. It is geographically dispersed, spread over 100 countries and all continents (led by Asia with 137 million tons; Eastern Europe and the former Soviet Union with 113 million tons; and North America with 117 million tons).

Current coal reserve/production rations confirm over 200 years of resource availability. The percent of the world reserves categorized by type and use is shown in Figure 13.2. Almost one half (48%) of the world's coal reserves are lignite and sub-bituminous coals, which are used primarily for power generation.

13.4 Clean Coal Technology Development

At an increasing rate in the last few years, innovations have been developed and tested aimed at reducing emissions through improved combustion and environmental control in the near term, and in the longer term by fundamental changes in the way coal is preprocessed before converting its chemical energy to electricity. Such technologies are referred to as "clean coal technologies" and described by a family of precombustion, combustion–conversion, and postcombustion technologies. They are designed to provide the coal user with added technical capabilities and flexibility and the world with an opportunity to exploit its most abundant fossil source. They can be categorized as:

- *Precombustion*: Sulfur and other impurities are removed from the fuel before it is burned.
- *Combustion*: Techniques to prevent pollutant emissions are applied in the boiler while the coal burns.

- *Postcombustion*: The flue gas released from the boiler is treated to reduce its content of pollutants.
- *Conversion*: Coal, rather than being burned, is changed into a gas or liquid that can be cleaned and used as a fuel.

13.4.1 Coal Cleaning

Cleaning of coal to remove sulfur and ash is well established in the U.S., with more than 400 operating plants, mostly at mines. Coal cleaning removes primarily pyritic sulfur (up to 70% SO_2 reduction is possible) and in the process increases the heating value of the coal, typically about 10% but occasionally 30% or higher. The removal of organic sulfur, chemically part of the coal matrix, is more difficult, but may be possible using micro-organisms or through chemical methods; research is underway (Couch 1991). Heavy metal trace elements can be removed also; conventional cleaning can remove (typically) 30%–80% of arsenic, mercury, lead, nickel, antimony, selenium, and chromium.

13.5 Pulverized-Coal Plants

In the late 1950s the first units operating at supercritical pressures were introduced, initially in the U.S. and Germany. American Electric Power put the Philo supercritical unit in service in 1957 and Philadelphia Electric soon followed with Eddystone 1, a unit still in active service. Today, 159 supercritical units operate in the U.S., and worldwide more than 500 supercritical units are operating with ratings from 200 to 300 MW. Steam pressures for these units are typically 240 Bar (3500 psi), most of them single reheat designs. Steam temperatures are usually limited to about 594°C (1100°F), in order to utilize all-ferritic materials for thick wall components. A few (for example, Eddystone 1) utilize higher steam temperatures. The increased pressures and temperatures provide significant efficiency improvements over subcritical units, with attendant reductions in environmental emissions: SO_x, NO_x, CO_2, and particulates.

The greatest concentration of installed supercritical units is in the countries of the former U.S.S.R. where the 232 units in operation provide 40% of all electric power. These units are designed at specific sizes of 300, 500, 800, or 1200 MW, and have steam conditions typically 24 MPa/565°C/565°C. The former U.S.S.R. has also manufactured 18 supercritical units of 300- and 500-MW output, with shaft speeds of 3000 rpm, for China and Cuba (Oliker and Armor 1992).

In Japan more than 60 supercritical plants are in operation; 25 are coal fired with another nine due to start up in the next 2 years. Until the early 1990s, these plants had steam conditions of 24.6 MPa/538°C/566°C but, starting in 1993, the steam temperatures of the newer plants have climbed to the ultrasupercritical range, approaching 600°C.

There are about 60 supercritical units in Western Europe, largely in Germany, Italy (mostly oil fired), and Denmark. In the U.S., it is notable that the original Eddystone 1 unit, constructed in 1960, is still operating with the highest steam pressures and temperatures in the world, 4800 psi, 1150°F (322 Bar, 610°C), more than 35 years after commissioning. After the first supercritical units were installed in the late 1950s, early problems with these first-generation units began to surface. The majority of the problems were related to specific designs, changes in operating conditions, or plant malfunctions unrelated to higher pressure and temperature levels. Nevertheless, a few areas needed improvement, as noted later, to push the envelope of operating conditions as well as to improve the reliability and availability of these plants.

The design and material improvements that have been achieved now offset any deleterious effect of higher steam conditions on the cyclic fatigue life of components. The net result is that current designs do not push design margins on materials beyond the traditional margin of safety used for subcritical units; thus, new units based on their designs should have the same, if not better, availability.

13.5.1 Materials Advances

Higher steam temperatures (to 1150°F) and **supercritical steam pressures** are an important aspect of the modern pulverized coal plant. They are possible now because of advances in ferritic materials technology that extend life, provide greater creep and fatigue strength, and offer resistance to **temper embrittlement** and, in the boiler, to coal ash corrosion (Armor et al. 1999). Of particular note in those units operating today are:

- **Coextruded tubing** or monotubing for superheaters and reheaters, resistant to coal ash corrosion
- Improved 9-chrome steel (P91) for steam piping, valves, headers, and casings
- Improved creep-resistant 12-chrome forgings for HP/IP turbines
- "Superclean" 3.5 NiCrMoV rotors for LP turbines that are resistant to temper embrittlement.

Because of the severe temperatures and stresses that exist throughout the boiler and turbine, alloys have been developed that mitigate creep and creep-fatigue problems. Particularly in the headers, steam lines, and HP and IP rotors, the impact of start–stop cycling is a concern as steam temperatures advance. To date, it is reasonable to view 1100°F (593°C) as a steam temperature for which ferritic steels for boilers and turbines are well established using P91 steels. It is likely that 1150°F (620°C) will be possible with improved ferritic steels in the near future, and perhaps even 1200°F (650°C) with the addition of tungsten as an alloying agent of steels such as P92 and P122 (Armor et al. 2003).

Additional long-term creep data may be necessary for these advanced steels. Beyond 1200°F (650°C), it is anticipated that superalloys will replace the traditional ferritic steels for HP and IP rotors. Nickel-based superalloys will be required for turbine forgings, as well as for superheater tubes, boiler outlet headers, steam piping, valve bodies, and turbine casings (Dalton et al. 2001). Because of the increased thermal expansion coefficients of these materials over ferritic steels, forging and casting thermal stresses become an important issue in start-ups and load cycling. For this reason, it is thought that such high temperature designs might be more suitable for base load duty in which thermal stress-caused fatigue damage is limited, rather than for on–off or load-change operation.

13.5.2 Cycle Selection

The selection of a supercritical versus a subcritical cycle is dependent on many site-specific factors, including fuel cost; emissions regulations; capital cost; load factor; duty; local labor rates; and perceived reliability and availability. As to the reliability and availability of the supercritical cycle, it can match or better the subcritical cycle for base-loaded operation because early problems in first-and second-generation supercritical boilers and steam turbines have been overcome.

In fact, the use of subcritical cycles for the limited number of plants built in the U.S. in the last 20 years has been mainly due to relatively low fuel costs, which eliminated the cost justification for the somewhat higher capital costs of the higher efficiency cycles. However, in some international markets in which fuel cost is a higher fraction of the total cost, the higher efficiency cycles offer a favorable cost-of-electricity comparison and provide lower emissions compared to a subcritical plant. This is true in Europe and Japan.

For future plants though, the issue on cycle selection is likely to be decided based on efficiency considerations. The reduction of CO_2 emissions due to the supercritical cycle could be a deciding factor as ways are sought to reduce global warming concerns.

13.5.3 Supercriticals

13.5.3.1 Designs in U.S.

A survey of 159 supercritical units operating in the U.S. (EPRI 1986) showed significant efficiency advantages (up to 3%), compared to typical subcritical units, and outage rates comparable to drum units after an initial learning period. Further studies were carried out in the early 1980s on the optimum steam

TABLE 13.1 Advanced Supercritical Cycles at U.S. Locations With Double Reheat

Unit Name and Company	Steam Conditions (MPa/°C/°C/°C)	Design Capacity (MW)
Eddystone 1, PECO	34.3/649/565/565	325
Breed 1, AEP	24/565/565/565	450
Sporn 5, AEP	24/565/565/565	450
Eddystone 2, PECO	24/565/565/565	325
Tanners Creek 4, AEP	24/538/552/565	580
Muskingum River 5, AEP	24/538/552/565	590
Cardinal 1&2, AEP	24/538/552/565	600
Hudson 1, PSEG	24/538/552/565	400
Brayton Point 3, NEP	24/538/552/565	600
Hudson 2, PSEG	24/538/552/565	600
Big Sandy 2, AEP	24/538/552/565	760
Chalk Point 1&2, PEPCO	24/538/552/565	355
Haynes 5&6, LADWP	24/538/552/565	330
Mitchell 1&2, AEP	24/538/552/565	760
Amos 1&2	24/538/552/565	760

Note: More than 150 supercriticals operate in the U.S., although few have been installed in recent years.

pressures and temperatures for supercritical cycles (EPRI 1982) and on the materials of choice for boiler and turbine components. As noted earlier, better materials have now been adopted worldwide for new supercritical units and include the use of P91 (super 9 chrome) for thick wall headers, steam lines, valves, and turbine casings.

The optimum design for a new supercritical cycle was recommended (EPRI 1985) as a 700-MW double-reheat unit with steam conditions of 309 Bar/594/594/594°C (4500 psi, 1100/1100/1100°F). Such units have been constructed or are in planning in Japan and Denmark. As mentioned, it is notable that the original Eddystone 1 unit, constructed in 1960, is still operating with the highest steam pressures and temperatures in the world (322 Bar, 610°C), more than 35 years after commissioning. Table 13.1 notes other double-reheat units operating in the U.S.

13.5.3.2 Designs in Japan

In Japan more than 60 supercritical plants are in operation; 25 are coal fired with another nine due to start up by 2004. Until the early 1990s, these plants had steam conditions of 24.6 MPa/538°C/566°C (3500 psi/960°F/1000°F); however, starting in 1993, the steam temperatures of the newer plants have climbed to the ultrasupercritical range, approaching 600°C (1100°F). The more recent of the large-scale, coal-fired supercritical plants to come on line and those planned for commissioning by 2004 are shown in Table 13.2 and Table 13.3, respectively. It is notable that new USC plants such as Tachibana will use advanced ferritic steels NF616 (P92) and HCM 12A (P122).

TABLE 13.2 Recent Coal-Fired Ultrasupercritical Units in Japan

Power Plant	Company	Output (MW)	Steam Conditions (MPa/°C/°C)	Start-Up Date
Hekinann #3	Chubu	700	24.6/538/593	Apr. 1993
Noshiro #2	Tohoku	600	24.6/566/593	Dec. 1994
Nanao-Ohta #1	Hokuriku	500	24.6/566/593	Mar. 1995
Reihoku #1	Kyushu	700	24.1/566/566	Jul. 1995
Haramachi #1	Tohoku	1000	25/566/593	Jul. 1997
Maatsuura #2	EPDC	1000	24.6/593/593	Jul. 1997
Misumi #1	Chugoku	1000	25/600/600	Jun. 1998
Haramachi #2	Tohoku	1000	25/600/600	Jul. 1998
Nanoa-Ohta #2	Hokuriku	700	24.6/593/593	Jul. 1998

Note: Several other supercriticals are fired by oil or LNG.

TABLE 13.3 New and Upcoming Coal-Fired Ultrasupercritical Units in Japan

Power Plant	Company	Output (MW)	Steam Conditions (MPa/°C/°C)	Start-Up Date
Hekinann #4	Chubu	1000	24.6/566/593	Nov. 2001
Hekinann #5	Chubu	1000	24.6/566/593	Nov. 2002
Tsuruga #2	Hokuriku	700	24.6/593/593	Oct. 2000
Tachibana-wan	Shikoku	700	24.6/566/566	Jul. 2000
Karita #1 (PFBC)	Kyushu	350	24.6/566/593	Jul. 2000
Reihoku #2	Kyushu	700	24.6/593/593	Jul. 2003
Tachibana-wan #1	EPDC	1050	25/600/610	Jul. 2000
Tachibana-wan #2	EPDC	1050	25/600/610	Jul. 2001
Isogo (New #1)	EPDC	600	25.5/600/610	Apr. 2002
Hitachinaka #1	Tokyo	1000	24.5/600/600	2002
Maizuni #1	Kansai	900	24.1/593/593	2003
Maizuni #2	Kansai	900	24.1/593/593	2003

Note: Noticeably, these are all single reheat units; double reheat is reserved for the more expensive oil- and gas-fired units.

With few natural resources, Japan depends on efficient fossil generation plants burning coal, LNG, and heavy oil (as well as nuclear). Efficiency has always been a key issue for resource use minimization as well as environmental control. The Japanese supercriticals have adopted the best in European and U.S. technologies for their plants, including the European spiral-wound boiler and the U.S. partial arc admission turbine. Japanese steel-making is among the best in the world and the original EPRI advanced plant studies involved the leading Japanese steel makers. The first superclean low-pressure turbine steels (an EPRI innovation to avoid temper embrittlement) were first deployed in units at the Chubu Kawagoe station. Material advances for ferritic steels beyond P91 have been stimulated by formulations using tungsten as an alloying element (P92, for example).

It is noticeable in the tables that the units are of large size—sometimes a deterrent to cycling operation because of the thermal stresses involved. However, careful adaptation of automated start-up systems and the use of turbine by-pass systems have essentially solved today's start-up and cycling problems. The next generation, however, with the likely use of austenitic steels or of nickel-based superalloys will present new design challenges.

13.5.3.3 Designs in Europe

There are more than 100 supercritical units in Europe, largely in Germany, Italy (mostly oil fired), Holland, and Denmark. The most recent European coal-fired units with advanced supercritical steam conditions are listed in Table 13.4. Pioneering work on supercritical machines was carried out in Germany in the late 1950s and early 1960s, parallel with the U.S. advances. Particularly of note was the development of the spiral-wound boiler that permitted the pressure to slide up and down without concerns related to any departure from nucleate boiling, a situation that would severely damage boiler tubing. These boilers are now routinely used in Japan and Europe for full sliding pressure supercritical operation.

It is also significant that supercritical units in Germany, with double reheat, were built in unit sizes down to 220 MW—a size that would have appeal in the U.S., where smaller sizes are often sought by generating companies. Unit sizes are climbing in Germany, particularly as shown by the big lignite units at Schwarze Pumpe, Lippendorf, and Niederaussem. Lignite is a major resource in Germany and in several other European counties, such as Greece, where a lignite-fired, supercritical, district heating plant is in construction at Florina. New advanced plants for steam conditions are seen in Denmark, where the 411-MW, double-reheat supercritical units at Skaerbeck and Nordjyllands have steam temperatures of 580°C (1050°F), and at Avedore, a single-reheat design, 600°C (1110°F).

TABLE 13.4 Recent European Supercritical Units with Advanced Steam Conditions

Power Plant	Fuel	Output (MW)	Steam Conditions (MPa/°C/°C/°C)	Start-Up Date
Skaerbaek 3	Gas	411	29/582/580/580	1997
Nordjyllands 3	Coal	411	29/582/580/580	1998
Avedore	Oil, biomass	530	30/580/600	2000
Schopau A,B	Lignite	450	28.5/545/560	1995–1996
Schwarze Pumpe A,B	Lignite	800	26.8/545/560	1997–1998
Boxberg Q,R	Lignite	818	26.8/545/583	1999–2000
Lippendorf R,S	Lignite	900	26.8/554/583	1999–2000
Bexbach II	Coal	750	25/575/595	1999
Niederaussem K	Lignite	1000	26.5/576/599	2002

Note: The trend in Europe appears to be in the direction of larger unit sizes.

13.6 Emissions Controls for Pulverized Coal Plants (Armor and Wolk 2002)

Today, worldwide, about 40% of electricity is generated from coal and the total installed coal-fired generating capacity is more than 1000 GW, largely made up of 340 GW in North America; 220 GW in Western Europe, Japan, and Australia; 250 GW in Eastern Europe and the former U.S.S.R.; and 200 GW in China and India. More than 200 GW of new coal capacity has been added since 1990. Thus, together with the potential impact of carbon dioxide emissions contributing to global warming, the control of particulates, sulfur dioxides and nitrogen oxides from those plants is one of the most pressing needs of today and the future. To combat these concerns, a worldwide move toward environmental retrofitting of older fossil-fired power plants is underway, focused largely on sulfur dioxide scrubbers and combustion or postcombustion optimization for nitrogen oxides. Carbon dioxide control and sequestration options are now under study worldwide.

13.6.1 Conventional Lime/Limestone Wet Scrubber

The dominant SO_2 scrubbing system is the wet limestone design because limestone costs one quarter that of lime as a reagent. In this system (Figure 13.3), the limestone is ground and mixed with water in a reagent preparation area. It is then conveyed to a spray tower, called an absorber, as a slurry of 90% water and 10% solids, and sprayed into the flue gas stream. The SO_2 in the flue gas is absorbed in the slurry and collected in a reaction tank where it combines with the limestone to produce water and calcium sulfate or calcium sulfate crystals. A portion of the slurry is then pumped to a thickener where these solids/crystals settle out before going to a filter for final dewatering. Mist eliminators installed in the system ductwork at the spray tower outlet collect slurry/moisture entrained in the flue gas stream. Calcium sulfate is typically mixed with flyash (1:1) and lime (5%) and disposed of in a landfill.

Various improvements can be made to this basic process, including the use of additives for performance enhancement and the use of a hydrocyclone for dewatering, replacing the thickener and leading to a saleable gypsum byproduct. The Chiyoda-121 process (Figure 13.4) reverses the classical spray scrubber and bubbles the gas through the liquid. This eliminates the need for spray pumps, nozzle headers, separate oxidation towers, and thickeners. The Chiyoda process was demonstrated in a DOE Clean Coal Technology project on a 100-MW unit at the Yates plant of Georgia Power Company.

13.6.2 Spray Drying

Spray drying (Figure 13.5) is the most advanced form of dry SO_2 control technology. Such systems tend to be less expensive than wet FGD but remove typically a smaller percentage of the sulfur (90% compared

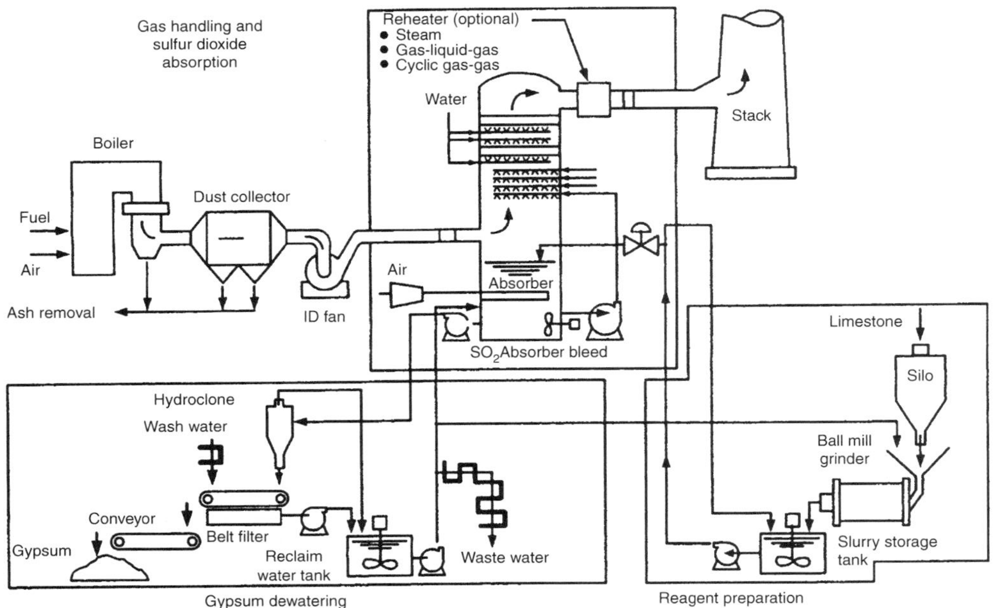

FIGURE 13.3 The conventional lime/limestone wet scrubber is the dominant system in operation in the U.S. With recent refinements, this system can be 98%–99% effective in removing SO_2.

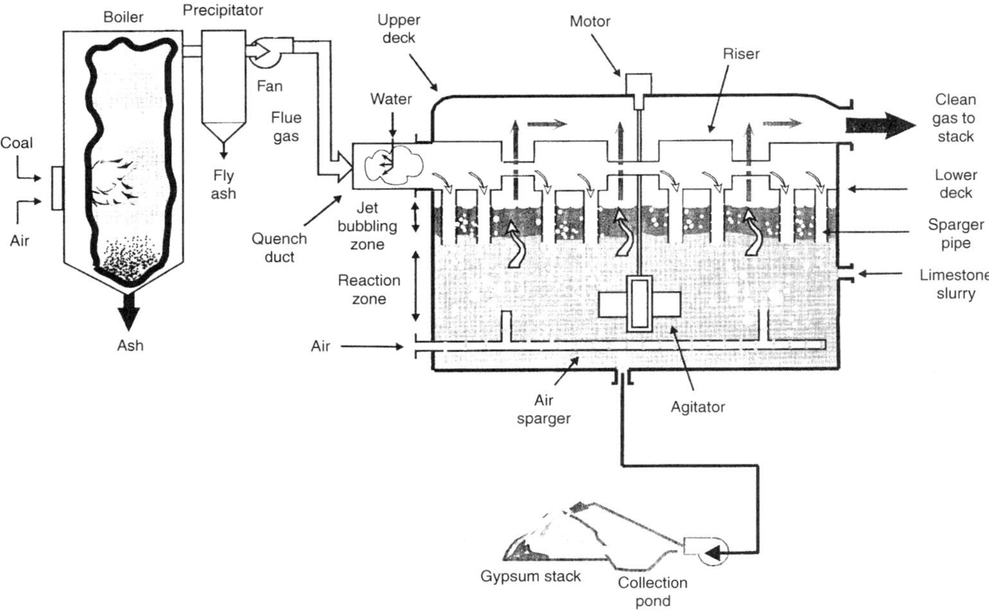

FIGURE 13.4 The Chioda-121 scrubber simplifies the process by bubbling the flue gas through the liquid, eliminating some equipment needs.

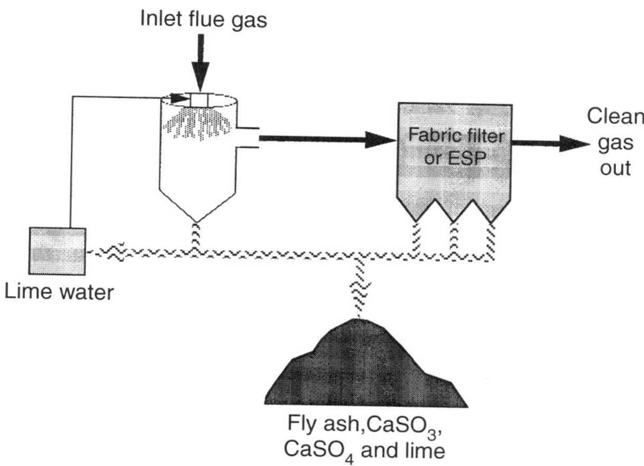

FIGURE 13.5 Spray driers use a calcium oxide reagent mixed with water, which is dried by the flue gas. A dry product is collected in a fabric filter.

with 98%). They are usually used when burning low-sulfur coals and utilize fabric filters for particle collection, although recent tests have shown applicability to high-sulfur coals also.

Spray driers use a calcium oxide reagent (quicklime) that, when mixed with water, produces a calcium hydroxide slurry. This slurry is injected into the spray drier where it is dried by the hot flue gas. As the drying occurs, the slurry reacts to collect SO_2. The dry product is collected at the bottom of the spray tower and in the downstream particulate removal device where further SO_2 removal may take place. It may then be recycled to the spray drier to improve SO_2 removal and alkali utilization.

For small, older power plants with existing **electrostatic precipitators** (ESPs), the most cost effective retrofit spray dry configuration locates the spray dryer and fabric filter downstream of the ESP, separating in this manner the spray dryer and fly ash waste streams. The fly ash can then be sold commercially.

13.6.3 Control of Nitrogen Oxides

Nitrogen oxides can be removed during or after coal combustion. The least expensive option, and the one generating the most attention in the U.S., is combustion control, first through adjustment of the fuel–air mixture, and second through combustion hardware modifications. Postcombustion processes seek to convert NO_x to nitrogen and water vapor through reactions with amines such as ammonia and urea. Selective catalytic reduction (SCR) injects ammonia in the presence of a catalyst for greater effectiveness. The options (Figure 13.6) can be summarized as:

- *Operational changes*: Reduced excess air, and biased firing, including taking burners out of service
- *Hardware combustion modifications*: Low NO_x burners, air staging, and fuel staging (reburning)
- *Postcombustion modifications*: Injection of ammonia or urea into the convection pass, selective catalytic reduction (SCR), and wet or dry NO_x scrubbing (usually together with SO_2 scrubbing)

Low NO_x burners can reduce NO_x by 50% and SCR by 80%, but the low NO_x burner option is much more cost effective in terms of cost per ton of NO_x removed. Reburning is intermediate in cost per removed ton and can reduce NO_x 50%, or 75% in conjunction with low NO_x burners.

13.6.3.1 Postcombustion Options

Selective catalytic reduction is used widely in Europe (especially in Germany, where it is installed on more than 30,000 MW of coal-fired boilers) and in Japan, but sparingly to date in the U.S., although applications of SCR are increasing. In an SCR, ammonia is injected into the boiler exhaust gases

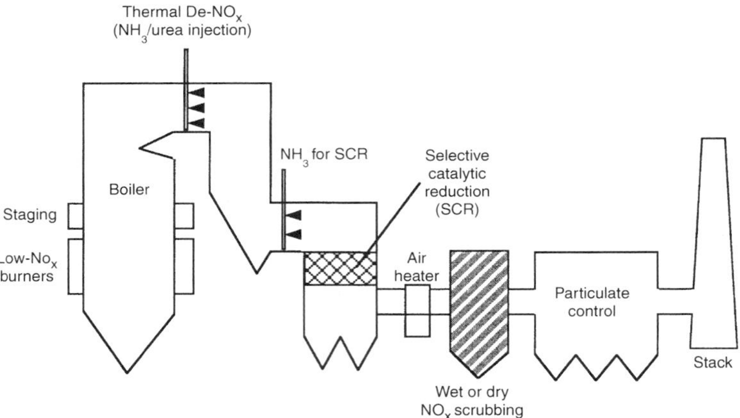

FIGURE 13.6 Control options for NO$_x$ include operational, hardware, and postcombustion modifications.

before the catalyst bank (at about 550°F–750°F). NO$_x$ and NH_3 then react to produce nitrogen and water; the chemical reactions are:

$$4NO + 4NH_3 + O_2 \rightarrow 4N_2 + 6H_2O$$

$$6NO_2 + 8NH_3 \rightarrow 7N_2 + 12H_2O$$

$$2NO_2 + 4NH_3 + O_2 \rightarrow 3N_2 + 6H_2O$$

The reaction can result in a potential NO$_x$ removal capability of more than 90%, though practical limitations include ineffective mixing of NO$_x$ and NH_3; velocity and temperature variations; NH_3 slip; and gradual loss of catalyst activity. Retrofit installation of an SCR system can require considerable space, although the reactor can be placed inside the original ductwork if NO$_x$ reduction levels are modest. This would be difficult for coal-fired systems due to the high gas velocities. A separate reactor allows more flexibility in design.

In general, SCR has been successful in current installations, although impacts on the boiler system have included air heater deposition and plugging due to the formation of ammonium sulfate and bisulfate; ammonia contamination of fly ash; and ammonia slip. SCR systems operate at significantly lower temperatures than SNCR and are much more flexible in achieving the desired degree of NO$_x$ reduction. These systems utilize catalyst coated metal plates to react flue gas. Inlet temperatures to the catalyst panels are controlled by bypassing flue gas around the catalyst zone to meet the required level of NO$_x$ removal. Clearly, an option with relatively high operational cost, it remains an effective proven method for NO$_x$ reductions.

Selective noncatalytic combustion (SNCR) is a promising lower capital cost alternative to SCR ($10/kW versus more than $50/kW), but with lower performance (20%–35% reduction compared with 50 to as high as 80% for SCR). In SNCR, the injection of a reagent like urea or ammonia into the upper part of the furnace converts NO$_x$ from combustion into nitrogen; this conversion is a direct function of furnace temperature and reagent injection rate. Usual practice is to inject the reagent into a region of the boiler or convective pass where the temperature is in the range of 1600°F–2100°F (871°C–1149°C). NO$_x$ reductions can range from 20 to as high as 50%, as previously stated, but are typically in the range of 20%–35%. The level of reduction varies with the amount of nitrogen-containing chemical injected.

One major operating issue is that the system is not very flexible because the temperature of the flue gas at the point of ammonia injection is a function of boiler load. Additional operating issues of

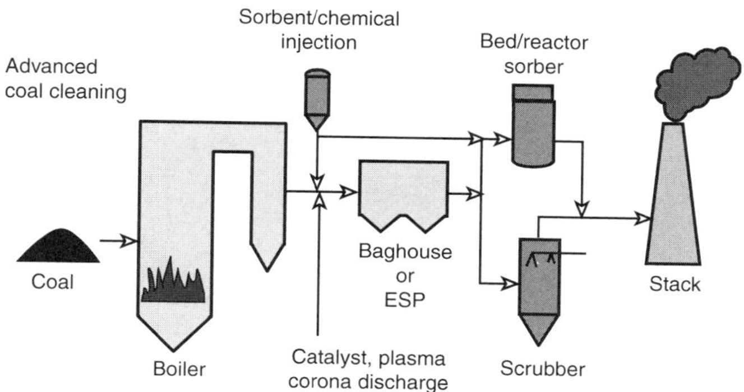

FIGURE 13.7 Options for the removal of mercury.

concern are the amount of unreacted ammonia (or slip) that exits the reaction zone; air heater fouling by sulfate compounds formed by reaction between SO_3 and excess ammonia; plume formation; and ammonia content of flyash. The operating cost is significant, although reagent consumption can be controlled.

13.6.4 Control of Mercury (EPRI 2000)

Trace amounts of mercury are present in coal. Consequently, whenever it is combusted, such as in the generation of electricity, some of this mercury is emitted into the air along with exhaust gases. Power plant emissions account for about one third of the mercury emitted to the air from industrial sources in the U.S., but this is only about 2%–3% of total global mercury emissions. After power plants release mercury, it becomes part of a global cycle.

The behavior of mercury in this cycle depends on its chemical form. Oxidized mercury (also called ionic mercury) may fall or wash out of the air and return to the Earth relatively near its source. Elemental mercury becomes part of the global inventory of mercury in the atmosphere, where it may remain for months to years. Eventually, it too returns to the Earth. This deposited mercury may enter bodies of water, directly or via runoff from surrounding soils, and enter the food chain. The ultimate environmental concern is humans' and animals' ingestion of fish containing elevated levels of mercury. Significant uncertainty and controversy remain about the contribution of individual sources to local or regional water bodies, as well as what constitutes safe levels of mercury in fish.

The removal of mercury from coal-fired units can be accomplished in several ways (Figure 13.7). Coal cleaning before combustion can remove some mercury and other heavy metals. After combustion, the injection of a sorbent such as an activated carbon can be very effective. Existing precipitators and SO_2 scrubbers can capture from 20 to 60% of mercury. Catalysts and certain chemicals can be injected that oxidize elemental mercury to enhance scrubber capture. Fixed beds, coated with materials such as gold, can form amalgams with mercury.

13.7 Fluidized Bed Plants

Introduced nearly 30 years ago, the **fluidized bed combustion** (FBC) boiler has found growing application for power generation. From the first FBC boiler, which generated 5000 lb/h of steam in 1967, the technology has matured to the 350-MW size units available today. In North America, more than 160 units now generate in excess of 9000 MW (EPRI 2002). Burning coal in a suspended bed with

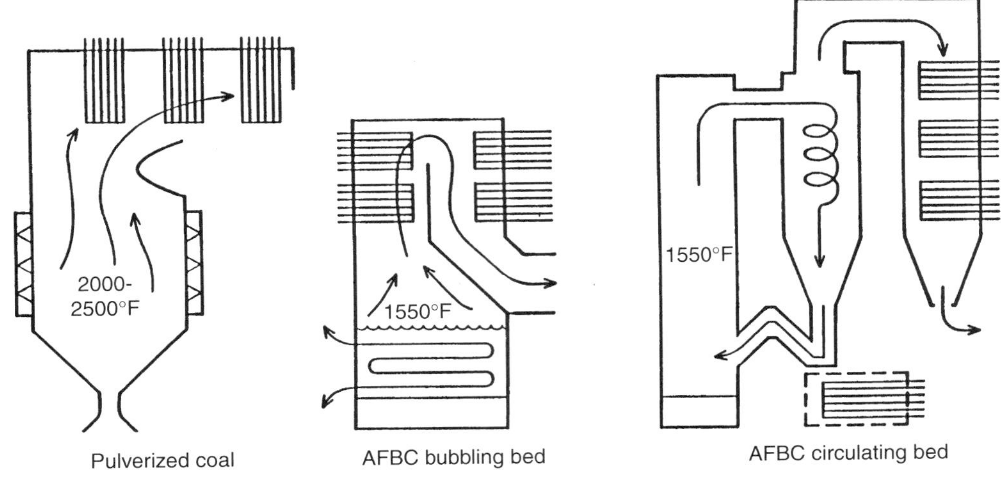

FIGURE 13.8 An illustration of the distinguishing features of pulverized coal and fluidized bed boilers. Noticeable in this diagram are the in-bed tubes characteristic of bubbling beds, and the cyclone separator of the circulating bed.

limestone or dolomite permits effective capture of sulfur and fuel flexibility allows a broad range of opportunity fuels. These fuels might include coal wastes (culm from anthracite, gob from bituminous coal), peat, petroleum coke, and a wide range of coals from bituminous to lignite. A low (1500°F) combustion temperature leads to low NO_x formation. The salient features of atmospheric fluidized bed boilers, compared with a pulverized coal boiler, are shown in Figure 13.8.

Utility size demonstration projects at the Tennessee Valley Authority in 1989 (Manaker 1992) (Shawnee, 160 MW) and Northern States Power in 1986 (Hinrichsen 1989) (Black Dog, 133 MW) are examples of successful atmospheric bubbling-bed units. The Black Dog unit has been dispatched in a daily cycling mode and has successfully fired a blend of coal and petroleum coke. However, the focus of AFBC in the U.S. is now on the circulating fluid bed (CFB). In fact, more than 70% of operating fluid bed boilers in the U.S. are of the circulating type. The CFB unit at Nucla (Tri-State G&T Association) (Blunden 1989) has been successful in demonstrating the technology at the 110-MW level and commercial CFB plants have now reached 250 MW in size. Most fluidized bed units for electricity generation are installed by independent power producers in the 50- to 100-MW size range; here, the inherent SO_2 and NO_x advantages over the unscrubbed PC plant have encouraged installations even in such traditional noncoal arenas as California (Melvin and Friedman 1994). Worldwide, the AFBC boiler is employed largely for steam heat; hundreds of them operate in Russia and India, and thousands in China.

13.7.1 Atmospheric Fluidized Bed Combustion (AFBC)

In the bubbling bed version of the AFBC, the fuel and inert matter, together with limestone or dolomite for SO_2 capture, is suspended through the action of fluidizing air, which flows at a velocity of 3–8 ft/s in essentially a one-pass system. Circulating fluid beds (CFB) differ from bubbling beds in that much of the bed material passes through a cyclone separator before being circulated back to the boiler (Figure 13.9). In-bed tubes are generally not used for CFB units permitting a much higher fluidizing velocity of 16–26 ft/s. Since the early AFBC designs, attention has been directed toward increasing unit efficiency, and reheat designs are now usual in large units. When SO_2 capture is important, a key parameter is the ratio of calcium in the limestone to sulfur in coal. Typical calcium–sulfur ratios for 90% SO_2 reduction

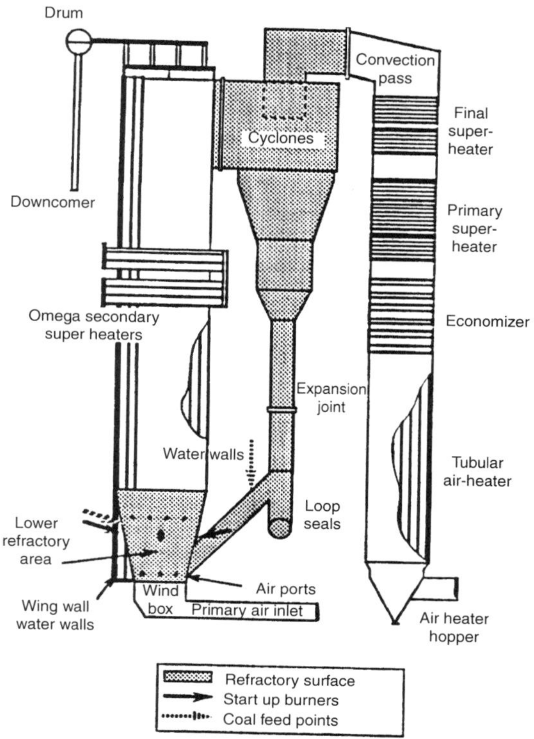

FIGURE 13.9 A circulating fluid bed boiler installed at the ACE Cogeneration Company at Trona, California. This 108-MW unit burns low-sulfur, western bituminous coal with limestone in a bed, which circulates back to the boiler after passing through a cyclone separator.

are in the range of 3.0–3.5 for bubbling beds and 2.0–2.5 for circulating beds. This depends on the fuel, however, and the 200-MW CFB plant at the NISCO cogeneration plant, which burns 100% **petroleum coke** (4.5% S), has a Ca–S ratio of below 1.7 for more than 90% sulfur capture. NO_x levels in AFBCs are inherently low and nominally less than 0.2 lb/MMBtu.

It is important to note that for CFBs, boiler efficiencies can be as high as a pulverized coal unit (Table 13.5). In fact, designs now exist for AFBCs with supercritical steam conditions, with prospects for cycles up to 4500 psia, 1100°F with **double reheat** (Skowyra et al. 1995). Large CFB units in the Americas are shown in Table 13.6.

13.7.2 Pressurized Fluidized Bed Combustion (PFBC)

In a PFBC combined cycle unit (Figure 13.10), coal in a fluid bed is burned with dolomite or limestone in a pressurized steel chamber, raising steam for a steam turbine generator. The pressurized flue gases are expanded through a gas turbine. Commercial plants at about the 80-MW level in Sweden, the U.S., and Spain have demonstrated that bubbling-bed PFBC plants, with a calcium–sulfur molar ratio of about 1.5, offer sulfur capture up to 95%, together with inherently low NO_x emissions due to low combustion temperatures. Cleanup of the flue gas before entry to the gas turbines is a key technical objective, and first-generation units have used cyclones together with gas turbines ruggedized with special blade coatings. For more advanced, higher efficiency PFBC systems, hot-gas cleanup technology, where the gas is directed through large ceramic filter units, will likely be needed.

TABLE 13.5 Typical Boiler Efficiencies, Pulverized Coal (PC), and Fluidized Beds

Loss/Gain Parameter	PC	Calculated Heat Loss (%)		
		Highest Efficiency CFB	Lowest Efficiency CFB	Bubbling Bed
Moisture in limestone	NA	0.06	0.10	0.10
Calcination	NA	1.02	1.69	2.70
Sulfation credit	NA	−1.60	−1.60	−1.60
Unburned carbon	0.25	0.50	2.0	4.0
Heat in dry flue gas	5.28	5.57	5.60	5.75
Moisture in fuel	1.03	1.03	1.03	1.03
Moisture from burning H_2	41.9	4.19	4.19	4.19
Radiation and convection	0.30	0.30	0.80	0.30
Moisture in air	0.13	0.14	0.14	0.14
Sensible heat in boiler ash	0.03	0.09	0.76	0.50
Bottom ash	0.05	NA	NA	NA
Fan-power credit	−0.25	−0.75	−0.40	−0.50
Pulverizer/crusher power gain	−0.20	NA	NA	NA
Total losses/gains	10.81	10.55	14.31	16.51
Overall boiler efficiency, %	89.19	89.45	85.69	83.49

Source: Fluidized Bed Boilers, *Power Magazine*, January 1987.

Early designs included the 80-MW units at Vaertan (Sweden) and Escatron (Spain) and the 70-MW unit at Tidd (AEP), which operated satisfactorily. The modular aspect of the PFBC unit is a particularly attractive feature, leading to short construction cycles and low-cost power. One promising use for PFBC units is for small, in-city cogeneration plants in which the inherent size advantages, high efficiencies, and effective coal gas cleanup approach permits compact plants to be retrofitted in place of heating boilers; the small steam turbines can be easily be adapted to electricity and hot water supply (Olesen 1985).

A 250-MW subcritical unit based on the Hitachi bubbling bed technology was built at Osaka for Chugoku Electric and started commercial operation in late 2000. Steam conditions are 16.7 MPa/566°C/593°C (386 psig/1050°F/1100°F). A 360-MW supercritical unit based on the ABB technology was constructed in Japan at Karita for Kyushu Electric Power Company. The boiler is designed for steam conditions of 24.1 MPa/565°C/593°C (3500 psig/1050°F/1100°F). The plant started up in late 1999.

In early 2000, ABB decided that it would no longer market PFBC but would confine its role to that of supplying the gas turbine. Alstom subsequently acquired ABB's power business. Thus, it appears that the future of PFBC will depend on the market in Japan However, in the Japanese market, PFBC faces strong competition from the several 800- to 1000-MW ultrasupercritical (USC) PC plants that have entered service. These have been performing very well, so the economies of scale represented by these large USC PC plants will make it very difficult for PFBC to compete. At this time (2003), commercialization of PFBC in the U.S is not currently actively progressing (Courtright et al. 2003).

13.8 Gasification Plants

One option of growing interest to coal-burning utilities is that of coal gasification. After the EPRI Cool Water demonstration in 1984 at the 100-MW level, the technology has moved ahead in the U.S. largely through demonstrations under the Clean Coal Technology (CCT) program (U.S. DOE 2002).

Gasification-based plants have among the lowest emissions of pollutants of any central station fossil technology. Using the efficiency advantages of combined cycles, CO_2 emissions are also low. Fuel flexibility is an additional benefit because the gasifier can accommodate a wide range of coals, plus

TABLE 13.6 U.S., Canadian, and Latin American CFB Units Larger Than 75 MW

Plant/Location (Vendor)	Start-Up	Capacity, MW (Net)	Fuels
Tri-State Generation & Transmission/Colorado	1987	1×100	Bit. coal
AES Shady Point/Oklahoma	1989	4×75	Bit. coal
AES Thames/Connecticut	1989	2×90	Bit. coal
Schuylkill Energy/Pennsylvania	1989	1×80	Culm
ACE Cogeneration/California	1990	1×97	Low-S bit. coal
Texas-New Mexico Power/Texas	1990	2×150	Lignite
AES Barbers Point/Hawaii	1992	2×90	Bit. coal
Nelson Industrial Steam Co. (NISCO)/Louisiana	1992	2×110	Coke
Cedar Bay Generating Co./Florida	1993	3×90	Bit. coal
Nova Scotia Power/Nova Scotia	1993	1×165	30% bit. coal and 70% coke
Colver Power/Pennsylvania	1995	1×105	Gob
Northampton Generating Co./Pennsylvania	1995	1×112	Culm
ADM/Illinois	1996/2000	2×132	Bit. coal and up to 5% TDF
ADM/Iowa	2000	1×132	Bit. coal
AES Warrior Run/Maryland	1999	1×180	Bit. coal
Choctaw Generation-the Red Hills project/Mississippi	2001	2×220	Lignite
Bay Shore Power—First Energy/Ohio	2001	1×180	Coke
AES Puerto Rico/Puerto Rico	2002	2×227	Bit. coal
JEA/Florida	2002	2×265	Bit. coal and coke
Southern Illinois Power Cooperative/Illinois	2002	1×113	Waste bit. coal
Termoelectrica del Golfo/Mexico	2002	2×115	Coke
Termoelectrica de Penoles/Mexico	2003	2×115	Coke
Reliant Energy Seward Station/Pennsylvania (ALSTOM)	2004	2×260	Gob & bit. coal
East Kentucky Power Cooperative/Kentucky	2004	1×268	Unwashed high-sulfur bit. coals
Figueira/Brazil	2004	1×128	Bit. coal

Source: EPRI, *Atmospheric Fluidized-Bed Combustion Handbook*, EPRI Report 1004493, December 2002.

petroleum coke. Integrated gasification combined cycle (IGCC) plants permit a hedge against long-term increases in natural gas prices because natural gas-fired combustion turbines can be installed initially, and gasifiers at a later time, when a switch to coal becomes prudent (Douglas 1986).

The pioneering Cool Water plant, the first of its kind in the world, operated for more than 4 years, gasifying 1.1 million tons of coal and producing 2.8 million MWh of electricity. The project was a collaborative effort of the industry involving the utility (Southern California edition); equipment manufacturers (Texaco, General Electric); and consultants/research consortia (Bechtel, EPRI, and others). Particularly notable was the achievement of exceptionally low levels of emissions of SO_2, NO_x, and particulates, as shown in Figure 13.11.

Basically, IGCC plants replace the traditional coal combustor with a gasifier and gas turbine. Ultralow emissions are realized; over 99% of the coal's sulfur is removed before the gas is burned in the gas turbine. A gasification cycle can take advantage of all the technology advances made in combustion turbines and steam turbines, so as to enhance overall cycle efficiency.

There are currently two coal-based IGCC commercial sized demonstration plants operating in the U.S. and two in Europe (Table 13.7). The U.S. projects were all supported under the U.S. Department of

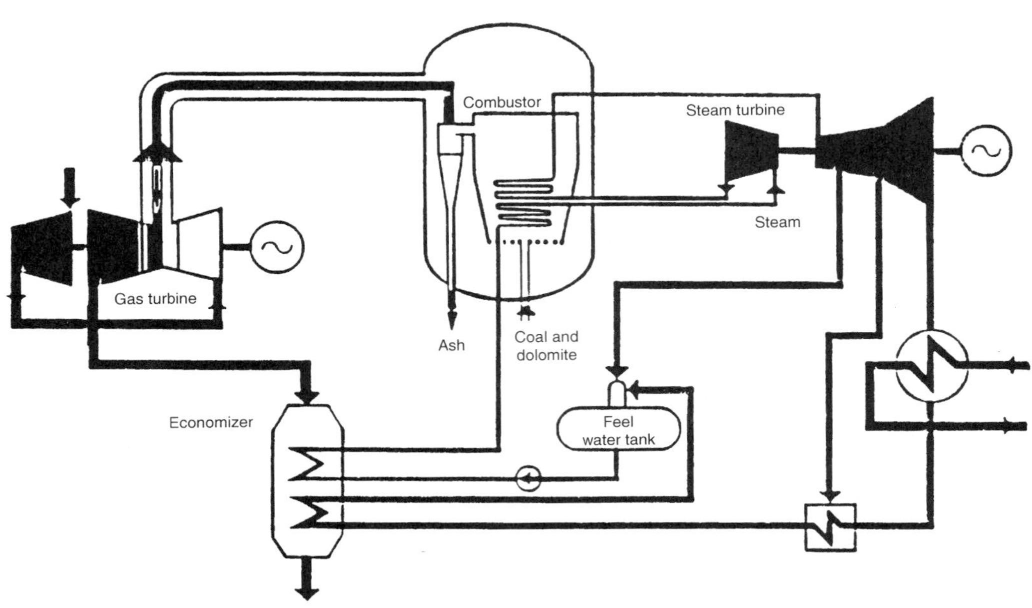

FIGURE 13.10 Pressurized, fluidized-bed combustor with combined cycle. This 70-MW system has operated at the Tidd plant of American Electric Power.

Energy's (DOE) Clean Coal Technology (CCT) demonstration program. The 262-MW Wabash River IGCC repowering project in Indiana started up in October 1995 and uses the E-GAS™ (formerly Destec) gasification technology. The 250-MW Tampa Electric Company (TEC) IGCC project in Florida started up in September 1996 and is based on the Texaco gasification technology. The first of the European IGCC plants, the SEP/Demkolec project at Buggenum, the Netherlands, uses the Shell gasification technology and started operations in early 1994. The second European project, the ELCOGAS project in Puertollano, Spain, which uses the Prenflo gasification technology, started coal-based operations in December 1997.

13.8.1 Polk County IGCC

Texaco's pressurized, oxygen-blown, entrained-flow gasifier is used at the Tampa Electric Polk County plant to produce a medium-Btu fuel gas (Figure 13.12). Coal/water slurry and oxygen are combined at high temperature and pressure to produce a high-temperature syngas. Molten coal ash flows out of the bottom of the vessel and into a water-filled quench tank where it is turned into a solid slag. The syngas from the gasifier moves to a high-temperature heat-recovery unit, which cools the gases.

The cooled gases flow to a particulate-removal section before entering gas-cleanup trains. A portion of the syngas is passed through a moving bed of metal oxide absorbent to remove sulfur. The remaining syngas is further cooled through a series of heat exchangers before entering a conventional gas-cleanup train where sulfur is removed by an acid-gas removal system. These cleanup systems combined are expected to maintain sulfur levels below 0.21 lb/million Btu (96% capture). The cleaned gases are then routed to a combined-cycle system for power generation. A gas turbine generates about 192 MW. Thermally generated NO_x is controlled to below 0.27 lb/MM Btu by injecting nitrogen as a dilutent in the turbine's combustion section. A heat-recovery steam generator uses heat from the gas-turbine exhaust to reduce high-pressure steam. This steam, along with the steam generated in the gasification process, is routed to the steam turbine to generate an additional 120 MW. The IGCC heat rate for this demonstration is approximately 8600 Btu/kWh (40% efficient). Coals used in the demonstration are Illinois 6 and Pittsburgh 8 bituminous coals with sulfur contents ranging from 2.5 to 3.5%.

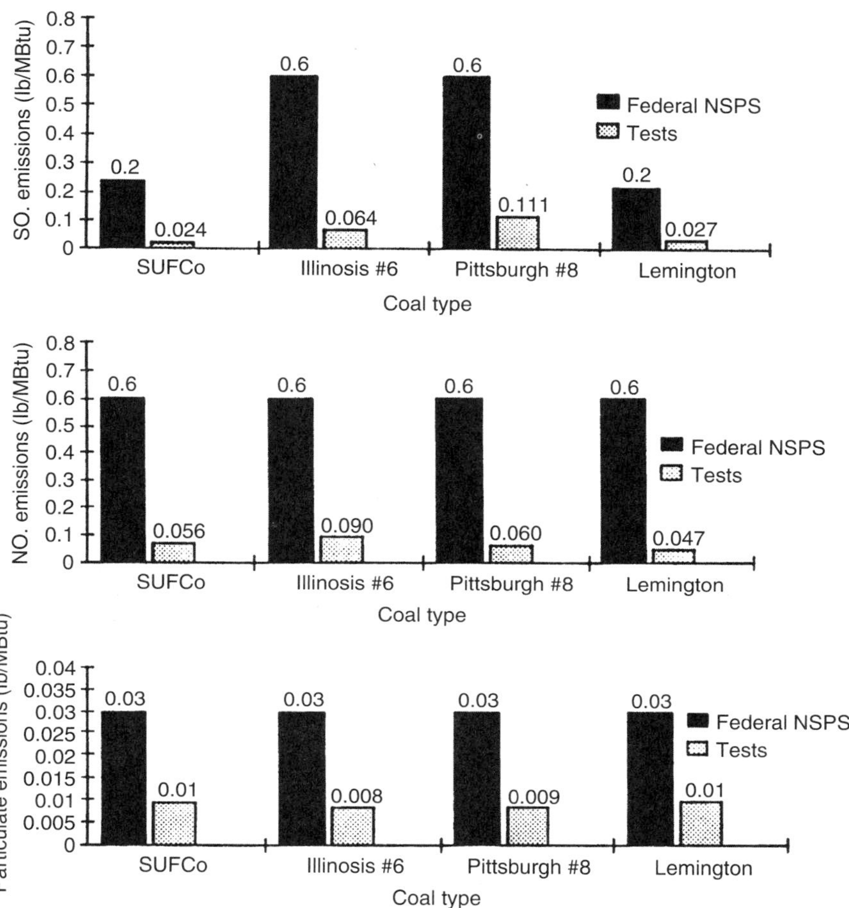

FIGURE 13.11 Tests at Cool Water on four coals show emissions of SO_2, NO_x, and particulates substantially below the federal New Source Performance Standards.

By-products from the process—sulfuric acid and slag—can be sold commercially, sulfuric acid by-products as a raw material to make agricultural fertilizer and the nonleachable slag for use in roofing shingles and asphalt roads and as a structural fill in construction projects.

13.8.2 Buggenum IGCC

Tests are in progress on the 250-MW IGCC plant in Buggenum, Netherlands. After successful operations running on natural gas, a switch was made to coal gas using Columbia coals. Buggenum comprises a 2000

TABLE 13.7 Coal-Based Commercial Size IGCC Plants

Operating IGCC Plants	Gasification Technology	Plant Size (MW)	Start-Up Date
Wabash River, Indiana	Destec	262	Oct. 1995
Tampa Electric, Florida	Texaco	250	Sep. 1996
SEP/Demkolec, Buggenum, The Netherlands	Shell	253	Early 1994
ELCOGAS, Puertollano, Spain	Krupp-Uhde Prenflo	310	Dec. 1997 on coal

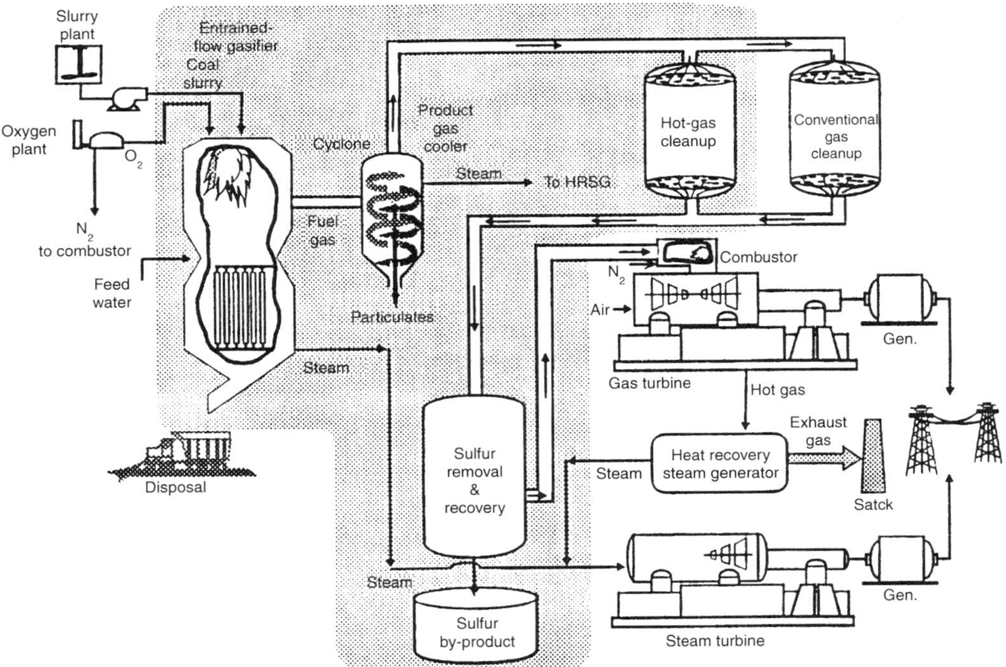

FIGURE 13.12 Integrated gasification, combined cycle at Tampa Electric, Polk County plant. A Texaco oxygen-blown gasifier is used. Total net generation is 250 MW.

ton/day single reactor coal gasification unit and an air separation plant able to produce 1700 ton/day of 95% pure oxygen. Syngas drives a Siemens combined cycle power unit, including a 156-MW, V94.2 gas turbine and a 128-MW steam turbine. The gasifier, operating at 28-bar pressure and 2700°F is designed to produce syngas containing 42% nitrogen, 25% carbon monoxide, and 12% hydrogen, with a combustion value of 4.3 MJ/kg. The environmental constraints are defined by permit requirements fixing upper limits of SO_2 at 0.22 g/kWh, NO_x at 0.62 g/kWh, and particulates at 0.007 g/kWh.

Key steps for limiting emissions include:

- Removing fly ash with cyclone and candle filters after gas cooling
- Removing halogens and other soluble pollutants with water scrubbing
- Desulfurizing gas by catalytic and chemical processes; sulfur is fixed in sulfinol-*M* solvent, which is further treated to produce elemental sulfur
- Desulfurized gas is mixed with nitrogen from the air separation units and saturated with water vapor to reduce its lower heating value from about 11,000 to 4300 kJ/kg, thus greatly reducing NO_x production.

13.9 Combustion Turbine Plants

Combustion turbine (CT)-based plants comprise the fastest growing technology in power generation. Almost all of these CT and CC plants will be gas fired, leading to a major expansion of gas for electricity generation. It is likely that combustion turbines and combined cycles will grow steadily more important in all generation regimes—peaking, mid-range and base load. The present 2300°F firing temperature machines operate reliably and durably and CT and CC plants are beginning to replace older steam plants.

TABLE 13.8 Modern Gas Turbine Specifications

Turbine		Large Heavy Frame Machines			
		Simple Cycle		Combined Cycle	
		MW	Efficiency % (LVH)	MW	Plant Efficiency % (LHV)
		Current			
GE	GE7FA/9FA	172/256	36.0	262/367	56.0/55.3
MHI	M501F/701F	185/270	38.1	280/—	56.7/—
Siemens/W	W501F	187	38.1	273	55.5
Siemens/W	V94.2A	190	36.2	293	55.2
		New			
Alstom	GT24/26	179/262	—	260/378	56.5/57.0
Siemens/W	V94.3A	265	38.0	385	57.1
MHI	M501G/701G	254/334	38.5	500/—	58.0/—
GE	GE7G/9G	240/282	39.5	350/420	58.0
GE	GE7H/9H	—	—	400/480	60.0/60.0

Source: *Gas Turbine World 2000–2001* Handbook, Vol. 21, Pequot Publishing Inc.

Combustion turbine plants will be a competitive choice for new fossil generation, and advanced CT cycles, with intercooling, reheat, and possibly chemical recuperation and humidification; they will spearhead the drive to higher efficiencies and lower capital costs. Gasification, which guarantees a secure bridge to coal in the near term, will come into its own as natural gas prices rise under demand pressure.

Modern gas turbines for power generation are mostly heavy-frame machines, with 60 Hz ratings in a simple cycle configuration around 170–190 MW for the high firing temperatures (~2300°F) of the "**F-class" machines.** Efficiencies (lower heating value) are 36%–38% in simple cycles. In combined cycles, the units are 260–380 MW in size and 53%–56% efficient. The next generation of CTs, with efficiencies from 57 to 60% is now emerging (Table 13.8). Smaller scale aeroderivative machines have benefited from turbofan engines designed for wide-body aircraft and today are available in ratings of 35–65 MW and with efficiencies of 40% or more for turbine inlet temperatures around 2250°F.

13.9.1 Advanced Combustion Turbines

Under the Department of Energy Advanced Turbine System (ATS) program (DOE 2001), development work was carried out with two manufacturers to enhance the efficiency and environmental performance of utility-scale gas turbines. The goals were to achieve 60% efficiency or more in a combined-cycle mode; NO_x emission levels less than 9 ppm; and a 10% reduction in the cost of electricity. To achieve the efficiency objective required significantly higher turbine inlet temperatures. These higher temperatures required advancements in materials, cooling systems, and low-NO_x combustion techniques.

The focus of General Electric work for DOE was the "H" series gas turbine. To accommodate turbine inlet temperatures of 2800°F, General Electric is employing closed loop steam cooling for the first and second stages to reduce the differential between combustion (2800°F) and firing (2600°F) temperatures; the company is also developing new single-crystal (nickel superalloy) turbine blades with better thermal fatigue and creep characteristics. Thermal barrier coatings protect the metal substrate from the combustion gases using a ceramic top coat for thermal resistance and a metal bond coat for oxidation resistance. An MS9001H unit is being deployed in the U.K., and an MS7001H in Scriba, New York, in 2004.

Siemens–Westinghouse used its 501G turbine as a test bed for the ATS design. Computer modeling has allowed design refinements contributing to capital cost reduction and efficiency enhancement. These include a piloted ring combustor, which uses a lean, premixed multistage design to produce ultralow pollutant emissions and stable turbine operation. Siemens–Westinghouse has also developed brush and

TABLE 13.9 Comparison of "F" and "H" Class Machines

	GE Advanced Machines			
	7FA	7H	9FA	9H
		Characteristics		
Firing temperature F©	2350 (1300)	2600/1430	2350 (1300)	2600 (1430)
Air flow, lb/s (kg/s)	974 (442)	1230/558	1327 (602)	1510 (685)
Pressure ratio	15	23	15	23
Specific work, MW/lb/s (MW/kg/s)	0.26 (0.57)	0.33 (0.72)	0.26 (58)	0.32 (70)
		Performance		
Simple cycle output, MW	168	—	227	—
Simple cycle efficiency, %	36	—	36	—
Combined cycle net output, MW	253	400	349	480
Combined cycle net efficiency, %	55	60	55	60
NO_x (ppmvd at 15% O2)	9	9	25	25

Source: GE Power Systems, Power system for the 21st century: H gas turbine combined cycle, 1995.

abradable coating seals to reduce internal leakage and new thermal barrier coatings for turbine blades to permit higher temperatures. These ATS developments will be incorporated into the commercial 501G turbine at the Lakeland Electric McIntosh station in Florida and demonstrated in 2005. A comparison of the "F" and "H" class machines for GE is shown in Table 13.9.

13.9.2 Humidified Air Power Plants (Cohn 1994)

A new class of combustion turbines has been designed based on humidifying the combustion air. In these combustion turbine cycles, the compressor exit air is highly humidified prior to combustion. This reduces the mass of dry air needed and the energy required to compress it, thus raising plant efficiency.

The continuous plant cycle for this concept is termed the **humid air turbine** (HAT). This cycle has been calculated to have a heat rate for natural gas about 5% better than current high-technology combined cycles. The HAT cycle is adaptable to coal gasification—leading to the low-emissions and high-efficiency characteristics of gasification combined cycle plants—but at a low capital cost because the steam turbine bottoming cycle is eliminated. A simple humidified air turbine cycle is shown in Figure 13.13. The addition of moisture means that perhaps 25% more mass flow goes through the

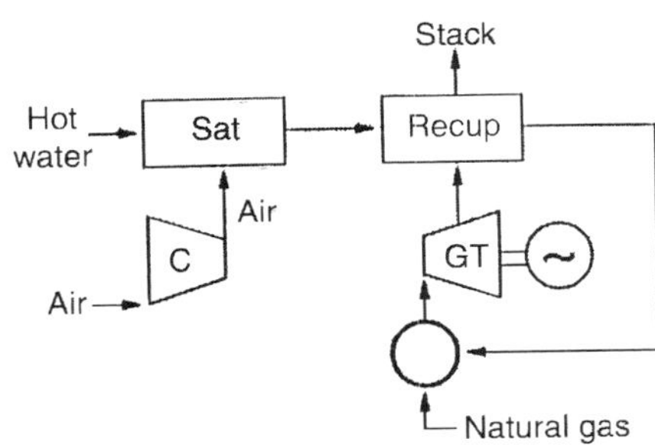

FIGURE 13.13 The humid air turbine (HAT) cycle adds moisture to the compressor exit air, reducing the air mass flow needed and increasing cycle efficiency.

turbine than through the compressor. This suggests the use of separate spools for the turbine and compressor. Using today's 2350°F firing temperatures, it is reasonable to expect a HAT heat rate of about 6100 Btu/kWh from this cycle.

The ideal natural-gas-fired HAT plant has been calculated to have higher efficiency (about 2 points higher) than a combined cycle for the same turbine cooling technology. Thus, it would provide the lowest heat rate for a natural-gas-fired thermal plant and would be utilized in baseload or long intermediate dispatch. The capital cost of this power plant has been calculated to be only slightly higher than that of a combined cycle. However, the anticipated development cost for the ideal turbo machinery has been estimated to be very high, in excess of $250 million.

In contrast, the CHAT (cascaded humid air turbine) plant utilizes turbine components, which are now available, with few exceptions, in a cascade arrangement that allows them to match together. The development cost of the CHAT equipment is currently estimated to be only in the $5–$10 million range, making its development much more practical.

The HAT and CHAT cycles can be integrated with gasification. Because these cycles directly incorporate humidification, they can make direct use of hot water generated in the gasification plant, but cannot readily utilize steam. Thus, they match well with the lower capital cost, but lower-efficiency, quench types of gasifier. This provides an overall power plant with efficiency about the same as an IGCC. Moreover, the capital cost of the IGHAT plant has been calculated to be about $150/kW less than an IGCC plant. These humidification cycles have yet to be offered commercially. The main obstacle is the need to demonstrate low-emission, high-efficiency, full-scale combustors utilizing very humid air experimentally.

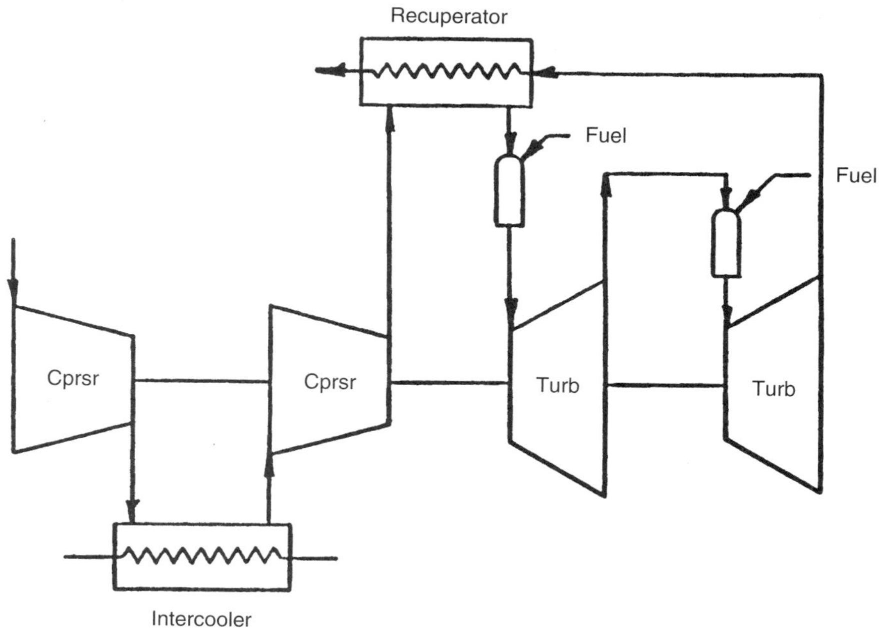

FIGURE 13.14 Improvement in combustion turbine performance is illustrated in this schematic, which combines an intercooler for the compressor with a recuperator using combustion turbine exhaust heat, and a reheat cycle for the turbine to improve efficiency.

13.9.3 Other Combustion Turbine Cycle Enhancements (Lukas 1986)

Several variants of the combustion turbine-based Brayton cycle increase plant efficiency and capacity. **Regenerative cycles** use storage type heat exchangers, where porous or honeycomb wall structures store energy from the hot gases. This is released later to the cold gases. A **recuperative cycle** uses a heat exchanger where the hot and cold streams are separated by walls through which heat transfer occurs. This is the approach commonly used in combustion turbines, allowing gains in efficiency and reduced fuel consumption, but no specific output increase.

Intercooling between compressor stages increases useful output by about 30% for a given air mass flow, by reducing the volume flow and increasing available energy to the power turbine. It has minimal effect on efficiency because heat removed must be added back in the combustion chamber, but it is commonly used in conjunction with recuperation. In a reheat cycle the fuel is introduced at two locations, increasing the total energy available to produce work. A combination of intercooling, reheat, and recuperation is shown in Figure 13.14.

Steam injection, in which the steam is injected directly into the combustion chamber, increases the mass flow through the turbine and results in increased output power. Steam-injected gas turbine (SIGT) cycles have been compared from the viewpoints of efficiency, power generation, capital and operating costs, and environmental impacts with combined cycle systems (Esposito 1989). Above 50-MW size, it was found that combined cycle plants were more economical and achieved significantly better heat rates, although cooling tower fog, visible plumes, and drift deposition favored SIGT plants for a flat site.

13.10 Central Station Options for New Generation

Coal and gas fuels are expected to continue to dominate U.S. central stations in the next decade, with gas-fired combined cycles supplanting several older fossil steam stations. This process is already underway, and by the end of 1999, 118 gas-fired combined cycles were in planning, totaling 56 GW of new power.

The U.S. central station generation options for fossil fuels may be described as follows:

- *Coal, oil, and gas-fired plants of conventional design with typical plant efficiencies of 35% (coal and oil) and 33% (gas); mostly Rankine cycles.* These represent the majority of plants currently in operation. On average, they are 30 years old, many (more than 70,000 MW) equipped with SO_2 scrubbers; most are facing NO_x control additions and perhaps other environmental upgrades. Yet they provide for the bulk of electricity needs; are extremely reliable; and are increasingly in demand as evidenced by an average capacity factor at an all-time high of nearly 70%.
- *Repowered plants, based on gas-firing and combined cycle operation, with efficiencies up to 45%.* Many of the gas-fired steam plants are now targeted for repowering, i.e., combustion turbines will be added to provide exhaust heat for producing steam for the existing steam turbines. This combination of gas and steam turbine cycles adds megawatts, reduces emissions, and improves efficiencies 5% or more.
- *New combined cycles based on gas-firing (about 45% efficiency with today's gas turbines and 50% with advanced gas turbines), and on coal-firing with gasification, utilizing advanced gas and steam turbine technology (50% efficiency).* Gas-fired combined cycles are currently the new plants of choice. Although relatively few are in operation today, more than 50 GW are planned. The massive deployment of these plants in the future raises questions of gas and gas pipeline availability, gas prices, and a potential retreat from coal that could have serious future energy consequences. Coal plants, whether pulverized or gasified, also lend themselves to combined cycles, although none are planned domestically outside the DOE Clean Coal demonstrations.
- *Coal-fired Rankine cycles with advanced steam conditions and up to 50% plant efficiency.* Advancing steam temperatures and pressures in pulverized coal steam plants greatly improve overall efficiency. Such ultrasupercritical cycles are already in operation outside the U.S. Advancing

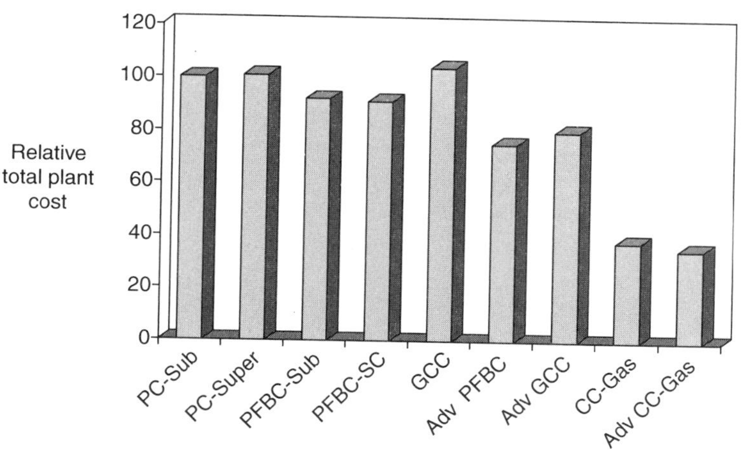

FIGURE 13.15 Capital costs for coal and gas-fired central stations. These costs assume mature technologies and fully functional equipment manufacturing lines for the newer technologies.

steam temperatures to 750°C from current levels of about 590°C permits plants to rival the best gas combined cycles. When used in coal combined cycles and with temperatures increased to 850°C or beyond, a coal plant approaching 60% efficiency is attained. Significant challenges still exist in materials technology.

- *Integrated coal gasification fuel cells, perhaps combined with gas turbines with efficiencies of 60% or more.* The fuel cell is an exciting advance that will change the energy picture in the long term. Shorter term, and in small sizes, great advances are being made in mobile as well as stationary applications. If the fuel cell can be used as a combuster for a gas turbine, efficiencies can be raised above 60%. Clearly, this is a power source of great promise for the second half of this century.

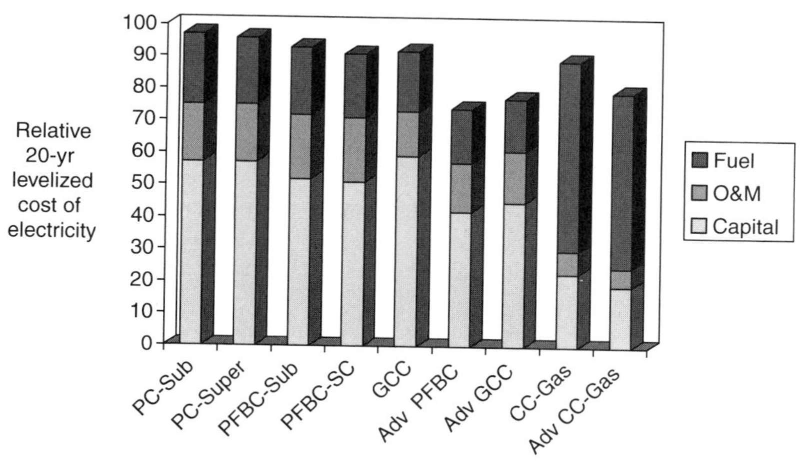

FIGURE 13.16 Levelized cost of electricity comparisons based on mean fuel prices for the U.S. and a 2010 start-up date. Coal is assumed at \$1.06/MMBtu and gas at \$3.08/MMBtu, a price differential about 40% greater than in 2000.

13.10.1 Overall Plant Performance and Cost Estimates

As shown in Figure 13.15, mature versions of IGCC plants are expected to have capital costs slightly higher than the capital costs for PC plants, while the capital costs for PFBC plants are expected to be slightly lower than for PC plants. More advanced coal plants, when they become commercially available, are expected to have capital costs that are 20%–25% lower than today's PC plants (Booras 1999). These calculations include all engineering and construction costs. They assume a mature technology in which the plant achieves rated performance as designed (Figure 13.16).

13.11 Summary

The preceding section has described how the future for electric power generation will increasingly be dominated by environmental control needs, putting an emphasis on the base efficiency of new generation, and on heat rate recovery for existing units. The pulverized coal-fired power plant with flue gas desulfurization will remain a focus of most near-term activity related to upgrades and retrofits. However, new technology, based on coal gasification, is under development and being tested in a growing number of demonstration plants that promise extremely low emissions.

The future for many nations will be based on exploiting the opportunities offered by clean and efficient use of coal. This implies access to the range of new technologies now being tested at large scale in the U.S. and other developed nations. This strategy is timely and prudent on a global basis as the world increasingly voices concerns related to carbon combustion.

New, base-load, central generation plants will largely be focused in the immediate future on the rapidly developing areas of the world: Asia (particularly China and India) and Latin America. In these areas, the fuel of choice will likely be coal, particularly in Asia, and the generating unit most often will be a conventional pulverized coal unit; such units will increasingly deploy supercritical steam at temperatures of 1100°F and above. In North America, Europe, and Japan, gas-fired central plants using combustion turbines, often in a combined cycle, will continue to be built through 2010 due to the short construction time and lower carbon dioxide emissions (compared with conventional PC plants).

As the cost of natural gas, relative to coal, increases, this will then encourage the installation of coal gasification units enabling the enormous world coal reserves to be utilized fully. Then, as the 21st century progresses, smaller distributed generating sources will begin to emerge, based on gas-fired fuel cells, small combustion turbines, or possibly photovoltaics. As the economics for the distributed option become favorable, these smaller generating units will encourage broad electrification of the developing countries of the world, avoiding the immediate need for large high-voltage transmission systems. Such distributed generation plants will also be added at strategic locations in the developed world as ways are sought to relieve the growing congestion on transmission networks.

Defining Terms

Lower heating value: Fuels containing hydrogen produce water vapor as a product of combustion. Fuel heating value is said to be "lower" if the combustion process leaves all products in the gaseous state, but "higher" if the fuel heating value includes the latent heat of vaporization. Practice in the U.S. is to use the higher value.

Combined cycle: Power stations that employ combustion turbines (Brayton cycle) and condensing steam turbines (Rankine cycle) where the waste heat from the CTs generate steam for the STs are called "combined" cycle plants. Overall plant efficiency improves.

Coal gasification: Coal can be converted into a mixture of carbon monoxide and hydrogen by burning it with a controlled deficiency of oxygen. Use of pure oxygen produces a medium calorific value gas, and air a low calorific value gas. This "syngas" can then be used to power a combustion turbine.

Flue gas desulfurization: Removal of sulfur dioxide, SO_2, from combustion gases is accomplished in a number of flue gas desulfurization methods. Most of these involve wet "scrubbing" of the gas using lime or limestone, and result in a calcium sulfate waste product. A 95% removal efficiency, or higher, is possible.

Slagging and fouling: The mineral matter in coal can attach itself following combustion to the boiler walls and heat exchanger surfaces. Oxides of silicon, aluminum, iron, calcium, and magnesium can foul all boiler surfaces, requiring soot blowers for cleaning. Hot ash can melt, becoming sticky and sometimes coalescing in the furnace to cause slagging problems.

On–off cycling capability: Generating units are often not required on a 24-h basis. Some are shut down during low demand times and started up perhaps hours later. This form of "on–off cycling" imposes thermal stresses on the equipment, leading to premature equipment failure unless special measures are taken to deal with this.

Temper embrittlement: Tempering of steel in the manufacturing process removes some of the brittleness and is carried out by a heating and cooling process. During operation, though, it is possible that ductility can worsen close to specific "tempering temperatures." The material is then said to be temper embrittled and premature cracking may follow.

Coextruded tubing: Tubing for superheaters and reheaters must be strong enough to withstand the pressures and temperatures expected, and also corrosion resistant to depositions of fly ash. By making tubing with a strong inner layer and corrosion-resistant outer layer through an extrusion process, both concerns can be dealt with.

Spray drying: Spray dryers, for desulfurization, used typically when burning lower sulfur coals, use a spray of quicklime, which is dried by the hot flue gas and results in a dry solid product. A 90% removal efficiency is typical.

Electrostatic precipitators: Flue gas particles, when electrically charged in an ionized gas flow, collect on electrodes in the presence of a strong electrostatic field. Collected dust is discharged by rapping into hoppers. A collection efficiency above 99% is possible.

Fluidized bed: A process of burning solid fuels, particularly coal, by combustion suspending the fuel within a column of air supplied from below the furnace. This method permits effective combustion of poor-quality fuels; lowers NO_x emissions due to low combustion temperatures; and captures sulfur in the bed by mixing limestone or dolomite in with the fuel.

Petroleum coke: Petroleum coke is a residual product of the oil refining process, and in its fuel grade form is an almost pure carbon by-product. About 19 million tons of fuel grade pet coke is produced each year in the U.S. It is inexpensive, although it may have high sulfur and vanadium content.

Double reheat: Modern designs of fossil steam-generating units remove a portion of the steam before full expansion through the turbine and reheat it in the boiler before returning it to the turbine. This enhances the thermal efficiency of the cycle by up to 5%. For supercritical cycles, two stages of reheat can be justified—double reheat.

Cogeneration: Cogeneration refers to the production of multiple products from a power plant. Typically, process steam, or hot water for heating, are produced in addition to electricity. This approach leads to high plant utilization, the "effective" heat rate being 70% or more.

Ash-softening temperature: The tendency for fly ash to adhere to tube banks is increased as the ash softens and melts. The point at which the ash begins to soften is dependent on the type of coal and is difficult to predict, depending on the many coal constituents. Slagging and fouling of tubes can lead to severe tube corrosion.

Fuel cell: Fuel cells convert gaseous or liquid fuels directly to electricity without any combustion process. Like a continuous battery, the fuel cell has electrodes in a electrolyte medium. Typically, hydrogen and air are supplied and DC electricity, water, and carbon dioxide are produced. They are currently high-cost, low-size devices, but with minimum environmental emissions.

Hot gas cleanup: Cycles that use gas from the combustion of coal, typically pressurized fluidized bed or gasification cycles, need to clean up the ash particles before passing them through a gas turbine. This prevents severe erosion of the turbine blades and other components. Hot gas cleanup can involve the application of hanging particulate traps, using ceramic filters.

F-class machines: Recent designs of combustion turbines have increased efficiencies resulting from increased firing temperatures. The first generation of these machines has firing temperatures of about 2300°F. They have been termed "F class" machines (for example the GE 7F). Even higher temperatures have now been incorporated into "G-class" turbines. Aeroderivative turbine: in the 1960s, gas turbines derived from military jet engines formed a source of utility peaking capacity. Now modern airline fan jets are being converted to utility service. These lighter combustion turbines are highly efficient and can have low NO_x emissions, high pressure ratios, and low capital cost.

Humid air turbine: A new type of combustion turbine uses humidified compressor exit air for the combustor. The mass of dry air needed is thus lessened for a given mass flow, and turbine efficiency increases. Several applications of this "HAT" appear attractive in gasification and compressed air storage cycles.

Regenerative cycles: Combustion turbine cycles using heat exchangers to store and transfer heat from hot gases to cold gases are termed regenerative cycles.

Recuperative cycle: Recuperative cycles for combustion turbines use walls between the hot and cold streams through which heat is transferred. This improves efficiency and reduces fuel consumption.

Intercooling: Increased output from a combustion turbine can be obtained by cooling the air between compressor stages. This reduces volume flow and increases energy to the power turbine.

Steam injection: Injecting steam directly into the combustion chamber of a combustion turbine increases turbine mass flow and thus increases the output power.

References

Armor, A. F., Bakker, W. T., Holt, N. H., and Viswanathan, R. 1999. Supercritical fossil-fired power plants: designs and materials for the new millennium, Proc. PowerGen International Conf., New Orleans, LA, November 1.

Armor, A. F. and Wolk, R. H. 2002. *Productivity Improvement Handbook for Fossil Steam Plants.* 3rd Ed., EPRI Report 1006315, October.

Armor, A. F., Viswanathan, R., and Dalton, S. M. 2003. Ultrasupercritical steam turbines: design and materials issues for the next generation, coal utilization and fuel systems, DOE Annu. Conf., Clearwater, FL, March 10–13.

Blunden, W. E. 1989. Colorado-UTE's Nucla circulating AFBC demonstration project, EPRI Report CS-5831, February.

Booras, G. 1999. Overview of the economics of clean coal technologies as compared with alternatives for power generation, Proc. DOE Clean Coal Technol. Conf., Knoxville, TN, May.

Carpenter, L. K. and Dellefield, R. J. 1994. The U.S. Department of Energy PFBC perspective, EPRI fluidized bed combustion for power generation conference, Atlanta, May 17–19.

Cohn, A. 1994. Humidified power plant options, AFPS developments, Spring, Electric Power Research Institute.

Couch, G. 1991. Advanced coal cleaning technology, IEACR/44, London, IEA Coal Research, December.

Courtright, H. A., Armor, A. F., Holt, N. H., and Dalton, S. M. 2003. Clean coal technologies and their commercial development, POWER-GEN International, Conf. Proc., Las Vegas, NV, December 9–11.

Dalton, S. M., Viswanathan, R., Gehl, S. M., Armor, A. F., and Purgert, R. 2001. Ultrasupercritical materials, DOE Clean Coal Conf., Washington, D.C., November 20.

DOE Fossil energy—tomorrow's turbines, April 30, 2001, http://fossil.energy.gov/coal_power/turbines/index.shtml

Douglas, J. 1986. IGCC: phased construction for flexible growth. *EPRI J.*, September.

EPRI. 1982. Engineering assessment of an advanced pulverized-coal power plant, EPRI Report CS-2555, August.

EPRI. 1985. Development plan for advanced fossil fuel power plants, EPRI Report CS-4029, May.

EPRI. 1986. Assessment of supercritical power plant performance, EPRI Report CS-4968, December.

Esposito, N. T. 1989. A comparison of steam-injected gas turbine and combined cycle power plants, EPRI Report GS-6415, June.

EPRI. 2000. An assessment of mercury emissions from U.S. coal-fired power plants, EPRI Report 1000608, November.

EPRI. 2002. Atmospheric fluidized-bed combustion handbook, EPRI Report 1004493, December.

GE Power Systems. 1995. Power system for the 21st century: H gas turbine combined cycle.

Gas Turbine World 2000–2001 Handbook, Vol. 21, Pequot Publishing Inc.

Hinrichsen, D. 1989. AFBC conversion at Northern States Power Company, EPRI Report CS-5501, April.

Lucas, H. 1986. Survey of alternative gas turbine engineer and cycle design, EPRI Report AP-4450, February.

Manaker, A. M. 1992. TVA 160-MW atmospheric fluidized-bed combustion demonstration project, EPRI Report TR-100544, December.

Melvin, R. H. and Friedman, M. A. 1994. Successful coal-fired AFBC cogeneration in California: 108 MW ACE cogeneration facility, EPRI Fluidized Bed Combustion Conf., Atlanta, May 17–19.

Olesen, C. 1985. Pressurized fluidized bed combustion for power generation, in EPRI CS-4028, *Proc. Pressurized Fluidized-Bed Combustion Power Plants*, May.

Oliker, I. and Armor, A. F. 1992. Supercritical power plants in the USSR, EPRI Report TR-100364, February.

Skowyra et al. 1995. Design of a supercritical sliding pressure circulating fluidized bed boiler with vertical waterwalls, *Proc. 13th Int. Conf. Fluidized Bed Combustion*, ASME.

U.S. Department of Energy. 2002. Clean Coal Technology Demonstration Program, DOE/FE-0444, July.

For Further Information

Annual energy outlook. 2003. Energy Information Administration: www.eia.doe.gov
National Engineering Technology Laboratory: www.netl.doe.gov
EPRI: www.epri.com
Steam, Its Generation and Use, Babcock and Wilcox, New York.
Combustion: Fossil Power Systems, J. G. Singer, ed., Combustion Engineering, Inc. Windsor, CT.
Tapping global expertise in coal technology, EPRI J., Jan./Feb., 1986.
IGCC: new fuels, new players, EPRI J., Jul./Aug., 1994.
A brighter future for PFBC, EPRI J., Dec., 1993.
Fuel cells for urban power, EPRI J., Sept., 1991.
Distributed generation, EPRI J., Apr./May, 1993.
Plant repowering, EPRI J., Sep./Oct., 1995.
Smart materials, EPRI J., Jul./Aug., 1998.
Merchant plants, EPRI J., Summer, 1999.
Energy and air emissions, EPRI J., Summer, 2000.
Global coal initiative, EPRI J., Summer, 2001.

14

Combined-Cycle Power Plants

14.1 Combined-Cycle Concepts 14-1
14.2 Combined-Cycle Thermodynamics 14-2
14.3 Combined-Cycle Arrangements 14-4
14.4 Combined Heat and Power from Combined-Cycle Plants 14-7
14.5 Environmental Aspects 14-8
For Further Information 14-8

Alex Lezuo
Siemans Power Generation

In the decades since the early 1970s, power plant concepts featuring a combination of gas and steam turbines have been successfully commercialized. These combined-cycle power plants make very efficient use of fuel compared with other power plants. The first combined-cycle power plants in the 1970s achieved net efficiencies of about 40%, while the most recent ones attain net plant efficiencies of more than 58%.

14.1 Combined-Cycle Concepts

The considerable amount of energy available in the exhaust of a gas turbine also can be used in a secondary system, increasing overall efficiency considerably. Figure 14.1 shows three of the most common types of combined-cycle (CC) concepts.

The *natural-gas-fired CC* power plant, which offers the highest efficiency potential, is the most commonly used system today. However, to attain these high efficiencies, only fuels such as natural gas can be burned in the gas turbine. The use of other fuels results in a higher heat rate. The arrangement is relatively simple and therefore has the lowest specific investment costs among all power generation systems.

The *parallel-powered CC* power plant is mainly used for repowering existing coal-fired power plants. This is an efficient way of reducing the heat rate of older coal-fired power plants with a relatively small investment and short implementation schedules. In addition, this concept provides excellent part-load behavior, thus making it suitable for cycling duty. Instead of generating steam with the combustion turbine exhaust gas, it is also possible simply to heat the feedwater of a coal-fired power plant and thus reduce the amount of extraction steam required from the steam turbine for the feedwater heaters. This can increase the output and efficiency of the coal plant.

When no natural gas is available, the CC can also be used in connection with a coal gasification plant. Integrated systems (*integrated gasification combined cycle, IGCC*) increase system complexity and most likely decrease overall plant availability, but permit the use of lower cost fuels such as coal or petcoke in an

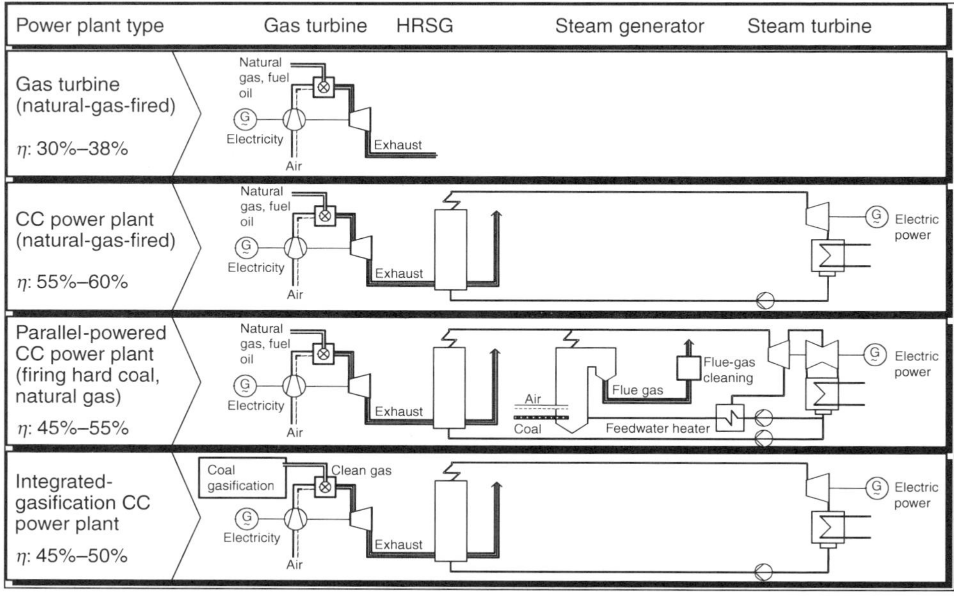

FIGURE 14.1 Combined-cycle power plant concepts.

environmentally acceptable manner. Several industrial-scale demonstration plants have proven the technical feasibility, but wider application is dependent on the price spread between coal and natural gas.

Other CC developments have been proposed but the natural-gas-fired CC has proven to be the most efficient and most economical under present technical and economical boundary conditions. Therefore, this CC will be described in greater detail.

14.2 Combined-Cycle Thermodynamics

The exhaust gas temperature of today's gas turbines varies between 500°C (932°F) for small and older gas turbines and 600°C (1112°F) for advanced GTs, and the specific exhaust gas stream amounts to 2–3 kg/s (4.4–6.6 lb/s) per megawatt, where the smaller figure relates to large, advanced gas turbines. By using the heat of this exhaust gas in a water–steam cycle, total cycle efficiency can be raised considerably because a very high total temperature difference can be utilized in the combined cycle (Figure 14.2) compared to a simple-cycle gas turbine.

With the improvement of gas turbines and the accompanying increase in exhaust temperature, main steam pressure and temperature of the steam cycle have been raised to as high as 165 bar (2400 psi) and 565°C (1050°F). In addition, multipressure heat-recovery steam generators (HRSGs) with reheat are used, which represents another improvement in utilization of the exhaust heat. In this way, the losses can be reduced as shown in Figure 14.3, which compares a single-pressure HRSG with a triple-pressure system. The ideal temperature/heat transfer diagram would be one in which the temperature difference in the HRSG between the steam and the exhaust gases is constant.

The exhaust gases leaving the gas turbine enter the HRSG at a temperature between approximately 550 and 600°C (1022 and 1112°F) and leave the HRSG typically between 80 and 100°C (176 and 212°F). The water in the steam cycle portion of the combined cycle enters the HRSG economizer as a subcooled liquid. The temperature of the water is increased in the economizer until the liquid becomes saturated. At this point, the minimum temperature difference between the water in the steam cycle and the

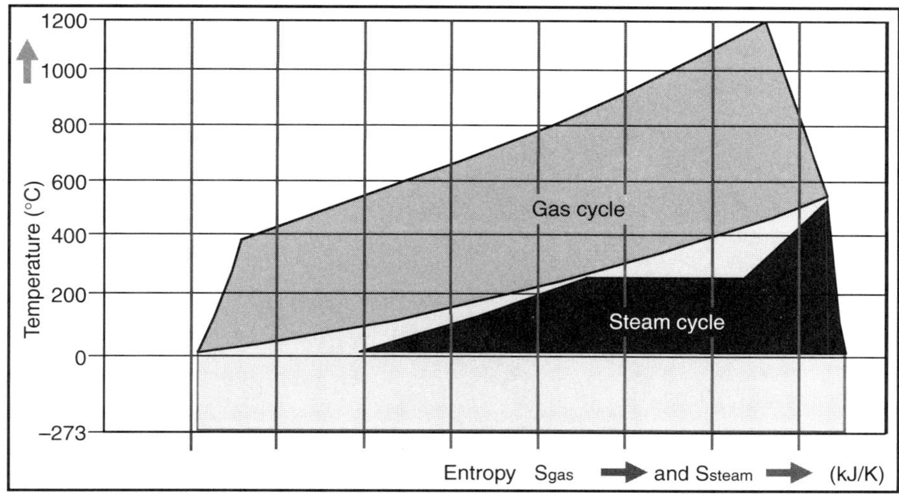

FIGURE 14.2 *T–S* diagram of a combined-cycle power plant with single-pressure steam cycle.

exhaust gases occurs and is called the "pinch point." Typical pinch-point values range from 8 to 30°C (46°F–86°F); the smaller the pinch point difference is, the larger the required heat-transfer surface area. After evaporation at constant temperature, the steam is superheated to the final temperature in the superheater section.

Figure 14.3 shows the improvement in heat transfer from the single-pressure cycle to the triple-pressure cycle. High-performance plant designs are today equipped with a triple-pressure HRSG with one reheat stage.

Table 14.1 lists net plant output and performance values for a selection of CCs from the four main suppliers worldwide. This is just a selection of available gas turbines and possible configurations because manufacturers are developing and improving their engines on a regular basis. Performance data are based on standard designs and configurations at ISO conditions of 15°C (59°F), which provides a normalized basis for comparison.

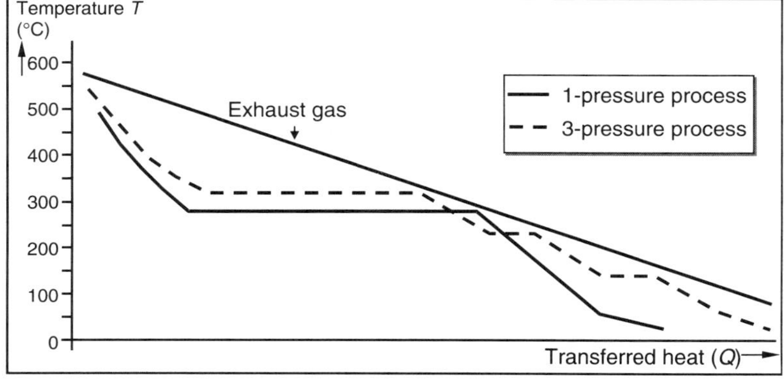

FIGURE 14.3 *T–Q* diagram for heat-recovery steam generators.

TABLE 14.1 Net Plant Output and Performance Data for Selected Combined Cycles

Supplier Model	Frequency (Hz)	Number of GTs	Net Plant Output (MW)	Heat Rate, LHV (Btu/kWh)	Net Plant Efficiency (%)	Configuration
Alstom Power (AP)						
KA8C-2	50	2	80.0	6965	49.0	2-P
KA 13E2-2	50	2	480.0	6450	52.9	2-P
KA 24-1	60	1	250.8	6129	55.7	2-P/RH
KA 26-1	50	1	392.5	6061	56.3	3-P/RH
GE Power Systems						
S106B	50/60	1	64.3	7020	49.0	3-P
S206FA	50/60	2	218.7	6930	54.1	3-P/RH
S107EA	60	1	130.2	6380	50.2	3-P
S209EC	50	2	522.6	6415	54.4	3-P/RH
S107FA	60	1	262.6	6170	56.0	3-P/RH
Mitsubishi Heavy Industries (MHI)						
M501F	60	1	279.0	6074	56.2	3-P/RH
M701F	50	1	399.0	5994	56.9	3-P/RH
M501G	60	1	371.0	5879	58.0	3-P/RH
M701G	50	1	484.0	5879	58.0	3-P/RH
Siemens PG//Siemens Westinghouse						
CC 1S.V64.3A	60	1	99.8	6541	52.2	2-P
CC 1.V94.2	50	1	239.4	6533	52.2	2-P
CC 1S.V943A	50	1	392.2	5946	57.4	3-P/RH
CC 2.W501F	60	2	568.5	6060	56.3	3-P/RH
CC 1S.W501G	60	1	365.0	5880	58.0	3-P/RH

Note: 2-P, double-pressure HRSG; 3-P, triple-pressure HRSG; RH, reheat.
Source: From company Web pages, August 2003. (Alstom.com, GEPower.com, MHI.co.jp, powergeneration.siemens.com)

14.3 Combined-Cycle Arrangements

Today, several designs for natural-gas-fired CCs are available. All consist of the following main components (Figure 14.4):

- Gas turbine
- Steam turbine
- Generator
- HRSG
- Stack
- Condenser with heat removal system
- Condensate pumps
- Feedwater pumps
- Auxiliary systems for gas and steam turbine
- Main and auxiliary transformers
- Fuel supply system
- Electrical equipment
- Instrumentation and control systems.

Figure 14.5 shows the configuration of a *single-shaft CC* arrangement with the steam turbine at one end, the common generator in the center, and a synchronous self-shifting (SSS) clutch between the generator and the steam turbine. The gas turbine can be started independently without any restrictions arising from the actual condition (hot, warm, or cold) of the steam turbine and the cooling system. The clutch acts as a coupling to enable unrestricted axial expansion of the steam turbine shaft relative to the generator. This allows optimized axial clearances for the steam turbine blading, resulting in improved efficiency.

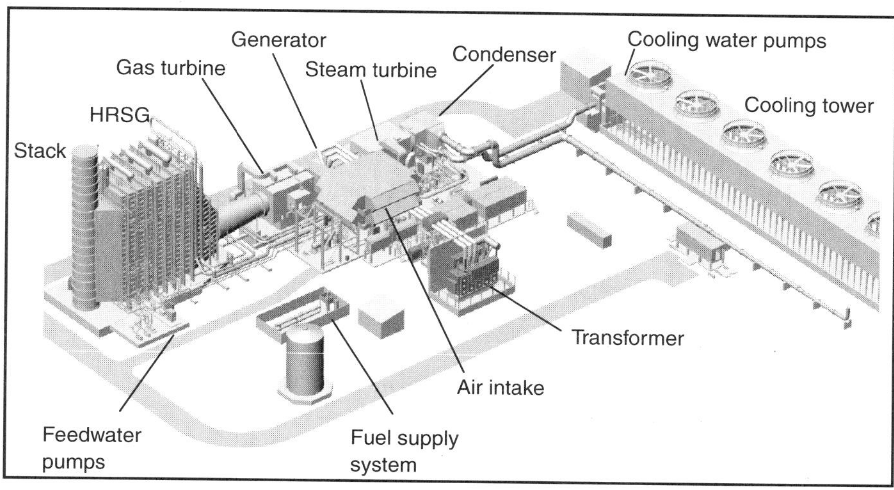

FIGURE 14.4 Arrangement and main components of a single-shaft CC plant.

The clutch engages automatically once the steam turbine approaches the operational rotating speed of the generator (3000 or 3600 rpm). It automatically disconnects the steam turbine whenever the steam turbine slows down relative to the generator. The clutch allows the gas turbine to be started and operated independently of the steam turbine. The gas turbine can be started up relatively quickly and the steam turbine can be accelerated at a suitable rate once the gas turbine is loaded. The steam turbine can be shut down at any time, leaving the gas turbine operating in open-cycle mode. In this mode, the steam produced is dumped to the condenser via the bypass station. The single-shaft configuration requires less space and can be built in efficient and independent units.

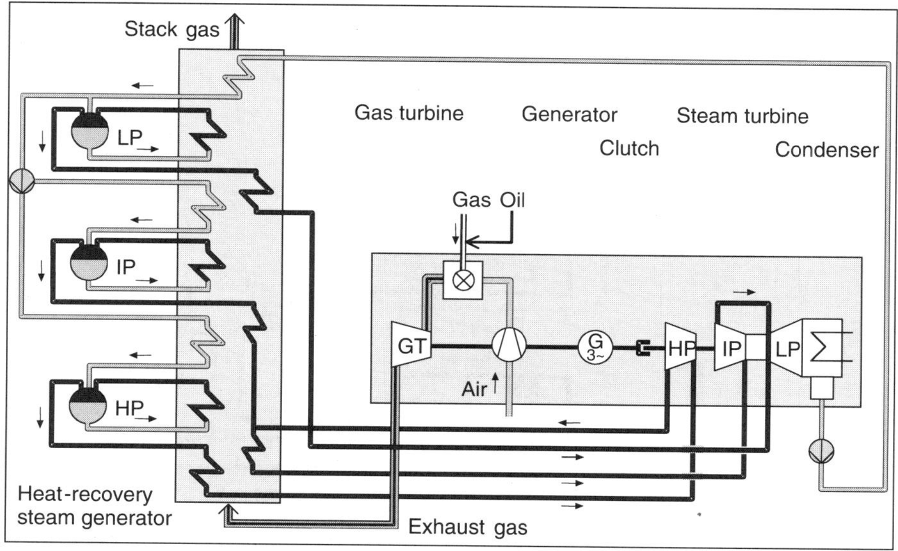

FIGURE 14.5 Single-shaft CC plant schematic with triple-pressure reheat.

In the *multishaft CC* arrangement, the gas turbine and the steam turbine are separated; each engine has its own generator and transformer. They are built in 1 + 1 configuration, which means one gas turbine, one HRSG and one steam turbine as the schematic in Figure 14.6 shows, or in the very common 2 + 1 configuration (two gas turbines, two HRSGs, and one common steam turbine). Sometimes a 3 + 1 arrangement and even a 4 + 1 arrangement are used.

The multishaft arrangement offers the advantage that the plant can be built in what is known as "phased construction." The gas turbine with its very fast construction time can be installed first and operated in a simple cycle (of course, with low efficiency); the plant can be extended later by adding an HRSG and steam turbine along with all the other necessary equipment. In this way, the customer gets not only a gradual increase in power output but also a deferment of the necessary investment. In such an unfired configuration, about two thirds of the total power output is generated in the gas turbine and one third in the steam turbine. This is due to the fixed amount of exhaust energy from the gas turbine.

Natural-gas-fired CC plants have relatively small auxiliary systems and do not require major stack gas clean-up systems. Consequently, the parasitic power of CC plants is low compared to steam power plants—amounting to approximately 1.5% of total rated power output.

One way to increase the plant output is to introduce supplementary firing in the exhaust duct between the gas turbine exit and HRSG inlet. This increases temperature and available heat transferred to the steam cycle (Figure 14.7).

The advantages of adding supplementary or duct firing are:

- The total output from the combined cycle will increase with a higher fraction of the output coming from the steam turbine cycle.
- The temperature at the HRSG inlet can be controlled. This is important because the temperature and mass flow rate at the exit from the gas turbine are highly dependent on ambient temperature.
- The implementation of supplementary firing requires relatively low investment.
- Duct firing can be easily turned on or off, making it an excellent choice for peaking capacity.

The disadvantage is that a drop in overall plant efficiency occurs when the duct firing is employed.

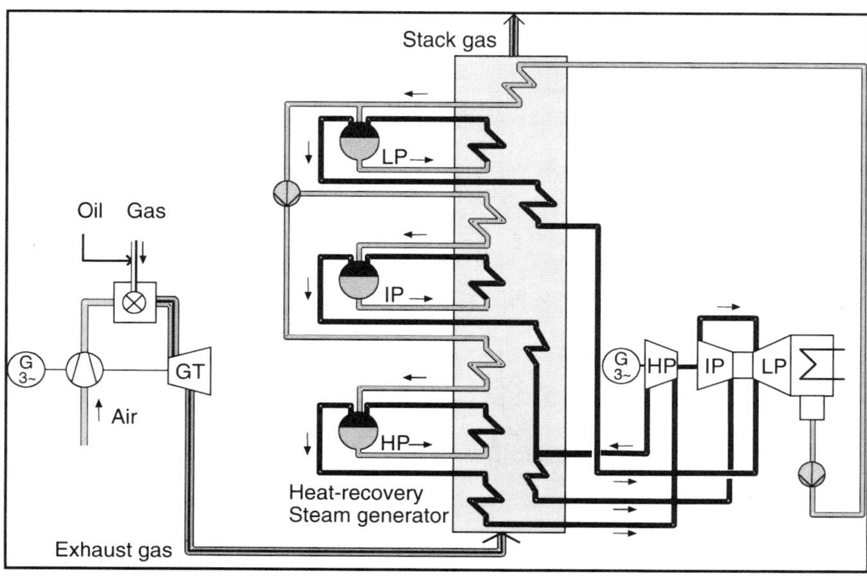

FIGURE 14.6 Multishaft CC plant arrangement with triple-pressure reheat.

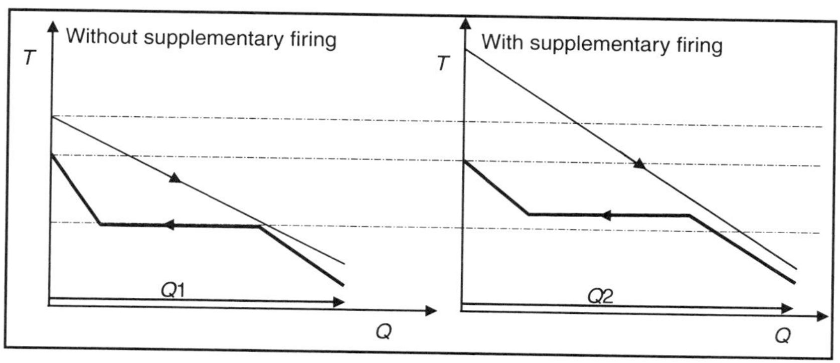

FIGURE 14.7 *T–Q* diagrams for CC without/with supplementary firing.

Duct firing can be employed in varying amounts, from small firing rates to compensate for high ambient temperature reductions in CT output to firing levels that can double the steam cycle output. For this reason, an economic analysis is recommended before applying this option.

14.4 Combined Heat and Power from Combined-Cycle Plants

As in other power plants, not only electricity but also usable heat can be produced in CC plants and thus a fuel utilization factor $((P_{\text{electricity}}+P_{\text{heat}})/Q_{\text{fuel}})$ up to 90% can be achieved. Depending on specific requirements, several different gas turbine power plant configurations can be utilized in cogeneration applications (Figure 14.8). A plant with a backpressure turbine has a relatively inflexible operation profile, while the use of an extraction condensing turbine makes it possible to vary the electric power-to-heat ratio depending on the demand. The required temperature and amount of heated steam are the primary factors for determining the design concept.

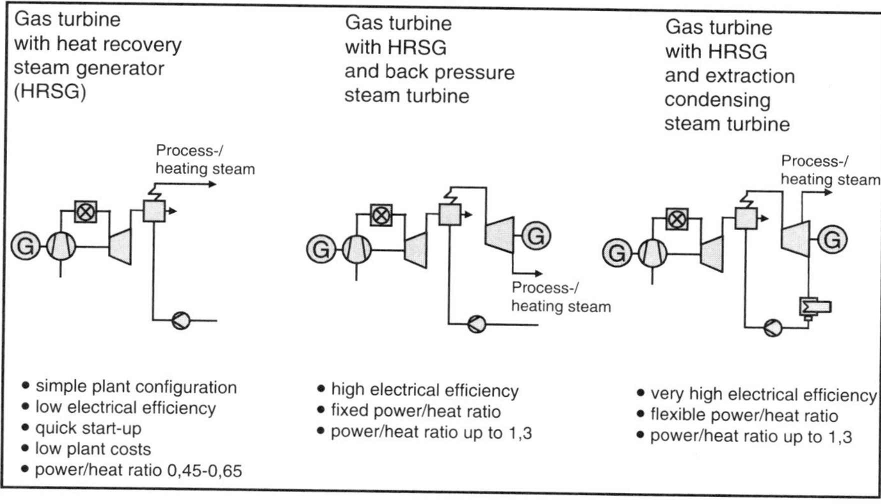

FIGURE 14.8 CHP concepts using gas turbines.

14.5 Environmental Aspects

From an environmental perspective, the natural-gas-fired CC is one of the most environmentally benign fossil-fueled power plant designs. As a result of its high efficiency and the use of clean natural gas, emissions of the greenhouse gas carbon dioxide (CO_2) are very low. Depending on plant design, a value of approximately 320–380 g/kWh (0.70 and 0.83 lb/kWh) can be obtained. The very high efficiency is also the reason for the relatively small amount of waste heat that must be dumped into the atmosphere.

With the development of dry low-NO_x (DLN) burners, the standard NO_x emissions are today <25 ppmvd with a tendency to even lower values. With the help of selective catalytic reduction (SCR) values of <5 ppmvd can be achieved. CO emissions in the exhaust gas are usually <10 ppmvd in the upper load range. CO catalysts can lower this value to <4 ppmvd.

For Further Information

Barclay, F. J. 1998. Combined power and process. An exergy approach. *Prof. Eng.*

Bonzani, G. et al. 1991. Technical and economic optimization of a 450-MW combined cycle plant. In *1991 ASME Cogen-Turbo, 5th International Symposium and Exposition on Gas Turbines in Cogeneration, Repowering and Peak-Load Power Generation*, S. van der Linden et al. ed., pp. 131–143. ASME, New York.

Dechamps, P. J. et al. 1993. Advanced combined cycle alternatives with advanced gas turbines. In *ASME Cogen-Turbo Power93, 7th Congress and Exposition on Gas Turbines in Cogeneration and Utility*, H. W. Holland et al. ed., pp. 387–396. ASME, New York.

Gyarmathy, G. and Ortmann, P. 1991. The off design of single- and dual-pressure steam cycles in CC plants. In *1991 ASME Cogen-Turbo, 5th International Symposium and Exposition on Gas Turbines in Cogeneration, Repowering and Peak-Load Power Generation*, S. van der Linden et al. ed., pp. 271–280. ASME, New York.

Hannemann, F. et al. 2002. V94.2 Buggenum experience and improved concepts for syngas application. Gasification Technology Council (GTC), San Francisco, 2002.

Horlock, J. H. 1992. *Combined Power Plants Including Combined Cycle Gas Turbine (CCGT) Plants.* Pergamon Press, New York.

Huettenhofer, K. and Lezuo, A. *Cogeneration Power Plant Concepts Using Advanced Gas Turbines.* VGB PowerTech 6/2001, pp. 50–56.

Kehlhofer, W. 1991. *Combined-Cycle Gas & Steam Turbine Power Plants.* Fairmont Press, Englewood Cliffs, NJ.

Kehlhofer, R., Bachmann, R., Nielson, H., and Warner, J. 1999. *Combined Cycle Gas and Steam Turbine Power Plants. 2nd Ed.* PennWell Books, Tulsa, OK.

Kiameh, P. and McCombs, K. 2002. *Power Generation Handbook. Selection, Applications, Operation, and Maintenance.* McGraw-Hill, New York.

Maurer, R. 1992. Destec's successes and plans for coal gasification combined cycle (CGCVC) power systems. In *1992 ASME Cogen-Turbo, 6th International Conference in Cogeneration and Utility*, D. H. Cooke et al. ed., pp. 75–85. ASME, New York.

15

Energy Storage Technologies

15.1 Overview of Storage Technologies 15-1
15.2 Principal Forms of Stored Energy 15-3
15.3 Applications of Energy Storage 15-3
15.4 Specifying Energy Storage Devices 15-4
15.5 Specifying Fuels 15-6
15.6 Direct Electric Storage 15-7
Ultracapacitors • Superconducting Magnetic Energy Storage
15.7 Electrochemical Energy Storage 15-8
Secondary Batteries • Lead–Acid • Lithium-Ion • Nickel–Cadmium • Nickel–Metal Hydride • Sodium–Sulfur • Zebra • Flow Batteries • Electrolytic Hydrogen
15.8 Mechanical Energy Storage 15-13
Pumped Hydro • Compressed Air • Flywheels
15.9 Direct Thermal Storage 15-15
Sensible Heat • Latent Heat
15.10 Thermochemical Energy Storage 15-18
Biomass Solids • Ethanol • Biodiesel • Syngas
References 15-21

Roel Hammerschlag
Institute for Lifecycle Environmental Assessment

Christopher P. Schaber
Institute for Lifecycle Environmental Assessment

15.1 Overview of Storage Technologies

Energy storage will play a critical role in an efficient and renewable energy future; much more so than it does in today's fossil-based energy economy. There are two principal reasons that energy storage will grow in importance with increased development of renewable energy:

- Many important renewable energy sources are intermittent, and generate when weather dictates, rather than when energy demand dictates.
- Many transportation systems require energy to be carried with the vehicle.[1]

[1]This is almost always true for private transportation systems, and usually untrue for public transportation systems, which can rely on rails or overhead wires to transmit electric energy. However, some public transportation systems such as buses do not have fixed routes and also require portable energy storage.

Energy can be stored in many forms: as mechanical energy in rotating, compressed, or elevated substances; as thermal or electrical energy waiting to be released from chemical bonds; or as electrical charge ready to travel from positive to negative poles on demand.

Storage media that can take and release energy in the form of electricity have the most universal value, because electricity can efficiently be converted either to mechanical or heat energy, whereas other energy conversion processes are less efficient. Electricity is also the output of three of the most promising renewable energy technologies: wind turbines, solar thermal, and photovoltaics. Storing this electricity in a medium that naturally accepts electricity is favored, because converting the energy to another type usually has a substantial efficiency penalty.

Still, some applications can benefit from mechanical or thermal technologies. Examples are when the application already includes mechanical devices or heat engines that can take advantage of the compatible energy form; lower environmental impacts that are associated with mechanical and thermal technologies; or low cost resulting from simpler technologies or efficiencies of scale.

In this chapter, the technologies are grouped into five categories: direct electric, electrochemical, mechanical, direct thermal, and thermochemical. Table 15.1 is a summary of all of the technologies covered. Each is listed with indicators of appropriate applications that are further explained in Section 15.1.3.

TABLE 15.1 Overview of Energy Storage Technologies and Their Applications

	Utility Shaping	Power Quality	Distributed Grid	Automotive
Direct electric				
Ultracapacitors		✓		✓
SMES		✓		
Electrochemical				
Batteries				
Lead–acid	✓	✓	✓	
Lithium-ion	✓	✓	✓	✓
Nickel–cadmium	✓	✓		
Nickel–metal hydride				✓
Zebra				✓
Sodium–sulfur	✓	✓		
Flow Batteries				
Vanadium redox	✓			
Polysulfide bromide	✓			
Zinc bromide	✓			
Electrolytic hydrogen				✓
Mechanical				
Pumped hydro	✓			
Compressed air	✓			
Flywheels		✓		✓
Direct Thermal				
Sensible Heat				
Liquids			✓	
Solids			✓	
Latent Heat				
Phase change	✓		✓	
Hydration–dehydration	✓			
Chemical reaction	✓		✓	
Thermochemical				
Biomass solids	✓		✓	
Ethanol	✓			✓
Biodiesel				✓
Syngas	✓			✓

15.2 Principal Forms of Stored Energy

The storage media discussed in this chapter can accept and deliver energy in three fundamental forms: electrical, mechanical, and thermal. Electrical and mechanical energy are both considered high-quality energy because they can be converted to either of the other two forms with fairly little energy loss (e.g., electricity can drive a motor with only about 5% energy loss, or a resistive heater with no energy loss).

The quality of thermal energy storage depends on its temperature. Usually, thermal energy is considered low quality because it cannot be easily converted to the other two forms. The theoretical maximum quantity of useful work W_{max} (mechanical energy) extractable from a given quantity of heat Q is

$$W_{max} = \frac{T_1 - T_2}{T_1} \times Q,$$

where T_1 is the absolute temperature of the heat and T_2 is the surrounding, ambient absolute temperature.

Any energy storage facility must be carefully chosen to accept and produce a form of energy consistent with either the energy source or the final application. Storage technologies that accept and/or produce heat should, as a rule, only be used with heat energy sources or with heat applications. Mechanical and electric technologies are more versatile, but in most cases electric technologies are favored over mechanical because electricity is more easily transmitted, because there is a larger array of useful applications, and because the construction cost is typically lower.

15.3 Applications of Energy Storage

In Table 15.1 above, each technology is classified by its relevance in one to four different, principal applications:

- *Utility shaping* is the use of very large capacity storage devices to answer electric demand, when a renewable resource is not producing sufficient generation. An example would be nighttime delivery of energy generated by a solar thermal plant during the prior day.
- *Power quality* is the use of very responsive storage devices (capable of large changes in output over very short timescales) to smooth power delivery during switching events, short outages, or plant run-up. Power-quality applications can be implemented at central generators, at switchgear locations, and at commercial and industrial customers' facilities. Uninterruptible power supplies (UPS) are an example of this category.
- *Distributed grid technologies* enable energy generation and storage at customer locations, rather than at a central (utility) facility. The distributed grid is an important, enabling concept for photovoltaic technologies that are effective at a small scale and can be installed on private homes and commercial buildings. When considered in the context of photovoltaics, the energy storage for the distributed grid is similar to the utility shaping application in that both are solutions to an intermittent, renewable resource, but distributed photovoltaic generation requires small capacities in the neighborhood of a few tens of MJ, while utility shaping requires capacities in the TJ range.[2] Renewable thermal resources (solar, geothermal) can also be implemented on a distributed scale, and require household-scale thermal storage tanks. For the purposes of this chapter, district-heating systems are also considered a distributed technology.
- *Automotive applications* include battery-electric vehicles (EVs), hybrid gasoline–electric vehicles, plug-in hybrid electric vehicles (PHEVs), and other applications that require mobile batteries larger than those used in today's internal combustion engine cars. A deep penetration of

[2]Storage capacities in this chapter are given in units of MJ, GJ, and TJ: 1 MJ = 0.28 kWh, 1 GJ = 280 kWh, and 1 TJ = 280 MWh.

automotive batteries also could become important in a distributed grid. Large fleets of EVs or PHEVs that are grid connected when parked would help enable renewable technologies, fulfilling utility shaping and distributed grid functions as well as their basic automotive function.

Additional energy storage applications exist, most notably portable electronics and industrial applications. However, the four applications described here make up the principal components that will interact in a significant way with the global energy grid.

15.4 Specifying Energy Storage Devices

Every energy storage technology, regardless of category, can be roughly characterized by a fairly small number of parameters. Self-discharge time, unit size, and efficiency serve to differentiate the various categories. Within a category, finer selections of storage technology can be made by paying attention to cycle life, specific energy, specific power, energy density, and power density.

Self-discharge time is the time required for a fully charged, noninterconnected storage device to reach a certain depth of discharge (DOD). DOD is typically described as a percentage of the storage device's useful capacity, so that, for instance, 90% DOD means 10% of the device's energy capacity remains. The relationship between self-discharge time and DOD is rarely linear, so self-discharge times must be measured and compared at a uniform DOD. Acceptable self-discharge times vary greatly, from a few minutes for some power-quality applications, to years for devices designed to shape annual power production.

Unit size describes the intrinsic scale of the technology, and is the least well-defined of the parameters listed here. If the unit size is small compared to the total required capacity of a project, complexity and supply shortages can increase the cost relative to technologies with a larger unit size. Some technologies have a fairly large unit size that prohibits small-scale energy storage.

Figure 15.1 maps all of the technologies discussed in this chapter, according to their unit size and 10% self-discharge time. The gamut of technologies available covers many orders of magnitude on each axis, illustrating the broad choice available. Utility shaping applications require a moderate self-discharge time and a large unit size; power-quality applications are much less sensitive to self-discharge time but require a moderate unit size. Distributed grid and automotive applications both require a moderate self-discharge time and a moderate unit size.

Efficiency is the ratio of energy output from the device, to the energy input. Like energy density and specific energy, the system boundary must be carefully considered when measuring efficiency. It is particularly important to pay attention to the form of energy required at the input and output interconnections, and to include the entire system necessary to attach to those interconnections. For instance, if the system is to be used for shaping a constant-velocity, utility wind farm, then presumably both the input and output will be AC electricity. When comparing a battery with a fuel cell in this scenario, it is necessary to include the efficiencies of an AC-to-DC rectifier for the battery, an AC-powered hydrogen generation system for the fuel cell system, and DC-to-AC converters associated with both systems.

Efficiency is related to self-discharge time. Technologies with a short self-discharge time will require constant charging to maintain a full charge; if discharge occurs much later than charge in a certain application, the apparent efficiency will be lower because a significant amount of energy is lost in maintaining the initial, full charge.

Cycle life is the number of consecutive charge–discharge cycles a storage installation can undergo while maintaining the installation's other specifications within certain, limited ranges. Cycle-life specifications are made against a chosen DOD depending on the application of the storage device. In some cases, for example pressurized hydrogen storage in automobiles, each cycle will significantly discharge the hydrogen canister and the appropriate DOD reference might be 80% or 90%. In other cases, for example a battery used in a hybrid electric vehicle, most discharge cycles may consume only 10% or 20% of the energy stored in the battery. For most storage technologies, cycle life is significantly larger for shallow discharges than deep discharges, and it is critical that cycle-life data be compared across a uniform DOD assumption.

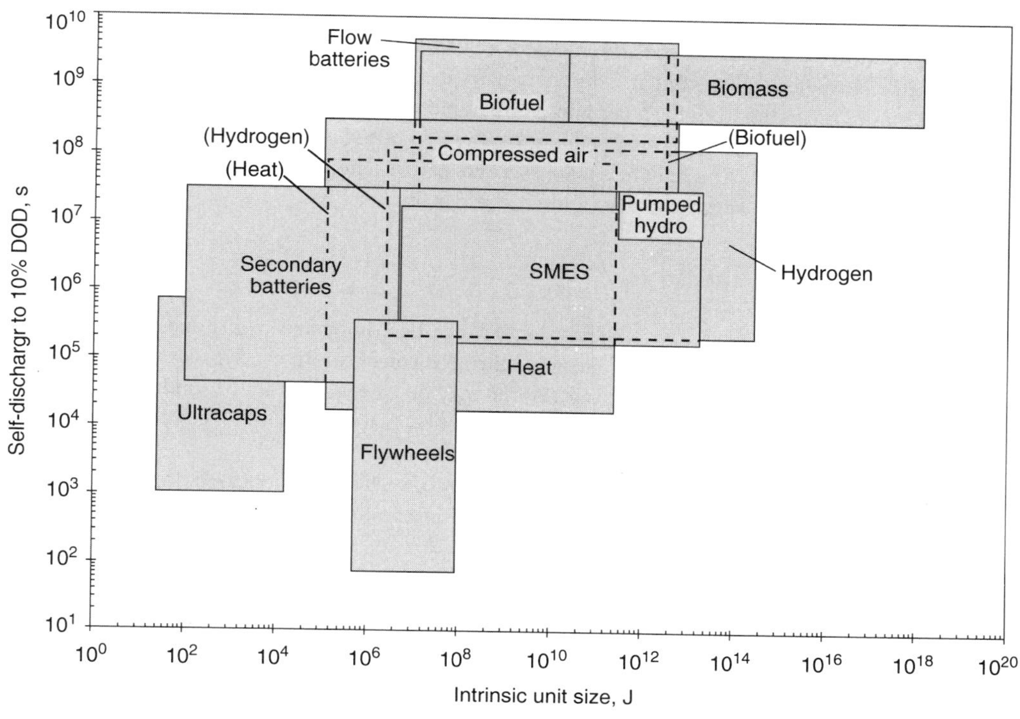

FIGURE 15.1 All storage technologies, mapped by self-discharge time and unit size. Not all hidden lines are shown. Larger self-discharge times are always more desirable, but more or less important depending on the application. Intrinsic unit size does not have a desirability proportional to its value, but rather must be matched to the application.

Specific energy is a measure of how heavy the technology is. It is measured in units of energy per mass, and in this chapter this quantity will always be reported in MJ/kg. The higher the specific energy, the lighter the device. Automotive applications require high specific energies; for utility applications, specific energy is relatively unimportant, except where it impacts construction costs.

Energy density is a measure of how much space the technology occupies. It is measured in units of energy per volume, and in this chapter we will always report this quantity in MJ/L. The higher the energy density, the smaller the device. Again, this is most important for automotive applications, and rarely important in utility applications. Typical values for energy density associated with a few automotive-scale energy technologies are listed in Table 15.2, together with cycle-life and efficiency data.

TABLE 15.2 Nominal Energy Density, Cycle Life and Efficiency of Automotive Storage Technologies

	Energy Density MJ/L	Cycle Life at 80% DOD[a]	Electric Efficiency %
Ultracapacitors	0.2	50,000	95
Li-ion batteries	1.8	2,000	85
NiMH batteries	0.6	1,000	80
H_2 at 350 bar	3.0	n/a[b]	47
H_2 at 700 bar	5.0	n/a	45
Air at 300 bar	<0.1	n/a	37
Flywheels	<0.1	20,000	80
Ethanol	23.4	n/a	n/a

Electric efficiencies are calculated for electric-to-electric conversion and momentary storage.
[a]Depth of discharge.
[b]Not applicable.

Energy-density and specific-energy estimates are dependent on the system definition. For example, it might be tempting to calculate the specific energy of a flow battery technology by dividing its capacity by the mass of the two electrolytes. But it is important to also include the mass of the electrolyte storage containers, and of the battery cell for a fair and comparable estimate of its specific energy. Therefore, the energy density and specific energy are dependent on the size of the specific device; large devices benefit from efficiency of scale with a higher energy density and specific energy. *Specific power* and *power density* are the power correlates to specific energy and energy density.

15.5 Specifying Fuels

A fuel is any (relatively) homogenous substance that can be combusted to produce heat. Though the energy contained in a fuel can always be extracted through combustion, other processes may be used to extract the energy (e.g., reaction in a fuel cell). A fuel may be gaseous, liquid, or solid. All energy storage technologies in the thermochemical category store energy in a fuel. In the electrochemical category, electrolytic hydrogen is a fuel.

A fuel's lower heating value (LHV) is the total quantity of sensible heat released during combustion of a designated quantify of fuel. For example, in the simplest combustion process, that of hydrogen,

$$2\,H_2 + O_2 \rightarrow 2\,H_2O(\text{vapor}) + \text{LHV},$$

or for the slightly more complex combustion of methane,

$$CH_4 + 2\,O_2 \rightarrow CO_2 + 2\,H_2O(\text{vapor}) + \text{LHV}.$$

In this chapter, the quantity of fuel is always expressed as a mass, so that LHV is a special case of specific energy. Like specific energy, LHV is expressed in units of MJ/kg in this chapter.

Higher heating value (HHV) is the LHV, plus the latent heat contained in the water vapor resulting from combustion.[3] For the examples of hydrogen and methane, this means

$$2\,H_2 + O_2 \rightarrow 2\,H_2O(\text{liquid}) + \text{HHV},$$

and

$$CH_4 + 2\,O_2 \rightarrow CO_2 + 2\,H_2O(\text{liquid}) + \text{HHV}.$$

The latent heat in the water vapor can be substantial, especially for the hydrogen-rich fuels typical in renewable energy applications. Table 15.3 lists LHVs and HHVs of fuels discussed in this chapter; in the

TABLE 15.3 Properties of Fuels

	Chemical Formula	Density g/L	LHV MJ/kg	HHV MJ/kg
Methanol	CH_3OH	794	19.9	22.7
Ethanol	C_2H_5OH	792	26.7	29.7
Methane	CH_4	0.68	49.5	54.8
Hydrogen	H_2	0.085	120	142
Dry syngas, airless process[a]	$40H_2 + 21CO + 10CH_4 + 29CO_2$	0.89	11.2	12.6
Dry syngas, air process[a]	$25H_2 + 16CO + 5CH_4 + 15CO_2 + 39N_2$	0.99	6.23	7.01

[a]Chemical formulae and associated properties of syngas are representative; actual composition of syngas will vary widely according to manufacturing process.

Source: From All except syngas from U.S. Department of Energy, *Properties of Fuels*, Alternative Fuels Data Center 2004.

[3]The concepts of sensible and latent heat are explained further in Section 15.1.9.

most extreme case of molecular hydrogen, the HHV is some 18% higher than the LHV. Recovery of the latent heat requires controlled condensation of the water vapor.

In this chapter, all heating values are reported as HHV rather than LHV. HHV is favored for two reasons: (1) its values allow easier checking of energy calculations with the principle of energy conservation, and (2) when examining technologies for future implementation, it is wise to keep an intention of developing methods for extracting as much of each energy source's value as possible.

15.6 Direct Electric Storage

15.6.1 Ultracapacitors

A capacitor stores energy in the electric field between two oppositely charged conductors. Typically, thin conducting plates are rolled or stacked into a compact configuration with a dielectric between them. The dielectric prevents arcing between the plates and allows the plates to hold more charge, increasing the maximum energy storage. The ultracapacitor—also known as supercapacitor, electrochemical capacitor, or electric double layer capacitor (EDLC)—differs from a traditional capacitor in that it employs a thin electrolyte, on the order of only a few angstroms, instead of a dielectric. This increases the energy density of the device. The electrolyte can be made of either an organic or an aqueous material. The aqueous design operates over a larger temperature range, but has a smaller energy density than the organic design. The electrodes are made of a porous carbon that increases the surface area of the electrodes and further increases energy density over a traditional capacitor.

Ultracapacitors' ability to effectively equalize voltage variations with quick discharges make them useful for power-quality management and for regulating voltage in automotive systems during regular driving conditions. Ultracapacitors can also work in tandem with batteries and fuel cells to relieve peak power needs (e.g., hard acceleration) for which batteries and fuel cells are not ideal. This could help extend the overall life and reduce lifetime cost of the batteries and fuel cells used in hybrid and electric vehicles. This storage technology also has the advantage of very high cycle life of greater than 500,000 cycles and a 10- to 12-year life span.[1] The limitations lie in the inability of ultracapacitors to maintain charge voltage over any significant time, losing up to 10% of their charge per day.

15.6.2 Superconducting Magnetic Energy Storage

An superconducting magnetic energy storage (SMES) system is well suited to storing and discharging energy at high rates (high power.) It stores energy in the magnetic field created by direct current in a coil of cryogenically cooled, superconducting material. If the coil were wound using a conventional wire such as copper, the magnetic energy would be dissipated as heat due to the wire's resistance to the flow of current. The advantage of a cryogenically cooled, superconducting material is that it reduces electrical resistance to almost zero. The SMES recharges quickly and can repeat the charge/discharge sequence thousands of times without any degradation of the magnet. A SMES system can achieve full power within 100 ms.[2] Theoretically, a coil of around 150–500 m radius would be able to support a load of 18,000 GJ at 1000 MW, depending on the peak field and ratio of the coil's height and diameter.[3] Recharge time can be accelerated to meet specific requirements, depending on system capacity.

Because no conversion of energy to other forms is involved (e.g., mechanical or chemical), the energy is stored directly and round-trip efficiency can be very high.[2] SMES systems can store energy with a loss of only 0.1%; this loss is due principally to energy required by the cooling system.[3] Mature, commercialized SMES is likely to operate at 97%–98% round-trip efficiency and is an excellent technology for providing reactive power on demand.

15.7 Electrochemical Energy Storage

15.7.1 Secondary Batteries

A secondary battery allows electrical energy to be converted into chemical energy, stored, and converted back to electrical energy. Batteries are made up of three basic parts: a negative electrode, positive electrode, and an electrolyte (Figure 15.2). The negative electrode gives up electrons to an external load, and the positive electrode accepts electrons from the load. The electrolyte provides the pathway for charge to transfer between the two electrodes. Chemical reactions between each electrode and the electrolyte remove electrons from the positive electrode and deposit them on the negative electrode. This can be written as an overall chemical reaction that represents the states of charging and discharging of a battery. The speed at which this chemical reaction takes place is related to the internal resistance that dictates the maximum power at which the batteries can be charged and discharged.

Some batteries suffer from the "memory effect" in which a battery exhibits a lower discharge voltage under a given load than is expected. This gives the appearance of lowered capacity but is actually a voltage depression. Such a voltage depression occurs when a battery is repeatedly discharged to a partial depth and recharged again. This builds an increased internal resistance at this partial depth of discharge and the battery appears as a result to only be dischargeable to the partial depth. The problem, if and when it occurs, can be remedied by deep discharging the cell a few times. Most batteries considered for modern renewable applications are free from this effect, however.

15.7.2 Lead–Acid

Lead–acid is one of the oldest and most mature battery technologies. In its basic form, the lead–acid battery consists of a lead (Pb) negative electrode, a lead dioxide (PbO_2) positive electrode and a separator to electrically isolate them. The electrolyte is dilute sulfuric acid (H_2SO_4), which provides the sulfate ions for the discharge reactions. The chemistry is represented by:

$$PbO_2 + Pb + 2\,H_2SO_4 \leftrightarrow 2\,PbSO_4 + 2\,H_2O.$$

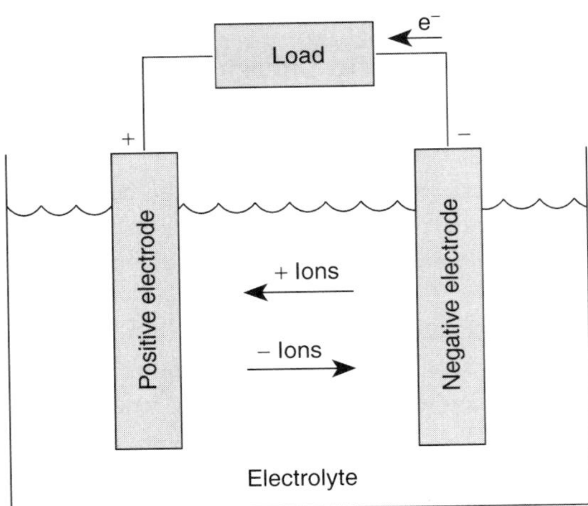

FIGURE 15.2 Schematic of a generalized secondary battery. Directions of electron and ion migration shown are for discharge, so that the positive electrode is the cathode and the negative electrode is the anode. During charge, electrons and ions move in the opposite directions and the positive electrode becomes the anode while the negative electrode becomes the cathode.

(In all battery chemistries listed in this chapter, left-to-right indicates battery discharge and right-to-left indicates charging.)

There are three main types of lead–acid batteries: the flooded cell, the sealed gel cell, and the sealed absorbed glass mat (AGM) lead–acid battery. The wet cell has a liquid electrolyte that must be replaced occasionally to replenish the hydrogen and oxygen that escape during the charge cycle. The sealed gel cell has a silica component added to the electrolyte to stiffen it. The AGM design uses a fiberglass-like separator to hold electrolyte in close proximity to the electrodes, thereby increasing efficiency. For both the gel and AGM configurations, there is a greatly reduced risk of hydrogen explosion and corrosion from disuse. These two types do require a lower charging rate, however. Both the gel cells and the AGM batteries are sealed and pressurized so that oxygen and hydrogen produced during the charge cycle are recombined into water.

The lead–acid battery is a low-cost and popular storage choice for power-quality applications. Its application for utility shaping, however, has been very limited due to its short cycle life. A typical installation survives a maximum of 1500 deep cycles.[4] Yet, lead–acid batteries have been used in a few commercial and large-scale energy management applications. The largest one is a 140-GJ system in Chino, California, built in 1988. Lead–acid batteries have a specific energy of only 0.18 MJ/kg and would therefore not be a viable automobile option apart from providing the small amount of energy needed to start an engine. It also has a poor energy density at around 0.25 MJ/L. The advantages of the lead–acid battery technology are low cost and high power density.

15.7.3 Lithium-Ion

Lithium-ion and lithium polymer batteries, although primarily used in the portable electronics market, are likely to have future use in many other applications. The cathode in these batteries is a lithiated metal oxide ($LiCoO_2$, $LiMO_2$, etc.) and the anode is made of graphitic carbon with a layer structure. The electrolyte consists of lithium salts (such as $LiPF_6$) dissolved in organic carbonates; an example of Li-ion battery chemistry is

$$Li_xC + Li_{1-x}CoO_2 \leftrightarrow LiCoO_2 + C.$$

When the battery is charged, lithium atoms in the cathode become ions and migrate through the electrolyte toward the carbon anode where they combine with external electrons and are deposited between carbon layers as lithium atoms. This process is reversed during discharge. The lithium polymer variation replaces the electrolyte with a plastic film that does not conduct electricity but allows ions to pass through it. The 60°C operating temperature requires a heater, reducing overall efficiency slightly.

Lithium-ion batteries have a high energy density of about 0.72 MJ/L and have low internal resistance; they will achieve efficiencies in the 90% range and above. They have an energy density of around 0.72 MJ/kg. Their high energy efficiency and energy density make lithium-ion batteries excellent candidates for storage in all four applications considered here: utility shaping, power quality, distributed generation, and automotive.

15.7.4 Nickel–Cadmium

Nickel–cadmium (NiCd) batteries operate according to the chemistry:

$$2\,NiOOH + 2\,H_2O + Cd \leftrightarrow 2\,Ni(OH)_2 + Cd(OH)_2.$$

NiCd batteries are not common for large stationary applications. They have a specific energy of about 0.27 MJ/kg, an energy density of 0.41 MJ/L and an efficiency of about 75%. Alaska's Golden Valley Electric Association commissioned a 40-MW/290-GJ nickel–cadmium battery in 2003 to improve reliability and to supply power for essentials during outages.[5] Resistance to cold and relatively low cost were among the deciding factors for choosing the NiCd chemistry.

Cadmium is a toxic heavy metal and there are concerns relating to the possible environmental hazards associated with the disposal of NiCd batteries. In November 2003, the European Commission adopted a proposal for a new battery directive that includes recycling targets of 75% for NiCd batteries. However, the possibility of a ban on rechargeable batteries made from nickel–cadmium still remains and hence the long-term viability and availability of NiCd batteries continues to be uncertain. NiCd batteries can also suffer from "memory effect," where the batteries will only take full charge after a series of full discharges. Proper battery management procedures can help to mitigate this effect.

15.7.5 Nickel–Metal Hydride

The nickel–metal hydride (Ni*M*H) battery operates according to the chemistry:

$$MH + NiOOH \leftrightarrow M + Ni(OH)_2,$$

where M represents one of a large variety of metal alloys that serve to take up and release hydrogen. Ni*M*H batteries were introduced as a higher energy density and more environmentally friendly version of the nickel–cadmium cell. Modern nickel–metal hydride batteries offer up to 40% higher energy density than nickel–cadmium. There is potential for yet higher energy density, but other battery technologies (lithium-ion, in particular) may fill the same market sooner.

Nickel–metal hydride is less durable than nickel–cadmium. Cycling under heavy load and storage at high temperature reduces the service life. Nickel–metal hydride suffers from a higher self-discharge rate than the nickel–cadmium chemistry. Nickel–metal hydride batteries have a specific energy of 0.29 MJ/kg, an energy density of about 0.54 MJ/L and an energy efficiency of about 70%. These batteries have been an important bridging technology in the portable electronics and hybrid automobile markets. Their future is uncertain because other battery chemistries promise higher energy storage potential and cycle life.

15.7.6 Sodium–Sulfur

A sodium–sulfur (NaS) battery consists of a liquid (molten) sulfur positive electrode and liquid (molten) sodium negative electrode, separated by a solid beta-alumina ceramic electrolyte (Figure 15.3). The chemistry is as follows:

$$2\,Na + x\,S \leftrightarrow Na_2S_x.$$

When discharging, positive sodium ions pass through the electrolyte and combine with the sulfur to form sodium polysulfides. The variable x in the equation is equal to 5 during early discharging, but after free sulfur has been exhausted a more sodium-rich mixture of polysulfides with lower average values of x develops. This process is reversible as charging causes sodium polysulfides in the positive electrode to release sodium ions that migrate back through the electrolyte and recombine as elemental sodium. The battery operates at about 300°C. NaS batteries have a high energy density of around 0.65 MJ/L and a specific energy of up to 0.86 MJ/kg. These numbers would indicate an application in the automotive sector, but warm-up time and heat-related accident risk make its use there unlikely. The efficiency of this battery chemistry can be as high as 90% and would be suitable for bulk storage applications while simultaneously allowing effective power smoothing operations.[6]

15.7.7 Zebra

Zebra is the popular name for the sodium–nickel-chloride battery chemistry:

$$NiCl_2 + 2\,Na \leftrightarrow Ni + 2\,NaCl.$$

Zebra batteries are configured similarly to sodium–sulfur batteries (see Figure 15.3), and also operate at about 300°C. Zebra batteries boast a greater than 90% energy efficiency, a specific energy of up to

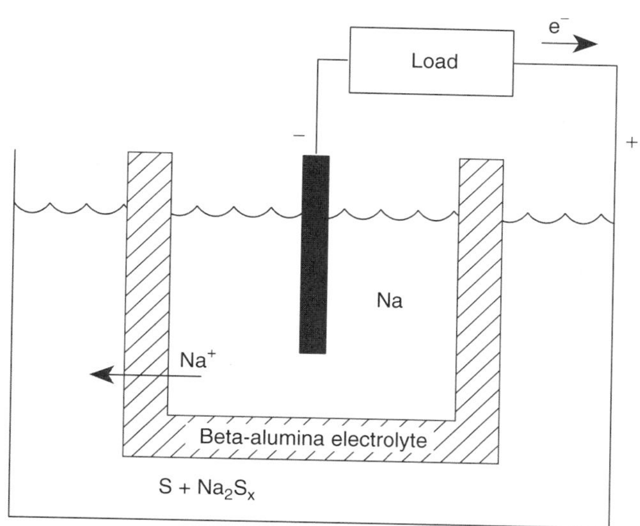

FIGURE 15.3 Sodium–sulfur battery showing discharge chemistry. The sodium (Na) and sulfur (S) electrodes are both in a liquid state and are separated by a solid, beta-alumina ceramic electrolyte that allows only sodium ions to pass. Charge is extracted from the electrolytes with metal contacts; the positive contact is the battery wall.

0.32 MJ/kg and an energy density of 0.49 MJ/L.[7] Its tolerance for a wide range of operating temperature and high efficiency, coupled with a good energy density and specific energy, make its most probable application the automobile sector, and as of 2003 Switzerland's MES-DEA is pursuing this application aggressively.[8] Its high energy efficiency also makes it a good candidate for the utility sector.

15.7.8 Flow Batteries

Most secondary batteries use electrodes both as an interface for gathering or depositing electrons, and as a storage site for the products or reactants associated with the battery's chemistry. Consequently, both energy and power density are tied to the size and shape of the electrodes. Flow batteries store and release electrical energy by means of reversible electrochemical reactions in two liquid electrolytes. An electrochemical cell has two compartments—one for each electrolyte—physically separated by an ion exchange membrane. Electrolytes flow into and out of the cell through separate manifolds and undergo chemical reaction inside the cell, with ion or proton exchange through the membrane and electron exchange through the external electric circuit. The chemical energy in the electrolytes is turned into electrical energy and vice versa for charging. They all work in the same general way but vary in chemistry of electrolytes.[9]

There are some advantages to using the flow battery over a conventional secondary battery. The capacity of the system is scaleable by simply increasing the amount of solution. This leads to cheaper installation costs as the systems get larger. The battery can be fully discharged with no ill effects and has little loss of electrolyte over time. Because the electrolytes are stored separately and in large containers (with a low surface area to volume ratio), flow batteries show promise to have some of the lowest self-discharge rates of any energy storage technology available.

Poor energy densities and specific energies remand these battery types to utility-scale power shaping and smoothing, although they might be adaptable for distributed-generation use. There are three types of flow batteries that are closing in on commercialization: vanadium redox, polysulfide bromide, and zinc bromide.

15.7.8.1 Vanadium Redox

The vanadium redox flow battery (VRB) was pioneered at the University of New South Wales, Australia, and has shown potentials for long cycle life and energy efficiencies of over 80% in large installations.[10] The VRB uses compounds of the element vanadium in both electrolyte tanks. The reaction chemistry at the positive electrode is:

$$V^{5+} + e^{-} \leftrightarrow V^{4+},$$

and at the negative electrode,

$$V^{2+} \leftrightarrow V^{3+} + e^{-}.$$

Using vanadium compounds on both sides of the ion-exchange membrane eliminates the possible problem of cross-contamination of the electrolytes and makes recycling easier.[11] As of 2005, two small, utility-scale VRB installations are operating, one 2.9-GJ unit on King Island, Australia and one 7.2-GJ unit in Castle Valley, Utah.

15.7.8.2 Polysulfide Bromide

The polysulfide bromide battery (PSB) utilizes two salt solution electrolytes, sodium bromide (NaBr) and sodium polysulfide (Na_2S_x). PSB electrolytes are separated in the battery cell by a polymer membrane that only passes positive sodium ions. The chemistry at the positive electrode is

$$NaBr_3 + 2\,Na^{+} + 2\,e^{-} \leftrightarrow 3\,NaBr,$$

and at the negative electrode,

$$2\,Na_2S_2 \leftrightarrow Na_2S_4 + 2\,Na^{+} + 2\,e^{-}.$$

The PSB battery is being developed by Canada's VRB Power Systems, Inc.[12] This technology is expected to attain energy efficiencies of approximately 75%.[13] Although the salt solutions themselves are only mildly toxic, a catastrophic failure by one of the tanks could release highly toxic bromine gas. Nevertheless, the Tennessee Valley Authority released a finding of no significant impact for a proposed 430-GJ facility and deemed it safe.[14]

15.7.8.3 Zinc Bromide

In each cell of a zinc bromide (ZnBr) battery, two different electrolytes flow past carbon-plastic composite electrodes in two compartments separated by a microporous membrane. Chemistry at the positive electrode follows the equation:

$$Br_2(aq) + 2\,e^{-} \leftrightarrow 2\,Br,$$

and at the negative electrode:

$$Zn \leftrightarrow Zn^{2+} + 2e^{-}.$$

During discharge, Zn and Br combine into zinc bromide. During charge, metallic zinc is deposited as a thin-film on the negative electrode. Meanwhile, bromine evolves as a dilute solution on the other side of the membrane, reacting with other agents to make thick bromine oil that sinks to the bottom of the electrolytic tank. During discharge, a pump mixes the bromine oil with the rest of the electrolyte. The zinc bromide battery has an energy efficiency of nearly 80%.[15]

Exxon developed the ZnBr battery in the early 1970 s. Over the years, many GJ-scale ZnBr batteries have been built and tested. Meidisha demonstrated a 1-MW/14-GJ ZnBr battery in 1991 at Kyushu Electric Power Company. Some GJ-scale units are now available preassembled, complete with plumbing and power electronics.

15.7.9 Electrolytic Hydrogen

Diatomic, gaseous hydrogen (H_2) can be manufactured with the process of electrolysis; an electric current applied to water separates it into components O_2 and H_2. The oxygen has no inherent energy value, but the HHV of the resulting hydrogen can contain up to 90% of the applied electric energy, depending on the technology.[16] This hydrogen can then be stored and later combusted to provide heat or work, or to power a fuel cell (see Chapter 23).

The gaseous hydrogen is low density and must be compressed to provide useful storage. Compression to a storage pressure of 350 bar, the value usually assumed for automotive technologies, consumes up to 12% of the hydrogen's HHV if performed adiabatically, although the loss approaches a lower limit of 5% as the compression approaches an isothermal ideal.[17] Alternatively, the hydrogen can be stored in liquid form, a process that costs about 40% of HHV using current technology, and that at best would consume about 25%. Liquid storage is not possible for automotive applications, because mandatory boil-off from the storage container cannot be safely released in closed spaces (i.e., garages).

Hydrogen can also be bonded into metal hydrides using an absorption process. The energy penalty of storage may be lower for this process, which requires pressurization to only 30 bar. However, the density of the metal hydride can be between 20 and 100 times the density of the hydrogen stored. Carbon nanotubes have also received attention as a potential hydrogen storage medium.[18]

15.8 Mechanical Energy Storage

15.8.1 Pumped Hydro

Pumped hydro is the oldest and largest of all of the commercially available energy storage technologies, with existing facilities up to 1000 MW in size. Conventional pumped hydro uses two water reservoirs, separated vertically. Energy is stored by moving water from the lower to the higher reservoir, and extracted by allowing the water to flow back to the lower reservoir. Energy is stored according to the fundamental physical principle of potential energy. To calculate the stored energy, E_s, in joules, use the formula:

$$E_s = Vdgh,$$

where V is the volume of water raised (m^3), d is the density of water (1000 kg/m^3), g is the acceleration of gravity (9.8 m/s^2), and h is the elevation difference between the reservoirs (m) often referred to as the *head*.

Though pumped hydro is by nature a mechanical energy storage technology, it is most commonly used for electric utility shaping. During off-peak hours electric pumps move water from the lower reservoir to the upper reservoir. When required, the water flow is reversed to generate electricity. Some high dam hydro plants have a storage capability and can be dispatched as pumped hydro storage. Underground pumped storage, using flooded mine shafts or other cavities, is also technically possible but probably prohibitively expensive. The open sea can also be used as the lower reservoir if a suitable upper reservoir can be built at close proximity. A 30-MW seawater pumped hydro plant was first built in Yanbaru, Japan in 1999.

Pumped hydro is most practical at a large scale with discharge times ranging from several hours to a few days. There is over 90 GW of pumped storage in operation worldwide, which is about 3% of global electric generation capacity.[19] Pumped storage plants are characterized by long construction times and high capital expenditure. Its main application is for utility shaping. Pumped hydro storage has the limitation of needing to be a very large capacity to be cost-effective, but can also be used as storage for a number of different generation sites.

Efficiency of these plants has greatly increased in the last 40 years. Pumped storage in the 1960s had efficiencies of 60% compared with 80% for new facilities. Innovations in variable speed motors have helped these plants operate at partial capacity, and greatly reduced equipment vibrations, increasing plant life.

15.8.2 Compressed Air

A relatively new energy storage concept that is implemented with otherwise mature technologies is compressed air energy storage (CAES). CAES facilities must be coupled with a combustion turbine, so are actually a hybrid storage/generation technology.

A conventional gas turbine consists of three basic components: a compressor, combustion chamber, and an expander. Power is generated when compressed air and fuel burned in the combustion chamber drive turbine blades in the expander. Approximately 60% of the mechanical power generated by the expander is consumed by the compressor supplying air to the combustion chamber.

A CAES facility performs the work of the compressor separately, stores the compressed air, and at a later time injects it into a simplified combustion turbine. The simplified turbine includes only the combustion chamber and the expansion turbine. Such a simplified turbine produces far more energy than a conventional turbine from the same fuel, because there is potential energy stored in the compressed air. The fraction of output energy beyond what would have been produced in a conventional turbine is attributable to the energy stored in compression.

The net efficiency of storage for a CAES plant is limited by the heat energy loss occurring at compression. The overall efficiency of energy storage is about 75%.[20]

CAES compressors operate on grid electricity during off-peak times, and use the expansion turbine to supply peak electricity when needed. CAES facilities cannot operate without combustion because the exhaust air would exit at extremely low temperatures causing trouble with brittle materials and icing. If 100% renewable energy generation is sought, biofuel could be used to fuel the gas turbines. There might still be other emissions issues but the system could be fully carbon neutral.

The compressed air is stored in appropriate underground mines, caverns created inside salt rocks or possibly in aquifers. The first commercial CAES facility was a 290-MW unit built in Hundorf, Germany in 1978. The second commercial installation was a 110-MW unit built in McIntosh, Alabama in 1991. The third commercial CAES is a 2,700-MW plant under construction in Norton, Ohio. This nine-unit plant will compress air to about 100 bar in an existing limestone mine 2200 ft. (766 m) underground.[21] The natural synergy with geological caverns and turbine prime movers dictate that these be on the utility scale.

15.8.3 Flywheels

Most modern flywheel energy storage systems consist of a massive rotating cylinder (comprised of a rim attached to a shaft) that is supported on a stator by magnetically levitated bearings that eliminate bearing wear and increase system life. To maintain efficiency, the flywheel system is operated in a low vacuum environment to reduce drag. The flywheel is connected to a motor/generator mounted onto the stator that, through some power electronics, interact with the utility grid.

The energy stored in a rotating flywheel, in joules, is given by

$$E = \frac{1}{2} I \omega^2$$

where I is the flywheel's moment of inertia (kg m^2), and ω is its angular velocity (s^{-2}). I is proportional to the flywheel's mass, so energy is proportional to mass and the square of speed. In order to maximize energy capacity, flywheel designers gravitate toward increasing the flywheel's maximum speed rather than increasing its moment of inertia. This approach also produces flywheels with the higher specific energy.

Some of the key features of flywheels are low maintenance, a cycle life of better than 10,000 cycles, a 20-year lifetime and environmentally friendly materials. Low-speed, high-mass flywheels (relying on I for energy storage) are typically made from steel, aluminum, or titanium; high-speed, low-mass flywheels (relying on ω for energy storage) are constructed from composites such as carbon fiber.

Flywheels can serve as a short-term ride-through before long-term storage comes online. Their low energy density and specific energy limit them to voltage regulation and UPS capabilities. Flywheels can have energy efficiencies in the upper 90% range depending on frictional losses.

15.9 Direct Thermal Storage

Direct thermal technologies, although they are storing a lower grade of energy (heat, rather than electrical or mechanical energy) can be useful for storing energy from systems that provide heat as a native output (e.g., solar thermal, geothermal), or for applications where the energy's commodity value is heat (e.g., space heating, drying).

Although thermal storage technologies can be characterized by specific energy and energy density like any other storage technology, they can also be characterized by an important, additional parameter: the delivery temperature range. Different end uses have more or less allowance for wide swings of the delivery temperature. Also, some applications require a high operating temperature that only some thermal storage media are capable of storing.

Thermal storage can be classified into two fundamental categories: sensible heat storage and latent heat storage. Applications that have less tolerance for temperature swings should utilize a latent heat technology.

Input to and output from heat energy storage is accomplished with heat exchangers. The discussion below focuses on the choice of heat storage materials; the methods of heat exchange will vary widely depending on properties of the storage material, especially its thermal conductivity. Materials with higher thermal conductivity will require a smaller surface area for heat exchange. For liquids, convection or pumping can reduce the need for a large heat exchanger. In some applications, the heat exchanger is simply the physical interface of the storage material with the application space (e.g., phase-change drywall, see below).

15.9.1 Sensible Heat

Sensible heat is the heat that is customarily and intuitively associated with a change in temperature of a massive substance. The heat energy, E_s, stored in such a substance is given by:

$$E_s = (T_2 - T_1)cM,$$

where c is the specific heat of the substance (J/kg °C) and M is the mass of the substance (kg); T_1 and T_2 are the initial and final temperatures, respectively (°C). The specific heat c is a physical parameter measured in units of heat per temperature per mass: substances with the ability to absorb heat energy with a relatively small increase in temperature (e.g., water) have a high specific heat, whereas those that get hot with only a little heat input (e.g., lead) have a low specific heat. Sensible heat storage is best accomplished with materials having a high specific heat.

15.9.1.1 Liquids

Sensible heat storage in a liquid is, with very few exceptions, accomplished with water. Water is unique among chemicals in having an abnormally high specific heat of 4,186 J/kg K, and furthermore has a reasonably high density. Water is also cheap and safe. It is the preferred choice for most nonconcentrating solar thermal collectors.

Liquids other than water may need to be chosen if the delivery temperature must be higher than 100°C, or if the system temperature can fall below 0°C. Water can be raised to temperatures higher than 100°C, but the costs of storage systems capable of containing the associated high pressures are usually prohibitive. Water can be mixed with ethylene glycol or propylene glycol to increase the useful temperature range and prevent freezing.

When a larger temperature range than that afforded by water is required, mineral, synthetic, or silicone oils can be used instead. The tradeoffs for the increased temperature range are higher cost, lower specific heat, higher viscosity (making pumping more difficult), flammability, and, in some cases, toxicity.

For very high temperature ranges, salts are usually preferred that balance a low specific heat with a high density and relatively low cost. Sodium nitrate has received the most prominent testing for this purpose in the U.S. Department of Energy's Solar Two Project located in Barstow, California.

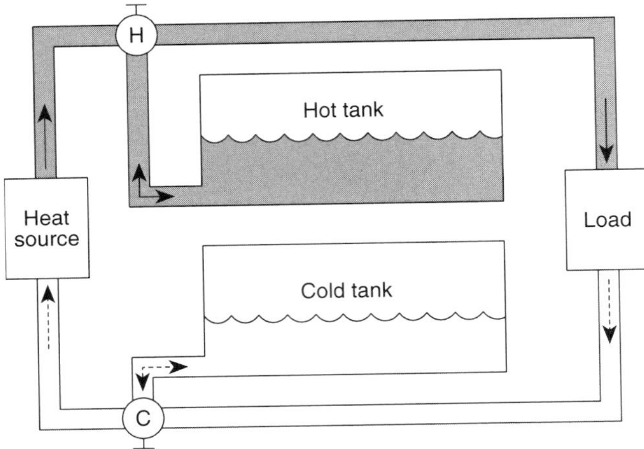

FIGURE 15.4 Two-tank thermal storage system; hot water is shown in gray and cold water is shown in white. When the heat source is producing more output than required for the load, valve H is turned to deposit hot liquid in the tank. When it is producing less than required for the load, the valve is turned to provide supplemental heat from the storage tank. Note that each tank must be large enough to hold the entire fluid capacity of the system.

Liquid sensible heat storage systems are strongly characterized not just by the choice of heat-transfer fluid, but also by the system architecture. Two-tank systems store the cold and hot liquids in separate tanks (Figure 15.4). Thermocline systems use a single tank with cold fluid entering or leaving the bottom of the tank and hot fluid entering or leaving the top (Figure 15.5). Thermocline systems can be particularly low cost because they minimize the required tank volume, but require careful design to prevent mixing of the hot and cold fluid.

One particularly interesting application of the thermocline concept is nonconvecting, salinity-gradient solar ponds that employ the concept in reverse. Solar ponds are both an energy collection and energy storage technology. Salts are dissolved in the water to introduce a density gradient, with the densest (saltiest) water on the bottom and lightest (freshest) on top. Solar radiation striking the dark bottom of the pond heats the

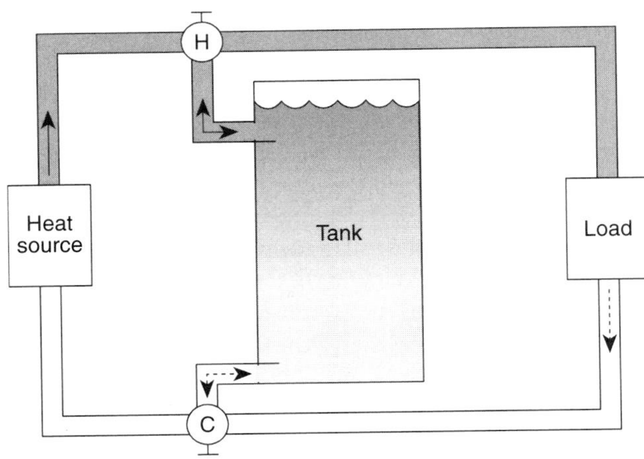

FIGURE 15.5 Thermocline storage tank. Thermocline storage tanks are tall and narrow to encourage the gravity-assisted separation of hot and cold fluid, and include design features (especially at the input/output connectors) to prevent mixing in the stored fluid.

densest water, but convection of the heated water to the top cannot occur because the density gradient prevents it. Salinity-gradient ponds can generate and store hot water at temperatures approaching 95°C.[22]

15.9.1.2 Solids

Storage of sensible heat in solids is usually most effective when the solid is in the form of a bed of small units, rather than a single mass. The reason is that the surface-to-volume ratio increases with the number of units, so that heat transfer to and from the storage device is faster for a greater number of units. Energy can be stored or extracted from a thermal storage bed by passing a gas (such as air) through the bed. Thermal storage beds can be used to extract and store the latent heat of vaporization from water contained in flue gases.

Although less effective for heat transfer, monolithic solid storage has been successfully used in architectural applications and solar cookers.

15.9.2 Latent Heat

Latent heat is absorbed or liberated by a phase change or a chemical reaction and occurs at a constant temperature. A phase change means the conversion of a homogenous substance among its various solid, liquid, or gaseous phases. One very common example is boiling water on the stovetop: though a substantial amount of heat is absorbed by the water in the pot, the boiling water maintains a constant temperature of 100°C. The latent heat, E_s, stored through a phase change is:

$$E_s = lM,$$

where M is the mass of material undergoing a phase change (kg), and l is the latent heat of vaporization (for liquid–gas phase changes) or the latent heat of fusion (for solid–liquid phase changes), in J/kg; l is measured in units of energy per mass. Conservation of energy dictates that the amount of heat absorbed in a given phase change is equal to the amount of heat liberated in the reverse phase change.

Although the term *phase change* is used here to refer only to straightforward freezing and melting, many sources use the term *phase-change materials* or *PCMs* to refer to any substance storing latent heat (including those described in Section 15.1.9.6 and Section 15.1.9.7, as well.)

15.9.2.1 Phase Change

Practical energy storage systems based on a material phase change are limited to solid–solid and solid–liquid phase changes. Changes involving gaseous phases are of little interest due to the expense associated with containing a pressurized gas, and difficulty of transferring heat to and from a gas.

Solid–solid phase changes occur when a solid material reorganizes into a different molecular structure in response to temperature. One particularly interesting example is lithium sulfate (Li_2SO_4) which undergoes a change from a monoclinic structure to a face-centered cubic structure at 578°C, absorbing 214 J/g in the process, more than most solid–liquid phase changes.[23]

Some common chemicals, their melting points and heats of fusion are listed in Table 15.4. Fatty acids and paraffins received particular attention in the 1990s as candidate materials for the heat storage component of phase-change drywall, a building material designed to absorb and release heat energy near room temperature for the purpose of indoor temperature stabilization.[24] In this application, solids in the drywall maintain the material's structural integrity even though the phase-change materials are transitioning between solid and liquid states.

15.9.2.2 Hydration–Dehydration

In this process, a salt or similar compound forms a crystalline lattice with water below a "melting-point" temperature, and at the melting point the crystal dissolves in its own water of hydration. Sodium sulfate (Na_2SO_4) is a good example, forming a lattice with ten molecules of water per molecule of sulfate ($Na_2SO_4 \cdot 10H_2O$) and absorbing 241 J/g at 32°C.[25]

TABLE 15.4 Melting Points and Heats of Fusion for Solid–Liquid Phase Changes

	Melting Point°C	Heat of Fusion J/g
Aluminum bromide	97	42
Aluminum iodide	191	81
Ammonium bisulfate	144	125
Ammonium nitrate	169	77
Ammonium thiocyanate	146	260
Anthracine	96	105
Arsenic tribromide	32	37
Beeswax	62	177
Boron hydride	99	267
Metaphosphoric acid	43	107
Naphthalene	80	149
Naphthol	95	163
Paraffin	74	230
Phosphoric acid	70	156
Potassium	63	63
Potassium thiocyanate	179	98
Sodium	98	114
Sodium hydroxide	318	167
Sulfur	110	56
Tallow	76	198
Water	0	335

Source: From Kreith, F. and Kreider J.F., *Principles of Solar Engineering*, Taylor & Francis, 1978. With permission

Hydration–dehydration reactions have not found significant application in renewable energy systems, although they, too, have been a candidate for phase-change drywall.

15.9.2.3 Chemical Reaction

A wide variety of reversible chemical reactions are available that release and absorb heat (see, for example, Hanneman, Vakil, and Wentorf[26]). The principal feature of this category of latent heat storage technologies is the ability to operate at extremely high temperatures, in some cases over 900°C. Extremely high temperature applications have focused primarily on fossil and advanced nuclear applications; to date, none of these chemical methods of heat storage have been deployed in commercial renewable energy applications.

15.10 Thermochemical Energy Storage

This section provides an overview of biomass storage technologies from an energetic perspective only. Additional details on biomass fuels are presented in Chapter 22.

15.10.1 Biomass Solids

Plant matter is a storage medium for solar energy. The input mechanism is photosynthesis conversion of solar radiation into biomass. The output mechanism is combustion of the biomass to generate heat energy.

Biologists measure the efficiency of photosynthetic energy capture with the metric net primary productivity (NPP), which is usually reported as a yield in units similar to dry Mg/ha-yr (dry metric tons per hectare per year). However, to enable comparisons of biomass with other solar energy storage technologies, it is instructive to estimate a solar efficiency by multiplying the NPP by the biomass heating value (e.g., MJ/dry Mg) and then dividing the result by the average insolation at the crop's location (e.g., MJ/ha-yr). The solar efficiency is a unitless value describing the fraction of incident solar energy ultimately available as biomass heating value. Most energy crops capture between 0.2 and 2% of the incident solar energy in heating value of the biomass; Table 15.5 shows examples of solar efficiencies estimated for a number of test crops.

The principal method for extracting useful work or electricity from biomass solids is combustion. Therefore, the solar efficiencies listed in Table 15.5 need to be multiplied by the efficiency of any associated combustion process to yield a net solar efficiency. For example, if a boiler-based electric generator extracts 35% of the feedstock energy as electricity, and the generator is sited at a switchgrass plantation achieving 0.30% solar capture efficiency on a mass basis, the electric plant has a net solar efficiency of $0.30\% \times 35\% = 0.11\%$. Because biomass is a low-efficiency collector of solar energy, it is very land intensive compared to photovoltaic or solar thermal collectors that deliver energy at solar efficiencies over 20% (see Chapter 19 and Chapter 20 for a full discussion). However, the capacity of land to store standing biomass over time is extremely high, with densities up to several hundred Mg/ha (and therefore several thousand GJ/ha), depending on the forest type. Standing biomass can serve as long-term storage, although multiple stores need to be used to accommodate fire risk. For short-term storage, woody biomass may be dried, and is frequently chipped or otherwise mechanically treated to create a fine and homogenous fuel suitable for burning in a wider variety of combustors.

TABLE 15.5 Primary Productivity and Solar Efficiency of Biomass Crops

Location	Crop	Yield (dry Mg/ha-yr)	Average Insolation (W/m^2)	Solar Efficiency(%)
Alabama	Johnsongrass	5.9	186	0.19
Alabama	Switchgrass	8.2	186	0.26
Minnesota	Willow and hybrid poplar	8–11	159	0.30–0.41
Denmark	Phytoplankton	8.6	133	0.36
Sweden	Enthropic lake angiosperm	7.2	106	0.38
Texas	Switchgrass	8–20	212	0.22–0.56
California	*Euphorbia lathyris*	16.3–19.3	212	0.45–0.54
Mississippi	Water hyacinth	11.0–33.0	194	0.31–0.94
Texas	Sweet sorghum	22.2–40.0	239	0.55–0.99
Minnesota	Maize	24.0	169	0.79
West Indies	Tropical marine angiosperm	30.3	212	0.79
Israel	Maize	34.1	239	0.79
Georgia	Subtropical saltmarsh	32.1	194	0.92
Congo	Tree plantation	36.1	212	0.95
New Zealand	Temperate grassland	29.1	159	1.02
Marshall Islands	Green algea	39.0	212	1.02
New South Wales	Rice	35.0	186	1.04
Puerto Rico	*Panicum maximum*	48.9	212	1.28
Nova Scotia	Sublittoral seaweed	32.1	133	1.34
Colombia	Pangola grass	50.2	186	1.50
West Indies	Tropical forest, mixed ages	59.0	212	1.55
California	Algae, sewage pond	49.3–74.2	218	1.26–1.89
England	Coniferous forest, 0–21 years	34.1	106	1.79
Germany	Temperate reedswamp	46.0	133	1.92
Holland	Maize, rye, two harvests	37.0	106	1.94
Puerto Rico	*Pennisetum purpurcum*	84.5	212	2.21
Hawaii	Sugarcane	74.9	186	**2.24**
Java	Sugarcane	86.8	186	2.59
Puerto Rico	Napier grass	106	212	2.78
Thailand	Green algae	164	186	4.90

Source: From Klass, D. L., Biomass for Renewable Energy, Fuels, and Chemicals, Academic Press, San Diego, CA, 1998. With permission.

15.10.2 Ethanol

Biomass is a more practical solar energy storage medium if it can be converted to liquid form. Liquids allow for more convenient transportation and combustion, and enable extraction on demand (through reciprocating engines) rather than through a less dispatchable, boiler- or turbine-based process. This latter property also enables its use in automobiles.

Biomass grown in crops or collected as residue from agricultural processes consists principally of cellulose, hemicellulose, and lignin. The sugary or starchy by-products of some crops such as sugarcane, sugar beet, sorghum, molasses, corn, and potatoes can be converted to ethanol through fermentation processes, and these processes are the principal source of ethanol today. Starch-based ethanol production is low efficiency, but does succeed in transferring about 16% of the biomass heating value to the ethanol fuel.[27]

When viewed as a developing energy storage technology, ethanol derived from cellulose shows much more promise than the currently prevalent starch-based ethanol.[28] Cellulosic ethanol can be manufactured with two fundamentally different methods: either the biomass is broken down to sugars using a hydrolysis process, and then the sugars are subjected to fermentation; or the biomass is gasified (see below), and the ethanol is subsequently synthesized from this gas with a thermochemical process. Both processes show promise to be far cheaper than traditional ethanol manufacture via fermentation of starch crops, and will also improve energy balances. For example, it is estimated that dry sawdust can yield up to 224 L/Mg of ethanol, thus recovering about 26% of the higher heating value of the sawdust.[29] Because the ethanol will still need to be combusted in a heat engine, the gross, biomass-to-useful-work efficiency will be well below this. In comparison, direct combustion of the biomass to generate electricity makes much more effective use of the biomass as an energy storage medium. Therefore, the value of ethanol as an energy storage medium lies mostly in the convenience of its liquid (rather than solid) state.

15.10.3 Biodiesel

As starch-based ethanol is made from starchy by-products, most biodiesel is generated from oily by-products. Some of the most common sources are rapeseed oil, sunflower oil, and soybean oil. Biodiesel yields from crops like these range from about 300 to 1000 kg/ha-yr, but the crop as a whole produces about 20 Mg/ha-yr, meaning that the gross solar capture efficiency for biodiesel from crops ranges between 1/20 and 1/60 the solar capture efficiency of the crop itself. Because of this low solar-capture efficiency, biomass cannot be the principal energy storage medium for transportation needs.[30]

Biodiesel can also be manufactured from waste vegetable or animal oils; however, in this case, the biodiesel is not functioning per se as a solar energy storage medium, so is not further treated in this work.

15.10.4 Syngas

Biomass can be converted to a gaseous state for storage, transportation, and combustion (or other chemical conversion).[31] Gasification processes are grouped into three different classes: *pyrolysis* is the application of heat in anoxic conditions; *partial oxidation* is combustion occurring in an oxygen-starved environment; *reforming* is the application of heat in the presence of a catalyst. All three processes form syngas, a combination of methane, carbon monoxide, carbon dioxide and hydrogen. The relative abundances of the gaseous products can be controlled by adjusting heat, pressure, and feed rates. The HHV of the resulting gas can contain up to 78% of the original HHV of the feedstock, if the feedstock is dry.[29] Compositions and heating values of two example syngases are listed in Table 15.3.

The equivalent of up to 10% of the gas HHV will be lost when the gas is pressurized for transportation and storage. Even with this loss, gasification is a considerably more efficient method than ethanol manufacture for transferring stored solar energy to a nonsolid medium.

References

1. Linden, D. and Reddy, T. B. 2002. *Handbook of Batteries, 3rd Ed.*, pp. 40–163. McGraw Hill, New York.
2. Luongo, C. A. 1996. Superconducting storage systems: An overview. *IEEE Transactions on Magnetics, Vol. 32*, 2214–2223.
3. Cheung, K. Y. C., Cheung, S. T. H. et al. 2003. *Large Scale Energy Storage Systems*, pp. 40–163. Imperial College London, London.
4. EA Technology, *Review of Electrical Energy Storage Technologies and Systems and of their Potential for the UK*, pp. 40–163. Department of Trade and Industry, London, UK.
5. DeVries, T. 2002. World's biggest battery helps stabilize Alaska. *Modern Power Systems*, 22, 40.
6. Nourai, A. 2003. NaS battery demonstration in the USA. *Electricity Storage Association Spring Meeting*, Electricity Storage Association, Morgan Hill, California.
7. Sudworth, J. L. 2001. Sodium/nickel chloride (Zebra) battery. *Journal of Power Sources*, 100, 149–163.
8. Fleets and Fuels. 2003. *The Zebra*, (February 17), 8.
9. Price, A. 2000. Technologies for energy storage, present and future: Flow batteries. *2000 Power Engineering Society Meeting*. IEEE Power Engineering Society, Piscataway, New Jersey.
10. Skyllas-Kazacos, M. 2000. *Recent Progress with the Vanadium Redox Battery*, pp. 27–148. University of New South Wales, Sydney.
11. Menictas, C. 1994. Status of the vanadium battery development program. *Proceedings of the Electrical Engineering Congress*, Sydney.
12. VRB Power Systems, Inc. http://www.vrbpower.com (accessed August 15, 2005).
13. Wilks, N. 2000. Solving current problems. *Professional Engineering*, 13, 27.
14. Scheffler, P. 2001. *Environmental Assessment for the Regenesys Energy Storage System*, pp. 142–148. Tennessee Valley Authority.
15. Lex, P. and Jonshagen, B. 1999. The zinc/bromide battery system for utility and remote applications. *Power Engineering Journal*, 13, 142–148.
16. Kruger, P. 2001. Electric power requirement for large-scale production of hydrogen fuel for the world vehicle fleet. *International Journal of Hydrogen Energy*, 26, 1137–1147.
17. Bossel, U., Eliasson, B., and Taylor, G. 2003. *The Future of the Hydrogen Economy: Bright or Bleak? Oberrohrdorf*, pp. 1863–1874. European Fuel Cell Forum, Switzerland.
18. Dillon, A. et al. Storage of hydrogen in single-walled carbon nanotubes. *Nature*, 386, 377–379.
19. Donalek, P. 2003. Advances in pumped storage. *Electricity Association Spring Meeting*, Chicago, IL.
20. Kondoh, J. et al. 2000. Electrical energy storage systems for energy networks. *Energy Conservation and Management*, 41, 1863–1874.
21. van der Linden, S. 2002. The case for compressed air energy system. *Modern Power Systems*, 22 19–21.
22. Hull, J. et al. 1988. *Salinity Gradient Solar Ponds*, pp. 393–403. CRC Press, Boca Raton, FL.
23. Sørensen, B. 2004. *Renewable Energy: Its Physics, Engineering, Environmental Impacts, Economics and Planning, 3rd Ed.*, pp. 393–403. Elsevier Academic Press, Burlington, MA.
24. Neeper, D. A. 2000. Thermal dynamics of wallboard with latent heat storage. *Solar Energy*, 68, 393–403.
25. Goswami, D. Y., Kreith, F., and Kreider, J. F. 2000. *Principles of Solar Engineering, 2nd Ed.*, pp. 631–653. Taylor & Francis, New York.
26. Hanneman, R., Vakil, H., and Wentorf, R. Jr. 1974. Closed loop chemical systems for energy transmission, conversion and storage, *Proceedings of the 9th Intersociety Energy Conversion Engineering Conference*, pp. 631–653. American Society of Mechanical Engineers, New York.
27. Shapouri, H., Duffield, J. A., Wang, M. 2002. *The Energy Balance of Corn Ethanol: An Update. Report 814*. USDA Office of Energy Policy and New Uses, Agricultural Economics, Washington, DC.

28. Hammerschlag, R. 2006. Ethanol's energy return on investment: A survey of the literature 1990–Present. *Environmental Science and Technology*, 40, 1744–1750.
29. Klass, D. L. 1998. *Biomass for Renewable Energy, Fuels, and Chemicals*, pp. 631–653. Academic Press, San Diego.
30. Bockey, D. and Körbitz, W. 2003. *Situation and Development Potential for the Production of Biodiesel—An International Study*, pp. 631–653. Union zur Forderung von Oel- und Proteinpflanzen, Berlin.
31. Bridgwater, A. V. 1995. The technical and economic feasibility of biomass gasification for power generation. *Fuel*, 74, 631–653.

16

Nuclear Power Technologies

16.1 Introduction 16-1
16.2 Development of Current Power-Reactor Technologies 16-2
Current Nuclear Power Plants Worldwide • Pressurized-Water Reactors • Boiling-Water Reactors • Pressurized Heavy-Water Reactor • Gas-Cooled Reactors • Other Power Reactors • Growth of Nuclear Power
16.3 Next-Generation Technologies 16-8
Light-Water Reactors • Heavy-Water Reactors • High-Temperature Gas-Cooled Reactors • Fast-Neutron Reactors • Summary of Generation-III Reactors
16.4 Generation-IV Technologies 16-11
Gas-Cooled Fast-Reactor System • Very-High-Temperature Reactor • Supercritical-Water-Cooled Reactor • Sodium-Cooled Fast Reactor • Lead-Cooled Fast Reactor • Molten-Salt Reactor
16.5 Fuel Cycle 16-20
Uranium and Thorium Resources • Mining and Milling • Conversion and Enrichment • Fuel Fabrication and Use • Reprocessing • Spent-Fuel Storage • Spent-Fuel Transportation
16.6 Nuclear Waste 16-26
Types of Radioactive Wastes
16.7 Nuclear Power Economics 16-29
16.8 Conclusions 16-29
References 16-30
For Further Information 16-31

Edwin A. Harvego
Idaho National Laboratory

Kenneth D. Kok
WSMS Mid-America

16.1 Introduction

Nuclear power is derived from the fission of heavy element nuclei or the fusion of light element nuclei. This chapter will discuss nuclear power derived from the fission process because fusion as a practical power source will not reach the stage of commercial development in the next 20–25 years. In a nuclear reactor, the energy available from the fission process is captured as heat that is transferred to working fluids that are used to generate electricity. Uranium-235 (^{235}U) is the primary fissile fuel currently used in nuclear power plants. It is an isotope of uranium that occurs naturally at about 0.72% of all natural uranium deposits. When ^{235}U is "burned" (fissioned) in a reactor, it provides about one megawatt day of energy for each gram of ^{235}U fissioned (3.71×10^{10} Btu/lb).

Nuclear power technology includes not only the nuclear power plants that produce electric power but also the entire nuclear fuel cycle. Nuclear power begins with the mining of uranium. The ore is processed and converted to a form that can be enriched in the ^{235}U isotope so that it can be used efficiently in today's light-water-moderated reactors. The reactor fuel is then fabricated into appropriate fuel forms for use in nuclear power plants. Spent fuel can then be either reprocessed or stored for future disposition. Radioactive waste materials are generated in all of these operations and must be disposed of. The transportation of these materials is also a critical part of the nuclear fuel cycle.

In this chapter, the development, current use, and future of nuclear power will be discussed. The second section of this chapter is a brief review of the development of nuclear energy as a source for production of electric power, and looks at nuclear power as it is deployed today both in the United States and worldwide. The third section examines the next generation of nuclear power plants that will be built. The fourth section reviews concepts being proposed for a new generation of nuclear power plants. The fifth section describes the nuclear fuel cycle, beginning with the availability of fuel materials and ending with a discussion of fuel reprocessing technologies. The sixth section discusses nuclear waste and the options for its management. The seventh section addresses nuclear power economics. Conclusions are presented in Section 16.2.8.

16.2 Development of Current Power-Reactor Technologies

The development of nuclear reactors for power production began following World War II when engineers and scientists involved in the development of the atomic bomb recognized that controlled nuclear chain reactions could provide an excellent source of heat for the production of electricity. Early research on a variety of reactor concepts culminated in President Eisenhower's 1953 address to the United Nations in which he gave his famous "Atoms for Peace" speech, in which he pledged the United States "to find the way by which the miraculous inventiveness of man shall not be dedicated to his death, but consecrated to his life." In 1954, President Eisenhower signed the 1954 Atomic Energy Act that fostered the cooperative development of nuclear energy by the Atomic Energy Commission (AEC) and private industry. This marked the beginning of the commercial nuclear power program in the United States.

The world's first large-scale nuclear power plant was the Shippingport Atomic Power Station in Pennsylvania, which began operation in 1957. This reactor was a pressurized-water reactor (PWR) nuclear power plant designed and built by the Westinghouse Electric Company and operated by the Duquesne Light Company. The plant produced 68 MWe and 231 MWt.

The first commercial-size boiling-water reactor (BWR) was the Dresden Nuclear Power Plant that began operation in 1960. This 200 MWe plant was owned by the Commonwealth Edison Company and was built by the General Electric Company at Dresden, Illinois, about 50 miles southwest of Chicago.

Although other reactor concepts, including heavy-water-moderated, gas-cooled and liquid-metal-cooled reactors, have been successfully operated, the PWR and BWR reactor designs have dominated the commercial nuclear power market, particularly in the U.S. These commercial power plants rapidly increased in size from the tens of MWe generating capacity to over 1000 MWe. Today, nuclear power plants are operating in 33 countries. The following section presents the current status of nuclear power plants operating or under construction around the world.

16.2.1 Current Nuclear Power Plants Worldwide

At the end of 2004 there were 439 individual nuclear power reactors operating throughout the world. More than half of these nuclear reactors are PWRs. The distribution of current reactors by type is listed in Table 16.1. As shown in Table 16.1, there are six types of reactors currently used for electricity generation

TABLE 16.1 Nuclear Power Units by Reactor Type, Worldwide

Reactor Type	Main Countries	# Units Operational	GWe	Fuel
Pressurized light-water reactors (PWR)	U.S., France, Japan, Russia	263	237	Enriched UO_2
Boiling light-water reactors (BWR and AWBR)	U.S., Japan, Sweden	92	81	Enriched UO_2
Pressurized heavy-water reactors—CANDU (PHWR)	Canada	38	19	Natural UO_2
Gas-cooled reactors (Magnox & AGR)	U.K.	26	11	Natural U (metal), enriched UO_2
Graphite-moderated light-water reactors (RBMK)	Russia	17	13	Enriched UO_2
Liquid-metal-cooled fast-breeder reactors (LMFBR)	Japan, France, Russia	3	1	PuO_2 and UO_2
		439	362	

Source: Information taken from World Nuclear Association Information Paper "Nuclear Power Reactors".

throughout the world. The following sections provide a more detailed description of the different reactor types shown in the table.

16.2.2 Pressurized-Water Reactors

Pressurized-water reactors represent the largest number of reactors used to generate electricity throughout the world. They range in size from about 400–1500 MWe. The PWR shown in Figure 16.1 consists of a reactor core that is contained within a pressure vessel and is cooled by water under high pressure. The nuclear fuel in the core consists of uranium dioxide fuel pellets enclosed in zircaloy rods that are held together in fuel assemblies. There are 200–300 rods in an assembly and

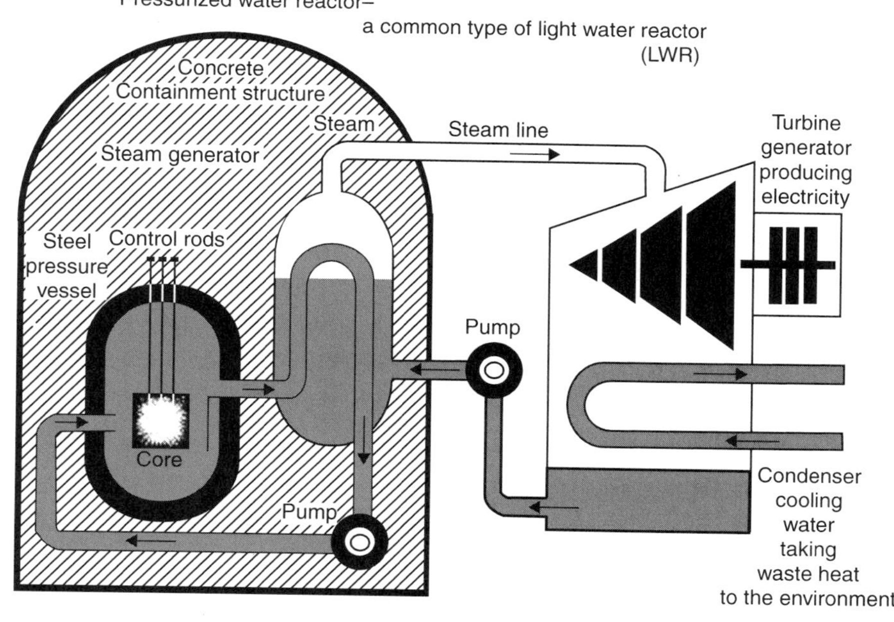

FIGURE 16.1 Sketch of a typical PWR power plant. (From World Nuclear Association, http://www.world-nuclear.org.)

100–200 fuel assemblies in the reactor core. The rods are arranged vertically and contain 80–100 tons of enriched uranium.

The pressurized water at 315°C is circulated to the steam generators. The steam generator is a tube- and shell-type of heat exchanger with the heated high-pressure water circulating through the tubes. The steam generator isolates the radioactive reactor cooling water from the steam that turns the turbine generator. Water enters the steam generator shell side and is boiled to produce steam that is used to turn the turbine generator producing electricity. The pressure vessel containing the reactor core and the steam generators are located in the reactor containment structure. The steam leaving the turbine is condensed in a condenser and returned to the steam generator. The condenser cooling water is circulated to cooling towers where it is cooled by evaporation. The cooling towers are often pictured as an identifying feature of a nuclear power plant.

16.2.3 Boiling-Water Reactors

The BWR power plants represent the second-largest number of reactors used for generating electricity. The BWRs range in size from 400 to 1200 MWe. The BWR, shown in Figure 16.2, consists of a reactor core located in a reactor vessel that is cooled by circulating water. The cooling water is heated to 285°C in the reactor vessel and the resulting steam is sent directly to the turbine generators. There is no secondary loop as there is in the PWR. The reactor vessel is contained in the reactor building. The steam leaving the turbine is condensed in a condenser and returned to the reactor vessel. The condenser cooling water is circulated to the cooling towers where it is cooled by evaporation.

16.2.4 Pressurized Heavy-Water Reactor

The so-called CANDU reactor was developed in Canada beginning in the 1950s. It consists of a large tank called a calandria containing the heavy-water moderator. The tank is penetrated horizontally by pressure tubes that contain the reactor fuel assemblies. Pressurized heavy water is passed over the fuel and heated

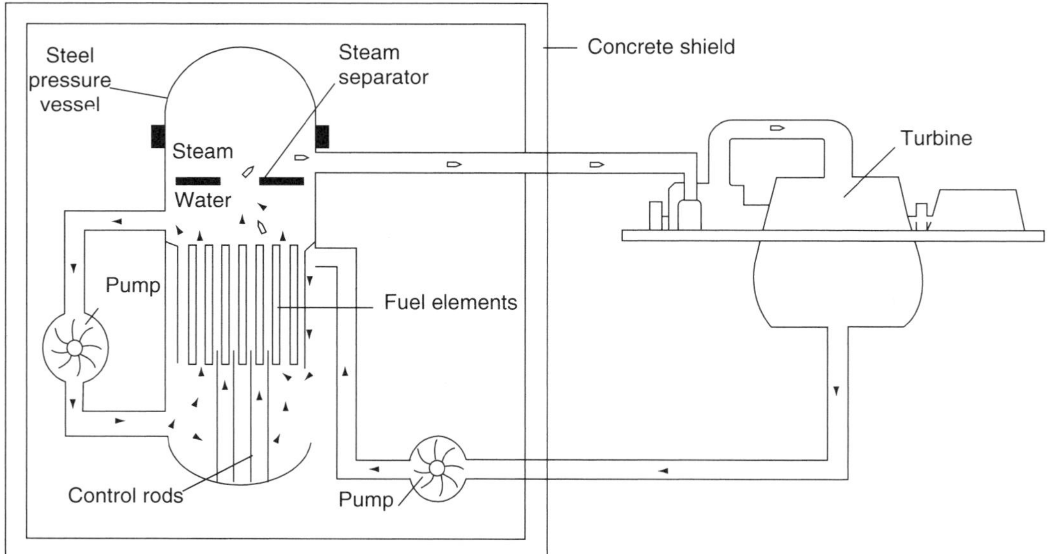

FIGURE 16.2 Sketch of a typical BWR power plant. (From World Nuclear Association, http://www.world-nuclear.org.)

to 290°C. As in the PWR, this pressurized water is circulated to a steam generator where light water is boiled, thereby forming the steam used to drive the turbine generators.

The pressure-tube design allows the CANDU reactor to be refueled while it is in operation. A single pressure tube can be isolated and the fuel can be removed and replaced while the reactor continues to operate. The heavy water in the calandria is also circulated and heat is recovered from it. The CANDU reactor is shown in Figure 16.3.

16.2.5 Gas-Cooled Reactors

Gas-cooled reactors were developed and implemented in the U.K. The first generation of these reactors was called Magnox, followed by the advanced gas-cooled reactor (AGR). These reactors are graphite moderated and cooled by CO_2. The Magnox reactors are fueled with uranium metal fuel, whereas the AGRs use enriched UO_2 as the fuel material. The CO_2 coolant is circulated through the reactor core and then to a steam generator. The reactor and the steam generators are located in a concrete pressure vessel. As with the other reactor designs, the steam is used to turn the turbine generator to produce electricity. Figure 16.4 shows the configuration for a typical gas-cooled reactor design.

16.2.6 Other Power Reactors

The remaining reactors listed in Table 16.5 are the light-water graphite-moderated reactors used in Russia, and the liquid-metal-cooled fast-breeder reactors (LMFBRs) in Japan, France, and Russia. In the

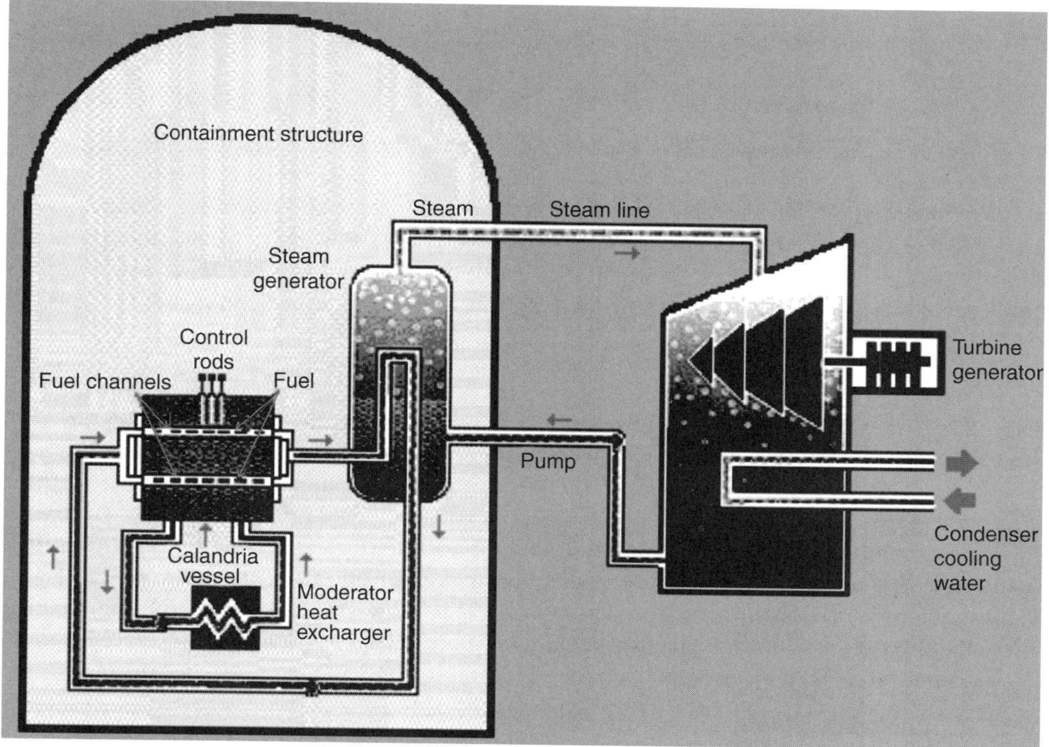

FIGURE 16.3 Sketch of a typical CANDU reactor power station. (From World Nuclear Association, http://www.world-nuclear.org.)

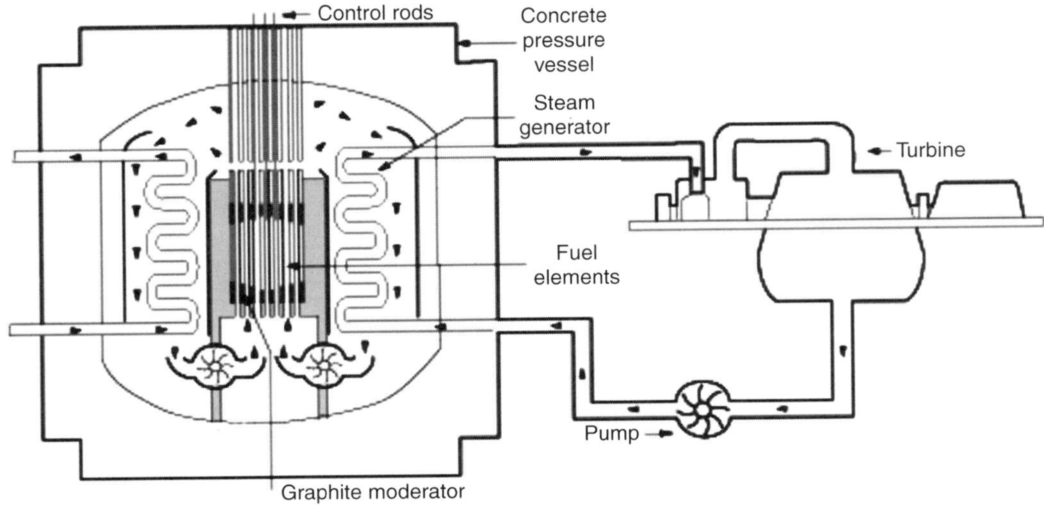

FIGURE 16.4 Sketch of a typical gasd-cooled reactor power station. (From World Nuclear Association, http://www.world-nuclear.org.)

light-water graphite-moderated reactors, the fuel is contained in vertical pressure tubes where the cooling water is allowed to boil at 290°C and the resulting steam is circulated to the turbine generator system as it is in a BWR. In the case of the LMFBR, sodium is used as the coolant and a secondary sodium cooling loop is used to provide heat to the steam generator.

16.2.7 Growth of Nuclear Power

The growth of nuclear power generation is being influenced by three primary factors. These factors are: (1) current plants are being modified to increase their generating capacity, (2) the life of old plants is being lengthened by life-extension practices that include relicensing, and (3) new construction is adding to the number of plants operating worldwide. According to the IAEA, in May 2005, there were 440 nuclear power plants in operation with a total net installed capacity of 367 GWe. They now anticipate that 60 new plants will be constructed in the next 15 years, increasing the installed capacity to 430 GWe by 2020.

16.2.7.1 Increased Capacity

Operating nuclear plants are being modified to increase their generating capacity. Reactors in the U.S., Belgium, Sweden, Germany, Switzerland, Spain, and Finland are being uprated. In the U.S., 96 reactors have been uprated since 1977, with some of them having capacity increased up to 20%. The number of operating reactors in the U.S. peaked in 1991 with a gross electrical generation of over 70,000 MW-years; however, in 2003, the net electrical generation approached 90,000 MW-years from six fewer reactors. The generating capacity increase was due to both power uprating and improvements in operation and maintenance practices to produce higher plant availability. Switzerland increased the capacity of its plants by over 12%, whereas in Spain, uprating has added 11% to that country's nuclear capacity. The uprating process has proven to be a very cost effective way to increase overall power production capacity while avoiding the high capital cost of new construction.

16.2.7.2 Plant-Life Extension

Life extension is the process by which the life of operating reactors is increased beyond the original planned and licensed life. Most reactors were originally designed and licensed for an operational life of 40 years. Without life extension, many of the reactors that were built in the 1970 s and 1980 s would reach the end of their operational lives during the years 2010–2030. If they were not replaced with new plant construction, there would be a significant decrease in nuclear-based electricity generation as these plants reached the end of their useful life.

Engineering assessments of current nuclear plants have shown that they are able to operate for longer than their original planned and licensed lifetime. Fifteen plants in the U.S. have been granted 20-year extensions to their operating licenses by the U.S. Nuclear Regulatory Commission (NRC). The operators of most of the remaining plants are also expected to apply for license extensions. This will give the plants an operating life of 60 years. In Japan, operating lifetimes of 70 years are envisaged.

The oldest nuclear power stations in the world were operated in Great Britain. Chalder Hall and Chaplecross were built in the 1950 s and were expected to operate for 20–25 years. They were authorized to operate for 50 years, but were shut down in 2003 and 2004 for economic reasons. In 2000, the Russian government extended the lives of their 12 oldest reactors by 15 years for a total of 45 years.

Although life extension has become the norm throughout the world, many reactors have been shut down due to economic, regulatory, and political reasons. Many of these reactors were built early in the development of nuclear power. They tended to be smaller in size and were originally built for

TABLE 16.2 Power Reactors under Construction

Start Operation	Country, Organization	Reactor	Type	MWe (net)
2005	Japan, Tohoku	Higashidori 1	BWR	1067
2005	India, NPCIL	Tarapur 4	PHWR	490
2005	China, CNNC	Tianwan 1	PWR	950
2005	Ukraine, Energoatom	Khmelnitski	PWR	950
2005	Russia, Rosenergoatom	Kalinin 3	PWR	950
2006	Iran, AEOI	Bushehr 1	PWR	950
2006	Japan, Hokuriku	Shika 2	ABWR	1315
2006	India, NPCIL	Tarapur 3	PHWR	490
2006	China, CNNC	Tianwan 2	PWR	950
2006	China, Taipower	Lungmen 1	ABWR	1300
2007	India, NPCIL	Rawatbhata 5	PHWR	202
2007	Romania, SNN	Cernavoda 2	PHWR	650
2007	India, NPCIL	Kudankulam 1	PWR	950
2007	India, NPCIL	Kaiga 3	PHWR	202
2007	India, NPCIL	Kaiga 4	PHWR	202
2007	USA, TVA	Browns Ferry 1	BWR	1065
2007	China, Taipower	Lungmen 2	ABWR	1300
2008	India, NPCIL	Kudankulam 2	PWR	950
2008	India, NPCIL	Rawatbhata 6	PHWR	202
2008	Russia, Rosenergoatom	Volgodonsk-2	PWR	950
2008	Korea, KHNP	Shin Kori 1	PWR	950
2009	Finland, TVO	Oikiluoto 3	PWR	1600
2009	Japan, Hokkaido	Tomari 3	PWR	912
2009	Korea, KHNP	Shin Kori 2	PWR	950
2009	Korea, KHNP	Shin Wolsong 1	PWR	950
2010	Russia, Rosenergoatom	Balakovo 5	PWR	950
2010	Russia, Rosenergoatom	Kalinin 4	PWR	950
2010	India, NPCIL	Kalpakkam	FBR	440
2010	Pakistan, PAEC	Chashma 2	PWR	300
2010	Korea, KHNP	Shin Wolsong 2	PWR	950
2010	North Korea, KEDO	Sinpo 1	PWR(KSNP)	950
2010	China, Guangdong	Lingao 3	PWR	950
2010	Russia, Rosenergoatom	Beloyarsk 4	FBR	750
2011	China, Guangdong	Lingao 4	PWR	950
2011	China, CNNC	Sanmen 1 and 2	PWR	?
2011	China, CNNC	Yangiang 1 and 2	PWR	?

Source: From World Nuclear Association, Plans for New Reactors Worldwide, http://www.world-nuclear.org, 2005.

demonstration purposes. However, the political and regulatory process in some countries has led to the termination of nuclear power programs and the shutdown of viable reactor plants.

16.2.7.3 New Nuclear Plant Construction

New nuclear power plants are currently being constructed in several countries. The majority of the new construction is in Asia. Plants currently under construction are listed in Table 16.2.

16.3 Next-Generation Technologies

The next generation, generation-III nuclear power reactors, are being developed to meet power production needs throughout the world. These reactors incorporate the lessons that have been learned by operation of nuclear power systems since the 1950 s. The reactors are designed to be safer, more economical, and more fuel efficient. The first of these reactors were built in Japan and began operation in 1996.

The biggest change in the generation-III reactors is the addition of passive safety systems. Earlier reactors relied heavily on operator actions to deal with a variety of operational upset conditions or abnormal events. The advanced reactors incorporate passive or inherent safety systems that do not require operator intervention in the case of a malfunction. These systems rely on such things as gravity, natural convection, or resistance to high temperatures.

Generation-III reactors also have:

- Standardized designs with many modules of the reactor being factory constructed and delivered to the construction site leading to expedited licensing, reduction of capital cost and reduced construction time
- Simpler designs with fewer components that are more rugged, easier to operate, and less vulnerable to operational upsets
- Longer operating lives of 60 years and designed for higher availability
- Reduced probability of accidents leading to core damage
- Higher fuel burnup reducing refueling outages and increasing fuel utilization with less waste produced

The following sections describe the different types of generation-III reactors being developed worldwide.

16.3.1 Light-Water Reactors

Generation-III advanced light-water reactors are being developed in several countries. These will be described below on a country by country basis.

16.3.1.1 United States

Even though no new reactors are being built in the U.S., U.S. companies have continued to design advanced systems in anticipation of sales both in the U.S. and other parts of the world. In the U.S., the commercial nuclear industry in conjunction with the U.S. Department of Energy (DOE) has developed four advanced light-water reactor designs.

Two of these are based on experience obtained from operating reactors in the U.S., Japan, and western Europe. These reactors will operate in the 1300-MW range. One of the designs is the advanced boiling-water reactor (ABWR). This reactor was designed in the U.S. and is already being constructed and operated in Asia. The NRC gave final design certification to the ABWR in 1997. It was noted that the design exceeded NRC "safety goals by several orders of magnitude." The other type, designated System 80+, is an advanced PWR. This reactor system was ready for commercialization, but the sale of this design is not being pursued.

The AP-600 (AP=advanced passive), designed by Westinghouse, was the second reactor system to receive NRC certification. The certification came in 1999. The reactor is designed with passive safety features that result in projected core damage frequencies nearly 1000 times less than current NRC licensing requirements.

The Westinghouse AP-1000 (a scaled up version of the AP-600) received final design approval from the NRC and is scheduled for full design certification in 2005. The passive safety systems in this reactor design lead to a large reduction in components including 50% fewer valves, 35% fewer pumps, 80% less pipe, 45% less seismic building volume, and 70% less cable.

Another aspect of the AP-1000 is the construction process. After the plant is ordered, the plant will be constructed in a modular fashion, with modules being fabricated in a factory setting and then transported to the reactor site. The anticipated design construction time for the plant is 36 months. The construction cost of an AP-1000 is expected to be $1200/kW and the generating costs are postulated to be less then 3.5 cents/kWh. The plant is designed to have a 60-year operating life. China, Europe, and the U.S. are considering purchases of the AP-1000.

General Electric has created a modification of the ABWR for the European market. The European simplified BWR is a 1300 MWe reactor with passive safety systems. It is now called the economic and simplified boiling-water reactor (ESBWR). General Electric has a 1500-MWe version of this reactor in the preapplication stage for design certification by the NRC.

An international project being led by Westinghouse is designing a modular 335-MWe reactor known as the international reactor innovative & secure (IRIS). This PWR is being designed with integral steam generators and a primary cooling system that are all contained in the reactor pressure vessel. The goal of this system is to reach an eight-year refueling cycle using 10% enriched fuel with an 80,000-MWd/t burn-up. U.S. Nuclear Regulatory Commission design certification of this plant is anticipated by 2010.

16.3.1.2 Japan

Japan has three operating ABWRs. The first two, Kashiwazaki Kariwa-6 and Kashiwazaki Kariwa-7, began operation in 1996, and the third, Hamaoka-5, started up in 2004. These plants are expected to have a 60-year life and produce power at about $0.07/kWh. Several of these plants are under construction in Japan and Taiwan.

Hitachi has completed systems design of three additional ABWRs. These are rated at 600, 900, and 1700 MWe and are based on the design of the 1350-MWe plants. The smaller versions are designed with standardized components that will allow construction times on the order of 34 months.

Westinghouse and Mitsubishi, in conjunction with four utilities, are developing a large, 1500-MWe advanced PWR. This design will have both active and passive cooling systems and will have a higher fuel burn-up of 55 GWd/t of fuel. Mitsubishi is also participating with Westinghouse in the design of the AP-1000.

16.3.1.3 South Korea

The South Koreans have the APR-1400 system that evolved from the U.S. System 80+ and is known as the Korean next generation reactor. The first of these will be Shin-Kori-3 and Shin-Kori-4. Capital cost for the first systems is estimated to be $1400/kW with future plants coming in at $1200/kW with a 48-month construction time.

16.3.1.4 Europe

Four designs are being developed in Europe to meet the European utility requirements that were derived from French and German requirements. These systems have stringent safety requirements.

Framatome ANP has designed a large (1600–1750 MWe) European pressurized-water reactor (EPR). This reactor is the new standard design in France and it received design approval in 2004. The first of these units is scheduled to be built at Olkiluoto in Finland and the second at Flamanville in France. It is capable of operating in a load following manner and will have a fuel burn-up of 65 GWd/t. It has the highest thermal efficiency of any light-water reactor at 36%.

Framatome ANP, in conjunction with German utilities and safety authorities, is developing the supercritical-water-cooled reactor (SWR), a 1000–1290-MWe BWR. This design was completed in 1999 and is ready for commercial deployment. Framatome ANP is seeking U.S. design certification for this system.

General Electric and Westinghouse are also developing designs for the European market. The General Electric system, known as the ESBWR, is 1390 MWe and is based on the ABWR. They are in the preapplication stage for a 1500-MWe version of this reactor for design certification by the U.S. NRC. Westinghouse is working with European and Scandinavian authorities on the 90+ PWR to be built in Sweden. These reactors all have passive safety systems.

16.3.1.5 Russia

Russia has also developed several advanced PWR designs with passive safety systems. The Gidropress 1000 MWe V-392 is being built in India with another planned for Novovoronezh. They are also building two VVER-91 reactors in China at Jiangsu Tainwan. The VVER-91 is designed with western control systems.

OKBM is developing the VVER-1500 for replacement of two plants each in Leningrad and Kursk. The design is planned to be complete in 2007 and the first units will be commissioned in 2012–2013.

Gidropress is developing a 640-MWe PWR with Siemans control systems which will be designated the VVER-640. OKBM is designing the VVER-600 with integral steam generators. Both of these designs will have enhanced safety systems.

16.3.2 Heavy-Water Reactors

Heavy-water reactors continue to be developed in Canada by AECL. They have two designs under development. The first, designated CANDU-9, is a 925–1300-MWe extension of the current CANDU-6. The CANDU-9 completed a two-year license review in 1997. The interesting design feature of this system is the flexible fuel requirements. Fuel materials include natural uranium, slightly enriched uranium, uranium recovered from the reprocessing of PWR fuel, mixed oxide (MOX) fuels, direct use of spent PWR fuel, and also thorium. The second design is the advanced CANDU Reactor (ACR). It uses pressurized light water as a coolant and maintains the heavy water in the calandria. The reactor is run at higher temperature and pressure, which gives it a higher thermal efficiency than earlier CANDU reactors.

The ACR-700 is smaller, simpler, cheaper, and more efficient than the CANDU-6. It is designed to be assembled from prefabricated modules that will cut the construction time to a projected 36 months. Heavy-water reactors have been plagued with a positive-void reactivity coefficient, which led some to question their safety. The ACR-700 will have a negative-void reactivity coefficient that enhances the safety of the system, as do the built-in passive safety features. AECL is seeking certification of this design in Canada, China, the U.S., and the U.K.

A follow-up to the ACR-700 is the ACR-1000, which will contain additional modules and operate in the range of 1100–1200 MWe. Each module of this design contains a single fuel channel and is expected to produce 2.5 MWe. The first of these systems is planned for operation in Ontario by 2014.

The long-range plan of AECL is to develop the CANDU-X, which will operate at a much higher temperature and pressure, yielding a projected thermal efficiency of 40%. The plan is to commercialize this plant after 2020 with a range of sizes from 350 to 1150 MWe.

India is also developing an advanced heavy-water reactor (AHWR). This reactor is part of the Indian program to utilize thorium as a fuel material. The AHWR is a 300-MWe heavy-water-moderated reactor. The fuel channels are arranged vertically in the calandria and are cooled by boiling light water. The fuel cycle will breed ^{233}U from ^{232}Th.

16.3.3 High-Temperature Gas-Cooled Reactors

The third generation of HTGRs is being designed to directly drive a gas turbine generating system using the circulating helium that cools the reactor core. The fuel material is a uranium oxycarbide in the form of small particles coated with multiple layers of carbon and silicon carbide. The coatings will contain the fission products and are stable up to 1600°C. The coated particles can be arranged in fixed graphite fuel elements or contained in "pebbles" for use in a pebble-bed-type reactor.

In South Africa, a consortium lead by the utility Eskom is developing the pebble-bed-modular reactor (PBMR). This reactor will have modules with power outputs of 165 MWe. It will utilize the direct gas turbine technology and is projected to have a thermal efficiency of 42%. The goal is to obtain a fuel burn-up of 90 GWd/t at the outset and eventually reach 200 GWd/t. The intent is to build a demonstration plant for operation in 2006 and obtain commercial operation in 2010.

In the U.S., a larger system is being design by General Atomics in conjunction with Minatom of Russia and Fuji of Japan. This reactor, designated the gas turbine-modular helium reactor (GT–MHR), utilizes hexagonal fuel elements of the kind that were used in the Fort St. Vrain reactor. The initial use of this reactor is expected to be to burn the weapons-grade plutonium at Tomsk in Russia.

16.3.4 Fast-Neutron Reactors

Several nations are working on developing improved fast-breeder reactors (FBRs). Fast-breeder reactors are fast-neutron reactors and about 20 of these reactors have operated since the 1950 s. They are able to use both ^{238}U and ^{235}U as reactor fuel, thus making use of all the uranium. These reactors use liquid metal as a coolant. In Europe, research work on the 1450-MWe European FBR has been halted.

In India, at the Indira Gandhi Centre for Atomic Research, a 40-MWt fast-breeder reactor has been operating since 1985. This reactor is used to research the use of thorium as reactor fuel by breeding ^{233}U. India has used this experience and began the construction of a 500-MWe prototype fast-breeder reactor in 2004. This unit at Kalpakkam is expected to be operating in 2010.

In Japan, the Joyo experimental reactor has been operating since 1977 and its power is now being raised to 140 MWt.

In Russia, the BN-600 FBR has been supplying electricity since 1981. It is considered to be the best operating reactor in Russia. The BN-350 FBR operated in Kazakhstan for 27 years and was used for water desalinization as well as electricity production. The BN-600 is being reconfigured to burn plutonium from the military stockpiles.

Russia has also begun construction of the BN-800 (880 MWe), which has enhanced safety features and improved fuel economy. This reactor will also be used to burn stockpiled plutonium. Russia has also experimented with lead-cooled reactor designs. A new Russian design is the BREST fast-neutron reactor. It will operate at 300 MWe or more and is an inherently safe reactor design. A pilot unit is being built at Beloyarsk. The reactor is fueled with plutonium nitride fuel and it has no blanket so no new plutonium is produced.

In the U.S., General Electric is involved in the design of a 150-MWe modular liquid-metal-cooled inherently safe reactor called PRISM. This design, along with a larger 1400-MWe design being developed jointly by GE and Argonne, has been withdrawn from NRC review.

16.3.5 Summary of Generation-III Reactors

As can be seen from the discussion above, there are many reactor systems of many types under development. The key feature of all of these reactors is the enhancement of safety systems. Some of these reactors have already been built and are in operation, whereas others are under construction. This activity indicates that there will be a growth of nuclear-reactor-generated electricity during the next 20 years. Table 16.3, taken from World Nuclear Association information on advanced nuclear power reactors, shows the advanced thermal reactors that are being marketed around the world.

16.4 Generation-IV Technologies

As discussed earlier, the development of nuclear power occurred in three general phases. The initial development of prototype reactor designs occurred in the 1950s and 1960s, development and

TABLE 16.3 Advanced Thermal Reactors Being Marketed

Country and Developer	Reactor	Size (MWe)	Design Progress	Main Features
U.S.-Japan (GE-Hitachi-Toshiba)	ABWR	1300	Commercial operation in Japan since 1996–1997, in U.S.: NRC certified 1997, first-of-a-kind engineering	Evolutionary design More efficient, less waste Simplified construction (48 months) and operation
South Korea (derived from Westinghouse)	APR-1400 (PWR)	1400	NRC certified 1997, Further developed for new S. Korean Shin Kori 3 and 4, expected to be operating in 2010	Evolutionary design Increased reliability Simplified construction and operation
U.S.A (Westinghouse)	AP-600	600	AP-600: NRC certified 1999, FOAKE	Passive safety features Simplified construction and operation 3 years to build 60-year plant life
Japan (Utilities, Westinghouse, Mitsubishi)	AP-1000 (PWR) APWR	1100 1500	AP-1000 NRC design approval 2004 Basic design in progress, planned at Tsuruga	Hybrid safety features Simpliified construction and operation
France–Germany (Framatome ANP)	EPR (PWR)	1600	Confirmed as future French standard, French design approval, to be built in Finland	Evolutionary design Improved safety features High fuel efficiency Low-cost electricity
U.S.A (GE)	ESBWR	1390	Developed from the ABWR, precertification in U.S.A	Evolutionary design Short construction time Enhanced safety features
Germany (Framatome ANP)	SWR-1000 (BWR)	1200	Under development, precertification in U.S.A	Innovative design High-fuel efficiency Passive safety features
Russia (OKBM)	V-448 (PWR)	1500	Replacement for Leningrad and Kursk plants	High-fuel efficiency Enhanced safety

Russia (Gidropress)	V-392 (PWR)	950	Two being build in India, likely bid for China	Evolutionary design 60-year plant life Enhanced safety features
Canada (AECL)	CANDU-9	925-1300	Licensing approval 1997	Evolutionary design Single stand-alone unit Flexible fuel requirements Passive safety features
Canada (AECL)	ACR	700	ACR-700: precertification in U.S.A	Evolutionary design Light-water cooling Low-enriched fuel Passive safe features
		1000	ACR-1000 proposed for U.K.	
South Africa (Eskom, BNFL)	PBMR	165 (module)	Prototype due to start building, precertification in U.S.A	Modular plant, low cost Direct cycle gas turbine High-fuel efficiency Passive safety features
U.S.A-Russia et al. (General Atomics, Minatom)	GT-MHR	285 (module)	Under development in the U.S.A. and Russia by multinational joint venture	Modular plant, low cost Direct-cycle gas turbine High-fuel efficiency Passive safety features

Source: From World Nuclear Association, Plans for New Reactors Worldwide, http://www.world-nuclear.org, 2005.

deployment of large commercial plants occurred in the 1970s and 1980s, and development of advanced light-water reactors occurred in the 1990s.

Although the earlier generations of reactors have effectively demonstrated the viability of nuclear power, the nuclear industry still faces a number of challenges that need to be overcome for nuclear power to achieve its full potential. Among these challenges are (1) public concern about the safety of nuclear power in the wake of the Three Mile Island accident in 1979 and the Chernobyl accident in 1986, (2) high capital costs and licensing uncertainties associated with the construction of new nuclear power plants, (3) public concern over potential vulnerabilities of nuclear power plants to terrorist attacks, and (4) issues associated with the accumulation of nuclear waste and the potential for nuclear material proliferation in an environment of expanding nuclear power production.

To address these concerns and to fully realize the potential contributions of nuclear power to future energy needs in the United States and worldwide, the development of a new generation of reactors, termed generation IV, was initiated in 2001. The intent or objective of this effort is to develop multiple generation-IV nuclear power systems that would be available for international deployment before the year 2030. The development of the generation-IV reactor systems is an international effort, initiated by the U.S. DOE with participation from 10 countries. These countries established a formal organization referred to as the Generation IV International Forum (GIF). The GIF countries included Argentina, Brazil, Canada, France, Japan, the Republic of Korea, the Republic of South Africa, Switzerland, the United Kingdom, and the United States. The intent of the GIF is "...to develop future-generation nuclear energy systems that can be licensed, constructed, and operated in a manner that will provide competitively priced and reliable energy products while satisfactorily addressing nuclear safety, waste, proliferation, and public perception concerns."

The process used by the GIF to identify the most promising reactor concepts for development (referred to as the Generation IV Technology Roadmap) consisted of three steps. These steps were (1) to develop a set of goals for new reactor systems, (2) solicit proposals from the worldwide nuclear community for new reactor systems to meet these goals, and (3) using experts from around the world, evaluate the different concepts to select the most promising candidates for further development.

The eight goals developed by the GIF for generation-IV nuclear systems were:

- Sustainability 1: Generation-IV nuclear energy systems will provide sustainable energy generation that meets clean air objective and promotes long-term availability of systems and effective fuel utilization for worldwide energy production.
- Sustainability 2: Generation-IV nuclear energy systems will minimize and manage their nuclear waste and notably reduce the long-term stewardship burden in the future, thereby improving protection for the public health and the environment.
- Economics 1: Generation-IV nuclear energy systems will have a clear life-cycle cost advantage over other energy sources.
- Economics 2: Generation-IV nuclear energy systems will have a level of financial risk comparable to other energy projects.
- Safety and reliability 1: Generation-IV nuclear energy systems operations will excel in safety and reliability.
- Safety and reliability 2: Generation IV nuclear energy systems will have a very low likelihood and degree of reactor core damage.
- Safety and reliability 3: Generation-IV nuclear energy systems will eliminate the need for offsite emergency response.
- Proliferation resistance and physical protection: Generation-IV nuclear energy systems will increase the assurance that they are a very unattractive and the least desirable route for diversion or theft of weapons-usable materials, and provide increased physical protection against acts of terrorism.

Over 100 generation-IV candidates were evaluated by experts from the GIF countries and six reactor systems were selected for further evaluation and potential development. The six reactor systems selected were:

16.4.1 Gas-Cooled Fast-Reactor System

The gas-cooled fast-reactor system (GFR) is a fast-neutron spectrum reactor that uses helium as the primary coolant. It is designed to operate at relatively high helium outlet temperatures, making it a good candidate for the high-efficiency production of electricity or hydrogen. As shown in Figure 16.5 below, a direct Brayton cycle is used for the production of electricity with the helium gas delivered from the reactor outlet to a high-temperature gas turbine connected to a generator that produces electricity. In alternative designs, the high-temperature helium can also be used to produce hydrogen using either a thermochemical process or high-temperature electrolysis, or for other high-temperature process heat applications.

The reference plant is designed to produce 288 MWe using the direct Brayton cycle with a reactor outlet temperature of 850°C. The fuel forms being considered for high-temperature operation include composite ceramic fuel, advanced fuel particles, or ceramic clad elements of actinide compounds. Alternative core configurations include prismatic blocks and pin- or plate-based assemblies. The GFR's

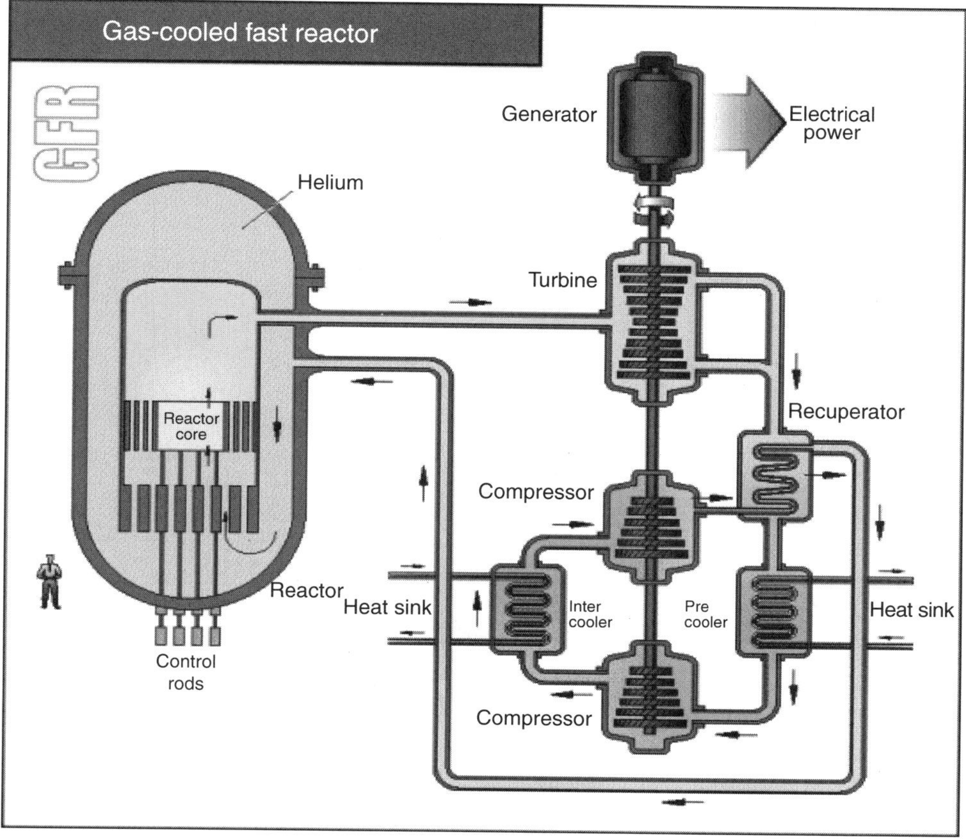

FIGURE 16.5 Gas-cooled fast reactor. (From US DOE Nuclear Energy Research Advisory Committee and the Generation IV International Forum. 2002. *A Technology Roadmap for the Generation IV Nuclear Energy Systems.*)

fast-neutron spectrum also makes it possible to efficiently use available fissile and fertile materials in a once-through fuel cycle.

16.4.2 Very-High-Temperature Reactor

The very-high-temperature reactor (VHTR) is a helium-cooled reactor designed to provide heat at very high temperatures, in the range of 1000°C for high-temperature process heat applications. In particular, the 1000°C reactor outlet temperature makes it a good candidate for the production of hydrogen using either thermochemical or high-temperature electrolysis processes. As shown in Figure 16.6 below, heat for the production of hydrogen is delivered through an intermediate heat exchanger that serves to isolate the reactor system from the hydrogen production process.

The reference design for the VHTR is a 600-MWt reactor with an outlet temperature of 1000°C. The reactor core uses graphite as a moderator to produce the thermal neutrons for the fission process. The core configuration can be either graphite blocks or pebbles about the size of billiard balls in which fuel particles are dispersed. For electricity production, either a direct Brayton cycle gas turbine using the primary helium coolant as the working fluid, or an indirect Rankine cycle using a secondary working fluid can be used. The high-temperature characteristics of this reactor concept also make it an ideal

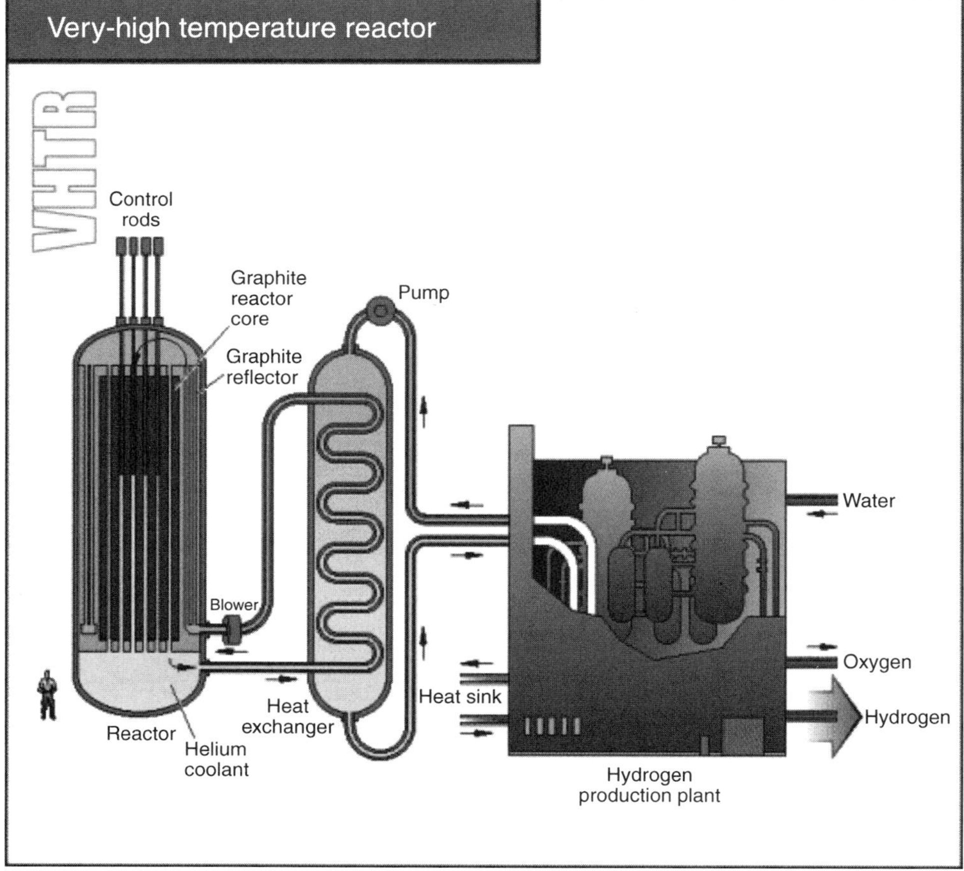

FIGURE 16.6 Very-high temperature reactor. (From US DOE Nuclear Energy Research Advisory committee and the Generation IV International Forum, *A Technology Roadmap for the Generation IV Nuclear Energy Systems*.)

candidate for cogeneration applications to meet both electricity and hydrogen production or other high-temperature process heat needs.

16.4.3 Supercritical-Water-Cooled Reactor

The supercritical-water-cooled reactor (SWR) is a relatively high-temperature, high-pressure reactor designed to operate above the thermodynamic critical point of water, which is 374°C and 22.1 MPa. Because there is no phase change in the supercritical coolant water, the balance of plant design, shown in Figure 16.7, utilizes a relatively simple direct-cycle power-conversion system. The reference design for this concept is a 1700-MWe reactor operating at a pressure of 25 MPa with a reactor outlet temperature ranging between 510 and 550°C. This reactor can be designed as either a fast-neutron-spectrum or thermal-neutron-spectrum reactor. The relatively simple design also allows for the incorporation of passive safety features similar to those of the simplified boiling-water reactor discussed earlier. However, unlike the previously discussed concepts, the lower reactor outlet temperature is not well suited for the efficient production of hydrogen, which requires minimum temperatures in the range of 850°C–900°C. Therefore, this reactor concept is primarily intended for the efficient, low-cost production of electricity.

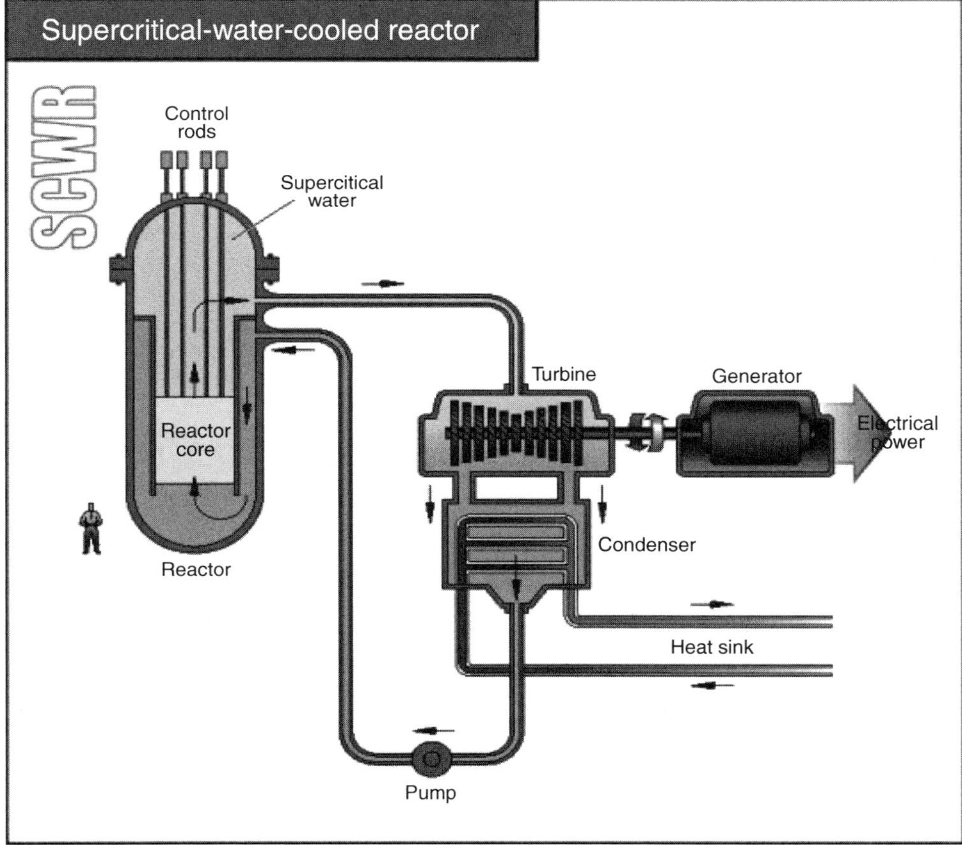

FIGURE 16.7 Supercritical-water-cooled reactor. (From US DOE Nuclear Energy Research Advisory Committee and the Generation IV International Forum, 2002. *A Technology Roadmap for the Generation IV Nuclear Energy Systems.*)

16.4.4 Sodium-Cooled Fast Reactor

The sodium-cooled fast reactor (SFR), shown in Figure 16.8, is a sodium-cooled fast-neutron-spectrum reactor designed primarily for the efficient management of actinides and conversion of fertile uranium in a closed fuel cycle. Two reference designs to support different fuel reprocessing options have been defined for this concept. The first is a medium-sized sodium-cooled reactor with a power output between 150 and 500 MWe that utilized uranium-plutonium-minor-actinide-zirconium metal alloy fuel. This reactor concept is supported by a fuel cycle based on pyrometallurgical processing in which the processing facilities are an integral part of the reactor plant design.

The second reactor reference design is a large sodium-cooled reactor with a power output capability between 500 and 1500 MWe that utilizes uranium-plutonium oxide fuel. This reactor design is supported by a fuel cycle based on an advanced aqueous process that would include a centrally located processing facility supporting a number of reactors.

Both versions of this reactor concept would operate at coolant outlet temperatures in the range of 550°C, and are intended primarily for the management of high-level waste and the production of electricity. In addition to design innovations to reduce capital costs, these reactors incorporate a number of enhanced safety features that include:

- Long thermal response time
- Large margin to coolant boiling
- Primary system that operates near atmospheric pressure
- Intermediate sodium system between the radioactive sodium in the primary system and the water and steam in the power plant.

16.4.5 Lead-Cooled Fast Reactor

The lead-cooled fast reactor (LFR) is a fast-neutron-spectrum reactor cooled by either molten lead or a lead-bismuth eutectic liquid metal. It is designed for the efficient conversion of fertile uranium and the management of actinides in a closed fuel cycle. The reactor core for this design, shown in Figure 16.9,

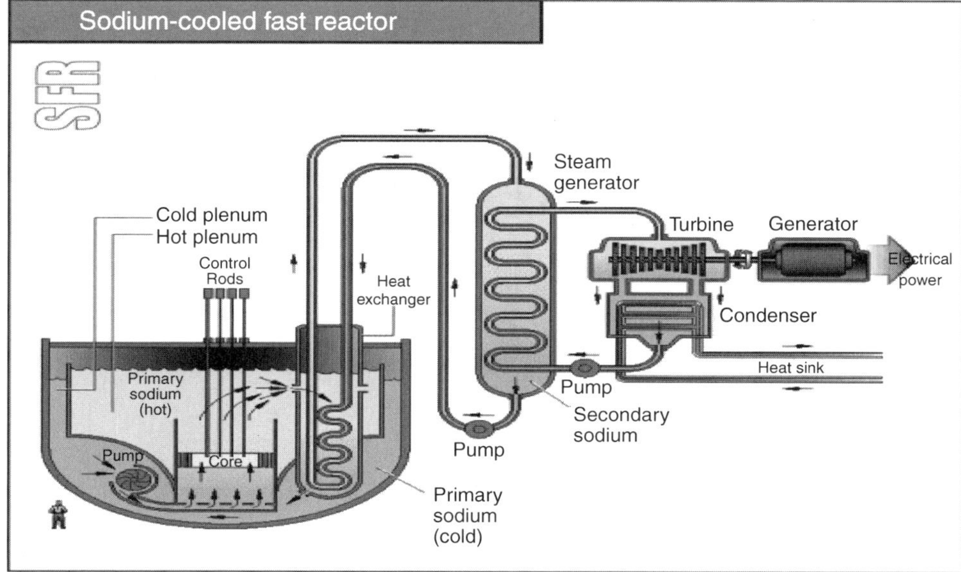

FIGURE 16.8 Sodium-cooled fast reactor. (From US DOE Nuclear Energy Research Advisory Committee and the Generation IV International Forum, 2002. *A Technology Roadmap for the Generation IV Nuclear Energy Systems.*)

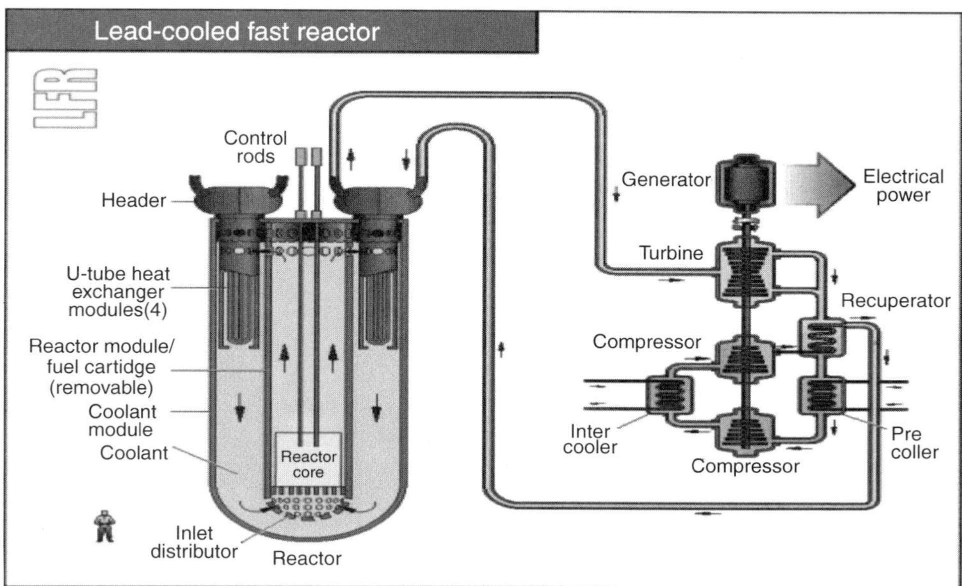

FIGURE 16.9 Lead-cooled fast reactor. (From US DOE Nuclear Energy Research Advisory Committee and the Generation IV International Forum, 2002. *A Technology Roadmap for the Generation IV Nuclear Energy Systems.*)

utilizes a metal or nitride-based fuel containing fertile uranium and transuranics. As shown in Figure 16.9, the LFR relies on natural convection to cool the reactor core. The outlet temperature for the current reactor concept is about 550°C, but with advanced materials, reactor outlet temperatures of 800°C may be possible. An indirect-gas Brayton cycle is used to produce electrical power.

There are currently three versions of the reference design for this concept. The smallest design, rated at 50–150 MWe is intended for distributed power applications or electricity production on small grids. This reactor design, referred to as a battery, features modular design with a factory fabrication "cassette" core. The reactor is designed for very long refueling intervals (15–20 years), with refueling accomplished by replacement of the cassette core or reactor module.

The other two versions of this design are a modular system rated at 300–400 MWe, and a large plant rated at 1200 MWe. The different power options for this design are intended to fill different needs or opportunities in the power market, and be economically competitive with comparable alternative power sources.

16.4.6 Molten-Salt Reactor

The molten-salt reactor (MSR), shown in Figure 16.10, produces power by circulating a molten salt and fuel mixture through graphite-core flow channels. The slowing down of neutrons by the graphite moderator in the core region provides the epithermal neutrons necessary to produce the fission power for sustained operation of the reactor. The heat from the reactor core is then transferred to a secondary system through an intermediate heat exchanger and then through a tertiary heat exchanger to the power conversion system that produces the electric power. The circulating coolant flow for this design is a mixture of sodium, uranium, and zirconium fluorides. In a closed fuel cycle, actinides such as plutonium can be efficiently burned by adding these constituents to the liquid fuel without the need for special fuel fabrication. The reference design for this concept is a 1000 MWe power plant with a coolant outlet temperature of 700°C. To achieve higher thermal efficiencies for this concept, coolant outlet temperatures as high as 800°C may also be possible.

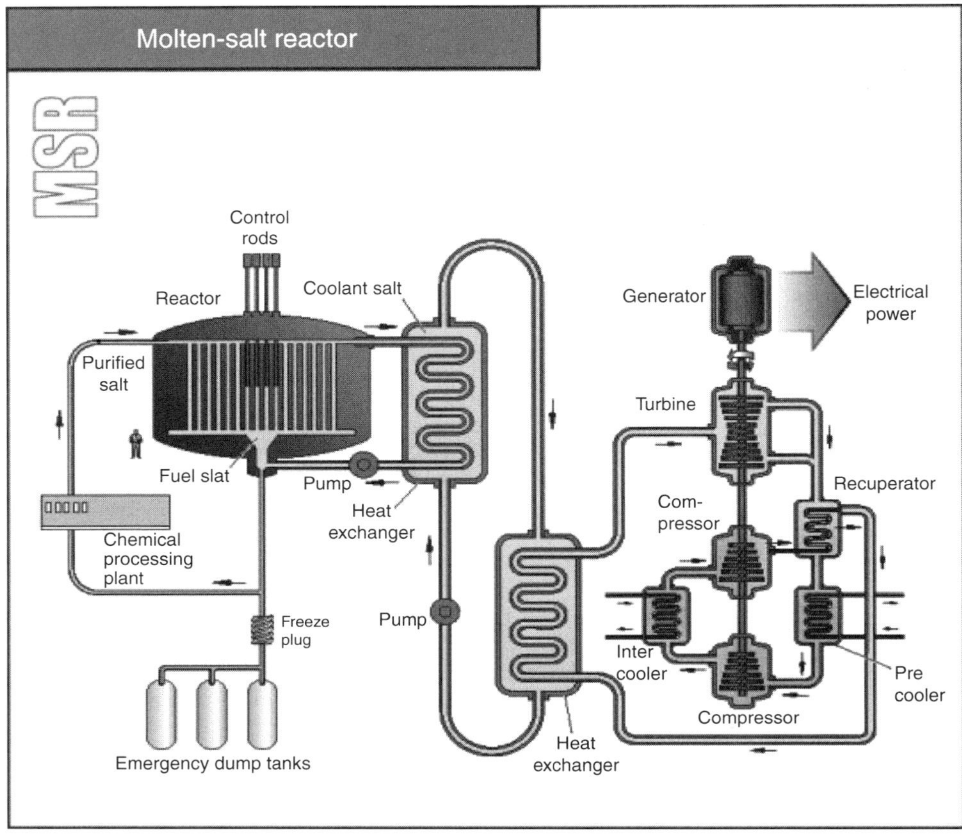

FIGURE 16.10 Molten-salt reactor. (From US DOE Nuclear Energy Research Advisory Committee and the Generation IV International Forum, 2002. *A Technology Roadmap for the Generation IV Nuclear Energy Systems.*)

16.5 Fuel Cycle

The process of following the fuel material from the uranium or thorium mine through processing and reactor operation until it becomes waste is called the fuel cycle for nuclear systems. After a discussion of the fuel cycle in general, the fuel cycle will be examined by looking at uranium and thorium resources, mining and milling, enrichment, reactor fuel use, spent fuel storage, nuclear materials transportation, and reprocessing. Nuclear waste will be addressed in a separate section.

General discussion of the fuel cycle will often include the terms "open" or "closed." The open fuel cycle is also called the *once-through* cycle. In the once-through fuel cycle, the uranium fuel is fabricated and run through the reactor once and then disposed of as waste. There is no reprocessing of the fuel. In the closed cycle, the fuel is reprocessed after leaving the reactor so that it can be reused to improve overall fuel utilization.

In the open cycle, the fuel is introduced into the reactor for one to two years. It is then removed and placed into long-term storage for eventual disposal. The impact of this cycle is the waste of about 95% of the energy contained in the fuel. The U.S. adopted the open cycle in 1977 when President Carter issued an executive order to stop reprocessing as a part of the fuel cycle. Canada has also adopted the open cycle.

The closed cycle was envisioned when the development of nuclear power began. The uranium and plutonium removed from reactors would be reprocessed and returned to reactors as fuel. Breeder reactors would be used to breed additional plutonium for use in thermal reactors. Thorium could also be used as a breeding material to generate ^{233}U as a reactor fuel. The intent of the closed fuel cycle was to maximize the use of available reactor fuel resources while minimizing waste generated by operating reactors.

Currently, reprocessing is used in Europe and Japan, but the benefits of the closed cycle have not been fully realized because there has only been limited use of the separated plutonium. As discussed above, the U.S. and Canada, for reasons described later, have not pursued closed cycle reprocessing of spent fuel. As a result, only a small fraction of the available fuel resources are utilized, and disposal of large quantities of potentially usable spent fuels has become a major issue for the U.S. nuclear industry.

16.5.1 Uranium and Thorium Resources

Uranium is a common material in the earth's crust. It is also present in sea water. Thorium is about three times more plentiful then uranium. Typical concentrations of uranium measured in parts per million (ppm) are shown in Table 16.4.

The amount of recoverable uranium is dependent upon the price. As the price increases, more material is economically recoverable. Also, more exploration will occur and it is likely that additional orebodies will be discovered. An orebody is defined as an occurrence of mineralization from which the metal, in this case uranium, can be recovered economically. Because of the uncertainties of price and its impact on exploration, any statement of recoverable amounts of uranium is simply a picture at an instant in time and is likely to change many times in the future. There is also a store of highly enriched uranium that is being recovered as nuclear weapons are dismantled. In addition, there are millions of tons of ^{238}U that are the results of previous enrichment activities around the world. The ^{238}U can be blended with highly enriched uranium or plutonium to make fuel for nuclear power plants. The ^{238}U can also be used to breed plutonium in FBR fuel cycles.

Table 16.5 presents a list of recoverable resources of uranium. The table is taken from information gathered by the World Nuclear Association from other sources and was generated in 2004.

The 3.5 Mt is enough to fuel the world's current reactors for 50 years assuming the same fuel cycles currently in use. IAEA estimates the world supply at over 14 Mt, which provides a supply exceeding 200 years at the current rate of use. This estimate does not include the uranium in phosphate deposits estimated at 22 Mt or the uranium available in seawater estimated at 1400 Mt. In addition, the ability of nuclear reactors to achieve higher burn-ups (utilize more of the uranium in the fuel) has also increased. This increases the efficiency of uranium use. Because thorium is not included in these fuel supply numbers, and as noted above is about three times as plentiful as uranium, there does not appear to be a fuel supply limitation for nuclear power in the foreseeable future.

16.5.2 Mining and Milling

Uranium is being mined using traditional underground and open-pit excavation technologies, and also using in situ leaching or solution-mining techniques.

TABLE 16.4 Typical Concentrations of Uranium

Source	Uranium Concentration (ppm)
High-grade ore: 2% U	20,000
Low-grade ore: 0.1% U	1000
Granite	4
Sedimentary rock	2
Earth's continental crust (avg)	2.8
Seawater	0.003

Source: From World Nuclear Association, Supply of Uranium, http://www.world-nuclear.org, 2004.

TABLE 16.5 Known Recoverable Resources of Uranium

Country	Tons of Uranium	Percentage of Total
Australia	989,000	28
Kazakhstan	622,000	18
Canada	439,000	12
South Africa	298,000	8
Namibia	213,000	6
Brazil	143,000	4
Russian Federation	158,000	4
U.S.A	102,000	3
Uzbekistan	93,000	3
World total	3,537,000	—

Source: From World Nuclear Association, Supply of Uranium, http://www.world-nuclear.org, 2004.

Underground mining is used when the orebody is deep underground, usually greater than 120 m deep. In underground mines, only the orebody material is extracted. Underground mining is hazardous and made more so by high concentrations of radon from the radioactive decay of the uranium. Once mined, the extracted ore is sent to a mill where the uranium in the ore is concentrated.

Open-pit technology is used when the orebody is near the surface. This leads to the excavation of large amounts of material that does not contain the ore itself. The ore that is recovered is also sent to a mill for further processing.

Solution mining involves the introduction of an aqueous solution into the orebody. The solution, oxygenated ground water, is pumped into the porous orebody and the uranium is dissolved. The uranium-rich solution is then extracted and sent to the mill for further processing.

The milling process for the solid ore material involves crushing the ore and then subjecting it to a highly acidic or alkaline solution to dissolve the uranium. Mills are normally located close to the mining activity and a single mill will often support several mines. The solution containing the uranium goes through a precipitation process that yields a material called *yellow cake.* The yellow cake contains about 80% uranium oxide. The yellow cake is packaged and sent to a conversion and enrichment facility for further processing.

16.5.3 Conversion and Enrichment

Prior to entering the enrichment process, the impure U_3O_8 is converted through a series of chemical processing steps to UF_6. During these processes, the uranium is purified. Conversion facilities are operating commercially in the U.S., Canada, France, the U.K., and Russia. UF_6 is a solid at room temperature but converts to its gaseous form at moderate temperature levels, making the compound suitable for use in the enrichment process. UF_6 is very corrosive and reacts readily with water. It is transported in large cylinders in the solid state.

Conversion of the U_3O_8 to UO_2 is also done at conversion facilities. The natural UO_2 is used in reactors such as the CANDU that do not require enriched uranium as fuel.

The first enrichment facilities were operated during the 1940 s. The electromagnetic isotope-separation process was used to separate the ^{235}U used in the first atomic bomb. The process used a magnetic field to separate the ^{235}U from the ^{238}U. As the ions were accelerated and turned, they moved differently because of the difference in their masses. Multiple stages were required and the process was very difficult to run efficiently; it was therefore soon abandoned.

Today, only two processes—gaseous diffusion and gas centrifugation—are used commercially. The capacity of enrichment plants is measured in separative work units (SWU). The SWU is a complex term that is dependent on the amount of uranium that is processed, and the concentration of ^{235}U in the product and in the tails. It is a measure of the amount of energy used in the process.

The first commercial enrichment was carried out in large gaseous diffusion plants in the U.S. It has also been used in Russia, the U.K., France, China, and Argentina. Today, operating plants remain in the U.S., France, and China, with a total nominal capacity of 30 million SWU.

In the gaseous diffusion process, UF_6 is pumped through a series of porous walls or membranes that allow more of the light ^{235}U to pass through. Because the lighter ^{235}U particles travel faster then the heavier ^{238}U particles, more of them penetrate the membrane. This process continues through a series of membranes with the concentration of ^{235}U increasing each time. For commercial reactor fuel, the process continues until the ^{235}U concentration is 3%–5%. The slower ^{238}U particles are left behind and collect as a product referred to as *tails*. The tails have a reduced concentration of ^{235}U and are commonly referred to as *depleted uranium*. This process uses a very large amount of energy and thus is very expensive to operate.

In the centrifuge enrichment process, the gaseous UF_6 is placed in a high-speed centrifuge. The spinning action forces the heavier ^{238}U particles to the outside while the lighter ^{235}U particles remain closer to the center. To obtain the enrichment required for power reactor fuel, many stages of separation are required. The arrangement is know as a *cascade*. Again, the process is continued until the ^{235}U concentration is 3%–5%. The centrifuge process uses only about 2% of the energy required by gaseous diffusion.

Table 16.6 shows the location and size of enrichment facilities around the world.

16.5.4 Fuel Fabrication and Use

Following enrichment, the UF_6 is shipped to a fuel fabrication facility. Here, the UF_6 is converted to UO_2 and pressed into cylindrical ceramic pellets. The pellets are sintered, heated to high temperature, and inserted in the fuel cladding tubes. The tubular material is zircaloy, an alloy of zirconium. The tubes are sealed forming fuel rods that are assembled into fuel assemblies and shipped to a reactor for use. All of the dimensions of the pellets and fuel rods are very carefully controlled to assure uniformity throughout the fuel assemblies.

The primary hazard in the fabrication facility is the potential for an accidental criticality because they are working with enriched uranium. Therefore, all of the processing quantities and the dimensions of the processing vessels must be controlled. This must be done even with low-enriched uranium.

A typical 1000-MWe reactor will use about 27 tons of UO_2 each year. Typical burn-up in current reactors is 33 GWd/t of uranium fed to the reactor. The energy available from the fission of uranium is 1 MW/g of uranium or 1000 GW/t. Using these numbers, the actual amount of uranium burned is only 3%–5%. This means that the unused energy available from the spent fuel, if it could be completely burned, is over 95%. During the operation of the reactor, some of the ^{238}U is converted to plutonium, which also contributes to the thermal energy of the reactor.

Advanced fuel use in reactors is estimated to be up to 200 GWd/t. In this case, about 80% of the energy available from the uranium remains in the spent fuel. These facts are the driving force behind the questions regarding reprocessing. In the once-through fuel cycle, the spent fuel will be disposed of as waste. In the closed cycle, the spent fuel is reprocessed and the remaining uranium and also the plutonium are recovered.

16.5.5 Reprocessing

In the 1940 s, reactors were operated solely for the production of plutonium for use in weapons. The fuels from the production reactors were reprocessed to recover the plutonium. The chemical processes were developed to separate the fission products and the uranium from the plutonium. The most common process was the PUREX process. This is the process that is used today by countries that reprocess power reactor fuels.

The purpose of reprocessing is to recover the uranium and plutonium in the spent fuel. As discussed above, these materials contain a large amount of potential energy if they are reused as reactor fuel. Plutonium separated in the PUREX process can be mixed with uranium to form a MOX fuel. Plutonium from the dismantlement of weapons can be used in the same way.

TABLE 16.6 Location, Size, and Type of Enrichment Facilities Around the World

Country	Owner/Controller	Plant Name/Location	Capacity (1000 SWU)
	Gaseous Diffusion Plants		
China	CNNC	Lanzhou	900
France	EURODIF	Tricastin	10,800
		Paducah, KY	11,300
United States	U.S. Enrichment Corporation	Portsmouth, OH (Closed since May, 2001)	7400
Subtotal			30,400
	Centrifuge Plants		
		Hanzhong	500
China	CNNC	Lanzhou	500
Germany	Urenco	Gronau	1462.5
	JNC	Ningyo Toge	200
Japan	Japan Nuclear Fuel Limited (JNFL)	Rokkasho-mura	1050
The Netherlands	Urenco	Almelo	1950
Pakistan	Pakistan Atomic Energy Commission (PAEC)	Kahuta	5
		Ural Electrochemical Integrated Enterprise (UEIE), Novouralsk	7000
		Siberian Chemical Combine (SKhK), Seversk	4000
Russia	Minatom	Electrochemical Plant (ECP), Zelenogorsk	3000
		Angarsk Electrolytic Chemical Combine (AEKhK), Angarsk	1000
United Kingdom	Urenco	Capenhurst	2437.5
Subtotal			23,105
Total			53,505

Source: From WISE Uranium Project, World Nuclear Fuel Facilities, http://www.wise-uranium.org, 2005.

The potential availability of separated plutonium is seen by some as a potential mechanism for the proliferation of nuclear weapons. This was the basis of the U.S. decision to halt reprocessing. In the 1970 s, research began into methods for modifying the chemical process so that the plutonium and uranium would remain together at the end of the process. In this method, called coprocessing, the short-lived fission products would be separated and the remaining uranium, plutonium, and other actinide elements would remain together. This remaining mixture would be highly radioactive, but could be remotely processed into new reactor fuel. A blend of fast neutron and thermal reactors could be used to maximize the use of this material.

The current worldwide reprocessing capability is shown in Table 16.7. These facilities all use the PUREX technology. More then 80,000 tons of commercial fuel have been reprocessed in these facilities.

Three processes are considered to be mature options for reprocessing fuel: PUREX, UREX+, and pyroprocessing. Each of these processes has certain advantages and disadvantages.

16.5.5.1 PUREX

The PUREX process is the oldest and most common reprocessing option. It uses liquid–liquid extraction to process light-water reactor spent fuel. The spent fuel is dissolved in nitric acid, and then the acid solution is mixed with an organic solvent consisting of tributyl phosphate in kerosene. The uranium and plutonium are extracted in the organic phase and the fission products remain in the aqueous phase. Further processing allows the separation of the uranium and plutonium. The advantage of this process is the long-term experience with the process. The disadvantage is that it cannot separate fission products such as technetium, cesium, and strontium, nor can it separate actinides such as neptunium, americium, and curium.

16.5.5.2 UREX+

The UREX+ process is a liquid–liquid extraction process like PUREX. It can be used for light-water reactor fuels and it includes additional extraction steps that allow separation of neptunium/plutonium, technetium, uranium, cesium/strontium, americium, and curium. The advantage of this process is that it meets the requirements for continuous recycle in light-water reactors and it builds on current technology. The disadvantage is that it cannot be used to process short-cooled fuels and it cannot be used for some specialty fuels being developed for advanced reactors.

16.5.5.3 Pyroprocessing

This process was developed and tested at Experimental Breeder Reactor-2 (EBR-2) by Argonne National Laboratory in the U.S. It is an electrochemical process rather then a liquid–liquid extraction process. Oxide fuels are first converted to metals to be processed. The metallic fuel is then treated to separate uranium and the transuranic elements from the fission products. The advantage of this process is the ability to process short-cooled and specialty fuels designed for advanced reactors. The disadvantage is

TABLE 16.7 World Commercial Reprocessing Capacity

Type of Fuel	Location	Tons/year
LWR fuel	France, La Hague	1700
	U.K., Sellafield (THORP)	900
	Russia, Ozersk (Mayak)	400
	Japan	14
	Subtotal	3000
Other nuclear fuels	U.K., Sellafield	1500
	India	275
	Subtotal	1750
Civilian capacity	Total	4750

Source: From Uranium Information Centre, Nuclear Issues Briefing Paper 72, Processing of Used Nuclear Fuel, http://www.uic.com.au, 2005.

that it does not meet the requirements for continuous recycle from thermal reactors; however, it is ideal for fuel from fast-neutron reactors.

16.5.6 Spent-Fuel Storage

Spent fuel is routinely discharged from operating reactors. As it is discharged, it is moved to the spent-fuel storage pool that is an integral part of the reactor facility. Reactors are built with storage pools that will hold fuel from many years of operation. The pools are actively cooled by circulating cooling water. The fuel stored at many of the older reactors is reaching the capacity of the on-site storage pools. At this point, the fuel is being transferred to dry storage. Dry storage takes place in large metal or concrete storage facilities. These dry facilities are passively cooled by the air circulating around them.

16.5.7 Spent-Fuel Transportation

Spent fuel is transported in large engineered containers designated as type-B containers (casks). The casks provide shielding for the highly radioactive fuel so that they can be safely handled. They are constructed of cast iron or steel. Many of them use lead as the shielding material. They are also designed to protect the environment by maintaining their integrity in the case of an accident. They are designed to withstand severe accidents, including fires, impacts, immersion, pressure, heat and cold, and are tested as part of the design certification process.

Casks have been used to transport radioactive materials for over 50 years. The IAEA has published advisory regulations for safe transportation of radioactive materials since 1961. Casks are built to standards designed to meet the IAEA advisory regulations specified by licensing authorities such as the NRC in the U.S.

Spent fuel is shipped from reactor sites by road, rail, or water. The large casks can weigh up to 110 tons and hold about 6 tons of spent fuel. Since 1971, about 7000 shipments of spent fuel (over 35,000 tons) have been transported over 30 million km with no property damage or personal injury, no breach of containment, and a very low dose rate to the personnel involved.

16.6 Nuclear Waste

Radioactive wastes are produced throughout the reactor fuel cycle. The costs of managing these wastes are included in the costs of the nuclear fuel cycle and thus are part of the electricity cost. Because these materials are radioactive, they decay with time. Each radioactive isotope has a half life, which is the time it takes for half of the material to decay away. Eventually, these materials decay to a stable nonradioactive form.

The process of managing radioactive waste involves the protection of people from the effects of radiation. The longer lived materials tend to emit alpha and beta particles. It is relatively easy to shield people from this radiation but if these materials are ingested the alpha and beta radiation can be harmful. The shorter lived materials usually emit gamma rays. These materials require greater amounts of shielding.

16.6.1 Types of Radioactive Wastes

The strict definitions of types of radioactive waste may vary from country to country. In the following discussion, the more generally accepted terminology will be used.

16.6.1.1 Mine Tailings

Mining and milling of uranium produces a sandy type of waste that contains the naturally occurring radioactive elements that are present in uranium ore. The decay of these materials produces radon gas that must be contained. This is often accomplished by covering the tailings piles with clay to contain the radon gas. Technically, tailings are not classified as radioactive waste.

16.6.1.2 Low-Level Wastes

Low-level wastes (LLW) is generated from medical and industrial uses of radioactive materials as well as from the nuclear fuel cycle. In general, these wastes include materials such as paper, clothing, rags, tools, filters, soils, etc., that contain small amounts of radioactivity. The radioactivity tends to be short-lived. These materials generally do not have to be shielded during transport and they are suitable for shallow land burial. The volume of these materials may be reduced by compacting or incinerating prior to disposal. They make up about 90% of the volume of radioactive waste but contain only about 1% of the radioactivity of all the radioactive waste.

16.6.1.3 Intermediate-Level Wastes

Intermediate-level wastes (ILW) are generated during the operation of nuclear reactors, in the reprocessing of spent fuel, and from the decommissioning of nuclear facilities. These materials contain higher amounts of radioactivity and generally require some shielding during storage and transportation. Intermediate-level wastes is generally made up of resins, chemical sludges, fuel cladding, and contaminated materials from decommissioned nuclear facilities. Some of these materials are processed before disposal by solidifying them in concrete or bitumen. They make up about 7% of the volume and have about 4% of the radioactivity of all the radioactive waste.

16.6.1.4 High-Level Wastes

High-level wastes (HLW) is generated in the operation of a nuclear reactor. This waste consists of fission products and transuranic elements generated during the fission process. This material is highly radioactive and it is also thermally hot so that it must be both shielded and cooled. It accounts for 95% of the radioactivity produced by nuclear power reactors.

16.6.1.5 Managing HLW from Spent Fuel

The form of HLW from spent fuel is either the spent fuel itself or the waste products from reprocessing. The level of radioactivity from spent fuel falls to about one thousandth of the level it was when removed from the reactor in 40–50 years. This means the heat generated is also greatly reduced.

Currently, 270,000 tons of spent fuel are in storage at reactor sites around the world. An additional 12,000 tons are generated each year and about 3,000 tons of this are sent for reprocessing.

When spent fuel reprocessing is used, the uranium and plutonium are first removed during reprocessing, and then the much smaller volume of remaining HLW is solidified using a vitrification process. In this process, the fission products are mixed in a glass material, vitrified in stainless steel canisters and stored in shielded facilities for later disposal.

High-level waste will eventually be disposed of in deep geologic facilities. Several countries have selected sites for these facilities and they are expected to be commissioned for use after 2010.

16.6.1.6 Managing Other Radioactive Wastes

Generally, ILW and LLW are disposed of by burial. Intermediate-level wastes generated from fuel reprocessing will be disposed of in deep geological facilities. Some low-level liquid wastes from reprocessing plants are discharged to the sea. These liquids include some distinctive materials such as ^{99}Tc that can be discerned hundreds of kilometers away. Such discharges are tightly controlled and regulated so that the maximum dose any individual receives is a small fraction of natural background radiation.

Nuclear power stations and reprocessing facilities release small quantities of radioactive gases to the atmosphere. Gases such as ^{85}Kr and ^{133}Xe are chemically inert, and gases such as ^{131}I have short half-lives. The net effect of these gases is too small to warrant further consideration.

Table 16.8 provides a summary of waste management adopted by countries throughout the world.

TABLE 16.8 Waste Management Policies for Spent Fuel for Countries Throughout the World

Country	Policy	Facilities and Progress Toward Final Disposition
Belgium	Reprocessing	Central waste storage and underground laboratory established Construction of repository to begin about 2035
Canada	Direct disposal	Underground repository laboratory established Repository planned for use 2025
China	Reprocessing	Central spent fuel storage in LanZhou
Finland	Direct disposal	Spent fuel storages in operation Low and intermediate-level repositories in operation since 1992 Site near Olkiluoto selected for deep repository for spent fuel, from 2020
France	Reprocessing	Two facilities for storage of short-lived wastes Site selection studies underway for deep repository for commissioning 2020
Germany	Reprocessing but moving to direct disposal	Low-level waste sites in use since 1975 Intermediate-level wastes stored at Ahaus Spent fuel storage at Ahaus and Gorleben High-level repository to be operational after 2010
India	Reprocessing	Research on deep geological disposal for HLW
Japan	Reprocessing	Low-level waste repository in operation High-level waste storage facility at Rokkasho-mura since 1995 Investigations for deep geological repository begun, operation from 2035
Russia	Reprocessing	Sites for final disposal under investigation Central repository for low and intermediate-level wastes planned from 2008 Central interim HLW store planned for 2016
South Korea	Direct disposal	Central low- and ILW repository planned from 2008 Investigating deep HLW repository sites Low and intermediate-level waste repository in operation
Spain	Direct disposal	Final HLW repository site selection program for commissioning 2020 Central spent fuel storage facility in operation since 1985
Sweden	Direct disposal	Final repository for low to intermediate waste in operation since 1988 Underground research laboratory for HLW repository Site selection for repository in two volunteered locations
Switzerland	Reprocessing	Central interim storage for high-level wastes at Zwilag since 2001 Central low and intermediate-level storages operating since 1993 Underground research laboratory for high-level waste repository with deep repository to be finished by 2020
United Kingdom	Reprocessing	Low-level waste repository in operation since 1959 High-level waste is vitrified and stored at Sellafield Underground HLW repository planned
U.S.A	Direct disposal	Three low-level waste sites in operation 2002 decision to proceed with geological repository at Yucca Mountain

Source: From World Nuclear Association, Waste Management in the Nuclear Fuel Cycle, http://www.world-nuclear.org, 2004.

16.7 Nuclear Power Economics

Any discussion of the economics of nuclear power involves a comparison with other competitive electric generation technologies. The competing technologies are usually coal and natural gas.

Nuclear power costs include capital costs, fuel cycle costs, waste management costs and the cost of decommissioning after operation. The costs vary widely depending on the location of the generating plant. In countries such as China, Australia and the U.S. coal remains economically attractive because of large accessible coal resources. This advantage could be changed if a charge is made on carbon emissions. In other areas nuclear energy is competitive with fossil fuels even though nuclear costs include the cost of all waste disposal and decommissioning.

As previously stated, nuclear power costs include spent fuel management, plant decommissioning, and final waste disposal. These costs are not generally included in the costs of other power generation technologies.

Decommissioning costs are estimated to be 9%–15% of the initial cost of a nuclear plant. Because these costs are discounted over the life of the plant, they contribute only a few percent to the investment cost of the plant and have an even lower impact on the electricity generation cost. This impact in the U.S. is about 0.1–0.2 cent/kWh or about 5% of the cost of electricity produced.

Spent-fuel interim storage and ultimate disposal in a waste repository contribute another 10% to the cost of electricity produced. This cost is reduced if the spent fuel is disposed of directly. This does not account for the energy that could be extracted from the fuel if it was reprocessed.

Costs for nuclear-based electricity generation have been dropping over the last decade. This reduction in the cost of nuclear-generated electricity is a result of reductions in nuclear plant fuel, operating costs, and maintenance costs. However, the capital construction costs for nuclear plants are significantly higher than coal- and gas-fired plants. Because the capital cost of nuclear plants contribute more to the cost of electricity than coal- or gas-fired generation, the impact of changes in fuel, operation costs, and maintenance costs on the cost of electricity generation is less than those for coal- or gas-fired generation.

One of the primary contributors to the capital cost of nuclear plants has been the cost of money used to finance nuclear plant construction. The financing costs increase when the time required to license and construct a plant increases. Two factors are leading to the reduction in this portion of the cost. First, especially in the U.S., the licensing process is changing so that a plant receives both the construction permit and the operating license prior to the start of construction. Under this process, there is no large investment in plant hardware prior to completion of a significant portion of the licensing process, leading to a reduction in time required for the plant to begin producing revenue. Second, the new generation of nuclear plants will be highly standardized and modularized. This will allow a significant reduction in the time required to construct a new plant. It is estimated that the time from the start of construction to the start of operation will be reduced from nearly 10 years to 4–5 years. This will have a significant impact on capital costs.

The reduced capital costs associated with the licensing and construction of new nuclear power plants, and the fact that nuclear power is inherently less susceptible to large fluctuations in fuel costs, have made nuclear power an attractive energy option for many countries seeking to diversify their energy mix in the face of rising fossil fuel costs.

16.8 Conclusions

The development of nuclear power began after World War II and continues today. The first power-generating plants were constructed in the late 1950 s. During the 1960 s and 1970 s, there was a large commitment to nuclear power until the accidents occurred at Three Mile Island in 1979 and then at Chernobyl in 1986. The new safety requirements and delays caused by these accidents drove up the costs and at the same time caused a loss of public acceptance. In the U.S., many plant orders were canceled; in other countries, entire nuclear programs were canceled.

The ability of nuclear reactors to produce electricity economically and safely without the generation of greenhouse gasses has revitalized the interest in nuclear power as an alternative energy source. Many lessons have been learned from the operation of current power plants that have allowed the safety of newly designed plants to be improved. This, coupled with the desire of many nations to develop secure energy sources and a diversity of energy options, have resulted in the continuing development of a whole new generation of nuclear plants to meet future energy needs.

Nuclear power is also not as susceptible to fluctuation in fuel costs as petroleum and natural gas. As shown, the supply of uranium is very large, and if it is supplemented with thorium, the fuel supply is seemingly unlimited. This drives many other aspects of the fuel cycle, such as the choice between closed and open fuel cycles discussed earlier. For example, because of the large uranium resource and the fears of nuclear proliferation, the once-through (open) fuel cycle is favored by many. This will require large deep geologic waste repositories for the disposal of large quantities of spent fuel. However, when reprocessing is included in the closed fuel cycle, the amount of needed repository space is greatly reduced, but the expense of operation is increased. Finally, it may be possible to essentially eliminate the need for repositories by utilizing advanced fuel cycles that utilize almost all of the energy available in the uranium and the other transuranic products of reactor operation.

The need for energy and the use of electricity as the primary energy source for the end user will drive the increase in electricity generation around the world. The drive to reduce the production of greenhouse gases will contribute to a wider use of nuclear power for electricity generation. The recognition that nuclear power can safely provide large base-load generating capacity at a reasonable cost using known technologies will also be a major factor in its future development.

References

1. CRC Press. 1997. Generation technologies through the year 2005. In *CRC handbook of energy efficiency.* CRC Press.
2. World Coal Institute. 2000. *Coal Power for Progress. 4th Ed.* World Coal Institute.
3. EPRI. 1999. *The Electricity Technology Roadmap: Powering Progress.* 1999 Summary and Synthesis. EPRI Report C1-112677-V1.
4. Armor, A. F. and Wolk, R. H. 2002. *Productivity Improvement Handbook for Fossil Steam Plants.* 3rd Ed. EPRI Report 1006315.
5. Armor, A. F. and Wolk, R. H. 2005. *Productivity Improvement for Fossil Steam Power Plants 2005: One Hundred Case Studies.* EPRI Report 1012098.
6. EPRI. 2000. *An Assessment of Mercury Emissions from U.S. Coal-Fired Power Plants.* EPRI Report 1000608.
7. Dalton, S. M., Viswanathan, R., Gehl, S. M., Armor, A. F., Purgert, R. 2001. Ultrasupercritical materials. In *DOE Clean Coal Conference.*
8. Armor, A. F., Viswanathan, R., and Dalton, S. M. 2003. Ultrasupercritical steam turbines: Design and materials issues for the next generation, coal utilization and fuel systems. In *DOE Annual Conference.*
9. US Department of Energy. 2001. DOE fossil energy-Tomorrow's turbines. http://fossil.energy.gov/coal_power/turbines/index.shtml (accessed on).
10. EPRI. 2002. *Atmospheric Fluidized—Bed Combustion Handbook.* EPRI Report 1004493.
11. Skowyra, R. S. et al. 1995. Design of a supercritical sliding pressure circulating fluidized bed boiler with vertical waterwalls. In *Proceedings of 13th International Conference on Fluidized Bed Combustion*
12. EPRI. 2002. *Technical Status, Operating Experience, and Risk Assessment of Clean Coal Technologies-2002.* EPRI Report 1004480.

13. Courtright, H. A., Armor, A. F., Holt, N. H., and Dalton, S. M. 2003. Clean coal technologies and their commercial development. In *POWER-GEN International, Conference Proceedings*.
14. EPRI. 2004. Decommissioning Handbook for Coal-Fired Power Plants. EPRI Report 1011220.
15. US Department of Energy. 2002. Clean Coal Technology Demonstration Program. DOE/FE-0444, US DOE.
16. El-Wakil, M. M. 1962. Nuclear Power Engineering. McGraw-Hill, New York.
17. Nuclear News. 2005.
18. American Nuclear Society. 2005. Nuclear News.
19. Deutch, J. et al. 2003. The Future of Nuclear Power. Massachusetts Institute of Technology, Cambridge, MA.
20. Sutherland, J. K. 2003. Nuclear power comparisons and perspectives, http://www.energypulse.net (accessed on April 26, 2004).
21. World Nuclear Association. 2005. Plans for new reactors worldwide. World Nuclear Association information paper, http://www.world-nuclear.org, (accessed on June 16, 2005).
22. World Nuclear Association. 2005. Advanced nuclear power reactors. World Nuclear Association information paper, http://www.world-nuclear.org, (accessed on June 3, 2005).
23. World Nuclear Association. 2001. The nuclear fuel cycle. World Nuclear Association information paper. http://www.world-nuclear.org, (accessed on June 22, 2005).
24. WISE Uranium Project. 2005. World nuclear fuel facilities. http://www.wise-uranium.org, (accessed on June 22, 2005).
25. World Nuclear Association. 2004. Waste management in the nuclear fuel cycle. World Nuclear Association information paper, http://www.world-nuclear.org, (accessed on June 29, 2005).
26. World Nuclear Association. 2005. Nuclear waste disposal concepts. World Nuclear Association information paper, http://www.world-nuclear.org (accessed on June 29, 2005).
27. Uranium Information Centre. Processing of used nuclear fuel. UIC Nuclear Issues Briefing Paper 72, http://www.uic.com.au (accessed on June 29, 2005).
28. Sutherland, J. K. 2003. Nuclear cycles and nuclear resources, http://www.energypulse.net, (accessed on April 26, 2004).
29. World Nuclear Association. 2003. Uranium enrichment. World Nuclear Association information paper, http://www.world-nuclear.org (accessed on June 22, 2005).
30. World Nuclear Association. 2003. Transport of nuclear materials. World Nuclear Association information paper, http://www.world-nuclear.org (accessed on June 22, 2005).
31. Finck, P. J. 2005. Congressional testimony on nuclear fuel reprocessing. http://www.anl.gov/Media_-Center/News/2005/testimony050616.html (accessed on June 29, 2005).
32. World Nuclear Association. 2004. Supply of uranium. World Nuclear Association information paper, http://www.world-nuclear.org (accessed on June 22, 2005).
33. World Nuclear Association. 2005. The economics of nuclear power. World Nuclear Association information paper, http://www.world-nuclear.org (accessed on June 3, 2005).
34. World Nuclear Association. 2004. Energy analysis of power systems. World Nuclear Association information paper, http://www.world-nuclear.org (accessed on June 16, 2005).
35. Bruschi, H. J. The Westinghouse AP1000—Final design approved. Nuclear News, (November 2004).
36. US DOE Nuclear Energy Research Advisory Committee and the Generation IV International Forum. 2002. A technology roadmap for the generation IV nuclear energy systems. GIF-002-00.

For Further Information

Energy Information Administration. Annual energy outlook. 2003. http://www.eia.doe.gov.

National Engineering Technology Laboratory. http://www.netl.doe.gov.

EPRI. http://www.epri.com.

Babcock and Wilcox Co. *Steam, its generation and use*. Babcock and Wilcox, New York.

Combustion Engineering, Inc. *Combustion: Fossil power systems.* Combustion Engineering, Inc, Windsor, CT.
EPRI Journal. 1986. Tapping global expertise in coal technology. *EPRI Journal.* (Jan/Feb).
IGCC. 1994. New fuels, new players. *EPRI Journal.* (July/August).
EPRI Journal. 1993. A brighter future for PFBC. *EPRI Journal.* (December).
EPRI Journal. 1991. Fuel cells for urban power. *EPRI Journal.* (September).
EPRI Journal. 1993. Distributed generation. *ERPI Journal.* (April/May).
EPRI Journal. 1995. Plant repowering. *EPRI Journal.* (September/October).
EPRI Journal. 1998. Smart materials. *EPRI Journal.* (July/August).
EPRI Journal. 1999. Merchant plants. *EPRI Journal.* (Summer).
EPRI Journal. 2000. Energy and air emissions. *EPRI Journal.* (Summer).
EPRI Journal. 2001. Global coal initiative. *EPRI Journal.* (Summer).

17
Nuclear Fusion

Thomas E. Shannon
University of Tennessee

17.1 Introduction ... 17-1
17.2 Fusion Fuel ... 17-1
17.3 Confinement Concepts ... 17-2
17.4 Tokamak Reactor Development ... 17-2
17.5 Fusion Energy Conversion and Transport ... 17-4
Defining Terms ... 17-4
References ... 17-4
For Further Information ... 17-4

17.1 Introduction

Nuclear fusion holds the promise of providing almost unlimited power for future generations. If the process can be commercialized as envisioned by reactor design studies (Najmabadi et al. 1994), many of the problems associated with central electric power stations could be eliminated. Fusion power plants would not produce the pollution caused by the burning of fossil fuel and would eliminate the concern for meltdown associated with nuclear fission. The amount of radioactive waste material produced by a fusion reactor will be much less than that of a fission reactor since there is essentially no radioactive ash from the fusion reaction. If **low activation advanced materials** such as silicon carbide composites can be developed for the reactor structural material, the problem of disposal of activated components can also be eliminated.

17.2 Fusion Fuel

Although a number of different atomic nuclei can combine to release net energy from fusion, the reaction of **deuterium and tritium** (D–T) is the basis of planning for the first generation of fusion reactors. This choice is based on considerations of reactor economy. The D–T reaction occurs at the lowest temperature, has the highest probability for reaction, and provides the greatest output of power per unit of cost (Shannon 1989). The disadvantages of D–T as a fusion fuel are twofold. Tritium does not occur naturally in nature and must be bred in the fusion reactor or elsewhere. Second, tritium is a radioactive isotope of hydrogen with a relatively long **half-life** of 12.3 years. Since tritium can readily combine with air and water, special safety procedures will be required to handle the inventory necessary for a fusion reactor. There is hope that a less reactive fuel, such as deuterium alone (D–D) will eventually prove to be an economically acceptable alternative (Shannon 1989).

17.3 Confinement Concepts

Magnetic fusion, based on the tokamak concept, has received the majority of research funding for fusion energy development. However, other magnetic fusion concepts, such as the stellarator, the spherical torus, reversed-field pinch, and field-reversed configurations, are being developed as possible alternatives to the tokamak (Sheffield 1994). It may also be possible to develop fusion power reactors by inertial confinement concepts (Waganer et al. 1992). Research on these concepts has been done primarily in support of weapons development; therefore, the level of scientific understanding for power reactor applications is significantly less than that of magnetic fusion. The remainder of this discussion on reactor development, fusion energy conversion, and transport will consider only the tokamak magnetic fusion concept.

17.4 Tokamak Reactor Development

The tokamak device has proved to be the most effective means of producing the conditions necessary for magnetic fusion energy production. In 1994, researchers at the Princeton Plasma Physics Laboratory achieved in excess of 10 MW of D–T fusion power in a research tokamak, the Tokamak Fusion Test Reactor (TFTR). This accomplishment, coupled with worldwide progress in 40 years of magnetic fusion research, has established the scientific feasibility of the tokamak concept. The next major step, the International Thermonuclear Experimental Reactor (ITER) is being carried out under an international agreement among Europe, Japan, Russia, and the United States (Conn et al. 1992). A drawing of the ITER tokamak is shown in Figure 17.1. If the project is approved for construction, it will be in operation around 2005. The ITER is being designed to produce a fusion power in excess of 1000 MW. This will be a significant step on the path to commercial fusion power.

The U.S. Department of Energy has proposed a strategy, shown in Figure 17.2, which will lead to a demonstration power reactor by the year 2025. Supporting research and development programs necessary to achieve this goal are shown in this figure.

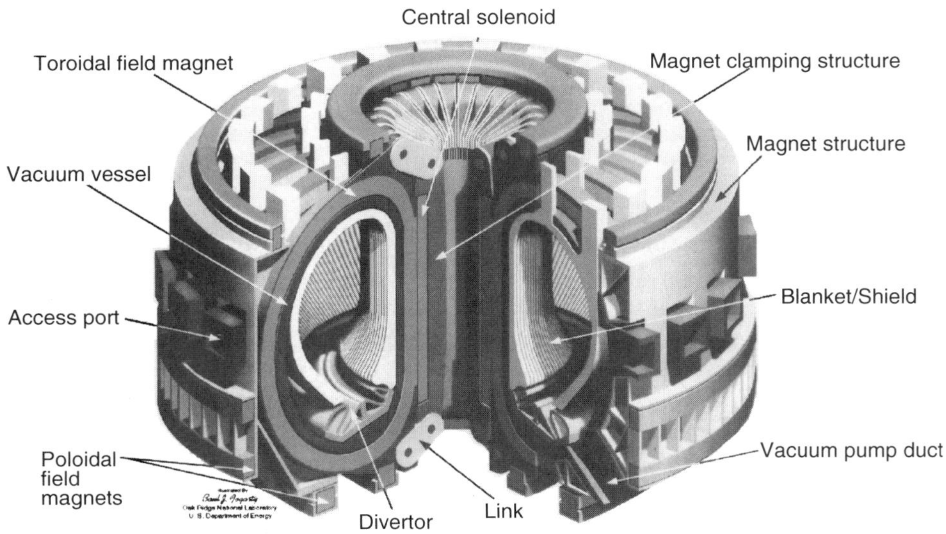

FIGURE 17.1 The International Thermonuclear Experimental Reactor (ITER).

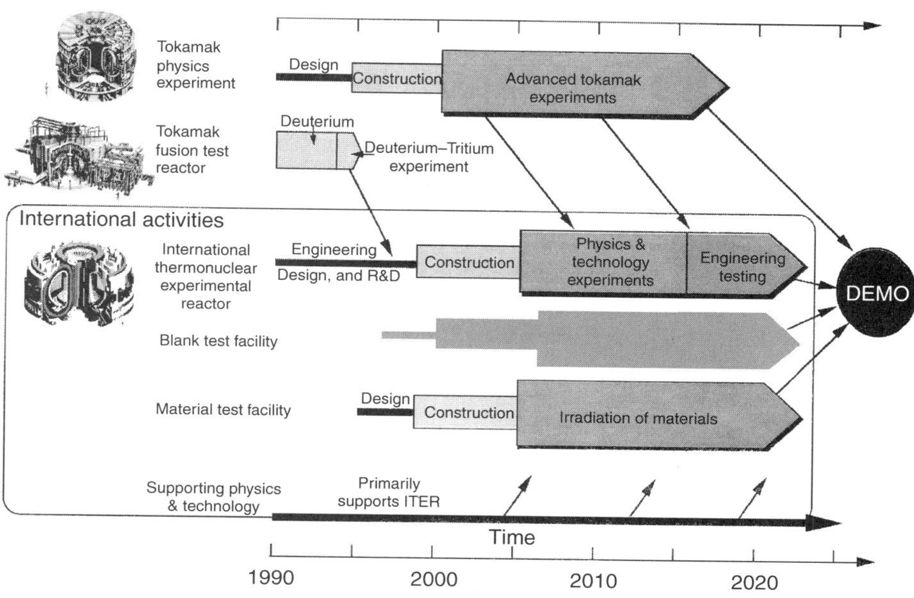

FIGURE 17.2 The U.S. Department of Energy magnetic fusion energy strategy.

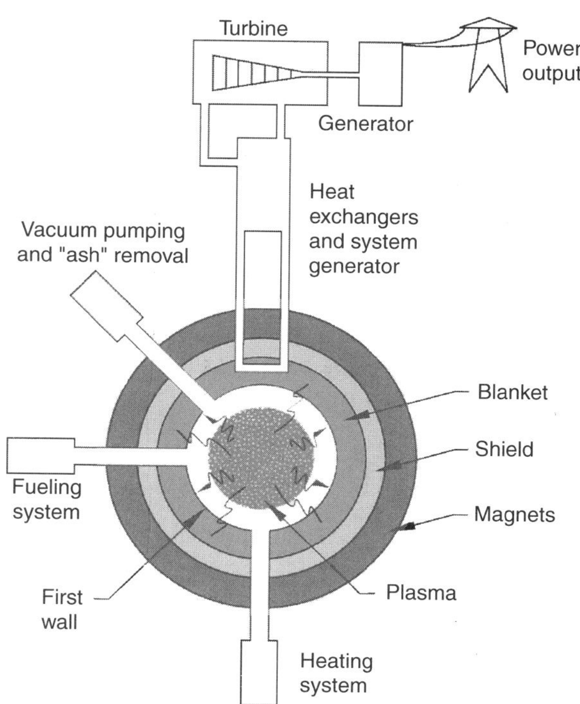

FIGURE 17.3 Schematic diagram of a magnetic fusion reactor power plant.

17.5 Fusion Energy Conversion and Transport

The energy from fusion is created in the form of charged particles and neutrons. The D–T reaction produces a neutron with an energy of 14.1 MeV and an alpha particle (helium) with an energy of 3.5 MeV in the reaction

$$D + T \rightarrow He^4\ (3.5\ \text{MeV}) + n\ (14.1\ \text{MeV}).$$

In the tokamak device, the reaction will take place in the toroidal vacuum vessel as previously shown in the ITER drawing, Figure 17.1. The D–T fuel, in the form of a **plasma**, will absorb energy from the positively charged alpha particles to sustain the temperature necessary for the reaction to continue. The neutron, having no charge, will escape from the plasma and pass through the wall of the vessel and penetrate into the surrounding blanket/shield structure. The kinetic energy of the alpha particles from the fusion reaction is eventually deposited on the wall of the vacuum vessel by radiation and conduction heat transfer from the plasma while the neutron deposits most of its energy within the cross section of the blanket/shield. The resulting thermal energy is transferred by a coolant such as water to a steam generator where a conventional steam to electric generator system may be used to produce electricity. An overall schematic diagram of the energy conversion and heat-transport system is shown in Figure 17.3.

Defining Terms

Deuterium and tritium: Isotopes of hydrogen as the fuel for fusion reactors.

Half-life: The time required for half of the radioactive material to disintegrate.

Low activation advanced materials: Structural materials that significantly reduce the radioactivity induced by exposure to fusion neutrons.

Plasma: A gas such as a mixture of deuterium and tritium raised to a very high temperature at which the electrons and the nuclei of the atoms separate. The plasma, consisting of electrons and ions, can conduct electricity and react to magnetic fields.

References

Conn, R. W., Chuyanov, V. A., Inoue, N., and Sweetman, D. R. 1992. The International Thermonuclear Experimental Reactor. *Sci. Am.*, 266 (4).

Najmabadi, F. et al. 1994. The ARIES-II and -IV Second Stability Tokamak Reactors, University of California, Los Angeles, report UCLA-PPG-1461.

Shannon, T. E. 1989. Design and cost evaluation of a generic magnetic fusion reactor using the D–D fuel cycle. *Fusion Technol.*, 15 (2), 1245–1253 (Part 2B).

Sheffield, J. 1994. The physics of magnetic fusion reactors. *Rev. Mod. Phys.*, 66 (3).

Waganer, L. et al. 1992. Inertial Fusion Energy Reactor Design Studies. U.S. Department of Energy Report. Vol. I, II, and III, DOE/ER-54101 MDC 92E0008.

For Further Information

The U.S. Department of Energy, Office of Fusion Energy maintains a home page on the World Wide Web. The address http://wwwofe.er.doe.gov provides an excellent source of up-to-date information and access to information from most institutions involved in fusion research.

18

Solar Thermal Energy Conversion

18.1 Active Solar Heating Systems 18-1
Introduction • Solar Collectors • Long-Term Performance of Solar Collectors • Solar Systems • Controls • Thermal Storage Systems • Solar System Simulation • Solar System Sizing Methodology • Solar System Design Methods • Design Recommendations and Costs

References 18-48

18.2 Solar Heat for Industrial Processes 18-49
The Potential for Solar Process Heat • Solar Thermal Systems in Industrial Processes: Integration and Basic Design Guidelines • Overview of Existing Solar Process Heat Plants

References 18-59

18.3 Passive Solar Heating, Cooling, and Daylighting 18-59
Introduction • Solar Thermosyphon Water Heating • Passive Solar Heating Design Fundamentals • Passive Space Cooling Design Fundamentals • Daylighting Design Fundamentals

Glossary 18-119

References 18-119

For Further Information 18-121

18.4 Solar Cooling 18-121
Vapor Compression Cycle • Absorption Air Conditioning • Solar Desiccant Dehumidification • Liquid-Desiccant Cooling System

References 18-133

T. Agami Reddy
Drexel University

Riccardo Battisti
University of Rome "La Sapienza"

Hans Schweiger
Active Solar System Group

Werner Weiss
AEE INTEC

Jeffrey H. Morehouse
University of South Carolina

Sanjay Vijayaraghavan
Intel Technology India Pvt. Ltd.

D. Yogi Goswami
University of South Florida

18.1 Active Solar Heating Systems

T. Agami Reddy

18.1.1 Introduction

This section defines the scope of the entire chapter and presents a brief overview of the types of applications that solar thermal energy can potentially satisfy.

18.1.1.1 Motivation and Scope

Successful solar system design is an iterative process involving consideration of many technical, practical, reliability, cost, code, and environmental considerations (Mueller Associates 1985). The success of a project involves identification of and intelligent selection among trade-offs, for which a proper understanding of goals, objectives, and constraints is essential. Given the limited experience available in the solar field, it is advisable to keep solar systems as simple as possible and not be lured by the promise of higher efficiency offered by more complex systems. Because of the location-specific variability of the solar resource, solar systems offer certain design complexities and concerns not encountered in traditional energy systems.

The objective of this chapter is to provide energy professionals which a fundamental working knowledge of the scientific and engineering principles of solar collectors and solar systems relevant to both the prefeasibility study and the feasibility study of a solar project. Conventional equipment such as heat exchangers, pumps, and piping layout are but briefly described. Because of space limitations, certain equations/correlations had to be omitted, and proper justice could not be given to several concepts and design approaches. Effort has been made to provide the reader with pertinent references to textbooks, manuals, and research papers.

A detailed design of solar systems requires in-depth knowledge and experience in (i) the use of specially developed computer programs for detailed simulation of solar system performance, (ii) designing conventional equipment, controls, and hydronic systems, (iii) practical aspects of equipment installation, and (iv) economic analysis. These aspects are not addressed here, given the limited scope of this chapter. Readers interested in acquiring such details can consult manuals such as Mueller Associates (1985) or SERI (1989).

The lengthy process outlined above pertains to large solar installations. The process is much less involved when a small domestic hot-water system, or unitary solar equipment or single solar appliances such as solar stills, solar cookers, or solar dryers are to be installed. Not only do such appliances differ in engineering construction from region to region, there are also standardized commercially available units whose designs are already more or less optimized by the manufacturers, normally as a result of previous experimentation, both technical or otherwise. Such equipment is not described in this chapter for want for space.

The design concepts described in this chapter are applicable to domestic water heating, swimming pool heating, active space heating, industrial process heat, convective drying systems, and solar cooling systems.

18.1.2 Solar Collectors

18.1.2.1 Collector Types

A solar thermal collector is a heat exchanger that converts radiant solar energy into heat. In essence this consists of a receiver that absorbs the solar radiation and then transfers the thermal energy to a working fluid. Because of the nature of the radiant energy (its spectral characteristics, its diurnal and seasonal variability, changes in diffuse to global fraction, etc.), as well as the different types of applications for which solar thermal energy can be used, the analysis and design of solar collectors present unique and unconventional problems in heat transfer, optics, and material science. The classification of solar collectors can be made according to the type of working fluid (water, air, or oils) or the type of solar receiver used (nontracking or tracking).

Most commonly used working fluids are water (glycol being added for freeze protection) and air. Table 18.1 identifies the relative advantages and potential disadvantages of air and liquid collectors and associated systems. Because of the poorer heat transfer characteristics of air with the solar absorber, the air collector may operate at a higher temperature than a liquid-filled collector, resulting in greater thermal losses and, consequently, a lower efficiency. The choice of the working fluid is usually dictated by the application. For example, air collectors are suitable for space heating and convective drying applications, while liquid collectors are the obvious choice for domestic and industrial hot-water applications. In certain high-temperature applications, special types of oils are used that provide better heat transfer characteristics.

The second criterion of collector classification is according to the presence of a mechanism to track the sun throughout the day and year in either a continuous or discreet fashion (see Table 18.2). The

TABLE 18.1 Advantages and Disadvantages of Liquid and Air Systems

Characteristics	Liquid	Air
Efficiency	Collectors generally more efficient for a given temperature difference	Collectors generally operate at slightly lower efficiency
System configuration	Can be readily combined with service hot-water and cooling systems	Space heat can be supplied directly but does not adapt easily to cooling. Can preheat hot-water
Freeze protection	May require antifreeze and heat exchangers that add cost and reduce efficiency	None needed
Maintenance	Precautions must be taken against leakage, corrosion and boiling	Low maintenance requirements. Leaks repaired readily with duct tape, but leaks may be difficult to find
Space requirements	Insulated pipes take up nominal space and are more convenient to install in existing buildings	Duct work and rock storage units are bulky, but ducting is a standard HVAC installation technique
Operation	Less energy required to pump liquids	More energy required by blowers to move air; noisier operation
Cost	Collectors cost more	Storage costs more
State of the art	Has received considerable attention from solar industry	Has received less attention from solar industry

Source: From SERI, *Engineering Principles and Concepts for Active Solar Systems*, Hemisphere Publishing Company, New York, 1989.

stationary flat-plate collectors are rigidly mounted, facing toward the equator with a tilt angle from the horizontal roughly equal to the latitude of the location for optimal year-round operation. The compound parabolic concentrators (CPCs) can be designed either as completely stationary devices or as devices that need seasonal adjustments only. On the other hand, Fresnel reflectors, paraboloids, and heliostats need two-axis tracking. Parabolic troughs have one axis tracking either along the east–west direction or the north–south direction. These collector types are described by Kreider (1979a, 1979b) and Rabl (1985).

A third classification criterion is to distinguish between nonconcentrating and concentrating collectors. The main reason for using concentrating collectors is not that *more energy* can be collected but that the thermal energy is obtained at higher temperatures. This is done by decreasing the area from which heat losses occur (called the receiver area) with respect to the aperture area (i.e., the area that intercepts the solar radiation). The ratio of the aperture to receiver area is called the *concentration ratio.*

18.1.2.2 Flat-Plate Collectors

18.1.2.2.1 Description

The flat-plate collector is the most common conversion device in operation today, since it is most economical and appropriate for delivering energy at temperatures up to about 100°C. The construction of flat-plate collectors is relatively simple, and many commercial models are available.

Figure 18.1 shows the physical arrangements of the major components of a conventional flat-plate collector with a liquid working fluid. The blackened absorber is heated by radiation admitted via the transparent cover. Thermal losses to the surroundings from the absorber are contained by the cover,

TABLE 18.2 Types of Solar Thermal Collectors

Nontracking Collectors	Tracking Collectors
Basic flat-plate	Parabolic troughs
Flat-plate enhanced with side reflectors or V-troughs	Fresnel reflectors
Tubular collectors	Paraboloids
Compound parabolic concentrators (CPCs)	Heliostats with central receivers

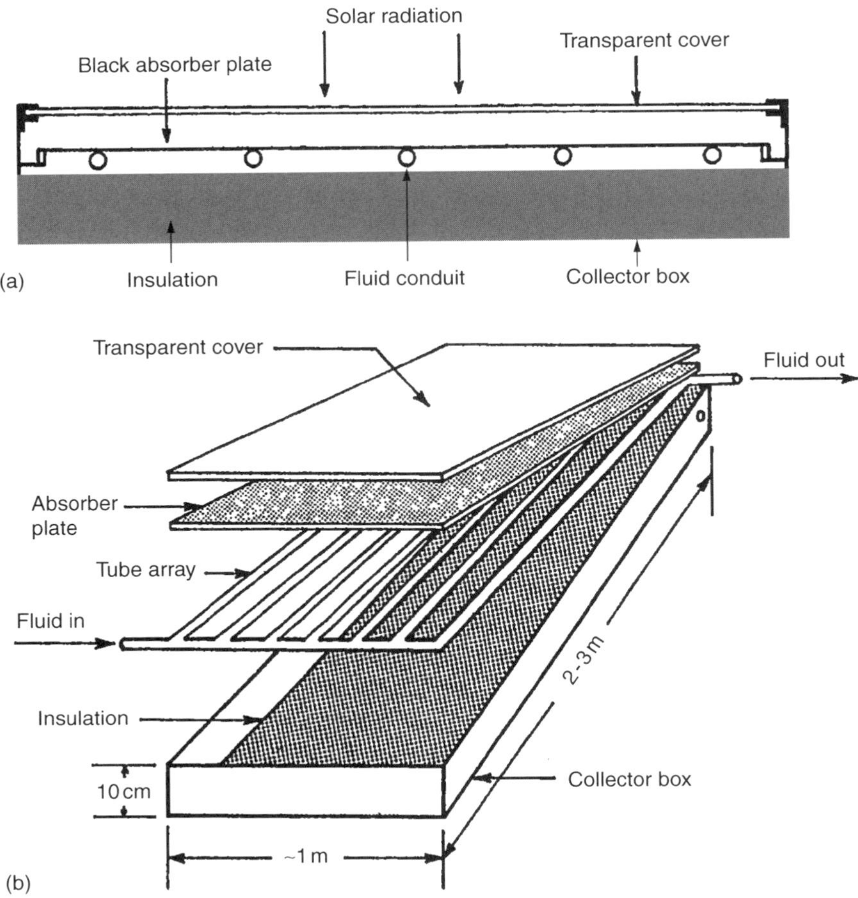

FIGURE 18.1 Cross-section and isometric view of a flat-plate collector.

which acts as a black body to the infrared radiation (this effect is called the *greenhouse* effect), and by insulation provided under the absorber plate. Passages attached to the absorber are filled with a circulating fluid, which extracts energy from the hot absorber. The simplicity of the overall device makes for long service life.

The absorber is the most complex portion of the flat-plate collector, and a great variety of configurations are currently available for liquid and air collectors. Figure 18.2 illustrates some of these concepts in absorber design for both liquid and air absorbers. Conventional materials are copper, aluminum, and steel. The absorber is either painted with a dull black paint or can be coated with a *selective surface* to improve performance (see "Improvements to Flat-Plate Collector Performance" for more details). Bonded plates having internal passageways perform well as absorber plates because the hydraulic passageways can be designed for optimal fluid and thermal performance. Such collectors are called *roll-bond* collectors. Another common absorber consists of tubes soldered or brazed to a single metal sheet, and mechanical attachments of the tubes to the plate have also been employed. This type of collector is called a *tube-and-sheet* collector. Heat pipe collectors have also been developed, though these are not as widespread as the previous two types. The so-called *trickle type* of flat-plate collector, with the fluid flowing directly over the corrugated absorber plate, dispenses entirely with fluid passageways. Tubular collectors have also been used because of the relative ease by which air can be evacuated from such collectors, thereby reducing convective heat losses from the absorber to the ambient air.

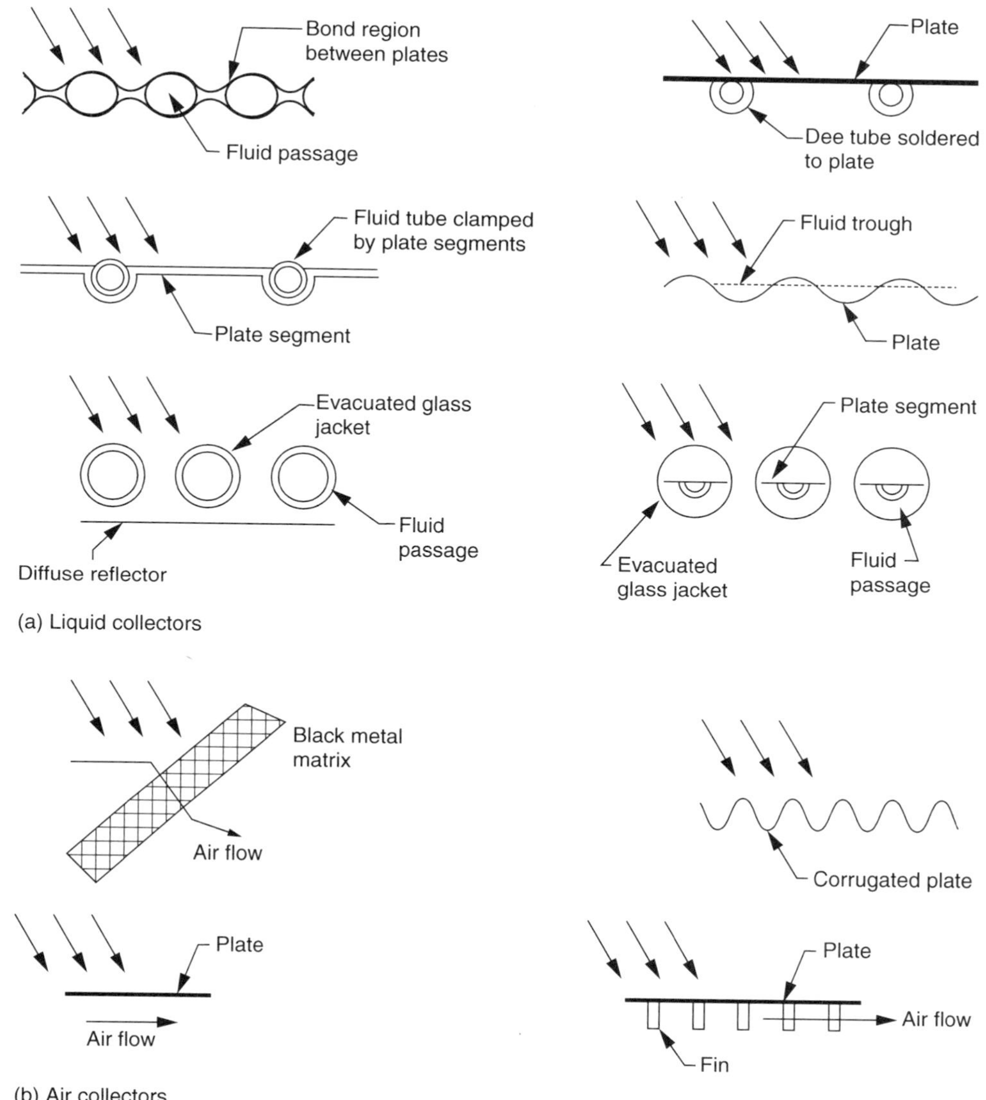

FIGURE 18.2 Typical flat-plate absorber configurations.

The absorber in an air collector normally requires a larger surface than in a liquid collector because of the poorer heat transfer coefficients of the flowing air stream. Roughness elements and producing turbulence by way of devices such as expanded metal foil, wool, and overlapping plates have been used as a means for increasing the heat transfer from the absorber to the working fluid. Another approach to enhance heat transfer is to use packed beds of expanded metal foils or matrices between the glazing and the bottom plate.

18.1.2.2.2 Modeling

A particular modeling approach and the corresponding degree of complexity in the model are dictated by the objective as well as by experience gained from past simulation work. For example, it has been found that transient collector behavior has insignificant influence when one is interested in determining the long-term performance of a solar thermal system. For complex systems or systems meant for nonstandard applications, detailed modeling and careful simulation of system operation are a must initially, and simplifications in component models and system operation can subsequently be made. However, in the case of solar thermal systems, many of the possible applications have been studied to date and a backlog of experience is available not only concerning system configurations but also with reference to the degree of component model complexity.

Because of low collector time constants (about 5–10 min), heat capacity effects are usually small. Then the instantaneous (or hourly, because radiation data are normally available in hourly time increments only) steady-state useful energy q_C in watts delivered by a solar flat-plate collector of surface area A_C is given by

$$q_C = A_C F'[I_T \eta_0 - U_L(T_{Cm} - T_a)]^+ \tag{18.1}$$

where F' is the plate efficiency factor, which is a measure of how good the heat transfer is between the fluid and the absorber plate; η_0 is the optical efficiency, or the product of the transmittance and absorptance of the cover and absorber of the collector; U_L is the overall heat loss coefficient of the collector, which is dependent on collector design only and is normally expressed in W/(m^2°C); T_{Cm} is the *mean* fluid temperature in the collector (in °C); and I_T is the radiation intensity on the plane of the collector (in W/m^2). The + sign denotes that negative values are to be set to zero, which physically implies that the collector should not be operated when q_C is negative (i.e., when the collector loses more heat than it can collect).

However, because T_{Cm} is not a convenient quantity to use, it is more appropriate to express collector performance in terms of the fluid *inlet* temperature to the collector (T_{Ci}). This equation is known as the classical Hottel–Whillier–Bliss (HWB) equation and is most widely used to predict instantaneous collector performance:

$$q_C = A_C F_R[I_T \eta_0 - U_L(T_{Ci} - T_a)]^+ \tag{18.2}$$

where F_R is called the heat removal factor and is a measure of the solar collector performance as a heat exchanger, since it can be interpreted as the ratio of actual heat transfer to the maximum possible heat transfer. It is related to F' by

$$\frac{F_R}{F'} = \frac{(mc_p)_C}{A_C F' U_L}\left\{1 - \exp\left[-\frac{A_C U_L F'}{(mc_p)_C}\right]\right\} \tag{18.3}$$

where m_C is the total fluid flow rate through the collectors and c_{pc} is the specific heat of the fluid flowing through the collector. The variation of (F_R/F') with $[(mC_p)_C/A_C U_L F']$ is shown graphically in Figure 18.3. Note the asymptotic behavior of the plot, which suggests that increasing the fluid flow rate more than a certain amount results in little improvement in F_R (and hence in q_C) while causing a quadratic increase in the pressure drop.

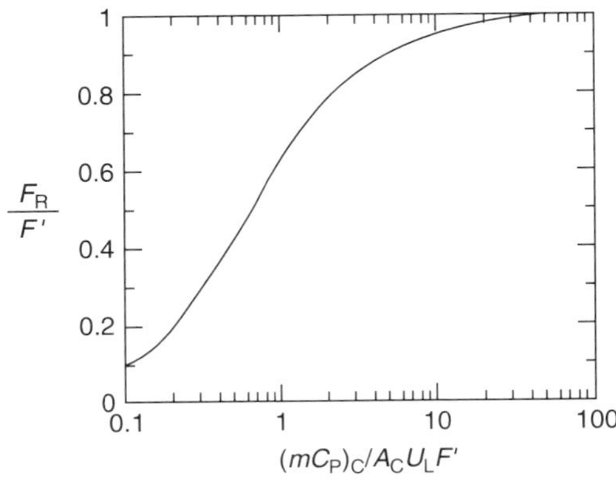

FIGURE 18.3 Variation of F_R/F' as a function of $[(mc_p)_c/(A_c U_L F')]$. (From Duffie, J. A. and Beckman, W. A., *Solar Engineering of Thermal Processes*, Wiley Interscience, New York, 1980.)

Factors influencing solar collector performance are of three types: (i) constructional, that is, related to collector design and materials used, (ii) climatic, and (iii) operational, that is, fluid temperature, flow rate, and so on. The plate efficiency factor F' is a factor that depends on the physical constructional features and is essentially a constant for a given liquid collector. (This is not true for air collectors, which require more careful analysis.) Operational features involve changes in m_C and T_{Ci}. While changes in m_C affect F_R as per Equation 18.3, we note from Equation 18.2 that to enhance q_C, T_{Ci} needs to be kept as low as possible. For solar collectors that are operated under more or less constant flow rates, specifying $F_R\eta_0$ and F_RU_L is adequate to predict collector performance under varying climatic conditions.

There are a number of procedures by which collectors have been tested. The most common is a *steady-state procedure*, where transient effects due to collector heat capacity are minimized by performing tests only during periods when radiation and ambient temperature are steady. The procedure involves simultaneous and accurate measurements of the mass flow rate, the inlet and outlet temperatures of the collector fluid, and the ambient conditions (incident solar radiation, air temperature, and wind speed). The most widely used test procedure is the ASHRAE Standard 93–77 (1978), whose test setup is shown in Figure 18.4. Though a solar simulator can be used to perform indoor testing, outdoor testing is always more realistic and less expensive. The procedure can be used for nonconcentrating collectors using air or liquid as the working fluid (but not two phase mixtures) that have a single inlet and a single outlet and contain no integral thermal storage.

Steady-state procedures have been in use for a relatively long period and though the basis is very simple the engineering setup is relatively expensive (see Figure 18.4). From an overall heat balance on the collector fluid and from Equation 18.2, the expressions for the instantaneous collector efficiency under normal solar incidence are

$$\eta_C \equiv \frac{q_C}{A_C I_T} = \frac{(mc_p)_C(T_{Co} - T_{Ci})}{A_C I_T} \tag{18.4}$$

$$= \left[F_R\eta_n - F_RU_L\left(\frac{T_{Ci} - T_a}{I_T}\right)\right] \tag{18.5}$$

where η_n is the optical efficiency at normal solar incidence.

From the test data, points of η_c against reduced temperature $[(T_{Ci} - T_a)/I_T]$ are plotted as shown in Figure 18.5. Then a linear fit is made to these data points by regression, from which the values of $F_R\eta_n$ and F_RU_L are easily deduced. It will be noted that if the reduced term were to be taken as $[(T_{Cm} - T_a)/I_T]$, estimates of $F'\eta_n$ and $F'U_L$ would be correspondingly obtained.

18.1.2.2.3 Incidence Angle Modifier

The optical efficiency η_0 depends on the collector configuration and varies with the angle of incidence as well as with the relative values of diffuse and beam radiation. The incidence angle modifier is defined as $K_\eta \equiv (\eta_0/\eta_n)$. For flat-plate collectors with 1 or 2 glass covers, K_η is almost unchanged up to incidence angles of 60°, after which it abruptly drops to zero.

A simple way to model the variation of K_η with incidence angle for flat-plate collectors is to specify η_n, the optical efficiency of the collector at normal beam incidence, to assume the entire radiation to be beam, and to use the following expression for the angular dependence (ASHRAE 1978)

$$K_\eta = 1 + b_0\left(\frac{1}{\cos\theta} - 1\right) \tag{18.6}$$

where θ is the solar angle of incidence on the collector plate (in degrees) and b_0 is a constant called the incidence angle modifier coefficient. Plotting K_η against $[(1/\cos\theta) - 1]$ results in linear plots (see Figure 18.6), thus justifying the use of Equation 18.6. We note that for one-glass and two-glass covers, approximate values of b_0 are -0.10 and -0.17, respectively.

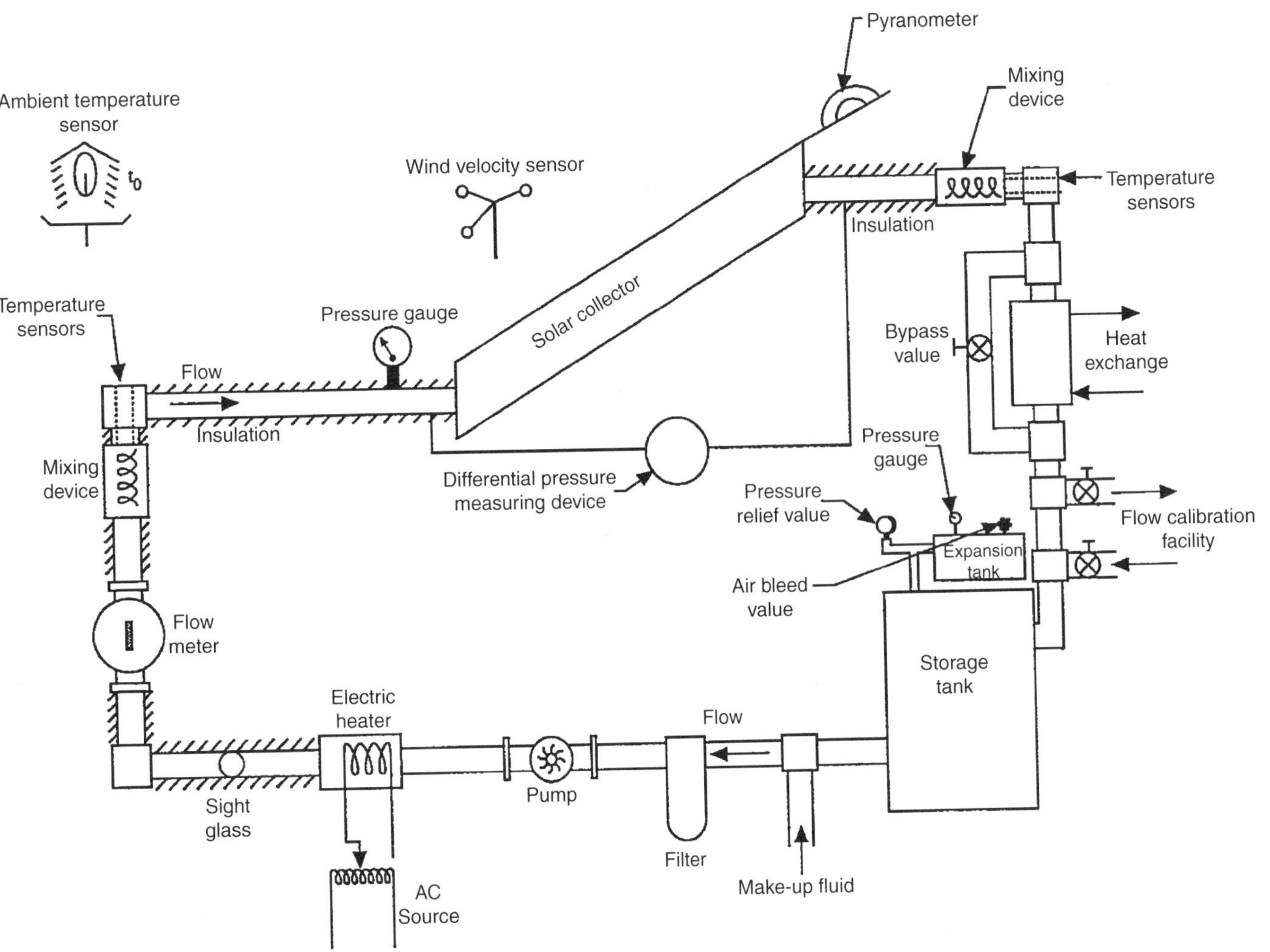

FIGURE 18.4 Set up for testing liquid collectors according to ASHRAE Standard 93–72.

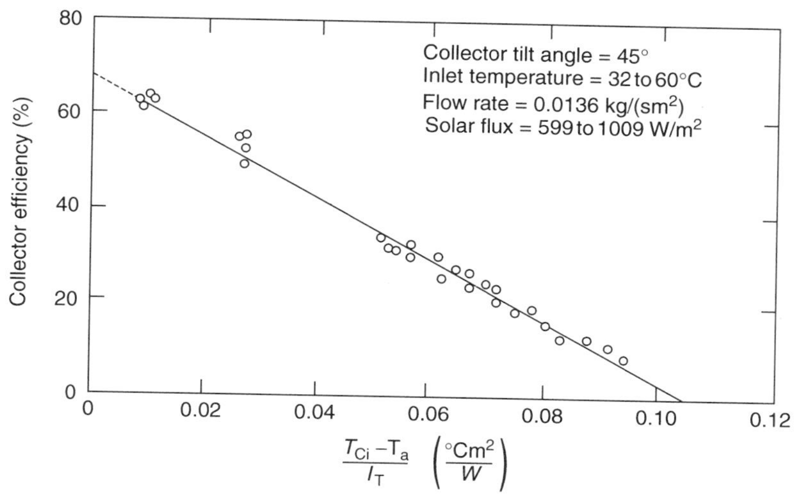

FIGURE 18.5 Thermal efficiency curve for a double glazed flat-plate liquid collector. Test conducted outdoors on a 1.2 m by 1.25 m panel with 10.2 cm of glass fiber back insulation and a flat copper absorber with black coating of emissivity of 0.97. (From ASHRAE Standard 93–77, *Methods of Testing to Determine the Thermal Performance of Solar Collectors*, American Society of Heating, Refrigeration and Air Conditioning Engineers, New York, 1978.)

In case the diffuse solar fraction is high, one needs to distinguish between beam, diffuse, and ground-reflected components. Diffuse radiation, by its very nature, has no single incidence angle. One simple way is to assume an equivalent incidence angle of 60° for diffuse and ground-reflected components. One would then use Equation 18.6 for the beam component along with its corresponding value of θ and account for the contribution of diffuse and ground reflected components by assuming a value of $\theta = 60°$ in Equation 18.6. For more accurate estimation, one can use the relationship between the effective diffuse solar incidence angle versus collector tilt given in Duffie and Beckman (1980). It should be noted that the preceding equation gives misleading results with incidence angles close to 90°. An alternative functional form for the incidence angle modifier for both flat-plate and concentrating collectors has been proposed by Rabl (1981).

Example 18.1.1

From the thermal efficiency curve given in Figure 18.5 determine the performance parameters of the corresponding solar collector.

Extrapolating the curve yields y-intercept $= 0.69$, x-intercept $= 0.105$ ($m^2°C/W$). Since the reduced temperature in Figure 18.5 is in terms of the inlet fluid temperature to the collector, Equation 18.5 yields $F_R\eta_n = 0.69$ and $F_RU_L = 0.69/0.105 = 6.57$ W/($m^2°C$). Alternatively, the collector parameters in terms of the plate efficiency factor can be deduced. From Figure 18.5, the collector area $= 1.22 \times 1.25 = 1.525$ m^2, while the flow rate $(m/A_C) = 0.0136$ kg/(s m^2). From Equation 18.3,

$$F'/F_R = -(0.0136 \times 4190/6.57)\ln[-6.57/(0.0136 \times 4190)] = 1.0625$$

Thus $F'U_L = 6.57 \times 1.0625 = 6.98$ W/($m^2°C$) and $F'\eta_n = 0.69 \times 1.0625 = 0.733$.

Example 18.1.2

How would the optical efficiency be effected at a solar incidence angle of 60° for a flat-plate collector with two glass covers?

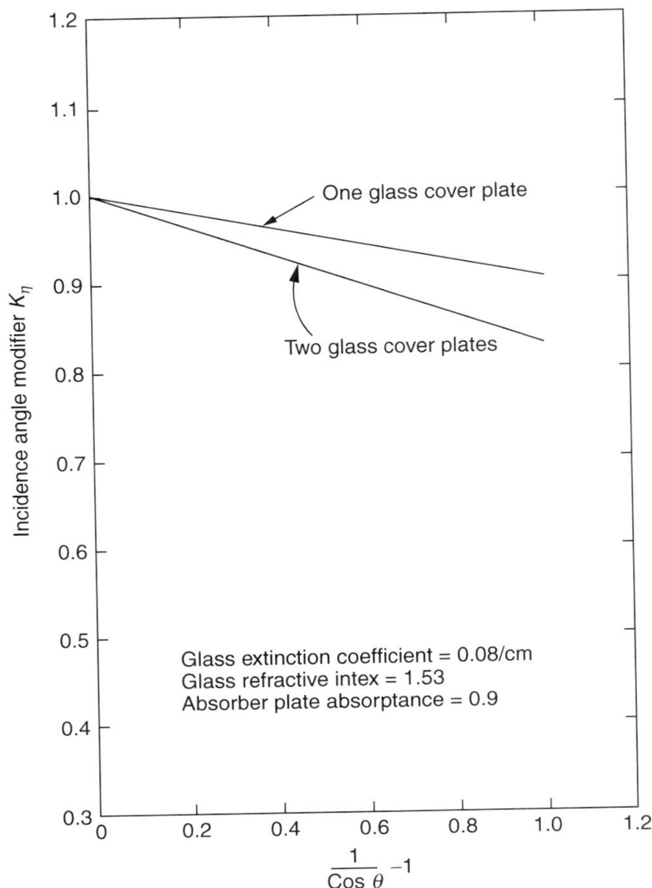

FIGURE 18.6 Incidence angle modifiers for two flat-plate collectors with nonselective coating on the absorber. (Adapted From ASHRAE Standard 93–77, *Methods of Testing to Determine the Thermal Performance of Solar Collectors*, American Society of Heating, Refrigeration and Air Conditioning Engineers, New York, 1978.)

Assume a value of $b_0 = -0.17$. From Equation 18.6, $K_\eta = 0.83$. Thus

$$F_R \eta_0 = F_R \eta_n K_\eta = 0.69 \times 0.83 = 0.57$$

18.1.2.2.4 *Other Collector Characteristics*

There are three collector characteristics that a comprehensive collector testing process should also address. The collector *time constant* is a measure that determines how intermittent sunshine affects collector performance and is useful in defining an operating control strategy for the collector array that avoids instability. Collector performance is usually enhanced if collector time constants are kept low. ASHRAE 93–77 also includes a method for determining this value. Commercial collectors usually have time constants of about 5 min or less, and this justifies the use of the HWB model (see Equation 18.2).

Another quantity to be determined from collector tests is the collector *stagnation temperature*. This is the equilibrium temperature reached by the absorber plate when no heat is being extracted from the collector. Determining the maximum stagnation temperature, which occurs under high I_T and T_a values, is useful in order to safeguard against reduced collector life due to thermal damage to collectors (namely irreversible thermal expansion, sagging of covers, physical deterioration, optical changes, etc.) in the field when not in use. Though the stagnation temperature could be estimated from Equation 18.2 by setting $q_C=0$ and solving for T_{Ci}, it is better to perform actual tests on collectors before field installation.

The third collector characteristic of interest is the *pressure drop* across the collector for different fluid flow rates. This is an important consideration for liquid collectors, and more so for air collectors, in order to keep parasitic energy consumption (namely electricity to drive pumps and blowers) to a minimum in large collector arrays.

18.1.2.3 Improvements to Flat-Plate Collector Performance

There are a number of ways by which the performance of the basic flat-plate collectors can be improved. One way is to enhance optical efficiency by treatment of the glass cover thereby reducing reflection and enhancing performance. As much as a 4% increase has been reported (Anderson 1977). Low-iron glass can also reduce solar absorption losses by a few percent.

These improvements are modest compared to possible improvements from reducing losses from the absorber plate. Essentially, the infrared upward reradiation losses from the heated absorber plate have to be decreased. One could use a second glass cover to reduce the losses, albeit at the expense of higher cost and lower optical efficiency. Usually for water heating applications, radiation accounts for about two-thirds of the losses from the absorber to the cover with convective losses making up the rest (conduction is less than about 5%). The most widely used manner of reducing these radiation losses is to use selective surfaces whose emissivity varies with wavelength (as against matte-black painted absorbers, which are essentially gray bodies). Note that 98% of the solar spectrum is at wavelengths less than 3.0 μm, whereas less than 1% of the black body radiation from a 200°C surface is at wavelengths less than 3.0 μm. Thus selective surfaces for solar collectors should have high-solar absorptance (i.e., low reflectance in the solar spectrum) and low long-wave emittance (i.e., high reflectance in the long-wave spectrum). The spectral reflectance of some commonly used selective surfaces is shown in Figure 18.7. Several commercial collectors for water heating or low-pressure steam (for absorption cooling or process heat applications) are available that use selective surfaces.

Another technique to simultaneously reduce both convective and radiative losses between the absorber and the transparent cover is to use honeycomb material (Hollands 1965). The honeycomb material can be reflective or transparent (the latter is more common) and should be sized properly. Glass honeycombs have had some success in reducing losses in high-temperature concentrating receivers, but plastics are

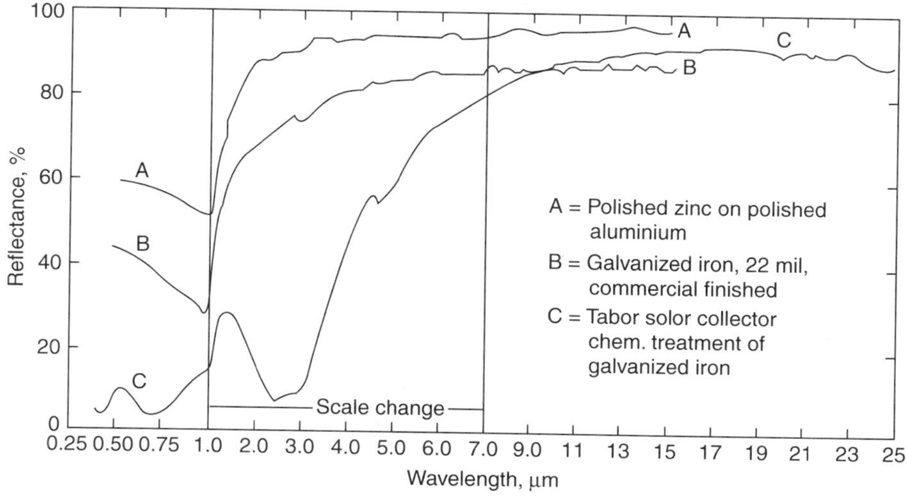

FIGURE 18.7 Spectral reflectance of several surfaces. (From Edwards, D. K., Nelson, K. E., Roddick, R. D., and Gier, J. T., Basic Studies on the Use of Solar Energy, Report no. 60–93, Department of Engineering. University of California at Los Angeles, CA, 1960.)

usually recommended for use in flat-plate collectors. Because of the poor thermal aging properties, honeycomb flat-plate collectors have had little commercial success. Currently the most promising kind seems to be the simplest (both in terms of analysis and construction), namely collectors using horizontal rectangular slats (Meyer 1978). Convection can be entirely suppressed provided the slats with the proper aspect ratio are used.

Finally, collector output can be enhanced by using side reflectors, for instance a sheet of anodized aluminum. The justification in using these is their low cost and simplicity. For instance, a reflector placed in front of a tilted collector cannot but increase collector performance because losses are unchanged and more solar radiation is intercepted by the collector. Reflectors in other geometries may cast a shadow on the collector and reduce performance. Note also that reflectors would produce rather nonuniform illumination over the day and during the year, which, though not a problem in thermal collectors, may drastically penalize the electric output of photovoltaic modules. Whether reflectors are cost-effective depends on the particular circumstances and practical questions such as aesthetics and space availability. The complexity involved in the analysis of collectors with planar reflectors can be reduced by assuming the reflector to be long compared to its width and treating the problem in two dimensions only. How optical performance of solar collectors are affected by side planar reflectors is discussed in several papers, for example Larson (1980), Chiam (1981).

18.1.2.4 Other Collector Types

18.1.2.4.1 Evacuated Tubular Collectors

One method of obtaining temperatures between 100 and 200°C is to use evacuated tubular collectors. The advantage in creating and being able to maintain a vacuum is that convection losses between glazing

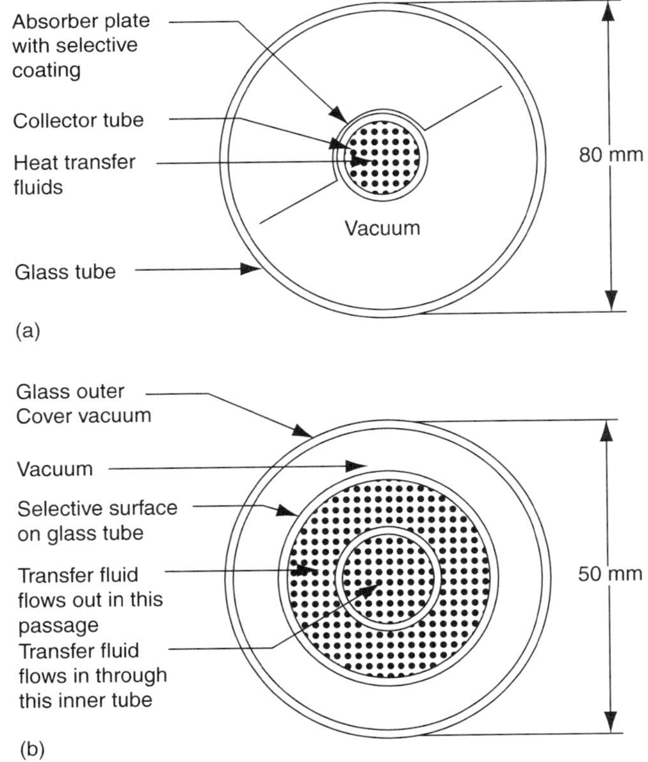

FIGURE 18.8 Evacuated tubular collectors. (From Charters, W. W. S. and Pryor, T. L., *An Introduction to the Installation of Solar Energy Systems*, Victoria Solar Energy Council, Melbourne, Australia, 1982.)

and absorber can be eliminated. There are different possible arrangements of configuring evacuated tubular collectors. Two designs are shown in Figure 18.8. The first is like a small flat-plate collector with the liquid to be heated making one pass through the collector tube. The second uses an all-glass construction with the glass absorber tube being coated selectively. The fluid being heated passes up the middle of the absorber tube and then back through the annulus. Evacuated tubes can collect both direct and diffuse radiation and do not require tracking. Glass breakage and leaking joints due to thermal expansion are some of the problems which have been experienced with such collector types. Various reflector shapes (like flat-plate, V-groove, circular, cylindrical, involute, etc.) placed behind the tubes are often used to usefully collect some of the solar energy, which may otherwise be lost, thus providing a small amount of concentration.

18.1.2.4.2 Compound Parabolic Concentrators

The CPC collector, discovered in 1966, consists of parabolic reflectors that funnel radiation from aperture to absorber rather than focusing it. The right and left halves belong to different parabolas (hence the name *compound*) with the edges of the receiver being the foci of the opposite parabola (see Figure 18.9). It has been proven that such collectors are *ideal* in that any solar ray, be it beam or diffuse, incident on the aperture within the acceptance angle will reach the absorber while all others will bounce back to and fro and reemerge through the aperture. CPCs are also called *nonimaging* concentrators because they do not form clearly defined images of the solar disk on the absorber surface as achieved in classical concentrators. CPCs can be designed both as low-concentration devices with large acceptance angles or as high-concentration devices with small acceptance angles. CPCs with low-concentration ratios (of about 2) and with east–west axes can be operated as stationary devices throughout the year or at most

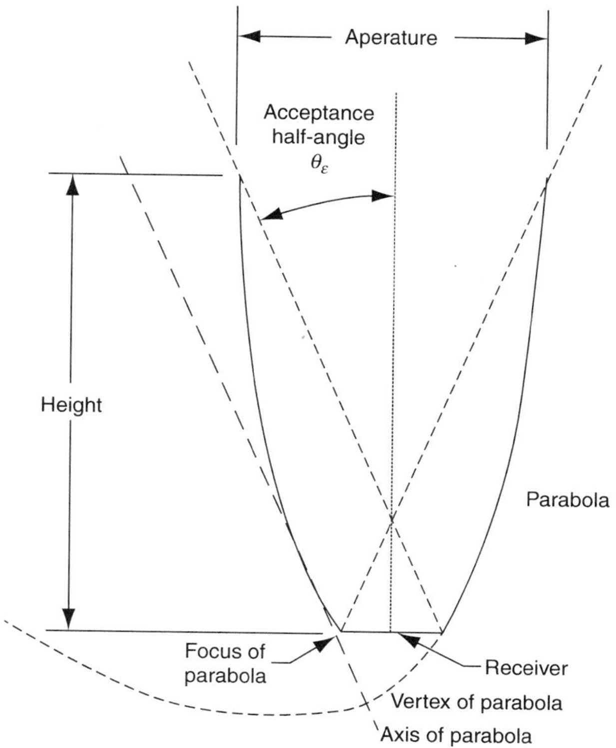

FIGURE 18.9 Cross-section of a symmetrical nontruncated CPC. (From Duffie, J. A. and Beckman, W. A., *Solar Engineering of Thermal Processes*, Wiley Interscience, New York, 1980.)

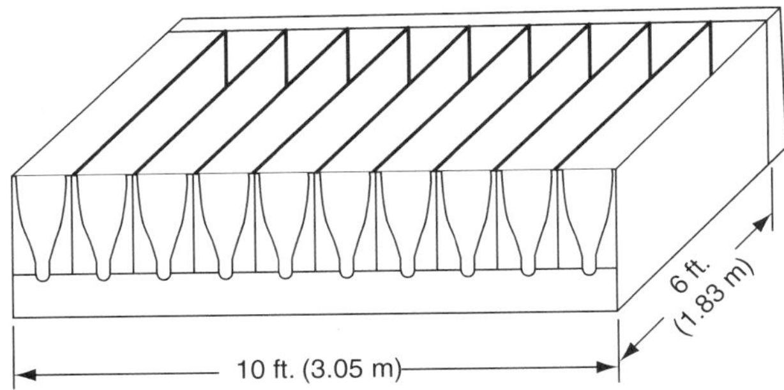

FIGURE 18.10 A CPC collector module. (From SERI, *Engineering Principles and Concepts for Active Solar Systems*, Hemisphere Publishing Company, New York, 1989.)

with seasonal adjustments only. CPCs, unlike other concentrators, are able to collect all the beam and a large portion of the diffuse radiation. Also they do not require highly specular surfaces and can thus better tolerate dust and degradation. A typical module made up of several CPCs is shown in Figure 18.10. The absorber surface is located at the bottom of the trough, and a glass cover may also be used to encase the entire module. CPCs show considerable promise for water heating close to the boiling point and for low-pressure steam applications. Further details about the different types of absorber and receiver shapes used, the effect of truncation of the receiver and the optics, can be found in Rabl (1985).

18.1.3 Long-Term Performance of Solar Collectors

18.1.3.1 Effect of Day-to-Day Changes in Solar Insolation

Instantaneous or hourly performance of solar collectors has been discussed in "Flat-Plate Collectors." For example, one would be tempted to use the HWB Equation 18.2 to predict long-term collector performance at a prespecified and constant fluid inlet temperature T_{Ci} merely by assuming average hourly values of I_T and T_a. Such a procedure would be erroneous and lead to underestimation of collector output because of the presence of the control function, which implies that collectors are turned on only when $q_C > 0$, that is, when radiation I_T exceeds a certain critical value I_C. This critical radiation value is found by setting q_C in Equation 18.2 to zero:

$$I_C = U_L(T_{Ci} - T_a)/\eta_0 \tag{18.7a}$$

To be more rigorous, a small increment δ to account for pumping power and stability of controls can also be included if needed by modifying the equation to

$$I_C = U_L(T_{Ci} + \delta - T_a)/\eta_0 \tag{18.7b}$$

Then, Equation 18.2 can be rewritten in terms of I_C as

$$q_C = A_C F_R \eta_0 [I_T - I_C]^+ \tag{18.8}$$

Why one cannot simply assume a mean value of I_T in order to predict the mean value of q_C will be illustrated by the following simple concept (Klein 1978). Consider the three identical day sequences shown in sequence A of Figure 18.11. If I_{C1} is the critical radiation intensity and if it is constant over the whole day, the useful energy collected by the collector is represented by the sum of the shaded areas. If a higher critical radiation value shown as I_{C2} in Figure 18.11 is selected, we note that no useful energy is collected at all. Actual weather sequences would not look like that in sequence A but rather like that in sequence B, which is comprised of an excellent, a poor, and an average day. Even if both sequences have

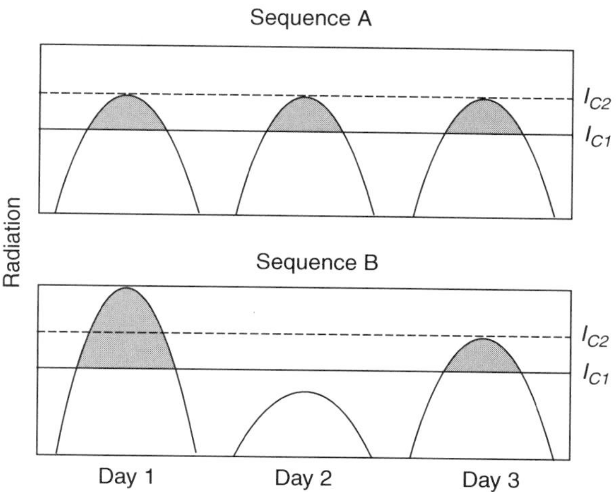

FIGURE 18.11 Effect of radiation distribution on collector long-term performance. (From Klein, S. A., Calculation of flat-plate collector utilizability, *Solar Energy*, 21, 393, 1978.)

the same average radiation over 3 days, a collector subjected to sequence B will collect useful energy when the critical radiation is I_{C2}. Thus, neglecting the variation of radiation intensity from day-to-day over the long term and dealing with mean values would result in an underestimation of collector performance.

Loads are to a certain extent repetitive from day-to-day over a season or even the year. Consequently, one can also expect collectors to be subjected to a known diurnal repetitive pattern or mode of operation, that is, the collector inlet temperature T_{Ci} has a known repetitive pattern.

18.1.3.2 Individual Hourly Utilizability

In this mode, T_{Ci} is assumed to very over the day but has the same variation for all the days over a period of N days (where $N=30$ days for monthly and $N=365$ for yearly periods). Then from Equation 18.8, *total* useful energy collected over N days during individual hour i of the day is

$$q_{CN}(i) = A_C F_R \bar{\eta}_0 \bar{I}_{Ti} \sum_{i=1}^{N} \frac{[I_{Ti} - I_C]^+}{\bar{I}_{Ti}} \tag{18.9}$$

Let us define the radiation ratio

$$X_i = I_{Ti}/\bar{I}_{Ti} \tag{18.10}$$

and the critical radiation ratio

$$X_C = I_C/\bar{I}_{Ti}$$

The modified HWB Equation 18.8 can be rewritten as

$$q_{CN}(i) = A_C F_R \bar{\eta}_0 \bar{I}_{Ti} N \phi_i(x_c) \tag{18.11}$$

where the individual hourly utilizability factor ϕ_i is identified as

$$\phi_i(X_C) = \frac{1}{N} \sum^{N} (X_i - X_C)^+ \tag{18.12}$$

Thus ϕ_i can be considered to be the fraction of the incident solar radiation that can be converted to useful heat by an ideal collector (i.e., whose $F_R\eta_0=1$). The utilizability factor is thus a *radiation statistic* in the sense that it depends solely on the radiation values at the specific location. As such, it is in no way dependent on the solar collector itself. Only after the radiation statistics have been applied is a collector dependent significance attached to X_C.

Hourly utilizability curves on a *monthly* basis that are independent of location were generated by Liu and Jordan (1963) over 30 years ago for flat-plate collectors (see Figure 18.12). The key climatic parameter which permits generalization is the *monthly clearness index* $\bar{K}$ of the location defined as

$$\bar{K} = \bar{H}/\bar{H}_0 \tag{18.13}$$

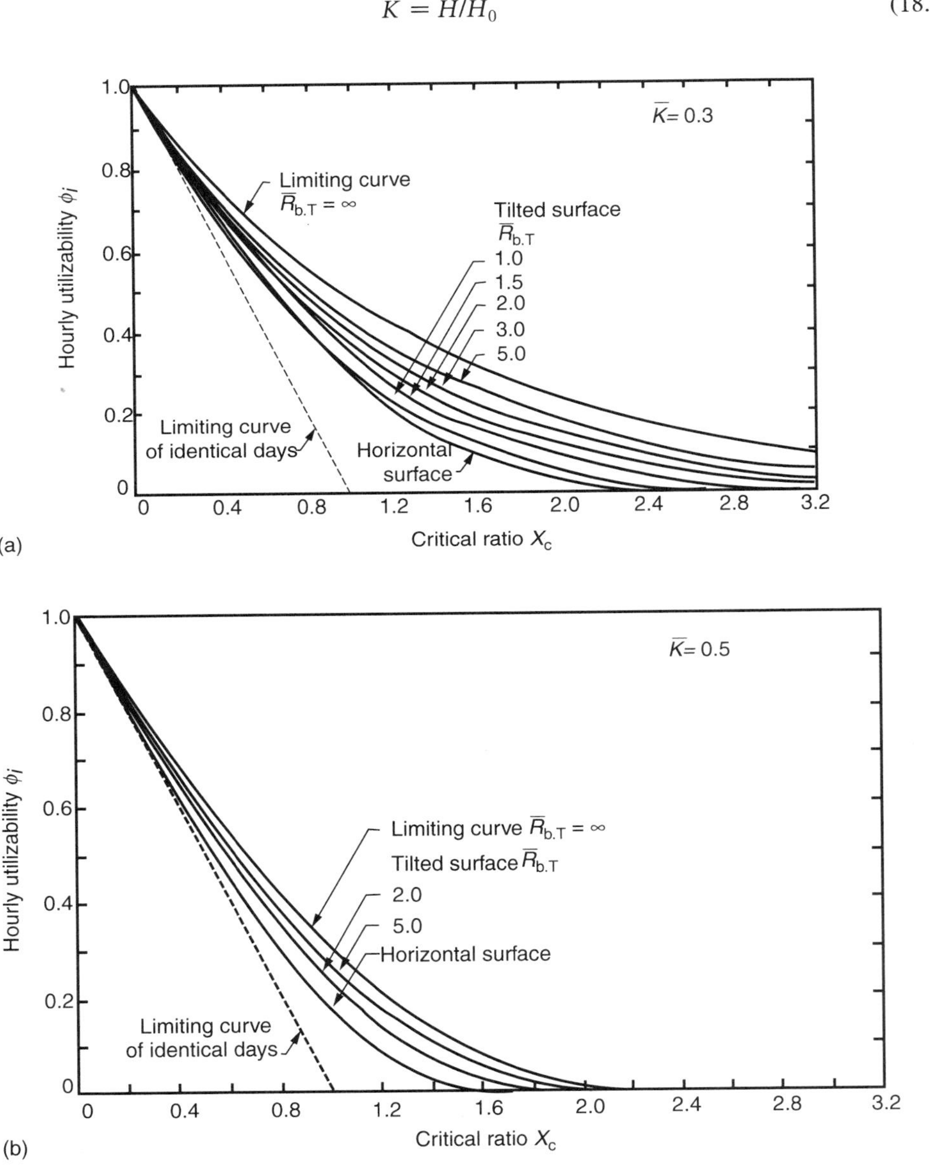

FIGURE 18.12 Generalized hourly utilizability curves of Liu and Jordan (1963) for three different monthly mean clearness indices K. (a) $\bar{K} = 0.3$, (b) $\bar{K} = 0.5$, (c) $\bar{K} = 0.7$. (From Liu, B. Y. H. and Jordan, R. C., A rational procedure for predicting the long-term average performance of flat-plate solar energy collectors, *Solar Energy*, 7, 53, 1963.)

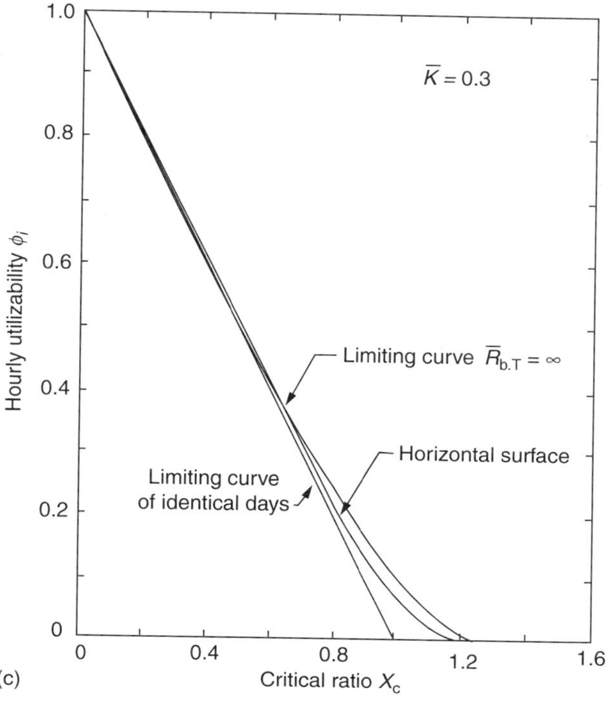

FIGURE 18.12 *(continued)*

where $\bar{H}$ is the monthly mean daily global radiation on the horizontal surface and $\bar{H}_0$ is the monthly mean daily extraterrestrial radiation on a horizontal surface.

Extensive tables giving monthly values of $\bar{K}$ for several different locations worldwide can be found in several books, for example, Duffie and Beckman (1980) or Reddy (1987). The curves apply to equator-facing tilted collectors with the effect of collector tilt accounted for by the factor $\bar{R}_{b,T}$ which is the ratio of the monthly mean daily extraterrestrial radiation on the tilted collector to that on a horizontal surface. Monthly mean daily calculations can be made using the 15th of the month, though better accuracy is achieved using slightly different dates (Reddy 1987). Clark, Klein, and Beckman (1983), working from measured data from several U.S. cities, have proposed the following correlation for individual hourly utilizability over monthly time scales applicable to flat-plate collectors only:

$$\begin{aligned} \phi_i &= 0 \quad \text{for } X_C \geq X_{max} \\ &= (1 - X_C/X_{max})^2 \quad \text{for } X_{max} = 2 \\ &= \left| |a| - [a^2 + (1 + 2a)(1 - X_C/X_{max})^2]^{1/2} \right| \quad \text{otherwise} \end{aligned} \tag{18.14}$$

where

$$a = (X_{max} - 1)(2 - X_{max}) \tag{18.15}$$

and

$$X_{max} = 1.85 + 0.169(\bar{r}_T/\bar{k}^2) - 0.0696 \cos \beta/\bar{k}^2 - 0.981\bar{k}/(\cos \delta)^2 \tag{18.16}$$

where $\bar{k}$ is the monthly mean *hourly* clearness index for the particular hour, δ is the solar declination, β is the tilt angle of the collector plane with respect to the horizontal, and $\bar{r}_T$ is the ratio of monthly average hourly global radiation on a tilted surface to that on a horizontal surface for that particular hour. For an isotropic sky assumption, $\bar{r}_T$ is given by

$$\bar{r}_T = (1 - \bar{I}_d \bar{I}) r_{b,T} + \left(\frac{1 + \cos\beta}{2}\right) \bar{I}_d \bar{I} + \left(\frac{1 - \cos\beta}{2}\right) \rho \tag{18.17}$$

where $\bar{I}_d$ and $\bar{I}$ are the hourly diffuse and global radiation on the horizontal surface, $r_{b,T}$ is the ratio of hourly beam radiation on the tilted surface to that on a horizontal surface (this is a purely astronomical quantity and can be calculated accurately from geometric considerations), and ρ is the ground albedo.

Example 18.1.3

Compute the total energy collected during 11:30–12:30 for the month of September in New York, NY (latitude: 40.75°N, T_a=20°C) by a flat-plate solar collector of 5 m^2 area having zero tilt. The collector performance parameters are $F_R\eta_0$=0.54 and F_RU_L=3.21 W/(m$^{2\circ}$C) and the collector inlet temperature is 80°C. The corresponding hourly mean clearness index $\bar{k}$ is 0.44, and the monthly mean hourly radiation on a horizontal surface $\bar{I}_{Ti}$ (11:30–12:30) is 6.0 MJ/(m^2 h).

From Equation 18.7a, critical radiation I_C=3.21×(80−20)/0.54=356.7 W/m^2=1.28 MJ/(m^2 h). For the average day of September, solar declination δ=2.2°. Also, because the collector is horizontal $\bar{r}_T$ = 1 and β=0. Thus from Equation 18.16

$$X_{max} = 1.85 + 0.169/0.44^2 - 0.0696/0.44^2 - 0.981 \times 0.44/(\cos 2.2)^2 = 1.93.$$

Also from Equation 18.15, a=(1.93−1)/(2−1.93)=13.29.
The critical radiation ratio X_C=1.28/1.93=0.663.
Because $X_C < X_{max}$, from Equation 18.14 we have

$$\phi_i(X_C) = \left|13.29 - [13.29^2 + (1 + 2 \times 13.29)(1 - 0.663/1.93)^2]^{1/2}\right| = |13.29 - 13.73| = 0.44.$$

Finally, the total energy collected is given by Equation 18.11

$$q_{CN}(11:30\text{–}12:30) = 5 \times 0.54 \times 60 \times 30 \times 0.44 = 214 \text{ MJ/h}$$

18.1.3.3 Daily Utilizability

18.1.3.3.1 Basis

In this mode, T_{Ci}, and hence the critical radiation level, is assumed constant during all hours of the day. The *total* useful energy over N days that can be collected by solar collectors operated all day over n hours is given by

$$Q_{CN} = A_C F_R \bar{\eta}_0 \bar{H}_T N \bar{\phi} \tag{18.18}$$

where $\bar{H}_T$ is the average daily global radiation on the collector surface, and $\bar{\phi}$ (called Phibar) is the daily utilizability factor, defined as

$$\bar{\phi} = \sum^N \sum^n (I_T - I_C)^+ / \sum^N \sum^n I_T = \frac{1}{Nn} \sum^N \sum^n (X_i - X_C)^+ \tag{18.19}$$

Generalized correlations have been developed both at monthly time scales and for annual time scales based on the parameter $\bar{K}$. Generalized (i.e., location and month independent) correlations for $\bar{\phi}$ on a *monthly* time scale have been proposed by Theilacker and Klein (1980). These are strictly applicable for flat-plate collectors only. Collares-Pereira and Rabl (1979) have also proposed generalized correlations for $\bar{\phi}$ on a monthly time scale which, though a little more tedious to use are applicable to concentrating collectors as well. The reader may refer to Rabl (1985) or Reddy (1987) for complete expressions.

18.1.3.3.2 Monthly Time Scales

The Phibar method of determining the daily utilizability fraction proposed by Theilacker and Klein (1980) correlates $\bar{\phi}$ to the following factors:

1. A geometry factor $\bar{R}_T/\bar{r}_{T,noon}$, which incorporates the effects of collector orientation, location, and time of year. $\bar{R}_T$ is the ratio of monthly average global radiation on the tilted surface to that on a horizontal surface. $\bar{r}_{T,noon}$ is the ratio of radiation at noon on the tilted surface to that on a horizontal surface for the average day of the month. Geometrically, $\bar{r}_{T,noon}$ is a measure of the maximum height of the radiation curve over the day, whereas $\bar{R}_T$ is a measure of the enclosed area. Generally the value $(\bar{R}_T/\bar{r}_{T,noon})$ is between 0.9 and 1.5.
2. A dimensionless critical radiation level $\bar{X}_{C,K}$ where

$$\bar{X}_{C,K} = I_C/\bar{I}_{T,noon} \tag{18.20}$$

with $\bar{I}_{T,noon}$, the radiation intensity on the tilted surface at noon, given by

$$\bar{I}_{T,noon} = \bar{r}_{noon}\bar{r}_{T,noon}\bar{H} \tag{18.21}$$

where $\bar{r}_{noon}$ is the ratio of radiation at noon to the daily global radiation on a horizontal surface during the mean day of the month which can be calculated from the following correlation proposed by Liu and Jordan (1960):

$$r(W) = \frac{I(W)}{H} = \frac{\pi}{24}(a + b\cos W)\frac{(\cos W - \cos W_S)}{\left(\sin W_S - \frac{\pi}{180}W_S\cos W_S\right)} \tag{18.22}$$

with

$$a = 0.409 + 0.5016\sin(W_S - 60)$$
$$b = 0.6609 - 0.4767\sin(W_S - 60)$$

where W is the hour angle corresponding to the midpoint of the hour (in degrees) and W_S is the sunset hour angle given by

$$\cos W_S = -\tan L \tan\delta \tag{18.23}$$

where L is the latitude of the location. The fraction r is the ratio of hourly to daily global radiation on a horizontal surface. The factors $\bar{r}_{T,noon}$ and $\bar{r}_{noon}$ can be determined from Equation 18.17 and Equation 18.22, respectively, with $W = 0°$.

The Theilacker and Klein correlation for the daily utilizability for equator-facing flat-plate collectors is

$$\bar{\phi}(X_{C,K}) = \exp\{[a' + b'(\bar{r}_{T,noon}/\bar{R}_T)][X_{C,K} + c'X_{C,K}^2]\} \tag{18.24}$$

where

$$\begin{aligned} a' &= 7.476 - 20.00\bar{K} + 11.188\bar{K}^2 \\ b' &= -8.562 + 18.679\bar{K} - 9.948\bar{K}^2 \\ c' &= -0.722 + 2.426\bar{K} + 0.439\bar{K}^2 \end{aligned} \tag{18.25}$$

How $\bar{\phi}$ varies with the critical radiation ratio $\bar{X}_{C,K}$ for three different values of $\bar{K}$ is shown in Figure 18.13.

Example 18.1.4

A flat-plate collector operated horizontally at Fort Worth, Texas ($L = 32.75°N$), has a surface area of 20 m^2. It is used to heat 10 kg/min. of water entering the collector at a constant temperature of 80°C each day from 6 a.m. to 6 p.m. The collector performance parameters are $F_R\eta_0 = 0.70$ and $F_R U_L = 5.0$ W/(m$^{2\circ}$C). Use Klein's correlation to compute the energy collected by the solar collectors during September.

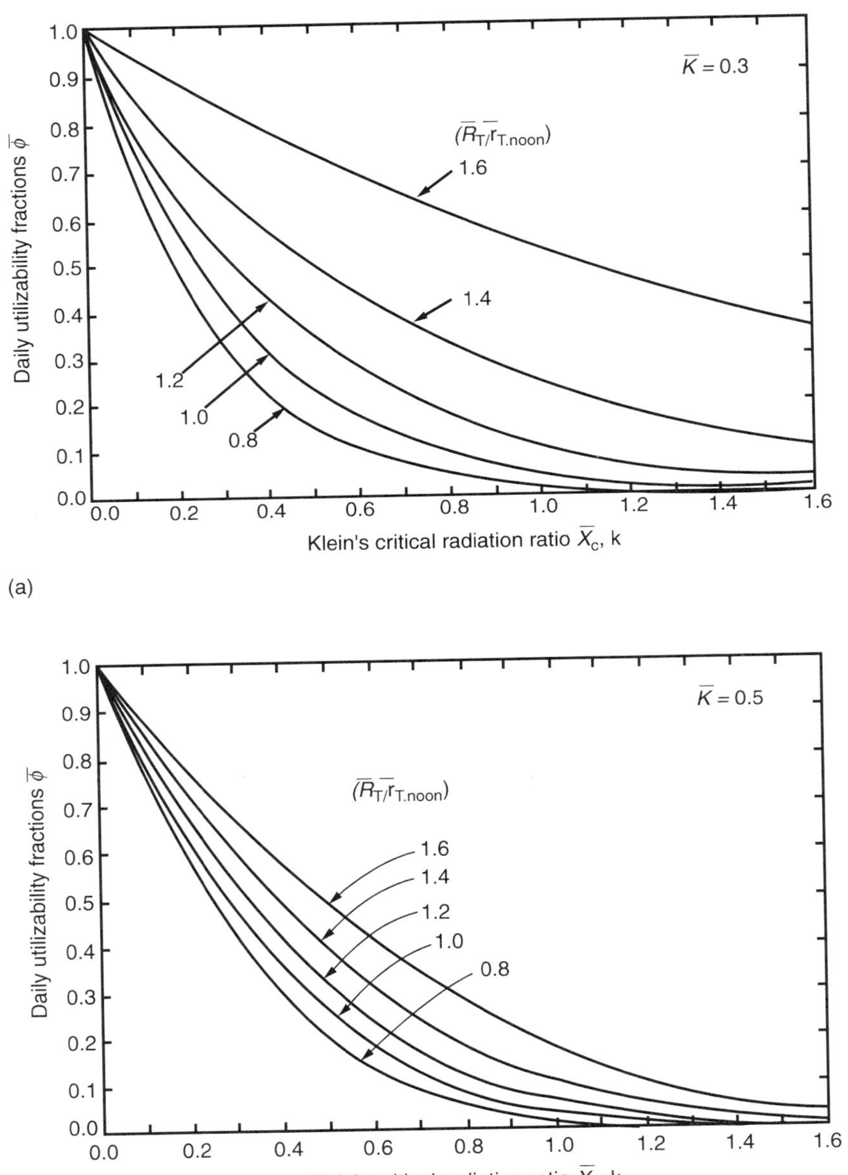

FIGURE 18.13 Generalized daily utilizability curves of Theilacker and Klein (1980) for three different K values. (a) $\bar{K} = 0.3$, (b) $\bar{K} = 0.5$, (c) $\bar{K} = 0.7$. (From Theilacker, J. C. and Klein, S. A., Improvements in the utilizability relationships, *American Section of the International Solar Energy Society Meeting Proceedings*, p. 271. Phoenix, AZ, 1980.)

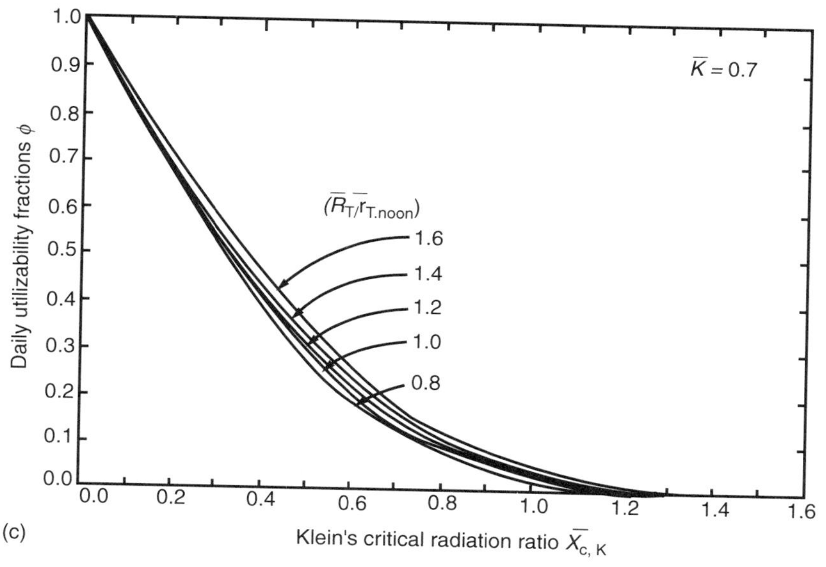

FIGURE 18.13 *(continued)*

Assume $\bar{H} = 18.28$ MJ/(m^{2}d), $\bar{K} = 0.57$ and $\bar{T}_a = 25°$C. Assume the mean sunset hour angle for September to be 90°.

The critical radiation is calculated first:

$$I_C = (5/0.7)(80 - 25) = 393 \text{ W/m}^2 = 1.414 \text{ MJ/(m}^2\text{h)}$$

For a horizontal surface, $\bar{R}_T = \bar{r}_{T,noon} = 1$. From Equation 18.22, $r(W=0) = \pi/24(a+b) = 0.140$. Klein's critical radiation ratio (17.1.20) $\bar{X}_{C,K} = 1.414/(18.28 \times 0.140) = 0.553$. From Equation 18.24, $\bar{\phi} = 0.318$. Finally, from Equation 18.18, the total monthly energy collected by the solar collectors is $Q_{CM} = 20 \times 0.7 \times 30 \times 0.318 \times 18.28 = 2.44$ GJ/month.

18.1.3.3.3 Annual Time Scales

Generalized expressions for the *yearly* average energy delivered by the principal collector types with constant radiation threshold (i.e., when the fluid inlet temperature is constant for all hours during the day over the entire year) have been developed by Rabl (1981) based on data from several U.S. locations. The correlations are basically quadratic of the form

$$\frac{Q_{CY}}{A_C F_R \eta_n} = \tilde{a} + \tilde{b} I_C + \tilde{c} I_C^2 \qquad (18.26)$$

where the coefficients $\tilde{a}$, $\tilde{b}$, and $\tilde{c}$ are functions of collector type and/or tracking mode, climate, and in some cases, latitude. The complete expressions as revised by Gordon and Rabl (1982) are given in Reddy (1987). Note that the yearly *daytime* average value of T_a should be used to determine I_C. If this is not available, the yearly mean *daily* average value can be used. Plots of Q_{CY} versus I_C for flat-plate collectors that face the equator with tilt equal to the latitude are shown in Figure 18.14. The solar radiation enters these expressions as $\tilde{I}_{bn}$, the annual average beam radiation at normal incidence. This can be estimated from the following correlation

$$\tilde{I}_{bn} = 1.37\tilde{K} - 0.34 \qquad (18.27)$$

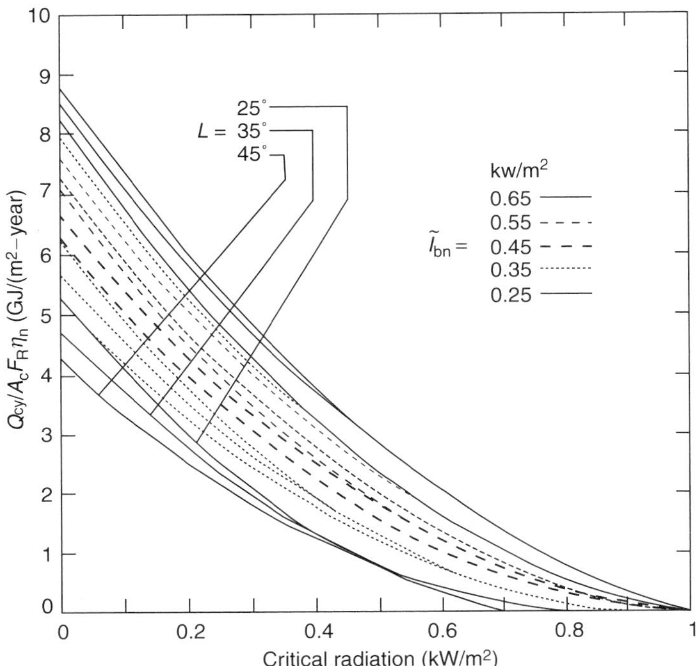

FIGURE 18.14 Yearly total energy delivered by flat-plate collectors with tilt equal to latitude. (From Gordon, J. M. and Rabl, A., Design, analysis and optimization of solar industrial process heat plants without storage, *Solar Energy*, 28, 519, 1982.)

where $\tilde{I}_{bn}$ is in kW/m^2 and $\tilde{K}$ is the annual average clearness index of the location. Values of $\tilde{K}$ for several locations worldwide are given in Reddy (1987).

This correlation is strictly valid for latitudes ranging from 25 to 48°. If used for lower latitudes, the correlation is said to lead to overprediction. Hence, it is recommended that for such lower latitudes a value of 25° be used to compute Q_{Cy}.

A direct comparison of the yearly performance of different collector types is given in Figure 18.15 (from Rabl 1981). A latitude of 35°N is assumed and plots of Q_{Cy} U.S. $(T_{Ci}-T_a)$ have been generated in a sunny climate with $l_{bn}=0.6$ kW/m^2. Relevant collector performance data are given in Figure 18.15. The crossover point between flat-plate and concentrating collectors is approximately 25°C above ambient temperature whether the climate is sunny or cloudy.

18.1.4 Solar Systems

18.1.4.1 Classification

Solar thermal systems can be divided into two categories: standalone or solar supplemented. They can be further classified by means of energy collection as active or passive, by their use as residential or industrial. Further, they can be divided by collector type into liquid or air systems, and by the type of storage they use into seasonal or daily systems.

18.1.4.1.1 Standalone and Solar Supplemented Systems

Standalone systems are systems in which solar energy is the only source of energy input used to meet the required load. Such systems are normally designed for applications where a certain amount of tolerance is permissible concerning the load requirement; in other words, where it is not absolutely imperative that

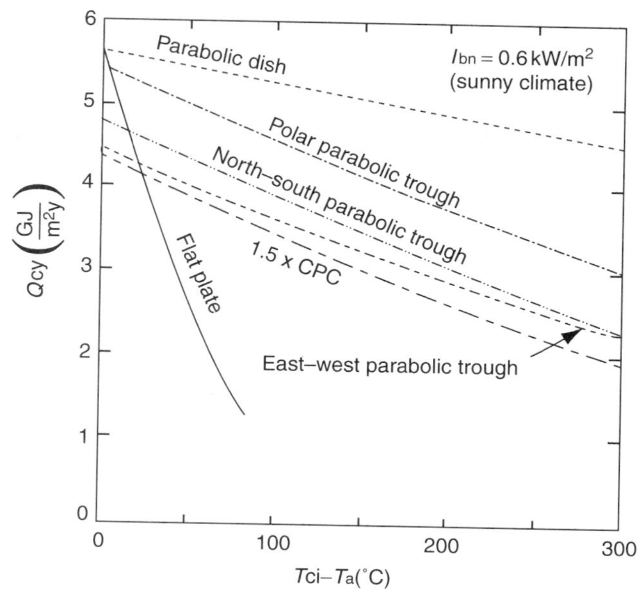

FIGURE 18.15 Figure illustrating the comparative performance (yearly collectible energy) of different collector types as a function of the difference between collector inlet temperature and ambient collector performance parameters $F'\eta_0$ and $F'U_L$ in W/(m^2 °C) are: flat plate (0.70 and 5.0), CPC (0.60 and 0.75), parabolic trough (0.65 and 0.67), and parabolic dish (0.61 and 0.27). (From Rabl, A., Yearly average performance of the principal solar collector types, *Solar Energy*, 27, 215, 1981.)

the specified load be met each and every instant. This leniency is generally admissible in the case of certain residential and agricultural applications. The primary reasons for using such systems are their low cost and simplicity of operation.

Solar-supplemented systems, widely used for both industrial and residential purposes, are those in which solar energy supplies part of the required heat load, the rest being met by an auxiliary source of heat input. Due to the daily variations in incident solar radiation, the portion of the required heat load supplied by the solar energy system may vary from day-to-day. However, the auxiliary source is so designed that at any instant it is capable of meeting the remainder of the required heat load. It is normal practice to incorporate an auxiliary heat source large enough to supply the entire heat load required. Thus, the benefit in the solar subsystem is not in its capacity credit (i.e., not that a smaller capacity conventional system can be used), but rather that a part of the conventional fuel consumption is displaced. The solar subsystem thus acts as a fuel economizer.

Solar-supplemented energy systems will be the primary focus of this chapter. Designing such systems has acquired a certain firm scientific rationale, and the underlying methodologies have reached a certain maturity and diversity, which may satisfy professionals from allied fields. On the other hand, unitary solar apparatus are not discussed here, since these are designed and sized based on local requirements, material availability, construction practices, and practical experience. Simple rules of thumb based on prior experimentation are usually resorted to for designing such systems.

18.1.4.1.2 Active and Passive Systems

Active systems are those systems that need electric pumps or blowers to collect solar energy. It is evident that the amount of solar energy collected should be more than the electrical energy used. Active systems are invariably used for industrial applications and for most domestic and commercial applications as well. *Passive systems* are those systems that collect or use solar energy without direct recourse to any source of conventional power, such as electricity, to aid in the collection. Thus, either such systems operate by natural thermosyphon (for example, domestic water heating systems) between collector, storage, and load or, in the case of space heating, the architecture of the building is such as to favor optimal use of solar energy. Use of a passive system for space heating applications, however, in no way precludes the use of a backup auxiliary system. This chapter deals with active solar systems only.

18.1.4.1.3 Residential and Industrial Systems

Basically, the principles and the components used in these two types of systems are alike, the difference being in the load distribution, control strategies, and relative importance of the components with respect to each other. Whereas *residential* loads have sharp peaks in the early morning or in the evening and have significant seasonal variations, industrial loads tend to be fairly uniform over the year. Constant loads favor the use of solar energy because good equipment utilization can be achieved. Because of differences in load distribution, the role played by the storage differs for both applications. Residential loads often occur at times when solar radiation is no longer available. Thus the collector and the storage subsystems interact in a mode without heat withdrawal from the storage. Finally, for economic reasons, many residential systems are designed to operate by natural thermosyphon, in which case no pumps or controls are needed.

On the other hand, for *industrial and commercial* applications, there is no a priori relationship between the time dependence of the load and the period of sunshine. Moreover, a high reliability has to be assured, so the solar system will have to be combined with a conventional system. Very often, a significant portion of the load can be directly supplied by the solar system even without storage. Another option is to use buffer storage for short periods, on the order of a few hours, in case of discontinuous batch process loads. Thus, the proper design of the storage component has to be given adequate consideration. At present, due to economic constraints as well as the fact that proper awareness of the various installations and operational difficulties associated with larger solar thermal systems is still lacking, solar thermal systems are normally designed either (i) with the no-storage option, or (ii) with buffer storage where a small fraction of the total heat demand is only supplied by the solar system.

18.1.4.1.4 Liquid and Air Collectors

Although air has been the primary fluid for space heating and drying applications, solar air heating systems have until recently been relegated to second place, mainly as a result of the engineering difficulties associated with such systems. Also, applications involving hot air are probably less common than those needing hot water. Air systems for space heating are well described by Löf (1981).

Even with liquid solar collectors, various configurations are possible, and these can be classified basically as *nontracking* (which include flat-plate collectors and CPCs) or *tracking* collectors (which include various types of concentrating collectors). For low-grade thermal heat, for which solar energy is most suited, flat-plate collectors are far more appropriate than concentrating collectors, not only because of their lower cost but also because of their higher thermal efficiencies at low temperature levels. Moreover, their operation and maintenance costs are lower. Finally, for locations having a high fraction of diffuse radiation, as in the tropics, flat-plate collectors are considered to be thermally superior because they can make use of diffuse radiation as well as beam radiation. Although the system design methodologies presented in this chapter explicitly assume flat-plate collector systems, these design approaches can be equally used with concentrating collectors.

18.1.4.1.5 Daily and Seasonal Storage

By *daily storage* is meant systems having capacities equivalent to at most a few days of demand (i.e., just enough to tide over day-to-day climatic fluctuations). In *seasonal storage*, solar energy is stored during the summer for use in winter. Industrial demand loads, which are more or less uniform over the year, are badly suited for seasonal-storage systems. This is also true of air-conditioning for domestic and commercial applications because the load is maximum when solar radiation is also maximum, and vice versa. The present-day economics of seasonal storage units do not usually make such systems an economical proposition except for community heating in cold climates.

18.1.4.2 Closed-Loop and Open-Loop Systems

The two possible configurations of solar thermal systems with daily storage are classified as closed-loop or open-loop systems. Though different authors define these differently, we shall define these as follows. A *closed-loop system* has been defined as a circuit in which the performance of the solar collector is directly dependent on the storage temperature. Figure 18.16 gives a schematic of a closed-loop system in which the fluid circulating in the collectors does not mix with the fluid supplying thermal energy to the load. Thus, these two subsystems are distinct in the sense that any combination of fluids (water or air) is theoretically feasible (a heat exchanger, as shown in the figure, is of course imperative when the fluids are different). However, in practice, only water-water, water-air, or air-air combinations are used. From the point of system performance, the storage temperature normally varies over the day and, consequently, so does collector performance. Closed-loop system configurations have been widely used to date for domestic hot water and space heating applications. The flow rate per unit collector area is generally around 50 kg/(h m^2) for liquid collectors. The storage volume makes about 5–10 passes through the collector during a typical sunny day, and this is why such systems are called *multipass* systems. The temperature rise for each pass is small, of the order of 2°C–5°C for systems with circulating pumps and about 10°C for thermosyphon systems. An expansion tank and a check valve to prevent reverse thermosyphoning at nights, although not shown in the figure, are essential for such system configurations.

Figure 18.17 illustrates one of the possible configurations of *open-loop systems*. Open-loop systems are defined as systems in which the collector performance is independent of the storage temperature. The working fluid may be rejected (or a heat recuperator can be used) if contaminants are picked up during its passage through the load. Alternatively, the working fluid could be directly recalculated back to the entrance of the solar collector field. In all these open-loop configurations, the collector is subject to a given or known inlet temperature specified by the load requirements.

If the working fluid is water, instead of having a continuous flow rate (in which case the outlet temperature of the water will vary with isolation), a solenoid valve can be placed just at the exit of the collector, set so as to open when the desired temperature level of the fluid in the collector is reached. The water is then discharged into storage, and fresh water is taken into the collector. The solar collector will thus operate in a discontinuous manner, but this will ensure that the temperature in the storage is always at the desired level. An alternative way of ensuring uniform collector outlet temperature is to vary the flow rate according to the incident radiation. One can collect a couple of percent more energy

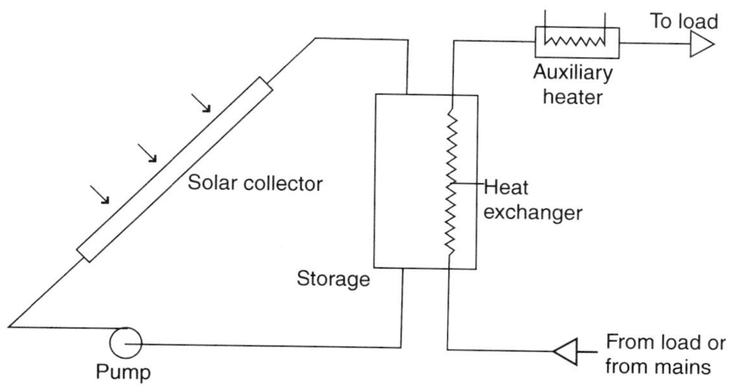

FIGURE 18.16 Schematic of a closed-loop solar system.

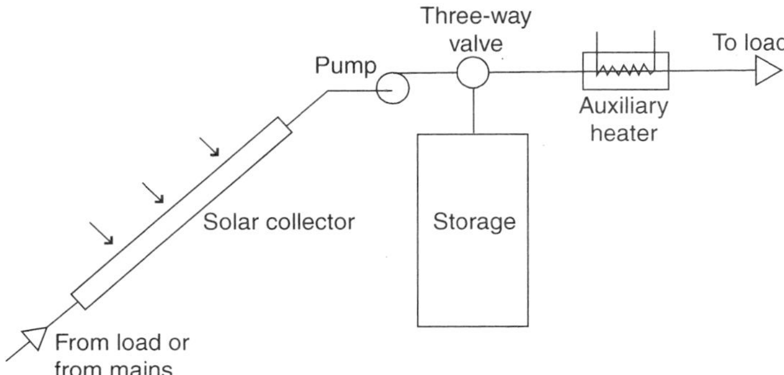

FIGURE 18.17 Schematic of an open-loop solar system.

than with constant rate single-pass designs (Gordon and Zarmi 1985). However, this entails changing the flow rate of the pump more or less continuously, which is injurious to the pump and results in reduced life. Of all the three variants of the open-loop configuration, the first one, namely the single-pass open-loop solar thermal system configuration with constant flow rate and without a solenoid valve, is the most common.

As stated earlier, closed-loop systems are appropriate for domestic applications. Until recently, industrial process heat systems were also designed as large solar domestic hot-water systems with high collector flow rates and with the storage tank volume making several passes per day through the collectors. Consequently, the storage tank tends to be fairly well mixed. Also the tank must be strong enough to withstand the high pressure from the water mains. The open-loop single-pass configuration, wherein the required average daily fluid flow is circulated just once through the collectors with the collector inlet temperature at its lowest value, has been found to be able to deliver as much as 40% more yearly energy for industrial process heat applications than the multipass designs (Collares-Pereira et al. 1984). Finally, in a closed-loop system where an equal amount of fresh water is introduced into storage whenever a certain amount of hot water is drawn off by the load, it is not possible to extract the entire amount of thermal energy contained in storage since the storage temperature is continuously reduced due to mixing. This *partial depletion effect* in the storage tank is not experienced in open-loop systems. The penalty in yearly energy delivery ranges typically from about 10% for daytime-only loads to around 30% for nighttime-only loads compared to a closed-loop multipass system where the storage is depleted every day. Other advantages of open-loop systems are (i) the storage tank need not be pressurized (and hence is less costly), and (ii) the pump size and parasitic power can be lowered.

A final note of caution is required. The single-pass design is not recommended for *variable* loads. The tank size is based on yearly daily load volumes, and efficient use of storage requires near-total depletion of the daily collected energy each day. If the load draw is markedly lower than its average value, the storage would get full relatively early the next day and solar collection would cease. It is because industrial loads tend to be more uniform, both during the day and over the year, than domestic applications that the single-pass open-loop configuration is recommended for such applications.

18.1.4.2.1 Description of a Typical Closed-Loop System

Figure 18.18 illustrates a typical closed-loop solar-supplemented liquid heating system. The useful energy is often (but not always) delivered to the storage tank via a collector-heat exchanger, which separates the collector fluid stream and the storage fluid. Such an arrangement is necessary either for antifreeze protection or to avoid corrosion of the collectors by untreated water containing gases and salts. A safety relief valve is provided because the system piping is normally nonpressurized, and any steam produced in the solar collectors will be let off from this valve. When this happens, energy dumping is said to take place.

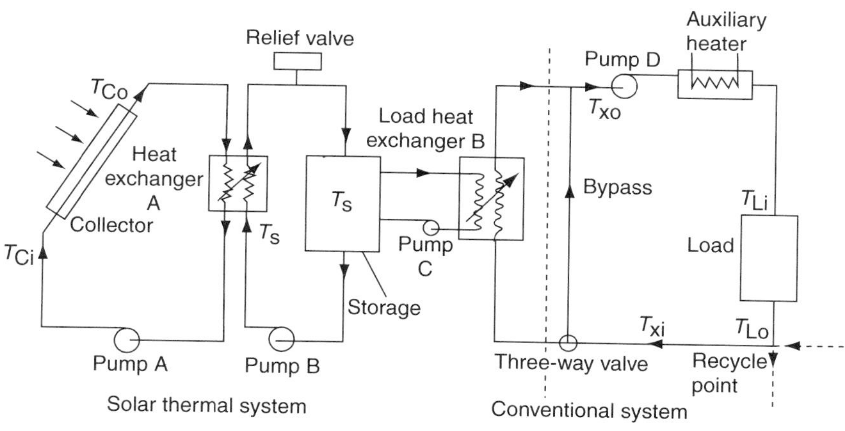

FIGURE 18.18 Schematic of a typical closed-loop system with auxiliary heater placed in series (also referred to as a topping-up type).

Fluid from storage is withdrawn and made to flow through the load-heat exchanger when the load calls for heat. Whenever possible, one should withdraw fluid directly from the storage and pass it through the load, and avoid incorporating the load-heat exchanger, since it introduces additional thermal penalties and involves extra equipment and additional parasitic power use. Heat is withdrawn from the storage tank at the top and reinjected at the bottom in order to derive maximum benefit from the thermal stratification that occurs in the storage tank. A bypass circuit is incorporated prior to the load heat exchanger and comes into play

1. when there is no heat in the storage tank (i.e., storage temperature T_S is less than the fluid temperature entering the load heat exchanger T_{Xi})
2. when T_S is such that the temperature of the fluid leaving the load heat exchanger is greater than that required by the load (i.e., $T_{Xo} > T_{Li}$, in which case the three-way valve bypasses part of the flow so that $T_{Xo} = T_{Li}$). The bypass arrangement is thus a differential control device which is said to modulate the flow such that the above condition is met. Another operational strategy for maintaining $T_{Xo} = T_{Li}$ is to operate the pump in a "bang-bang" fashion (i.e., by short cycling the pump). Such an operation is not advisable, however, since it would lead to premature pump failure.

An auxiliary heater of the *topping-up type* supplies just enough heat to raise T_{Xo} to T_{Li}. After passing through the load, the fluid (which can be either water or air) can be recirculated or, in case of liquid contamination through the load, fresh liquid can be introduced. The auxiliary heater can also be placed in parallel with the load (see Figure 18.19), in which case it is called an *all-or-nothing type.* Although such an arrangement is thermally less efficient than the topping-up type, this type is widely used during the solar retrofit of heating systems because it involves little mechanical modifications or alterations to the auxiliary heater itself.

It is obvious that there could also be solar-supplemented energy systems that do not include a storage element in the system. Figure 18.20 shows such a system configuration with the auxiliary heater installed in series. The operation of such systems is not very different from that of systems with storage, the primary difference being that whenever instantaneous solar energy collection exceeds load requirements (i.e., $T_{Co} > T_{Li}$), energy dumping takes place. It is obvious that by definition there cannot be a closed-loop, no-storage solar thermal system. Solar thermal systems without storage are easier to construct and operate, and even though they may be effective for 8–10 h a day, they are appropriate for applications such as process heat in industry.

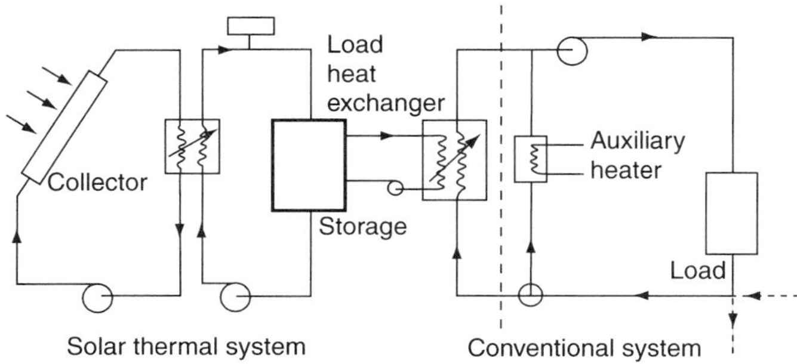

FIGURE 18.19 Schematic of a typical closed-loop system with auxiliary heater placed in parallel (also referred to as an all-or-nothing type).

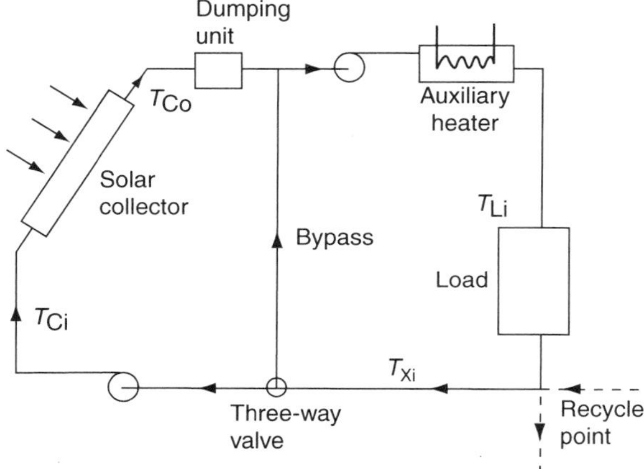

FIGURE 18.20 Simple solar thermal system without storage.

Active closed-loop solar systems as described earlier are widely used for service hot-water systems, that is, for domestic hot water and process heat applications as well as for space heat. There are different variants to this generic configuration. A system without the collector-heat exchanger is referred to having collectors *directly coupled* to the storage tank (as against *indirect coupling* as in Figure 18.16). For domestic hot-water systems, the system can be simplified by placing the auxiliary heater (which is simply an electric heater) directly inside the storage tank. One would like to maintain stratification in the tank so that the coolest fluid is at the bottom of the storage tank, thereby enhancing collection efficiency. Consequently, the electric heater is placed at about the upper third portion of the tank so as to assure good collection efficiency while assuring adequate hot water supply to the load. A more efficient but expensive option is widely used in the United States: the *double tank system*, shown in Figure 18.21. Here the functions of solar storage and auxiliary heating are separated, with the solar tank acting as a preheater for the conventional gas or electric unit. Note that a further system simplification can be achieved for domestic applications by placing the load heat exchanger directly inside the storage tank. In certain cases, one can even eliminate the heat exchanger completely.

Another system configuration is the *drain-back* (also called drain-out) system, where the collectors are emptied each time the solar system shuts off. Thus the system invariably loses collector fluid at least once, and often several times, each day. No collector-heat exchanger is needed, and freeze protection is inherent in such a configuration. However, careful piping design and installation, as well as a two-speed pump, are

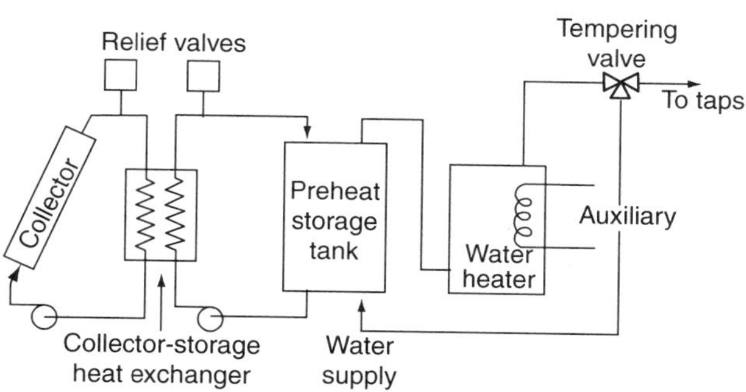

FIGURE 18.21 Schematic of a standard domestic hot-water system with double tank arrangement. (From Duffie, J. A. and Beckman, W. A., *Solar Engineering of Thermal Processes*, Wiley Interscience, New York, 1980.)

needed for the system to work properly (Newton and Gilman 1981). The drain-back configuration may be either open (vented to atmosphere) or closed (for better corrosion protection). Long-term experience in the United States with the drain-back system has shown it to be very reliable if engineered properly. A third type of system configuration is the *drain-down* system, where the fluid from the collector array is removed only when adverse conditions, such as freezing or boiling, occur. This design is used when freezing ambient temperatures are only infrequently encountered.

Active solar systems of the type described above are mostly used in countries such as the United States and Canada. Countries such as Australia, India, and Israel (where freezing is rare) usually prefer thermosyphon systems. No circulating pump is needed, the fluid circulation being driven by density difference between the cooler water in the inlet pipe and the storage tank and the hotter water in the outlet pipe of the collector and the storage tank. The low fluid flow in thermosyphon systems enhances thermal stratification in the storage tank. The system is usually fail-proof, and a study by Liu and Fanney (1980) reported that a thermosyphon system performed better than several pumped service hot-water systems. If operated properly, thermosyphon and active solar systems are comparable in their thermal performance. A major constraint in installing thermosyphon systems in already existing residences is the requirement that the bottom of the storage tank be at least 20 cm or more higher than the top of the solar collector in order to avoid reverse thermosyphoning at night. To overcome this, spring-loaded one-way valves have been used, but with mixed success.

18.1.5 Controls

There are basically five categories to be considered when designing automatic controls (Mueller Associates 1985): (i) collection to storage, (ii) storage to load, (iii) auxiliary energy to load, (iv) miscellaneous (i.e., heat dumping, freeze protection, overheating, etc.), and (v) alarms. The three major control system components are sensors, controllers, and actuating devices. Sensors are used to detect conditions (such as temperatures, pressures, etc.). Controllers receive output from the sensors, select a course of action, and signal a system component to adjust the condition. Actuated devices are components such as pumps, valves, and alarms that execute controller commands and regulate the system.

The sensors for the controls must be set, operated, and located correctly if the solar system is to collect solar energy effectively, reduce operating time, wear and tear of active components, and minimize auxiliary and parasitic energy use. Moreover, sensors also need to be calibrated frequently. For diagnostic purposes, it may be advisable to add extra sensors and data acquisition equipment in order to verify system operation and keep track of long-term system operation. Potential problems can be then rectified in time. The reader may refer to manuals by Mueller Associates (1985) or by SERI (1989) for more details on controls pertaining to solar energy systems.

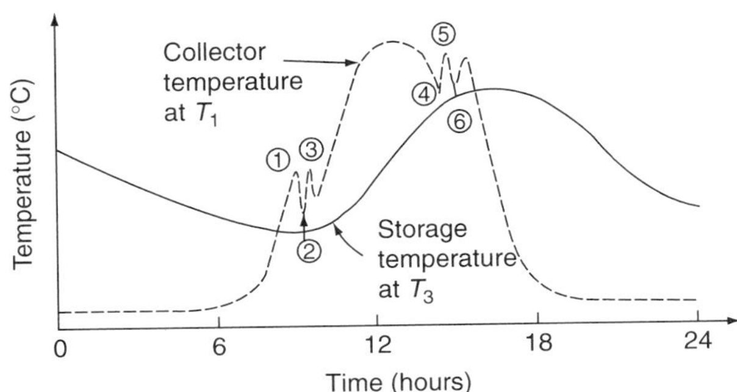

FIGURE 18.22 Typical diurnal variation of collector and storage temperatures. (From CSU, Solar Heating and Cooling of Residential Buildings—Design of Systems, manual prepared by the Solar Energy Applications Laboratory, Colorado State University, 1980.)

Though single-point temperature controllers or solar-cell-activated controls have been used for activated solar collectors, the best way to do so is by differential temperature controllers. Temperature sensors are used to measure the fluid temperature at collector outlet and at the bottom of the storage tank. When the difference is greater than a set amount, say 5°C, then the controller turns the pump on. If the pump is running and the temperature difference falls below another preset value, say 1°C, the controller stops the pump. The temperature deadband between switching-off and reactivating levels should be set with care, since too high a deadband would adversely affect collection efficiency and too low a value would result in short cycling of the collector pump. Figure 18.22 taken from CSU (1980), shows typical diurnal temperature variations of the liquids at collector exit T_1 and in the storage bottom T_3 as a result of heat withdrawal and/or heat losses from the storage. At about 8:30 a.m., $T_1 > T_3$ and, since there is no flow in the collector, T_1 increases rapidly until the difference $(T_1 - T_3)$ reaches the preset activation level (shown as point 1). The collector pump A comes on, and liquid circulation through the collector begins. Because of this cold water surge, T_1 decreases, resulting in a drop of $(T_1 - T_3)$ to the preset deactivating level (shown as point 2). The pump switches off, and liquid flow through the collectors stops. Gradually T_1 increases again, and so on. The number of on–off cycles at system start-up depends on solar intensity, fluid flow rate, volume of water in the collector loop, and the differential controller setting. A similar phenomenon of cycling also occurs in the afternoon. However, the error introduced in solar collector long-term performance predictions by neglecting this cycling effect in the modeling equations is usually small.

18.1.5.1 Corrections to Collector Performance Parameters

18.1.5.1.1 Combined Collector-Heat Exchanger Performance

The use of the heat exchanger A in Figure 18.18 imposes a penalty on the performance of the solar system because T_{Ci} is always higher than T_S, thereby decreasing q_C (see Figure 18.23). The collector-heat exchanger can be implicitly accounted for by suitably modifying the collector performance parameters. Recall from basic heat transfer the concept of heat exchanger effectiveness E defined as the ratio of the actual heat transfer rate to the maximum possible heat transfer rate, that is,

$$E = (mc_p)_a(T_{ai} - T_{ao})/(mc_p)_{min}(T_{ai} - T_{bi}) \quad (18.28a)$$

$$= (mc_p)_b(T_{bo} - T_{bi})/(mc_p)_{min}(T_{ai} - T_{bi}) \quad (18.28b)$$

where $(mC_p)_x$ = capacitance rate of fluid X (with $X = a$ for the warmer fluid, or $X = b$ for the cooler fluid) and $(mC_p)_{min}$ is the lower heat capacitance value of either stream. The advantage of this modeling approach is that, to a good approximation, E can be considered constant in spite of variations in

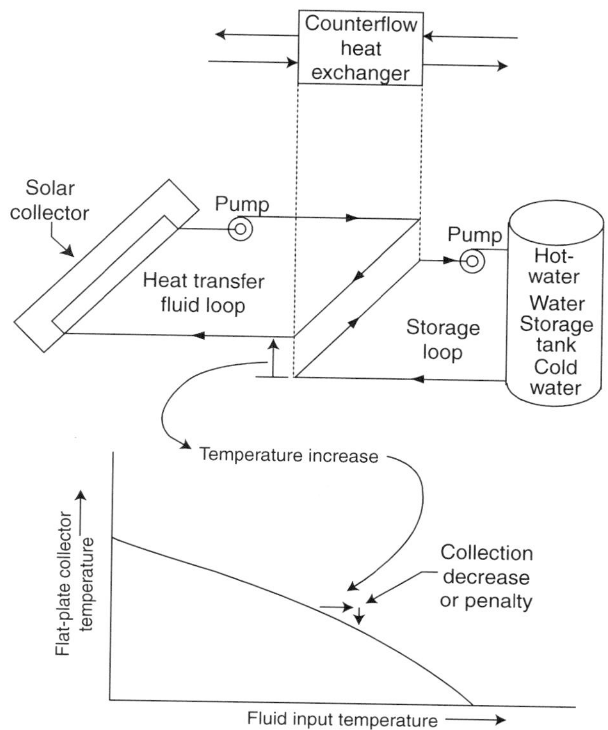

FIGURE 18.23 Heat collection decrease caused by double-loop heat exchangers. (From Cole, R. L., Nield, K. J., Rohde, R. R., and Wolosewicz, R. M. eds., *Design and Installation Manual for Thermal Energy Storage*, ANL-79-15, Argonne National Laboratory, Argonne, IL, 1979.)

temperature levels provided the mass flow rates of both fluids remain constant. Thus knowing the two flow rates, E, T_{ai}, and T_{bi}, both the exit fluid temperatures can be conveniently deduced. De Winter (1975) has shown that the combined performance of the solar collector and the heat exchanger can be conveniently modeled by replacing the collector heat removal factor F_R by a combined collector-exchanger heat removal factor F'_R such that

$$\frac{F'_R}{F_R} = \left[1 + \frac{F_R U_L A_C}{(mc_p)_C}\left\{\frac{(mc_p)_C}{E_A(mc_p)\mathrm{min}} - 1\right\}\right]^{-1} \tag{18.29}$$

where $(mC_p)_c$ is the capacitance rate of the fluid through the collector and E_A is the effectiveness of heat exchanger A. The variation of F'_R/F_R is shown in Figure 18.24. The plots exhibit the same type of asymptotic behavior with mass flow rate as in Figure 18.3.

The design of the collector-heat exchanger also requires care if the penalty imposed by it on the solar collection is to be minimized. Using a large heat exchanger increases the effectiveness and lowers this penalty; that is, the ratio (F'_R/F_R) is high, but the associated initial and operating costs may be higher. Both these considerations need to be balanced for optimum design (see Figure 18.25). Optimum heat exchanger area A_X can be found from the following equation proposed by Cole et al. (1979):

$$A_X = A_C\left[\frac{F_R U_L C_C}{U_X C_X}\right]^{1/2} \tag{18.30}$$

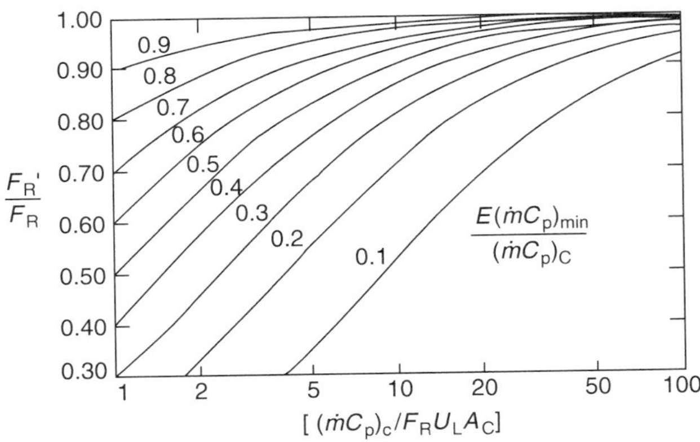

FIGURE 18.24 Variation of collector-heat exchanger correction factor. (From Duffie, J. A. and Beckman, W. A., *Solar Engineering of Thermal Processes*, Wiley Interscience, New York, 1980.)

where A_C is the collector area, C_C is the cost per unit collector area, C_X is the cost per unit heat exchanger area, and U_X is the heat loss per unit area of the heat exchanger.

18.1.5.1.2 Collector Piping and Shading Losses

Other corrections that can be applied to collector performance parameters include those for thermal losses from the piping (or from ducts) between the collection subsystem and the storage unit. Beckman (1978) has shown that these losses can be conveniently taken into consideration by suitably modifying the η_n and U_L terms of the solar collectors as follows:

$$\frac{\eta'_n}{\eta_n} = \left[1 + \frac{u_d A_0}{(mc_p)_C}\right]^{-1} \quad \text{and} \quad \frac{U'_L}{U_L} = \frac{1 - \frac{U_d A_i}{(mc_p)_C} + \frac{U_d (A_i + A_0)}{A_C F_R U_L}}{1 + \frac{U_d A_0}{(mc_p)_C}} \qquad (18.31)$$

where U_d is the heat coefficient from the pipe or duct, A_0 is the heat loss area of the outlet pipe or duct, and A_i is the heat loss area of the inlet pipe or duct.

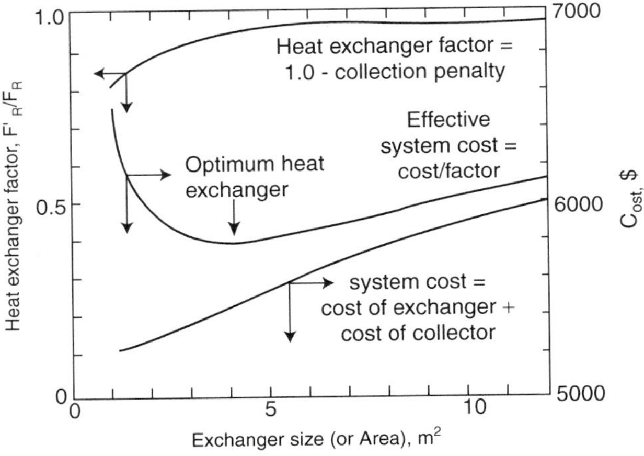

FIGURE 18.25 Typical heat-exchanger optimization plot. (From Cole, R. L., Nield, K. J., Rohde, R. R., and Wolosewicz, R. M. eds., *Design and Installation Manual for Thermal Energy Storage*, ANL-79-15, Argonne National Laboratory, Argonne, IL, 1979.)

When large collector arrays are mounted on flat roofs or level ground, multiple rows of collectors are usually arranged in a sawtooth fashion. These multiple rows must be spaced so that they do not shade each other at low sun angles. Unlimited space is rarely available, and it is desirable to space the rows as close as possible to minimize piping and to keep land costs low. Some amount of shading, especially during early mornings and late evenings during the winter months is generally acceptable. Detailed analysis of shading losses is cumbersome though not difficult and equations presented in standard text books such as Duffie and Beckman (1980) can be used directly.

18.1.6 Thermal Storage Systems

Low-temperature solar thermal energy can be stored in liquids, solids, or phase change materials (PCMs). Water is the most frequently used liquid storage medium because of its low cost and high specific heat. The most widely used solid storage medium is rocks (usually of uniform circular size 25–40 mm in diameter). PCM storage is much less bulky because of the high-latent heat of the PCM material, but this technology has yet to become economical and safe for widespread use.

Water storage would be the obvious choice when liquid collectors are used to supply hot water to a load. When hot air is required (for space heat or for convective drying), one has two options: an air collector with a pebble-bed storage or a system with liquid collectors, water storage, and a load heat exchanger to transfer heat from the hot water to the working air stream. Though a number of solar air systems have been designed and operated successfully (mainly for space heating), water storage is very often the medium selected. Water has twice the heat capacity of rock, so water storage tanks will be smaller than rock-bed containers. Moreover, rock storage systems require higher parasitic energy to operate, have higher installation costs, and require more sophisticated controls. Water storage permits simultaneous charging and discharging while such an operation is not possible for rock storage systems. The various types of materials used as containers for water and rock-bed storage and the types of design, installation, and operation details one needs to take care of in such storage systems are described by Mueller Associates (1985), SERI (1989).

Sensible storage systems, whether water or rock-bed, exhibit a certain amount of thermal stratification. Standard textbooks present relevant equations to model such effects. In the case of active closed-loop multipass hot-water systems, storage stratification effects can be neglected for long-term system performance with little loss of accuracy. Moreover, this leads to conservative system design (i.e., solar contribution is underpredicted if stratification is neglected). A designer who wishes to account for the effect of stratification in the water storage can resort to a formulation by Phillips and Dave (1982), who showed that this effect can be fairly well modeled by introducing a *stratification coefficient* (which is a system constant that needs to be determined only once) and treating the storage subsystem as fully mixed. However, this approach is limited to the specific case of no (or very little) heat withdrawal from storage during the collection period. Even when water storage systems are highly stratified, simulation studies seem to indicate that modeling storage as a one-dimensional plug-flow three-node heat transfer problem yields satisfactory results of long-term solar system performance.

The thermal losses q_w from the storage tank can be modeled as

$$q_w = (UA_S)(T_S - T_{env}) \tag{18.32}$$

where (UA_S) is the storage overall heat loss per unit temperature difference and T_{env} is the temperature of the air surrounding the storage tank. Note that (UA_S) depends (i) on the storage size, which is a parameter to be sized during system design, and (ii) on the configuration of the storage tank (i.e., on the length by diameter ratio in case of a cylindrical tank). For storage tanks, this ratio is normally in the range of 1.0–2.0.

18.1.7 Solar System Simulation

A system model is nothing but an assembly of appropriate component modeling equations that are to be solved over time subject to certain forcing functions (i.e., the meteorological data and load data). The resulting set of simultaneous equations can be solved either analytically or numerically.

The analytical method of resolution is appropriate, or possible, only for simplified system configurations and operating conditions. This approach has had some success in the analysis and design of open-loop systems (refer to Reddy 1987, and Gordon and Rabl 1986, for more details). On the other hand, numerical simulation can be performed for any system configuration and operating strategy, however, complex. However, this is time-consuming and expensive in computer time and requires a high level of operator expertise.

We shall illustrate the approach of numerical simulation by considering the simple solar system shown in Figure 18.18. Assuming a fully mixed storage tank, the instantaneous energy balance equation is

$$(Mc_p)_S(dT_S/dt) = q_C - q_u - q_w \tag{18.33}$$

where

q_C is the useful energy delivered by the solar collector (given by Equation 18.2.)
q_w is the thermal loss from the storage tank (given by Equation 18.32)
q_u is the useful heat transferred through the load heat exchanger, which can be determined as follows:

The maximum hourly rate of energy transfer through the load heat exchanger is

$$q_{max} = E_B(mc_p)_{min}(T_S - T_{Xi})\delta_L \tag{18.34}$$

where δ_L is a control function whose value is either 1 or 0 depending upon whether there is a heat demand or not. Since q_{max} can be greater than the amount of thermal energy q_L actually required by the load, the bypass arrangement can be conveniently modeled as

$$q_u = \min(q_{max}, q_L) \tag{18.35}$$

where

$$q_L = (mc_p)_L(T_{Li} - T_{Xi}) \tag{18.36}$$

for water heating and industrial process heat loads. Space heating and cooling loads can be conveniently determined by one of the several variants of the bin-type methods (ASHRAE 1985).

The amount of energy q_{max} supplied by a topping-up type of auxiliary heater is

$$q_{max} = q_L - q_u \tag{18.37}$$

Assuming $T_{env} = T_a$, Equation 18.33 can be expanded into

$$(Mc_p)_S \frac{dT_S}{dt} - A_C F_R[I_T\eta_0 - U_L(T_S - T_a)]^+ - (mc_p)_s(T_S - T_{Xi})\delta_L - (UA)_S(T_S - T_a) \tag{18.38}$$

The presence of control functions and time dependence of I_T and T_a prevent a general analytical treatment, though, as mentioned earlier, specific cases can be handled. The numerical approach involves expressing this differential equation in finite difference form. After rearranging, one gets

$$= T_{S,b} + \frac{\Delta t}{(Mc_p)_S}\{A_C F_R[I_T\eta_0 - U_L(T_{S,b} - T_a)]^+ - (mc_p)_S(T_{S,b} - T_{Xi})\delta_L - (UA)_S(T_{S,b} - T_a)\}T_{S,f} \tag{18.39}$$

where $T_{S,b}$ and $T_{S,f}$ are the storage temperatures at the beginning and the end of the time step Δt. The time step is sufficiently small (say 1 h) that I_T and T_a can be assumed constant. This equation is repeatedly used over the time period in question (day, month, or year), and the total energy supplied by the collector or to the load can be estimated.

Such methods of simulation, referred as stepwise steady-state simulations, implicitly assume that the solar thermal system operates in a steady-state manner during one time step, at the end of which it undergoes an abrupt change in operating conditions as a result of changes in the forcing functions, and thereby attains a new steady-state operating level. Although in reality, the system performance varies smoothly over time and is consequently different from that outlined earlier, it has been found that, in most cases, taking time steps of the order of 1 h yields acceptable results of long-term performance.

The objective of solar-supplemented energy systems is to displace part of the conventional fuel consumption of the auxiliary heater. The index used to represent the contribution of the solar thermal system is the *solar fraction*, which is the fraction of the total energy required by the load that is supplied by the solar system. The solar fraction could be expressed over any time scale, with month and year being the most common. Two commonly used definitions of the monthly solar fraction are

1. Thermal solar fraction:

$$f_Y = Q_{UM}/Q_{LM} = 1 - Q_{aux,M}/Q_{LM} \tag{18.40}$$

   where
   Q_{UM} is the monthly total thermal energy supplied by the solar system
   Q_{LM} is the monthly total thermal requirements of the load
   $Q_{aux,M}$ is the monthly total auxiliary energy consumed
2. Energy solar fraction (i.e., thermal plus parasitic energy):

$$f'_M = Q'_{UM}/Q'_{LM} \tag{18.41}$$

   where Q'_{UM} is Q_{UM} *minus* the parasitic energy consumed by the solar system and Q'_{UM} is Q_{LM} *plus* the parasitic energy consumed by the load.

Example 18.1.5

Simulate the closed-loop solar thermal system shown in Figure 18.18 for each hour of a day assuming both collector and load heat exchangers to be absent (i.e., $E_A = E_B = 1$). Assume the following data as input for the simulation: $A_C = 10\ m^2$, $F_R U_L = 5.0\ W/m^{2\circ}C)$, $F_R \eta_0 = 0.7$, $(Mc_p) = 2.0\ MJ/^\circ C$ and $(UA)_S = 3\ W/^\circ C$. Water is withdrawn to meet a load from 9 a.m. to 7 p.m. (solar time) at a constant rate of 60 kg/h and is replenished from the mains at a temperature of 25°C. The storage temperature at the start (i.e., at 6 a.m.) is 40°C, and the environment temperature is equal to the ambient temperature. The temperature of the water entering the load should not exceed 55°C. The hourly values of the solar radiation on the plane of the collector are given in column 2 of Table 18.3 and the ambient temperature is assumed constant over the day and equal to 25°C. The variation of the optical efficiency with angle of incidence can be neglected.

The results of the simulation are given in Table 18.3. The following equations should permit the reader to verify for himself the results obtained. Simulating the system entails solving the following equations in the sequence given here:

Column 4. Useful energy delivered by the collector (Equation 18.2)

$$q_C = 10[0.7 I_T - 5(3{,}600/10^6) T_{S,b} - 25]^+ \text{ (MJ/h)}$$

The term $(3600/10^6)$ is introduced to convert W/m^2 (the units in which I_T is expressed) into $MJ/(h\ m^2)$. Note that $T_{S,b}$ is taken to be equal to $T_{S,f}$ of the final hour.

Column 5. Thermal losses from the storage tank (Equation 18.32)

$$q_w = 3(3{,}600/10^6)(T_{S,b} - 25) \text{ (MJ/h)}$$

TABLE 18.3 Simulation Results of Example 18.1.5

(1) Solar Time (h)	(2) I_T (MJ/m²h)	(3) $T_{S,f}$ (°C)	(4) q_C (MJ/h)	(5) q_w (MJ/h)	(6) q_{max} (MJ/h)	(7) q_L (MJ/h)	(8) q_U (MJ/h)	(9) q_{aux} (MJ/h)
Start		40.00						
6–7	0.37	39.92	0.00	0.16	0.00	0.00	0.00	0.00
7–8	0.95	41.82	3.96	0.16	0.00	0.00	0.00	0.00
8–9	1.54	45.61	7.75	0.18	0.00	0.00	0.00	0.00
9–10	2.00	48.05	10.29	0.22	5.18	754	5.18	2.36
10–11	2.27	50.90	11.74	0.25	5.79	7.54	5.79	1.75
11–12	2.46	53.78	12.56	0.28	6.51	7.54	6.51	1.03
12–13	2.50	56.17	12.32	0.31	7.24	7.54	7.24	0.31
13–14	2.24	57.26	10.07	0.34	7.84	7.54	7.54	0.00
14–15	2.12	57.84	9.03	0.35	8.11	7.54	7.54	0.00
15–16	1.37	55.73	3.68	0.35	8.25	7.54	7.54	0.00
16–17	0.76	51.79	0.00	0.33	7.72	7.54	7.54	0.00
17–18	0.23	48.28	0.00	0.29	6.73	7.54	6.73	0.81
18–19	0.00	45.23	0.00	0.25	5.85	7.54	5.85	1.69
Total	18.81	—	81.41	3.48	69.24	75.40	67.48	7.94

Column 6. The maximum rate of energy that can be transferred from the load can be calculated from Equation 18.34

$$q_{max} = 60(4{,}190/10^6)(T_{S,b} - 25)\ (\text{MJ/h})$$

Column 7. The thermal energy required by the load (from Equation 18.36)

$$q_L = 60(4{,}190/10^6)(55 - 25) = 7.54\ \text{MJ/h}$$

Column 8. The actual amount of heat withdrawn from storage (Equation 18.35)

$$q_w = \min[\text{column 6, column 7}]$$

Column 9. The amount of energy supplied by the auxiliary heater (Equation 18.2.37)

$$q_{aux} = \text{column 7} - \text{column 8}$$

The final storage temperature $T_{S,f}$ is now calculated from Equation 18.39

$$T_{S,f} = T_{S,b} + [\text{column 4} - \text{column 8} - \text{column 5}]/2.0$$

From Table 18.3, we note that the solar collector efficiency over the entire day is $[81.41/(18.81 \times 10)] = 0.43$. The corresponding daily solar fraction $= (67.48/75.40) = 0.895$.

18.1.8 Solar System Sizing Methodology

Sizing of solar systems primarily involves determining the collector area and storage size that are most cost effective. Standalone and solar-supplemented systems have to be treated separately since the basic design problem is somewhat different. The interested reader can refer to Gordon (1987) for sizing standalone systems.

18.1.8.1 Solar-Supplemented Systems

18.1.8.1.1 Production Functions

Because of the annual variation of incident solar radiation, it is not normally economical to size a solar subsystem such that it provides 100% of the heat demand. Most solar energy systems follow the *law of diminishing returns.* This implies that increasing the size of the solar collector subsystem results in a less than proportional increase in the annual fuel savings (or alternatively, in the annual solar fraction).

Any model has two types of variables: exogenous and endogenous. The *exogenous parameters* are also called the input variables, and these in turn may be of two kinds. *Variable exogenous parameters* are the collector area A_C, the collector performance parameters $F_R\eta_n$ and F_RU_L, the collector tilt, the thermal storage capacity $(Mc_p)_S$, the heat exchanger size, and the control strategies of the solar thermal system. On the other hand, the climatic data specified by radiation and the ambient temperature, as well as the end use thermal demand characteristics, are called *constrained exogenous parameters* because they are imposed externally and cannot be changed. The *endogenous parameters* are the output parameters whose values are to be determined, the annual solar fraction being one of the parameters most often sought.

Figure 18.26 illustrates the law of diminishing results. The annual solar fraction f_Y is seen to increase with collector area but at a decreasing rate and at a certain point will reach saturation. Variation of any of the other exogenous parameters also exhibits a similar trend. The technical relationship between f_Y and one or several variable exogenous parameters for a given location is called the *yearly production function.*

It is only for certain simple types of solar thermal systems that an analytical expression for the production can be deduced directly from theoretical considerations. The most common approach is to carry out computer simulations of the particular system (solar plus auxiliary) over the complete year for several combinations of values of the exogenous parameters. The production function can subsequently be determined by an empirical curve fit to these discrete sets of points.

Example 18.1.6

Kreider (1979a, 1979b) gives the following expression for the production function of an industrial solar water heater for a certain location:

$$f_Y = Q_{UY}/Q_{LY} = \left(0.35 - \frac{F_RU_L}{100F_R\eta_n}\right)\ln\left(1 + \frac{20F_R\eta_nA_C}{Q_{LY}}\right) \tag{18.42}$$

where Q_{UY} is the thermal energy delivered by the solar thermal system over the year in GJ/y; Q_{LY} is the yearly thermal load demand, also in GJ/y; and F_RU_L is in W/(m^{2}°C). Note that only certain solar system exogenous parameters figure explicitly in this expression, thereby implying that other exogenous parameters (for example, storage volume) have not been varied during the study. As an illustration,

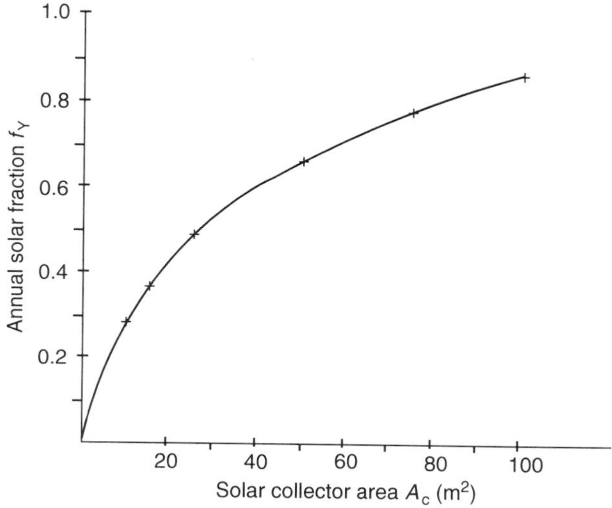

FIGURE 18.26 A typical solar system production function (see Example 18.3.6).

let us assume the following nominal values: $Q_{LY} = 100$ GJ/y, $F_R U_L = 2.0$ W/(m^2°C), and $F_R \eta_n = 0.7$. For a 1% increase in collector area A_C, the corresponding percentage increase in Q_{UY} (called elasticity) can be determined:

$$\frac{dQ_{UY}}{Q_{UY}} = \frac{dA_C}{A_C}\left[\left(\frac{Q_{LY}}{20F_R\eta_n A_C}+1\right)\ln\left(1+\frac{20F_R\eta_n A_C}{Q_{LY}}\right)\right]^{-1} \tag{18.43}$$

From this, we obtain the expression for *marginal productivity*

$$\frac{dQ_{UY}}{dA_C} = \frac{Q_{UY}}{A_C}\left[\left(\frac{Q_{LY}}{20F_R\eta_n A_C}+1\right)\ln\left(1+\frac{20F_R\eta_n A_C}{Q_{LY}}\right)\right]^{-1} \tag{18.44}$$

Numerical values can be obtained from the preceding expression. Though Q_{UY} increases with A_C, the marginal productivity of Q_{UY} goes on decreasing with increasing A_C, thus illustrating the law of diminishing returns. A qualitative explanation of this phenomenon is that as A_C increases, the mean operating temperature level of the collector increases, thus leading to decreasing solar collection rates. Figure 18.26 illustrates the variation of f_Y with A_C as given by Equation 18.42 when the preceding numerical values are used.

The objective of the sizing study in its widest perspective is to determine, for a given specific thermal end use, the size and configuration of the solar subsystem that results in the most economical operation of the entire system. This economical optimum can be determined using the production function along with an appropriate economic analysis. Several authors—for example, Duffie and Beckman (1980) or Rabl (1985)—have presented fairly rigorous methodologies of economic analysis, but a simple approach is adequate to illustrate the concepts and for preliminary system sizing.

18.1.8.1.2 Simplified Economic Analysis

It is widely recognized that *discounted cash flow analysis* is most appropriate for applications such as sizing an energy system. This analysis takes into account both the initial cost incurred during the installation of the system and the annual running costs over its entire life span.

The economic objective function for optimal system selection can be expressed in terms of either the energy cost incurred or the energy savings. These two approaches are basically similar and differ in the sense that the objective function of the former has to be minimized while that of the latter has to be maximized. In our analysis, we shall consider the latter approach, which can further be subdivided into the following two methods:

1. Present worth or life cycle savings, wherein all running costs are discounted to the beginning of the first year of operation of the system.
2. Annualized life cycle savings, wherein the initial expenditure incurred at the start as well as the running costs over the life of the installation are expressed as a yearly mean value.

18.1.9 Solar System Design Methods

18.1.9.1 Classification

Design methods may be separated into three generic classes. The *simple* category, usually associated with the prefeasibility study phase involves quick manual calculations of solar collector/system performance and rule-of-thumb engineering estimates. For example, the generalized yearly correlations proposed by Rabl (1981) and described in Section 18.1.2 could be conveniently used for year-round, more or less constant loads. The approach is directly valid for open-loop solar systems, while it could also be used for closed-loop systems if an *average* collector inlet temperature could be determined. A simple manner of selecting this temperature $\bar{T}_m$ for domestic closed-loop multipass systems is to assume the following empirical relation:

$$\bar{T}_{\mathrm{m}} - T_{\mathrm{mains}}/3 + (2/3)T_{\mathrm{set}} \tag{18.45}$$

where T_{mains} is the average annual supply temperature and T_{set} is the required hot-water temperature (about 60°C–80°C in most cases).

These manual methods often use general guidelines, graphs, and/or tables for sizing and performance evaluation. The designer should have a certain amount of knowledge and experience in solar system design in order to make pertinent assumptions and simplifications regarding the operation of the particular system.

Mid-level design methods are resorted to during the feasibility phase of a project. The main focus of this chapter has been toward this level, and a few of these design methods will be presented in this section. A personal computer is best suited to these design methods because they could be conveniently programmed to suit the designer's tastes and purpose (spreadsheet programs, or better still one of the numerous equation-solver software packages, are most convenient). Alternatively, commercially available software packages such as f-chart (Beckman, Klein, and Duffie 1977) could also be used for certain specific system configurations.

Detailed design methods involve performing hourly simulations of the solar system over the entire year from which accurate optimization of solar collector and other equipment can be performed. Several simulation programs for active solar energy systems are available, TRNSYS (Klein et al. 1975, 1979) developed at the University of Wisconsin–Madison being perhaps the best known. This public-domain software has technical support and is being constantly upgraded. TRNSYS contains simulation models of numerous subsystem components (solar radiation, solar equipment, loads, mechanical equipment, controls, etc.) that comprise a solar energy system. A user can conveniently hook up components representative of a particular solar system to be analyzed and then simulate that system's performance at a level of detail that the user selects. Thus TRNSYS provides the design with large flexibility, diversity, and convenience of usage.

As pointed out by Rabl (1985), the detailed computer simulations approach, though a valuable tool, has several problems. Judgment is needed both in the selection of the input and in the evaluation of the output. The very flexibility of big simulation programs has drawbacks. So many variables must be specified by the user that errors in interpretation or specification are common. Also, learning how to use the program is a time-consuming task. Because of the numerous system variables to be optimized, the program may have to be run for numerous sets of combinations, which adds to expense and time. The inexperienced user can be easily misled by the second-order details while missing first-order effects. For example, uncertainties in load, solar radiation, and economic variables are usually very large, and long-term performance simulation results are only accurate to within a certain degree. Nevertheless, detailed simulation programs, if properly used by experienced designers, can provide valuable information on system design and optimization aspects at the final stages of a project design.

There are basically three types of mid-level design approaches: the empirical correlation approach, the analytical approach, and the one-day repetitive methods (described fully in Reddy 1987). We shall illustrate their use by means of specific applications.

18.1.9.2 Active Space Heating

The solar system configuration for this particular application has become more or less standardized. For example, for a liquid system, one would use the system shown in Figure 18.27. One of the most widely used design methods is the f-chart method (Beckman, Klein, and Duffie 1977; Duffie and Beckman 1980), which is applicable for standardized liquid and air heating systems as well as for standardized domestic hot-water systems. The f-chart method basically involves using a simple algebraic correlation that has been deduced from numerous TRNSYS simulation runs of these standard solar systems subject to a wide range of climates and solar system parameters (see Figure 18.28). Correlations were

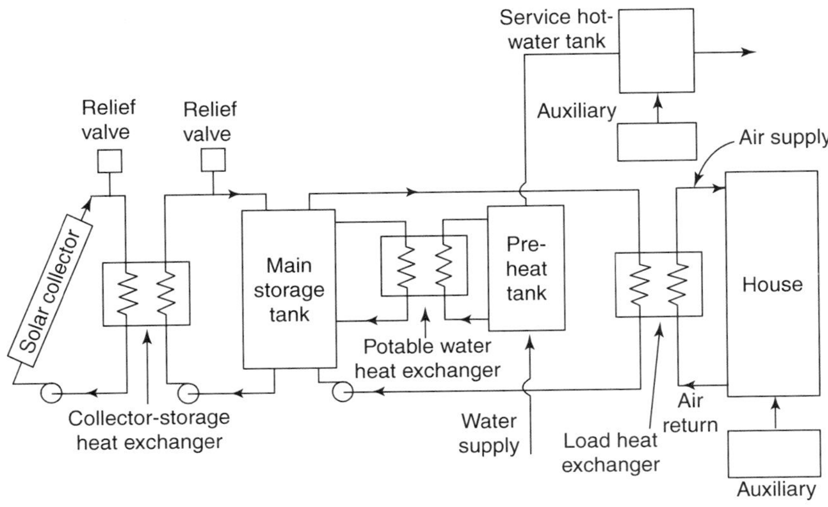

FIGURE 18.27 Schematic of the standard space heating liquid system configuration for the f-chart method. (From Duffie, J. A. and Beckman, W. A., *Solar Engineering of Thermal Processes*, Wiley Interscience, New York, 1980.)

developed between monthly solar fractions and two easily calculated dimensionless variables X and Y, where

$$X = (A_C F'_R U_L (T_{Ref} - \bar{T}_a)\Delta t)/Q_{LM} \tag{18.46}$$

$$Y = A_C F'_R \bar{\eta}_0 \bar{H}_T N / Q_{LM} \tag{18.47}$$

where

- A_C collector area (m^2)
- F'_R collector-heat exchanger heat removal factor (given by Equation 18.29)
- U_L collector overall loss coefficient ($W/(m^{2\circ}C)$)
- Δt total number of seconds in the month $= 3600 \times 24 \times N = 86{,}400 \times N$

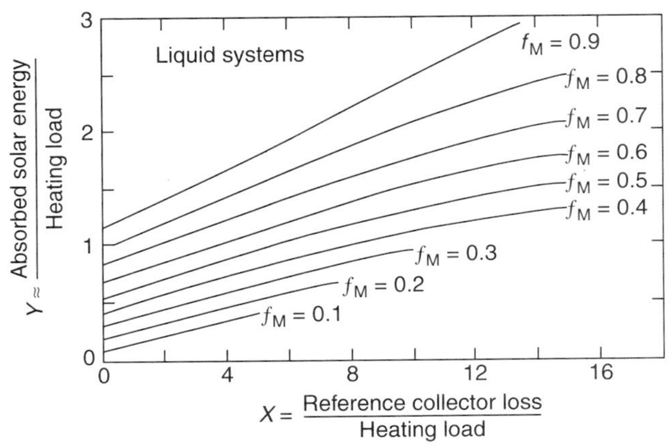

FIGURE 18.28 The f-chart correlation for liquid system configuration. (From Duffie, J. A. and Beckman, W. A., *Solar Engineering of Thermal Processes*, Wiley Interscience, New York, 1980.)

$\bar{T}_a$ monthly average ambient temperature (°C)
T_{Ref} an empirically derived reference temperature, taken as 100°C
Q_{LM} monthly total heating load for space heating and/or hot water (J)
$\bar{H}_T$ monthly average daily radiation incident on the collector surface per unit area (J/m^2)
N number of days in the month
$\bar{\eta}_0$ monthly average collector optical efficiency

The dimensionless variable X is the ratio of reference collector losses over the entire month to the monthly total heat load; the variable Y is the ratio of the monthly total solar energy absorbed by the collectors to the monthly total heat load. It will be noted that the collector area and its performance parameters are the predominant exogenous variables that appear in these expressions. For changes in secondary exogenous parameters, the following corrective terms X_C and Y_C should be applied for liquid systems:

1. for changes in storage capacity:

$$X_C/X = (\text{actual storage capacity/standard storage capacity})^{-0.25} \tag{18.48}$$

where the standard storage volume is 75 L/m^2 of collector area.

2. for changes in heat exchanger size:

$$Y_C/Y = 0.39 + 0.65\exp[-(0.139(UA)_B/(E_L(mc_p)_{min}))] \tag{18.49}$$

The monthly solar fraction for liquid space heating can then be determined from the following empirical correlation:

$$f_M = 1.029Y - 0.065X - 0.245Y^2 + 0.0018X^2 + 0.0215Y^3 \tag{18.50}$$

subject to the conditions that $0 \le X \le 15$ and $0 \le Y \le 3$. This empirical correlation is shown graphically in Figure 18.28.

A similar correlation has also been proposed for space heating systems using air collectors and pebble-bed storage. The procedure for exploiting the preceding empirical correlations is as follows. For a predetermined location, specified by its 12 monthly radiation and ambient temperature values, Equation 18.50 is repeatedly used for each month of the year for a particular set of variable exogenous parameters. The monthly solar fraction f_M and thence the annual thermal energy delivered by the solar thermal system are easily deduced. Subsequently, the entire procedure is repeated for different values and combinations of variable exogenous parameters. Finally, an economic analysis is performed to determine optimal sizes of various solar system components. Care must be exercised that the exogenous parameters considered are not outside the range of validity of the f-chart empirical correlations.

Example 18.1.7

(Adapted from Duffie and Beckman 1980). A solar heating system is to be designed for Madison, Wisconsin (latitude 43°N) using one-cover collectors with $F_R\eta_n = 0.74$ and $F_RU_L = 4$ W/(m$^{2\circ}$C). The collector faces south with a slope of 60° from the horizontal. The average daily radiation on the tilted surface in January is 12.9 MJ/m^2, and the average ambient temperature is -7°C. The heat load is 36 GJ for space heating and hot water. The collector-heat exchanger correction factor is 0.97 and the ratio of monthly average to normal incidence optical efficiency is 0.96. Calculate the energy delivered by the solar system in January if 50 m^2 of collector area is to be used.

From Equation 18.46 and Equation 18.47, with $A_C = 50\ \text{m}^2$,

$$X = 4.0 \times 0.97[100 - (-7)]31 \times 86{,}400 \times 50/(36 \times 10^9) = 1.54$$

$$Y = 0.74 \times 0.97 \times 0.96 \times 12.9 \times 10^6 \times 31 \times 50/(36 \times 10^9) = 0.38$$

From Equation 18.50, the solar fraction for January is $f_M = 0.26$. Thus the useful energy delivered by the solar system $= 0.26 \times 36 = 9.4$ GJ.

In an effort to reduce the tediousness involved in having to perform 12 monthly calculations, two analogous approaches that enable the annual solar fraction to be determined directly have been developed by Barley and Winn (1978), Lameiro and Bendt (1978). These involve the computation of a few site-specific empirical coefficients, thereby rendering the approach less general. For example, the *relative-area* method suggested by Barley and Winn enable the designer to directly calculate the annual solar fraction of the corresponding system using four site-specific empirical coefficients. The approach involves curve fits to simulation results of the f-chart method for specific locations in order to deduce a correlation such as:

$$f = c_1 + c_2 \ln(A/A_{0.5}) \tag{18.51}$$

where c_1 and c_2 are location-specific parameters that are tabulated for several United States locations, and $A_{0.5}$ is the collector area corresponding to an annual solar fraction of 0.5 given by

$$A_{0.5} = A_S(UA)/(F'_R\eta_0 - F'_R U_L Z) \tag{18.52}$$

where A_S and Z are two more location specific parameters, UA is the overall heat loss coefficient of the building, and $F'_R\eta_0$ and $F'_R U_L$ are the corresponding solar collector performance parameters corrected for the effect of the collector-heat exchanger.

Barley and Winn also proposed a simplified economic life-cycle analysis whereby the optimal collector area could be determined directly. Another well-known approach is the *Solar Load Radio* (SLR) method for sizing residential space heating systems (Hunn 1980).

18.1.9.3 Domestic Water Heating

The f-chart correlation (Equation 18.50) can also be used to predict the monthly solar fraction for domestic hot-water systems represented by Figure 18.21 provided the water mains temperature T_{mains} is between 5 and 20°C and the minimum acceptable hot-water temperature drawn from the storage for end use (called the set water temperature T_w) is between 50 and 70°C. Further, the dimensionless parameter X must be corrected by the following ratio

$$X_w/X = (11.6 + 1.8T_w + 3.86T_{\text{mains}} - 2.32\bar{T}_a)/(100 - \bar{T}_a) \tag{18.53}$$

In case the domestic hot-water load is much smaller than the space heat load, it is recommended that Equation 18.50 be used without the above correction.

18.1.9.4 Industrial Process Heat

As discussed in "Description of a Typical Closed-Loop System," two types of solar systems for industrial process heat are currently used: the closed-loop multipass systems (with an added distinction that the auxiliary heater may be placed either in series or in parallel (see Figure 18.18 and Figure 18.19) and the open-loop singlepass system. How such systems can be designed will be described next.

18.1.9.4.1 Closed-Loop Multipass Systems

Auxiliary Heater in Parallel. The Phibar-f chart method (Klein and Beckman 1979; Duffie and Beckman 1980; Reddy 1987) is a generalization of the f-chart method in the sense that no restrictions need be imposed on the temperature limits of the heated fluid in the solar thermal system. However, three basic criteria for the thermal load have to be satisfied for the Phibar-f chart method to be applicable: (i) the thermal load must be constant and uniform over each day and for at least a month, (ii) the thermal energy supplied to the load must be above a minimum temperature that completely specifies the temperature level of operation of the load, and (iii) either there is no conversion efficiency in the load (as in the case of hot water usage) or the efficiency of conversion is constant (either because the load temperature level is constant or because the conversion efficiency is independent of the load temperature level). The approach is strictly applicable to solar systems with the auxiliary heater in parallel (Figure 18.19).

A typical application for the Phibar-f chart method is absorption air-conditioning. The hot water inlet temperature from the collectors to the generator must be above a minimum temperature level (say, 80°C) for the system to use solar heat. If the solar fluid temperature is less (even by a small amount), the entire energy to heat up the water to 80°C is supplied by the auxiliary system.

As a result of continuous interaction between storage and collector in a closed-loop system, the variation of the storage temperature and hence the fluid inlet temperature to the collectors) over the day and over the month is undetermined. The Phibar-f chart method implicitly takes this into account and reduces these temperature fluctuations down to a monthly mean equivalent storage temperature $\bar{T}_S$ The determination of this temperature in conjunction with the daily utilizability approach is the basis of the design approach.

The basic empirical correlation of the Phibar-f chart method, shown graphically in Figure 18.29, is as follows:

$$f_M = Y\bar{\phi} - a[\exp(bf_M) - 1][1 - \exp(cX)] \tag{18.54}$$

with $0<X<20$ and $0<Y<1.6$, and $\bar{\phi}$ is the Klein daily utilizability fraction described in "Daily Utilizability" and given by Equation 18.24. Y is given by Equation 18.47, and X is now slightly different from Equation 18.46 and is defined as:

$$X = A_C F_R U_L \Delta t(100°C)/Q_{LM} \tag{18.55}$$

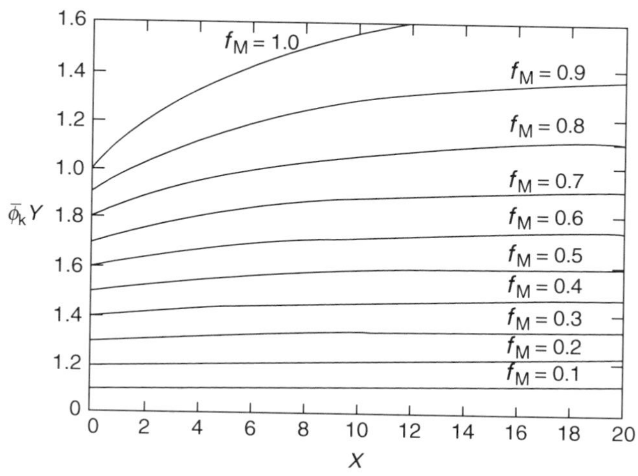

FIGURE 18.29 The Phibar-f chart correlation for a storage capacity of 350 kJ/m^2 and for a 12 h per day thermal load. (From Duffie, J. A. and Beckman, W. A., *Solar Engineering of Thermal Processes*, Wiley Interscience, New York, 1980.)

The values of the constants a, b, and c are given by the following:

1. for an end use load operating between 6 a.m. and 6 p.m. every day of the month,

$$a = 0.015[(Mc_p)_S/350\ \text{kJ}/(\text{m}^2\,°\text{C})]^{-0.76} \quad \text{for} \quad 175 \leq [(Mc_p)_S/A_C] \leq 1{,}400\ \text{kJ}/(\text{m}^2\,°\text{C}) \tag{18.56}$$

and $b = 3.85$ and $c = -0.15$

2. for an end use load operating 24 h per day over the entire month,

$$a = 0.043 \quad \text{only for}[(Mc_p)_S/A_C] = 350\text{kJ}/(\text{m}^2\,°\text{C}),\ b = 2.81,\ \text{and}\ c = -0.18 \tag{18.57}$$

It will be noted that $(Y\bar{\phi})$ denotes the maximum solar fraction that would have resulted had T_{Ci}, the inlet temperature to the collector, been equal to T_{Li} throughout the month. The term in Equation 18.54 that is subtracted from $(Y\bar{\phi})$ represents the decrease in the solar fraction as a result of $T_{Ci} > T_{Xi}$. The solar fraction computed from Equation 18.54 has to be corrected for the effect of thermal losses from the storage as well as the presence of the load-heat exchanger, both of which will decrease the solar fraction. For complete details, refer to Duffie and Beckman (1980) or Reddy (1987). Note that Equation 18.54 needs to be solved for f_m in an iterative manner.

Auxiliary Heater in Series. The Phibar-f chart method has also been modified to include solar systems with the auxiliary heater in series as shown in Figure 18.18. This configuration leads to higher solar fractions but retrofit to existing systems may be more costly.

In this case, the empirical correlation given by Equation 18.54 has been modified by Braun, Klein, and Pearson (1983) as follows:

$$f_M = Y\bar{\phi} - a[\exp(bf_M) - 1][1 - \exp(cX)]\exp(-1.959Z) \tag{18.58}$$

with $Z = Q_{LM}/(C_L \times 100°\text{C})$ and (i) when there is no load-heat exchanger, C_L is the monthly total load heat capacitance, which is the product of the monthly total mass of water used and the specific heat capacity of water, and (ii) when there is a load-heat exchanger present $C_L = E_L \times C_{min}$, where E_L is the effectiveness of the load-heat exchanger and C_{min} is the monthly total heat capacitance, which is the lesser of the two fluids rates across the load heat exchanger.

The modified Phibar-f chart is similar to the original method in respect to load uniformity on a day-to-day basis over the month and in assuming no conversion efficiency. The interested may refer to Braun, Klein, and Pearson (1983) or Reddy (1987) for complete details.

18.1.9.4.2 Open-Loop Single-Pass Systems

The advantages offered by open-loop single-pass systems over closed-loop multipass systems for meeting constant loads has been described in "Closed-Loop and Open-Loop Systems." Because industrial loads operate during the entire sun-up hours or even for 24 h daily, the simplest solar thermal system is one with no heat storage (Figure 18.20). A sizable portion (between 25 and 70%) of the daytime thermal load can be supplied by such systems and consequently, the sizing of such systems will be described below (Gordon and Rabl 1982). We shall assume that T_{Li} and T_{Xi} are constant for all hours during system operation. Because no storage is provided, excess solar energy collection (whenever $T_{Ci} > T_{Li}$) will have to be dumped out.

The maximum collector area $\hat{A}_C$ for which energy dumping does not occur at any time of the year can be found from the following instantaneous heat balance equation:

$$P_L = \hat{A}_C \hat{F}_R[I_{max}\eta_n - U_L(T_{Ci} - T_a)] \tag{18.59}$$

where P_L, the instantaneous thermal heat demand of the load (say, in kW) is given by

$$P_L = m_L c_p(T_{Li} - T_{Xi}) \tag{18.60}$$

and F_R is the heat removal factor of the collector field when its surface area is $\hat{A}_C$. Since $\hat{A}_C$ is as yet unknown, the value of $\hat{F}_R$ is also undetermined. (Note that though the *total* fluid flow rate is known, the flow rate per unit collector area is not known.) Recall that the plate efficiency factor F' for liquid collectors can be assumed constant and independent of fluid flow rate per unit collector area. Equation 18.59 can be expressed in terms of critical radiation level I_C:

$$P_L = \hat{A}_C \hat{F}_R \eta_n (I_{max} - I_C) \tag{18.61a}$$

or

$$\hat{A}_C \hat{F}_R \eta_n = P_L/(I_{max} - I_C) \tag{18.61b}$$

Substituting Equation 18.3 in lieu of F_R and rearranging yields

$$\hat{A}_C = -(m_L c_p/F' U_L)\ln[1 - P_L U_L/(\eta_n(I_{max} - I_C)m_L c_p)] \tag{18.62}$$

If the actual collector area A_C exceeds this value, dumping will occur as soon as the radiation intensity reaches a value I_D, whose value is determined from the following heat balance:

$$P_L = A_C F_R \eta_n (I_D - I_C) \tag{18.63a}$$

Hence

$$I_D = I_C + P_L/(A_C F_R \eta_n) \tag{18.63b}$$

Note that the value of I_D decreases with increasing collector area A_C, thereby indicating that increasing amounts of solar energy will have to be dumped out.

Since the solar thermal system is operational during the entire sunshine hours of the year, the yearly total energy collected can be directly determined by the Rabl correlation given by Equation 18.26. Similarly, the yearly total solar energy collected by the solar system which has got to be dumped out is

$$Q_{DY} = A_C F_R \eta_n (\tilde{a} + \tilde{b} I_D + \tilde{c} I_D^2) \tag{18.64}$$

The yearly total solar energy delivered to the load is

$$\begin{aligned} Q_{UY} &= Q_{CY} - Q_{DY} \\ &= A_C F_R \eta_n [\tilde{b}(I_C - I_D) + \tilde{c}(I_C^2 - I_D^2)] \\ &= -(\tilde{b} + 2\tilde{c} I_C) P_L - \tilde{c} P_L^2/(A_C F_R \eta_n) \end{aligned} \tag{18.65}$$

$$= -(\tilde{b} + 2\tilde{c} I_C) P_L - \tilde{c} P_L^2/(A_C F_R \eta_n) \tag{18.66}$$

Replacing the value of F_R given by Equation 18.3, the annual production function in terms of A_C is

$$Q_{UY} = -(\tilde{b} + 2\tilde{c} I_C) P_L - \frac{\tilde{c} P_L^2}{\left(\frac{F'\eta_n}{F' U_L}\right)(m_L c_p)\left[1 - \exp\left(-\frac{F' U_L A_C}{m_L c_p}\right)\right]} \tag{18.67}$$

subject to the condition that $A_C > \hat{A}_C$. If the thermal load is not needed during all days of the year due to holidays or maintenance shutdown, the production function can be reduced proportionally. This is illustrated in the following example.

Example 18.1.8

Obtain the annual production function of an open-loop solar thermal system without storage that is to be set up in Boston, Massachusetts according to the following load specifications: industrial hot water load for 12 h a day (6 a.m.–6 p.m.) and during 290 days a year, mass flow rate $m_L = 0.25$ kg/s, required inlet temperature $T_{Li} = 60°C$. Contaminants are picked up in the load, so that all used water is to be rejected and fresh water at ambient temperature is taken in. Flat-plate collectors with tilt equal to latitude with the following parameters are used $F'\eta_n = 0.75$ and $F'U_L = 5.5$ W/(m^{2}°C). The latitude of Boston is 42.36°N = 0.739 radians. The yearly $\tilde{K} = 0.45$ and $\tilde{T}_a = 10.9°C$. Use the following Gordon and Rabl (1981) correlation:

$$\begin{aligned} Q_{CY}/A_c F_R \eta_n &= [(5.215 + 6.973 I_{bn}) + (-5.412 + 4.293 I_{bn})L + (1.403 - 0.899 I_{bn})L^2] + \\ &[(-18.596 - 5.931 I_{bn}) + (15.468 + 18.845 I_{bn})L + (-0.164 - 35.510 I_{bn})L^2] I_C + \\ &[(-14.601 - 3.570 I_{bn}) + (13.675 - 15.549 I_{bn})L + (-1.620 + 30.564 I_{bn})L^2] I_C^2 \end{aligned}$$

From Equation 18.27, $\tilde{I}_{bn} = 1.37 \times 0.45 - 0.34 = 0.276$ kW/m^2. The critical radiation level $I_C = 0$, since $T_{Ci} = T_a$. Consequently, Equation 18.26, using the above expression reduces to

$$\begin{aligned} Q_{CY}/(A_C F_R \eta_n) &= 5.215 + 6.973 \times 0.276 + (-5.412 + 4.293 \times 0.276)0.739 + (1.403 - 0.899 \\ &\quad \times 0.276)0.739^2 \\ &= 4.646 \text{ GJ/(m}^2\text{y)}. \end{aligned}$$

The expression for the dumped out energy is found from Equation 18.64 and the previous expression by replacing I_C by I_D:

$$\begin{aligned} Q_{CY,dump}/(A_C F_R \eta_n) &= 4.646 + [(-18.596 - 5.931 \times 0.276) + (15.468 + 18.845 \times 0.276)0.739 + \\ &(-0.164 - 35.510 \times 0.276)0.739^2] I_D + [(14.601 - 3.57 \times 0.276) + \\ &(-13.675 - 15.549 \times 0.276)0.739 + (1.620 + 30.564 \times 0.276)0.739^2] I_D^2 \\ &= 4.646 - 10.40 I_D + 5.83 I_D^2 (\text{GJ/m}^2\text{y}) \end{aligned}$$

The thermal energy demand $P_L = 0.25 \times 4.19(60 - 10.9) = 51.43$ kW.

The annual production function is

$$\begin{aligned} Q_{UY}(365/290) &= -(-10.40 + 2\tilde{c} \times 0)51.43 - (5.83 \times 51.43^2)/\{(0.75/5.5)(0.25 \times 4.19) \\ &\quad [1 - \exp[-(5.5 \times A_C)/(0.25 \times 4190)]]\} \\ \text{or } Q_{UY} &= 424.96 - 85.78/[1 - \exp(-A_C/190.45)](\text{GJ/y}) \end{aligned}$$

Complete details as well as how this approach can be extended to solar systems with storage (see Figure 18.30) can be found in Rabl (1985), Gordon and Rabl (1986) or Reddy (1987).

18.1.10 Design Recommendations and Costs

18.1.10.1 Design Recommendations

As mentioned earlier, design methods reduce computational effort compared to detailed computer simulations. Even with this decrease, the problem of optimal system design and sizing remains formidable because of

a. The presence of several solar thermal system configuration alternatives.
b. The determination of optimal component sizes for a given system.

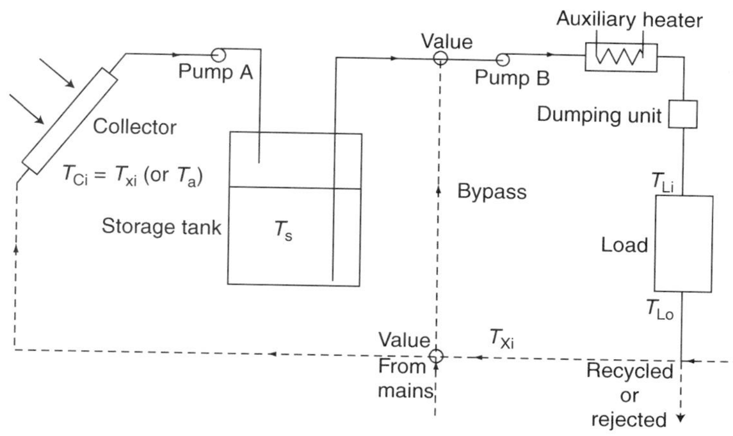

FIGURE 18.30 Open-loop solar industrial hot-water system with storage.

c. The presence of certain technical and economic constraints.
d. The choice of proper climatic, technical, and economic input parameters.
e. The need to perform sensitivity analysis of both technical and economic parameters.

For most practical design work, a judicious mix of theoretical expertise and practical acumen is essential. Proper focus right from the start on the important input variables as well as the restriction of the normal range of variation would lead to a great decrease in design time and effort several examples of successful case studies and system design recommendations are described in the published literature (see, for example, Kutcher et al. 1982).

18.1.10.2 Solar System Costs

How the individual components of the solar system contribute to the total cost can be gauged from Table 18.4. We note that collectors constitute the major fraction (from 15 to 30%), thus suggesting that collectors should be selected and sized with great care. Piping costs are next with other collector-related costs like installation and support structure being also important.

TABLE 18.4 Percentage of Total System Cost by Component

Cost Component	Percentage Range
Collectors	15–30
Collector installation	5–10
Collector support structure	5–20[a]
Storage tanks	5–7
Piping and specialties	10–30
Pumps	1–3
Heat exchangers	0–5[b]
Chiller	5–10
Miscellaneous	2–10
Instrumentation	1–3
Insulation	2–8
Control subsystem	4–9
Electrical	2–6

[a] For collectors mounted directly on a tilted roof.
[b] For systems without heat exchangers.
Source: From Mueller Associates. 1985. Active Solar Thermal Design Manual, funded by U.S. DOE (no. EG-77-C-01-4042), SERI (XY-2-02046-1) and ASHRAE (project no. 40). Baltimore, MD.

References

Anderson, B. 1977. *Solar Energy: Fundamentals in Building Design*, McGraw-Hill, New York.

ASHRAE Standard 93-77. 1978. *Methods of Testing to Determine the Thermal Performance of Solar Collectors*. American Society of Heating, Refrigeration and Air Conditioning Engineers, New York.

ASHRAE. 1985. *Fundamentals*, American Society of Heating, Refrigeration and Air Conditioning Engineers, New York.

Barley, C. D. and Winn, C. B. 1978. Optimal sizing of solar collectors by the method of relative areas. *Solar Energy*, 21, 279.

Beckman, W. A. 1978. Technical note: Duct and pipe losses in solar energy systems. *Solar Energy*, 21, 531.

Beckman, W. A., Klein, S. A., and Duffie, J. A. 1977. *Solar Heating Design by the f-Chart Method*, Wiley Interscience, New York.

Braun, J. E., Klein, S. A., and Pearson, K. A. 1983. An improved design method for solar water heating systems. *Solar Energy*, 31, 597.

Charters, W. W. S. and Pryor, T. L. 1982. In *An Introduction to the Installation of Solar Energy Systems*, pp. 503–127. Victoria Solar Energy Council, Melbourne, Australia.

Chiam, H. F. 1981. Planar concentrators for flat-plate solar collectors, *Solar Energy*, 26, 503.

Clark, D. R., Klein, S. A., and Beckman, W. A. 1983. Algorithm for evaluating the hourly radiation utilizability function. *ASME Journal of Solar Energy Engineering*, 105, 281.

Cole, R. L., Nield, K. J., Rohde, R. R., and Wolosewicz, R. M. eds. 1979. *Design and Installation Manual for Thermal Energy Storage*, ANL-79-15, pp. 223–397. Argonne National Laboratory, Argonne, IL.

Collares-Pereira, M. and Rabl, A. 1979. Derivation of method for predicting the long-term average energy delivery of solar collectors. *Solar Energy*, 23, 223.

Collares-Pereira, M., Gordon, J. M., Rabl, A., and Zarmi, Y. 1984. Design and optimization of solar industrial hot water systems with storage. *Solar Energy*, 32, 121.

CSU. 1980. *Solar Heating and Cooling of Residential Buildings*—Design of Systems, manual prepared by the Solar Energy Applications Laboratory, Colorado State University. Colorado State University, Fort Collins, CO.

de Winter, F. 1975. Heat exchanger penalties in double loop solar water heating systems. *Solar Energy*, 17, 335.

Duffie, J. A. and Beckman, W. A. 1980. *Solar Engineering of Thermal Processes*, Wiley Interscience, New York.

Edwards, D. K., Nelson, K. E., Roddick, R. D., and Gier, J. T. 1960. Basic Studies on the Use of Solar Energy, Report no. 60-93. Department of Engineering. University of California, Los Angeles, CA.

Gordon, J. M. and Rabl, A. 1982. Design, analysis and optimization of solar industrial process heat plants without storage. *Solar Energy*, 28, 519.

Gordon, J. M. and Zarmi, Y. 1985. An analytic model for the long-term performance of solar thermal systems with well-mixed storage. *Solar Energy*, 35, 55.

Gordon, J. M. and Rabl, A. 1986. Design of solar industrial process heat systems, In *Reviews of Renewable Energy Sources* chapter 15. Sodha, M. S. Mathur, S. S. Malik, M. A.S and Kandpal, T. C. eds., pp. 55–177. Wiley Eastern, New Delhi.

Gordon, J. M. 1987. Optimal sizing of stand-alone photovoltaic systems. *Solar Cells*, 20, 295.

Hollands, K. G. T. 1965. Honeycomb devices in flat-plate solar collectors. *Solar Energy*, 9, 159.

Hunn, B. D. 1980. A simplified method for sizing active solar space heating systems, *In Solar Energy Technology Handbook, Part B: Applications, System Design and Economics*, Dickinson, W. C. and Cheremisinoff, P. N. eds., pp. 639–255. Marcel Dekker, New York.

Klein, S. A., Cooper, P. I., Freeman, T. L., Beekman, D. M., Beckman, W. A. and Doffie, J. A. 1975. A method of simulation of solar processes and its applications, *Solar Energy*, 17, 29.

Klein, S. A. 1978. Calculation of flat-plate collector utilizability. *Solar Energy*, 21, 393.

Klein, S. A. et al. 1979. TRNSYS-A Transient System Simulation User's Manual, University of Wisconsin-Madison Engineering Experiment Station Report. University of Wisconsin-Madison Engineering Experiment Station Report 38-10.

Klein, S. A. and Beckman, W. A. 1979. A general design method for closed-loop solar energy systems. *Solar Energy*, 22, 269.

Kreider, J. F. 1979. Medium and High Temperature Solar Energy Processes. Academic Press, New York.

Kutcher, C. F., Davenport, R. L., Dougherty, D. A., Gee, R. C., Masterson, P. M., and May, E. K. 1982. Design Approaches for Solar Industrial Process Heat Systems, SERI/TR-253-1356. Solar Energy Research Institute, Golden, CO.

Lameiro, G. F. and Bendt, P., 1978. The GFL method for designing solar energy space heating and domestic hot water systems, In Proceedings of American Solar Energy Society Conference, Vol. 2, p. 113. Boulder, CO.

Larson, D. C. 1980. Optimization of flat-plate collector flat mirror system. *Solar Energy*, 24, 203.

Larson, R. W., Vignola, F., and West, R. 1992a. *Economics of Solar Energy Technologies*. American Solar Energy Society Report, Boulder, CO.

Liu, B. Y. H. and Jordan, R. C. 1960. The inter-relationship and characteristic distribution of direct, diffuse and total solar radiation. *Solar Energy*, 4, 1.

Liu, B. Y. H. and Jordan, R. C. 1963. A rational procedure for predicting the long-term average performance of flat-plate solar energy collectors. *Solar Energy*, 7, 53.

Liu, S. T. and Fanney, A. H. 1960. Comparing experimental and Computer-predicted performance for solar hot water systems, *ASHRAE Journal*, 22, 5, 34.

Löf, G. O. G. 1981. Air based solar systems for space heating, In *Solar Energy Handbook*, J. F. Kreider and F. Kreith, eds., pp. 39–44. McGraw-Hill, New York.

Meyer, B. A. 1978. Natural convection heat transfer in small and moderate aspect ratio enclosures—An application to flat-plate collectors, In *Thermal Storage and Heat Transfer in Solar Energy Systems*, Kreith, F. Boehm, R. Mitchell, J. and Bannerot, R. eds., pp. 555–558. American Society of Mechanical Engineers, New York.

Mitchell, J. C., Theilacker, J. C., and Klein, S. A. 1981. Technical note: Calculation of monthly average collector operating time and parasitic energy requirements. *Solar Energy*, 26, 555–558.

Mueller Associates. 1985. *Active Solar Thermal Design Manual*, funded by U. S. DOE (no. EG-77-C-01-4042). SERI(XY-2-02046-l) and ASHRAE (project no. 40). Baltimore, MD.

Newton, A. B. and Gilman, S. H. 1981. *Solar Collector Performance Manual*, funded by U. S. DOE (no. EG-77-C-01-4042), SERI(XH-9-8265-1) and ASHRAE (project no. 32, Task 3).

Phillips, W. F. and Dave, R. N. 1982. Effect of stratification on the performance of liquid-based solar heating systems. *Solar Energy*, 29, 111.

Rabl, A. 1981. Yearly average performance of the principal solar collector types. *Solar Energy*, 27, 215.

Rabl, A. 1985. *Active Solar Collectors and Their Applications*. Oxford University Press, New York.

Reddy, T. A. 1987. *The Design and Sizing of Active Solar Thermal Systems*. Oxford University Press, Oxford, U. K.

SERI 1989. Engineering Principles and Concepts for Active Solar Systems. Hemisphere Publishing Company, New York.

Theilacker, J. C. and Klein, S. A. 1980. Improvements in the utilizability relationships. *American Section of the International Solar Energy Society Meeting Proceedings*, p. 271. Phoenix, AZ.

Liu, S. T. and Fanney, A. H. 1980. Comparing experimental and computer-prodicted performance for solar hot water systems. *ASHRAE Journal*, 22, 5, 34.

Klein, S. A., Cooper, P. I., Freeman, T. L., Beekman, D. M., Beckman, W. A. and Doffie, J. A. 1975. A method of simulation of solar processes and its applications, *Solar Energy*, 17, 29.

18.2 Solar Heat for Industrial Processes

Riccardo Battisti, Hans Schweiger, and Werner Weiss

18.2.1 The Potential for Solar Process Heat

Currently, the widespread use of residential solar thermal energy has focused almost exclusively on swimming pools, domestic hot water preparation, and space heating, while its use in the service sector and in industrial applications is insignificant. On the other hand, the industrial sector accounts for a large share of the total energy consumption in the OECD (Organization for Economic Co-operation and Development) countries at approximately 30%, and a significant portion of industrial heat demand falls within a temperature range compatible with solar thermal collectors.

As a matter of fact, 30%–50% of the thermal energy needed in commercial and industrial companies for production processes is below 250°C (Kreider 1979; European Commission 2001). In this temperature range, the heat demand in the European Union (EU) for industrial processes can be estimated with about 300 TWh, or 7% of the total energy demand. The total potential for industrial process heat at below 150°C was estimated to be 202.8 TWh for the 12 countries that formed the EU in 1994 (Laue and Reichert 1994).

Studies for the application potential for industrial solar thermal systems were carried out in Austria (Müller 2004), Germany and Greece (PROCESOL 2000), Italy (IEA 2005), Netherlands (KWA 2001), Portugal and Spain (European Commission 2001). The potential for solar low temperature heat ranges between 3% and 4% of the total industrial heat demand in Italy, Spain, Portugal, and Austria. These studies primarily showed that solar thermal plants can readily provide the required low- and medium-temperature process heat.

The most promising industrial sectors and processes are shown in Table 18.5, and the distribution of heat demand by temperature range is shown in Figure 18.31.

Another important result showed by the studies is that the available surface area on factory roofs is often a limiting factor for the installation of a solar plant for industrial use.

TABLE 18.5 Industrial Sectors and Processes Suitable for Solar Thermal Use

Sector	Processes	Temperature (°C)
Brewing and malting	Wort boiling	100
	Bottle washing	60
	Drying	90
	Cooling	60
Milk	Pasteurization	60–85
	Sterilization	130–150
Food preservation	Pasteurization	110–125
	Sterilization	<80
	Cooking	70–100
	Scalding	95–100
	Bleaching	<90
Meat	Washing, sterilization, cleaning	<90
	Cooking	90–100
Wine and beverage	Bottle washing	60–90
	Cooling (single effect absorption cooling)	85
Textile	Washing, bleaching, dyeing	<90
	Cooking	140–200
Automobile	Paint drying	160–220
	Degreasing	35–55
Paper	Paper pulp: cooking	170–180
	Boiler feed water	<90
	Bleaching	130–150
	Drying	130–160
Tanning	Water heating for damp processes	165–180 (steam)
Cork	Drying, cork baking	40–155

Source: From European Commission, *The Potential of Solar Heat for Industrial Processes*, Final report, EC Project, Contract No. NNE5-1999-0308, 2001, http://www.solarpaces.org. With permission.

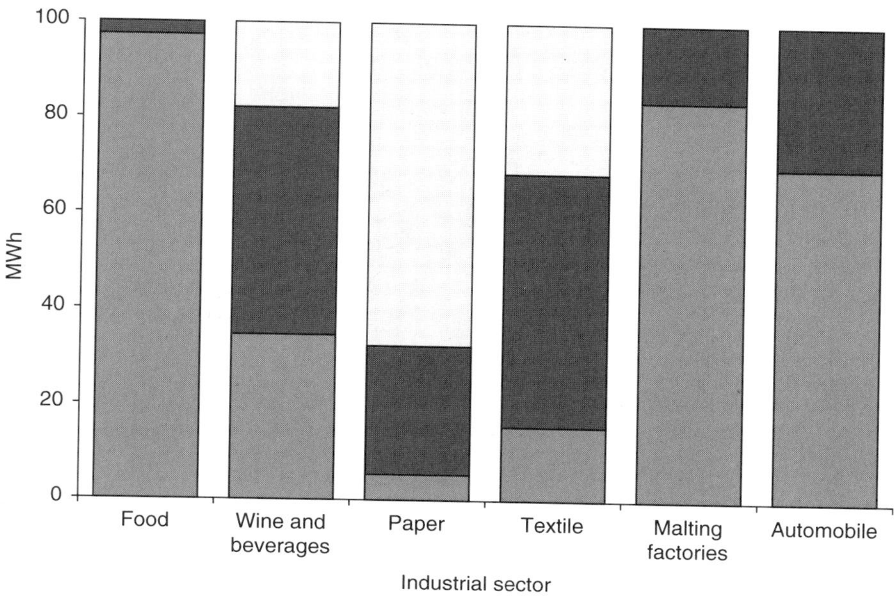

FIGURE 18.31 Distribution of the heat demand by temperature range in some selected companies studied within the POSHIP project, grouped by industrial sectors. (From European Commission, *The Potential of Solar Heat for Industrial Processes*, Final Report, EC Project, Contract No. NNE5-1999-0308, 2001, http://www.aiguasol.com/poship.htm. With permission.)

18.2.2 Solar Thermal Systems in Industrial Processes: Integration and Basic Design Guidelines

18.2.2.1 Which Solar Collectors?

For process temperatures up to about 60°C, flat-plate collectors with selective absorbers would be most appropriate and could be the most economical solution even up to a temperature range of 90°C. For temperatures above this range, other collector types should be considered: evacuated tubes, high efficiency flat plate, CPC or line-axis concentrating collectors. In the framework of IEA Task 33/IV, "Solar Heat for Industrial Processes," which is carried out within the Solar Heating and Cooling Program of the IEA, new "medium temperature collectors" (i.e., with operating temperatures between 80 and 250°C) are being developed, including:

- Improved flat-plate collectors
- Stationary low-concentration collectors
- Small parabolic trough collectors

Different collector technologies already available on the market can be used for applications in the 80°C–120°C temperature range. For example, flat-plate collectors with double antireflection glazings and hermetically sealed collectors with inert-gas fillings, or even a combination of both, reduce collector heat losses without significantly sacrificing the optical performance. Figure 18.32 shows estimated efficiency curves of single-, double-, and triple-glazed flat-plate collectors when newly developed antireflection glazings (AR-glass) are used.

Another solution for medium temperature collectors is to reduce heat losses by concentration by, for example, using stationary CPC collector without vacuum and with a low concentration factor (in the range of two).

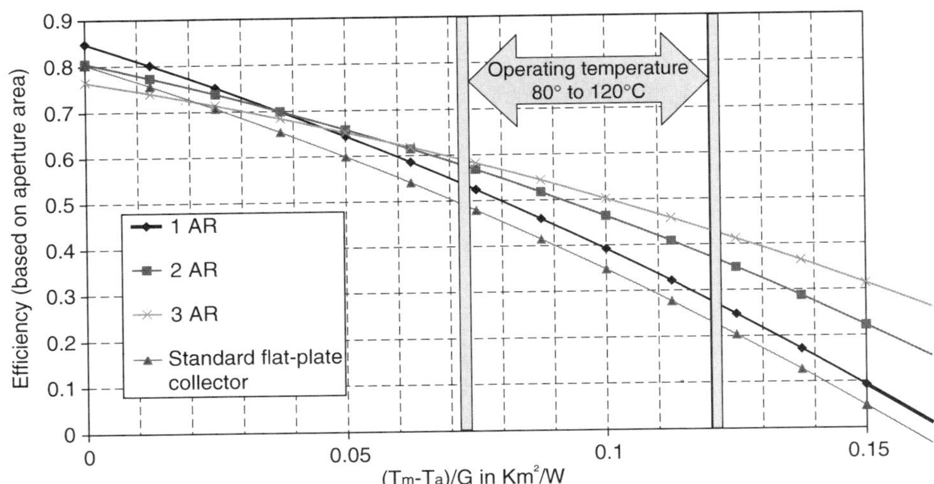

FIGURE 18.32 Efficiency curves of a single-, double-, and triple-glazed antireflection collector in comparison with a standard flat-plate collector with ordinary solar glass. (From IEA., *Task 33/IV, Solar Heat for Industrial Processes*, International Energy Agency, 2005.)

Between 150 and 250°C, it is appropriate to consider the parabolic trough collector technology. Much is known about high-temperature applications (400°C–600°C) using parabolic trough collectors for electric power production, but adjustments must be made for the medium temperature range. Current developments involved in Task 33/IV have been carried out in Spain, Austria, and Germany. The first results showed that parabolic troughs could even be an appropriate alternative for large systems at low temperatures (about 60°C). It is noteworthy that small parabolic trough are readily available in the market and are a reliable technology (e.g., in the U.S., some plants have been operating since the early 1990s).

18.2.2.2 Coupling the Solar Thermal System with the Processes

The integration of solar heat into industrial production processes is challenging for both the process engineer and the solar designer. Existing heating systems based on steam or hot water from boilers are normally designed for much higher temperatures (150°C–180°C) compared to those that the processes need (100°C or lower) to keep temperature differences small. On the contrary, the solar thermal system should always be coupled to the existing heat supply at the lowest possible temperature. Nevertheless, for fluid preheating, solar heat should be introduced only after preheating by waste heat recovery systems, and not as an alternative to these systems (Figure 18.33).

Even if the waste heat recovery raises the working temperature in the solar thermal system, the combination of both systems yields better results than a solar thermal system at lower temperature but without heat recovery.

Today, one of the most used heat recovery assessment methodologies is the "pinch" analysis. The discovery of the heat recovery "pinch" was a major breakthrough in the development of design methods for energy efficient industrial processes. Based on the analysis of hot and cold streams within the process, the pinch methodology gives fundamental hints about the possibility and the right position of heat exchangers for waste heat recovery. This allows developing integral solutions for solar thermal energy applications in given industrial processes. From the point of view of energy, the streams that constitute a process flow sheet may be classified into two groups: "hot" streams, i.e., streams that must be cooled, and "cold" streams, i.e., streams that must be heated.

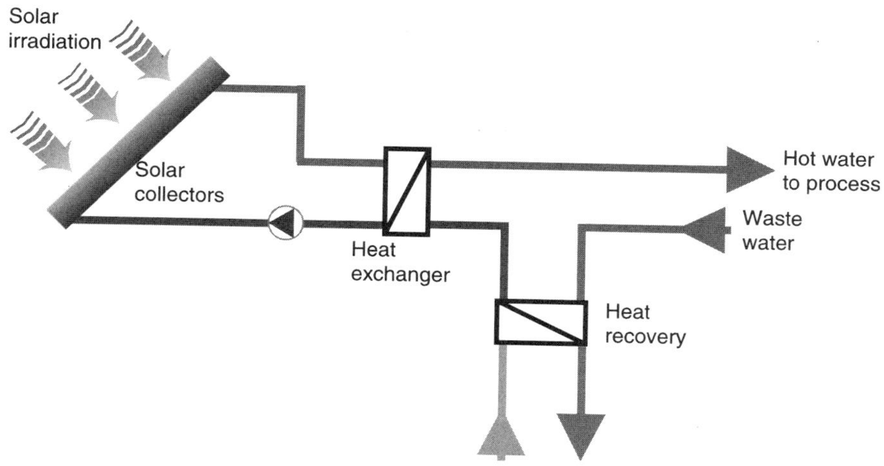

FIGURE 18.33 Combination of solar thermal system and waste heat recovery. (From European Commission, *The Potential of Solar Heat for Industrial Processes*, Final Report, EC Project, Contract No. NNE5-1999-0308, 2001, http://www.aiguasol.com/poship.htm. With permission.)

When there are a large number of streams, the selection of the best match between these streams is not obvious. The pinch methodology provides a systematic way to find the optimal solutions for the implementation of heat recovery techniques.

The solar thermal system may be coupled with the conventional heat supply system in several ways, including direct coupling to a specific process, preheating of water, and steam generation in the central system (Figure 18.34).

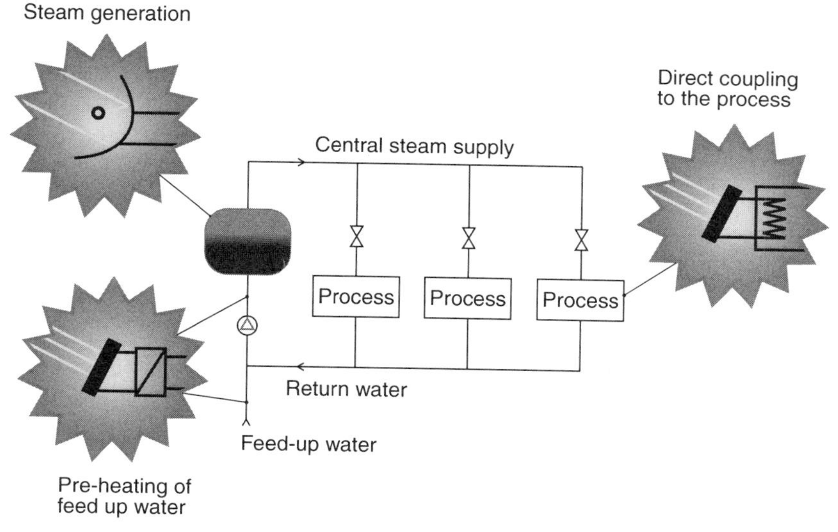

FIGURE 18.34 Coupling the solar thermal system with the conventional heat supply. (From European Commission, *The Potential of Solar Heat for Industrial Processes*, Final Report, EC Project, Contract No. NNE5-1999-0308, 2001, http://www.aiguasol.com/poship.htm. With permission.)

Whenever possible, a direct coupling of the solar thermal systems to one or several processes is preferred because the working temperatures are lower. Direct coupling to a process can mainly be carried out in the following two ways:

- Preheating of a circulating fluid (e.g., feed-up water, return of closed circuits, air preheating). This solution is feasible if fluid circulation is either continuous or periodic (e.g., periodic replacement of bath water). If circulation is discontinuous, a storage tank must be introduced. The mean working temperature of the solar thermal system is lower than the required final process temperature. The smaller the solar fraction, the lower the mean working temperature. For very low solar fractions, the mean working temperature may be close to the fluid inlet (or return) temperature.
- Heating of liquid baths or hot (e.g., drying) chambers. The energy demand is both for heating-up at the operational start-up, either concentrated in the early morning hours or periodically each time the used fluid is replaced with fresh fluid, and for maintaining the operating temperature, which is generally a nearly constant load.

The existing heat exchangers for bath heating generally require steam at temperatures that are too high for a solar thermal system. The introduction of additional heat exchangers with a larger exchange area into existing baths is not always possible due to lack of space or other technical restrictions. In some cases, an external heat exchanger in combination with a circulation pump can be used.

If the process baths are well-insulated, they can be used for solar heat storage. For example, maintaining the temperature of the solar thermal system during a weekend without operation can reduce the heat demand for start-up on Monday morning.

In almost all industries, coupling of a solar thermal system to the central heat supply system is possible. This can be done either by preheating the feed-up water for the steam boilers (the temperature level rises with increasing condensate recovery) or by a solar steam generator. The latter is only recommended at sites with a high level of solar radiation and if concentrating collectors are used.

18.2.2.3 Heat Demand and Storage

Another challenge in applying solar thermal energy to industrial production processes is the time dependency of the solar energy supply and the heat demand of the processes (Figure 18.35). Very few

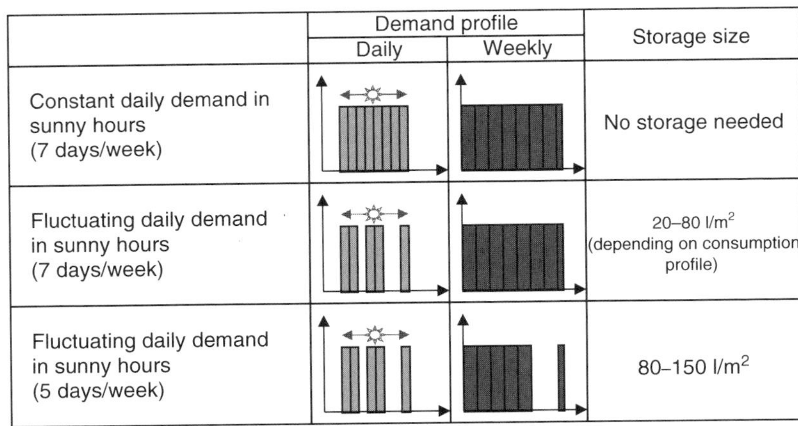

	Demand profile		Storage size
	Daily	Weekly	
Constant daily demand in sunny hours (7 days/week)			No storage needed
Fluctuating daily demand in sunny hours (7 days/week)			20–80 l/m^2 (depending on consumption profile)
Fluctuating daily demand in sunny hours (5 days/week)			80–150 l/m^2

FIGURE 18.35 Heat storage size depending on daily and weekly heat demand profiles.

production lines run at constant load throughout the day. Most processes in smaller companies run for one or two shifts per day and show a batch operation mode.

When the process heat demand is continuous during sunny hours with no weekend breaks, the load is always higher than the solar gains. Therefore, the solar thermal system can be designed without storage allowing the solar heat to be fed directly to the process or to the heat supply system. This enables building lower cost solar thermal systems by eliminating storage related costs.

In case the total weekly demand is constant, but there are strong fluctuations in the daily demand during operational periods (e.g., demand peaks, short breaks of operation), storage of 20–80 l/m^2 of collectors is necessary, depending on the process heat demand profile.

If the daily fluctuations come together with weekend breaks (five days of operation per week), then the storage size is recommended to be 80–150 l/m^2. Weekend storage is generally not recommended for small systems. The larger the system size, the more effective the heat storage over longer periods (e.g., weekends). A weekend storage becomes economic for systems with an installed capacity of about 350 kW_{th}, corresponding to 500 m^2 or more of collector array.

Storage for longer periods (seasonal storage) can only be considered for very large systems (greater than 3.5 MW_{th}), but systems with only seasonal utilization (less than six months of operation a year) are generally not economically viable.

From the previous considerations, key criteria for the feasibility of a solar process heat plant can be drawn:

- Temperature level: solar heat at temperatures above 150°C is technically feasible but economically reasonable only at favorable locations; applications at low temperature (less than 60°C) offer the best economics
- Continuous or quasicontinuous demand (otherwise storage is needed and plant costs increase)
- Technical possibility of introducing a heat exchanger in the existing equipment or heat supply circuit for the solar thermal system

The different possible schemes of solar thermal plants for process heat production are summarized in Figure 18.36. This classification scheme, currently under development and improvement, allows choosing the most suitable plant scheme depending on key process features (open/closed, collector fluid same as or different from heat distribution fluid, need for storage). The scheme is divided into three parts: energy supply (both solar and conventional), heat transfer/storage, and process load. For each column corresponding to a class of processes, two possibilities are given depending on the heat distribution medium: water/air or steam. These concepts are currently being developed and tested in demonstration plants.

18.2.3 Overview of Existing Solar Process Heat Plants

Since the 1980s, several solar thermal systems for industrial applications have been developed and are currently operating. At present, 84 plants are reported worldwide, with a total installed thermal power of about 24 MW_{th} (34,000 m^2).

The majority of the operational plants are in the sectors of food and beverage, textile, and transport (e.g., washing and painting of car components). The main applications differ among countries. In Greece, for instance, several "solar dairies" are operating, whereas in Germany the most common application is for car-washing facilities.

Most of the reported plants supply heat at temperature levels between 60 and 100°C and therefore, standard flat-plate collectors are suitable. As a matter of fact, flat-plate collectors are used in about 65% of the operational systems. Some plants are working at temperatures above 160°C, whereas there is only one project operating in the intermediate range from 100 to 160°C. Analysis of the reported plants reveals that they appear to encompass a broad range of working temperatures, but with no significant correlation with the solar field size. Figure 18.37 through Figure 18.40 show some examples of solar industrial process heat plants.

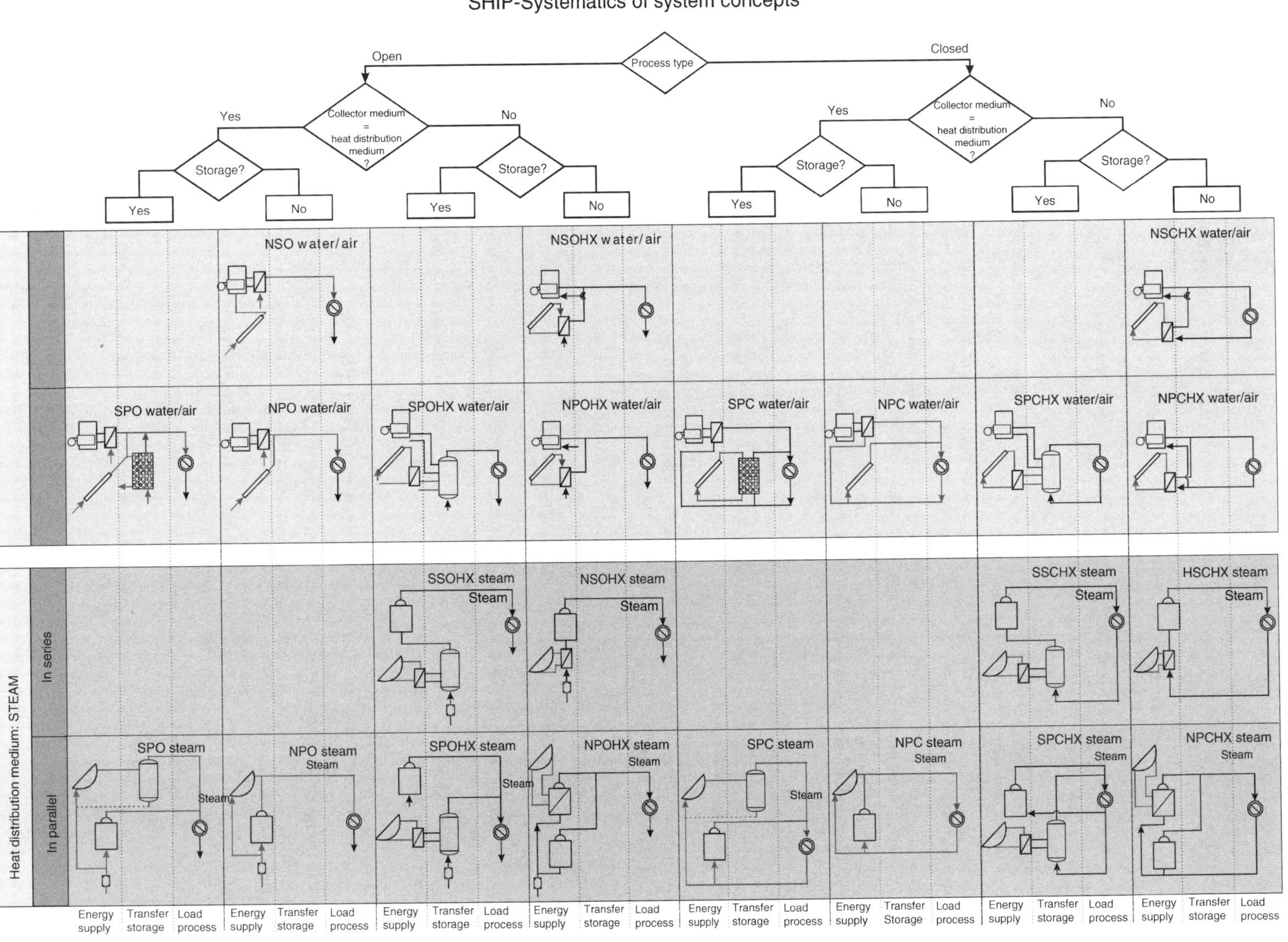

FIGURE 18.36 Systematic of systems concepts. (From IEA., *Task 33/IV, Solar Heat for Industrial Processes*, International Energy Agency, 2005.)

FIGURE 18.37 El NASR Pharmaceutical Chemicals, Egypt. Installed capacity: 1.33 MW_{th}. (From Fichtner Solar GmbH, Germany. With permission.)

FIGURE 18.38 Alpino SA, dairy industry, Greece. Installed capacity: 518 kW_{th}. (From Alpino SA, Greece. With permission.)

FIGURE 18.39 Wine cooling and bottle washing, Austria. Installed capacity: 70 kW_{th}. (From S.O.L.I.D. GmbH, Austria. With permission.)

FIGURE 18.40 Parking service Castellbisbal SA, container washing, Spain. Installed capacity: 357 kW_{th}. (From Aiguasol Engineering, Spain. With permission.)

References

European Commission. 2001. The Potential of Solar Heat for Industrial Processes. Final Report, EC Project, Contract No. NNE5-1999-0308, 2001, http://www.aiguasol.com/poship.htm (accessed on 2006/10/11).

IEA (International Energy Agency). 2005. *Task 33/IV. Solar heat for industrial processes*, International Energy Agency.

Kreider, J. F. 1979. *Medium and High Temperature Solar Energy Processes*, Academic Press, New York.

KWA. 2001. Bedrijfsadviseurs B. V., *Onderzoek naar het potentieel van zonthermishce energie in de industrie*. Relatienummer 8543.00, Rapportnummer 2009740DR01.DOC.

Laue, H. J. and Reichert, J. 1994. *Potential for medium and large sized industrial heat pumps in Europe*. Contract No. XVII/7001/90-8, Final report, European Commission, Directorate General for Energy (DGXII).

Müller, T. et al. 2004. *PROMISE-Produzieren mit Sonnenenergie, Projekt im Rahmen der Programmlinien "Fabrik der Zukunft" des Bundesministeriums für Verkehr*. Innovation und Technologie, Endbericht, Gleisdorf.

PROCESOL. 2000. *Solar Thermal Process Heating in Industrial Applications: A Stimulate Plan*, Final Report, EC ALTENTER Project, Contract No 4.1030/Z/98–205.

18.3 Passive Solar Heating, Cooling, and Daylighting

Jeffrey H. Morehouse

18.3.1 Introduction

Passive systems are defined, quite generally, as systems in which the thermal energy flow is by natural means: by conduction, radiation, and natural convection. A *passive heating system* is one in which the sun's radiant energy is converted to heat upon absorption by the building. The absorbed heat can be transferred to thermal storage by natural means or used to directly heat the building. *Passive cooling systems* use natural energy flows to transfer heat to the environmental sinks: the ground, air, and sky.

If one of the major heat transfer paths employs a pump or fan to force flow of a heat transfer fluid, then the system is referred to as having an active component or subsystem. Hybrid systems—either for heating or cooling—are ones in which there are both passive and active energy flows. The use of the sun's radiant energy for the natural illumination of a building's interior spaces is called *daylighting*. Daylighting design approaches use both solar beam radiation (referred to as *sunlight*) and the diffuse radiation scattered by the atmosphere (referred to as *skylight*) as sources for interior lighting, with historical design emphasis on utilizing skylight.

18.3.1.1 Distinction Between a Passive System and Energy Conservation

A distinction is made between energy conservation techniques and passive solar measures. Energy conservation features are designed to reduce the heating and cooling energy required to thermally condition a building: the use of insulation to reduce either heating and cooling loads, and the use of window shading or window placement to reduce solar gains, reducing summer cooling loads. Passive features are designed to increase the use of solar energy to meet heating and lighting loads, plus the use of ambient "coolth" for cooling. For example, window placement to enhance solar gains to meet winter heating loads and/or to provide daylighting is passive solar use, and the use of a thermal chimney to draw air through the building to provide cooling is also a passive cooling feature.

18.3.1.2 Key Elements of Economic Consideration

The distinction between passive systems, active systems, or energy conservation is not critical for economic calculations, as they are the same in all cases: a trade-off between the life-cycle cost of the energy saved (performance) and the life-cycle cost of the initial investment, operating, and maintenance costs (cost).

18.3.1.2.1 Performance: Net Energy Savings

The key performance parameter to be determined is the net annual energy saved by the installation of the passive system. The basis for calculating the economics of any solar energy system is to compare it against a "normal" building; thus, the actual difference in the annual cost of fuel is the difference in auxiliary energy that would be used with and without solar. Therefore, the energy saved rather than energy delivered, energy collected, useful energy, or some other energy measure, must be determined.

18.3.1.2.2 Cost: Over and Above "Normal" Construction

The other significant part of the economic trade-off involves determining the difference between the cost of construction of the passive building and of the "normal" building against which it is to be compared. The convention, adopted from the economics used for active solar systems, is to define a "solar add-on cost." Again, this may be a difficult definition in the case of passive designs because the building can be significantly altered compared to typical construction since, in many cases, it is not just a one-to-one replacement of a wall with a different wall, but it is more complex and involves assumptions and simulations concerning the "normal" building.

18.3.1.2.3 General System Application Status and Costs

Almost 500,000 buildings in the U.S. were constructed or retrofitted with passive features in the 20 years after 1980. Passive heating applications are primarily in single-family dwellings and secondarily in small commercial buildings. Daylighting features that reduce lighting loads and the associated cooling loads are usually more appropriate for large office buildings.

A typical passive heating design in a favorable climate might supply up to one-third of a home's original load at a cost of \$5 to \$10 per million Btu net energy saved. An appropriately designed daylighting system can supply lighting at a cost of 2.5–5 ¢ per kWh (Larson, Vignola, and West 1992a, 1992b, 1992c).

18.3.2 Solar Thermosyphon Water Heating

Solar hot-water heating systems are composed of a collector and a storage tank. When the flow between the collector and tank is by natural circulation, these passive solar hot-water systems are referred to as *thermosyphon systems*. This ability of thermosyphon systems to heat water without an externally powered pump has spurred its use in both regions where power is unavailable and where power is very expensive.

18.3.2.1 Thermosyphon Concept

The natural tendency of a less dense fluid to rise above a more dense fluid can be used in a simple solar water heater to cause fluid motion through a collector. The density difference is created within the solar collector where heat is added to increase the temperature and decrease the density of the liquid. This collection concept is called a *thermosyphon*, and Figure 18.41 schematically illustrates the major components of such a system.

The flow pressure drop in the fluid loop (ΔP_{FLOW}) must equal the buoyant force "pressure difference" ($\Delta P_{\mathrm{BUOYANT}}$) caused by the differing densities in the "hot" and "cold" legs of the fluid loop:

$$\Delta P_{\mathrm{FLOW}} = \Delta P_{\mathrm{BUOYANT}}$$

$$= \rho_{\mathrm{stor}} g H - \left[\int_{O}^{L} \rho(x) g \, \mathrm{d}x + \rho_{\mathrm{out}} g (H - L) \right], \tag{18.68}$$

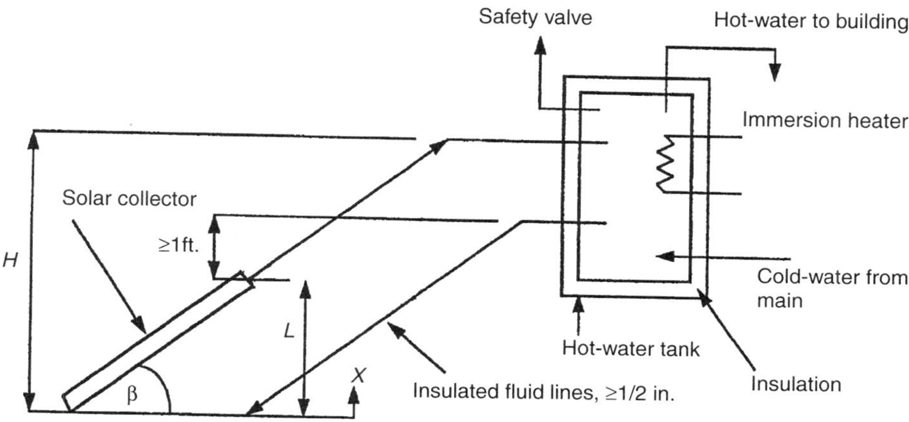

FIGURE 18.41 Schematic diagram of thermosyphon loop used in a natural circulation, service water-heating system. The flow pressure drop in the fluid loop must equal the bouyant force "pressure" $[\int_0^L g\rho(x)dx - \rho_{stor}gL]$ where $\rho(x)$ is the local collector fluid density and ρ_{stor} is the tank fluid density, assumed uniform.

where H is the height of the "legs," L is the height of the collector (see Figure 18.41), $\rho(x)$ is the local collector fluid density, ρ_{stor} is the tank fluid density, and ρ_{out} is the collector outlet fluid density; the latter two densities are assumed to be uniform. The flow pressure term, ΔP_{FLOW}, is related to the flow loop system headloss that is in turn directly connected to friction and fitting losses and the loop flow rate:

$$\Delta P_{FLOW} = \oint_{LOOP} \rho \, d(h_L), \tag{18.69}$$

where $h_L = KV^2$, with K being the sum of the component loss "velocity" factors (see any fluid mechanics text), and V is the flow velocity.

18.3.2.2 Thermo-Fluid System Design Considerations

Because the driving force in a thermosyphon system is only a small density difference and not a pump, larger than normal plumbing fixtures must be used to reduce pipe friction losses. In general, one pipe size larger than would be used with a pump system is satisfactory. Under no conditions should piping smaller than 1/2-in (12-mm) national pipe thread (NPT) be used. Most commercial thermosyphons use 1-in (25-mm) NPT pipe. The flow rate through a thermosyphon system is about 1 gal/ft.^{2} h (40 L/m^2 h) in bright sun, based on collector area.

Because the hot-water system loads vary little during a year, the best angle to tilt the collector is that equal to the local latitude. The temperature difference between the collector inlet water and the collector outlet water is usually 15–20°F (8–11°C) during the middle of a sunny day (Close 1962). After sunset, a thermosyphon system can reverse its flow direction and lose heat to the environment during the night. To avoid reverse flow, the top header of the absorber should be at least 1 ft. (30 cm) below the cold leg fitting on the storage tank, as shown.

To provide heat during long cloudy periods, an electrical immersion heater can be used as a backup for the solar system. The immersion heater is located near the top of the tank to enhance stratification and so that the heated fluid is at the required delivery temperature at the delivery point. Tank stratification is desirable in a thermosyphon to maintain flow rates as high as possible. Insulation must be applied over

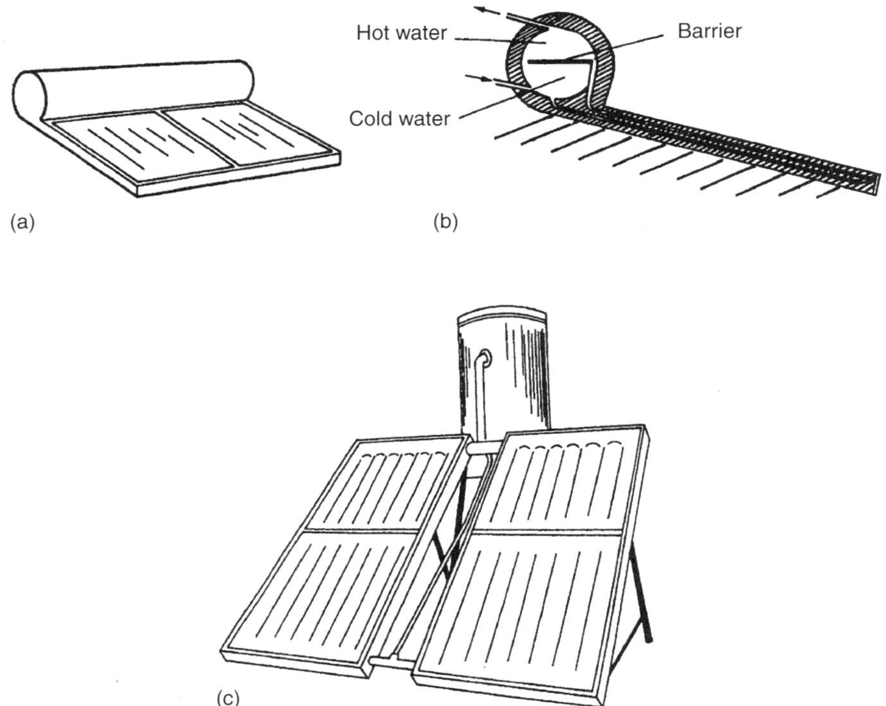

FIGURE 18.42 Passive solar water heaters; (a) compact model using combined collector and storage, (b) section view of the compact model, and (c) tank and collector assembly.

the entire tank surface to control heat loss. Figure 18.42 illustrates two common thermosyphon system designs.

Several features inherent in the thermosyphon design limit its utility. If it is to be operated in a freezing climate, a nonfreezing fluid must be used, which in turn requires a heat exchanger between collector and potable water storage. (If potable water is not required, the collector can be drained during cold periods instead.) Heat exchangers of either the shell-and-tube type or the immersion-coil type require higher flow rates for efficient operation than a thermosyphon can provide. Therefore, the thermosyphon is generally limited to nonfreezing climates. A further restriction on thermosyphon use is the requirement for an elevated tank. In many cases structural or architectural constraints prohibit raised-tank locations. In residences, collectors are normally mounted on the roof, and tanks mounted above the high point of the collector can easily become the highest point in a building. Practical considerations often do not permit this application.

Example 18.3.1

Determine the "pressure difference" available for a thermosyphon system with 1-m high collector and 2-m high "legs." The water temperature input to the collector is 25°C and the collector outputtemperature is 35°C. If the overall system loss velocity factor (K) is 15.6, estimate the system flow velocity.

Solution. Equation 18.68 is used to calculate the pressure difference, with the water densities being found from the temperatures (in steam tables):

$$\rho_{\text{stor}}(25°\text{C}) = 997.009 \text{ kg/m}^3;$$

$$\rho_{\text{out}}(35°\text{C}) = 994.036\ \text{kg/m}^3;$$

$$\rho_{\text{coll.ave.}}(30°\text{C}) = 996.016\ \text{kg/m}^3$$

(note: average collector temperature used in "temperature") and with $H=2$ m and $L=1$ m,

$$\Delta P_{\text{BUOYANT}} = (997.009)9.81(2) - [(996.016)9.81(1) + (994.036)9.81(1)] = 38.9\ \text{N/m}^2(\text{Pa}).$$

The system flow velocity is estimated from the "system K" given, the pressure difference calculated above, taking the average density of the water around the loop (at 30°C), and substituting into Equation 18.69:

$$\Delta P_{\text{BUOYANT}} = (\rho_{\text{loop.ave}})(h_L)_{\text{loop}} = (\rho_{\text{loop.ave}})KV^2,$$

$$V^2 = 38.9/(996.016)(15.6),$$

$$V = 0.05\ \text{m/s}.$$

18.3.3 Passive Solar Heating Design Fundamentals

Passive heating systems contain the five basic components of all solar systems, as described in the previous chapter on Active Solar Systems. Typical passive realizations of these components are:

1. Collector: windows, walls and floors
2. Storage: walls and floors, large interior masses (often these are integrated with the collector absorption function)
3. Distribution system: radiation, free convection, simple circulation fans
4. Controls: moveable window insulation, vents both to other inside spaces or to ambient
5. Backup system: any nonsolar heating system

The design of passive systems requires the strategic placement of windows, storage masses, and the occupied spaces themselves. The fundamental principles of solar radiation geometry and availability are instrumental in the proper location and sizing of the system's "collectors" (windows). Storage devices are usually more massive than those used in active systems and are frequently an integral part of the collection and distribution system.

18.3.3.1 Types of Passive Heating Systems

A commonly used method of cataloging the various passive system concepts is to distinguish three general categories: direct, indirect, and isolated gain. Most of the physical configurations of passive heating systems are seen to fit within one of these three categories.

For direct gain (Figure 18.43), sunlight enters the heated space and is converted to heat at absorbing surfaces. This heat is then distributed throughout the space and to the various enclosing surfaces and room contents.

For indirect gain category systems, sunlight is absorbed and stored by a mass interposed between the glazing and the conditioned space. The conditioned space is partially enclosed and bounded by this thermal storage mass, so a natural thermal coupling is achieved. Examples of the indirect approach are the thermal storage wall, the thermal storage roof, and the northerly room of an attached sunspace.

In the thermal storage wall (Figure 18.44), sunlight penetrates the glazing and is absorbed and converted to heat at a wall surface interposed between the glazing and the heated space. The wall is usually masonry (Trombe wall) or containers filled with water (water wall), although it might contain phase-change material. The attached sunspace (Figure 18.45) is actually a two-zone combination of direct gain and thermal storage wall. Sunlight enters and heats a direct gain southerly "sunspace" and also heats a

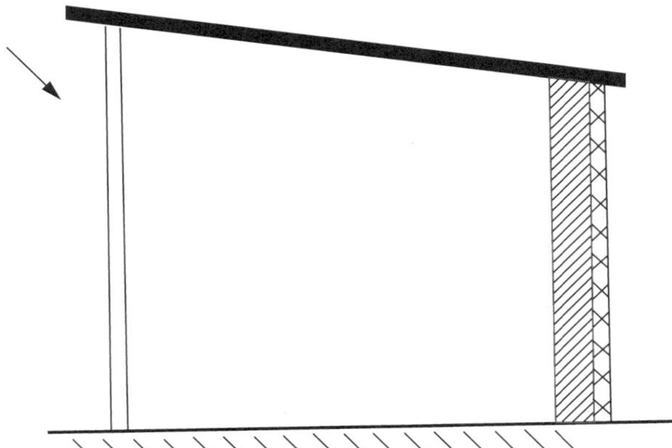

FIGURE 18.43 Direct gain.

mass wall separating the northerly buffered space, which is heated indirectly. The "sunspace" is frequently used as a greenhouse, in which case, the system is called an "attached greenhouse." The thermal storage roof (Figure 18.46) is similar to the thermal storage wall except that the interposed thermal storage mass is located on the building roof.

The isolated gain category concept is an indirect system, except that there is a distinct thermal separation (by means of either insulation or physical separation) between the thermal storage and the heated space. The convective (thermosyphon) loop, as depicted in Figure 18.41, is in this category and, while often used to heat domestic water, is also used for building heating. It is most akin to conventional active systems in that there is a separate collector and separate thermal storage. The thermal storage wall, thermal storage roof, and attached sunspace approaches can also be made into isolated systems by insulating between the thermal storage and the heated space.

18.3.3.2 Fundamental Concepts for Passive Heating Design

Figure 18.47 is an equivalent thermal circuit for the building illustrated in Figure 18.44, the Trombe wall-type system. For the heat transfer analysis of the building, three temperature nodes can be identified: room temperature, storage wall temperature, and the ambient temperature. The circuit responds to

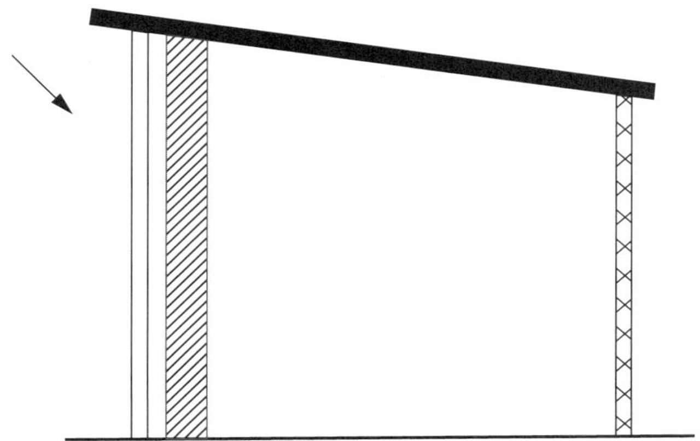

FIGURE 18.44 Thermal storage wall.

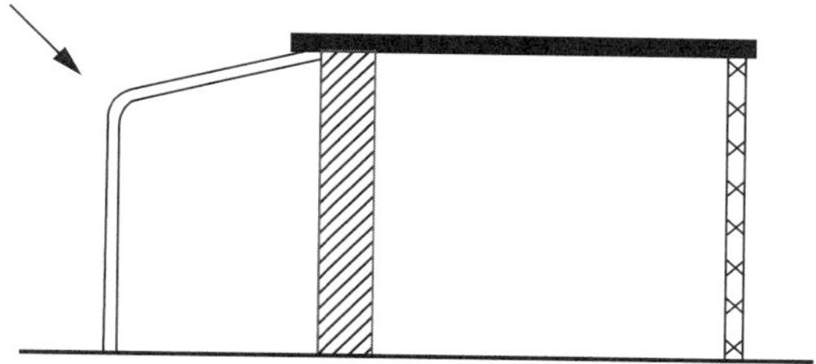

FIGURE 18.45 Attached sunspace.

climatic variables represented by a current injection I_s (solar radiation) and by the ambient temperature T_a. The storage temperature, T_s, and room temperature, T_r, are determined by current flows in the equivalent circuit. By using seasonal and annual climatic data, the performance of a passive structure can be simulated and the results of many such simulations correlated to give the design approaches described below.

18.3.3.3 Passive Design Approaches

Design of a passive heating system involves selection and sizing of the passive feature type(s), determination of thermal performance, and cost estimation. Ideally, a cost/performance optimization would be performed by the designer. Owner and architect ideas usually establish the passive feature type, with general size and cost estimation available. However, the thermal performance of a passive heating system has to be calculated.

There are several "levels" of methods that can be used to estimate the thermal performance of passive designs. First-level methods involve a rule of thumb and/or generalized calculation to get a starting estimate for size and/or annual performance. A second-level method involves climate, building, and passive system details, which allow annual performance determination, plus some sensitivity to passive system design changes. Third-level methods involve periodic calculations (hourly, monthly) of performance and permit more detailed variations of climatic, building, and passive solar system design parameters.

These three levels of design methods have a common basis in that they all are derived from correlations of a multitude of computer simulations of passive systems (PSDH 1980, 1984). As a result, a similar set of defined terms is used in many passive design approaches:

- A_p, solar projected area, m^2 (ft.^{2}): the net south-facing passive solar glazing area projected onto a vertical plane

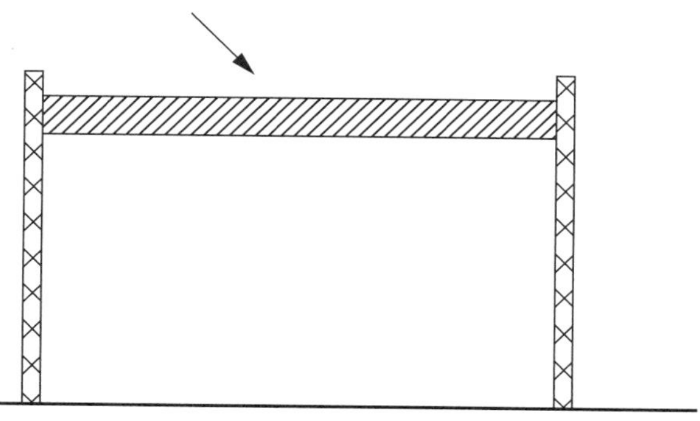

FIGURE 18.46 Thermal storage roof.

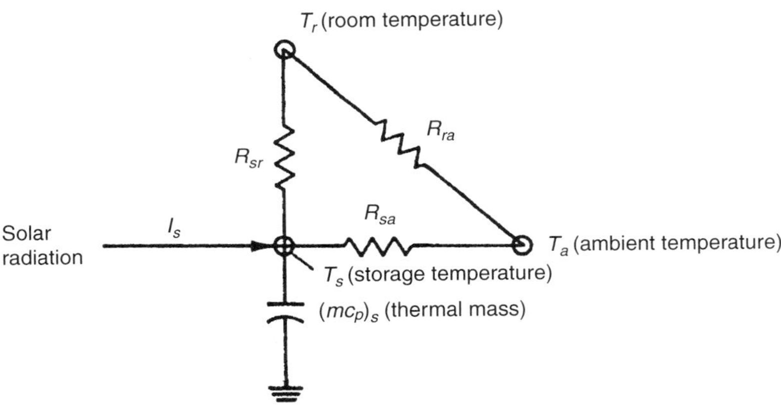

FIGURE 18.47 Equivalent thermal circuit for passively heated solar building in Figure 18.44.

- NLC, net building load coefficient, kJ/CDD (Btu/FDD): net load of the nonsolar portion of the building per degree day of indoor–outdoor temperature difference. The CDD and FDD terms refer to Celsius and Fahrenheit degree days, respectively
- Q_{net}, net reference load, Wh (Btu): heat loss from nonsolar portion of building as calculated by

$$Q_{net} = \text{NLC} \times (\text{Number of degree days}). \tag{18.70}$$

- LCR, load collector ratio, kJ/m^2 CDD (Btu/ft.^{2} FDD): ratio of NLC to A_p,

$$\text{LCR} = \text{NLC}/A_p \tag{18.71}$$

- SSF, solar savings fraction, %: percentage reduction in required auxiliary heating relative to net reference load,

$$\text{SSF} = 1 - \frac{\text{Auxiliary heat required } (Q_{aux})}{\text{Net reference load } (Q_{net})} \tag{18.72}$$

Therefore, using Equation 18.70, the auxiliary heat required, Q_{aux}, is given by

$$Q_{aux} = (1 - \text{SSF}) \times \text{NLC} \times (\text{Number of degree days}). \tag{18.73}$$

The amount of auxiliary heat required is often a basis of comparison between possible solar designs as well as being the basis for determining building energy operating costs. Thus, many of the passive design methods are based on determining SSF, NLC, and the number of degree days in order to calculate the auxiliary heat required for a particular passive system by using Equation 18.73.

18.3.3.4 The First Level: Generalized Methods

A first estimate or starting value is needed to begin the overall passive system design process. Generalized methods and rules of thumb have been developed to generate initial values for solar aperture size, storage size, solar savings fraction, auxiliary heat required, and other size and performance characteristics. The following rules of thumb are meant to be used with the defined terms presented above.

18.3.3.5 Load

A rule of thumb used in conventional building design is that a design heating load of 120–160 kJ/CDD per m^2 of floor area (6–8 Btu/FDD ft.^{2}) is considered an energy conservative design. Reducing these non-solar values by 20% to solarize the proposed south-facing solar wall gives rule-of-thumb NLC values per unit of floor area:

$$\text{NLC/Floor area} = 100\text{–}130\ \text{kJ/CDD m}^2\ (4.8\text{–}6.4\ \text{Btu/FDD ft.}^2). \tag{18.74}$$

18.3.3.6 Solar Savings Fraction

A method of getting starting-point values for the solar savings fraction is presented in Figure 18.48 (PSDH 1984). The map values represent optimum SSF (in percent) for a particular set of conservation and passive-solar costs for different climates across the United States. With the Q_{net} generated from the NLC rule of thumb (see above) and the SSF read from the map, the Q_{aux} can be determined.

18.3.3.7 Load Collector Ratio (LCR)

The A_p can be determined using the NLC from above if the LCR is known. The rule of thumb associated with "good" values of LCR (PSDH 1984) differs depending on whether the design is for a "cold" or "warm" climate:

$$\text{"Good" LCR} = \begin{cases} \text{For cold climate}: 410\ \text{kJ/m}^2\ \text{CDD}\ (20\ \text{Btu/ft.}^2\ \text{FDD}) \\ \text{For warm climate}: 610\ \text{kJ/m}^2\ \text{CDD}\ (30\ \text{Btu/ft.}^2\ \text{FDD}) \end{cases} \tag{18.75}$$

18.3.3.8 Storage

Rules of thumb for thermal mass storage relate storage material total heat capacity to the solar projected area (PSDH 1984). The use of the storage mass is to provide for heating on cloudy days and to regulate sunny day room air temperature swing. When the thermal mass directly absorbs the solar radiation, each

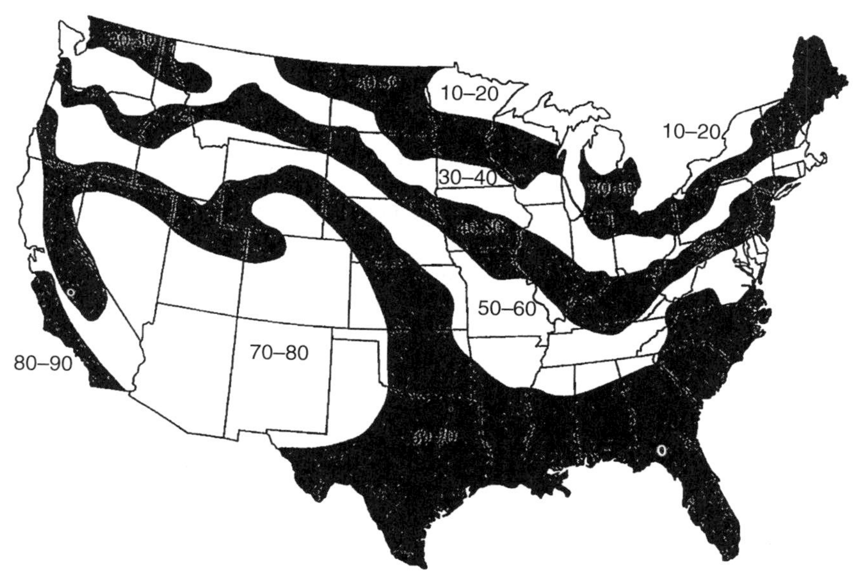

FIGURE 18.48 Starting-point values of solar savings fraction (SSF) in percent. (From PSDH, *Passive Solar Design Handbook*. Part One: Total Environmental Action, Inc., Part Two: Los Alamos Scientific Laboratory, Part Three: Los Alamos National Laboratory. Van Nostrand Reinhold, New York, 1984.)

square meter of the projected glazing area requires enough mass to store 613 kJ/°C. If the storage material is not in direct sunlight, but heated from room air only, then four times as much mass is needed. In a room with a directly sunlight-heated storage mass, the room air temperature swing will be approximately one-half the storage mass temperature swing. For room air heated storage, the air temperature swing is twice that of the storage mass.

Example 18.3.2

A Denver, Colorado, building is to have a floor area of 195 m^2 (2100 ft.^{2}). Determine rule-of-thumb size and performance characteristics.

Solution. From Equation 18.72, the NLC is estimated as

$$\begin{aligned} \text{NLC} &= (115\ \text{kJ/CDD m}^2) \times (195\ \text{m}^2) \\ &= 22,400\ \text{kJ/CDD}\ (11,800\ \text{Btu/FDD}). \end{aligned}$$

Using the "cold" LCR value and Equation 18.71, the passive solar projected area is

$$\begin{aligned} A_\text{p} &= \text{NLC/LCR} = (22,400\ \text{kJ/CDD})/(410\ \text{kJ/m}^2\ \text{CDD}) \\ &= 54.7\ \text{m}^2 (588\ \text{ft.}^2) \end{aligned}$$

Locating Denver on the map of Figure 18.48 gives an SSF value in the 70%–80% range (use 75%). An annual °C-degree-day value can be found in city climate tables (PSDH 1984; NCDC 1992), and is 3491 CDD (6283 FDD) for Denver. Thus, the auxiliary heat required, Q_aux, is found using Equation 18.73:

$$\begin{aligned} Q_\text{aux} &= (1-0.75)(22,400\ \text{kJ/CDD})(3491\ \text{CDD}) \\ &= 19,600\ \text{MJ}\ (18.5 \times 10^6\ \text{Btu})\ \text{annually}. \end{aligned}$$

The thermal storage can be sized using directly solar-heated and/or room air heated mass by using the projected area. Assuming brick with a specific heat capacity of 840 J/kg°C, the storage mass is found by

$$\begin{aligned} A_p \times (613\ \text{kJ/C}) &= m \times (840\ \text{J/kg°C}); \\ m_\text{d} &= 40,000\ \text{kg}\ (88,000\ \text{lbm})\ [\text{Direct sun}] \\ \text{or} \quad m_\text{a} &= 160,000\ \text{kg}\ (351,000\ \text{lbm})\ [\text{Air heated}] \end{aligned}$$

A more location-dependent set of rules of thumb is presented in PSDH (1980). The first rule of thumb relates solar projected area as a percentage of floor area to solar savings fraction, with and without night insulation of the solar glazing:

"A solar projected area of (B1)% to (B2)% of the floor area can be expected to produce a SSF in (location) of (S1)% to (S2)%, or, if R9 night insulation is used, of (S3)% to (S4)%."

The values of B1, B2, S1, S2, S3 and S4 are found using Table 18.6 for the location. The thermal storage mass rule of thumb is again related to the solar projected area:

"A thermal storage wall should have 14 kg×SSF (%) of water or 71 kg×SSF (%) of masonry for each square meter of solar projected area. For a direct gain space, the mass above should be used with a surface area of at least three times the solar projected area, and masonry no thicker than 10–15 cm. If the mass is located in back rooms, then four times the above mass is needed."

TABLE 18.6 Values to Be Used in the Glazing Area and SSF Relations Rules of Thumb

City	B1	B2	S1	S2	S3	S4
Birmingham, Alabama	0.09	0.18	22	37	34	58
Mobile, Alabama	0.06	0.12	26	44	34	60
Montgomery, Alabama	0.07	0.15	24	41	34	59
Phoenix, Arizona	0.06	0.12	37	60	48	75
Prescott, Arizona	0.10	0.20	29	48	44	72
Tucson, Arizona	0.06	0.12	35	57	45	73
Winslow, Arizona	0.12	0.24	30	47	48	74
Yuma, Arizona	0.04	0.09	43	66	51	78
Fort Smith, Arkansas	0.10	0.20	24	39	38	64
Little Rock, Arkansas	0.10	0.19	23	38	37	62
Bakersfield, California	0.08	0.15	31	50	42	67
Baggett, California	0.07	0.15	35	56	46	73
Fresno, California	0.09	0.17	29	46	41	65
Long Beach, California	0.05	0.10	35	58	44	72
Los Angeles, California	0.05	0.09	36	58	44	72
Mount Shasta, California	0.11	0.21	24	38	42	67
Needles, California	0.06	0.12	39	61	49	76
Oakland, California	0.07	0.15	35	55	46	72
Red Bluff, California	0.09	0.18	29	46	41	65
Sacramento, California	0.09	0.18	29	47	41	66
San Diego, California	0.04	0.09	37	61	46	74
San Francisco, California	0.06	0.13	34	54	45	71
Santa Maria, California	0.05	0.11	31	53	42	69
Colorado Springs, Colorado	0.12	0.24	27	42	47	74
Denver, Colorado	0.12	0.23	27	43	47	74
Eagle, Colorado	0.14	0.29	25	35	53	77
Grand Junction, Colorado	0.13	0.27	29	43	50	76
Pueblo, Colorado	0.11	0.23	29	45	48	75
Hartford, Connecticut	0.17	0.35	14	19	40	64
Wilmington, Delaware	0.15	0.29	19	30	39	63
Washington, District Of Columbia	0.12	0.23	18	28	37	61
Apalachicola, Florida	0.05	0.10	28	47	36	61
Daytona Beach, Florida	0.04	0.08	30	51	36	63
Jacksonville, Florida	0.05	0.09	27	47	35	62
Miami, Florida	0.01	0.02	27	48	31	54
Orlando, Florida	0.03	0.06	30	52	37	63
Tallahassee, Florida	0.05	0.11	26	45	35	60
Tampa, Florida	0.03	0.06	30	52	36	63
West Palm Beach, Florida	0.01	0.03	30	51	34	59
Atlanta, Georgia	0.06	0.17	22	36	34	58
Augusta, Georgia	0.06	0.16	24	40	35	60
Macon, Georgia	0.07	0.15	25	41	35	59
Savannah, Georgia	0.06	0.13	25	43	35	60
Boise, Idaho	0.14	0.28	27	38	48	71
Lewiston, Idaho	0.15	0.29	22	29	44	65
Pocatello, Idaho	0.13	0.26	25	35	51	74
Chicago, Illinois	0.17	0.35	17	23	43	67
Moline, Illinois	0.20	0.39	17	22	46	70
Springfield, Illinois	0.15	0.30	19	26	42	67
Evansville, Indiana	0.14	0.27	19	29	37	61
Fort Wayne, Indiana	0.16	0.33	13	17	37	60
Indianapolis, Indiana	0.14	0.28	15	21	37	60
South Bend, Indiana	0.18	0.35	12	15	39	61
Burlington, Iowa	0.18	0.36	20	27	47	71
Des Moines, Iowa	0.21	0.43	19	25	58	75

(continued)

TABLE 18.6 ***(Continued)***

City	B1	B2	S1	S2	S3	S4
Mason City, Iowa	0.22	0.44	18	19	56	79
Sioux City, Iowa	0.23	0.46	20	24	53	76
Dodge City, Kansas	0.12	0.23	27	42	46	73
Goodland, Kansas	0.13	0.27	26	39	47	74
Topeka, Kansas	0.14	0.26	24	35	45	71
Wichita, Kansas	0.14	0.26	26	41	45	72
Lexington, Kentucky	0.13	0.27	17	26	35	58
Louisville, Kentucky	0.13	0.27	18	27	35	59
Baton Rouge, Louisiana	0.06	0.12	26	43	34	59
Lake Charles, Louisiana	0.06	0.11	24	41	32	57
New Orleans, Louisiana	0.05	0.11	27	46	35	61
Shreveport, Louisiana	0.08	0.15	26	43	36	61
Caribou, Maine	0.25	0.30	NR	NR	53	74
Portland, Maine	0.17	0.34	14	17	45	69
Baltimore, Maryland	0.14	0.27	19	30	38	62
Boston, Massachusetts	0.15	0.29	17	25	40	64
Alpena, Michigan	0.21	0.42	NR	NR	47	69
Detroit, Michigan	0.17	0.34	13	17	39	61
Flint, Michigan	0.15	0.31	11	12	40	62
Grand Rapids, Michigan	0.19	0.38	12	13	39	61
Sault Ste. Marie, Michigan	0.25	0.50	NR	NR	50	70
Traverse City, Michigan	0.18	0.36	NR	NR	42	62
Duluth, Minnesota	0.25	0.50	NR	NR	50	70
International Falls, Minnesota	0.25	0.50	NR	NR	47	66
Minneapolis-St. Paul, Minnesota	0.25	0.50	NR	NR	55	76
Rochester, Minnesota	0.24	0.49	NR	NR	54	76
Jackson, Mississippi	0.06	0.15	24	48	34	59
Meridian, Mississippi	0.08	0.15	23	39	34	58
Columbia, Missouri	0.13	0.26	20	30	41	66
Kansas City, Missouri	0.14	0.29	22	32	44	70
Saint Louis, Missouri	0.15	0.29	21	33	41	65
Springfield, Missouri	0.13	0.26	22	34	40	65
Billings, Montana	0.16	0.32	24	31	53	76
Cut Bank, Montana	0.24	0.49	22	23	62	81
Dillon, Montana	0.16	0.32	24	32	54	77
Glasgow, Montana	0.25	0.50	NR	NR	55	75
Great Falls, Montana	0.18	0.37	23	26	56	77
Helena, Montana	0.20	0.39	21	25	55	77
Lewistown, Montana	0.19	0.38	21	25	54	76
Miles City, Montana	0.23	0.47	21	23	60	80
Missoula, Montana	0.18	0.36	15	16	47	68
Grand Island, Nebraska	0.18	0.36	24	33	51	76
North Omaha, Nebraska	0.20	0.48	21	29	51	76
North Platte, Nebraska	0.17	0.34	25	36	50	76
Scottsbluff, Nebraska	0.16	0.31	24	36	49	74
Elko, Nevada	0.12	0.25	27	39	52	76
Ely, Nevada	0.12	0.23	27	41	50	77
Las Vegas, Nevada	0.09	0.18	35	56	48	75
Lovelock, Nevada	0.13	0.25	32	48	53	78
Reno, Nevada	0.11	0.22	31	48	49	76
Tonopah, Nevada	0.11	0.23	31	48	51	77
Winnemucca, Nevada	0.13	0.26	28	42	49	75
Concord, New Hampshire	0.17	0.34	13	15	45	68
Newark, New Jersey	0.13	0.25	19	29	39	64
Albuquerque, New Mexico	0.11	0.22	29	47	46	73

(continued)

TABLE 18.6 (*Continued*)

City	B1	B2	S1	S2	S3	S4
Clayton, New Mexico	0.10	0.20	28	45	45	73
Farmington, New Mexico	0.12	0.24	29	45	49	76
Los Alamos, New Mexico	0.11	0.22	25	40	44	72
Roswell, New Mexico	0.10	0.19	30	49	45	73
Truth or Consequences, New Mexico	0.09	0.17	32	51	46	73
Tucumcari, New Mexico	0.10	0.20	30	48	45	73
Zuni, New Mexico	0.11	0.21	27	43	45	73
Albany, New York	0.21	0.41	13	15	43	66
Binghamton, New York	0.15	0.30	NR	NR	35	56
Buffalo, New York	0.19	0.37	NR	NR	36	57
Massena, New York	0.25	0.50	NR	NR	50	71
New York (Central Park), New York	0.15	0.30	16	25	36	59
Rochester, New York	0.18	0.37	NR	NR	37	58
Syracuse, New York	0.19	0.38	NR	NR	37	59
Asheville, North Carolina	0.10	0.20	21	35	36	61
Cape Hatteras, North Carolina	0.09	0.17	24	40	36	60
Charlotte, North Carolina	0.08	0.17	23	38	36	60
Greensboro, North Carolina	0.10	0.20	23	37	37	63
Raleigh-Durham, North Carolina	0.09	0.19	22	37	36	61
Bismarck, North Dakota	0.25	0.50	NR	NR	56	77
Fargo, North Dakota	0.25	0.50	NR	NR	51	72
Minot, North Dakota	0.25	0.50	NR	NR	52	72
Akron-Canton, Ohio	0.15	0.31	12	16	35	57
Cincinnati, Ohio	0.12	0.24	15	23	35	57
Cleveland, Ohio	0.15	0.31	11	14	34	55
Columbus, Ohio	0.14	0.28	13	18	35	57
Dayton, Ohio	0.14	0.28	14	20	36	59
Toledo, Ohio	0.17	0.34	13	17	38	61
Youngstown, Ohio	0.16	0.32	NR	NR	34	54
Oklahoma City, Oklahoma	0.11	0.22	25	41	41	67
Tulsa, Oklahoma	0.11	0.22	24	38	40	65
Astoria, Oregon	0.09	0.19	21	34	37	60
Burns, Oregon	0.13	0.25	23	32	47	71
Medford, Oregon	0.12	0.24	21	32	38	60
North Bend, Oregon	0.09	0.17	25	42	38	64
Pendleton, Oregon	0.14	0.27	22	30	43	64
Portland, Oregon	0.13	0.26	21	31	38	60
Redmond, Oregon	0.13	0.27	26	38	47	71
Salem, Oregon	0.12	0.24	21	32	37	59
Allentown, Pennsylvania	0.15	0.29	16	24	39	63
Erie, Pennsylvania	0.17	0.34	NR	NR	35	55
Harrisburg, Pennsylvania	0.13	0.26	17	26	38	62
Philadelphia, Pennsylvania	0.15	0.29	19	29	38	62
Pittsburgh, Pennsylvania	0.14	0.28	12	16	33	55
Wilkes-Barre-Scranton, Pennsylvania	0.16	0.32	13	18	37	60
Providence, Rhode Island	0.15	0.30	17	24	40	64
Charleston, South Carolina	0.07	0.14	25	41	34	59
Columbia, South Carolina	0.08	0.17	25	41	36	61
Greenville-Spartanburg, South Carolina	0.08	0.17	23	38	36	60
Huron, South Dakota	0.25	0.50	NR	NR	58	79
Pierre, South Dakota	0.22	0.43	21	23	58	80
Rapid City, South Dakota	0.15	0.30	23	32	51	76
Sioux Falls, South Dakota	0.22	0.45	18	19	57	79
Chattanooga, Tennessee	0.09	0.19	19	32	33	56
Knoxville, Tennessee	0.09	0.18	20	33	33	56

(continued)

TABLE 18.6 *(Continued)*

City	B1	B2	S1	S2	S3	S4
Memphis, Tennessee	0.09	0.19	22	36	36	60
Nashville, Tennessee	0.10	0.21	19	30	33	55
Abilene, Texas	0.09	0.18	29	47	41	68
Amarillo, Texas	0.11	0.22	29	46	45	72
Austin, Texas	0.06	0.13	27	46	37	63
Brownsville, Texas	0.03	0.06	27	46	32	57
Corpus Christi, Texas	0.05	0.09	29	49	36	63
Dallas, Texas	0.08	0.17	27	44	38	64
Del Rio, Texas	0.06	0.12	30	50	39	66
El Paso, Texas	0.09	0.17	32	53	45	72
Forth Worth, Texas	0.09	0.17	26	44	38	64
Houston, Texas	0.06	0.11	25	43	34	59
Laredo, Texas	0.05	0.09	31	52	39	64
Lubbock, Texas	0.09	0.19	30	49	44	72
Lufkin, Texas	0.07	0.14	26	43	35	61
Midland-Odessa, Texas	0.09	0.18	32	52	44	72
Port Arthur, Texas	0.06	0.11	26	44	34	60
San Angelo, Texas	0.08	0.15	29	48	40	67
San Antonio, Texas	0.06	0.12	28	48	38	64
Sherman, Texas	0.10	0.20	25	41	38	64
Waco, Texas	0.06	0.15	27	45	38	64
Wichita Falls, Texas	0.10	0.20	27	45	41	67
Bryce Canyon, Utah	0.13	0.25	26	39	52	78
Cedar City, Utah	0.12	0.24	28	43	48	75
Salt Lake City, Utah	0.13	0.26	27	39	48	72
Burlington, Vermont	0.22	0.43	NR	NR	46	68
Norfolk, Virginia	0.09	0.19	23	38	37	62
Richmond, Virginia	0.11	0.22	21	34	37	61
Roanoke, Virginia	0.11	0.23	21	34	37	61
Olympia, Washington	0.12	0.23	20	29	38	59
Seattle-Tacoma, Washington	0.11	0.22	21	30	39	59
Spokane, Washington	0.20	0.39	20	24	48	68
Yakima, Washington	0.18	0.36	24	31	49	70
Charleston, West Virginia	0.13	0.25	16	24	32	54
Huntington, West Virginia	0.13	0.25	17	27	34	57
Eau Claire, Wisconsin	0.25	0.50	NR	NR	53	75
Green Bay, Wisconsin	0.23	0.46	NR	NR	53	75
La Crosse, Wisconsin	0.21	0.43	NR	NR	52	75
Madison, Wisconsin	0.20	0.40	15	17	51	74
Milwaukee, Wisconsin	0.18	0.35	15	18	48	71
Casper, Wyoming	0.13	0.26	27	39	53	78
Cheyenne, Wyoming	0.11	0.21	25	39	47	74
Rock Springs, Wyoming	0.14	0.28	26	38	54	79
Sheridan, Wyoming	0.16	0.31	22	30	52	75
		Canada				
Edmonton, Alberta	0.25	0.50	NR	NR	54	72
Suffield, Alberta	0.25	0.50	28	30	67	85
Nanaimo, British Columbia	0.13	0.26	26	35	45	66
Vancouver, British Columbia	0.13	0.26	20	28	48	60
Winnipeg, Manitoba	0.25	0.50	NR	NR	54	74
Dartmouth, Nova Scotia	0.14	0.28	17	24	45	70
Moosonee, Ontario	0.25	0.50	NR	NR	48	67
Ottawa, Ontario	0.25	0.50	NR	NR	59	80
Toronto, Ontario	0.18	0.36	17	23	44	68
Normandie, Quebec	0.25	0.50	NR	NR	54	74

Note: NR, not recommended.
Source: From PSDH, *Passive Solar Design Handbook*, U.S. Department of Energy, Washington, DC, 1980.

Example 18.3.3

Determine size and performance passive solar characteristics with the location-dependent set of rules of thumb for the house of the previous example.

Solution. Using Table 18.6 with the 195 m^2 house in Denver yields:

$$\text{Solar projected area} = 12\% \text{ to } 23\% \text{ of floor area}$$
$$= 23.4\ m^2 \text{ to } 44.9\ m^2.$$
$$\text{SSF (no night insulation)} = 27\% \text{ to } 43\%.$$
$$\text{SSF (R9 night insulation)} = 47\% \text{ to } 74\%.$$

Using the rule of thumb for the thermal storage mass:

$$m = 17\ \text{kg} \times 43\% \times 44.9\ m^2$$
$$= 33,000\ \text{kg } (72,000\ \text{lbm})[\text{Thermal wall or direct gain}]$$

Comparing the results of this example to those of the previous example, the two rules of thumb are seen to produce "roughly" similar answers. General system cost and performance information can be generated with results from rule-of-thumb calculations, but a more detailed level of information is needed to determine design-ready passive system type (direct gain, thermal wall, sunspace), size, performance, and costs.

18.3.3.9 The Second Level: LCR Method

The LCR method is useful for making estimates of the annual performance of specific types of passive system(s) combinations. The LCR method was developed by calculating the annual SSF for 94 reference passive solar systems for 219 U.S. and Canadian locations over a range of LCR values. Table 18.7 includes the description of these 94 reference systems for use both with the LCR method and with the SLR method described below. Tables were constructed for each city with LCR versus SSF listed for each of the 94 reference passive systems. (Note that the solar load ratio (SLR) method was used to make the LCR calculations, and this SLR method is described in the next section as the third-level method.) Although the complete LCR tables (PSDH 1984) include 219 locations, Table 18.8 only includes six "representative" cities (Albuquerque, Boston, Madison, Medford, Nashville, Santa Maria), purely due to space restrictions. The LCR method consists of the following steps (PSDH 1984):

1. Determine the building parameters:
   a. Building load coefficient, NLC
   b. Solar projected area, A_p
   c. Load collector ratio, $\text{LCR} = \text{NLC}/A_p$
2. Find the short designation of the reference system closest to the passive system design (Table 18.7)
3. Enter the LCR Tables (Table 18.8)
   a. Find the city
   b. Find the reference system listing
   c. Determine annual SSF by interpolation using the LCR value from above
   d. Note the annual heating degree days (Number of degree days)
4. Calculate the annual auxiliary heat required:

$$\text{Auxiliary heat required} = (1 - \text{SSF}) \times \text{NLC} \times (\text{Number of degree days}).$$

If more than one reference solar system is being used, then find the "aperture area weighted" SSF for the combination. Determine each individual reference system SSF using the total aperture area LCR, then take the "area weighted" average of the individual SSFs.

TABLE 18.7 Designations and Characteristics for 94 Reference Systems

(a) Overall System Characteristics

Masonry Properties	
Thermal conductivity (k)	
Sunspace floor	0.5 Btu/h/ft./°F
All other masonry	1.0 Btu/h/ft./°F
Density (Q)	150 lb/ft.^{3}
Specific heat (c)	0.2 Btu/lb/°F
Infrared emittance of normal surface	0.9
Infrared emittance of selective surface	0.1
Solar Absorptances	
Waterwall	1.0
Masonry, Trombe wall	1.0
Direct gain and sunspace	0.8
Sunspace: water containers	0.9
Lightweight common wall	0.7
Other lightweight surfaces	0.3
Glazing Properties	
Transmission characteristics	Diffuse
Orientation	Due south
Index of refraction	1.526
Extinction coefficient	0.5 in.$^{-1}$
Thickness of each pane	⅛ in.
Gap between panes	½ in.
Ared emittance	0.9
Control Range	
Room temperature	65°F–75°F
Sunspace temperature	45°F–95°F
Internal heat generation	0
Thermocirculation Vents (when used)	
Vent area/projected area (sum of both upper and lower vents)	0.06
Height between vents	8 ft.
Reverse flow	None
Nighttime Insulation (when used)	
Thermal resistance	R9
In place, solar time	5:30 P.M. to 7:30 A.M.
Solar Radiation Assumptions	
Shading	None
Ground diffuse reflectance	0.3

(b) Direct-Gain (DG) System Types

Designation	Thermal Storage Capacity[a] (Btu/ft.^{2}/°F)	Mass Thickness[a] (in.)	Mass-Area-to-Glazing-Area Ratio	No. of Glazings	Nighttime Insulation
A1	30	2	6	2	No
A2	30	2	6	3	No
A3	30	2	6	2	Yes
B1	45	6	3	2	No
B2	45	6	3	3	No
B3	45	6	3	2	Yes
C1	60	4	6	2	No
C2	60	4	6	3	No
C3	60	4	6	2	Yes

(continued)

TABLE 18.7 (***Continued***)

(c) Vented Trombe Wall (TW) System Types

Designation	Thermal Storage Capacity[a] (Btu/ft.^{2}/°F)	Wall Thickness[a] (in.)	ρck (Btu^2/h/ft.^{4}/°F^2)	No. of Glazings	Wall Surface	Nighttime Insulation
A1	15	6	30	2	Normal	No
A2	22.5	9	30	2	Normal	No
A3	30	12	30	2	Normal	No
A4	45	18	30	2	Normal	No
B1	15	6	15	2	Normal	No
B2	22.5	9	15	2	Normal	No
B3	30	12	15	2	Normal	No
B4	45	18	15	2	Normal	No
C1	15	6	7.5	2	Normal	No
C2	22.5	9	7.5	2	Normal	No
C3	30	12	7.5	2	Normal	No
C4	45	18	7.5	2	Normal	No
D1	30	12	30	1	Normal	No
D2	30	12	30	3	Normal	No
D3	30	12	30	1	Normal	Yes
D4	30	12	30	2	Normal	Yes
D5	30	12	30	3	Normal	Yes
E1	30	12	30	1	Selective	No
E2	30	12	30	2	Selective	No
E3	30	12	30	1	Selective	Yes
E4	30	12	30	2	Selective	Yes

(d) Unvented Trombe Wall (TW) System Types

Designation	Thermal Storage Capacity[a] (Btu/ft.^{2}/°F)	Wall Thickness[a] (in.)	ρck (Btu^2/h/ft.^{4}/°F^2)	No. of Glazings	Wall Surface	Nighttime Insulation
F1	15	6	30	2	Normal	No
F2	22.5	9	30	2	Normal	No
F3	30	12	30	2	Normal	No
F4	45	18	30	2	Normal	No
G1	15	6	15	2	Normal	No
G2	22.5	9	15	2	Normal	No
G3	30	12	15	2	Normal	No
G4	45	18	15	2	Normal	No
H1	15	6	7.5	2	Normal	No
H2	22.5	9	7.5	2	Normal	No
H3	30	12	7.5	2	Normal	No
H4	45	18	7.5	2	Normal	No
I1	30	12	30	1	Normal	No
I2	30	12	30	3	Normal	No
I3	30	12	30	1	Normal	Yes
I4	30	12	30	2	Normal	Yes
I5	30	12	30	3	Normal	Yes
J1	30	12	30	1	Selective	No
J2	30	12	30	2	Selective	No
J3	30	12	30	1	Selective	Yes
J4	30	12	30	2	Selective	Yes

(continued)

TABLE 18.7 (*Continued*)

(e) Waterwall (WW) System Types

Designation	Thermal Storage Capacity[a] (Btu/ft.^{2}/°F)	Wall Thickness (in.)	No. of Glazings	Wall Surface	Nighttime Insulation
A1	15.6	3	2	Normal	No
A2	31.2	6	2	Normal	No
A3	46.8	9	2	Normal	No
A4	62.4	12	2	Normal	No
A5	93.6	18	2	Normal	No
A6	124.8	24	2	Normal	No
B1	46.8	9	1	Normal	No
B2	46.8	9	3	Normal	No
B3	46.8	9	1	Normal	Yes
B4	46.8	9	2	Normal	Yes
B5	46.8	9	3	Normal	Yes
C1	46.8	9	1	Selective	No
C2	46.8	9	2	Selective	No
C3	46.8	9	1	Selective	Yes
C4	46.8	9	2	Selective	Yes

(f) Sunspace (SS) System Types

Designation	Type	Tilt (°)	Common Wall	End Walls	Nighttime Insulation
A1	Attached	50	Masonry	Opaque	No
A2	Attached	50	Masonry	Opaque	Yes
A3	Attached	50	Masonry	Glazed	No
A4	Attached	50	Masonry	Glazed	Yes
A5	Attached	50	Insulated	Opaque	No
A6	Attached	50	Insulated	Opaque	Yes
A7	Attached	50	Insulated	Glazed	No
A8	Attached	50	Insulated	Glazed	Yes
B1	Attached	90/30	Masonry	Opaque	No
B2	Attached	90/30	Masonry	Opaque	Yes
B3	Attached	90/30	Masonry	Glazed	No
B4	Attached	90/30	Masonry	Glazed	Yes
B5	Attached	90/30	Insulated	Opaque	No
B6	Attached	90/30	Insulated	Opaque	Yes
B7	Attached	90/30	Insulated	Glazed	No
B8	Attached	90/30	Insulated	Glazed	Yes
C1	Semienclosed	90	Masonry	Common	No
C2	Semienclosed	90	Masonry	Common	Yes
C3	Semienclosed	90	Insulated	Common	No
C4	Semienclosed	90	Insulated	Common	Yes
D1	Semienclosed	50	Masonry	Common	No
D2	Semienclosed	50	Masonry	Common	Yes
D3	Semienclosed	50	Insulated	Common	No
D4	Semienclosed	50	Insulated	Common	Yes
E1	Semienclosed	90/30	Masonry	Common	No
E2	Semienclosed	90/30	Masonry	Common	Yes
E3	Semienclosed	90/30	Insulated	Common	No
E4	Semienclosed	90/30	Insulated	Common	Yes

[a] The thermal storage capacity is per unit of projected area, or, equivalently, the quantity ρck. The wall thickness is listed only as an appropriate guide by assuming $\rho c = 30$ Btu/ft.^{3}/°F.

Source: From PSDH, Passive Solar Design Handbook. Part One: Total Environmental Action, Inc., Part Two: Los Alamos Scientific Laboratory, Part Three: Los Alamos National Laboratory. Van Nostranal Laboratory. Van Nostranal Reinhold, New York, 1984.

TABLE 18.8 LCR Tables for Six Representative Cities (Albuquerque, Boston, Madison, Medford, Nashville, and Santa Maria)

SSF	0.10	0.20	0.30	0.40	0.50	0.60	0.70	0.80	0.90
Santa Maria, California									3053 DD
WW A1	1776	240	119	73	50	35	25	18	12
WW A2	617	259	154	103	74	54	39	28	19
WW A3	523	261	164	114	82	61	45	33	22
WW A4	482	260	169	119	87	65	48	35	24
WW A5	461	263	175	125	92	69	52	38	26
WW A6	447	263	177	128	95	72	54	40	27
WW B1	556	220	128	85	60	43	32	23	15
WW B2	462	256	168	119	88	66	49	36	25
WW B3	542	315	211	151	112	85	64	47	32
WW B4	455	283	197	144	109	83	63	47	32
WW B5	414	263	184	136	103	79	60	45	31
WW C1	569	330	221	159	118	89	67	49	33
WW C2	478	288	197	143	107	81	61	45	31
WW C3	483	318	228	170	130	100	77	57	40
WW C4	426	280	200	149	114	88	68	51	35
TW A1	1515	227	113	70	48	34	24	17	11
TW A2	625	234	134	89	63	46	33	24	16
TW A3	508	231	140	95	68	50	37	27	18
TW A4	431	217	137	95	69	51	38	28	19
TW B1	859	212	112	71	49	35	25	18	12
TW B2	502	209	124	83	59	43	32	23	15
TW B3	438	201	123	84	60	44	33	24	16
TW B4	400	184	112	76	55	40	30	22	14
TW C1	568	188	105	69	48	35	25	18	12
TW C2	435	178	105	70	50	36	27	19	13
TW C3	413	165	97	64	46	33	25	18	12
TW C4	426	146	82	54	38	27	20	14	10
TW D1	403	170	101	67	48	35	25	18	12
TW D2	488	242	152	105	76	57	42	31	21
TW D3	509	271	175	123	90	67	50	36	25
TW D4	464	266	177	127	94	71	53	39	27
TW D5	425	250	169	122	91	69	52	38	26
TW E1	581	309	199	140	102	76	57	42	28
TW E2	512	283	186	132	97	73	55	40	27
TW E3	537	328	225	164	123	94	71	53	36
TW E4	466	287	199	145	109	83	63	47	32
TW F1	713	198	107	68	47	34	25	18	12
TW F2	455	199	120	81	58	42	31	22	15
TW F3	378	190	120	83	60	45	33	24	16
TW F4	311	169	110	77	57	42	32	23	16
TW G1	450	170	98	65	46	33	24	17	12
TW G2	331	163	102	70	51	38	28	20	14
TW G3	278	147	94	66	48	36	27	20	13
TW G4	222	120	78	55	40	30	22	16	11
TW H1	295	137	84	57	41	30	22	16	11
TW H2	226	118	75	52	38	28	21	15	10
TW H3	187	99	64	44	33	24	18	13	9
TW H4	143	75	48	33	24	18	14	10	7
TW I1	318	144	88	59	42	31	23	16	11
TW I2	377	203	132	93	68	51	38	28	19
TW I3	404	226	149	106	78	58	44	32	22
TW I4	387	230	156	113	84	64	48	36	24
TW I5	370	226	155	113	85	65	49	36	25

(continued)

TABLE 18.8 (*Continued*)

SSF	0.10	0.20	0.30	0.40	0.50	0.60	0.70	0.80	0.90
TW J1	483	271	179	127	94	71	53	39	26
TW J2	422	246	165	119	88	67	50	37	25
TW J3	446	283	199	146	111	85	65	48	33
TW J4	400	254	178	132	100	77	58	43	30
DG A1	392	188	117	79	55	38	26	16	7
DG A2	389	190	121	85	61	45	32	22	14
DG A3	443	220	142	102	77	58	44	31	19
DG B1	384	191	122	86	64	48	35	24	13
DG B2	394	196	127	91	69	53	40	29	19
DG B3	445	222	145	105	80	62	49	37	25
DG C1	451	225	146	104	78	61	47	34	21
DG C2	453	226	148	106	80	63	49	37	25
DG C3	509	254	167	121	92	73	58	45	31
SS A1	1171	396	220	142	98	69	49	34	22
SS A2	1028	468	283	190	135	98	71	50	33
SS A3	1174	380	209	133	91	64	45	31	20
SS A4	1077	481	289	193	136	98	71	50	32
SS A5	1896	400	204	127	86	60	42	29	18
SS A6	1030	468	283	190	135	97	71	50	32
SS A7	2199	359	178	109	72	50	35	24	15
SS A8	1089	478	285	190	133	96	69	48	31
SS B1	802	298	170	111	77	55	40	28	18
SS B2	785	366	224	152	108	79	57	41	27
SS B3	770	287	163	106	74	52	37	26	17
SS B4	790	368	224	152	108	78	57	40	26
SS B5	1022	271	144	91	62	44	31	22	14
SS B6	750	356	219	149	106	77	56	40	26
SS B7	937	242	127	80	54	38	27	19	12
SS B8	750	352	215	146	103	75	55	39	25
SS C1	481	232	144	99	71	52	39	28	19
SS C2	482	262	170	120	88	66	49	36	24
SS C3	487	185	107	71	50	36	27	19	13
SS C4	473	235	147	102	74	55	41	30	20
SS D1	1107	477	282	188	132	95	68	48	31
SS D2	928	511	332	232	169	125	92	66	43
SS D3	1353	449	248	160	110	78	56	39	25
SS D4	946	500	319	222	160	117	86	61	40
SS E1	838	378	227	153	108	78	56	40	26
SS E2	766	419	272	190	138	102	75	54	36
SS E3	973	322	178	115	79	56	40	28	18
SS E4	780	393	247	170	122	89	65	47	31
Albuquerque, New Mexico									4292 DD
WW A1	1052	130	62	38	25	18	13	9	6
WW A2	354	144	84	56	39	29	21	15	10
WW A3	300	146	90	62	45	33	24	18	12
WW A4	276	146	93	65	47	35	26	19	13
WW A5	264	148	97	69	50	38	28	21	14
WW A6	256	148	99	70	52	39	30	22	15
WW B1	293	111	63	41	28	20	15	11	7
WW B2	270	147	96	67	49	37	28	20	14
WW B3	314	179	119	84	62	47	35	26	18
WW B4	275	169	116	85	64	49	37	28	19
WW B5	252	159	110	81	61	47	36	27	19
WW C1	333	190	126	89	66	50	38	28	19
WW C2	287	171	115	83	62	47	36	27	18
WW C3	293	191	136	101	77	59	46	34	24

(*continued*)

TABLE 18.8 ***(Continued)***

SSF	0.10	0.20	0.30	0.40	0.50	0.60	0.70	0.80	0.90
WW C4	264	172	122	91	69	54	41	31	22
TW A1	900	124	60	37	25	17	12	9	6
TW A2	361	130	73	48	33	24	18	13	8
TW A3	293	129	77	52	37	27	20	15	10
TW A4	249	123	76	52	38	28	21	15	10
TW B1	502	117	60	38	26	18	13	9	6
TW B2	291	118	68	45	32	23	17	12	8
TW B3	254	114	68	46	33	24	18	13	9
TW B4	233	104	63	42	30	22	16	12	8
TW C1	332	106	58	37	26	19	14	10	6
TW C2	255	101	58	39	27	20	15	11	7
TW C3	243	94	54	36	25	18	13	10	7
TW C4	254	84	46	30	21	15	11	8	5
TW D1	213	86	50	33	23	17	12	9	6
TW D2	287	139	86	59	43	32	24	17	12
TW D3	294	153	97	68	49	37	27	20	14
TW D4	281	158	104	74	55	41	31	23	16
TW D5	260	151	101	73	54	41	31	23	16
TW E1	339	177	113	78	57	43	32	23	16
TW E2	308	168	109	77	56	42	32	23	16
TW E3	323	195	133	96	72	55	42	31	21
TW E4	287	175	120	88	66	50	38	28	20
TW F1	409	108	57	36	24	17	13	9	6
TW F2	260	110	65	43	31	22	17	12	8
TW F3	216	106	66	45	33	24	10	13	9
TW F4	178	95	61	42	31	23	17	13	9
TW G1	256	93	53	34	24	17	13	9	6
TW G2	189	91	56	38	27	20	15	11	7
TW G3	159	82	52	36	26	20	15	11	7
TW G4	128	68	43	30	22	16	12	9	6
TW H1	168	76	45	31	22	16	12	9	6
TW H2	130	66	41	29	21	15	11	8	6
TW H3	108	56	35	25	8	13	10	7	5
TW H4	83	42	27	19	13	10	7	5	4
TW I1	166	73	43	29	20	15	11	8	5
TW I2	221	117	75	52	30	28	21	16	11
TW I3	234	128	83	59	43	32	24	10	12
TW I4	234	137	92	66	49	37	28	21	14
TW I5	226	136	93	67	50	38	29	22	15
TW J1	282	156	102	72	53	40	30	22	15
TW J2	254	146	97	69	51	39	29	22	15
TW J3	269	169	118	86	65	50	38	29	20
TW J4	247	155	106	80	60	46	35	26	18
DG A1	211	97	57	36	22	13	5	—	—
DG A2	227	107	67	46	32	23	16	10	5
DG A3	274	131	83	59	44	34	25	18	10
DG B1	210	97	60	42	30	21	13	6	—
DG B2	232	110	69	49	37	28	21	14	8
DG B3	277	134	85	61	47	37	28	21	14
DG C1	253	120	74	53	39	30	22	14	—
DG C2	271	130	82	59	45	35	26	19	12
DG C3	318	155	96	71	54	43	34	26	18
SS A1	591	187	101	64	44	31	22	16	10
SS A2	531	232	137	92	65	47	34	25	16
SS A3	566	170	90	56	38	27	19	13	8

(continued)

TABLE 18.8 (*Continued*)

SSF	0.10	0.20	0.30	0.40	0.50	0.60	0.70	0.80	0.90
SS A4	537	230	135	89	63	45	33	23	15
SS A5	980	187	92	56	37	26	18	13	8
SS A6	529	231	136	91	64	47	34	24	16
SS A7	1103	158	74	44	29	20	14	10	6
SS A8	540	226	131	87	61	44	32	23	15
SS B1	403	141	78	50	35	25	18	13	8
SS B2	412	186	111	75	53	39	28	20	14
SS B3	372	130	71	46	31	22	16	11	7
SS B4	403	181	106	72	51	37	27	20	13
SS B5	518	127	65	40	27	19	13	9	6
SS B6	390	179	106	73	52	38	28	20	13
SS B7	457	108	54	33	22	16	11	8	5
SS B8	379	171	102	69	49	35	26	19	12
SS C1	270	126	77	52	37	27	20	15	10
SS C2	282	150	97	68	49	37	28	20	14
SS C3	276	101	57	37	26	19	14	10	7
SS C4	277	135	83	57	41	31	23	17	11
SS D1	548	225	130	85	59	43	31	22	14
SS D2	474	253	162	113	82	61	45	33	22
SS D3	683	212	113	72	49	35	25	17	11
SS D4	484	248	156	107	77	57	42	30	20
SS E1	410	176	103	68	48	35	25	18	12
SS E2	390	208	133	92	67	50	37	27	18
SS E3	487	151	80	51	35	25	18	12	8
SS E4	400	195	120	82	59	43	32	23	15
Nashville, Tennessee									3696 DD
WW A1	588	60	24	13	8	5	3	2	1
WW A2	192	70	38	23	15	11	7	5	3
WW A3	161	72	42	27	18	13	9	6	4
WW A4	148	72	43	29	20	14	10	7	5
WW A5	141	74	46	31	22	16	11	8	5
WW A6	137	74	47	32	22	16	12	8	5
WW B1	135	41	19	10	6	3	2	—	—
WW B2	152	78	48	33	23	17	12	9	6
WW B3	179	97	61	42	30	22	16	12	8
WW B4	164	97	65	46	34	25	19	14	9
WW B5	153	93	63	45	33	25	19	14	9
WW C1	193	105	67	46	33	24	18	13	8
WW C2	169	97	63	44	32	24	18	13	8
WW C3	181	115	79	58	43	33	25	18	12
WW C4	164	104	72	53	39	30	23	17	11
TW A1	509	59	25	13	8	5	3	2	1
TW A2	199	64	33	20	13	9	6	4	3
TW A3	160	65	36	23	15	11	8	5	3
TW A4	136	62	36	23	16	11	8	6	4
TW B1	282	57	26	15	9	6	4	3	2
TW B2	161	59	32	20	13	9	6	4	3
TW B3	141	58	32	21	14	10	7	5	3
TW B4	131	54	30	19	13	9	7	5	3
TW C1	188	53	27	16	10	7	5	3	2
TW C2	144	52	28	18	12	8	6	4	2
TW C3	139	49	27	17	11	8	5	4	2
TW C4	149	45	23	14	9	7	5	3	2
TW D1	99	33	16	9	5	3	2	1	—
TW D2	164	75	44	29	20	14	10	7	5

(*continued*)

TABLE 18.8 (*Continued*)

SSF	0.10	0.20	0.30	0.40	0.50	0.60	0.70	0.80	0.90
TW D3	167	82	49	33	23	17	12	8	5
TW D4	168	91	58	40	29	21	15	11	7
TW D5	160	89	58	40	29	22	16	12	8
TW E1	198	98	59	40	28	20	15	10	7
TW E2	182	95	59	40	29	21	15	11	7
TW E3	197	115	76	54	39	29	22	16	11
TW E4	178	105	70	50	37	27	20	15	10
TW F1	221	50	23	13	8	5	4	2	1
TW F2	139	53	29	18	12	8	6	4	2
TW F3	116	52	30	19	13	9	7	5	3
TW F4	96	47	28	19	13	9	7	5	3
TW G1	137	44	22	13	9	6	4	3	2
TW G2	101	44	25	16	11	8	5	4	2
TW G3	86	41	24	16	11	8	6	4	2
TW G4	69	34	21	14	10	7	5	3	2
TW H1	89	36	20	13	8	6	4	3	2
TW H2	69	33	19	12	9	6	4	3	2
TW H3	59	28	17	11	8	5	4	3	2
TW H4	46	22	13	9	6	4	3	2	1
TW I1	74	26	13	7	4	2	1	—	—
TW I2	125	62	38	25	18	13	9	7	4
TW I3	133	69	43	29	20	15	11	8	5
TW I4	139	78	51	35	26	19	14	10	7
TW I5	137	80	53	37	27	20	15	11	7
TW J1	164	86	54	36	26	19	14	10	6
TW J2	150	82	53	36	26	19	14	10	7
TW J3	165	101	68	49	36	27	20	15	10
TW J4	153	93	63	46	34	25	19	14	10
DG A1	98	34	—	—	—	—	—	—	—
DG A2	130	55	31	19	11	6	—	—	—
DG A3	173	78	47	32	23	16	11	7	2
DG B1	100	36	17	—	—	—	—	—	—
DG B2	134	58	33	22	15	10	6	—	—
DG B3	177	81	49	33	24	18	14	10	6
DG C1	131	52	28	17	9	—	—	—	—
DG C2	161	71	42	28	20	14	10	6	—
DG C3	205	94	57	39	29	22	17	12	8
SS A1	351	100	50	29	19	13	9	6	4
SS A2	328	135	76	49	33	24	17	12	8
SS A3	330	87	41	24	15	10	6	4	2
SS A4	331	133	74	47	32	22	16	11	7
SS A5	595	98	43	24	15	10	7	4	2
SS A6	324	132	75	48	32	23	16	11	7
SS A7	668	79	32	17	10	6	4	2	1
SS A8	330	129	71	45	30	21	15	10	6
SS B1	236	74	38	23	15	10	7	5	3
SS B2	258	110	63	41	28	20	14	10	6
SS B3	212	65	32	19	12	8	5	3	2
SS B4	251	105	60	39	27	19	13	9	6
SS B5	307	65	30	17	10	7	4	3	2
SS B6	241	104	60	39	27	19	14	10	6
SS B7	264	52	23	12	7	5	3	2	—
SS B8	233	98	56	36	25	17	12	9	5
SS C1	141	60	33	21	14	10	7	5	3
SS C2	161	81	50	33	23	17	12	9	6
SS C3	149	48	25	15	10	7	4	3	2

(continued)

TABLE 18.8 ***(Continued)***

SSF	0.10	0.20	0.30	0.40	0.50	0.60	0.70	0.80	0.90
SS C4	160	73	43	28	19	14	10	7	5
SS D1	317	119	64	39	26	18	13	8	5
SS D2	287	147	90	61	43	31	23	16	10
SS D3	405	113	55	33	21	14	10	6	4
SS D4	295	144	87	58	40	29	21	15	10
SS E1	229	89	48	29	19	13	9	6	4
SS E2	233	118	72	48	34	24	18	12	8
SS E3	283	77	37	22	14	9	6	4	2
SS E4	242	111	65	43	29	21	15	11	7
Medford, Oregon									4930 DD
WW A1	708	64	24	11	—	—	—	—	—
WW A2	212	73	38	22	13	7	3	—	—
WW A3	174	75	41	25	16	9	5	2	—
WW A4	158	74	43	27	17	11	6	3	1
WW A5	149	75	45	29	19	12	7	4	2
WW A6	144	75	46	30	20	13	8	4	2
WW B1	154	43	16	—	—	—	—	—	—
WW B2	162	80	48	31	21	14	9	6	3
WW B3	190	100	62	41	28	19	13	8	5
WW B4	171	99	65	45	32	23	16	11	7
WW B5	160	95	63	45	32	23	17	12	7
WW C1	205	108	67	45	31	21	15	10	6
WW C2	178	99	63	43	30	22	15	10	6
WW C3	189	117	80	57	42	31	23	16	10
WW C4	170	106	72	52	38	28	21	15	9
TWAI	607	63	25	12	5	—	—	—	—
TW A2	222	68	33	19	11	6	2	—	—
TW A3	175	67	36	21	13	8	4	2	—
TW A4	147	64	36	22	14	9	5	3	1
TW B1	327	61	27	14	7	3	—	—	—
TW B2	178	62	32	19	12	7	4	2	—
TW B3	154	60	33	20	12	8	4	2	1
TW B4	143	56	31	19	12	8	5	2	1
TW C1	212	56	27	15	9	5	2	—	—
TW C2	159	55	28	17	11	7	4	2	—
TW C3	154	52	27	16	10	6	4	2	1
TW C4	167	48	24	14	9	5	3	2	—
TW D1	112	34	14	—	—	—	—	—	—
TW D2	177	77	44	28	18	12	8	5	3
TW D3	180	85	50	32	21	14	9	6	3
TW D4	177	93	58	39	27	19	13	9	5
TW D5	168	92	58	40	28	20	14	10	6
TW E1	213	101	60	39	26	18	12	8	4
TW E2	194	98	59	39	27	19	13	9	5
TW E3	208	118	77	53	38	27	20	13	8
TW E4	186	108	71	49	36	26	19	13	8
TW F1	256	53	23	12	5	—	—	—	—
TW F2	153	56	29	17	10	5	2	—	—
TW F3	125	54	30	18	11	7	3	1	—
TW F4	102	48	28	18	11	7	4	2	1
TW G1	153	46	22	12	7	—	—	—	—
TW G2	109	46	25	15	9	5	3	1	—
TW G3	92	42	24	15	9	6	3	2	—
TW G4	74	35	20	13	8	5	3	2	—
TW H1	97	38	20	12	7	4	1	—	—

(continued)

TABLE 18.8 (*Continued*)

SSF	0.10	0.20	0.30	0.40	0.50	0.60	0.70	0.80	0.90
TW H2	75	34	19	12	7	5	3	1	—
TW H3	63	29	17	10	7	4	3	1	—
TW H4	49	23	13	8	5	3	2	1	—
TW I1	83	27	10	—	—	—	—	—	—
TW I2	133	64	38	24	16	11	7	4	2
TW I3	142	71	43	28	19	13	9	5	3
TW I4	146	80	51	35	25	17	12	8	5
TW I5	144	82	53	37	26	19	13	9	6
TW J1	175	89	54	36	24	17	11	7	4
TW J2	158	85	53	36	25	18	12	8	5
TW J3	173	103	69	48	35	26	18	13	8
TW J4	160	96	64	45	33	24	17	12	8
DG A1	110	35	—	—	—	—	—	—	—
DG A2	142	58	32	18	9	—	—	—	—
DG A3	187	82	48	32	22	15	9	5	—
DG B1	110	40	15	—	—	—	—	—	—
DG B2	146	61	35	21	13	7	—	—	—
DG B3	193	84	51	34	24	17	12	7	3
DG C1	144	57	29	13	—	—	—	—	—
DG C2	177	75	44	28	19	12	6	—	—
DG C3	224	98	60	41	29	21	14	10	5
SS A1	415	110	51	28	16	9	4	2	—
SS A2	372	146	79	48	31	21	14	8	5
SS A3	397	96	42	21	10	—	—	—	—
SS A4	379	144	76	46	29	19	12	7	4
SS A5	732	111	45	23	12	5	—	—	—
SS A6	368	143	77	47	30	20	13	8	4
SS A7	846	90	33	14	—	—	—	—	—
SS A8	379	140	73	44	27	17	11	6	3
SS B1	274	81	38	21	12	6	3	—	—
SS B2	288	117	65	40	26	18	12	7	4
SS B3	249	71	33	17	8	—	—	—	—
SS B4	282	113	62	38	25	16	11	7	4
SS B5	368	72	30	15	7	—	—	—	—
SS B6	269	111	62	30	25	17	11	7	4
SS B7	323	58	23	10	—	—	—	—	—
SS B8	262	106	57	35	23	15	9	6	3
SS C1	153	62	33	19	11	5	—	—	—
SS C2	172	83	50	32	22	15	10	6	3
SS C3	166	51	24	13	7	3	—	—	—
SS C4	173	76	43	27	18	12	8	5	3
SS D1	367	129	65	37.	22	13	7	3	1
SS D2	318	156	92	60	40	27	18	12	7
SS D3	480	124	57	31	18	10	5	2	—
SS D4	328	153	89	57	38	26	17	11	6
SS E1	262	95	48	27	15	7	—	—	—
SS E2	257	124	73	47	31	21	14	9	5
SS E3	334	84	38	20	10	4	—	—	—
SS E4	269	118	67	42	27	18	12	7	4
Boston, Massachusetts									5621 DD
WW A1	368	28	9	—	—	—	—	—	—
WW A2	119	41	20	12	7	5	3	2	—
WW A3	101	43	24	15	10	6	4	3	1
WW A4	93	44	26	16	11	7	5	3	2
WW A5	89	45	27	18	12	8	6	4	2

(*continued*)

TABLE 18.8 *(Continued)*

SSF	0.10	0.20	0.30	0.40	0.50	0.60	0.70	0.80	0.90
WW A6	87	46	28	19	13	9	6	4	3
WW B1	59	—	—	—	—	—	—	—	—
WW B2	103	52	31	21	15	10	7	5	3
WW B3	123	66	41	28	20	14	10	7	5
WW B4	118	70	46	33	24	18	13	9	6
WW B5	113	69	46	33	25	18	14	10	7
WW C1	135	72	46	31	22	16	12	8	5
WW C2	121	68	44	31	22	16	12	9	6
WW C3	136	86	60	44	33	25	19	14	9
WW C4	124	78	54	40	30	23	17	12	8
TW A1	324	30	11	4	—	—	—	—	—
TW A2	126	37	18	10	6	4	2	1	—
TW A3	102	39	21	13	8	5	3	2	1
TW A4	88	38	22	14	9	6	4	3	2
TW B1	180	32	13	7	4	2	—	—	—
TW B2	104	36	19	11	7	5	3	2	1
TW B3	92	36	19	12	8	5	3	2	1
TW B4	86	34	19	12	8	5	4	2	i
TW C1	122	32	15	9	5	3	2	1	—
TW C2	95	33	17	10	7	4	3	2	1
TW C3	93	31	16	10	6	4	3	2	1
TW C4	102	29	15	9	6	4	3	2	1
TW D1	45	—	—	—	—	—	—	—	—
TW D2	112	49	28	18	12	9	6	4	3
TW D3	113	54	32	21	15	10	7	5	3
TW D4	121	64	41	28	20	15	11	8	5
TW D5	118	66	42	30	21	16	12	8	6
TW E1	138	67	40	27	18	13	9	7	4
TW E2	130	66	41	28	20	14	10	7	5
TW E3	146	84	56	39	29	21	16	11	8
TW E4	133	78	52	37	27	20	15	11	7
TW F1	134	25	10	4	—	—	—	—	—
TW F2	86	30	16	9	5	3	2	1	—
TW F3	72	31	17	11	7	4	3	2	1
TW F4	61	29	17	11	7	5	3	2	1
TW G1	83	24	11	6	3	2	—	—	—
TW G2	63	26	14	9	5	4	2	1	—
TW G3	54	25	14	9	6	4	3	2	1
TW G4	45	21	12	8	5	4	3	2	1
TW H1	54	21	11	6	4	2	1	—	—
TW H2	44	20	11	7	5	3	2	1	—
TW H3	38	17	10	6	4	3	2	1	—
TW H4	30	14	8	5	3	2	2	1	—
TW I1	30	—	—	—	—	—	—	—	—
TW I2	84	41	24	16	11	8	6	4	2
TW I3	91	46	28	19	13	9	7	5	3
TW I4	100	56	36	25	18	13	10	7	5
TW I5	101	58	38	27	20	15	11	8	5
TW J1	114	59	37	25	17	12	9	6	4
TW J2	107	58	37	25	18	13	10	7	4
TW J3	123	75	51	36	27	20	15	11	7
TW J4	115	70	47	34	25	19	14	10	7
DG A1	43	—	—	—	—	—	—	—	—
DG A2	85	34	18	9	—	—	—	—	—
DG A3	125	56	33	22	16	11	7	4	—

(continued)

TABLE 18.8 ***(Continued)***

SSF	0.10	0.20	0.30	0.40	0.50	0.60	0.70	0.80	0.90
DG B1	44	—	—	—	—	—	—	—	—
DG B2	87	36	20	12	7	—	—	—	—
DG B3	129	58	35	24	17	13	9	6	3
DG C1	71	23	—	—	—	—	—	—	—
DG C2	109	47	27	17	12	8	4	—	—
DG C3	151	68	41	28	21	16	12	8	5
SS A1	230	61	29	16	10	6	4	2	1
SS A2	231	93	52	33	22	15	11	7	5
SS A3	205	48	20	10	4	—	—	—	—
SS A4	229	90	49	31	20	14	9	6	4
SS A5	389	58	23	11	6	3	—	—	—
SS A6	226	91	50	32	21	15	10	7	4
SS A7	420	40	12	—	—	—	—	—	—
SS A8	226	86	46	28	19	12	8	6	3
SS B1	151	44	21	12	7	4	2	1	—
SS B2	183	77	43	28	19	13	9	6	4
SS B3	129	36	16	8	3	—	—	—	—
SS B4	176	73	41	26	17	12	8	6	4
SS B5	193	36	15	7	3	—	—	—	—
SS B6	169	72	41	26	18	12	9	6	4
SS B7	157	25	7	—	—	—	—	—	—
SS B8	160	66	37	23	16	11	7	5	3
SS C1	84	33	17	10	6	4	2	1	—
SS C2	110	54	33	22	15	11	8	5	3
SS C3	91	26	12	7	4	2	—	—	—
SS C4	109	48	28	18	12	9	6	4	3
SS D1	206	73	38	22	14	9	5	3	2
SS D2	203	103	63	42	29	21	15	10	6
SS D3	264	69	32	18	10	6	4	2	1
SS D4	208	100	60	39	27	19	14	9	6
SS E1	140	51	25	14	8	4	2	—	—
SS E2	161	80	48	32	22	15	11	7	5
SS E3	177	44	19	10	5	2	—	—	—
SS E4	166	75	43	28	19	13	9	6	4
Madison, Wisconsin									7730 DD
WW A1	278	—	—	—	—	—	—	—	—
WW A2	91	27	12	—	—	—	—	—	—
WW A3	77	30	15	8	3	—	—	—	—
WW A4	72	32	17	10	5	—	—	—	—
WW A5	69	33	19	11	7	4	—	—	—
WW A6	67	34	19	12	7	4	2	—	—
WW B1	—	—	—	—	—	—	—	—	—
WW B2	84	41	24	15	10	7	5	3	2
WW B3	102	53	32	21	15	10	7	5	3
WW B4	101	59	39	27	19	14	10	7	5
WW B5	98	59	39	28	20	15	11	8	5
WW C1	113	59	37	25	17	12	8	6	3
WW C2	103	57	37	25	18	13	9	6	4
WW C3	119	75	51	37	28	21	15	11	7
WW C4	109	68	47	34	25	19	14	10	7
TW A1	249	16	—	—	—	—	—	—	—
TW A2	97	26	11	4	—	—	—	—	—
TW A3	79	28	13	7	3	—	—	—	—
TW A4	69	28	15	9	5	3	—	—	—

(continued)

TABLE 18.8 ***(Continued)***

SSF	0.10	0.20	0.30	0.40	0.50	0.60	0.70	0.80	0.90
TW B1	139	20	5	—	—	—	—	—	—
TW B2	81	26	12	6	3	—	—	—	—
TW B3	72	27	13	7	4	2	—	—	—
TW B4	69	26	13	8	5	3	1	—	—
TW C1	96	23	10	4	—	—	—	—	—
TW C2	76	25	12	7	4	2	—	—	—
TW C3	75	24	12	7	4	2	1	—	—
TW C4	84	23	11	6	4	2	1	—	—
TW D1	—	—	—	—	—	—	—	—	—
TW D2	91	39	22	13	9	6	4	2	1
TW D3	93	43	25	16	10	7	5	3	1
TW D4	103	54	34	23	16	12	8	6	4
TW D5	102	56	36	25	18	13	10	7	4
TW E1	115	54	32	21	14	10	7	4	3
TW E2	110	55	34	22	16	11	8	5	3
TW E3	126	72	47	33	24	18	13	9	6
TW E4	116	68	45	32	23	17	13	9	6
TW F1	99	13	—	—	—	—	—	—	—
TW F2	65	20	8	—	—	—	—	—	—
TW F3	55	22	11	5	—	—	—	—	—
TW F4	47	21	11	7	4	2	—	—	—
TW G1	61	14	—	—	—	—	—	—	—
TW G2	47	18	8	4	—	—	—	—	—
TW G3	42	18	9	5	3	—	—	—	—
TW G4	35	16	9	5	3	2	—	—	—
TW H1	41	13	6	—	—	—	—	—	—
TW H2	34	14	7	4	2	—	—	—	—
TW H3	29	13	7	4	2	1	—	—	—
TW H4	24	10	6	3	2	1	—	—	—
TW I1	—	—	—	—	—	—	—	—	—
TW I2	68	32	18	12	8	5	3	2	1
TW I3	75	37	22	14	10	7	4	3	2
TW I4	85	47	30	21	15	11	8	5	3
TW I5	87	50	33	23	16	12	9	6	4
TW J1	95	48	29	19	13	9	6	4	3
TW J2	91	48	30	21	14	10	7	5	3
TW J3	106	65	43	31	23	17	12	9	6
TW J4	100	61	41	29	21	16	12	9	6
DG A1	—	—	—	—	—	—	—	—	—
DG A2	68	25	11	—	—	—	—	—	—
DG A3	109	47	28	18	12	8	5	—	—
DG B1	—	—	—	—	—	—	—	—	—
DG B2	70	27	14	6	—	—	—	—	—
DG B3	114	50	30	20	14	10	7	4	—
DG C1	47	—	—	—	—	—	—	—	—
DG C2	91	37	21	13	7	—	—	—	—
DG C3	133	59	35	24	17	13	9	6	3
SS A1	192	47	20	9	3	—	—	—	—
SS A2	200	78	42	26	17	12	8	5	3
SS A3	166	32	—	—	—	—	—	—	—
SS A4	197	74	39	23	15	10	6	4	2
SS A5	329	42	13	—	—	—	—	—	—
SS A6	195	75	40	25	16	11	7	5	3
SS A7	349	22	—	—	—	—	—	—	—
SS A8	192	69	36	21	13	8	5	3	2

(continued)

TABLE 18.8 (*Continued*)

SSF	0.10	0.20	0.30	0.40	0.50	0.60	0.70	0.80	0.90
SS B1	122	32	13	5	—	—	—	—	—
SS B2	158	64	36	22	15	10	7	5	3
SS B3	100	22	—	—	—	—	—	—	—
SS B4	150	60	33	29	13	9	6	4	2
SS B5	156	24	—	—	—	—	—	—	—
SS B6	145	59	33	20	13	9	6	4	2
SS B7	122	—	—	—	—	—	—	—	—
SS B8	136	54	29	18	11	7	5	3	2
SS C1	61	20	7	—	—	—	—	—	—
SS C2	90	43	25	16	11	7	5	3	2
SS C3	67	16	—	—	—	—	—	—	—
SS C4	90	38	22	13	9	6	4	2	1
SS D1	169	56	26	13	6	—	—	—	—
SS D2	175	86	51	34	23	16	11	7	5
SS D3	221	52	21	10	—	—	—	—	—
SS D4	179	84	49	32	21	15	10	7	4
SS E1	108	34	12	—	—	—	—	—	—
SS E2	135	65	38	24	16	11	7	5	3
SS E3	141	29	8	—	—	—	—	—	—
SS E4	140	61	34	21	14	9	6	4	2

Source: From PSDH, *Passive Solar Design Handbook*, Los Alamos National Laboratory, Van Nostrand Reinhold, New York, 1984.

The LCR method allows no variation from the 94 reference passive designs. To treat off-reference designs, sensitivity curves have been produced that illustrate the effect on SSF of varying one or two design variables. These curves were produced for the six "representative" cities, chosen for their wide geographical and climatological ranges. Several of these SSF "sensitivity curves" are presented in Figure 18.49 for storage wall (a, b, c) and sunspace (d) design variations.

Example 18.3.4

The previously used 2100 ft.^{2} building with NLC = 11,800 Btu/FDD is preliminarily designed to be located in Medford, Oregon, with 180 ft.^{2} of 12-in thick vented Trombe wall and 130 ft.^{2} of direct gain, both systems with double glazing, nighttime insulation, and 30 Btu/ft.^{2} thermal storage capacity. Determine the annual auxiliary energy needed by this design.

Solution. Step 1 yields:

$$\text{NLC} = 11{,}800 \text{ Btu/FDD}.$$

$$A_p = 180 + 130 = 320 \text{ ft.}^2$$

$$\text{LCR} = 11{,}800/320 = 36.8 \text{ Btu/FDD ft.}^2$$

Step 2 yields: From Table 18.7 the short designations for the appropriate systems are

TWD4 (Trombe wall)

DGA3 (Direct gain)

Step 3 yields: From Table 18.8 for Medford, Oregon, with LCR = 36.8,

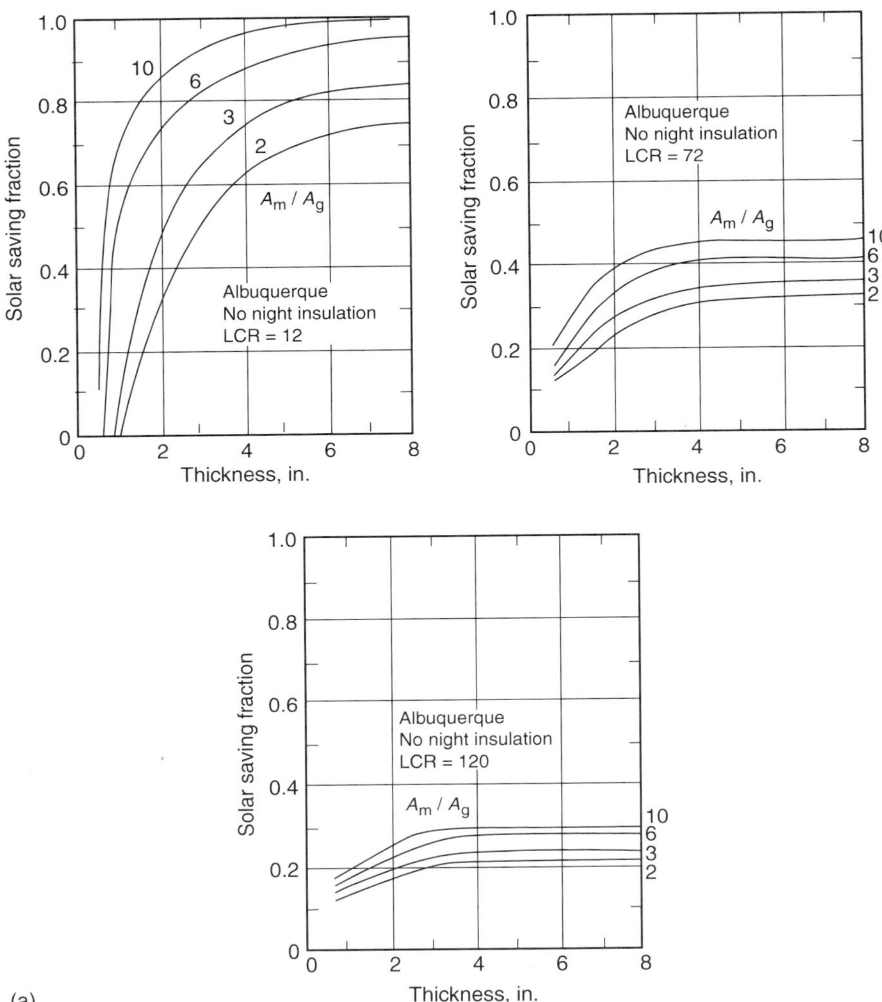

FIGURE 18.49 (a) Storage wall: Mass thickness. Sensitivity of SSF to Off-Reference conditions. (b) Storage wall: ρck product. (c) Storage wall: Number of glazings. (d) Sunspace : Storage volume to projected area ratio. (From PSDH. *Passive Solar Design Handbook*. Volume One: Passive Solar Design Concepts, DOE/CS-0127/1, March 1980. Prepared by Total Environmental Action, Inc. (B. Anderson, C. Michal, P. Temple, and D. Lewis); Volume Two: *Passive Solar Design Analysis*, DOE/CS-0127/2, January 1980. Prepared by Los Alamos Scientific Laboratory (J. D. Balcomb, D. Barley, R McFarland, J. Perry, W. Wray and S. Noll). U.S. Department of Energy, Washington, DC, 1980.)

$$\text{TWD4} : \text{SSF(TW)} = 0.42$$

$$\text{DGA3} : \text{SSF(DG)} = 0.37$$

Determine the "weighted area" average SSF:

$$\text{SSF} = \frac{180(0.42) + 130(0.37)}{320} = 0.39.$$

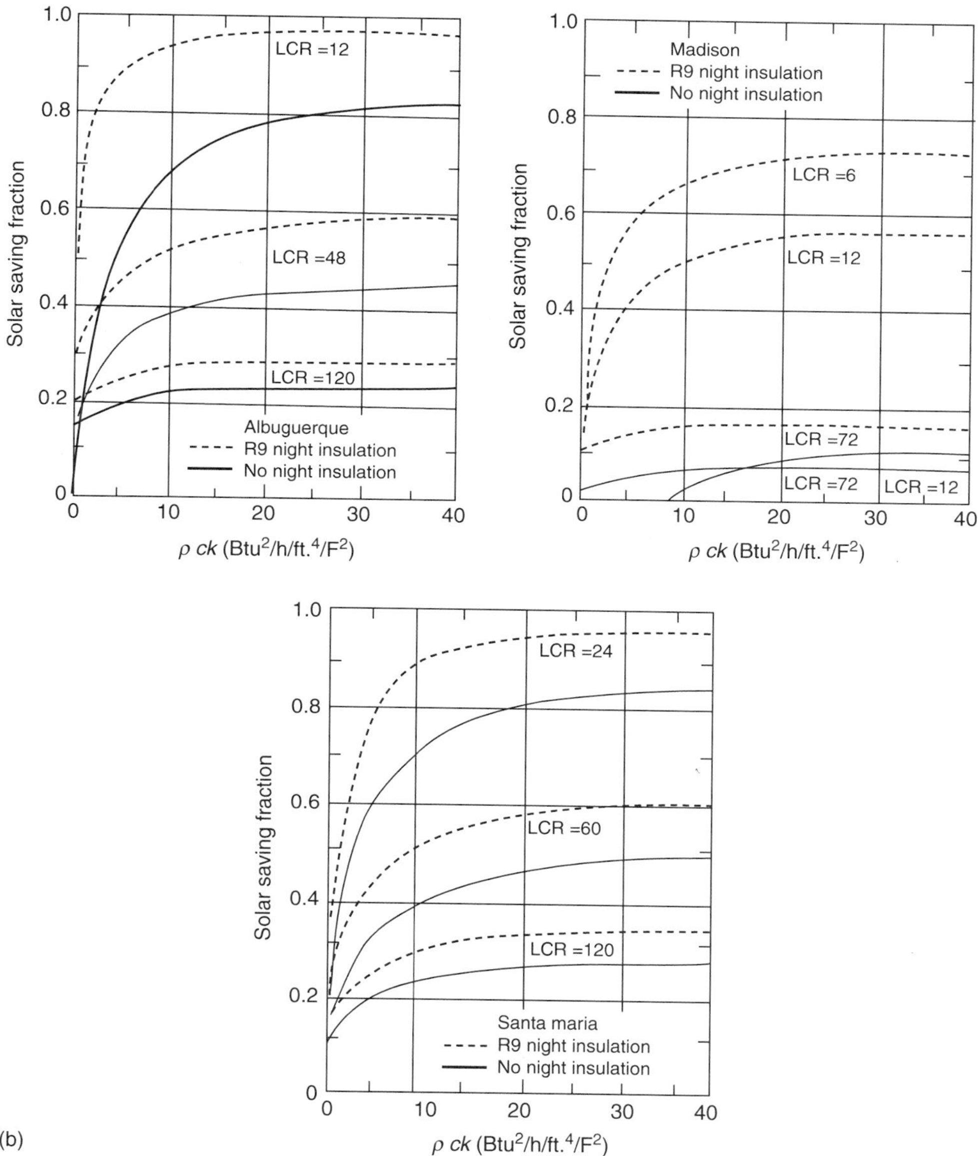

FIGURE 18.49 *(continued)*

Step 4 yields: Using Equation 18.73 and reading 4930 FDD from Table 18.8,

$$Q_{\text{aux}} = (1 - 0.39) \times 11,800 \text{ Btu} \times 4,930 \text{ FDD} = 35.5 \times 106 \text{ Btu annually.}$$

Using the reference system characteristics yields the thermal storage size: Trombe wall ($\rho ck = 30$, concrete properties from Table 18.7c):

$$\begin{aligned} m\,(\text{TW}) &= \text{density} \times \text{area} \times \text{thickness} \\ &= 150 \text{ lbm/ft.}^3 \times 180 \text{ ft.}^2 \times 1 \text{ ft.} \\ &= 27{,}000 \text{ lbm.} \end{aligned}$$

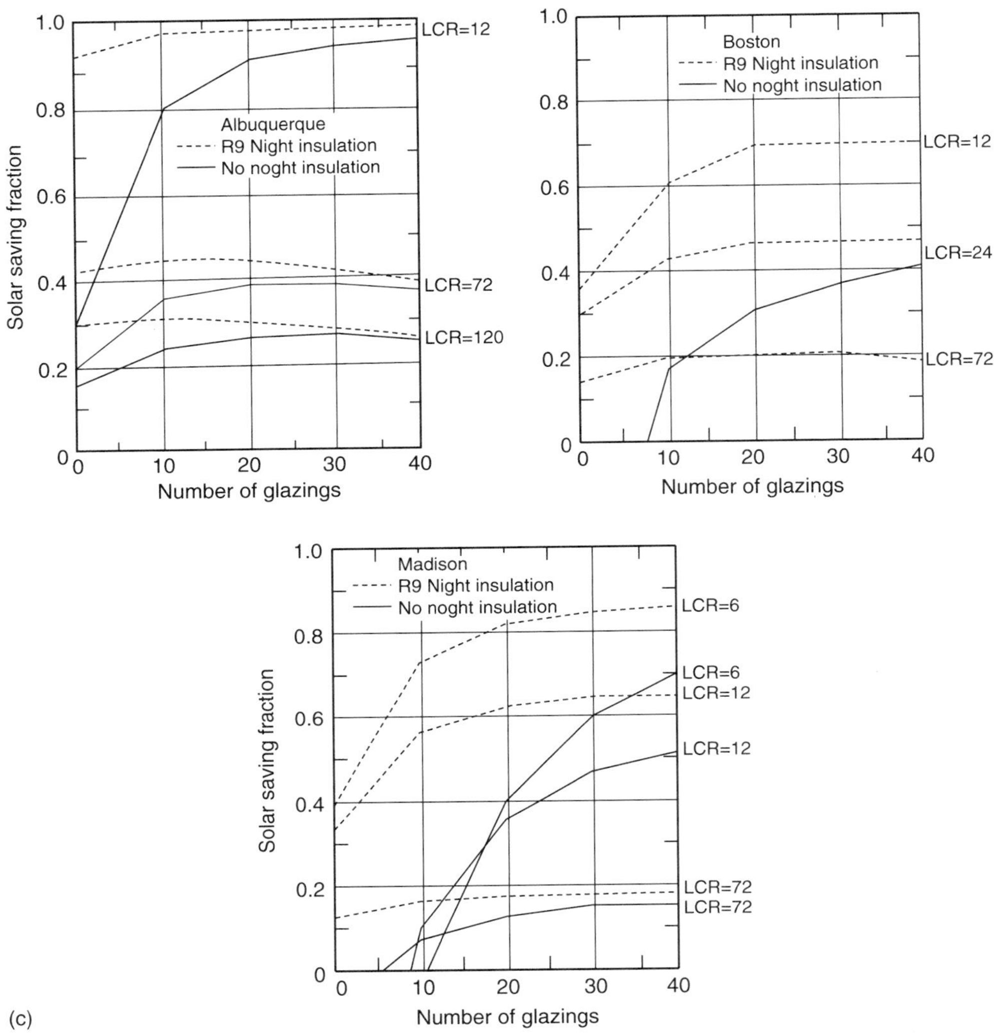

FIGURE 18.49 *(continued)*

Direct Gain (ρck = 30, concrete properties), using mass area to glazing area ratio of 6:

$$\text{mass area} = 6 \times 130 = 780 \text{ ft.}^2 \text{ of } 2'' \text{ thick concrete}$$

$$m(\text{DG}) = 150 \text{ lbm/ft.}^3 \times 780 \text{ ft.}^2 \times 1/6 \text{ ft.}$$
$$= 19,500 \text{ lbm}$$

Using the LCR method allows a basic design of passive system types for the 94 reference systems, and the resulting annual performance. A bit more design variation can be obtained by using the sensitivity curves of Figure 18.49 to modify the SSF of a particular reference system. For instance, a direct gain system SSF of 0.37 would increase by approximately 0.03 if the mass-glazing-area ratio (assumed 6) were increased to 10, and would decrease by about 0.04 if the mass-glazing-area ratio were decreased to 3. This information provides a designer with quantitative information for making trade-offs.

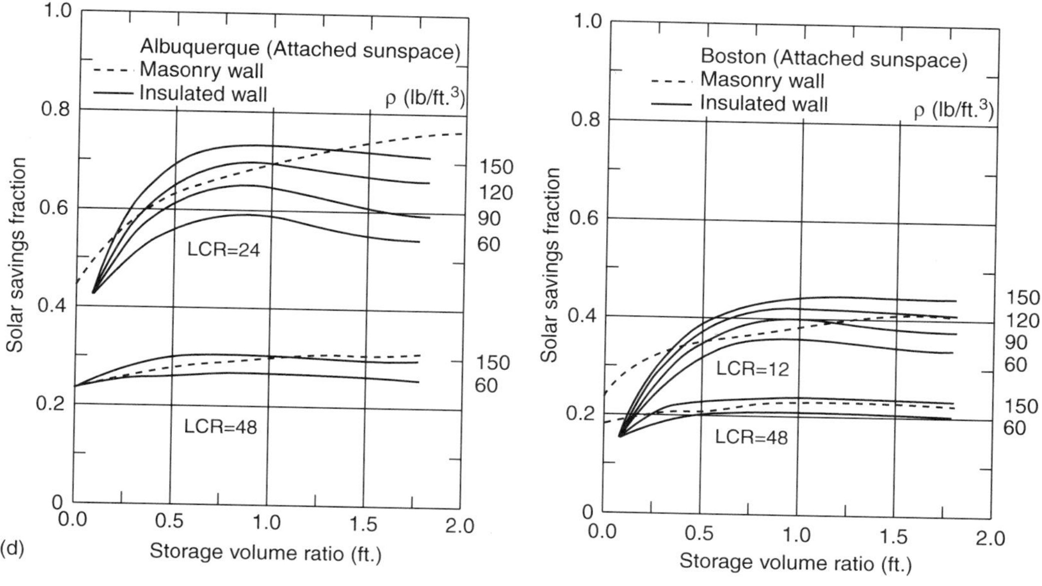

FIGURE 18.49 *(continued)*

18.3.3.10 The Third Level: SLR Method

The solar-load ratio (SLR) method calculates monthly performance, and the terms and values used are monthly based. The method allows the use of specific location weather data and the 94 reference design passive systems (Table 18.7). In addition, the sensitivity curves (Figure 18.49) can again be used to define performance outside the reference design systems. The result of the SLR method is the determination of the monthly heating auxiliary energy required that is then summed to give the annual requirement for auxiliary heating energy. Generally, the SLR method gives annual values within $\pm$ 3% of detailed simulation results, but the monthly values may vary more (PSDH 1984; Duffie and Beckman 1991). Thus, the monthly SLR method is more "accurate" than the rule-of-thumb methods, plus providing the designer with system performance on a month-by-month basis.

The SLR method uses equations and correlation parameters for each of the 94 reference systems combined with the insolation absorbed by the system, the monthly degree days, and the system's LCR to determine the monthly SSF. These correlation parameters are listed in Table 18.9 as *A*, *B*, *C*, *D*, *R*, *G*, *H*, and LCRs for each reference system (PSDH 1984). The correlation equations are

$$\text{SSF} = 1 - K(1 - F), \tag{18.76}$$

where

$$K = 1 + G/\text{LCR}, \tag{18.77}$$

$$F = \begin{cases} AX, \text{ when } X < R \\ B - C\exp(-DX), \text{ when } X > R \end{cases} \tag{18.78}$$

$$X = \frac{S/DD - (LCRs)\,H}{(LCR)K}. \tag{18.79}$$

and X is called the *generalized solar load ratio*. The term S is the monthly insolation absorbed by the system per unit of solar projected area. Monthly average daily insolation data on a vertical south facing

TABLE 18.9 SLR Correlation Parameters for the 94 Reference Systems

Type	A	B	C	D	R	G	H	LCRs	STDV
WW A1	0.0000	1.0000	0.9172	0.4841	−9.0000	0.00	1.17	13.0	0.053
WW A2	0.0000	1.0000	0.9833	0.7603	−9.0000	0.00	0.92	13.0	0.046
WW A3	0.0000	1.0000	1.0171	0.8852	−9.0000	0.00	0.85	13.0	0.040
WW A4	0.0000	1.0000	1.0395	0.9569	−9.0000	0.00	0.81	13.0	0.037
WW A5	0.0000	1.0000	1.0604	1.0387	−9.0000	0.00	0.78	13.0	0.034
WW A6	0.0000	1.0000	1.0735	1.0827	−9.0000	0.00	0.76	13.0	0.033
WW B1	0.0000	1.0000	0.9754	0.5518	−9.0000	0.00	0.92	22.0	0.051
WW B2	0.0000	1.0000	1.0487	1.0851	−9.0000	0.00	0.78	9.2	0.036
WW B3	0.0000	1.0000	1.0673	1.0087	−9.0000	0.00	0.95	8.9	0.038
WW B4	0.0000	1.0000	1.1028	1.1811	−9.0000	0.00	0.74	5.8	0.034
WW B5	0.0000	1.0000	1.1146	1.2771	−9.0000	0.00	0.56	4.5	0.032
WW C1	0.0000	1.0000	1.0667	1.0437	−9.0000	0.00	0.62	12.0	0.038
WW C2	0.0000	1.0000	1.0846	1.1482	−9.0000	0.00	0.59	8.7	0.035
WW C3	0.0000	1.0000	1.1419	1.1756	−9.0000	0.00	0.28	5.5	0.033
WW C4	0.0000	1.0000	1.1401	1.2378	−9.0000	0.00	0.23	4.3	0.032
TW A1	0.0000	1.0000	0.9194	0.4601	−9.0000	0.00	1.11	13.0	0.048
TW A2	0.0000	1.0000	0.9680	0.6318	−9.0000	0.00	0.92	13.0	0.043
TW A3	0.0000	1.0000	0.9964	0.7123	−9.0000	0.00	0.85	13.0	0.038
TW A4	0.0000	1.0000	1.0190	0.7332	−9.0000	0.00	0.79	13.0	0.032
TW B1	0.0000	1.0000	0.9364	0.4777	−9.0000	0.00	1.01	13.0	0.045
TW B2	0.0000	1.0000	0.9821	0.6020	−9.0000	0.00	0.85	13.0	0.038
TW B3	0.0000	1.0000	0.9980	0.6191	−9.0000	0.00	0.80	13.0	0.033
TW B4	0.0000	1.0000	0.9981	0.5615	−9.0000	0.00	0.76	13.0	0.028
TW C1	0.0000	1.0000	0.9558	0.4709	−9.0000	0.00	0.89	13.0	0.039
TW C2	0.0000	1.0000	0.9788	0.4964	−9.0000	0.00	0.79	13.0	0.033
TW C3	0.0000	1.0000	0.9760	0.4519	−9.0000	0.00	0.76	13.0	0.029
TW C4	0.0000	1.0000	0.9588	0.3612	−9.0000	0.00	0.73	13.0	0.026
TW D1	0.0000	1.0000	0.9842	0.4418	−9.0000	0.00	0.89	22.0	0.040
TW D2	0.0000	1.0000	1.0150	0.8994	−9.0000	0.00	0.80	9.2	0.036
TW D3	0.0000	1.0000	1.0346	0.7810	−9.0000	0.00	1.08	8.9	0.036
TW D4	0.0000	1.0000	1.0606	0.9770	−9.0000	0.00	0.85	5.8	0.035
TW D5	0.0000	1.0000	1.0721	1.0718	−9.0000	0.00	0.61	4.5	0.033
TW E1	0.0000	1.0000	1.0345	0.8753	−9.0000	0.00	0.68	12.0	0.037
TW E2	0.0000	1.0000	1.0476	1.0050	−9.0000	0.00	0.66	8.7	0.035
TW E3	0.0000	1.0000	1.0919	1.0739	−9.0000	0.00	0.61	5.5	0.034
TW E4	0.0000	1.0000	1.0971	1.1429	−9.0000	0.00	0.47	4.3	0.033
TW F1	0.0000	1.0000	0.9430	0.4744	−9.0000	0.00	1.09	13.0	0.047
TW F2	0.0000	1.0000	0.9900	0.6053	−9.0000	0.00	0.93	13.0	0.041
TW F3	0.0000	1.0000	1.0189	0.6502	−9.0000	0.00	0.86	13.0	0.036
TW F4	0.0000	1.0000	1.0419	0.6258	−9.0000	0.00	0.80	13.0	0.032
TW G1	0.0000	1.0000	0.9693	0.4714	−9.0000	0.00	1.01	13.0	0.042
TW G2	0.0000	1.0000	1.0133	0.5462	−9.0000	0.00	0.88	13.0	0.035
TW G3	0.0000	1.0000	1.0325	0.5269	−9.0000	0.00	0.82	13.0	0.031
TW G4	0.0000	1.0000	1.0401	0.4400	−9.0000	0.00	0.77	13.0	0.030
TW H1	0.0000	1.0000	1.0002	0.4356	−9.0000	0.00	0.93	13.0	0.034
TW H2	0.0000	1.0000	1.0280	0.4151	−9.0000	0.00	0.83	13.0	0.030
TW H3	0.0000	1.0000	1.0327	0.3522	−9.0000	0.00	0.78	13.0	0.029
TW H4	0.0000	1.0000	1.0287	0.2600	−9.0000	0.00	0.74	13.0	0.024
TW I1	0.0000	1.0000	0.9974	0.4036	−9.0000	0.00	0.91	22.0	0.038
TW I2	0.0000	1.0000	1.0386	0.8313	−9.0000	0.00	0.80	9.2	0.034
TW I3	0.0000	1.0000	1.0514	0.6886	−9.0000	0.00	1.01	8.9	0.034
TW I4	0.0000	1.0000	1.0781	0.8952	−9.0000	0.00	0.82	5.8	0.032
TW I5	0.0000	1.0000	1.0902	1.0284	−9.0000	0.00	0.65	4.5	0.032
TW J1	0.0000	1.0000	1.0537	0.8227	−9.0000	0.00	0.65	12.0	0.037
TW J2	0.0000	1.0000	1.0677	0.9312	−9.0000	0.00	0.62	8.7	0.035

(continued)

TABLE 18.9 (*Continued*)

Type	A	B	C	D	R	G	H	LCRs	STDV
TW J3	0.0000	1.0000	1.1153	0.9831	−9.0000	0.00	0.44	5.5	0.034
TW J4	0.0000	1.0000	1.1154	1.0607	−9.0000	0.00	0.38	4.3	0.033
DG A1	0.5650	1.0090	1.0440	0.7175	0.3931	9.36	0.00	0.0	0.046
DG A2	0.5906	1.0060	1.0650	0.8099	0.4681	5.28	0.00	0.0	0.039
DG A3	0.5442	0.9715	1.1300	0.9273	0.7068	2.64	0.00	0.0	0.036
DG B1	0.5739	0.9948	1.2510	1.0610	0.7905	9.60	0.00	0.0	0.042
DG B2	0.6180	1.0000	1.2760	1.1560	0.7528	5.52	0.00	0.0	0.035
DG B3	0.5601	0.9839	1.3520	1.1510	0.8879	2.38	0.00	0.0	0.032
DG C1	0.6344	0.9887	1.5270	1.4380	0.8632	9.60	0.00	0.0	0.039
DG C2	0.6763	0.9994	1.4000	1.3940	0.7604	5.28	0.00	0.0	0.033
DG C3	0.6182	0.9859	1.5660	1.4370	0.8990	2.40	0.00	0.0	0.031
SS A1	0.0000	1.0000	0.9587	0.4770	−9.0000	0.00	0.83	18.6	0.027
SS A2	0.0000	1.0000	0.9982	0.6614	−9.0000	0.00	0.77	10.4	0.026
SS A3	0.0000	1.0000	0.9552	0.4230	−9.0000	0.00	0.83	23.6	0.030
SS A4	0.0000	1.0000	0.9956	0.6277	−9.0000	0.00	0.80	12.4	0.026
SS A5	0.0000	1.0000	0.9300	0.4041	−9.0000	0.00	0.96	18.6	0.031
SS A6	0.0000	1.0000	0.9981	0.6660	−9.0000	0.00	0.86	10.4	0.028
SS A7	0.0000	1.0000	0.9219	0.3225	−9.0000	0.00	0.96	23.6	0.035
SS A8	0.0000	1.0000	0.9922	0.6173	−9.0000	0.00	0.90	12.4	0.028
SS B1	0.0000	1.0000	0.9683	0.4954	−9.0000	0.00	0.84	16.3	0.028
SS B2	0.0000	1.0000	1.0029	0.6802	−9.0000	0.00	0.74	8.5	0.026
SS B3	0.0000	1.0000	0.9689	0.4685	−9.0000	0.00	0.82	19.3	0.029
SS B4	0.0000	1.0000	1.0029	0.6641	−9.0000	0.00	0.76	9.7	0.026
SS B5	0.0000	1.0000	0.9408	0.3866	−9.0000	0.00	0.97	16.3	0.030
SS B6	0.0000	1.0000	1.0068	0.6778	−9.0000	0.00	0.84	8.5	0.028
SS B7	0.0000	1.0000	0.9395	0.3363	−9.0000	0.00	0.95	19.3	0.032
SS B8	0.0000	1.0000	1.0047	0.6469	−9.0000	0.00	0.87	9.7	0.027
SS C1	0.0000	1.0000	1.0087	0.7683	−9.0000	0.00	0.76	16.3	0.025
SS C2	0.0000	1.0000	1.0412	0.9281	−9.0000	0.00	0.78	10.0	0.027
SS C3	0.0000	1.0000	0.9699	0.5106	−9.0000	0.00	0.79	16.3	0.024
SS C4	0.0000	1.0000	1.0152	0.7523	−9.0000	0.00	0.81	10.0	0.025
SS D1	0.0000	1.0000	0.9889	0.6643	−9.0000	0.00	0.84	17.8	0.028
SS D2	0.0000	1.0000	1.0493	0.8753	−9.0000	0.00	0.70	9.9	0.028
SS D3	0.0000	1.0000	0.9570	0.5285	−9.0000	0.00	0.90	17.8	0.029
SS D4	0.0000	1.0000	1.0356	0.8142	−9.0000	0.00	0.73	9.9	0.028
SS E1	0.0000	1.0000	0.9968	0.7004	−9.0000	0.00	0.77	19.6	0.027
SS E2	0.0000	1.0000	1.0468	0.9054	−9.0000	0.00	0.76	10.8	0.027
SS E3	0.0000	1.0000	0.9565	0.4827	−9.0000	0.00	0.81	19.6	0.028
SS E4	0.0000	1.0000	1.0214	0.7694	−9.0000	0.00	0.79	10.8	0.027

Source: From PSDH, *Passive Solar Design Handbook*, Los Alamos National Laboratory, Van Nostrand Reinhold, New York, 1984.

surface can be found and/or calculated using various sources (PSDH 1984; McQuiston and Parker 1994) and the S term can be determined by multiplying by a transmission and an absorption factor and the number of days in the month. Absorption factors for all systems are close to 0.96 (PSDH 1984), whereas the transmission is approximately 0.9 for single glazing, 0.8 for double glazing, and 0.7 for triple glazing.

Example 18.3.5

For a vented, 180 ft.^{2}, double-glazed with night insulation, 12-in thick Trombe wall system (TWD4) in a NLC=11,800 Btu/FDD house in Medford, Oregon, determine the auxiliary energy required in January.

Solution. Weather data for Medford, Oregon (PSDH 1984) yields for January (N=31, days): daily vertical surface insolation=565 Btu/ft.^{2} and 880 FDD, so $S=(31)(565)(0.8)(0.96)=$ 13,452 Btu/ft.^{2} month.

$$\text{LCR} = \text{NLC}/A_p = 11,800/180 = 65.6 \text{ Btu/FDD ft.}^2.$$

From Table 18.9 at TWD4: $A=0$, $B=1$, $C=1.0606$, $D=0.977$, $R=-9$, $G=0$, $H=0.85$, LCRs= 5.8 Btu/FDD ft.^{2}.

Substituting into Equation 18.77 gives

$$K = 1 + 0/65.6 = 1.$$

Equation 18.79 gives

$$X = \frac{(13,452/880) - (5.8 \times 0.85)}{65.6 \times 1} = 0.16.$$

Equation 18.78 gives

$$F = 1 - 1.0606\ e^{-0.977 \times 0.16} = 0.09,$$

and Equation 18.76 gives

$$\text{SSF} = 1 - 1(1 - 0.09) = 0.09.$$

The January auxiliary energy required can be calculated using Equation 18.73:

$$\begin{aligned} Q_{aux}(\text{Jan}) &= (1 - \text{SSF}) \times \text{NLC} \times (\text{Number of degree days}) \\ &= (1 - 0.09) \times 11,800 \times 880 \\ &= 9,450,000 \text{ Btu}. \end{aligned}$$

As mentioned, the use of sensitivity curves (PSDH 1984) as in Figure 18.49 will allow SSF to be determined for many off-reference system design conditions involving storage mass, number of glazings, and other more esoteric parameters. Also, the use of multiple passive system types within one building would be approached by calculating the SSF for each type system individually using a "combined area" LCR, and then a weighted-area (aperture) average SSF would be determined for the building.

18.3.4 Passive Space Cooling Design Fundamentals

Passive cooling systems are designed to use natural means to transfer heat from buildings, including convection/ventilation, evaporation, radiation, and conduction. However, the most important element in both passive and conventional cooling design is to prevent heat from entering the building in the first place. Cooling conservation techniques involve building surface colors, insulation, special window glazings, overhangs and orientation, and numerous other architectural/engineering features.

18.3.4.1 Solar Control

Controlling the solar energy input to reduce the cooling load is usually considered a passive (versus conservation) design concern because solar input may be needed for other purposes, such as daylighting throughout the year and/or heating during the winter. Basic architectural solar control is normally "designed in" via the shading of the solar windows, where direct radiation is desired for winter heating and needs to be excluded during the cooling season.

The shading control of the windows can be of various types and "controllability," ranging from drapes and blinds, use of deciduous trees, to the commonly used overhangs and vertical louvers. A rule of thumb design for determining proper south-facing window overhang for both winter heating and summer

TABLE 18.10 South-Facing Window Overhang Rule of Thumb

$$\text{Length of the Overhang} = \frac{\text{Window Height}}{F}$$

(a) Overhang Factors		(b) Roof Overhang Geometry
North Latitude	F^a	
28	5.6–11.1	
32	4.0–6.3	
36	3.0–4.5	
40	2.5–3.4	
44	2.0–2.7	
48	1.7–2.2	
52	1.5–1.8	
56	1.3–1.5	Properly sized overhangs shade out hot summer sun but allow winter sun (which is lower in the sky) to penetrate windows

[a] Select a factor according to your latitude. Higher values provide complete shading at noon on June 21; lower values, until August 1.

Source: From Halacy, 1984.

shading is presented in Table 18.10. Technical details on calculating shading from various devices and orientations are found in Olgyay and Olgyay (1976) and ASHRAE (1989, 1993, 1997).

18.3.4.2 Natural Convection/Ventilation

Air movement provides cooling comfort through convection and evaporation from human skin. ASHRAE (1989) places the comfort limit at 79°F for an air velocity of 50 ft./min (fpm), 82°F for 160 fpm, and 85°F for 200 fpm. To determine whether or not comfort conditions can be obtained, a designer must calculate the volumetric flow rate, Q, which is passing through the occupied space. Using the cross-sectional area, A_x, of the space and the room air velocity, V_a, required, the flow is determined by

$$Q = A_x V_a. \tag{18.80}$$

The proper placement of windows, "narrow" building shape, and open landscaping can enhance natural wind flow to provide ventilation. The air flow rate through open windows for wind-driven ventilation is given by ASHRAE (1989, 1993, 1997):

$$Q = C_v V_w A_w, \tag{18.81}$$

where Q is air flow rate (m^3/s), A_w is free area of inlet opening (m^2), V_w is wind velocity (m/s), and C_v is effectiveness of opening that is equal to 0.5–0.6 for wind perpendicular to opening, and 0.25–0.35 for wind diagonal to opening.

The stack effect can induce ventilation when warm air rises to the top of a structure and exhausts outside, while cooler outside air enters the structure to replace it. Figure 18.50 illustrates the solar

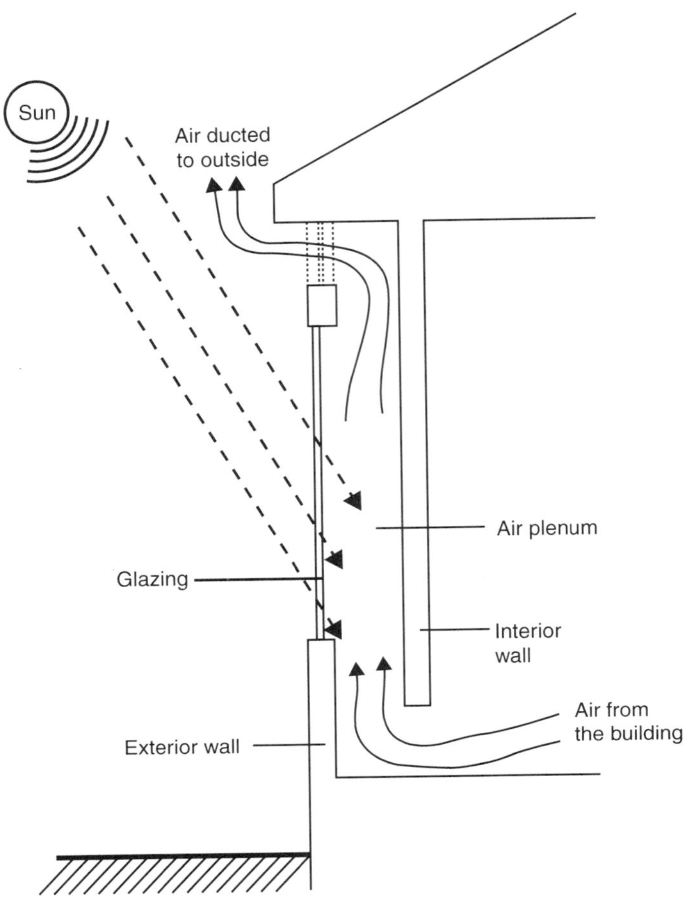

FIGURE 18.50 The stack-effect/solar chimney concept to induce convection/ventilation. (From PSDH. *Passive Solar Design Handbook*. Volume One: Passive Solar Design Concepts, DOE/CS-0127/1, March 1980. Prepared by Total Environmental Action, Inc. (B. Anderson, C. Michal, P. Temple, and Lewis); Volume Two: *Passive* Solar Design Analysis, DOE/CS-0127/2, January 1980. Prepared by Los Alamos Scientific Laboratory (J. D. Balcomb, D. Barley, R McFarland, J. Perry, W. Wray and S. Noll). U.S. Department of Energy, Washington, DC, 1980.)

chimney concept, which can easily be adapted to a thermal storage wall system. The greatest stack-effect flow rate is produced by maximizing the stack height and the air temperature in the stack, as given by

$$Q = 0.116\, A_{\mathrm{j}} \sqrt{h(T_{\mathrm{s}} - T_{\mathrm{o}})} \tag{18.82}$$

where Q is stack flow rate (m^3/s), A_j is the area of inlets or outlets, whichever is smaller (m^2), h is the inlet to outlet height (m), T_s is the average temperature in stack (°C), and T_o is the outdoor air temperature (°C).

If inlet or outlet area is twice the other, the flow rate will increase by 25%, and by 35% if the areas' ratio is 3:1 or larger (Table 18.11).

Example 18.3.6

A two-story (5-m) solar chimney is being designed to produce a flow of 0.25 m^3/s through a space. The preliminary design features include a 25 cm × 1.5 m inlet, a 50 cm × 1.5 m outlet, and an estimated 35°C average stack temperature on a sunny 30°C day. Can this design produce the desired flow?

TABLE 18.11 Ground Reflectivities

Material	ρ (%)
Cement	27
Concrete	20–40
Asphalt	7–14
Earth	10
Grass	6–20
Vegetation	25
Snow	70
Red brick	30
Gravel	15
White paint	55–75

Source: From Murdoch, J. B., *Illumination Engineering—From Edison's Lamp to the Laser*, Macmillan, New York, 1985.

Solution. Substituting the design data into Equation 18.82,

$$Q = 0.116(0.25 \times 1.5)[5(5)]^{1/2}$$
$$= 0.2\ \text{m}^3/\text{s}.$$

Because the outlet area is twice the inlet area, the 25% flow increase can be used:

$$Q = 0.2(1.25) = 0.25\text{m}^3/\text{s}$$

(answer: Yes, the proper flow rate is obtained).

18.3.4.3 Evaporative Cooling

When air with less than 100% relative humidity moves over a water surface, the evaporation of water causes both the air and the water itself to cool. The lowest temperature that can be reached by this direct evaporative cooling effect is the wet-bulb temperature of the air, which is directly related to the relative humidity, with lower wet-bulb temperature associated with lower relative humidity. Thus, dry air (low relative humidity) has a low wet-bulb temperature and will undergo a large temperature drop with evaporative cooling, while humid air (high relative humidity) can only be slightly cooled evaporatively. The wet-bulb temperature for various relative humidity and air temperature conditions can be found via the "psychrometric chart" available in most thermodynamic texts. Normally, an evaporative cooling process cools the air only part of the way down to the wet-bulb temperature. To get the maximum temperature decrease, it is necessary to have a large water surface area in contact with the air for a long time, and interior ponds and fountain sprays are often used to provide this air-water contact area.

The use of water sprays and open ponds on roofs provides cooling primarily via evaporation. The hybrid system involving a fan and wetted mat, the "swamp cooler," is by far the most widely used evaporative cooling technology. Direct, indirect, and combined evaporative cooling system design features are described in ASHRAE (1989, 1991, 1993, 1995, 1997).

18.3.4.4 Nocturnal and Radiative Cooling Systems

Another approach to passive convective/ventilative cooling involves using cooler night air to reduce the temperature of the building and/or a storage mass. Thus, the building/storage mass is prepared to accept part of the heat load during the hotter daytime. This type of convective system can also be combined with evaporative and radiative modes of heat transfer, utilizing air and/or water as the convective fluid. Work in Australia (Close et al. 1968) investigated rock storage beds that were chilled using evaporatively cooled night air. Room air was then circulated through the bed during the day to provide space cooling. The use of encapsulated roof ponds as a thermal cooling mass has been tried by

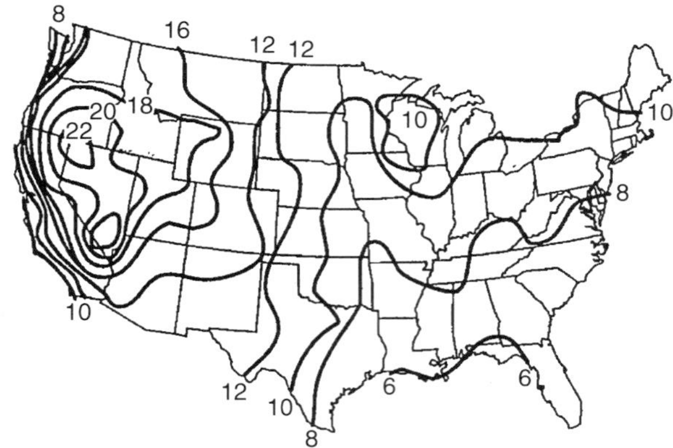

FIGURE 18.51 Average monthly sky temperature depression ($I_{AIR} - I_{SKY}$) for July in °F. (Adapted from Martin, M. and Berdahl, P., *Solar Energy*, 33(314), 321–336, 1984.

several investigators (Hay and Yellot 1969; Marlatt, Murray, and Squire 1984; Givoni 1994) and is often linked with nighttime radiative cooling.

All warm objects emit thermal infrared radiation; the hotter the body, the more energy it emits. A passive cooling scheme is to use the cooler night sky as a sink for thermal radiation emitted by a warm storage mass, thus chilling the mass for cooling use the next day. The net radiative cooling rate, Q_r, for a horizontal unit surface (ASHRAE 1989, 1993, 1997) is

$$Q_r = \varepsilon\sigma\left(T_{\text{body}}^4 - T_{\text{sky}}^4\right), \tag{18.83}$$

where Q_r is the net radiative cooling rate, W/m^2 (Btu/h ft.^{2}), ε is the surface emissivity fraction (usually 0.9 for water), σ is 5.67×10^{-8} W/m^2 K^4 (1.714×10^{-9} Btu/h ft.^{2} R^4), T_{body} is the warm body temperature, Kelvin (Rankine), and T_{sky} is the effective sky temperature, Kelvin (Rankine).

The monthly average air–sky temperature difference has been determined (Martin and Berdahl 1984) and Figure 18.51 presents these values for July (in °F) for the United States.

Example 18.3.7

Estimate the overnight cooling possible for a 10 m^2, 85°F water thermal storage roof during July in Los Angeles.

Solution. Assume the roof storage unit is black with $\varepsilon = 0.9$. From Figure 18.51, $T_{\text{air}} - T_{\text{sky}}$ is approximately 10°F for Los Angeles. From weather data for LA airport, (PSDH 1984; ASHRAE 1989), the July average temperature is 69°F with a range of 15°F. Assuming night temperatures vary from the average (69°F) down to half the daily range (15/2), then the average nighttime temperature is chosen as $69-(1/2)(15/2)=65$°F. Therefore, $T_{\text{sky}}=65-10=55$°F. From Equation 18.83,

$$\begin{aligned} Q_r &= 0.9(1.714\times10^{-9})[(460+85)^4 - (460+55)^4] \\ &= 27.6 \text{ Btu/h ft.}^2. \end{aligned}$$

For a 10-h night and 10 m^2 (107.6 ft.^{2}) roof area,

$$\begin{aligned} \text{Total radiative cooling} &= 27.6(10)(107.6) \\ &= 29,700 \text{ Btu.} \end{aligned}$$

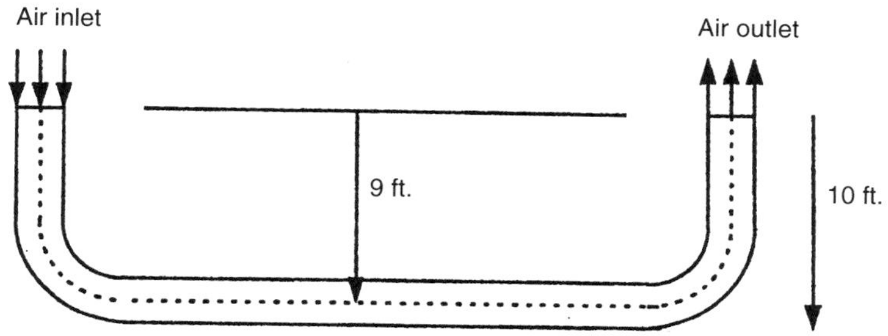

FIGURE 18.52 Open loop underground air tunnel system.

Note that this does not include the convective cooling possible, which can be approximated (at its maximum rate) for still air (ASHRAE 1989, 1993, 1997) by

$$\begin{aligned}\text{Maximum total } Q_{\text{conv}} &= hA(T_{\text{roof}} - T_{\text{air}})(\text{Time}) \\ &= 5(129)(85-55)(10) \\ &= 161{,}000 \text{ Btu}.\end{aligned}$$

This is a maximum since the 85°F storage temperature will drop as it cools; this is also the case for the radiative cooling calculation. However, convection is seen to usually be the more dominant mode of nighttime cooling.

18.3.4.5 Earth Contact Cooling (or Heating)

Earth contact cooling or heating is a passive summer cooling and winter heating technique that utilizes underground soil as the heat sink or source. By installing a pipe underground and passing air through the pipe, the air will be cooled or warmed depending on the season. A schematic of an open loop system and a closed loop air-conditioning system are presented in Figure 18.52 and Figure 18.53, respectively (Goswami and Biseli 1994).

The use of this technique can be traced back to 3000 BC when Iranian architects designed some buildings to be cooled by natural resources only. In the nineteenth century, Wilkinson (USDA 1960) designed a barn for 148 cows where a 500-ft. long underground passage was used for cooling during the summertime. Since that time, a number of experimental and analytical studies of this technique have

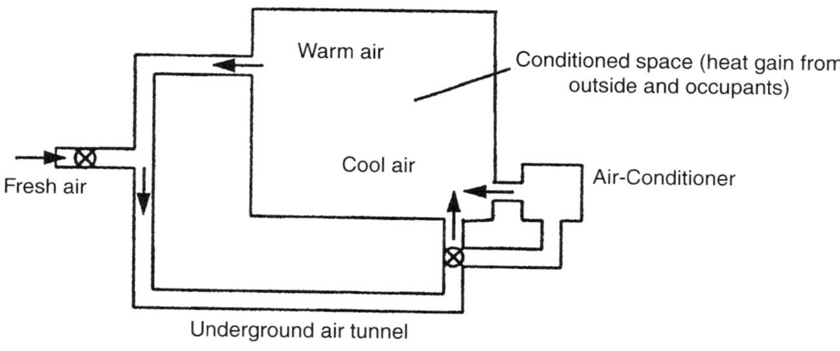

FIGURE 18.53 Schematic of closed loop air-conditioning system using air-tunnel.

continued to appear in the literature (Krarti and Kreider 1996; Hollmuller and Lachal 2001; De Paepe and Janssens 2003). Goswami and Dhaliwal (1985) have given a brief review of the literature, as well as presenting an analytical solution to the problem of transient heat transfer between the air and the surrounding soil as the air is made to pass through a pipe buried underground.

18.3.4.5.1 Heat Transfer Analysis

The transient thermal analysis of the air and soil temperature fields (Goswami and Dhaliwal 1985) is conducted using finite elements with the convective heat transfer between the air and the pipe and using semi-infinite cylindrical conductive heat transfer to the soil from the pipe. It should be noted that the thermal resistance of the pipe (whether of metal, plastic or ceramic) is negligible relative to the surrounding soil.

Air and Pipe Heat Transfer—The pipe is divided into a large number of elements and a psychrometric energy balance written for each, depending on whether the air leaves the element (1) unsaturated, or (2) saturated.

1. If the air leaves an element as unsaturated, the energy balance on the element is

$$mC_p(T_1 - T_2) = hA_p(T_{air} - T_{pipe}). \qquad (18.84)$$

T_{air} can be taken as $(T_1 + T_2)/2$. Substituting and simplifying,

$$T_2 = \left[\left(1 - \frac{U}{2}\right)T_1 + UT_{pipe}\right] / \left(1 + \frac{U}{2}\right), \qquad (18.85)$$

where U is defined as

$$U = \frac{A_p h}{mC_p}$$

2. If the air leaving the element is saturated, the energy balance is

$$mC_pT_1 + m(W_1 - W_2)H_{fg} = mC_pT_2 + hA_p(T_{air} - T_{pipe}). \qquad (18.86)$$

Simplifying gives:

$$T_2 = \left(1 - \frac{U}{2}T_1\right) + \frac{W_1 - W_2}{C_p}H_{fg} + UT_{pipe}/\left(1 + \frac{U}{2}\right). \qquad (18.87)$$

The convective heat transfer coefficient h in the preceding equations depends on Reynolds number, the shape, and roughness of the pipe.

Using the exit temperature from the first element as the inlet temperature for the next element, the exit temperature for the element can be calculated in a similar way. Continuing this way from one element to the next, the temperature of air at the exit from the pipe can be calculated.

Soil Heat Transfer—The heat transfer from the pipe to the soil is analyzed by considering the heat flux at the internal radius of a semi-infinite cylinder formed by the soil around the pipe. For a small element the problem can be formulated as

$$\frac{\partial^2 T(r,t)}{\partial r^2} + \frac{1}{r}\frac{\partial T(r,t)}{\partial r} = \frac{1}{\alpha}\frac{\partial T(r,t)}{\partial t}, \qquad (18.88)$$

with initial and boundary conditions as

$$T(r,0) = T_e,$$

$$T(\infty,t) = T_e,$$

$$-K\frac{\partial T}{\partial r}(r,t) = q'',$$

where T_e is the bulk earth temperature and q'' is also given by the amount of heat transferred to the pipe from the air by convection, i.e., $q^{11} = h(T_{air} - T_{pipe})$.

18.3.4.5.2 Soil Temperatures and Properties

Kusuda and Achenback (1965) and Labs (1981) studied the earth temperatures in the United States. According to both of these studies, temperature swings in the soil during the year are dampened with depth below the ground. There is also a phase lag between the soil temperature and the ambient air temperature, and this phase lag increases with depth below the surface. For example, the soil temperature for light dry soil at a depth of about 10 ft. (3.05 m) varies by approximately ± 5°F (2.8°C) from the mean temperature (approximately equal to mean annual air temperature) and has a phase lag of approximately 75 days behind ambient air temperature (Labs 1981).

The thermal properties of the soil are difficult to determine. The thermal conductivity and diffusivity both change with the moisture content of the soil itself, which is directly affected by the temperature of and heat flux from and to the buried pipe. Most researchers have found that using constant property values for soil taken from standard references gives reasonable predictive results (Goswami and Ileslamlou 1990).

18.3.4.5.3 Generalized Results from Experiments

Figure 18.54 presents data from Goswami and Biseli (1994) for an open system, 100-ft. long, 12-in diameter pipe, buried 9 ft. deep. The figure shows the relationship between pipe inlet-to-outlet temperature reduction ($T_{in} - T_{out}$) and the initial soil temperature with ambient air inlet conditions of 90°F, 55% relative humidity for various pipe flow rates.

Other relations from this same report which can be used with the Figure 18.54 data include: (1) the effect of increasing pipe/tunnel length on increasing the inlet-to-outlet air temperature difference is fairly

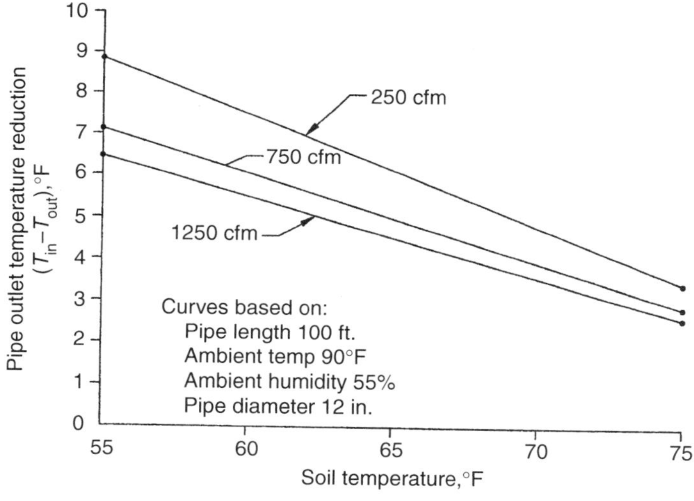

FIGURE 18.54 Air temperature drop through a 100-ft. long, 12-in. diameter pipe buried 9 ft. underground.

linear up to 250 ft.; and (2) the effect of decreasing pipe diameter on lowering the outlet air temperature is slight, and only marginally effective for pipes less than 12-in. in diameter.

Example 18.3.8

Provide the necessary 12-in diameter pipe length(s) that will deliver 1500 cfm of 75°F air if the ambient temperature is 85°F and the soil at 9 ft. is 65°F.

Solution. From Figure 18.54, for 100 ft. of pipe at 65°F soil temperature, the pipe temperature reduction is

$$\begin{aligned} T_{in} - T_{out} &= 6°\text{F (at 250 cfm)} \\ &= 5°\text{F (at 750 cfm)} \\ &= 4.5°\text{F (at 1250 cfm)}. \end{aligned}$$

Because the "length versus temperature reduction" is linear (see text above), the 10°F reduction required (85 down to 75) would be met by the 750 cfm case (5°F for 100 ft.) if 200 ft. of pipe is used. Then, two 12-in diameter pipes would be required to meet the 1500-cfm requirement.

Answer: Two 12-in diameter pipes, each 200 ft. long. (Note: see what would be needed if the 250 cfm or the 1250 cfm cases had been chosen. Which of the three flow rate cases leads to the "cheapest" installation?)

18.3.5 Daylighting Design Fundamentals

Daylighting is the use of the sun's radiant energy to illuminate the interior spaces in a building. In the 19th century, electric lighting was considered an alternative technology to daylighting. Today the situation is reversed, primarily due to the economics of energy use and conservation. However, there are good physiological reasons for using daylight as an illuminant. The quality of daylight matches the human eye's response, thus permitting lower light levels for task comfort, better color rendering, and clearer object discrimination (Robbins 1986; McCluney 1998; Clay 2001).

18.3.5.1 Lighting Terms and Units

Measurement of lighting level is based on the "standard candle", where the lumen (lm), the unit of luminous flux (ϕ), is defined as the rate of luminous energy passing through a 1-m^2 area located 1 m from the candle. Thus, a standard candle generates 4π lumens, which radiate away in all directions. The illuminance (E) on a surface is defined as the luminous flux on the surface divided by the surface area, $E=\phi/A$. Illuminance is measured in either lux (lx), as lm/m^2, or footcandles (fc), as lm/ft.^{2}.

Determination of the daylighting available at a given location in a building space at a given time is important to evaluate the reduction possible in electric lighting and the associated impact on heating and cooling loads. Daylight provides about 110 lm/W of solar radiation, fluorescent lamps about 75 lm/W of electrical input, and incandescent lamps about 20 lm/W; thus daylighting generates only 1/2–1/5 the heating that equivalent electric lighting does, significantly reducing the building cooling load.

18.3.5.2 Approach to Daylighting Design

Aperture controls such as blinds and drapes are used to moderate the amount of daylight entering the space, as are the architectural features of the building itself (glazing type, area, and orientation; overhangs and wingwalls; lightshelves; etc.). Many passive and "active" reflective, concentrating, and diffusing devices are available to specifically gather and direct both the direct and diffuse components of

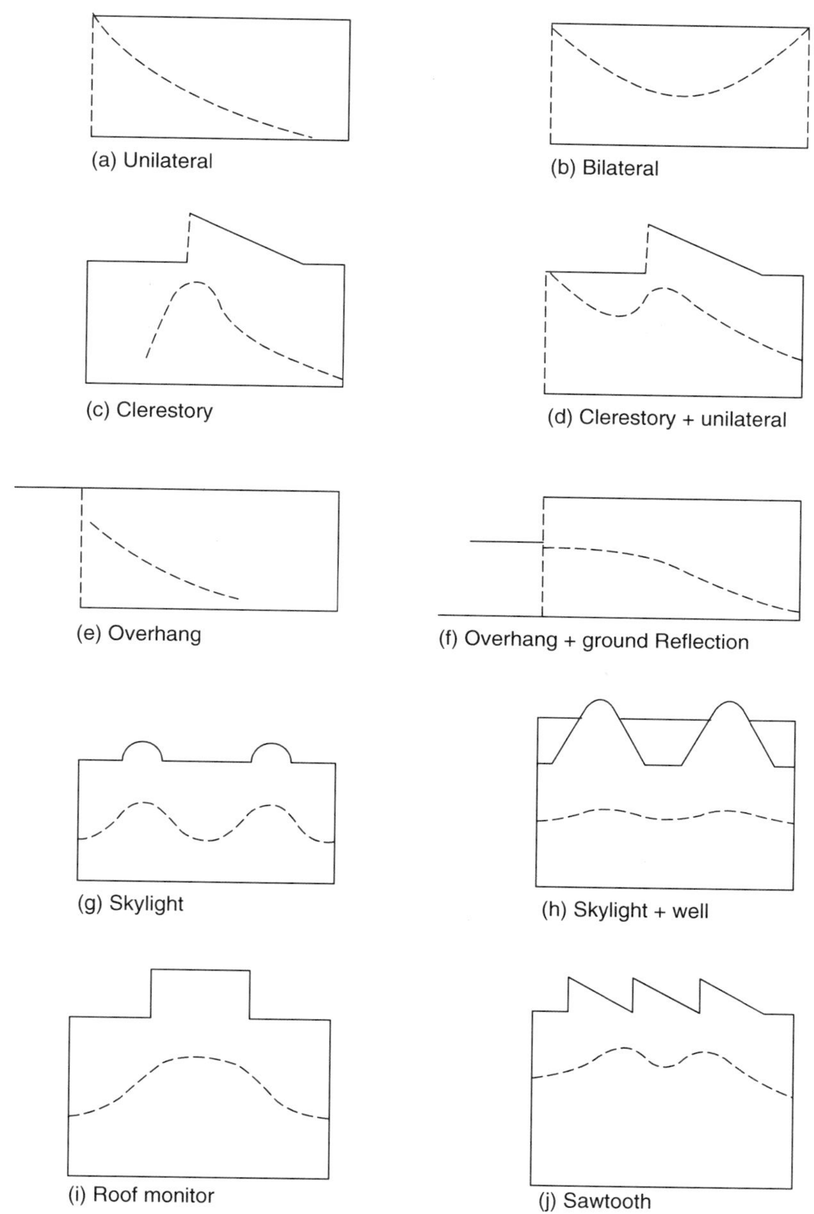

FIGURE 18.55 Examples of sidelighting and toplighting architectural features (dashed lines represent illuminance distributions). (From Murdoch, J. B., *Illumination Engineering from Edison's Lamp to the Laser*. Macmillan, New York, 1985)

daylight to areas within the space (Kinney et al. 2005). Electric-lighting dimming controls are used to adjust the electric light level based on the quantity of the daylighting. With these two types of controls (aperture and lighting), the electric lighting and cooling energy use and demand, as well as cooling system sizing, can be reduced. However, the determination of the daylighting position and time illuminance value within the space is required before energy usage and demand reduction calculations can be made.

Daylighting design approaches use both solar beam radiation (referred to as sunlight) and the diffuse radiation scattered by the atmosphere (referred to as skylight) as sources for interior lighting, with historical design emphasis being on utilizing skylight. Daylighting is provided through a variety of glazing features, which can be grouped as sidelighting (light enters via the side of the space) and toplighting (light enters from the ceiling area). Figure 18.55 illustrates several architectural forms producing sidelighting and toplighting, with the dashed lines representing the illuminance distribution within the space. The calculation of work-plane illuminance depends on whether sidelighting and/or toplighting features are used, and the combined illuminance values are additive.

18.3.5.3 Sun-Window Geometry

The solar illuminance on a vertical or horizontal window depends on the position of the sun relative to that window. In the method described here, the sun and sky illuminance values are determined using the sun's altitude angle (α) and the sun-window azimuth angle difference (Φ). These angles need to be determined for the particular time of day, day of year, and window placement under investigation.

18.3.5.3.1 Solar Altitude Angle

The solar altitude angle, α, is the angle swept out by a person's arm when pointing to the horizon directly below the sun and then raising the arm to point at the sun. The equation to calculate solar altitude, α, is

$$\sin \alpha = \cos L \cos \delta \cos H + \sin L \sin \delta, \tag{18.89}$$

where L is the local latitude (degrees), δ is the earth-sun declination (degrees) given by $\delta = 23.45 \sin[360(n-81)/365]$, n is the day number of the year, and H is the hour angle (degrees) given by

$$H = \frac{(12 \text{ noon} - \text{time})(\text{in minutes})}{4}; \ (+\text{ morning}, \ -\text{afternoon}). \tag{18.90}$$

18.3.5.3.2 Sun-Window Azimuth Angle Difference

The difference between the sun's azimuth and the window's azimuth, Φ, needs to be calculated for vertical window illuminance. The window's azimuth angle, γ_w, is determined by which way it faces, as measured from south (east of south is positive, westward is negative). The solar azimuth angle, γ_s, is calculated:

$$\sin \gamma_s = \frac{\cos \delta \sin H}{\cos \alpha}. \tag{18.91}$$

The sun-window azimuth angle difference, Φ, is given by the absolute value of the difference between γ_s and γ_w:

$$\Phi = |\gamma_s - \gamma_w|. \tag{18.92}$$

18.3.5.4 Daylighting Design Methods

To determine the annual lighting energy saved (ES_L), calculations using the lumen method described below should be performed on a monthly basis for both clear and overcast days for the space under investigation. Monthly weather data for the site would then be used to prorate clear and overcast lighting energy demands monthly. Subtracting the calculated daylighting illuminance from the design illuminance leaves the supplementary lighting needed, which determines the lighting energy required.

The approach in the method below is to calculate the "sidelighting" and the "skylighting" of the space separately, and then combine the results. This procedure has been computerized (Lumen II/ Lumen Micro) and includes many details of controls, daylighting technologies, and weather. ASHRAE (1989, 1993, 1997) lists many of the methods and simulation techniques currently used with daylighting and its associated energy effects.

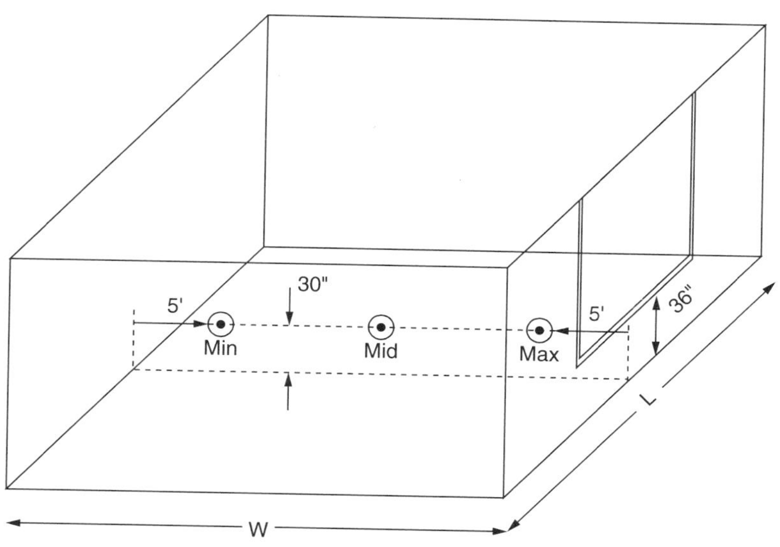

FIGURE 18.56 Location of illumination points within the room (along centerline of window) determined by lumen method of sidelighting.

18.3.5.4.1 Lumen Method of Sidelighting (Vertical Windows)

The lumen method of sidelighting calculates interior horizontal illuminance at three points, as shown in Figure 18.56, at the 30-in (0.76-m) work-plane on the room-and-window centerline. A vertical window is assumed to extend from 36 in (0.91 m) above the floor to the ceiling. The method accounts for both direct and ground-reflected sunlight and skylight, so both horizontal and vertical illuminances from sun and sky are needed. The steps in the lumen method of sidelighting are presented next.

As mentioned, the incident direct and ground-reflected window illuminance are normally calculated for both a cloudy and a clear day for representative days during the year (various months), as well as for clear or cloudy times during a given day. Thus, the interior illumination due to sidelighting and skylighting can then be examined for effectiveness throughout the year.

Step 1: Incident Direct Sky and Sun Illuminances—The solar altitude and sun-window azimuth angle difference are calculated for the desired latitude, date, and time using Equation 18.89 and Equation 18.92, respectively. Using these two angles, the total illuminance on the window (E_{sw}) can be determined by summing the direct sun illuminance (E_{uw}) and the direct sky illuminance (E_{kw}), each determined from the appropriate graph in Figure 18.57.

Step 2: Incident Ground-Reflected Illuminance—The sun illuminance on the ground (E_{ug}), plus the overcast or clear sky illuminance (E_{kg}) on the ground, make up the total horizontal illuminance on the ground surface (E_{sg}). A fraction of the ground surface illuminance is then considered diffusely reflected onto the vertical window surface (E_{gw}), where gw indicates from the ground to the window.

The horizontal ground illuminances can be determined using Figure 18.58, where the clear sky plus sun case and the overcast sky case are functions of solar altitude. The fractions of the ground illuminance diffusely reflected onto the window depends on the reflectivity (ρ) of the ground surface (see Table 18.11.) and the window-to-ground surface geometry.

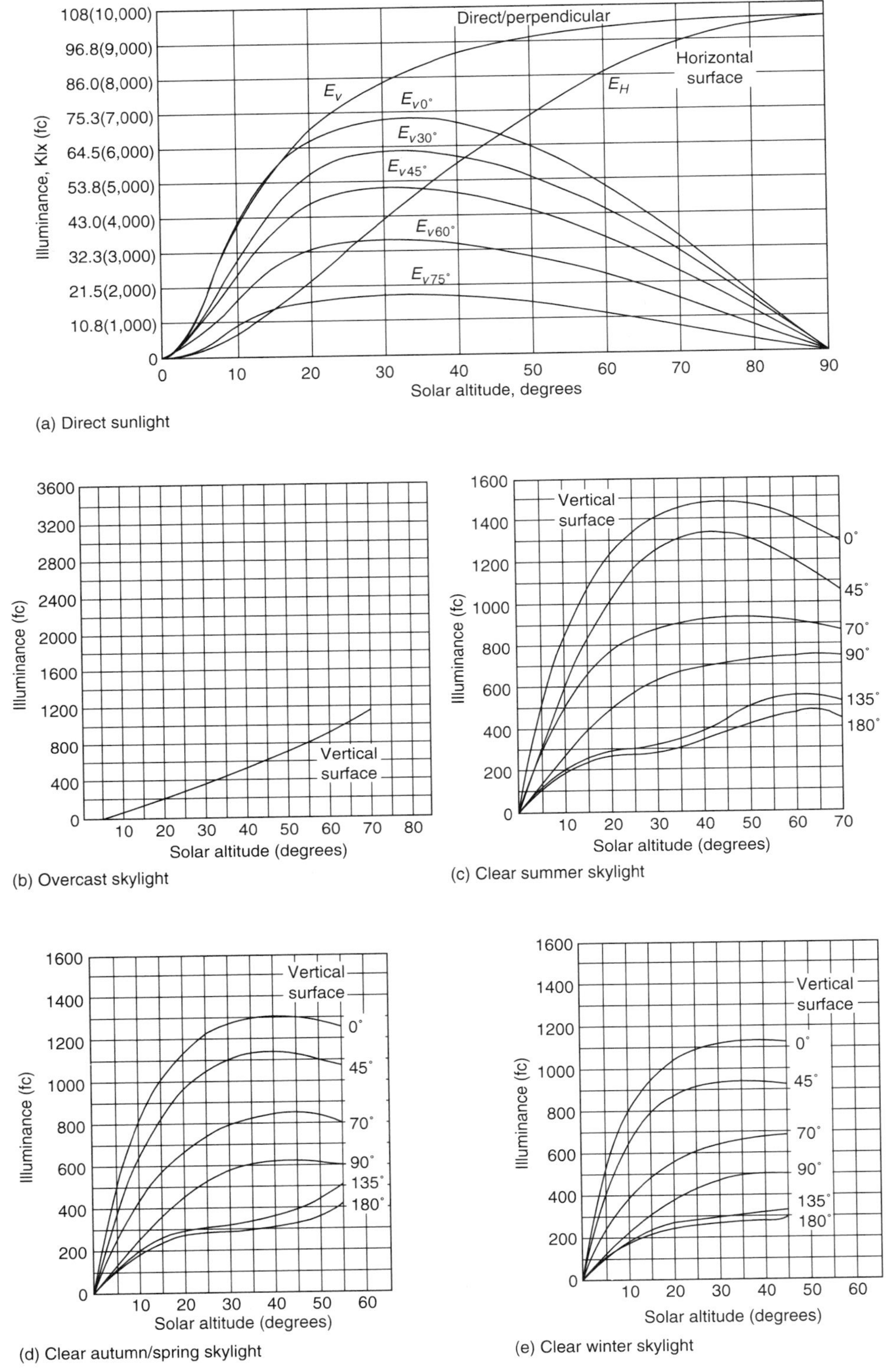

FIGURE 18.57 Vertical illuminance from (a) direct sunlight and (b-c) skylight, for various sun-window azimuth angle differences. (From IES (Illumination Engineering Society), *Lighting Handbook, Applications Volume*. Illumination Engineering Society, New York, 1987.)

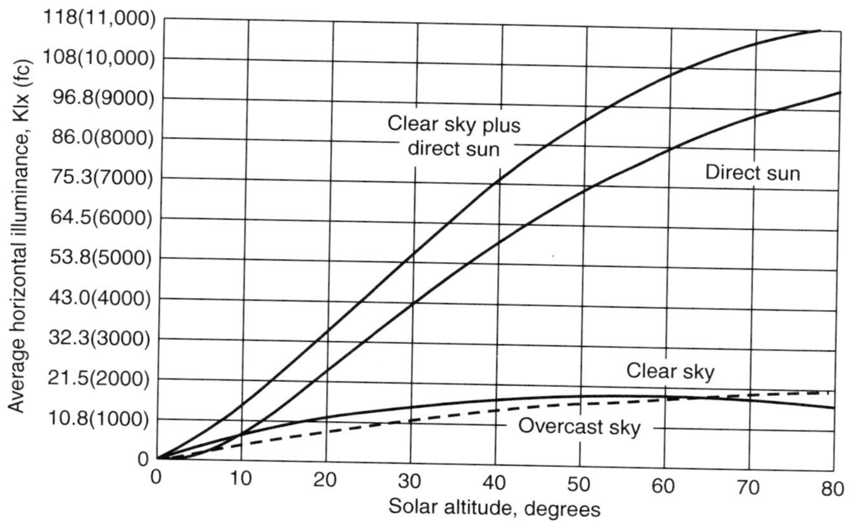

FIGURE 18.58 Horizontal illuminance for overcast sky, clear sky, direct sun, and clear sky plus direct sun. (From Murdoch, J. B., *Illumination Engineering — From Edison's Lamp to the Laser*, Mac Millan, New York, 1985.)

If the ground surface is considered uniformly reflective from the window outward to the horizon, then the illuminance on the window from ground reflection is

$$E_{gw} = \frac{\rho E_{sg}}{2}. \tag{18.93}$$

A more complicated ground-reflection case is illustrated in Figure 18.59, where multiple "strips" of differently reflecting ground are handled using the angles to the window where a strip's illuminance on a window is calculated,

$$E_{gw(strip)} = \frac{\rho_{strip} E_{sg}}{2}(\cos\theta_1 - \cos\theta_2). \tag{18.94}$$

And the total reflected onto the window is the sum of the strip illuminances:

$$E_{gw} = \frac{E_{sg}}{2}[\rho_1(\cos 0 - \cos\theta_1) + \rho_2(\cos\theta_1 - \cos\theta_2) + \cdots + \rho_n(\cos\theta_{n-1} - \cos 90)]. \tag{18.95}$$

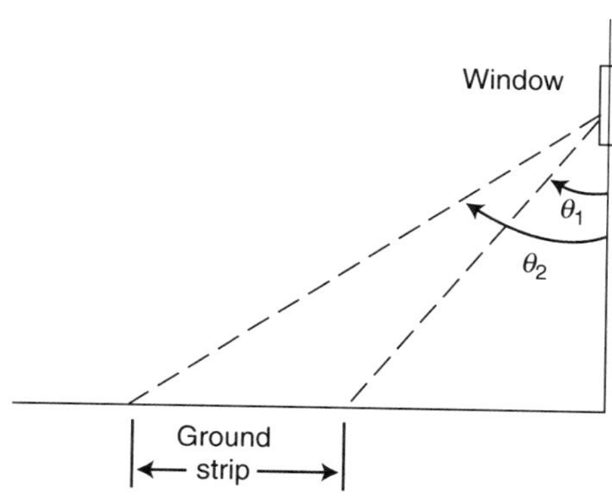

FIGURE 18.59 Geometry for ground "strips". (From Murdoch, J. B., *Illumination Engineering — From Edison's Lamp to the Laser*, Mac Millan, New York, 1985.)

TABLE 18.12 Glass Transmittances

Glass	Thickness (in.)	τ (%)
Clear	1/3	89
Clear	3/16	88
Clear	1/4	87
Clear	5/16	86
Grey	1/8	61
Grey	3/16	51
Grey	1/4	44
Grey	5/16	35
Bronze	1/8	68
Bronze	3/16	59
Bronze	1/4	52
Bronze	5/16	44
Thermopane	1/8	80
Thermopane	3/16	79
Thermopane	1/4	77

Source: From Murdoch, J. B., *Illumination Engineering—From Edison's Lamp to the Laser*, Macmillan, New York, 1985.

Step 3: Luminous Flux Entering Space—The direct sky–sun and ground-reflected luminous fluxes entering the building are attenuated by the transmissivity of the window. Table 18.12 presents the transmittance fraction (τ) of several window glasses. The fluxes entering the space are calculated from the total sun–sky and the ground-reflected illuminances by using the area of the glass, A_w:

$$\begin{aligned}\phi_{sw} &= E_{sw}\tau A_w,\\ \phi_{gw} &= E_{gw}\tau A_w,\end{aligned} \tag{18.96}$$

Step 4: Light Loss Factor—The light loss factor (K_m) accounts for the attenuation of luminous flux due to dirt on the window (WDD, window dirt depreciation) and on the room surfaces (RSDD, room surface dirt depreciation). WDD depends on how often the window is cleaned, but a 6-month average for offices is 0.83 and for factories is 0.71 (Murdoch 1985).

The RSDD is a more complex calculation involving time between cleanings, the direct-indirect flux distribution, and room proportions. However, for rooms cleaned regularly, RSDD is around 0.94 and for once-a-year-cleaned dirty rooms, the RSDD would be around 0.84.

The light loss factor is the product of the preceding two fractions:

$$K_m = (\text{WDD})(\text{RSDD}). \tag{18.97}$$

Step 5: Work-Plane Illuminances—As discussed earlier, Figure 18.56 illustrates the location of the work-plane illuminances determined with this lumen method of sidelighting. The three illuminances (max, mid, min) are determined using two coefficients of utilization, the *C* factor and *K* factor. The *C* factor depends on room length and width and wall reflectance. The *K* factor depends on ceiling-floor height, room width, and wall reflectance. Table 18.13 presents *C* and *K* values for the three cases of incoming fluxes: sun plus clear sky, overcast sky, and ground-reflected. Assumed ceiling and floor reflectances are given for this case with no window controls (shades, blinds, overhangs, etc.). These further window control complexities can be found in Libbey-Owens-Ford Company (1976); IES (1987), and others. A reflectance of 70% represents light-colored walls, with 30% representing darker walls.

The work-plane max, mid, and min illuminances are each calculated by adding the sun–sky and ground-reflected illuminances, which are given by

$$E_{sp} = \phi_{sw} C_s K_s K_m,$$
$$E_{gp} = \phi_{gw} C_g K_g K_m, \tag{18.98}$$

where the "sp" and "gp" subscripts refer to the sky-to-work-plane and ground-to-work-plane illuminances.

Example 18.3.9

Determine the clear-sky illuminances for a 30-ft.-long, 30-ft.-wide, 10-ft.-high room with a 20-ft.-long window with a 3-ft. sill. The window faces 10°E of south, the building is at 32°N latitude, and it is January 15 at 2 p.m. The ground cover outside is grass, the glass is 1/4-in clear, and the walls are light colored.

Solution. Following the steps in the "sidelighting" method:
Step 1: With $L=32$, $n=15$, $H=(12-14)60/4=-30$,

$$\delta = 23.45 \sin[360(15-81)/365] = -21.3°.$$

Then, Equation 18.89 yields $\alpha=41.7°$, Equation 18.91 yields $\gamma_s=-38.7°$, and Equation 18.92 yields

$$\Phi = |-38.7-(+10)| = 48.7°.$$

From Figure 18.57 with $\alpha=41.7°$ and $\Phi=48.7°$:

(a) For clear sky (winter, no sun): $E_{kw}=875$ fc.
(b) For direct sun: $E_{uw}=4{,}100$ fc.
(c) Total clear sky plus direct: $E_{sw}=4{,}975$ fc.

(Note: A high E_{uw} value probably indicates a glare situation!)
Step 2: Horizontal illuminances from Figure 18.58: $E_{sg}=4007$ fc.
Then Equation 18.93 yields, with $\rho_{grass}=0.06$, $E_{gw}=222$ fc.
Step 3: From Equation 18.96, with $\tau=0.87$ and $A_w=140$ ft.^{2},

$$\Phi_{sw} = 4975(0.87)(140) = 605{,}955 \text{ lm},$$

$$\Phi_{gw} = 222(0.87)(140) = 27{,}040 \text{ lm}.$$

Step 4: For a clean office room:

$$K_m = (0.83)(0.94) = 0.78.$$

Step 5: From Table 18.13, for 30′ width, 30′ length, 10′ ceiling, and wall reflectivity 70%,

(a) Clear sky:

$$C_{s,\,max} = 0.0137;\ K_{s,\,max} = 0.125,$$

$$C_{s,\,mid} = 0.0062;\ K_{s,\,mid} = 0.110,$$

$$C_{s,\,min} = 0.0047;\ K_{s,\,min} = 0.107.$$

TABLE 18.13a *C* and *K* Factors for No Window Controls for Overcast Sky

Illumination by Overcast Sky

C: Coefficient of Utilization

Room Length (ft.)		20		30		40	
Wall Reflectance (%)		70	30	70	30	70	30
	Room Width (ft.)						
Max	20	0.0276	0.0251	0.0191	0.0173	0.0143	0.0137
Max	30	0.0272	0.0248	0.0188	0.0172	0.0137	0.0131
Max	40	0.0269	0.0246	0.0182	0.0171	0.0133	0.0130
Mid	20	0.0159	0.0177	0.0101	0.0087	0.0081	0.0071
Mid	30	0.0058	0.0050	0.0054	0.0040	0.0034	0.0033
Mid	40	0.0039	0.0027	0.0030	0.0023	0.0022	0.0019
Min	20	0.0087	0.0053	0.0063	0.0043	0.0050	0.0037
Min	30	0.0032	0.0019	0.0029	0.0017	0.0020	0.0014
Min	40	0.0019	0.0009	0.0016	0.0009	0.0012	0.0008

K: Coefficient of Utilization

Ceiling Height (ft.)		8		10		12		14	
Wall Reflectance (%)		70	30	70	30	70	30	70	30
	Room Width (ft.)								
Max	20	0.125	0.129	0.121	0.123	0.111	0.111	0.0991	0.0973
Max	30	0.122	0.131	0.122	0.121	0.111	0.111	0.0945	0.0973
Max	40	0.145	0.133	0.131	0.126	0.111	0.111	0.0973	0.0982
Mid	20	0.0908	0.0982	0.107	0.115	0.111	0.111	0.105	0.122
Mid	30	0.156	0.102	0.0939	0.113	0.111	0.111	0.121	0.134
Mid	40	0.106	0.0948	0.123	0.107	0.111	0.111	0.135	0.127
Min	20	0.0908	0.102	0.0951	0.114	0.111	0.111	0.118	0.134
Min	30	0.0924	0.119	0.101	0.114	0.111	0.111	0.125	0.126
Min	40	0.111	0.0926	0.125	0.109	0.111	0.111	0.133	0.130

Source: From IES, 1979.

TABLE 18.13b *C* and *K* Factors for No Window Controls for Clear Sky

Illumination by Clear Sky

C: Coefficient of Utilization

Room Length (ft.)		20		30		40	
Wall Reflectance (%)		70	30	70	30	70	30
	Room Width (ft.)						
Max	20	0.0206	0.0173	0.0143	0.0123	0.0110	0.0098
Max	30	0.0203	0.00.173	0.0137	0.0120	0.0098	0.0092
Max	40	0.0200	0.0168	0.0131	0.0119	0.0096	0.0091
Mid	20	0.0153	0.0104	0.0100	0.0079	0.0083	0.0067
Mid	30	0.0082	0.0054	0.0062	0.0043	0.0046	0.0037
Mid	40	0.0052	0.0032	0.0040	0.0028	0.0029	0.0023
Min	20	0.0106	0.0060	0.0079	0.0049	0.0067	0.0043
Min	30	0.0054	0.0028	0.0047	0.0023	0.0032	0.0021
Min	40	0.0031	0.0014	0.0027	0.0013	0.0021	0.0012

K: Coefficient of Utilization

Ceiling Height (ft.)		8		10		12		14	
Wall Reflectance (%)		70	30	70	30	70	30	70	30
	Room Width (ft.)								
Max	20	0.145	0.155	0.129	0.132	0.111	0.111	0.101	0.0982
Max	30	0.141	0.149	0.125	0.130	0.111	0.111	0.0954	0.101
Max	40	0.157	0.157	0.135	0.134	0.111	0.111	0.0964	0.0991
Mid	20	0.110	0.128	0.116	0.126	0.111	0.111	0.103	0.108
Mid	30	0.106	0.125	0.110	0.129	0.111	0.111	0.112	0.120
Mid	40	0.117	0.118	0.122	0.118	0.111	0.111	0.123	0.122
Min	20	0.105	0.129	0.112	0.130	0.111	0.111	0.111	0.116
Min	30	0.0994	0.144	0.107	0.126	0.111	0.111	0.107	0.124
Min	40	0.119	0.116	0.130	0.118	0.111	0.111	0.120	0.118

Source: From IES, 1979.

TABLE 18.13c *C* and *K* Factors for No Window Controls for Ground Illumination (Ceiling Reflectance, 80%; Floor Reflectance, 30%)

Ground Illumination

C: Coefficient of Utilization

	Room Length (ft.)	20		30		40	
	Wall Reflectance (%)	70	30	70	30	70	30
	Room Width (ft.)						
Max	20	0.0147	0.0112	0.0102	0.0088	0.0081	0.0071
Max	30	0.0141	0.0012	0.0098	0.0088	0.0077	0.0070
Max	40	0.0137	0.0112	0.0093	0.0086	0.0072	0.0069
Mid	20	0.0128	0.0090	0.0094	0.0071	0.0073	0.0060
Mid	30	0.0083	0.0057	0.0062	0.0048	0.0050	0.0041
Mid	40	0.0055	0.0037	0.0044	0.0033	0.0042	0.0026
Min	20	0.0106	0.0071	0.0082	0.0054	0.0067	0.0044
Min	30	0.0051	0.0026	0.0041	0.0023	0.0033	0.0021
Min	40	0.0029	0.0018	0.0026	0.0012	0.0022	0.0011

K: Coefficient of Utilization

	Ceiling Height (ft.)	8		10		12		14	
	Wall Reflectance (%)	70	30	70	30	70	30	70	30
	Room Width (ft.)								
Max	20	0.124	0.206	0.140	0.135	0.111	0.111	0.0909	0.0859
Max	30	0.182	0.188	0.140	0.143	0.111	0.111	0.0918	0.0878
Max	40	0.124	0.182	0.140	0.142	0.111	0.111	0.0936	0.0879
Mid	20	0.123	0.145	0.122	0.129	0.111	0.111	0.100	0.0945
Mid	30	0.0966	0.104	0.107	0.112	0.111	0.111	0.110	0.105
Mid	40	0.0790	0.0786	0.0999	0.106	0.111	0.111	0.118	0.118
Min	20	0.0994	0.108	0.110	0.114	0.111	0.111	0.107	0.104
Min	30	0.0816	0.0822	0.0984	0.105	0.111	0.111	0.121	0.116
Min	40	0.0700	0.0656	0.0946	0.0986	0.111	0.111	0.125	0.132

Source: From IES, 1979.

(b) Ground reflected:

$$C_{g,\ max} = 0.0098;\ K_{g,\ max} = 0.140.$$

$$C_{g,\ mid} = 0.0062;\ K_{,\ mid} = 0.107.$$

$$C_{g,\ min} = 0.0041;\ K_{g,\ min} = 0.0984.$$

Then using Equation 18.98,

$$E_{sp,\ max} = 605,955(0.0137)(0.125)(0.78) = 809\ \text{fc},$$

$$E_{sp,\ mid} = 605,955(0.0062)(0.110)(0.78) = 322\ \text{fc},$$

$$E_{sp,min} = 605,955(0.0047)(0.107)(0.78) = 238\ \text{fc},$$

$$E_{gp,\ max} = 27,040(0.0098)(0.140)(0.78) = 29\ \text{fc},$$

$$E_{gp,\ mid} = 27,040(0.0062)(0.107)(0.78) = 14\ \text{fc},$$

$$\text{E}_{gp,\ min} = 27,040(0.0041)(0.984)(0.78) = 9\ \text{fc}.$$

Thus,

$$E_{max} = 838\ \text{fc},$$

$$E_{mid} = 336\ \text{fc},$$

$$E_{min} = 247\ \text{fc}.$$

18.3.5.4.2 Lumen Method of Skylighting

The lumen method of skylighting calculates the average illuminance at the interior work-plane provided by horizontal skylights mounted on the roof. The procedure for skylighting is generally the same as that described above for sidelighting. As with windows, the illuminance from both overcast sky and clear sky plus sun cases are determined for specific days in different seasons and for different times of the day, and a judgment is then made as to the number and size of skylights and any controls needed.

The procedure is presented in four steps: (1) finding the horizontal illuminance on the outside of the skylight; (2) calculating the effective transmittance through the skylight and its well; (3) figuring the interior space light loss factor and the utilization coefficient; and finally, (4) calculating illuminance on the work-plane.

Step 1: Horizontal Sky and Sun Illuminances—The horizontal illuminance value for an overcast sky or a clear sky plus sun situation can be determined from Figure 18.58 knowing only the solar altitude.

Step 2: Net Skylight Transmittance—The transmittance of the skylight is determined by the transmittance of the skylight cover(s), the reflective efficiency of the skylight well, the net-to-gross skylight area, and the transmittance of any light-control devices (lenses, louvers, etc.).

The transmittance for several flat-sheet plastic materials used in skylight domes is presented in Table 18.14. To get the effective dome transmittance (T_D) from the flat-plate transmittance (T_F) value (AAMA 1977), use

TABLE 18.14 Flat-Plate Plastic Material Transmittance for Skylights

Type	Thickness (in.)	Transmittance (%)
Transparent	1/8–3/16	92
Dense translucent	1/3	32
Dense translucent	3/16	24
Medium translucent	1/8	56
Medium translucent	3/16	52
Light translucent	1/8	72
Light translucent	3/16	68

Source: From Murdoch, J. B., *Illumination Engineering—From Edison's Lamp to the Laser*, Macmillan, New York, 1985.

$$T_D = 1.25T_F(1.18 - 0.416T_F). \tag{18.99}$$

If a double-domed skylight is used, then the single-dome transmittances are combined as follows (Pierson 1962):

$$T_D = \frac{T_{D_1}T_{D_2}}{T_{D_1}T_{D_2} - T_{D_1}T_{D_2}} \tag{18.100}$$

If the diffuse and direct transmittances for solar radiation are available for the skylight glazing material, it is possible to follow this procedure and determine diffuse and direct dome transmittances separately. However, this difference is usually not a significant factor in the overall calculations.

The efficiency of the skylight well (N_w) is the fraction of the luminous flux from the dome that enters the room from the well. The well index (WI) is a geometric index (height, h; length, l; width, w) given by

$$\mathrm{WI} = \frac{h(w + l)}{2wl}, \tag{18.101}$$

and WI is used with the well–wall reflectance value in Figure 18.60 to determine well efficiency, N_w.

With T_D and N_w determined, the net skylight transmittance for the skylight and well is given by:

$$T_n = T_D N_w R_A T_C, \tag{18.102}$$

where R_A is the ratio of net to gross skylight areas and T_C is the transmittance of any light-controlling devices.

Step 3: Light Loss Factor and Utilization Coefficient—The light loss factor (K_m) is again defined as the product of the room surface dirt depreciation (RSDD) and the skylight direct depreciation (SDD) fractions, similar to Equation 18.97. Following the reasoning for the sidelighting case, the RSDD value for clean rooms is around 0.94 and 0.84 for dirty rooms. Without specific data indicating otherwise, the SDD fraction is often taken as 0.75 for office buildings and 0.65 for industrial areas.

The fraction of the luminous flux on the skylight that reaches the work-plane (K_u) is the product of the net transmittance (T_n) and the room coefficient of utilization (RCU). Dietz et al. (1981) developed RCU equations for office and warehouse interiors with ceiling, wall, and floor reflectances of 75%, 50%, and 30%, and 50%, 30%, and 20%, respectively.

$$\mathrm{RCU} = \frac{1}{1 + A(\mathrm{RCR})^B} \text{ if } \mathrm{RCR} < 8, \tag{18.103}$$

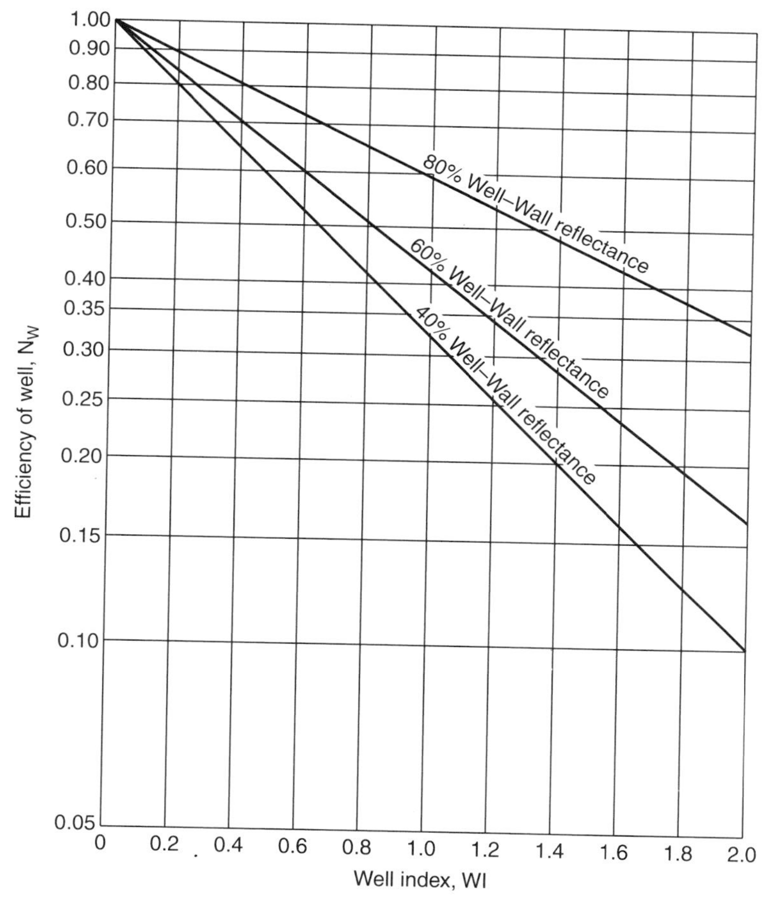

FIGURE 18.60 Efficiency of well versus well index. (From IES (Illumination Engineering Society), *Lighting Handbook, Applications Volume*. Illumination Engineering Society, New York, 1987.)

where A is 0.0288 and B is 1.560 for offices, and A is 0.0995 and B is 1.087 for warehouses. Room cavity ratio (RCR) is given by

$$RCR = \frac{5h_c(l + w)}{lw}, \tag{18.104}$$

where h_c is the ceiling height above the work-plane and l and w are the room length and width, respectively.

The RCU is then multiplied by the previously determined T_n to give the fraction of the external luminous flux passing through the skylight and incident on the workplace:

$$K_u = T_n(\text{RCU}). \tag{18.105}$$

Step 4: Work-Plane Illuminance—The illuminance at the work-plane (E_{TWP}) is given by

$$E_{TWP} = E_H\left(\frac{A_T}{A_{WP}}\right)K_u K_m, \tag{18.106}$$

where E_H is the horizontal overcast or clear sky plus sun illuminance from Step 1, A_T is total gross area of the skylights (number of skylights times skylight gross area), and A_{WP} is the work-plane area (generally room length times width). Note that in Equation 18.106, it is also possible to fix the E_{TWP} at some desired value and determine the required skylight aread.

Rules of thumb for skylight placement for uniform illumination include 4%–8% of roof area and spacing less than 1.5 times ceiling-to-work-plane distance between skylights (Murdoch 1984).

Example 18.3.10

Determine the work-plane "clear sky plus sun" illuminance for a $30\times30\times10$ ft.^{3} office with 75% ceiling, 50% wall, and 30% floor reflectance with four 4×4 ft.^{2} double-domed skylights at 2:00 p.m. on January 15 at 32° latitude. The skylight well is 1 ft. deep at with 60% reflectance walls, and the outer- and inner-dome flat-plastic transmittances are 0.85 and 0.45, respectively. The net skylight area is 90%.

Solution. Follow the four steps in the lumen method for skylighting.

Step 1: Use Figure 18.58 with the solar altitude of 41.7° (calculated from Equation 18.93) for the clear sky plus sun curve to get horizontal illuminance:

$$E_H = 7400 \text{ fc.}$$

Step 2: Use Equation 18.99 to determine domed transmittances from the flat plate plastic transmittances given,

$$T_{D1} = 1.25(0.85)[1.18 - 0.416(0.85)] = 0.89,$$

$$T_{D_2}(T_F = 0.45) = 0.56,$$

and use Equation 18.100 to get total dome transmittance from the individual dome transmittances:

$$T_D = \frac{(0.89)(0.56)}{(0.89) + (0.56) - (0.89)(0.56)} = 0.52.$$

To determine well efficiency, use WI=0.25 from Equation 18.101 with 60% wall reflectance in Figure 18.60 to give $N_w=0.80$. With $R_A=0.90$, use Equation 18.102 to calculate net transmittance:

$$T_n = (0.52)(0.80)(0.90)(1.0) = 0.37.$$

Step 3: The light loss factor is assumed to be from "typical" values in Equation 18.97: $K_m = (0.75)(0.94) = 0.70$. The room utilization coefficient is determined using Equation 18.103 and Equation 18.104:

$$\text{RCR} = \frac{5(7.5)(30 + 30)}{(30)(30)} = 2.5,$$

$$\text{RCU} = [1 + 0.0288(2.5)^{1.560}]^{-1} = 0.89.$$

Equation 18.104 yields $K_u = (0.37)(0.89) = 0.33$.

Step 4: The work-plane illuminance is calculated by substituting the above values into Equation 18.106:

$$E_{TWP} = 7400\left[\frac{4(16)}{30(30)}\right]0.33(0.70),$$
$$E_{TWP} = 122 \text{ fc.}$$

18.3.5.5 Daylighting Controls and Economics

The economic benefit of daylighting is directly tied to the reduction in lighting electrical energy operating costs. Also, lower cooling-system operating costs are possible due to the reduction in heating caused by the reduced electrical lighting load. The reduction in lighting and cooling system electrical power during peak demand periods could also beneficially affect demand charges.

The reduction of the design cooling load through the use of daylighting can also lead to the reduction of installed or first-cost cooling system dollars. Normally, economics dictate that an automatic lighting control system must take advantage of the reduced lighting/cooling effect, and the control system cost minus any cooling system cost savings should be expressed as a "net" first cost. A payback time for the lighting control system ("net" or not) can be calculated from the ratio of first costs to yearly operating savings. In some cases, these paybacks for daylighting controls have been found to be in the range of 1–5 years for office building spaces (Rundquist 1991).

Controls, both aperture and lighting, directly affect the efficacy of the daylighting system. As shown in Figure 18.61, aperture controls can be architectural (overhangs, light shelves, etc.) and/or window shading devices (blinds, automated louvers, etc.). The aperture controls generally moderate the sunlight entering the space to maximize/minimize solar thermal gain, permit the proper amount of light for visibility, and prevent glare and beam radiation onto the workplace. Photosensor control of electric lighting allows the dimming (or shutting off) of the lights in proportion to the amount of available daylighting illuminance.

In most cases, increasing the solar gain for daylighting purposes, with daylighting controls, saves more in electrical lighting energy and the cooling energy associated with the lighting than is incurred with the added solar gain (Rundquist 1991). In determining the annual energy savings total from daylighting, ES_T, the annual lighting energy saved from daylighting, ES_L, is added with the reduction in cooling system energy, ΔES_C, and with the negative of the heating system energy increase ΔES_H:

$$ES_T = ES_L + \Delta ES_C - \Delta ES_H. \tag{18.107}$$

A simple approach to estimating the heating and cooling energy changes associated with the lighting energy reduction is by using the fraction of the year associated with the cooling or heating season (f_C, f_H) and the seasonal COP of the cooling or heating equipment. Thus, Equation 18.107 can be

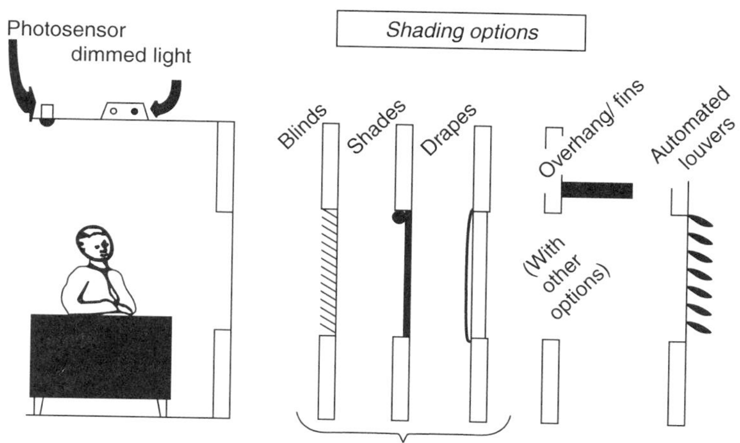

FIGURE 18.61 Daylighting system controls. (From Rundquist, R. A., Daylighting controls: Orphan of HVAC design. *ASHRAE Journal*, 11 (November), 30–340, 1991.)

expressed as

$$ES_T = ES_L + \frac{f_C ES_L}{COP_c} - \frac{f_H ES_L}{COP_H},$$

$$ES_T = ES_L\left(1 + \frac{f_C}{COP_C} - \frac{f_H}{COP_H}\right). \tag{18.108}$$

It should be noted that the increased solar gain due to daylighting has not been included here but would reduce summer savings and increase winter savings. If it is assumed that the increased wintertime daylighting solar gain approximately offsets the reduced lighting heat gain, then the last term in Equation 18.108 becomes negligible.

To determine the annual lighting energy saved (ES_L), calculations using the lumen method described earlier should be performed on a monthly basis for both clear and overcast days for the space under investigation. Monthly weather data for the site would then be used to prorate clear and overcast lighting energy demands monthly. Subtracting the calculated (controlled) daylighting illuminance from the design illuminance leaves the supplementary lighting needed, which determines the lighting energy required.

This procedure has been computerized and includes many details of controls, daylighting methods, weather, and heating and cooling load calculations. ASHRAE (1989) lists many of the methods and simulation techniques currently used with daylighting and its associated energy effects.

Example 18.3.11

A 30×20 ft.^{2} space has a photosensor dimmer control with installed lighting density of 2.0 W/ft.^{2}. The required workplace illuminance is 60 fc and the available daylighting illuminance is calculated as 40 fc on the summer peak afternoon. Determine the effect on the cooling system (adapted from Rundquist 1991).

Solution. The lighting power reduction is (2.0 W/ft.^{2}) (30×20) ft.$^2 \times$(40 fc/60 fc)$=$800 W. The space cooling load would also be reduced by this amount (assuming CLF$=$1.0):

$$\frac{800\text{ W} \times 3.413\text{ Btu h/W}}{12,000\text{ Btu h/tn.}} = 0.23\text{ tn.}$$

Assuming 1.5 tn. nominally installed for 600 ft.^{2} of space at \$2200/tn., the 0.23-tn. reduction is "worth" $0.23 \times$\$2200/tn.$=$\$506. The lighting controls cost about \$1/ft.^{2} of controlled area, so the net installed first cost is

$$\text{Net first cost} = \$\ 600\text{ controls} - \$\ 500\text{ A/C savings} = \$\ 100.$$

Assuming the day-to-monthly-to-annual illuminance calculations gave a 30% reduction in annual lighting, the associated operating savings can be determined. Lighting energy savings are

$$ES_L = 0.30 \times 2.0\text{ W/ft.}^2 \times 600\text{ ft.}^2 \times 2500\text{ h/year} = 900\text{ kWh}.$$

Using Equation 18.108 to also include cooling energy saved due to lighting reduction (with $COP_c=$ 2.5, $f_c=0.5$, and neglecting heating) gives

$$ES_T = 900(1 + 0.5/2.5 - 0) = 1080\text{ kWh}.$$

At \$0.10 per kWh, the operating costs savings are \$0.10/kWh$\times$1080 kWh$=$\$108/year.

Thus, the simple payback is approximately 1 year (100/108) for the "net" situation, and a little over 5.5 years (600/108) against the "controls" cost alone. It should also be noted that the 800 W lighting electrical reduction at peak hours, with an associated cooling energy reduction of 800 W/2.5 COP=320 W, provides a peak demand reduction for the space of 1.1 kW, which can be used as a "first-cost savings" to offset control system costs.

Glossary

Active system: A system employing a forced (pump or fan) convection heat transfer fluid flow.

Daylighting: The use of the sun's radiant energy for illumination of a building's interior space.

Hybrid system: A system with parallel passive and active flow systems or one using forced convection flow to distribute from thermal storage.

Illuminance: The density of luminous flux incident on a unit surface. Illuminance is calculated by dividing the luminous flux (in lumens) by the surface area (m^2, $ft.^2$). Units are lux (lx) (lumens/m^2) in SI and footcandles (fc) (lumens/$ft.^2$) in English systems.

Luminous flux: The time rate of flow of luminous energy (lumens). A lumen (lm) is the rate which luminous energy from a 1 candela (cd) intensity source is incident on a 1-m^2 surface 1 m from the source.

Passive cooling system: A system using natural energy flows to transfer heat to the environmental sinks (ground, air, and sky).

Passive heating system: A system in which the sun's radiant energy is converted to heat by absorption in the system, and the heat is distributed by naturally occurring processes.

Sidelighting: Daylighting by light entering through the wall/side of a space.

Skylight: The diffuse solar radiation from a clear or overcast sky, excluding the direct radiation from the sun.

Sunlight: The direct solar radiation from the sun.

Toplighting: Daylighting by light entering through the ceiling area of a space.

References

AAMA (Architectural Aluminum Manufacturers Association). 1977. Publication 1602.1.1977, *Voluntary Standard Procedure for Calculating Skylight Annual Energy Balance.* AAMA, Chicago, IL.

ASHRAE (American Society of Heating, Refrigerating and Air-Conditioning Engineers). 1989. Fundamentals. In *ASHRAE Handbook.* ASHRAE, Atlanta, GA.

ASHRAE (American Society of Heating, Refrigerating and Air-Conditioning Engineers), 1993. Fundamentals. In *ASHRAE Handbook.* ASHRAE, Atlanta, GA

ASHRAE (American Society of Heating, Refrigerating and Air-Conditioning Engineers). 1991. Heating, ventilating, and air-conditioning applications, In *ASHRAE Handbook.* ASHRAE, Atlanta, GA.

ASHRAE (American Society of Heating, Refrigerating and Air-Conditioning Engineers), 1995. Heating, ventilating, and air-conditioning applications, In *ASHRAE Handbook*, pp. 279–127. ASHRAE, Atlanta, GA.

Clay, R. A. 2001. Green is good for you. *Monitor on Psychology*, 32, 4, 40–42.

Close, D. J., Dunkle, R. V., and Robeson, K. A. 1968. Design and performance of a thermal storage air conditioning system. *Mechanical and Chemical Engineering Transactions*, MC4, 45.

De Paepe, M. and Janssens, A. 2003. Thermo-hydraulic design of earth-air heat exchanger. *Energy and Buildings*, 35, 389–397.

Dietz, P., Murdoch, J., Pokoski, J., and Boyle, J. 1981. A skylight energy balance analysis procedure. *Journal of the Illuminating Engineering Society*, 11, October, 27–34.

Duffie, J. A. and Beckman, W. A. 1991. *Solar Engineering of Thermal Processes. 2nd Ed.*, Wiley, New York.

Goswami, D. Y. and Biseli, K. M. 1994. Use of underground air tunnels for headting and coooling agricultural residential buildings. Report EES-78, Florida Energy Extension service, University of Florida, Gainesville, Fl. August.

Goswami, D. Y. and Dhaliwal, A. S. 1985. Heat transfer analysis in environmental control using an underground air tunnel. *J. of Solar Energy Eng.*, 107 (May): 141–45.

Goswami, D. Y. and Ileslamlou, S. 1990. Performance analysis of a closed-loop climate control system using undergroung air tunnel. *J. of Solar Energy Eng.* 112 (May): 76–81.

Givoni, B. 1994. *Passive and Low Energy Cooling of Buildings*, Van Nostrand Reinhold, New York.

Hay, H. and Yellott, J. 1969. Natural air conditioning with roof ponds and movable insulation. *ASHRAE Transactions*, 75 (1), 165–177.

Hollmuller, P. and Lachal, B. 2001. Cooling and preheating with buried pipe systems: Monitoring, simulation and economic aspects. *Energy and Buildings*, 33, 509–518.

IES (Illumination Engineering Society). 1987. *Lighting Handbook, Applications Volume.* Illumination Engineering Society, New York.

Kinney, L., McCluney, R., Cler, G., and Hutson, J. 2005. New designs in active daylighting: Good ideas whose time has (finally) come. In *Proceedings of the 2005 Solar World Congress.* August 6–12, 2005, ISES, Orlando, FL.

Krarti, M. and Kreider, J. F. 1996. Analytical model for heat transfer in an underground air tunnel. *Energy Conversion Management*, 37, 10, 1561–1574.

Kusuda, T. and Achenbach, P. R. 1965. Earth temperature and thermal diffusivity at selected stations in the United States. ASHRAE Transactions, 71 (1), 965.

Labs, K. 1981. Regional analysis of ground and above ground climate. Report ORNL/Sub-81/40451/1, Oak Ridge National Laboratory, Oak Ridge, TN.

Larson, R., Vignola, F., and West, R. eds. 1992. *Economics of Solar Energy Technologies.* American Solar Energy Society, Orlando, FL.

Libbey-Owens-Ford Company 1976. *How to Predict Interior Daylight Illumination.* Libbey-Owens-Ford Company, Toledo, OH.

Martin, M. and Berdahl, P. 1984. Characteristics of infrared sky radiation in the United States. *Solar Energy*, 33, 314, 321–336.

Marlatt,W., Murray, C., and Squire, S. 1984. Roof Pond Systems Energy Technology Engineering Center. Report No. ETEC 6, April. Rockwell International, New York.

McCluney, R. 1998. Advanced fenestration daylighting systems. *In International Daylighting Conference '98*, Natural Resources Canada/CETC, Ottawa, Canada.

McQuiston, P. C. and Parker, J. D. 1994. *Heating, Ventilating, and Air Conditioning. 4th Ed.*, Wiley, New York.

Murdoch, J. B. 1985. *Illumination Engineering—From Edison's Lamp to the Laser.* Macmillan, New York.

NCDC (National Climactic Data Center). 1992. *Climatography of the U.S. #81.* NCDC, Asheville, NC.

Olgyay, A. and Olgyay, V. 1967. *Solar Control and Shading Devices*, Princeton University Press, Princeton, NJ.

Pierson, O. 1962. *Acrylics for the Architectural Control of Solar Energy.* Rohm and Haas, Philadelphia, PA.

Robbins, C. L. 1986. *Daylighting—Design and Analysis*, Van Nostrand, New York.

Rundquist, R. A. 1991. Daylighting controls: Orphan of HVAC design. *ASHRAE Journal*, 11 (November), 30–34.

PSDH. 1980. *Passive Solar Design Handbook.* Volume One: Passive Solar Design Concepts, DOE/CS-0127/1, March 1980. Prepared by Total Environmental Action, Inc. (B. Anderson, C. Michal, P. Temple, and D. Lewis); Volume Two: *Passive Solar Design Analysis*, DOE/CS-0127/2, January 1980. Prepared by Los Alamos Scientific Laboratory (J.D. Balcomb, D. Barley, R. McFarland, J. Perry, W. Wray and S. Noll). U.S. Department of Energy, Washington, DC.

PSDH. 1984. *Passive Solar Design Handbook.* Part One: Total Environmental Action, Inc., Part Two: Los Alamos Scientific Laboratory, Part Three: Los Alamos National Laboratory. Van Nostrand Reinhold, New York.

For Further Information

The most complete basic reference for passive system heating design is still the Passive Solar Design Handbook, all three parts. Solar Today magazine, published by the American Solar Energy Society, is the most available source for current practice designs and economics, as well as a source for passive system equipment suppliers.The ASHRAE Handbook of Fundamentals is a general introduction to passive cooling techniques and calculations, with an emphasis on evaporative cooling. Passive Solar Buildings and Passive Cooling, both published by MIT Press, contain a large variety of techniques and details concerning passive system designs and economics. All the major building energy simulation codes (DOE-2, EnergyPlus, TRNSYS, TSB13, etc.) now include passive heating and cooling technologies.The Illlumination Engineering Society's Lighting Handbook presents the basis for and details of daylighting and artificial lighting design techniques. However, most texts on illumination present simplified format day lighting procedures. Currently used daylighting computer programs include various versions of Lumen Micro, Lightscape, and Radiance Passive Solar Design Strategies: Guidelines for Homebuilders (Passive Solar Industries Council, Washington, DC, 1989) presents a user-friendly approach to passive solar design.

18.4 Solar Cooling

D. Yogi Goswami and Sanjay Vijayaraghavan

In some ways, solar energy is better suited to space cooling and refrigeration than to space heating. The seasonal variation of solar energy is extremely well-suited to the space-cooling requirements of buildings. The principal factors affecting the temperature in a building are the average quantity of radiation received and the environmental air temperature. Because the warmest seasons of the year correspond to periods of high insolation, solar energy is most available when comfort cooling is most needed. Moreover, the efficiency of solar collectors increases with increasing insolation and increasing environmental temperature. Consequently, in the summer, the amount of energy delivered per unit surface area of collector can be larger than that in winter.

Solar cooling using various refrigeration cycles is technically feasible and has been demonstrated several times over past few decades. However, application of these systems has not become popular due to the unfavorable economics. The most widely used methods applied to solar cooling and air conditioning are vapor compression cycles, absorption-cooling cycles, and desiccant cooling. The vapor compression refrigeration cycle is probably the most widely used refrigeration cycle. The vapor compression refrigeration cycle requires energy input into the compressor which may be provided as electricity from a photovoltaic system or as mechanical energy from a solar driven heat engine. Referring to Figure 18.62, the compressor raises the pressure of the refrigerant, which also increases the temperature. The compressed high-temperature refrigerant vapor then transfers heat to the ambient environment in the condenser, where it condenses to a high-pressure liquid at a temperature close to the environmental temperature. The liquid refrigerant is then passed through the expansion valve where the pressure is suddenly reduced, resulting in a vapor–liquid mixture at a much lower temperature. The low-temperature refrigerant is then used to cool air or water in the evaporator where the liquid refrigerant evaporates by absorbing heat from the medium being cooled. The cycle is completed by the vapor returning to the compressor. If water is cooled in the evaporator, the device is usually called a *chiller*. The chilled water could then be used to cool air in a building.

In an absorption system, the refrigerant is evaporated or distilled from a less volatile absorbent, the vapor is condensed in a water- or air-cooled condenser, and the resulting liquid is passed through a pressure-reducing valve to the cooling section (evaporator) of the unit. The refrigerant from the evaporator flows into the absorber, where it is reabsorbed in the stripped absorbing liquid and

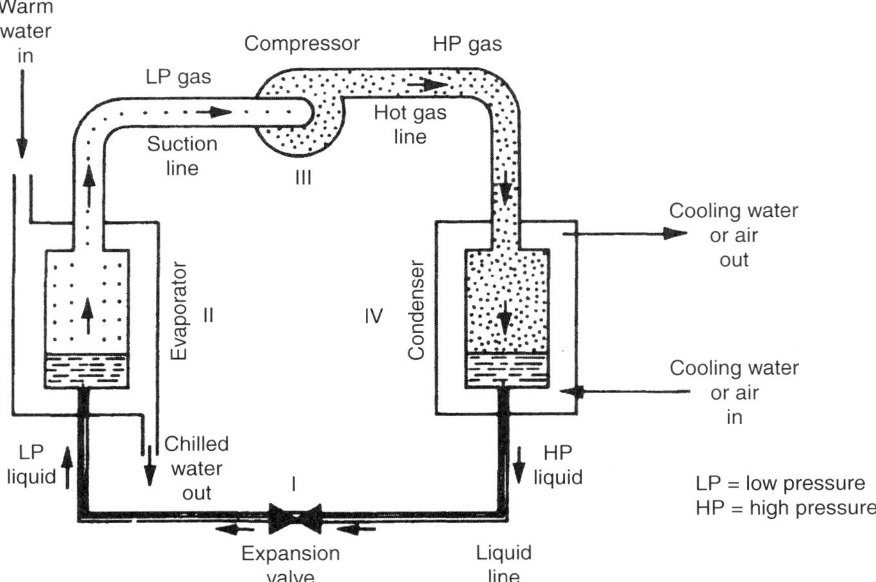

FIGURE 18.62 A schematic diagram showing a typical vapor compression refrigeration cycle.

pumped back to the heated generator. The heat required to evaporate the refrigerant in the generator can be supplied directly from solar energy as shown in Figure 18.63.

In humid climates, removal of moisture from the air represents a major portion of the air-conditioning load. In such climates, desiccant systems can be used for dehumidification, in which solar energy can provide most of the energy requirements. There are several passive space cooling techniques that are

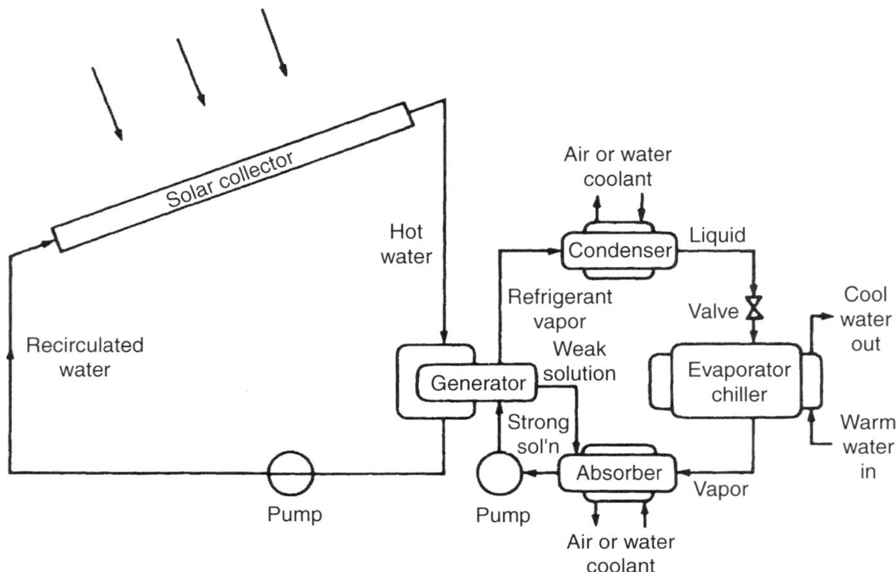

FIGURE 18.63 Figure shows the basic arrangement of a solar driven absorption cycle.

described elsewhere. The present section covers the active solar cooling techniques based on vapor compression and vapor-absorption refrigeration cycles and desiccant humidification.

18.4.1 Vapor Compression Cycle

The principle of operation of a vapor compression refrigeration cycle can be illustrated conveniently with the aid of a pressure–enthalpy diagram as shown in Figure 18.64. The ordinate is the pressure of the refrigerant in N/m^2 absolute, and the abscissa its enthalpy in kJ/kg. The roman numerals in Figure 18.64 correspond to the physical locations in the schematic diagram of Figure 18.62. Process I is a throttling process in which hot liquid refrigerant at the condensing pressure p_c passes through the expansion valve, where its pressure is reduced to the evaporator pressure, p_e. This is an isenthalpic (constant enthalpy) process in which the temperature of the refrigerant decreases. In this process, some vapor is produced and the state of the mixture of liquid refrigerant and vapor entering the evaporator is shown by point A. Because the expansion process is isenthalpic, the following relation holds:

$$h_{ve}f + h_{le}(1-f) = h_{lc},$$

where f is the fraction of mass in vapor state, subscripts "v" and "l" refer to vapor and liquid states, respectively, and "c" and "e" refer to states corresponding to condenser and evaporator pressures, respectively.

Process II represents the vaporization of the remaining liquid. This is the process during which heat is removed from the chiller. Thus, the specific refrigeration effect (per kilogram of refrigerant flow), q_r, is

$$q_r = h_{ve} - h_{lc}, \text{ in kJ/kg}.$$

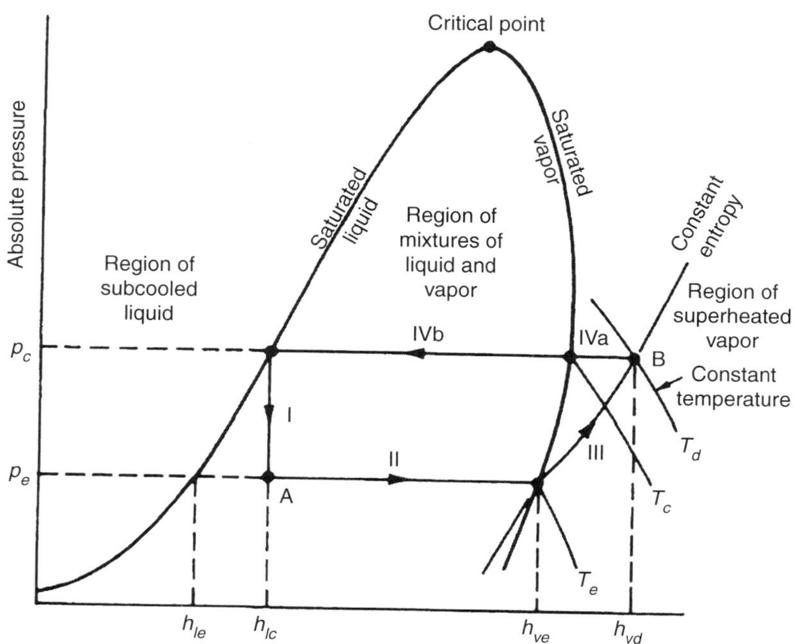

FIGURE 18.64 The thermodynamic state processes of the vapor compression refrigeration cycle shown on a pressure–enthalpy (p–h) diagram.

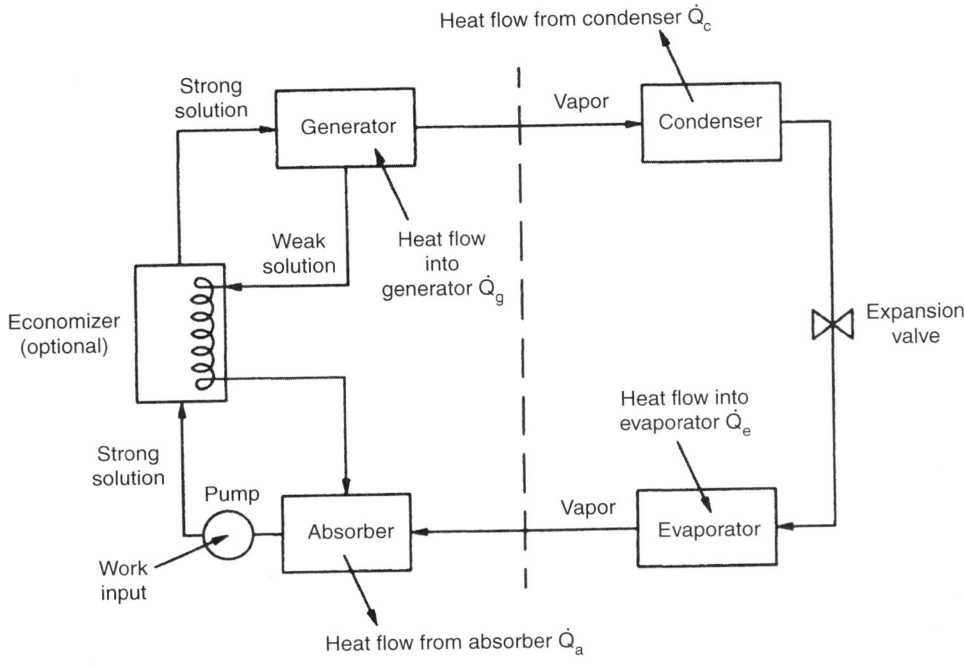

FIGURE 18.65 A typical absorption refrigeration cycle.

In the United States, it is still common practice to measure refrigeration in terms of tons. One ton is the amount of cooling produced if 1 ton of ice is melted over a period of 24 h. One ton of cooling is equivalent to 3.516 kW, or 12,000 Btu/h.

Process III in Figure 18.64 represents the compression of refrigerant from pressure p_e to pressure p_c. The process requires work input from an external source, which may be obtained from a solar-driven expander turbine or a solar electrical system. In general, if the heated vapor leaving the compressor is at the condition represented by point B in Figure 18.64, the work of compression is

$$W_c = \dot{m}_r(h_{vd} - h_{ve}).$$

In an idealized cycle analysis, the compression process is usually assumed to be isentropic.

Process IV represents the condensation of the refrigerant. Actually, sensible heat is first removed in subprocess IVa as the vapor is cooled at constant pressure from T_d to T_c and latent heat is removed at the condensation temperature T_c corresponding to the saturation pressure, p_c, in the condenser. The heat transfer rate in the condenser, $\dot{Q}_c$, is

$$\dot{Q}_c = \dot{m}_r(h_{vd} - h_{lc}).$$

This heat must be rejected to the environment, either to cooling water or to the atmosphere if no water is available.

The overall performance of a refrigeration machine is usually expressed as the coefficient of performance that is defined as the ratio of the heat transferred in the evaporator, $\dot{Q}_r$, to the shaft work supplied by the compressor:

$$\text{COP} = \frac{\dot{Q}_r}{W_c} = \frac{h_{ve} - h_{lc}}{h_{vd} - h_{ve}}.$$

18.4.2 Absorption Air Conditioning

Absorption air conditioning is compatible with solar energy because a large fraction of the energy required is thermal energy at temperatures that solar collectors can easily provide. Low- and medium-temperature solar collectors have been used to drive several absorption air conditioning systems (Macriss and Zawacki 1989; Mathur 1989; Manrique 1991; Chinnappa and Wijeysundera 1992; Siddiqui 1993; Thornbloom and Nimmo 1994; Hewett 1995). Although single-effect absorption refrigeration systems can be run using solar heated hot water at 80°C, higher temperatures are preferred for better refrigeration cycle performance. The key difference between a conventional gas-fired absorption chiller and one used for solar applications is the larger heat transfer area required to make the cycle work using the lower driving temperatures available in solar applications. Figure 18.65 shows a schematic of an absorption refrigeration system. Absorption refrigeration differs from vapor compression air conditioning only in the method of compressing the refrigerant (left of the dashed line in Figure 18.65). In absorption air conditioning systems, the pressurization is accomplished by first dissolving the refrigerant in a liquid (the absorbent) in the absorber section, then pumping the solution to a high pressure with a liquid pump. The low-boiling refrigerant is then driven from solution by the addition of heat in the generator. By this means, the refrigerant vapor is compressed without the large input of high-grade shaft work that the vapor compression cycle demands.

The effective performance of an absorption cycle depends on the two materials that comprise the refrigerant–absorbent pair. Desirable characteristics for the refrigerant–absorbent pair are as follows:

1. Absence of a solid-phase sorbent
2. A refrigerant more volatile than the absorbent so that separation from the absorbent occurs easily in the generator
3. An absorbent that has a strong affinity for the refrigerant under conditions in which absorption takes place
4. A high degree of stability for long-term operations
5. Nontoxic and nonflammable fluids for residential applications; this requirement is less critical in industrial refrigeration
6. A refrigerant that has a large latent heat so that the circulation rate can be kept low
7. A low fluid viscosity that improves heat and mass transfer and reduces pumping power
8. Fluids that do not have long-term environmental effects

Lithium bromide–water (LiBr–H_2O) and ammonia–water (NH_3–H_2O) are the two pairs that meet most of the requirements and have been used commercially in several applications. In the LiBr–H_2O system, water is the refrigerant and LiBr is the absorbent, whereas in the ammonia–water system, ammonia is the refrigerant and water is the absorbent. Because the LiBr–H_2O system has a high-volatility ratio, it can operate at lower pressures and therefore, at the lower generator temperatures achievable by flat-plate collectors. A disadvantage of this system is that LiBr has a tendency to crystallize in the stream returning from the generator. Crystallization is avoided by careful system design and by the use of additives. Furthermore, because the refrigerant is water, the system evaporator cannot be operated at or below the freezing point of water. Therefore, the LiBr–H_2O system is operated at evaporator temperatures of 5°C or higher. Using a mixture of LiBr with some other salt as the absorbent can overcome the crystallization problem. The ammonia–water system has the advantage that the evaporator can be maintained at very low temperatures. However, for temperatures much below 0°C, water vapor must be removed from ammonia as much as possible to prevent ice crystals from forming. This requires a rectifying column after the boiler. Also, ammonia is a safety code group B2 fluid (ASHRAE Standard 34-1992) that restricts its use indoors (ASHRAE 1997). Consequently, the ammonia–water system cannot use a direct expansion (DX) evaporator. Other refrigerant–absorbent pairs include (Macriss and Zawacki 1989):

- Ammonia–salt
- Methylamine–salt

- Alcohol–salt
- Ammonia–organic solvent
- Sulfur dioxide–organic solvent
- Halogenated hydrocarbons–organic solvent
- Water–alkali nitrate
- Ammonia–water–salt

If the pump work is neglected, the COP of an absorption air conditioner can be calculated from Figure 18.65:

$$\mathrm{COP} = \frac{\text{cooling effect}}{\text{heat input}} = \frac{\dot{Q}_e}{\dot{Q}_g}.$$

The COP values for absorption air conditioning range from 0.5 for a small, single-stage unit to 0.85 for a double-stage, steam-fired unit. Another figure of merit for absorption systems is the ratio of the cooling effect to work supplied to the system (circulation pumps, fans, etc.).

Explicit procedures for the mechanical and thermal design as well as the sizing of the heat exchangers are presented in standard heat transfer texts. In large commercial units, it may be possible to use higher concentrations of LiBr, operate at a higher absorber temperature, and thus save on heat-exchanger costs. In a solar-driven unit, this approach would require concentrator-type or high-efficiency flat-plate solar collectors.

18.4.2.1 Ammonia–Water Systems

The main difference between an ammonia–water system and a water–lithium bromide system is that a small amount of absorbent (water) also evaporates along with the refrigerant (ammonia) in the vapor generator. Therefore, ammonia–water systems use a rectifier (also called a *dephlagmator*) after the generator to condense as much water vapor out of the vapor mixture as possible. Figure 18.66 shows a schematic of an NH_3–H_2O absorption refrigeration system. Because ammonia has a much lower boiling point than water, a very high fraction of ammonia and a very small fraction of water are boiled off in the

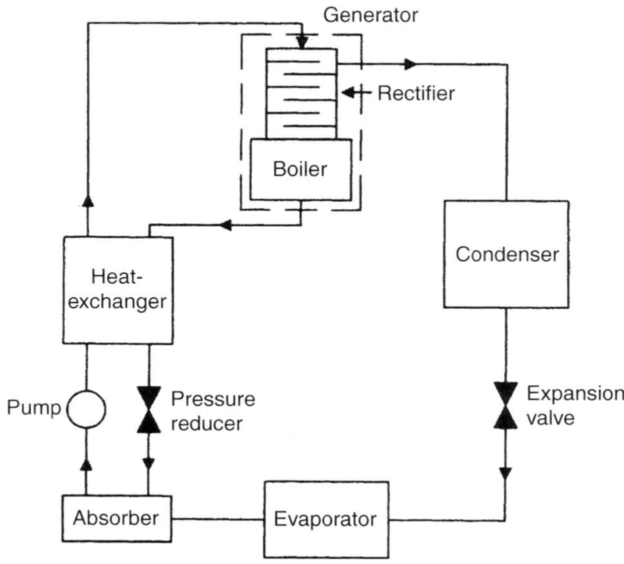

FIGURE 18.66 A diagram showing the arrangement of components for an ammonia-absorption cycle.

boiler. The vapor is cooled as it rises in the rectifier by the countercurrent flow of the strong NH_3–H_2O solution from the absorber; therefore, some moisture is condensed. The weak ammonia–water solution from the boiler goes through a pressure-reducing valve to the absorber, where it absorbs the ammonia vapor from the evaporator. The high-pressure ammonia from the rectifier is condensed by rejecting heat to the atmosphere. It may be further subcooled before expanding in a throttle valve. The two-phase low-temperature ammonia from the throttle valve provides refrigeration in the evaporator. The vapor from the evaporator is recombined with the weak ammonia solution in the absorber. Operating pressures are primarily controlled by the ambient air temperature for an air-cooled condenser, the evaporator temperature, and the concentration of the ammonia solution in the absorber.

18.4.2.2 Multieffect Systems

A major price component in designing absorption systems is the solar collector field. To improve the economics of a solar absorption system, the efficiency of the solar collectors must be improved in addition to the COP of the absorption system, thus reducing the required collector area. A single-effect absorption system has a typical efficiency of around 0.7. For higher COP, double-effect systems are used. Double-effect systems typically operate at higher temperatures than single-effect systems requiring higher concentration solar collectors to provide the heat input. An example of a typical double-effect lithium bromide system is shown in Figure 18.67.

A double-effect lithium bromide cycle has two generators at two different pressure levels. Vapor is generated using the solar heat source in the first generator. This vapor is condensed in the second generator and the heat of condensation is used to produce more vapor (this arrangement is known as a *condenser-coupled system*). Thus the double-effect absorption cycle is a triple pressure cycle. A double-effect ammonia–water system is configured slightly differently (absorber-coupled), but still uses the same

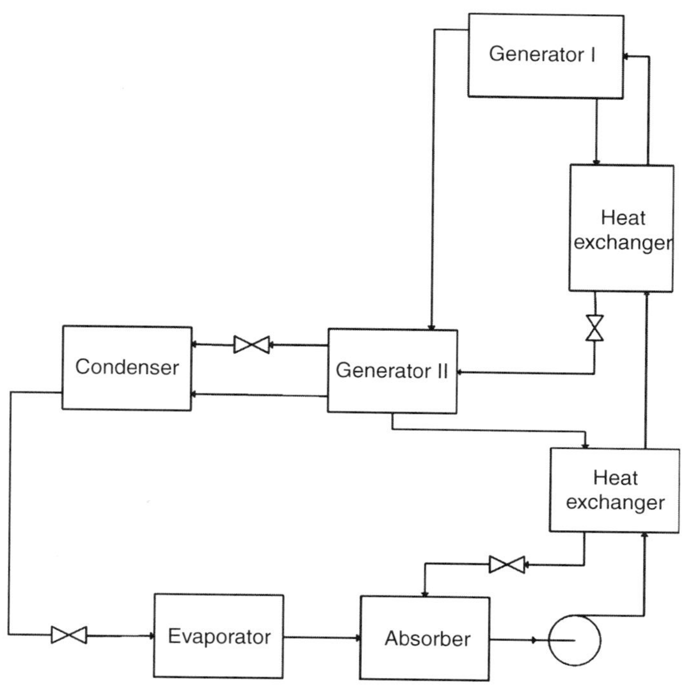

FIGURE 18.67 A schematic diagram of a condenser coupled double-effect absorption cooling cycle. (Adapted from Wahlig, M., In *Active Solar Systems*, Vol. 6 of Solar Heat Technologies: Fundamentals and Applications, MIT Press, Cambridge, MA, 747–854, 1988. With permission.)

principle of internal heat recovery to produce more refrigerant vapor than is possible in a single-effect system. However, it requires a much higher driving temperature (140°C or higher) to operate efficiently.

18.4.3 Solar Desiccant Dehumidification

In hot and humid regions of the world experiencing significant latent cooling demand, solar energy may be used for dehumidification using liquid or solid desiccants. Rangarajan, Shirley, and Raustad (1989) compared a number of strategies for ventilation air conditioning for Miami, Florida, and found that a conventional vapor compression system could not even meet the increased ventilation requirements of ASHRAE Standard 62-1989. By pretreating the ventilation air with a desiccant system, proper indoor humidity conditions could be maintained and significant electrical energy could be saved. A number of researchers have shown that a combination of a solar desiccant and a vapor compression system can save from 15 to 80% of the electrical energy requirements in commercial applications such as supermarkets (Meckler 1988; Meckler, Parent, and Pesaran 1993; Meckler 1994; Meckler 1995; Spears and Judge 1997; Oberg and Goswami 1998a, 1998b).

In a desiccant air conditioning system, moisture is removed from the air by bringing it in contact with the desiccant, followed by sensible cooling of the air by a vapor compression cooling system, vapor-absorption cooling system, or evaporative cooling system. The driving force for the process is the water vapor pressure. When the vapor pressure in air is higher than on the desiccant surface, moisture is transferred from the air to the desiccant until an equilibrium is reached (see Figure 18.67). To regenerate the desiccant for reuse, the desiccant is heated, which increases the water vapor pressure on its surface. If air with lower vapor pressure is brought into contact with this desiccant, the moisture passes from the desiccant to the air (Figure 18.68). Two types of desiccants are used: solids such as silica gel and lithium chloride, or liquids such as salt solutions and glycols.

The two solid desiccant materials that have been used in solar systems are silica gel and molecular sieves, a selective absorber. Figure 18.69 shows the equilibrium absorption capacity of several substances. Note that molecular sieves has the highest capacity up to 30% humidity, and silica gel is optimal between 30 and 75%, the typical humidity range for buildings.

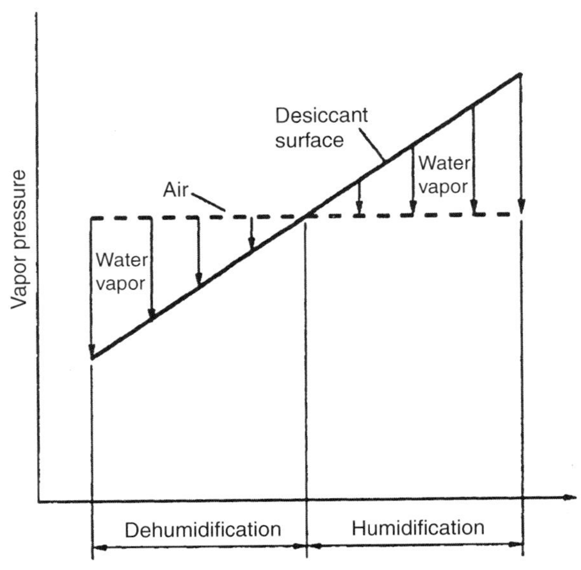

FIGURE 18.68 Vapor pressure vs. temperature and water content for desiccant and air.

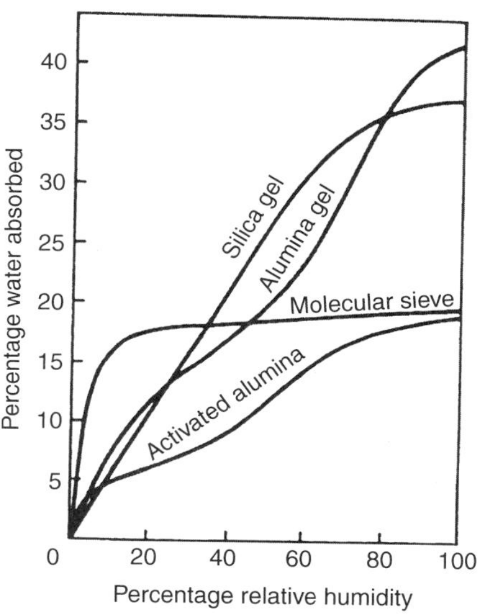

FIGURE 18.69 Equilibrium capacities of common water absorbents.

Figure 18.70 is a schematic diagram of a desiccant cooling ventilation cycle (also known as the Pennington cycle), which achieves both dehumidification and cooling. The desiccant bed is normally a rotary wheel of a honeycomb-type substrate impregnated with the desiccant. As the air passes through the rotating wheel, it is dehumidified while its temperature increases (processes 1 and 2) due to the latent heat of condensation. Simultaneously, a hot air stream passes through the opposite side of the rotating wheel, which removes moisture from the wheel. The hot and dry air at state 2 are cooled in a heat exchanger wheel to condition 3 and further cooled by evaporative cooling to condition 4. Air at condition 3 may be further cooled by vapor compression or vapor absorption systems instead of evaporative cooling. The return air from the conditioned space is cooled by evaporative cooling (processes 5 and 6), which in turn cools the heat exchanger wheel. This air is then heated to condition 7. Using solar heat, it is further heated to condition 8 before going through the desiccant wheel to regenerate the desiccant. A number of researchers have studied this cycle, or an innovative variation of it, and have found thermal COPs in the range of 0.5–2.58 (Pesaran, Penney, and Czandema 1992).

18.4.4 Liquid-Desiccant Cooling System

Liquid desiccants offer a number of advantages over solid desiccants. The ability to pump a liquid desiccant makes it possible to use solar energy for regeneration more efficiently. It also allows several small dehumidifiers to be connected to a single regeneration unit. Because a liquid desiccant does not require simultaneous regeneration, the liquid may be stored for regeneration later when solar heat is available. A major disadvantage is that the vapor pressure of the desiccant itself may be enough to cause some desiccant vapors to mix with the air. This disadvantage, however, may be overcome by proper choice of the desiccant material.

A schematic of a liquid desiccant system is shown in Figure 18.71. Air is brought into contact with concentrated desiccant in a countercurrent flow in a dehumidifier. The dehumidifier may be a spray column or packed bed. The packings provide a very large area for heat and mass transfer between the air and the desiccant. After dehumidification, the air is sensibly cooled before entering the conditioned space. The dilute desiccant exiting the dehumidifier is regenerated by heating and exposing it to a countercurrent flow of a moisture-scavenging air stream. Liquid desiccants commonly used are aqueous solutions of lithium bromide, lithium chloride, calcium chloride, mixtures of these solutions, and

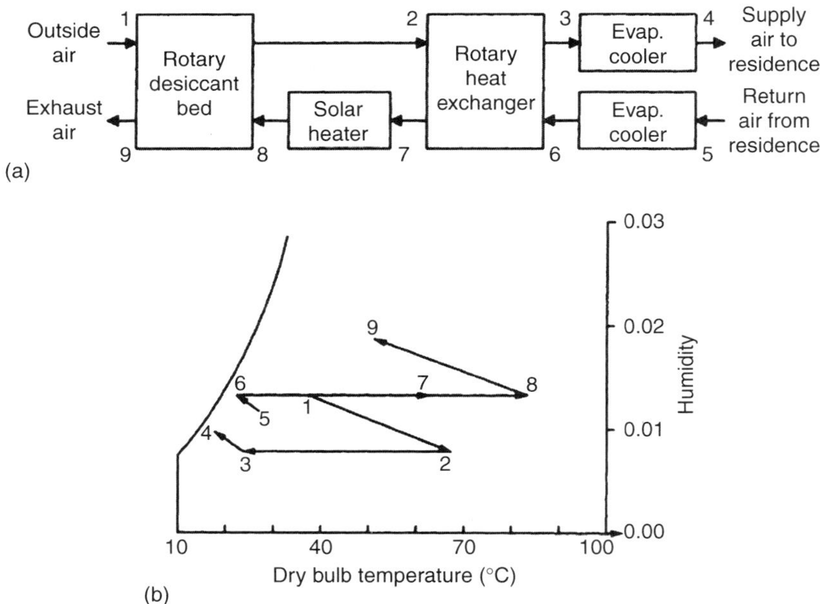

FIGURE 18.70 Schematic of a desiccant cooling ventilation cycle, (a) schematic of air flow, (b) the process on a psychrometric chart.

triethylene glycol (TEG). (See Oberg and Goswami 1998b). Vapor pressures of these common desiccants are shown in Figure 18.72 as a function of concentration and temperature, based on a number of references (Cyprus Foote Mineral Company; Dow Chemical Company 1992, 1996; Ertas, Anderson, and Kiris 1992; Zaytsev and Aseyev 1992). Other physical properties important in the selection of desiccant

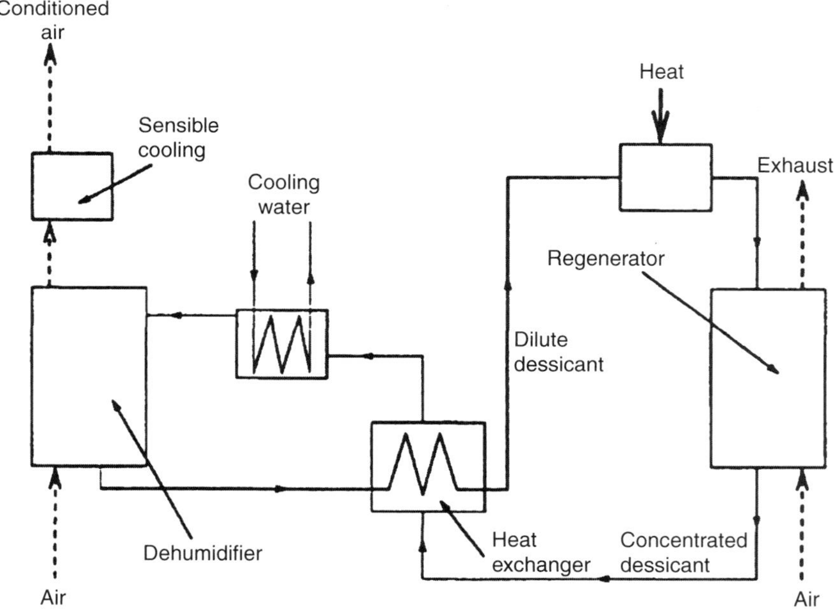

FIGURE 18.71 A conceptual liquid-desiccant cooling system.

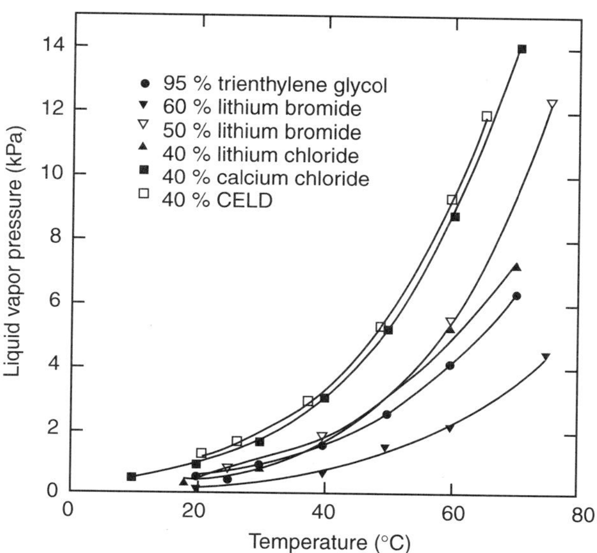

FIGURE 18.72 Vapor pressures of liquid desiccants.

materials are listed in Table 18.15. Although salt solutions and TEG have similar vapor pressures, the salt solutions are corrosive and have higher surface tension. The disadvantage of TEG is that it requires higher pumping power because of its higher viscosity.

Oberg and Goswami (1998b) have presented an in-depth review of liquid-desiccant cooling systems. Based on an extensive numerical modeling and on experimental studies, they have presented correlations for the performance of a packed-bed liquid-desiccant dehumidifier and a regenerator.

The performance of a packed-bed dehumidifier or a regenerator may be represented by a humidity effectiveness, ε_y, defined as the ratio of the actual change in humidity of the air to the maximum possible for the operating conditions (Ullah, Kettleborough, and Gandhidasan 1988; Chung 1989; Khan 1994):

$$\varepsilon_y = \frac{Y_{in} - Y_{out}}{Y_{in} - Y_{eq}},$$

where Y_{in} and Y_{out} are the humidity ratios of the air inlet and outlet, respectively, and Y_{eq} is the humidity ratio in equilibrium with the desiccant solution at the local temperature and concentration (Figure 18.73).

In addition to the humidity effectiveness, an enthalpy effectiveness, ε_H, is also used as a performance parameter (Khan 1994; Kettleborough and Waugaman 1995):

$$\varepsilon_H = \frac{H_{a,in} - H_{a,out}}{H_{a,in} - H_{a,eq}},$$

where $H_{a,in}$, $H_{a,out}$, and $H_{a,eq}$ are the enthalpies of the air at the inlet and outlet, and in equilibrium with the desiccant, respectively.

Oberg and Goswami (1998a) found the following correlation for ε_y and ε_H:

$$\varepsilon_y, \varepsilon_H = 1 - C_1 (L/G)^a (H_{a,in}/H_{L,out})^b (aZ)^c,$$

TABLE 18.15 Physical Properties of Liquid Desiccants at 25°C

Desiccant	Density, $\rho \times 10^{-3}$ (kg/m^3)	Viscosity, $\mu \times 10^3$ (Ns/m^2)	Surface Tension, $\gamma \times 10^3$ (N/m)	Specific Heat, c_p (kJ/kg °C)	Reference
95% by weight triethylene glycol	1.1	28	46	2.3	Thornbloom and Nimmo (1996)
55% by weight lithium bromide	1.6	6	89	2.1	Gordon and Rabl (1986); Oberg and Goswami (1998a, 1998b)
40% calcium chloride	1.4	7	93	2.5	Gordon and Rabl (1986); Siddiqui (1993); Spears and Judge (1997)
40% by weight lithium chloride	1.2	9	96	2.5	Gordon and Rabl (1986)
40% by weight CELD	1.3	5	—	—	Cyprus Foote Mineral Company

Source: From Oberg, V. and Goswami, D. Y., *Advances in Solar Energy*, 12, 431–470, 1998.

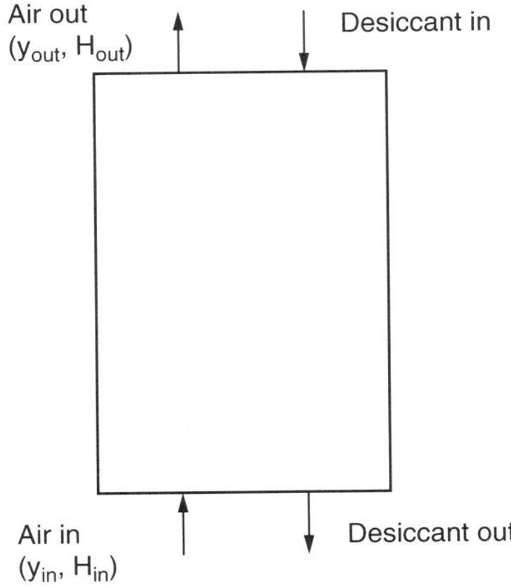

FIGURE 18.73 Exchange of humidity and moisture between desiccant and air in the tower.

TABLE 18.16 Constants for Performance Correlations

	C_1	B	k_1	M_1	k_2	m_2
ε_y	48.345	−0.751	0.396	−1.573	0.033	−0.906
ε_H	3.766	−0.528	0.289	−1.116	0.004	−0.365

where

$$a = k_1 \frac{\gamma_L}{\gamma_c} + m_1,$$

$$c = k_2 \frac{\gamma_L}{\gamma_c} + m_2.$$

Here, C_1, b_1, k_1, m_1, and m_2 are constants listed in Table 18.16. L and G are the liquid and air mass-flow rates, respectively; a is the packing surface area per unit volume for heat and mass transfer in m^2/m^3; Z is the tower height in meters; γ_L is the surface tension of the liquid desiccant; and γ_c is the critical surface tension for the packing material.

Although liquid-desiccant cooling systems are not an off-the-shelf variety currently on the market, there are a number of examples of their use, especially in hybrid combinations with conventional vapor compression systems. In a hybrid system, the liquid desiccant would remove moisture from the air, allowing the vapor compression system to be downsized. Hybrid systems are especially useful in their ability to maintain comfort conditions in hot and humid climates, where conventional high-efficiency systems usually have trouble maintaining low humidity. Mago and Goswami (2003) and Mago (2003a, 2003b) did a simple cost analysis of a hybrid system for a residential building and a supermarket. They found that a house that typically requires a 5-tn. conventional air conditioning unit in Florida could use a hybrid system consisting of a desiccant tower of height 1.1 m, and a vapor compression system of 2 tn.

The total cost of the conventional system was estimated at $4000, whereas the hybrid system would be $3250 ($1000 for the desiccant system + $2250 for a 2-tn. vapor compression system). A solar system for regeneration of the desiccant was estimated at $6500, making the total costs of a solar hybrid liquid-desiccant system at $9780. They estimated that based on the electrical savings, a simple payback period would be 9.5 years. The payback period is high mainly because the solar regeneration system is not utilized throughout the year. For applications where the system can be used year-around, the payback period would be much shorter.

For a commercial application in a supermarket where the hybrid desiccant system and the solar regeneration system would be used throughout the year, the payback period would be less than two years (Mago and Goswami 2003; Mago 2003a). In this application, Mago also estimated that a conventional 13-tn. system, at an estimated cost of $9300, would be replaced by a hybrid system consisting of a 2-m-high desiccant tower at a cost of $1500 and a 5-tn. vapor compression system at an estimated cost of $4000. A solar regeneration system for this desiccant system was estimated at $9300. The two-year payback period confirmed the estimate of the U.S. Department of Energy (1996) that a simple payback is typically less than five years.

References

ASHRAE (American Society of Heating, Refrigerating and Air-Conditioning Engineers). 1997. Fundamentals. In *ASHRAE Handbook*, ASHRAE, Atlanta, GA.

Chinnappa, J. C. V., and Wijeysundera, N. E. 1992. Simulation of solar-powered ammonia-water integrated hybrid cooling system. *Journal of Solar Energy Engineering*, 114, 125–127.

Chung, T. -W. 1989. Predictions of the moisture removal efficiencies for packed-bed dehumidification systems. In *Solar engineering—1989, Proceedings of the 11th Annual ASME Solar Energy Conference*, 371–377, ASMF, New York.

Cyprus Foote Mineral Company. Technical data on lithium bromide and lithium chloride. Bulletins 145 and 151, Cyprus Foote Mineral Company, Kings Mountain, NC.

Dow Chemical Company. 1996. *Calcium chloride handbook*, Dow Chemical Company, Midland, MI.

Dow Chemical Company. 1992. *A guide to glycols*, Dow Chemical Company, Midland, MI.

Ertas, A., Anderson, E. E. and Kiris, I. 1992. Properties of a new liquid desiccant solution—Lithium chloride and calcium chloride mixture. *Solar Energy*, 49, 205–212.

Hewett, R. 1995. Solar absorption cooling: An innovative use of solar energy. In *AlChE Symposium Series, No. 306*, Vol. 91, AICHE, New Year.

Kettleborough, C. F., and Waugaman, D. G. 1995. An alternative desiccant cooling cycle. *Journal of Solar Energy Engineering*, 117, 251–255.

Khan, A. Y. 1994. Sensitivity analysis and component modeling of a packed-type liquid desiccant system at partial load operating conditions. *International Journal of Energy Research*, 18, 643–655.

Macriss, R. A., and Zawacki, T. S. 1989. Absorption fluid data survey: 1989 update. Oak Ridge National Laboratory Report, ORNL/Sub84-47989/4.

Mago, P. J. 2003a. Sistema Hibrido de Enfriamiento como alternative al sistema convencional de aire acondicionado. *Revista SABER*, 15(1), 39–44.

Mago, P. J. 2003b. Analisis economico de la utilización de un sistema hibrido enfriamientote liquido secante en aplicaciones residenciales y comerciales. *Revista SABER*, 15(1), 45–50.

Mago, P. J., and Goswami, D. Y. 2003. Study of the performance of a hybrid liquid desiccant system using lithium chloride. *ASME Journal of Solar Energy Engineering*, 125(1), 129–131.

Manrique, J. A. 1991. Thermal performance of ammonia-water refrigeration system. *International Communications in Heat and Mass Transfer*, 19(6), 779–789.

Mathur, G. D. 1989. Solar-operated absorption coolers. *Heating/Piping/Air Conditioning*. (November). 61, 11, 103–108.

Meckler, M. 1988. Off-peak desiccant cooling and cogeneration combine to maximize gas utilization. *ASHRAE Transactions*, 94(Part 1), 575–596.

Meckler, H. 1994. Desiccant-assisted air conditioner improves IAQ and comfort. *Heating, Piping & Air Conditioning*, 66(10), 75–84.

Meckler, M. 1995. Desiccant outdoor air preconditioners maximize heat recovery ventilation potentials. *ASHRAE Transactions*, 101(Part 2), 992–1000.

Meckler, M., Parent, Y. O. and Pesaran, A. A. 1993. Evaluation of dehumidifiers with polymeric desiccants. Gas Institute Report, Contract No. 5091-246-2247, Gas Research Institute, Chicago, IL.

Oberg, V., and Goswami, D. Y. 1998a. Experimental study of heat and mass transfer in a packed bed liquid desiccant air dehumidifier. In *Solar engineering*, Morehouse J. H. and Hogan, R. E. ed., ASMF, New York, pp. 155–166.

Oberg, V., and Goswami, D. Y. 1998b. A review of liquid desiccant cooling. *Advances in Solar Energy*, 12, 431–470.

Pesaran, A. A., Penney, T. R. and Czandema, A. W. 1992. Desiccant cooling: State-of-the-art assessment. Report No. NREL/TP-254-4147, National Renewable Energy Laboratory, Golden, CO.

Rangarajan, K., Shirley, III, D. B. and Raustad, R. A. 1989. Cost-effective HVAC technologies to meet ASHRAE Standard 62-1989 in hot and human climates. *ASHRAE Transactions*. (Part 1), 166–182.

Siddiqui, A. M. 1993. Optimum generator temperatures in four absorption cycles using different sources of energy. *Energy Conversion and Management*, 34(4), 251–266

Spears, J. W., and Judge, J. 1997. Gas-fired desiccant system for retail super center. *ASHRAE Journal*, 39, 65–69.

Thornbloom, M. and Nimmo, B. 1994. Modification of the absorption cycle for low generator firing temperatures. Solar Engineering 1994, In *Proceedings of the Joint Solar Energy Engineering Conference ASME* 1994, ASME, New York, 367–372.

Thornbloom, M. and Nimmo, B. 1996. Impact of design parameters on solar open cycle liquid desiccant dehumidification system, In *Solar '96 Proceedings of the 1996 Annual Conference of the American Solar Energy Society*, American Solar Energy Society, Boulder, Colorado, pp. 107–111.

Ullah, M. R., Kettleborough, C. F., and Gandhidasan, P. 1988. Effectiveness of moisture removal for an adiabatic counterflow packed tower absorber operating with CaCl-air contact system. *Journal of Solar Energy Engineering*, 110, 98–101.

US Department of Energy. 1996. Desiccant cooling programs. What's new in building energy research. Report DOE/GO-10096-084, Washington, DC.

Zaytsev, 1.0. and Aseyev, G. G. 1992. *Properties of aqueous solutions of electrolytes*, CRC Press, Boca Raton, FL.

19

Concentrating Solar Thermal Power

19.1 Introduction and Context **19**-2
19.2 Solar Concentration and CSP Systems **19**-6
Why Use Concentrating Solar Energy Systems? Dependence of Efficiency on Temperature
19.3 Solar Concentrator Beam Quality **19**-9
19.4 Solar Concentration Ratio: Principles and Limitations of CSP Systems **19**-13
19.5 Solar Thermal Power Plant Technologies **19**-15
19.6 Parabolic Trough Solar Thermal Power Plants **19**-18
The Operational Principles and Components of the PTC • Performance Parameters and Losses in a PTC • Efficiencies and Energy Balance in a PTC • Industrial Applications for PTCs • Sizing and Layout of Solar Fields with PTCs • Electricity Generation with PTCs • Thermal Storage Systems for PTCs • Direct Steam Generation • The SEGS Plants and State-of-the-Art of Solar Power Plants with PTCs
19.7 Central Receiver Solar Thermal Power Plants.............. **19**-50
Technology Description • Water/Steam Plants: The PS10 Project • Molten Salt Systems: Solar Two
19.8 Volumetric Atmospheric Receivers: PHOEBUS and Solair **19**-80
19.9 Solar Air Preheating Systems for Combustion Turbines: The SOLGATE Project.................................. **19**-82
19.10 Dish/Stirling Systems.. **19**-85
System Description • Dish/Stirling Developments
19.11 Market Opportunities **19**-91
19.12 Conclusions **19**-92
References **19**-92

Manuel Romero-Alvarez and Eduardo Zarza

Plataforma Solar de Almeria-CIEMAT

19.1 Introduction and Context

Solar energy has a high exergetic value since it originates from processes occurring at the sun's surface at a blackbody equivalent temperature of approximately 5777 K. Due to this high exergetic value, more than 93% of the energy may be theoretically converted to mechanical work by thermodynamic cycles (Winter, Sizmann, and Vant-Hull 1991), or to Gibbs free energy of chemicals by solarized chemical reactions (Kodama 2003), including promising hydrogen production processes (Seinfeld 2005). According to thermodynamics and Planck's equation, the conversion of solar heat to mechanical work or Gibbs free energy is limited by the Carnot efficiency, and therefore to achieve maximum conversion rates, the energy should be transferred to a thermal fluid or reactants at temperatures close to that of the sun.

Even though solar radiation is a source of high temperature and energy at origin, with a high radiosity of 63 MW/m^2, sun–earth geometrical constraints lead to a dramatic dilution of flux and to irradiance available for terrestrial use only slightly higher than 1 kW/m^2, and consequently, supply of low temperatures to the thermal fluid. It is therefore an essential requisite for solar thermal power plants and high-temperature solar chemistry applications to make use of optical concentration devices that enable the thermal conversion to be carried out at high solar flux and with relatively little heat loss. A simplified model of a concentrating solar thermal power (CSP) plant is depicted in Figure 19.1.

The optimum CSP system design combines a relatively large, efficient optical surface (e.g., a field of high-reflectivity mirrors), harvesting the incoming solar radiation and concentrating it onto a solar receiver with a small aperture area. The solar receiver is a high-absorptance and transmittance, low-reflectance, radiative/convective heat exchanger that emulates as closely as possible the performance of a radiative black body. An ideal solar receiver would thus have negligible convection and conduction losses. In the case of a solar thermal power plant, the solar energy is transferred to a thermal fluid at an outlet temperature high enough to feed a heat engine or a turbine that produces electricity. The solar thermal element can be a parabolic trough field, a linear Fresnel reflector field, a central receiver system or a field of parabolic dishes, commonly designed for a normal incident radiation of 800–900 W/m^2. Annual

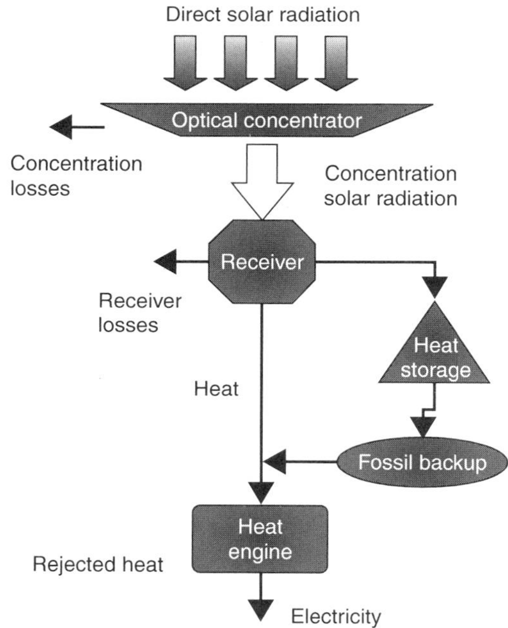

FIGURE 19.1 Flow diagram for a typical solar thermal power plant.

normal incident radiation varies from 1600 to 2800 kWh/m^2 depending on the available radiation at the particular site. This rate assumes 2000–3500 annual full-load operating hours with the solar element.

Solar transients and fluctuation in irradiance can be mitigated by using an oversized mirror field (solar multiple higher than 1) and then making use of the excess energy to load a thermal or chemical storage system. Hybrid plants with fossil backup burners connected in series or in parallel are also possible. The use of heat storage systems and fossil backup makes CSP systems highly flexible for integration with conventional power plant design and operation and for blending the thermal output with fossil fuel, biomass and geothermal resources (Mancini, Kolb, and Prairie 1997). The use of large solar multiples with low-cost heat storage systems of up to 12 h (equivalent at nominal power), facilitates the design of firm capacity plants supplying between 2000 and 6000 h of operation (equivalent at full load). In addition, hybridization is possible in power booster and fuel saver modes with natural gas combined cycles and coal-fired Rankine plants, and may accelerate near-term deployment of projects due to improved economics and reduced overall project risk (Kolb 1998). As a consequence, CSP can currently supply transferable power and meet peaking and intermediate loads at the lowest electricity costs of any grid-connected solar technology (Tyner et al. 2000).

Additional advantages of CSP are (Morse 2000):

- Proven capabilities, e.g., 354 MW of trough plants in operation in California since 1985 have selectively demonstrated excellent performance, availability, a reduction in investment cost of almost 50%, and significant reductions in O&M cost.
- Modular, and thus suitable for large central facilities in the 100 s of MW down to distributed generation in the 10 s of kW.
- Can be rapidly deployed using entirely domestic resources and existing infrastructure.
- Scale can be significant enough to impact climate change targets.
- Suitable for both IPP (Independent Power Producer) and turnkey projects.
- Proven potential for further cost reduction, including those resulting from economies of scale, i.e., from mass production of glass, steel, etc.

Electricity production with concentrating solar thermal technologies is not an innovation of the last few years. The French mathematician Augustin Mouchot built a machine able to convert concentrating solar power into mechanical work to run a printing press by generating steam (Pifre 1882). A showcase with a parabolic dish connected to an engine and producing electricity was presented in Paris in 1878. Other remarkable pioneers that deserve recognition are Ericsson (1888), Eneas (1901), Shuman (1913), and Francia (1968). However solar thermal concentrating technologies were not sufficiently developed for industrial use till the 1970s. The oil crisis triggered R&D on CSP, and several pilot plants were built and tested around the world during the 1980s (Winter, Sizmann, and Vant-Hull 1991). Nevertheless, most of these experiences ended without having reached the final goal of making the concentrating solar thermal technologies commercial. The only exception is the experience accumulated by the LUZ International Company in the nine 354 MW total capacity solar electric generating system (SEGS) plants, which it built between 1984 and 1991. These plants have created more than 12,000 GWh for the Southern California grid since their construction. All SEGS plants were developed, financed, built, and are still operated on a purely private basis.

However, no new commercial solar thermal power plants have been built since the last two 80 MW$_e$ parabolic trough plants (SEGS VIII and IX) were connected to the grid in 1991 and 1992, respectively, for the following main reasons (Becker et al. 2002):

- Financial uncertainties caused by delayed renewal of favorable tax provisions for solar systems in California
- Financial problems and subsequent bankruptcy of the U.S./Israeli LUZ group, the first commercial developer of private solar power projects
- Rapid drop in fossil energy prices followed by years of worldwide stability at those low levels

- The large solar thermal power station unit capacities required to meet competitive conditions for the generation of bulk electricity, resulting in financial constraints due to their inherently large share of capital costs
- Rapidly decreasing depreciation times of capital investments in power plants due to the deregulation of the electricity market and the worldwide shift to private investor ownership of new plant projects
- Drops in cost and enhanced efficiencies of installed conventional power plants, particularly in combined cycle power plants
- Lack of a favorable financial and political environment for the development of solar thermal power plant project initiatives in sunbelt countries

In spite of the aforementioned factors, the Cost Reduction Study for Solar Thermal Power Plants prepared for the World Bank in early 1999 (Enermodal 1999) concludes that the potential CSP market could reach an annual installation rate of 2000 MW of electricity. In the foregoing scenario, this rate is reached by 2020. Advanced low-cost CSP systems are likely to offer energy output at an annual capacity factor of 0.22 or more. So the contribution of CSP would be about 24–36 TWh of electricity by 2020 and 1600–2400 TWh by 2050. However the same study estimated that the current CSP capital cost is 2.5–3.5 times higher than the capital cost of a conventional fossil-fueled thermal power plant, and showed that the price of electricity generation is from 2 to 4 times the conventional generation price. But the potential learning curve is enormous, and technology roadmaps predict that over a 60% cost reduction is possible by 2020. Production costs for solar-only plants could drop below 8¢/kWh, by combining innovation, mass production and scaling-up factors (Sargent & Lundy 2003; Pitz-Paal et al. 2005a, 2005b).

Given the huge solar resource on earth and the abovementioned cost scenarios, CSP is foreseen to have enormous impact on the world's bulk power supply by the middle of the century. In Southern Europe alone, the technical potential of CSP is estimated at 2,000 TWh and in Northern Africa it is beyond any quantifiable guess (Nitsch and Krewitt 2004). Worldwide, the exploitation of less than 1% of the total solar thermal power plant potential would be enough to meet the recommendations of the United Nations' Intergovernmental Panel on Climate Change for long-term climate stabilization (Aringhoff et al. 2003; Philibert 2004). In contrast to conventional fossil plants, CSP plants do not produce CO_2 during operation, and are therefore suitable to meet the challenge of keeping standards of living without compromising environmental issues. A MW of installed CSP avoids 688 t of CO_2 compared to a combined cycle conventional plant and 1360 t of CO_2 compared to a conventional coal/steam plant. A 1 m^2 mirror in the primary solar field produces 400 kWh of electricity per year, avoids 12 t of CO_2 and contributes to a 2.5 t savings of fossil fuels during its 25-year operation lifetime. The energy payback time of concentrating solar power systems is less than one year. In addition, most solar field materials and structures can be recycled and used again for further plants.

In terms of electricity grid and quality of bulk power supply, it is the ability to provide dispatchability on demand that makes CSP stand out from other renewable energy technologies like PV or wind. Even though the sun is an intermittent source of energy, CSP systems offer the advantage of being able to run the plant continuously at a predefined load. Thermal energy storage systems store excess thermal heat collected by the solar field. A typical storage concept consists of two storage tanks filled with a liquid storage medium at different temperatures (Falcone 1986). When charging the storage, the medium is pumped from the "cold" to the "hot" tank being heated up (directly or indirectly) by the solar heat collected. When discharging, the storage medium is pumped from the "hot" to the "cold" tank extracting the heat in a steam generator that drives the power cycle. Storage systems, alone or in combination with some fossil fuel backup, keep the plant running under full-load conditions. Storing thermal energy allows the managers of the plant to know when the plant must stop supplying energy. Figure 19.2 shows how stable operation can be extended for several hours after sunset. With an appropriate weather forecast, a 24–48 h prediction of solar capacity appears to be feasible. Keep in mind that thermal energy storage systems are only designed to shift the energy a few hours (e.g., from daytime to evening) or days. It cannot compensate for the seasonal difference in the solar input, which comes from the changing

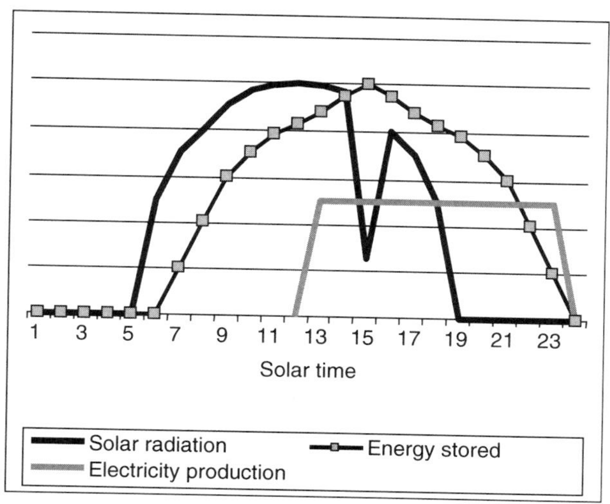

FIGURE 19.2 Extended operation with an solar-only CSP plant by using some hours of thermal energy storage.

duration of sunlight from summer to winter, but with an appropriately small percentage of fossil fuel hybridization, a firm capacity can be ensured and investment can be reduced by 30% (Kolb 1998). Figure 19.3 shows a solar thermal power plant providing firm capacity. The thermal storage system supplies most of the energy required by the turbine after sunset. Overnight and early in the morning, operation is ensured by fossil fuel backup. The example shown is a "fuel saver" scheme in which solar energy is used to save fossil fuel during the daytime. There are also other options, like the "power booster" scheme, in which the fossil burner is kept constant all the time and solar energy is fed into the turbine for peaks during solar hours, in which case the power block can absorb power increments to some extent.

This capability of storing high-temperature thermal energy leads to economically competitive design options, since only the solar part has to be oversized. This fact means that there is an incremental cost for the storage system and additional solar field, while the size of the conventional part of the plant (power block) remains the same. Furthermore, storage system efficiencies are over 95% which is high. Specific investment costs of less than \$10–30/kWh$_{th}$ resulting in \$25–75/kWh$_{el}$ are possible today. This CSP plant feature is tremendously relevant, since penetration of solar energy into the bulk electricity market is possible only when substitution of intermediate-load power plants

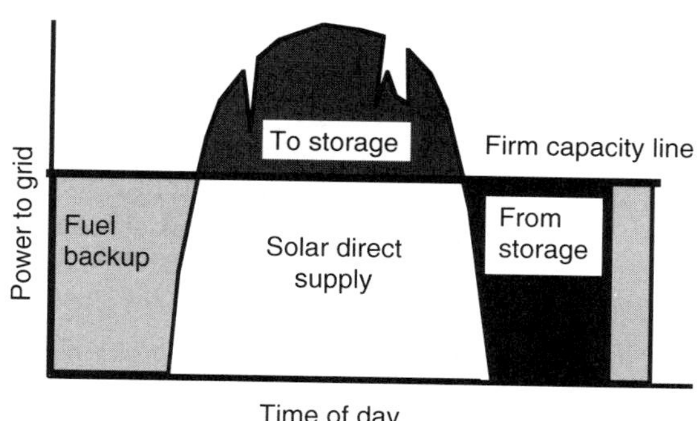

FIGURE 19.3 Example of operational strategy for a CSP plant with thermal storage and fuel backup to maintain a constant firm capacity supply round the clock.

of about 4000–5000 h/year is achieved. This is the situation in Europe, where 50% of the electricity demand by 2020 is supposed to be supplied by new intermediate-load plants to be built in the forthcoming years (Pitz-Paal et al. 2005b).

Fortunately, the context for deployment of solar thermal power plants has changed for the better. New opportunities are opening up for CSP as a result of the global search for clean energy solutions and new plants are being constructed after more than two decades of interruption (IEA 2003). Feed-in tariffs, green portfolios and other environmentally related incentives are helping new projects to become reality in Spain, the United States, Australia, Algeria, and elsewhere (Aringhoff et al. 2005).

19.2 Solar Concentration and CSP Systems

19.2.1 Why Use Concentrating Solar Energy Systems? Dependence of Efficiency on Temperature

Explained simply, solar concentration allows "higher-quality" energy to be collected because higher temperatures, and thereby greater capacity for generating mechanical work, can be achieved. According to the second law of thermodynamics, the higher the operating temperature, Top, is, the better the efficiency of a heat engine (for example, the one in a CSP plant). The heat engine operating temperature is directly dependent on the solar receiver, or absorber, outlet temperature.

Moreover, with solar concentration, the receiver–absorber aperture area can be reduced, minimizing infrared losses. Finally, concentration of solar radiation allows us to develop at a smaller absorber, which has a larger cost reduction potential.

Solar concentrating systems are characterized by the use of devices, such as mirrors or lenses, which are able to redirect the incident solar radiation received onto a particular surface, collector surface A_C, and concentrate it onto a smaller surface, absorber surface A_{abs}, or absorber. The quotient of these two areas is called the geometric concentration ratio, $C_{onc}=A_{abs}/A_c$.

Assume a simplified model of a solar thermal power plant, like the one represented in Figure 19.1, made up of an ideal optical concentrator, a solar receiver performing as a blackbody and therefore having only emission losses (cavity receivers and volumetric receivers theoretically approach this condition), and a turbine or heat engine with Carnot ideal efficiency. System efficiency will depend on the balance of radiative and convective losses in the solar receiver, as shown in (19.1), where α, τ, and ε are absorber absorbance, transmittance, and emittance. When the concentrated solar flux impinges on the absorber, its temperature augments and, simultaneously, radiation losses from the absorber surface to the ambient increase. With a thermal fluid cooling the absorber, when equilibrium, or steady state, is reached, the solar radiation gain equals the sum of infrared emission losses plus the useful energy gained by the cooling fluid.

$$\frac{Q^*_{gain}}{A} = \alpha C\phi - \sigma\varepsilon(T^4_{abs} - T^4_{amb}), \tag{19.1}$$

where Q_{gain} is the power gain or useful power outlet from solar receiver [W], A is the absorber aperture area [m^2], α is the hemispherical absorptivity of the absorber, C is the geometrical concentration ratio,σ is the Stefan–Boltzmann constant [5.67e-08 W/m^{2}K^4], ε is the hemispherical emissivity of the absorber, ϕ is the direct normal irradiance [W/m^2], T_{abs} is the temperature (homogeneous) of the absorber [K], and T_{amb} is the effective temperature of ambient or atmosphere "viewed" by the absorber [K].

Solar receiver efficiency, defined as the quotient of power gain flux and concentrated solar radiation flux incident on the receiver (absorber), can be formulated as:

$$\eta_{rec} = \frac{Q^*_{gain}/A}{C\phi}. \tag{19.2}$$

By substituting Equation 19.1 into Equation 19.2, the dependence of thermal efficiency versus parameters and variables of the receiver is observed.

$$\eta_{rec} = \alpha - \sigma\varepsilon \frac{(T_{abs}^4 - T_{amb}^4)}{C\phi}. \tag{19.3}$$

Equation 19.3 is plotted for different concentration ratios in Figure 19.4. This graphic representation, valid for flat absorbers, leads to the following conclusions:

- The maximum theoretical optical efficiency (when $T_{abs} \geq T_{amb}$) is the effective absorptivity of the receiver, α.
- The higher the incident solar flux ($C\phi$), the better the optical efficiency.
- The higher the absorber temperature, the higher the radiative loss and, therefore, optical efficiency is lower.
- The higher the effective emissivity, ε, the lower the optical efficiency.

Figure 19.4 shows the evolution of optical efficiency versus temperature and concentration ratio. It also includes the Carnot cycle efficiency, defined as:

$$\eta_{Carnot} = \frac{T_{abs} - T_{amb}}{T_{abs}}. \tag{19.4}$$

The Carnot cycle efficiency is the ideal efficiency (for reversible processes) that increases with temperature and sets the thermodynamic limit of the conversion efficiency.

As ideal absorber temperature increases, thermal radiation losses increase as well. When losses and gains are equal, the net useful heat is zero and the receiver should have achieved the maximum temperature or stagnation temperature. The stagnation temperature is described in Equation 19.5 and Figure 19.5.

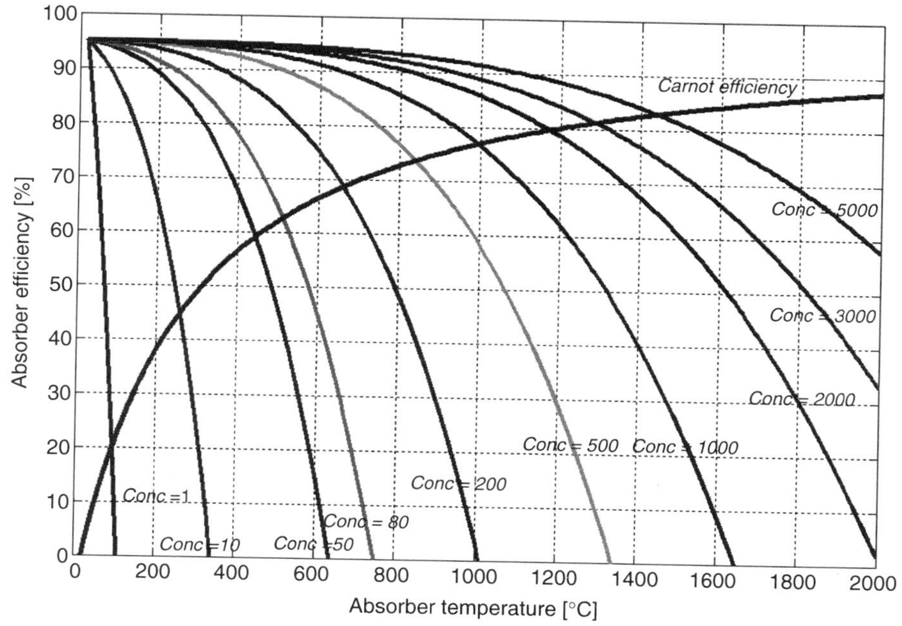

FIGURE 19.4 Efficiency of the solar receiver versus T_{abs} and versus solar concentration ratio, assuming $T_{amb} = 20°C$, $\phi = 770$ W/m^2 and $\alpha = \varepsilon = 0.95$.

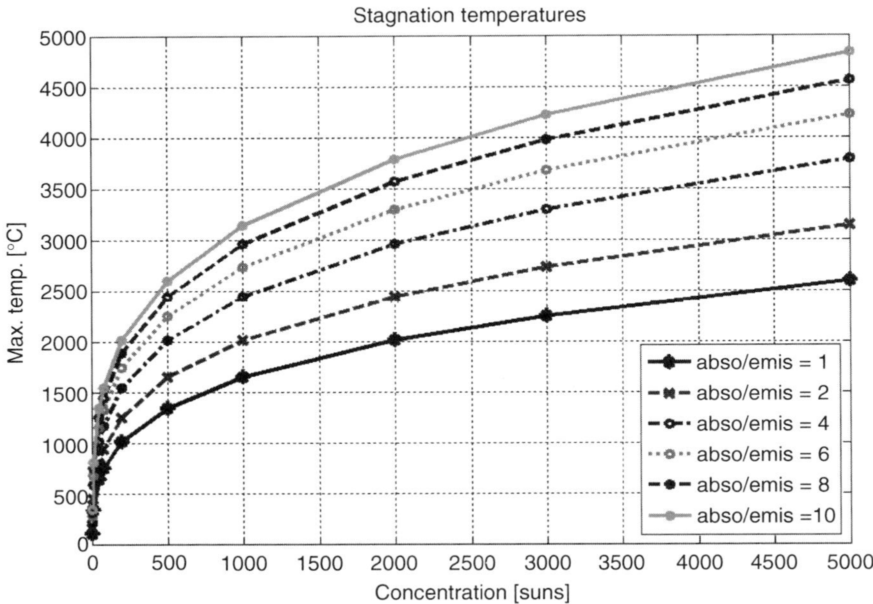

FIGURE 19.5 Stagnation temperatures for a given solar absorber as a function of concentration ratio, for a direct normal irradiance of 770 W/m², ambient temperature of 20°C, $\alpha = 1$ and different values of emissivity, ε.

$$T_{\text{abs\_max}} = \left[\frac{\alpha C\phi}{\varepsilon\sigma} + T_{\text{amb}}^4\right]^{1/4}. \tag{19.5}$$

It should be noted that:

- A nonselective absorber ($\alpha = \varepsilon = 1$) reaches 95°C in stagnation conditions, without concentration and for the given solar irradiance.
- A selective coating can enable much higher stagnation temperatures to be reached. For instance, at $C = 1000$, the maximum temperature is higher than 1600°C, for $\alpha/\varepsilon = 1$, and about 3200°C, for $\alpha/\varepsilon = 10$ ($\alpha = 1$; $\varepsilon = 0.1$).

From the abovementioned correlations, it may clearly be concluded that, in terms of solar receiver efficiency, high solar concentrations and low temperatures are the best compromise. For a given concentration ratio there is an absorber threshold temperature at which radiation losses increase dramatically. But when analyzing a theoretical CSP system, the convolution of the solar receiver and the heat engine should also be taken into consideration. What is the optimum temperature for a complete system including receiver and Carnot cycle? The combined efficiency of both systems can easily be visualized by multiplying the optical efficiency of the absorber (19.2) and the Carnot cycle efficiency (19.4). The result would represent the ideal conversion efficiency of our system from solar radiation to work.

$$\eta_{\text{tot\_rec}} = \eta_{\text{rec}}\eta_{\text{Carnot}}. \tag{19.6}$$

Figure 19.6 depicts the combined efficiency of the receiver/heat engine system versus concentration and temperature. It may be observed that for each concentration the efficiency increases with temperature up to a maximum (Carnot term prevails). Once this peak is achieved, a temperature increment represents a decrement in efficiency (infrared losses at receiver prevail).

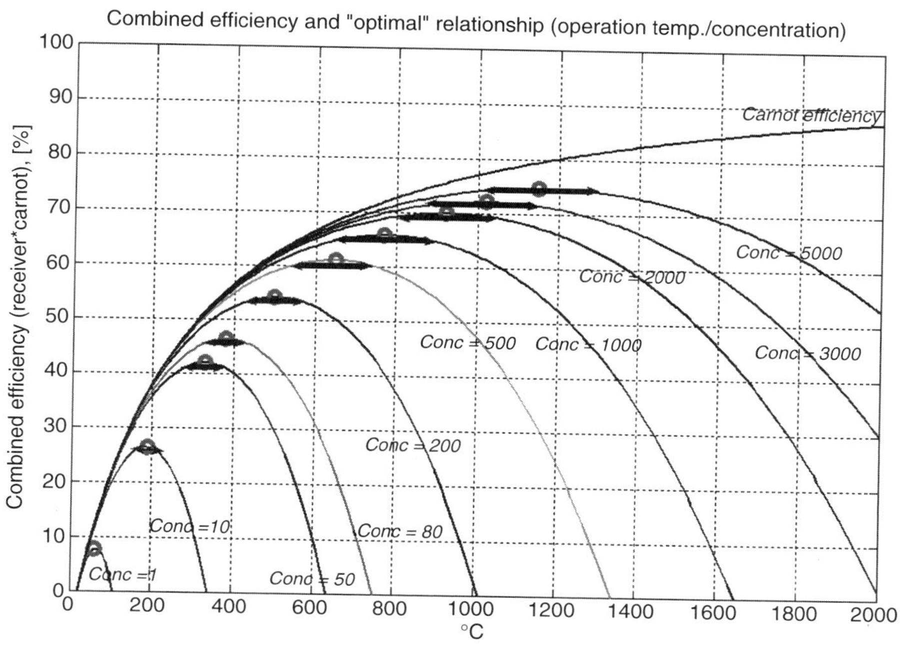

FIGURE 19.6 Combined efficiency of the solar receiver/heat engine system for different solar concentration factors and operation temperatures of the absorber. Direct normal irradiance of 770 W/m^2, ambient temperature of 20°C, $\alpha = \varepsilon = 1$.

As a result, it may be concluded that for any ideal receiver working at a given concentration there is an optimum temperature, and this temperature can obtained by:

$$\frac{d\eta_{\text{tot\_rec}}}{dT} = 0. \tag{19.7}$$

Substituting Equation 19.1, Equation 19.2, Equation 19.4, and Equation 19.5 into Equation 19.6 and finding the derivative yields a polynomial expression in T_{abs}, and its real (positive) roots give the optimum temperature:

$$4\sigma T_{\text{abs}}^5 - 3\sigma\varepsilon T_{\text{amb}} T_{\text{abs}}^4 - (\sigma T_{\text{amb}}^5 + \alpha C\phi T_{\text{amb}}) = 0 \tag{19.8}$$

Figure 19.6 includes the optimum temperature for different solar concentrations as calculated from Equation 19.8.

In conclusion, solar concentration is necessary to convert solar energy into mechanical work; furthermore, for each geometrical concentration, there is a theoretical optimum absorber operating temperature.

19.3 Solar Concentrator Beam Quality

Optical concentration leads to two significant limitations on the practical use of solar radiation that are intrinsic to the characteristics of the radiation source. On one hand, the nonnegligible diffuse solar radiation cannot be concentrated on the absorber (energy spillage), therefore only direct solar radiation from the solar disc can be used. Secondly, costly mechanical devices are required to track the sun. Consequently, there are practical physical limitations to the concentration level depending on the application (Sizmann 1991). Sun tracking and use of beam radiation are not the only restrictions.

It should also be taken into account that solar rays coming directly from the solar disc are not completely collimated, but have a certain solid angle. The subtended solid angle is 32′; this means an angular radius of 4.653 mrad or 16′ of arc, therefore even an ideal parabolic concentrator would reflect the image of the sun on a spot having the same target-to-mirror solid angle. This means that for an ideal heliostat located 500 m from the optical target or focal point, the theoretical diameter of the spot would be 4.7 m. The effect of the size of the sun on the reflected cone for a heliostat and the size of the spot on the target can be observed in Figure 19.7.

Therefore, when designing a real solar concentrator and the aperture of a solar receiver, it is necessary to take into account the minimum size of the spot at a given distance. One additional characteristic of the sun must be considered: the sunshape. Dispersion and absorption effects on the solar photosphere modify the uniform distribution of the expected irradiance from an ideal blackbody. Because of this modification, it is more realistic to substitute a "limb-darkened" distribution for the ideal uniform distribution because the sun is darker near the rim than at the center (Vant-Hull 1991). Assuming that the sun is an ideal Lambertian emitter, a uniform distribution of radiance (pillbox) with a constant value of $L_0 = 13.23\ \mathrm{MW/m^2}$ would be required over the entire solar disc, providing the integrated value of the solar irradiance as $E = \int L\, d\Omega = \pi\, \theta^2 L_0$, where $\theta = R/D_{ES} = 4.653$ mrad is the ratio of the radius of the solar disc and the earth–sun distance and Ω is the solid angle. For the "limb-darkened" distribution, the following expression of radiance is obtained:

$$\frac{L}{L_0} = 0.36 + 0.84\sqrt{1 - \frac{\xi^2}{\theta_S^2}} - 0.20\left(1 - \frac{\xi^2}{\theta_S^2}\right), \tag{19.9}$$

where L is the radiance ($\mathrm{MW/(m^2 \cdot sr)}$), L_0 is the radiance at the center of the disc ($13.23\ \mathrm{MW/m^2}$); for $\xi = 0$, $\xi = r/D_{ES} \leq \theta$, radial coordinate normalized to D_{ES}, and $\theta_S = R/D_{ES} = 4.653$ mrad.

The extraterrestrial irradiance is modified as it enters the atmosphere because of absorption and multiple dispersions, producing the well-known aureole. That is why better sunshape fit is obtained if the previous radiance distribution is separated into two regions, the central solar disc and the circumsolar

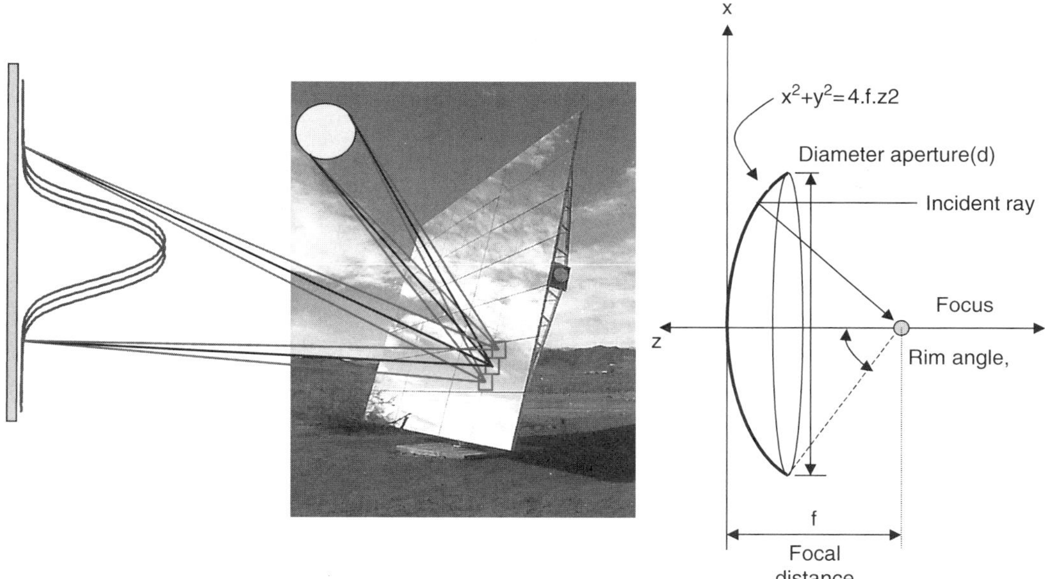

FIGURE 19.7 Configuration of an ideal parabolic concentrador (right) and effect of the size of the sun on the reflected image with a real reflectant surface, heliostat (left).

region (Rabl 1985). The ratio between the circumsolar irradiance and direct irradiance varies depending on atmospheric conditions, but in typical sites with good solar radiation, the monthly average does not exceed 5%, provided the operating threshold for system start-up is above 300 W/m^2. A detailed "ray-tracing" analysis reproducing reflection on our solar concentrating surface should take into account the sunshape. To this effect, other factors, like curvature and waviness errors of the reflecting surface, as well as the possible tracking errors in the drive mechanism, must be added. It is relatively simple to approximate all these nonsystematic errors of the concentrator to a standard deviation, $\sigma^2 = \sum \sigma_i^2$, to quantify the beam quality of the reflector. The consequence of the convolution of all the above-mentioned errors from the sun, tracking system and reflecting surface, leads to the fact that instead of an ideal point focus parabolic concentrator, Figure 19.7 (right), the spot and energy profile obtained on a flat absorber can be approximated to a Gaussian shape, Figure 19.7 (left). The real image obtained is also known as *degraded sunshape*. Subsequently, the designer of a solar receiver should take into account the beam quality of the solar concentrator and the concentrated flux distribution to optimize the heat transfer process.

However, the main interest of a solar concentrator is the energy flux and not the quality of the image. This fact means 4.65 mrad is a good reference for comparing the extent of optical imperfections. Those errors deflecting the reflected ray significantly less than 4.65 mrad are of minor importance, while deviations over 6 mrad contribute drastically to the reduction of concentration and energy spillage at the receiver aperture.

Solar concentrators follow the basic principles of Snell's law of reflection (Rabl 1985), as depicted in Figure 19.8. In a specular surface like the mirrors used in solar thermal power plants, the reflection angle equals the angle of incidence. In a real mirror with intrinsic and constructional errors, the reflected ray distribution can be described with "cone optics." The reflected ray direction has an associated error that can be described with a normal distribution function. The errors of a typical reflecting solar concentrator may be either microscopic (specularity) or macroscopic (waviness of the mirror and error of curvature). All the errors together end up modifying the direction of the normal compared to the reference reflecting element. However, it is necessary to discriminate between microscopic and macroscopic errors. Microscopic errors are intrinsic to the material and depend on the fabrication process, and can be measured at the lab with mirror samples. Macroscopic errors are characteristic of the concentrator and the erection process. Therefore, they should be measured and quantified with the final system in operation (Biggs and Vittitoe 1979).

The parameter best defining the "macroscopic" quality of a reflective concentrator is the slope error (β). The slope error is the angle between the normal to the reference surface ($\vec{N}_0$) and the normal to the real reflecting surface ($\vec{N}$), as represented in Figure 19.9. The root mean square (RMS), a statistical mean distribution of slope errors, is used to specify the distribution of β on a real surface. For a given differential element of surface (dA), the RMS is obtained as:

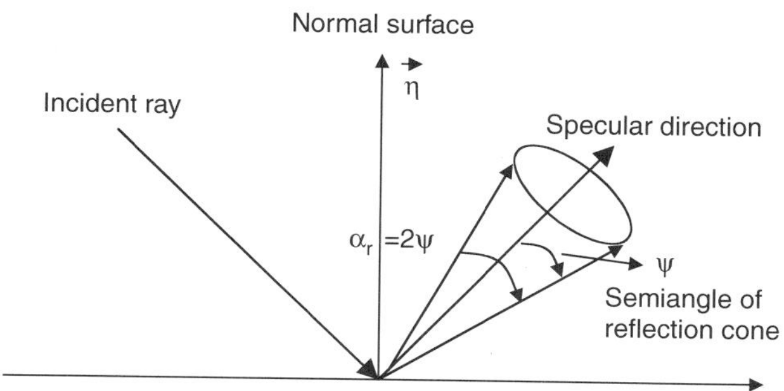

FIGURE 19.8 Geometry of reflection according to the principles of Snell.

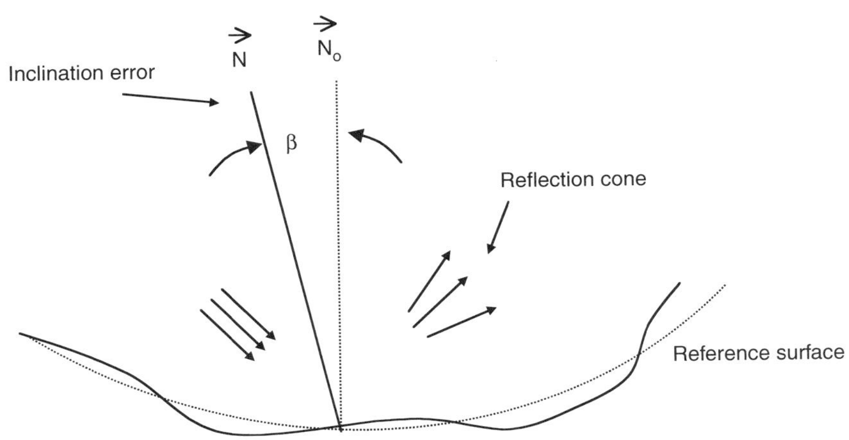

FIGURE 19.9 Normal error produced by grainy texture of the material and deficient curvature of the concentrator.

$$\text{RMS} = \langle \beta^2 \rangle^{1/2} = \left[\frac{\int \beta^2 \text{d}A}{\int \text{d}A} \right]^{1/2}. \tag{19.10}$$

RMS is a deterministic value of the surface errors, but it can be expressed with a probabilistic value as the standard deviation. Because it is more practical to determine the probabilistic error in the reflected image, it is appropriate to translate the RMS of the normals on the reflector to the standard deviation of the reflected rays. Assuming a new reference plane *rs* placed at a unitary distance, the probability of $\vec{N}$ intersecting the element of surface $\text{d}r \cdot \text{d}s$ is $F(r, s) \cdot \text{d}r \cdot \text{d}s$ where $F(r, s)$ is a probability density function normalized to 1 when integrated over the entire plane *rs* (Figure 19.10).

In the aforementioned case, the probability function can be approximated to a normal distribution function, since further convolution with other errors like specularity, solar tracking, or sunshape, leads to a damping effect according to the central limit theorem. In addition, since the total error is the convolution of a series of random surface errors, the distribution is circular normal.

If the new coordinates defined for the plane as depicted in Figure 19.10 are used,

$$\rho = \tan\beta = (r^2 + s^2)^{1/2}, \tag{19.11}$$

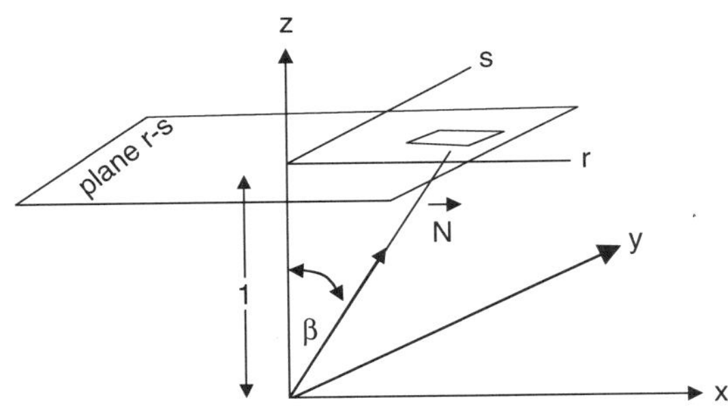

FIGURE 19.10 Translation of the normal error to a new reference plane at a distance 1.

and for those values of ρ close to β, which is the case for actual solar concentrators used in CSP plants, the function F may be expressed as:

$$F(\rho) = \frac{1}{2\pi\sigma^2}\exp\left[-\frac{\rho^2}{2\sigma^2}\right], \tag{19.12}$$

where the parameter σ is the standard deviation of the reflected ray, and for circular symmetry: $\sigma_r = \sigma_s = \sigma$.

By integrating the previous expression to obtain the RMS of ρ, σ is correlated with β, and with the RMS, by the following equation (Biggs and Vittitoe 1979; Vant-Hull 1991):

$$\text{RMS} = \langle\rho^2\rangle^{1/2} = \sigma\sqrt{2} = \langle\beta^2\rangle^{1/2}. \tag{19.13}$$

Summarizing, the beam quality of the concentrating reflector may be expressed by means of three parameters related to the inclination error of the surface elements: the RMS of β, the dispersion σ of the normal, or the dispersion σ of the reflected beam. The total standard deviation of a solar concentrator or beam quality would be the sum of several sources of error:

$$\sigma_C^2 = \sigma_{sp+wav}^2 + \sigma_{curvature}^2 + \sigma_{tracking}^2. \tag{19.14}$$

The total error of the image, also known as degraded sun, would be the convolution of the beam quality of the concentrator with the sunshape.

$$\sigma_D^2 = \sigma_{sunshape}^2 + \sigma_C^2, \tag{19.15}$$

where $\sigma_{Sunshape}$ is the beam standard deviation due to the sunshape effect (approximately 2.19 mrad), σ_{sp+wav} is the beam standard deviation due to specularity and waviness (measured with reflected rays from material samples using a reflectometer), $\sigma_{curvature}$ is the beam standard deviation due to curving (and should be measured on the concentrator itself), and $\sigma_{tracking}=$ is the beam standard deviation due to aiming point and other drive mechanism-related sources of error.

The association of the flux profile on the target with a Gaussian shape and the determination of the beam quality of the concentrator are useful in identifying the optimum aperture area of the receiver for a specific fraction of intercepted power. For a given receiver aperture radius length (ρ_A), the probability of $\rho < \rho_A$ may be obtained by integrating Equation 19.12:

$$P(\rho < \rho_A) = 2\pi\int_0^{\rho_A} \rho F(\rho)\mathrm{d}\rho = 1 - \exp\left[-\frac{\rho_A^2}{2\sigma^2}\right]. \tag{19.16}$$

With this simple correlation, the beam standard deviation (σ) and cone radius (ρ_A) can be correlated by intercepting a certain percentage of reflected power or the probability of $\rho < \rho_A$ (P).

19.4 Solar Concentration Ratio: Principles and Limitations of CSP Systems

The most practical and simplest primary geometrical concentrator typically used in CSP systems is the parabola. Even though there are other concentrating devices like lenses or compound parabolic concentrators (CPCs) (Welford and Winston 1989), the reflective parabolic concentrators and their analogues are the systems with the greatest potential for scaling up at a reasonable cost. Parabolas are imaging concentrators able to focus all incident paraxial rays onto a focal point located on the optical axis (see Figure 19.7). The paraboloid is a surface generated by rotating a parabola around its axis. The parabolic dish is a truncated portion of a paraboloid. For optimum sizing of the parabolic dish and absorber geometries, the geometrical ratio between the focal distance, f, the aperture diameter of the concentrator, d, and the rim angle, Θ must be taken into account. The ratio can be deducted from the equation describing the geometry of a truncated paraboloid, $x^2+y^2=4fz$, where x and y are the coordinates on the aperture plane and z is the distance from the plane to the vertex. For small rim angles, the paraboloid tends to be a sphere, and in many cases spherical facets are used. Therefore, in most solar concentrators the following correlation is valid:

$$f/d = \frac{1}{4\tan(\Theta/2)}. \tag{19.17}$$

For example, a paraboloid with a rim angle of 45° has an f/d of 0.6 (see Figure 19.11). The ratio f/d increases as the rim angle decreases. A parabolic concentrator with a very small rim angle has very little curvature and the focal point far from the reflecting surface. Because of this postitioning, CSP systems making use of cavity receivers with small apertures should use small rim angles. Conversely, those CSP systems using external or tubular receivers will make use of large rim angles and short focal lengths.

The equation describing the minimum concentration ratio as a function of the rim angle for a given beam quality, σ, is:

$$Cmin_{\text{conc}} = \frac{\sin^2\Theta\cos^2(\Theta+\sigma)}{\sin^2\sigma}. \tag{19.18}$$

From Equation 19.18, it can be concluded that 45° is the optimum rim angle for any beam quality, in terms of solar concentration (Figure 19.12). Therefore, $f/d=0.6$ is the optimum focal length-to-diameter ratio in a parabolic concentrator.

The thermodynamic limit or maximum concentration ratio for an ideal solar concentrator would be set by the size of the sun and not by the beam quality. By applying the geometrical conservation of energy in a solar concentrator, the following expressions are obtained for 3D and 2D systems (for a refraction index $n=1$):

$$C\max,3\text{D} = \frac{1}{\sin^2\theta_{\text{S}}} \leq 46{,}200; \tag{19.19}$$

$$C\max,2\text{D} = \frac{1}{\sin\theta_{\text{S}}} \leq 215. \tag{19.20}$$

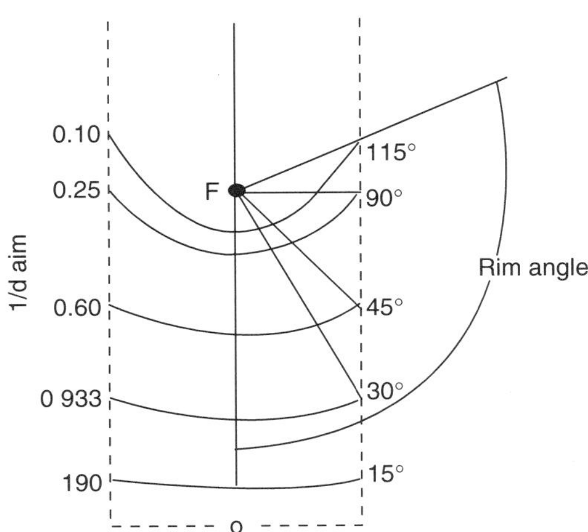

FIGURE 19.11 Visualization of dependence of f/d vs. rim angle for a parabolic concentrator.

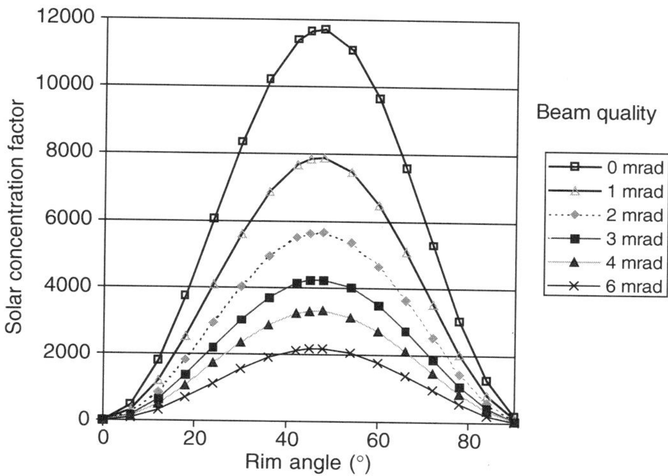

FIGURE 19.12 Dependence of minimum concentration achieved for a parabola versus rim angle and beam quality.

Then, the semi-angle subtended by the sun is $\theta_S = 4.653 \times 10^{-3}$ rad (16′), and the maximum concentration values are 46,200 for 3D and 215 for 2D. For real concentrators, the maximum ratios of concentration are much lower, because of microscopic and macroscopic, tracking and mechanical, sunshape and other errors. Engineers designing a specific CSP plant should give special attention to the expected real beam quality and rim angle of the reflecting system to obtain an appropriate sizing of the solar receiver.

19.5 Solar Thermal Power Plant Technologies

Solar thermal power plants with optical concentration technologies are important candidates for providing the bulk solar electricity needed within the next few decades. Four concentrating solar power technologies are today represented at pilot and demonstration-scale (Mills 2004): parabolic trough collectors (PTC), linear Fresnel reflector systems (LF), power towers or central receiver systems (CRS), and dish/engine systems (DE). All the existing pilot plants mimic parabolic geometries with large mirror areas and work under real operating conditions. Reflective concentrators are usually selected since they have better perspectives for scale-up (Figure 19.13).

PTC and LF are 2D concentrating systems in which the incoming solar radiation is concentrated onto a focal line by one-axis tracking mirrors. They are able to concentrate the solar radiation flux 30–80 times, heating the thermal fluid up to 393°C, with power conversion unit (PCU) sizes of 30–80 MW, and therefore, they are well suited for centralized power generation with a Rankine steam turbine/generator cycle in dispatchable markets. CRS optics are more complex, since the solar receiver is mounted on top of a tower and sunlight is concentrated by means of a large paraboloid that is discretized into a field of heliostats. This 3D concentrator is therefore off-axis and heliostats require two-axis tracking. Concentration factors are between 200 and 1000 and unit sizes are between 10 and 200 MW. Therefore, they are well suited for dispatchable markets and integration into advanced thermodynamic cycles. A wide variety of thermal fluids, like saturated steam, superheated steam, molten salts, atmospheric air or pressurized air, can be used, and temperatures vary between 300°C and 1000°C. Finally, DE systems are small modular units with autonomous generation of electricity by Stirling engines or Brayton miniturbines located at the focal point. Dishes are parabolic 3D concentrators with high concentration ratios (1000–4000) and unit sizes of 5–25 kW. Their current market niche is in both distributed on-grid and remote/off-grid power applications (Becker et al. 2002).

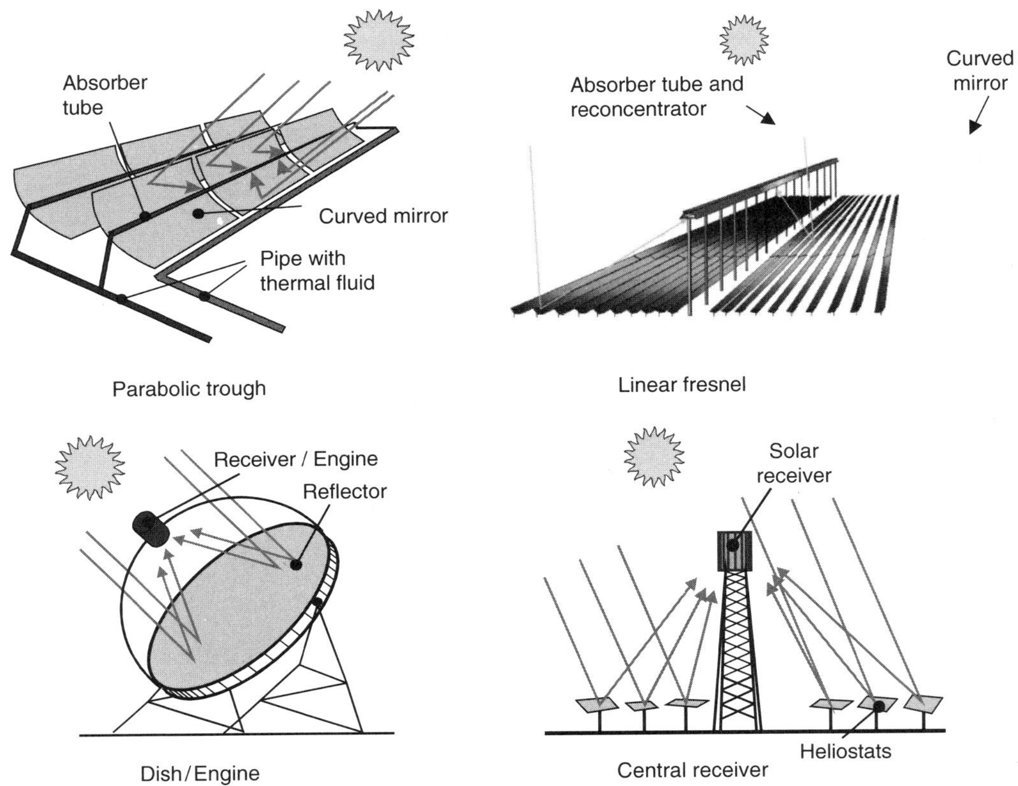

FIGURE 19.13 Schematic diagrams of the four CSP systems scaled up to pilot and demonstration sizes.

Typical solar-to-electric conversion efficiencies and annual capacity factors, as compiled by IEA SolarPACES for the four technologies, are listed Table 19.1 (DeMeo and Galdo 1997; Tyner et al. 2000). The values for parabolic troughs, by far the most mature technology, have been demonstrated commercially. Those for linear Fresnel, dish, and tower systems are, in general, projections based on component and large-scale pilot plant test data and the assumption of mature development of current technology. With current investment costs, all current CSP technologies are generally thought to require a public financial support strategy for market deployment. An independent study promoted by the World Bank (Enermodal 1999), confirms CSP as the most economical technology for solar production of bulk electricity. However, its analysis finds the direct capital costs of CSP to be 2.5–3.5 times those of a fossil-fueled power plant and, therefore, generation costs of electricity 2–4 times higher.

Every square meter of CSP field can produce up to 1200 kWh thermal energy per year or up to 500 kWh of electricity per year. That means a cumulative savings of up to 12 tn. of carbon dioxide and

TABLE 19.1 Characteristics of Concentrating Solar Power Systems

System	Peak Efficiency (%)	Annual Efficiency (%)	Annual Capacity Factor (%)[a]
Trough/linear Fresnel	21	10–12 (d) 14–18 (p)	24 (d)
Power tower	23	14–19 (p)	25–70 (p)
Dish/engine	29	18–23 (p)	25 (p)

[a] (d) = demonstrated, (p) = projected, based on pilot-scale testing. Annual capacity factor refers to the fraction of the year the technology can deliver solar energy at rated power.

2.5 tn. of fossil fuel per square meter of CSP system over its 25-year lifetime (Geyer 2002). In spite of their promising applications and environmental benefit, to date only parabolic troughs have operated commercially (in nine SEGS plants totaling 354 MW, built by LUZ in California in the 1980s and 1990s). The high initial investment required by early commercial plants ($3000–$4000/kW) and the restricted modularity generally motivated by their expensive thermodynamic cycle, combined with the lack of appropriate power purchase agreements and fair taxation policies, have led to a vicious circle in which the first generation of commercial grid-connected plants becomes difficult to implement without market incentives. After two decades of frozen or failed projects, approval in the past few years of specific financial incentives in Europe, the United States, Australia, and elsewhere, is now paving the way for the launching of the first commercial ventures.

The parabolic trough is today considered a fully mature technology, ready for deployment (Price et al. 2002). Early costs are expected to be in the range of $0.15–0.18/kWh (and even up to 30% less in hybrid systems), and technological and financial risks are expected to be low. In recent years, the five plants at the Kramer Junction site (SEGS III–VII) have achieved a 30% reduction in operating and maintenance costs, a record annual plant efficiency of 14% and a daily solar-to-electric efficiency near 20%, as well as peak efficiencies up to 21.5%. Annual plant availability exceeds 98% and collector field availability is over 99% (Cohen, Kearney, and Kolb 1999). In view of this advanced state of development, several utilities are already pursuing opportunities to build plants similar to the California SEGS plants using existing technology, with various hybridization options, including integrated solar combined cycle systems (ISCCS).

LF reflector systems are conceptually simple, using inexpensive, compact optics, that can produce saturated steam at 150–360°C with land use of less than 1 Ha/MW. It is therefore a CSP technology best suited for integration with combined cycle recovery boilers, to replace the steam bled in regenerative Rankine power cycles or for saturated steam turbines. The most extensive prototype-scale experience has been the very compact designs with multitower aiming of the mirror slabs developed at the University of Sydney in Australia (Mills and Morrison 2000). Despite their simplicity, LF systems lack scale-up experience for eventual electricity production. This situation is hopefully changing, since a first 36 MW commercial project was initiated in Australia in 2003 by the Solar Heat & Power Company (Hu et al. 2003). This proof-of-technology project is located in an existing coal-fired power plant, with the 132,500 m^2 compact LF field supplying 270°C preheat thermal energy to replace the steam bled in the regenerative Rankine power cycle.

Power tower technology, after a proof-of-concept stage, is today on the verge of commercialization, although less mature than the parabolic trough technology. To date, more than 10 different CRS experimental plants have been tested worldwide, generally, small demonstration systems of between 0.5 and 10 MW, and most of them operated in the 1980s (Romero et al. 2002). That experience demonstrated the technical feasibility of the CRS power plants, and their capability of operating with large heat storage systems. The most extensive operating experience has been in the European pilot projects located in Spain on the premises of the Plataforma Solar de Almería (PSA), and in the United States at the 10-MW Solar One and Solar Two facilities located in California. Continuous technological improvement now places current predictions for CRS design point efficiency at 23% and annual at 20%, but a first generation of commercial demonstration plants still remains an essential requisite to validate the technology under real operating and market conditions. The three most promising power tower technologies that are expected to lead to commercial plants in the next few years are: molten salt technology, open or closed loop volumetric air technologies, and saturated steam technology (Romero et al. 2002). PS10 is the only commercial solar tower project currently in operation. The project, led by Solúcar Energía, consists of an 11-MW system, with the solar receiver producing 40 bar/250°C saturated steam (Osuna et al. 2004).

DE systems are modular and ideal for sizes between 5 and 25 kW. Two decades ago, dish/Stirling systems had already demonstrated their high conversion efficiency, concentration of more than 3000 suns, and operating temperatures of 750°C at annual efficiencies of 23% and 29% peak (Stine and Diver 1994). Unfortunately, DE have not yet surpassed the proof-of-reliability operation phase. Only a limited number of prototypes have been tested worldwide, and annual availability above 90% still remains a challenge. Given the fact that autonomous operation and off-grid markets are the first priorities of this technology, more long-endurance test references must be accumulated. DE technology investment costs, which are twice as high as those of parabolic troughs, would have to be dramatically reduced by mass production of specific components, such as the engine and the concentrator. However, these systems, because of their modular nature, are targeted toward much higher-value markets. DE system industries and initiatives are basically confined to the United States and Europe (Mancini et al. 2003).

19.6 Parabolic Trough Solar Thermal Power Plants

19.6.1 The Operational Principles and Components of the PTC

PTCs are linear focus concentrating solar devices suitable for working in the 150–400°C temperature range (Price et al. 2002).

A PTC is basically made up of a parabolic trough-shaped mirror that reflects direct solar radiation, concentrating it onto a receiver tube located in the focal line of the parabola. Concentration of the direct solar radiation reduces the absorber surface area with respect to the collector aperture area and thus significantly reduces the overall thermal losses. The concentrated radiation heats the fluid that circulates through the receiver tube, thus transforming the solar radiation into thermal energy in the form of the sensible heat of the fluid. Figure 19.14 shows a typical PTC and its components.

PTCs are dynamic devices because they have to rotate around an axis, the so-called *tracking axis*, to follow the apparent daily movement of the sun. Collector rotation around its axis requires a drive unit. One drive unit is usually sufficient for several parabolic trough modules connected in series and driven together as a single collector. The type of drive unit assembly depends on the size and dimensions of the collector. Drive units composed of an electric motor and a gearbox combination are used for small collectors (aperture area less than 100 m^2), whereas powerful hydraulic drive units are required to rotate large collectors. A drive unit placed on the central pylon is commanded by a local control unit in order to track the sun.

Local control units currently available on the market can be grouped into two categories: (1) control units based on sun sensors and (2) control units based on astronomical algorithms.

Control units in group 1 use photocells to detect the sun position, whereas those in group 2 calculate the sun vector using very accurate mathematical algorithms that find the sun elevation and azimuth every second and measure the angular position of the rotation axis by means of electronic devices (angular encoders or magnetic coded tapes attached to the rotation axis).

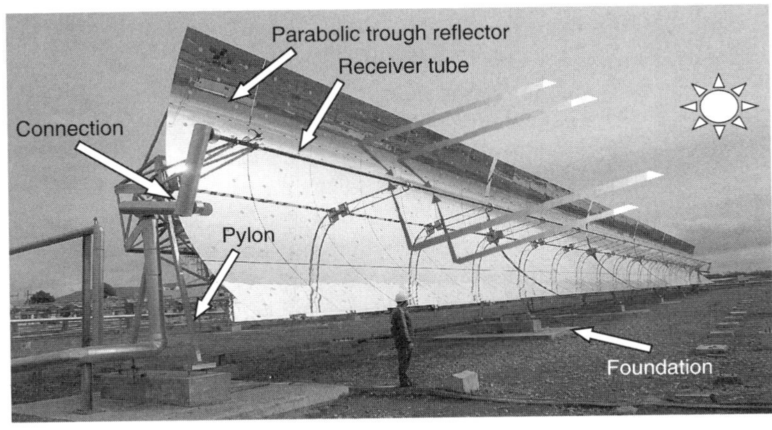

FIGURE 19.14 A typical parabolic trough collector.

Shadow band and flux line trackers are in group 1. Shadow band trackers are mounted on the parabolic concentrator and face the sun when the collector is in perfect tracking (i.e., the sun vector is within a plane that includes the receiver tube and is perpendicular to the concentrator aperture plane). Two photosensors, one on each side of a separating shadow wall, detect the sun's position. When the collector is correctly pointed, the shadow wall shades both sensors equally, and their electric output signals are identical.

Flux line trackers are mounted on the receiver tube. Two sensors are also placed on both sides of the absorber tube to detect the concentrated flux reaching the tube. The collector is correctly pointed when both sensors are equally illuminated and their electrical signals are of the same magnitude.

At present, all commercial PTC designs use a single-axis sun-tracking system. Though PTC designs with two-axis sun-tracking systems have been designed, manufactured, and tested in the past, evaluation results show that they are less cost-effective. Though the existence of a two-axis tracking system allows the PTC to permanently track the sun with an incidence angle equal to 0° (thus reducing optical losses while increasing the amount of solar radiation available at the PTC aperture plane) the length of passive piping (i.e., connecting pipes between receiver pipes of adjacent parabolic troughs on the same collector) and the associated thermal losses are significantly higher than in single-axis collectors. Furthermore, their maintenance costs are higher and their availability lower because they require a more complex mechanical design.

Thermal oils are commonly used as the working fluid in these collectors for temperatures above 200°C, because at these high operating temperatures normal water would produce high pressures inside the receiver tubes and piping. This high pressure would require stronger joints and piping, and thus raise the price of the collectors and the entire solar field. However, the use of demineralized water for high temperatures/pressures is currently under investigation at the PSA and the feasibility of direct steam generation (DSG) at 100 bar/400°C in the receiver tubes of PTC has already been proven in an experimental stage. For temperatures below 200°C, either a mixture of water/ethylene glycol or pressurized liquid water can be used as the working fluids because the pressure required in the liquid phase is moderate.

When choosing a thermal oil to act as working fluid, the main limiting factor to be taken into consideration for stability is the maximum oil bulk temperature. Above this temperature, oil cracking and rapid degradation occur.

The oil most widely used in PTCs for temperatures up to 395°C is VP-1, which is a eutectic mixture of 73.5% diphenyl oxide and 26.5% diphenyl. The main problem with this oil is its high solidification temperature (12°C) that requires an auxiliary heating system when oil lines run the risk of cooling below this temperature. Because the boiling temperature at 1013 mbar is 257°C, the oil circuit must be pressurized with nitrogen, argon, or some other inert gas when oil is heated above this high temperature. Blanketing of the entire oil circuit with an oxygen-free gas is a must when working at high temperatures because high pressure mists can form an explosive mixture with air. Though there are other suitable thermal oils for slightly higher working temperatures with lower solidification temperatures, they are too expensive for large solar plants.

The typical PTC receiver tube is composed of an inner steel pipe surrounded by a glass tube to reduce convective heat losses from the hot steel pipe. The steel pipe has a selective high-absorptivity (greater than 90%), low-emissivity (less than 30% in the infrared) coating that reduces radiative thermal losses. Receiver tubes with glass vacuum tubes and glass pipes with an antireflective coating achieve higher PTC thermal efficiency and better annual performance, especially at higher operating temperatures. Receiver tubes with no vacuum are usually for working temperatures below 250°C, because thermal losses are not so critical at these temperatures. Due to manufacturing constraints, the maximum length of single receiver pipes is less than 6 m, so the complete receiver tube of a PTC is composed of a number of single receiver pipes welded in series up to the total length of the PTC. The total length of a PTC is usually within 25–150 m.

Figure 19.15 shows a typical PTC vacuum receiver pipe. The outer glass tube is attached to the steel pipe by means of flexible metal differential expansion joints which compensate for the different thermal expansion of glass and steel when the receiver tube is working at nominal temperature. At present there are only two manufacturers of PTC vacuum absorber tubes: the German company Schott and the Israeli

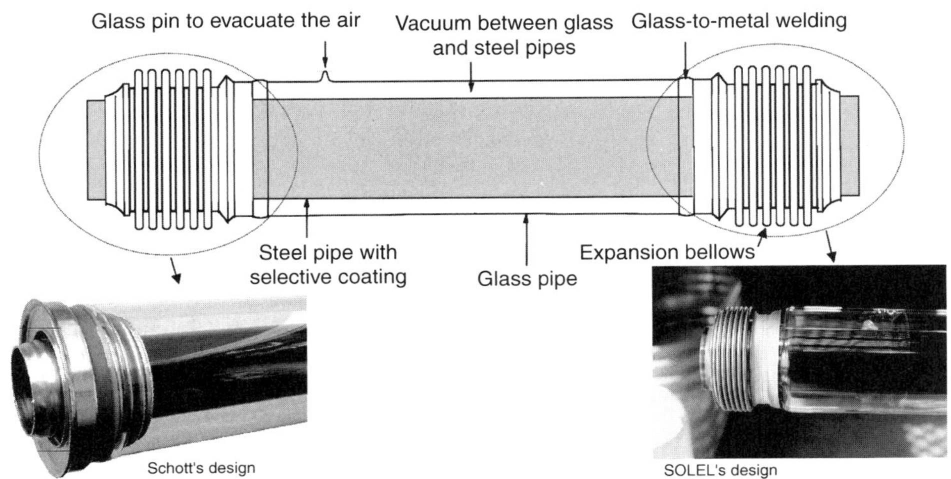

FIGURE 19.15 A typical receiver tube of a PTC.

company SOLEL. The flexible expansion joint used by these two manufacturers is shown in Figure 19.15. The glass-to-metal welding used to connect the glass tube and the expansion joint is a weak point in the receiver tube and has to be protected from the concentrated solar radiation to avoid high thermal and mechanical stress that could damage the welding. An aluminum shield is usually placed over the joint to protect the welds.

As seen in Figure 19.15, several chemical getters are placed in the gap between the steel receiver pipe and the glass cover to absorb gas molecules from the fluid that get through the steel pipe wall to the annulus.

PTC reflectors have a high specular reflectance (greater than 88%) to reflect as much solar radiation as possible. Solar reflectors commonly used in PTC are made of back-silvered glass mirrors, since their durability and solar spectral reflectance are better than the polished aluminum and metallized acrylic mirrors also available on the market. Solar spectral reflectance is typically 0.93 for silvered glass mirrors and 0.87 for polished aluminum. Low-iron glass is used for the silvered glass reflectors and the glass tube of the receiver pipes to improve solar transmission.

The parabolic trough reflector is held by a steel support structure on pylons in the foundation. At present, there are several commercial PTC designs. Large solar thermal power plant designs are much larger than those developed for industrial process heat (IPH) applications in the range of 125–300°C. Examples of PTC designs for IPH applications are the designs developed by the American company IST (Industrial Solar Technology, http://www.industrialsolartech.com) and the European companies Solitem (http://www.solitem.de) and SOLEL Solar Systems (http://www.solel.com). IST and Solitem designs are very similar in size (approx 50 m total length and 2 m wide) and have aluminum reflectors.

Two PTC designs specially conceived for large solar thermal power plants are the LS-3 and EuroTrough (ET-100) (Table 19.2), both of which have a total length of 100 m and a width of 5.76 m, with back-silvered thick glass mirrors and vacuum absorber pipes. The main difference between these two collector designs is their steel structure; EuroTrough's mechanical rigidity to torsion is assured by a steel torque box of trusses and beams, while the LS-3 steel structure is based on two "V-trusses" held together by endplates. Figure 19.16 shows the steel structure of these two PTC designs, and Figure 19.17 shows the overall dimensions of an LS-3 collector that was installed in the later SEGS plants. The main constraint when developing the mechanical design of a PTC is the maximum torsion at the collector ends, because high torsion would lead to a smaller intercept factor and lower optical efficiency.

TABLE 19.2 Parameters of the ET-100 Parabolic Trough Collector

Overall length of a single collector (m)	98.5
Number of parabolic trough modules per collector	8
Gross length of every concentrator module (m)	12.27
Parabola width (m)	5.76
Outer diameter of steel absorber pipe (m)	0.07
Inner diameter of steel absorber pipe (m)	0.055
Number of ball joints between adjacent collectors	4
Net collector aperture per collector (m^2)	548.35
Peak optical efficiency	0.765
Cross section of the steel absorber pipes (m^2)	2.40×10^{-3}
Inner roughness factor of the steel absorber pipes (m)	4.0×10^{-5}
Relative roughness of the steel absorber pipes (m)	7.23×10^{-4}

PTCs are usually installed with the rotation axis oriented either north–south or east–west, however, any other orientation would be feasible too. The orientation of this type of solar collector is sometimes imposed by the shape and orientation of the site where they are installed. Solar collector orientation influences the sun incidence angle on the aperture plane which, in turn, affects collector performance. Seasonal variations in north–south-oriented trough collector output can be quite wide. Three to four times more energy is delivered daily during summer months than in the winter, depending on the geographical latitude and local weather conditions. Seasonal variations in energy delivery are much smaller for an east–west orientation, usually less than 50%. Nevertheless, a north–south sun-tracking axis orientation provides more energy on a yearly basis. Daily variation in the incidence angle is always greater for the east–west orientation, with maximum values at sunrise and sunset and a minimum of 0° every day at solar noon (Rabl 1985).

Therefore, the orientation of the rotation axis is very important for performance, and selection of the best orientation depends on the answers to the following questions:

- Which season of the year should the solar field produce the most energy? If more energy is needed in summer than in winter, the most suitable orientation is north–south.
- Is it better for energy to be evenly distributed during the year, although in winter the production is significantly less than in summer? If the answer is "yes," the best orientation is east–west.

FIGURE 19.16 LS-3 collector with flexible hose and ET-100 collector with ball–joint connections to allow collector rotation and linear thermal expansion of receiver tubes.

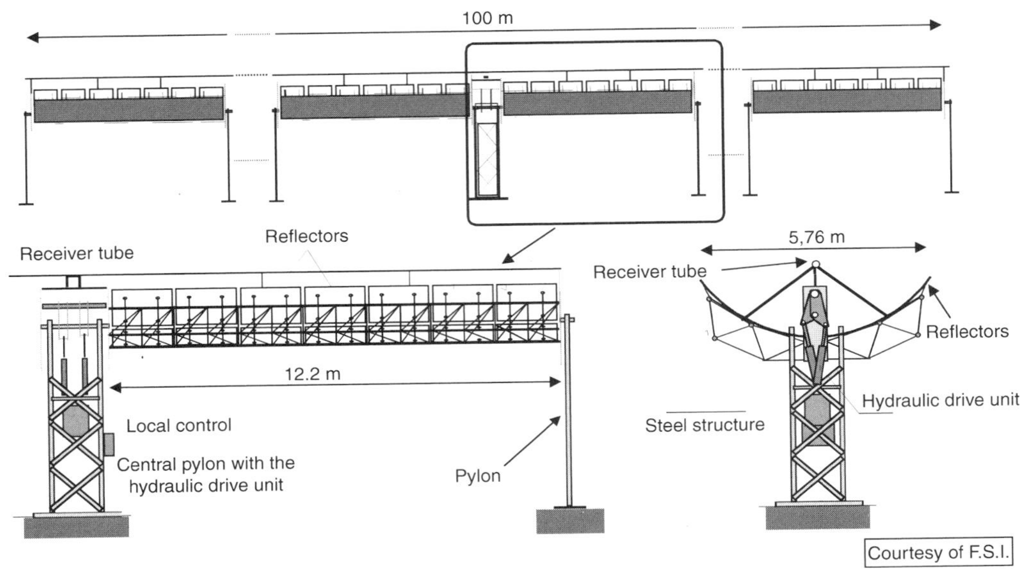

FIGURE 19.17 LS-3 collector dimensions.

- Is the solar field expected to supply similar thermal power in summer and winter? If the answer is yes, the proper orientation is east–west.

In a typical PTC field, several collectors connected in series make a row, and a number of rows are connected in parallel to achieve the required nominal thermal power output at design point. The number of collectors connected in series in every row depends on the temperature increase to be achieved between the row inlet and outlet. In every row of collectors, the receiver tubes in adjacent PTCs have to be connected by flexible joints to allow independent rotation of both collectors as they track the sun during the day. These flexible connections are also necessary to allow the linear thermal expansion of the receiver tubes when their temperature increases from ambient to nominal temperature during system start-up. Two main types of flexible connections are available: flexible hoses and ball joints.

Flexible hoses for temperatures below 300°C are composed of an inner hose that can withstand this maximum temperature and an outer metal braid shield protecting the inner hose. The outer braid is thermally insulated to reduce thermal loss. For higher temperatures, stainless steel bellows are usually used. This type of hose is not as flexible, and causes a significant pressure drop in the circuit because of its high friction coefficient. The minimum bending radius defined by the manufacturer must be taken into consideration to prevent overstressing of the bellows.

Ball joints are another option for flexible connection between the receiver tubes of adjacent collectors. The main benefit of this option is a significantly lower pressure drop because pressure drop for one ball joint is equivalent to a 90° elbow. Another advantage of ball joints is the connected pipes have two degrees of freedom of movement, because the connected pipes can rotate freely (360°) simultaneously and with a maximum pivot angle of about $\pm 7.5°$. Ball joints are also provided with an inner graphite sealing to reduce friction and avoid leaks.

Today's PTCs working at temperatures above 300°C are connected by ball joints instead of flexible hoses. Furthermore, the flexible hoses initially installed in the solar power plants in California between 1985 and 1990 are being replaced by ball joints because they are more reliable and have lower maintenance costs. Flexible hoses are likely to suffer from fatigue failures resulting in a leak, whereas ball joints only require the graphite sealing to be refilled after many thousands of hours of operation.

Flexible connections are also needed to connect the row to the main field pipe header inlet and outlet. Figure 19.16 shows typical main pipe headers with flexible hose and ball joint connections.

19.6.2 Performance Parameters and Losses in a PTC

Three of the design parameters required for a PTC are the geometric concentration ratio, the acceptance angle, and the rim angle (see Figure 19.18). The concentration ratio is the ratio between the collector aperture area and the total area of the absorber tube, whereas the acceptance angle is the maximum angle that can be formed by two rays in a plane transversal to the collector aperture so that they intercept the absorber pipe after being reflected by the parabolic mirrors. The concentration ratio, C, is given by

$$C = \frac{l_a l}{\pi d_0 l} = \frac{l_a}{\pi d_0}, \tag{19.21}$$

where d_0 is the outer diameter of the receiver pipe, l is the collector length, and l_a is the parabola width.

The wider the collector acceptance angle, the less accurate the tracking system has to be, as the collector will not need to update its position as frequently. Usual values of the concentration ratio of PTCs are about 20, although the maximum theoretical value is on the order of 70. High concentration ratios are associated with very small acceptance angles, which require very accurate sun-tracking systems and consequently, higher costs.

The minimum practical acceptance angle is 32′ (0.53°), which is the average solid angle at which the sun sphere is seen from the earth. This means that any PTC with an acceptance angle smaller than 32′ would always loose a fraction of the direct solar radiation. In fact, recommended acceptance angles for commercial PTCs are between 1° and 2°. Smaller angles would demand very accurate sun-tracking system and frequent updating of the collector position, whereas higher values would lead to small concentration ratios and, therefore, lower working temperatures. Therefore, acceptance angle values between 1° and 2° are the most cost-effective.

The rim angle, ϕ, is directly related to the concentrator arc length and its value can be calculated from Equation 19.22 as a function of the parabola focal distance, f, and aperture width, l_a:

$$\frac{l_a}{4f} = \tan\frac{\phi}{2}. \tag{19.22}$$

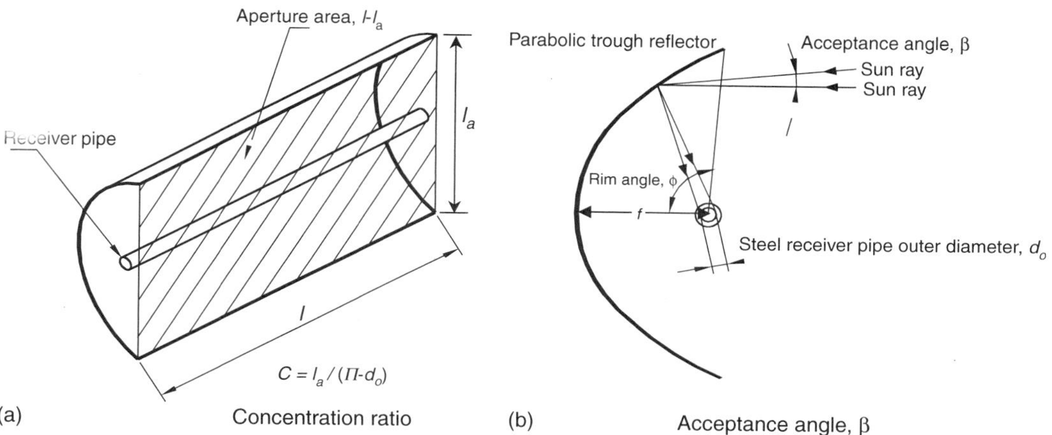

FIGURE 19.18 Concentration ratio and acceptance angle of a parabolic trough collector.

Usual values for rim angles in a PTC are between 70° and 110°. Smaller rim angles are not advisable because they reduce the aperture surface. Rim angles over 110° are not cost-effective because they increase the total reflecting surface without effectively increasing the aperture width.

When direct solar radiation reaches the surface of a PTC, a significant amount of it is lost due to several different factors. The total loss can be divided into three types, which in descending order of importance are:

- Optical losses
- Thermal losses from the absorber pipe to the ambient
- Geometrical losses

The optical losses are associated with four parameters (see Figure 19.19), which are:

- Reflectivity, ρ, of the collector reflecting surface: because the reflectivity of the parabolic trough concentrator is less than 1, only a fraction of the incident radiation is reflected towards the receiver tube. Typical reflectivity values of clean silvered glass mirrors are around 0.93. After washing the mirrors, their reflectivity continuously decreases as dirt accumulates until the next washing. Commercial parabolic trough mirrors are washed when their reflectivity is of about 0.9.
- Intercept factor, Υ: a fraction of the direct solar radiation reflected by the mirrors does not reach the glass cover of the absorber tube due to either microscopic imperfections of the reflectors or macroscopic shape errors in the parabolic trough concentrators (e.g., imprecision during assembly). These errors cause reflection of some rays at the wrong angle, and therefore they do not intercept the absorber tube. These losses are quantified by an optical parameter called the intercept factor,Υ, that is typically 0.95 for a collector properly assembled.
- Transmissivity of the glass tube, τ: the metal absorber tube is placed inside an outer glass tube in order to increase the amount of absorbed energy and reduce thermal losses. A fraction of the direct solar radiation reflected by the mirrors and reaching the glass cover of the absorber pipe is not able to pass through it. The ratio between the radiation passing through the glass tube and the total incident radiation on it, gives transmissivity, τ, which is typically $\tau = 0.93$.
- Absorptivity of the absorber selective coating, α: this parameter quantifies the amount of energy absorbed by the steel absorber pipe, compared with the total radiation reaching the outer wall of the steel pipe. This parameter is typically 0.95 for receiver pipes with a cermet coating, whereas it is slightly lower for pipes coated with black nickel or chrome.

Multiplication of these four parameters (reflectivity, intercept factor, glass transmissivity, and absorptivity of the steel pipe) when the incidence angle on the aperture plane is 0° gives what is called the *peak optical efficiency* of the PTC, $\eta_{\text{opt},0°}$:

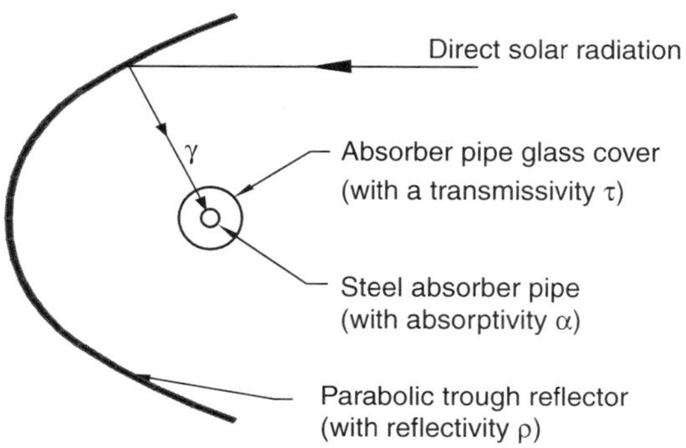

FIGURE 19.19 Optical parameters of a parabolic trough collector.

$$\eta_{\text{opt},0°} = \rho \times \gamma \times \tau \times \alpha|_{\varphi=0°}. \tag{19.23}$$

$\eta_{\text{opt},0°}$ is usually in the range of 0.70–0.76 for clean, good-quality PTCs.

The total thermal loss in a PTC, $P_{\text{Q,collector}\rightarrow\text{ambient}}$, is due to radiative heat loss from the absorber pipe to ambient, $P_{\text{Q,absorber}\rightarrow\text{ambient}}$, and convective and conductive heat losses from the absorber pipe to its outer glass tube, $P_{\text{Q,absorber}\rightarrow\text{glass}}$. Although this heat loss is governed by the well-known mechanisms of radiation, conduction, and convection, it is a good practice to calculate them all together using the thermal loss coefficient, $U_{\text{L)abs}}$, according to

$$P_{\text{Q,collector}\rightarrow\text{ambient}} = U_{\text{L)abs}}\pi d_0 l(T_{\text{abs}} - T_{\text{amb}}), \tag{19.24}$$

where T_{abs} is the mean absorber pipe temperature, T_{amb} is the ambient air temperature, d_0 is the outer diameter of the absorber pipe, and l is the absorber pipe length (PTC length).

In Equation 19.24, the thermal loss coefficient is given in $(\text{W/m}^2_{\text{abs}}\cdot\text{K})$ units per square meter of the steel absorber pipe surface. Equation 19.25 can be used to find the value of the thermal loss coefficient per square meter of aperture surface of the PTC, $U_{\text{L)col}}$:

$$U_{\text{L)col}} = U_{\text{L)abs}}/C[\text{W/m}^2_{\text{col}}\ \text{K}] \tag{19.25}$$

The heat loss coefficient depends on absorber pipe temperature and is found experimentally by performing specific thermal loss tests with the PTC operating at several temperatures within its typical working temperature range. Variation in the thermal loss coefficient versus the receiver pipe temperature can usually be expressed with a second-order polynomial equation like Equation 19.26, with coefficients a, b, and c experimentally calculated:

$$U_{\text{L)abs}} = a + b(T_{\text{abs}} - T_{\text{amb}}) + c(T_{\text{abs}} - T_{\text{amb}})^2 \quad \left(\frac{\text{W}}{\text{m}^2_{\text{abs}}\ \text{K}}\right). \tag{19.26}$$

It is sometimes difficult to find values for coefficients a, b, and c valid for a wide temperature range. When this happens, different sets of values are given for smaller temperature ranges. Table 19.3 gives the values of coefficients a, b, and c that have been experimentally calculated by Ajona (1999) for LS-3 collectors installed at SEGS VIII and IX.

A typical value of $U_{\text{L)abs}}$ for absorber tubes with vacuum in the space between the inner pipe and the outer glass tube is lower than 5 $\text{W/m}^2{}_{\text{abs}}$ K. High-vacuum conditions are not needed to significantly reduce the convective heat losses. However, the low thermal stability in hot air of the cement coatings currently used in these receiver tubes requires a high vacuum to assure good coating durability.

The optimum space between the absorber pipe and the outer glass tube for receivers without vacuum to minimize the convective heat loss is calculated as a function of the Rayleigh number (Ratzel and Simpson 1979). The possible bowing of the steel pipe has to be considered also in determining the minimum gap, because the possibility of contact with the glass cover has to be avoided.

The third group of losses in a PTC are the geometrical losses that are due to the incidence angle, ϕ, of direct solar radiation on the aperture plane of the collector. The incidence angle is the angle between the normal to the aperture plane of the collector and the sun's vector, both contained on a plane perpendicular to the collector axis. This angle depends on the day of the year and the time of day.

TABLE 19.3 Values of Coefficients a, b, and c for a LS-3 Collector

$T_{\text{abs}}/(°\text{C})$	a	b	c
<200	0.687257	0.001941	0.000026
>200; <300	1.433242	−0.00566	0.000046
>300	2.895474	−0.0164	0.000065

The incidence angle of direct solar radiation is a very important factor, because the fraction of direct radiation that is useful to the collector is directly proportional to the cosine of this angle, which also reduces the useful aperture area of the PTC (see Figure 19.20). The incidence angle reduces the aperture area of a PTC in an amount A_e called the *collector geometrical end losses* and is calculated by:

$$A_e = l_a l_\phi = l_a f_m \tan\varphi, \tag{19.27}$$

$$f_m = f + [(f\, l_a^2)/(48\, f^2)], \tag{19.28}$$

where l_a is the parabola width, l is the collector length, f is the focal distance of the parabolic trough concentrator, f_m is the mean focal distance in a cross section of the parabolic trough concentrator, and ϕ is the incidence angle of the direct solar radiation.

The incidence angle also affects PTC optical parameters (i.e., mirror reflectivity, selective coating absorptivity, intercept factor, and glass transmissivity) because these parameters are not isotropic. The

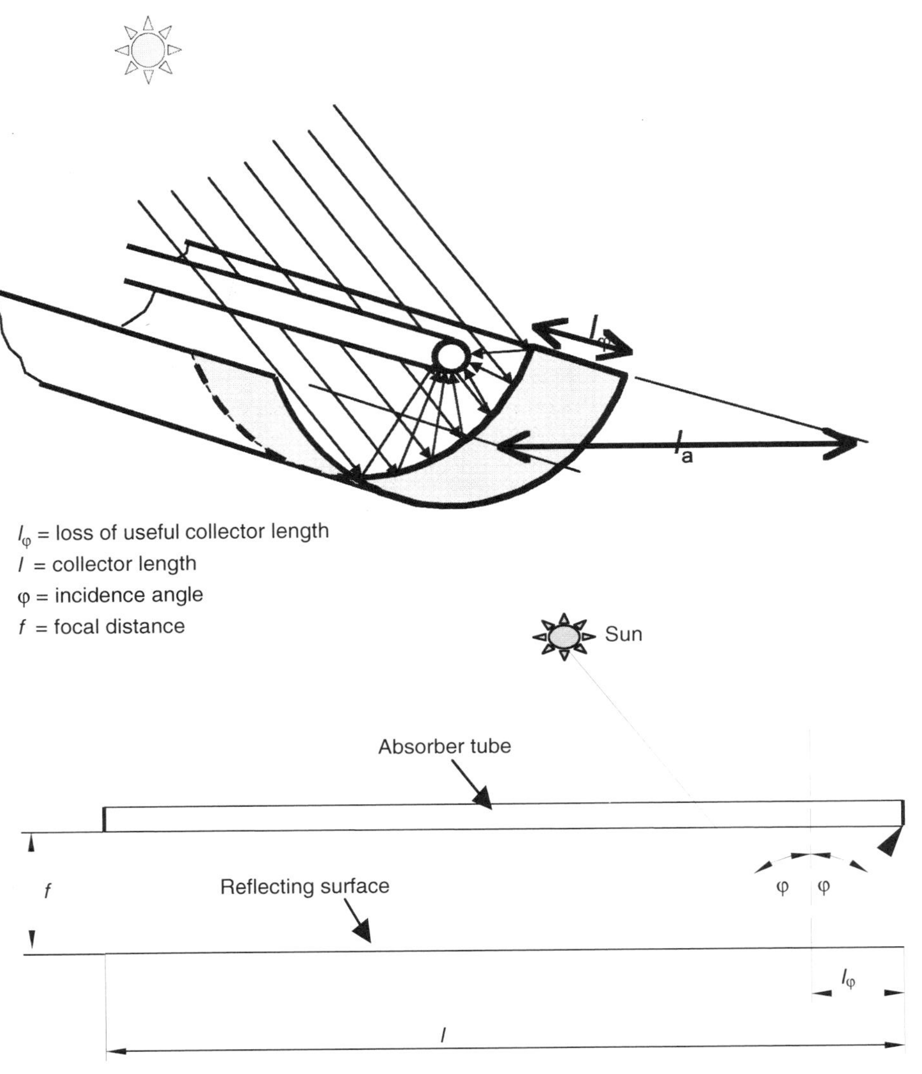

FIGURE 19.20 Geometrical losses at the end of a parabolic trough collector.

effect of the incidence angle on the optical efficiency and useful aperture area of a PTC is quantified by the incidence angle modifier, $K(\phi)$, because this parameter includes all optical and geometric losses due to an incidence angle greater than 0°.

The incidence angle modifier, which directly depends on the incidence angle, is usually given by a polynomial equation so that it is equal to 0 for $\phi = 90°$, and equal to 1 for $\phi = 0°$. Therefore, for instance, the incidence angle modifier for an LS-3 collector is given by

$$K(\varphi) = 1 - 2{,}23073E - 4 \times \varphi - 1{,}1E - 4 \times \varphi^2 + 3{,}18596E - 6 \times \varphi^3 - 4{,}85509E - 8 \times \varphi^4 \quad (0° < \varphi < 80°)$$

$$K(\varphi) = (85° < \varphi < 90°) \tag{19.29}$$

Coefficients of Equation 19.29 are calculated experimentally by means of tests performed with different incidence angles (González, Zarza, and Yebra 2001). The incidence angle of direct solar radiation depends on PTC orientation and Sun position, which can be easily calculated by means of the azimuth, AZ, and elevation, EL, angles. For horizontal north–south and east–west PTC orientations, the incidence angle is given by Equation 19.30 and Equation 19.31, respectively. The sun elevation angle is measured with respect to the horizon (positive upwards), while azimuth is 0° to the south and positive clockwise.

$$\varphi = \arccos[1 - \cos^2(\mathrm{EL})\sin^2(\mathrm{AZ})]^{1/2}, \tag{19.30}$$

$$\varphi = \arccos[1 - \cos^2(\mathrm{EL})\cos^2(\mathrm{AZ})]^{1/2}. \tag{19.31}$$

19.6.3 Efficiencies and Energy Balance in a PTC

The combination of three different efficiencies,

η_{global} = global efficiency
$\eta_{opt,0°}$ = peak optical efficiency (optical efficiency with an incidence angle of 0°)
η_{th} = thermal efficiency

and one parameter,

$K(\phi)$ = incidence angle modifier,

describe the performance of a PTC. Their definition is graphically represented in the diagram shown in Figure 19.21, which clearly shows that a fraction of the energy flux incident on the collector aperture plane is lost due to the optical losses accounted for by the peak optical efficiency, while another fraction is lost because of an incidence angle $\phi > 0°$, which is taken into account by the incidence angle modifier, $K(\phi)$. The remaining PTC losses are thermal losses at the absorber tube.

As explained above, the peak optical efficiency, $\eta_{opt,0°}$, considers all optical losses that occur with an incidence angle of $\phi = 0°$ (reflectivity of the mirrors, transmissibility of the glass tube, absorptivity of the steel absorber pipe and the intercept factor). The incidence angle modifier, $K(\phi)$, considers all optical and geometrical losses that occur in the PTC because the incidence angle is greater than 0° (collector end losses, collector center losses, blocking losses due to absorber tube supports, angle dependence of the intercept factor, angle dependence of reflectivity, transmissivity, and absorptivity). Thermal efficiency, η_{th}, includes all absorber tube heat losses from conduction, radiation, and convection.

Global efficiency, η_{global}, includes the three kinds of losses that occur in the PTC (optical, geometrical, and heat) and can be calculated as a function of the peak optical efficiency, incidence angle modifier, and thermal efficiency using Equation 19.32.

$$\eta_{global} = \eta_{opt,0°} K(\varphi) \eta_{th}. \tag{19.32}$$

The global efficiency can be also calculated as the ratio between the net thermal output power delivered by the collector, $P_{Q,collector \rightarrow fluid}$, and the solar energy flux incident on the collector aperture plane,

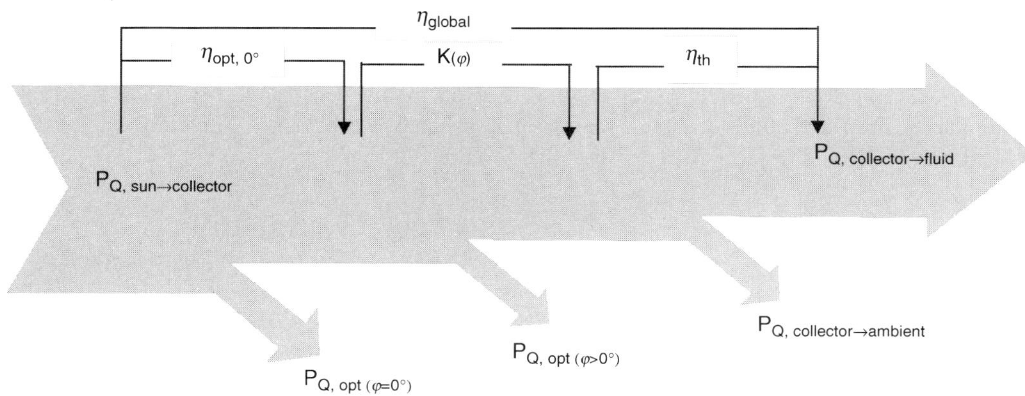

FIGURE 19.21 Diagram of efficiencies and losses in a parabolic trough collector.

$P_{Q,sun \rightarrow collector}$, by Equation 19.33, Equation 19.34, and Equation 19.35:

$$\eta_{global} = \frac{P_{Q,collector \rightarrow fluid}}{P_{Q,sun \rightarrow collector}}, \tag{19.33}$$

$$P_{Q,sun \rightarrow collector} = A_c E_d \cos(\varphi), \tag{19.34}$$

$$P_{Q,collector \rightarrow fluid} = q_m (h_{out} - h_{in}), \tag{19.35}$$

where A_c = collector aperture surface, E_d = direct solar irradiance, ϕ = incidence angle, q_m = fluid mass flow through the absorber tube of the collector, h_{in} = fluid specific mass enthalpy at the collector inlet, and h_{out} fluid specific mass enthalpy at the collector outlet

The net output thermal power delivered by a PTC can be calculated by means of Equation 19.35 if the fluid mass flow and the inlet and outlet temperatures are known when the collector is in operation. However, these data are not known during the solar field design phase and the expected net thermal output has to be calculated starting from the values of the direct solar irradiance, ambient air temperature, incidence angle, and PTC optical, thermal, and geometrical parameters. Equation 19.36 can be used for this purpose:

$$P_{Q,collector \rightarrow fluid} = P_{Q,sun \rightarrow collector} \eta_{global} = A_c E_d \cos(\varphi) \eta_{opt,0°} K(\varphi) \eta_{th} F_e \tag{19.36}$$

From a practical standpoint, calculation of the net thermal output power during the design phase is easier if thermal losses in the PTC, $P_{Q,collector \rightarrow ambient}$, are used instead of the thermal efficiency, η_{th}. In this case, the net thermal output power is given by Equation 19.37 that must be used in combination with Equation 19.23:

$$\begin{aligned} P_{Q,collector \rightarrow fluid} &= A_c E_d \cos(\varphi) \eta_{opt,0°} K(\varphi) F_e - P_{Q,collector \rightarrow ambient} \\ &= A_c E_d \cos(\varphi) \eta_{opt,0°} K(\varphi) F_e - U_{L)abs} \pi d_0 l (T_{abs} - T_{amb}) \end{aligned} \tag{19.37}$$

All the parameters used in Equation 19.37 have been explained previously, with the exception of the soiling factor, F_e, which is $0 < F_e < 1$ and takes into account the progressive soiling of mirrors and glass tubes after washing. This means that the reflectivity and transmissivity are usually lower than nominal values and the peak optical efficiency is also lowered. Usual values of F_e are around 0.97, which is equivalent to a mirror reflectivity of 0.90 for mirrors with a nominal reflectivity of 0.93.

19.6.4 Industrial Applications for PTCs

The large potential market existing for solar systems with PTCs can be clearly seen in the statistical data. U.S. industry consumes about 40% of the total energy demand in that country. Of this usage approximately half (about 20% of the total energy consumption) involves IPH suitable for solar applications with PTCs, which are internationally known as *IPH* applications. As an example of the situation in other countries with a good level of direct solar radiation, industry is also the biggest energy consumer in Spain (more than 50% of the total energy demand) and 35% of the industry demand is in the mid-temperature range (80–300°C) for which PTCs are very suitable.

Besides this large potential market for parabolic trough systems, there is also an environmental benefit that is taken more and more into consideration: contrary to fossil fuels, solar energy does not contaminate and it is independent of political or economic interruptions of supply (due to war, trade boycott, etc).

Because industrial process energy requirements in the mid-temperature range are primarily met by steam systems, representative configurations of solar steam generation systems are presented in this section with simple diagrams to facilitate their understanding.

Steam is the most common heat transport medium in industry for temperatures below 250°C where there is a great deal of experience with it. Compact steam generators have proven to be extremely reliable. Integration of a solar steam generation system for a given industrial process involves a simple plant interface to feed steam directly into the existing process, with no major facility changes. Medium-temperature steam can be supplied with PTCs in three different ways:

- Using a high-temperature, low-vapor pressure working fluid in the solar collectors and transferring the heat to an unfired boiler where steam is produced. Oil is widely used for this purpose.
- Circulating pressurized hot water in the solar collectors and flashing it to steam in a flash tank. This method is suitable for temperatures that are not too high (below 200°C), because of the high pressure required in the absorber pipes and flash tank for higher temperatures.
- Boiling water directly in the collectors, (the so-called *DSG* process).

A brief description of each of these methods follows.

19.6.4.1 Unfired Boiler System

Figure 19.22 shows the schematic diagram for an unfired steam boiler system with PTCs. A heat transfer fluid is circulated through the collector field and steam is generated in an unfired boiler. A variation of the system shown in Figure 19.22 incorporates a preheater in the water makeup line that increases system cost but also reduces the inlet temperature to the solar field. Water could be circulated in the collector loop, but the fluid generally selected is a low-vapor pressure, nonfreezing hydrocarbon, or silicone oil. The use of oil overcomes the disadvantages associated with water (high vapor pressure and risk of freezing) and accommodates energy storage, but certain characteristics of these oils cause other problems. Generally, precautions must be taken to prevent the oil from leaking out of the system, which could cause fire. Oil is also expensive and has poorer heat transport properties than water. They are extremely viscous when cold, and a positive displacement pump is sometimes needed to start the system after it has cooled down. The use of a fluid to transfer thermal energy from the solar field to an unfired boiler where steam is produced is internationally known as *heat transfer fluid* (HTF) technology.

The unfired boiler itself is an expensive item requiring alloy tubes for corrosion protection, and it is an additional resistance to heat flow. As in the flash steam system, the collectors must operate at a temperature of approximately 20°C above the steam delivery temperature. The process steam must be maintained at a certain temperature, so the solar-generated steam is delivered at a variable flow rate depending on the solar radiation available at the aperture of the solar collectors. The collector outlet temperature can be held constant by varying the oil flow rate through the collectors as the collected solar energy varies due to cloud passage or any other reason.

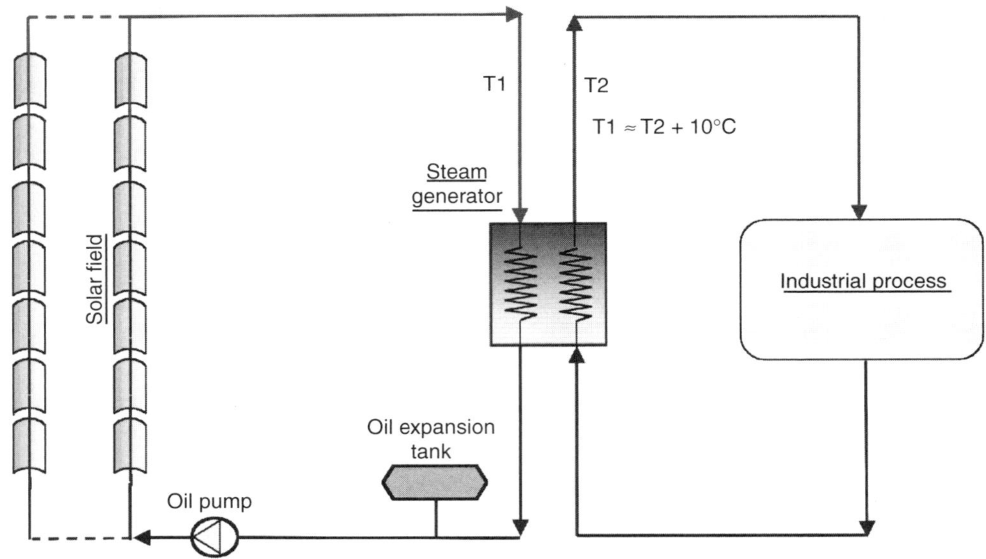

FIGURE 19.22 Unfired boiler steam generating system.

19.6.4.2 Flash Steam Systems

A diagram of a flash steam system is shown in Figure 19.23. Water at a high enough pressure to prevent boiling is circulated through the collector field and flashed to steam across a throttling valve into a separator. This constant enthalpy process converts the sensible heat of the water into the latent heat of a two-phase mixture at the conditions prevailing in the separator. The maximum steam quality (i.e., the fraction of total flow that is converted into steam) is less than 10% due to thermodynamic constraints. The steam thus produced is fed into the industrial process, while the water remaining in the flash tank is recirculated to the solar field inlet. Feedwater makeup is injected from the flash tank into the pump suction to maintain the liquid level in the tank.

Using water as a heat transfer fluid simplifies the construction of a flash steam system. However, although water is an excellent heat transport medium, freezing problems can occur. Therefore, the freeze protection mechanism must be carefully designed and controlled to ensure that a minimum amount of heat is supplied to the water to prevent freezing due to low ambient temperature.

The disadvantages of the flash steam system are associated with the steam generation mechanism. Collector temperatures must be significantly higher than the steam delivery temperature to obtain reasonable steam qualities downstream of the throttling valve and to limit the water recirculation rate. But higher temperatures reduce the collector's efficiency. In addition, the circulating pump must overcome the pressure drop across the flash valve, which can be important.

Moreover, the rapid rise in the water vapor pressure at temperatures above 175°C limits the steam pressure that can be achieved by this method to approximately 2 MPa (305 psig) at acceptable levels of electrical power required for pumping. For higher pressures, the electricity consumption of the recirculation pump would be excessive and would jeopardize the efficiency of the whole system.

19.6.4.3 Direct Steam Generation

DSG in the absorber tubes of PTC is an attractive concept because the average collector operating temperature would be near the steam delivery temperature and because the phase change reduces the required water flow through the circulating pump. The system diagram would be similar to that of the flash steam system but without a flash valve. The disadvantages of this concept are associated with the thermo-hydraulic problems associated with the two-phase flow existing in the evaporating section of the

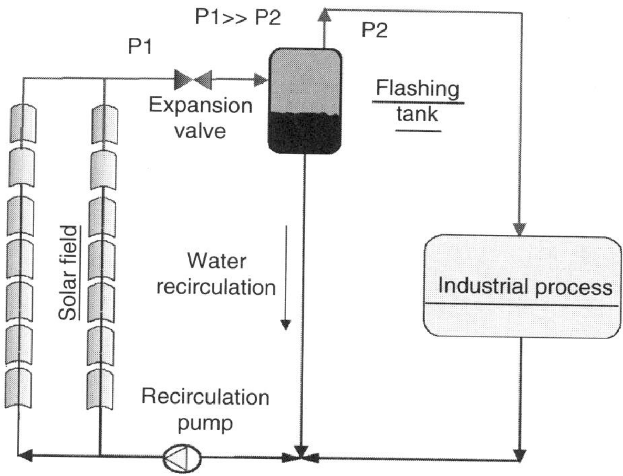

FIGURE 19.23 Flash steam solar system.

solar field. Nevertheless, experiments performed at the PSA in Spain have proven the technical feasibility of DSG with horizontal PTCs at 100 bar/400°C (Zarza et al. 2002).

19.6.5 Sizing and Layout of Solar Fields with PTCs

A typical parabolic trough solar collector field (Figure 19.24) is composed of a number of parallel rows of several collectors connected in series so that the working fluid circulating through the absorber pipe is heated as it passes from the inlet to the outlet of each row.

The first step in the design of a parabolic trough solar field is the definition of the so-called *design point*, which is a set of parameters that determine solar field performance. Parameters to be defined for the design point are:

- The collector orientation
- The date (month and day) and time of design point
- The direct solar irradiance and ambient air temperature for the selected date and time
- The geographical location of the plant site (latitude and longitude)
- The total thermal output power to be delivered by the solar field
- The solar collector soiling factor
- The solar field inlet/outlet temperatures
- The working fluid for the solar collectors
- The nominal fluid flow rate

If oil is used in the solar field to transfer the energy to an unfired boiler (HTF technology), then the selected temperature of the fluid at the solar field outlet must be at least 15°C higher than the steam temperature demanded by the process to be fed. For example, if the industrial process to be fed by the solar system requires 300°C steam, the oil temperature at the solar field outlet must be about 315°C. This difference is necessary to compensate for thermal losses between the solar field outlet and the steam generator inlet and the boiler pinch point, which is on the order of 5–7°C.

After the design point has been defined, the number of collectors to be connected in series in each row can be calculated using the parameters of the selected PTC (peak optical efficiency, incidence angle modifier, heat loss coefficient and aperture area) and fluid (density, heat capacity, and dynamic viscosity). The number of collectors in each row depends on the nominal temperature difference between solar field

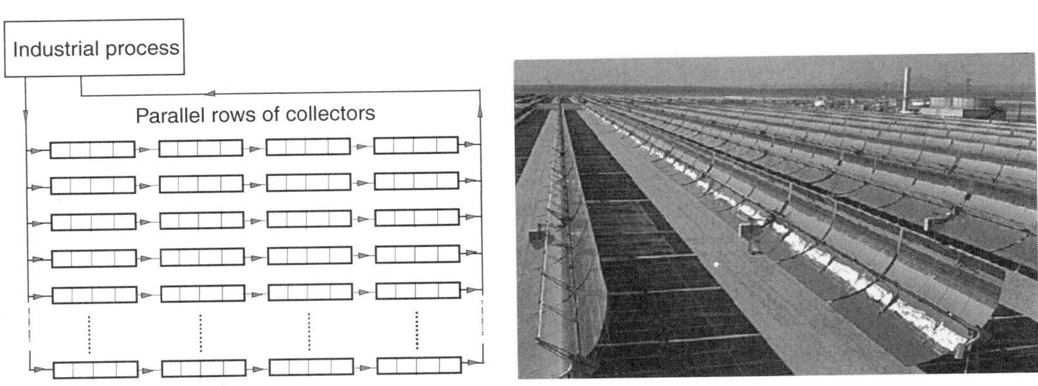

FIGURE 19.24 A typical solar field with parabolic trough collectors.

inlet and outlet, ΔT, and the single collector temperature step, ΔT_c. Therefore, if a collector field is intended to supply thermal energy to an unfired boiler which requires a temperature step of 70°C between inlet and outlet, with a nominal inlet temperature of 220°C, the inlet and outlet temperatures in each row will be 220°C and 290°C, respectively, with a $\Delta T = 70$°C. After this ΔT has been determined, the number of collectors required in each row, N, is given by the ratio:

$$N = \frac{\Delta T}{\Delta T_c},$$

where N is the number of collectors to be connected in series in a row, ΔT is the temperature difference required by the industrial process, and ΔT_c is the difference between the single collector nominal inlet and outlet working temperatures.

After the number of collectors to be connected in series in each row has been calculated, the next step is to determine the number of rows to be connected in parallel. This number depends on the thermal power demanded by the industrial process. The procedure for determining the number of rows is calculated as: the ratio between the thermal power demanded by the industrial process and the thermal power delivered by a single row of collectors at design point.

After sizing the solar field, the designer has to layout the piping. Three basic layouts are used in solar fields with PTCs. These layouts (direct-return, reverse-return, and center-feed) are shown schematically in Figure 19.25. In all three options, the hot outlet piping is shorter than the cold inlet piping to minimize thermal losses. The advantages and disadvantages in each of these three configurations are explained in following paragraphs.

The direct-return piping configuration is the simplest and probably the most extensively used in small solar fields. Its main disadvantage is that there is a much greater pressure difference between the inlets in parallel rows, so that balancing valves must be used to keep flow rates the same in each row. These valves cause a significant pressure drop at the beginning of the array, and thus their contribution to the total system pressure loss is also significant. The result is higher parasitic energy consumption than for the reverse-return layout, where the fluid enters the collector array at the opposite end. Pipe headers with different diameters are used in this configuration to balance the array flow. The use of larger pipe headers also results in lower parasitic power requirements, but these could be offset by increases in initial investment costs and thermal energy losses.

The reverse-return layout has an inherently more balanced flow. While balancing valves may still be required, the additional system pressure loss is much lower than in a direct-return configuration. (Alternatively, header pipes can be stepped down in size on the inlet side and stepped up on the outlet

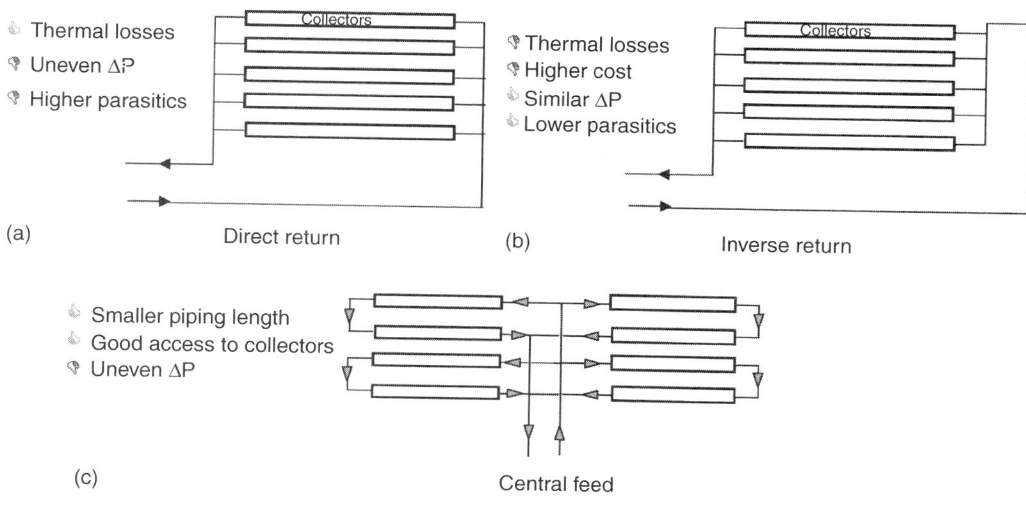

FIGURE 19.25 Solar field layouts for parabolic trough collectors.

side to keep flow rate in the headers constant, thereby providing uniform flow.) The extra length of piping at the solar field inlet is a disadvantage in the reverse-return configuration because of the additional heat loss, although this greatly depends on the solar field inlet temperature. If this temperature is low, additional heat loss is negligible. Adding to pipe length, however, results in higher piping, insulation, and fluid inventory costs.

The center-feed configuration is the most widely used layout for large solar fields. As in the direct-return design, pressure loss in the solar field is greater if balancing valves are installed at the row inlets. This layout minimizes the total amount of piping because there is no pipe running the length of the collector row. There is also direct access to each collector row without buried pipes.

19.6.6 Electricity Generation with PTCs

The suitable PTC temperature range and their good solar-to-thermal efficiency up to 400°C make it possible to integrate a parabolic trough solar field in a Rankine water/steam power cycle to produce electricity. The simplified scheme of a typical solar thermal power plant using parabolic troughs integrated in a Rankine cycle is shown in Figure 19.26. So far, all the solar thermal power plants with PTCs use the HTF technology because steam production by flashing is not suitable for 100 bar steam pressure and commercial DSG has not yet been proven.

A parabolic trough power plant is basically composed of three elements: the solar system, the steam generator, and the power conversion system (PCS) (see Figure 19.26). The solar system is composed of a parabolic trough solar collector field and the oil circuit. The solar field collects the solar energy available in the form of direct solar radiation and converts it into thermal energy as the temperature of the oil circulating trough the receiver tubes of the collectors increases.

Once heated in the solar field, the oil goes to the steam generator, which is an oil–water heat exchanger where the oil transfers its thermal energy to the water that is used to generate the superheated steam required by the turbine. The steam generator is, therefore, the interface between the solar system (solar field + oil circuit) and the PCS itself. Normally, the steam generator used in these solar power plants consists of three stages:

- Preheater: where water is preheated to a temperature close to evaporation
- Evaporator: where the preheated water is evaporated and converted into saturated steam

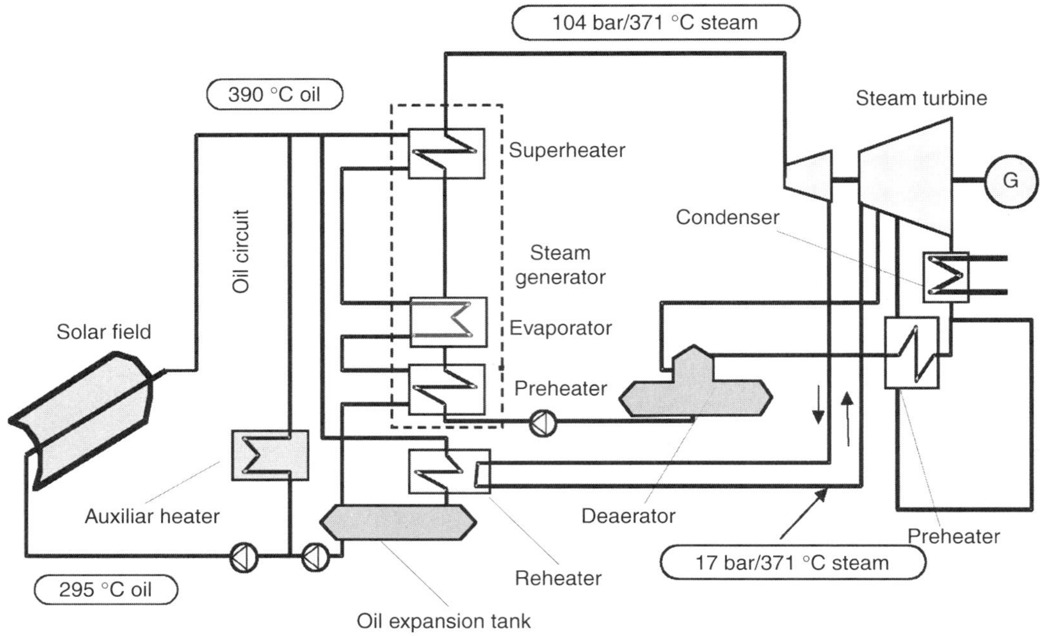

FIGURE 19.26 Simplified scheme of a solar thermal power plant with parabolic trough collectors.

- Superheater: the saturated steam produced in the evaporator is heated in the superheater to the temperature required by the steam turbine

The PCS transforms the thermal energy delivered by the solar field into electricity, using the superheated steam delivered by the steam generator. This PCS is similar to that of a conventional Rankine power plant, except for the heat source.

The superheated steam delivered by the steam generator is then expanded in a steam turbine that drives an electricity generator. The steam turbine is usually composed of two consecutive stages, for high- and low-pressure steam. Steam leaving the turbine high-pressure stage goes to a reheater where its temperature rises before entering the low-pressure turbine stage. After this stage, the steam is condensed and the condensate goes to a water deaerator to remove oxygen and gases dissolved in the water. The steam leaving the turbine low-pressure stage can be condensed either in a wet cooling system (evaporative cooling towers) or in a dry cooling system (air-cooled condenser). The selection of the best cooling system is strongly influenced by the on-site availability of water resources. The main pump takes feed water for the steam generator from the deaerator, thus starting the Rankine thermodynamic cycle again.

Though parabolic trough power plants usually have an auxiliary gas-fired heater to produce electricity when direct solar radiation is not available, the amount of electricity produced with natural gas is always limited to a reasonable level. This limit changes from one country to another: 25% in California (U.S.A.), 15% in Spain, and no limit in Algeria. Figure 19.27 shows what a solar thermal power plant with PTCs looks like. The PCS is located at the center of the plant, surrounded by the solar field. The plant shown in Figure 19.27 is provided with a wet cooling system and the steam leaving the cooling towers is clearly shown.

Parabolic trough power plants can play an important role in achieving sustainable growth because they save about 2000 tn. of CO_2 emissions per MW of installed power yearly. Typical solar-to-electric efficiencies of a large solar thermal power plant (greater than 30 MW_e) with PTCs is between 15% and 22%, with an average value of about 17%. The yearly average efficiency of the solar field is about 50%.

Although not included in Figure 19.26, a thermal energy storage system can be implemented in parabolic trough power plants to allow operation of the PCS when direct solar radiation is not available. In this case, the solar field has to be oversized so that it can simultaneously feed the PCS and charge the storage system during sunlight hours. Thermal energy from the storage system is then used to keep the steam turbine running and producing electricity after sunset or during cloudy periods. Yearly hours of operation can be significantly increased and plant amortization is thus enhanced when a storage system is implemented. However, the required total investment cost is also higher.

19.6.7 Thermal Storage Systems for PTCs

The main problem with using solar radiation is its intermittency and the fact that it is only possible to collect it during sunlight hours. There is an additional limitation when dealing with concentrating solar systems as these systems can only collect the direct solar radiation, so they need clear sky conditions, because clouds block direct solar radiation. Thermal storage systems are implemented to solve these limitations.

When a solar system does not have to supply thermal energy during the night or during cloudy periods, a storage system is not necessary. On the other hand, if the industrial process has to be supplied during periods without direct solar radiation, a storage system has to be implemented to store part of the thermal energy supplied by the solar collectors during the sunlight hours.

Thermal storage systems have two main advantages:

- Thermal energy can be supplied during hours when direct solar radiation is not available, so that solar energy collection and thermal energy supply do not have to be simultaneous.
- The solar field inlet can be isolated from possible disturbances at the outlet, because the storage system behaves as a good thermal cushion and avoids feedback of the disturbances affecting the solar field outlet temperature.

This second advantage is important because it enhances solar field operation on days with frequent cloud transients. No matter how effective solar field control is, the fluid temperature at the outlet is affected by cloud transients and temperature fluctuations are likely. These fluctuations would immediately affect the working fluid temperature at the inlet if there were not a thermal storage system in between.

A hot water storage system such as the one used in low-temperature solar conversion systems (i.e., flat-plate collectors) is not suitable for parabolic trough systems because the high pressure in the storage tank would make the system too expensive. For this reason, PTCs require a different storage medium.

FIGURE 19.27 Overall view of a solar power plant with parabolic trough collectors.

Depending on the medium where the thermal energy is stored, there are two types of systems: single-medium storage systems and dual-medium storage systems.

19.6.7.1 Single-Medium Storage Systems

Single-medium storage systems are those in which the storage medium is the same fluid circulating through the collectors. The most common is thermal oil as both the working fluid and the storage medium. The efficiency of these systems is over 90%. Oil storage systems can be configured in two different ways:

Systems With a Single Oil Tank. For low-capacity storage systems, thermal energy can be stored in a single tank, in which the oil is stratified by temperature. Figure 19.28a shows this type of system configuration, in which the solar field can supply hot oil to the tank by means of three-way valves installed at the solar field inlet and outlet, so that all of it or only part of it enters. The density of the thermal oils commonly used as working fluids in these systems strongly varies with temperature. Therefore, for instance, the density of Santotherm 55 oil at 90°C is 842.5 kg/m^3, while at 300°C it is 701.4 kg/m^3. Due to its lower density, the hot oil entering the storage tank through the top inlet manifold remains in the upper layers inside the tank, while the cold oil always remains at the bottom of the tank. As seen in Figure 19.28a, the boiler supplying the thermal energy demanded by the industrial process can be fed from either the storage tank or the solar field, depending on the position of the three-way valve. When discharging the storage system, the hot oil leaves the tank through the top outlet manifold and returns to the bottom after leaving the boiler. Pump B2 is used exclusively to feed the boiler from the storage tank, when the solar field is not in operation. Cold oil leaves from the bottom of the storage tank and goes to the solar field during daylight hours to be heated and then returned to the top of the storage tank. The storage system is fully charged when all the oil stored in the tank is hot. As already mentioned, the use of a single oil storage tank is feasible only for small storage systems. For high-capacity systems, two oil tanks (i.e., one tank for cold oil and another for hot oil) are needed.

Systems With Two Oil Tanks. There are two oil tanks in these systems (see Figure 19.28b), one hot tank and one cold tank. The boiler is always fed from the hot tank and once the oil has transferred heat to the water in the unfired boiler, it goes to the cold tank. This tank supplies the solar field, which at the same time feeds the hot tank with the oil heated by the collectors.

One of the drawbacks of using oil as the storage medium is the need to keep the oil in the storage tank(s) pressurized and in an inert atmosphere. Thermal oil has to be kept pressurized above the vapor pressure corresponding to the maximum temperature in the oil circuit to prevent the oil from changing into gas. Fortunately, the vapor pressure of the thermal oils used in these systems is usually low for the 100–400°C temperature range, and pressurization is easily maintained by injecting argon or nitrogen. This inert atmosphere also avoids the risk of explosion in the tank from pressurized mists which are explosive in air.

Another disadvantage of oil systems is the need for appropriate firefighting systems, as well as a concrete oil sump to collect any leaks and avoid contamination. All additional equipment increases the cost of the storage system. Thermal oil storage systems usually have two safety systems to avoid excessive overpressure inside the tank when the temperature increases:

- A relief valve to discharge the N_2 into the atmosphere when the pressure inside the tank is over a predefined value; this valve usually works when the pressure inside the tank increases slowly (for instance, during charging)
- Due to the small section of the relief valve, it cannot dissipate sudden overpressures. There is an additional security device for this purpose: a pressure-rated ceramic rupture disc. This system allows gases to be rapidly evacuated into the atmosphere. The rupture disc is destructive, because it consists of a ceramic membrane that breaks if the pressure in the tank is higher than calibrated. The rupture disc only works in case of an emergency when overpressure occurs so quickly that the relief valve cannot keep the pressure inside the tank below the limit.

An additional auxiliary system required in thermal oil storage tanks is a small vessel where gas and volatile compounds produced by oil cracking are condensed and evacuated.

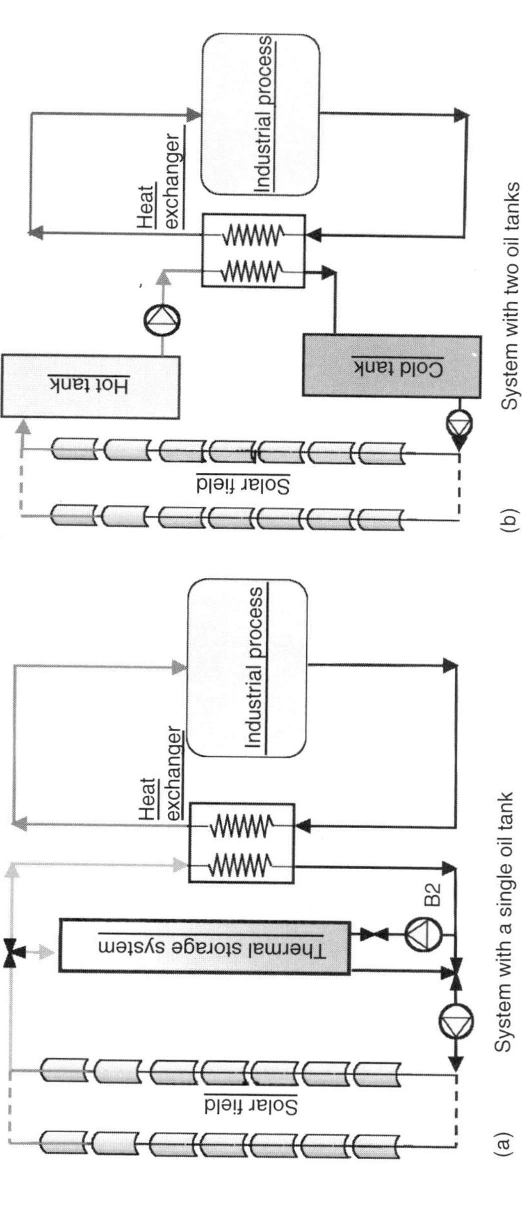

FIGURE 19.28 Thermal storage systems with one and two oil tanks.

19.6.7.2 Dual-Medium Storage Systems

Dual-medium storage systems are those in which the heat is stored in a medium other than the working fluid heated in the solar collectors. Iron plates, ceramic materials, molten salts, or concrete can be used as the storage medium. In these systems, the oil is commonly used as the heat transfer medium between the solar field and the material where the thermal energy is stored in the form of sensible heat. In the case of thermal storage in iron plates, the oil circulates through channels between cast iron slabs placed inside a thermally insulated vessel, transferring thermal energy to them (charging process) or taking it from them (discharging process).

Molten salts (an eutectic mixture of sodium and potassium nitrates) can also be used for dual-medium thermal storage systems in parabolic trough solar plants. In this case, two tanks are needed: one for cold molten salt and another to store the hot molten salt. Obviously, the lowest temperature is always above the melting point of the salt (approximately 250°C). In this case a heat exchanger is needed to transfer energy from the oil used in the solar field (heat transfer medium) to the molten salt used for energy storage (storage medium). Figure 19.29 shows a simplified scheme of a parabolic trough power plant with a molten salt thermal energy storage system. This type of thermal storage system is claimed to be the most cost-effective option for large commercial solar power plants with large solar shares.

19.6.8 Direct Steam Generation

All solar power plants with PTCs implemented to date use thermal oil as the working fluid in the solar field, and they usually follow the general scheme depicted in Figure 19.26, with only slight differences from one plant to another. The technology of these plants has been improved since the implementation of the first commercial plant in 1984 (Lotker 1991; Price et al. 2002). However, though the collector design and connection between the solar system and the PCS have been improved, some further improvements could still be implemented to reduce costs and increase efficiency. The main limitation to improving their competitiveness is the technology itself: the use of oil as a heat carrier medium between the solar field and the PCS which entails a high pressure drop in the oil circuit, limitation of the maximum temperature of the Rankine cycle and O&M costs of the oil-related equipment. If the superheated steam required to feed the steam turbine in the power block were produced directly in the receiver tubes of the PTCs (i.e., DSG), the oil would be no longer necessary and temperature limitation and environmental risks associated with the oil would be avoided (Ajona and Zarza 1994). Figure 19.30 shows the overall scheme of a parabolic trough power plant with DSG in the solar field. Simplification of overall plant configuration is evident when comparing Figure 19.26 and Figure 19.30.

DSG has technical advantages that must be considered (Zarza et al. 1999):

- No danger of pollution or fire due to the use of thermal oil at temperatures of about 400°C
- Opportunity to increase the maximum temperature of the Rankine cycle above 400°C, the limit imposed by the thermal oil currently used
- Reduction in the size of the solar field, thus reducing the investment cost
- Reduction in operation and maintenance-related costs, as thermal oil-based systems require a certain amount of the oil inventory to be changed every year, as well as antifreeze protection when the air temperature is below 14°C

However, DSG presents certain challenges as a way to improve the current technology of parabolic trough solar power plants, due to the two-phase flow (liquid water + steam) existing in the absorber tubes of the solar field evaporating section. The existence of this two-phase flow involves some uncertainties that must be clarified before a commercial plant making use of this technology can be built. Some of these uncertainties are:

- Solar field control
- Process stability

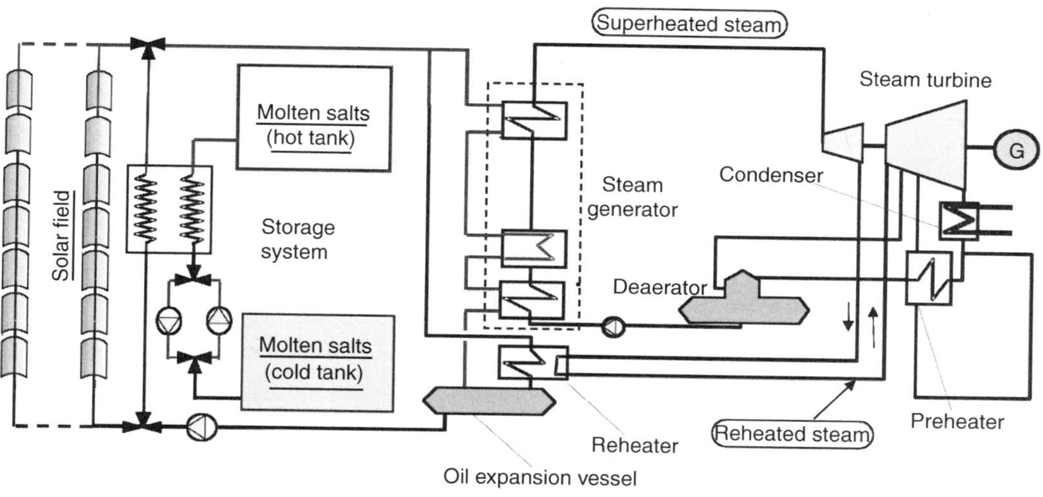

FIGURE 19.29 Scheme of a parabolic trough power plant with molten salt storage system.

- Stress in the receiver pipes
- Higher steam loss (leaks) than oil-based systems

Figure 19.31 shows the typical two-phase flow pattern in a horizontal pipe. As observed in Figure 19.31, four main flow patterns are possible, depending on the superficial velocities in the liquid and steam phases: bubbly, intermittent, stratified, and annular. The borders between adjacent flow patterns are not as well defined as they appear in Figure 19.31, but are rather separated by transition zones.

In bubbly and intermittent flow, the steel absorber pipe inner wall is well-wetted, thus avoiding dangerous temperature gradients between the bottom and the top of the pipe when it is heated from one side. The result is a good heat transfer coefficient all the way around the pipe because the liquid phase is not stratified.

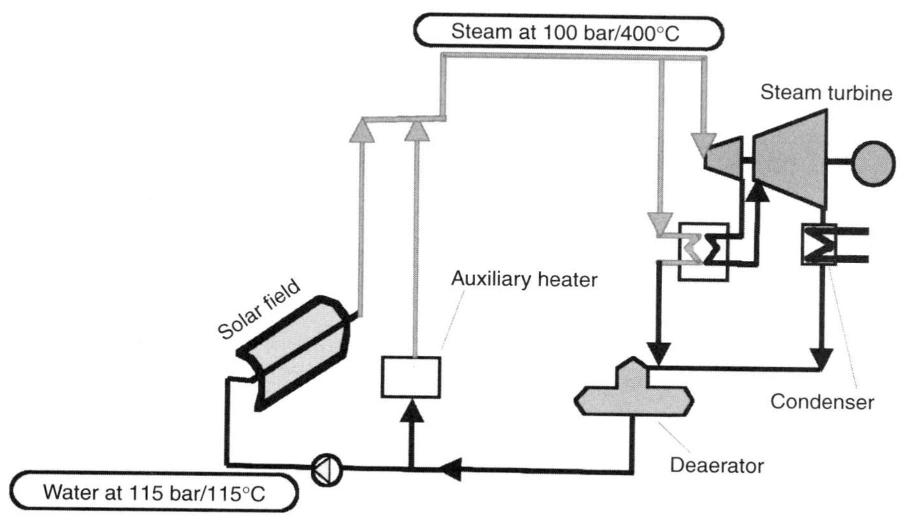

FIGURE 19.30 Simplified scheme of a parabolic trough power plant with direct steam generation.

In the stratified region, the liquid water is in the bottom of the absorber pipe while the steam remains above the surface of the liquid water. The result of this stratification is an uneven heat transfer coefficient around the pipe. Wetting of the bottom of the pipe is still very good and so is the heat transfer coefficient. But the cooling effect of the steam is poorer and the heat transfer coefficient in the top section of the absorber pipe can be very low, resulting in a wide temperature difference of more than 100°C between the bottom and the top of the pipe in a given cross section when it is heated from one side. The thermal stress and bending from this steep temperature gradient can destroy the pipe. Figure 19.32 shows what happens in a cross section of the steel absorber pipe when it is heated underneath (Figure 19.32b, parabolic trough concentrator looking upwards) and from one side (Figure 19.32a, parabolic trough concentrator looking at the horizon). The figure clearly shows how stratified flow can cause steep temperature gradients only when the vector normal to the aperture plane of the concentrator is almost horizontal (Figure 19.32a).

In the annular region, though there is partial stratification of water at the bottom of the pipe, there is a thin-film of water wetting the upper part of the pipe. This film is enough to ensure a good heat transfer coefficient all the way around the pipe, thus avoiding dangerous thermal gradients that could destroy it. Typical absorber pipe cross sections in the stratified and annular regions are also shown in Figure 19.31.

Nevertheless, the technical problems due to water stratification inside the absorber pipes can be avoided. There are three basic processes, called *once-through*, *injection*, and *recirculation*, that can be used without facing dangerous temperature gradients in the receiver pipes. These three options require a solar field composed of long rows of PTCs connected in series to perform the complete DSG process: water preheating, evaporation, and steam superheating. Figure 19.33 summarizes the advantages and disadvantages of the three basic DSG options.

In the once-through process, stratification can be avoided by keeping a high water mass flow rate throughout the absorber pipes. All the feed water is introduced at the collector row inlets and converted into superheated steam as it circulates through the collector rows.

In the injection process, small fractions of feed water are injected along the collector row. The main advantage of this process is the good controllability of the superheated steam parameters at the field outlet. On the downside, this makes the system more complex and increases its cost.

The third option, the so-called *recirculation process*, is the most conservative one. In this case, a water steam separator is placed at the end of the evaporating section of the collector row. Feed water enters the solar field inlet at a higher flow rate than the steam to be produced by the system. Only a fraction of this water is converted into steam as it circulates through the collectors of the preheating and evaporating

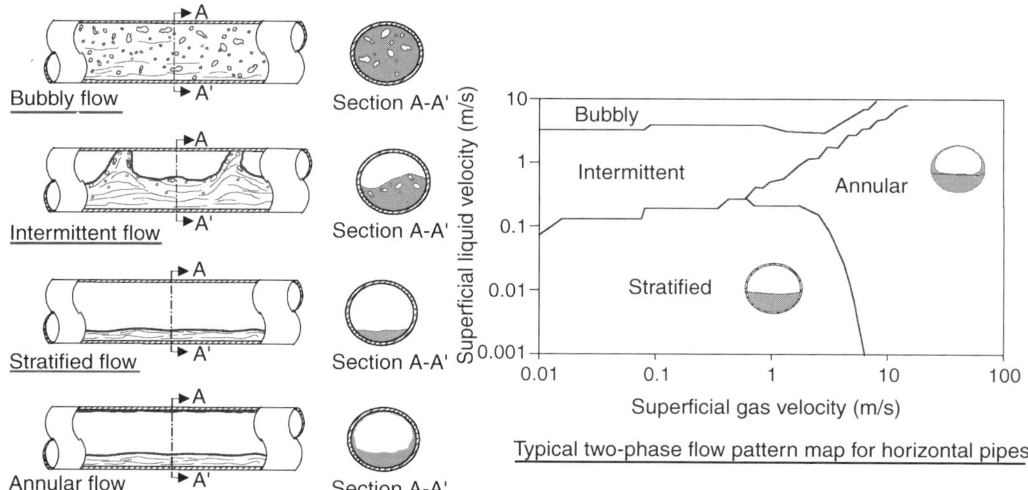

FIGURE 19.31 Two-phase flow configurations and typical flow pattern map for a horizontal receiver pipe.

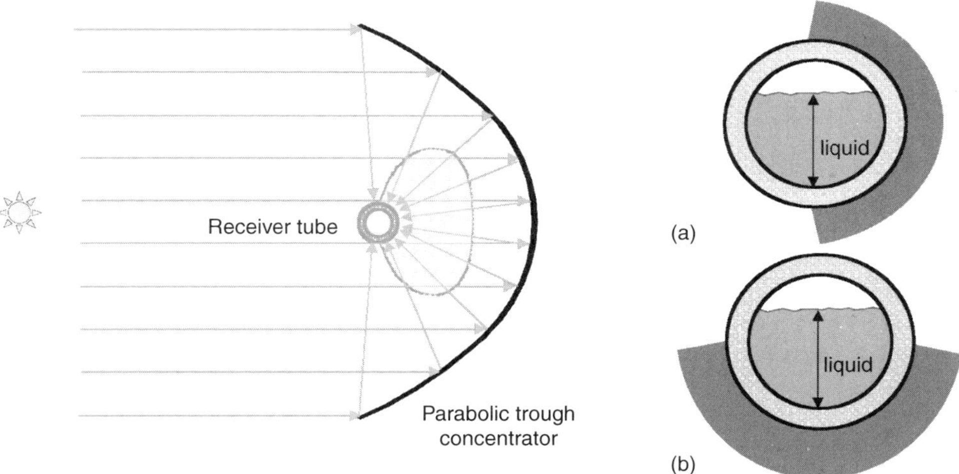

FIGURE 19.32 Liquid phase stratification and concentrated incident solar flux onto the receiver pipe.

sections. At the end of the evaporating section, the saturated steam is separated from the liquid water by the separator, and this remaining liquid water is then recirculated to the solar field inlet by a recirculation pump. The excess water in the evaporating section guarantees good wetting of the receiver pipes and makes stratification of liquid water impossible. Good controllability is the main advantage of this DSG option, but the need for a recirculation pump and the excess water that has to be recirculated from the water/steam separator to the solar field inlet increase system parasitic loads, penalizing overall efficiency.

Figure 19.33 shows that, compared to each other, each of the three basic DSG options has advantages and disadvantages (Zarza et al. 1999) and only their evaluation under real solar conditions can show which option is the best. This is why in 1996 a European consortium launched an R&D project to investigate all the technical questions concerning the DSG process. The name of that project was DISS

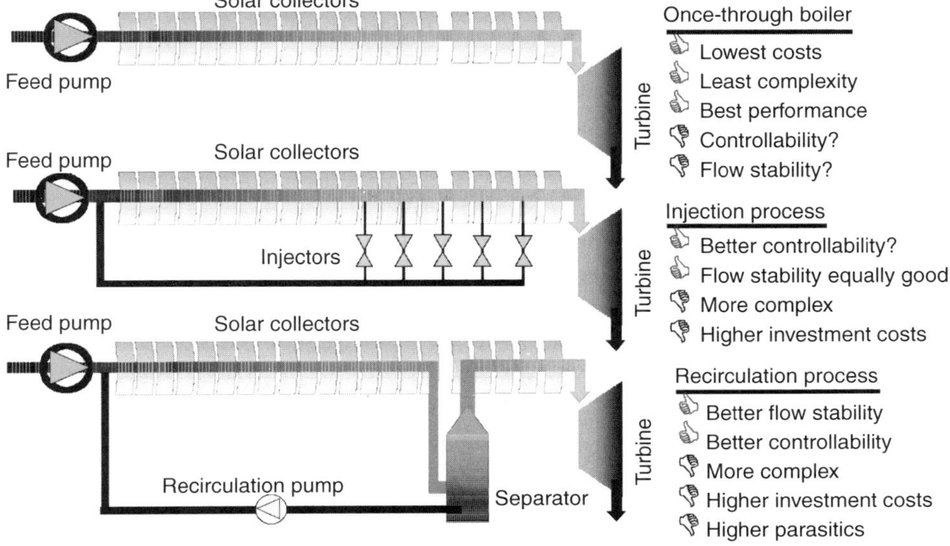

FIGURE 19.33 The three basic DSG processes.

(direct solar steam), and it was developed by partners belonging to all the sectors involved in this technology (i.e., electric utilities, industry, engineering companies, and research centers) in efficient collaboration with other European projects related to PTCs and solar thermal power plants (e.g., STEM, ARDISS, GUDE, EuroTrough, PRODISS, etc.).

The only DSG life-sized test facility currently available in the world was designed and implemented at the PSA during the first phase of the DISS project to investigate the feasibility of DSG in PTCs under real solar conditions. DISS-phase I started in January 1996 and ended in November 1998, with the financial support of the European Commission. The second phase of the project, also partly funded by the European Commission, started in December 1998 and lasted 37 months.

Although the DISS solar field can be operated over a wide temperature/pressure range, the three main operating modes are:

Solar Field Conditions	Inlet (°C)	Outlet (°C)
Mode 1	40 bar/210	30 bar/300
Mode 2	68 bar/270	60 bar/350
Mode 3	108 bar/300	100 bar/375

The DISS test facility accumulated more than 3500 h of operation in 1999, 2000, and 2001. The experimental data gathered in the project and the simulation tools that were developed on the basis of test data provided enough information to evaluate and compare the three DSG basic operation modes (i.e., recirculation, injection, or once-through) or any combination of them. The facility is provided with water injectors at the inlet of the solar collectors and water pumps are provided with frequency converters for smooth, efficient speed control from 10% to 100%. The DISS facility is still in operation and delivering very useful information for the short-term commercial deployment of DSG.

The PSA DISS test facility solar field was initially composed of a single row of 11 PTCs connected in series, with a total length of 550 meters and 2700 m^2 of aperture surface. The collector row is connected to the balance of plant where the superheated steam delivered by the solar field is condensed and used as feed water for the solar field (closed-loop operation).

The DISS test facility implemented at the PSA in DISS-phase I was improved during the subsequent years and it has become a very flexible and powerful life-size test facility, suitable not only for investigating thermo-hydraulic aspects of DSG under real solar conditions, but also evaluating optimized components and O&M procedures for commercial DSG solar plants. Figure 19.34 shows the present schematic diagram of the PSA DISS facility. A water/steam separator (marked TK-4 in Figure 19.34) connects the end of the evaporating section to the inlet of the steam superheating section when the facility is operated in recirculation mode. Two additional collectors were added to the first row in 2003, so it is currently 750 m long with an aperture area of 3822 m^2, with a nominal thermal power of 1.8 MW_{th} and a maximum 400°C/100 bar superheated steam production of 1 kg/s.

The DISS project results proved the feasibility of the DSG process in horizontal PTCs and important know-how was acquired by the project partners regarding the thermo-hydraulic parameters of the liquid water/steam flow in DSG solar fields within a wide range of pressures (30–100 bar). Experimental results were evaluated and complemented with results from lab-scale experiments and simulation tools. Additionally, several models developed in the past to calculate the pressure drop in pipes with two-phase flow were compared with the project's experimental results, and the models with the most accurate simulation results were identified. The proposed Chisholms (Chisholms 1980) and Friedel models proved to be the best match (Zarza 2004).

The good match between experimental data and simulation results obtained with finite element models for the temperature profile in the DSG absorber pipes was another conclusion achieved in the DISS project. It was proven that the temperature gradients in the steel absorber pipes are within safe limits for a wide range of mass fluxes. Results concerning pressure drop and temperature gradients in the absorber pipes of the DISS collector row enabled accurate simulation and design tools for large DSG

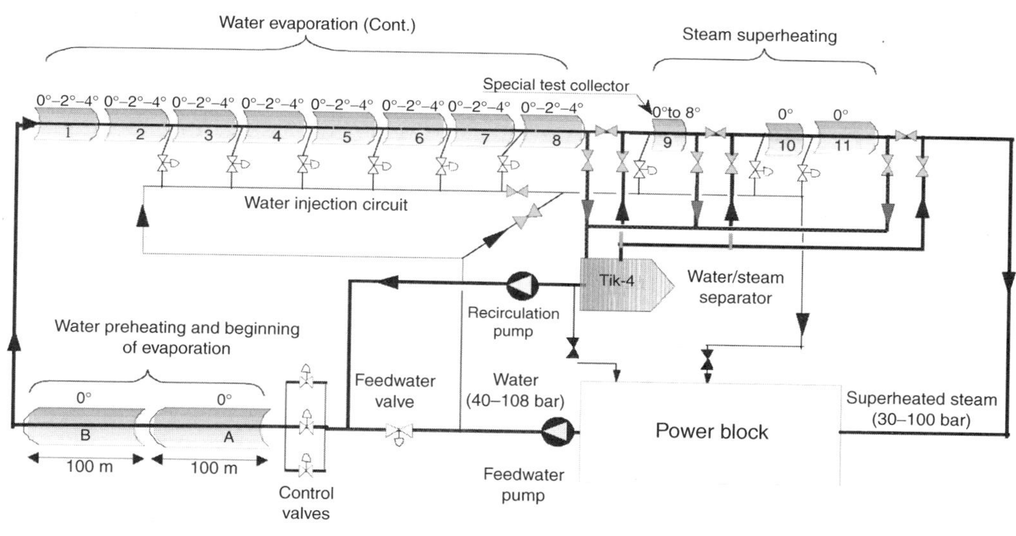

FIGURE 19.34 Schematic diagram of the PSA DISS test facility.

commercial solar fields to be developed (Eck et al. 2003), and was one of the major contributions of DISS to the development of the DSG technology.

Different possibilities (i.e., inner capillary structures and displacers) for enhancing heat transfer in the DSG absorber pipes were also investigated, however, experimental results showed that the absorber pipes do not need these devices to keep the temperature gradients within safe limits.

Several control schemes for once-through and recirculation operation modes were developed and evaluated with good results in the project. As initially expected, it was found that the temperature and pressure of the superheated steam produced by the solar field can be controlled more easily in the recirculation mode, though an efficient control was also achieved for the once-through mode.

The influence of the inclination of the absorber pipes on the thermo-hydraulic parameters of the two-phase flow was investigated also. Tests at 30, 60, and 100 bar were performed in two positions: horizontal and tilted 4°. Though inclination of the absorber pipe reduced the stratified region, the test results obtained without inclination clearly showed that inclination of the absorber pipes is not required to guarantee sufficient cooling for the wide range of operating conditions investigated in the project.

Of the three DSG basic processes (i.e., injection, recirculation, and once-through), experimental results showed that recirculation is the most feasible option for financial, technical and O&M-related parameters of commercial application. Test results showed good stability of the recirculation process even with a low recirculation ratio, thus making possible the use of small, inexpensive water/steam separators in the solar field.

The experience and know-how acquired in DISS was applied to the INDITEP project (2002–2005) in designing the first precommercial DSG solar power plant, in which the thermal energy delivered by a DSG solar field is used to feed a superheated-steam Rankine cycle (Zarza et al. 2004). The INDITEP project was the logical continuation of the DISS project, because the design and simulation tools developed in it for DSG solar fields were used in INDITEP, and most of the partners had also been involved in the previous project. INDITEP was promoted by a Spanish–German Consortium of engineering companies, power equipment manufacturers, research centers, and businesses involved in the energy market: Iberdrola Ingeniería Consultoría (project Coordinator), CIEMAT, DLR, FLAGSOL GmbH, FRAMATONE, GAMESA Energía Servicios S.A., INITEC Tecnología S.A., Instalaciones Inabensa S.A., and ZSW. The European Commission also provided financial assistance.

Three basic requirements were defined for the design of this first precommercial DSG solar power plant:

1. The power block must be robust and operable under flexible conditions in order to assure durability and reliability. Higher priority was therefore given to power block robustness and flexibility, while efficiency was considered less critical for this first DSG plant.
2. The plant must be small to limit the financial risk. Operating stability of a multirow DSG solar field under uneven distribution of solar radiation and solar radiation transients must be proven.
3. The solar field must operate in recirculation mode because the DISS project demonstrated that the recirculation mode is the best option for commercial DSG solar fields (Eck and Zarza 2002).

A 5.47-MW_e power block was selected for this plant to meet the first two requirements. Though due to its small size, this power block is not very highly efficient, its robustness was guaranteed by the manufacturer with references from facilities already in operation. Figure 19.35 shows a schematic diagram of the power block design and its main parameters (Table 19.4).

The DSG solar field design consists of 70 ET-100 PTCs developed by the European EuroTrough Consortium with financial support from the European Commission. This collector was chosen over the LUZ company's LS-3 collector, which is in use at the most recent SEGS plants erected in California (U.S.A.), because of its improved optical efficiency and lighter steel structure (Luepfert et al. 2003).

The seven parallel rows of ET-100 PTC axes are oriented north–south to collect the largest amount of solar radiation per year, even though the differences in solar field thermal energy output in winter and summer are more significant than with east–west orientation. Figure 19.36 shows the schematic diagram of a typical solar field collector row at the design point. Each row is made up of 10 collectors: three collectors for preheating water, five collectors for evaporating water, and two collectors for superheating steam. The end of the boiling section and the inlet of the superheating steam section in every row are connected by a compact water/steam separator, which in turn drains into a larger shared vessel. Water from the separator in every row goes to the final vessel from which it is then recirculated to the solar field inlet by the recirculation pump.

The temperature of the superheated steam produced in each row of collectors is controlled by means of a water injector placed at the inlet of the last collector. The amount of water injected at the inlet of the last

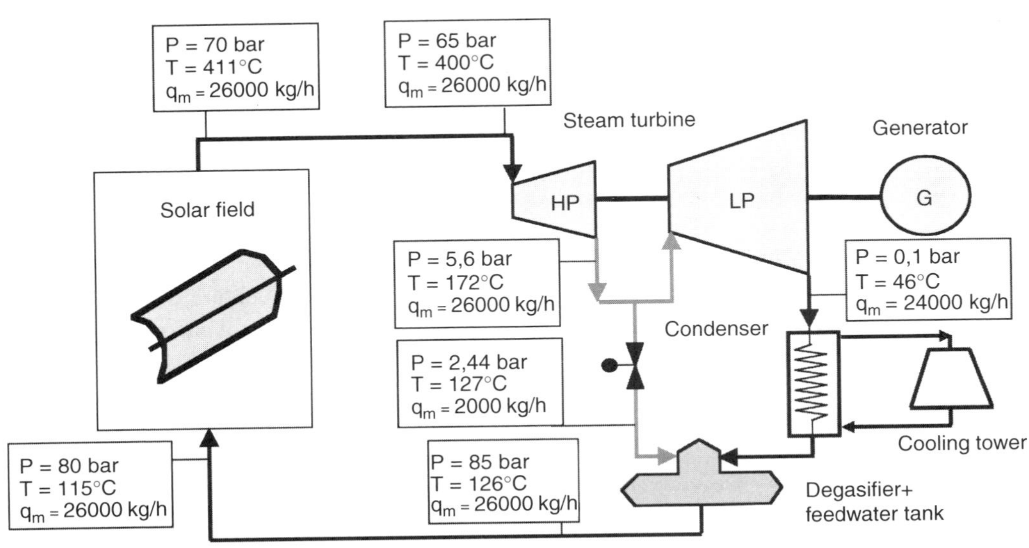

FIGURE 19.35 Schematic diagram of the power block designed for the DSG plant INDITEP.

TABLE 19.4 INDITEP Project 5-MW_e DSG Solar Power Plant Power Block Parameters

Manufacturer	KKK
Gross power (kW_e)	5472
Net power (kW_e)	5175
Net heat rate (kJ/kWh)	14,460
Gross efficiency (%)	26.34
Net efficiency (%)	24.9

collector in every row is increased or decreased by the control system to keep the superheated steam temperature at the outlet of the row as close as possible to the set point defined by the operator.

19.6.9 The SEGS Plants and State-of-the-Art of Solar Power Plants with PTCs

From a commercial standpoint, PTCs are the most successful technology for generating electricity with solar thermal energy. Eight commercial solar thermal power plants with PTCs were in operation in 2005. These plants, called the SEGS (solar electricity generating systems) II to IX, which use thermal oil as the working fluid (HTF technology), were designed and implemented by the LUZ International Limited company from 1985 to 1990. All the SEGS plants are located in the Mojave desert, northwest of Los Angeles (California, U.S.A.). With their daily operation and over 2,200,000 m^2 of PTCs, SEGS plants are this technology's best example of commercial maturity and reliability. Their plant availability is over 98% and their solar-to-electric annual efficiency is in the range of 14–18%, with a peak efficiency of 22% (DeMeo and Galdo 1997) (Price et al. 2002). All SEGS plant configurations are similar to that shown in Figure 19.26, with only slight differences between one plant and another. Though LUZ implemented nine SEGS plants, the first one, SEGS I, was destroyed by a fire in February 1999 (Table 19.5).

The SEGS plants configuration was modified by LUZ from one plant to another in order to achieve higher efficiencies and lower O&M costs, so the only plant with a thermal energy storage system was SEGS I, which had a two-tank oil storage system. LUZ came to the conclusion that the auxiliary gas heater had to be implemented in the oil circuit because plant operating procedures were more difficult with the auxiliary heater installed in the water/steam circuit. LUZ research and development in design and improvement of the solar collector field components and integration of the power block achieved

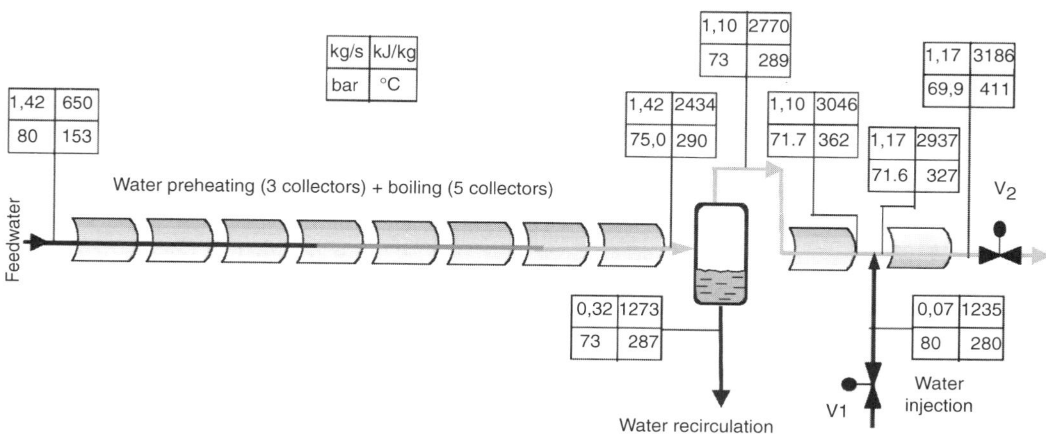

FIGURE 19.36 Scheme of a typical row of collectors in the DSG plant designed in INDITEP.

TABLE 19.5 SEGS Plants Installed by LUZ Internacional

Plant	Net Capacity (MW_e)	Location	Inauguration
SEGS I	14	Dagget, CA	1984
SEGS II	30	Dagget, CA	1985
SEGS III	30	Kramer Jn, CA	1986
SEGS IV	30	Kramer Jn, CA	1986
SEGS V	30	Kramer Jn, CA	1987
SEGS VI	30	Kramer Jn, CA	1988
SEGS VII	30	Kramer Jn, CA	1988
SEGS VIII	80	Harper Lake, CA	1989
SEGS IX	80	Harper Lake, CA	1990

higher plant efficiencies and provided a technology for large solar power plants, leading to a reduction in the cost of the solar field per square meter to about 75% of the first SEGS plant, and solar system thermal efficiency was increased 8% [Table 19.6 (Harats and Kearney 1989)].

Figure 19.37 shows an aerial view of the SEGS III and IV plants. The layout chosen by LUZ for the solar fields is the so-called *central feed*, because it allows easy washing and maintenance access to all the collector rows. The power block (steam generator, steam turbine, electricity generator, condenser, etc.) building is located in the center of the collector field.

The SEGS IX plant, put into operation in 1990, was the last one installed by LUZ before its bankruptcy in 1991. In SEGS plants VIII and IX (see Figure 19.26), there is no gas-fired steam reheater and the steam turbine has two stages working with steam at 371°C/104 bar and 371°C/17 bar, respectively. The thermal oil is heated in the solar collectors up to a temperature of approximately 390°C, and split into two parallel circuits, the steam generator and reheater. In the steam generating circuit, the oil passes through a steam superheater, boiler, and preheater, generating steam at 371°C and 104 bar. This steam expands in the first stage of the turbine, passing to a reheater thermally fed by the other oil circuit, reheating steam to 371°C and 17 bar. During June, July, August, and September, the auxiliary gas boiler is put into operation to keep the turbine working at full load during peak demand hours. For the rest of the time, the steam turbine is mainly driven by the solar system only. The electric generator is cooled by air and water, with a gross nominal power of 108 MVA, 13,8 kV, three-phase at 60 Hz and 3600 rpm.

LUZ Industries developed three generations of PTCs, called LS-1, LS-2, and LS-3 (Table 19.7), The LS-1 and LS-2 designs are conceptually very similar. The main differences are the overall dimensions. The parabolic reflectors are glass panels simply screwed to helicoidal steel tube frames that provide the assembly with the required integrity and structural stiffness.

The LS-3 collector is twice as long as the LS-2, with an aperture area 14% wider, which reduces the number of flexible connections, local control units, temperature sensors, hydraulic drives, and similar equipment by more than half. However, the LS-3 design represents a change in collector philosophy, more than a change in scale. Although the mechanical components of the LS-2 model were designed with wide tolerances and assembled onsite to obtain the required optical behavior, the LS-3 is made of a central steel space frame that is assembled on site with precision jigs. The result of this innovation is a lighter and more resistant structure, with highly accurate operation in heavy winds.

The LS-3 hydraulic drive supplies twice as much torque as the one required by the LS-2 collectors: 7400 kg-meters (peak) for the inner collector rows, and higher in the outer rows.

The electricity produced by the SEGS plants is sold to the local utility, Southern California Edison, under individual 30-year contracts for every plant. To optimize the profitability of these plants, it is essential to produce the maximum possible energy during peak demand hours, when the electricity price is the highest. The gas boilers can be operated for this, either to supplement the solar field or alone. Nevertheless, the total yearly electricity production using natural gas is limited by the Federal Commission for Energy Regulation in the United States to 25% of the overall yearly production.

TABLE 19.6 Basic Characteristics of the SEGS I–IX Plants

	SEGS 1	SEGS 2	SEGS 3	SEGS 4	SEGS 5	SEGS 6	SEGS 7	SEGS 8	SEGS 9
Starting-up:	Dec'84	Dec'85	Dec'86	Dec'86	Oct'87	Dec'88	Dec'88	Dec'89	Sep'90
Investment (M$)	62	96	101	104	122	116	117	231	
Electricity yearly production (MWh)	30,100	80,500	92,780	92,780	91,820	90,850	92,646	2,52,750	2,56,125
Estimated life (years)	20	25	30	30	30	30	30	30	30
Number of stages of the turbine	1	2	2	2	2	2	2	2	2
Solar steam (P,T) (bar,C):	248/38	300/27	327/43.4	327/43.4	327/43.4	371/100	371/100	371/104	371/104
Steam with gas (P,T) (bar,C)	417/37	510/104	510/104	510/104	510/100	510/100	510/100	371/104	371/104
Efficiency in solar mode (%)	31.50	29.40	30.60	30.60	37.70	37.50	37.50	37.60	37.60
Efficiency with gas (%)	—	37.30	37.40	37.40	37.40	39.50	39.50	37.60	37.60
Solar field									
Collector type	LS1/LS2	LS1/LS2	LS2	LS2	LS2	LS2	LS3	LS3	LS3
Aperture area (m^2)	82,960	1,88,990	2,30,300	2,30,300	2,33,120	1,88,000	1,94,280	4,64,340	4,83,960
Working temperature	279	321	349	349	349	390	390	390	390
Inlet/outlet collector field temperature©	241/307	248/320	248/349	248/349	248/349	293/393	293/390	293/390	293/390
Type of oil	ESSO 500	M-VP1	M-VP1	M-VP1	M.VP1	M-VP1	M-VP1	M-VP1	M-VP1
Oil volume (m^3)	3217	379	403	404	461	372	350		

FIGURE 19.37 Aerial view of the SEGS III and IV plants.

Peak demand hours are when there is the most electricity consumption and, therefore, the tariff is the highest. Off-peak and super off-peak hours are when electricity consumption is low, and the electricity price is therefore also lower. At present, 16% of the SEGS plants' annual net production is generated during summer peak demand hours, and the revenues from this are on the order of 55% of the annual total. These figures show how important electricity generated during peak demand hours is for the profitability of these plants.

Thanks to the continuous improvements in the SEGS plants, the total SEGS I cost of \$0.22/kWh$_e$ for electricity produced was reduced to \$0.16/kWh$_e$ in the SEGS II and down to \$0.09/kWh$_e$ in SEGS IX (Kearney and Cohen 1997).

The other important contribution to the profitability of the SEGS plants was the favorable tax laws it took advantage of. The large tax rebate was a crucial factor in the economic feasibility of the SEGS plants. The eventual reduction in these tax exemptions was overcome by the considerable cost reduction from one plant to another achieved by LUZ. The significant reduction in both fossil fuel prices and tax

TABLE 19.7 Characteristics of the LS-1, LS-2, and LS-3 Collectors

	LS-1	LS-2	LS-3
Solar tracking accuracy (°)	0.10	0.10	0.10
Maximum wind velocity to operate (km/h)	56	56	56
Steel structure based on	Steel tube	Steel tube	Space frame
Selective coating of absorber tubes	Black Cr	Black Cr	Cermet
Absorptivity/transmissivity (%)	94/94	94/95	96/95
Emissivity (%)	39 (300°C)	24 (300°C)	18 (350°C)
Focal distance of the solar concentrator	0.68	1.49	1.71
Aperture angle	85	80	80
Reflectivity (%)	94	94	94
Trough aperture (m)	2.5	5	5.76
Absorber steel pipe outer diameter (mm)	42	70	70
Geometric concentration	19	23	26
Overall length (m)	50.2	47.1	99.0
Distance between supports (m)	6.3	8.0	12.0
Mirrors surface per collector (m^2)	128	235	545
Maximum working temperature (°C)	307	350	390
Distance between parallel rows (m)	7	12.5/15	17
Intercept factor (%)	87	89	93

exemptions made it impossible for the SEGS plants to maintain the profit margin they had had at the beginning, resulting in the bankruptcy of the company and the cancellation of the planned erection of more SEGS plants. Another factor that led to the bankruptcy of LUZ was the short time in which one SEGS plant had to be fully implemented to obtain tax benefits. (Lotker 1991).

Though LUZ achieved a significant cost reduction in the SEGS plants, the current cost of electricity produced by parabolic trough solar power plants must still be further reduced before it can become competitive with conventional power plants (Pitz-Paal et al. 2005a, 2005b). Without tax incentives or an adequate premium for electricity generated by this type of plants, they are not profitable, because current investment and O&M costs demand a public support strategy for commercial development. At present, the direct capital cost is about 3 times that of a fossil-fueled power plant and the generating cost is also about 3 times higher. Annual thermal efficiency (solar-to-useful thermal energy delivered) of commercial parabolic trough solar fields is about 50%. This means a yearly savings of about 0.45 tn. of CO_2 emissions and 0.1 tn. of fossil fuel per square meter of PTC (Geyer 2002).

In spite of their environmental benefits, there are some barriers to the commercial use of this technology. The main barriers at present are the high investment cost (2500–4000 \$ kW^{-1}, depending on plant size and thermal storage capacity) and the minimum size of the power block required for high thermodynamic efficiency. However, these barriers are shared by all the solar thermal power technologies currently available (i.e., CRS, dish/Stirling systems, and linear Fresnel concentrator systems) (Romero, Martinez, and Zarza 2004).

The technology commercially available at present for parabolic trough power plants is the HTF technology that uses oil as the heat carrier between the solar field and the power block. At the time of writing, there were three solar collector designs available for large solar power plants, the LS-3 design (owned by the Israeli company SOLEL Solar Systems), the EuroTrough design (owned by the EuroTrough Consortium), and the American Solargenix design, which has an aluminum structure. However, other collector designs under development are expected to be commercially available in the short-to-medium term.

At present, there are several solar power plants initiatives with PTCs using the HTF technology. Some of these initiatives integrate a parabolic trough solar field in the bottoming cycle of a combined cycle gas-fired power plant. This configuration is called the integrated solar combined cycle system (ISCCS). Though the contribution of the solar system to the overall plant power output is small (approximately 10–15%) in the ISCCS configuration, it seems to be a good approach to market penetration in some countries, which is why the World Bank, through its Global Environment Facility (GEF), is promoting ISCCS plants in India, Egypt, Morocco, and Mexico (Geyer et al. 2003). Figure 19.38 shows the schematic diagram of a typical ISCCS plant.

There are also plans to install parabolic trough power plants similar to SEGS in the U.S.A. (Romero, Martinez, and Zarza 2004) and Spain. The new Spanish regulation approved in 2004 (Royal Decree No. 436/204, March 12, 2004) defines a premium of about 0.18 €/kWh for electricity produced by 50 MW_e capacity or smaller solar thermal power plants. This new premium has leveraged the construction of several 50 MW_e parabolic trough power plant projects in Spain. The most mature project in Spain is the AndaSol plant (Aringhoff et al. 2002a, 2002b), a SEGS-like plant with a 6-h molten salt thermal storage system.

Studies performed in recent years have pointed out the possibility of achieving a significant midterm cost reduction in parabolic trough technology (Enermodal 1999; Sargent & Lundy 2003). Mass production and component improvement through R&D could lead to an electricity cost of \$0.10/kWh for 100 MW_e plants in the midterm, and \$0.05/kWh in advanced 200 MW_e-capacity plants or five adjacent plants located at the same site (Price et al. 2002).

It is therefore clear that parabolic trough technology must seek ways to become more competitive with conventional power plants. Within the potential improvements for cost reduction, DSG of high-pressure/high-temperature superheated steam in the receiver pipes seems to be the most promising way to achieve this goal, because the thermal oil currently used in the HTF technology as the working fluid

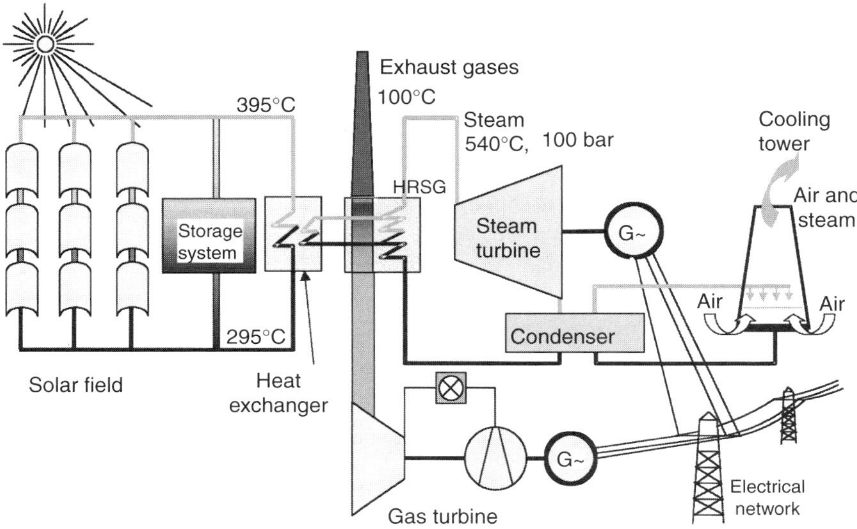

FIGURE 19.38 Schematic diagram of a typical ISCCS (integrated solar combined cycle system) plant.

and the associated equipment (oil/water heat exchanger required to produce the superheated steam, oil circuit, expansion vessel) would no longer be needed.

Several R&D projects have been developed in Europe to investigate the DSG technology under real solar conditions. Technical feasibility of the DSG process was proven in the DISS project (Zarza et al. 1999, 2002), whereas DSG components have been optimized in the INDITEP project (Rueda 2003), which also undertook the detailed design of a first precommercial 5 MW_e DSG plant. When this small DSG power plant is put into operation and its O&M procedures and components optimized, this new technology is likely to become the best bet for commercially available lower-cost parabolic trough power plants.

19.7 Central Receiver Solar Thermal Power Plants

CRS with large heliostat fields and solar receivers located on top of a tower are now in a position for deployment of the first generation of grid-connected commercial plants. The CRS power plant technology can be considered as sufficiently mature after the pioneering experience of several 0.5–10 MW pilot plants in the early 1980s, and the subsequent improvement of such key components as heliostats and solar receivers in many later projects merging international collaboration during the past 20 years. Solar-only plants like Solar Tres and PS10 and hybrid configurations like Solgas, ConSolar, or SOLGATE have provided a portfolio of alternatives which have led to the first scaled up plants for the period 2005–2010. Those small 10–15 MW projects, still nonoptimized, already reveal a dramatic cost reduction over previous estimates and provide a path for a realistic LEC milestone of $0.08/kWh by 2015.

In power towers, incident sunrays are tracked by large mirrored collectors (heliostats), which concentrate the energy flux onto radiative/convective heat exchangers called solar receivers, where energy is transferred to a thermal fluid. After energy collection by the solar subsystem, the thermal energy conversion to electricity is quite similar to fossil-fueled thermal power plants and the above-described parabolic trough system power block.

Reflective solar concentrators are employed to reach the temperatures required for thermodynamic cycles (Mancini, Kolb, and Prairie 1997). In power towers or CRS the solar receiver is mounted on top of a tower and sunlight is concentrated by means of a large paraboloid that is discretized into a field of heliostats (Figure 19.39). CRS have a high potential for midterm cost reduction of electricity produced since there are many intermediate steps between their integration in a conventional Rankine cycle up to the higher exergy cycles using gas turbines at temperatures above 1300°C, leading to higher efficiencies and throughputs.

The typical optical concentration factor ranges from 200 to 1000. Plant sizes of 10–200 MW are chosen because of economy of scale, even though advanced integration schemes are claiming the economics of smaller units as well (Romero et al. 2000a, 2000b). The high solar flux incident on the receiver (averaging between 300 and 1000 kW m^{-2}) enable operation at relatively high temperatures of up to 1000°C and integration of thermal energy into more efficient cycles in a step-by-step approach. CRS can be easily integrated in fossil plants for hybrid operation in a wide variety of options and has the potential for generating electricity with high annual capacity factors through the use of thermal storage. With storage, CRS plants are able to operate over 4500 h per year at nominal power (Kolb 1998). Their main characteristics are summarized in Table 19.8.

19.7.1 Technology Description

A solar power tower, or central receiver system, plant may be described in terms of the following subsystems:

- Collector system, or heliostat field, created with a large number of two-axis tracking units distributed in rows
- Solar receiver, where the concentrated flux is absorbed. It is the key element of the plant and serves as the interface between the solar portion of the plant and the more conventional power block
- Heat exchanger system, where a heat transfer fluid may be used to carry the thermal energy from the receiver to the turbine
- Heat storage system, with which system dispatchability is ensured during events like cloud passages, and can adapt to demand curves
- Fossil fuel backup for hybrid systems with a more stable output
- Power block, including steam generator and turbine-alternator
- Master control, UPS, and heat rejection systems

A detailed description and historical perspective of all the subsystems would be excessive, given the large number of configurations and components tested to date. Because of their higher temperatures,

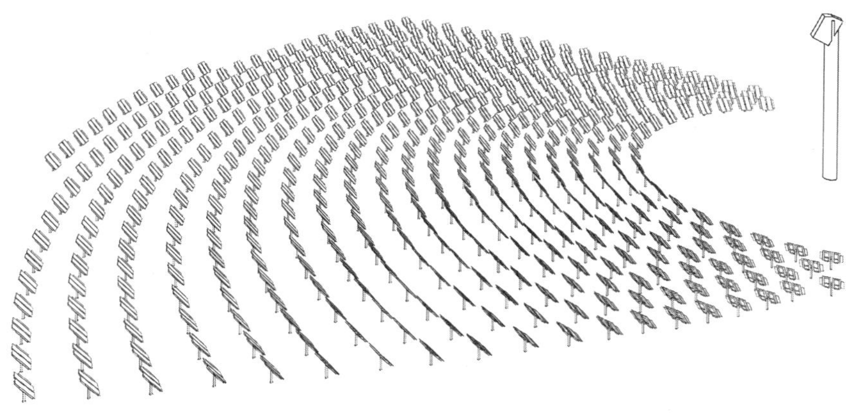

FIGURE 19.39 Artist's view of a heliostat field focusing sunlight onto a receiver/tower system.

TABLE 19.8 Characteristics of Central Receiver Solar Thermal Power Systems

Typical size	10–200 MW[a]
Operating temperature	
Rankine	565°C
Brayton	800°C
Annual capacity factor	20–77%[a]
Peak efficiency	16–23%[a]
Annual net efficiency	12–20%[a]
Commercial status	Scale-up demonstration (10–30 MW)
Technology development risk	Medium
Storage available	Nitrate salt for molten salt receivers Ceramic bed for air receivers
Hybrid designs	Yes
Investment cost	
$\$\cdot W^{-1}$	4.4–2.5[a]
$\$\cdot W_p^{-1b}$	2.4–0.9[a]

[a] Figure indicate expected progress from present to 2030.
[b] $\$/W_p$ removes the effect of energy storage or solar multiple, as in PV.
Source: Adapted From DeMeo, E. A. and Galdo, J. F. 1997. TR-109496 Topical Report. U.S. Department of Energy, Washington, DC.

CRS have been able to make use of a diversity of thermal fluids, such as air, water/steam, molten nitrate salt, and liquid sodium. In general, the components that impact the most on investment cost are the heliostat field, the tower–receiver system, and the power block. The heliostat field and solar receiver systems distinguish solar thermal tower power plants from other CSP plants, and are therefore given more attention below. In particular, the heliostat field is the single factor with the most impact on plant investment, as seen in Figure 19.40. Collector field and power block together represent about 72% of the typical solar-only plant (without fossil backup) investment, of which heliostats represent 60% of the solar share. Even though the solar receiver impacts the capital investment much less (about 14%), it can be considered the most critical subsystem in terms of performance, since it centralizes the entire energy flux exchange. The largest heliostat investment is the drive mechanism and reflecting surface, which alone are almost 70% of the total.

19.7.1.1 Heliostat and Collector Field Technology

The collector field consists of a large number of tracking mirrors, called heliostats, and a tracking control system to continuously focus direct solar radiation onto the receiver aperture area. During cloud passages and transients the control system must defocus the field and react to prevent damage to the receiver and tower structure.

Heliostats fields are characterized by their off-axis optics. Since the solar receiver is located in a fixed position, the entire collector field must track the sun in such a way that each and every heliostat individually places its surface normal to the bisection of the angle subtended by sun and the solar receiver. Figure 19.41 shows the variability of elevation angles in a heliostat field and identifies the elevation angle. The geometrical definition of the inclination angle n of a single heliostat is a function of the tower height, its distance from the tower and the incidence angle of the sun. Assuming Z_S is the heliostat vertical dimension and Z_T is the geometrical tower height above ground, the so-called *optical tower height* may be defined as the elevation of the center of the receiver aperture area above the pivot point of the heliostat ($Z_T - Z_S/2$).

$$\psi = h + n - 90°(\text{degrees}), \tag{19.38}$$

$$90° - n = \arctan\left[\frac{z_T - z_S/2}{x}\right] + \psi(\text{degrees}), \tag{19.39}$$

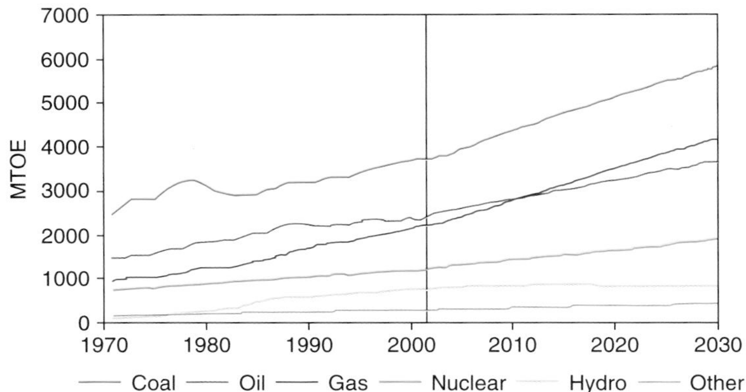

COLOR FIGURE 1.10 World primary energy demand by fuel types. (According to IEA, *World Energy Outlook*, International Energy Agency, Paris, France, 2004.)

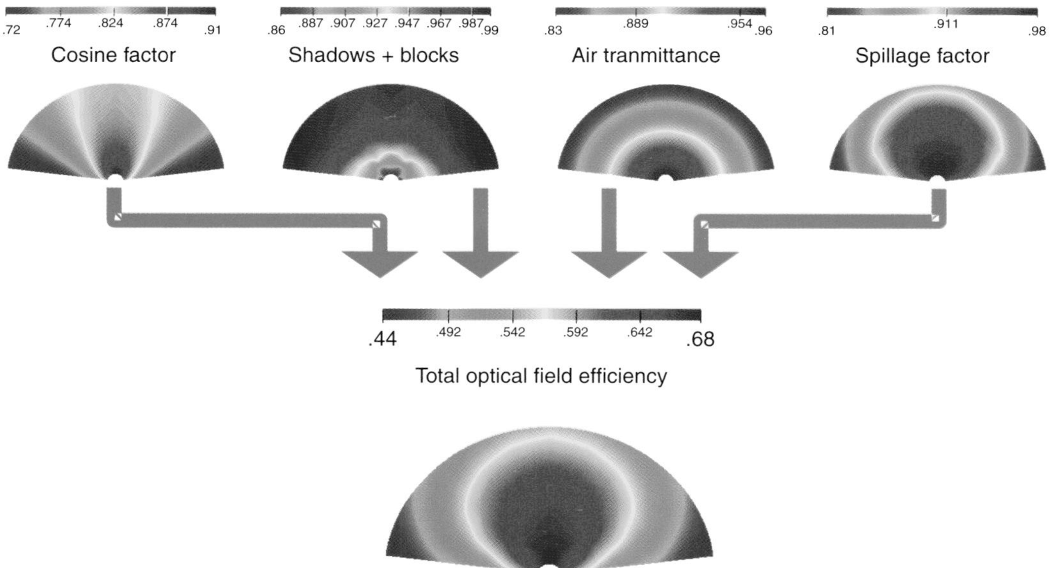

COLOR FIGURE 19.44 Mapping of total optical efficiency of a north field area of heliostats and its breakdown into cosine factor, shadowing and blocking, air transmittance, and receiver spillage.

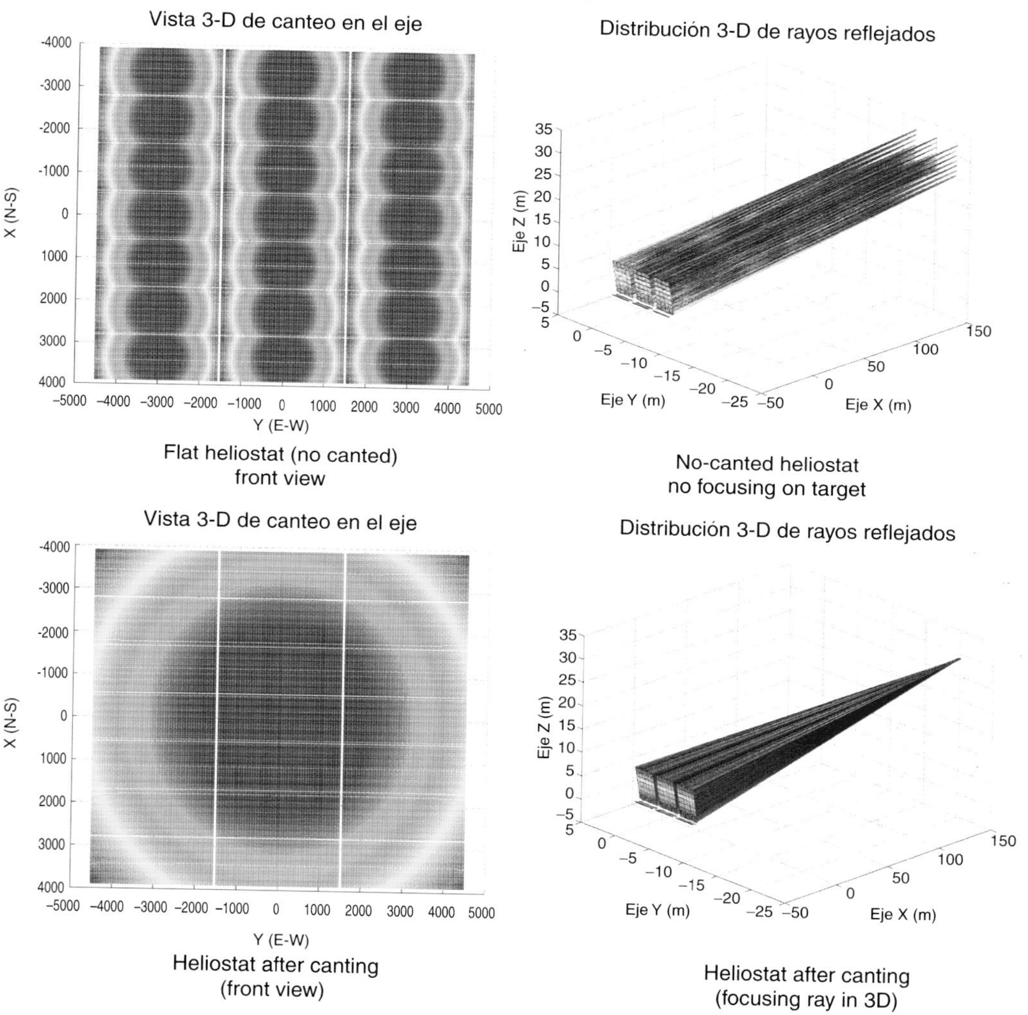

COLOR FIGURE 19.47 Effect of canting facets in a glass-metal heliostat. In the upper part it can be observed that a flat heliostat with curved facets is not focusing in a single image. In the lower part a canted heliostat is depicted.

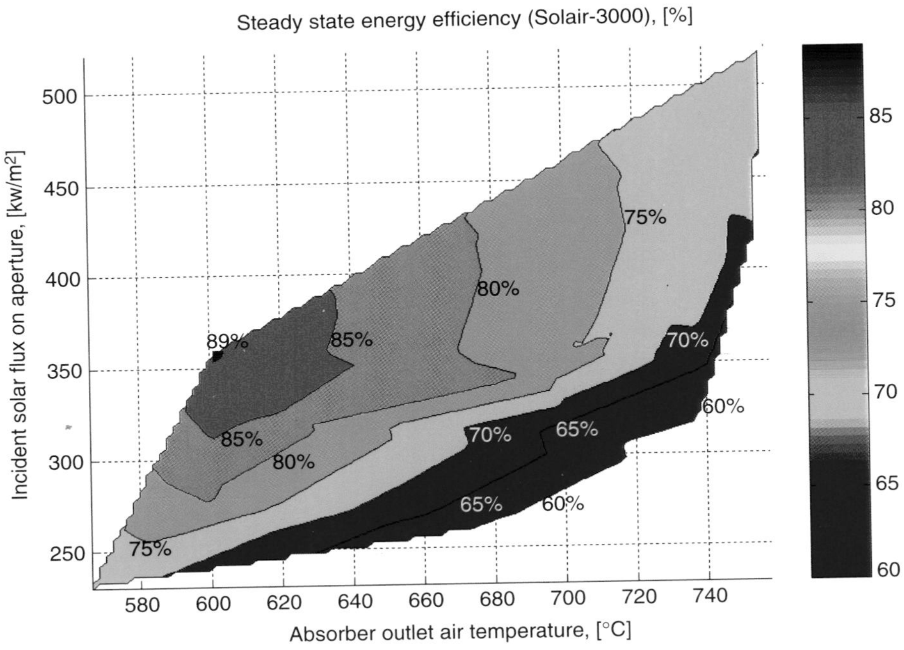

COLOR FIGURE 19.57 Map of Solair-3 MW volumetric receiver showing thermal efficiency versus incident solar flux on aperture and air outlet temperature.

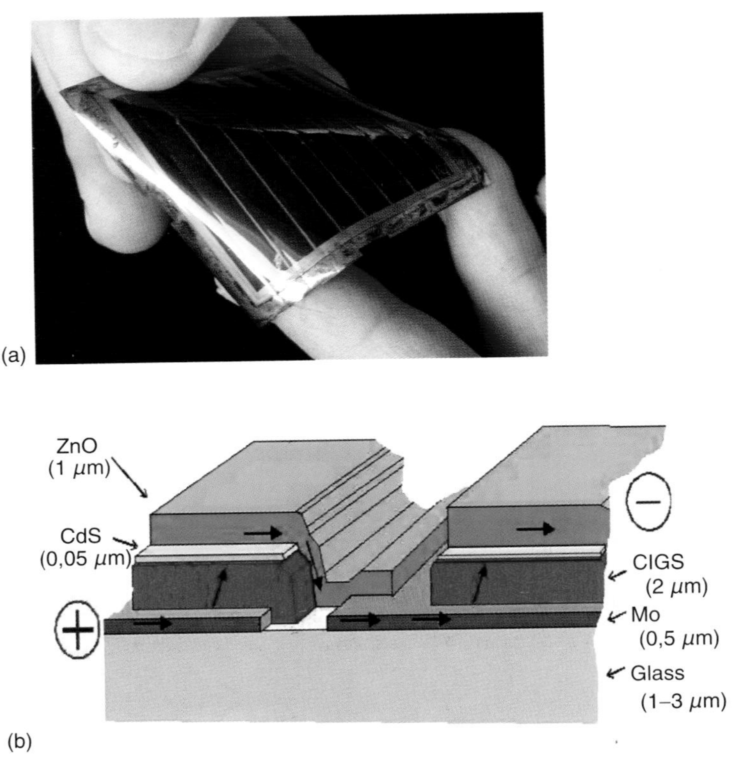

COLOR FIGURE 20.37 Flexible monolithic CIGS modules showing a prototype mini-module on a polymer foil (a) and schematic cross-section of the module (image taken from the website of ZSW) exhibiting the material component layers and their interconnect patterns (b).

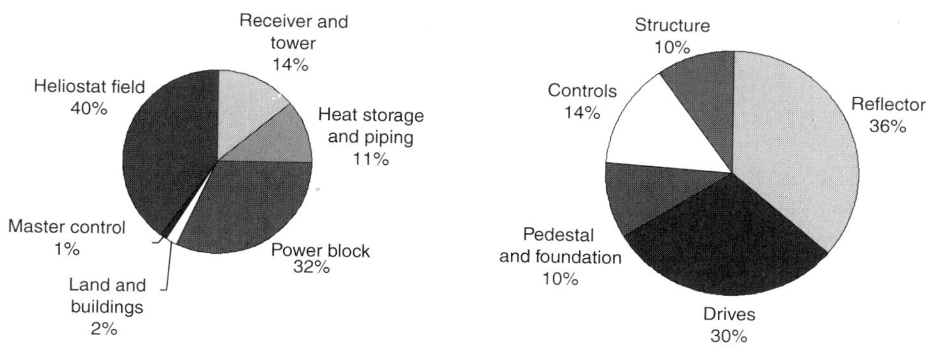

FIGURE 19.40 (Left) Investment costs breakdown for a CRS plant, only solar. As it can be observed the heliostat field and the power block are the most impacting subsystems on plant investment. (Right) Breakdown of production cost for a single heliostat distributed among its main components. Reflector and tracking mechanism are in this case the most capital intensive components.

$$n = \frac{180^\circ - h - \arctan\left[\frac{z_T - z_S/2}{x}\right]}{2} \text{ (degrees)}. \tag{19.40}$$

Heliostat field performance is defined in terms of the optical efficiency, which is equal to the ratio of the net power intercepted by the receiver to the product of the direct insolation and the total mirror area. The optical efficiency includes the cosine effect, shadowing, blocking, mirror reflectivity, atmospheric attenuation, and receiver spillage (Falcone 1986).

Because of the large area of land required, complex optimization algorithms are used to optimize the annual energy produced by unit of land, and heliostats must be packed as close as possible so the receiver can be small and concentration high. However, the heliostats are individual tracking reflective Fresnel segments subject to complex performance factors, which must be optimized over the hours of daylight in the year, by minimizing the cosine effect, shadowing and blocking, and receiver spillage. Because the reflective surface of the heliostat is not normal to the incident rays, its effective area is reduced by the cosine of the angle of incidence ψ; the annual average $\cos\psi$ varies from about 0.9 at two tower heights north of the tower to about 0.7 at two tower heights south of the tower. Of course, annual average cosine is highly dependent on site latitude. Consequently, in places close to the equator a surround field would be the best option to make best use of the land and reduce the tower height. North fields improve performance as latitude increases (south fields in the southern hemisphere), in which case, all the

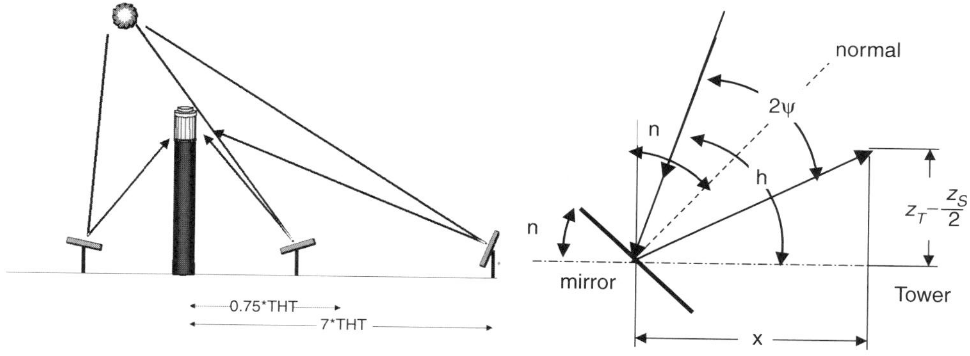

FIGURE 19.41 (Left) Visualization of the off-axis optics of heliostats representing different inclination angles of mirrors located in a heliostat field. (Right) Geometrical definition of elevation angle n.

heliostats are arranged on the north side of the tower. Representative surround and north collector field configurations are depicted in Figure 19.42.

Another point to be considered in the layout is the heliostat dimensions. If the heliostats are spaced too close together, their corners could collide, so mechanical limits preclude pedestal spacing closer than the maximum dimension of the heliostat (i.e., the diagonal or diameter of the heliostat, D_m). This causes a significant disadvantage when the aspect ratio (height/width) differs greatly from one.

Blocking of reflected rays is also an important limitation on spacing heliostats. Blocking is produced by neighboring heliostats. To avoid blocking losses, the distance Δx between the heliostat rows must be calculated according to the following equation:

$$\Delta x = x \frac{z_S}{z_T}. \tag{19.41}$$

Shading produced by neighboring heliostats also has to be taken into account. This occurs mostly at low sun angles and in the middle of the field where blocking conditions would allow close spacing. The shadows move during the day and year, as does the heliostat orientation, so there is no simple rule. In addition, the tower or other objects may also cast a shadow over part of the heliostat field. Usually shadowing in the field is calculated by projecting the outlines of the heliostats aligned, the tower and anything else that casts a shadow onto a plane perpendicular to the center sunray. Shadowed portions of

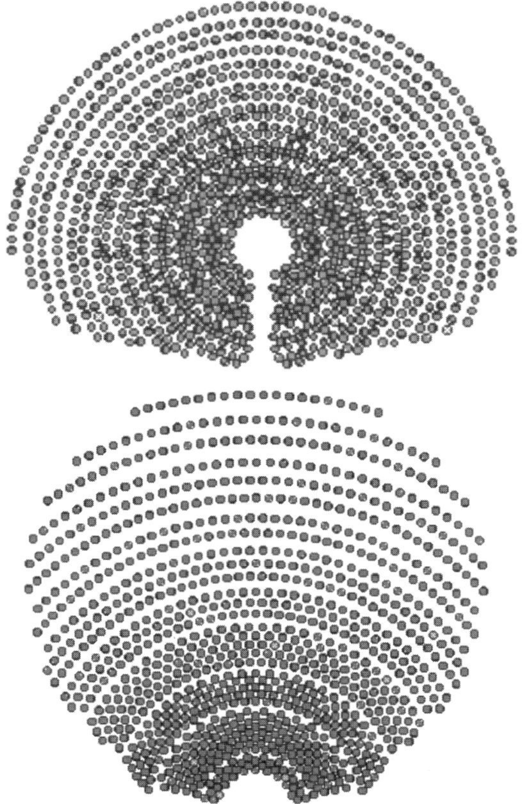

FIGURE 19.42 Representation of optimized fields for latitude of 36° with surround field (left) and north field (right) configurations.

any heliostat appear in the overlapping areas in this projection. Classical computer codes like HELIOS provide this calculation (Biggs and Vittitoe 1979).

Not all the sunlight that clears the heliostats reaches the vicinity of the receiver. Some of the energy is scattered and absorbed by the atmosphere. This effect is referred to as the attenuation loss (Falcone 1986). This factor increases when water vapor or aerosol content in the atmosphere is high, and is typically anywhere between 5–15% in a solar field. Atmospheric attenuation is usually expressed as a function or experimental correlation depending on the range of heliostat inclination.

The size of the image formed by each heliostat depends on mirror focusing and canting and on the size of the heliostat and errors, as expressed by its beam quality. Because of that, there is an intercept factor for a given receiver aperture area. Some of the energy spills over around the receiver. Although spillage can be eliminated by increasing the size of the receiver, at some point, increased size becomes counterproductive because of the resulting increased receiver losses and receiver costs.

The combination of all the above-mentioned factors influencing the performance of the heliostat field should be optimized to determine an efficient layout. There are many optimization approaches to establish the radial and azimuthal spacing of heliostats and rows. One of the most classic, effective, and widespread procedures is the *radial staggered* pattern, as shown in Figure 19.43, originally proposed by the University of Houston in the 1970s (Lipps and Vant-Hull 1978). Typical radially staggered field spacing at 35° latitude using square low-cost heliostats can be expressed by:

$$\frac{\Delta R}{\Delta D_\mathrm{m}} = \frac{1.009}{\Theta} - 0.063 + 0.4803\Theta, \tag{19.42}$$

$$\frac{\Delta Z}{\Delta D_\mathrm{m}} = 2.170 - 0.6589\Theta + 1.247\Theta^2, \tag{19.43}$$

where Θ represents the receiver elevation angle in radians:

$$\Theta = \arctan\left[\frac{z_\mathrm{T} - z_\mathrm{S}/2}{x}\right].$$

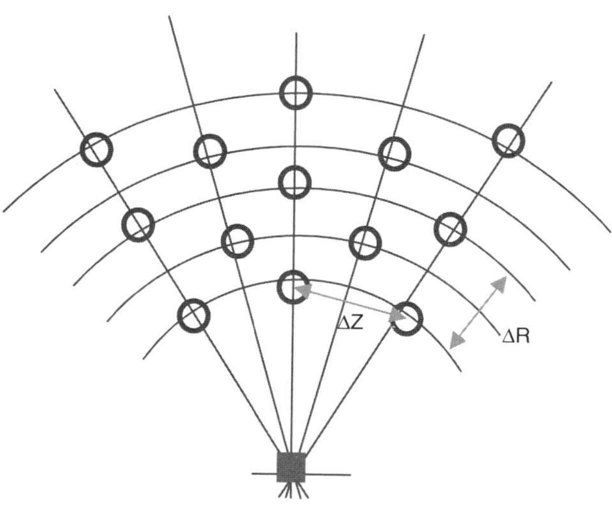

FIGURE 19.43 Radial staggered field layout, where ΔZ is representing the azimuthal spacing and ΔR represents the radial spacing.

Integral optimization of the heliostat field is decided by a trade-off between cost and performance parameters. Heliostats, land, and cabling network must be correlated with costs. Cost and performance also often have reverse trends, so that when heliostats are packed closer together, blocking and shadowing penalties increase, but related costs for land and wiring decrease. A classical code in use since the 1980s for optimization of central receiver subsystems is DELSOL3 (Kistler 1986). In DELSOL3, heliostat field layout is optimized for tower height and solar receiver geometry. Figure 19.44 shows a breakdown of efficiency maps for the different performance factors in a typical one-sided north heliostat field. It may be deduced that the heliostat density is greatest at the inner boundary and decreases with increasing radial distance from the tower. The average land coverage ratio is typically 0.20–0.25.

The radially staggered distribution clearly creates "prearranged" grids based on the tower height vs. row-to-radius ratio. This geometrical procedure provides a smart solution to the problem with good optimization of computing resources. However, with today's computers, it is possible to calculate the yearly energy available at any point in a site for a given tower height, "the yearly normalized energy surface" YNES. Yearly efficiency maps can be generated based on the cosine factor, the spillage factor, and the site atmospheric attenuation coefficient using real direct normal irradiance data (DNI) within a reasonable computing time. It is therefore easy to find the place where the yearly energy available is the highest for location of the first heliostat. It is also possible to calculate the effect of shadowing and blocking by this heliostat on the YNES, so YNES can be recalculated and the best position for the next heliostat can be found. Although this iterative method is time consuming, it is worthwhile if either the efficiency of the solar plant can be increased or the capital cost reduced. This YNES-based layout method enables better flexibility than predetermined gridding strategies (Sánchez and Romero 2006).

Mature low-cost heliostats consist of a reflecting surface, a support structure, a two-axis tracking mechanism, pedestal, foundation, and control system (Figure 19.45). The development of heliostats shows a clear trend from the early first-generation prototypes, with a heavy, rigid structure, second-surface mirrors and reflecting surfaces of around 40 m^2 (Mavis 1989), to the more recent designs with large 100–120 m^2 reflecting surfaces, lighter structures, and lower-cost materials (Romero, Conejero, and Sánchez 1991). Since the first-generation units, heliostats have demonstrated beam qualities below

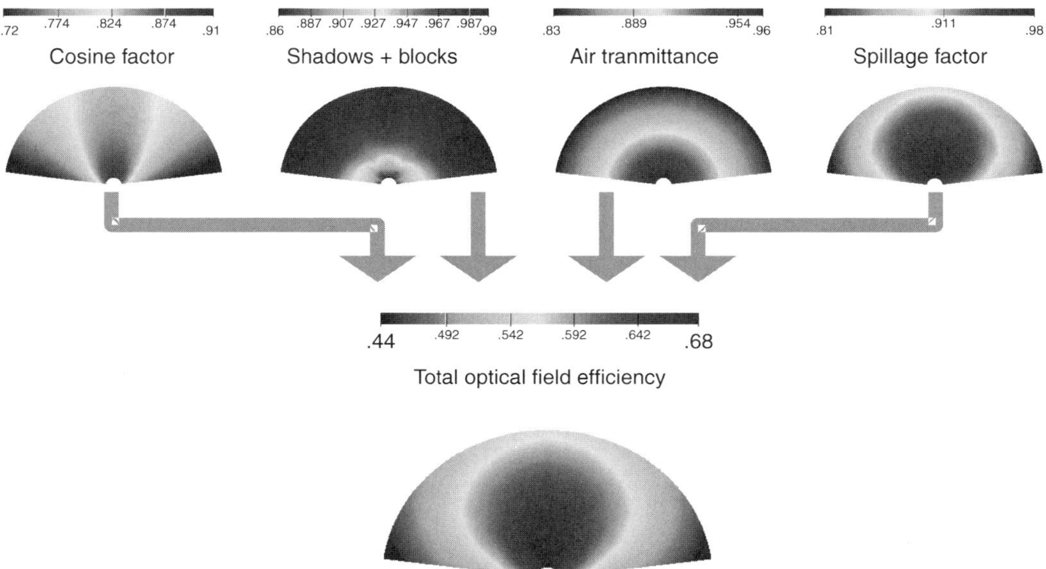

FIGURE 19.44 **(See color insert following page 19-52.)** Mapping of total optical efficiency of a north field area of heliostats and its breakdown into cosine factor, shadowing and blocking, air transmittance, and receiver spillage.

2.5 mrad that are good enough for practical applications in solar towers, so the main focus of development is directed at cost reduction. Two basic approaches are being pursued to reduce per-m^2 installed cost.

The first approach is devoted to increasing the reflective area by employing curved surfaces made up of several mirror facets. Each facet surface typically goes from 3 to 6 m^2. This increment in optical surface results in a cost reduction, since some components, like the drive mechanism, pedestal and control, do not increase linearly. However there is a limit to this advantage, since the larger the area, the higher the optical errors and washing problems are also.

The second line of development is the use of new light reflective materials like polymer reflectors and composites in the supporting structure, such as in the stretched-membrane heliostats. The stretched-membrane drum consists of a metal ring to which prestressed 0.4-mm stainless steel membranes are welded. One of the membranes is glued to a polymer reflector or thin mirrors. A vacuum is created inside the plenum with a controlled blower to ensure curvature.

In Spain, some developments worthy of mention are the 40 m^2 COLON SOLAR prototype (Osuna et al. 1999), the 105 m^2 GM-100 (Monterreal et al. 1997) and more recently, the 90 and 120 m^2 Sanlúcar heliostats (Osuna et al. 2004). In the U.S.A., the latest development in glass/metal technology is the 150 m^2 ATS heliostat (Alpert and Houser 1990). Estimated production costs for sustainable market scenarios are around \$130/m^2–\$200/m^2. The stretched-membrane milestone is the 150 m^2 Steinmuller heliostat ASM-150 with an excellent beam quality of 2 mrad (Weinrebe, Schmitz-Goeb, and Schiel 1997). In spite of the good quality achieved by stretched membranes, projected costs are higher than the more mature glass/metal units.

The drive mechanism is in charge of independent azimuthal and elevation movement, in such a way that the specular surface follows the sun position and reflects the beam onto the focal point. The ratio between the angle of incidence and the reflected ray leads to angular errors doubling at the target. It is therefore crucial for a tracking system to be highly accurate. Heliostat drives should have the following characteristics:

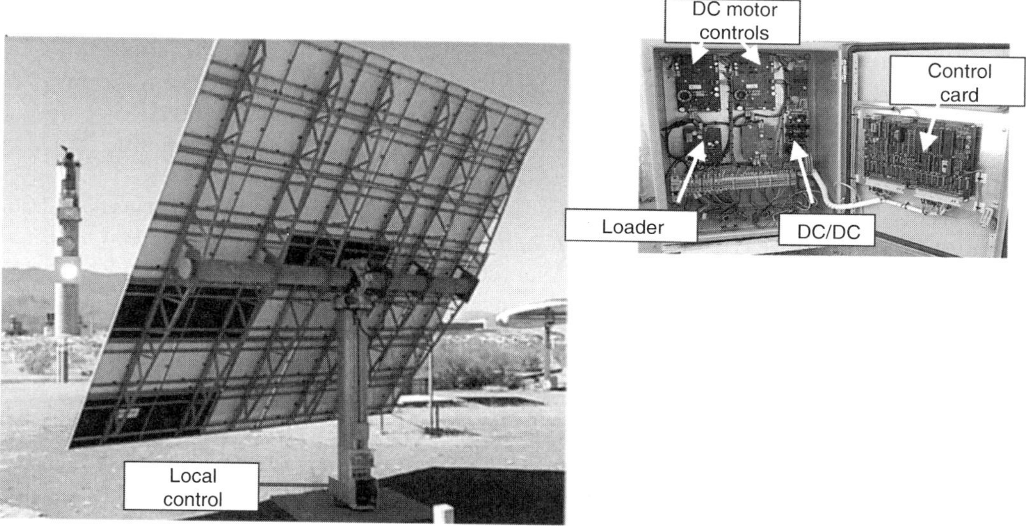

FIGURE 19.45 Rear view of the heliostat COLON SOLAR of 70 m^2 with typical T-shape structure and torque tube. At the bottom of the pedestal it is located the control box. Reflected image is shown at the Lambertian target located on the tower.

- Sufficiently robust to support their own weight, the movable structure and wind loads, and be rigid enough to avoid low frequency vibrations
- Able to generate extremely slow movement, with high reduction ratios (up to 40,000:1)
- Highly accurate positioning (use of encoders) and no free movement
- Able to ensure relatively fast return to stow position in case of high winds or other dangerous weather conditions, and other events
- Resistance to outdoor exposure
- Easy maintenance
- Low-cost manufacture and operation

The most common drive mechanism configuration makes use of worm-gear systems for both elevation and azimuth axes (Figure 19.46). Both gears are essentially analogous in terms of tooth shape and reduction ratio. In many cases there is a first planetary reduction step and then a second worm-gear reduction step at the outlet. The advantage of the planetary system is the high ratio of reduction in a limited space. The worm gear provides high reduction ratios at high momentums. However worm gears are less efficient because of the stress of high friction. This stress has a positive reaction, since the self-locking worm comes to a halt whenever the angle of friction between the worm thread and the gear teeth is higher than the nominal design angle.

Mature glass/metal faceted heliostats report availabilities over 95%, and beam qualities of 2.4–2.8 mrad. Yearly average reflectivity of a heliostat field reaches 85–92%. Faceted heliostats with curved mirrors require canting of the facets to form a large paraboloid. In Figure 19.47, the effect of canting facets can be seen.

Control of CRS is more complicated than other types of solar thermal power plants since optics are off-axis and each and every heliostat individually tracks the sun. The control system in a central receiver system is naturally separated into the heliostat field control system (HFCS) and the receiver and power system control system (RPSCS) (Yebra et al. 2004). The main purpose of the HFCS is to keep each heliostat positioned at the desired coordinates at all times, depending on power system demand. The general purpose of the HFCS is to generate a uniform time-spatial distribution of the temperature on the volumetric receiver by controlling the timed insertion of an associated group of heliostats at predefined aiming points on the receiver by modifying the aiming point coordinates and changing from one heliostat group to another during operation. This is accomplished by an HFCS aiming point strategy (García-Martín et al. 1999). The current trend in control systems is a distributed control, with a hard real-time operating system (RTOS), integrating heterogeneous hardware and software platforms in real time, that guarantee a deterministic response to external (physical environment) and internal (operator interface) events. The RPSCS regulates the pressure and temperature of the heat transfer fluid, and steam generator.

The heliostat local control is responsible for all the emergency and security maneuvers and sun-tracking calculations, as well as communication with the control room. The current trend is to increase the heliostat intelligence and autonomy. In addition, some drive mechanism options consider the use of wireless communications and PV power supply, eliminating the need for cabling and trenching. This is the case of the stand-alone heliostat developed at the PSA in Spain, where a field of 92 such heliostats is in operation (Garcia, Egea, and Romero 2004).

19.7.1.2 Solar Receiver

In a solar power tower plant, the receiver is the heat exchanger where the concentrated sunlight is intercepted and transformed into thermal energy useful in thermodynamic cycles. Radiant flux and temperature are substantially higher than in parabolic troughs, and therefore, high technology is involved in the design, and high-performance materials should be chosen. The solar receiver should mimic a blackbody by minimizing radiation losses. To do so cavities, black-painted tube panels or porous absorbers able to trap incident photons are used. In most designs, the solar receiver is a single unit that

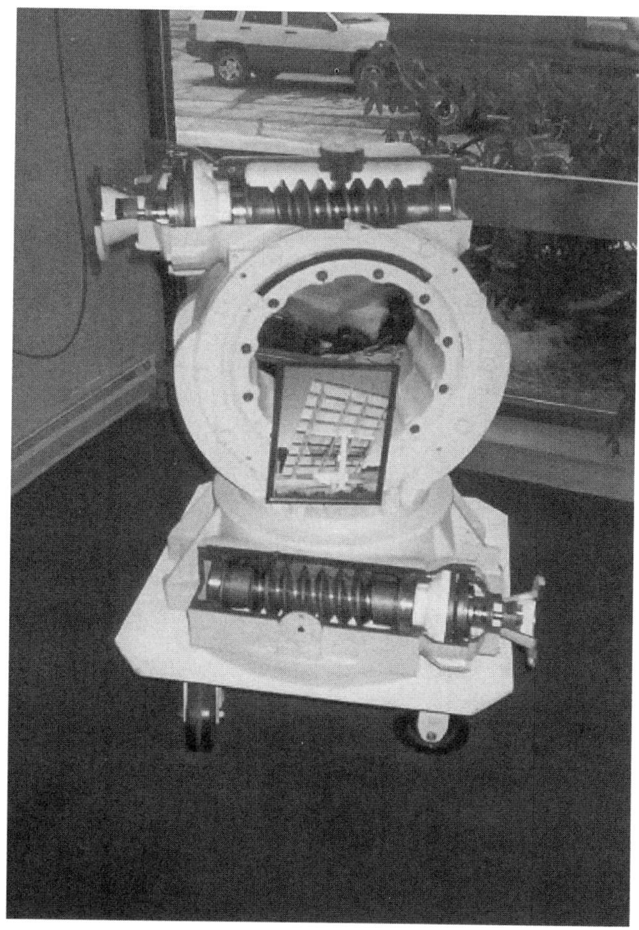

FIGURE 19.46 Detail of a typical worm-gear drive mechanism. At the bottom it is shown the azimuthal actuator located on top of the pedestal. The upper worm actuates against the gear connected to the torque tube.

centralizes all the energy collected by the large mirror field, and therefore high availabilities and durability are a must. Just as cost reduction is the priority for further development in the collector field, in solar receivers, the priorities are thermal efficiency and durability. Typical receiver-absorber operating temperatures are between 500°C and 1200°C and incident flux covers a wide range between 300 and over 1000 kW/m^2. The picture in Figure 19.48 clearly shows the high flux to be withstood by the receiver.

Thermal and optical losses are the key parameters for quantifying the efficiency of a solar receiver.

$$\eta_{\mathrm{REC}} = (\alpha\tau_{\mathrm{W}}) + (\alpha\varepsilon_{\mathrm{W}})\frac{\sigma T_{\mathrm{W}}^4}{C\phi} - (\varepsilon\bar{\rho})\frac{\sigma T^4}{C\phi} - U\frac{(T - T_a)}{C\phi}. \tag{19.44}$$

This equation is the result of the energy balance of gains and losses in the receiver, with an absorber surface at temperature T. In some cases, the receiver has a transparent window that absorbs part of the incident radiation at a temperature, T_{W}, higher than the ambient, T_{a}. The concentrated solar irradiance $(C\phi)$ is absorbed with an efficiency of $(\alpha\tau_{\mathrm{W}})$. The radiation must often go through a transparent window, where it is partially absorbed, before reaching the absorber surface, reflected and transmitted onto the absorber. These not-so-simple effects are represented by the apparent window transmittance, τ_{W}.

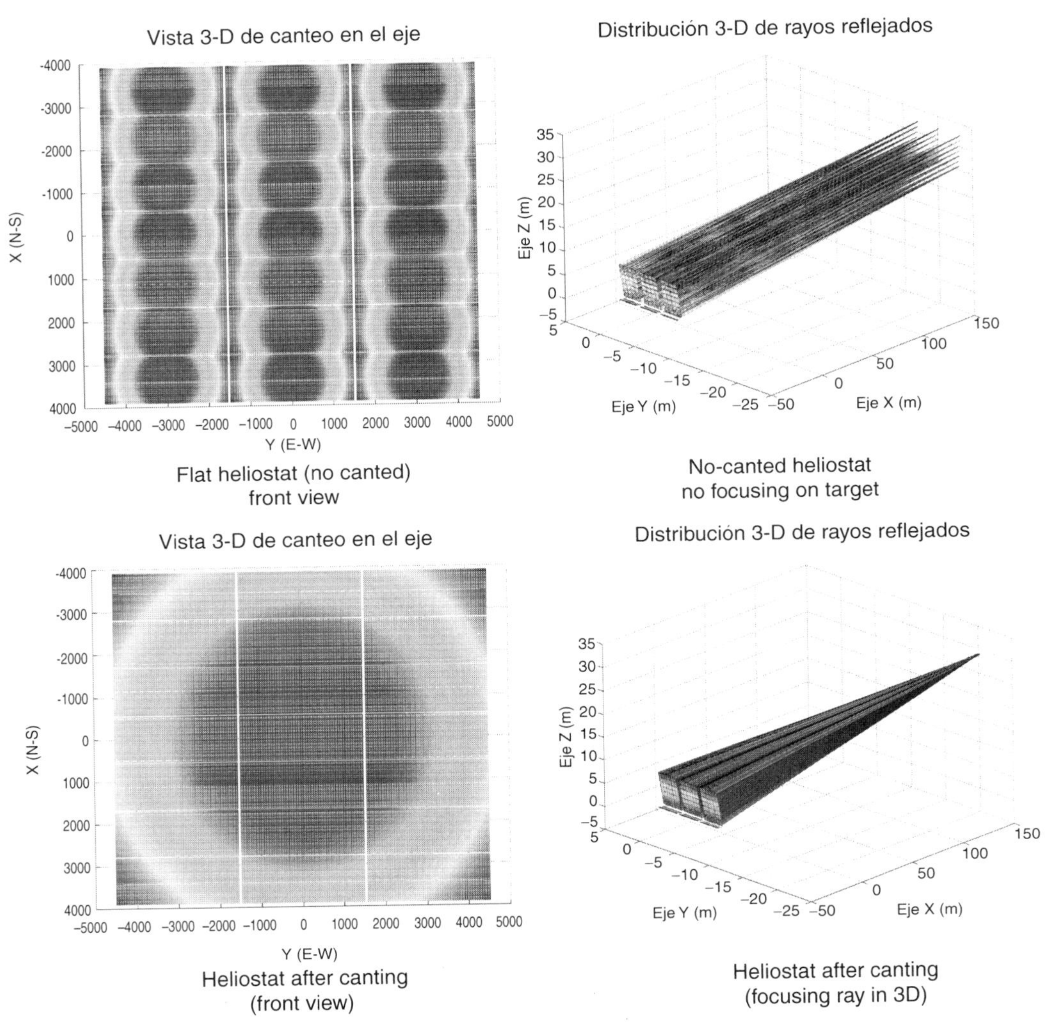

FIGURE 19.47 **(See color insert following page 19-52.)** Effect of canting facets in a glass-metal heliostat. In the upper part it can be observed that a flat heliostat with curved facets is not focusing in a single image. In the lower part, a canted heliostat is depicted.

Likewise, the second term in Equation 19.44 expresses the energy emitted of temperature T_W from the hot window toward the absorber. Therefore, this second term adds gains to the first. The loss terms are of two different types. The most important in a central receiver represents energy thermally radiated by the absorber through the receiver aperture. These radiation losses depend on emissivity of the absorber and on thermal radiation reflectivity ρ_W of the window. In the product $(\varepsilon\bar{\rho})$, $\bar{\rho}$ indicates $(1-\rho_W)$ or $(\alpha_W+\tau_W)$.

The absorber convective, or conductive loss, to the ambient is determined by the heat loss coefficient U that depends on the temperature and the forced convection due to wind. In good central receiver designs, U can be sufficiently reduced by thermal insulation and decreased aperture area, therefore U is basically expressed as a convective loss heat transfer coefficient. Generally, this coefficient is obtained from the dimensionless Nusselt number and subsequently as a function of numbers like Reynolds (Re), Prandtl (Pr) and Grashoff (Gr). Forced convection is determined by combinations of the Re and Pr numbers, whereas natural convection is characterized by Pr and Gr numbers.

FIGURE 19.48 Lateral view of the 80-m concrete tower of the CESA-1 facility located at the Plataforma Solar de Almería in Spain. The concentrated flux is aiming a standby position previous to start operation.

In the solar tower, the convective heat loss is calculated differently depending on whether the receiver is a cavity or a cylindrical external receiver. A typical simple mixed convection coefficient for an external receiver can be obtained calculated (Siebers and Kraabel, 1984) as

$$U_{\text{mix}} = (h_{\text{forced}}^{3.2} + h_{\text{nat}}^{3.2})^{1/3.2}, \tag{19.45}$$

where $h_{\text{nat}} = 9.09$ (W/m^2°C) for an average absorber temperature of 480°C, and h_{forced} is separated into three cases depending on the receiver diameter (Kistler 1986). In all cases, the Reynolds number is $\text{Re} = (1.751 \times 10^5)D$

Case (1): $D \leq 4.0$ m	$h_{\text{forced}} = \left(\frac{1}{D}\right) \cdot \left[0.3 + 0.488 \cdot \text{Re}^{0.5} \cdot \left(1.0 + \left(\frac{\text{Re}}{282{,}000}\right)^{0.625}\right)^{0.8}\right] \cdot 0.04199,$
Case (2): $4.0 < D \leq 125.0$	$h_{\text{forced}} = 14.0,$
Case (3): $D > 125.0$ m	$h_{\text{forced}} = 33.75 \cdot D^{-0.19}.$

For a cavity receiver, the convective heat loss can be directly calculated as (Kistler 1986):

$$Q_{\text{conv}} = Q_{\text{forced}} + Q_{\text{nat}}(W),$$
$$Q_{\text{forced}} = 7631 \frac{A}{W_{\text{ap}}^{0.2}},$$
$$Q_{\text{nat}} = 5077\, A_{\text{cav}},$$

where A is the aperture area (m^2), W_{ap} is the aperture width (m), and A_{cav} is the approximation to total area inside cavity (m^2).

For more detailed correlations applicable to convection losses in different kinds of solar receivers, Becker and Vant-Hull (1991) is recommended.

There are different solar receiver classification criteria depending on the construction solution, the use of intermediate absorber materials, the kind of thermal fluid used, or heat transfer mechanisms. According to the geometrical configuration, there are basically two design options, external and cavity-type receivers. In a cavity receiver, the radiation reflected from the heliostats passes through an aperture into a box-like structure before impinging on the heat transfer surface. Cavities are constrained angularly and subsequently used in north field (or south field) layouts. External receivers can be designed with a flat-plate tubular panel or a cylindrically shaped. Cylindrical external receivers are the typical solution adopted for surround heliostat fields. Figure 19.49 shows examples of cylindrical external, billboard external, and cavity receivers.

Receivers can be directly or indirectly irradiated depending on the absorber materials used to transfer the energy to the working fluid (Becker and Vant-Hull 1991). Directly irradiated receivers make use of fluids or particle streams able to efficiently absorb the concentrated flux. Particle receiver designs make use of falling curtains or fluidized beds. Darkened liquid fluids can use falling films. In many applications, and to avoid leaks to the atmosphere, direct receivers should have a transparent window. Windowed receivers are excellent solutions for chemical applications as well, but they are strongly limited by the size of a single window, and therefore clusters of receivers are necessary.

The key design element if indirectly heated receivers is the radiative/convective heat exchange surface or mechanism. Basically, two heat transfer options are used, tubular panels and volumetric surfaces. In tubular panels, the cooling thermal fluid flows inside the tube and removes the heat collected by the external black panel surface by convection. It is therefore operating as a recuperative heat exchanger. Depending on the heat transfer fluid properties and incident solar flux, the tube might undergo thermomechanical stress. Because heat transfer is through the tube surface, it is difficult to operate at an incident flux above 600 kW/m^2 (peak). Table 19.9 shows how only with high thermal conductivity liquids like sodium is it possible to reach operating fluxes above 1 MW/m^2. Air-cooled receivers have difficulties working with tubular receivers because of the lower heat transfer coefficients, as already found in the German–Spanish GAST project where two tubular receivers, one metal and one ceramic, were tested at the PSA in Spain (Becker and Boehmer 1989). To improve the contact surface, a different approach based on wire, foam or appropriately shaped materials within a volume are used. In volumetric receivers, highly porous structures operating as convective heat exchangers absorb the concentrated solar radiation. The solar radiation is not absorbed on an outer surface, but inside the structure "volume." The heat transfer medium (mostly air) is forced through the porous structure and is heated by convective heat transfer. Figure 19.50 shows a comparison of the two absorber principles. Volumetric absorbers are usually made of thin heat-resistant wires (in knitted or layered grids) or either metal or ceramic (reticulated foams, etc.) open-cell matrix structures. Good volumetric absorbers are very porous, allowing the radiation to penetrate deeply into the structure. Thin substructures (wires, walls or struts) ensure good convective heat transfer. A good volumetric absorber produces the so-called "volumetric effect," which means that the irradiated side of the absorber is at a lower temperature than the medium leaving the absorber. Under specific operating conditions, volumetric absorbers tend to have an unstable mass flow distribution. Receiver arrangements with mass flow adaptation elements (e.g., perforated plates) located behind the absorber can reduce this tendency, as well as appropriate selection of the operating conditions and the absorber material.

Selection of a particular receiver technology is a complex task, since operating temperature, heat storage system and thermodynamic cycle influence the design. In general, tubular technologies allow either high temperatures (up to 1000°C) or high pressures (up to 120 bar), but not both (Kribus 1999). Directly irradiated or volumetric receivers allow even higher temperatures but limit pressures to below 15 bar.

19.7.1.3 Tubular Receivers

The most common systems used in the past have been tubular receivers where concentrated radiation is transferred to the cooling fluid through a metal or ceramic wall. Conventional panels with darkened metal tubes have been used with steam, sodium and molten salts for temperatures up to 500–600°C. Much less experience is available on tubular receivers with gas, though temperatures in the range of 800–900°C are possible. Cavity receivers have been tested in France (Themis) and Spain (IEA–SSPS–CRS project and CESA-1 plant). External tubular receivers were used in Solar One (U.S.A.), IEA–SSPS–CRS

FIGURE 19.49 Different configurations of solar receivers. From left to right and top to bottom: (a) external tubular cylindrical, (b) cavity tubular, (c) billboard tubular, and (d) volumetric.

(Spain) and Solar Two (U.S.A.) (Grasse, Hertlein, and Winter 1991; Pacheco and Gilbert 1999). Other experimental systems using water/steam and tubular receivers were Nio in Japan, Eurelios in Italy, and SES-5 in the former Soviet Union.

Solar One in Barstow, California, U.S.A., used a once-through superheated water–steam receiver. It operated from 1984 to 1988 and has been the largest central receiver in the world for two decades. It was an external cylindrical receiver made up of 24 1-m-wide by 14-m-long rectangular panels (Figure 19.51). The six panels on the south side were feedwater preheat units. Preheated water was then transferred to once-through boilers and superheaters on the north side. The tubes in each panel were welded throughout their length and painted black with Pyromark® paint. The design specifications were for steam at 516°C and a pressure of 100 bar. Up to 42 MW_t could be absorbed by the receiver. The initial

TABLE 19.9 Operating Temperature and Flux Ranges of Solar Tower Receivers

Fluid	Water/Steam	Liquid Sodium	Molten Salt (nitrates)	Volumetric Air
Flux (MW/m^2)				
Average	0.1–0.3	0.4–0.5	0.4–0.5	0.5–0.6
Peak	0.4–0.6	1.4–2.5	0.7–0.8	0.8–1.0
Fluid outlet temperature (°C)	490–525	540	540–565	700–800 (>800)

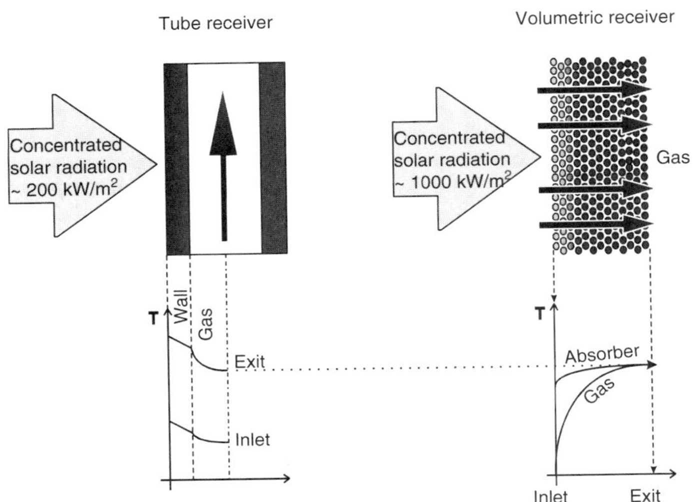

FIGURE 19.50 Heat transfer principles in tubular and volumetric receivers.

thermal efficiency was 77% for an absorbed power of 34 MW_t. After painting it black and curing the surface it increased to 82% with $\alpha = 0.97$. Almost constant thermal losses have been verified (4.5–5 MW_t) for this kind of receiver, due basically to radiation losses and operating the receiver at constant temperature by regulating the mass flow rate. One of the main problems found during testing was overheating and deformation in the superheating section because of solar transients and poor heat transfer (Radosevich and Skinrood 1989). Cracking and leaking was observed at the top of the boiler tubes after 18 months of operation. The temperature gradient between the edges and the center of the tubes can go up to 111°C during start-ups and shutdowns.

CESA-1, which operated 1631 h between 1983 and 1986 in Almería, Spain, was a north-facing water–steam cavity receiver. The 3.4-m square aperture was tilted 20° to the heliostat field. The receiver panel configuration and position of the steam drum are shown in Figure 19.52. The boiler consisted of three panels of A-106 Gr B carbon steel tubes, with an effective surface of 48.6 m^2, and a superheater made of X-20 Cr Mo V 121 steel. The 6.7 MW maximum incident power on the receiver produced superheated steam at 110 kg/cm^2 and 525°C. As in Solar One, operating problems and noticeable deformations were found in the superheating zone, requiring it to be operated with lower flux. Also because of that, operation had to proceed slowly during start-up and transients, penalizing efficiency. More than 45 min were required to reach nominal conditions (Sánchez 1986).

Molten salt tubular receivers are represented by the Themis system (cavity) and Solar Two (cylindrical external). In a molten salt system, cold salt at about 290°C is pumped from a tank at ground level to the receiver mounted atop a tower where it is heated by concentrated sunlight to 565°C. Using molten salts as receiver coolant provides a number of benefits because there is no phase change and it is possible to heat up to 565°C without the problems associated in tubes with superheating sections. Mixtures of 60% sodium nitrate and 40% potassium nitrate have been extensively tested with satisfactory results in France and the U.S.A.. Molten nitrates provide good thermal conductivity (0.52 W/mK) and heat capacity (1.6 kJ/kgK) at relatively low prices. Molten nitrate salt, though an excellent thermal storage medium, can be a troublesome fluid to deal with because of its relatively high freezing point (220°C). To keep the salt molten, a fairly complex heat trace system must be employed. (Heat tracing is composed of electric wires attached to the outside surface of pipes. Pipes are kept warm by way of resistance heating.) Problems were experienced during the start-up of Solar Two due to the improper installation of the heat trace. Though

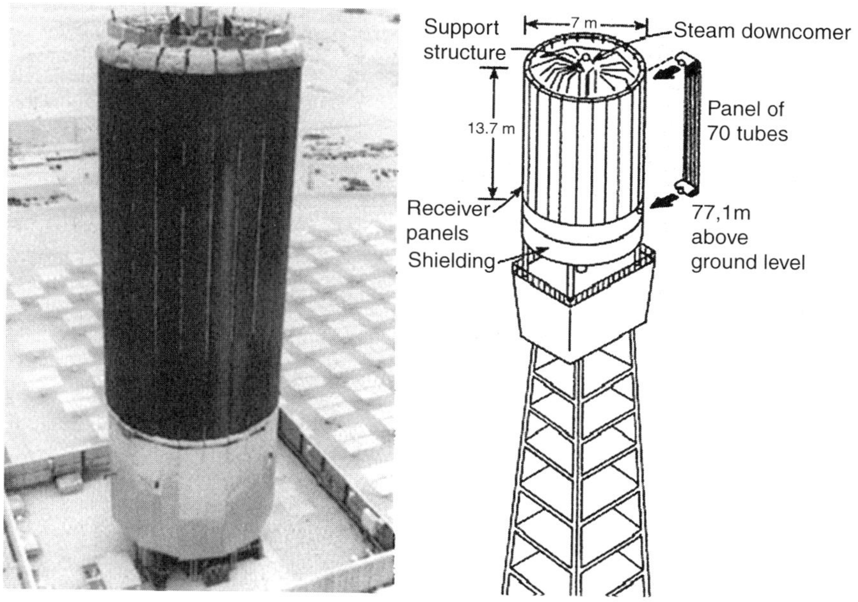

FIGURE 19.51 External cylindrical tubular receiver used in Solar One power plant, Barstow, California.

this problem has been addressed and corrected, research is needed to reduce the reliance on heat tracing in these plants. Also, valves can be troublesome in molten salt service. Special packings must be used, oftentimes with extended bonnets, and leaks are not uncommon. Furthermore, freezing in the valve or packing can prevent it from operating correctly. While today's valve technology is adequate for molten salt power towers, design improvements and standardization would reduce risk and ultimately reduce O&M costs (DeMeo and Galdo 1997).

Solar Two, tested between 1996 and 1999 (Pacheco et al. 2000), is still the technical reference for molten salt tubular receivers. The 42-MW absorber consisted of a 6.2-m high and 5.1-m diameter cylinder, with 768 2-cm diameter tubes. Reported efficiency with no wind was 88% for 34 MW absorbed (86% with wind under 8 km/h). Peak concentrated solar flux was 800 kW/m^2 and average 400 kW/m^2. Though reported efficiency was close to nominal, during the three years of operation there were many incidents, and modifications and repairs to avoid freezing and obstructions in tubes, downcomers, manifolds, valves, and pipelines. The consequence was a very limited experience in long-term testing. Some attempts were made in the early nineties to simplify molten salt receivers by removing the salt-in-tube heat exchanger and introducing open-air falling film flat-plates (Romero et al. 1995). Table 19.10 summarizes typical operating temperatures, incident flux, pressures, and efficiencies in tubular water/steam and molten salt receivers.

19.7.1.4 Volumetric Receivers

As already mentioned, volumetric receivers use highly porous structures for the absorption of the concentrated solar radiation deep inside (in the "volume") of the structure. Volumetric receivers can work open to the ambient or enclosed by a transparent window. With metal absorbers it is possible to achieve air outlet temperatures up to 850°C, and with ceramic fibers, foams, or monoliths (SiC), the temperature can surpass 1000°C.

The main advantages of an air-cooled volumetric receiver are:

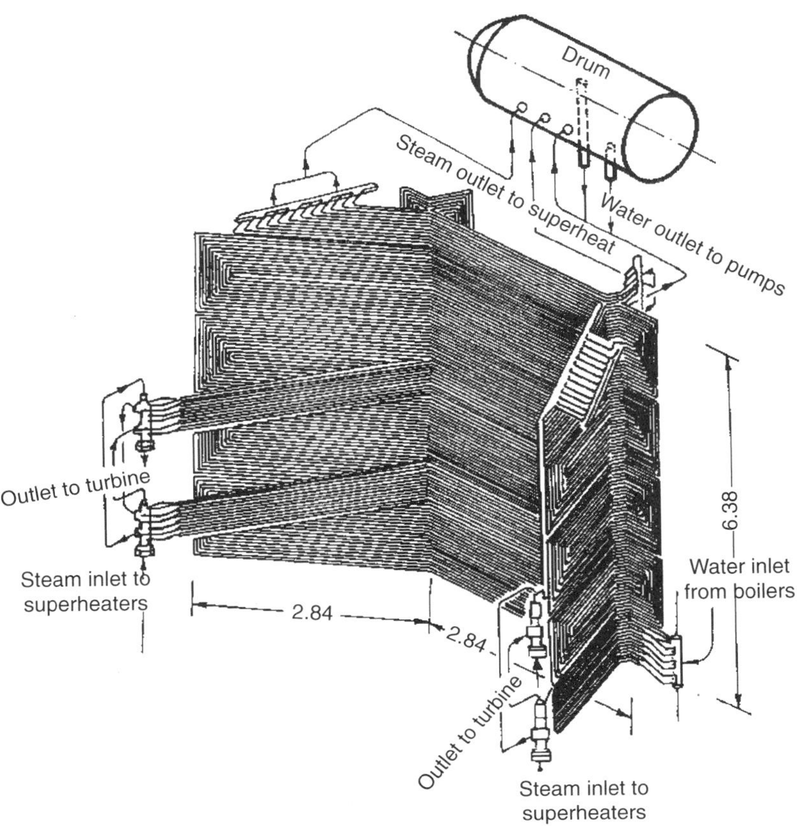

FIGURE 19.52 Inner view of tubular panels and drum of the water/steam cavity receiver used in the CESA-1 project in Spain.

- The air is free and fully available at the site
- No risk of freezing
- Higher temperatures are possible and therefore the integration of solar thermal energy into more efficient thermodynamic cycle looks achievable
- No phase change
- Simpler system
- Fast response to transients or changes in incident flux
- No special safety requisites
- No environmental impact

TABLE 19.10 Summary of Operational Range for Tubular Water/Steam and Molten Salt Receivers

Receivers Water/Steam	Molten Salt Receivers
Temperature fluid outlet 250/525°C	Temperature outlet 566°C
Incident flux 350 Kw/m^2	Incident flux 550 kW/m^2
Peak flux 700 kW/m^2	Peak flux 800 kW/m^2
Pressure 100–135 bar	Efficiency 85–90%
Efficiency 80–93%	

Open volumetric receivers have made dramatic progress since the pioneering experiences of the late 1970s (Sanders 1979) and early 1980s (Fricker et al. 1988). More than 20 absorbers and prototypes in the 200–300 kW_{th} range have been tested in the Sulzer test bed at the PSA (Becker et al. 1989; Becker, Cordes, and Böhmer 1992; Hoffschmidt et al. 2001). Wire mesh, knitted wire, foam, metal and ceramic monolith volumetric absorbers have been developed worldwide. The relatively large number of volumetric prototypes tested has demonstrated the feasibility of producing hot air at temperatures of 1000°C and upward and with aperture areas similar to those used in molten salt or water/steam receivers. Average flux of 400 kW/m^2 and peaks of 1000 kW/m^2 have been proven, and their low inertia and quick sun-following dispatchability are excellent. Comparative assessments have demonstrated that wire mesh has the lowest thermal losses (Table 19.11). This can mainly be explained by the very high porosity of the absorber, which permits a large portion of the irradiation to penetrate deep into its volume. The choice of ceramics as the absorber material, makes higher gas outlet temperatures possible. In particular, siliconized SiC been has revealed as a good option because of its high thermal conductivity. Even though ceramic absorbers have lower efficiencies at 680°C (this is the reference temperature for applications where hot air is used as the heat transfer fluid to produce superheated steam in a heat exchanger), they have demonstrated efficiencies about 80% at temperatures of 800°C. With higher solar flux and temperatures, more compact designs and smaller receivers can be developed.

A milestone in the experimental scale-up of open volumetric receivers was the TSA (Technology Program Solar Air Receiver) project, under the leadership of the German Steinmüller Company. A 2.5 MWt air-cooled receiver was tested on top of the PSA CESA-1 tower late in 1991. The TSA experimental setup was a small-scale PHOEBUS-type receiver (Schmitz-Goeb and Keintzel 1997), in which atmospheric air is heated up through a wire mesh receiver to temperatures on the order of 700°C to produce steam at 480–540°C and 35–140 bar, in a heat recovery steam generator with a separate superheater, reheater, evaporator, and economizer feeding a Rankine turbine–generator system (Figure 19.65). All the hardware, including receiver, steam generator, and heat storage, was located atop the CESA-1 tower. Average solar flux at the receiver aperture was 0.3 MW/m^2. The heliostat field control system was implemented with heuristic algorithms to obtain an automatic aiming strategy able to maintain stationary flux and air outlet temperature (García-Martín et al. 1999). One of two blowers controlled the air mass flow rate through the receiver and the other the air through the steam generator. The 1000 kWh capacity thermocline storage system (alumina pellets) was enough to allow 30 min of nominal off-sun operation. Depending on the load, it could be charged, bypassed or discharged. The 3m-diameter absorber was made of hexagonal wire mesh modules (Figure 19.53). The TSA receiver was successfully operated by DLR and CIEMAT for a total of nearly 400 h between April and December 1993 (Haeger 1994), and for shorter periods in 1994 and 1999, demonstrating that a receiver outlet temperature of 700°C could easily be achieved within 20 minutes of plant start-up and receiver thermal efficiencies up to 75% were obtained as shown in Figure 19.54.

A significant number of volumetric prototypes have been reported unable to reach nominal design conditions because of local cracks and structural damages. Those failures were, in many cases, caused by thermal shock, material defects or improper operation. In the middle 1980s, some projects promoted by the German Aerospace Center (DLR) studied the fluid dynamics and thermal mechanisms inherent in volumetric absorbers (Hoffschmidt et al. 2001). One of the conclusions of these studies was that in highly porous absorber materials the airflow through the absorber structure is unstable under high solar flux, which leads to the destruction (cracks or melting) of the absorber structure by local overheating (Pitz-Paal et al. 1996). As a consequence of this analysis, a new approach developed monolithic ceramic absorbers able to work at high temperatures and fluxes due to their high thermal conductivity and geometric modularity, which were put to use first in the German–Spanish HiTRec project (Hoffschmidt et al. 1999) and afterwards in the European Solair project (Hoffschmidt et al. 2002).

The HiTRec–Solair receiver principle is shown in Figure 19.55. A stainless steel support structure on the back of a set of ceramic absorber modules forms the base of the receiver. Similar to ceramic burner

TABLE 19.11 Properties and Efficiencies Reported for Several Absorber Materials Tested at the Plataforma Solar de Almería

Type of Receiver	Designed by	Absorber Structure	Porosity (V_p/V_{tot})	*Dm*,p[a] mm	Absorber Thickness mm	Material	Thermal Conductivity (500°C) W/m.K	Efficiency 680°C (%)
Metallic wire	SULZER	Wire mesh	0.95	2.5	35	Stainless steel	20	75
Ceramic foam	SANDIA	Amorphous foam	0.8	1.0	30	Al_2O_3	25	54
Metallic foils	Interatom/Emitec	Prismatic channels	0.9	1.0	90	$X_5CrAl_2O_5$	20	57
Ceramic foils	DLR/Ceramtec	Prismatic channels	0.4	3.0	92	SiSiC	80	60
Ceramic cups	DLR/STOBBE	Prismatic channels	0.5 + 0.12 apert.	2.1	80	SiSiC	80	60

[a] Mean pore diameter.

FIGURE 19.53 Front view of the TSA volumetric receivers. The absorber was composed of hexagonal pieces of wire mesh. The outer ring includes the air return system.

tubes, the absorber modules are separated from the back and allowed to expand both axially and radially with thermal expansion of the modules or movement of the stainless steel construction behind them during start-up or shutdown. The absorber modules are spaced to avoid touching adjacent modules. The support structure is a double-sheet membrane that may be cooled by either ambient or returned air. Tubes attached to the absorber cups pass through holes in the front sheet and are welded to the rear sheet. The cooling air circulates between the two sheets and, as it leaves through the sides of the segments, also cools the support structure. The air reaches the absorber aperture through the spaces between the segments. Outgoing air and ambient air are mixed and sucked back into the segments. As they penetrate the absorber structure, the air is heated up by convection. On leaving the absorber structure, the hot air is ducted to the bottom of the cup. There an orifice, previously sized according to solar flux simulations, adjusts the air mass flow rate to compensate the expected solar flux profile over the aperture and to

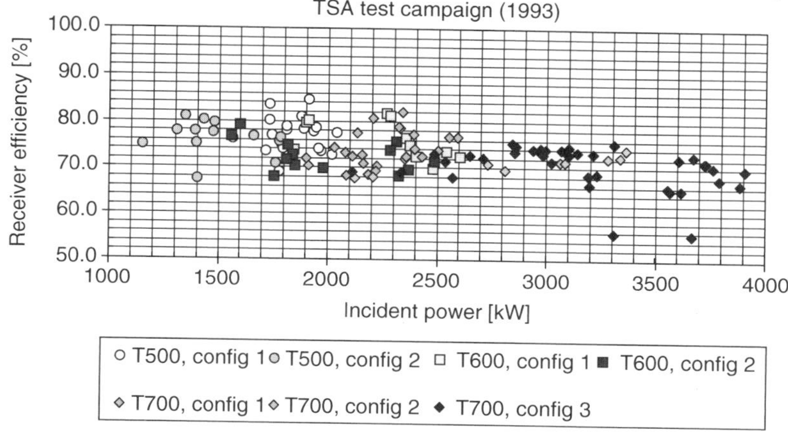

FIGURE 19.54 TSA receiver efficiency at different temperatures and incident powers during test campaign carried out in 1993.

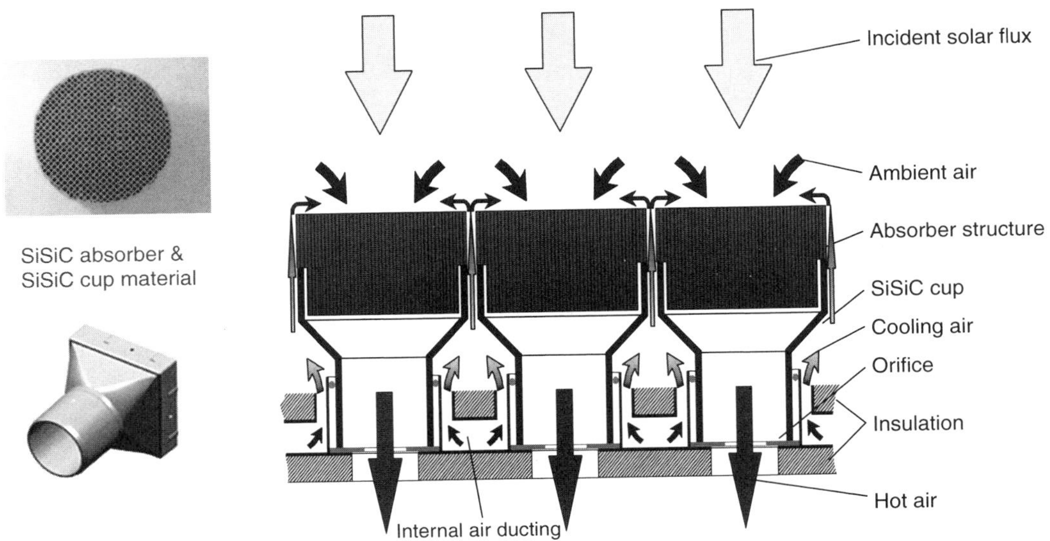

FIGURE 19.55 Principle of volumetric receivers used in HiTRec and Solair projects.

provide homogeneous outlet air temperatures from the cups. Then the air passes through the tubes hosting the cups and across the holes in the rear sheet of the membrane.

A first milestone in the development of this type of receiver was the qualification of the 200 kW HiTRec-II receiver (Hoffschmidt et al. 2003). Testing lasted from November 2000 through May 2001. During the course of the test program, the HiTRec-II receiver was operated over a period of 38 days, accumulating a total of 155 operating hours with a thermal efficiency of up to 76±7% at 700°C. Eventually, a Solair 3 MW prototype was developed, installed, and tested in the TSA test bed in the PSA and therefore connected to the heat storage thermocline, the steam generator and the air circuit. The receiver was designed as a real modular absorber, which was assembled from 270–140-mm square ceramic modules. Each module consisted of a square absorber structure glued to a SiSiC cup. The cups were square at the aperture, but round at the back. The honeycomb absorber structure was made of recrystallized SiC with a normal open porosity of 49.5%. The Solair-3000 receiver had a total aperture surface of 5.67 m^2 (Figure 19.56). During testing, the incident solar radiation is reflected by the heliostat field and concentrated on the ceramic volumetric absorber with an average flux density of 0.5 MW/m^2. The air leaves the absorber outlet at 700–750°C. The system was evaluated during 2003 and 2004 presenting an air return ratio (ARR) of 0.5, efficiency of (72±9)% at a temperature of 750°C and efficiency of (74±9)% at 700°C. The wide margin of uncertainty (of 9%) was due to the accumulation of uncertainties in the measurement process, mainly due to air mass flow and incident solar power measurements. For the same outlet air temperature, absorber efficiency increases with incident solar flux. This is because thermal losses are quite the same and thermal gains are larger. Absorber efficiency is therefore better represented by a three-dimensional plot (Figure 19.57). Efficiencies were estimated at over 85% (and up to 89%) for outlet air temperatures in the range of 590–630°C and mean incident solar fluxes of 310–370 kW/m^2.

Although simplicity and operating results are satisfactory, it is obvious that open volumetric receiver thermal efficiencies must be improved to achieve cost-effective plant designs able to replace tubular receivers. In addition, there have not yet been any long-term endurance tests, radiation losses must be further reduced and the ARR should be improved in open receiver designs. The need for improving the ARR for the next generation of open volumetric receivers is clearly justified when the integration of the solar receiver and PCU must be optimized. The impact of the ARR on PHOEBUS-type central receiver system performance was already quantified in a comparative assessment carried out in 1992 by a joint

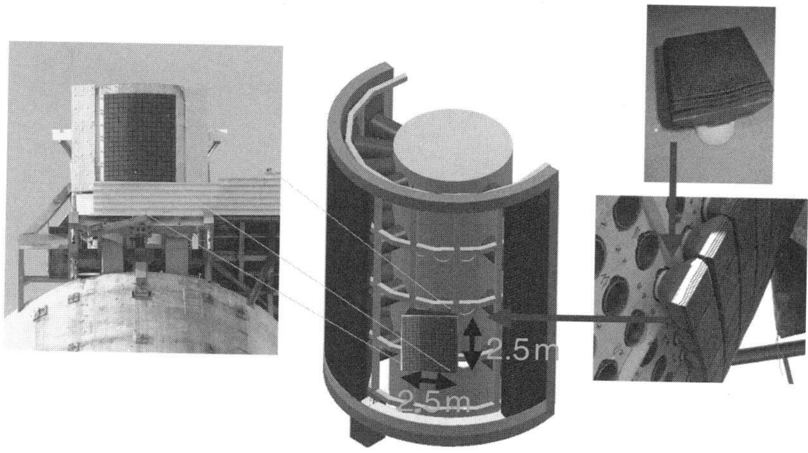

FIGURE 19.56 On the left the 3-MW Solair receiver tested at PSA. The modularity of the concept allows designing any size by replication. At the center a proposal of 55-MW receiver made of a cluster of 3-MW modules.

U.S.A./German team that identified the potential for molten salt and air-cooled plants (Chavez, Kolb and Meinecke 1993). The study revealed that one of the weak points of air-cooled plants is related to the losses in the open volumetric receiver, especially mixing losses at the receiver inlet. The Rankine cycle forces air from the steam generator to return at temperatures between 110–170°C. Plant performance analysis leads to the conclusion that ARR in a PHOEBUS-type receiver should be close to 70% to keep air-cooled solar plants in the same efficiency range as other power tower plants cooled with molten salt or water/steam (Marcos, Romero, and Palero 2004). For an optimistic $ARR = 0.7$ and $T_{rec.in} = 150°C$, the air mixing

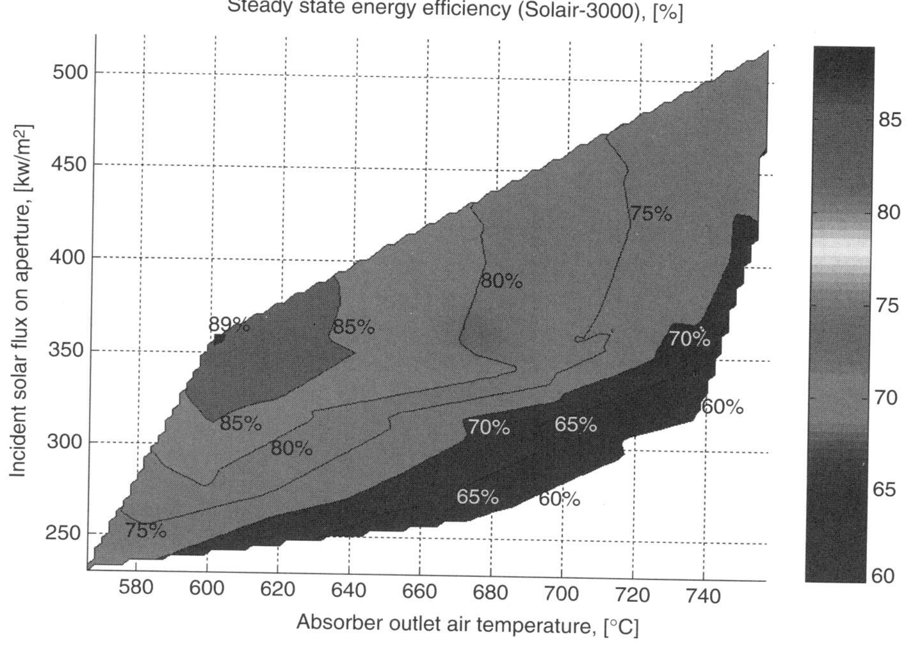

FIGURE 19.57 **(See color insert following page 19-52.)** Map of Solair-3 MW volumetric receiver showing thermal efficiency versus incident solar flux on aperture and air outlet temperature.

efficiency is 0.93. Air mixing efficiency can be defined as:

$$\eta_{mix} = \frac{T_{Rec.out} - T_{Rec.in}}{T_{Rec.out} - T_{mix}}, \tag{19.46}$$

where $T_{rec.in}$ is the receiver air inlet temperature (i.e., 150°C), $T_{rec.out}$ is the receiver air outlet temperature (i.e., 700°C), and T_{mix} is the receiver air mixing temperature (i.e., 111°C)

$$T_{mix} = xT_{Rec.in} + (1-x)T_{amb}. \tag{19.47}$$

$$x = \text{ARR} = 0.7.$$

$$T_{amb} = \text{ambient temperature} = 20°\text{C}.$$

The future use of open volumetric receivers in more efficient thermodynamic cycles, with air return temperatures up to 500°C and more, may lead to even more stringent requirements for the ARR as depicted in Figure 19.58. Assuming a target air mixing efficiency of at least 0.93 or higher it means that for air return temperatures of 200°C ARR should be over 0.8 and for temperatures above 400°C the ARR should raised to 0.9.

Another option for working with closed-loop air circuits is the use of windowed volumetric receivers. One attractive application is the use of solar receivers as the preheating chamber of a gas turbine (Romero, Buck, and Pacheco 2002). In 1996, the German Aerospace Center (DLR) initiated a specific development program called REFOS for the purpose of producing an optimum 350 kW windowed module design able to work at temperatures up to 1000°C and with pressures up to 15 bar. The aim of the REFOS project was to develop, build, and test modular pressurized volumetric receivers under representative operating conditions for coupling to gas turbines. Emphasis was on testing solar preheating of air, accompanied by basic materials research. The REFOS receiver consists of a cylindrical vessel containing a curved knitted absorber. A quartz dome is used to pressurize the air cycle. A hexagonal secondary concentrator with a 1.2 m inner diameter is used to increment the flux density and protect the window flange. The hexagonal shape was selected to optimize the layout of cluster packing, in such a way that the REFOS modules can be used in either small or large power plants (Figure 19.59). Typical REFOS module specifications are:

- Absorbed thermal power (design point): 350 kW per unit.
- Absolute pressure (operation): 15 bar

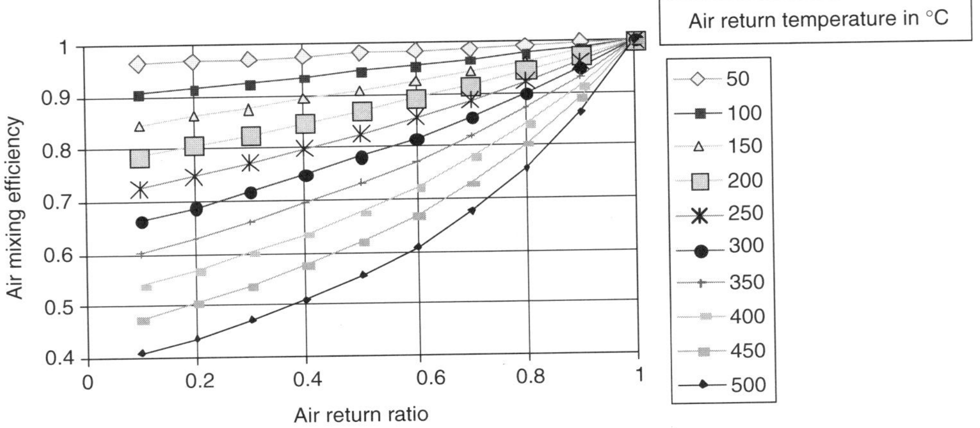

FIGURE 19.58 Influence of air return inlet temperature and ARR on air mixing efficiency. The values are obtained for an outlet receiver temperature of 800°C and ambient temperature of 20°C. Air return inlet range covers temperatures between 50°C and 500°C.

- Air outlet temperature: 800°C for metal absorber and up to 1000°C for ceramics.
- Temperature increment per module: 150°C
- Receiver efficiency (including secondary): 80% at design point.

Several modules have been tested at the PSA under cooperation agreements between DLR and CIEMAT. In the most extensive tests performed in 1999, the design conditions were demonstrated with a single-module operating at air outlet temperatures of 800°C at 15 bar, at power levels up to 400 kW_t. More than 247 h of testing proved the feasibility of the receiver, which reached a maximum temperature of 1050°C without incurring damage (Buck et al., 2002). Thermal efficiencies was between 63% and 75%.

19.7.1.5 Experience in CRS

Although there have been a large number of STPP tower projects, only a few have culminated in the construction of an entire experimental system. Table 19.12 lists systems that have been tested all over the world along with new plants that are likely to be built. In general terms, as observed, they are characterized as being small demonstration systems between 0.5 and 10 MW, and most of them were operated in the 80 s (Entropie, 1982; Falcone, 1986; Grasse, Hertlein, and Winter 1991). The thermal fluids used in the receiver have been liquid sodium, saturated or superheated steam, nitrate-based molten salts, and air. All of them can easily be represented by flow charts, where the main variables are determined by working fluids, with the interface between power block and the solar share.

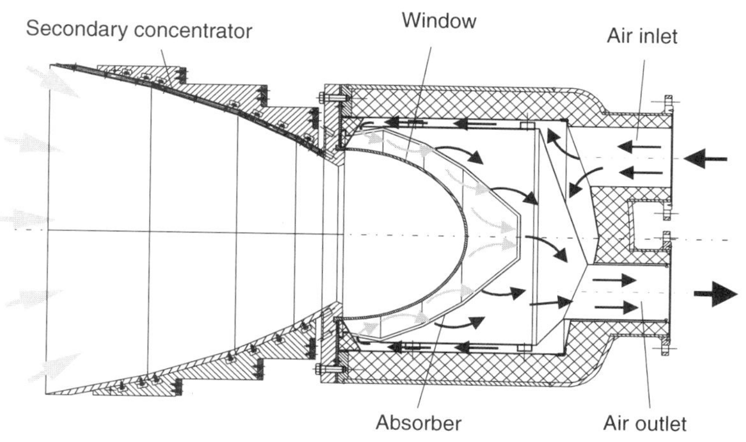

FIGURE 19.59 Example of 350 kW REFOS module. (From DLR, Germany.)

The set of experiences referred to has served to demonstrate the technical feasibility of the CRS power plants, whose technology is sufficiently mature. The most extensive experience has been collected by several European projects located in Spain at the premises of the PSA (Grasse, Hertlein, and Winter 1991), and the 10 MW Solar One (Radosevich and Skinrood 1989) and Solar Two plants (Pacheco and Gilbert 1999) in the U.S. annual efficiencies of 7% have been demonstrated, and the predictions are to reach efficiencies of 23% at design point and 20% annual by 2030, with investment costs of $0.9/W$_p$ (DeMeo and Galdo 1997), but the first generation of commercial demonstration plants have to be built to validate the technology under real operating and market conditions. Since the early 1990s, most proposals for the first generation of commercial plants have focused on nitrate salt and air as the receiver heat transfer fluids because a joint U.S.A/German study identified their potential and economics in power tower plants (Chavez, Kolb, and Meinecke 1993), though in Spain parallel initiatives have developed saturated steam designs (Silva, Blanco, and Ruiz 1999). Several penetration strategies have been proposed since then, and many more may be developed in the future, since solar towers have the great advantage of admitting very open integration designs depending on the dispatching scenarios, annual capacity factors and hybridization schemes. Three of the most promising power tower technologies ready for implementation are: (1) molten salt, (2) open or closed-loop volumetric air, and (3) saturated steam. The best developed technology in the U.S.A. is based on solar-only power plants with large thermal storage capacity based on molten nitrate salts as the working fluid. The use of volumetric receivers, both with closed air loops for efficient integration into gas turbine cycles or open air for intermediate storage and/or hybridization solutions, have been promoted in Europe and Israel in projects like SOLGATE, Solair, and ConSolar. Finally, a more conservative approach in Spain makes use of solar saturated steam receivers for electricity generation, in such projects as Solgas, COLON SOLAR, and PS10. PS10 is the first solar tower plant operating and selling electricity to the public distribution grid in the World (Osuna et al. 2004).

19.7.2 Water/Steam Plants: The PS10 Project

Production of superheated steam in the solar receiver has been demonstrated in several plants, such as Solar One, Eurelios, and CESA-1, but operatining experience showed critical problems related to the control of zones with dissimilar heat transfer coefficients like boilers and superheaters (Grasse, Hertlein, and Winter 1991). Better results regarding absorber panel lifetime and controllability have been reported

TABLE 19.12 Experimental Power Towers in the World

Project	Country	Power (MW_e)	Heat Transfer Fluid	Storage Media	Beginning Operation
SSPS	Spain	0.5	Liquid sodium	Sodium	1981
Eurelios	Italy	1	Steam	Nitrate salt/water	1981
Sunshine	Japan	1	Steam	Nitrate salt/water	1981
Solar One	U.S.A.	10	Steam	Oil/rock	1982
CESA-1	Spain	1	Steam	Nitrate salt	1982
MSEE/Cat B	U.S.A.	1	Nitrate salt	Nitrate salt	1983
Themis	France	2.5	Hitec salt	Hitech salt	1984
SPP-5	Russia	5	Steam	Water/steam	1986
TSA	Spain	1	Air	Ceramic	1993
Solar two	U.S.A.	10	Nitrate salt	Nitrate salt	1996
ConSolar	Israel	0.5[a]	Pressurized air	Fossil hybrid	2001
SOLGATE	Spain	0.3	Pressurized air	Fossil hybrid	2002
PS10	Spain	11	Water/steam	Saturated steam	2006
PS20[b]	Spain	20	Water/steam	Saturated steam	2008
Solar Tres[b]	Spain	17	Nitrate salt	Nitrate salt	2009

[a] Thermal.
[b] Projects under development.

for saturated steam receivers. In particular, the STEOR (solar thermal enhanced oil recovery) pilot plant, which proved to be highly reliable for oil extraction using direct injection of steam, was successfully operated in Kern County, California, for 345 days in 1983 (Blake, Gorman, and McDowell 1985). The good performance of saturated steam receivers was also qualified at the 2 MW Weidman receiver that produced steam at 15 bar for 500 h in 1989 (Epstein et al. 1991). Even though technical risks are reduced by saturated steam receivers, the outlet temperatures are significantly lower than those of superheated steam, making applications where heat storage is replaced by fossil fuel backup necessary.

At present, estimated costs of electricity production from the solar share of hybrid systems are \$0.08–\$0.15 per kWh, whereas expect costs for solar-only plants are in the range of \$0.15–0.20 per kWh. The implementation of hybridized systems is one of the paths leading to a breakthrough in the financial barriers to deployment of solar electric technologies as it reduces the initial investment (Kolb 1998). The use of hybrid plants with the low technological risk of a central receiver system with saturated steam as the working fluid is the starting point. Two projects subsidized by the European Commission, the Solgas project promoted by SODEAN and the COLON SOLAR project promoted by the Spanish utility, SEVILLANA (Ruiz, Silva, and Blanco 1999), established the strategy of market penetration on the basis of the integration of saturated steam receivers in cogeneration systems and repowering of combined cycles. The size of the cavity receiver was optimized to supply 19.8 MW_t to the fluid at 135 bar and 332.8°C outlet temperature. The collector subsystem consisted of 489 heliostats (each with a 70-m^2 reflective surface) and a 109 m tower. As observed in Table 19.13, the use of low temperature receivers and phase-change saturated steam yields a much higher thermal efficiency of up to 92% at nominal load. The table shows a theoretical comparison between a typical volumetric air-cooled receiver working at 700°C air outlet temperature and saturated steam receiver at 250°C thermal outlet. Both are cavity receivers with an incident power of 45 MW (Osuna et al. 2004).

Integrating power towers into existing combined cycle plants can create issues with respect to heliostat field layout, since the solar field is forced to make use of sites near gas pipelines and industrial areas. Land becomes a nonnegligible share of plant cost and site constraints lead to layout optimization and subsequent optical performance problems. This was the case of the COLON heliostat field that represented a real design challenge because of the significant restrictions imposed by the available site (Romero, Fernández, and Sánchez 1999). The hybrid solar gas scheme predicts solar production costs below \$0.11/kWh and annual solar shares in the range of 8–15%. The lack of public support schemes for hybrid solar thermal power plants in Spain at that time led to project abortion and the plant was never built.

The follow-up of the COLON SOLAR project was finally a solar-only saturated steam plant called PS10 (Planta Solar 10 MW). PS10 came about as a consequence of the Spanish legal framework for a special regime of feed-in tariffs for renewable electricity, issued in March 2004. PS10 is located on the Casa Quemada estate (37.2° latitude), 15 km west of the city of Seville, Spain. The 11 MW plant was designed to achieve an annual electricity production of 23 GWh at an investment cost of less than 3500 €/kW. The project made use of available, well-proven technologies like the glass–metal heliostats developed by the Spanish INABENSA Company and the saturated steam cavity receiver developed by the TECNICAL Company to produce steam at 40 bar and 250°C (Osuna et al. 2004). The plant is a solar-only system with

TABLE 19.13 Comparison of Thermal Losses and Efficiency in Air Volumetric and Saturated Steam Receivers for the PS10 Project

Losses	Air (%)	Steam (%)
Reflection	7.9	2.0
Radiation	8.6	0.8
Convection	0.0	2.6
Spillage	5.0	2.1
Air return	3.7	0.0
Total efficiency	74.8	92.4

saturated steam heat storage able to supply 50 min of plant operation at 50% load. The system makes use of 624 121 m^2 heliostats, distributed in a north field configuration, a 90 m high tower, a 15 MWh heat storage system and a cavity receiver with four 4.8×12 m^2 tubular panels. The basic flow diagram selected for PS10 is shown in Figure 19.60. Though the system makes use of a saturated steam turbine working at extremely low temperature, the nominal efficiency of 30.7% is relatively good. This efficiency is the result of optimized management of waste heat in the thermodynamic cycle. At the turbo generator exit, the steam is sent to a water-cooled condenser, working at 0.06 bar. The condenser outlet is preheated with 0.8 and 16 bar steam bled from the turbine. The output of the first preheater is sent to the deaerator that is fed with steam again bled from the turbine. A humidity separator is installed between the high- and low-pressure sections of the turbine to increase steam quality in the last stages of expansion. A third and last preheater fed with steam from the receiver increases water temperature to 245°C. When mixed with water returned from the drum, 247°C receiver feedwater is obtained. As summarized in Table 19.14, the combination of optical, receiver, and power block efficiencies leads to a total nominal efficiency at design point of 21.7%. Total annual efficiency decreases to 15.4%, including operational losses and outages. The subsystem energy efficiency, as in any solar thermal power plant, can be viewed with an energy cascade diagram, as presented in Figure 19.61. PS10 is a milestone in the CRS deployment process because it is the first solar power tower plant developed for commercial exploitation. Commissioning and start-up is planned for September 2006. The construction of a second, 20 MW_e plant with the same technology as PS10 is planned to begin in April 2006.

19.7.3 Molten Salt Systems: Solar Two

The Solar One pilot plant successfully demonstrated operation of a utility-scale power tower plant. The Solar One receiver heated subcooled water to superheated steam, which drove a turbine. The superheated steam was also used to charge an oil–rock thermocline storage system. Solar One operated for 6 years from 1982 to 1988, the last three of which were devoted to power production (Radosevich and Skinrood 1989). Although Solar One successfully demonstrated the feasibility of the power tower concept, the thermal storage system was inadequate for operating the turbine at peak efficiency, because the storage system operated only between 220°C and 305°C, whereas the receiver outlet (and design turbine inlet) temperature was 510°C. The primary mode of operation was directly from receiver outlet to turbine input, bypassing the thermal storage system, and storage provided auxiliary steam during offline periods.

For high annual capacity factors, solar-only power plants must have an integrated cost-effective thermal storage system. One such thermal storage system employs molten nitrate salt as the receiver heat transfer fluid and thermal storage media. To be usable, the operating range of the molten nitrate salt, a mixture of 60% sodium nitrate and 40% potassium nitrate, must match the operating temperatures of modern Rankine cycle turbines. In a molten salt power tower plant, cold salt at 290°C is pumped from a tank at ground level to the receiver mounted atop a tower where it is heated by concentrated sunlight to 565°C. The salt flows back to ground level into another tank. To generate electricity, hot salt is pumped from the hot tank through a steam generator to make superheated steam. The superheated steam powers a Rankine cycle turbine. A diagram of a molten salt power tower is shown in Figure 19.62. The collector field can be sized to collect more power than is demanded by the steam generator system, and the excess salt is accumulated in the hot storage tank. With this type of storage system, solar power tower plants can be built with annual capacity factors up to 70%. As molten salt has a high energy storage capacity per volume (500–700 kWh/m^3), as shown in Figure 19.63, they are excellent candidates for solar thermal power plants with large capacity factors. Even though nitrate salt has a lower specific heat capacity per volume than carbonates, they still store 250 kWh/m^3. The average heat conductivity of nitrates is 0.52 W/mK and their heat capacity is about 1.6 kJ/kgK. Nitrates are a cheap solution for large storage systems. The cost of the material is \$0.70/kg or \$5.20/kWht. Estimates for large systems including vessels are in the range of \$13/kWht.

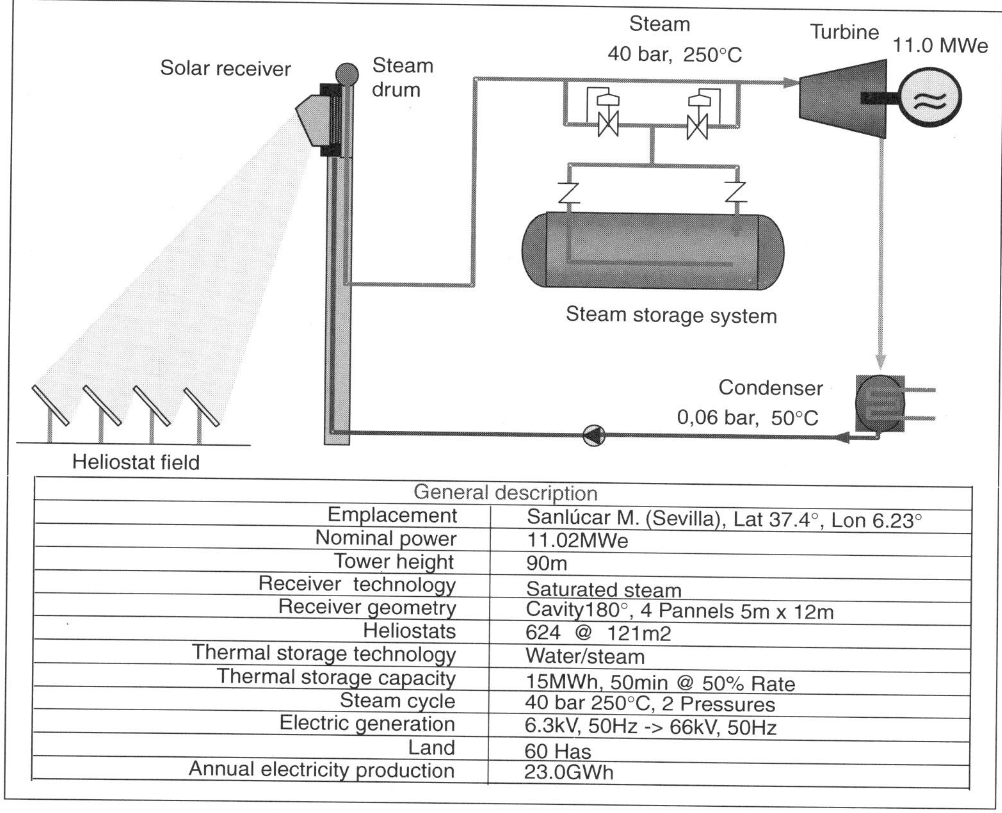

General description	
Emplacement	Sanlúcar M. (Sevilla), Lat 37.4°, Lon 6.23°
Nominal power	11.02MWe
Tower height	90m
Receiver technology	Saturated steam
Receiver geometry	Cavity180°, 4 Pannels 5m x 12m
Heliostats	624 @ 121m2
Thermal storage technology	Water/steam
Thermal storage capacity	15MWh, 50min @ 50% Rate
Steam cycle	40 bar 250°C, 2 Pressures
Electric generation	6.3kV, 50Hz -> 66kV, 50Hz
Land	60 Has
Annual electricity production	23.0GWh

FIGURE 19.60 Basic scheme of PS10 solar thermal power plant with saturated steam as thermal fluid.

Several molten salt development and demonstration experiments have been conducted over the past two and half decades in the U.S.A. and Europe to test entire systems and develop components. The largest demonstration of a molten salt power tower was the Solar Two project, a 10 MW power tower located near Barstow, CA. A picture of Solar Two is shown in Figure 19.64.

TABLE 19.14 Annual Energy Balance for the PS10 Plant at Nominal Conditions

Nominal Rate Operation		
Optical efficiency	77.0%	67.5 MW → 51.9 MW
Receiver and heat handling efficiency	92.0%	51.9 MW → 47.7 MW
Thermal power to storage		11.9 MW
Thermal power to turbine		35.8 MW
Thermal pow. → electric pow. efficiency	30.7%	35.8 MW → 11.0 MW
Total efficiency at nominal rate	21.7%	
Energetical balance in annual basis		
Mean annual optical efficiency	64.0%	148.63 GWh(useful) → 95.12 GWh
Mean annual receiver & heat handling efficiency	90.2%	95.12 GWh → 85.80 GWh
Operational efficiency (starts up/stops)	92.0%	85.80 GWh → 78.94 GWh
Operational efficiency (breakages, O&M)	95.0%	78.94 GWh → 75.00 GWh
Mean annual thermal ener. → electric efficiency	30.6%	75.00 GWh → 23.0 GWh
Total annual efficiency	15.4%	

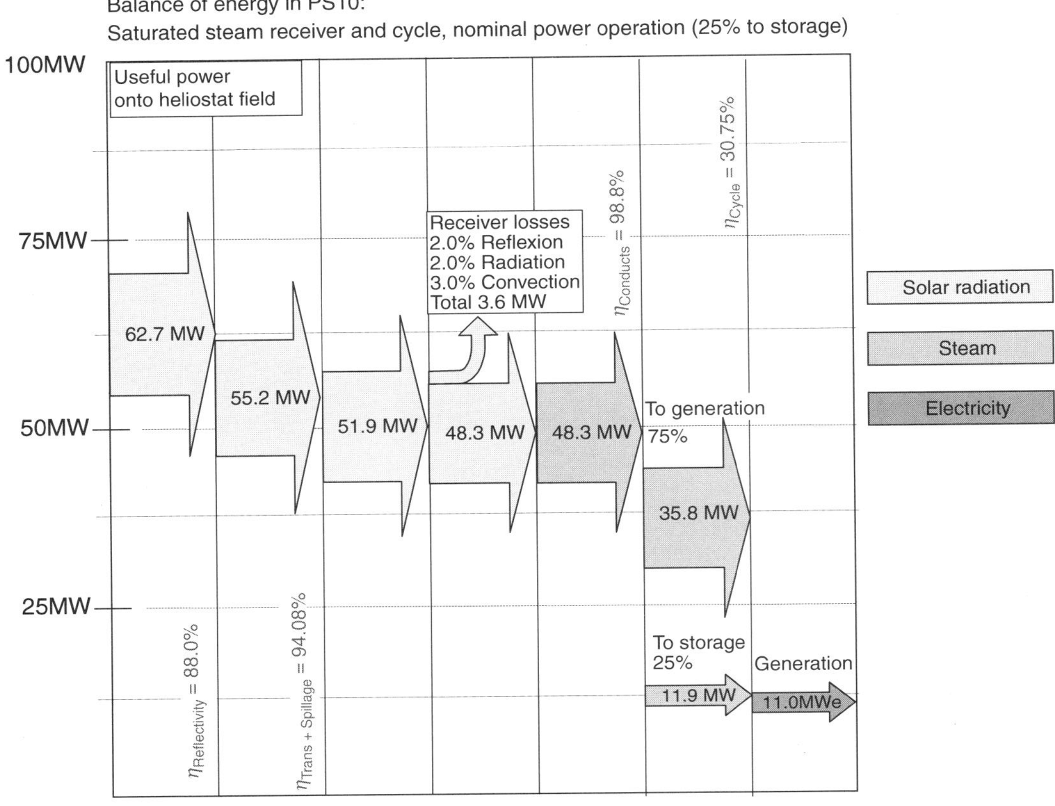

FIGURE 19.61 Energy cascade for PS10 subsystems.

The purpose of the Solar Two project was to validate the technical characteristics of the molten salt receiver, thermal storage, and steam generator technologies, improve the accuracy of economic projections for commercial projects by increasing the capital, operating, and maintenance cost database, and distribute information to utilities and the solar industry to foster a wider interest in the first commercial plants. The Solar Two plant was built at the same site as the Solar One pilot plant and reused much of the hardware including the heliostat collector field, tower structure, 10 MW turbine, and balance of plant. A new, 110 MWh_t two-tank molten salt thermal storage system was installed, as well as a new 42 MW_t receiver, a 35 MW_t steam generator system (535°C, 100 bar), and master control system (Kelly and Singh 1995).

The plant began operating in June 1996. The project successfully demonstrated the potential of nitrate salt technology. Some of the key results were: receiver efficiency was measured at 88%, the thermal storage system had a measured round-trip efficiency of over 97%, and the gross Rankine turbine cycle efficiency was 34%, all of which matched performance projections. The collector field performance was less than predicted, primarily due to the low availability of the heliostats (85–95% versus 98% expected), the degradation of the mirrored surfaces, and poor heliostat canting. Most of the heliostat problems were attributed to the fact that the heliostat field had sat idle and unmaintained for six years between Solar One shutdown and Solar Two startup. The overall peak conversion efficiency of the plant was measured at 13.5%. The plant successfully demonstrated its ability to dispatch electricity independently from solar collection. On one occasion, the plant operated around the clock for 154 h straight (Pacheco et al. 2000). The plant met daily performance projections when

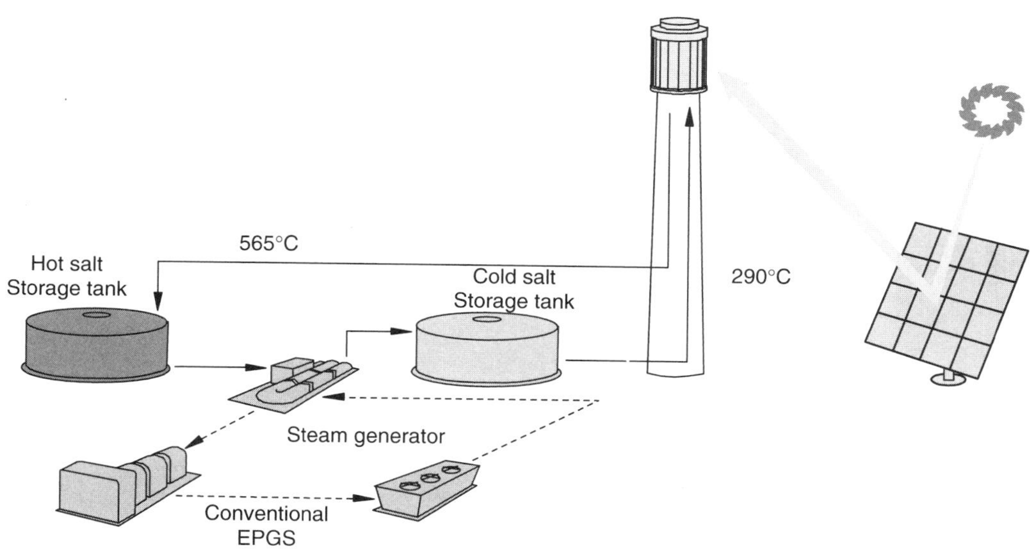

FIGURE 19.62 Schematic of a molten salt central receiver system with cylindrical tubular receiver. (From SunLab, U.S. Department of Energy, USA.)

the actual heliostat availability was accounted for. Although there were some plant start-up issues and it did not run long enough to establish annual performance or refine operating and maintenance procedures, the project identified several areas where the technology could be simplified and its reliability improved. On April 8, 1999, testing and evaluation of this demonstration project was completed and it was shut down.

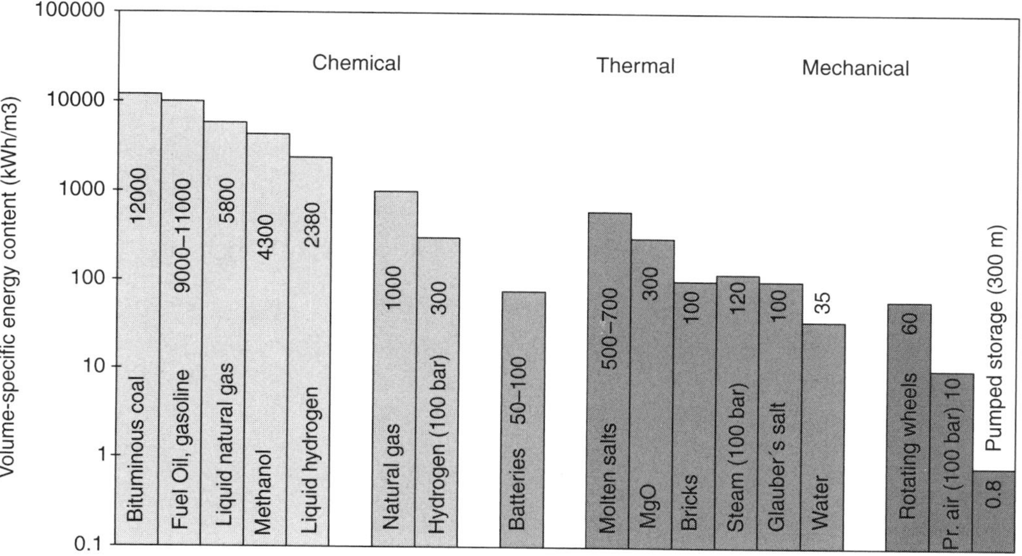

FIGURE 19.63 Comparison of volume energy storage capacity for several chemical, thermal and, mechanical media.

FIGURE 19.64 Aerial view of Solar Two power plant located in Barstow, California. Cylindrical 42-MW receiver appears illuminated. Hot and cold molten storage tanks are located close to the tower.

Because Solar Two was a demonstration project and was quite small by conventional power plant standards, it could not compete economically with fossil-fired power plants without special subsidies. Solar-only commercial power tower plants must be larger to take advantage of economies of scale, more efficient designs, and distribute the costs of the maintenance crew over more energy production.

To reduce the risks associated with scaling-up hardware, the first commercial molten salt power tower should be approximately three times the size of Solar Two (Zavoico et al. 2001). One attempt to prove scaled-up molten salt technology is the Solar Tres project being promoted by the Spanish company, SENER, (Table 19.15) and partially funded by the European Commission.

The investment cost foreseen for the Solar Tres plant is 147 M€. Its design efficiently optimizes levelized electricity cost reduction and required investment. With only 17 MW_e, the plant is predicted to produce 105 GWh_e/year. A large heliostat field of 264,825 m^2 is oversized to supply 15-h equivalent heat storage capacity. The plant is designed to operate around the clock in summertime, leading to an annual capacity factor of 71%. Fossil backup corresponding to 15% of annual production will be added. The levelized energy costs are estimated to be approximately $0.16/kWh.

19.8 Volumetric Atmospheric Receivers: PHOEBUS and Solair

The use of air as the working fluid in CRS plants has been demonstrated since the early 1980s. The selection of air is based on its intrinsic advantages, such as its availability from the ambient, environmentally friendly characteristics, no troublesome phase change, higher working temperatures, easy operation and maintenance, and high dispatchability. Several project initiatives were shifted to air-cooled receivers after the operating problems identified in producing superheated steam in CRS plants like Solar One, CESA-I, or Eurelios. A German–Spanish project called the GAST coordinated by the Interatom and Asinel Companies proposed the construction of a 20 MW plant in Southern Spain using a tube panel air-cooled receiver (Becker and Boehmer 1989). Several plant components, such as heliostats and receiver panels, were tested at the PSA, in particular, a metal tube receiver in 1985–1986, producing 2.45 kg/s hot air at 9.5 bar and 800°C outlet temperature. A second panel with ceramic SiC tubes was

TABLE 19.15 Technical Specifications and Design Performance of the Solar Tres Project

Technical Specifications	
Heliostat field reflectant surface	264,825 m^2
Number heliostats	2750
Land area	142.31 Ha
Receiver thermal power	120 MWt
Tower height	120 m
Heat storage capacity	15 h
Power at turbine	17 MWe
Power NG burner	16 MWt
Operation	
Solar irradiance onto heliostats	2062 kWh/m^2
Electricity dispatched	105,566 MWhe
Production from fossil (annual)	15%
Capacity factor	71%

tested in 1987, with a mass flow rate of 0.48 kg/s at 9.3 bar and 1000°C. The high estimated investment cost and low incident solar flux admitted by the tubes (less than 200 kW/m^2) made it impractical to pursue construction of the plant.

With the advent of volumetric receivers, interest in the air-cooled solar towers was renewed. The relatively large number of volumetric prototypes tested has demonstrated the feasibility of producing hot air at temperatures of 1000°C and upward and with aperture areas similar to those used in molten salt or water/steam receivers (Becker, Cordes, and Böhmer 1992; Hoffschmidt et al. 2001). An average flux of 400 kW/m^2 and peak of 1000 kW/m^2 have been proven, and their low inertia and quick sun-following dispatchability are excellent. Even though long endurance experience is lacking, radiation losses need to be further reduced and air recirculation ratios in open designs should be improved, several plants initiatives have been promoted in the last 15 years.

The first study for a CRS solar plant with an atmospheric air heat transfer circuit was the METAROZ project in the Swiss Alps in the early 1980s (Fricker 1985), followed by a second study performed by the Swiss SOTEL Consortium. These pioneering studies laid the foundations for what was later called the PHOEBUS scheme (Schmitz-Goeb and Keintzel 1997), in which atmospheric air is heated up through a metal wire mesh receiver to temperatures on the order of 700°C and used to produce steam at 480–540°C and 35–140 bar, in a heat recovery steam generator with separate superheater, reheater, evaporator, and economizer sections feeding a Rankine turbine–generator system. The PHOEBUS scheme integrates several equivalent hours of ceramic thermocline thermal storage able to work in charge and discharge modes by reversing the air flow with two axial blowers. Current heat storage capacity restrictions lead to designs with a limited number of hours (between 3 and 6 h maximum), and therefore, for higher annual capacity factors, hybrid designs with burner backup from a duct located in the downcomer of the receiver have been proposed.

In 1986, under a SOTEL-DLR initiative, the study of a 30 MW plant for Jordan was begun. The international PHOEBUS Consortium was formed by companies from Germany, Switzerland, Spain, and the U.S.A., and a feasibility study was completed in March 1990 (Grasse 1991). Unfortunately, the project was unable to acquire the necessary grants and financial support and it was never built. Technology development of key components followed through the German TSA Consortium (Technology Program Solar Air Receiver), under the leadership of the Steinmüller company. A 2.5 MW_t air-receiver facility comprising the complete PHOEBUS power plant cycle that included air recirculation loop, thermal storage, and steam generator was assembled on top of the CESA-1 tower at PSA in Spain at the end of 1991. The plant was successfully operated by DLR and CIEMAT for nearly 400 h between April and

December 1993, and for shorter periods in 1994 and 1999, demonstrating that a receiver outlet temperature of 700°C could easily be achieved within 20 minutes of plant start-up (Haeger 1994).

During the development phase of the PS10 project described above, a feasibility study and conceptual design were performed to analyze the potential use of a 10 MW commercial plant using PHOEBUS-type technology. The results of that study are presently the most reliable information for assessment of the use of open-air volumetric receivers for electricity generation (Romero et al. 2000). A schematic and main design and performance characteristics are summarized in Figure 19.65 and Table 19.16. The left blower regulates air through the receiver and the right blower to the steam generator. Receiver inlet temperature is 110°C and outlet 680°C. The hot air ducts are thermally insulated on the inside permitting the use of lower-cost carbon steel for the ducts. Because of this inner insulation, the air speed may not exceed 33 m/s. Heat storage is not a high technical risk. It is a technology well developed in blast furnaces and other industries, like cement or textiles, for waste heat recovery from stacks. Alumina pellets have demonstrated an excellent thermocline effect and temperature gradient in the 1000 kWht TSA project ceramic checker storage system. The steam generator is a meander-type tube bundle with natural circulation. Steam (10.73 kg/s) will be produced at 460°C and 65 bar. The turbine generator will produce 11 MW_e gross and 10 MW_e net with 30% efficiency.

Parametric material analyses showed that it is possible to design and manufacture a relatively economical storage system, based on an Al_2O_3 ceramic saddle storage core material. The size of the heat storage system has been reduced to half an hour (18 MWh total capacity with a useful storage capacity of 14.4 MWh) by running from storage at a reduced air mass flow rate corresponding to approximately 70% of the nominal value.

Compact designs are required to avoid long air ducts. An example of how the air loop is integrated in the solar tower is shown in Figure 19.66. Though air-cooled open volumetric receivers are a promising way of producing superheated steam, the modest thermal efficiency (74% nominal and 61.4% annual average) must still be improved. At present, all the benefits from using higher outlet temperatures are sacrificed by radiation losses at the receiver, leading to low annual electricity production, so it is clear that volumetric receiver improvements must reduce losses.

19.9 Solar Air Preheating Systems for Combustion Turbines: The SOLGATE Project

Introducing solar energy into the gas turbine of combined cycles (CC) offers significant advantages over other solar hybrid power plant concepts. A very promising way to do it is by solar preheating of the

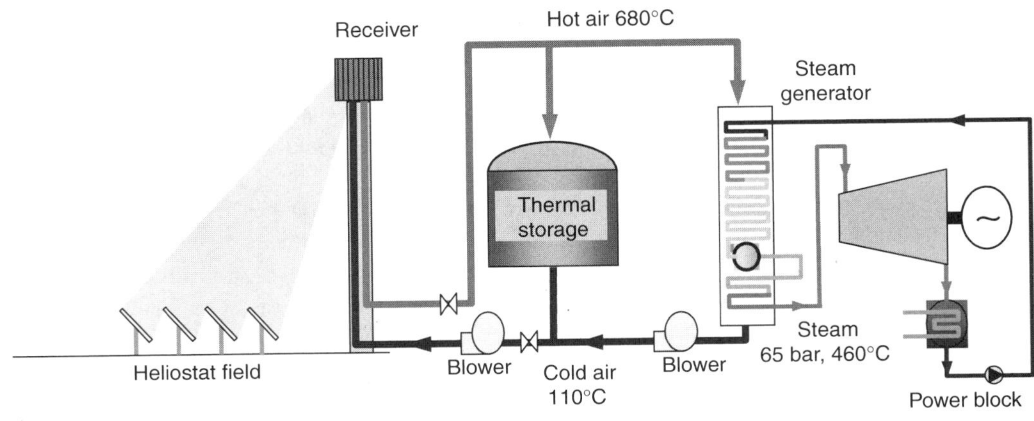

FIGURE 19.65 Process flow diagram of a 10-MW open-air volumetric receiver solar tower power plant.

TABLE 19.16 Design Parameters for a 10-MW Solar Plant

Annual irradiation [kWh/m²]		2063
Design point day		355 (noon)
Design point irradiance [W/m²]		860
Design point power [MWe]		10
Solar multiple/heat storage capacity [MWh]		1.15/18
Tower height [m]		90
Number heliostats/heliostat reflective surface [m²]		981/91
Annual average heliostat reflectivity		0.90
Focal length [m]		500
Receiver shape		Half cylinder
Receiver diameter [m]/receiver height [m]		10.5/10.5
	Design Point	Annual Balance
Power/energy onto reflective surface	75.88 MW	183.50 GWh
Heliostat field optic efficiency	0.729	0.647
Gross power/energy onto receiver	55.27 MW	118.72 GWh
Receiver and air circuit efficiency	0.740	0.614
Power/energy to working fluid	40.92 MW	72.90 GWh
Power/energy to storage	5.34 MW	
Power/energy to turbine	35.58 MW	72.90 GWh
Thermal → Electric Efficiency	0.309	0.303
Gross electric power/energy	11.00 MW	22.09 GWh
Parasitic losses	1.00 MW	2.89 GWh
Net electric power/energy	10.00 MW	19.20 GWh

compressor discharge air before it enters the gas turbine combustor. A diagram of the concept is shown in Figure 19.67.

Solar air preheating offers superior performance, as the solar energy absorbed in the heated air is directly converted with the high CC plant efficiency. For a certain annual solar share, this results in reduced heliostat field size and thus lower overall investment cost for the solar part compared to solar steam generation. Solar air preheating has a high potential for cost reduction of solar thermal power. In

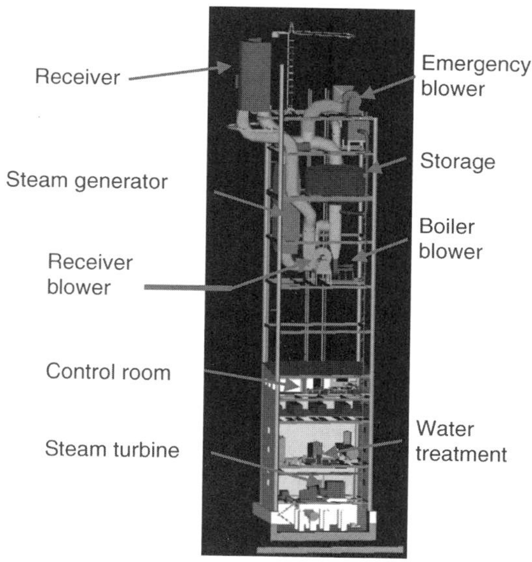

FIGURE 19.66 Compact design of the receiver, air loop, storage, and steam generator for a 10-MW carried out by the German company KAM for Abengoa, during the PS10 feasibility study.

addition, this concept could be applied to a wide range of power levels (1–100 MW_e). At lower power levels, highly efficient heat recovery gas turbine cycles can be used instead of CC. The solar share can be chosen quite flexibly by the receiver outlet temperature, which could be significantly higher than with other hybrid concepts (e.g., ISCCS with parabolic troughs).

Air can be preheated by molten salt solar receivers (up to 560°C) (Price, Whitney, and Beebe 1996) or with pressurized volumetric receivers (Kribus et al. 1997; Buck, Lüpfert and Téllez 2000), in which, due to the limited size of the quartz window, a number of receiver modules are placed on the tower. Each module consists of a pressurized receiver unit with a secondary concentrator in front. The secondary concentrator with a hexagonal aperture (located in the focal plane of the heliostat field) reconcentrates the solar radiation onto the aperture of the pressure vessel which is enclosed by a domed quartz window to maintain pressure. After passing through the window, the radiation is absorbed in the volumetric absorber which transfers the heat by forced convection to the airstream flowing through it. Power is upscaled by installing many modules in a honeycomb formation to cover the entire focal spot. The modules are then interconnected in parallel and to a serial connection.

This was the configuration tested in 2001 in a European project for integrating the receivers with a gas turbine in a complete solar hybrid power system of about 250 kW_e. The project (called SOLGATE) also included further development of the receiver technology, especially to increase the temperature to 1000°C and reduce cost (Buck et al. 2004). The system was erected in the CESA-1 solar tower test facility at the PSA, Spain, at the 60 m level. The test setup consisted of three receiver modules connected in series, a helicopter engine with modified combustor, and generator. In the receiver cluster, the air from the turbine compressor is heated from 290°C to 1000°C with solar energy (Figure 19.68). A bypass lowered the temperature to 800°C at the combustor inlet, due to limitations of the combustor design. The combustor could also be fed with gas. The two-shaft turbine drove the compressor and the generator which was connected to the grid.

For high temperatures, a highly porous SiC ceramic foam absorber with a pore size of 20 ppi was used. The pressure-resistant, domed quartz window was cooled on the atmospheric side by air jets. For the low-temperature receiver the aim was to achieve an overall cost reduction at the first, low-temperature stage of the receiver cluster by employing simple, less expensive modules. The concept selected was a multitube coil attached to a hexagonal secondary concentrator, in which the air was heated convectively while flowing through the tubes. The coiled tubes were very flexible and thus reduced mechanical stresses from thermal expansion of the tube material. The final layout consisted of 16 2.3 m long, 28 mm diameter tubes connected in parallel (Buck et al. 2004). According to the design calculations, the absorber should have a temperature increase of about 200 K and an associated pressure drop of 100 mbar.

After the new ceramic absorber, the window cooling system and the IR scanner had been installed, operation was resumed in June 2003. A maximum receiver temperature of 960°C was achieved. From the experimental data the receiver module efficiencies were derived based on incident flux measurements.

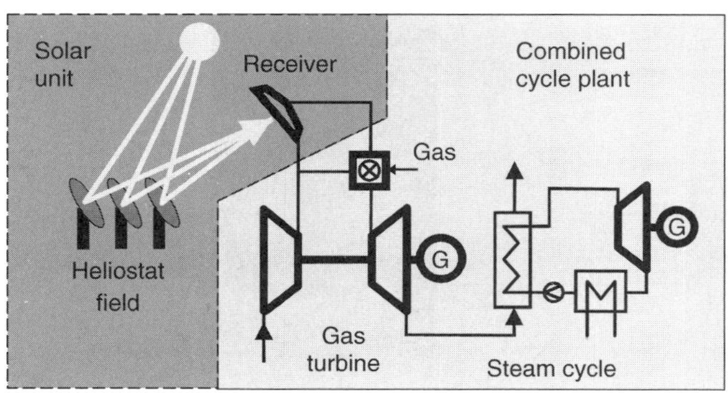

FIGURE 19.67 Solar air preheating system for gas turbines. (From DLR, Germany.)

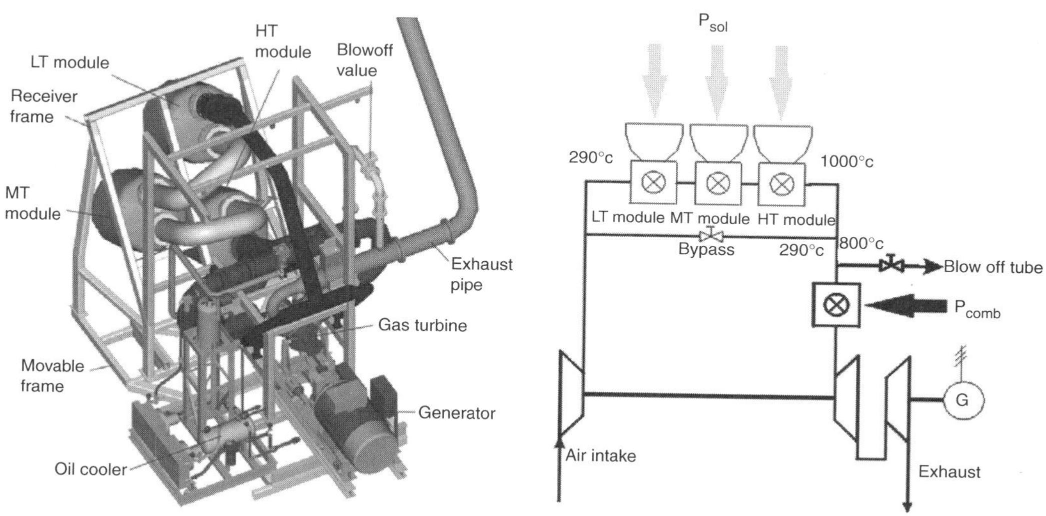

FIGURE 19.68 Test setup used at the SOLGATE Project where three modules of high, medium, and low temperature were used to produce pressurized air at 1000°C.

Measured efficiencies ranged between 68% and 79%. The pressure drop through the receiver cluster was about 120 mbar. The solar fractions reached 70%, and the remaining contribution was from fuel combustion. SOLGATE solarized turbine efficiency, at nominal conditions of 230 kWe, was about 20% (Heller et al. 2004). Gas turbine system performance for design conditions is shown in Figure 19.69. Power production at 960°C was about 190 kW_e (130 kW_e of which was from the solar contribution). The estimated overall thermal efficiency of the receiver cluster was 77% at 800°C and 70% at 950°C.

ConSolar, another similar project, was started by an Israeli consortium in 1995 and was supported by the Israeli Ministry of Industry and Trade. Part of the project was the development of the *reflective tower* concept, intended for solar power plants with introduction of solar energy into gas turbines. This concept foresees an additional reflector at the top of a tower which redirects the concentrated solar radiation to a receiver located on the ground. The receiver is also a modular arrangement of pressurized units with secondary concentrators in front (Kribus et al. 1997). The volumetric absorber is composed of thin convectively cooled ceramic pins. In a 50 kW_t prototype receiver, temperatures up to 1200°C were achieved at pressures up to 20 bar.

Cost predictions indicate potential competitive applications in the green power market with solar LEC of below 10 cent/kWh for an equivalent solar electric power level of 12.6 MW. However, to exploit the full potential of high combined cycle efficiencies (greater than 50%) power levels must be above 50 MW_e. This means a very high investment cost which is not realistic for the introduction of a new technology. At power levels below 10 MW_e, gas turbine systems are mainly used for decentralized power generation with cogeneration of heat or cooling power. First cost assessments for such cogeneration units have indicated potential for solar hybrid gas turbine units (Sugarmen et al. 2003). It is a logical first step to prove the technical–economic feasibility of the technology for cogeneration. This is the goal of the European project SOLHYCO.

19.10 Dish/Stirling Systems

Solar thermal power plants can also be applied to distributed generation through parabolic dishes in which a PCU is attached by an arm directly to the concentrator. Although there have been other modular PCU system initiatives in the past, like the dish/Brayton tested by Cummins and DLR (Buck, Heller, and

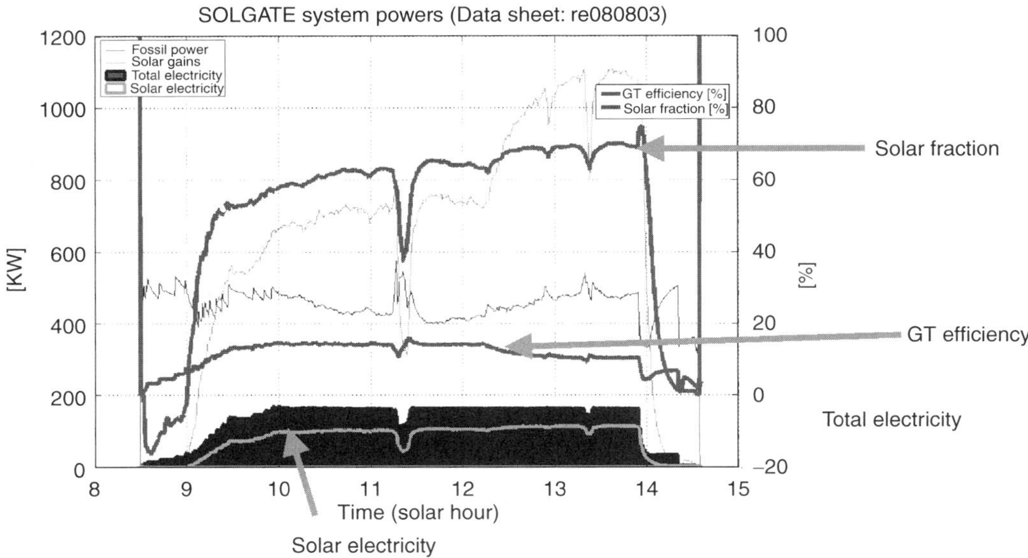

FIGURE 19.69 Example for gas turbine system performance.

Koch 1996) or the use of dish farms to produce superheated steam designed to feed a centralized Rankine cycle in Georgia (Alvis 1984), it is the dish/Stirling system that has demonstrated from the earliest prototypes the highest peak conversion efficiencies of above 30% solar to electric and a daily average of up to 25%. Dish/Stirling systems are considered the most efficient way to convert solar energy into electricity of any large solar power technology. This record is due to their high concentration ratios (up to 3000×) and high working temperatures of above 750°C (Stine and Diver 1994).

19.10.1 System Description

Dish/Stirling systems track the sun and focus solar energy onto a cavity receiver, where it is absorbed and transferred to a heat engine/generator. An electrical generator, directly connected to the crankshaft of the engine, converts the mechanical energy into electricity (AC). To constantly keep the reflected radiation at the focal point during the day, a sun-tracking system continuously rotates the solar concentrator on two axes following the daily path of the sun. With current technologies, a 5 kW_e dish/Stirling system would require 5.5 m diameter concentrator, and for 25 kW_e, the diameter would have to increase up to 10 m. Stirling engines are preferred for these systems because of their high efficiencies (40% thermal-to-mechanical), high power density (40–70 kW/L), and potential for long-term, low-maintenance operation. Dish/Stirling systems are modular, i.e., each system is a self-contained power generator, allowing their assembly in plants ranging in size from a few kilowatts to tens of megawatts (Mancini et al. 2003).

Global efficiency of the system can be defined as:

$$\eta = \eta_C \eta_R \eta_{Stir} \eta_{Gen} = \frac{P}{A_C I}, \quad (19.48)$$

where η_c is the concentrator efficiency, η_R is the receiver efficiency, η_{Stir} is the Stirling engine efficiency, η_{Gen} is the generator efficiency, P is the gross power generated, A_c is the projected concentrator area, and I is the direct normal irradiance.

19.10.1.1 Concentrator

The concentrator is a key element of any dish/Stirling system. The curved reflective surface can be manufactured by attached segments, by individual facets or by a stretched membranes shaped by a continuous plenum. In all cases, the curved surface should be coated or covered by aluminum or silver reflectors. Second-surface glass mirrors, front surface thin glass mirrors or polymer films have been used in various different prototypes.

First-generation parabolic dishes developed in the 1980s were shaped with multiple, spherical mirrors supported by a trussed structure (Lopez and Stone 1992). Though extremely efficient, this structure concept was costly and heavy. Large monolithic reflective surfaces can be obtained by using stretched membranes in which a thin reflective membrane is stretched across a rim or hoop. A second membrane is used to close off the space behind forming a partially evacuated plenum between them, giving the reflective membrane an approximately spherical shape. This concept was developed by the German SBP Company in the 1990s, and several prototypes have been tested at the PSA in Spain (Schiel, Schweizer, Stine 1994b; Schiel et al. 1994a). An alternative to the single stretched membrane in large units is a composition of a number of small circular rings, each with their corresponding membrane (Beninga et al. 1997).

A good example of the latest developments in dish/Stirling systems is the EuroDish prototype (Figure 19.70). The concentrator consists of 12 single segments made of fiberglass resin. When mounted, the segments form an 8.5 m diameter parabolic shell. The shell rim is stiffened by a ring truss to which bearings and the Stirling support structure are later attached. Thin 0.8 mm thick glass mirrors, are glued onto the front of the segments for durable, high reflectivity of around 94% (Keck, Heller, and Weinrebe 2003).

Solar tracking is usually done by two different methods (Adkins 1987):

- Azimuth elevation tracking by an orientation sensor or by calculated coordinates of the sun performed by the local control.
- Polar tracking, where the concentrator rotates about an axis parallel to the earth's axis rotation. The rate of rotation is constant and equal to 15°/h. Declination angle movement is only ± 23.5° per year (0.016°/h) and therefore adjusted from time to time.

19.10.1.2 Receiver

As in central receivers and parabolic trough absorbers, the receiver absorbs the light and transfers the energy as heat to the engine's working gas, usually helium or hydrogen. Thermal fluid working temperatures are between 650°C and 750°C. This temperature strongly influences the efficiency of the engine. Because of the high operating temperatures, radiation losses strongly penalize the efficiency of the receiver; therefore, a cavity design is the optimum solution for this kind of system.

Two different heat transfer methods are commonly used in parabolic dish receivers (Diver 1987). In directly illuminated receivers, the same fluid used inside the engine is externally heated in the receiver through a pipe bundle. Although this is the most conventional method, a good high-pressure, high-velocity, heat transfer gas like helium or hydrogen (Figure 19.71 left) must be used. In indirect receivers, an intermediate fluid is used to decouple solar flux and working temperature from the engine fluid. One such method is heat pipes, which employ a metal capillary wick impregnated with a liquid metal heated up through the receiver plate and vaporized. The vapor then moves across the receiver and condenses in a cooler section, transferring the heat to the engine. The phase change guarantees good temperature control, providing uniform heating of the Stirling engine (Moreno et al. 2001).

19.10.1.3 Stirling Engine

Stirling engines solarized for parabolic dishes are externally heated gas-phase engines in which the working gas is alternately heated and cooled in constant-temperature, constant-volume processes (Figure 19.71 right). This possibility of integrating additional external heat in the engine is what makes it an ideal candidate for solar applications. Because the Stirling cycle is very similar to the Carnot

FIGURE 19.70 Example of segmented parabolic concentrators. Rear and front view of two EuroDish prototype tested at the Plataforma Solar de Almería, Spain.

cycle, the theoretical efficiency is high. High reversibility is achieved since work is supplied to and extracted from the engine at isothermal conditions. The clever use of a regenerator that collects the heat during constant-volume cooling and heating substantially enhances the final system efficiency. For most engine designs, power is extracted kinematically by rotating a crankshaft connected to the pistons by a connecting rod. An example of a kinematic Stirling engine is shown in Figure 19.71-left. Though theoretically, Stirling engines may have a high life cycle projection, the actual fact is that today their availability is still not satisfactory, as an important percentage of operating failures and outages are caused by pistons and moving mechanical components. Availability is therefore one of the key issues, since it must operate for more than 40,000 h in 20-year lifetime, or 10 times more than an automobile engine. One option to improve availability is the use of free-piston designs. Free-piston engines make use of gas or a mechanical spring so that mechanical connections are not required to move reciprocating pistons. Apparently, they are better than kinematic engines in terms of availability and reliability. The most relevant program in developing dishes with free-piston technology was promoted by Cummins in

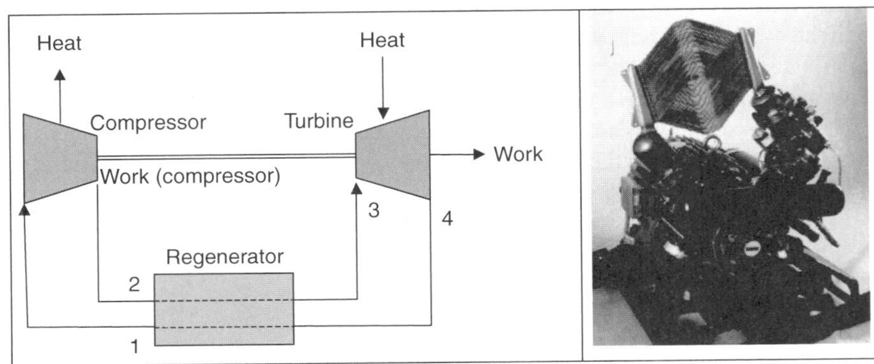

FIGURE 19.71 On the left representation of the ideal Stirling cycle. On the right kinematic Stirling engine V-160 of 10 kWe manufactured by Solo Kleinmotoren situated in a V and connected to a tubular array heat exchanger.

the U.S.A. in 1991. Unfortunately, there were technical problems with the PCU and the project was cancelled (Bean and Diver 1995). Due to the flexibility of the heat source, a Stirling engine can also be operated with a solar/fossil or solar/biomass hybrid receiver (Laing and Trabing 1997), making the system available during cloudy periods and at night.

19.10.2 Dish/Stirling Developments

Like the other CSP technologies, practical dish/Stirling development started in the early 1980s. Most development has concentrated in the U.S.A. and Germany, and though developed for commercial markets they have been tested in a small number of units (Stine and Diver 1994).

The first generation of dishes was a facet-type concentrator with second-surface mirrors that already established concentration records ($C=3000$), and had excellent performances, though their estimated costs for mass production were above \$300/m^2. Their robust structures were extremely heavy, weighing in at 100 kg/m^2 (Grasse, Hertlein, and Winter 1991). The 25 kW Vanguard-1 prototype built by Advanco was operated at Rancho Mirage, California, in the Mojave desert in production mode for 18 months (February 1984 to June 1985) and results were published by EPRI (Droher and Squier 1986).

This system was 10.7 m in diameter with a reflecting surface of 86.7 m^2 and a 25 kW PCU made by United Stirling AB (USAB) model 4-95 Mark II. This engine had four cylinders with a 95 cm^3 cylinder displacement. Cylinders were distributed in parallel and assembled in a square. They were connected to the regenerator and cooler and had double-acting pistons. The working gas was hydrogen at a maximum pressure of 20 MPa and temperature of 720°C. Engine power was controlled by varying the working gas pressure (Schiel 1999). The Advanco/Vanguard system, with a net conversion efficiency (including ancillary systems) of more than 30%, still holds the world's conversion record.

McDonnell Douglas later developed another somewhat improved dish system making use of the same technology and the same engine. The dish was 10.5 m and 25 kW. The 88 m^2 parabolic dish consisted of 82 spherically curved glass facets. Six of these units were produced and installed at sites around the U.S.A. for testing in operation. Southern California Edison continued to evaluate the system later. Reported performances and efficiencies were similar to those of Advanco/Vanguard (Lopez and Stone 1992). The project was frozen for several years until in 1996, Stirling Energy Systems (SES) acquired the intellectual and technology rights to the concentrator and the U.S. manufacturing rights to what is now called the Kockums, 4–95 Stirling engine-based PCU (Mancini et al. 2003). Under a DOE industry cost-sharing project to commercialize the dish/Stirling system for emerging markets, SES started testing and improvement of several units in different locations in the U.S.A. and South Africa. More than 100,000 h of operation accumulated for all the systems have been reported (Stone et al. 2001; Mancini et al. 2003). Daily efficiency has been found to be 24–27% and the annual average 24%, and what is even more important, they have achieved availabilities of 94% at irradiances of just over 300 W/m^2. SES is at present engaged in the promotion of several commercial projects for large plants in some Western states like Nevada and Arizona.

Since the pioneering Vanguard dish records, with the exception of SES, most of the design options have been directed at the development of lowering costs by such strategies as less demanding temperatures, thereby penalizing efficiency, and introducing lighter and less expensive reflectors made of polymers or thin glass glued onto resin-based structures. These dishes, which have a lower optical performance, were first used in non-Stirling applications, with lower operating temperatures, such as the Shenandoah (Solar Kinetics) and Solarplant 1 in Warner Springs (LaJet), (Grasse, Hertlein, and Winter 1991). Typical concentrations were in the range of 600–1000 and working temperatures were in the range of 650°C. Several prototypes were developed by Acurex, Lajet, GE, SKI, and SBP. These developments were followed up in the U.S.A. by SAIC (Mayette, Davenport, and Forristall 2001) and WGA (Diver et al. 2001) under the DOE industry R&D program, Dish Engine Critical Components—DECC (Mancini et al. 2003).

The most extensive testing of this light material concept has been done with the stretched-membrane concentrator developed in Germany by Schalich, Bergermann und Partner (SBP). More than 50,000 h of testing have been accumulated in the six-prototype field, promoted by SBP and Steinmüller, and evaluated at the PSA in Spain (Schiel, Schweizer, Stine 1994b; Schiel et al. 1994a). The concentrator is a single 7.5 m-diameter facet made of a single 0.23 mm thick preformed stainless steel stretched membrane. Thin glass mirrors are bonded to the stainless steel membrane. The membrane is pre-stretched beyond its elastic limit using a combination of the weight of water on the front and vacuum on the back, to form a nearly ideal paraboloid. Then, a slight active vacuum within the membrane drum preserves the optical shape. The V-160 engine, originally produced by Stirling Power Systems, is at present manufactured by the German company Solo Kleinmotoren (Figure 19.71). The engine sweeps 160 cm^3 of helium with two pistons. The engine has an efficiency of 30% and reported overall conversion efficiency of 20.3% (Figure 19.72). The figure demonstrates the high dispatchability of dish/Stirling systems at part loads.

Though stretched membranes had excellent optical results, the economics revealed production costs higher than expected. The successor of the SBP membrane dishes is the EuroDish system. The EuroDish project is a joint venture undertaken by the European Community, German/Spanish industry (SBP, MERO, Klein + Stekl, Inabensa), and research institutions DLR and CIEMAT. The new design replaces the stretched-membrane concentrator with a glass–fiber composite shell onto which glass mirrors are bonded with an adhesive. The engine used in the EuroDish is the next generation SOLO Kleinmotoren 161. Two new 10 kW EuroDish units, shown in Figure 19.70, were installed at the Plataforma Solar de Almeria, Spain, early in 2001 for testing and demonstration. In a follow-up project called EnviroDish, additional units were deployed in France, India, Italy, and Spain to accumulate operating experience at different sites. The peak solar-to-net-electric energy conversion efficiency of the system is expected to be 21–22%, based on the experience of former projects with the same engine. The peak system efficiency was first measured at 20%. The estimated annual production of a EuroDish system operating in Albuquerque, New Mexico, is 20,252 kWh of electricity with 90% availability and an annual efficiency of 15.7% (Mancini et al. 2003). SBP and the associated EuroDish industry have performed cost estimates for a yearly production rate of 500 units per year (5 MW/year) and 5000 units per year, which corresponds to 50 MW/year. The actual cost of the 10 kW unit without transportation and installation cost and excluding foundations is approximately U.S.$10,000/kW. The cost projections at production rates of 500 and 5000 units per year are U.S.$2,500/kW and U.S.$1,500/kW, respectively.

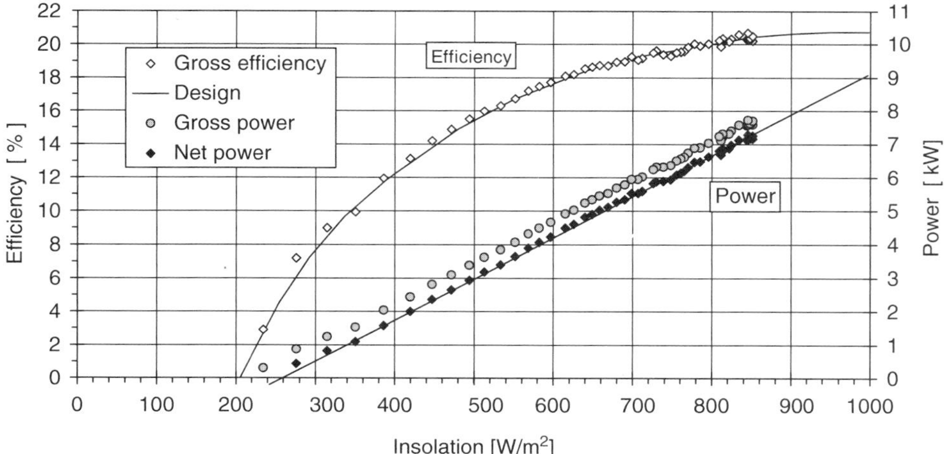

FIGURE 19.72 Input/output parameters and performance measured for a 9-kW SBP stretched-membrane dish/Stirling system tested in Almería, Spain (DISTAL).

19.11 Market Opportunities

World energy outlooks issued by international organizations and government departments, such as the WEC (World Energy Council), the IEA (International Energy Agency), the U.S.-DOE (U.S. Department of Energy), and the EC (European Commission) anticipate steady growth of renewables for the period 2000–2030. In spite of this development, and because initial levels were very low, electricity from renewable sources is not expected to exceed 3% of the world total electricity production in 2030 (16% if large hydro and geothermal plants are included). This is, for example, the EC WETO (World Energy, Technology and Climate Policy Outlook) prediction published in 2003 (European Commission 2003). Solar technologies play a very small role according to these general studies and their share will be very limited. CSP technologies have been poorly publicized, and the absence of commercial references leads to the real fact that most of these energy predictions simply obviate solar thermal power as a credible option for the next two decades. It is, therefore, in more renewable-oriented studies and technical roadmaps elaborated in close cooperation with solar engineers where detailed information about potential deployment of CSP plants can be found. Studies sponsored by the EPRI (DeMeo and Galdo 1997), the World Bank (Enermodal 1999), the International Solar Energy Society—ISES (Aitken 2003), IEA-SolarPACES/Greenpeace/ESTIA (Aringhoff et al. 2005), EUREC—European Renewable Energy Centers Agency (Becker et al. 2002), SunLab in U.S.A. (Sargent & Lundy 2003), UNDP (Turkenburg 2000), and IEA (IEA 2003) consider the implementation of 2–8 GW by the year 2010 highly feasible, rising to between 20–45 GW by 2020. The United Nations Development Program (UNDP), United Nations Department of Economic and Social Affairs (UNDESA), and World Energy Council (WEC) published the World Energy Assessment (WEA) in 2000. The WEA forecasts for beyond 2020 are for a growth rate of 20–25% after 2010, and an average 15% a year after 2020, which would result in 800–1200 GW of CSP electricity by 2050. The Cost Reduction Study for Solar Thermal Power Plants, prepared for the World Bank in early 1999 concludes that the large potential market of CSP could reach an annual installation rate of 2000 MW of electricity (Enermodal 1999) between 2015 and 2020.

The need to find global joint strategies for removal of the nontechnological barriers to concentrating solar power and to devise a bold and effective plan to greatly expand the international market for it in an accelerated time frame were the subject of two international executive conferences (June 2002 in Berlin, Germany, and October 2003 in Palm Spring, U.S.A.). The result of those conferences, attended by senior executives from the energy, financial, and policy sectors of many countries, was the definition and launching of the global market initiative (GMI), a coordinated global market initiative for concentrating solar power (CSP) aimed at erecting 5000 MW of large-scale CSP power projects in prime areas around the world on an aggressive timescale (Geyer 2002).

CSP technologies are capable of meeting the requirements of two major electric power markets, large-scale dispatchable markets comprised of grid-connected peaking and base load power, and rapidly expanding distributed markets, including both on-grid and remote/off-grid applications. Dispatchable markets, where power must be produced on demand, can be served with large trough and tower plants using storage and hybridization. The most appropriate CSP technology for distributed applications is the dish/engine system, however, its marketing requires that the price of electricity drop to below 12¢/kWh in industrialized countries and to below 30¢/kWh in developing countries, for which capital costs of $1–2/W (IC) and $4/W (DC) and about 100 MW installed capacity would be required. To market PT/CRS, the electricity price would have to drop below 8¢/kWh (low enough to compete in large-scale peaking and green markets), capital costs would have to be 2–4 times lower and 1000 MW installed capacity would be required (Morse, 2000).

The reduction in electricity production costs should be a consequence, not only of mass production, but also of scaling-up and R&D (Pitz-Paal et al. 2005a, 2005b). According to the most recent parabolic trough analyses conducted at SunLab, the cost reduction projected for 2020 will be 37% from plant scale-up, 42% from technology development and 21% from mass production (Jones 2003).

19.12 Conclusions

Solar thermal power plants are excellent candidates for supply of a significant share of solar bulk electricity by the year 2020. Their strong point is their flexibility for adapting to both dispatchable and distributed markets. A near-term target of 5000 MW of grid-connected systems is considered feasible according to the global market initiative (GMI) strategy, which will significantly influence the competitiveness of STPP technologies. The portfolio of concentrating solar power technologies includes mature technologies, like the market-ready parabolic troughs, technologies like solar towers and linear Fresnel systems, which are ready to start up in early 10–30 MW commercial/demonstration plants, and finally dish-engine systems, which have high conversion efficiencies, but have still only been tested in a few 5–25 kW prototypes.

Ten years from now, STPP may already have reduced production costs to ranges competitive with intermediate-load fossil power plants. An important portion of this reduction (up to 42%) will be from R&D and technology advances in materials and components, efficient integration with thermodynamic cycles, highly automated control, and low-cost heat storage systems.

References

Adkins, D. R. 1987. Control Strategies and Hardware Used in Solar Thermal Applications, SAND86-1943. Sandia National Laboratories, Albuquerque, NM.

Aitken, D. W. 2003. *White Paper: Transitioning to a Renewable Energy Future.* International Solar Energy Society (ISES), Freiburg, Germany.

Ajona, J. I. 1999. Electricity generation with distributed collector systems. *Solar Thermal Electricity Generation. Lectures from the Summer School at the Plataforma Solar de Almería*, pp. 7–77, CIEMAT, Madrid.

Ajona, J. I. and Zarza, E. 1994. Benefits potential of electricity production with direct steam generation in parabolic troughs, *Proceedings of the 7th International Symposium on Solar Thermal Concentrating Technologies*, O. Popel, S. Fris, and E. Shchedrova, eds., pp. 300–314. Institute for High Temperature of Russian Academy of Science, Moscú.

Alpert, D. J. and Houser, R. M. 1990. Performance evaluation of large-area glass-mirror heliostats. *Libro: Research, Development and Applications of Solar Thermal Technology*, B. P. Gupta and W. H. Traugott, eds., pp. 91–100. Hemisphere, New York.

Alvis, R. L. 1984. *Some Solar Dish/Heat Engine Design consideration*, Technical Report SAND84-1698, Sandia National Laboratories, Albuquerque, NM.

Aringhoff, R., Geyer, M., Herrmann, U., Kistner, R., Nava, P., and Osuna, R. 2002. AndaSol-50 MW solar plants with 9 hour storage for southern Spain. In *Proceedings of the 11th SolarPACES International Symposium on Concentrated Solar Power and Chemical Energy Technologies*, A. Steinfeld, ed., pp. 37–42.

Aringhoff, R., Geyer, M., Herrmann, U., Kistner, R., Nava, P., and Osuna, R. 2002. 50 MW solar plants with 9 hour storage for southern Spain. *Proceedings of the 11th International Symposium on Concentrating Solar Power and Chemical Energy Technologies.*

Aringhoff, R., Aubrey, C., Brakmann, G., and Teske, S. 2003. *Solar Thermal Power 2020.* Greenpeace International/European Solar Thermal Power Industry Association, Amsterdam/Birmingham.

Aringhoff, R., Brakmann, G., Geyer, M., and Teske, S. 2005. *Concentrated Solar Thermal Power-Now!.* Greenpeace, ESTIA and IEA/SolarPACES.

Bean, J. R. and Diver, R. B. 1995. Technical status of the dish/stirling joint venture program. *Proceedings of the 30th IECEC*, 2.497-2.504. Report SAND95-1082C, Sandia National Laboratories, Albuquerque, NM.

Becker, M. and Boehmer, M. 1989. *GAST: The Gas Cooled Solar Tower Technology Program. Proceedings of the Final Presentation, Springer, Berlin.*

Becker, M. and Vant-Hull, L. L. 1991. Thermal receivers, *Solar Power Plants*, C. J. Winter, R. L. Sizmann, and L. L. Vant-Hull, eds., pp. 163–197. Springer, Berlin.

Becker, M., Böhmer, M., Meinecke, W., and Unger, E. 1989. *Volumetric Receiver Evaluation*. SSPS Technical Report Number 3/89.DLR, Cologne.

Becker, M., Cordes, S., and Böhmer, M. 1992. The development of open volumetric receivers. *Proceedings of the 6th International Symposium on Solar Thermal Concentrating Technology, Vol. 2*, pp. 945–952, CIEMAT, Madrid.

Becker M., Meinecke, W., Geyer, M., Trieb, F., Blanco, M., Romero, M., and Ferriere, A. 2002. Solar thermal power plants. In *The Future for renewable Energy 2: Prospects and Directions*, pp. 115–137, Eurec Agency, James & James Science Publishers, London.

Beninga, K., Davenport, R., Sellars, J., Smith, D., and Johansson, S. 1997. Performance results for the SAIC/STM prototype dish/Stirling system. *ASME International Solar Energy Conference*, pp. 77–80.

Biggs, F. and Vittitoe, C. 1979. The helios model for the optical behavior of reflecting solar concentrators. Report SAND76-0347.

Blake, F. A., Gorman, D. N., and McDowell, J. H 1985. *ARCO Power Systems Littleton, CO. ARCO Central Receiver Solar Thermal Enhanced Oil Recovery Project.*

Buck, R., Heller, P., and Koch, H. 1996. Receiver development for a Dish-Brayton system. In *Solar Engineering*, ASME 1996, pp. 91–96.

Buck, R., Lüpfert, E., and Téllez, F. 2000. Receiver for solar-hybrid gas turbine and CC ystems (REFOS). In *Proceedings, 10th SolarPACES International Symposium 'Solar Thermal 2000'*, pp. 95–100.

Buck, R., Bräuning, T., Denk, T., Pfänder, M., Schwarzbözl, P., and Tellez, F. 2002. Solar-hybrid gas turbine-based power tower systems (REFOS). *Journal of Solar Energy Engineering*, 124, 2–9.

Buck, R., Heller, P., Schwarzbozl, P., Sugarmen, C., Ring, A., Téllez, F., and Enrile, J. 2004. Solar hybrid gas turbine plants: Status and perspective. *Proceedings EuroSun2004, Vol.* 1, pp. 822–831.

Chavez, J. M., Kolb, G. J., and Meinecke, W. 1993. *Second Generation Central Receiver Technologies—A Status Report*, M. Becker and P. C. Klimas, eds., pp. 363–367. Verlag C.F. Müller, Karlsruhe, Germany.

Chisholms, D. 1980. Two-phase flow in bends. *International Journal of Multiphase Flow*, 6, 363–367.

Cohen, G. E., Kearney, D. W., and Kolb, G. J., 1999. Final report on the operation and maintenance improvement program for concentrating solar power plants, SAND99-1290, Sandia National Laboratories, Albuquerque, NM.

DeMeo, E. A. and Galdo, J. F. 1997. Renewable energy technology characterizations, TR-109496 Topical Report, U.S. Department of Energy, Washington, DC.

Diver, R. B. 1987. *Journal of Solar Energy Engineering*, 109 (3), 199–204.

Diver, R., Andraka, C., Rawlinson, K., Thomas, G., and Goldberg, V. 2001. The advanced dish development system project, *Proceedings of Solar Forum 2001 Solar Energy: The Power to Choose*, S. J. Kleis and C. E. Bingham, eds., pp. 341–351. ASME (CD-Rom), New York.

Droher, J. J. and Squier, S. E. 1986. Performance of the Vanguard solar dish-Stirling engine module, Technical Report EPRI AP-4608, Electric Power Research Institute, Palo Alto, CA.

Eck, M. and Zarza, E. 2002. Assessment of operation modes for direct solar steam generation in parabolic troughs. In *Proceedings of the 11th SolarPACES International Symposium on Concentrated Solar Power and Chemical Energy Technologies*, A. Steinfel, ed., pp. 591–598.

Eck, M., Zarza, E., Eickhoff, M., Rheinländer, J., and Valenzuela, L. 2003. Applied research concerning the direct steam generation in parabolic troughs. *Solar Energy.*, 74 341–351.

Eneas, A. 1901. U.S. Patent 670,917.

Enermodal Engineering, Ltd. 1999. Cost reduction study for solar thermal power plants. Final report.

Entropie. 1982. Centrales à tour: Conversion thermodynamique de l'energie solaire. *Entropie*, 103 (Special Issue), 1–115.

Epstein, M., Liebermann, D., Rosh, M., and Shor, A. J. 1991. Solar testing of 2 MW (th) water/steam receiver at the Weidman Institute solar tower. *Solar Energy Materials*, 24, 265–278.

Ericsson, J. 1888. The sun motor. *Nature*, 38, 321.

European Commission 2003. *World Energy, Technology and Climate Policy Outlook 2030-WETO. Directorate-General for Research; Energy. EUR 20366,* Office for Official Publications of the European Communities, Luxembourg.

Falcone, P. K. 1986. *A Handbook for Solar Central Receiver Design. SAND86-8009,* Sandia National Laboratories, Livermore, CA.

Francia, G. 1968. Pilot plants of solar steam generation systems. *Solar Energy,* 12, 51–64.

Fricker, H. 1985. Studie über die möglichkeiten eines alpenkraftwerkes. *Bulletin SEV/VSE,* 76, 10–16.

Fricker, H. W., Silva, M., García, C., Winkler, C., and Chavez, J. 1988. Design and test results of the wire receiver experiment Almeria, *Solar Thermal Technology—Proceedings of the 4th International Workshop,* B. K. Gupta, ed., pp. 265–277. Hemisphere Publishing, New York.

Garcia, G., Egea, A., and Romero, M. 2004. Performance evaluation of the first solar tower operating with autonomous heliostats: PCHA project. In *Proceedings of the 12th SolarPACES International Symposium,* C. Ramos and J. Huacuz, eds.

García-Martín, F. J., Berenguel, M., Valverde, A., and Camacho, E. F. 1999. Heuristic knowledge-based heliostat field control for the optimization of the temperature distribution in a volumetric receiver. *Solar Energy,* 66 (5), 355–369.

Geyer, M. 2002. Panel 1 briefing material on status of major project opportunities. The current situation, issues, barriers and planned solutions. In *International Executive Conference on Expanding the Market for Concentrating Solar Power (CSP)—Moving Opportunities into Projects.*

Geyer, M., Romero, M., Steinfeld, A., and Quaschning, V. 2003. *International Energy Agency–Solar Power and Chemical Energy Systems. Annual Report 2002,* DLR, ed., http://www.solarpaces.org (accessed).

González, L., Zarza, E., and Yebra, L. 2001. Determinación del Modificador por Angulo de Incidencia de un colector solar LS-3, incluyendo las pérdidas geométricas por final de colector. Technical report DISS-SC-SF-30, Plataforma Solar de Almería, Almería, Spain.

Grasse, W. 1991. PHOEBUS- international 30 MWe solar tower plant. *Solar Energy Materials,* 24, 82–94.

Grasse, W., Hertlein, H. P., and Winter, C. J. 1991. Thermal solar power plants experience, *Solar Power Plants,* C. J. Winter, R. L. Sizmann, and L. L. Vant-Hull, eds., pp. 215–282. Springer, Berlin.

Haeger, M. 1994. Phoebus technology program: Solar air receiver (TSA), PSA Technical Report PSA-TR02/94.

Harats, Y. and Kearney, D. 1989. Advances in parabolic trough technology in the SEGS plants, *Proceedings of the 11th Annual American Society of Mechanical Engineers Solar Energy Conference,* A. H. Fanney and K. O. Lund, eds., pp. 471–476. American Society of Mechanical Engineers, New York.

Heller, P., Pfaender, M., Denk, T., Téllez, F., Valverde, A., Fernández, J., and Ring, A. 2004. Test and evaluation of a solar powered gas turbine system. In *Proceedings 12th SolarPACES International Symposium,* C. Ramos and J. Huacuz, eds.

Hoffschmidt, B., Pitz-Paal, R., Böhmer, M., Fend, T., and Rietbrock, P. 1999. 200 kWth open volumetric air receiver (HiTRec) of DLR reached 1000°C average outlet temperature at PSA. *Journal of physics,* 9 Pr3-551–Pr3-556. IV France.

Hoffschmidt, B., Fernández, V., Konstandopoulos, A. G., Mavroidis, I., Romero, M., Stobbe, P., and Téllez, F. 2001. Development of ceramic volumetric receiver technology, *Proceedings of 5th Cologne Solar Symposium,* K.-H. Funken and W. Bucher, eds., pp. 51–61. Forschungsbericht, Cologne, Germany.

Hoffschmidt, B., Fernandez, V., Pitz-Paal, R., Romero, M., Stobbe, P., and Téllez, F. 2002. The development strategy of the HitRec volumetric receiver technology—Up-scaling from 200 kWth via 3 MWth up to 10 MWel. *11th SolarPACES International Symposium on Concentrated Solar Power and Chemical Energy Technologies,* pp. 117–126.

Hoffschmidt, B., Téllez, F. M., Valverde, A., Fernández-Reche, J., and Fernández, V. 2003. Performance evaluation of the 200-kWth HiTRec-II open volumetric air receiver. *Journal of Solar Energy Engineering,* 125 87–94.

Hu, E. J., Mills, D. R., Morrison, G. L., and le Lievre, P. 2003. Solar power boosting of fossil fuelled power plants. *Proceedings ISES Solar World Congress 2003.*

IEA 2003. *Renewables for Power Generation. Status and Prospects IEA Publications, Paris.*

Jones, S. 2003. U.S. trough and tower systems, *Proceedings of SolarPACES Tasks Meetings*, M. Romero, A. Steinfeld, and V. Quaschning, eds., pp. 217–224. CIEMAT, Madrid, Spain.

Kearney, D. W. and Cohen, G. E. 1997. Current experiences with the SEGS parabolic trough plants, In *Proceedings of the 8th International Symposium on Solar Thermal Concentrating Technologies, Vol. 1, Becker and M. Böhmer, eds, pp. 217–224.* C. F. Möller, Heidelberg, Alemania.

Keck, T., Heller, P., and Weinrebe, G. 2003. Envirodish and EuroDish—system and status. In *Proceedings of ISES Solar World Congress.*

Kelly, B. and Singh, M. 1995. Summary of the final design for the 10 MWe solar two central receiver project. *Solar Engineering*, 1, 575.

Kistler, B. L. 1986. A user's manual for DELSOL3: A computer code for calculating the optical performance and optimal system design for solar thermal central receiver plants, Sandia Report Sand-86-8018, Sandia National Laboratories, Livermore, CA.

Kodama, T. 2003. High-temperature solar chemistry for converting solar heat to chemical fuels. *Progress in Energy and Combustion Science*, 29, 567–597.

Kolb, G. J. 1998. Economic evaluation of solar-only and hybrid power towers using molten-salt technology. *Solar Energy*, 62, 51–61.

Kribus, A. 1999. Future directions in solar thermal electricity generation. *Solar Thermal Electricity Generation. Colección documentos CIEMAT*, pp. 251–285. CIEMAT, Madrid, Spain.

Kribus, A., Zaibel, R., Carrey, D., Segal, A., and Karni, J. 1997. A solar-driven combined cycle power plant. *Solar Energy*, 62, 121–129.

Lipps, F. W. and Vant-Hull, L. L. 1978. A cellwise method for the optimization of large central receiver systems. *Solar Energy*, 20, 505–516.

Laing, D. and Trabing, C. 1997. Second generation sodium heat pipe receivers for USAB V-160 Stirling engine: Evaluation of on-Sun test results using the proposed IEA guidelines and analysis of heat pipe damage. *Journal of Solar Energy Engineering*, 119 (4), 279–285.

Lopez, C. and Stone, K. 1992. Design and performance of the Southern California Edison Stirling dish. *Solar Engineering, Proceedings of ASME International Solar Energy Conference*, pp. 945–952.

Lotker, M. 1991. *Barriers to Commercialization of Large-Scale Solar Electricity: Lessons Learned from LUZ experience.* Informe Técnico SAND91-7014. *Sandia National Laboratories, Albuquerque, NM.*

Luepfert, E., Zarza, E., Schiel, W., Osuna, R., Esteban, A., Geyer, M., Nava, P., Langenkamp, J., and Mandelberg, E. 2003. Collector qualification complete—Performance test results from PSA. *Proceedings of the ISES 2003 Solar World Congress.*

Mancini, T., Kolb, G. J., and Prairie, M. 1997. Solar thermal power, *Advances in Solar Energy: An Annual Review of Research and Development, Vol. 11, K. W. Boer, ed.,*, pp. 135–151. American Solar Energy Society, Boulder, CO.

Mancini, T., Heller, P., Butler, B., Osborn, B., Schiel, W., Goldberg, V., Buck, R., Diver, R., Andraka, C., and Moreno, J. 2003. Dish-Stirling systems: An overview of development and status. *International Journal of Solar Energy Engineering*, 125, 135–151.

Marcos, M. J., Romero, M., and Palero, S. 2004. Analysis of air return alternatives for CRS-type open volumetric receiver. *Energy*, 29 677–686.

Mavis, C. L. 1989. *A Description and Assessment of Heliostat Technology*, SAND87-8025, Sandia National Laboratories, Enero.

Mayette, J., Davenport, R., and Forristall, R. 2001. The salt river project sundish dish-stirling system, *Proceedings of Solar Forum 2001 Solar Energy: The Power to Choose*, S. J. Kleis and C. E. Bingham, eds., pp. 19–31. ASME (CD-Rom), New York.

Mills, D. 2004. Advances in solar thermal electricity technology. *Solar Energy*, 76, 19–31.

Mills, D. R. and Morrison, L. G. 2000. Compact linear Fresnel reflector solar thermal power plants. *Solar Energy*, 68, 263–283.

Mills, D., Le Lièvre, P., and Morrison, G. L., Lower temperature approach for very large solar power plants. *Proceedings of the 9th Solar PACES, Symposium*, IEA/Solar PACES International Energy Agency, Paris, France.

Monterreal, R., Romero, M., García, G., and Barrera, G. 1997. Development and testing of a 100 m^2 glass-metal heliostat with a new local control system, *Solar Engineering 1997*, D. E. Claridge and J. E. Pacheco, eds., pp. 251–259. ASME, New York.

Moreno, J. B., Modesto-Beato, M., Rawlinson, K. S., Andraka, C. E., Showalter, S. K., Moss, T. A., Mehos, M., and Baturkin, V. 2001. Recent progress in heat-pipe solar receivers. *SAND2001-1079, 36th Intersociety Energy Conversion Engineering Conference*, pp. 565–572.

Morse, F. H. 2000. The commercial path forward for concentrating solar power technologies: A review of existing treatments of current and future markets.

Nitsch, J. and Krewitt, W. 2004. *Encyclopedia of Energy., Vol. 5*, Elsevier, San Diego, CA.

Osuna, R., Cerón, F., Romero, M., and García, G. 1999. Desarrollo de un prototipo de helióstato para la planta colón solar. *Energía*, 25 (6), 71–79.

Osuna, R., Fernández, V., Romero, S., Romero, M., and Sanchez, M. 2004. PS10: A 11-MW solar tower power plant with saturated steam receiver. In *Proceedings 12th SolarPACES International Symposium*, C. Ramos and J. Huacuz, eds.

Pacheco, J. E. and Gilbert, R. 1999. Overview of recent results of the Solar Two test and evaluations program. In *Renewable and advanced energy systems for the 21st century RAES'99*, R., Hogan, Y., Kim, S., Kleis, D., O'Neal, and T. Tanaka, eds., RAES99-7731. ASME, New York.

Pacheco, J. E., Reilly, H. E., Kolb, G. J., and Tyner, C. E. 2000. Summary of the solar two test and evaluation program. *Proceeding of the Renewable Energy for the New Millennium*, pp. 1–11.

Philibert, C. 2004. International energy technology collaboration and climate change mitigation, case study 1: Concentrating solar power technologies. OECD/IEA Information Paper.

Pifre, A. 1882. A solar printing press. *Nature*, 21, 503–504.

Pitz-Paal, R., Hoffschmidt, B., Böhmer, M., and Becker, M. 1996. Experimental and numerical evaluation of the performance and flow stability of different types of open volumetric absorbers under non-homogeneous irradiation. *Solar Energy*, 60 (3/4), 135–159.

Pitz-Paal, R., Dersch, J., Milow, B., Ferriere, A., Romero, M., Téllez, F., Zarza, E., Steinfeld, A., Langnickel, U., Shpilrain, E., Popel, O., Epstein, M., and Karni, J. 2005. ECOSTAR roadmap document for the European Commission; SES-CT-2003-502578. R. Pitz-Paal, J. Dersch and B. Milow eds., Deutsches Zentrum für Luft- und Raumfahrt e.V, Cologne, Germany.

Pitz-Paal, R., Dersch, J., Milow, B., Téllez, F., Zarza, E., Ferriere, A., Langnickel, U., Steinfel, A., Karni, J., and Popel, O. 2005. A European roadmap for concentrating solar power technologies (ECOSTAR). In *Proceedings submitted to ISEC2005, 2005 International Solar Energy Conference.*

Price, H. W., Whitney, D. D., and Beebe, H. I. 1996. SMUD Kokhala power tower study. *Proceedings of the 1996 International Solar Energy Conference.*

Price, H., Luepfert, E., Kearney, D., Zarza, E., Cohen, G., Gee, R., and Mahoney, R. 2002. Advances in parabolic trough solar power technology. *International Journal of Solar Energy Engineering*, 124 109–125.

Pye, J. D., Morrison, G. L., and Behnia, M. 2003. Transient modeling of cavity receiver heat transfer for the compact linear Fresnel reflector. *ANZSES Annual Conference*, pp. 69–78.

Rabl, A. 1985. *Active Solar Collectors and their Applications,* Oxford University Press, New York.

Radosevich, L. G. and Skinrood, C. A. 1989. The power production operation of solar one, the 10 MWe solar thermal central receiver pilot plant. *Journal of Solar Energy Engineering*, 111 144–151.

Ratzel, C. A. and Simpson, C. E. 1979. Heat loss reduction techniques for annular receiver design. Informe técnico SAND78-1769, Sandia National Laboratories, Albuquerque, NM.

Romero, M., Conejero, E., and Sánchez, M. 1991. Recent experiences on reflectant module components for innovative heliostats. *Solar Energy Materials*, 24, 320–332.

Romero, M., Sanchez-Gonzalez, M., Barrera, G., Leon, J., and Sanchez-Jimenez, M. 1995. Advanced salt receiver for solar power towers, *Libro: Solar Engineering 1995, Vol. 1*, W. B. Stine, T. Tanaka, and D. E. Clardige, eds., pp. 657–664. ASME, New York.

Romero, M., Fernández, V., and Sánchez, M. 1999. Optimization and performance of an optically asymmetrical heliostat field. *Journal of Physics*, 9, Pr3-71–Pr3-76. IV France.

Romero, M., Marcos, M. J., Osuna, R., and Fernández, V. 2000. Design and implementation plan of a 10 MW solar tower power plant based on volumetric-air technology in Seville (Spain), *Solar Engineering 2000-Proceedings of the ASME International Solar Energy Conference*, J. D. Pacheco and M. D. Thornbloom, eds., pp. 249–264. ASME, New York.

Romero, M., Marcos, M. J., Téllez, F. M., Blanco, M., Fernández, V., Baonza, F., and Berger, S. 2000. Distributed power from solar tower systems: A MIUS approach. *Solar Energy*, 67 (4–6), 249–264.

Romero, M., Buck, R., and Pacheco, J. E. 2002. An update on solar central receiver systems, projects, and technologies. *International Journal of Solar Energy Engineering*, 124, 98–108.

Romero, M., Martinez, D., and Zarza, E. 2004. Terrestrial solar thermal power plants: On the verge of commercialization. *Proceedings of the 4th International Conference on Solar Power from Space-SPS '04*, pp. 81–89. European Space Agency, Noordwijk, The Netherlands.

Rueda, F. 2003. Project INDITEP. First yearly report. IBERDROLA Ingeniería y consultoría, Madrid, Spain.

Ruiz, V., Silva, M., and Blanco, M. 1999. Las centrales energéticas termosolares. *Energía*, 25 (6), 47–55.

Sánchez, F. 1986. Results of Cesa-1 plant, *Proceedings of the 3rd International Workshop on Solar Thermal Central Receiver Systems*, B. M. Konstanz, ed., pp. 46–64. Springer, Berlin.

Sánchez, M. and Romero, M. 2006. Methodology for generation of heliostat field layout in central receiver systems based on yearly normalized energy surfaces. *Solar Energy*, 80(7), 861–874.

Sanders Associates, Inc. 1979. 1/4-Megawatt solar receiver. Final report. DOE/SF/90506-1.

Sargent & Lundy, LLC. 2003. Assessment of parabolic trough and power tower solar technology cost and performance forecasts, Report NREL/SR-550-34440, National Renewable Energy Laboratory, Golden, CO.

Schiel, W. 1999. Dish/stirling systems. *Solar Thermal Electricity Generation. Colección Documentos CIEMAT*, pp. 209–250, CIEMAT, Madrid, Spain.

Schiel, W., Keck, T., Kern, J., and Schweitzer, A. 1994. Long term testing of three 9 kW dish/Stirling systems. *Solar Engineering 1994, ASME 1994 Solar Engineering Conference*, pp. 541–550.

Schiel, W., Schweizer, A., and Stine, W. 1994. Evaluation of the 9-kW dish/Stirling system of Schlaich Bergermann und Partner using the proposed IEA dish/Stirling performance analysis guidelines. *Proceedings of 29th IECEC*, pp. 1725–1729.

Schmitz-Goeb, M. and Keintzel, G. 1997. The Phoebus solar power tower. In *Proceedings of the 1997 ASME International Solar Energy Conference*, D. E. Claridge and J. E. Pacheco, eds., pp. 47–53. ASME, NY.

Seinfeld, A. 2005. Solar thermochemical production of hydrogen—a review. *Solar Energy*, 78, 603–615.

Shuman, F. 1913. The most rational source of energy: Tapping the sun's radiant energy directly. *Scientific American*, 109, 350.

Siebers, D. L. and Kraabel, J. S. 1984. Estimating convective energy losses from solar central receivers. SAND84-8717, Sandia National Laboratories, Livermore, CA.

Silva, M., Blanco, M., and Ruiz, V. 1999. Integration of solar thermal energy in a conventional power plant: The COLON SOLAR project. *Journal de Physique IV Symposium Series*, 9, 189–194.

Sizmann, R. L. 1991. Solar radiation conversion, *Solar Power Plants*, C. J. Winter, R. L. Sizmann, and L. L. Vant-Hull, eds., pp. 17–83. Springer, Berlin.

Stine, W. and Diver, R. B. 1994. A compendium of solar dish/Stirling technology, Report SAND93-7026, Sandia National Laboratories, Albuquerque, NM.

Stone, K., Leingang, E., Rodriguez, G., Paisley, J., Nguyen, J., Mancini, T., and Nelving, H. 2001. Performance of the SES/Boeing dish Stirling system, *Proceedings of Solar Forum 2001 Solar Energy: The Power to Choose*, S. J. Kleis and C. E. Bingham, eds., pp. 21–27. ASME (CD-Rom), New York.

Sugarmen, C., Ring, A., Buck, R., Heller, P., Schwarzbözl, P., Téllez, F., Marcos, M. J., and Enrile, J. 2003. Solar-hybrid gas turbine power plants—Test results and market perspectives. *Proceedings of ISES Solar World Congress.*

Turkenburg, W. C. 2000. Renewable energy technologies. *World Energy Assessment*, pp. 219–272, United Nations Development Programme, New York.

Tyner, C. E., Kolb, G. J., Geyer, M., and Romero, M. 2000. Concentrating solar power in 2001- An IEA/SolarPACES summary of status and future prospects, edited by IEA-SolarPACES. http://www.solarpaces.org (accessed).

Vant-Hull, L. L. 1991. Solar radiation conversion, *Solar Radiation Conversion*, C. J. Winter, R. L. Sizmann, and L. L. Vant-Hull, eds., pp. 21–27. Springer, Berlin.

Weinrebe, G., Schmitz-Goeb, M., and Schiel, W. 1997. On the performance of the ASM150 stressed membrane heliostat. *1997 ASME/JSME/JSES International Solar Energy Conference.*

Welford, W. T. and Winston, R. 1989. *High Collection Non-Imaging Optics.* Academic Press, New York.

Winter, C. J., Sizmann, R. L., and Vant-Hull, L. L. eds. 1991. *Solar Power Plants*, pp. 21–27. Springer, Berlin.

Yebra, L., Berenguel, M., Romero, M., Martínez, D., and Valverde, A. 2004. Automation of solar plants. *Proceedings EuroSun2004, Vol. 1*, pp. 978–984.

Zarza, E. 2004. *Generación directa de vapor con colectores solares cilindro parabólicos. Proyecto direct solar steam (DISS)*, CIEMAT, Madrid, Spain.

Zarza, E., Hennecke, K., Hermann, U., Langenkamp, J., Goebel, O., Eck, M., Rheinländer, J., Ruiz, M., Valenzuela, L., Zunft, S., and Weyers, D.-H. 1999. DISS-phase I project, Final project report, Editorial CIEMAT, Madrid, Spain.

Zarza, E., Valenzuela, L., Leon, J., Hennecke, K., Eck, M., Weyers, D.-H., and Eickhoff, M. 2002. Direct steam generation in parabolic troughs. Final results and conclusions of the DISS project, *Book of Proceedings of 11th SolarPACES International Symposium on Concentrated Solar Power and Chemical Energy Technologies*, A. Steinfeld, ed., pp. 21–27. Paul Scherrer Institute, Villigen, Switzerland.

Zarza, E., Rojas, M. E., González, L., Caballero, J. M., and Rueda, F. 2004. INDITEP: The first DSG pre-commercial solar power plant. In *Proceeding submitted to the 12th Solar PACES International Symposium, held in Oaxaca (México) in October 6th–8th.*

Zavoico, A. B., Gould, W. R., Kelly, B. D., Grimaldi, I., and Delegado, C. 2001. Solar power tower (SPT) design innovations to improve reliability and performance—Reducing technical risk and cost. *Proceedings of Forum 2001 Conference.*

20

Photovoltaics Fundamentals, Technology and Application

Roger Messenger
Florida Atlantic University

D. Yogi Goswami
University of South Florida

Hari M. Upadhyaya
Loughborough University

Takhir M. Razykov
Uzbek Academy of Sciences

Ayodhya N. Tiwari
Loughborough University

Roland Winston
University of California

Robert McConnell
National Renewable Energy Laboratory

20.1 Photovoltaics 20-1
Introduction • The PV Cell • Manufacture of Solar Cells • PV Modules and PV Arrays • The Sun and PV Array Orientation • System Configurations • PV System Components • PV System Examples • Latest Developments in PV • Future Challenges for PV Systems

20.2 Thin-Film PV Technology 20-28
Introduction • Thin-Film Silicon • Cadmium Telluride Solar Cells • Cu(In Ga)Se_2 Solar Cells • Environmental Concerns and Cd Issue • Conclusions

20.3 Concentrating PV Technologies 20-54
Introduction • CPV Market Entry • Future Growth • Energy Payback • The Need for Qualification Standards

Nomenclature 20-58

Symbols 20-58

Acronyms 20-59

References 20-59

20.1 Photovoltaics

Roger Messenger and D. Yogi Goswami

20.1.1 Introduction

Photovoltaic conversion is the direct conversion of sunlight into electricity with no intervening heat engine. Photovoltaic devices are solid state; therefore, they are rugged and simple in design and require very little maintenance. Perhaps the biggest advantage of solar photovoltaic devices is that they can be constructed as standalone systems to give outputs from microwatts to megawatts. That is why they have been used as the power sources for calculators, watches, water pumps, remote buildings, communications, satellites and space vehicles, and even megawatt-scale power plants. Photovoltaic panels can be

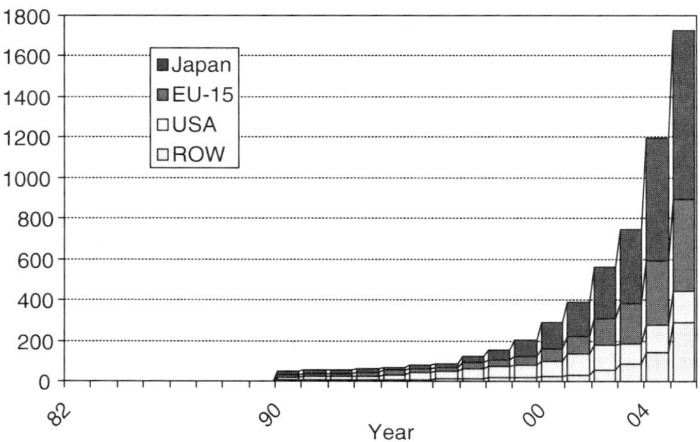

FIGURE 20.1 Worldwide production of photovoltaic panels. (From Maycock, P. EPIA.)

made to form components of building skin, such as roof shingles and wall panels. With such a vast array of applications, the demand for photovoltaics is increasing every year. In 2003, 750 MW_p (peak MW or MW under peak solar radiation of 1 kW/m^2) of photovoltaic panels were sold for the terrestrial markets and the market is growing at a phenomenal rate of 30% per year worldwide (see Figure 20.1).

In the early days of solar cells in the 1960s and 1970s, more energy was required to produce a cell than it could ever deliver during its lifetime. Since then, dramatic improvements have taken place in the efficiencies and manufacturing methods. In 1996, the energy payback periods were reduced to about 2.5 to 5 years, depending on the location of use (Nijs 1997) while panel lifetimes were increased to over 25 years. The costs of photovoltaic panels have come down from about $30 to $3 per peak watt over the last three decades and are targeted to reduce to around $1 per peak watt in the next ten years. Even the $3/W costs of solar panels results in system costs of $5–$7/W, which is very high for on-grid applications.

To reduce the costs further, efficiency of PV cells must be increased and the manufacturing costs will have to be decreased. At present, module efficiencies are as high as 15% (Hamakawa 2005). The main constraint on the efficiency of a solar cell is related to the bandgap of the semiconductor material of a PV cell. As explained later in this chapter, a photon of light with energy equal to or greater than the bandgap of the material is able to free-up one electron when absorbed into the material. However, the photons that have energy less than the bandgap are not useful for this process. When absorbed on the cell, they just produce heat. And for the photons with more energy than the bandgap, the excess energy above the bandgap is not useful in generating electricity. The excess energy simply heats up the cell. These reasons account for a theoretical maximum limit on the efficiency of a conventional single-junction PV cell to less than 25%. The actual efficiency is even lower because of reflection of light from the cell surface, shading of the cell due to current collecting contacts, internal resistance of the cell, and recombination of electrons and holes before they are able to contribute to the current.

The limits imposed on solar cells due to bandgap can be partially overcome by using multiple layers of solar cells stacked on top of each other, each layer with a bandgap higher than the layer below it. For example (Figure 20.2), if the top layer is made from a cell of material A (bandgap corresponding to λ_A), solar radiation with wavelengths less than λ_A would be absorbed to give an output equal to the hatched area A.

The solar radiation with wavelength greater than λ_A would pass through A and be converted to give an output equal to the hatched area B. The total output and therefore the efficiency of this tandem cell would be higher than the output and the efficiency of each single cell individually. The efficiency of a multijunction cell can be about 50% higher than a corresponding single cell. The efficiency would increase with the number of layers. For this concept to work, each layer must be as thin as possible, which puts a very difficult if not an insurmountable constraint on crystalline and polycrystalline cells to be made

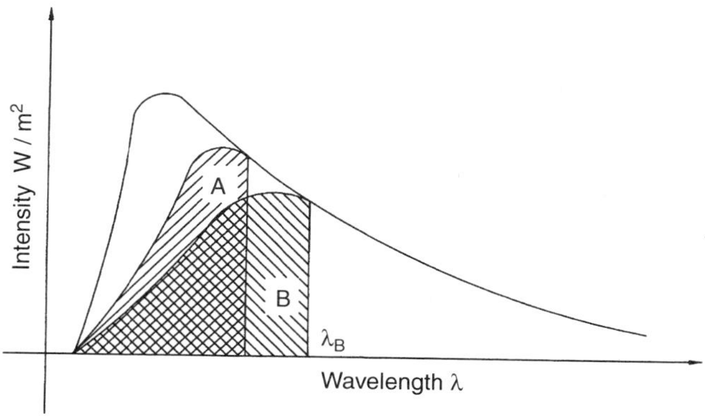

FIGURE 20.2 Energy Conversion from a two layered stacked PV cell. (From Goswami, D. Y., Kreith, F. and Kreider, J. F. *Principles of Solar Engineering, 2nd Ed.*, Taylor & Francis, Philadelphia, PA, 2000.)

multijunction. As a result, this concept is being investigated mainly for thin-film amorphous or microcrystalline solar cells. Efficiencies as high as 24.7% have been reported in the literature (Hamakawa 2005).

In this section, the physics of PV electrical generation will be briefly reviewed, followed by a discussion of the PV system design process. Several PV system examples will be presented, then a few of the latest developments in crystalline silicon PV will be summarized, and, finally, some of the present challenges (2004–2005) facing the large-scale deployment of PV energy sources will be explored. Emphasis will be on nonconcentrating, crystalline or multicrystalline silicon, terrestrial PV systems because such systems represent nearly 95% of systems currently being designed and built. However, the design procedures outlined at the end of the section also can be applied to other PV technologies, such as thin-films. Thin-film solar cells and concentrating PV cells are described in sections 20.2 and 20.3 respectively.

20.1.2 The PV Cell

20.1.2.1 The *p–n* Junction

PV cells have been made with silicon (Si), gallium arsenide (GaAs), copper indium diselenide (CIS), cadmium telluride (CdTe), and a few other materials. The common denominator of PV cells is that a *p–n* junction, or the equivalent, such as a Schottky junction, is needed to enable the photovoltaic effect. Understanding the *p–n* junction is thus at the heart of understanding how a PV cell converts sunlight into electricity. Figure 20.3 shows a Si *p–n* junction.

The junction consists of a layer of *n*-type Si joined to a layer of *p*-type Si, with an uninterrupted Si crystal structure across the junction. The *n*-layer has an abundance of free electrons and the *p*-layer has an abundance of free holes. Under thermal equilibrium conditions, meaning that temperature is the only external variable influencing the populations of free holes and electrons, the relationship between hole density, *p*, and electron density, *n*, at any given point in the material, is given by

$$np = n_i^2, \tag{20.1}$$

where n_i is approximately the density of electrons or holes in intrinsic (impurity-free) material. When impurities are present, then $n \cong N_d$ and $p \cong N_a$, where N_d and N_a are the densities of donor and acceptor impurities. For Si, $n_i \cong 1.5\times10^{10}$ cm^{-3} at $T=300$ K, while N_d and N_a can be as large as 10^{21} cm^{-3}. Hence, for example, if $N_d=10^{18}$ on the *n*-side of the junction, then $p=2.25\times10^2$ cm^{-3}.

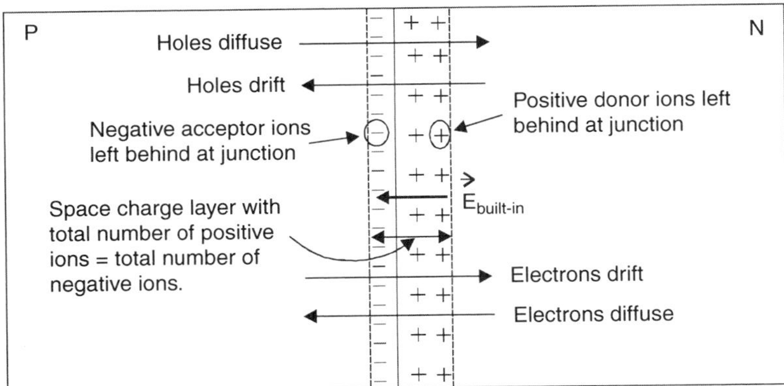

FIGURE 20.3 The *p–n* junction showing electron and hole drift and diffusion. (From Messenger, R. and Ventre, G. *Photovoltaic Systems Engineering, 2nd Ed.*, CRC Press, Boca Raton, FL, 2004.)

Both electrons and holes are subject to random diffusion within the Si crystalline structure, so each tends to diffuse from regions of high concentration to regions of low concentration. The enormous concentration differences of hole and electron concentrations between the *n*-side and the *p*-side of the junction cause large concentration gradients across the junction. The net result is that the electrons diffuse across the junction into the *p*-region and the holes diffuse across the junction into the *n*-region, as shown in Figure 20.3.

Before formation of the junction, both sides of the junction are electrically neutral. Each free electron on the *n*-side of the junction comes from a neutral electron donor impurity atom, such as arsenic (As), whereas each free hole on the *p*-side of the junction comes from a neutral hole donor (acceptor) impurity atom, such as boron (B). When the negatively charged electron leaves the As atom, the As atom becomes a positively charged As ion. Similarly, when the positively charged hole leaves the B atom, the B atom becomes a negatively charged B ion. Thus, as electrons diffuse to the *p*-side of the junction, they leave behind positively charged electron donor ions that are covalently bound to the Si lattice. As holes diffuse to the *n*-side of the junction, they leave behind negatively charged hole-donor ions that are covalently bound to the Si lattice on the *p*-side of the junction. The diffusion of charge carriers across the junction thus creates an electric field across the junction, directed from the positive ions on the *n*-side to the negative ions on the *p*-side, as shown in Figure 20.3. Gauss's law requires that electric field lines originate on positive charges and terminate on negative charges, so the number of positive charges on the *n*-side must be equal to the number of negative charges on the *p*-side.

Electric fields exert forces on charged particles according to the familiar $f = qE$ relationship. This force causes the charge carriers to drift. In the case of the positively charged holes, they drift in the direction of the electric field, i.e., from the *n*-side to the *p*-side of the junction. The negatively charged electrons drift in the direction opposite the field, i.e., from the *p*-side to the *n*-side of the junction. If no external forces are present other than temperature, then the flows of holes are equal in both directions and the flows of electrons are equal in both directions, resulting in zero net flow of either holes or electrons across the junction. This is called the *law of detailed balance*, which is consistent with Kirchoff's current law.

Carrying out an analysis of electron and hole flow across the junction ultimately leads to the development of the familiar diode equation,

$$I = I_o\left(e^{\frac{qV}{kT}} - 1\right), \tag{20.2}$$

where q is the electronic charge, k is the Boltzmann constant, T is the junction temperature in K, and V is the externally applied voltage across the junction from the *p*-side to the *n*-side of the junction.

20.1.2.2 The Illuminated *p–n* Junction

Figure 20.4 illustrates the effect of photons impinging upon the junction area.

The energy of a photon is given by Equation 20.3:

$$e = h\nu = \frac{hc}{\lambda}, \tag{20.3}$$

where λ is the wavelength of the photon, h is Planck's constant (6.625×10^{-34} J·s), and c is the speed of light (3×10^8 m/s).

The energy of a photon in electron-volts (eV) becomes $1.24/\lambda$, if λ is in μm (1 eV $= 1.6 \times 10^{-19}$ J). If a photon has an energy that equals or exceeds the semiconductor bandgap energy of the *p-n* junction material, then it is capable of creating an electron-hole pair (EHP). For Si, the bandgap is 1.1 eV, so if the photon wavelength is less than 1.13 μm, which is in the near infrared region, then the photon will have sufficient energy to generate an EHP.

Although photons with energies higher than the bandgap energy can be absorbed, one photon can create only one EHP. The excess energy of the photon is wasted as heat. As photons enter a material, the intensity of the beam depends upon a wavelength-dependent absorption constant, α. The intensity of the photon beam as a function of penetration depth into the material is given by $F(x) = F_o e^{-\alpha x}$, where x is the depth of penetration into the material. Optimization of photon capture, thus, suggests that the junction should be within $(1/\alpha)$ of the surface to ensure transmission of photons to within a diffusion length of the *p–n* junction, as shown in Figure 20.4.

If an EHP is created within one minority carrier diffusion length, D_x, of the junction, then, on the average, the EHP will contribute to current flow in an external circuit. The diffusion length is defined to be $L_x = \sqrt{D_x \tau_x}$, where D_x and τ_x are the minority carrier diffusion length and lifetime for electrons in the *p*-region if $x = n$, and D_x and τ_x are the minority carrier diffusion length and lifetime for holes in the *n*-region if $x = p$. So the idea is to quickly move the electron and hole of the EHP to the junction before either has a chance to recombine with a majority charge carrier. In Figure 20.4, points A, B, and C represent EHP generation within a minority carrier diffusion length of the junction. But if an EHP is generated at point D, it is highly unlikely that the electron will diffuse to the junction before it recombines.

The amount of photon-induced current flowing across the junction and into an external circuit is directly proportional to the intensity of the photon source. Note that the EHPs are swept across the junction by the builtin E-field, so the holes move to the *p*-side and continue to diffuse toward the *p*-side external contact. Similarly, the electrons move to the *n*-side and continue to diffuse to the *n*-side external contact. Upon reaching their respective contacts, each contributes to external current flow if an external

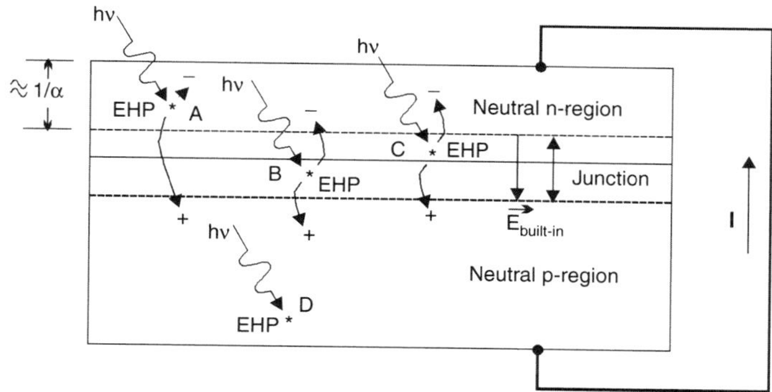

FIGURE 20.4 The illuminated *p–n* junction showing desirable geometry and the creation of electron-hole pairs. (From Messenger, R. and Ventre, G. *Photovoltaic Systems Engineering, 2nd Ed.*, CRC Press, Boca Raton, FL, 2004.)

path exists. In the case of holes, they must recombine at the contact with an electron that enters the material at the contact. Electrons, on the other hand, are perfectly happy to continue flowing through an external copper wire.

At this point, an important observation can be made. The external voltage across the diode that results in significant current flow when no photons are present, is positive from p to n. The diode current and voltage are defined in this direction, and the diode thus is defined according to the passive sign convention. In other words, when no photons impinge on the junction, the diode dissipates power. But when photons are present, the photon-induced current flows opposite to the passive direction. Therefore, current leaves the positive terminal, which means that the device is generating power. This is the photovoltaic effect. The challenge to the manufacturers of PV cells is to maximize the capture of photons and, in turn, maximize the flow of current in the cell for a given incident photon intensity. Optimization of the process is discussed in detail in Messenger and Ventre (2004). When the photocurrent is incorporated into the diode equation, the result is

$$I = I_l - I_o\left(e^{\frac{qV}{kT}} - 1\right) \cong I_l - I_o e^{\frac{qV}{kT}}. \tag{20.4}$$

Note that in Equation 20.4, the direction of the current has been reversed with respect to the cell voltage. With the active sign convention implied by Equation 20.4, the junction device is now being defined as a cell, or PV cell. Figure 20.5 shows the I–V curves for an ideal PV cell and a typical PV cell, assuming the cell has an area of approximately 195 cm^2.

It is evident that the ideal curve closely represents that of an ideal current source for cell voltages below 0.5 V, and it closely represents that of an ideal voltage source for voltages near 0.6 V. The intersection of the curve with the $V=0$ axis represents the short circuit current of the cell. The intersection of the curve with the $I=0$ axis represents the open circuit voltage of the cell. To determine the open circuit voltage of the cell, simply set $I=0$ and solve Equation 20.4 for V_{OC}. The result is

$$V_{oc} = \frac{kT}{q}\ln\frac{I_l}{I_O}. \tag{20.5}$$

The direct dependence of I on I_l and the logarithmic dependence of V_{OC} on I_l is evident from Equation 20.4 and Equation 20.5, as well as from Figure 20.5.

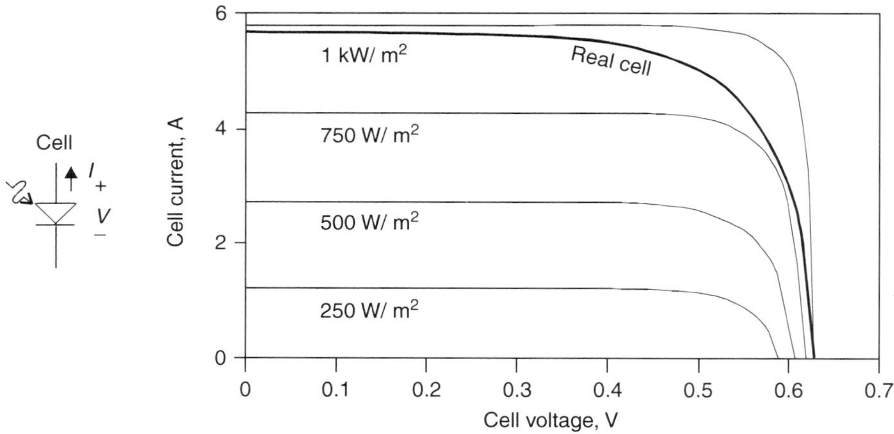

FIGURE 20.5 I–V characteristics of real and ideal PV cells under different illumination levels. (From Messenger, R. and Ventre, G. *Photovoltaic Systems Engineering, 2nd Ed.*, CRC Press, Boca Raton, FL, 2004.)

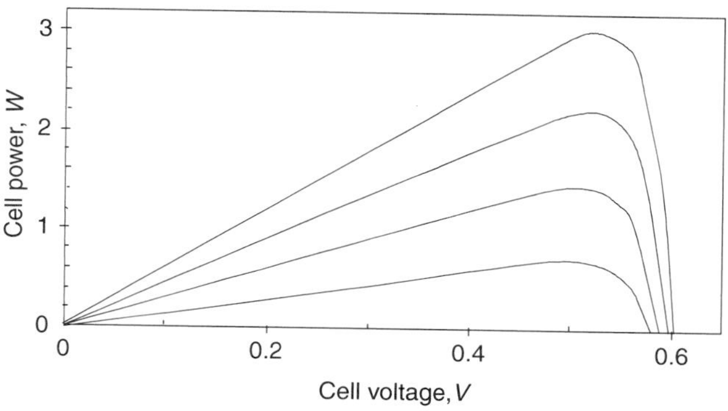

FIGURE 20.6 Power vs. voltage for a PV cell for four illumination levels. (From Messenger, R. and Ventre, G. *Photovoltaic Systems Engineering, 2nd Ed.*, CRC Press, Boca Raton, FL, 2004.)

The departure of the real curve from the ideal prediction is primarily due to unavoidable series resistance between the cell contacts and the junction.

20.1.2.3 Properties of the PV Cell

Another property of the *I–V* curves of Figure 20.5 is the presence of a single point on each curve at which the power delivered by the cell is a maximum. This point is called the *maximum power point of the cell*, and is more evident when cell power is plotted vs. cell voltage, as shown in Figure 20.6. Note that the maximum power point of the cell remains at a nearly constant voltage as the illumination level of the cell changes.

Not shown in Figure 20.5 or Figure 20.6 is the temperature dependence of the photocurrent. It turns out that I_o increases rapidly with temperature. Thus, despite the KT/q multiplying factor, the maximum available power from a Si PV cell decreases at approximately 0.47%/°C, as shown in Figure 20.7.

Furthermore, the maximum power voltage also decreases by approximately this same factor. An increase of 25°C is not unusual for an array of PV cells, which corresponds to a decrease of approximately 12% in maximum power and in maximum power voltage. Because of this temperature degradation of the performance of a PV cell, it is important during the system design phase to endeavor to keep the PV cells as cool as possible.

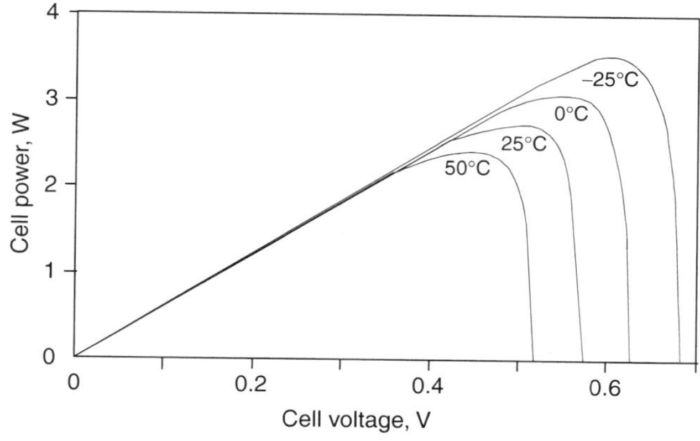

FIGURE 20.7 Temperature dependence of the power vs. voltage curve for a PV cell. (From Messenger, R. and Ventre, G. *Photovoltaic Systems Engineering, 2nd Ed.*, CRC Press, Boca Raton, FL, 2004.)

20.1.3 Manufacture of Solar Cells

20.1.3.1 Manufacture of Crystalline and Multicrystalline Silicon PV Cells

Although crystalline and multicrystalline silicon PV cells require highly purified, electronic-grade silicon, the material can be about an order of magnitude less pure than semiconductor grade silicon and still yield relatively high performance PV cells. Recycled or rejected semiconductor-grade silicon is often used as the feedstock for PV-grade silicon. Once adequately refined silicon is available, a number of methods have been devised for the production of single-crystal and multicrystalline PV cells. Single-crystal Si PV cells have been fabricated with conversion efficiencies just over 20%, while conversion efficiencies of champion multicrystalline Si PV cells are about 16% (Hanoka 2002; Rosenblum et al. 2002).

Single-crystal Si cells are almost exclusively fabricated from large single crystal ingots of Si that are pulled from molten, PV-grade Si. These ingots, normally *p*-type, are typically on the order of 200 mm in diameter and up to 2 m in length. The Czochralski method (Figure 20.8a) is the most common method of growing single-crystal ingots.

A seed crystal is dipped in molten silicon doped with a *p*-material (boron) and drawn upward under tightly controlled conditions of linear and rotational speed and temperature. This process produces cylindrical ingots of typically 10-cm diameter, although ingots of 20-cm diameter and more than 1 m long can be produced for other applications. An alternative method is called the *float zone method* (Figure 20.8b). In this method a polycrystalline ingot is placed on top of a seed crystal and the interface is melted by a heating coil around it. The ingot is moved linearly and rotationally, under controlled conditions. This process has the potential to reduce the cell cost. Figure 20.9 illustrates the process of manufacturing a cell from an ingot.

The ingots are sliced into wafers that are approximately 0.25 mm thick. The wafers are further trimmed to a nearly square shape, with only a small amount of rounding at the corners. Surface degradation from the slicing process is reduced by chemically etching the wafers. To enhance photon absorption, it is common practice to use a preferential etching process to produce a textured surface finish. An *n*-layer is then diffused into the wafer to produce a *p–n* junction, contacts are attached, and the cell is then encapsulated into a module (Figure 20.10).

Detailed accounts of cell and module fabrication processes can be found in Hanoka (2002); Messenger and Ventre (2004); Hamakawa (2005), and Saitoh (2005).

Growing and slicing single-crystal Si ingots is highly energy intensive, and, as a result, imposes a relatively high energy cost on this method of cell fabrication. This high-energy cost imposes a lower limit on the cost of production of a cell and, although the cell will ultimately generate more energy than was used to produce it, the energy payback time is longer than desirable. Reducing the energy cost of cell and module fabrication has been the subject of a great deal of research over the past 40 years. The high-energy cost of crystalline Si led to the work on thin-films of amorphous Si, CdTe, and other materials that is

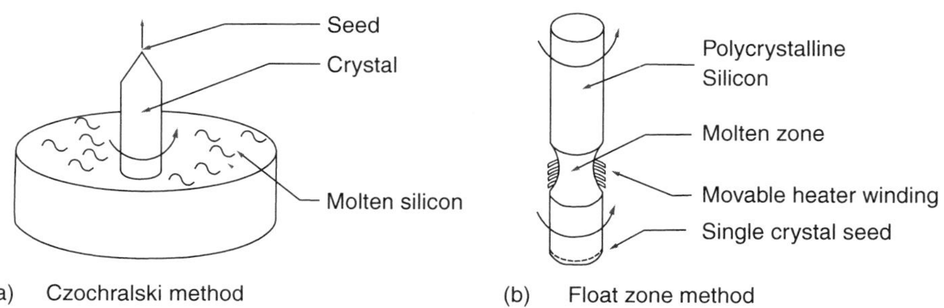

FIGURE 20.8 Crystalline silicon ingot production methods. (From Goswami, D. Y., Kreith, F. and Kreider, J. F. *Principles of Solar Engineering, 2nd Ed.*, Taylor & Francis, Philadelphia, PA, 2000.)

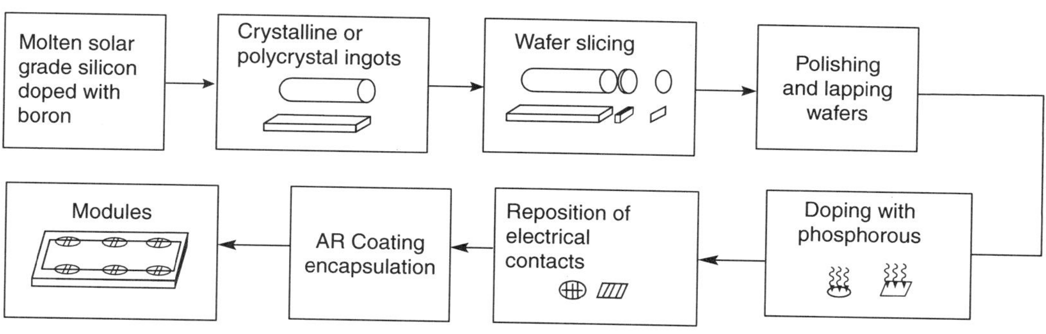

FIGURE 20.9 Series of processes for the manufacture of crystalline and polycrystalline cells. (From Goswami, D. Y., Kreith, F. and Kreider, J. F. *Principles of Solar Engineering, 2nd Ed.*, Taylor & Francis, Philadelphia, PA, 2000.)

described later in this handbook. A great deal of work has also gone into developing methods of growing Si in a manner that will result in lower-energy fabrication costs.

Three methods that are less energy intensive are now commonly in use—crucible growth, the EFG process, and string ribbon technology. These methods, however, result in the growth of multicrystalline Si, which, upon inspection, depending upon the fabrication process, has a speckled surface appearance, as opposed to the uniform color of single crystal Si. Multicrystalline Si has electrical and thermodynamic characteristics that match single crystal Si relatively closely, as previously noted.

The crucible growth method involves pouring molten Si into a quartz crucible and carefully controlling the cooling rate (Figure 20.11).

A seed crystal is not used, so the resulting material consists of a collection of zones of single crystals with an overall square cross-section. It is still necessary to saw the ingots into wafers, but the result is square wafers rather than round wafers that would require additional sawing and corresponding loss of material. Wafers produced by this method can achieve conversion efficiencies of 15% or more (Hamakawa 2005).

The edge-defined film-fed growth (EFG) process is another method currently being used to produce commercial cells (ASE International). The process involves pulling an octagon tube, 6-m long, with a wall thickness of 330 μm, directly from the Si melt. The octagon is then cut by a laser along the octagonal edges into individual cells. Cell efficiencies of 14% have been reported for this fabrication method (Rosenblum et al. 2002). Figure 20.12 illustrates the process.

A third method of fabrication of multicrystalline Si cells involves pulling a ribbon of Si, or dendritic web, from the melt (Figure 20.13).

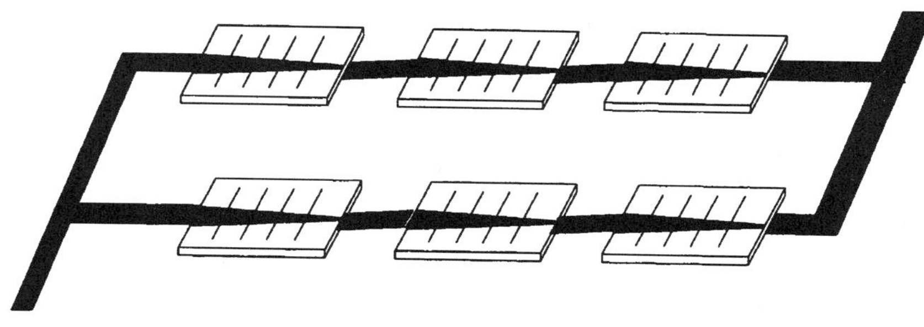

FIGURE 20.10 Assembly of solar cells to form a module. (From Goswami, D. Y., Kreith, F. and Kreider, J. F. *Principles of Solar Engineering, 2nd Ed.*, Taylor & Francis, Philadelphia, PA, 2000.)

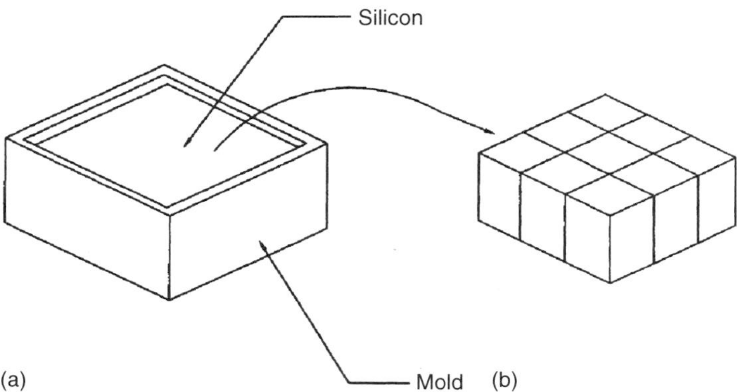

FIGURE 20.11 Polycrystalline ingot production. (From Goswami, D. Y., Kreith, F. and Kreider, J. F. *Principles of Solar Engineering, 2nd Ed.*, Taylor & Francis, Philadelphia, PA, 2000.)

Controlling the width of the ribbon is the difficult part of this process. High-temperature string materials are used to define the edges of the ribbon. The string materials are pulled through a crucible of molten Si in an Ar atmosphere after the attachment of a seed crystal to define the crystal structure of the ribbon. The nonconducting string material has a coefficient of thermal expansion close to that of Si, so during the cooling process, the string material will not affect the Si crystallization process (Hanoka 2002). The ribbons of Si are then cut into cells, typically rectangular in shape, as opposed to the more common square configuration of other multicrystalline technologies. Once the multicrystalline wafers have been fabricated, further processing is the same as that used for single crystal cells.

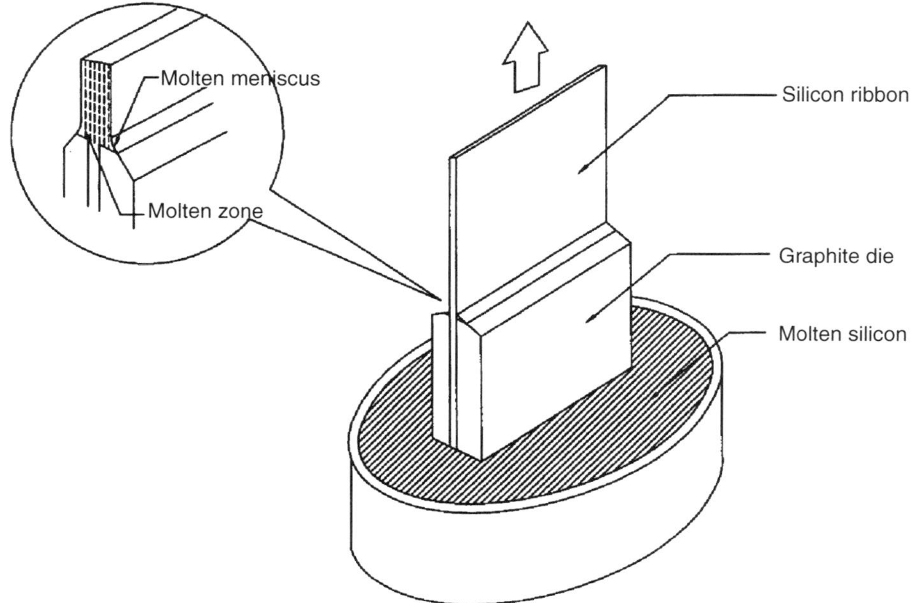

FIGURE 20.12 Thin-film production by edge-defined film-fed growth (EFG). (From Goswami, D. Y., Kreith, F. and Kreider, J. F. *Principles of Solar Engineering, 2nd Ed.*, Taylor & Francis, Philadelphia, PA, 2000.)

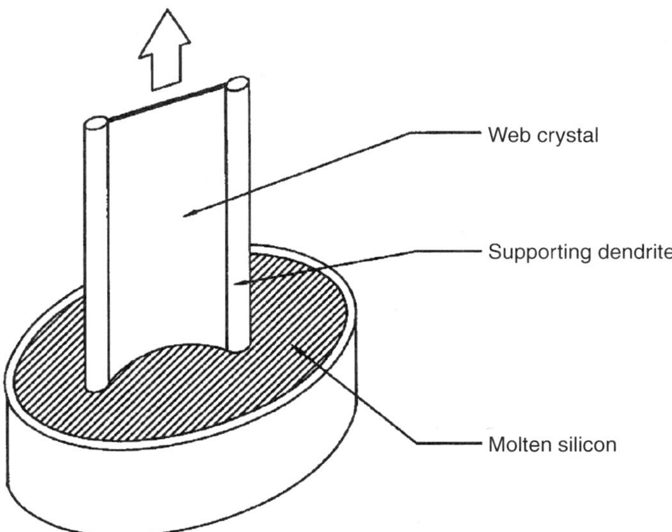

FIGURE 20.13 Thin-film production by dendritic web growth. (From Goswami, D. Y., Kreith, F. and Kreider, J. F. *Principles of Solar Engineering, 2nd Ed.*, Taylor & Francis, Philadelphia, PA, 2000.)

20.1.3.2 Amorphous Silicon and Multijunction-Thin-Film Fabrication

Amorphous silicon (*a*-Si) cells are made as thin-films of *a*-Si:H alloy doped with phosphorous and boron to make *n* and *p* layers respectively. The atomic structure of an *a*-Si cell does not have any preferred orientation. The cells are manufactured by depositing a thin layer of *a*-Si on a substrate (glass, metal or plastic) from glow discharge, sputtering or chemical vapor deposition (CVD) methods. The most common method is by an RF glow discharge decomposition of silane (SiH_4) on a substrate heated to a temperature of 200–300°C. To produce *p*-silicon, diborane (B_2H_6) vapor is introduced with the silane vapor. Similarly phosphene (PH_3) is used to produce *n*-silicon. The cell consists of an *n*-layer, and intermediate undoped *a*-Si layer, and a *p*-layer on a substrate. The cell thickness is about 1 μm. The manufacturing process can be automated to produce rolls of solar cells from rolls of substrate. Figure 20.14 shows an example of roll-to-roll *a*-Si cell manufacturing equipment using a plasma CVD method. This machine can be used to make multijunction or tandem cells by introducing the appropriate materials at different points in the machine.

The four previously mentioned cell fabrication techniques require contacts on the front surface and on the back surface of the cells. Front-surface contacts need to cover enough area to minimize series resistance between cell and contact, but if too much area is covered, then photons are blocked from entering the crystal. Thus, it is desirable, if possible, to design cells such that both contacts are on the back of the cell. Green and his PV team at the University of South Wales have devised a buried contact cell (Green and Wenham 1994) that has both contacts on the back and also is much thinner and therefore much less material-intensive than conventional Si cells. In conventional cells, charge carrier flow is perpendicular to the cell surface, while in the buried contact cell, even though the multiple *p*–*n* junctions are parallel to the cell surfaces, charge carrier flow is parallel to the cell surfaces. The fabrication process involves depositing alternate *p*-type and *n*-type Si layers, each about 1-μm thick, on an insulating substrate or superstrate. Grooves are laser cut in the layers and contacts are deposited in the grooves. Elimination of the ingot and wafer steps in processing, along with the reduced amount of material used, reduces correspondingly the energy overhead of cell production. Conversion efficiencies in excess of 20% and high cell fill factors have been achieved with this technology.

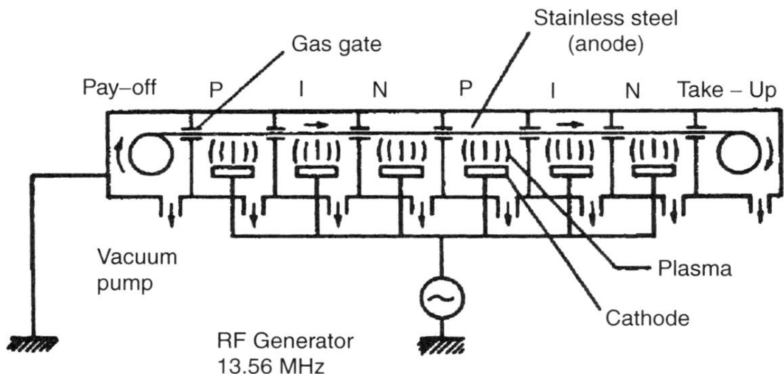

FIGURE 20.14 A schematic diagram of a roll-to-roll plasma CVD machine. (Adapted from ASE International, http://www.ase-international.com)

20.1.4 PV Modules and PV Arrays

Because individual cells have output voltages limited to approximately 0.5 V and output currents limited to approximately 7 A, it is necessary to combine cells in series and parallel to obtain higher voltages and currents. A typical PV module consists of 36 cells connected in series in order to produce a maximum power voltage of approximately 17 V, with a maximum power current of approximately 7 A at a temperature of 25°C. Such a module will typically have a surface area of about 10 ft^2. Modules also exist with 48 or more series cells so that three modules in series will produce the same output voltage and current as four 36-cell modules in series. Other larger modules combine cells in series and in parallel to produce powers up to 300 W per module.

Modules must be fabricated so the PV cells and interconnects are protected from moisture and are resistant to degradation from the ultraviolet component of sunlight. Since the modules can be expected to be exposed to a wide range of temperatures, they must be designed so that thermal stresses will not cause delamination. Modules must also be resistant to blowing sand, salt, hailstones, acid rain and other unfriendly environmental conditions. And, of course, the module must be electrically safe over the long term. A typical module can withstand a pressure of 50 psf and large hailstones and is warranteed for 25 years. Details on module fabrication can be found in Messenger and Ventre (2004); Saitoh (2005), and Bohland (1998).

It is important to realize that when PV cells with a given efficiency are incorporated into a PV module, the module efficiency will be less than the cell efficiency, unless the cells are exactly identical electrically. When cells are operated at their maximum power point, this point is located on the cell *I–V* curve at the point where the cell undergoes a transition from a nearly ideal current source to a nearly ideal voltage source. If the cell *I–V* curves are not identical, since the current in a series combination of cells is the same in each cell, each cell of the combination will not necessarily operate at its maximum power point. Instead, the cells operate at a current consistent with the rest of the cells in the module, which may not be the maximum power current of each cell.

When modules are combined to further increase system voltage and/or current, the collection of modules is called an *array*. For the same reason that the efficiency of a module is less than the efficiencies of the cells in the module, the efficiency of an array is less than the efficiency of the modules in the array. However, because a large array can be built with subarrays that can operate essentially independently of each other, in spite of the decrease in efficiency at the array level, PV arrays that produce in excess of 1 MW are in operation at acceptable efficiency levels. The bottom line is that most efficient operation is achieved if modules are made of identical cells and if arrays consist of identical modules.

20.1.5 The Sun and PV Array Orientation

As explained in detail in Chapter 5, total solar radiation is composed of components, direct or beam, diffuse and reflected. In regions with strong direct components of sunlight, it may be advantageous to have a PV array mount that will track the sun. Such tracking mounts can improve the daily performance of a PV array by more than 20% in certain regions. In cloudy regions, tracking is less advantageous.

The position of the sun in the sky can be uniquely described by two angles—the azimuth, γ, and the altitude, α. The azimuth is the deviation from true south. The altitude is the angle of the sun above the horizon. When the altitude of the sun is 90°, the sun is directly overhead.

Another convenient, but redundant, angle, is the hour angle, ω. Because the earth rotates 360° in 24 h, it rotates 15°/h. The sun thus appears to move along its arc 15° toward the west each hour. The hour angle is 0° at solar noon, when the sun is at its highest point in the sky during a given day. In this handbook, we have a sign convention such that the hour angle and the solar azimuth angle are negative before noon and positive after noon. For example, at 10:00 a.m. solar time, the hour angle will be −30°.

A further important angle that is used to predict sun position is the declination, δ. The declination is the apparent position of the sun at solar noon with respect to the equator. When $\delta=0$, the sun appears overhead at solar noon at the equator. This occurs on the first day of fall and on the first day of spring. On the first day of northern hemisphere summer (June 21), the sun appears directly overhead at a latitude, L, of 23.45° north of the equator. On the first day of winter (December 21), the sun appears directly overhead at a latitude of 23.45° south of the equator. At any other latitude, the altitude, $\alpha=90°-|L-\delta|$ when the sun is directly south, (or north) i.e., at solar noon. At solar noon, the sun is directly south for $L>\delta$ and directly north for $L<\delta$. Note that if L is negative, it refers to the southern hemisphere.

Several important formulas for determining the position of the sun (Messenger and Ventre 2004; Markvart 1994) include the following, where n is the day of the year with January 1 being day 1:

$$\delta = 23.45°\sin\frac{360[n-80]}{365}, \tag{20.6}$$

$$\omega = \pm 15°(\text{hours from local solar noon}) \tag{20.7}$$

$$\sin\alpha = \sin\delta\sin L + \cos\delta\cos L\cos\omega, \tag{20.8}$$

and

$$\cos\gamma = \frac{\cos\delta\sin\omega}{\cos\alpha}. \tag{20.9}$$

Solution of Equation 20.6 through Equation 20.9 shows that for optimal annual performance of a fixed PV array, it should face directly south and should be tilted at an angle approximately equal to the latitude, L. For best summer performance, the tilt should be at $L-15°$ and for best winter performance, the array should be tilted at an angle of $L+15°$.

Although Equation 20.6 through Equation 20.9 can be used to predict the location of the sun in the sky at any time on any day at any location, they cannot be used to predict the degree of cloud cover. Cloud cover can only be predicted on a statistical basis for any region, and thus the amount of sunlight available to a collector will also depend upon cloud cover. The measure of available sunlight is the peak sun hour (psh). If sunlight intensity is measured in kW/m^2, then if the sunlight intensity is integrated from sunrise to sunset over 1 m^2 of surface, the result will be measured in kWh. If the daily kWh/m^2 is divided by the peak sun intensity, which is defined as 1 kW/m^2, the resulting units are hours. Note that this hour figure multiplied by 1 kW/m^2 results in the daily kWh/m^2. Hence, the term *peak sun hours*, because the psh is the number of hours the sun would need to shine at peak intensity to produce the daily sunrise to sunset kWh. Obviously the psh is also equivalent to kWh/m^2/day. For locations in the United States, the National Renewable Energy Laboratory publishes psh for fixed and single-axis tracking PV arrays at tilts of horizontal, latitude $-15°$, latitude, latitude $+15°$, and vertical. NREL also tabulates data for double axis trackers. These tables are extremely useful for determining annual performance of a PV array.

20.1.6 System Configurations

Figure 20.15 illustrates four possible configurations for PV systems.

Perhaps the simplest system is that of Figure 20.15a, in which the output of the PV module or array is connected directly to a DC load. This configuration is most commonly used with a fan or a water pump, although it is likely that the water pump will also use a linear current booster (LCB) between the array and the pump motor. Operation of the LCB will be explained later.

The configuration of Figure 20.15b includes a charge controller and storage batteries so the PV array can produce energy during the day that can be used day or night by the load. The charge controller serves a dual function. If the load does not use all the energy produced by the PV array, the charge controller prevents the batteries from overcharge. While flooded lead acid batteries require overcharging about once per month, frequent overcharging shortens the life of the batteries. As the batteries become discharged, the charge controller disconnects the load to prevent the batteries from overdischarge. Normally PV systems incorporate deep discharge lead-acid batteries, but the life of these batteries is reduced significantly if they are discharged more than 80%. Modern charge controllers typically begin charging as constant current sources. In the case of a PV system, this simply means that all array current is directed to the batteries. This is called the "bulk" segment of the charge cycle. After the battery voltage reaches the bulk voltage, which is an owner programmable value, as determined by the battery type and the battery temperature, the charging cycle switches to a constant voltage mode, commonly called the *absorption* mode. During the absorption charge mode, the charge controller maintains the bulk charge voltage for a preprogrammed time, again depending upon manufacturers' recommendations. During the absorption charge, battery current decreases as the batteries approach full charge. At the end of the absorption charge period, the charging voltage is automatically reduced to the "float" voltage level, where the charging current is reduced to a "trickle" charge. Because quality charge controllers are microprocessor controlled, they have clock circuitry so that they can be programmed to automatically subject the batteries to an "equalization" charge approximately once a month. The equalization mode applies a voltage higher than the bulk voltage for a preset time to purposely overcharge the batteries. This process causes the electrolyte to bubble, which helps to mix the electrolyte as well as to clean the battery plates. Equalization is recommended *only* for flooded lead-acid batteries. Sealed varieties can be seriously damaged if they are overcharged.

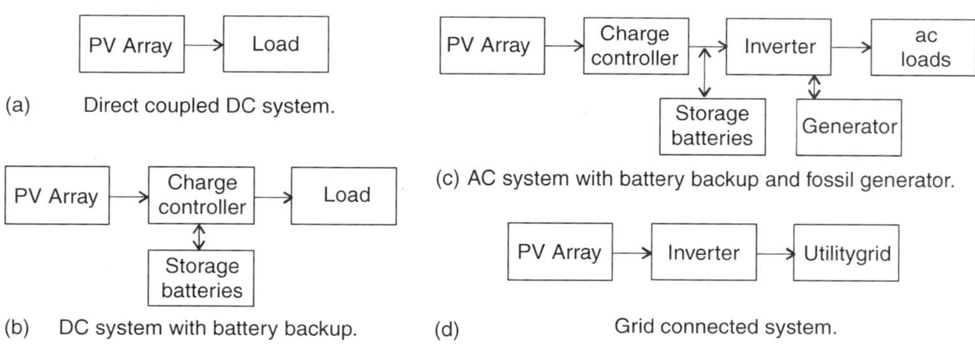

FIGURE 20.15 Several examples of PV systems.

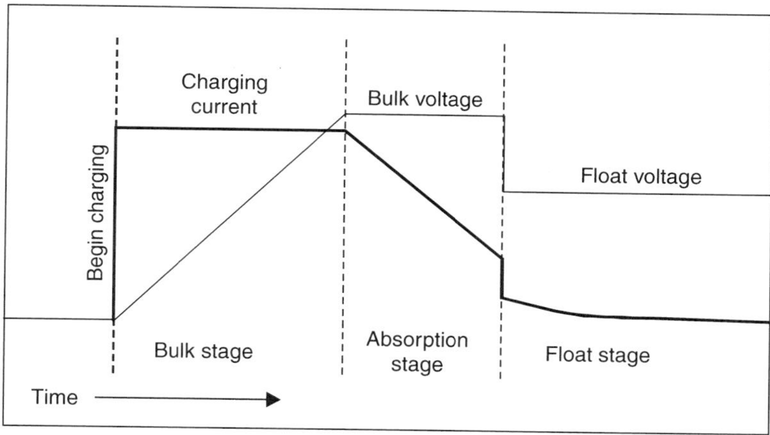

FIGURE 20.16 Charging cycle for typical PV charge controller.

Figure 20.16 shows the currents and voltages during the bulk, absorption and float parts of the charging cycle. Note that all settings are programmable by the user in accordance with manufacturers' recommendations. Some charge controllers incorporate maximum power tracking as a part of their charge control algorithm. Because the maximum power voltage of a module or an array is generally higher than needed to charge the batteries, the array will not normally operate at its maximum power point when it is charging batteries, especially if the array temperature is low. For example, if it takes 14.4 V to charge a 12.6-V battery, and if a module maximum power voltage is 17 V, then the charging current can be increased by a factor of 17/14.4, or approximately 18%, assuming close to 100% efficiency of the MPT.

The configuration of Figure 20.15c incorporates an inverter to convert the DC PV array output to AC and a backup generator to supply energy to the system when the supply from the sun is too low to meet the needs of the load. Normally the backup generator will be a fossil-fueled generator, but it is also possible to incorporate wind or other renewable generation into the system. In this case, the charge controller prevents overcharge of the batteries. The inverter is equipped with voltage sensing circuitry so that if it detects the battery voltage going too low, it will automatically start the generator so the generator will provide power for the load as well as provide charging current for the batteries. This system is called a *hybrid system* because it incorporates the use of more than one energy source.

The first three configurations are standalone systems. The fourth system, shown in Figure 20.15d, is a grid-connected, or utility-interactive, system. The inverter of a utility interactive system must meet more stringent operational requirements than the standalone inverter. The inverter output voltage and current must be of "utility-grade" quality. This means that it must have minimal harmonic content. Furthermore, the inverter must sense the utility and, if utility voltage is lost, the inverter must shut down until utility voltage is restored to within normal limits.

20.1.7 PV System Components

20.1.7.1 Maximum Power Trackers and Linear Current Boosters

The linear current booster (LCB) was mentioned in conjunction with the water pumping example. The function of the LCB is to match the motor *I–V* characteristic to the maximum power point of the PV array, so that at all times the array delivers maximum power to the load. Note that the LCB acts as a DC-to-DC transformer, converting a higher voltage and lower current to a lower voltage and higher current, with minimal power loss in the conversion process. A more general term that includes the

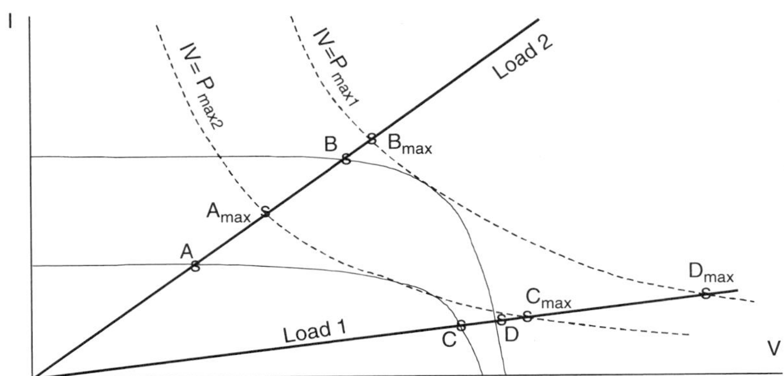

FIGURE 20.17 Operation of the LCB or MPT. (From Messenger, R. and Ventre, G. *Photovoltaic Systems Engineering, 2nd Ed.*, CRC Press, Boca Raton, FL, 2004.)

possibility of converting voltage upward defines the *maximum power tracker* (MPT). Figure 20.17 shows the operating principle of the LCB and MPT.

Note that normally the *I–V* characteristic of the load will not intersect the *I–V* characteristic of the PV array at the maximum power point of the array, as shown by points A, B, C, and D for the two loads and the two sunlight intensity levels. For the lower intensity situation, the characteristic of load 1 intersects the array characteristic at point C and the characteristic of load 2 intersects the array characteristic at point A. The two hyperbolas are the loci of points where the *IV* product is equal to the maximum available power from the array at the particular sunlight intensity. Hence, the intersection of these hyperbolas with the load characteristics represents the transfer of all available power from the array to the load. Although the increase in power for points B and C is not particularly impressive, as shown by points B_{max} and C_{max}, the increase in power for points A and D is considerably greater, as shown by points A_{max} and D_{max}. The assumption here, of course, is 100% efficiency in the transformation. In fact, efficiencies in excess of 95% are not unusual for quality MPT and LCB devices.

The final observation for Figure 20.17 is that points A_{max} and B_{max} occur at voltages below the maximum power voltages of the array, while C_{max} and D_{max} occur at voltages above the maximum power voltages of the array. Because the input voltage and current of the MPT or LCB is the maximum power voltage and current of the array, the MPT or LCB output voltage and current points A_{max} and B_{max} represent down conversion of the array voltage and points C_{max} and D_{max} represent up conversion of the array voltage. These forms of conversion are discussed in power electronics books, such as that by Krein (1998). The difference between the MPT and the LCB is that the LCB only performs a down conversion, so the operating voltage of the load is always below the maximum power voltage point of the array. The terms LCB and MPT are often used interchangeably for down conversion, but normally LCB is limited to the description of the black box that optimizes performance of pumps, whereas MPT is used for more general applications.

20.1.7.2 Inverters

Inverters convert DC to AC. The simplest inverter converts DC to square waves. Although square waves will operate many AC loads, their harmonic content is very high, and, as a result, there are many situations where square waves are not satisfactory. Other more suitable inverter output waveforms include the quasi-sine wave and the utility-grade sine wave. Both are most commonly created by the use of multilevel H-bridges controlled by microprocessors. There are three basic configurations for inverters: standalone, grid-tied, and UPS. The standalone inverter must act as a voltage source that delivers a prescribed amplitude and frequency rms sine wave without any external synchronization. The grid-tied inverter is essentially a current source that delivers a sinusoidal current waveform to the grid that is synchronized by the grid voltage. Synchronization is typically sufficiently close to maintain a power factor in excess of 0.9. The UPS inverter combines the features of both the standalone and the grid-tied inverter, so that if grid power is lost, the unit will act as a standalone inverter while supplying power to emergency loads. IEEE Standard 929 (IEEE 2000) requires that any inverter that is connected to the grid must monitor the utility grid voltage, and, if the grid voltage falls outside prescribed limits, the inverter must stop delivering current to the grid. Underwriters Laboratory (UL) Standard 1741 (UL 1999) provides the testing needed to ensure compliance with IEEE 929.

Although it may seem to be a simple matter to shut down if the utility shuts down, the matter is complicated by the possibility that additional utility-interactive PV systems may also be on line. Hence, it may be possible for one PV system to "fool" another system into thinking that it is really the utility. To prevent this "islanding" condition, sophisticated inverter control algorithms have been developed to ensure that an inverter will not appear as the utility to another inverter. Some PV system owners do not want their PV system to shut down when the utility shuts down. Such a system requires a special inverter that has two sets of AC terminals. The first set, usually labeled *AC IN*, is designed for connection to the utility. If the utility shuts down, this set of terminals disconnects the inverter output from the utility, but continues to monitor utility voltage until it is restored. When the utility connection is restored, the inverter will first meet the needs of the emergency loads and then will feed any excess output back to the main distribution panel.

The second set of terminals is the emergency output. If the utility shuts down, the inverter almost instantaneously transfers into the emergency mode, in which it draws power from the batteries and/or the PV array to power the emergency loads. In this system, the emergency loads must be connected to a separate emergency distribution panel. Under emergency operation, the loads in the main distribution panel are without power, but the emergency panel remains energized. Such a system is shown in Figure 20.18.

The reader is referred to the book by Messenger and Ventre (2004) and to that by Krein (1998) for detailed explanations of the operation of inverters, including the methods used to ensure that utility interactive inverters meet UL 1741 testing requirements.

20.1.7.3 Balance of System Components (BOS)

Aside from the array, the charge controller, and the inverter, a number of other components are needed in a code-compliant PV system. For example, if a PV array consists of multiple series-parallel connections, as shown in Figure 20.19, then it is necessary to incorporate fuses or circuit breakers in series with each series string of modules, defined as a source circuit.

This fusing is generally accomplished by using a source circuit combiner box as the housing for the fuses or circuit breakers, as shown in Figure 20.19. The combiner box should be installed in a readily accessible location. The PV output circuit of Figure 20.19 becomes the input to the charge controller, if a charge controller is used. If multiple parallel source circuits are used, it may be necessary to use more than one charge controller, depending upon the rating of the charge controller. When more than one charge controller is used, source circuits should be combined into separate output circuits for each charge controller input. In a utility-interactive circuit with no battery backup, a charge controller is not

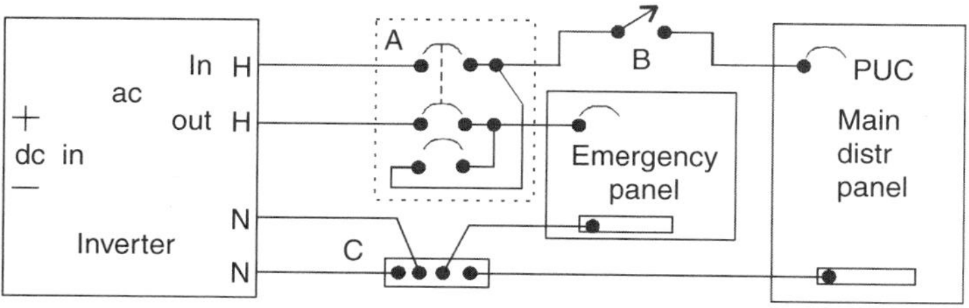

FIGURE 20.18 Utility-interactive PV system connections to emergency loads and to utility.

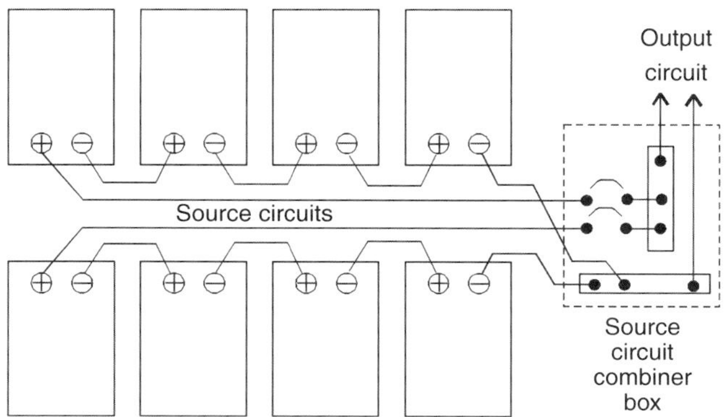

FIGURE 20.19 Example of PV source and output circuits.

necessary. The PV output circuit connects directly to the inverter through either a DC disconnect or a DC ground fault detection and interruption device (GFDI).

A GFDI device is required by the *National Electrical Code* (NEC) (NFPA 2002) whenever a PV array is installed on a residential rooftop. The purpose of the device is to detect current flow on the grounding conductor. The grounding conductor is used to ground all metal parts of the system. In a properly installed and operating system, no current will flow on the grounding conductor. Normally the negative conductor of the PV array is grounded, but this ground, if properly installed, will be attached to the grounding conductor at only one point, as shown in Figure 20.20, where the negative PV output conductor is connected to the equipment grounding bus through the 1A circuit breaker. The 1A circuit breaker is ganged to the 100 A circuit breaker so that if the current through the 1A circuit breaker exceeds 1A, both breakers will trip. When the two circuit breakers are open, current flow on both the PV output circuit conductors as well as the grounding conductor is interrupted. If the fault current on the grounding conductor was the result of an arcing condition between one of the PV circuit conductors and ground, the arc will be extinguished, thus preventing a fire from starting.

The NEC also requires properly rated disconnects at the inputs and outputs of all power conditioning equipment. An additional disconnect will be needed at the output of a charge controller as well as between any battery bank and inverter input or DC load center. If the disconnect is to disconnect DC, then the NEC requires that it be rated for DC. Additional disconnects are needed at the output of any inverter. If the inverter is utility interactive with battery backup for emergency loads, it is desirable to include an inverter bypass switch at the inverter output in case inverter maintenance is required without interruption of power to emergency loads. In addition to the inverter bypass switch, many utilities

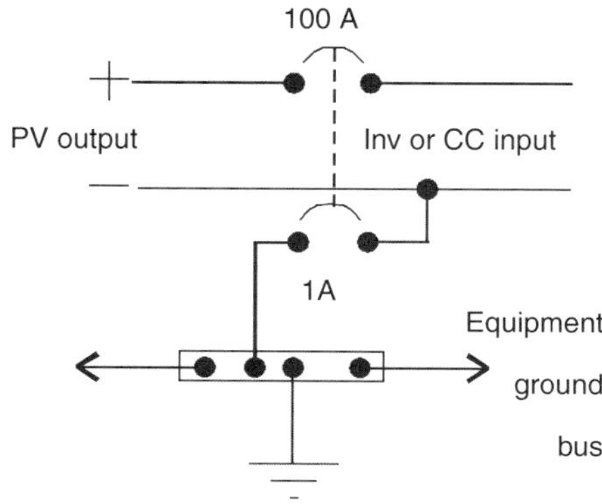

FIGURE 20.20 Use of GFDI.

require a visible, lockable, accessible, load break, disconnect between the inverter output and the point of utility connection. This switch is for use by the utility if they deem it necessary to disconnect the inverter from the line for any reason.

The point of utility connection for a utility-interactive system will normally be a backfed circuit breaker in a distribution panel. This circuit breaker is to be labeled so maintenance workers will recognize it as a source of power to the distribution panel. Figure 20.18 shows the connections for an inverter bypass switch (A), the utility disconnect switch (B) and the point of utility connection circuit breaker (PUC). The figure also shows a neutral bus (C) for connection of neutrals for the main distribution panel, the emergency panel and the inverter. Operation of the inverter bypass switch is as follows: The two-pole unit and the one-pole unit are ganged together so that either both are off or only one is on. Under normal operation, the two-pole unit is on and the one-pole unit is off. This connects the utility to the inverter and the inverter emergency output to the emergency panel. When the two-pole is off and the one-pole is on, the utility is connected to the emergency panel and the inverter is bypassed. When both are off, the utility is disconnected from both the inverter and the emergency panel. It is interesting to note that if the PUC circuit breaker in the main distribution panel is turned off, the inverter will interpret this as an interruption in utility power and will shut down the feed from the inverter to the main distribution panel. Thus, the energized portions of the circuit breaker will be the same as the energized portions of the other circuit breakers in the panel. When it is on, both sides of the circuit breaker will be energized. When it is off, only the line side will be energized.

Article 690 of the NEC governs the sizing of conductors in the PV system. The serious designer should carefully review the requirements of this article, especially because many PV systems use low-voltage DC where the voltage drop in the connecting wiring can be a problem. Sizing of conductors must be done carefully.

Chapter 18 provides information about storage batteries.

20.1.8 PV System Examples

20.1.8.1 A Standalone PV Well Pump System

As long as the depth of the well, the well replenishment rate, and the necessary flow rate are known, a PV pumping system can be designed. PV pumping systems are so common, in fact, that they often come in kits that include PV modules, a pump controller (LCB), and a pump. Pump manufacturers generally provide specifications that indicate, for a given pumping height, the amount of water pumped and the current drawn by the pump for specified pump voltages.

As an example, consider a system designed to pump 2000 gal/day from a well that is 200 ft deep and has a replenishment rate that exceeds the desired pumping rate. Assume the location for the pumping system has a minimum of 5 peak sun hours per day. This means that the 2000 gal must be pumped in 5 h, which corresponds to a pumping rate of 2000 gal/300 min $=6.67$ gpm. One pump that meets this requirement is a 1.0 HP pump that will pump 7.6 gpm to a height of 200 ft. Under these pumping conditions, the pump will draw 6.64 A at a DC voltage of 105 V. An 875-W PV array is recommended for the operation of this system by the distributor. Note that (6.64 A) $\times$ (105 V) $=697$ W, indicating that the recommended PV array is rated at 125% of the system requirements.

Before committing to this system, however, it should be compared with a system that uses battery storage and a smaller pump. The cost of the 1.0-hp pump is close to \$1800, while a 0.25-hp pump that will pump 2.15 gpm while consuming 186 W can be purchased for about \$500. This pump will need to pump for 15.5 h to deliver the 2000 gal, so the energy consumption of the pump will be (186 W) $\times$ (15.5 h) $=$ 2884 Wh. If the pump runs at 24 V DC, this corresponds to $2884 \div 24 = 120$ Ah per day. For PV storage, deep-discharge lead–acid batteries are normally used, and it is thus necessary to ensure that the batteries will provide adequate storage for the pump without discharging to less than 20% of full charge. Thus, the battery rating must be at least $120 \div 0.8 = 150$ Ah for each day of storage. If the water is pumped into a tank, then the water itself is a form of energy storage, and if the tank will hold several days supply of water, then the batteries will only need to store enough energy to operate the pump for a day. If it is less expensive to use more batteries than to use a larger water tank, then additional batteries can be used.

So, finally, a sensible system will probably consist of a 0.25-hp pump, an MPT charge controller for the batteries and a minimum of 150 Ah at 24 V of battery storage. With the MPT controller, the array size, assuming 5 psh minimum per day, becomes (2884 Wh)×1.25÷(5 h)=721 W, where the 1.25 factor compensates for losses in the array due to operation at elevated temperatures, battery charging and discharging losses, MPT losses, and wiring losses. This array can be conveniently achieved with 120-W modules configured in an array with two in series and three in parallel, as shown in Figure 20.21.

As a final note on the pumping system design, it is interesting to check the wire sizes. The NEC gives wire resistance in terms of Ω/kft. It is good design practice, but not an absolute requirement, to keep the voltage drop in any wiring at less than 2%. The overall system voltage drop must be less than 5%. The wire size for any run of wire can thus be determined from

$$\Omega/\text{kft} \le \frac{(\%VD)V_S}{0.2\cdot I\cdot d}, \tag{20.10}$$

where $\%VD$ is the allowed voltage drop in the wiring expressed as a percentage, V_S is the circuit voltage, I is the circuit current, and d is the one-way length of the wiring.

For the PV source circuit wiring, V_S will be about 34 V and I will be about 7 A. If the one-way source circuit length is 40 ft, then, for a 2% voltage drop, Equation 20.10 evaluates to Ω/kft =1.2143. NEC Chapter 9, Table 8 shows that #10 solid Cu wire has 1.21 Ω/kft, whereas #10 stranded Cu has 1.24 Ω/kft. So either type of #10 will keep the $\%VD$ very close to 2%. Because #10 THWN-2 is rated to carry 40 A at 30°C, it is adequate for the job even under most derating conditions. Because the pump will be submersed, it will need 200 ft of wire just to get out of the well. If the controller is close to the well, then d will be approximately 210 ft. Thus, for I=(186 W)÷(24 V)=7.75 A, and V_S=24 V, Equation 20.10 yields Ω/kft =0.1475, which requires #1/0 Cu according to NEC Chapter 9 Table 8. A 3% voltage drop would allow the use of #2 Cu. In either case, this is a good example of how wire size may need to be increased to keep voltage drop at acceptable levels when relatively low-voltage DC is used. The 30°C ampacity of #2 Cu, for example is 130 A. Therefore, even the small, low-voltage DC pump may not be the best choice. With inverter price decreasing and reliability increasing, and with AC motors generally requiring less maintenance than DC motors, at the time this article is being read it may be more cost effective to consider a 120-V or 240-V AC pump for this application.

20.1.8.2 A Standalone System for a Remote Schoolhouse

Standalone system design requires a tabulation of the system loads, generally expressed in ampere hours (Ah) at the battery voltage. Suppose, for example, it is desired to provide power for 400 W of lighting, 400 W of computers and 200 W of refrigeration, all at 120 V AC. Suppose that all of the loads operate for 8 h/day. This means the load to be met is 8 kWh/day at 120 V AC. If this load is supplied by an inverter that operates with 92% efficiency, then the batteries must supply the inverter

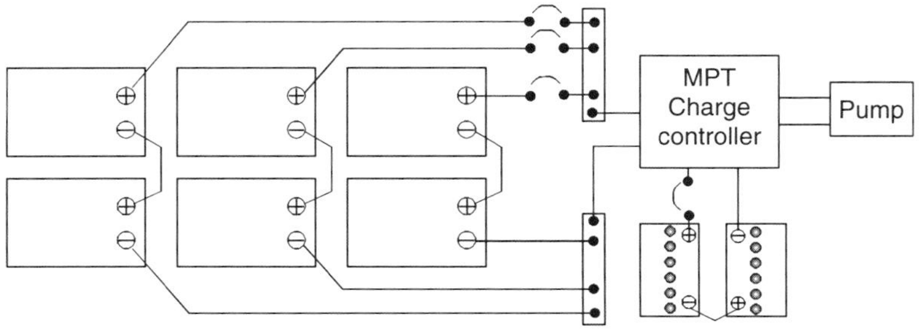

FIGURE 20.21 Water pumping system with battery storage and MPT charge controller.

with 8 ÷ 0.92 = 8.7 kWh/day. If the inverter input is 48 V DC, then the daily load in Ah is (8700 Wh) ÷ (48 V) = 181 Ah. To meet the needs for one day of operation, the batteries should thus be rated at 125% of 181 Ah = 226 Ah. But for a standalone system, it is usually desirable to provide more than one day of storage. For this system, three days would be more common, so a total of 678 Ah at 48 V should be used.

If an MPT charge controller is used, then the array can be sized based upon the daily system Wh and the available daily psh, taking losses into account. First, battery charging and discharging is only about 90% efficient. Therefore, to get 181 Ah out of the batteries, it is necessary to design for 181 ÷ 0.9 = 201 Ah into the batteries. At 48 V, this is 9648 Wh. Next, it is necessary to include a 10% degradation factor for array maintenance, mismatch and wiring losses, and another 15% factor for elevated array operating temperature. Therefore, the array should be designed to produce 9648 ÷ 0.9 ÷ 0.85 = 12,612 Wh/day. Assuming a worst-case psh = 5 h/day, this means an array size of 12,612 ÷ 5 = 2522 W will be needed.

This can be achieved with 20 125-W modules in a four-series-by-five-parallel array, or with with 15 167-W modules in a three-series-by-five-parallel array, or by any number of other module combinations that do not exceed the charge controller maximum input voltage limit when running open circuit. It must be remembered that achieving the nominal 48-V source circuit output may require two, three, or four modules in series. Thus, each additional parallel module will require additional series modules to achieve the system voltage. Figure 20.22 shows the block diagram of the schoolhouse system.

To this point, wind loading of the array has not been mentioned. In areas of high-wind loads, the size of the module and the mounting method may or may not be adequate to meet high wind-loading conditions. The final check on any PV system design must be a determination of whether the system will blow away in a high wind. This is especially undesirable considering the cost of a system as well as the fact that many systems are installed to provide power during emergencies, one of which might be a hurricane.

As a final note on system design, if an MPT charge controller is not used, then the array should be sized to provide 110% of the daily battery input Ah, using the maximum power current of the array. In this case, the daily battery input Ah is 201, so the array should be designed for 221 Ah. For 5 psh, this converts to an array current of 44.2 A. The 125-W modules have a 7-A maximum power current, so this means six in parallel will produce 42 A, which is close to the required amount. Therefore, the MPT controller saves four modules, or 500 W. At $4/W, this is a savings of $2,000, which more than pays for the additional cost of the MPT charge controller.

It is interesting to look at a life-cycle cost of the schoolhouse system. Using a discount rate of 5% and an inflation rate of 3%, an LCC cost estimate can be tabulated. In Table 20.1, it is assumed that the batteries are 12 V, 110 Ah, sealed, AGM lead-acid deep-cycle batteries with a rated lifetime of 8 years and a cost of $150 each.

To compare this system with a gasoline generator, note that the generator would need to generate 8 kWh/day for 20 years by operating 8 h/day over this period. A typical small gasoline generator will

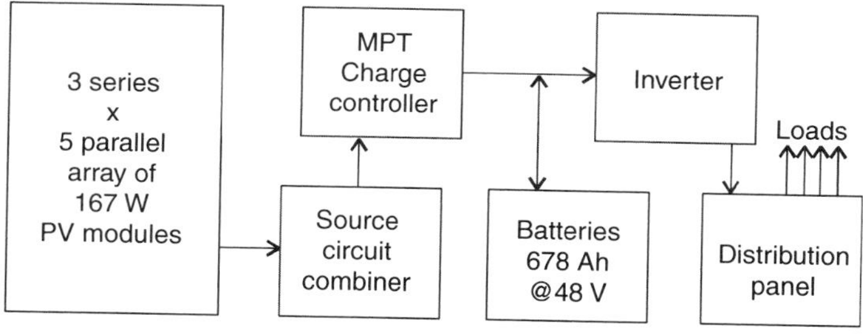

FIGURE 20.22 Block diagram of schoolhouse PV system.

TABLE 20.1 Life-Cycle Costs of a Standalone Photovoltaic System for a Schoolhouse

Item	Cost	Present Worth	% LCC
Capital Costs			
Array	$10,000	$10000	38.1
Batteries	3600	3600	13.7
Array mount	1250	1250	4.8
Controller	500	500	1.9
Inverter	500	500	1.9
BOS	1000	1000	3.8
Installation	2000	2000	7.6
Recurring Costs			
Annual Inspection	50	839	3.2
Replacement Costs			
Batteries −8 yr	3600	3087	11.8
Batteries −16 yr	3600	2646	10.1
Controller −10 yr	500	413	1.6
Inverter −10 yr	500	413	1.6
Totals		$26,247	100

generate 4 kWh/gal (Messenger and Ventre 2004), so will require 2 gal of gasoline per day. The generator will require an oil change every 25 h, a tune-up every 300 h and a rebuild every 3000 h. The LCC analysis for the generator is shown in Table 20.2. Clearly, the PV system is the preferred choice. And this does not account for the noise-free, pollution-free performance of the PV system.

20.1.8.3 A Straightforward Utility-Interactive PV System

Because utility-interactive PV systems are backed up by the utility, they do not need to be sized to meet any particular load. Sometimes they are sized to meet emergency loads, but if the system does not have emergency backup capabilities, then they may be sized to fit on a particular roof, to meet a particular budget, or to incorporate a particular inverter. Suppose the sizing criteria is the inverter, which has the following specifications:

DC Input	Input Voltage Range	250–550 V
	Maximum Input Current	11.2 A
AC Output	Voltage	240 V 1ϕ
	Nominal Output Power	2200 W
	Peak Power	2500 W
	Total Harmonic Distortion	< 4%
	Maximum Efficiency	94%

Note that if the input voltage is 550 V and the input current is 11.2 A, the input power would be 6160 W. Because the peak output power of this inverter is 2500 W, it would not make sense to use a 6160-W array

TABLE 20.2 Life-Cycle Costs of a Gasoline Generator for a Schoolhouse

Item	Cost	Present Worth	%LCC
Capital Costs			
Generator	$750	$750	2
Recurring Costs			
Annual fuel	1825	30,593	73
Annual oil changes	235	3939	9
Annual tune-ups	345	4107	10
Annual rebuilds	146	2447	6
		$41,836	100

since most of the output would be wasted, and it might be easier to damage the inverter. Therefore, an array size of about 2500 W would make better sense.

Because cloud focusing can increase the short circuit output current of a module by 25%, the array rated short circuit current should be kept below $11.2 \div 1.25 = 8.96$ A. The number of modules in series will depend upon the maximum voltage of the array at low temperatures remaining less than 550 V and the minimum array voltage at high array operating temperatures remaining greater than 250 V.

NEC Table 690.7 specifies multipliers for open-circuit voltages for different low-temperature ranges. For design purposes, suppose the coldest array temperature will be -25°C and the hottest array temperature will be 60°C. NEC Table 690.7 requires a multiplier of 1.25 for the array open-circuit voltage, so the maximum rated array open circuit voltage must be less than $550 \div 1.25 = 440$ V. If the open-circuit voltage of a module decreases by 0.47%/°C, then it will decrease by $35 \times 0.47 = 16.45\%$ when the module is operated at 60°C. Thus, the 25°C-rated array open-circuit voltage needs to be greater than $250 \div 0.8355 = 299$ V.

The next step is to look at PV module specifications. One module has $P_{max} = 125$ W, $V_{OC} = 21.0$ V and $I_{SC} = 7.2$ A. Thus, the maximum number of these modules in series will be $440 \div 21.0 = 20.95$, which must be rounded down to 20. The minimum number in series will be $299 \div 21.0 = 14.24$, which must be rounded up to 15. Checking power ratings gives 2500 W for 20 modules and 1875 W for 15 modules. Because 2500 W does not exceed the inverter rated maximum output power, and because the array will normally operate below 2500 W, it makes sense to choose 20 modules, as long as the budget can afford it and as long as there is room for 20 modules wherever they are to be mounted. Figure 20.15d shows the block diagram for this system.

The life-cycle cost of this type of system is usually looked at somewhat differently than that of the schoolhouse system. In this case, the cost of electricity generated is usually compared with the cost of electricity from the utility, neglecting pollution and other externalities. In regions with an abundance of trained installers, it is currently possible to complete a grid-connected installation for less than $7/W. The installed cost of the 2500-W system would thus be approximately $17,500. It is reasonable to expect an average daily output of 10 kWh for this system in an area with an average of 5 peak sun hours. The value of the annual system output will thus be approximately $365 at $0.10/kWh. This amounts to a simple payback period of 48 years—almost double the expected lifetime of the system.

Of course, what is not included in the analysis is the significant amount of CO_2 production that is avoided, as well as all the other pollutants associated with nonrenewable generation. Also not included are the many subsidies granted to producers of nonrenewable energy that keep the price artificially low. For that matter, it assumes that an abundance of fossil fuels will be available at low cost over the lifetime of the PV system. Finally, it should be remembered that the energy produced by the PV system over the lifetime of the system will be at least four times as much as the energy that went into the manufacture and installation of the system.

If the cost of the system could be borrowed at 3% over a period of 25 years, the annual payments would be $1005. Thus, if a grid-connected system is considered, unless there is a subsidy program, it could not be justified with simple economics. It would be purchased simply because it is the right thing to do for the environment. Of course, if the installation cost were less, the value of grid electricity were more and the average sunlight were higher, then the numbers become more and more favorable. If the values of externalities, such as pollution, are taken into account, then the PV system looks even better.

Because of the cost issue, as well as local PUC regulations or lack of them, ill-defined utility interface requirements, etc., few grid-connected PV systems are installed in areas that do not provide some form of incentive payments. In some cases, PV system owners are paid rebates based on dollars per watt. The problem with this algorithm is that there is no guarantee that the system will operate properly. In other cases, PV system owners are guaranteed a higher amount per kWh for a prescribed time, which guarantees that the system must work to qualify for incentive payments. In fact, at present, more kW of PV are installed annually in grid-connected systems than are installed in standalone systems (Maycock, 2003).

20.1.9 Latest Developments in PV

The PV field is developing rapidly. The latest developments of 2006 will be historic developments in 2008, so it is almost presumptuous to claim what are the "latest" developments when it is likely that by the time the reader sees this paragraph, the development will be history. Not only are significant developments being made in PV cell technology, but significant developments in PV system component technology are also being made. Several recent developments, however, are particularly notable, and other recent developments may also become notable. Rather than attempt to identify all of the latest important developments in PV, a few will be described here to foster sufficient curiosity in the reader to encourage further reading of those publications, such as the *Proceedings of the IEEE PV Specialists Conferences*, in which the latest developments are reported. Several additional publications that track new developments in PV are listed in the reference section (Maycock; American Solar Energy Society; Renewable Energy World; Goswami and Boer, International Solar Energy Society). *Cost*, *efficiency*, and *reliability* are the key words that drive the PV system development process, whether applied to PV cells, modules, inverters, array mounts, or other BOS components.

The current buzzword in crystalline silicon technology is "thin silicon." The buried contact cell of the University of New South Wales has been described earlier. Other processes that may show promise are thin Si on ceramics (Green 2004), thin crystalline Si on glass (CSG) (Basore 2002a; Basore 2002b), and epitaxially grown Si on existing crystalline Si with subsequent removal of the epitaxially grown cell from the existing substrate (the PSI process) (Brendel et al. 2005). Combinations of crystalline and amorphous Si are also showing very promising results. For example, the heterojunction with intrinsic thin-layer (HIT) process has been used to fabricate a 21% efficient cell that has been incorporated into an 18.4% module (Hamakawa 2005).

The interconnections of conventional Si cells add an additional processing step in module production. Perhaps more importantly, these interconnects have led to cell failures due to interconnect failure. Elimination of soldered interconnects through incorporating monolithic technology can lead to an expedited production process, greater module reliability, and, perhaps more importantly, less sensitivity of the module to shading because the cells extend the length of the module and have a relatively narrow width. It is thus more difficult to entirely shade a single cell of a module.

Historically, inverters have been the weakest link of a PV system. Recently, the US Department of Energy (2005) has sponsored the development of inverters designed to have at least a 10-year mean time to failure. Thus, as the industry continues to mature, PV systems continue to come closer to simple "plug-and-play" configurations that will be relatively simple to install and will require very little maintenance over the projected 25-year lifetime of the system.

Recently, Green (2006) compiled a listing of the highest confirmed efficiencies for a range of photovoltaic cell and module technologies. Table 20.3a was compiled from this reference and shows the efficiency values of cells and the independent test centers where efficiencies were confirmed. Table 20.3b shows similar information for modules. As expected, the module efficiencies are less than the cell efficiencies.

20.1.10 Future Challenges for PV Systems

The PV industry is currently engaged in an effort to ensure quality control at all levels of system deployment, including manufacturing, distribution, design, installation, inspection and maintenance. At this point in time (2005), most of the technical challenges for PV system components have been overcome. PV modules are very reliable, and most are warranted for 25 years. Reliability of other system components continues to improve and, at this point, PV system power conditioning equipment has proven to operate very reliably. Therefore, once installed properly, modern PV systems require very little maintenance. In fact, systems without batteries require almost no maintenance.

To meet the design challenge, many manufacturers and distributors are offering "kits" with standardized components and installation instructions. Although these systems meet many needs,

TABLE 20.3a Confirmed Terrestrial Cell and Submodule Efficiencies Measured Under the Global AM1.5 Spectrum (1,000 Wm^{-2})

Classification[a]	Effic. Description (%)[b]	Area (cm^2)[c]	V_{oc} (V)	J_{sc} (mA/cm^2)	FF (%)[d]	Test Center (and Date)[e]
Silicon						
Si (crystalline)	24.7±0.5 UNSW PERL	4.00 (da)	0.706	42.2	82.8	Sandia (3/99)
Si (multicrystalline)	20.3±0.5 FhG-ISE	1.002 (ap)	0.664	37.7	80.9	NREL (5/04)
Si (thin-film transfer)	16.6±0.4 U. Stuttgart (45 μm thick)	4.017 (ap)	0.645	32.8	78.2	FhG-ISE (7/01)
Si (amorphous)[f]	9.5±0.3 U. Neuchatel	1.070(ap)	0.859	17.5	63.0	NREL (4/03)
Si (nanocrystalline)	10.1±0.2 Kaneka (2 μm on glass)	1.199(ap)	0.539	24.4	76.6	JQA (12/97)
III–V Cells						
GaAs (crystalline)	25.1±0.8 Kopin, AlGaAs window	3.91(t)	1.022	28.2	87.1	NREL (3/90)
GaAs (thin-film)	24.5±0.5 Radboud U., NL	1.002(t)	1.029	28.8	82.5	FhG-ISE (5/05)
GaAs (multicrystalline)	18.2±0.5 RTI, Ge substrate	4.011(t)	0.994	23.0	79.7	NREL (11/95)
InP (crystalline)	21.9±0.7 Spire, epitaxial	4.02(t)	0.878	29.3	85.4	NREL (4/90)
Thin-film chalcogenide						
CIGS (cell)	18.4±0.5[g] NREL, CIGS on glass	1.04(ap)	0.669	35.7	77.0	NREL (2/01)
CIGS (submodule)	16.6±0.4 U. Uppsala, 4 serial cells	16.0(ap)	2.643	8.35	75.1	FhG-ISE (3/00)
CdTe (cell)	16.5±0.5[g] NREL, mesa on glass	1.032(ap)	0.845	25.9	75.5	NREL (9/01)
Photochemical						
Nanocrystalline dye Sharp	10.4±0.3	1.004 (ap)	0.729	21.8	65.2	AIST(8/05)
Nanocrystalline dye INAP (submodule)	4.7±0.2	141.4 (ap)	0.795	11.3	59.2	FhG-ISE (2/98)

(continued)

TABLE 20.3a (*Continued*)

Classification[a]	Effic. Description (%)[b]	Area (cm^2)[c]	V_{oc} (V)	J_{sc} (mA/cm^2)	FF (%)[d]	Test Center (and Date)[e]
Multijunction devices						
GaInP/GaAs/Ge	32.0 ± 1.5 Spectrolab (monolithic)	3.989(t)	2.622	14.37	85.0	NREL (1/03)
GaInP/GaAs	30.3 Japan Energy (monolithic)	4.0(t)	2.488	14.22	85.6	JQA (4/96)
GaAs/CIS (thin-film)	25.8 ± 1.3 Kopin/Boeing (4 terminal)	4.00(t)	–	–	–	NREL (11/89)
a-Si/CIGS (thin-film)[h]	14.6 ± 0.7 ARCO (4 terminal)	2.40(ap)	–	–	–	NREL (6/88)
a-Si/μc-Si (thin submodule)[i]	11.7 ± 0.4	14.23(ap)	5.462	2.99	71.3	AIST (9/04)

[a] CIGS = $CuInGaSe_2$; a-Si = amorphous silicon/hydrogen alloy.
[b] Effic. = efficiency.
[c] (ap) = aperture area; (t) = total area; (da) = designated illumination area.
[d] FF = fill factor.
[e] FhG-ISE = Fraunhofer-Insitut für Solare Energiesysteme; JQA = Japan Quality Assurance; AIST = Japanese National Institute of Advanced Industrial Science and Technology.
[f] Stabilized by 800 h, 1 sun AM1.5 illumination at a cell temperature of 50°C.
[g] Not measured at an external laboratory.
[h] Unstabilized results.
[i] Stabilized by 174 h, 1-sun illumination after 20 h, 5-sun illumination at a sample temperature of 50°C.

Source: From Deng, X. and Schiff, E. A., Amorphous silicon-based solar cells, In *Handbook of Phtovoltaic Science and Engineering*, Luque, A. and Hegedus, S., eds., Wiley, New York, 2003.

TABLE 20.3b Confirmed Terrestrial Module Efficiencies Measured Under the Global AM1.5 Spectrum (1,000 W/m^2) and at 25°C

Classification[a]	Effic. Description (%)[b]	Area (cm^2)[c]	V_{oc} (V)	I_{sc} (A)	FF (%)[d]	Test Center (and Date)
Si (crystalline)	22.7±0.6 UNSW/Gochermann	778 (da)	5.60	3.93	80.3	Sandia (9/96)
Si (multicrystalline)	15.3±0.4[e] Sandia/HEM	1017 (ap)	14.6	1.36	78.6	Sandia (10/94)
Si (thin-film polycrystalline)	8.2±0.2 Pacific Solar (1–2 μm on glass)	661 (ap)	25.0	0.318	68.0	Sandia (7/02)
CIGSS	13.4±0.7 Showa Shell (Cd free)	3459(ap)	31.2	2.16	68.9	NREL (8/02)
CdTe	10.7±0.5 BP Solarex	4874(ap)	26.21	3.205	62.3	NREL (4/00)
a-Si/a-SiGe/a-SiGe (tandem)[f]	10.4±0.5 USSC (a-Si/a-Si/a-Si:Ge)	905(ap)	4.353	3.285	66.0	NREL (10/98)

[a] CIGSS = CuInGaSSe; a-Si = amorphous silicon/hydrogen alloy; a-SiGe = amorphous silicon/germanium/hydrogen alloy.
[b] Effic. = efficiency.
[c] (ap) = aperture area; (da) = designated illumination area.
[d] FF = fill factor.
[e] Not measured at an external laboratory.
[f] Light soaked at NREL for 1000 h at 50°C, nominally 1-sun illumination.

there are still needs for custom systems, and design professionals are still generally unfamiliar with PV system design. In some areas where PV systems have been encouraged, there are properly trained installers and maintenance personnel. But in most areas, it is difficult to find a contractor who is familiar with PV systems. It is not unusual for a general contractor to oppose the installation of a PV system out of fear that it may compromise structural integrity. Furthermore, once a system is installed, it is usually difficult to find someone who understands PV systems well enough to be able to inspect a system for code compliance. It is also difficult to find someone who can troubleshoot a nonfunctional system, or, for that matter, to determine that a system is not working up to specifications. Therefore, in the case of design, installation, inspection and maintenance, there is a lot of education and training to accomplish.

There are also some nontechnical challenges. One is cost. The installed cost of a PV system is still three to six times as high as the installed cost of conventional nonrenewable generation. As fuel costs rise and as PV installed costs decrease, it is likely that within a decade, the two costs may become equal, especially if fuel and environmental costs are taken into account. If the cost of a PV system were to be measured in Btus, then perhaps the importance of PV systems would become more obvious. By the early 1980s, Odum and Odum (1981); Henderson (1981) had proposed using the Btu as the international monetary standard. Recently, Messenger and Ventre (2004) report a conservative factor of 4:1 return on the Btus expended in the deployment of a PV system, and Battisti and Corrado (2005) report energy payback times (EPBT) of approximately three years and CO_{2eq} payback times of about four years for PV systems. Because the life expectancy of a PV system exceeds 20 years, this equates to about a 7:1 return on the Btu's invested and a 5:1 return on avoidance of CO_2 and equivalent greenhouse gases released during the production and installation phase of a PV system. These positive return on investment figures make PV a very attractive energy source and suggest that the dollar cost of a PV system does not adequately portray the true environmental value of a PV system.

In a recent review paper, Hamakawa (2005) reported that in 2000, commercial silicon PV module efficiency was 13%–14% and the cost about 300 Yen/W (approximately \$3–\$4/W). He predicted that by 2010 the module efficiency would go up to 17% and the cost would come down to 100 Yen/W (approximately \$1/W) and by 2015–2020 the module efficiency will increase to 20% with the costs down to 75 Yen/W (approximately 75 cents/W). Based on the progress in research as reported by Green (2004), where thin-film cell efficiencies of 24.7% were achieved in 2003, it is not too difficult to imagine that Hamakawa's predictions about module efficiencies may be on target. Only the future will tell whether the cost predictions will be realized or not; however, it is clear that the costs are moving in the right direction.

20.2 Thin-Film PV Technology

Hari M. Upadhyaya, Takhir M. Razykov, and Ayodhya N. Tiwari

20.2.1 Introduction

20.2.1.1 Historical and Current Developments

Crystalline silicon (c-Si) technology has a lion's share in the present photovoltaic (PV) industry, contributing more than 95% through the cells and modules based on mono- and multicrystalline wafer technology. The recent growth rate of the photovoltaic industry and market is phenomenal, with a substantial surge of over 30%–40% recorded globally during last few years. During the early developmental phase of c-Si PV technology, the continuous feedstock support offered by the Si-based electronics industry played a key role in its growth. The high-purity and even second-grade wafer materials obtained at a relatively cheaper price proved favorable for the PV industry, as they led to a reasonable efficiency (average $\eta > 15\%$) and extremely good performance stability (more than 25 years) that are two essential requirements for any technology to successfully demonstrate its potential (Chopra et al. 2004; Jäger-Waldau 2004). However, continuously increasing demand for PV modules and the need for low-cost PV options have stretched these advantages to the limit and have exposed some inherent disadvantages of c-Si technology, such as the scarcity of feedstock material and costly processing of materials and device fabrication steps, as well as the inability for monolithic interconnections. These, in turn, restrict the potential of Si wafer technology and make it difficult to achieve PV module production cost below € 1/W (1€ is about US$1.20), which is considered essential for cost-competitive generation of solar electricity (Hegedus and Luque 2003; von Roedern et al. 2005; Zweibel 2000). The PV module cost depends on the total manufacturing cost of the module per square area, conversion efficiency, and long-term performance stability. Figure 20.23 gives an estimate of achievable cost with c-Si technology and a comparison with projected achievable costs with other PV technologies. It is generally agreed that c-Si wafer technology would be unable to meet the low cost targets, whereas thin-film technologies have the potential to provide a viable alternative in the near future.

In addition, Si wafers are fragile, solar cell area is limited by the wafer size, and the modules are bulky, heavy, and nonflexible. To overcome some of the problems of c-Si wafer technology, efforts are being made to develop monocrystalline and polycrystalline thin-film silicon solar cells, as reviewed by Bergmann (1999).

It has been realized that application of thin-film technologies to grow solar cells and modules based on other materials can solve some of the problems of the silicon-wafer technologies. Thin-film deposition of materials of required quality and suitable properties depends on the processes used and the control of several parameters. However, once optimized, these methods provide over an order of magnitude cheaper processing cost and low-energy payback time, which is certainly a big advantage.

Research on alternatives to c-Si started about four decades back. It has a low-absorption coefficient ($\sim 10^3$ cm^{-1}) and its bandgap ($E_g \sim 1.1$ eV) is far below than the optimum $E_g \sim 1.5$ eV required for high solar-electric conversion using a single-junction solar cell. Some of the most interesting semiconducting materials that have received considerable attention are: cadmium telluride (CdTe), gallium arsenide (GaAs), indium phosphide (InP), zinc phosphide (Zn_3P_2), copper sulfide (Cu_2S), copper indium diselenide (CIS), and copper–indium gallium diselenide (CIGS). These all have the electronic and optical properties suitable for the efficient utilization of the sun's spectrum (see Table 20.4 for bandgaps). Multijunction solar cells based on III–V materials (GaAs, InP, GaSb, GaInAs, GaInP, etc.) show high efficiency, exceeding 35%, but due to the high production costs and low availability of their constituents, these solar cells are not considered suitable for cost-effective terrestrial applications, although they are still very important for space PV applications.

Thin-film technology has answers and potential to eliminate many existing bottlenecks of c-Si PV programs, which are experienced at different levels from module production to its application. Thin-film

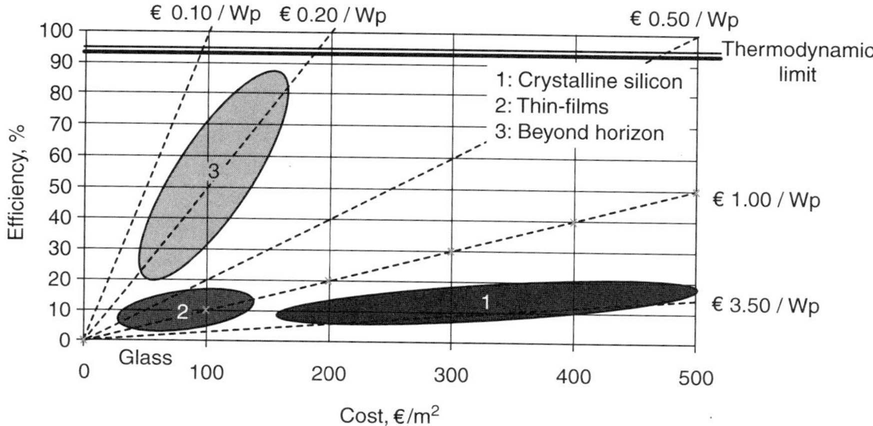

FIGURE 20.23 Comparison of estimated costs achievable with different PV technologies as a function of manufacturing cost (€/m^2) and conversion efficiency. (Courtesy M. Green, UNSW.)

TABLE 20.4 Bandgaps of Different Semiconductor Materials Suitable as Light Absorber in Solar Cells

Compound/Alloys	Bandgap (eV)	Compound/Alloys	Bandgap (eV)
c-Si	1.12	Zn_3P_2	1.50
a-Si	1.70	$CuInSe_2$	1.04
GaAs	1.43	$CuGaSe_2$	1.68
InP	1.34	$Cu(In,Ga)Se_2$	1.20
Cu_2S	1.20	$CuInS_2$	1.57
CdTe	1.45	$Cu(In,Ga)(S,Se)_2$	1.36

PV modules are manufactured on either rigid glass substrates or flexible substrates (thin metallic or plastic foils). Some of the advantages of the thin-film technologies are:

- The high-absorption coefficient ($\sim 10^5\ cm^{-1}$) of the absorber materials is about 100 times higher than c-Si (Figure 20.24); thus about 1–2 μm of material thickness is sufficient to harness more than 90% of the incident solar light. This helps in reducing the material mass significantly to make modules cost-effective. The estimated energy payback time of the thin-film PV is 3–5 times lower than that of c-Si PV.
- The formation of heterojunction and better device engineering for reduction of photon absorption losses and enhanced collection of photogenerated carriers are possible.
- Large-area deposition (on the order of m^2), along with the monolithic integration (interconnection of cells during processing of rigid and flexible devices) is possible which minimizes area losses, handling, and packaging efforts.
- Roll-to-roll manufacturing of flexible solar modules is possible. This gives high throughput and thus can reduce the energy payback time significantly.
- Tandem/multijunction devices could be realized to utilize the full solar spectrum to achieve higher-efficiency (greater than 50%) devices.
- Flexible and lightweight PV facilitates several attractive applications.

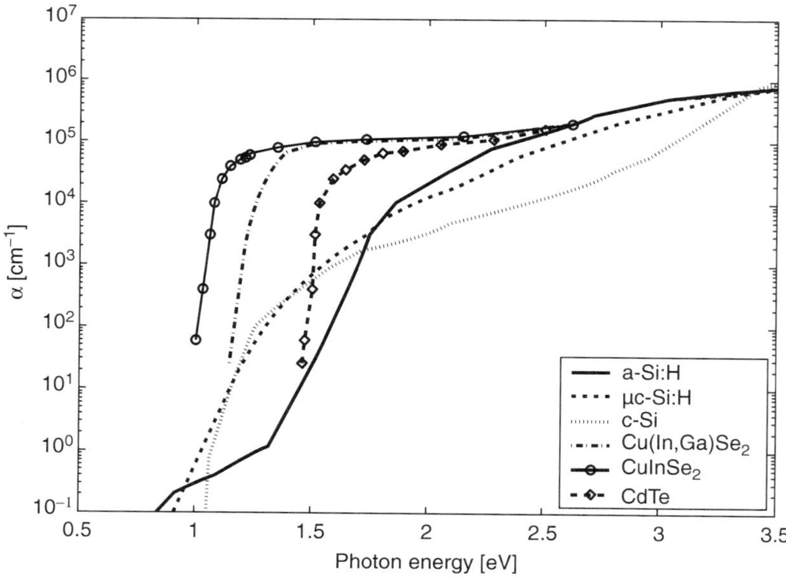

FIGURE 20.24 The optical absorption (α) versus bandgap (E_g) spectra of the c-Si and other prominent light-absorbing materials that are used in thin-film solar cells.

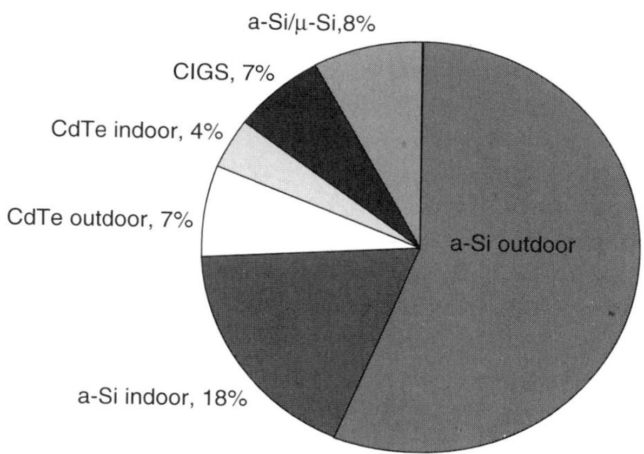

FIGURE 20.25 Worldwide sales of thin-film solar cells in 2002 were 43.8 MW. (From Maycock, P. D. *Renewable Energy World*, 6, 84–101, 2003.)

Historically, cadmium sulfide and copper sulfide (CdS/Cu_2S) heterojunctions were one of the early technologies, almost contemporary to c-Si, but could not survive due to the astounding success of c-Si technology that was efficient, stable, backed by electronic industries, and widely supported by space programs. Despite Cu_xS having stability problems, the technology went to pilot production in the late 1970s or early 1980s, but with the advent of a-Si and CdTe thin-film solar cells, it was abandoned before the mid 1980s.

Currently, a-Si, CdTe, and CIGS are considered the mainstream thin-film technologies. Out of these, a-Si currently has the highest market share (Figure 20.25), as with the involvement of several industries worldwide during the last two decades, this technology has attained a significant level of maturity for large volume production. In addition, CdTe and CIGS have proven their potential to become leading technologies but they lacked adequate R&D and industrial and investors' support to make a mark. However, recently several companies have started production-related activities and production volume is expected to increase during the next few years. Recent developments on a new technology for organic and dye-sensitized solar cells appear to be quite interesting with their low-temperature and cheaper processing cost on flexible substrates (Gratzel 2000; Durr et al. 2005). However, there are potential issues of instability and degradation. They are in the infancy of their development at the moment and would require considerable efforts and time to see them become a commercially viable option.

Figure 20.26 (von Roedern et al. 2005) summarizes the best laboratory-scale (cell area less than 1 cm^2) efficiencies of some prominent thin-film technologies. Out of these, an efficiency of 19.5% achieved at NREL, USA with CIGS makes it the most efficient thin-film PV device, which has narrowed the gap further between existing c-Si and thin-film technologies. Table 20.5 summarizes the current status of thin-film solar cell and module development at laboratory and industrial levels on rigid substrates (with one exception). These contain some of the latest developments and global efforts to make thin-film PV technology a viable option for cost-effective and competitive electrical energy production.

20.2.1.2 Cost Potentials and Material Availability Issues

Estimates suggest that solar cell production costs below 1 €/W to up to 0.30 €/W are possible depending on the volume of production (Bruton et al. 1997; Keshner and Arya 2004; Woodcock et al. 1997). At present, there is a significantly large gap between the efficiencies of the solar cells at the R&D level, and those at the industrial modules level, because of the process-scalability-related problems during the early phase of industrial development. The current production cost of the thin-film PV modules is rather high because of low volume of production through pilot-production lines or small-capacity (less than 30 MW) plants. However, with larger production volumes and higher module efficiencies, the cost is

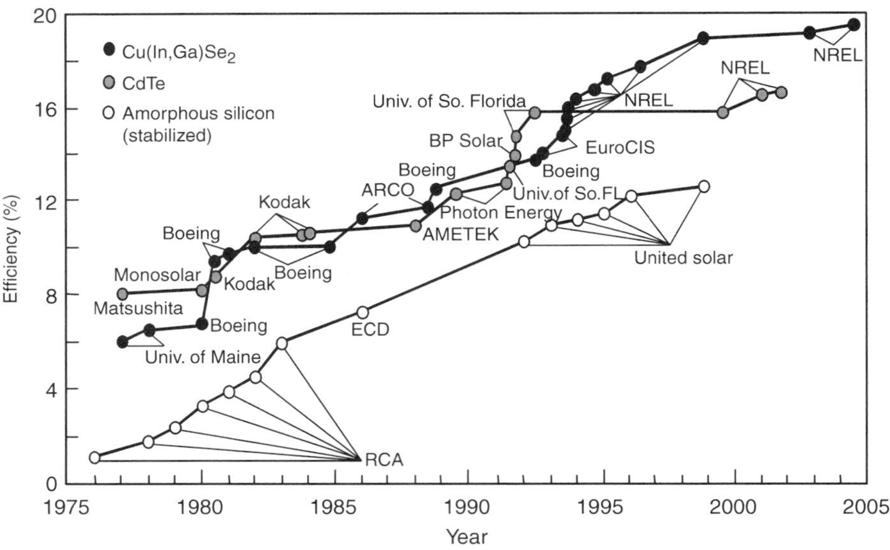

FIGURE 20.26 Development of "champion" thin-film solar cell efficiencies. CIGS and CdTe are single junction cells while a-Si is a multijunction device. (From von Roedern, B., et al., In *Proceedings of 31st IEEE Photovoltaic Specialists Conference*, pp. 183–188, 2005.)

expected to decrease significantly in the near future. On the basis of the trend derived from the past experiences of cumulative production vs. price variations (learning curve), it is estimated that module prices decrease by 20% for every doubling of the cumulative production (Surek 2003). These trends predict that with a growth rate of 25%, the cumulative production would reach ~75 GW by the year 2020, and the target cost below approximately 0.2€/W_p is achievable with existing thin-film technologies. c-Si counterpart technology will have to accelerate its growth rate many folds higher than the existing rate to achieve this target, which would be a difficult situation considering the balance in the cost of the feedstock and mature nature of the technology.

Currently, the conventional module assembly procedures used by the c-Si industries and adopted by thin-film industries restricts the overall cost of the thin-film modules. Thus, the advantage of using "less material" in thin-film technology under "back end" schemes of assembling is drastically reduced. Recently, some novel frameless mounting schemes (PV laminates on membranes) have been developed and pursued for a-Si technology, which bring significant reduction in cost potential and are very encouraging (Arya 2004). Figure 20.27 shows a promising potential with frameless modules. As can be seen, the cost of packaging and framing components has been drastically lowered from 25.5% to only 4%, which is significant; consequently, the cost reduction in materials and processing will have more meaning with further advancements in technologies of all the components.

Tuttle et al. (in press) have given cost estimates of CIGS solar modules from 2 to 2,000 MW/year capacity production plants including cost of equipment and materials. Similarly, there are several reviews and reports available that summarize the cost potential of thin-film modules in terms of different component costs and address some future concerns related with the cost and availability of materials, i.e., In, Ga, Se, and Te used in the leading thin-film technologies. An estimate in the subcontractor's report at NREL (Keshner and Arya 2004) suggests that the current production of these elements will have to be accelerated to fulfil the demand of 2–3 GW per year generation of electricity.

An important report (out of the DOE Solar Energy Technologies Program and PV FAQ, www.nrel.gov/ncpv) also exists (US Department of Energy 2004) that brings the judgment in favor of In, Ga, and Se. This report speculates that Ga and Se will not be constrained by supply, and a steady production growth rate of 0.16%

TABLE 20.5 Efficiencies of Solar Cells and Modules Based on Rigid Substrates (With the Exception of USSC Cells on Steel) with Their Configuration for Various Technologies

Material/Cell configuration	Companies Institutions/	Laboratory Cell/Modules Efficiency (%)	Area (cm^2)	Power (Watts)	Test Center/Date	Remarks/References
a-Si/a-SiGe/nc-SiGe	USSC	13.0(14.6)[a]	0.25		03/05	(Stainless steel substrate)
a-Si/a-SiGe	USSC	12.4	0.25		10/04	(Stainless steel substrate)
a-Si	Uni. Neuchatel	9.5	1.07		NREL (1/03)	Stabilized at 800 h
a-Si/μ-Si (submodule)	IPV Juelich	10.1(10.7)[a]	100		06/04	Stabilized at 1000 h
a-Si/a-SiGe/a-SiGe/SS	United Solar	7.6	9276	70.8	NREL (9/97)	
a-Si/a-SiGe/a-SiGe	BP Solarex	10.4	905		NREL(10/98)	
a-Si	Mitsubishi Heavy	6.4	15625	100	(7/ 05)	
a-Si	Kaneka	6.3	8100	51	NREL (7/04)	
a-Si/a-Si	EPV	5.7	7432	42.3	NREL (10/02)	
a-Si/ nano-Si	Sharp	11	4770	52.5	(7/05)	
a-Si/μ-Si (submodule)	Kaneka	11.7	14.23		AIST(9/04)	
μc Si (polycrystalline)	Pacific Solar	8.2	661		Sandia (7/02)	
CdS/CdTe	NREL	16.5	1.032		NREL (9/01)	
CdS/CdTe	BP Solarex	10.7	4874		NREL (4/00)	
CdS/CdTe	First Solar	10.2	6624	67.4	(2/04)	
CdS/CdTe	Antec Solar	7.3	6633	52.3	(6/04)	
CdS/CdTe	BP Solar	11.0	8390	92.5	NREL (9/01)	
CdS/CdTe	Matsushita	11.0	5413	59	NREL (6/00)	
CdS/CIGS	NREL	19.5	0.41		FhG-ISE (9/04)	
CdS/CIGS (submodule)	Univ. of Uppsala	16.6	16.0		FhG-ISE (3/00)	
CIGSS	Showa Shell	13.4	3459		NREL (8/02)	(Cd free)
CdS/ CIGS/SS	Global Solar	10.2	8709	88.9	NREL (5/05)	
CdS/CIGS/Glass	Wurth Solar	13.0	6500	84.6	NREL (6/04)	
CdS/ CIS-alloy/Glass	Shell Solar GmbH	13.1	4938	64.8	NREL (6/04)	
CdS/ CIS-alloy/Glass	Shell Solar Ind.	12.8	3626	46.5	NREL (3/03)	
Showa shell	13.4	3459	46.45	NREL (8/02)		ZnS(O.OH)CIGS/Glass
DSSC	EPFL, Lausanne	11.0	0.25		FhG-ISE (9/04),	Nanocrystalline TiO_2 (Gratzel, 2000)
DSSC	ECN	8.2	2.36		FhG-ISE (7/01)	
DSSC(submodule)	INAP	4.7	141.4		FhG-ISE (2/98)	

[a] Measured efficiency before light-induced degradation.

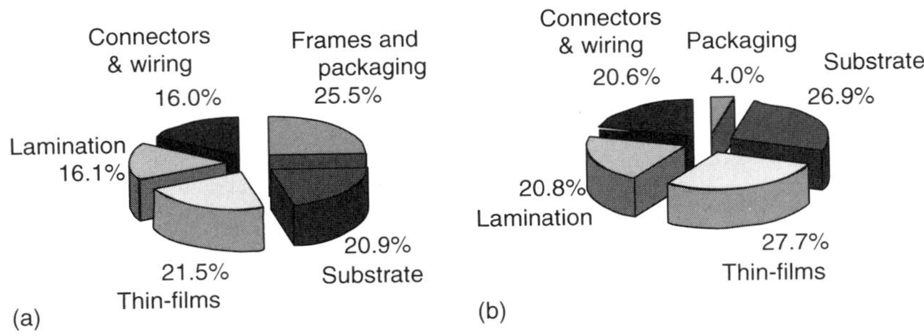

FIGURE 20.27 Distribution and break-up of the materials cost in (a) framed and (b) unframed a- Si based modules.

per year for Se would be sufficient to meet the demand of annual PV production of 20 GW/year by 2050. Furthermore, it is also estimated that there will be a steady supply of In as it is a by-product of Zn, which is relatively higher in abundance and more in demand in various industries. There could be some concerns about the availability of Te for multi-GW plants, but with the advancement in thin-film technology, the thickness of the material may be reduced by implementation of light trapping schemes, thus relieving the burden on production of materials.

In the following sections, the front-line thin-film technologies such as a-Si, CdTe, and CIGS are discussed in detail relating to their material and device aspects, current status, and issues related to environmental concerns.

20.2.2 Thin-Film Silicon

20.2.2.1 Material and Properties

The oil crisis of the early 1970s gave an impetus to the PV R&D activities and several concepts for thin-film PV started emerging. The first report on thin-film a-Si-based solar cells appeared in 1976 (Carlson and Wronski 1976). It took only five years to see the indoor consumer products appearing in the market with this technology in the 1980s, although it took a significant amount of time to understand the basics of the material and device properties and their inherent bottlenecks. Amorphous-Si is now the most studied and applied material for thin-film solar cells as compared to its counterparts. Silicon has the advantage of the material being in abundance in the earth's crust; therefore, following the trends of c-Si technology, a-Si has developed over the years into an industrially mature technology (Rech and Wagner 1999). In addition, a-Si has several other non-PV applications that provide it an additional standing.

Crystalline-Si has long-range atomic ordering extending up to a few cm in single crystals, whereas a-Si exhibits short-range atomic ordering of less than 1 nm, and thus the material is not a crystal. Amorphous silicon has a disordered lattice showing localized tetrahedral bonding schemes but with broken Si-Si bonds of random orientation, as shown in Figure 20.28. These broken (or unsaturated) bonds are called "dangling bonds" and contribute to the defect density in the material. Because of disorder, the momentum conservation rules are relaxed and a higher absorption coefficient (α) is observed in a-Si materials. The absorption coefficient of a-Si is about two orders of magnitude higher than c-Si, thus it only requires a couple of microns of thickness for effective absorption and utilization of the solar spectrum. However, due to its predominant disordered structure, a high density ($\sim 10^{19}\ \text{cm}^{-3}$) of localized defect states are created within the energy gap that causes the Fermi-level pinning. Hence, the material cannot be doped because the defects states act as a trap for all free carriers generated in the material.

One effective way to overcome this problem is to passivate the unsaturated bonds of a-Si with the help of small atoms that could get into the crystal and attach themselves with the available bonds. This is precisely done by adding 5%–10% atomic hydrogen into a-Si, which attaches itself to the uncoordinated bonds due to its high activity; this reduces the dangling bonds density from $\sim 10^{19}$ to $\sim 10^{15}\ \text{cm}^{-3}$.

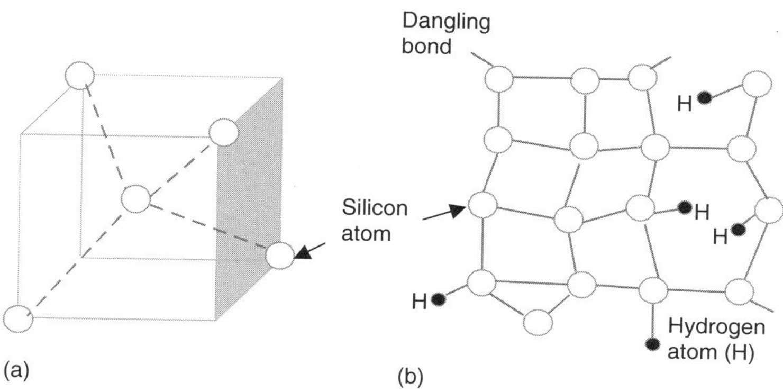

FIGURE 20.28 A schematic presentation of the tetrahedral bonding scheme in crystalline-Si (a) and network of a-Si:H exhibiting broken silicon atom dangling bonds that are passivated by hydrogen atoms (b).

At this order of defect density, the doping of material is possible, and the material can be made as *p*- or *n*-type using boron and phosphorous as dopants. However, the defect density still remains so high that even with high doping, the Fermi level is not significantly shifted; it remains mostly within the donor and defect levels at the center of the gap in the case of *n*-doping.

Amorphous silicon may be considered as an alloy of silicon with hydrogen. The distortion of the bond length and bond angle after passivation with hydrogen modifies the defect distribution and consequently changes the optical and electronic properties. By changing the deposition conditions (Guha et al. 1981; Guha et al. 1981; Guha et al. 1986; Vetterl 2000), hydrogen-diluted microcrystalline silicon (μc- Si) can be obtained that has rather different properties. Figure 20.24 compares the absorption coefficient of the a-Si:H, c-Si, and μc-Si:H, along with other photovoltaic materials. The absorption bands (plateau) appearing at low energy values for a-Si:H and μc-Si:H are ascribed to the presence of a large density of midgap defects and band-tail states. The absorption coefficient (α) of c-Si (monocrystalline silicon wafers) and microcrystalline thin-film Si have more or less the same onset of transition, but μc-Si has a higher α in the low-wavelength (λ) region. However, α for μc-Si is lower than that of a-Si, therefore thicker μc-Si layers are required compared to a-Si for the absorption of the solar spectrum. A stacked combination of the two—microcrystalline and amorphous Si layers—is attractive for absorption of the most useful part of the spectrum in thin layers. This has been successfully employed, first by IMT Neuchatel, Switzerland and later by several other groups, to develop a-Si/μc-Si tandem (also often called *micromorph*) solar cells.

The bandgap of a-Si can also be tailored by addition of O, C, and Ge to produce amorphous materials of wider or narrower bandgaps, e.g., with the addition of C and Ge in a-Si:H, bandgaps of 2.2–1.1 eV are achievable but with inferior electronic properties. Table 20.6 provides the list of these alloys with their respective bandgaps. Suitable a-SiC:H and a-SiGe:H for solar-cell devices have bandgaps of approximately 2.0 and 1.3 eV, respectively.

20.2.2.2 Deposition Techniques

Perhaps the most important feature of a-Si material is that a wide range of temperatures, from room temperature to 400°C, can be used for its deposition. Room-temperature deposition allows the use of a variety of substrates such as glass, metal, and plastic; furthermore, the possibility to use low-cost plastic PET (polyethylene terephthalate) could be of significant advantage in reducing the cost of the modules. There are various processes used for the deposition of the a-Si:H material. Silane (SiH_4), which is the basic precursor gas, is used in nearly all processes using the chemical vapor deposition (CVD) method, but excluding sputtering which is not preferred for active semiconductor layers in a-Si:H. Typical deposition temperatures for a-Si:H must be below 500°C, otherwise the incorporation of hydrogen in the film is not possible. At low-substrate temperatures, the predissociation of SiH_4 does not take place easily.

TABLE 20.6 Energy Bandgaps (E_g) of Certain Alloys of a-Si:H with Germanium and Carbon Used in Multiple Junction Solar Cell Structures

Material (Semiconductor/Alloy)	E_g min (eV)	E_g max (eV)
c-Si	1.1	1.1
μc-Si:H	1.0	1.2
a-Si:H	1.7	1.8
a-SiC:H	2.0 (in 20% C)	2.2
a-SiGe:H	1.3 (in 60% Ge)	1.7
a-Ge:H	1.1	

Hence, room-temperature-deposited layers give rise to inferior quality and efficiency. Therefore, plasma is used for dissociation of silane gas. Two of the most commonly used methods are plasma-enhanced chemical vapor deposition (PECVD) and glow-discharge CVD. Typically, 13.56-MHz plasma excitation frequencies with optimal plasma excitation power at 0.1–1 mbar pressure are used. SiH_4 diluted with hydrogen ($\sim$10%) is used in the deposition of a-Si:H, whereas increasing hydrogen dilution results in μc-Si:H layers, but with lower growth rates. The typical deposition rates for a-Si cells (in R&D) is $\sim$1 Å/s and results in fairly long deposition times (50 min for 0.3-μm-thick a-Si:H cell, and 5 h for a 1.8-μm-thick μc-Si:H cell), while 3 Å/s or higher deposition rates are generally preferred in production plants. For high deposition rates, the deposition technologies based on very-high-frequency (VHF), microwave, and high-pressure plasma are currently being pursued at the R&D level. Rates as high as 10 Å/s have been achieved at laboratory scale.

Alternative deposition methods using the hot-wire CVD (HWCVD) technique, electron cyclotron resonance reactor (ECR), and the combination of HWCVD and PECVD are also being carried out to increase the deposition rate. A detailed account of some of these techniques can be found in the following references: Carabe and Gandia (2004); Deng and Schiff (2003); Klein et al. (2004); Lechner and Schade (2002); Shah et al. (2004), and Sopori (2003).

20.2.2.3 Amorphous-Silicon Solar Cells and Configurations

The conventional *p–n* junction configuration for a-Si:H-based solar cells suffers from inherent limitations due to the presence of a large number of defect states even after H-passivation. The doping of a-Si:H further increases this concentration, which reduces the average lifetime of the free carriers as a result of very high recombination probabilities and low-diffusion lengths ($\sim$0.1 micron). Thus solar cells in the *p–n* configuration do not work and are not considered suitable. The basic structure of a-Si solar cell configuration is a *p–i–n* junction shown in Figure 20.29a, which illustrates qualitatively the thickness of different layers grown in the device in the "superstrate" configuration, along with applied texturing (roughness) of the transparent conducting electrodes for enhanced light trapping in the a-Si layer, as will be described later.

The *p–i–n* type configuration for an a-Si solar cell was introduced by Carlson et al. at RCA Laboratories, U.S.A. (Carlson and Wronski 1976), where an intrinsic (*i*) layer of a-Si:H is sandwiched between the *n*- and *p*-type doped layers of a-Si:H or its alloys. Because of the very short lifetime (or high recombination) of the carriers, the doped layers do not contribute to the photocurrent generation; i.e., the photons absorbed in these layers contribute to optical losses, but these *p*- and *n*-layers build up the electrical field across the *i*-layers. This electrical field drives the electrons and holes in the opposite direction. They are then photogenerated in the *i*-layer, which essentially acts as the absorber layer in a-Si:H solar cells. The electrical field depends on the doping concentration of the *p*- and *n*-layers, as well as the thickness, of the *i*-layer. Because the *p*- and *n*-doped layers do not contribute to the photocurrents and can cause further recombination of the generated carriers before sweeping across the layer, it is essential to minimize their thickness which is typically 10–30 nm. There is an upper limit to the thickness of the *i*-layer ($\sim$0.5 micron), because charge defects reduce the effective field; thus, if the width of the

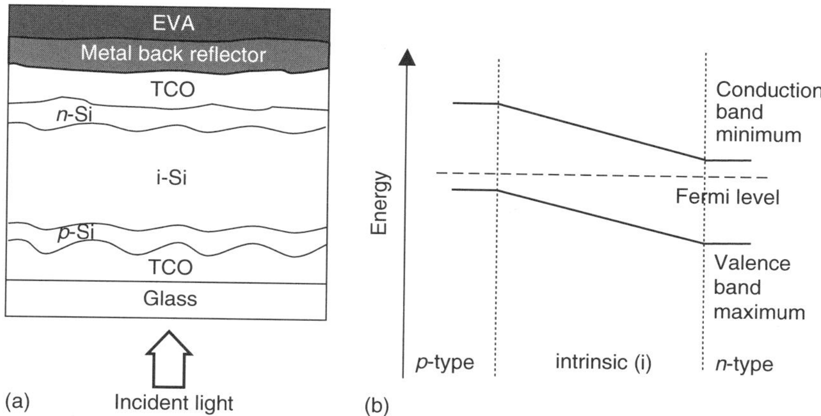

FIGURE 20.29 The schematic representation of single junction a-Si solar cell in superstrate configuration (a) and the energy band diagram of the *p–i–n* solar cell structure (b). EVA (Vinyl Acetate) is the polymer coating used for encapsulation of the solar cell.

i-layer exceeds the space charge width, then the extra width would act as a "dead" layer without actually contributing to the photocurrent.

Stability and Recombination Issues in a-Si Solar Cells. The initial results of a-Si cells in the 1970s indicated very promising potential for attaining efficiencies well above 10%. However, it was soon observed and realized that a-Si solar cells suffer from an inherent problem of light-induced degradation on their performance under continuous light exposure, later attributed to the Staebler-Wronsky effect (SWE) (Staebler and Wronski 1977). It was observed that in a time scale of a few months, the performance of the cells dropped about 30%–40% then stabilized at an efficiency lower than the initial values. This initial drop in performance is significant and takes the edge off the promise shown by this cheaper alternative to c-Si technology, keeping it much below the efficiency threshold limit (which is generally accepted to be ~10% module efficiency for thin-film vacuum-based technologies). An explanation for the light-induced degradation (SWE) is that with light exposure, the Si-H bonds break and further increase the density of the dangling bonds. Thus, the system is driven into an excited or higher energy state with active defect centers leading to higher recombination of the free carriers and hence a reduction in efficiency. The efficiency drop depends on the illumination level and operating temperature of the solar cell. It has been observed that efficiencies may be partially recovered on heating the cells. The heating recovery depends on the temperature of the cells, i.e., annealing at 70°C helps stabilize the system better than annealing at room temperature.

The control of defect (trap) states and dangling-bonds passivation for effective doping of the a-Si:H layers are significant in the a-Si cell's overall design. The first inherent problem of the technology is that SWE cannot be eliminated, but can be reduced by engineering of the device structure, e.g., by employing an a-Si thinner *i*-layer at the expense of absorption loss. It has been verified by several groups that efficiency degradation in a-Si solar cells and modules is lower with thinner *i*-layers. The second problem is that the doping of the a-Si leads to an increase of the trap density, leading to pronounced recombination effects in the device; therefore, limiting the thickness of the doped layers to 10–30 nm is needed for minimized recombination effects. The limits on *i*-layer and *n*- and *p*-doped layer thicknesses together have a direct bearing on the overall device structure and performance stability.

A-Si Solar Cell Configurations. An advantage of a-Si is that the solar cells can be grown in both "superstrate" and "substrate" configurations as shown in Figure 20.30. In the superstrate configuration, the cell is grown in the *p–i–n* sequence (starting with the *p*-layer followed by the *i* and *n* layers) onto a substrate that must be transparent (such as glass), and hence this configuration is not suitable for metal or highly

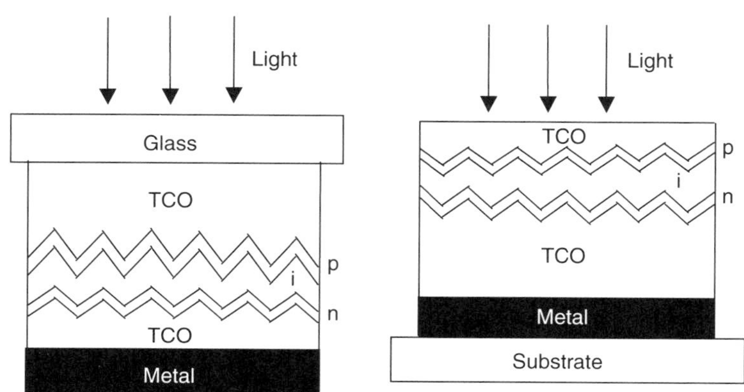

FIGURE 20.30 Schematic presentation of a-Si solar cell in "superstrate" (*p–i–n*) configuration (left) and "substrate" (*n–i–p*) configuration (right).

opaque polymeric substrates. The substrate configuration, however, can be grown on any type of substrate, such as rigid glass, flexible metal, or polymer foil. It bears an *n–i–p* configuration (cell growth starting with the *n*-layer followed by *i* and *p* layers) and the light enters through the last grown *p*-layer.

Generally, a-Si solar cells on glass are available in the superstrate configuration, starting with the TCOs (transparent conducting oxides) window with *p–i–n* layers grown on it, followed by another TCO layer and a metallic back reflector layer, as shown in Figure 20.30. One of the leading US-based companies, United Solar (USOC, formerly USSC) has been using substrate configuration for roll-to-roll production of cells on stainless steel (SS) and polymer foils. The layers can be grown in *n–i–p* or *p–i–n* sequences. However, irrespective of the substrate or superstrate configuration, incident light is allowed through the *p*-side, as it has a higher bandgap than the *i* or *n* parts. Also, because the mobility of holes is smaller as compared to electrons, a thin front *p*-layer supports hole collection in the device (Rech and Wagner 1999).

The choice of TCO material and its electrical and optical properties is not only important for electrical contacts, but also for efficient light trapping through the device. Light trapping is essential for efficient performance of a-Si solar cells, where device thickness is limited by several inimical factors, e.g., a thinner *i*-layer is desired for minimizing light-induced performance degradation. Consequently, the thickness of the intrinsic layer that acts as an absorber is generally limited to only ~300 nm, which is not sufficient for absorption of a large part of the solar spectrum. To effectively utilize the incident photons, the applied strategies are to reduce the reflection through refractive index grading structures for the entire spectral wavelength-range cell response, and allow multiple scattering of light for enhanced absorption of photons in the *i*-layer. These are achieved by an antireflection coating used on the glass where the light enters into the PV module, and also through suitable surface texture of TCO with the feature sizes comparable to the wavelength and application of metal reflectors. For detailed description of TCOs and light scattering, refer to the publications by Goetzberger et al. (2003); Granqvist (2003); Muller et al. (2004), and Shah et al. (2004).

TCOs such as SnO_x:F, ITO, and ZnO:Al have been extensively used in a-Si solar cells. Some of the requirements for a good a-Si:H solar cell are:

- Glass and front TCO should have a high (greater than 80%) transparency over the whole spectral range.
- TCO with a sheet resistance of at most 10–15 Ω/square (high conductivity) to be obtained by enhancing carrier mobility rather than the carrier concentration to minimize free carrier absorption over the near infrared region.

- TCO layers and doped silicon layers, which do not contribute to photogeneration and collection, should be kept as thin as possible and have very low absorption coefficients.
- TCO layers should not degrade by chemical reduction during a-Si:H deposition.
- Use of back reflectors with as little absorption as possible.

The properties of doped and intrinsic layers have been widely studied and layers are employed in optimized conditions. However, light trapping through various structures and patterns are relatively recent advancements and thus open up more possibilities in the improvement of the device performance (Granqvist 2003; Muller et al. 2004). There are other innovations related to device architecture by making use of tandem or multiple-junction and heterojunction cells for efficiency and stability improvements; these are discussed in the following sections.

Multiple-Junction or Tandem Solar Cells. The light-induced degradation (SWE) has become the biggest bottleneck of the a-Si technology and it has serious implications. In addition, the general effects of high density of trap and recombination centers have put restrictions on the thickness of the device layers that consequently limit the absorption of the incoming light. The clever *p–i–n* configuration, once considered to have great promise for good efficiency at a cheaper cost, was also masked by this instability component. To work within the limits of intrinsic layer thickness of ~300 nm and to make use of different light trapping arrangements, the concept of tandem cells using double and triple junctions have been thoroughly pursued worldwide. Single-junction a-Si solar cells are hardly used these days because of low efficiency and stability problems. Multijunction solar cells are used for better utilization of the solar spectrum and to improve the stability of the solar cells. The state-of-the-art cell-stabilized efficiency (small area) for a single junction is 9.3%, whereas 12.4% for a double-junction and 13.0% for a triple-junction have been achieved with a-Si:H and its alloys (Guha 2004). Figure 20.31 schematically presents different multijunction structures. The developments of multijunction solar cells are based on the following strategies:

1. The first strategy for tandem design is based on the use of only a-Si:H intrinsic layers shown in Figure 20.31a. Such double-junction devices have been developed by Fuji Electric & Co., Japan, Phototronics (part of RWE Schott), Germany and others. A stabilized laboratory efficiency of ~8.5% and module efficiency at about 5.5% are commercially available (Ichikawa et al. 1993; Diefenbach 2005).
2. The second strategy includes the use of a-Si and Ge alloys with different bandgap (lower than 1.7 eV) combinations to form tandem junctions, where the top cell is 1.7 eV a-Si:H-based, and bottom cells have a-SiGe:H alloy layers of lower (1.5–1.3 eV) bandgaps, as shown in Figure 20.31b. United Solar (USSC) has developed a 13% stabilized efficiency triple-junction solar cell (small area) in the substrate configuration and is selling their triple-junction modules on stainless steel at a stabilized efficiency of about 6.5%.
3. The third strategy introduced in 1994 by IMT Neuchatel, Switzerland is based on a novel concept of combining μc-Si:H (with a bandgap of 1.1 eV) and a-Si:H (with a bandgap of 1.7 eV) based solar cells. This has a promising potential because of the significantly reduced light-induced degradation effect in a tandem solar cell (Meier et al. 1994; Shah et al. 2004). The only degradation observed comes from the a-Si part that is optimized at 0.2–0.3 μm, whereas the μc-Si:H layer is kept around 1–2 μm. Figure 20.31c shows the schematic of the design in a *p–i–n* / *p–i–n* configuration on a rigid glass substrate. Kaneka Corporation, Japan has recently achieved large-area modules (910×455 mm^2) of initial efficiency ~13.2% and with stabilized efficiency approaching 10%. Using the concept of an intermediate TCO reflector layer for novel light trapping, Yamamoto et al. (2004)) have shown an initial efficiency of 14.7% for a test cell. The reason for the good efficiency of the cells lies in the spectral response of the combination of 1.7 eV a-Si:H-based cells and a 1.1 eV μc-Si:H-based second part. The superposition of the two results in a quantum efficiency spreading of around 80% between 500 nm and 800 nm, covering a large part of the solar spectrum.

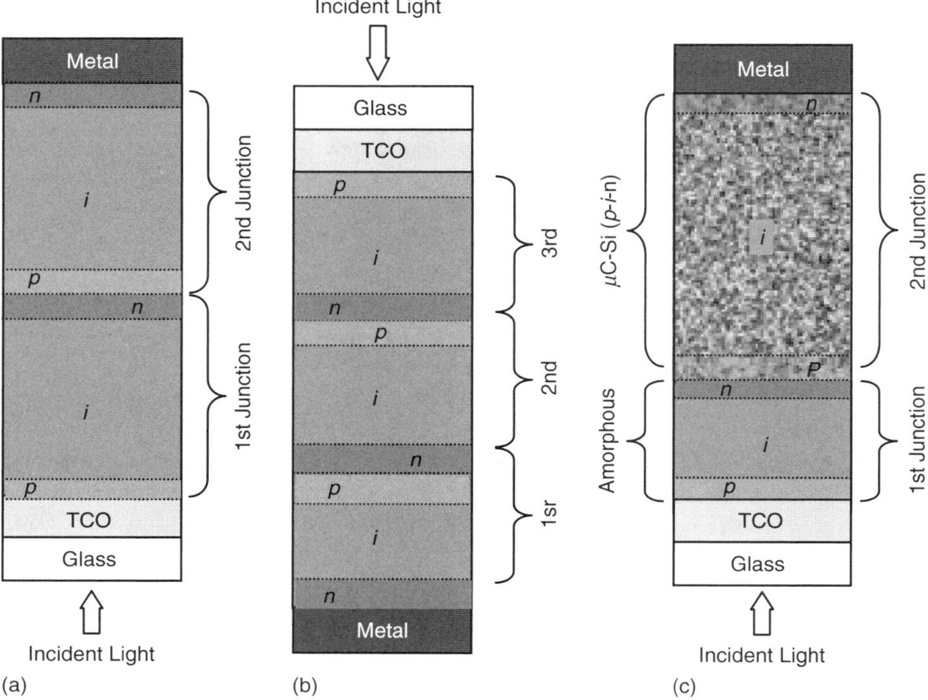

FIGURE 20.31 The schematics of multijunction cell architecture showing double-junction "superstrate" configuration (a), triple-junction in "substrate" configuration (b) and "micromorph" junction in "superstrate" configuration, i.e., glass/TCO/*p–i–n* (a-Si:H) /*p–i–n* (μc-Si:H) (c).

Hybrid Solar Cells. Amorphous silicon cells have been combined with nanocrystalline silicon junctions and the cells of other materials, e.g., junction with CIGS (Mitchell et al. 1988; Yamamoto et al. 1998). Another significant development in the design is the formation of a thick/thin-type of interface structure (heterostructure) between an a-Si:H layer and a c-Si wafer referred to as HIT (heterojunction with intrinsic thin-film layer) cells developed recently by Sanyo, Japan. Efficiency close to 21% over a cell area of 101 cm^2 has been reported (Sakata et al. 2000). This technology uses an n-type of Cz-silicon wafer as the base (light absorber) and low-temperature processes with the device structure being: a-Si(p^+)/a-Si(i) /c-Si(n)/a-Si(i)/ a-Si(n^+). The intrinsic a-Si layer is important because it contacts c-Si at both ends and provides passivation as well as extra stability to the system. As per reports of Sanyo, the cells have exhibited excellent stability. A pilot-plant production by Sanyo is already underway.

20.2.2.4 Flexible a-Si Solar Cells and Modules

Monolithic Modules. All solar modules require a number of solar cells to be electrically connected in series to provide power, depending on size and cell efficiency. Additional processing steps such as attachment of leads and encapsulation for protection against external influences are done to finalize the module structure. The superstrate configuration has advantages for monolithic electrical interconnection of solar cells to form solar modules because substrates (glass, polymer) are insulating; whereas in the substrate configuration, individual large area solar cells are mechanically connected, cell to cell, as is done in c-Si technology. USSC, USA follows this electrical contacting approach for triple-junction a-Si solar cells on steel foils.

Figure 20.32 illustrates the monolithic interconnection scheme to develop solar modules in the superstrate the configuration. For interconnection of solar cells, layers are laser scribed in three stages: first to separate the front TCO contact, then scribing a-Si layers to connect individual cells, and finally to isolate the

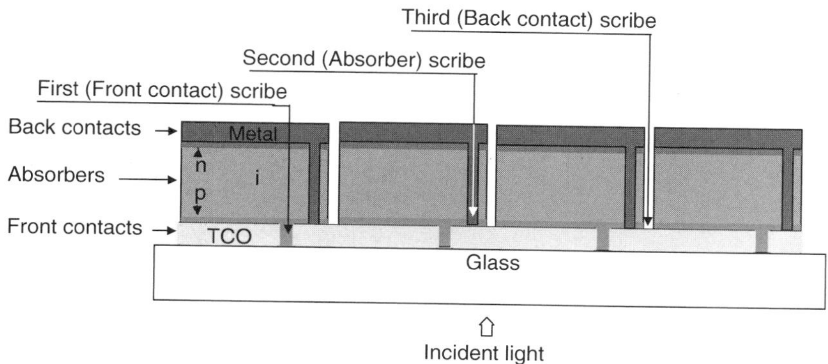

FIGURE 20.32 Schematic presentation of a monolithic module design with a single-junction a-Si cell in a "superstrate" configuration. Similar strategy can also be used for modules on polymer foils or multijunction solar cells, which may have a different sequence of layers.

conducting back electrode to obtain series interconnection. This approach removes the need for separating and connecting of cells, which is time-consuming, costly, and complex in conventional c-Si technology.

Flexible a-Si Solar Cells. Another important perspective of thin-film PV technology are the flexible modules. These are quite lucrative because of their applications for strategic space and military use, integration in roofs and facades of buildings, portable power sources, automobiles, consumer electronics, and value-added products, etc. Therefore, the technology has recently attracted major players in the thin-film PV industry to venture into this area. Because they can be available in different shades (even semi transparent), shapes, and sizes, these flexible a-Si solar cells are likely to be very popular and in demand for low- to medium-range power applications.

Some of the prominent companies that are currently involved in the production and development of flexible a-Si based modules are listed in Table 20.7. Some significant record efficiencies may also be found in Table 20.5. Of these, US-based United Solar (Iowa) and Japanese companies such as Sanyo and Fuji have entered into relatively large-scale production, whereas European companies such as Flexcell (Switzerland) and Akzo Nobel (the Netherlands) are starting pilot production plants for the consumer-oriented market. a-Si offers cheaper processing and materials costs. However, for it to remain as a front-line thin-film technology would mainly depend on its efficiency for larger modules with high throughput, which is the biggest bottleneck for the technology. A module efficiency above ~12% and a long-term (more than 30 years) stable performance of the large-area modules would compete effectively with c-Si and allow a-Si to remain the leader amongst its thin-film counterparts if low-cost modules could be developed with production processes that give high throughput and yield.

20.2.3 Cadmium Telluride Solar Cells

20.2.3.1 Material and Properties

Cadmium telluride (CdTe) is a direct-bandgap material and is of the II–VI family of materials referred to as *metal chalcogenides*. Its bandgap energy of ~1.45 eV is quite favorable for the requirement of the highest conversion of the solar spectrum into electricity with a single-junction solar cell. In addition, very high-optical absorption (10^5 cm^{-1}, see Figure 20.24) and *p*-type conductivity make it an ideal material for PV applications. It assumes the zincblende crystal structure and features a simple phase diagram because the constituents exhibit high vapor pressures. The layers of CdTe can be deposited using a number of processes and the compound can easily be grown in stoichimetric form at temperatures over 400°C. Like other II–VI materials, the electronic doping is controlled by substitution of atoms at vacancy sites.

TABLE 20.7 Summary of the Companies Active in Flexible a-Si Thin-Film Si PV[a]

Companies	Technology	Production	Remarks
United Solar	Triple junction SiGe alloys on stainless steel RF-PECVD	30 MW line, RF-PECVD, modules and consumer products	modules between 6.5–6.8%. Cutting-assembling process modules with 8% for military and space application
VHF-Technology	a-Si on PEN, polyimide VHF-PECVD	Pilot production and consumer products	Product efficiency ~4%
Iowa Thin-Film Technology (ITFT)	a-Si on polymide, RF-PECVD	Pilot production and consumer products	Product efficiency ~4%
Fuji Electric	a-Si/a-Si on polymide RF-PECVD	Pilot production and consumer products	Feed-through contacts through the substrates, 7% module, 8% active area stable efficiency
Sanyo	a-Si on plastic RF-PECVD	Consumer products	
Canon	μc-Si:H/a-Si:H on stainless steel (VHF-PECVD)	Development	
Akzo Nobel	Amorphous *p–i–n* on Al RF-PECVD	Development, pilot-line	Al sacrificial substrate dissolved after cell deposition

[a] Module efficiencies are generally in the 3–8% range.

Although *n*-type doping control is relatively easier, it is difficult to vary the doping concentration in *p*-type CdTe because of compensation effects. The most common CdTe thin-film solar cell structure comprises a *p*-type CdTe absorber layer and *n*-type CdS window layer forming a heterojunction, which has an intermixed interface region.

The first development of CdTe heterojunction solar cells started around 1962 using *n*-CdTe in combination with Cu_xTe (Bonnet 2004; Cusano 1963; Lebrun 1966;). Although 5%–6% efficiencies were achieved, those cells suffered from degradation due to Cu diffusion from Cu_xTe (similar to CdS/Cu_xS solar cells). A breakthrough in 1972 was achieved when a new heterojunction structure between CdTe

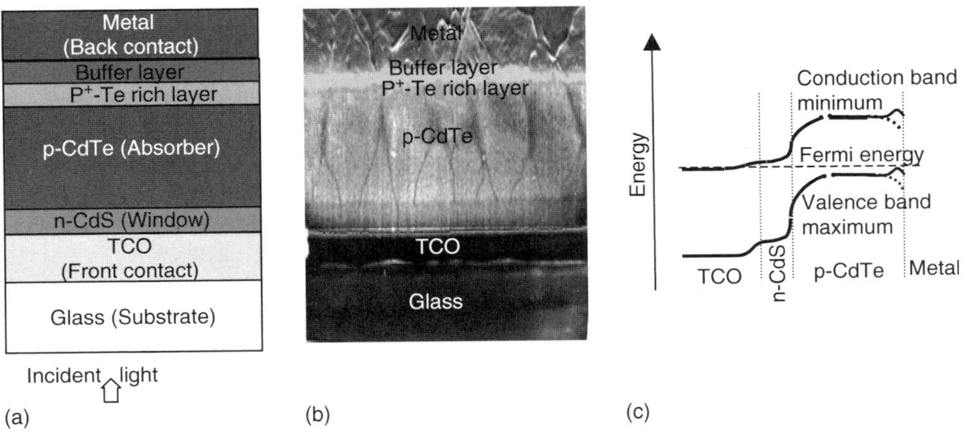

FIGURE 20.33 Schematic presentation of CdTe/ CdS solar cell in "superstrate" configuration showing different layers and their nomenclature (a), a corresponding SEM picture illustrating different layers (b), and the energy band diagram of the solar cell (c).

and CdS was developed by Bonnet and Rabenhorst (1972) and 6% efficiency was reported. Current CdTe solar cell structures are based on this device configuration as shown in Figure 20.33 (the details of this device structure are discussed later in the section 20.2.3.2), meanwhile, significant progress in CdTe devices has resulted in an efficiency rise up to 16.5% (Wu et al. 2001). Historical developments of CdTe PV technology have been reviewed by McCandless and Sites (2003).

The most attractive features of the CdTe compound are its chemical simplicity and robust stability. Because of its highly ionic nature, the surfaces and grain boundaries tend to passivate and are not detrimental (this is why polycrystalline CdTe solar cells are more efficient than their single-crystal counterparts). CdTe material is inherently stable under the solar spectrum because its bonding strength, ~5.7 eV, is much higher than the incident photon energy of the solar spectrum. This accounts for the excellent material stability and device performance. CdTe is not only stable for terrestrial applications; recent work has demonstrated the excellent stability of CdTe, superior to Si, GaAs, CIGS etc., under high-energy photon and electron irradiation for space applications (Bätzner et al. 2004).

Table 20.5 lists the efficiency of the CdTe solar cells reported by laboratories and prominent manufacturers. An efficiency of 16.5% has been reported by NREL USA (Wu et al. 2001) for an approximate area of 1 cm^2, whereas cell efficiencies of more than 12% are reported by several groups using different deposition methods and substrates. The theoretical efficiency of CdTe solar cell is estimated to be ~29%, (J_{sc} ~30 mA/cm^{-2} and V_{oc} ~1.0 V) which is higher than c-Si. There is therefore plenty of room for improvement in the conversion efficiency in the future. However, similar to other thin-film PV technologies, current CdTe technology also suffers from some problems, namely reproducibility of the device performance when small-area cells are scaled to a larger area. Other important issues are formation of good ohmic contacts on *p*-CdTe, and reduction of optical absorption losses from the CdS and TCO front-contact layers. These are some inherent critical issues requiring further investigation to achieve greater than 20% cell efficiency with single-junction devices. Progress of CdTe technology has also been hampered because of environmental concerns due to cadmium; these issues are addressed in Section 20.2.5.

20.2.3.2 CdTe Solar Cell Device Structure

The CdTe solar cells can be grown in both substrate and superstrate configurations. However, the highest efficiencies have been achieved in the superstrate configuration. Figure 20.33a gives the schematics of a CdTe solar cell grown on a TCO-coated glass substrate in a superstrate configuration. The glass substrate can be a low-cost soda-lime glass for processing temperatures below 500°C, or alkali-free glass (generally borosilicate) for high-temperature processing above 500°C. Substrate configuration would require CdTe to be deposited on metal foils or metal-deposited glass substrates. However, stability of the back contact, in particular, is a limiting factor in this configuration because CdTe/ CdS layers need to be grown at such high temperatures that interdiffusion degrades the CdTe back contact interface and cells are shunted. Therefore, substrate configuration is not actively pursued in CdTe solar cells (Figure 20.34).

In the superstrate configuration, layers of TCO, CdS, CdTe, and the metal back contact are sequentially grown on glass substrates. There are some intermediate processing steps that will be described later. An antireflection coating on the glass rear surface is often applied to reduce the reflection at the air-glass interface. The incident light passes through the glass, TCO, and CdS and gets absorbed in the CdTe layer, where it creates electron-hole pairs that contribute to the photovoltaic power.

TCO Front Electrical Contact. A highly transparent and *n*-type conducting TCO layer with an electron affinity below 4.5 eV is required to form an ohmic contact and to have good band alignment with *n*-CdS. Most of the TCOs, such as SnO_x:F, ITO, ZnO:Al, and Cd_2SnO_4 (cadmium stannate), have been used to grow high-efficiency (greater than 12%) solar cells, but the highest efficiency (16.5%) cells have been achieved on a bilayer stack of highly resistive Zn_2SnO_4 on conductive Cd_2SnO_4 by Wu et al. (2001). Though a single layer of TCO provides high-efficiency cells, a bilayer (combination of high and low resistive TCO stack) is often used to protect against shunts caused by pinholes in thin CdS layers. The most commonly used TCO in industrial production is SnO_x:F, and often with a thin ITO layer. ITO front contacts are often

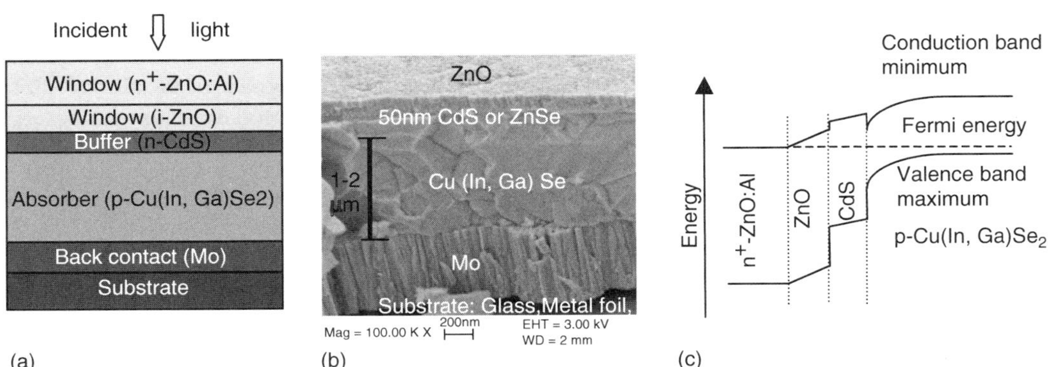

FIGURE 20.34 Schematic presentation of CIGS solar cell in "substrate" configuration (a), SEM cross-section image of CIGS device showing microstructure of layers (b), and the qualitative energy band diagram of CIGS solar cell (c).

sensitive to annealing treatment. An increase in electron affinity from around 4 to 5 eV can be caused by oxidation or postdeposition treatment; moreover, ITO is rather expensive and should be avoided if possible.

CdS Window Layer. CdS has bandgap of 2.4 eV with *n*-type electrical conductivity and it forms a heterojunction with a CdTe layer. The typical thickness of a CdS layer used in solar cells is in the range of 10–500 nm under as-deposited conditions. During the high-temperature steps of cell processing, this thickness can be effectively reduced because of interdiffusion with CdTe. A thin CdS layer (10–50 nm) is desired to minimize the photon-absorption losses so that the maximum number of photons can reach the CdTe layer. However, there must be a compromise, because very thin CdS may lead to a lower open-circuit voltage and fill factor through shunting in the device. CdS layers can be grown by different methods, such as chemical bath deposition (CBD), evaporation, sublimation, vapor transport, MOCVD, and sputtering (Ferekides et al. 1993; McCandless and Sites 2003; Romeo et al. 1999; Romeo et al. 2004).

CdTe Absorber Layer. The CdTe thin-film is the most important component because it absorbs the incident solar light and contributes to the photogenerated current. Because of its direct bandgap properties, only about a 2-μm-thick material is sufficient to absorb most of the useful part of the solar spectrum. CdTe layers may be grown by a variety of vacuum and nonvacuum methods classified into high-temperature and low-temperature processes. Some of the commonly used high temperature methods are closed space sublimation (CSS), vapor transport (VT), or vapor transport deposition (VTD) with deposition temperatures above 500°C; whereas methods such as electrodeposition (ED), screen printing (SP), chemical spraying (CS), high-vacuum evaporation (HVE), and sputtering with deposition temperatures below 450°C are classified under low-temperature processes (McCandless and Sites 2003; Romeo et al. 2004). Depending on deposition methods, the typical thickness of the CdTe layer in solar cells is in the range of 2–6 μm.

Junction Activation Treatment. The as-deposited CdTe/CdS solar cells always exhibit poor photovoltaic properties and thus require a special annealing treatment that improves the cell efficiency considerably (by a factor of 3–5). This is done by subjecting the CdTe/CdS stacks to a heat treatment under Cl-O ambient between 350–600°C. This is known as *$CdCl_2$ treatment* or *junction activation treatment.* After this annealing treatment, a significant enlargement of grain size, by a factor of 5–20, is observed in CdTe grown by low-temperature deposition methods, as can be seen in Figure 20.33b. The grain size of HVE CdS is in the range of 0.1 to 0.3 μm and the layers are rough. If the CdS is grown by a chemical-bath deposition (CBD), then it consists of small grains of about 0.1-μm widths. A treatment with $CdCl_2$ recrystallizes the CdS layers so that some of the small grains coalesce together and form bigger grains of 0.5-μm width (Romeo et al. 2004). In the case of high-temperature-grown CdTe, this annealing treatment recrystallization is observed, but the grain size near the top surface does not increase because the grains are already a few microns large even in the as-deposited condition.

For high-temperature growth processes, there is a tendency of CdS_xTe_{1-x} formation by the conversion of small CdS grains into CdTe due to interdiffusion at the interface, and little or low grain growth is noticed after the activation treatment. A stable CdS/CdTe interface can be obtained for 6% diffusion of sulfur atoms. However, under nonequilibrium conditions, the diffusion of S decreases the thickness of the CdS films, causing pinholes and eventually leading to shorting paths across the junction. This is a critical problem in CdTe solar cells that restricts the application of thinner CdS as desired to minimize the optical-absorption losses. Nevertheless, a thermal treatment of a CdS layer prior to CdTe deposition is frequently applied to restrict the interdiffusion of S. The formation of CdS_xTe_{1-x} after activation actually helps in reducing the lattice mismatch between CdS and CdTe, but only marginally as compared to improvements in electrical changes induced by Cl, O, and S. Apart from the reduction in density of stacking faults and misfit dislocation, there is an overall increase in the shallow-acceptor concentration in CdTe, leading to enhanced p-doping in CdTe after annealing. In particular, the grain-boundary regions become more p-doped, owing to preferred grain boundary diffusion and segregation of Cl and O. As a result, increased charge-carrier collection efficiency is measured and efficiency increases by a factor of 3–5.

Problems of Electrical Back Contact and Stability. An important issue in CdTe solar-cell technology is the formation of efficient and stable ohmic contacts on p-CdTe layers. For an ohmic contact to form on a p-type semiconductor, the work function of the metal should be higher than the sum of the bandgap and the electron affinity of the semiconductor, otherwise a Schottky contact is formed. For a p-CdTe layer, a metal with a work function higher than 5.7 eV is needed because CdTe has a bandgap of 1.45 eV and electron affinity is 4.3 eV. Metals with such high work functions are not available. To overcome this problem, a heavily doped p-CdTe surface is created with the help of chemical etching and a buffer layer of high carrier concentration is often applied. Subsequent postdeposition annealing diffuses some buffer material into CdTe, where it changes the band edges as a result of a change in the interface state density. A lowering in interface barrier height and width results that enables a quasi-ohmic or tunnelling contact between the metal and CdTe, as shown in Figure 20.33c. Commonly used buffer layer/metallization combinations are Cu/Au, Cu/graphite, or graphite pastes doped with Hg and Cu. However, back contacts containing Cu in any form are often not stable with time, as Cu migration from the back contact leads to efficiency degradation. However, alternate processes are being developed. Among them Sb_2Te_3/Mo and Sb/Mo contacts have provided high efficiency and long-term stable solar cells (Abken and Bartelt 2002; Bätzner et al. 2001; Bätzner et al. 2004; Romeo et al. 2004).

20.2.3.3 Deposition Techniques

The CdTe deposition process and substrate temperature have strong influence on the microstructure of CdTe layer and solar cell efficiency. Low-temperature processes (less than 450°C) yield lower efficiency (maximum is 14.5%), whereas high-temperature (500–600°C) processes yield 15–16.5% efficiency cells.

HVE is a simple deposition method where CdTe and CdS are congruently evaporated from crucibles/boats on substrates at 150°C for CdS and ~300°C for CdTe. In this process, the layers are grown in a high vacuum (~10^{-5} torr), but the distance between the source material and substrate is kept in the range of 10–30 cm. This allows the use of substrates at relatively lower temperature. Typical deposition rates vary between 2–10 μm/h. Solar cells of approximately 12–13% efficiency have been achieved by Stanford University, USA, ETH Zurich, Switzerland, and IEC, USA (Fahrenbruch et al. 1992; McCandless et al. 1999; Romeo et al. 2004).

Sputtering is another process that involves a vacuum system with ionized gases forming plasma discharge, where a CdTe target (a few-mm-thick plate) attached to one electrode is used as the source material. The energetic ionized atoms from the plasma strike the target and remove the material atoms; consequently a CdTe layer is deposited on the substrate, which is placed on the counter electrode facing the target. Sputtering methods are suitable for large-area deposition in an industrial environment. However, because of the difficulty in maintaining stoichiometry, they are considered unsuitable for compound semiconductors. However, 14% efficiency has been recently achieved with this process at the University of Toledo, USA (Compaan et al. 2004), which is a very encouraging result.

CSS and VT are the prominent industrial processes used for CdTe deposition owing to their very high rate (2–5 μm/min) of deposition. The CSS process consists of an arrangement involving the placement of a graphite crucible with the source material (CdTe compound) in a high-vacuum chamber ($\sim 10^{-5}$ torr). The CdTe compound sublimes at around 600°C and is deposited onto the substrate, which is kept with a separation of 1 to 5 mm above the crucible and heated typically above 550°C. Antec Solar GmbH in Germany uses this method for industrial production of 60×120 cm^2 modules on a 10 MW capacity plant. Parma University, Italy, University of South Florida, USA and NREL, USA have also used this method and cells of 15.5% to 16.5% efficiency have been achieved. First Solar, USA uses a variant of CSVT, where instead of a compound, elemental vapors are used (Romeo et al. 2004; McCandless and Dobson 2004; First Solar 2005). First Solar is the most successful CdTe company to date, as they upscale the production from $\sim$15 MW to 75 MW modules on 60×120 cm^2 glass substrates.

Details of other alternative methods such as screen printing, spray pyrolysis, MOCVD, CVD, atomic layer deposition (ALD), and electrodeposition (ED) may be found in the literature (Bonnet and Rabenhorst 1972; Romeo et al. 2004). For flexible substrates such as polymers, low-temperature methods such as sputtering, HVE, and electrodeposition (ED) are suitable.

20.2.3.4 Flexible CdTe Solar Cells

Even though the technology of CdTe solar cells on glass substrates has matured and efficiencies exceeding 16% have been achieved, not much effort has been placed into developing these devices on flexible substrates. The R&D on flexible CdTe has been supported by NASA and defense agencies in the US. Because of high radiation tolerance (superior to conventional Si and GaAS solar cells) against high-energy electron and proton irradiation, these solar cells are also attractive for space applications in addition to terrestrial applications.

Although the CdTe solar cells can be grown in "substrate" configuration, one of the hurdles in the development of high-efficiency CdTe solar cells on metallic substrates is that most of the metal foils do not form efficient ohmic contact with CdTe and it is difficult to incorporate an additional buffer layer as an ohmic contact to increase the cell efficiency. The criteria of matching thermal expansion coefficients and work function limit the choice of available substrate materials. Another reason is that during $CdCl_2$ annealing treatment, diffusion of impurities changes the ohmic-contact properties. Flexible solar cells of up to 8% efficiency have been achieved on metal foils.

The choice of an appropriate substrate is a crucial factor for flexible solar cells in the superstrate configuration because the substrate should be optically transparent and should withstand the high processing temperatures. Most of the CdTe/CdS cell fabrication techniques require temperatures in the range 450–500°C. Therefore, low-temperature (less than 450°C) deposition processes are required. However, recently, ETH in Zurich, Switzerland has reported 11.4%-efficiency flexible cells on polymer foil and demonstrated the first monolithically interconnected mini-modules. These devices were grown with the HVE method, which is suitable for roll-to-roll deposition, but the concept for industrial production has yet to be demonstrated (Mathew et al. 2004; Romeo et al. 2004). The 11.4% efficiency of single-junction CdTe compares well with the 12% (initial, prior to degradation) best efficiency of triple-junction a-Si solar cells.

20.2.4 Cu(In Ga)Se_2 Solar Cells

20.2.4.1 Material and Properties

Compound semiconductors from the I–III–VI_2 series of the periodic table, such as copper-indium-diselenide (CIS), copper-gallium-diselenide (CGS), and their mixed alloys copper-indium-gallium-diselenide (CIGS) are often simply referred to as *chalcopyrites* because of their tetragonal crystal structure. These materials are easily prepared in a wide range of compositions and their corresponding phase diagrams have been intensively investigated. Changing the stoichiometry and extrinsic doping can vary their electrical conductivity. However, for the preparation of solar cells, only slightly Cu-deficient compositions of *p*-type conductivity are suitable. Depending on the [Ga] / [In + Ga] ratio, the bandgap of CIGS can be varied continuously between 1.04 and 1.68 eV. The current high-efficiency devices are prepared with bandgaps in the range of 1.20–1.25 eV; this corresponds to a [Ga] / [In + Ga] ratio between 20% and 30%. Layers with higher Ga content, as needed to increase the bandgap towards ~1.5 eV, are of inferior electronic quality and yield lower-efficiency cells.

Other chalcopyrites such as $CuInS_2$ and $CuInTe_2$ were also investigated, but cell efficiencies were rather low and the R&D focus was placed on CIS. Recently, interest in $CuInS_2$ has resurfaced with the development of ~11.4% efficient cells at HMI Berlin. A spin-off company, Sulfurcell, also based in Berlin, started setting up a pilot production line in 2003. The device structure of the $CuInS_2$ solar cell is quite similar to CIGS solar cells in terms of other constituent layers.

The first CIS solar cell was developed with single-crystal material and ~12% efficiency was reported in 1974 (Wagner et al. 1974). The first thin-film CIS solar cell was reported by Kazmerski in 1976 by developing ~4% cells obtained with the evaporation of $CuInSe_2$ material (Kazmerski et al. 1976). The real breakthrough in CIS thin-film technology came with the pioneering work of the Boeing Corp., USA, where they used three-source evaporation of Cu, In, and Se elements and raised the efficiency from 5.7% in 1980 to above 10% in 1982 (Mickelsen and Chen 1981; Mickelsen and Chen 1982). This success, despite the apparent complexity of the material system, clearly showed the promising potential of the material. Later, in 1987, Arco Solar, USA, raised the cell efficiency to ~14% by using a different CIS deposition process where a stacked metal layer was selenized under H_2Se ambient (Mitchell et al. 1988). Subsequent improvements in efficiency were attained by the EUROCIS consortium in Europe and later at NREL, USA, which holds the efficiency record of ~19.5%, the highest for any thin-film, single-junction solar cells (Ramanathan et al. 2003). Efficiency improvements over the Boeing process occurred due to addition of Ga and S for bandgap engineering, the addition of Na in the absorber layer, optimization of the *n*-CdS (heterojunction part of the cell), and transparent front electrical contact layers.

The first industrial production of CIS modules was started by the Siemens (later Shell Solar) based on the Arco solar technology, whereas other companies such as Wurth Solar and Global Solar started development of CIS solar modules using coevaporation methods. Several other companies have been investigating various other methods of deposition, such as paste printing, electrodeposition, etc., but up to now these technologies have been less successful as compared to vacuum-based technologies.

The phase diagram of the ternary compound is described by the pseudo-binary phase diagram of the binary analogue, e.g., Cu_2Se and In_2Se_3 phase for $CuInSe_2$ ternary. Single-phase chalcopyrite $CuInSe_2$ exists at small copper deficiency, whereas for Cu-rich compositions, a mixed phase of Cu_xSe with $CuInSe_2$ forms that is not suitable for PV devices. For In-rich compositions, the defect-chalcopyrite phase ($CuIn_3Se_5$) forms, that is generally *n*-type. Despite an apparent complicated crystal structure and multicomponent system, the material properties of the PV-relevant compounds are fault-tolerant and not particularly affected by minor deviations from stoichiometry in the Cu-deficient range. Further, surfaces and grain boundaries in CIGS compounds are easy to passivate, resulting in high-efficiency cells even with submicron grain-size materials.

The PV-grade Cu-deficient CIGS material has a tetragonal crystal structure having vacancies and interstitials that act favorably, especially because the material is self-healing, owing to the defect relaxation caused by highly mobile Cu ions and its interaction with vacancies. Defects created in CIGS by external influence (e.g., radiation) are immediately healed. This is an inherent advantage of the CIGS material, leading to highly stable CIGS solar cells. However, care must be taken for proper encapsulation of devices against highly damped conditions, otherwise the degradation of electrical contacts (TCO or Mo) in moisture can lead to minor degradations in performance. Therefore, stability of encapsulated CIGS solar cells is not a problem, as proven by field tests conducted by ZSW, Shell, and NREL. CIGS is also tolerant against space radiation, being superior to Si and GaAs single-crystal cells, but somewhat inferior to CdTe.

20.2.4.2 CIGS Solar-Cell Configuration

The CIGS solar cells can be grown in both substrate and superstrate configurations, but the substrate configuration gives the highest efficiency due to favorable processing conditions (Figure 20.34). However, it requires an additional encapsulation layer and/or glass to protect the cell surface, which is not required in the superstrate configuration. Superstrate structures were investigated in the early 1980s, but efficiency was below 5%. However, recent efforts have improved the efficiency to ~13%. This has been possible with the introduction of undoped ZnO instead of CdS as a buffer layer, and coevaporation of Na_xSe during the CIGS deposition. Furthermore, the developments in bifacial CIGS solar cells with both front and rear transparent conducting contacts, is making sound progress (Nakada et al. 2004).

Electrical Back Contact. CIGS solar cells in substrate configuration can be grown on glass as well as metal and polymer foils. Molybdenum (Mo), grown by sputtering or e-beam evaporation is the most commonly used electrical back contact material for CIGS solar cells. Growth of the solar cell starts with the deposition of Mo on the substrate, which forms an electrically conducting back electrode with CIGS. When CIGS is grown on Mo, an interface layer of $MoSe_2$ is automatically formed that helps in ohmic transport between CIGS and Mo. Recently, alternative back contact materials have been explored, but industrial production is still based on Mo layers.

CIGS Absorber Layer. Because of a high absorption coefficient, a 2-μm-thick layer is sufficient for absorption of maximum incident radiation. CIGS layers can be grown with a variety of deposition methods (as described later). Although the grain size and morphology (surface roughness) depend on deposition methods, efficiencies greater than 13% have been achieved with most of the methods, which indicates that grain boundaries in CIGS are benign and can be easily passivated. High-efficiency cells have *p*-type $Cu(In,Ga)Se_2$ bulk, while a defect-chalcopyrite $Cu(In,Ga)_3Se_5$ phase in the form of a thin layer segregates at the top surface that is *n*-type, especially when doped by cation atoms diffusing from the buffer layer. It is believed that this inverted surface, leading to a *p–n* homojunction in the CIGS absorber, is crucial for high-efficiency cells.

Buffer Layer. Several semiconductor compounds with *n*-type conductivity and bandgaps between 2.0 and 3.4 eV have been applied as a buffer to form a heterojunction in CIGS solar cells. However, CdS remains the most widely investigated buffer layer because it has continuously yielded high-efficiency cells on different types of absorber layers. CdS for highest-efficiency CIGS cells is commonly grown by chemical-bath deposition (CBD), which is a low-cost, large-area process. However, its incompatibility with in-line vacuum-based production methods is a matter of concern. Although CBD-grown CdS serves as a reference for the highest efficiency, physical vapor deposition (PVD)-grown CdS layers yield lower-efficiency cells. Thin layers grown by PVD often do not show uniform coverage of CIGS and are less effective in chemically engineering the interface properties between the buffer and the absorber. The recent trend in buffer layers is to substitute CdS with Cd-free wide-bandgap semiconductors and to replace the CBD technique with in-line-compatible processes. Alternative materials such as In_2S_3, ZnO, ZnS, ZnSe, etc., using different methods such as PVD, RF sputtering, metal organic chemical vapor deposition (MOCVD), atomic layer deposition (ALD), or a novel technique called *ion layer gas reaction* (ILGAR), are being explored (Bhattacharya and Ramanathan 2004; Chaisitsak et al. 2001; Ohtake et al. 1995; Olsen et al. 2002; Spiering et al. 2003). A record efficiency of 18.8% has recently been achieved with for CBD-ZnS (Nakada and Mizutani 2002). Most industries are currently using CBD-CdS but Showa Shell has shown 14.2% efficiency on 864 cm^2 submodule developed with CBD-grown ZnS(O,OH) buffer layers (Kushiya, 2004).

Front Electrical Contact. TCOs with bandgaps of above 3 eV are the most appropriate and have become the ultimate choice for front-contact due to their excellent optical transparency (greater than 85%) and reasonably good electrical conductivity. Today, CIGS solar cells employ either ITO or, more frequently, RF-sputtered Al-doped ZnO. A combination of an intrinsic and a doped ZnO layer is commonly used. Although this double layer yields consistently higher efficiencies, the beneficial effect of intrinsic ZnO is still under discussion. Doping of the conducting ZnO layer is achieved by group-III elements, particularly with aluminum. However, investigations show boron to be a feasible alternative because it yields a high mobility of charge carriers and a higher transmission in the long-wavelength spectral region, giving rise to higher currents. For high-efficiency cells, the TCO deposition temperature should be lower than 150°C to avoid the detrimental interdiffusion across CdS/CIGS interface. RF sputtering is not considered suitable for industrial production; therefore, alternative sputtering and CVD methods are investigated and used.

Sodium Incorporation into CIGS. One of the breakthroughs in CIGS PV technology occurred when the alumina or borosilicate glass substrate was replaced by soda-lime glass to match the thermal expansion coefficients, which resulted in substantial efficiency improvement (Hedström et al. 1993). It was subsequently realized that sodium (Na) plays an important role in high-efficiency CIGS solar cells because it affects the microstructure (grain size), and passivates the grain boundaries, leading to changes in electronic conductivity by up to two orders of magnitude. The overall effect is efficiency improvement primarily because of an increase in the open-circuit voltage (V_{oc}) and fill factor (FF) of the solar cells. Most commonly, Na is introduced into CIGS by diffusion from the soda-lime glass substrate during the absorber deposition. However, sodium incorporation from such an approach is neither controllable nor reliable; therefore, alternative methods to add Na from external sources are used either during or after the deposition of CIGS layers. These methods include the coevaporation or the deposition of a thin precursor of a Na compound such as NaF, Na_2Se, or Na_2S for CIGS on Na-free substrates that include soda-lime glass covered with barrier layers (Al_2O_3, Si_3N_4, etc., as used by Shell Solar). These barrier layers inhibit sodium diffusion from the glass substrate. CIGS on flexible substrates (metal and polyimide foils) also need controlled incorporation of sodium, which is provided from a precursor layer applied prior to or after the CIGS growth.

20.2.4.3 Deposition and Growth of CIGS Absorber

There are number of processes used for the deposition of the CIGS thin-films, some of them are briefly described:

Coevaporation Processes. As described earlier, vacuum evaporation is the most successful technique for deposition of CIGS absorber layers for highest-efficiency cells. The vacuum evaporation method involves simultaneous evaporation of the constituent elements from multiple sources in single or sequential processes during the whole absorber deposition process. While a variation of the In to Ga ratio during the deposition process leads to only minor changes in the growth kinetics, the variation of the Cu content strongly affects the film growth. Thus, different coevaporation growth procedures are classified by their Cu evaporation profile (Figure 20.35).

One variant of coevaporation is a bilayer process (also called the *Boeing process*), which originates from the work of Mickelsen and Chen (McCandless and Dobson 2004) and yields larger grain sizes compared to the constant-rate (single-stage) process. This is attributed to the formation of a Cu_xSe phase during the Cu-rich first stage that improves the mobility of group-III atoms during growth.

The highest efficiencies in laboratories are achieved with the so-called *three-stage process*, introduced by NREL, (Gabor et al. 1994). With this process, the CIGS layer is obtained by starting the deposition of an $(In,Ga)_xSe_y$ precursor, followed by the codeposition of Cu and Se until a Cu-rich overall composition is reached, and finally the overall Cu concentration is readjusted by subsequent deposition of In, Ga, and Se. CIGS films prepared by the three-stage process exhibit a large-grained, smooth surface, that reduces the junction area and is thereby expected to reduce the number of defects at the junction and yield high efficiency. Several groups have developed 16–19.5% efficiency cells using CIGS grown with the three-stage process.

Selenization of Precursor Materials. The selenization of precursors is another method of obtaining CIGS thin-films. This sequential process is favorable due to its suitability for large-area film deposition with good control of the composition and film thickness following the initial success of Arco Solar in 1987. Such processes consist of the deposition of a precursor material obtained by sputtering, evaporation, electrodeposition, paste printing, spray pyrolysis, etc., followed by thermal annealing in controlled reactive or inert atmosphere for optimum compound formation via the chalcogenization reaction as illustrated below in Figure 20.36. The precursor materials are either stacked metal layers or a

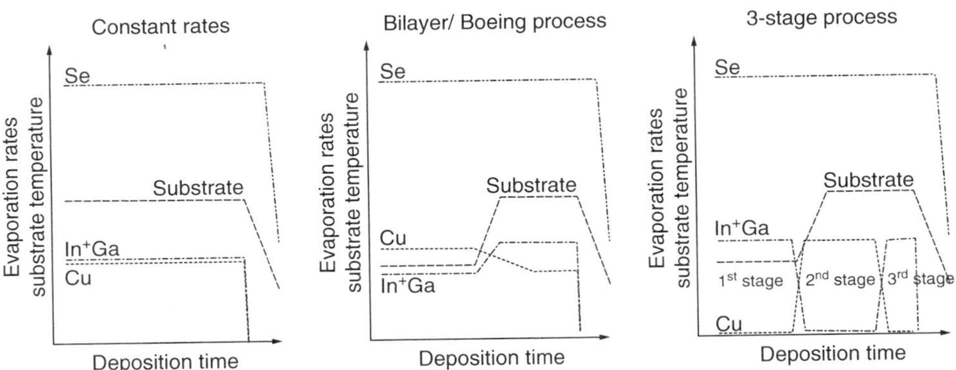

FIGURE 20.35 Diagram representing the recipes in the co-evaporation processing steps in the deposition of $Cu(In,Ga)Se_2$ used for constant rate, bilayer or Boeing, and three-stage process. (From Romeo, A., Terheggen, M., Abou-Ras, D., Bätzner, D. L., Haug, F.-J., Kälin, M., Rudmann, D., and Tiwari, A. N., *Prog. in Photovolt: Res. App.*, 12, 93–111.)

stack of their compounds and alloys. Shell Solar, USA and Showa Shell, Japan use sputtering techniques for precursor deposition and production of large-area solar modules up to 60 × 120 cm^2, yielding maximum efficiencies of 13% on 30 × 30 cm^2 modules (Karg 2001; Kushiya et al. 2003; Palm et al. 2004). Solar modules produced by Shell Solar, USA are commercially available in the market.

Alternative CIGS Growth Processes. There is substantial interest in using nonvacuum methods for CIGS deposition. An innovative approach utilizes the stability of the oxides to produce nanosized precursor particles (Eberspacher et al. 2001; Kapur et al. 2001; Kaelin et al. 2004). Nanosized metal oxides

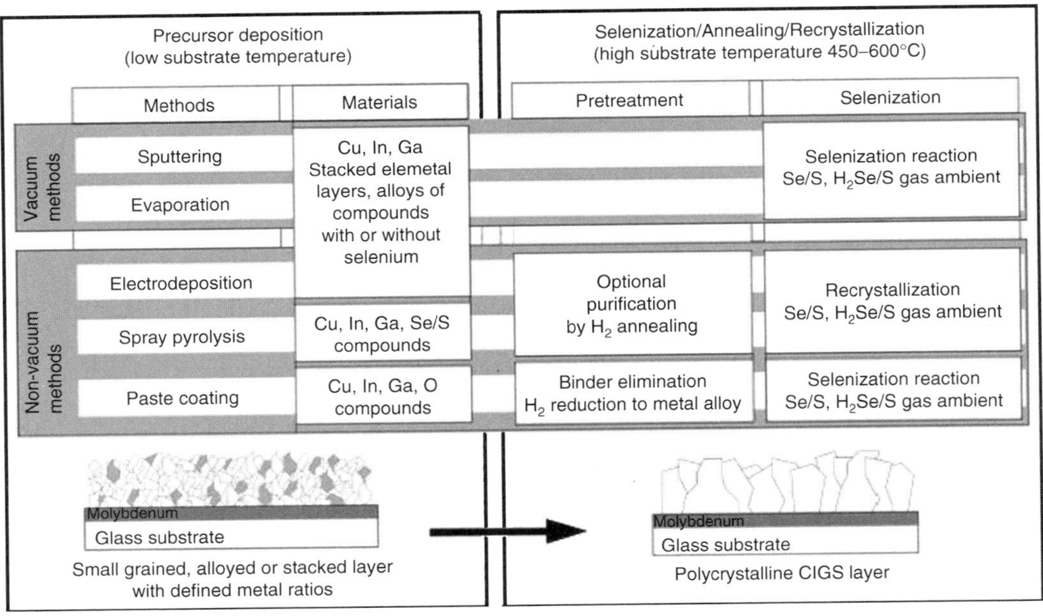

FIGURE 20.36 Schematic of the various processes for selenization of precursor materials. (From Romeo, A., Terheggen, M., Abou-Ras, D., Bätzner, D. L., Haug, F.-J., Kälin, M., Rudmann, D., and Tiwari, A. N., *Prog. in Photovolt: Res. App.*, 12, 93–111.)

are mixed in an ink suspension that allows low-cost, large-area deposition by doctor blading, screen printing, or spraying of the precursor. Such nonvacuum deposition of precursors allows a very efficient material utilization of almost 100% of the nonabundant metals indium and gallium. A selenization treatment converts the precursor into a CIGS layer and solar-cell efficiencies of over 13% have been achieved by ISET, USA. One of the drawbacks of the process is the toxicity of the H_2Se gas used for selenization. However, recent efforts are being made to selenize printed precursors with Se vapors (Kaelin et al. 2004).

The CIGS compound can also be formed directly by electrodeposition from a chemical bath (Hodes et al. 1986). Several groups, including EDF-CNRS, France, CIS Solar Technologies, Germany, and CISCuT, Germany, have been using such approaches and have obtained cells with efficiencies greater than 10%. With a hybrid approach that uses additional vacuum deposition on an electrodeposited precursor layer, efficiencies as high as 15.4% have been achieved at NREL, USA.

20.2.4.4 Flexible CIGS Solar Cells

Probably the ultimate advantage of thin-film technology is the application of roll-to-roll manufacturing for production of monolithically interconnected solar modules leading to low energy payback time because of high-throughput processing and low cost of the overall system. A large number of activities on highly efficient, stable, and flexible thin-film modules based on CIGS has recently drawn much interest for flexible solar cells on metal and plastic foils. Apart from the expected high efficiency and long-term stability for terrestrial applications, flexible CIGS has excellent potential for space application because of its space-radiation-tolerant properties that are 2–4 times superior to the conventional Si and GaAs cells. Lightweight and rollable solar array structures will not only reduce the overall cost of space-deployable solar modules, but also can substantially save on the launching cost of satellites.

Development and current status of flexible CIGS solar cells have been reviewed by Kessler and Rudmann (2004). Flexible CIGS cells can be grown on polyimide and on a variety of metals, e.g., stainless steel (SS), Mo, Ti, etc. Therefore, the choice of substrate is important because there are some advantages and disadvantages: (1) the density of usable metals is 4–8 times higher than that of polymers; therefore, cells on metals are heavier; (2) metals are conducting and have rough surfaces; therefore, monolithic module development is difficult, which, in contrast, is easier on polymer foils; (3) stainless steel foils need an extra barrier layer against detrimental impurity (e.g., Fe) diffusion of the metal into the CIGS during deposition; (4) metal foils can withstand high deposition temperatures (550°C–600°C), which leads to higher efficiency than that achieved with polymer foils that are not suitable for processing temperatures greater than 450°C.

High-record efficiencies of flexible CIGS solar cells are 17.5% on stainless steel by Daystar, U.S.A. (Tuttle et al. 2000) and 14.1% on polymer foil by ETH Zurich, Switzerland (Brémaud et al. 2005). Several research groups and industries are involved in the development of flexible solar cells, but Global Solar, U.S.A. is the only company manufacturing flexible CIGS on a pilot production line. They have reported a 13.17% cell of 68.8 cm^2 and modules of 11.13% (on 3898 cm^2) and 10.10% (on 7085 cm^2) on metal foils (Beck et al. 2005). The solar modules on SS foils are not monolithically connected; they are made by connecting individual cells into a large-area cell with an overlap method. ZSW, Germany is developing a scribing and patterning method for monolithically connected solar modules on metal and polymer foils (Kessler and Rudmann 2004). The basic schematic cross-section of a monolithic module on glass is shown in Figure 20.37 along with a flexible prototype mini-module developed on polymer foil by ZSW & ETH, Zurich, within a European collaborative project.

Table 20.8 gives an overview of different flexible solar cell technologies, including the organic and dye-sensitized TiO_2 based PV technologies. Because of late start in R&D, flexible CIGS and CdTe solar cells are industrially less mature compared to a-Si cells. However, high cell efficiencies and inherent stability advantages indicate a promising potential for these technologies.

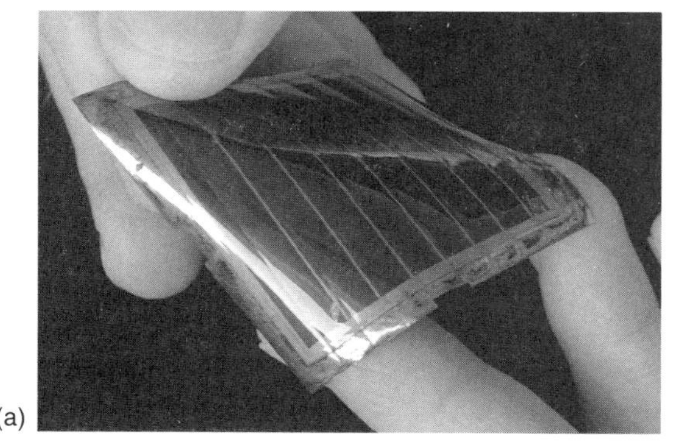
(a)

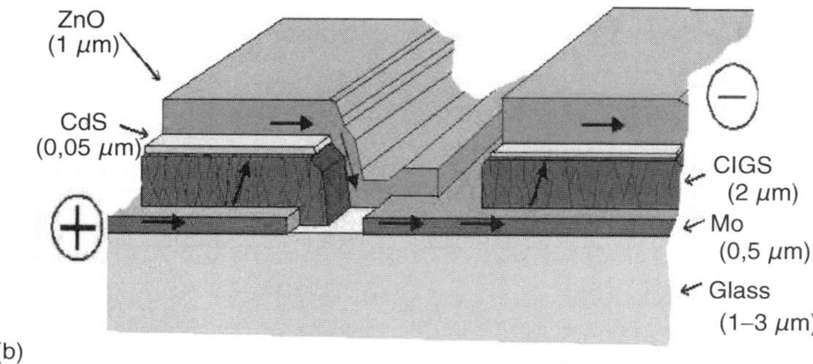

(b)

FIGURE 20.37 **(See color insert following page 19-52.)** Flexible monolithic CIGS modules showing a prototype mini-module on a polymer foil (a) and schematic cross-section of the module (image taken from the website of ZSW) exhibiting the material component layers and their interconnect patterns (b).

20.2.5 Environmental Concerns and Cd Issue

The CdTe and CIGS thin-film technologies have demonstrated excellent potential for the cost-effective production of solar electricity. However, these technologies, especially the CdTe, suffer from the perception of toxicity of the constituent element Cd, which is used in the form of a stable compound in thin-film modules. Often raised issues are the risks or hazards associated with the use of materials during the processing and fabrication of CdTe/CdS and CdS/CIGS solar cells and risks associated during the

TABLE 20.8 An Overview of Different Flexible Solar Cell Technologies

	CIGS	CdTe	Amorphous Silicon	Organic & Dye Solar Cells
Lab efficiency on plastic foil	14.1% (Single-junction cell)	11.4% (Single-junction cell)	8–12%[a] (multijunction cell)	5 to 8%
Lab efficiency on metal foil	17.5% (Single-junction cell)	8% (Single-junction cell)	14.6[a]/13% (multijunction cell)	7%
Industrial efficiency (typical values)	6–11% (On steel foil, not yet available on plastic foil)	Not yet demonstrated	4–8%[a] (Available on plastic and metal foils	Not yet demonstrated
Stability under light	Material stable	Material stable	Degrades	Stability not proven

[a] Initial values measured before light-induced degradation of solar cells.

cradle-to-grave operating lifetime of these modules. The environmental and health hazard (E&H) issues of CdTe solar modules have been extensively investigated by several independent agencies, including the national laboratories in Europe and the USA. (Fthenakis et al. 1999; Fthenakis and Zweibel 2003; Fthenakis et al. 2004). Although CdTe technology has no chance of eliminating Cd, there is some maneuverability in CIGS technology in the elimination of a very thin (typically ~50-nm) CdS buffer layer, and thus alternative buffer layers are being successfully pursued. Initial success has already been achieved as CIGS solar cells of 16%–18.8% efficiencies and modules of 13.4% efficiency have been developed with alternative, Cd-free buffers (Hedström et al. 1993; Kushiya 2004).

Referring to the perception and concerns on Cd issues, V. Fthenakis (Senior Chemical Engineer, Brookhaven National Laboratory) and K. Zweibel (Manager, Thin-film Partnership Program, NREL, USA) have presented a detailed account of their studies during their presentations at the 2003 NCPV program review meeting in the USA (Fthenakis et al. 2004) that confirmed that CdTe panels would be almost benign with zero emission and no associated health hazards and gave a clean chit to CdTe technology. The following points emerged out of the studies:

Cadmium is a by-product of zinc, lead, and copper mining. It constitutes only 0.25% of its main feedstock, ZnS (sphalerite). Cadmium is released into the environment from phosphate fertilizers, burning fuels, mining and metal-processing operations, cement production, and disposal of metal products. Releases from disposed Cd products, including Ni-Cd batteries are minimum contributors to human exposure because Cd is encapsulated in the sealed structures. Most human cadmium exposure comes from ingestion, and most of that stems from the uptake of cadmium by plants, through fertilizers, sewage sludge, manure, and atmospheric deposition. Although long-term exposure to elemental cadmium, a carcinogen, has detrimental effect on kidneys and bones, limited data exists in toxicology. However, the CdTe compound is more stable and less soluble than elemental Cd and is therefore likely to be less toxic.

Considering the electrolytic refinery production of CdTe powders (from Cd wastes from the zinc, iron, and steel industries) there would be 0.001% Cd gaseous emission. This would correspond to 0.01 g/GWh, which is significantly less as compared to the perceptions and hypes created that estimate it at 0.5 g/GWh based on other crude processes or unsubstantiated data. The only potential hazard would be a building fire. It has also been estimated quantitatively that the maximum temperature of a basement on fire is ~900°C, which is still less than the melting point of CdTe (1041°C). Furthermore, the vapor pressure at 800°C for CdTe is ~2.5 torr (0.003 atm), so this minimizes the risks further, and once sealed between glass plates, any Cd vapor emission is unlikely. The main conclusion of the studies was that the environmental risks associated with CdTe-based technology are minimal. Every energy source or product may present some direct or indirect environmental health and safety hazards, and those of CdTe should not be considered a problem. The following conclusions can be drawn:

- Cd is produced as a by-product of Zn and can either be put to beneficial uses or discharged to the environment posing another risk.
- CdTe in PV is much safer than other current Cd uses.
- CdTe PV uses Cd 2500 times more efficiently than Ni-Cd batteries.
- Occupational health risks are well managed.
- There is absolutely no emission of Cd during PV operation.
- A risk from fire emission is minimal.
- CdTe technology and modules are safe and do not pose significant risks.

20.2.6 Conclusions

1. The PV market is booming with a phenomenal surging growth rate. High demand for cost-effective PV installations with more consumer-oriented choices cannot be met by the c-Si wafer

technology and is not expected to achieve production cost targets of less than €1/W (Figure 20.23).

2. Thin-film PV has clearly demonstrated an excellent potential for cost-effective generation of solar electricity if modules are produced in high-capacity production plants. A mix of c-Si and thin-film PV technologies will cater to the market needs in the near- to midterm future, followed by the dominance of thin-film and other PV technologies in the long term.
3. Thin-film PV industries are growing fast; however, there are several issues:
   a. Reducing the gap between lab efficiency and larger area industrial production efficiency. This is achievable with the design of better equipment with in situ diagnostics. Nonavailability of a standard deposition system for thin-film PV has been a problem; therefore, effort is needed to develop large area equipment suitable for thin-film PV.
   b. For lower cost, high-throughput and high-yield efforts are needed that require further simplification and increased robustness of the processes and device structures.
   c. Further improvements are needed in the device structure for still higher efficiencies (greater than 25%) and enhanced material utilization (low wastage), along with thinner layers for less material consumption if multi-giga and terawatt PV facilities are to be successful for a safer and prosperous world.

20.3 Concentrating PV Technologies

Roland Winston, Robert McConnell, and D. Yogi Goswami

20.3.1 Introduction

The current state of solar cell development is illustrated in Figure 22.38. While single-junction silicon solar cells dominate today's solar industry, the rapid rise in efficiency vs. time (experience curve) of the multijunction cells makes this a particularly attractive technology path (NREL). The high efficiency, in comparison with single junction cells, such as silicon, is obtained by stacking several junctions in series, electrically isolated by tunnel diodes, as explained in section 22.1. These can be qualitatively viewed as adding the voltages of three junctions in series, while maintaining the current of a single junction. When compared with the very best silicon cells, the efficiency gain is about 2 today, but it is getting better with time as Figure 22.38 convincingly shows.

Under concentrated sunlight, multijunction (GaInP/GaAs/Ge) solar cells have demonstrated efficiencies twice (39.3%) those of most silicon cells (usually 15%–21%). This means that, in sunny areas, a multijunction concentrator system can generate almost twice as much electricity as a silicon panel with the same area. The concentrating optics focus the light onto a small area of cells, reducing the area of the solar cells by a factor of, typically 500–1000 times. The reduced cell area overcomes the increased cell cost. The cell cost is diminished in importance and is replaced by cost of optics. If the cost of the optics is comparable to the cost of the glass and support structure needed for silicon flat-plate modules, then the cost per unit area can remain fixed while the electricity production is essentially doubled. Thus, in high-direct insolation locations, the multijunction concentrator technology has the potential to reduce the cost of solar electricity by about a factor of two. As a side benefit, the cells are more efficient under concentration, provided a reasonable cell temperature can be maintained. Clearly, the technology may be extended to four and even five junctions if efficiency benefits justify added cost. The efficiency is a moving target; today's triple junction cell efficiency is nearly 40%. Thus, one may reasonably extrapolate that multijunction cells may reach 50% efficiency in the future. Using less cell material for a given power output has attraction to cell manufacturers that are having trouble producing sufficient material to keep up with demand. It is worth mentioning that this technology was first developed and proven in the space program, where specific power (power/mass) is a more important consideration than cost.

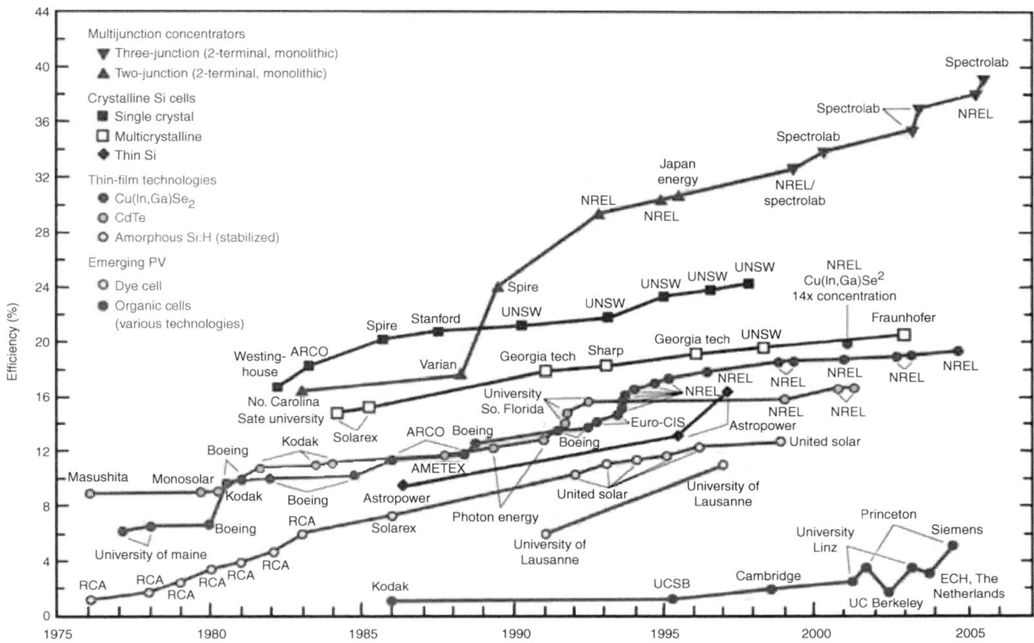

FIGURE 20.38 World-record conversion efficiencies for various PV technologies.

20.3.2 CPV Market Entry

In 2004, less than 1 MW of concentrator PV systems was installed, out of a total world PV market of 1200 MW. Admittedly, 1 MW does not constitute significant market entry. However, the significant development is an increased number of projects with sizes of several hundred kilowatts (kW), creating a market appropriate for CPV technology. Furthermore, CPV technologies are ready today for this market opportunity with high-efficiency solar cells and well-developed hardware. However, it is the near-term prospects for even better performance and lower-installed system costs that are leading to real market entry during 2005 and 2006.

Amonix, Inc., of Torrance, California, spent more than 15 years developing five generations of CPV prototypes that led to the products most recently installed in Arizona by Arizona Public Service (Figure 20.39) and in other southwestern U.S. locations (Amonix 2005).

Similarly, Solar Systems Pty, Ltd., of Hawthorn, Australia (Solar Systems 2005), spent more than 15 years developing several generations of their CPV prototypes that led to recent installations on aboriginal lands in Australia—part of a market served by large diesel generating plants with high transportation costs for the diesel fuel (Figure 20.40). The company plans to install more than 5 MW in 2006. Indeed, the growth from 1 MW of CPV in 2004 to some 18 MW in 2006 legitimately constitutes market entry for this technology.

20.3.3 Future Growth

With market entry comes the question of sustained market growth. Carlos Algora, from the Instituto de Energia Solar, Universidad Politecnica de Madrid, published a useful cost analysis in 2004 for concentrator PV technologies (Algora 2004). Many of the costs came from installed system costs for the 480-kW CPV project in Tenerife, but included technology improvements for production of 10 MW and incorporated learning curve reductions for cumulative production greater than 10 MW. The analysis

FIGURE 20.39 Several 35-kW Amonix modules installed in an Arizona Public Service power plant. The pickup truck underneath the module gives an idea of the scale.

included a wide range of parameters. Concentrations ranged from 400 to 1000 suns, with cell efficiencies ranging from 32%–40%. The presentday CPV-installed system cost was estimated at 2.34 €/W (US$3.00/W with the present exchange rate) using 32%-efficient multijunction solar cells in today's CPV technology with only 10 MW of product. This part of Algora's analysis did not incorporate the impact of a learning curve.

The lowest projected costs in Algora's analysis was 0.69 €/W (US$0.85/W) for 40%-efficient solar cells, 1000-sun concentration, and cumulative production of 1000 MW, using a conservative learning curve of 17.5% that started with 10 MW of product. Although 1000 MW may seem a distant cumulative production target for CPV, it is informative to realize that a total of about 1 MW of multijunction production capacity already exists at Spectrolab and other space satellite solar power providers. This 1 MW production capacity of high-efficiency multijunction solar cells corresponds to ~1000 MW of CPV plants at 1000-sun concentration.

20.3.4 Energy Payback

CPV system costs are much more sensitive to the price of steel than to the price of silicon. In this regard, CPV systems share similar concerns with the wind industry (McConnell 2002). Other

FIGURE 20.40 Several 25-kW solar systems modules. Again, note the people in the foreground for an idea of module scale.

FIGURE 20.41 An Amonix production facility in Los Angeles is strikingly different from a flat-plate PV manufacturing facility. This difference results in much lower capital costs for the facility.

technological similarities with wind systems include the low cost of production plants, suitability for distributed and large-scale utility generation, modularity, moving parts, and the need for a good resource, be it wind or solar.

Such observations suggest that CPV systems could follow in the footsteps of wind systems. It seems plausible that CPV system costs can approach wind system costs (typically US$0.80/W today—or about the same as projected CPV costs) if only because common materials (e.g., steel, glass, plastic) are dominant and because production plant costs are relatively low. Also consistent with these observations are recent estimates of energy payback for CPV technologies that are very close to values published for wind turbines at good wind sites. Specifically, the energy payback has been estimated as 8 months for a CPV system in a site having a good solar resource (Bett et al. 2005).

The cost of production plants can be a critical element in how rapidly a technology can expand to meet market demands (Figure 20.41). For manufacturing flat-plate modules of crystalline silicon, amorphous silicon, or polycrystalline thin-films, published costs range between $150 and $300 million for a production capacity of 100 MW per year. A 100-MW/year production facility for wind systems may cost between $10 and $15 million. Manufacturing of wind turbines or CPV systems resembles the assembly-line production of automobiles, including the dependence on suppliers and the sensitivity to steel prices (McConnell 2002).

It is much easier for an industry with low production-plant costs to sustain high growth rates because considerably less capital is needed to expand production facilities. However, uncertainty exists in the cost of a 100-MW/year CPV plant because one has yet to be built. Published estimates range between $15 and $50 million (McConnell 2002). Production-plant costs in this range are an asset for sustainable and augmented market growth of CPV technologies.

20.3.5 The Need for Qualification Standards

Qualification standards help developers design their new products by identifying weaknesses before production and project installation. They give customers the confidence that their project investments will pay off. In short, they can contribute immensely to a technology's successful market entry.

Fortunately, the CPV industry thought about this situation in the 1990s. Standards take years to develop because the process is based on consensus. Companies do not want standards that are unnecessarily strict or require expensive test procedures. Customers want standards that ensure good product performance. So, input from both groups—companies and customers—as well as from relatively objective research organizations leads to an accepted set of test procedures vital for successful entry of CPV into the market. The first CPV standard (IEEE 2001) was published in 2001. This standard, however, was most suitable for U.S. concentrator PV technologies using Fresnel lenses. The International Electrotechnical Commission (IEC), with input from engineers from more than 10 countries, has been developing a standard suitable for concentrators using mirrors or lenses with solar concentration ratios ranging from a couple of suns to 1000s of suns. The IEC group working on this standard hopes to complete their work and publish the standard in 2006 (McConnell and Symko-Davis 2005). They are also working on new standards for solar trackers and safety.

Nomenclature

c speed of light (3×10^8 m/s)
D_x minority carrier diffusion length
e energy of photon (eV)
$F(x)$ intensity of the photon beam as a function of penetration depth into the material
F_o initial intensity of the photon beam at the material surface
h Planck's constant (6.625×10^{-34} J·s)
I current (amperes)
I_ℓ short-circuit current (ampere)
k Boltzmann's constant (1.381×10^{-23} J/K)
L latitude angle
L_x diffusion length
n electron concentration
n day of year
N_a density of acceptor impurities
N_d density of donor impurities
n_i intrinsic carrier concentration
p hole concentration
P_{max} maximum power (watt)
q charge of an electron (1.602×10^{-19} Coulombs)
T junction temperature (K)
V voltage (volts)
V_{OC} open-circuit voltage (volt)
x depth of penetration into material

Symbols

λ wavelength (μm)
α absorption constant
α altitude angle of the sun
δ declination angle
(Azimuth angle of the sun
ω hour angle
Ω resistance (ohm)
τ_x minority carrier lifetime

Acronyms

a-Si amorphous silicon
CVD chemical vapor deposition
EHP electron-hole pair
GFDI ground fault detection and interruption device
HIT heterojunction with intrinsic thin-layer
LCB linear current booster
MPT maximum power tracker
psh peak sun hours

References

Abken, A. E. and Bartelt, O. J. 2002. *Thin Solid Films*, 403–404, 216–222.

Algora, C. 2004. The importance of the very high concentration in the third-generation solar cells, In *Next Generation Photovoltaics*, A. Marti and A. Luque, eds., pp. 4025–4027. Institute of Physics Publishing, Bristol and Philadelphia.

Matsuda, A., Yamaoka, T., Wolf, S., Koyama, M., Imanishi, Y., Kataoka, H., Matsumara, H., and Tanaka, K. 1986. *Journal of Applied Physics*, 60, 4025–4027.

American Solar Energy Society. *Solar Today*.

Amonix, Inc. 2005. http://www.amonix.com

Arya, R. R. 2004. In *Proceedings of the 19th European Photovoltaic Conference*.

ASE International. http://www.ase-international.com

Basore, P. A. 2002. Pilot production of thin-film crystalline silicon on glass modules. In *Proceedings of the 29th IEEE Pohotovoltaic Specialists Conference*.

Basore, P. A. 2002. Pilot production of thin-film crystalline silicon on glass modules. In *PV in Europe Conference*.

Battisti, R. and Corrado, A. 2005. Evaluation of technical improvements of photovoltaic systems through life cycle assessment methodology. *Energy*, 30, 952–967.

Bätzner, D. L., Romeo, A., Zogg, H., Wendt, R., and Tiwari, A. N. 2001. *Thin Solid Films*, 387, 151–154.

Bätzner, D. L., Romeo, A., Terheggen, M., Döbeli, M., Zogg, H., and Tiwari, A. N. 2004. *Thin Solid Films*, 451, 536–543.

Beck, M. E., Wiedeman, S., Huntington, R., Van Alsburg, J., Kanto, E., Butcher, R., and Britt, J. S. 2005. In *Proceedings of 31st IEEE Photovoltaic Specialists Conference*, pp. 211–214.

Bergmann, R. B. 1999. *Applied Physics A*, 69, 187–194.

Bett, A. W., Siefer, G., Baur, C. B., Riesen, S. V., Peharz, G., Lerchenmuller, H., and Dimroth, F. 2005. G. SFLATCON concentrator PV—Technology ready for the market. Presentation at the 20th European PV Solar Energy Conference and Exhibition.

Bhattacharya, R. N. and Ramanathan, K. 2004. *Solar Energy*, 77, 679–683.

BNL. PV EHS. http://www.pv.bnl.gov

Bohland, J. 1998. Accelerated aging of PV encapsulants by high intensity UV exposure. In *Proceedings of the 1998 Photovoltaic Performance and Reliability Workshop*.

Bonnet, D. 2004. In *Proceedings of the 19th European Photovoltaic Solar Energy Conference*, pp. 1657–1664.

Bonnet, D. and Rabenhorst, H. 1972. In *Proceedings of the 9th Photovoltaic Specialists Conference*, pp. 129–132.

Brémaud, D., Rudmann, D., Bilger, G., Zogg, H., and Tiwari, A. N. 2005. In *Proceedings of 31st IEEE Photovoltaic Specialists Conference*, pp. 223–226.

Brendel, R., Schmiga, C., Froitzheim, A., Ghosh, M., Motz, A., and Schmidt, J. 2005. In *Proceedings of the 29th IEEE Photovoltaic Specialists Conference*, p. 8689.

Bruton, T. M., Woodcock, J. M., Roy, K., Garrard, B., Alonso, J., Nijs, J., Räuber, A. et al. 1997. Final report European Commission Project RENA-CT94-0008 "MUSICFM".

Carabe, J. and Gandia, J. J. 2004. *Opto-electronics Reviews*, 12 (1), 1–6.

Carlson, D. and Wronski, C. 1976. *Applied Physics Letters*, 28, 671–673.

Chaisitsak, S., Yamada, A., and Konagai, M. 2001. In *Proceedings of the Materials Research Society Spring Meeting*, p. 668.

Chopra, K. L., Paulson, P. D., and Dutta, V. 2004. *Progress in Photovoltaics: Research and Applications*, 12, 69–92.

Compaan, A. D., Gupta, A. K., Lee, S., Wang, S., and Drayton, J. 2004. *Solar Energy*, 77, 815–822.

Cusano, D. A. 1963. *Solid State Electronics*, 6, 217–232.

Deng, X. and Schiff, E. A. 2003. Amorphous silicon-based solar cells, In *Handbook of Photovoltaic Science and Engineering*, A. Luque and S. Hegedus, eds., pp. 48–67. Wiley, New York.

Diefenbach, K. H. 2005. Wiped away. *Photon International*, February 48–67.

Durr, M., Schmid, A., Obermaier, M., Rosselli, S., Yasuda, A., and Nelles, G. 2005. *Nature Materials*, 4, 607–611.

Eberspacher, C., Fredric, C., Pauls, K., and Serra, J. 2001. *Thin Solid Films*, 387, 18–22.

Fahrenbruch, A., Bube, R., Kim, D., and Lopez-Otero, A. 1992. *International Journal of Solar Energy*, 12, 197–222.

Ferekides, C., Britt, J., Ma, Y., and Killian, L. 1993. In *Proceedings of the 23rd IEEE Photovoltaic Specialists Conference*, p. 389.

First Solar. 2005. http://www.firstsolar.com/index.html.

Fthenakis, V. and Zweibel, K. 2003. CdTe PV: Real and perceived EHS risks. Prepared for the NCPV and Solar Program Review Meeting.

Fthenakis, V., Morris, S., Moskowitz, P., and Morgan, D. 1999. Toxicity of cadmium telluride, copper indium diselenide, and copper gallium diselenide. *Progress in Photovoltaics*, 7, 489–497.

Fthenakis, V. M., Fuhrmann, M., Heiser, J., and Wang, W. 2004. Experimental investigation of emissions and redistribution of elements in CdTe PV modules during fires. Prepared for the 19th European PV Solar Energy Conference, Paris, France.

Gabor, A. M., Tuttle, J. R., Albin, D. S., Contreras, M. A., Noufi, R., and Hermann, A. M. 1994. *Applied Physics Letters*, 65, 198–200.

Goetzberger, A., Hebling, C., and Schock, H. W. 2003. *Materials Science and Engineering R*, 40, 1–46.

Goswami, D. Y. and Boer, K.W. eds. *Advances in solar energy*, Vol. 1–16. American Solar Energy Society, Boulder, CO.

Goswami, D. Y., Kreith, F., and Kreider, J. F. 2000. *Principles of Solar Engineering. 2nd Ed.* Taylor & Francis, Philadelphia, PA.

Granqvist, C. G. 2003. *Advanced Materials*, 15, 1789–1803.

Gratzel, M. 2000. *Progress in Photovoltaics: Research and Applications*, 8, 171–186.

Green, M. A. 2004. D. Y. Goswami and K. W. Boer, eds., In *Advances in Solar Energy, Vol. 15*, pp. 2907–51. American Solar Energy Society, Boulder, CO.

Green, M. A. and Wenham, S. R. 1994. Novel parallel multijunction solar cell. *Applied Physics Letters*, 65, 2907.

Green, M. A., Emery, K., Bocher, K., King, K. L., and Igari, S. 2006. Solar cell efficiency tables (version 27). *Progress in Photovoltaics: Research and Applications*, 14, 45–51.

Guha, S. 2004. *Solar Energy*, 77, 887–892.

Guha, S., Narsimhan, K. L., and Pietruszko, S. M. 1981. *Journal of Applied Physics*, 52, 859.

Guha, S., Yang, J., Nath, P., and Hack, M. 1986. *Applied Physics Letters*, 49, 218.

Hamakawa, Y. 2005. D. Y. Goswami and K. W. Boer, eds., In *Advances in Solar Energy, Vol. 16*, pp. C113–678. American Solar Energy Society, Boulder, CO.

Hanoka, J. I. 2002. In *Proceedings of the 29th IEEE Photovoltaics Specialists Conference*, pp. 66–69.

Hedström, J., Ohlsen, H., Bodegard, M., Kylner, A., Stolt, L., Hariskos, D., Ruckh, M., and Schock, H.W. 1993. In *Proceedings of the 23rd IEEE Photovoltaic Specialists Conference*, pp. 364–371.

Hegedus, S. S. and Luque, A. 2003. Status, trends, challenges and the bright future of solar electricity from photovoltaics, In *Handbook of Photovoltaic Science and Engineering*, A. Luque and S. Hegedus, eds., pp. C113–678. Wiley, New York.

Henderson, H. 1981. *The politics of the solar age: Alternatives to economics. Anchor Press/Doubleday, Garden City, NY.*

Hodes, G., Lokhande, C. D., and Cahen, D. 1986. *Journal of the Electrochemical Society*, 133 (3), C113.

Ichikawa, Y., Fujikake, S., Takayama, R., Saito, S., Ota, H., Yoshida, T., Ihara, T., and Sakai, H. 1993. In *Proceedings of the 23rd IEEE Photovoltaic Specialists Conference*, IEEE, Piscataway, NJ, p. 27.

IEEE. 2000. IEEE 929-2000. IEEE recommended practice for utility interface of residential and intermediate photovoltaic (PV) systems. IEEE Standards Coordinating Committee 21, Photovoltaics.

IEEE. 2001. IEEE recommended practice for qualification of concentrator photovoltaic receiver sections and modules. IEEE Standard, p. 1523–2001.

International Solar Energy Society, *Solar Energy Journal*, Elsevier.

Jäger-Waldau, A. 2004. *Solar Energy*, 77, 667–678.

James & James Publishers, Renewable energy world, James & James Publishers, London.

Kaelin, M., Rudmann, D., and Tiwari, A. N. 2004. *Solar Energy*, 77, 749–756.

Kapur, V. K., Fisher, M., and Roe, R. 2001. In *Proceedings of the 2001 MRS Spring Meeting*, p. 668, H2_6_1–7.

Karg, F. H. 2001. *Solar Energy Materials and Solar Cells*, 66, 645.

Kazmerski, L., White, F., and Morgan, G. 1976. *Applied Physics Letters*, 29, 268–269.

Keshner, M. S. and Arya, R. 2004. Study of potential cost reductions resulting from super-large-scale manufacturing of PV modules. NREL/SR-520-36846 (final subcontract report). http://www.osti.gov/bridge

Kessler, F. and Rudmann, D. 2004. *Solar Energy*, 77, 685–695.

Klein, S., Repmann, T., and Brammer, T. 2004. *Solar Energy*, 77, 893–908.

Krein, P. T. 1998. *Elements of power electronics. Oxford University Press, New York.*

Kushiya, K. 2004. *Solar Energy*, 77, 717–724.

Kushiya, K., Ohshita, M., Hara, I., Tanaka, Y., Sang, B., Nagoya, Y., Tachiyuki, M., and Yamase, O. 2003. *Solar Energy Materials and Solar Cells*, 75 (1–2), 171–178.

Lebrun, J. 1966. *Revue de Physique Appliquee*, 1, 204–210.

Lechner, P. and Schade, H. 2002. *Progress in Photovoltaics: Research and Applications*, 10, 85–97.

Lewis, N. S. and Crabtree, G., chairpersons. Report on the basic energy sciences: DoE Workshop on Solar Energy Utilization, 'Basic research needs for solar energy'.

Markvart, T. ed. 1994. Solar Electricity, pp. 831–838. Wiley, Chichester, UK.

Mathew, X., Enriquez, J. P., Romeo, A., and Tiwari, A. N. 2004. *Solar Energy*, 77, 831–838.

Maycock, P. D. 2003. *Renewable Energy World*, 6 (4), 84–101.

Maycock, P. D. Information about technology and business aspects of PV. Photovoltaic News.

McCandless, B. E. and Dobson, K. D. 2004. *Solar Energy*, 77, 839–856.

McCandless, B. E. and Sites, J. R. 2003. In *Handbook of Photovoltaic Science and Engineering*, A. Luque and S. Hegedus, eds., pp. 21–30. Wiley, Chichester, UK.

McCandless, B., Youm, I., and Birkmire, R. 1999. *Progress in Photovoltaics*, 7, 21–30.

McConnell, R. 2002. Large-scale deployment of concentrating PV: Important manufacturing and reliability issues. In *Proceedings of the First International Conference on Solar Electric Concentrators*, NREL/EL-590-32461.

McConnell, R. D. and Symko-Davies, M. 2005. High performance PV future: III–V multijunction concentrators. Plenary presentation at the 20th European PV Solar Energy Conference and Exhibition, Barcelona, Spain.

Meier, J., Dubail, S., Fluckinger, R., Fischer, D., Keppner, H., and Shah, A. 1994. In *Proceedings of the 1st World Conference on Photovoltaic Energy*, pp. 409–438.

Messenger, R. and Ventre, G. 2004. *Photovoltaic systems engineering. 2nd Ed.* CRC Press, Boca Raton, FL.

Mickelsen, R. and Chen, W. 1981. In *Proceedings of the 15th IEEE Photovoltaic Specialist Conference*, pp. 800–804.

Mickelsen, R. and Chen, W. 1982. In *Proceedings of the 16th IEEE Photovoltaic Specialist Conference*, pp. 781–785.

Mitchell, K., Eberspacher, C., Ermer, J., and Pier, D. 1988. In *Proceedings of the 20th IEEE Photovoltaic Specialists Conference*, pp. 1384–1389.

Muller, J., Rech, B., Springer, J., and Vanecek, M. 2004. *Solar Energy*, 77, 917–930.

Nakada, T. and Mizutani, M. 2002. *Japanese Journal of Applied Physics*, 41, 165–167.

Nakada, T., Hirabayashi, Y., Tokado, T., Ohmori, D., and Mise, T. 2004. *Solar Energy*, 77, 739–747.

NFPA. 2002. NFPA 70 national electrical code. National Fire Protection Association, Quincy, MA.

Nijs, J. and Morten, R. 1997. Energy payback time of crystalline silicon solar modules, Karl W. Boer et al., ed., In *Advances in Solar Energy, Vol. 11*, pp. 291–327. American Solar Energy Society, Boulder, CO.

NREL. National Renewable Energy Laboratory, 30-year average of monthly solar radiation, 1961–1990, spreadsheet portable data files. http://rredc.nrel.gov/solar/old_data/nsrdb/redbook/sum2/

CdTe and NREL. CdTe PV. http://www.nrel.gov/cdte

Odum, H. T. and Odum, E. C. 1981. *Energy basis for man and nature. 2nd Ed.* McGraw-Hill, New York.

Ohtake, Y., Kushiya, K., Ichikawa, M., Yamada, A., and Konagai, M. 1995. *Japanese of Journal Applied Physics*, 34, 5949–5955.

Olsen, L., Eschbach, P., and Kundu, S. 2002. In *Proceedings of the 29th IEEE Photovoltaic Specialists Conference*, pp. 652–655.

Palm, J., Probst, V., and Karg, F. H. 2004. *Solar Energy*, 77, 757–765.

Ramanathan, K., Contreras, M. A., Perkins, C. L., Asher, S., Hasoon, F. S., Keane, J., Young, D., Ward, J., Duda, A. et al. 2003. *Progress in Photovoltaics: Research and Applications*, 11, 225–230.

Rech, B. and Wagner, H. 1999. *Applied Physics A*, 69 155–167.

Romeo, A., Bätzner, D. L., Zogg, H., Tiwari, A. N., and Vignali, C. 1999. *Solar Energy Materials and Solar Cells*, 58 (2), 209–218.

Romeo, N., Bosio, A., Tedeschi, R., Romeo, A., and Canevari, V. 1999. *Solar Energy Materials and Solar Cells*, 58 (2), 209–218.

Romeo, N., Bosio, A., Canevari, V., and Podesta, A. 2004. *Solar Energy*, 77, 795–801.

Romeo, A., Khrypunov, G., Kurdesau, F., Bätzner, D. L., Zogg, H., and Tiwari, A. N. 2004. In *Technical Digest of the International PVSEC-14*, pp. 715–716.

Romeo, A., Terheggen, M., Abou-Ras, D., Bätzner, D. L., Haug, F.-J., Kälin, M., Rudmann, D., and Tiwari, A. N. 2004. *Progress in Photovoltaics: Research and Applications*, 12, 93–111.

Rosenblum, A. E. et al. 2002. In *Proceedings of the 29th IEEE Photovoltaic Specialists Conference*, pp. 58–61.

Saitoh, T. 2005. D. Y. Goswami and K. W. Boer, eds., In *Advances in Solar Energy, Vol. 16*, pp. 113–142. American Solar Energy Society, Boulder, CO.

Sakata, H., Kawamoto, K., Taguchi, M., Baba, T., Tsuge, S., Uchihashi, K., Nakamura, N., and Kiyam, S. 2000. In Proceedings of the 28th IEEE Photovoltaic Specialists Conference, p. 7.

Shah, A., Meier, J., Buechel, A., Kroll, U., Steinhauser, J., Meillaud, F., and Schade, H. 2004. Paper presented at the ICCG5 Conference in Saarbrücken.

Shah, A. V., Schade, H., Vanecek, M., Meier, J., Vallat-Sauvain, E., Wyrsch, N., Kroll, U., Droz, C., and Bailat, J. 2004. *Progress in Photovoltaics: Research and Applications*, 12, 113–142.

Solar Systems 2005. http://www.solarsystems.com.au

Sopori, B. 2003. In *Handbook of Photovoltaic Science and Engineering*, A. Luque and S. Hegedus, eds., pp. 359–363. Wiley, New York.

Spiering, S., Hariskos, D., Powalla, M., Naghavi, N., and Lincot, D. 2003. *Thin Solid Films*, 431–432 359–363.

Staebler, D. L. and Wronski, C. R. 1977. *Applied Physics Letters*, 31, 292.

Surek, T. 2003. In *Proceeding of 3rd World Conference on PV Energy Conversion*, p. 2507.

Tuttle, J. R., Szalaj, A., Keane, J. 2000. In *Proceedings of the 28th IEEE Photovoltaic Specialists Conference*, pp. 1042–1045.

Tuttle, J. R., Schuyler, T., Choi, E., and Freer, J. 2005. In *The in 20th European Photovoltaic Solar Energy Conference and Exhibition.*

Underwriters Laboratories, Inc 1999. *UL 1741. Standard for static inverters and charge controllers for use in photovoltaic power systems. Underwriters Laboratories, Northbrook, IL.*

US Department of Energy. 2004. Solar Energy Technologies Program, PV FAQ: Will we have enough materials for energy significant PV production. Availbale at http://www.nrel.gov/ncpv

Vetterl, O. et al. 2000. *Solar Energy Materials and Solar Cells*, 62, 97–108.

von Roedern, B., Zweibel, K., Ullal, H. S. 2005. In *Proceedings of 31st IEEE Photovoltaic Specialists Conference*, pp. 183–188.

Wagner, S., Shay, J., Migliorato, P., and Kasper, H. 1974. *Applied Physics Letters*, 25, 434–435.

Woodcock, J. M., Schade, H., Maurus, H., Dimmler, B., Springer, J., and Ricaud, A. 1997. In *Proceedings of the 14th European Photovoltaic Solar Energy Conference*, pp. 857–860, Stephens & Associates, UK.

Wu, X., Keane, J. C., Dhere, R. G., DeHart, C., Duda, A., Gessert, T. A., Asher, S., Levi, D. H., and Sheldon, P. 2001. In *Proceedings of the 17th European Photovoltaic Solar Energy Conference*, pp. 995–1000.

Yamamoto, K., Toshimi, M., Suzuki, T., Tawada, Y., Okamoto, T., Nakajima, A. 1998. A thin-film poly-Si solar cells on glass substrate fabricated at low temperature. Presented at MRS spring Meeting, San Francisco.

Yamamoto, K., Nakajima, A., Yoshimi, M., Sawada, T., Fukuda, S., Suezaki, T., Ichikawa, M., Matsuda, T., Kondo, M., Sasaki, T., Tawada, Y. et al. 2004. *Solar Energy*, 77, 939–949.

Zweibel, K. 2000. *Solar Energy Materials & Solar Cells.*, 63, 375–386.

21

Wind Energy Conversion

21.1 Introduction **21**-1

21.2 Wind Turbine Aerodynamics **21**-4
Aerodynamic Models

21.3 Wind Turbine Loads **21**-16

21.4 Wind Turbine Structural Dynamic Considerations **21**-16
Horizontal-Axis Wind Turbine Structural Dynamics • Vertical-Axis Wind Turbine Structural Dynamics

21.5 Peak Power Limitation **21**-18

21.6 Turbine Subsystems **21**-20
Electrical Power Generation Subsystem • Yaw Subsystem • Controls Subsystem

21.7 Other Wind-Energy Conversion Considerations **21**-23
Wind Turbine Materials • Wind Turbine Installations • Energy Payback Period • Wind Turbine Costs • Environmental Concerns

References **21**-27

Further Information **21**-30

Dale E. Berg
*Sandia National Laboratories**

21.1 Introduction

Wind energy is the most rapidly expanding source of energy in the world today. Over the past 10 years, the worldwide installed capacity of wind energy has grown at an average rate of over 28% per year, leading to an installed nameplate capacity at the end of 2004 of about 48,000 MW, enough to power about 16 million average American homes. As of January 2005, Germany was the world leader in wind-energy installations with about 16,600 MW installed, followed by Spain with 8300, the US with 6700, Denmark with 3100, India with 3000, Italy with 1100, The Netherlands with 1100, the United Kingdom with 900, Japan with 900, and China with 800. Although wind power supplies only about 0.6% of the world electricity demand today, the size of that contribution is growing rapidly. In Germany, the contribution of wind power to electricity consumption is over 5%, in Spain it is about 8%, and in Denmark it is approximately 20%. The cost to generate wind energy has decreased dramatically from more than 30 cents (U.S.) per kilowatt-hour (¢/kWh) in the early 1980s to under 4¢/kWh (at the best sites) in 2004. The cost has actually increased somewhat in the recent past, in spite of continuing technological improvements, as a result of worldwide increases in steel, concrete, and transportation

*Sandia is a multiprogram laboratory operated by Sandia Corporation, a Lockheed Martin Company, for the United States Department of Energy's National Nuclear Security Administration under contract DE-AC04-94AL85000.

costs that have led to increases in the prices of wind turbines. The large increases in the cost of natural gas and other fossil fuels have made wind-generated electricity a lower-cost option than natural gas for many utilities adding generating capacity. In fact, the demand for wind energy in the United States has been so strong that wind-power developers are demanding and receiving significantly higher prices in their new long-term power purchase agreements with utility companies than they did 2 years ago.

There is considerable anecdotal evidence that the first wind machines may have been built over 2000 years ago, perhaps in China, but there is no firm evidence to support this conjecture. However, there is considerable written evidence that the windmill was in use in Persia by AD 900, and, perhaps, as early as AD 640. Figure 21.1 illustrates the main features of this type of mill. The center vertical shaft was attached to a millstone and horizontal beams or arms were attached to the shaft above the millstone. Bundles of reeds attached vertically to the outer end of the arms acted as sails, turning the shaft when the wind blew. The surrounding structure was oriented so that the prevailing wind entered the open portion of the structure and pushed the sails downwind. The closed portion of the structure sheltered the sails from the wind on the upwind pass. The primary applications of these machines were to grind or mill grain and to pump water; they became generally known as *windmills*. The wind turbines of today may look much different than those first machines, but the basic idea remains the same—use the power in the wind to generate useful energy. Modern wind machines, called *wind turbines*, tend to have a small number of airfoil-shaped blades, in contrast to the older windmills that usually had several flat or slightly curved blades (such as the American multiblade water pumper shown in Figure 21.2). The reasons for this difference in blade number will be examined later in this chapter.

Although there are many different configurations of wind turbines, most of them can be classified as either horizontal-axis wind turbines (HAWTs), which have blades that rotate about a horizontal axis parallel to the wind, or vertical-axis wind turbines (VAWTs), which have blades that rotate about a vertical axis. Figure 21.3 illustrates the main features of these configurations. They both contain the same major components, but the details of those components differ significantly.

According to Shepard [1], the terms "horizontal" and "vertical" associated with these classifications are a potential source of confusion. Although they now refer to the driving shaft on which the rotor is mounted, in the past these terms referred to the plane in which the rotor turned. Thus, the ubiquitous multibladed water-pumper windmill shown in Figure 21.2, now referred to as a *horizontal-axis machine*,

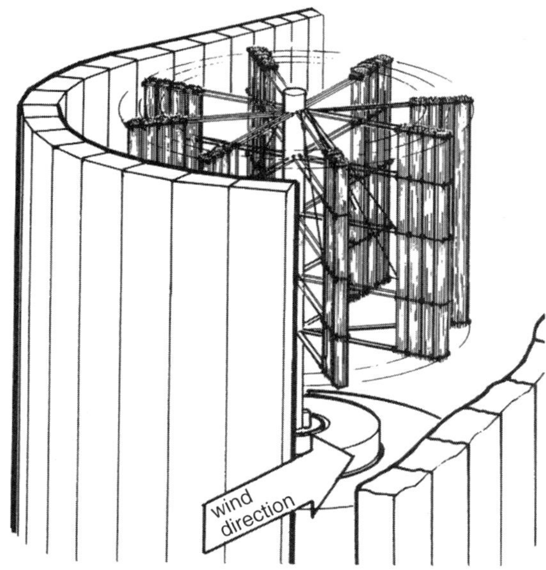

FIGURE 21.1 Illustration of ancient Persian wind mill.

FIGURE 21.2 Typical American multiblade windmill. (Courtesy of Nolan Clark, U.S. Department of Agriculture).

had a rotor that turned in a vertical plane, and was therefore at one point known as a *vertical* mill. Likewise, the earliest windmills, such as the one illustrated in Figure 21.1, had rotors that turned in a horizontal plane, and were known as *horizontal* windmills.

As shown in Figure 21.3, HAWTs and VAWTs have very different configurations. Each configuration has its own set of strengths and weaknesses. HAWTs usually have all of their drive train (the transmission, generator, and any shaft brake) equipment located in a nacelle or enclosure mounted on a tower, as shown. Their blades are subjected to cyclic stresses due to gravity as they rotate, and their rotors must be oriented (yawed) so the blades are properly aligned with respect to the wind. HAWTs may be readily placed on tall towers to access the stronger winds typically found at greater heights. The most common type of modern HAWT is the propeller-type machine, and these machines are generally classified according to the rotor orientation (upwind or downwind of the tower), blade attachment to the main shaft (rigid or hinged), maximum power control method (full or partial-span blade pitch or blade stall), and number of blades (generally two or three blades).

VAWTs, on the other hand, usually have most of their drive train on the ground; their blades do not experience cyclic gravitational stresses and do not require orientation with respect to the wind. However, VAWT blades are subject to severe alternating aerodynamic loading due to rotation, and VAWTs cannot readily be placed on tall towers to exploit the stronger winds at greater heights. The most common types of modern VAWTs are the Darrieus turbines, with curved, fixed-pitch blades, and the "H" or "box" turbines with straight fixed-pitch blades. All of these turbines rely on blade stall (loss of lift and increase in drag as the blade angle of attack increases) for maximum power control. Although there are still a few manufacturers of VAWTs, the overwhelming majority of wind turbine manufacturers devote their efforts to developing better (and usually larger) HAWTs.

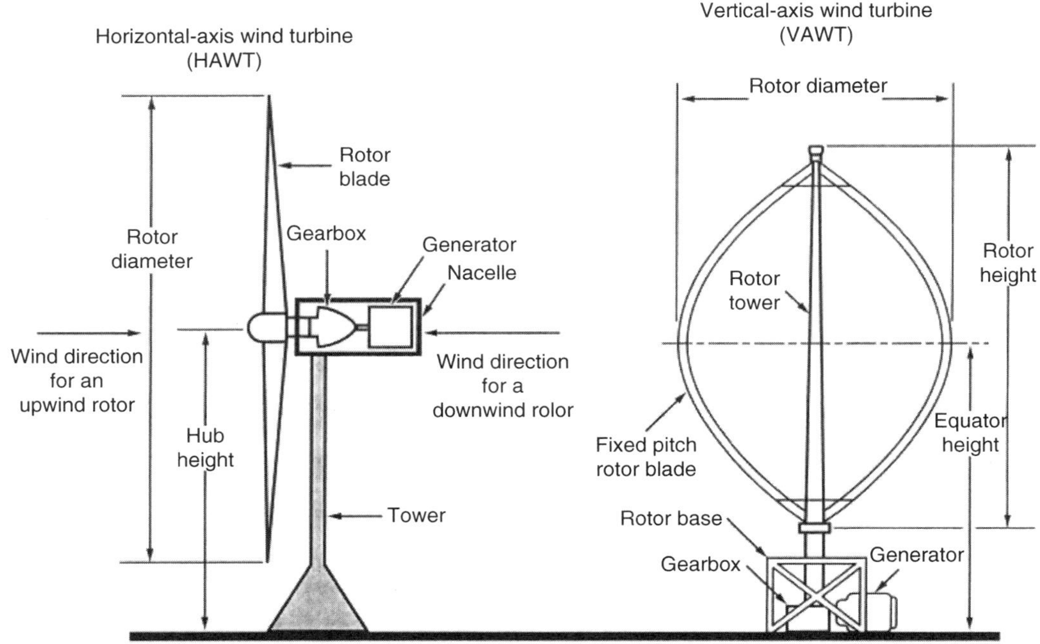

FIGURE 21.3 Schematic of basic wind turbine configurations.

Although the fuel for wind turbines is free, the initial cost of a wind turbine is a very large contributor to the cost of energy (COE) for that turbine. To minimize that COE, wind turbine designs must be optimized for the particular site or wind environment in which they will operate. Trial and error methods become very expensive and time-consuming when used to design and/or optimize turbines, especially larger ones. A large optimized wind turbine can be developed at a reasonable cost only if the designers can accurately predict the performance of conceptual machines and use modeling to investigate the effects of design alternatives. Over the past two decades, numerous techniques have been developed to accurately predict the aerodynamic and structural dynamic performance of wind turbines. These analytical models are not, in general, amenable to simple approximations, but must be solved with the use of computer codes of varying complexity. Several of these models are summarized below.

21.2 Wind Turbine Aerodynamics

Items exposed to the wind are subjected to forces in both the drag direction (parallel to the air flow) and the lift direction (perpendicular to the air flow). The earliest wind machines, known as *windmills*, used the drag on the blades to produce power, but many windmill designs over the last few centuries have made limited use of lift to increase their performance. For predominantly drag machines, such as those illustrated in Figure 21.1 and Figure 21.2, larger numbers of blades result in higher drag and produce more power; therefore, these machines tend to have many blades. The old Dutch windmills, such as the one shown in Figure 21.4, utilized lift as well as drag, and because lift devices must be widely separated to generate the maximum possible amount of power, those machines evolved with a small number of blades. The high-lift, low-drag shapes, referred to as *airfoils*, that were developed for airplane wings and propellers in the early part of the twentieth century were quickly incorporated into wind machines to produce the first modern wind machines, usually known as *wind turbines*. An example of a typical modern wind turbine is shown in Figure 21.5. Wind turbines use the lift generated by the blades to produce power. Because the blades must be widely separated to generate the maximum amount of lift,

FIGURE 21.4 Example of Dutch windmill. (Courtesy of Richard A. Neiser, Jr.).

lift-type machines have a small number of blades. The following paragraphs contrast the characteristics of the drag-type and lift-type machines.

Figure 21.6 illustrates the flow field about a moving drag device. The drag results from the relative velocity between the wind and the device, and the power that is generated by the device (the product of the drag force and the translation or blade velocity) may be expressed as

$$P = Dlv = [1/2\rho(U-v)^2]C_D clv, \tag{21.1}$$

where P is the power extracted from the wind, D is the drag force per unit length in the span direction (into the page), l is the length of device in the span direction (into the page), v is the translation (or blade) velocity, ρ is the air density, U is the steady free-stream wind velocity, $C_D = \text{Drag}/(1/2pclU^2)$ (drag coefficient, a function of device geometry), and c is the device width (perpendicular to the wind, in the plane of the page).

The translation (or blade) velocity of the device must always be less than the wind velocity or no drag is generated and no power is produced. The power extraction efficiency of the device may be expressed as the ratio of the power extracted by the device to the power available in the wind passing through the area occupied by the device (the projected area of the device), a ratio known as the *power coefficient*, C_P. From Equation 14.7, the power available is

$$P_A = \frac{1}{2}\rho U^3 A = \frac{1}{2}\rho U^3 cl,$$

where A is the area of the device projected perpendicular to the wind (cl).

Using Equation 21.1, the C_P for a drag machine is

$$C_P = \frac{P}{1/2\rho U^3 cl} = \frac{v}{U}\left[1 - \frac{v}{U}\right]^2 C_D. \tag{21.2}$$

FIGURE 21.5 General Electric 1.5-MW wind turbines near Lamar, Colorado.

Now consider a device that utilizes lift to extract power from the wind, i.e., an airfoil. Figure 21.7 depicts an airfoil that is moving at some angle relative to the wind and is subject to both lift and drag forces. The relative wind across the airfoil is the vector sum of the wind velocity, U, and the blade velocity, v. The angle between the direction of the relative wind and the airfoil chord (the straight line from the leading edge to the trailing edge of the airfoil) is termed the angle of attack, α. The power extracted by this device may be expressed as

$$P = 1/2\rho U^3 cl\frac{v}{U}\left[C_L - C_D\frac{v}{U}\right]\sqrt{1+\left(\frac{v}{U}\right)^2}, \quad (21.3)$$

where c is the chord length, and

$$C_L = \frac{\text{Lift}}{1/2\rho c l U^2} \text{(where lift is the coefficient, a function of airfoil shape and } \alpha\text{)},$$

$$C_D = \frac{\text{Drag}}{1/2\rho c l U^2} \text{(where drag is the coefficient, a function of airfoil shape and } \alpha\text{)}.$$

The other quantities are as defined for Equation 21.1. Lift and drag coefficients for some common airfoils may be found in the literature [2–7]. In this case, the projected area of the device is cl, and the power

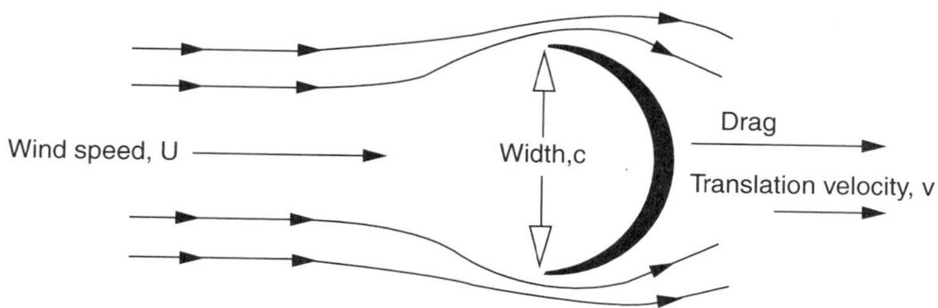

FIGURE 21.6 Schematic of translating drag device.

coefficient, using Equation 21.3, is

$$C_{\mathrm{P}} = \frac{v}{U}\left[C_{\mathrm{L}} - C_{\mathrm{D}}\frac{v}{U}\right]\sqrt{1+\left(\frac{v}{U}\right)^{2}}. \tag{21.4}$$

Keep in mind that Equation 21.2 and Equation 21.4 express the performance coefficients of these devices in terms of the projected area of the individual device. Figure 21.8 presents experimental lift and drag coefficient values for the S-809 airfoil designed by the National Renewable Energy Laboratory (NREL) for use on small HAWTs [8]. As the angle of attack increases beyond approximately 9°, the lift levels off and then drops slightly and the drag begins to rise rapidly. This is due to separation of the flow from the upper surface of the airfoil, a flow condition referred to as *stall*. Figure 21.9 compares Equation 21.2 and Equation 21.4 using $C_{\mathrm{L}}=10$ and $C_{\mathrm{D}}=0.10$ for the airfoil (conservative values for modern airfoils, as seen from Figure 21.8), and a drag coefficient of 2.0 (the maximum possible) for the drag device. The airfoil has a maximum power coefficient of about 15, compared with 0.3 for the drag device, i.e., it extracts 50 times more power per unit of device surface area. Of course, the airfoil must be translated across the wind to produce power, but this is easily achieved with rotating machines such as wind turbines.

As mentioned earlier, lift-type machines tend to have only a few blades, while drag-type machines tend to have many blades. Thus, the difference in the turbine performance coefficient (now based on the rotor

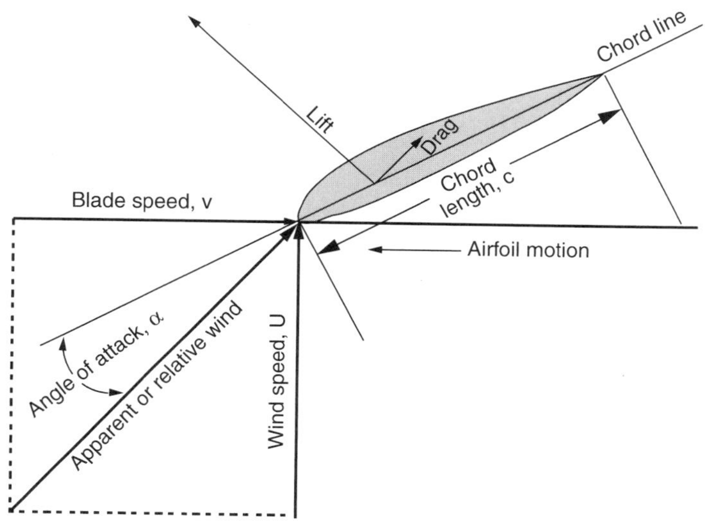

FIGURE 21.7 Schematic of translating lift device.

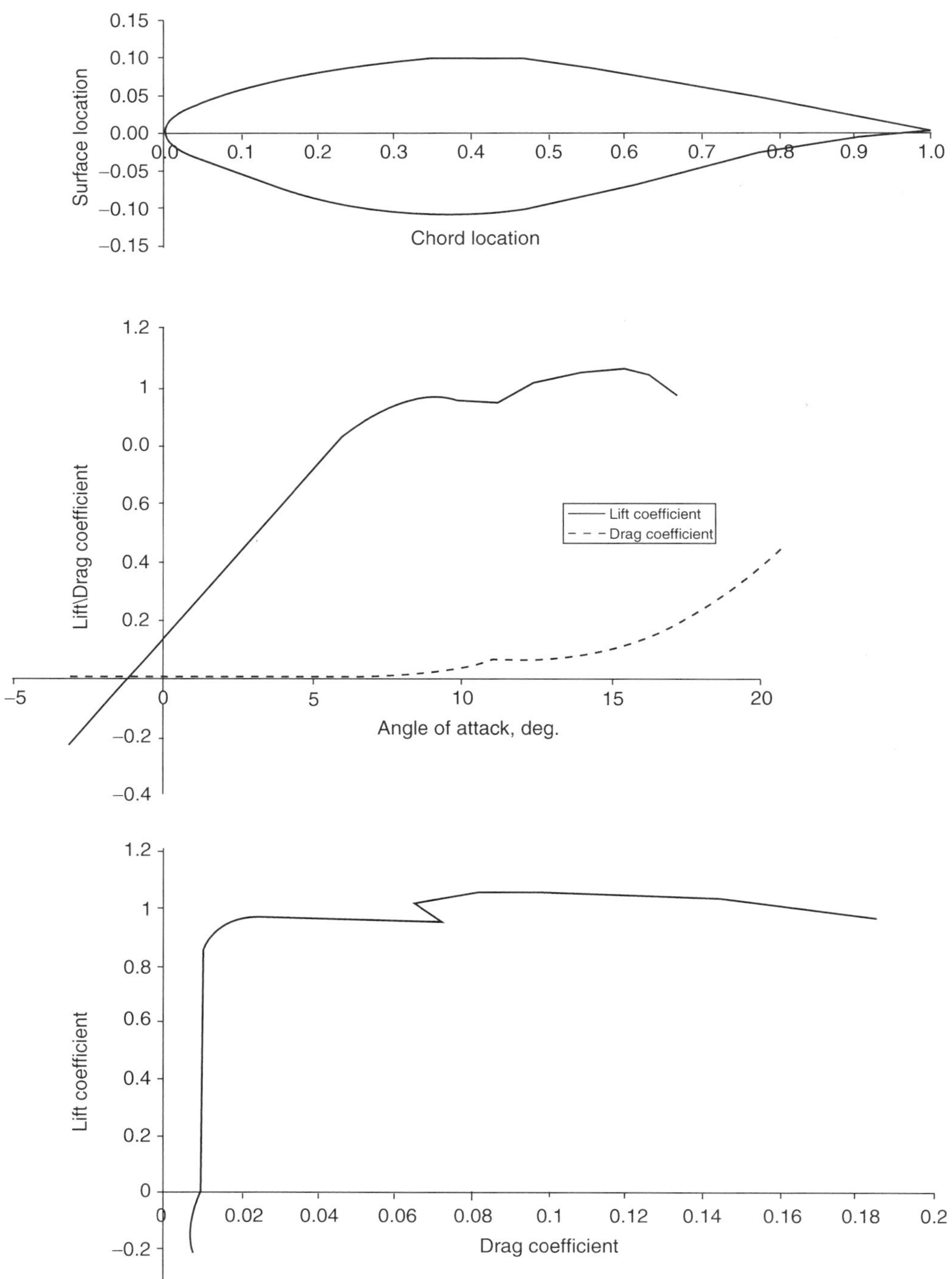

FIGURE 21.8 Profile and performance characteristics of the S-809 airfoil.

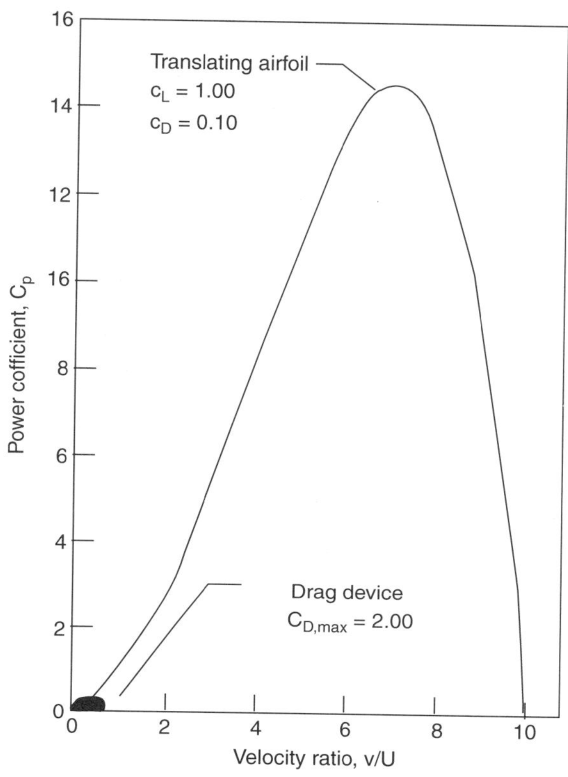

FIGURE 21.9 Comparison of power coefficients for drag-type and lift-type devices.

frontal area rather than the blade or bucket frontal area) of actual wind machines is much less than might be expected from the analysis presented above. A well-designed lift-type machine may achieve a peak power coefficient (based on the area covered by the rotating turbine blades) of 0.5–0.59, while a pure drag-type machine may achieve a peak power coefficient of no more than 0.2. Some of the multibladed drag-type windmills actually utilize a blade shape that creates some lift, and they may achieve power coefficients of 0.3 or slightly higher. The drag machines rotate slowly (the blade translation velocity cannot exceed the effective windspeed) and produce high torque, whereas the lift machines rotate quickly (to achieve a high translation velocity) and produce low torque. The slow-rotating, high-torque drag machines are very well-suited for mechanical power applications such as milling grain and pumping water. On the other hand, extensive experience has shown that fast-rotating, lift-type machines are much easier to adapt to electrical generators and can produce electricity at a significantly lower COE than drag-type machines. Because of their superior performance, only lift-type machines will be considered in the remainder of this chapter.

21.2.1 Aerodynamic Models

The aerodynamic analysis of a wind turbine has two primary objectives: (1) to predict the aerodynamic performance or power production of the turbine, and (2) to predict the detailed time-varying distribution of aerodynamic loads acting on the turbine rotor blades. In general, the same models are used to accomplish both objectives. Accurate prediction of turbine aerodynamic performance does not guarantee accurate prediction of the loading distribution—the performance predictions result from the integration of time-averaged aerodynamic lift and drag over the entire turbine, and significant errors may be present in the detailed lift and drag predictions but balance out in the performance predictions. Although there is a considerable body of data showing good agreement of predicted performance with measured performance, especially for codes that have been tailored to give good results for the particular configuration of interest, there are very few data available against which to compare detailed load-distribution predictions. The aerodynamics of wind turbines are far too complex to model with simple formulas that can be solved with hand-held calculators; computer-based models ranging from very simplistic to very complex are required. Several commonly used aerodynamic models are described below.

21.2.1.1 Momentum Models

The simplest aerodynamic model of a horizontal-axis wind turbine is the actuator disk or momentum model, in which the turbine rotor is modeled as a single porous disk. This analysis was originally adapted for wind turbine use by Betz [9], from the propeller theory developed by Froude [10] and Lanchester [11]. To develop the equations for this model, the axial force acting on the rotor is equated to the time rate of change of the momentum of the air stream passing through the rotor. It is assumed that the mass of air that passes through the rotor disk remains separate from the surrounding air and that only the air passing through the disk slows down. That mass of air with a boundary surface of a circular cross-section, extended upstream and downstream of the rotor disk, is shown in Figure 21.10. No air flows across the lateral boundary of this *stream-tube*, so the mass flow rate of air at any position along the stream-tube will be the same. Because the air is incompressible, the decrease in the velocity of the air passing through the disk must be accompanied by an increase in the cross-sectional area of the stream-tube to maintain the same mass flow. The presence of the turbine causes the air approaching from the upstream to gradually slow down; thus, the velocity of the air arriving at the rotor disk is already lower than the free-stream windspeed. As mentioned above, this causes the stream-tube to expand. In addition, the static pressure of the air rises to compensate for the decrease in kinetic energy.

As the air passes through the rotor disk, there is a drop in static pressure; the air immediately downstream of the disk is below the atmospheric pressure level, but there is no instantaneous change in velocity. As the air continues downstream, the static pressure gradually increases until it again comes into equilibrium with the surrounding atmosphere, and the velocity drops accordingly. This region of the flow is referred to as the *wake*. Thus, the difference in flow conditions between the far upstream and the far wake is a decrease in kinetic energy, but there is no change in static pressure.

Utilizing conservation of mass, conservation of axial momentum, the Bernoulli equation, and the first law of thermodynamics, and assuming isothermal flow, the power produced by the turbine (the product of the axial force and the air velocity at the rotor) may be readily derived. From the conservation of axial

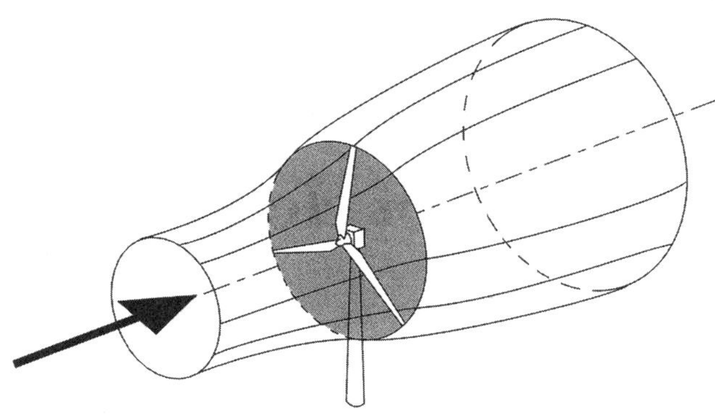

FIGURE 21.10 Schematic of stream-tube for horizontal-axis wind turbines.

momentum, the thrust on the rotor, T, may be expressed as

$$T = \dot{m}(U - v_w) = \rho A v (U - v_w), \tag{21.5}$$

where $\dot{m}$ is the mass flow rate of air ($\dot{m} = \rho A v$), v is the wind velocity at the rotor disk, v_w is the wind velocity far downstream of the rotor disk (in the wake), and A is the area of the rotor disk.

The thrust may also be expressed in terms of the pressure drop caused by the rotor:

$$T = A(p_u - p_d), \tag{21.6}$$

where p_u is the pressure just upwind of the rotor disk and p_d is the pressure just downwind of the rotor disk.

The Bernoulli equation, applied just upwind of the rotor and just downwind of the rotor, yields

$$p_\infty + 1/2\rho U^2 = p_u + 1/2\rho v_u^2, \tag{21.7}$$

$$p_d + 1/2\rho v_d^2 = p_w + 1/2\rho v_w^2. \tag{21.8}$$

Far upwind of the rotor and far downwind of the rotor, the pressures are equal ($p_\infty = p_w$), and the velocity is the same just upwind and just downwind of the rotor ($v_u = v_d$). Substituting Equation 21.7 and Equation 21.8 into Equation 21.6 and using the above equalities yields

$$T = A[(1/2\rho U^2 - 1/2\rho v_u^2) - (1/2\rho v_d^2 - 1/2\rho v_w^2)] = 1/2\rho A(U^2 - v_w^2). \tag{21.9}$$

Equating Equation 21.5 and Equation 21.9 then yields

$$v = 1/2(U + v_w). \tag{21.10}$$

That is, the velocity at the rotor disk is equal to the mean of the free-stream and wake velocities; thus, the velocity change between the free-stream and the wake is twice the change between the free-stream and the disk.

The power produced at the rotor, assuming isothermal flow and ambient pressure in the wake and utilizing Equation 21.10, is

$$P = Tv = 1/2\rho A(U^2 - v_w^2)v, = 1/2\rho A(U - v_w)(U + v_w)v, = 1/2\rho A[2(U - v)]2vv,$$

$$= 2\rho A(U - v)v^2. \tag{21.11}$$

Now, define

$$a = (U - v)/U$$

which is commonly known as the *axial interference factor*. Rearranging, Equation 21.11 becomes

$$P = 2\rho A(aU)(1 - a)^2 U^2 = 2\rho A U^3 a(1 - a)^2. \tag{21.12}$$

The power coefficient for the turbine, then, is

$$C_P = 4a(1 - a)^2. \tag{21.13}$$

This is maximized at $a = 1/3$, yielding $C_{p,max} = 16/27 = 0.593$ as the maximum possible performance coefficient for a lift-type machine, a maximum often referred to as the *Betz limit*. Expressed in slightly

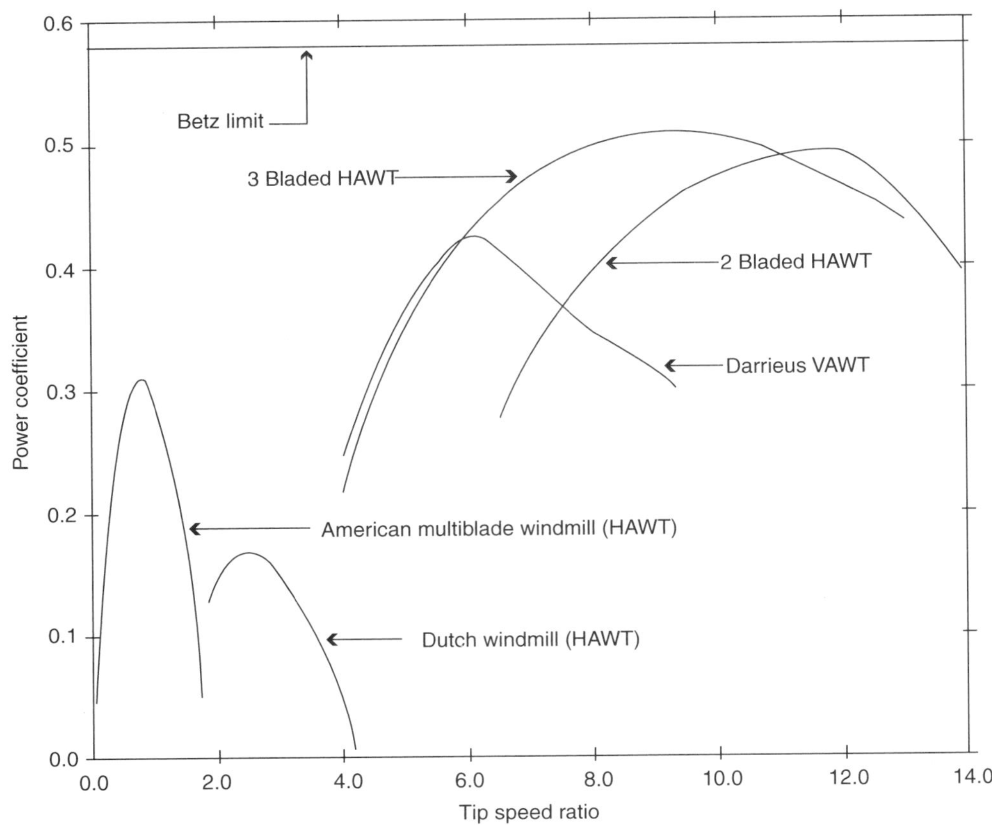

FIGURE 21.11 Typical performance of various types of wind turbines.

different terms, no lift-type turbine can extract more than 59.3% of the energy available in the wind passing through the rotor.

The typical performance of various types of wind machines is compared to the Betz limit in Figure 21.11, where the variations of the turbine power coefficients with the tip-speed ratio (the ratio of the speed of the blade tip to the free-stream windspeed) are presented. Even though the maximum performance of modern HAWTs and VAWTs is well above that of the older, drag-type machines such as the Dutch windmill and the American multiblade windmill, it is still somewhat below the Betz limit. Some HAWTs have demonstrated peak performance coefficients approaching 52%.

For horizontal-axis wind turbines, the momentum model can be expanded to the widely used blade element momentum (BEM) model in which the blades are divided into small radial elements and local flow conditions at each element are used to determine blade forces and loads on those elements. To obtain accurate predictions, this model typically incorporates numerous modifications to account for blade and turbine wake effects, the three-dimensional flow near blade tips, the thick blade sections near the root, and blade stall at high windspeeds. Additional information on these models may be found in Hansen and Butterfield [12], Wilson [13], and Snel [14].

A very similar derivation yields a momentum model for the vertical-axis wind turbine. This model may be expanded into the multiple stream-tube model (the turbine rotor is modeled as multiple actuator disks, rather than just one) and the double-multiple stream-tube model (multiple actuator disks, with separate ones modeling the upwind and downwind passes of the rotor blades) that are the VAWT equivalent of the HAWT blade element model. Additional information on these models may be found in Touryan et al. [15], Wilson [13], and Paraschivoiu [16].

Momentum-based models are extremely popular with wind turbine designers because they are simple, fast, and fairly accurate for performance prediction, especially after they are tuned for a particular configuration. However, they are approximate because they are based upon the assumptions of flow conditions that are fixed in time and space. They cannot predict the effects of yawed flow, unsteady aerodynamics, and other complex flows that are present on wind turbines, all of which can have large impacts on turbine performance and loads. In some cases, specialized codes based on experimental results are used to model some of these effects, but these codes are limited to specific turbine sizes and geometries. In other cases, more realistic models, such as vortex-based models, full computational fluid dynamics (CFD) models, and hybrid models, are used to estimate these effects.

21.2.1.2 Vortex Models

Vortex models are usually more properly referred to as lifting line or lifting surface models, depending on whether a lifting line or a lifting surface formulation is used to model the blades. In the lifting line method, each rotor blade is modeled as a series of segmented "bound line vortices" located at the blade ¼-chord line, as illustrated in Figure 21.12a. Line-vortex strengths, defined by the blade lift at each radial location, are associated with the vortex line segments. The lifting surface method represents the blade in more detail, as a distribution of vortex line segments over the blade surface, as illustrated in Figure 21.12b. Either of these models will generate both trailing vorticity (perpendicular to the span of the blade) due to the differences in vortex strength along the blade span and shed vorticity (parallel to the span of the blade) due to time-dependent changes in vortex strength, that are shed into the wake as the turbine rotates. These vortex methods lend themselves to the modeling of unsteady problems, as the shed vorticity models the time-dependent changes in the blade bound vortex strength. Solutions are achieved by impulsively starting the turbine in a uniform flow field and allowing the computational flow field to develop until it reaches a steady state or periodic condition.

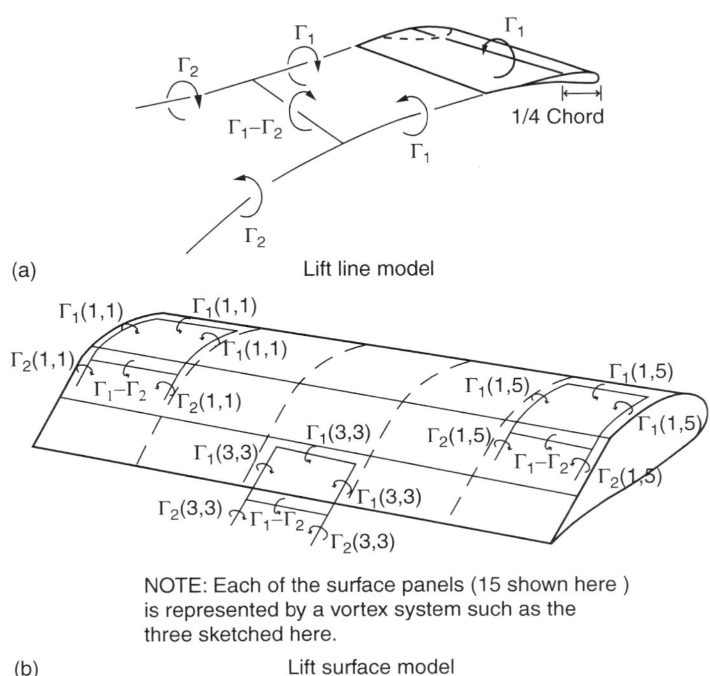

FIGURE 21.12 Schematic of lifting line and lifting surface models of turbine blades.

The manner in which the transport of the vorticity in the turbine wake is modeled depends on whether a free-wake or a fixed-wake (or prescribed-wake) model is used. In the free-wake model, the vorticity is allowed to convect, stretch, and rotate as it is transported through the wake. However, the movement of each line vortex is influenced by the presence of all of the other line vortices, including those on the blade. As the computation progresses in time, the number of vortices that must be followed and the time required to calculate the vortex interactions both grow very quickly. To minimize the need for large computer resources, the fixed or prescribed wake models have been developed. In these models, the geometry of the wake is either fixed or described by only a few parameters, and the vortex interactions in the wake are no longer directly calculated. The result is a much faster execution time, but the accuracy of the predicted power generation and blade loads depends very heavily upon the fidelity with which the specified wake approximates the actual physical wake. The three-dimensional, lifting-surface, free-wake formulation is the most physically realistic of the vortex models, but a computer program implementing such a model will require a large amount of computer resources and time. Experience has shown that the dramatic increase in computer resources required by such a model does not yield significantly more accurate predictions than what can be obtained with a three-dimensional, lifting-line, free-wake model. A major problem with vortex codes is the difficulty in finding a good balance between model simplification (and the associated limitations on fidelity), computation time, and desired accuracy. Vortex models are not widely used in the wind industry today; they are much more expensive in terms of computer resources than momentum models, and they are less accurate than computational fluid dynamics models. Additional information on vortex models for both HAWTs and VAWTs may be found in Kocurek [17], Snel [14] and Strickland et al. [18], among others.

21.2.1.3 Limitations Common to the Momentum and Vortex Models

Both momentum-based and vortex models normally utilize airfoil performance characteristic tables (lift and drag coefficients as functions of angle of attack, such as are shown in Figure 21.8), and air velocity to determine the blade lift and drag. These tables are generated from static two-dimensional wind tunnel test results or two-dimensional static airfoil design code predictions. The contents of the tables are modified with empirical, semiempirical, or analytic methods and used to estimate blade loads under the three-dimensional, dynamic conditions actually experienced by the turbine blades. The greatest difficulty in obtaining accurate load distribution predictions with either the momentum models or the vortex models is the challenge of accurately determining the appropriate airfoil performance characteristics.

21.2.1.4 Computational Fluid Dynamics Models

Computational fluid dynamics (CFD), in a broad sense, is the solution to the partial differential equations describing the flow field by approximating these equations with algebraic expressions (discretizing them), and then solving those expressions numerically with the aid of a computer. Within the wind energy community, the term CFD normally refers to the numerical solution of the unsteady Reynolds-averaged Navier–Stokes (RaNS) equations [19,20], often restricted to four partial differential equations (one conservation of mass equation, and conservation of momentum equations in three orthogonal directions) that describe general ideal-gas, incompressible, nonreacting fluid flow.

One might argue that the most detailed and physically realistic method of predicting the performance of and loads on a wind turbine is to utilize CFD to model the airflow around the turbine and through the rotor, calculating the airfoil lift and drag directly. The flow field in the vicinity of the wind turbine is approximated as a computational grid of variable density, and the discretized Navier–Stokes equations are applied to each element of that grid. The computational grid close to the turbine blades must be very, very dense to capture the details of the airflow around the blades; it becomes less dense as the distance from the blades becomes greater and the effect of the blades on the airflow decreases. The resulting set of simultaneous equations must be solved, frequently in a time-marching manner, to determine the time-dependent nature of the entire flow field.

Duque et al. [21] describe a fairly recent CFD investigation of a wind turbine in which they utilized a complex grid with 11.5 million points to model the flow around a 10-m diameter HAWT rotor and tower combination. A steady-flow solution (rotor facing directly into the wind) for that model at a single windspeed, utilizing eight PC processors, each operating at 1.4 GHz in a parallel processor computer configuration, required approximately 26 h. An unsteady-flow solution (rotor yawed at an angle to the wind) with the same computing resources required over 48 h for each rotor revolution. Sørensen et al. [22] and Johansen et al. [23] report other recent CFD modeling efforts.

At this point, CFD is suitable for research use or final design verification only—it is far too slow and requires far too many computer resources to be considered for use as a routine design tool. CFD also suffers from some shortcomings that limit its accuracy in performance and loads calculations. First, it does not consistently yield highly accurate results for airfoil lift and drag because it cannot adequately model the transition of flow over the airfoil surface from laminar, well-ordered, basically two-dimensional flow, to inherently three-dimensional and unsteady turbulent flow that is accompanied by rapid fluctuations in both velocity and pressure. Most wind turbines today utilize airfoils that will experience such a transition in the flow. Second, most CFD models today cannot adequately predict the effects of separated flow, especially three-dimensional separated flow, such as will occur near the hub of the wind turbine rotor under most operating conditions. In spite of these limitations, CFD calculations, when used wisely, can often yield much useful information. Efforts to improve the accuracy of CFD calculations continue.

21.2.1.5 Hybrid Models

The hybrid model approach typically approximates the airflow close to the turbine with the discretized Navier–Stokes equations, similar to the procedure used by the CFD models. However, away from the turbine, the model uses nonviscous or potential flow equations that are much less complex and that can be solved much faster than the Navier–Stokes equations. The two solutions must be merged at the boundary between the two regions. The result is a code with the accuracy of the CFD model, but one that requires an order of magnitude less computing resources to solve. Xu and Sankar [24] describe recent work on such a code. However, even this hybrid model requires a large amount of computer resources and is too slow and expensive to be considered a practical design tool.

21.2.1.6 Model Results

None of these aerodynamic models is capable of accurately predicting the performance of and detailed loads on an arbitrary wind turbine operating at a variety of windspeeds. To have high confidence in the code predictions for a turbine design, the code must be calibrated against the measured performance and loads obtained from turbines of similar size and shape. Simms et al. [25] report on the ability of 19codes based on the above models to predict the distributed loads on and performance of an upwind HAWT with a 10-m diameter rotor that was tested in the NASA/Ames 80 ft. $\times$120 ft. (24.4 m$\times$36.6 m) wind tunnel. Although the rotor was small compared to the 80-m diameter and larger commercial turbines that are being built today, a panel of experts from around the world concluded that it was large enough to yield results representative of what would be observed on the larger turbines. The comparisons of the code predictions and the experimental results were, in general, poor; turbine power predictions ranged from 30 to 275% of measured values, and blade-bending moment predictions ranged from 85 to 150% of measured values for what is considered to be the most easy-to-predict conditions of no yaw, steady state, and low windspeed. Many aerodynamic code developers have spent considerable effort over the years since that comparison attempting to identify the sources of the discrepancies and improving the accuracy of their various codes.

Additional information and references on wind turbine aerodynamics models may be found in Hansen and Butterfield [12], Wilson [13], and Snel [14] for HAWTs, and in Touryan et al. [15] and Wilson [13] for VAWTs.

Most wind turbine companies today continue to use the very fast momentum-based models for design purposes, in spite of the approximations and inaccuracies that are inherent in these models. The analysts doing the studies tweak these models to get good comparison with measurements from an existing turbine, and then use them to predict performance and loads for new turbines that are fairly close in size and geometry to the existing turbine. If the new turbine is significantly different in size and shape from the reference turbine, the performance predictions are considered to be subject to considerable error. Performance tests on prototype turbines are required to define the actual performance and loads.

21.3 Wind Turbine Loads

Wind turbines are fatigue-driven structures; they normally fail not as the result of a single application of a large-amplitude load, but as a result of the repeated application and removal (or cycling) of small-amplitude loads. Each load cycle causes microscopic damage to the structure, and the accumulated effect of many cycles of varying amplitude eventually leads to failure of the structure, a process referred to as *fatigue failure*. In general, the smaller the amplitude of the load, the larger the number of load cycles the structure can withstand before failing. Therefore, it is very important that the loads acting on a wind turbine be well understood.

The wind is random or stochastic in nature, with significant short-term variations or turbulence in both direction and velocity. Wind turbine aerodynamic loads may fall into one of two broad categories. One category consists of the deterministic loads occurring in narrow frequency bands resulting from the mean steady atmospheric wind, wind shear, rotor rotation, and other deterministic effects. The other category consists of the random loads occurring over all frequencies resulting from the wind turbulence. As recently as 10–12 years ago, the deterministic loads were frequently predicted with an aerodynamics code, such as those described earlier, utilizing a uniform wind input, while the random loads were estimated with empirical relations. However, turbine designers now recognize that this approach may lead to serious under prediction of both the maximum and random blade loads, resulting in costly short-term component failures. Most analysts today utilize an aerodynamics performance code with a wind model that includes a good representation of the turbulence of the wind in all three dimensions to predict long-term wind turbine loads. The appropriate method of determining the wind-induced extreme events and random loads that limit the lifetime of a turbine remains the subject of ongoing research. This lack of knowledge of loads is typically compensated for in the design process by incorporation of large safety margins, leading to excess material and cost. Civil engineers have spent decades developing statistical methods for predicting wind and wave loads on offshore drilling platforms to help reduce the cost and increase the reliability of those platforms. The wind turbine industry is just now starting to apply that technology to predict the wind loads on land-based wind turbines.

21.4 Wind Turbine Structural Dynamic Considerations

Input loads and dynamic interactions result in forces, moments, and motions in wind turbines, phenomena referred to as structural dynamics. By applying various analytical methods, the impact of changes in turbine configurations, controls, and subsystems on the behavior of the turbine can be predicted. General wind turbine structural dynamic concerns and methods of analysis are discussed in the following paragraphs.

21.4.1 Horizontal-Axis Wind Turbine Structural Dynamics

Small-horizontal-axis turbine designs usually use fairly rigid, high-aspect-ratio (the blade length is much greater than the blade chord) blades, cantilevered from a rigid hub and main shaft. As turbine size increases, the flexibility of the components tends to increase, even if the relative scales remain the same, so the blades on larger turbines tend to be quite flexible, and the hub and main shaft tend to be far less rigid than corresponding components on small turbines. The entire drivetrain assembly is mounted on and yaws about a tower that may also be flexible. These structures have many natural vibration modes and some of them may be excited by the wind or the blade rotation frequency to cause a resonance condition, amplifying vibrations and causing large stresses in one or more components. Operating at a resonance condition can quickly lead to component failure and result in the destruction of the turbine. Careful structural analysis during the turbine design may not guarantee that the turbine will not experience a resonance condition, as analysis techniques are not infallible, but ignoring the analysis altogether or failing to properly conduct parts of it may dramatically increase the probability that the turbine will experience one or more resonance conditions, leading to early failure. Although the relatively rigid small turbines are not likely to experience these resonance problems, the very flexible, highly dynamic, larger turbines may well experience resonance problems unless they are very carefully designed and controlled. Turbines that operate over a range of rotational speeds (variable-speed turbines) are especially challenging to design. The designer will usually try to minimize the number of resonances that occur within the operational speed range, and then implement a controller that will avoid operating at those resonance conditions. The actual resonances typically depend on the rotor speed, and the severity of the resonance depends on the windspeed, so the controller logic can become quite complicated.

The large relative motion between the rotor and the tower frequently precludes the use of standard commercial finite-element analysis codes and requires the use of a model constructed specifically for analysis of wind turbines. Development of such a model can be a rather daunting task, as it requires the formulation and solution of the full nonlinear governing equations of motion. The model must incorporate the yaw motion of the nacelle, the pitch control of the blades, any motion and control associated with hinged blades, the time-dependent interaction between the rotor and the supporting tower, etc. If the full equations of motion are developed with either finite-element or multibody dynamics formulations, the resultant models contain moderate numbers of elements and potential motions (or degrees of freedom, DOF), and significant computer resources are required to solve the problem. On the other hand, a modal formulation utilizing limited DOF may be able to yield an accurate representation of the wind turbine, resulting in models that do not require large computing resources. The development of the modal equations of motion may require somewhat more effort than do the finite-element or multibody equations, and the equations are apt to be more complex. The modal degrees of freedom must include, at a minimum, blade bending in two directions, blade motion relative to the main shaft, drive train torsion, tower bending in two directions, and nacelle yaw. Blade torsion (twisting about the long axis) is not normally included in current models, but it may become more important as the turbine sizes continue to increase and the blades become more flexible. The accuracy of some modal formulations is limited by their inability to model the direction-specific, nonlinear variation of airfoil lift with angle of attack that occurs as a result of aerodynamic stall. However, this is not an inherent limitation of the technique, and some formulations are free of these limitations. The modal formulation is the most widely used HAWT analysis tool today, but, as computer resources become more readily available, the more accurate finite-element (such as Abaqus [26]) and multibody dynamic codes (such as Adams [27]) will certainly become more widely used.

It is possible to develop techniques in the frequency domain to analyze many aspects of the turbine dynamics. The frequency domain calculations are fast, but they can only be applied to linear, time-invariant systems and, therefore, cannot deal with some important aspects of wind turbine operations such as aerodynamic stall, start-up and shutdown operations, variable speed operation, and nonlinear control system dynamics. In spite of these limitations, frequency domain solutions of modal formulations are frequently used in the preliminary design of a wind turbine, when quick analysis of many configurations is required. Regardless of the methods used in preliminary design, the state of the art in wind turbine design today is to use highly detailed modal, lumped-mass, or finite-element-based equations of motion, coupled with time-accurate solutions, to analyze the turbine behavior for the detailed final design calculations.

Malcolm and Wright [28] and Molenaar and Dijkstra [29] provide reviews of some of the available HAWT dynamics codes that have been developed, together with their limitations. Buhl et al. [30] compare some of the HAWT dynamics codes that have been extensively verified and that are widely used today, and Quarton [31] provides a good history of the development of HAWT wind turbine analysis codes. More general finite-element dynamics codes are described by others [32–35].

21.4.2 Vertical-Axis Wind Turbine Structural Dynamics

Darrieus turbine designs normally use relatively slender, high-aspect-ratio structural elements for the blades and supporting tower. As with large HAWTs, the result is a very flexible, highly dynamic structure, with many natural modes of vibration that again must be carefully analyzed to ensure that the turbine will avoid structural resonance conditions under all operating environments. The guy cables and turbine support structure can typically be analyzed with commercial or conventional finite-element codes, but the tower and blades require a more refined analysis, usually requiring the use of a finite-element code possessing options for analyzing rotating systems. With such a code, the blades and tower of a VAWT are modeled in a rotating coordinate frame, resulting in time-independent interaction coefficients. The equations of motion must incorporate the effects of the steady centrifugal and gravitational forces, the aerodynamic forces due to the turbulent wind, and the forces arising from rotating coordinate system effects. Detailed information on finite-element modeling of VAWTs may be found in Lobitz and Sullivan [36].

21.5 Peak Power Limitation

All turbines incorporate some method of regulating or limiting the peak power produced. The entire turbine, including the rotor, the transmission and the generator, must be sized to handle the loads associated with peak power production. While high winds (above, for example, 25 m/s) contain large amounts of available power, they do not occur very often, and the power that can be captured is very small. This is illustrated in Figure 21.13 for the Amarillo, Texas airport. In this figure, the power density is the power per unit of rotor area (normalized to yield a value under the curve of unity) that is available for capture by a wind turbine. This takes into account the amount of time that the wind actually blows at each windspeed (the probability density that is also shown on the figure). Amarillo occasionally

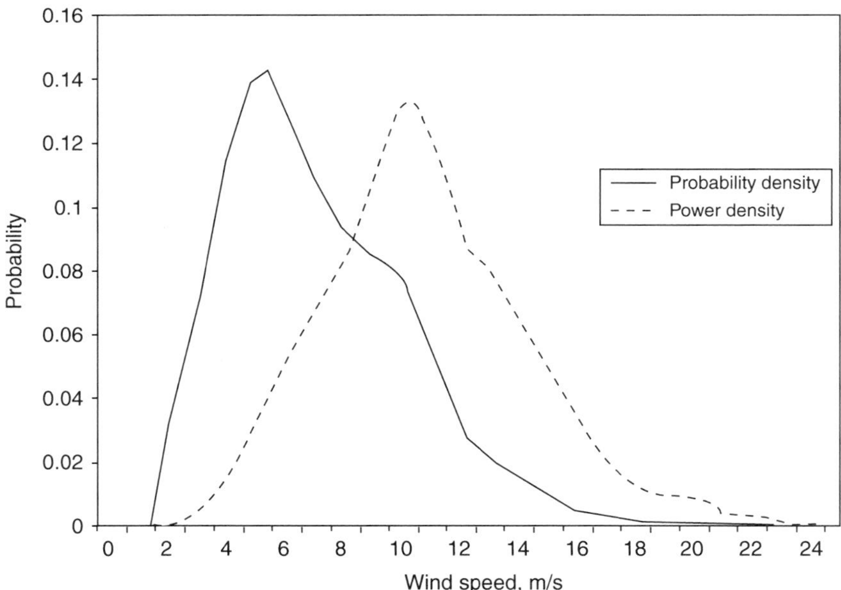

FIGURE 21.13 Windspeed and wind power probability densities for Amarillo, Texas airport.

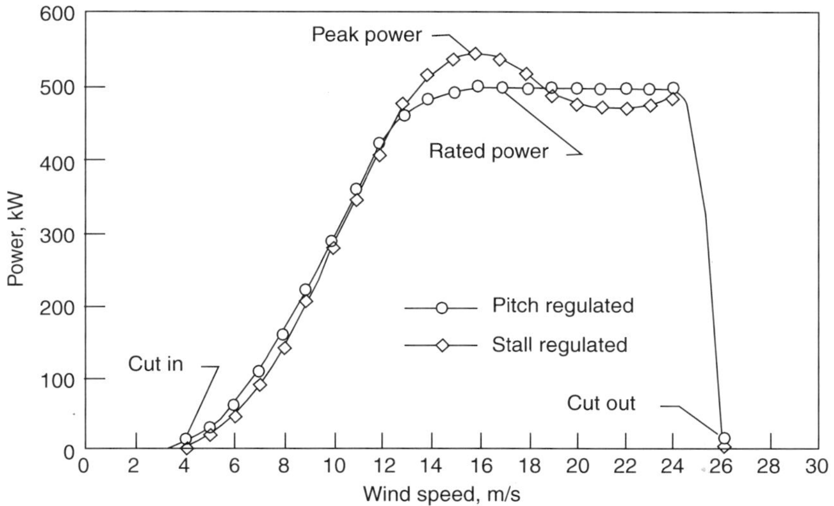

FIGURE 21.14 Sample power curves for stall-regulated and pitch-regulated wind turbines.

experiences very high winds, but as seen from this probability density, they do not occur very often, and the energy that could be captured from winds above 25 m/s is negligible.

Generators and transmissions operate most efficiently at their design conditions, typically close to their maximum capacity. These efficiencies drop off quickly at conditions below design. Cost trade-off studies reveal that it is far more cost effective to limit the maximum power level to that achieved at, for instance, 13–14 m/s and to shut the turbine down completely at a cutout windspeed of, for example, 26 m/s, as illustrated in Figure 21.14, than to try to capture the maximum amount of power at the higher windspeeds. Under these conditions, the transmission and generator are operating near design conditions for a significant portion of the time, and the turbine can be built with far less material than would be required for a turbine that generates peak power at 25 m/s. The additional energy captured due to the increase in generator and transmission efficiencies at the lower windspeeds is usually many times greater than that lost due to limiting the peak power at the rather infrequent winds above 14 m/s (refer to the windspeed distribution in Figure 21.13).

Nearly all modern large horizontal-axis turbines now use blade pitch control, where either the entire blade or a portion of it is rotated about the longitudinal axis to change the angle of attack and, therefore, the power output of the turbine, to limit peak power. Some turbines are designed with fixed-pitch blades, and rely on airfoil stall at high winds to limit the maximum power output of the machine. Small turbines frequently incorporate passive features, such as tail vanes and furling, that turn the rotor so the rotor axis is no longer aligned with the wind, to limit peak power production in high winds.

With full-span blade pitch control, the blade may be rotated about its longitudinal axis to decrease the effective angle of attack as the windspeed increases (commonly referred to as *pitch-to-feather*), causing decreased blade lift and limiting the peak power at the desired level. Alternatively, the blade may be rotated about that axis to increase the effective angle of attack (commonly referred to as *pitch-to-stall*), causing blade stall and limiting peak power. Although pitch-to-feather control results in decreased drag loads at high winds, a major disadvantage is poor peak power control during high-wind stochastic conditions—sudden increases in windspeed will result in corresponding increases in angle of attack, loads, and power generation. Power excursions can exceed twice the rated power levels before the high-inertia blade pitch system can compensate for the windspeed increases. Blade stall control (either fixed-pitch or pitch-to-stall), on the other hand, results in better peak power control at high winds. Major disadvantages of stall control include increased blade drag loads as the windspeed increases (even after stall) and possible large dynamic loads due to wind turbulence. A major advantage of fixed-pitch stall control is the lower cost compared to the active pitch control. Sample power curves for fixed-pitch stall- and pitch-regulated (pitch-to-feather) 500-kW turbines are shown in Figure 21.14. The average peak power cannot be as well regulated with fixed-pitch stall control as with either pitch-to-feather or pitch to stall control.

The use of only a portion of the airfoil surface (typically including the blade tip) as a control surface is referred to as "partial-span" control. Adjusting the pitch (and, thus, the lift and drag) of this portion of the blade independently of the remainder of the blade is used to control the peak power output of the turbine. These control surfaces are usually much smaller than the full blade, so they can respond to wind changes much faster than can the full-span blade. However, partial-span devices are very difficult to integrate into a blade, and the gaps around the devices tend to generate noise. Some recent research has investigated the use of very small load control devices that can generate large changes in lift and drag while experiencing small loads. Dimensions of these devices are typically on the order of 1% of the blade chord; their small size means they can be activated very quickly to alleviate excess loads and they are less likely to create large amounts of noise. Mayda et al. [37] discuss the use of CFD to investigate the effects of some of these devices.

Several other methods of pitch control have also been used, but on a limited basis. Passive pitch control techniques automatically adjust the blade pitch angle by using cams activated by centrifugal loads or by using tailored blade materials that permit the blade to twist toward feather or stall under high loads. These devices are very carefully tailored to maintain peak performance at lower windspeeds, but limit the peak power and blade loads at high windspeed.

Most vertical-axis wind turbines utilize stall regulation with fixed-pitch blades to control peak power, but some straight-bladed VAWTs are equipped with full-span pitch controls.

21.6 Turbine Subsystems

The wind turbine incorporates many subsystems, in addition to the actual turbine, to generate power. The electrical power generation, yaw, and controls systems are the only ones that will be discussed here.

21.6.1 Electrical Power Generation Subsystem

Once a wind turbine has converted the kinetic energy in the wind into rotational mechanical energy, that rotational energy is usually converted by a generator into electricity that can be readily transported to where it is needed. In most cases, the generator requires an input shaft speed that is much higher than the turbine shaft speed. These speeds are matched by a step-up transmission that increases the turbine shaft speed to that required by the generator.

While some wind turbines utilize permanent magnet alternators or generators, most grid-connected turbines today utilize either synchronous or induction generators. Synchronous generators are more complex and tend to be more expensive than induction generators, but they provide excellent voltage and frequency control of the generated power, and can deliver reactive power to the grid. However, these generators are not intrinsically self-starting (the blades are usually pitched to start the turbine), do not provide powertrain damping, and require sophisticated controls for connecting to the grid, as the output frequency (and, thus, the speed of the generator) must be precisely matched to the grid frequency before this may be accomplished.

Induction generators have a simple, rugged construction, they may readily be used as motors to spin a turbine up to speed, they are cheaper than synchronous generators, they may be connected to and disconnected form the grid relatively easily, and they provide some powertrain damping to smooth out the cyclic torque variations inherent in the wind turbine output. However, they require reactive power from either power electronics or the grid, and they can contribute to frequency and voltage instabilities in the grid to which they are connected. These adverse effects can usually be solved fairly quickly and at low cost with modern power electronics. Induction generators are, therefore, the most common type found on wind turbines.

Other types of generators, though less common, are also used on wind turbines. Permanent magnet generators eliminate the need for windings or winding current to provide the magnetic field, leading to a simple, rugged construction. The power produced by this type of generator is usually variable voltage and frequency AC that must be converted to DC or to fixed voltage and frequency AC with power electronics. Even with the addition of power electronics, these machines tend to achieve higher efficiency at low-power ratings than either induction or synchronous generators, leading to increased energy capture, but the technology is less mature for permanent magnet generators than it is for other types of generators. Although the cost of these generators has, historically, been somewhat higher than that of induction or synchronous generators, recent advances in technology have resulted in price decreases, making the technology quite competitive with the older technologies, and they are becoming more common. Fuchs et al. [38] describe the development of a permanent magnet generator for wind turbine use.

Direct drive generators are normally synchronous machines (with either conventional or permanent magnet excitation) of special design, built with a sufficient number of poles to permit the generator rotor to rotate at the same speed as the wind turbine rotor. This eliminates the need for a gearbox or transmission. These generators are usually used in conjunction with power electronic converters to decouple the generator from the network and provide flexibility in the voltage and frequency requirements of the generator.

Most older turbines operate at a single, fixed rotational speed, but many newer turbines, and especially the large ones, are variable-speed, operating within a fixed range of rotational speeds. Variable-speed turbine operation offers several major advantages over fixed-speed operation:

1. The aerodynamic efficiency of the rotor at low to moderate windspeeds may be improved by changing the rotational speed to keep the turbine operating close to the optimum tip-speed ratio, maximizing the power coefficient. As explained earlier, at higher windspeeds, the blades are either in stall or are pitched to limit peak power. The rotor speed may also be adjusted to fine-tune peak power regulation.
2. System dynamic loads may be attenuated by the inertia of the rotor as it speeds up and slows down in response to wind gusts.
3. The turbine may be operated in a variety of modes, including operation at maximum efficiency to maximize energy capture at lower windspeeds, and operation to minimize fatigue damage.

As mentioned earlier, for variable-speed operation, certain rotational speeds within the operating-speed range will likely excite turbine vibration modes, causing structural resonance and increased rates of fatigue damage. These rotational speeds must be avoided during operation, leading to complex control schemes.

Variable-speed operation is possible with any type of generator. In general, variable-speed operation results in variable-frequency/variable-voltage AC power. Interfacing the turbine to the power grid (a very common application) requires this power be converted to high quality, constant frequency power. Several methods have been developed for accomplishing this with sophisticated power electronics (see Smith [39], for example), but research to develop improved methods with higher efficiencies continues. Manwell et al. [40] give a brief overview of some common types of power converters and provide references for more extensive discussions of these devices.

Electrical generator efficiency tends to drop off rapidly as the generated power falls below the rated generator capacity, so single-generator wind turbine systems tend to be very inefficient at low windspeeds. Some turbine designs address this deficiency by incorporating multiple smaller generators. At low windspeed, only one generator might be attached to the drive train, with more being added as the windspeed increases. The net result is that each generator operates close to its rated power much of the time, increasing the overall generator efficiency. Similar increases in generator efficiency at low-power levels can be obtained with a single generator utilizing pole switching or multiple windings.

21.6.2 Yaw Subsystem

The rotor of a HAWT must be oriented so that the rotor axis is parallel to the wind direction for peak power production. While small turbines rely on passive systems, such as tail vanes, to accomplish this, most large, upwind HAWTs and a few downwind HAWTs incorporate active yaw control systems using a wind direction sensor and a drive motor/gear system to orient the rotor with respect to the wind direction. Some downwind HAWTs are designed to utilize the wind itself to automatically orient the rotor. Active yaw systems tended to be extremely problematical in early turbines, basically because the loads acting on them were not well understood. Yaw loads are much better understood today, and these systems are no longer a major problem area. VAWTs are not sensitive to wind direction and do not require yaw systems.

21.6.3 Controls Subsystem

One or more control systems are needed to integrate the operation of the many components of a wind turbine and to safely generate power. In a very general sense, a wind turbine control system consists of a number of sensors, a number of actuators, and a system of hardware and software that processes the signals from the sensors to generate signals to control the actuators. A turbine usually contains a minimum of two distinct controllers—a supervisory controller that handles the normal operation of the turbine, and a safety controller that will override the supervisory controller to bring the turbine to a safe state (usually stopped, with the brakes applied). Any particular turbine may incorporate many additional secondary or slave controllers, each of which handles only a limited number of tasks. An example of a secondary or slave controller would be a blade pitch controller that actually pitches the blade to follow a predetermined speed ramp during start-up or shutdown of the turbine and that regulates the power output of the turbine at the rated level in above-rated windspeed.

21.6.3.1 Supervisory Controller

The supervisory controller is usually computer or microprocessor-based and programmed with software to perform the controller functions. The basic turbine controller will start and stop the machine, connect or disconnect the generator output to the power grid, as needed, control the operation of the yaw and pitch systems (if present), perform diagnostics to monitor the operation of the machine, and perform normal or emergency shutdown of the turbine as required. For older turbines, the supervisory controller was frequently a fairly generic device with a minimum of machine-specific features that was simply added to an existing turbine. It would be programmed with either hardware or software modifications to implement the functions specific to that particular turbine. The controllers on newer turbines, especially large, variable-speed machines, incorporate much more intelligence than the old, generic type; they may be designed to utilize the pitch system to limit peak power and/or torque, to control the rotor speed, to maximize energy capture, to trade-off energy capture and loads mitigation, to reduce power fluctuations, to control power quality, and/or to actively control some turbine dynamics. Controllers with any or all of these capabilities are usually designed from scratch as an integral component of the wind turbine, and must be included in the models of system aerodynamics and structural dynamics to obtain accurate estimates of loads and motions.

21.6.3.2 Safety Controller

The safety controller acts as a backup to the supervisory controller, and takes over if the supervisory controller fails to maintain the turbine in a safe operating mode. It is normally triggered by the activation of certain safety sensors, such as excessive vibration, excessive rotor speed, or excessive generator power, that are independent of the sensors connected to the supervisory controller, but it may also be activated by an operator-controlled emergency stop button. This system must be as independent from the main control system as possible, and must be designed to be fail-safe and highly reliable, since it may be the last line of defense to save a turbine from self-destruction. The safety controller will normally consist of a hard-wired, fail-safe circuit monitoring a number of sensors. If any of the sensors indicates a problem, the safety controller takes control and ensures that the turbine is brought to a safe condition. This might include, for example, de-energizing all electrical systems, pitching the blades to the feather position, and engaging the spring-applied emergency brakes (that are held off in normal operation).

21.7 Other Wind-Energy Conversion Considerations

The actual turbine system is the single most important and most costly item of a wind energy conversion system, but there are many other aspects of the system that must be considered and carefully optimized before wind energy can be produced at a cost competitive price. These aspects include things like turbine siting, installation and foundations, operating and maintenance costs, manufacturing processes, transport of components to the site, turbine payback period, dependence of wind energy COE on turbine size, and environmental concerns. Turbine siting has already been discussed in Section 14.2. This chapter will only discuss materials, installations, costs, and environmental concerns. Additional information, including extensive reference lists, on all of these aspects can be found in Refs. [40–42].

21.7.1 Wind Turbine Materials

As mentioned earlier, wind turbines are fatigue-critical structures (their design is driven by consideration of the cyclic fatigue loads they must endure), and the number of fatigue cycles they experience in a 20–30 year design life is three orders of magnitude beyond the 10^6 cycles that has been the common limit of fatigue data for most materials. Most of the materials used in the construction of wind turbines are typical of those materials that are used in other rotating machinery and towers—relatively common structural materials such as metal, wood, concrete, and glass-fiber reinforced plastic (GFRP) composites. Towers are typically made of steel; a few have been built of concrete. Drive trains, generators, transmissions, and yaw drives are made of steel. These components can be readily designed so they experience very low stresses and will have a fatigue life of 20–30 years. Wind turbine blades, however, must be built with a minimum of lightweight material to minimize the gravity-induced cyclic loads on the blades, drivetrain, and tower. Over the past 15 years, high-cycle fatigue databases for many potential blade materials have been developed specifically for wind turbine applications. Mandell and Samborsky [43] and Mandell et al. [44] describe the main U.S. high-cycle fatigue database, while De Smet and Bach [45] and Kensche [46] describe the European counterpart. The material with the best all-around structural properties for wind turbine blades is carbon-fiber/epoxy composite. However, it is significantly more expensive than other potential materials. The blade material of choice today (and for the past decade or so) is GFRP, due to the high strength and stiffness that can be obtained, the ease of tailoring blades made of GFRP to handle the loads, and the relatively low cost of GFRP [47]. However, the trend towards larger and larger turbines, with the resultant increase in blade weight and flexibility, has created intense interest in utilizing some carbon fiber in the blades to decrease weight and add stiffness. The expense of carbon fiber, even in the cheapest form available today, means that turbine designers must incorporate it into blades in a cost-effective manner. According to Griffin [48] and Jackson et al. [49], one very efficient method of utilizing carbon fiber is to place it in the longitudinal spar caps, near the maximum thickness area of the blade, where its light weight and extra stiffness yield maximum benefits.

Obviously, the materials must be protected from the environment. Although paints and gel coats are adequate for most onshore installations, special care is required to protect turbine components from the extreme environment of offshore installations. This topic is beyond the scope of this chapter.

21.7.2 Wind Turbine Installations

Section 14.2 of this handbook discusses wind turbine siting considerations in some depth. Although individual turbines or small clusters of turbines may be used to provide power to small loads such as

FIGURE 21.15 Typical wind farm installation: New Mexico Wind Energy Center in Eastern New Mexico.

individual residences or businesses, the most common arrangement for producing large amounts of energy from the wind is to locate many wind turbines in close proximity to each other in a wind farm or wind park. Figure 21.15 is a photograph of several of the General Electric 1.5-MW turbines comprising the 204-MW New Mexico Wind Energy Center in eastern New Mexico (mentioned earlier in the case study in Section 14.2). Operating turbines in this manner leads to lower COE, as fixed construction costs, such as electrical grid interconnections and project development and management costs, and fixed maintenance costs, such as cranes, replacement parts and repair facilities, can be spread over a larger investment. The largest wind farms to date have capacity ratings in excess of 200 MW, and they consist of over 100 machines.

21.7.3 Energy Payback Period

A certain amount of energy is used in the manufacture, installation and eventual scrapping of any energy producing machine. The time required for the machine to generate as much energy as was used in its manufacture, installation and end of life scrapping is referred to as the energy payback period. Studies by Krohn [50] and Milborrow [51] show that the energy payback period of modern wind turbines ranges between 3 and 10 months, depending on the site windspeed and details of the turbine manufacture and installation. This payback period is among the shortest for any type of electricity production technology.

21.7.4 Wind Turbine Costs

Malcolm and Hansen [52], in a study performed for the U.S. Department of Energy, estimated the cost of wind turbines of four different sizes: 750 kW (46.6-m diameter rotor), 1.5 MW (66-m diameter rotor), 3.0 MW (93-m diameter rotor) and 5.0 MW (120-m diameter rotor). The hub height for each machine was 1.2 times the rotor diameter. These turbines were designed to operate in an 8.5-m/s annual average (at hub height) wind site, with a 50-year extreme 10-min windspeed of 42.5 m/s (International Electrotechnical Commission Class 2 site). The results are summarized in Table 21.1.

Although the intent of this study was strictly to look at the relative total turbine cost as a function of size, without regard to the actual turbine cost, the estimated cost for the 1.5-MW turbine was fairly close to actual turbine prices in 2002, the year of publication. Actual costs today are somewhat higher than those estimated by the study, due largely to increases in the cost of steel, the cost of concrete, the cost of

TABLE 21.1 Turbine Component Costs

Component\Rating	750 kW	1.5 MW	3.0 MW	5.0 MW
Rotor	**$101,897 (16%)**	**$247,530 (18%)**	**$727,931 (21%)**	**$1,484,426 (20%)**
Blades	10%	11%	13%	12%
Hub	3%	5%	6%	6%
Pitch mechanism and bearings	3%	3%	2%	2%
Drivetrain, nacelle	**$255,631 (39%)**	**$562,773 (40%)**	**$1,282,002 (37%)**	**$2,474,260 (33%)**
Low-speed shaft	1%	1%	2	2%
Bearings	1%	1%	1	1%
Gearbox	10%	11%	10	9%
Mechanical brake, HS coupling	<1%	<1%	<1%	<1%
Generator	7%	7%	6%	4%
Variable-speed electronics	8%	7%	6%	4%
Yaw drive and bearing	1%	1%	<1%	1%
Main frame	3%	5%	6%	6%
Electrical connections	5%	4%	4%	3%
Hydraulic systems	<1%	<1%	<1%	<1%
Nacelle cover	3%	3%	2%	2%
Control & safety system	**$10,000 (2%)**	**$10,200 (1%)**	**$10,490 (0.3%)**	**$10,780 (0.1%)**
Tower	**$66,660 (11%)**	**$183,828 (13%)**	**$551,415 (16%)**	**$1,176,152 (15%)**
Balance of station	**$217,869 (33%)**	**$388,411 (28%)**	**$873,312 (25%)**	**$2,458,244 (32%)**
Foundations	5%	4%	2%	1%
Transportation	4%	4%	7%	17%
Roads, civil works	7%	6%	4%	3%
Assembly and installation	4%	4%	3%	3%
Electrical interface/connections	11%	9%	7%	6%
Permits, engineering	2%	2%	2%	2%
Initial capital cost (ICC)	**$655,057**	**$1,392,741**	**$3,445,150**	**$7,603,862**
Normalized initial capital cost (ICC/Rating), ($/kW)	**873**	**928**	**1,148**	**1,520**
Annual energy production (AEP), (MWh)	**2,254**	**4,817**	**10,372**	**18,133**
Relative initial capital cost per kWh of energy produced	**1.01**	**1.00**	**1.15**	**1.45**

Source: From Malcolm, D. J. and Hansen, A. C. 2002. *WindPACT turbine rotor design study: June 2000–June 2002*, NREL/SR-500-32495, National Renewable Energy Laboratory, Golden, CO.

TABLE 21.2 Turbine Component Cost of Energy Contributions

Component\Rating	750 kW	1.5 MW	3.0 MW	5.0 MW
Rotor	11%	13%	16%	15%
Drive Train	27%	29%	28%	26%
Controls	1%	<1%	<1%	<1%
Tower	7%	9%	12%	12%
Balance of Station	23%	20%	19%	25%
Replacement Costs	11%	11%	9%	7%
O & M	18%	19%	17%	14%
Total COE (¢/kWh)	4.367	4.321	4.741	5.642

Source: From Malcolm, D. J. and A. C. Hansen. 2002. *WindPACT turbine rotor design study: June 2000–June 2002*, NREL/SR-500-32495, National Renewable Energy Laboratory, Golden, CO.

transportation, and the manufacturers' profit margins (resulting from the recent high demand for wind turbines).

Table 21.1, under "Balance of Station," shows that transportation costs increase dramatically as the size increases beyond 1.5 MW. This is due primarily to the difficulty of transporting the very large components over the highway. These costs would be significantly lower if the turbine could be shipped via ship or barge, but that is impossible for most onshore turbines. The table shows no advantage of scale on the initial capital cost, as that figure (in terms of $/kW) increases with turbine rating across the sizes studied. However, it does show that the annual energy production increases with rotor size to the power 2.2, due to the increase in tower height and the accompanying increase in windspeed arising from wind shear. The result is a minimum initial capital cost per generated kWh (ICC/AEP) for the 1.5-MW machine.

Malcolm and Hansen then estimated the cost of energy (COE) at a site with 5.8-m/s annual average winds at 10-m height, breaking the COE down by the contribution due to each major turbine component. The results are presented in Table 21.2. With current technology, it appears that the 1.5-MW turbine size actually has the lowest cost of energy. Keep in mind that the difference in COE for the three smaller turbines is relatively small; slight changes in the cost models could result in the lowest COE occurring for the 750-kW turbine or the 3.0-MW turbine rather than the 1.5-MW turbine. From these results, it is obvious that reductions in a single-component COE will not lead to dramatic changes in turbine COE. For example, the rotor contribution to the total COE is only 10%–16%, so a 20% decrease in the rotor contribution only yields a 2%–3% decrease in total COE. On the other hand, any improvement in a component that leads to increased turbine energy production may lead to a significant decrease in COE. A 20% increase in rotor energy capture at no additional capital or O&M cost, for example, would lead directly to a 20% decrease in COE.

21.7.5 Environmental Concerns

Although wind turbines generate electricity without causing any air pollution or creating any radioactive wastes, like all man-made structures, they do cause an impact on the environment. Wind turbines require a lot of land, but only about 5% of that land is used for turbine foundations, roads, electrical substations, and other wind-farm applications. The remaining 95% of the land is available for other uses such as farming or livestock grazing. Wind turbines do generate noise as well as electricity, but noise is seldom a problem with newer, large wind turbines. Some small turbines are quite noisy, but many of the newer ones are very quiet. Current industry standards call for characterization of turbine noise production and rate of decay with distance as part of the turbine testing process; therefore, noise information is readily available. Noise decreases quickly with distance from the source, so placing wind turbines appropriate distances from local homes has proven to be an effective means of eliminating noise as a problem. The noise level due to a typical, large, modern wind turbine, 300 m distant, is roughly comparable to the typical noise level in the reading room of a library.

21.7.5.1 Visual Impact

The visual impact of wind turbines is extremely subjective. What one person considers highly objectionable, another might consider as attractive or, at least, not objectionable. The relatively slow rotation of today's large wind turbines is viewed by most people as far less intrusive than the fast rotation of the early, small turbines. Visual impact can be minimized through careful design of a wind farm. The use of a single model of wind turbine in a wind farm and uniform spacing of the turbines helps alleviate concerns in this area. Computer simulation can be very helpful in evaluating potential visual impacts before construction begins.

21.7.5.2 Bird and Bat Collisions

One of the greatest environmental issues that the wind industry has had to face is the issue of bird deaths due to collisions with wind turbines. Concerns about this issue were, in large part, the result of relatively high numbers of raptor deaths in the Altamont Pass wind farms east of San Francisco, California in the 1980–1985 time frame. Dozens of studies of this issue have been conducted during the past 20 years. Sinclair and Morrison [53] and Sinclair [54] give overviews of the recent U.S. studies. One conclusion of the Altamont Pass studies is that the Altamont Pass situation is a worst-case scenario, due in large part to bad siting, and to the presence of overhead power lines that led to a large number of bird electrocutions. Colson [55] and Wolf [56] provide summaries of ways to minimize the impact of wind farms on birds. Among the specific recommendations are:

- Avoid bird migration corridors and areas of high bird concentrations(micro habitats or fly zones)
- Use fewer, larger turbines
- Minimize number of perching sites on turbine towers
- Bury electrical lines
- Conduct site-specific mitigation studies

In spite of the ongoing bird collision problems at Altamont Pass, the impact of wind energy on birds is very low compared with other human-related sources of bird deaths. According to the National Wind Coordinating Committee (NWCC), bird collisions with wind turbines caused the deaths of only 0.01%–0.02% of all the birds killed by collisions with man-made structures across the U.S. in 2001 [57]. Extrapolating their estimate of roughly two bird deaths per turbine per year to a scenario where 100% of U.S. electricity is provided by wind (and assuming a turbine size of 1.5 MW), yields an estimate of deaths due to collisions with wind turbines of 0.5%–1% of all bird deaths caused by collisions with structures. In contrast, bird collisions with buildings and windows account for about 55% of structure-related bird deaths, whereas collisions with vehicles, high-tension power lines, and communication towers account for about 17%.

Unexpectedly large numbers of bats have been killed by some wind farms in the eastern U.S. in the past few years. The wind industry has joined with Bat Conservation International, the U.S. Fish and Wildlife Service, and the National Renewable Energy Laboratory to identify and quantify the problem and to explore ways to mitigate these deaths. Several wind-energy companies are providing a portion of the funding for the cooperative effort that includes hiring a full-time biologist to spend three years coordinating the research effort. Efforts to resolve this issue are ongoing.

References

1. Shepherd, D. 1994. Historical development of the windmill, In *Wind Turbine Technology, Fundamental Concepts of Wind Turbine Engineering*, D. Spera, ed., ASME Press, New York, pp. 1–46.
2. Miley, S. J. 1982. A catalog of low Reynolds number airfoil data for wind turbine applications. RFP-3387. Department of Aerospace Engineering, Texas A&M University, College Station, TX.
3. Abbott, I. H. and von Doenhoff, A. E. 1959. *Theory of Wing Sections Including a Summary of Airfoil Data*. Dover Publications, New York.

4. Selig, M. S., Guglielmo, J. J., Broeren, A. P., and Giguere, P. 1995. *Summary of Low-Speed Airfoil Data., Vol. 1,* SoarTech Publications, Virginia Beach, VA.
5. Selig, M. S., Lyon, C. A., Giguere, P., Ninham, C. P., and Guglielmo, J. J. 1996. *Summary of Low-Speed Airfoil Data., Vol. 2,* SoarTech Publications, Virginia Beach, VA.
6. Lyon, C. A., Broeren, A. P., Giguere, P., Gopalarathnam, A., and Selig, M. S. 1997. *Summary of Low-Speed Airfoil Data, Vol. 3,*, SoarTech Publications, Virginia Beach, VA.
7. Selig, M. S. and McGranahan, B. D. 2004. Wind tunnel aerodynamic tests of six airfoils for use on small wind turbines; period of performance: October 31, 2002–January 31, 2003. NREL Report SR-500-34515. National Renewable Energy Laboratory, Golden, CO.
8. Somers, D. M. 1997. Design and experimental results for the S809 airfoil. SR-440-6918. National Renewable Energy Laboratory, Golden, CO.
9. Betz, A. 1920. Das maximum der theoretisch mőglichen ausnűtzung des windes durch windmotoren. *Z. Gesamte Turbinewesen.*, 26.
10. Froude, R. E. 1889. On the part played in propulsion by differences of fluid pressure. *Transactions of the Institute of Naval Architects*, 30, 390–405.
11. Lanchester, F. W. 1915. A contribution to the theory of propulsion and the screw propeller. *Transactions of the Institute of Naval Architects*, 57, 98–116.
12. Hansen, A. C. and Butterfield, C. P. 1993. Aerodynamics of horizontal-axis wind turbines. *Annual Review of Fluid Mechanics*, 25, 115–149.
13. Wilson, R. E. 1994. Aerodynamic behavior of wind turbines, In *Wind Turbine Technology, Fundamental Concepts of Wind Turbine Engineering*, D. Spera, ed., pp. 215–282. ASME Press, New York.
14. Snel, H. 1998. Review of the present status of rotor aerodynamics. *Wind Energy*, 1, 46–69.
15. Touryan, K. J., Strickland, J. H., and Berg, D. E. 1987. Electric power from vertical-axis wind turbines. *Journal of Propulsion and Power*, 3, 6, 481–493.
16. Paraschivoiu, I. 2002. *Wind Turbine Design with Emphasis on Darrieus Concept.* Polytechnic International Press, Montreal, Quebec.
17. Kocurek, D. 1987. Lifting surface performance analysis for horizontal axis wind turbines. SERI/STR-2 17 3163. Solar Energy Research Insitute, Golden, CO.
18. Strickland, J. H., Smith, T., and Sun, K. 1981. A vortex model of the Darrieus turbine: an analytical and experimental study. SAND81-7017. Sandia National Laboratories, Albuquerque, NM.
19. Xu, G. and Sankar, L.N. 1999. Computational study of horizontal axis wind turbines. *A Collection of the 1999 ASME Wind Energy Symposium Technical Papers*, pp. 192–199.
20. Sørensen, N. N. and Hansen, M. O. L. 1998. Rotor performance predictions using a Navier–Stokes method. *A Collection of the 1998 ASME Wind Energy Symposium Technical Papers*, pp. 52–59.
21. Duque, E. P. N., Burkland, M. D., and Johnson, W. 2003. Navier–Stokes and comprehensive analysis performance predictions of the NREL phase VI experiment. *A Collection of the 2003 ASME Wind Energy Symposium Technical Papers*, pp. 43–61.
22. Sørensen, N., Michelsen, J., and Schreck, S. 2002. Navier-Stokes predictions of the NREL phase VI rotor in the NASA Ames 80 ft. x 120 ft. wind tunnel. *Wind Energy*, 5, 2/3, 151–169.
23. Johansen, J., Sørensen, N., Michelsen, J., Schreck, S. 2003. Detached-eddy simulation of flow around the NREL phase-VI rotor. *Proceedings of the 2003 European Wind Energy Conference and Exhibition.*
24. Xu, G. and Sankar, L. N. 2002. Application of a viscous flow methodology to the NREL phase VI rotor. *A Collection of the 2002 ASME Wind Energy Symposium Technical Papers*, pp. 83–89.
25. Simms, D., Schreck, S., Hand, M., and Fingersh, L. 2001. NREL unsteady aerodynamics experiment in the NASA-Ames wind tunnel: a comparison of predictions to measurements. NREL/TP-500-29494. National Renewable Energy Laboratory, Golden, CO.
26. ABAQUS Analysis User's Manual, Version 6.5. 2005. ABAQUS, Inc. Providence, Rhode Island, http://www.abaqus.com

27. MSC Adams, 2006. http://www.mscsoftware.com
28. Malcolm, D. J. and Wright, A. D. 1994. The use of ADAMS to model the AWT-26 prototype. *Proceedings of 1994 ASME Wind Energy Symposium*, pp. 125–132.
29. Molenaar, D. P. and Dijkstra, S. 1999. State-of-the-art of wind turbine design codes: main features overview for cost-effective generation. *Wind Engineering*, 23, 5, 295–311.
30. Buhl, M. L., Jr., Wright, A. D., and Pierce, K. G. 2000. Wind turbine design codes: a comparison of the structural response. *A Collection of the 2000 ASME Wind Energy Symposium Technical Papers*, pp. 12–22.
31. Quarton, D. C. 1998. The evolution of wind turbine design analyses—a twenty year progress review. *Wind Energy*, 1, 5–24.
32. RCAS Theory Manual, Version 2.0. 2002. United States (US) Army Aviation and Missile Command/AeroFlightDynamics Directorate (USAAMCOM/AFDD) Technical Report (TR). USAAMCOM/AFDD TR 02-A-005. US Army Aviation and Missile Command, Moffett Field, CA.
33. RCAS User's Manual, Version 2.0. 2002. United States (US) Army Aviation and Missile COMmand/AeroFlightDynamics Directorate (USAAMCOM/AFDD) Technical Report (TR). USAAMCOM/AFDD TR 02-A-006. US Army Aviation and Missile Command, Moffett Field, CA.
34. RCAS Applications Manual, Version 2.0. 2002. United States (US) Army Aviation and Missile COMmand/AeroFlightDynamics Directorate (USAAMCOM/AFDD) Technical Report (TR) TR 02-A-007. US Army Aviation and Missile Command, Moffett Field, CA.
35. Bir, G. S. and Chopra, I. 1994. University of Maryland Advanced Rotorcraft Code (UMARC) theory manual. Technical Report UM-AERO 94–18, Center for Rotorcraft Education and Research, University of Maryland, College Park, MD.
36. Lobitz, D. W. and Sullivan, W. N. 1983. A comparison of finite element prediction and experimental data for forced response of DOE 100 kW VAWT. *Proceedings of the Sixth Biennial Wind Energy Conference and Workshop*, pp. 843–853.
37. Mayda, E. A., van Dam, C. P., and Yen-Nakafuji, D. 2005. Computational investigation of finite width microtabs for aerodynamic load control. *A Collection of the 2005 ASME Wind Energy Symposium Technical Papers*, pp. 424–436.
38. Fuchs, E. F., Erickson, R. W., Carlin, P. W., and Fardou, A. A. 1992. Permanent magnet machines for operation with large speed variations. *Proceedings of the 1992 American Wind Energy Association Annual Conference*, American Wind Energy Association, Washington, DC, pp. 291–297.
39. Smith, G. A. 1989. Electrical control methods for wind turbines. *Wind Engineering*, 13, 88–98.
40. Manwell, J. F., McGowan, J. G., and Rogers, A. L. 2003. *Wind Energy Explained: Theory, Design and Application*. Wiley, Chichester.
41. Spera, D. ed. 1994. *Wind Turbine Technology, Fundamental Concepts of Wind Turbine Engineering*, ASME Press, New York.
42. Burton, T., Sharpe, D., Jenkins, N., and Bossanyi, E. 2001. *Wind Energy Handbook*. Wiley, Chichester.
43. Mandell, J. F. and Samborsky, D. D. 1997. DOE/MSU composite material fatigue database: test methods, materials, and analysis. SAND97-3002. Sandia National Laboratories, Albuquerque, NM.
44. Mandell, J. F., Samborsky, D. D., and Cairns, D. S. 2002. Fatigue of composite material and substructures for turbine blades. SAND2002-077. Sandia National Laboratories, Albuquerque, NM.
45. De Smet, B. J. and Bach, P. W. 1994. Database FACT: fatigue of composites for wind turbines. ECN-C-94-045. ECN, Petten, The Netherlands.
46. Kensche, C. W. ed. 1996. Fatigue of materials and components for wind turbine rotor blades. EUR 16684. European Commission, Luxembourg, Belgium.
47. Sutherland, H. J. 2000. A summary of the fatigue properties of wind turbine materials. *Wind Energy*, 3, 1, 1–34.

48. Griffin, D. A. 2004. Blade system design studies volume II: preliminary blade designs and recommended test matrix. SAND2004-0073. Sandia National Laboratories, Albuquerque, NM.
49. Jackson, K. J., Zuteck, M. D., van Dam, C. P., Standish, K. J., and Berry, D. 2005. Innovative design approaches for large wind turbine blades. *Wind Energy*, 8, 2, 141–171.
50. Krohn, S. 1997. *The energy balance of modern wind turbines. WindPower Note No. 16*. Danish Wind Turbine Manufacturers Association, Copenhagen.
51. Milborrow, D. 1998. Dispelling the myths of energy payback time. *Wind Stats Newsletter*, 11, 2, 1–3.
52. Malcolm, D. J. and Hansen, A. C. 2002. WindPACT turbine rotor design study: June 2000-June 2002. NREL/SR-500-32495. National Renewable Energy Laboratory, Golden, CO.
53. Sinclair, K. C. and Morrison, M. L. 1997. Overview of the U.S. Department of Energy/National Renewable Energy Laboratory avian research program. *Proceedings of Windpower '97*, American Wind Energy Association, Washington, DC, pp. 273–279.
54. Sinclair, K. C. 1999. Status of the U.S. Department of Energy/National Renewable Energy Laboratory avian research program. *Proceedings of Windpower '99*, American Wind Energy Association, Washington, DC, pp. 273–279.
55. Colson, E. W. 1995. Avian interactions with wind energy facilities: a summary. *Proceedings of Windpower '95*, American Wind Energy Association, Washington, DC, pp. 77–86.
56. Wolf, B. 1995. Mitigating avian impacts: applying the wetlands experience to wind farms. *Proceedings of Windpower '95*, American Wind Energy Association, Washington, DC, pp. 109–116.
57. Erickson, W. P., Johnson, G. D., Strickland, M. D., Young, D. P., Sernka, K. J., and Good, R. E. 2001. Avian collisions with wind turbines: a summary of existing studies and comparisons to other sources of avian collision mortality in the United States. National Wind Coordinating Committee, Washington, DC http://www.nationalwind.org/publications/wildlife/avian_collisions.pdf.

Further Information

Excellent summaries of HAWT and VAWT aerodynamics, together with extensive reference lists, are presented by Hansen and Butterfield [12], and by Touryan, Strickland, and Berg [15], respectively. Volume 1, number 1 of *Wind Energy*, Wiley, 1998 contains a comprehensive set of review papers covering wind turbine rotor aerodynamics, design analysis, and overall system design.The latest developments in the field of wind energy in the United States and Europe may be found in the following annual conference proceedings: *A Collection of the ASME Wind Energy Symposium Technical Papers*, American Institute of Aeronautics and Astronautics, 59 John Street, 7th Floor, New York, NY 10038. *Proceedings of Windpower*, American Wind Energy Association (AWEA), 1101 14th St. NW, 12th Floor, Washington, DC 20005. *Proceedings of the European Wind Energy Association*, European Wind Energy Association, Renewable Energy House, Rue d'Arlon 63-65, BE-1040 Brussels. Belgium. *Proceedings of the British Wind Energy Association*, British Wind Energy Association, Renewable Energy House, 1 Aztec Row, Berners Road, London, N1 0PW, UK. The books by Manwell, et al. [40], Spera [41] and Burton et al. [42] contain a wealth of fairly current information on wind energy conversion, history, and technology, together with extensive reference lists. The National Wind Coordinating Committee [57] gives a good account of the studies that have been done of wind turbine/bird collisions and puts the results in context with bird deaths due to other causes. Extensive information on wind energy conversion technology may also be found on the World Wide Web. Excellent sites to start with include those of the U.S. National Renewable Energy Laboratory Wind Energy Technology Center at http://www.nwtc.nrel.gov, Sandia National Laboratories Wind Energy Technology Department at http://www.sandia.gov/wind, the Danish Risø National Laboratory at http://www.risoe.dk/vea/index.htm, the American Wind Energy Association at http://www.awea.org, the British Wind Energy Association at http://www.britishwindenergy.co.uk, the European Wind Energy Association at http://www.ewea.org, and the Danish Wind Industry Association at http://www.windpower.org/en/core.htm.

22

Biomass Conversion Processes For Energy Recovery

22.1 Energy Recovery by Anaerobic Digestion 22-2
Introduction • Organic Wastes and Biomass Used as Feedstocks in Anaerobic Digestion Process • The Issue of Biodegradability • Fundamental of Anaerobic Digestion • Monitoring of the Anaerobic Digestion Process • Reactor Types Used for Anaerobic Digestion • Modes of Operation for Anaerobic Digestion • Utilization of By-Products from In-Vessel Anaerobic Digestion Process • Commercial-Scale In-Vessel Anaerobic Digestion Technologies

References 22-34

22.2 Power Generation 22-37
Introduction • Direct Combustion • Combustion Equipment • Thermal Gasification • Stirling Cycle • Rankine Cycle • Brayton Cycle • Fuels Cells • Combined Cycles

References 22-50

22.3 Biofuels 22-51
Introduction • Ethanol • Ethanol from Starch Crops • Ethanol from Lignocellulosic Feedstocks • Biodiesel • Transportation Fuels from Biomass-Derived Syngas

References 22-65

Massoud Kayhanian and George Tchobanoglous
University of California at Davis

Robert C. Brown
Iowa State University

Transformation of waste materials into energy can generally be accomplished through biological, thermal, and chemical processes. The energy produced from these processes can be in the form of heat, gas, or liquid fuel.

This chapter is organized into three parts. Part 1 deals with anaerobic digestion process which is used to convert waste material into biogas composed of methane and carbon dioxide. Major emphasis is given to the recovery of energy from the organic fraction of MSW, as there are numerous commercial-scale digesters in operation throughout the world using processed MSW as a source of feedstock. In Part 2, the focus is on chemical conversion processing of biomass for the production of liquid fuel. Part 2 deals with combustion and gasification processes for the generation of electric power. In Part 3, the focus is on converting biomass into liquid fuels.

22.1 Energy Recovery by Anaerobic Digestion

Massoud Kayhanian and George Tchobanoglous

22.1.1 Introduction

The anaerobic digestion process, carried out in the absence of oxygen, involves the use of microorganisms for the conversion of biodegradable biomass material into energy, in the form of methane gas and a stable humus material. Anaerobic digestion can occur under control conditions in specially designed vessels (reactors), semi-control conditions such as in a landfill, or under uncontrolled conditions as it does in the environment. The focus in this part of Chapter 24 is on controlled anaerobic digestion process. It should be noted that anaerobic digestion is differentiated from anaerobic fermentation, which is the term usually applied to processes employing anaerobic microbes for the production of fermentation products such as alcohol, or lactic acid.

To describe anaerobic conversion of the biodegradable organic fraction of waste materials into energy, the following topics are examined in this chapter: (1) organic wastes and biomass used as feedstocks in anaerobic digestion process, (2) issue of organic waste biodegradability, (3) fundamental of anaerobic digestion process, (4) monitoring of anaerobic digestion process, (5) reactor types used in for anaerobic digestion, (6) modes of operation for anaerobic digestion, (7) utilization of an in-vessel anaerobic digestion process by-products, and (8) commercial-scale in-vessel anaerobic digestion technologies.

22.1.2 Organic Wastes and Biomass Used as Feedstocks in Anaerobic Digestion Process

The general scheme for a controlled anaerobic digestion process is shown in Figure 22.1. As shown, the recovery of energy involves feedstock preparation, methane gas generation, stabilization of digested solids, and the utilization of digester gas and humus as a source of energy and soil amendment, respectively. Major sources of waste materials considered as a feedstock for anaerobic digestion process are (1) municipal solid wate (MSW), (2) agricultural animal waste, (3) crop residues, biomass, and energy crops, and (4) wastewater treatment plant sludge (WWTPS). The typical composition of MSW in the U.S. is shown in Figure 22.2. The composition of MSW may vary greatly by season, geographical area, and community socio-economic level. The extrapolation of results from one location to another, therefore, may not be valid and should be done with caution.

As shown in Figure 22.2, paper, yard waste, and food waste are the principal biodegradable organic fractions. The biodegradability of these waste materials varies substantially as reported in Table 22.1. The issue of biodegradability is discussed further in the following section.

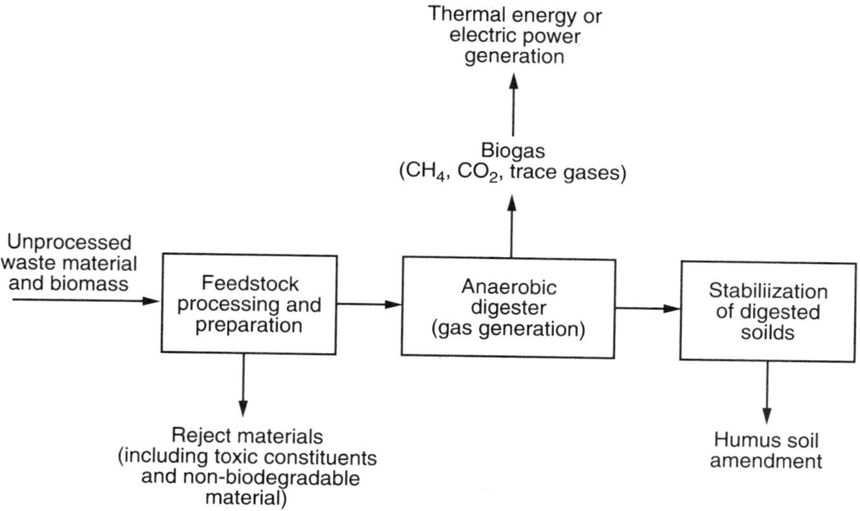

FIGURE 22.1 General flow diagram of bioenergy recovery system.

As illustrated in Figure 22.1, some waste material must be preprocessed to: (1) remove toxic constituents, (2) remove nonbiodegradable materials, and (3) prepare a balanced feedstock in terms of nutrient availability for methane recovery. Most livestock wastes and treatment plant sludge are relatively homogeneous, biologically active, and contain sufficient nutrients. Therefore, little or no preprocessing of feedstock may be needed. Also, because dairy and pig wastes contain a variety of anaerobic microorganisms, methane gas will be produced when proper conditions are provided. Thus, for the anaerobic digestion of organic wastes other than livestock wastes, the addition of dairy or pig manure is commonly used. The addition of livestock waste to activate a biological process is commonly known as "inoculation."

22.1.3 The Issue of Biodegradability

While each organic waste may contain a constant ultimate biodegradable fraction, practical biodegradability can be quite variable. Factors such as particle size, time, and environmental conditions

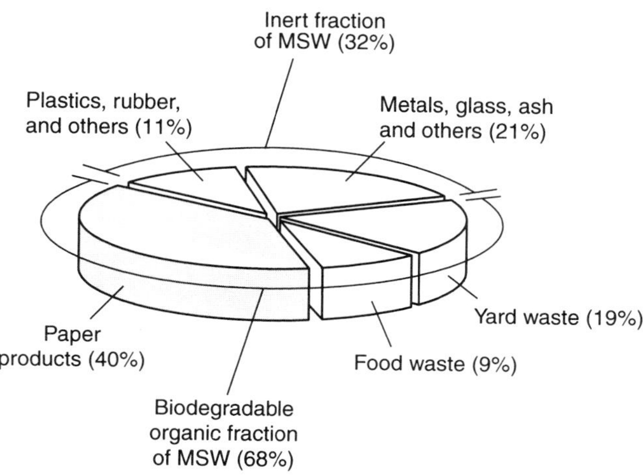

FIGURE 22.2 A typical MSW composition in the U.S.

TABLE 22.1 Weight Distribution and Biodegradability of the Organic Fraction of MSW Found in the U.S.

Component	Percent by Wet Weight		Biodegradability		
	Range	Typical	Low[a]	Medium[b]	High[c]
Paper					
Cardboard	3–10	10		X	
Magazines	2–8	6	X		
Newspaper	4–10	8	X		
Waxed cartoons	5–20	12			X
Other (mixed)	<1–5	3			X
Food waste	6–18	9			X
Yard wastes	5–20	15			X
Wood waste	1–4	1.5	X		

[a] Low-biodegradable materials are classified as having a biodegradability of less than 30%.
[b] Medium-biodegradable materials are classified as having a biodegradability of greater than 30% but lower than 75%.
[c] High-biodegradable materials are classified as having a biodegradability of greater than 75%.

(i.e., temperature, nutrient requirements, etc.) will influence the final outcome of biodegradation. For example, because favorable conditions do not exist in most landfills, the biodegradability estimated from analytical test will usually be greater than the actual biodegradation that occurs in landfills. For practical purposes, the discussion in this section deals with the biodegradability of the organic fraction of waste materials used in in-vessel anaerobic digestion processes. Substrate biodegradability is of special importance in an in-vessel anaerobic digestion processes, where the production of energy is of concern.[1,2]

22.1.3.1 Biodegradable Volatile Solids of an Organic Substrate

Dry organic substrates consist of volatile solids (VS) and ash. Taken together, these two components comprise the total solids (TS) of a substrate as illustrated in Figure 22.3. The fraction of ash typically depends on the nature of the organic substrate. VS is measured as loss on ignition. Only the biodegradable volatile solids (BVS) fraction of the VS has the potential for bioconversion largely because of the presence of refractory volatile solids (RVS) which, in most digester feedstocks, is mostly lignin. Lignin is a complex organic material which is difficult for anaerobic bacteria to degrade and normally requires a long period of time for complete degradation. It is clear from Figure 22.3 that organic substrates with high RVS and ash contents have a low biodegradability.

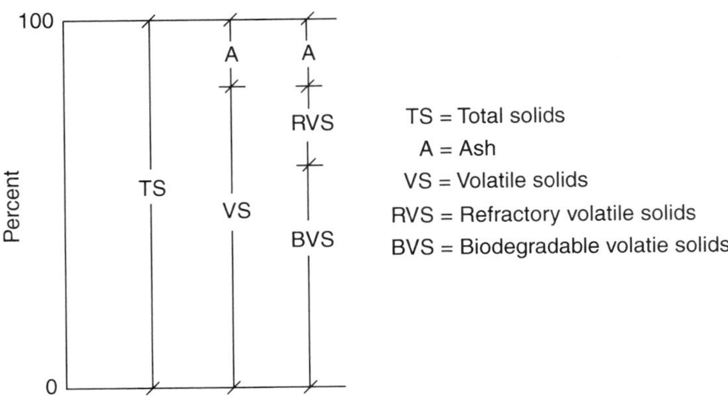

FIGURE 22.3 Characterization of a typical organic substrate.

The RVS in the organic fraction of MSW consists of the lignin content which is associated with cellulose in plant materials and thermoplastic materials. Lignin content of an unsorted organic fraction of MSW will fluctuate as the percentage of cellulose and thermoplastic materials will vary with season, socioeconomic conditions, and geographical locations. Knowing the BVS fraction of VS in the individual organic fractions of MSW will allow the biodegradability of any composite BOF/MSW to be estimated.

22.1.3.2 Methods to Estimate Biodegradability

Several methods can be used to estimate the BVS fraction of an organic substrate, including: (1) long-term batch digestion studies, (2) measurement of lignin content, and (3) chemostat studies. These methods are described briefly below.

22.1.3.2.1 Long-Term Batch Digestion

Batch digestion studies, designed to simulate a specific anaerobic digestion process, can be used to predict the biodegradable fraction of an organic substrate. One method involves graphical analysis of weight loss over time and is used commonly to predict the ultimate biodegradability of energy crops and agricultural wastes.[3] This method is based on a linear regression plot of the remaining VS concentration as the retention time approaches infinity. Regular weight measurements of the batch digester throughout the course of the study are required to apply this method.

An alternative method for determining the biodegradable fraction of an organic substrate, using batch digesters, is to determine and compare the initial and final dry masses. Initial measurements are made of the mass and the percent TS of active reactor mass of each mixture to be tested. Each mixture includes a percentage of seed, i.e., material taken from an active digester to provide suitable microorganisms. One batch mixture is always 100% seed. At the end of the batch study, the dry mass in each individual reactor is measured. The mass loss in each unit, corrected for seed biodegradation, is due to the conversion of the biodegradable portion of the substrate to biogas. The results of biodegradable fraction of municipal solid waste (BOF/MSW) based on long-term batch digestion studies are shown in Table 22.2.

22.1.3.2.2 Measurement of Lignin Content

An analytical method commonly used for determining the BVS fraction of an organic substrate is based on the measurement of the crude lignin content of the VS. Chandler et al.[1] correlated the biodegradability of various agricultural residues and newsprint, as determined by long-term batch digestion studies, with the lignin content of the substrate, as determined by sequential fiber analysis, developed by Robertson and Van Soest.[4] The following empirical relationship was developed to estimate the biodegradable fraction of an organic substrate from lignin test results:

$$\text{Biodegradable fraction} = 0.83\text{–}0.028 \times \text{LC} \tag{22.1}$$

TABLE 22.2 Estimated Biodegradable Fraction of Organic Waste Components of MSW Based on Long-Term Batch Studies

	BVS Fraction, %VS		
Organic Substrate	Reactor 1	Reactor 2	Average
Newspaper	22.5	24.8	23.7
Office paper	83.5	81.8	82.7
Food waste	83.0	82.5	82.8
Yard waste	73.0	70.5	71.8
Mix blend[a]	70.5	69.5	69.8[b]

[a] Mixed blend consists of 19% newsprint, 53% office paper, 15% yard waste, and 13% food waste (dry basis).
[b] The value reported for mixed blend is based on the actual organic waste and not on the value that can be computed from each individual component.

TABLE 22.3 Estimated Biodegradable Fraction of Organic Waste Components of MSW Based on the Lignin Content

Lignin Content Range	Average BVS Fraction	
Organic Substrate	%VS	%VS[a]
Newspaper	20–23	22
Office paper	0.2–1	82
Yard waste	4–10	72
Food waste	0.1–0.7	82
Mixed blend[b]	4–7	67.6

[a] Computed using Equation 22.1.
[b] Mixed blend consists of 19% newsprint, 53% office paper, 15% yard waste, and 13% food waste (dry basis).

where the biodegradable fraction is expressed as a fraction of the VS, and LC the lignin content, expressed as a percentage of the VS. The results of biodegradable fraction for BOF/MSW based on the lignin content are shown in Table 22.3.

22.1.3.2.3 *Chemostat Studies*

The BVS fraction of an organic substrate can also be estimated using chemostat techniques. The true digestible organic matter (TDOM) technique, developed by animal scientists, is used to assess the digestibility of animal feeds.[5] The test involves the digestion of a sample in vitro in rumen fluid for 48 h. A second chemostat technique, the biochemical methane potential (BMP) assay, is used to characterize the BMP and biodegradability of many organic substrates.[6] In the BMP assay, the substrate biodegradability is determined by monitoring the cumulative methane production from a slurry sample which is incubated anaerobically (typically for 30 days at 35°C) in a chemically defined media and inocula.

The biodegradability of energy crops as estimated using the BMP assay tends to be 3%–10% higher than the corresponding BVS value as estimated using the TDOM technique.[7,8] The BMP assay appears to be more suitable for determining the ultimate biodegradability.

22.1.3.3 Biodegradability of Various Organic Waste Materials

The estimates of BVS, as a percentage of VS, of a typical organic fraction of MSW based on long-term batch studies, lignin content, and a pilot study are compared in Figure 22.4. As can be seen, the average BVS fractions calculated using either the lignin content or the long-term batch study are essentially the same. The BVS fraction obtained from the pilot study, using a complete-mix reactor and mass retention time (MRT) of 30 days, is about 83 and 81% of the estimated values obtained from the lignin content or the batch study, respectively. Similarly, Richards et al.[7,8] were able to remove only 83% of BVS of sorghum, using a thermophilic, high rate, high-solids reactor with a MRT of 45 days.

The values of biodegradable fraction for various organic substrates using different methods are compared in Table 22.4. It is important to note that the estimated BVS values based on field operations are normally about 80%–85% of the values estimated by chemostat studies. Thus, the use of BMP values would result in an overestimation of biogas production rate that can be achieved in actual practice. It is important to note that the complete removal of BVS as estimated from batch digestion and analytical or a chemostat method at a practical range of MRT (20–40) is not possible. For an in-vessel high-solids anaerobic digestion process, BVS can be defined as that fraction of VS that can be biodegraded under optimum environmental and nutritional conditions in a period of 20–40 days.

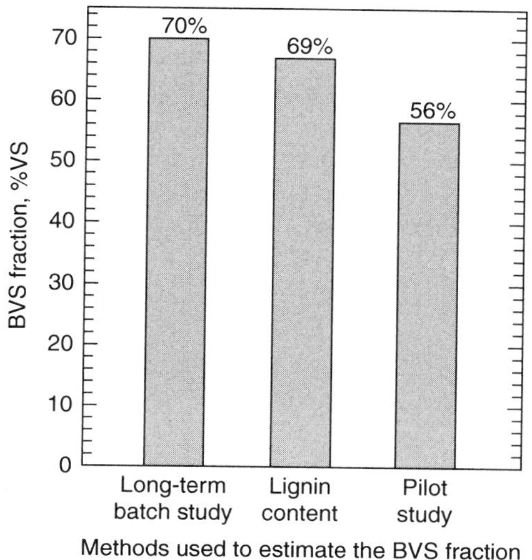

FIGURE 22.4 Comparison of the BVS fraction of a typical mixed BOF/MSW obtained using three methods of analysis.

22.1.3.4 Impact of BVS on Process Design and Performance

The common use of total or VS both in describing the organic loading rate and feedstock C/N ratio, and in estimating gas production from the BOF/MSW is misleading, as some of the organic compounds are low in biodegradability. The use of BVS instead allows meaningful comparisons to be made between the different anaerobic digestion processes reported in the literature. The impact of BVS on the organic loading rate, the computation of carbon to nitrogen (C/N) ratio, and the prediction of biogas production are discussed below.

Traditionally, the loading rates for biological processes have been based on the VS content of the substrate feedstock. When handling substrates with variable biodegradabilities, estimates of the methane production potential and digester performance cannot be made accurately using a VS loading rate. Loading rates based on the BVS fraction of the substrate should be used to estimate the conversion of organic wastes to methane. The BVS organic loading rate can be expressed as the input BVS mass per active reactor mass per day. The BVS mass of an organic substrate can be computed using total inflow VS mass multiplied by the BVS fraction of VS.

Customarily, the C/N ratio is determined based on the total dry mass of the organic matter and the corresponding percentage concentrations of carbon and nitrogen. This method of determining the C/N ratio may not be appropriate for the BOF/MSW because not all the organic carbon is biodegradable

TABLE 22.4 Biodegradability of Various Organic Substrate Using Batch Digestion, Chemostat, and Field Operation Methods

	Methods Used to Estimate Biodegradability				
	Average Value, %				
Substrate	Batch Digestion	TDOM	BMP	Field Operation[a]	Reference
Sorghum	60	79.1	82.0	72.6 (105)[b]	Richards et al. 1991a, 1991b[c]
Sorghum 1			90.9	75.0 (45)	Richards et al. 1991a
Sorghum 2			90.0	72.5 (29)	Richards et al. 1991a
Alpha-cellulose	81.6	90.6	99.8		Richards et al. 1991a
1:1 mix of sorghum/ Alpha-cellulose		70.8	87.6	90.7	Richards et al. 1991a
BOF/MSW	69.8			57 (30)	Kayhanian and Tchobanoglous 1993a

[a] All field operations were conducted in a thermophilic, high-rate, high-solids reactor.
[b] Numbers in parentheses are nominal MRT value in days.
[c] Data for field operations were obtained from Richards et al.[8]

and/or available for biological decomposition. However, it appears that almost all the nitrogen in the organic material is available for conversion to ammonia via microbial metabolism.[9] Because the available nitrogen in the organic feedstock can be converted to ammonia, the C/N ratio should be based on the nitrogen content of the total organic mass and the carbon content of the BVS mass.[10]

A more accurate biogas volume production can be made from the BVS mass, rather than the TS or VS mass. However, not all the BVS mass computed using the lignin content or the results of long-term batch studies will be converted into biogas in field operation. For design purpose, a correction factor of about 0.82, derived from lignin test or long-term batch digestion, must be used when using the BVS mass as a basis for theoretical gas production. Computation of BVS fraction of the biodegradable fraction of MSW is illustrated in the following example calculation.

22.1.3.4.1 *Example of BVS Calculation of MSW*

Estimate the total BVS weight (mass) of a 100 tn./d MSW material recovery facility.

Use the following information:

1. Assume that the composition of a typical MSW waste stream given in Figure 22.2 is valid:
2. Typical characteristics of MSW sub-fractions:

Waste Material	Total Solids, %	Volatile Solids, %TS	Lignin Content
Yard Waste	40	88	6.43
Food Waste	30	90	0.36
Paper	94	98	9.29

22.1.3.4.2 *Solution*

1. Start from the typical composition of MSW and prepare a table:

	Weight, lb/d	
Waste Material	Wet Weight	Dry Weight
Yard waste	38,000	15,200
Food waste	18,000	5400
Paper	80,000	75,200
Metal	42,000	42,000
Plastic	22,000	22,000

[a]Calculated using the percent TS.

2. Calculate the biodegradable organic fraction of MSW being serviced by the sorting facility:

$$\text{BOF/MSW per day} = (0.68)(100\ \text{tn. MSW/d}) = 68\ \text{tn.}$$

3. Compute the total weight of BVS. To compute the BVS content of a substrate:

$$\text{VS weight} = \text{dry weight} \times \text{VS\%}$$

$$\text{BVS weight} = \text{VS weight} \times \text{BF}$$

where

$$BF = 0.83-(0.028 \times LC)$$

and

$$LC = \text{lignin content.}$$

4. Prepare a table for the values calculated:

Biodegradable Waste	Dry wt, lb/d	VS, % TS	VS wt, lb/d	BF, % VS	BVS wt, lb/d
Yard Waste	15,200	88	13,376	0.65	8694
Food waste	5400	90	4860	0.82	3985
Paper (mixed)	75,200	98	73,696	0.57	42,005
Total	95,800		91,932		54,684

[a]The values in this column are calculated as VS = dry weight × VS%.[b]The BF fraction is computed using Equation 22.1 and the lignin content or using the biodegradability value determined by a long-term batch study.[c]The values in this column are calculated as BVS = VS weight × BF.

22.1.3.4.3 *Comments*

It is important to note that the BVS weight is about 27% of the total wet weight of MSW, 40% of the total wet weight of BOF/MSW, and 60% of the total VS fraction of the TS.

22.1.4 Fundamentals of Anaerobic Digestion

The purpose of this section is to familiarize the reader with the basic microbiology related to an anaerobic digestion process. Topics discussed include: (1) an introduction to anaerobic bacteria, (2) nutrient requirements for the anaerobic bacteria, (3) physical and chemical parameters that affect anaerobic bacteria, (4) basic biochemical reaction in anaerobic digestion process, and (5) pathway of complex organic substrate in anaerobic digestion process.

22.1.4.1 Anaerobic Bacteria

Effective anaerobic degradation of complex organic waste is a result of the combined and coordinated metabolic activity of the digester microbial population. This population of microorganisms is composed of several major trophic groups. At present, as reported in Table 22.5, four different groups of anaerobic bacteria are recognizable. With respect to methane recovery, the anaerobic bacteria are generally grouped

TABLE 22.5 Four Major Groups of Anaerobic Bacteria and their Function

Bacterial Group	Function
Hydrolytic Bacteria	Catabolize saccharides, protein, lipids, and other minor chemical constituents of biomass
Hydrogen-producing acetogenic Bacteria	Catabolize certain fatty acids and neutral end products
Homoacetogenic Bacteria	Catabolize unicarbon compounds (e.g., H_2/CO_2 or HCOOH) or hydrolyze multi-carbon compounds to acetic acid
Methanogenic Bacteria	Catabolize acetate and one carbon compound to methane

Source: From Holland, K. T., Knapp, J. S., and Shoesmith, J. G., *Anaerobic Bacteria*, Chapman and Hall, New York, 1987.

TABLE 22.6 Morphological Characteristics of Methane Bacteria

Organism	Morphology	Length, m	PH Optimum	Electron Donor
Methanobacterium formicium	Rods, single pairs or chains	2–15		Hydrogen and formate
M. strain MOH	Rods, single pairs or chains	2–4		Hydrogen
M. arborphilicum	Rods, single pairs or chains	2–3.5	7.5–8	Hydrogen
M. strain AZ	Rods, single pairs or chains	2–3	6.8–7.2	Hydrogen
Methanosarcina barkeri	Sarcina	1.5–5	7	Methanol and hydrogen
Methanobacterium rominantium	Coccus chains	1–2 diameters	6.8–7.2	Hydrogen and formate
Methanococcus vanniellii	Coccus	0.5–4 diameters	7.4–9.2	Formate
Methanobacterium mobile	Rod			Hydrogen and formate
Methanobacterium thermoautotrophium	Rod	5–10		Hydrogen
Methanobacterium hungatii	Spiral rods	50		Hydrogen and formate

Source: From Holland, K. T., Knapp, J. S., and Shoesmith, J. G., *Anaerobic Bacteria*, Chapman and Hall, New York, 1987.

as: (1) hydrolytics (hydrolyzing bacteria), (2) acetogens (acid-forming bacteria), and (3) methanogens (methane-forming bacteria).

Methane bacteria are among the most strictly anaerobic microorganisms known and occur naturally in the rumen of cows, marshes, and brackish waters, as well as in wastewater treatment plant digesters. Methanogenic bacteria have been isolated from anaerobic digesters, together with other obligate anaerobes, such as the *Propionibacter, Butyrobactor*, and *Lactobacillus*. The true methane bacteria include the following identified genera: *Methanococcus, Methanobacterium, Methanosarcina, Methanospirillium*, and *Methanobacillus*. The main group of methane bacteria along with some of their morphological and growth characteristics are given in Table 22.6.

As reported in Table 22.6, methane bacteria are rod, cocci, or sarcinate in shape. As a group, methane bacteria are nearly always Gram-positive (a stain used to identify bacterial types) and usually not motile. Many methane bacteria are pleomorphic; that is, they exhibit variable morphology when viewed under the microscope, which makes their identification very difficult. Their length can vary between 2 and 15 μm. Nutritionally, the methanogenic bacteria are said to be cemolihoheterotrophic, which means that they can build up their cell structures from either carbonate or organic compounds. Morphologically, methanogens are a diverse group; however, physiologically, they are quite similar as all share the common metabolic capacity to produce methane. The methanogens are possibly the most important group of anaerobic bacteria. They have scientific significance in their exclusive and distinctive properties among the bacteria.

22.1.4.2 Nutrient Requirements for Anaerobic Bacteria

To operate a high-solids anaerobic digestion process at a commercial level, attention must be focused on process stability. Successful operational parameters have been established for the high-solids process studies conducted at the University of California at Davis.[11] However, in the anaerobic digestion of BOF/MSW, bacterial nutritional requirements have often been overlooked. Nutritional deficiencies may result in reactor instability and incomplete bioconversion of the organic substrates. When the anaerobic digestion process is applied to the biodegradable organic fraction of MSW, bacterial nutritional requirements must be addressed and nutrient supplementation may be required.[12,13]

Methanogenic bacteria have a variety of mineral nutrient requirements for robust growth.[14,15] For a proper bacterial metabolism, a variety of nutrients must be present in the substrate. The nutrient requirements for anaerobic bacteria can generally be categorized as macro- and micronutrient. For a stable anaerobic digestion process, these nutrients must be present in the substrate in the correct ratios

TABLE 22.7 Representative Feedstock Nutrient Concentrations Required for the Robust Anaerobic Bioconversion of BOF/MSW

		Average Value (Dry Basis)	
Nutrient[a]	Unit	Range	Typical
C/N[a]		20–30	25
C/P		150–300	180
C/K		40–100	70
Co	mg/kg	<1–5	2
Fe	mg/kg	100–5000	1000
Mo	mg/kg	<1–5	2
Ni	mg/kg	5–20	10
Se	mg/kg	0–0.05	0.03
W	mg/kg	0.05–1	0.1

[a] C/N, C/P, and C/K ratios are based on the biodegradable organic carbon and total nitrogen, phosphorus, and potassium.
Source: From Kayhanian, M. and Rich, D., *Biomass and Bioenergy*, 8, 433–444, 1995.

and concentrations. Based on studies conducted at UC Davis, it was found that typical BOF/MSW used as a feedstock for the anaerobic digestion process is deficient in many essential nutrients.[16] To overcome feedstock nutrient deficiencies, supplementary nutrients must be added to stimulate the digestion process. The range of nutrient concentrations needed to stimulate the anaerobic treatment of BOF/MSW for gas production is not well known. The nutrient values reported in Table 22.7 are based on three years of experience at the UC Davis high-solids biogasification project.

22.1.4.3 Physical and Chemical Parameters that Affect Anaerobic Bacteria

The alteration of several physical and chemical environmental parameters in anaerobic digesters can influence microbial populations and digester performance. Major environmental factors are summarized and shown in Table 22.8. Changing temperature from the mesophilic range to (i.e., below 45°C) the thermophilic range (i.e., above 55°C) increases organic mineralization and establishes a different species composition (i.e., thermophilic bacteria). Zeikus[17] found that thermophilic bacteria have a limited species composition, but they possess all the major nutritional categories and metabolize the same substrate as mesophilic bacteria. The ability to proliferate at growth temperature optima well above 60°C is associated with extremely thermo-stable macromolecules. As a consequence of growth at high temperatures and unique micromolecular properties, thermophilic bacteria can possess high metabolic rates, physically and chemically stable enzymes, and lower growth but higher end product yields than similar mesophilic species. Thermophilic digesters are generally digester substrates within shorter retention time than mesophilic digesters.

TABLE 22.8 Physical and Chemical Effectors of Bacterial Populations in Anaerobic Digesters

Parameter	Population Response	Influence of Methanogenesis
Temperature (e.g., change to 65°C from 35°C)	Enrichment of thermophilic	Increase
pH (e.g., change to pH 5 from pH 7)	Enrichment of acidophiles	Decrease
Organic substrate composition (e.g., change soluble substrate (glucose) for particulate (lignoglucose))	Enrichment of biopolymer decomposers	Decrease
Substrate feed rate (e.g., change from slow to high glucose feed)	Enrichment of hydrolytic bacteria	Decrease
Inorganic composition (e.g., addition of excess sulfate)	Enrichment of sulfate reducers	Decrease

Source: From Holland, K. T., Knapp, J. S., and Shoesmith, J. G., *Anaerobic Bacteria*, Chapman and Hall, New York, 1987.

Values of pH well below neutral are indicative of digester failure. The pH ranges for growth of many bacterial species inherent to anaerobic methane digester are not known and may vary considerably. Nevertheless, important hydrogen-producing (e.g., *C. thermocellum*) and hydrogen-consuming (e.g., *M. thermoautitrophicum*) anaerobes do not grow at pH value below 6. Digester imbalances that favor more rapid growth of non-methanogens (e.g., high feed rate or chemical composition of feed) can lead to low pH and inhabitation of methanogenesis. Low pH values favor proton reduction to hydrogen, but not hydrogen oxidation to protons. Thus, methanogens may not function at low pH because they employ oxidoreductases for hydrogen oxidation and establishment of proton gradients during the catabolism of one carbon compound and acetate.

The organic composition of digester wastes affects species composition and methane yield. Most notably, methane yields from municipal wastes and animal manures are often limited by polymeric lignin surrounding cellulose and retarding its fermentation. High molecular weight lignin is not metabolized significantly by anaerobic bacteria. However, methane yield from digestion of lignocellulosics can be enhanced greatly by physico-chemical pretreatment that separates lignin from cellulose and/or solubilize lignin into anaerobically digestible substrates. To accomplish lignin solubilization, an adaptation period is required to obtain an active microbial population from untreated waste digesters that metabolized soluble chemical hydrolysis products of lignin. The organic composition modification caused by waste pretreatment is considered to be associated with change in species compositional change.

The most significant parameter affecting methanogenesis during anaerobic digestion processes is sulfate inhabitation. The addition of high levels of sulfate to decomposing organic matter can result in nearly complete inhabitation of methane formation as a result of species compositional change. The basis for this inhabitation is that carbon and electron flow during organic mineralization is diverted from methane formation to hydrogen sulfide production. The response of sulfide on many methanogens is concentration dependent. Sulfide is required by many methanogens as a sulfur source for growth. The addition of low sulfide concentrations often stimulates mixed culture methanogenesis, whereas addition of high concentrations can be inhibitory.

22.1.4.4 Pathway of a Complex Organic Substrate in Anaerobic Digestion Process

A generalized scheme for the anaerobic digestion of complex organic substrate is shown in Figure 22.5. Anaerobic digestion is generally considered to take place in three distinct stages. The three stages have been described as: (1) hydrolysis, (2) acetogenesis, and (3) methanogenesis. Each of the three stages has distinct bacterial groups and chemical reactions, and it proceeds in an assembly-line fashion.[18]

As depicted in Figure 22.5, the overall process begins with the hydrolysis of complex organic compounds into soluble components. Next, the acid-forming bacteria ferment the soluble components into a group of extracellular intermediates, including various volatile fatty acids (VFA's), hydrogen (H_2), and carbon dioxide (CO_2). The concentrations of the intermediate acids are usually small in proportion to their production and degradation rates, and they are quickly transformed to methanogenic substrates, including acetate, methanol, and formate. These products are then converted to methane by the methanogenic bacteria.

Methane bacteria can only use limited range of substrates for growth and energy production. During anaerobic digestion, a varied mixture of complex compounds is converted to a very narrow range of simple compounds, mainly methane and carbon dioxide. The ecology of the system is much more complicated and involves an interacting succession of microbes which influences each others' growth and metabolism. Nearly all anaerobic bacteria can use hydrogen and carbon dioxide, most can use formate, a few acetate, and fewer still methanol and methylamines. However, in the mixed population and with many organisms, the range of fermentation products is restricted. The restriction is due to the phenomenon of interspecies hydrogen transfer by which methanogens utilize hydrogen

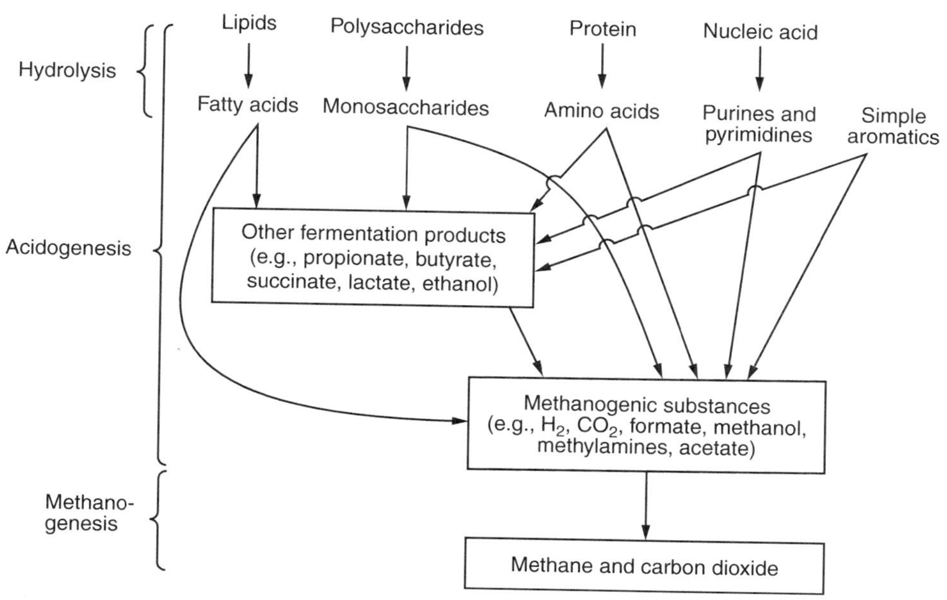

FIGURE 22.5 Stages of anaerobic digestion. (From Tchobanoglous, G., Theisen, H., and Vigil, S. A., *Integrated Solid Waste Management—Engineering Principles and Management Issues*, McGraw-Hill Book Company, New York, 1993.)

and thus benefit other organisms. Hydrogen removal is helpful as it allows some organisms to dispose of more electrons via hydrogen rather than via the production of more reduced carbon compounds, such as ethanol or butyrate. In general, fatty acids are the "key" intermediate products of the anaerobic fermentation of organic compounds prior to methane formation. Acetate is by far the predominant fatty acid in normally operated systems, being responsible for about 70% of the fatty acid present. The VFA measurements are usually expressed in mg/l, therefore, as being acetate equivalent.

22.1.4.5 Basic Biochemical Reaction in Anaerobic Digestion Process

The anaerobic bacteria described in previous sections are responsible for biochemical transformation of a wide variety of waste materials. These transformations are involved in the breakdown of complex polymers, such as cellulose, fats, and proteins, to long- and short-chain fatty acids and finally methane, carbon dioxide, and water. The basic biochemical reactions affected by anaerobic microbial population are oxidation/reduction reactions, where a number of organic compounds are oxidized by the removal of hydrogen. Carbon dioxide, is thereby, reduced in providing an oxidant for the methane bacteria. The hydrogen produced can be replaced by some of the organic acids and alcohols as direct reductants of carbon dioxide. The basic biochemical reactions utilizing hydrogen, carbon dioxide, carbon monoxide, alcohols, organic acids, methylamines, and other protein derivative compounds as a substrate are summarized in the energy-yielding equations given in Table 22.9.

The general anaerobic transformation of the biodegradable organic fraction of MSW can be described by the following equation:[19]

$$\begin{aligned}\text{Organic matter} + H_2O + \text{nutrients} \rightarrow \text{new cells} + \text{resistant organic matter} + CO_2 \\ + CH_4 + NH_3 + H_2S + \text{heat}\end{aligned} \qquad (22.2)$$

TABLE 22.9 Summary of Most Common Biochemical Reactions in Anaerobic Digestion Process

Substrate	Biochemical Reactions
Hydrogen and carbon dioxide	$4H_2+CO_2 \rightarrow CH_4+2H_2O$
Carbon monoxide	$4CO+2H_2O \rightarrow CH_4+3CO_2$
Alcohols	$4CH_3OH \rightarrow 3CH_4+CO_2+2H_2O$ $CH_3OH+H_2 \rightarrow CH_4+H_2O$
	$4HCOO^-+2H^+ \rightarrow CH_4+CO_2+2HCO_3^-$
Fatty acids	$HCOO^-+3H_2+H^+ \rightarrow CH_4+2H_2O$
	$CH_3COO^-+H_2O \rightarrow CH_4+HCO_3^-$
	$4CH_2NH2+2H_2O+4H^+ \rightarrow 3CH_4+CO_2+4NH_4^+$
Methylamines and other protein derivative compounds	$2(CH3)_2NH+2H_2O+2H^+ \rightarrow 3CH_4+CO_2+2NH_4^+$ $4(CH_3)_3N+6H_2O+4H^+ \rightarrow 9CH_4+3CO_2+4NH_4^+$ $2CH_3CH_2-N(CH_3)_2+2H_2O \rightarrow 3CH_4+CO_2+2CH_3CH_2NH_2$

For practical purposes, the overall conversion of the organic fraction of solid waste to methane, carbon dioxide, and ammonia can be represented by the following equation:[19]

$$C_aH_bO_cN_d \rightarrow nC_wH_xO_yN_z + mCH_4 + sCO_2 + rH_2O + (d\text{–}nx)NH_3 \tag{22.3}$$

where $s=a-nw-m$, $r=c-ny-2s$

The terms $C_aH_bO_cN_d$ and $C_wH_xO_yN_z$ are used to represent (on a molar basis) the composition of the organic material present at the start and the end of the process, respectively. If it is assumed that the organic wastes are stabilized completely, then the corresponding expression is

$$C_aH_bO_cN_d + \left(\frac{4a-b-2c+3d}{4}\right)H_2O \rightarrow \left(\frac{4a+b-2c-3d}{8}\right)CH_4 + \left(\frac{4a-b+2c+3d}{8}\right)CO_2 + dNH_3 \tag{22.4}$$

The above relationship can be used to estimate the theoretical biogas volume and composition. Estimation of the amount of gas produced by the complete anaerobic stabilization of the biodegradable VS fraction of MSW is illustrated in the following example calculation.

22.1.4.5.1 *Estimation of Gas Production*

Estimate the total theoretical amount of gas that could be produced from MSW under anaerobic conditions in an in-vessel anaerobic digestion process. Assume that 78% by weight of the MSW is organic material, including moisture. Further assume that the moisture content is 20%, the VS are 83.5% of the total organic solids, the BVS are 75% of the VS, and only 90% of the biodegradable VS will be converted to biogas. The overall chemical formula for the biodegradable organic material is $C_{60}H_{95}O_{38}N$.

22.1.4.5.2 *Solution*

1. Determine the BVS

$$BVS = 78 \times (1-0.2) \times 0.835 \times 0.75 = 39.1 \text{ lb}$$

2. Determine the total amount of BVS that will be converted to gas

$$BVS_{gas} = 39.1 \times 0.9 = 35.2$$

3. Using the chemical formula $C_{60}H_{95}O_{38}N$, estimate the amount of methane and carbon dioxide that can be produced using Equation 22.4.

$$C_aH_bO_cN_d + \left(\tfrac{4a-b-2c-3d}{4}\right)H_2O \rightarrow \left(\tfrac{4a+b-2c-3d}{8}\right)CH_4 + \left(\tfrac{4a-b+2c+3d}{8}\right)CO_2 + dNH_3$$

For the given chemical formula, $a=60$ $b=95$ $c=38$ $d=1$
The resulting equation is

$$\underset{(1437)}{C_{60}H_{94.3}O_{37.8}N} + \underset{(324)}{18.28H_2O} \rightarrow \underset{(512)}{31.96CH_4} + \underset{(1232)}{28CO_2} + \underset{(17)}{HN_3}$$

4. Determine the weight of methane and carbon dioxide from the equation derived in step 3 and the data from step 2.

$$\text{Methane} = \frac{512}{1437}(35.2\text{ lb}) = 12.5\text{ lb}(5.7\text{ kg})$$

$$\text{Carbondioxide} = \frac{1232}{1437}(35.2\text{ lb}) = 30.2\text{ lb}(13.7\text{ kg})$$

5. Convert the weight of gases, determined in step 4, to volume, assuming that the densities of methane and carbon dioxide are 0.0448 and 0.1235 lb/ft.^{3}, respectively.

$$\text{Methane} = \frac{12.5\text{ lb}}{0.0448\text{ lb/ft.}^3} = 279\text{ ft.}^3(7.9\text{ m}^3)$$

$$\text{Carbondioxide} = \frac{30.2\text{ lb}}{0.1235\text{ lb/ft.}^3} = 244\text{ ft.}^3(6.9\text{ m}^3)$$

6. Determine the percentage composition of the resulting gas mixture using the gas volume determined in step 5

$$\text{Methane}(\%) = [279/(279+244)] \times 100 = 53.35\%$$

$$\text{Carbondioxide}(\%) = 100 - 53.35\% = 46.65\%$$

7. Determine the total theoretical amount of gas generated per unit weight of dry BVS, as determined in step 1 and per ton of MSW

$$((279+244)/39.1) = 13.4\text{ ft.}^3/\text{lb BVS}(1.48\text{ m}^3/\text{kg BVS})$$

$$\text{Biogas/ton of MSW} = 13.4\text{ ft.}^3/\text{lb} \times 2000 \times 0.78 \times 0.8 \times 0.835 \times 0.75 = 10.470$$

22.1.4.5.3 *Comment*

The theoretical gas production value determined above has been achieved with the high-solids anaerobic decomposition process (see Section 25.1.6).

22.1.5 Monitoring of the Anaerobic Digestion Process

Although the high-solids anaerobic digestion process is generally robust, care must be taken to ensure balanced operation. To aid the prevention of unbalanced digester operation and to prevent digester failure, proper methods of monitoring the high-solids anaerobic digestion process are described, and possible operational problems are identified along with suggested remedial actions to be taken when these problems arise.

22.1.5.1 Balanced Digestion

A balanced digester is one in which anaerobic digestion proceeds with a minimum of control. Balanced operation means that the environmental parameters of the system remain within their optimum range, with only occasional fluctuations. When an imbalance does occur, the two main problems are: (1) identifying the beginning of an unbalanced condition and (2) identifying the cause of the imbalance. Unfortunately, there is no single parameter that will always indicate the commencement of an unhealthy anaerobic process. The parameters shown in Table 22.10 must all be monitored daily. None of these parameters can be used individually as a positive indicator of the development of digester imbalance.

The most immediate indication of impending operational problems is a significant decrease in the rate of gas production. If the growth of the microorganisms is being inhibited by one or more factors, it will be reflected in the total gas production. However, a decrease in the gas production rate may also be caused by a decrease in either the digester temperature or the rate at which the feed material is being added to the digester.

The most significant single indicator of a digester problem is a gradual decrease in pH. In an operating system, a decrease in pH is associated with an increase in organic acid concentration. Measurement of the increase in organic acids is also a good control parameter; however, proper laboratory facilities, equipment, and trained personnel are required to monitor this and most of the other control parameters affecting the anaerobic process. Gas production rate and pH are simple, quick measurements and are performed easily.

22.1.5.2 Common Problems and Solutions to Stabilize Digester Operation

In general, there are five major problems associated with the anaerobic digestion process: (1) increase in TS concentration, (2) organic overloading, (3) toxic overloading, (4) free ammonia toxicity, and (5) nutrient deficiency. Possible cures for these problems are summarized in Table 22.11.

22.1.6 Reactor Types Used for Anaerobic Digestion

In the past 50 years, a wide variety of in-vessel anaerobic digestion processes and reactors have been used for both waste stabilization and energy recovery. The anaerobic digestion is used most commonly in wastewater treatment plants for sludge stabilization with and without methane gas recovery plants.[20] In-vessel digesters have also been used by dairy and hog farmers to stabilize their waste as well as to produce methane gas for the production of electrical power to offset their power consumption during

TABLE 22.10 Indicators of Unbalanced Operation of the Anaerobic Digestion Process

Parameters	Warning Condition
Ammonia concentration	Increases
Percent of CH_4 in biogas	Decreases
Percent of CO_2 in biogas	Increases
Reactor pH	Decreases
Total gas production	Decreases
VFA concentration	Increases
Waste stabilization	Decreases

TABLE 22.11 Summary of the Most Common Problems Associated with the Anaerobic Digestion Process and Suggested Actions to Cure the Problems

Major Problem	Suggested Action
Free ammonia toxicity	1. Start feeding with an organic waste of higher C/N ratio 2. Dilute the active reactor mass with fresh water
Nutrient deficiency	1. Add chemical nutrient into the reactor 2. Add organic materials rich in the needed nutrients
Organic overloading	1. Do not feed 2. Add strong base to neutralize acids 3. Resume feeding at lower organic loading rate when pH reaches at least 6.8
Total solids build-up	1. Add water
Toxic overloading	1. Identify and remove the toxic element from the feedstock 2. If the population of methanogens are reduced (CH_4 concentration decreased), add proper methanogen seed 3. If pH decreases below 6.8, add strong base to neutralize acids 4. Resume feeding at lower organic loading rate when pH reaches at least 6.8

pick hours and on annual basis.[21,22] The principal types of in-vessel anaerobic digestion reactor designs that can be used for methane recovery and energy production are shown in Figure 22.6 and described below briefly.

22.1.6.1 Batch Reactor

A *batch reactor* is fed once, and then the biotransformation is allowed to proceed until completion before any material is added or removed. The evolution of compounds in the reactor can be monitored and a similar level of decomposition can be achieved by all the material in the reactor at one time. Additionally, the systems are generally simple, with less support equipment than continuous fed reactors (see below). Batch processes are necessary when the biotransformation being performed requires a long reaction time. Solids which are treated undiluted are often treated with batch reactors.

Usually, batch processes require more operator labor, for feeding and unloading, than continuous feed processes, and especially so for liquids and slurries, which are easily piped and pumped. For solids, the material handling needs are more similar for the two processes. Storage facilities are needed to contain waste received between batches. The need for storage can be inconvenient for large-scale, continuously produced wastes. A simplified diagram of a batch digestion is shown in Figure 22.6a.

22.1.6.2 Complete-Mix Continuous Flow Reactor

A reactor to which a waste stream is fed and a treated effluent stream is withdrawn continuously is known as a *continuous flow reactor*. Most municipal wastewater sludge treatment processes are continuous feed processes, as municipal sludge is produced continually. A process can be designed to fit the expected maximum inflow rate, although fluctuations in flow rate and process upsets may interfere with operation. If alternate storage is not available, discharge of untreated or partially treated wastes can occur.

The average residence time of material in a continuous feed reactor can be determined using the following equation:

$$t = V/Q \tag{22.5}$$

where t = residence time, V = reactor volume, and Q = flow rate (vol/time).

Reactors that are fed liquids or slurries continuously require less operator labor than batch reactors. Because storage space is not necessary to contain continuously produced wastes between batches, continuous feed processes are used commonly for industrial or municipal wastes. There are two types of

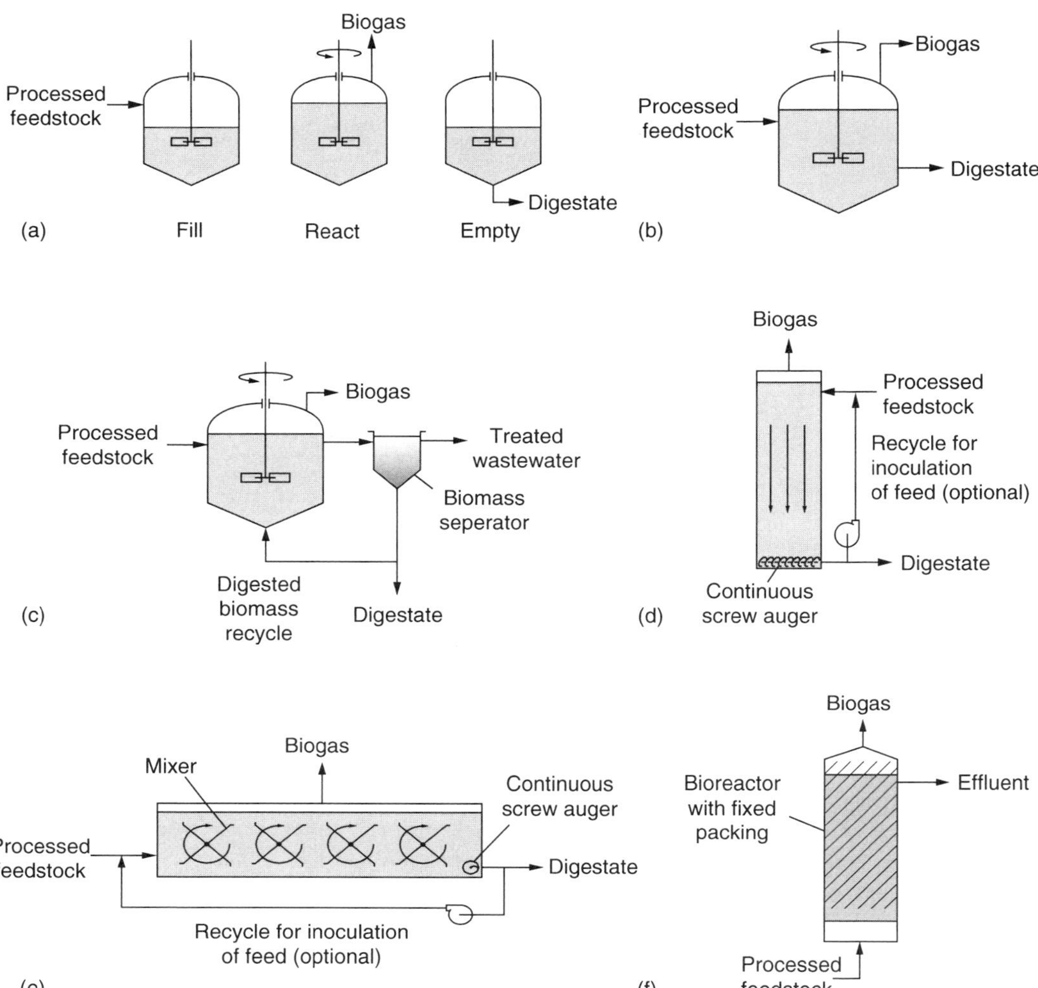

FIGURE 22.6 Typical reactor types used for anaerobic digestion: (a) batch reactor, (b) complete-mix continuous flow tractor, (c) anaerobic contact reactor, (d) vertical plug-flow reactor, (e) horizontal plug-flow reactor, and (f) attached growth reactor.

reactor which are intermediate between continuous feed and batch processes. A reactor which is fed once a day is a semi-continuous feed reactor. Also, a reactor which is fed continually, but only emptied when waste stabilization has been achieved, is a semi-batch reactor.

In a *complete-mix reactor*, the reactor contents are blended to homogeneity with a mixing device. The effluent leaving the reactor is exactly the same as the material in every part of the reactor. If fed continuously, the input waste is considered to be immediately mixed completely with the reactor contents. These reactors can also be described as well-mixed or as continuous stirred-tank reactors (CSTRs). A simplified diagram of a complete-mix reactor is shown in Figure 22.6b. One advantage of a complete-mix process is that fluctuations in feed concentration or composition are diluted into the larger reactor mass. Because of this dilution and the concentration of waste nutrient determines the rate of waste decomposition, complete-mix reactors have a slower decomposition rate.

22.1.6.3 Anaerobic Contact Reactor

The anaerobic contact process (ACP) is used to overcome the disadvantages of the complete-mix reactor without recycle. To enhance the rate of treatment, biomass is separated from the effluent and returned to the reactor. Biomass recycle can be used to reduce the reactor size and cost. A simplified ACP is shown in Figure 22.6c. Hydraulic retention times as low as a half-day have been achieved resulting in a significant reduction in plant size. ACP has been applied successfully for the treatment of meat-packing waste, where a retention time of several hours was measured. ACP is also used for the treatment of a high-strength waste material and is usually operated under low solids (TS less than 8%).

22.1.6.4 Plug-Flow Reactor

All *plug-flow reactors* are continuously or semi-continuously fed. In a plug-flow reactor, material passes through, ideally, without interacting with the material fed in before or after it. These reactors can be thought of as tubes through which independent batches of reacting material pass. The reactor can be either vertical or horizontal flow as shown in Figure 22.6d and e, respectively.

The retention time for a plug-flow reactor is the length of time it takes a mass introduced at the beginning of the reactor to pass through and be removed from the other end (see also Equation 22.5). Most horizontal plug-flow digesters are operated under low solids (usually with TS of less than 10%). Because the incoming, high-nutrient wastes are not diluted into the rest of the digester contents, the plug-flow digesters are more susceptible to system upsets due to sudden increases in waste concentration, called shock loadings. Fortunately, a certain degree of back mixing is unavoidable throughout the reactor due to longitudinal dispersion.

22.1.6.5 Anaerobic Attached Growth Reactor

Anaerobic attached growth reactors are used to prevent the depletion of the bacterial population within the reactor and, thus, improve digester efficiency. The retention of bacteria within the reactor is achieved by introducing some type of packing material on which the bacteria can grow. Various materials such as stone granules, wood chips, and plastic materials of various shapes and sizes have been used as packing material. The bacteria adhering to the filter particles bring about the treatment of the liquid as it passes through the reactor. Excess biomass dies and/or sloughs off and passes out as sludge. The anaerobic filtration process is usually used for the treatment of a high-strength waste material and usually operated under low solids (TS less than 8%). A simplified anaerobic filtration process is shown in Figure 22.6f.

22.1.6.6 Reactors with Recycle Flow

Often it is advantageous to recycle and mix effluent from an anaerobic reactor (digestate) with the inflowing waste stream (see Figure 22.6c, d, and e). Recycling is especially advantageous when it is necessary to inoculate the incoming waste with bacteria that have acclimated to the system. Recycling of biomass can also be used to increase average residence time in the reactor and to dilute high-concentration wastes. Recycling a portion of the biomass can also be used for process control, allowing the reactor to respond to fluctuations in the waste stream. Recycling systems, however, add costs, both capital and operational, to a system. Hence, unless it is necessary, the use of a simpler, *unrecycled* system is typically favored.

22.1.7 Modes of Operation for Anaerobic Digestion

To design an in-vessel anaerobic digestion, several parameters must be specified. These parameters may influence the physical characteristics of the reactor, the mode of operation, and the performance of a digestion system. Several modes of operation are discussed below.

22.1.7.1 Low-Solids Digestion

Anaerobic digestion systems for municipal wastewater operate under *low-solids* conditions; that is, the concentration of solids in the waste substrate is typically less than 10%. This mode of operation is appropriate because both municipal wastewater and the sludge collected for disposal from aerobic wastewater treatment processes are intrinsically low solids. The waste substrate in a low-solids system can be pumped and piped easily as it behaves like water. A low-solids system is also more able to handle feed composition fluctuations and higher ammonia levels, because the large volume of water serves to dilute the incoming waste compounds. Dilution of waste compounds also means dilution of nutrients, unfortunately, so that low-solids digestions can require more retention time if the nutrient level is too low. Also, the high dilution means that larger-sized tanks are required to accommodate wastes.

It is possible to digest substrates other than wastewater treatment sludge, however, and many of these waste substrates have a higher solids content. *High-solids* anaerobic digestion has recently been developed to accommodate these wastes. Originally developed to dispose of agricultural wastes, particularly manures, high-solids anaerobic digestion can be used to degrade food industry wastes, the biodegradable organic fraction of municipal solid waste (BOF/MSW), agricultural and forestry residues, and other high-solids wastes.

22.1.7.2 High-Solids Digestion

High-solids anaerobic digestion operates at solids contents from 20 to 32%. Typically, high-solids wastes, such as BOF/MSW, must be mixed with water or a low-solids waste, such as wastewater treatment sludge, to dilute the solids content to within the operating range. A high-solids system can operate with tanks that are up to 75% smaller than a low-solids system, for the same dry weight of waste. Because the nutrient in a high-solids reactor is more concentrated, the conversion rate is higher. Additionally, the high-solids process produces an end product that requires less dewatering to convert it into a landfillable waste or a usable compost material. High-solids anaerobic digestion systems are, however, more sensitive to micronutrient deficiencies and toxic inhibition, since there is more mass to digest and is diluted with less moisture. Especially, thermophilic systems are critically dependent on the maintenance of a proper carbon-to-nitrogen ratio, as explained below.

22.1.7.3 Thermophilic Digestion

Process in which the waste substrate is heated to between 120 and 135°F is characterized as *thermophilic*. Although heating the reactor adds additional equipment and energy costs, thermophilic process has a higher reaction rates as compared with ambient processes. Because bacterial metabolism is greater at the higher temperatures, a wider range of bacteria can colonize the reactor. Due to a decrease in water solubilization with increased temperature, thermophilic processes can be more vulnerable to inhibition effects due to the presence of some soluble compounds. Thermophilic methanogenic systems, for example, are more sensitive to high ammonia concentrations than are similar mesophilic systems.

22.1.7.4 Mesophilic Digestion

Processes in which the waste substrate has a temperature of between 68 and 98°F are characterized as *mesophilic*. Often no heating equipment at all is used. Although less costly than thermophilic systems, the time necessary for bioconversion may be significantly longer. If heating equipment is not used, the efficiency of the process will also depend on the weather.

22.1.7.5 Two-Stage Digestion

As discussed in previous sections, anaerobic digestion is principally carried out by three separate microbial populations under three distinct and interconnected stages. In the second stage of the digestion process, a group of bacteria convert soluble wastes to organic acids and, finally, at third stage methane bacteria convert organic acids to methane gas. Acid-forming and methane bacteria are distinct from each other in several features. These features include: their physiology, nutrient requirements, growth capabilities, and sensitivity to environmental stress. A single-stage digester can be most efficient only when a compromise is achieved between the relatively fast growing acid bacteria and the slower, more sensitive methane generators.

A two-stage digester (see Figure 22.7) is designed to isolate the acid and methane bacteria into separate reactors and optimize each environment for maximum reaction rate. Distinct features

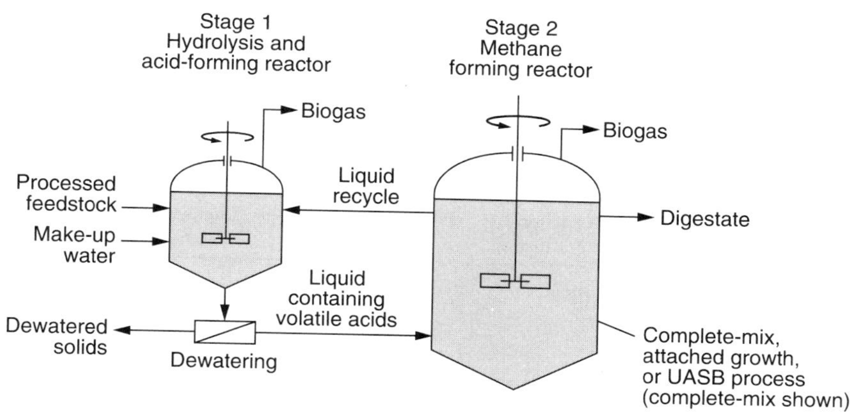

FIGURE 22.7 Two-stage anaerobic digestion.

of two-stage reactor are overall greater efficiency and system stability, and substantial reductions in total reactor volume. Several methods can be used to separate the non-methanogenic and methanogenic bacteria:

1. Various inhibitors may be introduced into the acid digester to prevent methanogenesis:
   a. adding chloroform
   b. adding carbon tetrachloride
   c. limiting oxygenation
   d. adjusting redox potential
2. Dialysis of methane bacteria by filtering the acids.
3. Adjusting the dilution rate and recycling of cell mass to each phase.

22.1.7.6 Co-Digestion of Wastewater Treatment Sludge and BOF/MSW

The co-digestion of WWTPS and BOF/MSW have been investigated by numerous researchers.[23–28] Most of these early studies concluded that high-quality source separated MSW could be digested without nutrient supplementation up to organic loading rates of about 2.6 kg VS/m^{3}d. They also reported that under conditions of nutrient addition, gas production could be enhanced and a more stable process could be achieved. Other studies conducted by Rivard et al.[12,13]; Babbitt et al.[29] and Cecchi et al.[30] confirmed the stabilizing effect of sludge to the digestion of MSW, with sludge comprising between 8 and 20% of feedstock VS.

One large-scale co-digestion of MSW and wastewater sludge has been demonstrated in the U.S. by the REFCOM project in Tampa, Florida.[31] The REFCOM system was based on a conventional low-solids digester design and operated as part of a total resource recovery process. The successful operation of the REFCOM process for mixed MSW at a rate of 50 tn./d or greater has confirmed the technical feasibility of the anaerobic digestion process. Reliable gas production was achieved in this facility for more than a year, indicating that MSW with wastewater sludge as a co-substrate can support an anaerobic biological conversion process.

Most co-digestion studies conducted prior to 1990 were low-solids processes, typically at a TS of 4%–8%. Only recently has the high-solids anaerobic digestion process been utilized for the co-digestion of MSW and WWTPS.[12,13,32] In these high-solids digestion studies, the sludge was mainly used to provide sufficient nutrients for microbial growth and metabolism. Nearly 10 years ago, the technical feasibility of the anaerobic composting process was demonstrated for co-management of BOF/MSW and WWTPS at UC Davis, as described below.[33]

TABLE 22.12 Comparison of Aerobic and Anaerobic Composting Processes

Characteristic	Aerobic Process	Anaerobic Process
Energy use	Net energy consumer	Net energy producer
End products	Humus, CO_2, H_2O	Sludge, CO_2, CH_4
Volume reduction	Up to 50%	Up to 50%
Processing time	21–30 days	20–40 days
Primary goal	Volume reduction	Energy production
Secondary goal	Compost production, waste stabilization	Volume reduction, waste stabilization

Source: From Kayhanian, M. and Tchobanoglous, G. 1993b. *Environmental Technology*, 14, 815–829.

22.1.7.7 Two-Stage Anaerobic Composting

In anaerobic composting, the focus is twofold: (1) production of methane as a source of energy and (2) complete waste stabilization to produce humus like material. Anaerobic digestion normally occurs at much higher solids content (e.g., 25%–32%). When a high-solids anaerobic digestion process is combined with a second stage aerobic biodryer, the two-stage process is termed "anaerobic composting".[34,35] The principal differences between the conventional aerobic and anaerobic composting processes are summarized in Table 22.12. The process flow diagram, physical characteristics of the reactors, and the energy recovery and volume reduction achieved are described below.

22.1.7.7.1 Process Flow Diagram

The process flow diagram for the anaerobic composting process studied at the University of California, Davis (UC Davis) is illustrated in Figure 22.8. As shown in Figure 22.8, the process involves two separate stages. The first stage of the two-stage process involves the high-solids anaerobic digestion of the biodegradable organic fraction of MSW to produce a gas, composed principally of CH_4 and CO_2, and digested solids. The second stage of the two-stage process is used to decrease the moisture content of the anaerobically digested solids. Typically, the solids content of the digested solids is increased from 25 to about 65%. Because the second stage aerobic process is used to dry the digested solids, the term *biodryer* is also used to describe the process. The physical and operational characteristics of the pilot anaerobic composting facilities used at UC Davis are briefly described below.

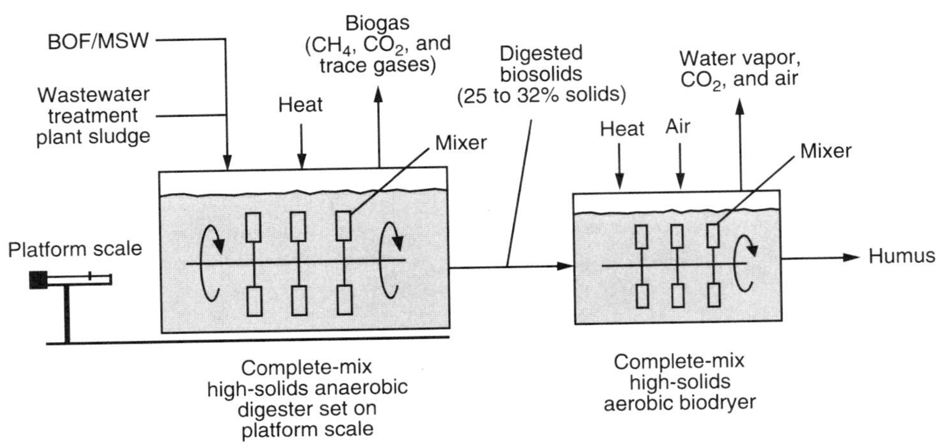

FIGURE 22.8 Flow diagram of the two-stage anaerobic composting process studied at UC Davis. The platform scale was used to obtain data for the preparation of materials mass balances.

TABLE 22.13 Summary of the Physical Characteristics of the UC Davis Pilot-Scale High-Solids Anaerobic Digestion/Aerobic Biodrying Processing Units

Item	Unit	Description/Value
Anaerobic digester		
Reactor type		Complete-mix
Mixing mechanism		Mechanical
Heating mechanism		Thermal blanket
Total reactor volume	l	2548
Volume of active reactor solids	l	1900
Aerobic biodrying unit		
Reactor type		Complete-mix
Mixing mechanism		Mechanical
Heating mechanism		Thermal blanket
Total rector volume	l	850
Volume of active reactor solids	l	765

Source: From Kayhanian, M., Lindenauer, K., Hardy, S., and Tchobanoglous, G., *BioCycle*, 32, 48–64, 1991.

22.1.7.7.2 *Physical Characteristics*

The physical characteristics of the pilot-scale anaerobic composting facilities are summarized in Table 22.14. As reported in Table 22.13, both the anaerobic digester and the biodryer are designed to operate as complete-mix reactors. Additionally, both units are equipped with thermal blankets and individual control panels so that they may be operated either manually or automatically. Further information on the UC Davis pilot-scale anaerobic digestion process design and performance can be obtained from Kayhanian and Tchobanoglous[10]; Kayhanian and Rich[33]; Kayhanian et al.[35]; Kayhanian[36]; Kayhanian and Hardy,[37] and Kayhanian et al.[38]

22.1.7.7.3 *Energy Recovery and Volume Reduction*

The results of the UC Davis co-digestion pilot study based on energy recovery and volume reduction are illustrated by the following two example calculations.

22.1.7.7.4 *Energy Recovery—Example Calculation*

Computation of the energy value that can be produced from the digestion of a 1000 tn./d MSW processing facility using primary sludge as a co-substrate.

1. Compute total organic mass
   Assume 75% of total mass is organic
   Organic mass $=0.75\times1000=750$ tn./d
2. Compute total dry organic mass
   Assume 20% of organic mass is moisture
   Dry organic mass $=750\times0.8=600$ tn./d.
3. Compute the biodegradable VS mass in MSW
   Assume 90% of dry organic mass is VS and
   70% of the VS mass is biodegradable
   BVS mass $=600\times0.9\times0.7=378$ tn./d
   BVS mass $=378$ tn./d $\times$ 1000 kg/tn. $=378\times10^3$ kg/d
4. Compute the amount of primary sludge required for co-digestion of the above BOF/MSW
   Assume that the TS of the primary sludge is 7% and that co-digestion occurs at 28% TS. Compute the amount of sludge needed considering the water already present in the biodegradable matter.
   Sludge required $=1170$ tn. $=1.17\times10^6$ kg/d

5. Compute the BVS mass in the primary sludge
   Assume that the VS of the primary sludge is about 70% and the biodegradable fraction of VS is about 60% of VS.
   BVS mass of primary sludge $=1.17\times10^6$ lb/d $\times0.07\times0.7\times0.6=34\times10^3$ kg/d
6. Compute the total BVS mass produced from the MSW and primary sludge
   Total BVS mass $=378\times10^3$ kg/d $+34\times10^3$ kg/d $=412\times10^3$ kg/d
7. Compute the biogas production rate
   Assume biogas production of about 0.86 m^3/kg BVS/d
   Total gas production rate $=(412\times10^3$ kg BVS/d$)\times(0.86$ m^3/kg BVS$)=354.32\times10^3$ m^3/d
8. Compute the energy value of the biogas volume produced
   Assume that the thermal energy value of biogas is about 16.23 MJ/m^3
   Thermal energy $=(354.32\times10^3$ m^3/d$)\times(16.23$ MJ/$m^3)=5.75\times10^6$ MJ/d
   Thermal energy $=(5.75\times10^6$ MJ/d$)\times365$ d/year $=2.098\times10^9$ MJ/yr
9. Compute the energy value relative to petroleum
   Assuming thermal energy value of one barrel of oil is about 6.218×10^6 MJ
   Energy equivalent in barrels of oil $=(2.098\times10^9$ MJ/yr$)/(6.218\times10^6$ MJ/barrel$)=337$ barrels of oil/yr

22.1.7.7.5 *Volume Reduction—Example Calculation*

Computation of volume reduction for BOF/MSW and WWTPS compared to a well compacted landfill

Case 1: Sludge cake fills the interstices of the compacted MSW in the landfill

1. Assumptions
   a. Input feedstock ratio (wet): 60% sludge to 40% BOF/MSW by weight
   b. Density of compacted MSW $=540$ kg/m^3
   c. Sludge TS $=5\%$
2. Compute the amount of sludge required for the anaerobic composting process
   For every unit volume of BOF/MSW (540 kg of MSW) added:
   $540\times(60/40)=810$ kg of sludge at 5% solids is required.
3. Compute the corresponding mass of sludge dewatered to 51% solids
   Dewatering the sludge to 51% solids would reduce the sludge weight to:
   810 kg $\times(51/5)\times0.01=82.6$ kg dewatered cake at 51% solids
4. Compute the combined weight of the MSW and sludge solids
   Combined weight of MSW and sludge cake $=540+82.6$ kg $=622.6$ kg
5. Compute the density of the combined MSW and sludge
   Assuming no change in MSW volume (1 m^3), combined density $=622.6$ kg/m^3

Case 2: Sludge cake remains unmixed from compacted MSW in landfill

1. Assumptions
   a. Input feedstock ratio (wet): 60% sludge to 40% BOF/MSW
   b. Density of compacted MSW $=540$ kg/m^3
   c. Sludge TS $=5\%$
   d. Density of sludge cake at 51% solids $=1240$ kg/m^3
2. Compute the amount of sludge required for the anaerobic composting process
   For every unit volume of BOF/MSW (540 kg of MSW) added:
   $540\times(60/40)=810$ kg of sludge at 5% solids is required.
3. Compute the corresponding mass of sludge dewatered to 51% solids
   Dewatering the sludge to 51% solids would reduce the sludge weight to:
   810 kg $\times(51/5)\times0.01=82.6$ kg dewatered cake at 51% solids

4. Compute the combined weight of the MSW and sludge solids
   Combined weight of MSW and sludge cake = 540 + 82.6 kg = 622.6 kg
5. Compute the combined volume of the MSW and sludge
   a. Volume of sludge = 82.62 kg/1240 kg/m^3 = 0.07 m^3
   b. Volume of MSW = 1 m^3
   c. Combined volume = 0.07 m^3 + 1.0 m^3 = 1.07 m^3
6. Compute the combined density of the MSW and sludge (unmixed)
   Combined density = 622.6 kg/1.07 m^3 = 583 kg/m^3

22.1.8 Utilization of By-Products from In-Vessel Anaerobic Digestion Process

The principal by-product of an anaerobic composting process is biogas and a stable digestate material with a low biodegradability and high lignin content. The digestate material can be processed further to produce a high-quality humus material for marketing. The characteristics of biogas and the humus material are presented in this section.

22.1.8.1 Characteristics of Biogas

The gas produced during anaerobic digestion of biodegradable organic material in a healthy fermentation system, called biogas, consists mainly of a mixture of methane (CH_4) and carbon dioxide (CO_2) with small amounts of other gases, including hydrogen sulfide (H_2S), hydrogen (H_2), nitrogen (N_2), and low-molecular weight hydrocarbons. Typically, digester gas has 50%–75% methane and 25%–50% carbon dioxide; the remaining gases are present in very small quantities. The composition of biogas, as obtained from various sources, is reported in Table 22.14. As reported in Table 22.14, a large variation exists in the composition of biogas, primarily due to differences in feedstocks and operating conditions.

Because biogas normally consists of a mixture of gases, biogas characteristics must be evaluated for each individual case. However, in many cases, the physical characteristics of the three main gas constituents, namely methane, carbon dioxide, and hydrogen sulfide, can be used to characterize biogas. Some physical and chemical characteristics of the principal gases found in biogas are presented in Table 22.15.

As a comparison, the weight of methane is roughly half that of air at 20°C (weight ratio = 1 m^3 of methane/1 m^3 of air = 0.716 kg/1.293 kg = 0.554).

Methane gas is not very soluble in water. Only three units of methane (by volume) can be dissolved in 100 units of water at 20°C and 1 atmosphere pressure. Methane is a very stable hydrocarbon compound and upon complete combustion it produces a blue flame and a large amount of heat. The complete combustion of 1 m^3 of methane can release 38 MJ or about 9500 kcal (1 kcal of heat will raise the temperature of 1 kg of water by 1°C). In comparison, a complete combustion of biogas yields a caloric value of about 20–26 MJ/m^3 (depending on the methane content), which represents a low fuel value compared with methane gas alone. In addition, biogas requires a pressure of about 34,450 kPa

TABLE 22.14 A Typical Composition of Biogas Produced from BOF/MSW

Constituent	Percent by Volume	
	Range	Typical[a]
Methane, CH_4	50–75	53.0
Carbon dioxide, CO_2	50–25	45.0
Hydrogen sulfide, H_2S	0.01–1.5	0.02
Hydrogen, H_2	Trace-3.5[b]	1.7
Nitrogen, N_2	Trace-8[b]	Trace
Other hydrocarbon	Trace-0.05	Trace

[a] Typical biogas composition from the biodegradable organic fraction of MSW.
[b] The hydrogen and nitrogen gases reported in this table are more commonly found in landfill gases.

TABLE 22.15 Physical Characteristics of Biogas

Characteristics	Unit	Average Value CO_2	CH_4	H_2S
Molecular weight	G	44.1	16.04	34.08
Vapor pressure at 21°C	kP	5719.0		1736.3
Specific volume at 21°C, 101 kP	M^3/kg	0.456	1.746	0.701
Boiling point at 101 kP	°C	−164.0	−161.61	−59.6
Freezing point at 101 kP	°C	−78.0	−182.5	−82.9
Specific gravity at 15°C (air = 1)		1.53	0.555	1.189
Density at 0 °C	kg/m^3	1.85	0.719	1.539
Critical temperature	°C	31.0	82.1	100.4
Critical pressure	kP	7386.0	4640.68	9007.0
Critical density	kg/m^3	0.468	0.162	0.349
Latent heat of vaporization at bp	kJ/kg	982.72	520.24	548.29
latent heat of fusion at mp	kJ/kg	189.0	58.74	69.78
Specific heat, Cp at 21°C, 101 kP	kJ/kg°C	0.83	2.206	1.06
Specific heat, Cv at 21°C, 101 kP	kJ/kg°C	0.64	1.688	0.803
Specific heat ratio, Cp/Cv		1.303	1.307	1.32
Thermal conductivity	W/m K	0.8323		0.0131
Flammable limits in air	% by volume		5.3–14	4.3–45
Solubility in water	kg/m^3	4.0	24.0	3.4
Viscosity	mPa s	0.0148	0.012	0.0116
Net heat of combustion at 25°C	MJ/m^3		36.71	
Gross heat of combustion at 25°C	MJ/m^3		37.97	
Ignition temperature	°C		650.0	
Octane rating			130.0	
Combustion equation			$CH_4 + 2O_2 \rightarrow CO_2 + 2H_2O$	$H_2S + 2O_2 \rightarrow SO_3 + H_2O$[a]

[a] Reaction is temperature dependent and the final product will be sulfuric acid ($SO_3 + H_2O \rightarrow H_2SO_4$).

(5,000 lb/in.^{2}) to liquefy it for storage. Therefore, biogas requires a larger storage volume for a given amount of energy than other fossil fuels.

22.1.8.2 Utilization of Biogas

To understand the potential use of biogas, it is important to gain a perspective on its energy potential by comparing it with more familiar uses of energy. The following are some examples of the use of 1 m^3 of biogas, at 60%–70% methane content, for common energy-consuming purposes (see Table 22.16):[39]

The potential commercial uses of biogas as a source of energy are summarized in Table 22.17. Some applications listed in Table 22.17 are presently practiced and others are in the process of research and development.

TABLE 22.16 Equivalent Uses of Biogas

Use	Equivalent Use
Cooking	Can cook three meals for a family of five to six
Lighting	Illumination equaling that of 60–100 W bulb for 6 h
Petroleum	Equivalent to 0.76 kg of petroleum
Car fuel	Can drive a 3 tn. truck 2.8 km
Motor power	Can run a 1 horse-power motor for 2 h
Electricity	Can generate 1.25 kW electricity

Source: Adapted from Barnett, A., Pyle, L., Subramanian, K. S., *Biogas Technology in the Third World: A Multidiciplinary Review*, International Development Research Center, Ottawa, Canada, 1978.

TABLE 22.17 Uses for Biogas Produced from the Biodegradation of the Organic Fraction of MSW

Applications	Comment
Fuel in an IC engine	The most common application of biogas, usually used to generate electricity. Modified engines available from Caterpillar, Cooper-Superior, and Waukesha. Modifications include corrosion resistance and proprietary designing. Specialized lubricating oils are necessary and must be changed more often. Power ratings are usually 5%–15% below natural gas ratings
Fuel in a gas turbine	Common application of biogas, used to produce electricity Modifications similar to those for IC engines are necessary. Biogas turbines are made by the Solar turbine division of Caterpillar. Power rating is 10–15% lower than for natural gas. Turbines must be checked often for corrosion or deposition. Water vapor should be removed from biogas before use
Boiler fuel	In moderate use, presently, requires little modification to present equipment. A mixture of biogas and natural gas may be used. Boilers appear to be less sensitive to gas contaminants than IC engines or gas turbines
Pipeline quality gas	Less common use. CO_2 and all contaminant gases, including water vapor, must be removed before gas will be accepted into a pipeline. Usually, not economical unless a large amount of gas is produced and a long-term contract for the sale is available
Fuel in a steam turbine	Presently in limited use. A large amount of gas is necessary before the system will be economical for the production of electricity. However, modifications are small
Fuel for space heating	Presently in limited use. Only minor modifications necessary to convert equipment which uses natural gas. Economical even on a small scale. Hampered only by unavailability of gas. Pipelines are not presently economical, so only used at production sites
Fuel for industrial heating	Presently in limited use. Can be used for lumber drying, kiln operations, and cement manufacturing. May be used in place of or with natural gas, using natural gas equipment
Fuel for fuel cell	Technology under development. These open fuel batteries can be designed to use biogas, or biogas can be converted to hydrogen in a catalytic pretreatment unit
Compressed vehicle fuel	Technology under development. Gas must be purified to near pipeline quality
Convert to methanol	Technology under development. Presently too costly
Synfuel or chemical feedstock	Technology under development. Liquid fuels and acetic acid production is being investigated

22.1.8.3 Characteristics of Digested Biosolids (Humus Material)

The physical, chemical, and biological characteristics of the humus produced from an anaerobic composting using the sorted biodegradable organic fraction of MSW are presented in this section. The data reported in this section were obtained from the UC Davis pilot demonstration project.[35,40]

TABLE 22.18 Physical Characteristics of Humus Produced from BOF/MSW by the Anaerobic Composting Process

Item	Unit	Value or Description
Bulk density	kg/m^3	560
Color		Dark brown
Moisture content	%	35
Odor		No offensive odor detected
Particle Size Distribution		
8 (2.362 mm)[a]	% of TM[b]	11.9
20 (0.833 mm)	% of TM	28.9
40 (0.351 mm)	% of TM	25.4
80 (0.175 mm)	% of TM	21.3
100 (0.147 mm)	% of TM	7.8
200 (0.074 mm)	% of TM	4.3
Pan	% of TM	0.4

[a] Sieve number (size, mm).
[b] TM = total mass (sample at 65% total solids).
Source: From Kayhanian, M. and Tchobanoglous, G., *Environmental Technology*, 14, 815–829, 1993.

The physical characteristics of the humus which are of interest are bulk density, color, moisture content, odor, and particle size distribution. The physical characteristics of the humus are summarized in Table 22.18. The chemical characteristics of the humus which are of interest can be determined by ultimate analysis, metal analysis, fiber analysis, nutrient analysis, energy content, and other tests. The chemical characteristics of the humus are summarized in Table 22.19. The biological characteristics of the humus which are of interest are the presence and the concentration of pathogenic bacterial, biodegradability, and phytotoxicity. The biological characteristics of the humus are summarized in Table 22.20.

22.1.8.4 Utilization of Humus Material

Representative applications for the humus produced from anaerobic digestion processes are summarized in Table 22.21. The most effective use of the humus material is as a soil amendment. Alternatively, because the humus is combustible, it appears that it can be fired directly in a boiler, when mixed with other fuels, or palletized for use as a fuel source. The application of the humus as a fuel source has been further studied and readers are referred to Jenkins et al.[41]

It is important to note that, depending on the final use, further aerobic composting of the digestate may be necessary to produce humus with no phytotoxic affect to be used as a soil amendment. For other applications specified in Table 22.21, no further stabilization may be needed.

22.1.9 Commercial-Scale In-Vessel Anaerobic Digestion Technologies

In the U.S., initial efforts to commercialize anaerobic digestion technology for converting biomass into energy and other products have been conducted primarily by livestock enterprises. The barrier to commercialization of anaerobic technology in the U.S. has been described by Lusk.[21] In addition, the commercial firms interested in the technology, the economics of the process, the statutes which apply, and the future prospects for application in the U.S. are identified and discussed. In general, the biggest barriers up to now have been financial rather than technical in nature. With rising energy costs and examples of successful anaerobic digester operation, more interest has been generated recently in these systems. Most large-scale anaerobic digestion applications for methane recovery in the U.S. are related to animal waste and wastewater treatment.[20,21] For instance, two large-scale farm digesters have been in operation in California since 1980.[21]

TABLE 22.19 Chemical Characteristics of the Humus Produced from BOF/MSW by the Anaerobic Composting Process

Chemical Analyses	Unit[a]	Range	Typical
Ultimate Analysis			
Carbon, C	%	30–35	32.4
Hydrogen, H	%	3.5–4	3.8
Chlorine, Cl	%	0.05–0.4	0.30
Nitrogen, N	%	1–2	1.9
Oxygen, O	%	30–35	31.4
Residue	%	25–35	30
Sulfur, S	%	0.1–0.4	0.25
Metal Analysis			
Aluminum, Al	mg/kg	30–194	54
Argon, Ar	mg/kg	<1–2	0.13
Arsenic, As	mg/kg	<1–3	1.04
Boron, B	mg/kg	12–64	18
Cadmium, Cd	mg/kg	<1–5	1.55
Calcium, Ca	%	0.8–1.7	1.08
Chromium, Cr	mg/kg	5–35	13
Cobalt, Co	mg/kg	<1–1	0.5
Copper, Cu	mg/kg	18–248	30
Iron, Fe	mg/kg	100–710	170
Lead, Pb	mg/kg	5–43	7
Magnesium, Mg	%	0.3–0.5	0.34
Manganese, Mn	mg/kg	120–175	133
Molybdenum, Mo	mg/kg	1–20	6
Nickel, Ni	mg/kg	18–186	28
Selenium, Se	mg/kg	<1	<1
Silicon, Si	mg/kg	0.1–4	0.21
Sodium, Na	%	0.1–0.3	0.3
Tungsten, W	mg/kg	<1–10	0.3
Zinc, Zn	mg/kg	98–376	176
Fiber Analysis			
Cellulose	%		35.3
Hemicellulose	%		3.9
Lipid	%		1.45
Protein	%		11.9
Lignin	%		26.5
Nutrient Analysis			
Nitrogen, N	%	1–2	1.9
Phosphorus, P	%	0.1–0.5	0.23
Phosphate, PO_4-P	mg/kg	50–200	170
Potassium, K	%	0.3–1	0.73
Sulfate, SO_4-S	mg/kg	300–800	547
Energy Content, HHV	MJ/kg	13–15	14.8
Others			
Cation exchange capacity, CEC meq/100 g dry		20–100	30
Electrical conductivity, EC	millimho/cm	5–15	9.4
PH		7–8.5	8.2

[a] % and mg/kg are based on dry mass.

Source: From Kayhanian, M. and Tchobanoglous, G., *Environmental Technology*, 14, 815–829, 1993.

A number of studies have been performed in the U.S. to utilize the anaerobic digestion process for methane gas recovery from solid waste materials.[13,23,31,35,42–46] Unfortunately, none of these projects have commercialized or implemented at full scale. Western Europe has been a leader in developing new solutions for managing MSW.[47] Germany and France, for example, have large-scale composting

TABLE 22.20 Biological Characteristics of the Humus Produced from the BOF/MSW by the Anaerobic Composting Process

Item	Unit	Value
Bacterial Concentration		
Total coliform	MPN/100 ml	Not detected[a]
Fecal coliform	MPN/100 ml	Not detected
Streptococcus and Enterococcus	MPN/100 ml	Not detected
Residual Biodegradability		
Biodegradable fraction	%VS	8.8[b]
Phytotoxicity		
Seed germinated with 100% leachate concentration	%	0[c]
Seed germinated with 20% leachate concentration	%	78
Seed germinated with 15% leachate concentration	%	95

[a] Detection limit is 0–6 organisms/10 ml at 95% confidence level.

[b] Biodegradable fraction (BF) is computed based on the lignin content (LC) presented in Table 22.19, using Equation 22.1: $BF = 0.83 - (0.028) \times LC$.

[c] Percent seed germinated is computed based on the ratio of seed germinated at each dilution, compared to the control.

Source: From Kayhanian, M. and Tchobanoglous, G., *Environmental Technology*, 14, 815–829, 1993.

plants in operation, and many of the recent innovations in the waste-to-energy industry have come from European countries. In 2005, De Baere, reviewed the state-of-the-art application of anaerobic digestion of MSW in Europe. Presently, there are 124 full-scale anaerobic digestion plants have been constructed in Europe by 27 suppliers.[48] Only three have constructed more than 15 plants based on the same concept. The suppliers include Dranco, Valorga, and Kompgas.

Two commercialized systems, the Dranco and Valorga, have been in operation at full scale in Europe since the late 1980s. The Kompgas process has been applied since 1990. All three of these European technologies employ high-solids digestion operated at TS concentrations of 20%–35%, thus achieving higher rates of gas production than have been reported from the conventional, low-solids anaerobic digestion of wastewater sludge or other natural organic materials. These three anaerobic digestion systems are the only commercial operations in the world that have performance data and a track

TABLE 22.21 Uses for Humus Material Produced from Anaerobic Digestate

Applications	Remarks
Use as soil amendment	The humus must be odorless, low in heavy metals, and free of pathogens. In addition, the amount of humus applied to the topsoil may be limited by the phytotoxic characteristics of the humus
Use as turf for sod production	Similar restrictions apply as specified for use of humus as a soil amendment
Use as topsoil for erosion control	If humus is to be used for the control of agricultural soil erosion, it must be of similar quality to that specified for soil amendments. For other erosion control applications, specifications may be less restrictive
Use as topsoil for landfills cover	Moisture content should be greater than about 35% to avoid wind erosion
Use as marsh restoration	Humus can be used to build up the organic material required for the growth of marsh plants in marsh restoration projects. A higher degree of humus stabilization may be required for this application
Use as absorbent agent for control and Humus may be used as an absorbent to limit	
Movement of hazardous wastes	Movement of liquid hazardous wastes
Use as fuel or fuel blend in power plants	Major characteristics of concern include: moisture content, foreign matter, heating value, and environmental impact on air quality

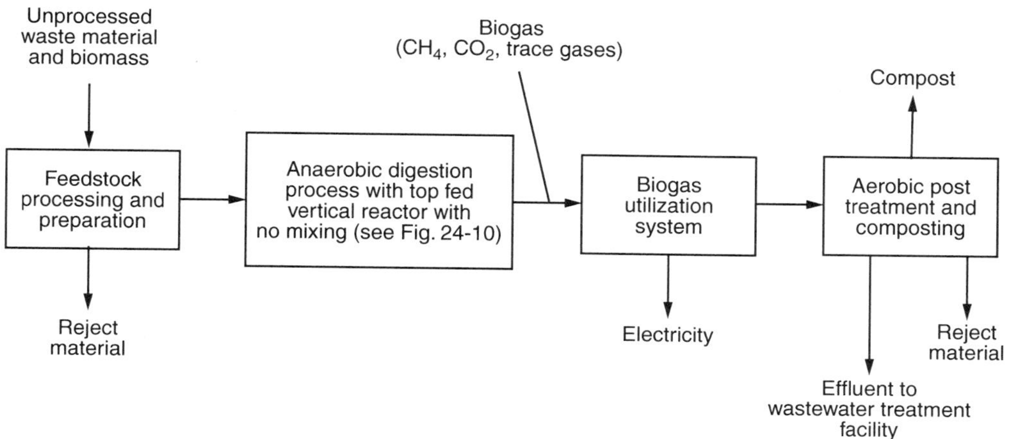

FIGURE 22.9 Schematic flow diagram of the DRANCO process.

record for over 15 years. The operational characteristics of these three systems are reviewed briefly below.

22.1.9.1 Dranco System

The Dranco (Dry ANaerobic COmposting) process (see Figure 22.9) was developed in Ghent, Belgium for the high-solids anaerobic digestion of refuse derived fuel (RDF). The RDF is prepared by completely sorting incoming urban solid waste. The Dranco system[49–51] is capable of removing most metals, glass, plastic, stone, recoverable paper, and other non-biodegradable items.

The anaerobic treatment takes place in a vertical fermenter (see Figure 22.10) for a period of 12–18 days, followed by a post-fermenter for a retention time of 2–3 days. The overall digestion time is, therefore, 14–21 days. The feed is mixed with recycled digester effluent and supernatant, and then pumped into the top of the reactor at a solids concentration of 355%–40%. Solids move downward in the digester in a plug-flow manner, without mechanical mixing. The digester operates under mesophilic (35°C) conditions with a VS reduction of about 55%. Presently, nine full-scale plants are operational in Europe and the capacity of each plant varies from 10,000 to 50,000 metric ton of MSW per year.

The solids are dewatered from about 30% to around 70% solids using a filter press and are dried further and marketed as compost, while the supernatant is used to dilute incoming feed. Additional information on the Dranco system can be obtained from the following web address: http://www.ows.be.

22.1.9.2 Valorga System

Valorga, a French company, was the first to use anaerobic technology for the treatment of MSW.[52,53] The complete Valorga system (see Figure 22.11) can be divided into five subunits: (1) preparation, (2) methanization, (3) biogas treatment, (4) combustion of combustible refuse, and (5) refining. The Valorga digester is distinguished by its unique design and is operated at a solids concentration of 35% under mesophilic (35°C) conditions with a retention time of 15 days. A typical Valorga reactor capacity is 500 m^3, and it can treat from 20 to 28 tn. of RDF per day. The VS removal rate is around 50% and the average gas production is about 140 m^3 biogas per ton of feed.

Two unique features of the Valorga process are: (1) the loading and (2) mixing mechanisms (see Figure 22.12). A piston pump is used to load feedstock into the system, supplying the digestion system with feed on a continuous basis. A control isolation valve system injects sufficient pressure to transfer the

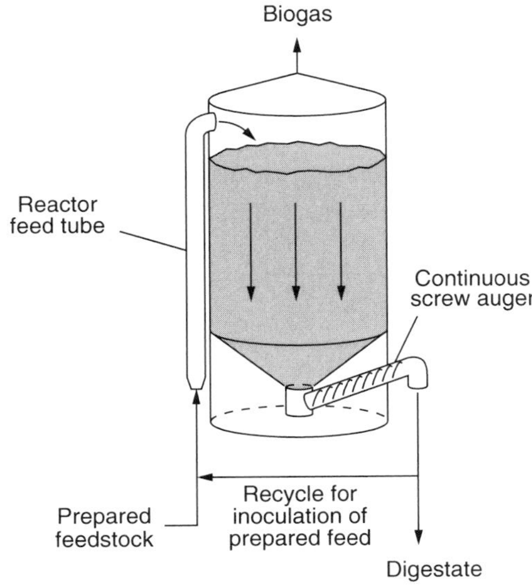

FIGURE 22.10 Schematic of the DRANCO anaerobic reactor. (Adapted from http://www.ows.be).

feedstock into the digester; at the same time, a similar quantity of digestate is extracted from the outlet. Mixing takes place through the programmed injection of pressurized biogas into a different section of the system.

Valorga has the most developed MSW anaerobic digestion technology in the world and over 15 full-scale plants are operational in various countries in Europe.[48] One large-scale Valorga system has been in operation in France since 1987. Three Valorga systems with annual capacity of over 200,000 metric ton became operational in various cities in Spain since 1999. Additional information on the Valorga system can be obtained from the following web address: http://www.valorgainternational.fr/index_en.php.

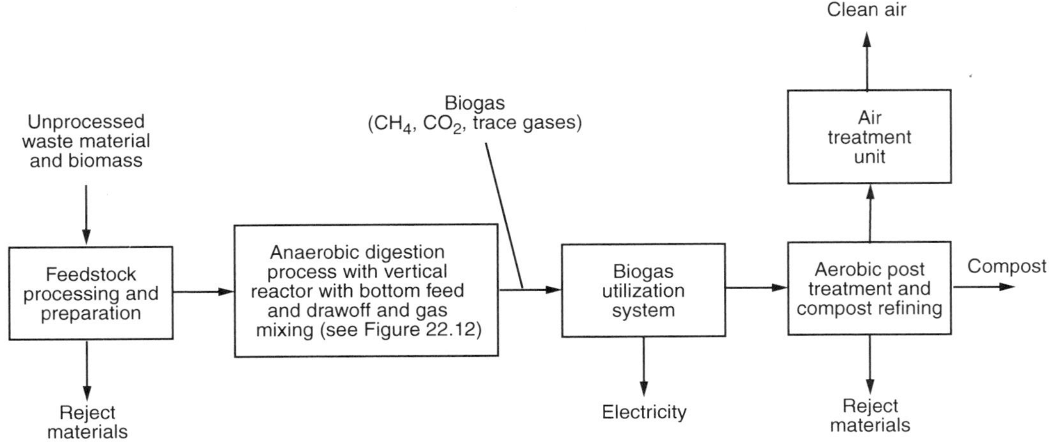

FIGURE 22.11 Schematic flow diagram of the VALORGA process.

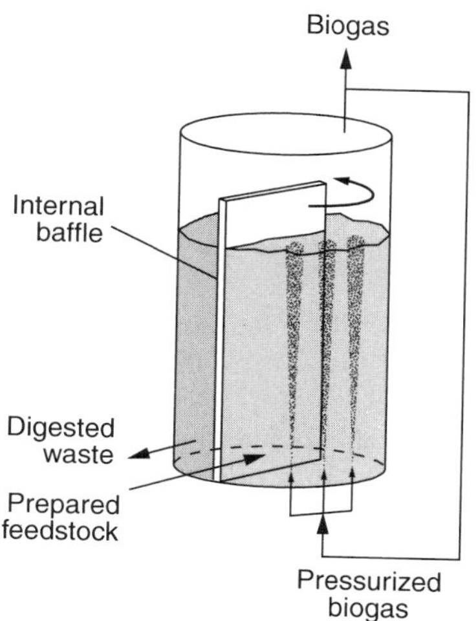

FIGURE 22.12 Schematic of the Valorga anaerobic reactor. (Adapted from http://www.valorgainternational.fr/index_en.php).

22.1.9.3 Kompogas System

Kompogas technology was developed in Switzerland and is generally composed of four major components (see Figure 22.13): (1) feedstock preparation, (2) digestion process, (3) biogas processing and utilization, and (4) digestate processing and compost production. Most Kompogas plants operate by using the source-separated mixture of easily biodegradable kitchen, food, and industrial wastes combined with wastes from gardens. To prepare the digester feed, a portion of the digestion slurry is mixed with the incoming biodegradable wastes.

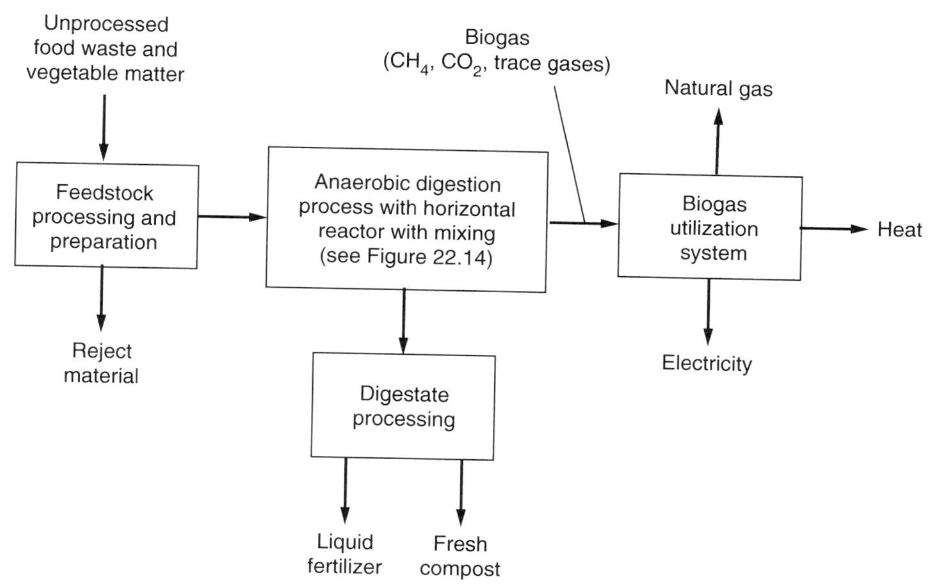

FIGURE 22.13 Schematic flow diagram of the Kompogas process.

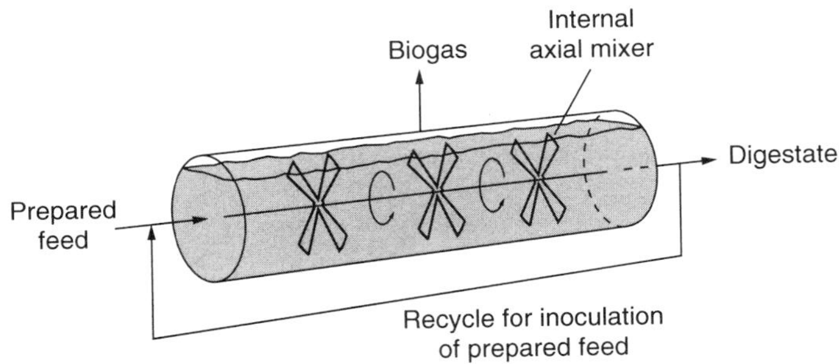

FIGURE 22.14 Schematic of the Kompogas anaerobic reactor. (Adapted from http://www.kompogas.com).

The Kompogas digester consists of a horizontal, cylindrical plug-flow reactor that operates under thermophilic temperature and solid content of about 20%. The reactor content is gently mixed along the longitudinal axis. A simplified diagram of Kompogas digester is shown in Figure 22.14. The horizontal design, with an agitation only right-angled to the flow direction, guarantees a good plug-flow behavior. The horizontal reactor design seems to be advantageous regarding clogging and sedimentation compared with vertical reactor designs.[54] With a few exceptions, nearly all Kompogas digesters are made out of steel.

At present, 22 large-scale Kompogas plants are in operation in Europe and elsewhere. Nine of these plants are located in Switzerland and the remaining thirteen plants are located in Japan, Germany, Denmark, Spain, and Astoria. Additional information on the Kompogas system can be obtained from the following web address: http://www.kompogas.com.

References

1. Chandler, J. A., Jewell, W. J., Gossett, J. M., Vansoest, P. J., and Robertson, B. J. 1980. Predicting methane fermentation biodegradability, *Biotechnology and Bioengineering Symposium,* 10, 93–107.
2. Kayhanian, M. 1995. Biodegradability of the organic fraction of municipal solid waste in a high-solids anaerobic digester. *Waste Management and Research,* 13, 2, 123–136.
3. Jewell, W. J., Kang, H., Herndon, F. G., Richards, B. K., and White, T. E. 1987. Engineering design considerations for methane fermentation of energy crops. pp. 105–106, Gas Research Institute.
4. Robertson, J. B. and Van Soest, P. J. 1981. The detergent system of analysis and its application to human foods, In *The Analysis of Dietary Fiber in Food,* W.P.T. James and O. Theander, eds., pp. 123–158. Marcel Dekker Inc., New York.
5. Van Soest, P. J. and Robertson, J. B. 1986. *Analysis of Forages and Fibrous Foods.* Cornel University, Itheca, NY.
6. Smith, P. H., Bordeaux, F. M., Wilkie, A., Yang, G., Boone, D., Mah, R. A., Chynoweth, D., and Jerger, D. 1988. Microbial aspects of biogas production, In *Methane from Biomass: A Systems Approach,* W. Smith and J.R. Frank, eds., Elsevier Applied Science, New York.
7. Richards, K. B., Cummings, R. J., Jewell, W. J., and Herenden, F. G. 1991. High-solid anaerobic methane fermentation of sorghum and cellulose. *Biomass and Bioenergy,* 1, 2, 47–53.
8. Richards, K. B., Cummings, R. J., White, E. T., and Jewell, W. J. 1991. Methods for kinetic analysis of methane fermentation in high solids biomass digesters. *Biomass and Bioenergy,* 1, 2, 65–73.
9. Mindermann, W. 1993. Conversion of nitrogen to ammonia in thermophilic anaerobic batch digestion of the organic fraction of municipal solid waste, M.S. thesis, University of California at Davis, Davis, CA.
10. Kayhanian, M. and Tchobanoglous, G. 1992. Computation of C/N ratios for various organic fractions. *BioCycle,* 33, 5, 42–45.

11. Kayhanian, M. and Tchobanoglous, G. 1993. Innovative two-stage process for the recovery of energy and compost from the organic fraction of municipal solid waste. *Water Science and Technology*, 27, 2, 133–143.
12. Rivard, J. C., Vinzant, T. B., Adney, W. S., and Grohmann, K. 1987. Nutrient requirements for aerobic and anaerobic digestion of processed municipal solid waste. *Journal of Environmental Health*, 5, 96–99.
13. Rivard, J. C., Himmel, M. E., Vinzant, T. B., Adney, W. S., Wyman, C. E., and Krohmann, K. 1987. *Anaerobic High Solids Fermentation of Processed Municipal Solid Wastes for the Production of Methane.* Biotechnology Research Branch, Solar Energy Research Institute, Golden, CO.
14. Speece, R. E. and McCarty, P. L. 1964. Nutrient requirements and biological solids accumulation in anaerobic digestion, In *Advances in Water Pollution Research*, W.W. Eckenfelder, ed., Vol. 2, pp. 305–322. Pergamon Press Ltd., Oxford, England.
15. Speece, R. E. and Parkin, G. F. 1985. Nutrient requirements for anaerobic digestion, In *Biotechnological Advances in Processing Municipal Waste for Fuels and Chemicals*, A.A. Antonopoulos, ed., , pp. 195–221. Argonne National Laboratory, Argonne, IL.
16. Kayhanian, M. and Rich, D. 1995. Pilot-scale high-solids anaerobic digestion of municipal solid waste with an emphasis on nutrient requirements. *Biomass and Bioenergy*, 8, 4, 433–444.
17. Zeikus, J. G. 1977. The biology of methanogenic bacteria. *Bacterial Review*, 41, 514–541.
18. Holland, K. T., Knapp, J. S., and Shoesmith, J. G. 1987. *Anaerobic Bacteria.* Chapman and Hall, New York.
19. Tchobanoglous, G., Theisen, H., and Vigil, S. A. 1993. *Integrated Solid Waste Management—Engineering Principles and Management Issues.* McGraw-Hill Book Company, New York.
20. Tchobanoglous, G., Burton, F. L., and Stensel, H. D. 2003. *Wastewater Engineering: Treatment, Disposal, Reuse. 4th Ed.* Metcalf & Eddy, Inc., McGraw-Hill Book Company, New York.
21. Lusk, P. 1994. Methane recovery from animal manures: A current opportunities casebook, Report prepared for the National Renewable Energy Laboratory, Golden, CO.
22. Markel, A. J. 1981. *Managing Livestock Wastes.* AVI Publishing Company, Inc., Westport, CN.
23. Diaz, L. F. and Trezek, G. T. 1977. Biogasification of a selected fraction of municipal solid wastes. *Compost Science*, 18, 2, 8–13.
24. Diaz, L. F., Savage, G. M., and Golueke, C. G. 1982. *Resource Recovery from Municipal Solid Wastes: Volume II Final Processing.* CRC Press, Boca Raton, FL.
25. Ghosh, S. and Klass, D. L. 1976. SNG from refuse and sewage sludge by the biogas process. *IGT Symposium on Clean Fuels from Biomass*, Sewage, Urban Refuse and Agricultural Wastes, Orlando, FL.
26. Klein, S. A. 1972. Anaerobic digestion of solid wastes. *Compost Science*, 13, 1, 6–13.
27. McFarland, J. M., Glassey, C. R., McGauhey, P. H., Brink, D. L., Klein, S. A., and Golueke, G. C. 1972. Coprehensive studies of solid waste management, Final Report, SERL Report No. 72-3, Berkeley, CA.
28. Pfeffer, J. T. and Liebman, J. C. 1974. Biological Conversion of organic refuse to methane, annual report NSF/RANN/SE/G1/39191/PR/75/2, Department of civil Engineering, University of Illiniois, Urbana, IL.
29. Babbitt, H. E., Leland, B. J., and Whitely, F. E. 1988. *The Biological Digestion of Garbage with Sewage Sludge.* p. 24, University of Illinois,.
30. Cecchi, F., Traverso, P. G., Mata-Alvarez, J., Clancy, J., and Zaror, C. 1988. State of the art of R&D in the anaerobic digestion process of municipal solid waste in europe. *Biomass*, 16, 257–284.
31. Isaacson, H. R., Pfeffer, J. P., Modij, A. J., and Geselbracht, J. J. 1987. RefCom-technical status, economic and market, In *Energy from Biomass and Wastes XI*, D.L. Klass, ed., pp. 1123–1163. Elsevier Applied Science Publishers and Institute of Gas Technology, Chicago, IL.
32. Poggi-Valardo, H. M. and Oleszkiewicz, J. A. 1992. Anaerobic co-composting of municipal solid waste and waste sludge at high total solids levels. *Environmental Technology*, 13, 96–105.
33. Kayhanian, M. and Rich, D. 1996. Sludge management using the biodegradable organic fraction of municipal solid waste as a primary substrate. *Water Environment Research*, 68, 2, 240–252.
34. Kayhanian, M. 1993. Anaerobic composting for MSW. *BioCycle*, 34, 5, 32–34.

35. Kayhanian, M., Lindenauer, K., Hardy, S., and Tchobanoglous, G. 1991. Two-stage process combines anaerobic and aerobic methods. *BioCycle*, 32, 3, 48–64.
36. Kayhanian, M. 1994. Performance of high-solids anaerobic digestion process under various ammonia concentrations. *Chemical Technology and Biotechnology*, 59, 5, 349–352.
37. Kayhanian, M. and Hardy, S. 1994. The impact of four design parameters on the performance of high-solids anaerobic digestion of municipal solid waste for fuel gas production. *Environmental Technology*, 15, 6, 557–567.
38. Kayhanian, M., Tchobanoglous, G., and Mata-Alvarez, J. 1996. Development of a mathematical model for the simulation of the biodegradation of municipal solid waste in high-solids anaerobic digestion process. *Chemical Technology and Biotechnology*, 66, 7, 312–322.
39. Barnett, A., Pyle, L., and Subramanian, K. S. 1978. *Biogas Technology in the Third World: A Multidiciplinary Review*. International Development Research Center, Ottawa, Canada.
40. Kayhanian, M. and Tchobanoglous, G. 1993. Characteristics of humus produced from the anaerobic composting process of the biodegradable organic fraction of municipal solid waste. *Environmental Technology*, 14, 9, 815–829.
41. Jenkins, B. M., Kayhanian, M., Baxter, L., and Salour, D. 1997. Combustion of residual biosolids from high-solids anaerobic digestion/aerobic composting process. *Biomass and Bioenergy*, 12, 5, 567–581.
42. Chynoweth, D. P., Earl, F. K., Bosch, G., and Lagrand, R. 1990. Biogasification of processed MSW. *Biocycle*, 31, 10, 50–51.
43. Ferrero, G. L., Feranti, M. P., and Naveau, H. eds. 1984. Anaerobic Digestion and Carbohydrate Hydrolysis of Waste, Elsevier Applied Science Publishers, New York.
44. Golueke, C. J. 1971. Comprehensive studies of solid waste management. Third Annual Report. University of California, Richmond Field Station, U.S. Environmental Protection Agency, Berkeley, CA.
45. Kispert, R. G., Sadek, S. E., and Wise, D. L. 1975. Fuel gas production from solid waste, *Final Report. NSF Contract C-827*, Dynatech Report No. 1258, Dynatech R/D Co., Cambridge, MA.
46. Pfeffer, J. P. 1987. Evaluation of the refcom proof-of-concept experiment, In *Energy from Biomass and Wastes X*, D.L. Glass, ed., pp. 1149–1171. Elsevier Applied Science Publishers and Institute of Gas Technology, Chicago, IL.
47. Mata-Alvarez, J., Cecchi, F., Pavan, P., and Fazzini, G. 1990. Performances of digesters treating the organic fraction of municipal solid waste differently sorted. *Biological Wastes*, 33, 181–199.
48. De Baere, L. 2005. Will anaerobic digestion of solid waste survive in the future?. In *Proceedings of 4th International Symposium on Anaerobic Digestion of Solid Waste*, Copenhagen, Denmark.
49. De Baere, L. 1984. High rate dry anaerobic composting process for the organic fraction of solid waste. In *7th Symposium on Biotechnology for Fuels and Chemicals*, Gatlinburg, TN.
50. De Baere, L., Verdonck, O., and Verstraete, W. 1985. High rate dry anaerobic composting process for the organic fraction of solid wastes. *Biotechnology Bioengineering Symposium*, 15, 321–332.
51. Six, W. and De Baere, L. 1992. Dry anaerobic conversion of municipal solid waste by means of the dranco process at brecht, belgium. In *The Proceedings of the International Symposium on Anaerobic Digestion of Solid Waste*, pp. 525–528. Venice, Italy.
52. Begouen, O., Thiebaut, E., Pavia, A., and Peillex, J. P. 1988. Thermophilic anaerobic digestion of municipal solid wastes by the VALORGA process, In *Proceedings of the Fifth International Symposium on Anaerobic Digestion*, May 22–26, Bologna, Italy, pp. 789–792.
53. Valorga. 1985. Waste Recovery as a Source of Methane and Fertilizer. In *2nd Annual International Symposium On Industrial Resource Management*, pp. 234–244. Philadelphia, PA.
54. Edelman, W. and Angeli, B. 2005. More than 12 years of experience with commercial anaerobic digestion of the organic fraction of municipal solid wastes in Switzerland. In *Proceedings of 4th International Symposium on Anaerobic Digestion of Solid Waste*, Copenhagen, Denmark.
55. Rich, D. and Kayhanian, M. 1994. Anaerobic co-composting yields quality humus. *BioCycle*, 35, 3, 82–87.

22.2 Power Generation

Robert C. Brown

22.2.1 Introduction

Although biomass encompasses many kinds of organic matter, fibrous plant material can be characterized as solid, carbonaceous fuel of high volatile content and heating value of about 18 MJ/kg.[1] Either direct combustion or thermal gasification can be used to transform this chemical energy into electric power. Direct combustion releases heat that can be used in Stirling engines or Rankine steam power cycles. Thermal gasification yields flammable gases suitable for firing in internal combustion engines, gas turbines, or fuel cells.

22.2.2 Direct Combustion

Combustion is the rapid oxidation of fuel to obtain energy in the form of heat. Since biomass fuels are primarily composed of carbon, hydrogen, and oxygen, the main oxidation products are carbon dioxide and water although fuel-bound nitrogen can be a source of significant nitrogen oxide emissions. Depending on the heating value and moisture content of the fuel, the amount of air used to burn the fuel, and the construction of the furnace, flame temperatures can exceed 1650°C.

Solid fuel combustion consists of four steps, which are illustrated in Figure 22.15: heating and drying, pyrolysis, flaming combustion, and char combustion.[2] Heating and drying of the fuel particle is normally not accompanied by chemical reaction. Water is driven from the fuel particle as the thermal front advances into the interior of the particle. As long as water remains, the temperature of the particle does not raise high enough to initiate pyrolysis, which is the second step in solid fuel combustion.

Pyrolysis is a complicated series of thermally driven chemical reactions that decompose organic compounds in the fuel. Pyrolysis proceeds at relatively low temperatures, which depend on the type of plant material. Hemicellulose begins to pyrolyze at temperatures between 150 and 300°C, cellulose pyrolyzes at 275°C–350°C and lignin pyrolysis is initiated between 250 and 500°C.[3]

The resulting decomposition yields a large variety of volatile organic and inorganic compounds; the types and amounts dependent on the fuel and the heating rate of the fuel. Pyrolysis products include

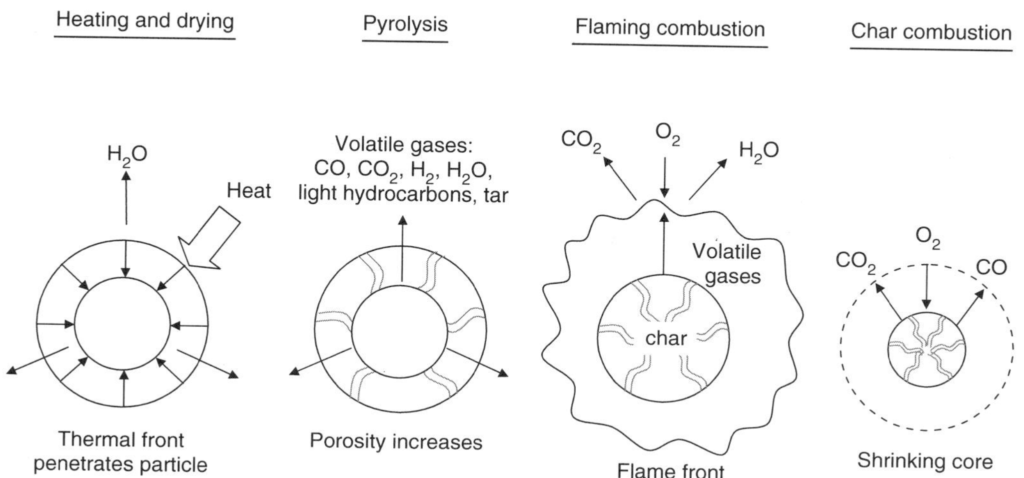

FIGURE 22.15 Processes of solid fuel combustion. (From Brown, R. C., *Biorenewable Resources: Engineering New Products from Agriculture*, Iowa State Press, Ames, IA, 2003.)

carbon monoxide (CO), carbon dioxide (CO_2), methane (CH_4), and high molecular weight compounds that condense to a tarry liquid if cooled before they are able to burn. Fine droplets of these condensable compounds represent much of the smoke associated with smoldering fires. Pyrolysis follows the thermal front through the particle, releasing volatile compounds and leaving behind pores that penetrate to the surface of the particle.

Pyrolysis is very rapid compared with the overall burning process and may be as short as a second for small particles of fuel but can extend to many minutes in wood logs. Although the net result of combustion is oxidation of fuel molecules and the release of heat, neither of these processes occurs to a significant extent during pyrolysis. Indeed, if pyrolysis is to proceed at all, heat must be added to the fuel. Oxygen is excluded from the pyrolysis zone by the large gaseous outflux of pyrolysis products from the surface of the fuel particle. Only after pyrolysis gases escape the particle and diffuse into the surrounding air, they are able to burn. There remains upon completion of pyrolysis a porous carbonaceous residue known as char.

Both the volatile gases and the char resulting from pyrolysis can be oxidized if sufficient oxygen is available to them. Oxidation of the volatile gases above the solid fuel results in flaming combustion. The ultimate products of volatile combustion are CO_2 and H_2O although a variety of intermediate chemical compounds can exist in the flame, including CO, condensable organic compounds, and long chains of carbon known as soot. Indeed, hot, glowing soot is responsible for the familiar orange color of wood fires.

Combustion intermediates will be consumed in the flame if sufficient temperature, turbulence, and time are allowed. High combustion temperature assures that chemical reactions will proceed at high rates. Turbulent or vigorous mixing of air with the fuel makes certain that every fuel molecule comes into contact with oxygen molecules. Long residence times for fuel in a combustor allow the fuel to be completely consumed. In the absence of good combustion conditions, a variety of noxious organic compounds can survive the combustion process including CO, soot, polycyclic aromatic hydrocarbons (PAH), and the particularly toxic families of chlorinated hydrocarbons known as furans and dioxins. In some cases, a poorly operated combustor can produce pollutants from relatively benign fuel molecules.

The next step in combustion of solid fuels is solid–gas reactions of char, also known as glowing combustion, familiar as red-hot embers in a fire. Char is primarily carbon with a small amount of mineral matter interspersed. Char oxidation is controlled by mass transfer of oxygen to the char surface rather than by chemical kinetics, which is very fast at the elevated temperatures of combustion. Depending on the porosity and reactivity of the char and the combustion temperature, oxygen may react with char at the surface of the particle or it may penetrate into the pores before oxidizing char inside the particle. The former situation results in a steadily shrinking core of char whereas the latter situation produces a constant diameter particle of increasing porosity. Both CO and CO_2 can form at or near the surface of burning char:[2]

$$C + \frac{1}{2}O_2 \rightarrow CO \tag{22.6}$$

$$CO + \frac{1}{2}O_2 \rightarrow CO_2 \tag{22.7}$$

These gases escape the immediate vicinity of the char particle where CO is oxidized to CO_2 if sufficient oxygen and temperature are available; otherwise, it appears in the flue gas as a pollutant.

22.2.3 Combustion Equipment

A combustor is a device that converts the chemical energy of fuels into high temperature exhaust gases. Heat from the high temperature gases can be employed in a variety of applications, including space heating, drying, and power generation. However, with the exception of kilns used by the cement industry, most solid fuel combustors today are designed to produce either low-pressure steam for process heat or high-pressure steam for power generation. Combustors integrated with steam-raising equipment are called boilers. In some boiler designs, distinct sections exist for combustion, high temperature heat transfer, and moderate temperature heat transfer: these are called the furnace, radiative, and convective sections of the boiler, respectively. In other designs, no clear separation between the processes of combustion and heat transfer exists.

Solid fuel combustors, illustrated in Figure 22.16, can generally be categorized as grate-fired systems, suspension burners, or fluidized beds.[4] Grate-fired systems were the first burner systems to be developed, evolving during the late nineteenth and early twentieth centuries into a variety of automated systems. The most common system is the spreader-stoker, consisting of a fuel feeder that mechanically or pneumatically flings fuel onto a moving grate where the fuel burns. Much of the ash falls off the end of the moving grate although some fly ash appears in the flue gas. Grate systems rarely achieve combustion efficiencies exceeding 90%.

Suspension burners were introduced in the 1920s, as a means of efficiently burning large quantities of coal pulverized to less than fifty micron particle sizes. Suspension burners suspend the fuel as fine powder in a stream of vertically rising air. The fuel burns in a fireball and radiates heat to tubes that contain water to be converted into steam. Suspension burners, also know as pulverized coal (PC) boilers, have dominated the U.S. power industry since World War II because of their high volumetric heat release rates and their ability to achieve combustion efficiencies often exceeding 99%. However, they are not well suited to burn coarse particles of biomass fuel and they are notorious generators of nitrogen oxides. Biomass is fed from a bunker through pulverizers designed to reduce fuel particle size enough to burn in suspension. The fuel particles are suspended in the primary airflow and fed to the furnace section of the boiler through burner ports where it burns as a rising fireball. Secondary air injected into the boiler helps to complete the combustion process. Heat is absorbed by steam tubes arrayed in banks of heat exchangers (waterwall, superheaters, and economizer) before exiting through a bag house designed to capture ash released from the fuel. Steam produced in the boiler is part of a Rankine power cycle.

Fluidized bed combustors are a recent innovation in boiler design. Air injected into the bottom of the boiler suspends a bed of sand or other granular refractory material producing a turbulent mixture of air and sand. The high rates of heat and mass transfer in this environment are ideal for efficiently burning a

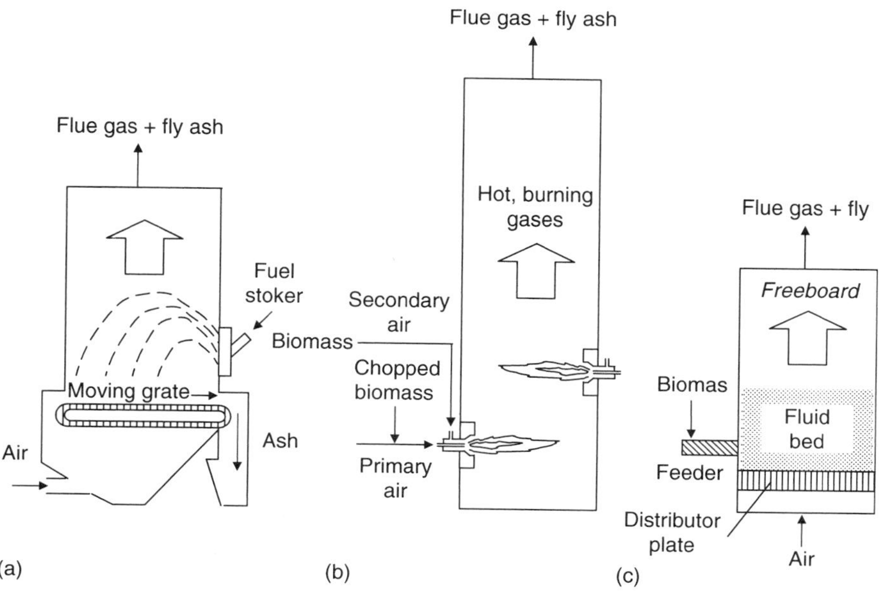

FIGURE 22.16 Common types of combustors: (a) grate-fired, (b) suspension, (c) fluidized bed. (From Brown, R. C., *Biorenewable Resources: Engineering New Products from Agriculture*, Iowa State Press, Ames, IA, 2003.)

variety of fuels. Furthermore, the large thermal mass of the sand bed allows the unit to be operated as low as 850°C, which lowers the emission of nitrogen compounds. A commercial market for fluidized bed boilers developed during the 1980s, especially for industrial applications.

Whole-tree burners have been proposed for electrical utility steam raising. The advantage of this approach is minimization of field processing and handling and the elimination of fuel chipping, which can save about 35% of the cost of harvesting and handling woody fuels.

A variety of biomass materials have proven suitable for direct combustion, including whole trees, wood chips, forestry residue, agricultural residue, pulp and paper refuse, food processing wastes, municipal solid waste, and straws and grasses. Wood and wood wastes dominate the biomass-to-power market in the U.S. and much of the rest of the world. However, a variety of agricultural wastes have been utilized through direct combustion in California, and Denmark has a number of commercial plants that burn straw.

Direct combustion has the advantage that it employs commercially well-developed technology. There are a number of vendors who supply turnkey systems and considerable operating experience and exists both in the U.S. and abroad. However, there are three prominent disadvantages with direct firing. These include penalties associated with burning high moisture fuels, agglomeration, and ash fouling due to alkali compounds in biomass, and relatively low thermodynamic efficiencies for steam power plants of the size appropriate to biomass power.

The moisture in biomass degrades boiler performance for two reasons. First, the energy required to vaporize fuel moisture can only be recovered if exhaust gases are cooled sufficiently to condense the water vapor in this gas. Although this procedure would result in very high thermal efficiencies for a boiler burning even very high moisture fuels, it is not commonly done in practice because the condensate is often corrosive to boiler tubes. Accordingly, most direct-combustion systems are penalized by moisture in the fuel. Second, high moisture fuels simply do not burn well because the process of fuel drying suppresses fuel temperatures below that required for ignition. Water contents exceeding 30% are unacceptable in most boilers for this reason. However, fluidized bed combustors, because of the enormous thermal mass associated with the hot bed material, can accept biomass with moisture content as high as 50% although even fluidized beds are penalized in thermal efficiency by the presence of this moisture. Depending on moisture content, as much as 15% of the heating value of biomass is required to dry the fuel. Obviously, field drying of biomass is desired to reduce both transportation costs and heating penalties within the boiler.

Alkali in biomass fuels presents a difficult problem for direct combustion systems. Compounds of alkali metals, such as potassium and sodium salts, are common in rapidly growing plant tissues. Annual biomass crops contain large quantities of alkali while the old-growth parts of perennial biomass contain relatively small quantities of alkali. These alkali compounds appear as oxides in the residue left after combustion of volatiles and char. Alkali vapors may combine with sulfur and silica to form low-melting point compounds.[5] These sticky compounds bind ash particles to fuel grates and heat exchanger surfaces. Boiler performance degrades as airflow and heat transfer are restricted by ash deposits. Boilers that fire straw, which contain both high silica and alkali, have experienced serious problems in slagging and fouling, as well as high temperature corrosion unless steam temperatures are kept relatively low (less than 500°C).

As an alternative to completely replace coal with biomass fuel in a boiler, mixtures of biomass and coal can be burned together in a process known as co-firing.[6] Co-firing offers several advantages for industrial boilers. Industries that generate large quantities of biomass wastes, such as lumber mills or pulp and paper companies, can use co-firing as an alternative to costly landfilling of wastes. Federal regulations also make co-firing attractive. The New Source Performance Standards, which limits particulate emissions from large coal-fired industrial boilers to 0.05 lb/MM Btu, doubles this allowance in co-fired boilers in which the capacity factor for biomass exceeds 10%. Adopting co-firing is a good option for companies that are slightly out of compliance with their coal-fired boilers. Similarly, a relatively inexpensive method for a company to reduce sulfur emissions from its boilers is to co-fire with biomass, which contains much less sulfur than coal. Co-firing capability also provides fuel flexibility and reduces ash-fouling problems associated with using only biomass as fuel.

The principal disadvantages of co-firing relate to the characteristics of biomass fuels. Because of the lower energy density and higher moisture content of biomass, the steam generating capacity of co-fired boilers is often reduced. Also, the elemental composition of fly ash from biomass is distinct from that of coal fly ash. Utilities are concerned that comingled biomass and coal ash will not meet the American Society for Testing and Materials (ASTM) definition of fly ash that is acceptable for concrete admixtures, thus eliminating an important market for this combustion by-product.

It is generally recommended that total fuel (single or mix) alkali content be limited to less than 0.17–0.34 kg/GJ (0.4–0.8 lb/MM Btu), which translates to only 5%–15% co-firing of biomass with coal. Also, furnace temperatures should be kept below 980°C to help to prevent the buildup of alkali-containing mineral combinations known as eutectics. At higher temperatures, molten eutectics bind dirt and other particulates to form slag and fouling deposits.

The best wood-fired power plants, which are typically 20–100 MW in capacity, have heat rates exceeding 12,500 Btu/kWh. In contrast, large, coal-fired power plants have heat rates of only 10,250 Btu/kWh. The relatively low thermodynamic efficiency of steam power plants at the sizes of relevance to biomass power systems may ultimately limit the use of direct combustion to convert biomass fuels to useful energy.

22.2.4 Thermal Gasification

Gasification is the high temperature (750°C–850°C) conversion of solid, carbonaceous fuels into flammable gas mixtures, sometimes known as producer gas, consisting of carbon monoxide (CO), hydrogen (H_2), methane (CH_4), nitrogen (N_2), carbon dioxide (CO_2), and smaller quantities of higher hydrocarbons. The overall process is endothermic and requires either the simultaneous burning of part of the fuel or the delivery of an external source of heat to drive the process.[7]

Gasification was placed into commercial practice as early as 1812, when coal was converted to gas for illumination (known as manufactured gas or town gas) in England. This technology was widely adopted in industrialized nations and was employed in the U.S. as late as the 1950s, when interstate pipelines made inexpensive supplies of natural gas available.[8] Some places, like China, still manufacture gas from coal.

The high volatile content of biomass (70–90 wt%) compared with coal (typically 30–40 wt%) and the high reactivity of its char make biomass an ideal gasification fuel. However, issues of cost and convenience of biomass gasification has limited its applications to special situations and niche markets. For example, in response to petroleum shortages during World War II, over one million small-scale wood gasifiers were built to supply low enthalpy gas for automobiles and steam boilers.[7] These were abandoned soon after the war.

Not only can producer gas be used for generation of heat and power, it can serve as feedstock for production of liquid fuels and chemicals. Because of this flexibility of application, gasification has been proposed as the basis for "energy refineries" that would provide a variety of energy and chemical products, including electricity and transportation fuels.

22.2.4.1 Fundamentals of Gasification

Figure 22.17 illustrates the several steps of gasification: heating and drying, pyrolysis, solid–gas reactions that consume char, and gas-phase reactions that adjust the final chemical composition of the producer gas.[9] Drying and pyrolysis are similar to the corresponding processes for direct combustion described in the previous section. Pyrolysis begins between 300 and 400°C and produces the intermediate products of char, gases (mainly CO, CO_2, H_2, and light hydrocarbons), and condensable vapor (including water, methanol, acetic acid, acetone, and heavy hydrocarbons). The distribution of these products depends on chemical composition of the fuel and the heating rate, and temperature achieved in the reactor. However, the total pyrolysis yield of pyrolysis products and the amount of char residue can be roughly estimated from the proximate analysis of the fuel. The fuel's volatile matter roughly corresponds to the pyrolysis yield, while the combination of fixed carbon and ash content can be used to estimate the char yield.

Heating and drying are endothermic processes, requiring a source of heat to drive them. This heat can be supplied by an external source in a process called indirectly heated gasification. More typically, a small amount of air or oxygen (typically not more than 25% of the stoichiometric requirement for complete combustion of the fuel) is admitted for the purpose of partial oxidation, which releases sufficient heat for drying and pyrolysis as well as for the subsequent endothermic chemical reactions described below.

The third step of gasification is solid–gas reactions. These reactions convert solid carbon into gaseous CO, H_2, and CH_4:[10]

$$\text{Carbon–oxygen reaction}: \quad C + \frac{1}{2}O_2 \leftrightarrow CO \quad \Delta H_R = -110.5 \text{ MJ/kmol} \tag{22.8}$$

$$\text{Boudouard reaction}: \quad C + CO_2 \leftrightarrow 2CO \quad \Delta H_R = 172.4 \text{ MJ/kmol} \tag{22.9}$$

$$\text{Carbon–water reaction}: \quad C + H_2O \leftrightarrow H_2 + CO \quad \Delta H_R = 131.3 \text{ MJ/kmol} \tag{22.10}$$

$$\text{Hydrogenation reaction}: \quad C + 2H_2 \leftrightarrow CH_4 \quad \Delta H_R = -74.8 \text{ MJ/kmol} \tag{25.11}$$

The first of these, known as the carbon–oxygen reaction, is strongly exothermic and is important in supplying the energy requirements for drying, pyrolysis, and endothermic solid–gas reactions. The hydrogenation reaction also contributes to the energy requirements of the gasifier, although significantly more char reacts with oxygen than hydrogen in the typical air-blown gasifier.

The fourth step of gasification is gas-phase reactions, which determine the final mix of gaseous products:[10]

$$\text{Water–gas shift reaction}: \quad CO + H_2O \leftrightarrow H_2 + CO_2 \quad \Delta H_R = -41.1 \text{ MJ/kmol} \tag{22.12}$$

$$\text{Methanation}: \quad CO + 3H_2 \leftrightarrow CH_4 + H_2O \quad \Delta H_R = -206.1 \text{ MJ/kmol} \tag{22.13}$$

The final gas composition is strongly dependent on the amount of oxygen and steam admitted to the reactor as well as the time and temperature of reaction. For sufficiently long reaction times, chemical equilibrium is attained and the products are essentially limited to the light gases CO, CO_2, H_2, and CH_4 (and nitrogen if air was used as a source of oxygen). Analysis of the chemical thermodynamics of these six gasification reactions reveals that low temperatures and high pressures favor the formation of CH_4, whereas high temperatures and low pressures favor the formation of H_2 and CO.

Often gasifier temperatures and reaction times are not sufficient to attain chemical equilibrium and the producer gas contains various amounts of light hydrocarbons such as C_2H_2 and C_2H_4 as well as up to 10 wt% heavy hydrocarbons that condense to a black, viscous liquid known as "tar."[12] This latter product is undesirable as it can block valves and filters and interferes with downstream conversion processes. Steam injection and addition of catalysts to the reactor are sometimes used to shift products toward lower-molecular weight compounds.

22.2.4.2 Gasification Equipment

Gasifiers are generally classified according to the method of contacting fuel and gas. The four classes of gasifiers are updraft (countercurrent), downdraft (concurrent), fluidized bed, and entrained flow.[12] These are illustrated in Figure 22.18 while their performance characteristics are summarized in Table 22.22.

Updraft gasifiers, illustrated in Figure 22.18a, are the simplest as well as the first type of gasifier developed. They were a natural evolution from charcoal kilns, which yield smoky yet flammable gas as a waste product, and blast furnaces, which generate product gas that reduces ore to metallic iron. Updraft gasifiers are little more than grate furnaces with chipped fuel admitted from above and insufficient air for complete combustion entering from below. Above the grate, where air first contacts the fuel, combustion

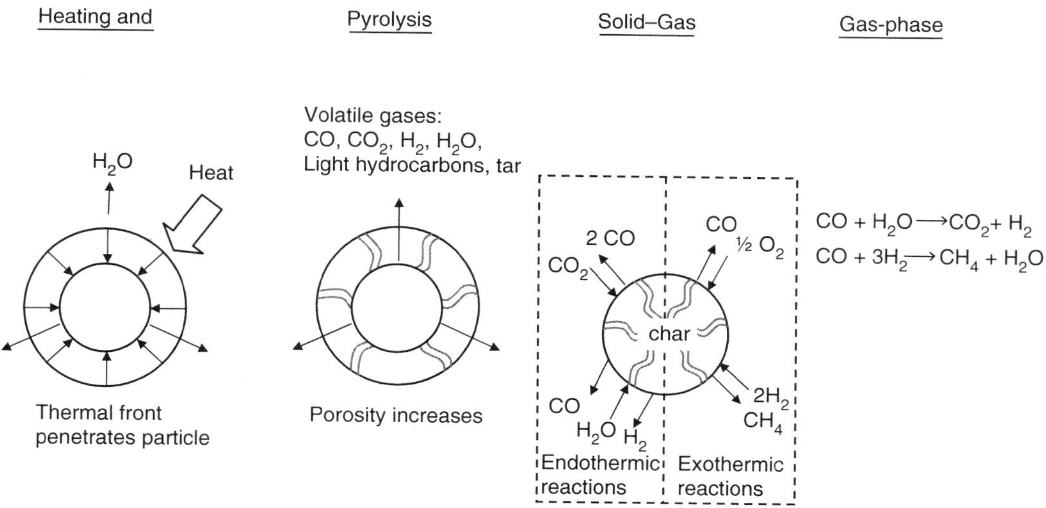

FIGURE 22.17 Processes of thermal gasification. (From Brown, R. C. 2003. *Biorenewable Resources: Engineering New Products from Agriculture*, Iowa State Press, Ames, IA.)

occurs and very high temperatures are produced. Although the gas flow is depleted of oxygen higher in the fuel bed, hot H_2O and CO_2 from combustion near the grate reduce char to H_2 and CO. These reactions cool the gas, but temperatures are still high enough to heat, dry, and pyrolyze the fuel moving down toward the grate. Of course, pyrolysis releases both condensable and non-condensable gases, and the producer gas leaving an updraft gasifier contains large quantities of tars on the order of 50 g/m^3. As a result, updraft gasifiers are generally not strong candidates for biomass energy applications.

In downdraft gasifiers, fuel and gas move in the same direction. Downdraft gasifiers appear to have been developed near the end of the nineteenth century, after the introduction of induced draft fans allowed air to be drawn downward through a gasifier in the same direction as the gravity-fed fuel. As shown in Figure 22.18b, contemporary designs usually add an arrangement of tuyeres that admit air or oxygen directly into a region known as the throat where combustion forms a bed of hot char. This design assures that condensable gases released during pyrolysis are forced to flow through the hot char bed, where tars are cracked. The producer gas is relatively free of tar ($<$1 g/m^3), making it a satisfactory fuel for engines. A disadvantage is the need for tightly controlled fuel properties (particles sized to between 1 and 30 cm, low ash content, and moisture less than 30%). Another disadvantage is a tendency for slagging or sintering of ash in the concentrated oxidation zone. Rotating ash grates or similar

TABLE 22.22 Producer Gas Composition from Various Kinds of Gasifiers

	Gaseous Constituents (vol% dry)						Gas Quality	
Gasifier type	H_2	CO	CO_2	CH_4	N_2	HHV (MJ/m^3)	Tars	Dust
Air-blown updraft	11	24	9	3	53	5.5	High (~10 g/m^3)	Low
Air-blown downdraft	17	21	13	1	48	5.7	Low (~1 g/m^3)	Medium
Air-blown fluidized bed	9	14	20	7	50	5.4	Medium (~10 g/m^3)	High
Oxygen-blown downdraft	32	48	15	2	3	10.4	Low (~1 g/m^3)	Low
Indirectly-heated fluidized bed	31	48	0	21	0	17.4	Medium (~10 g/m^3)	High

Source: From Milne, T. A., Abatzoglou, N., and Evans, R. J. 1998. Biomass Gasifier "Tars": Their Nature, Formation, and Conversion, National Renewable Energy Laboratory Report, NREL/TP-570-25357; Bridgwater, A. V., *Fuel*, 74, 631–653, 1995.

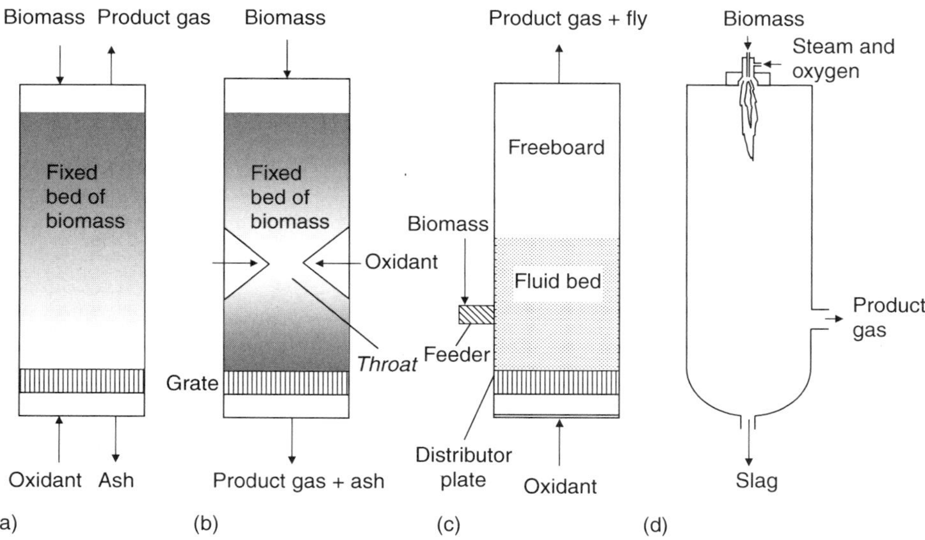

FIGURE 22.18 Types of biomass gasifiers: (a) updraft, (b) downdraft, (c) fluidized bed, (d) entrained flow. (From Brown, R. C., *Biorenewable Resources: Engineering New Products from Agriculture*, Iowa State Press, Ames, IA, 2003.)

mechanisms can solve this problem. Furthermore, incorporation of a throat region limits the maximum size of downdraft gasifiers to about 400 kg/h.

In a fluidized bed gasifier, illustrated in Figure 22.18c, a gas stream passes vertically upward through a bed of inert particulate material to form a turbulent mixture of gas and solid. Fuel is added at such a rate that it is only a few percent by weight of the bed inventory. Unlike the updraft and downdraft gasifiers, no segregated regions of combustion, pyrolysis, and tar cracking exist. The violent stirring action makes the bed uniform in temperature and composition with the result that gasification occurs simultaneously at all locations in the bed. Typically, a fluidized gasifier operates in the range of 700°C–850°C. By injecting fuel in the base of the bed, much of the tar can be cracked within the fluidized bed. However, a large insulated space above the bed, known as the freeboard, is usually included to promote additional tar cracking as well as more complete conversion of char. Nevertheless, tar production is intermediate between updraft and downdraft gasifiers (about 10 g/m^3). Fluidized beds are attractive for biomass gasification. They are able to process a wide variety of fuels including those with high moisture content and small particle size. They are easily scaled to large sizes suitable for electric power production. Disadvantages include relatively high power consumption to move gas through the fluidized bed; high exit gas temperatures, which complicates efficient energy recovery; and relatively high particulate burdens in the gas due to the abrasive forces acting within the fluidized bed.

A fourth kind of gasifier, illustrated in Figure 22.18d, employs finely pulverized fuel in an entrained flow reactor. This reactor was developed for steam-oxygen gasification of coal at temperatures of 1200°C–1500°C. These high temperatures assure excellent char conversion (approaching 100%) and low tar production and convert the ash to molten slag, which drains from the bottom of the reactor. The technology is attractive for advanced coal power plants but has not been widely explored for biomass because of the expense of finely dividing biomass, the difficulty of reaching high temperatures with biomass, and the presence of alkali in biomass, which leads to severe ash sintering. Despite these difficulties, entrained flow reactors are being considered for gasification of pretreated biomass (such as char and pyrolysis liquids).[13]

The fuel/air ratio is the single most important parameter for determining gasifier performance. Downdraft gasifiers can have better conversion efficiency and producer gas quality than fluidized bed gasifiers because they utilize a higher fuel/air ratio. Fluidized bed gasifiers, on the other hand, can generally handle a wider range of biomass feedstocks with higher moisture content.

Recent research in biomass gasification has focused on improving the heating value of product gas. Conventional gasification admits sufficient air or oxygen to the reactor to burn part of the fuel, thus releasing heat to support pyrolysis of the rest of the fuel. Gas produced in air-blown biomass gasifiers typically has heating value that is only 10%–20% than that of natural gas. This low heating value is largely the result of nitrogen diluting the fuel gas. Oxygen can be used as the fluidization agent, but high capital costs preclude this from consideration at the relatively small sizes envisioned for most biomass energy systems.

Indirectly heated gasification can improve gas-heating value by physically separating combustion and pyrolysis.[12] As a result, the products of combustion do not appear in the fuel gas. Higher heating values of 14,200 kJ/m^3 or higher are expected. Several schemes have been suggested for transporting heat from the combustion reactor to the pyrolysis reactor. These include transferring hot solids from the combustor to the pyrolyzer, transferring a chemically regenerative heat carrier between two reactors, transferring heat through a wall common to the reactors, and storing heat in high-temperature phase-change material.

Depending on the kind of gasifier, product gas can be contaminated by tar and particulate matter. Tar cannot be tolerated in many downstream applications, including internal combustion engines, fuel cells, and chemical synthesis reactors. Tar can deposit on surfaces in filters, heat exchangers, and engines where they reduce component performance and increase maintenance requirements. A well-designed gasifier has a high temperature zone through which pyrolysis products pass, allowing tars to thermally decompose ("crack") into the low molecular gases predicted by equilibrium theory. The production of particulate matter, a combination of ash and unreacted char from the biomass, varies considerably from one type of gasifier to another. Particulate matter can also deposit on critical power system components, degrading performance even at relatively low concentrations in some machinery.

Gas cleaning usually involves separate processes for removing solid particles and tar vapor or droplets. A gas cyclone operated above the condensation temperature for the least volatile tar constituent can remove most ash and char particles larger than about 5 μ in diameter. If removal of finer particles is required, then additional filtration such as ceramic candle filters or moving bed granular filters is required.

Two approaches are available for removing tar from product gas.[11] Scrubbing the gas stream with a fine mist of water or oil removes both tar and particles from the gas stream. Gas scrubbers are widely used in the chemical industry and are relatively inexpensive. However, cooling the gas and removing organic compounds reduce both sensible energy and chemical energy, which decreases overall energy conversion efficiency. Gas scrubbing also produces a toxic stream of tar, which complicates waste disposal.

The other approach for removing tar converts it into low molecular weight compounds, which has the advantages of increasing the heating value of the producer gas while eliminating a waste disposal problem. This "cracking" of tar into CO and H_2 can be accomplished by raising the gas to temperatures exceeding 1000°C. Such high temperatures are not easily achieved in most biomass gasifiers. The use of catalysts allows tar cracking to occur at much lower temperatures, in the range of 600°C–800°C. The gas is passed through a packed bed reactor containing catalyst with steam or oxygen sometimes used to promote reaction.

The thermodynamic efficiency of gasifiers is strongly dependent on the kind of gasifier and how the product gas is employed.[12] Some high-temperature, high-pressure gasifiers are able to convert 90% of the chemical energy of solid fuels into chemical and sensible heat of the product gas. However, these high efficiencies come at high capital and operating costs. Most biomass gasifiers have conversion efficiencies ranging between 70 and 80%. In some applications, such as process heaters or driers, both the chemical and sensible heat of the product gas can be utilized. In many power applications, though, the hot product gas must be cooled before it is utilized; thus, the sensible heat of the gas is lost. In this case, "cold gas" efficiency can be as low as 50%–60%. Whether the heat removed from the product gas can be recovered for other applications, like steam raising or fuel drying, ultimately determines which of these conversion efficiencies is most meaningful. Gas cleaning to remove tar and particulate matter also has a small negative impact on gasifier efficiency since it removes flammable constituents from the gas (tar and char particles) and generally requires a small amount of energy to run pumps.

22.2.5 Stirling Cycle

The Stirling cycle is an example of an external combustion engine; that is, the products of combustion do not come into contact with the fluid that undergoes the thermodynamic processes of the cycle.[14] In this respect, it resembles the Rankine steam cycle although its thermodynamic efficiency is theoretically higher than that of the Rankine cycle operating on the same fuel. In practice, the efficiencies of Stirling engines are relatively modest as are their output power, typically no more than a few kilowatts. They continue to attract attention mainly because of their low maintenance, high tolerance to contaminants, and relatively low pollution emissions. However, high costs have prevented significant market entry.

22.2.6 Rankine Cycle

The Rankine cycle is another example of an external combustion cycle.[15] As illustrated in Figure 22.19, fuel is burned in a furnace where the released heat is transferred to pressurized water contained within steel tubes. Steam generated in this process is expanded in a steam turbine, which drives an electric generator to produce electric power.

The Rankine steam power cycle has been the foundation of stationary power generation for over a century. Although Brayton cycles employing gas turbines and electrochemical cycles based on fuel cells will constitute much of the growth in power generation in the future, steam power plants will continue to supply the majority of electric power for decades to come and will find new applications in combination with advanced generation technologies. The reason for the Rankine cycle's preeminence has been its ability to directly fire coal and other inexpensive solid fuels. Constructed at scales of several hundred megawatts, the modern steam power plant can convert 35%–40% of chemical energy in fuel to electricity at a cost of \$0.02–\$0.05/kilowatt.

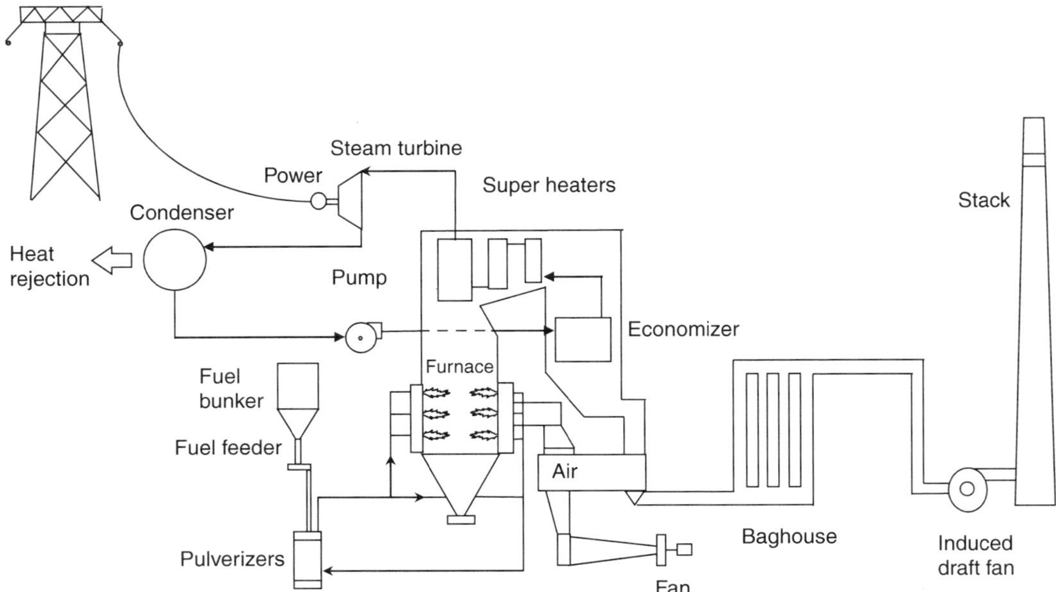

FIGURE 22.19 Rankine steam power plant. (From Brown, R. C., *Biorenewable Resources: Engineering New Products from Agriculture*, Iowa State Press, Ames, IA, 2003.)

Utility-scale steam power plants are not expected to dominate future growth in electric power infrastructure in the U.S. These giant plants take several years to plan and construct, which decreases their financial attractiveness in increasingly deregulated power markets. Coal and other fossil fuels burned in these plants are major sources of air pollution, including sulfur and nitrogen oxides, both of which are precursors to acid rain and the latter an important factor in smog formation; fine particulate matter, which is implicated in respiratory disease in urban areas; and heavy metals, the most prominent being mercury, which accumulates in the biosphere to toxic levels. Substitution of biorenewable resources such as wood and agricultural residues for coal in existing power plants could substantially reduce pollution emissions, although these plants are so large that the locally available biomass resources could supplant only a small fraction of the total energy requirement. Small-scale steam power plants sized for use of local biomass resources have low thermodynamic efficiencies, on the order of 25%, making them wasteful of energy resources.

22.2.7 Brayton Cycle

The Brayton cycle produces electric power by expanding hot gas through a turbine.[16] These gas turbines operate at temperatures approaching 1300°C compared with inlet temperatures of less than 650°C for steam turbines used in Rankine cycles. Although this difference in inlet temperature would suggest that Brayton cycles have much higher thermodynamic efficiencies than Rankine cycles, the Brayton cycle also has much higher exhaust temperature than does the Rankine cycle. Gas turbine exhaust temperatures are in the range of 400°C–600°C whereas steam turbine exhaust temperatures are on the order of 20°C. Furthermore, Brayton cycles, which contain both a gas compressor and gas turbine, have more sources of mechanical irreveribilities, further degrading thermodynamic efficiencies, which may only be marginally higher than the best Rankine steam cycles. However, improvements in gas turbine technology that allow operation at higher temperatures and pressures are expected to increase Rankine cycle efficiency for large power plants to greater than 50%, although 30% is more realistic for gas turbines sized appropriately for biomass power plants.

The two general classes of gas turbines for power generation are heavy-duty industrial turbines and lightweight aeroderivative gas turbines. The aeroderivatives are gas turbines originally developed for commercial aviation but adapted for stationary electric power generation. They are attractive for bioenergy applications because of their high efficiency and low unit capital costs at the modest scales required for biomass fuels.

Gas turbines are well suited to gaseous and liquid fuels that are relatively free of contaminants that rapidly erode or corrode turbine blades. In this respect, gas turbine engines are not suitable for directly firing most biomass fuels. Solid biomass releases significant quantities of alkali metals, chlorine, mineral matter, and lesser amounts of sulfur upon burning. These would be entrained in the gas flow entering the expansion turbine where they would quickly contribute to blade failure. Cleaning large quantities of hot fuel gas is not generally considered as an economical proposition. Even the gas released from anaerobic digestion contains too much hydrogen sulfide to be directly burned in a gas turbine without first chemically scrubbing the gas to remove this corrosive agent.

Nevertheless, gas turbine engines are considered one of the most promising technologies for bioenergy because of the relative ease of plant construction, cost-effectiveness in a wide range of sizes (from tens of kilowatts to hundreds of megawatts), and the potential for very high thermodynamic efficiencies when employed in advanced cycles. The key to their success in bioenergy applications is converting the biomass to clean-burning gas or liquid before burning it in the gas turbine combustor.

22.2.8 Fuels Cells

Among the most exciting new energy technologies are fuel cells, which directly convert chemical energy into work, thus, bypassing the restriction on efficiency imposed by the Carnot relationship.[17] This does not imply that fuel cells can convert 100% of chemical enthalpy of a fuel into work, as the process still must conform to the laws of thermodynamics. In practice, irreversibilities limit their conversion efficiencies to 35%–60%, depending upon the fuel cell design. Thus, fuel cells can produce significantly more work from a given amount of fuel than can heat engines. However, carbonaceous fuels must first be reformed to hydrogen before they are suitable for use in fuel cells. The energy losses associated with fuel reforming must be included when determining the overall fuel-to-electricity conversion efficiency of a fuel cell.

The gas mixture produced by a biomass gasifier contains dust and tar that must be removed or greatly reduced for most applications, including power generation in fuel cells. Removal of tar would ideally be performed at elevated temperatures. If the gas is to be used in fuel cells, further cleaning is required to remove ammonia (NH_3), hydrogen chloride (HCl), and hydrogen sulfide (H_2S).[18] To obtain high-energy efficiency, trace contaminant removal must be performed at elevated temperatures for fuel cells that operate at relatively high temperatures. Low temperature fuel cells cannot tolerate CO, which can be removed by the water–gas shift reaction. The catalysts that facilitate the shift reaction, however, are poisoned by trace contaminants, which must be removed prior to the shift reactors. One method for removing H_2S and HCl is the use of a fixed bed of calcined dolomite or limestone and zinc titanate at temperatures around 630° C. This is followed by steam reforming at high temperature (750°C–850°C) to destroy tar and ammonia.[19]

22.2.9 Combined Cycles

In an effort to enhance energy conversion efficiency, combined cycle power systems have been developed, which recognize that waste heat from one power cycle can be used to drive a second power cycle.[20] Combined cycles would be unnecessary if a single heat engine could be built to operate between the temperature extremes of burning fuel and the ambient environment. However, temperature and pressure limitations on materials of construction have prevented this realization. Combined cycles employ a topping cycle operating at high temperatures and a bottoming cycle operating on the rejected heat from the topping cycle. Most commonly, combined cycle power plants employ a gas turbine engine for the topping cycle and a steam turbine plant for the bottoming cycle, achieving overall efficiencies of 60% or higher.

Clean-burning fuel from biomass for use in a combined cycle can be obtained by thermal gasification. Integrated gasification/combined cycle (IGCC) power is illustrated in Figure 22.20. Compressed air enters an oxygen plant, which separates oxygen from the air. The oxygen is used to gasify biomass in a pressurized gasifier to produce medium heating-value producer gas. The producer gas passes through cyclones and a gas clean-up system to remove particulate matter, tar, and other contaminants that may adversely affect gas turbine performance (alkali and chloride being the most prominent among these). These clean-up operations are best performed at high temperature and pressure to achieve high cycle efficiency. The clean gas is then burned in air and expanded through a gas turbine operating as a "topping" cycle. The gas exits the turbine at temperatures ranging between 400 and 600°C. A heat recovery steam generator produces steam for a "bottoming" cycle that employs a steam turbine. Electric power is produced at two locations in this plant, yielding thermodynamic efficiencies exceeding 47%.

Integrated gasifier/combined cycle systems based on gas turbines are attractive for several reasons. These reasons include their relative commercial readiness and the expectation that they can generate electricity at the lowest cost of all possible biomass power options.

An alternative to IGCC is to generate steam for injection into the gas turbine combustor, which increases mass flow and power output from the turbine. This variation, called a steam-injected gas turbine (STIG) cycle[20,21] is less capital intensive than IGCC, since it does not employ a steam turbine. The STIG cycle is commercially developed for natural gas; lower flammability limits for producer gas make steam injection more problematic for biomass-derived producer gas. The intercooled steam-injected gas turbine (ISTIG) is an advanced version of the STIG. This cycle further improves thermodynamic efficiency by cooling gas flow between several stages of compression (intercooling).

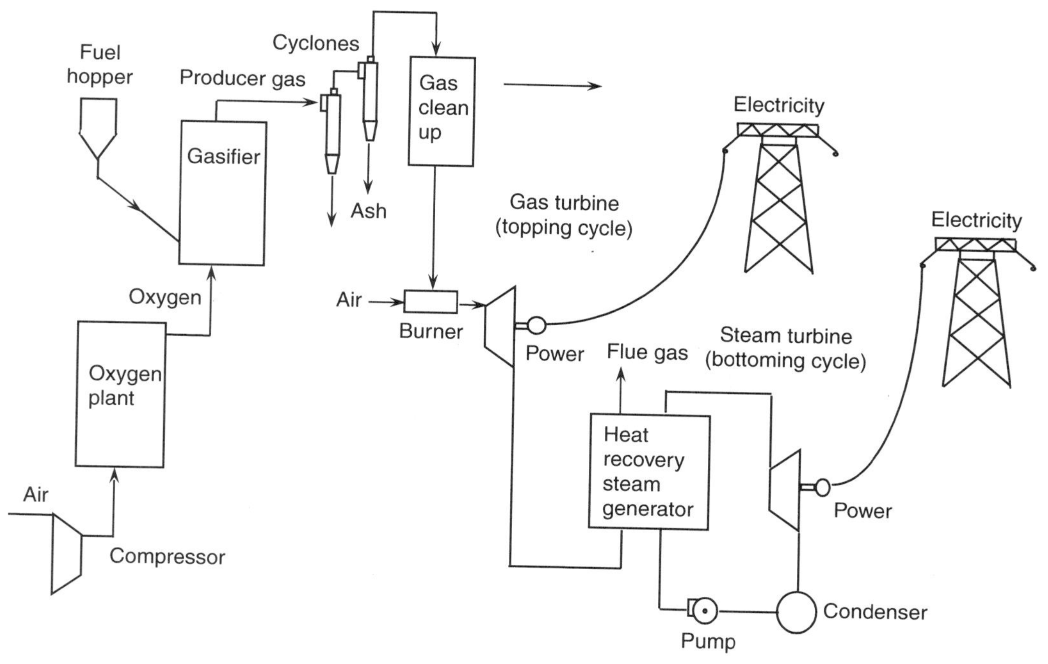

FIGURE 22.20 Integrated gasification/combined cycle power plant (IGCC) based on a gas turbine topping cycle. (From Brown, R. C., *Biorenewable Resources: Engineering New Products from Agriculture*, Iowa State Press, Ames, IA, 2003.)

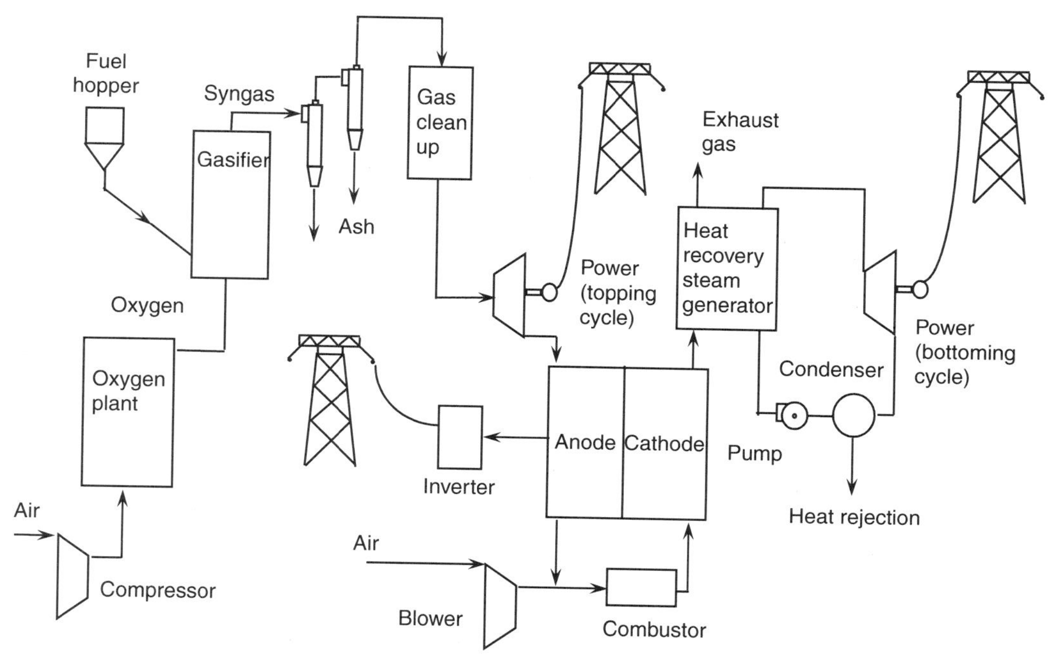

FIGURE 22.21 Combined cycle power plant based on high temperature fuel cells. (From Brown, R. C., *Biorenewable Resources: Engineering New Products from Agriculture*, Iowa State Press, Ames, IA, 2003.)

Figure 22.21 illustrates an IGCC power plant based on a molten carbonate fuel cell.[18] Biomass is gasified in oxygen to yield producer gas. Gasification occurs at elevated pressure to improve the yield of methane, which is important for proper thermal balance of this fuel cell. Hot-gas clean up to remove particulate matter, tar, and other contaminants is followed by expansion through a gas turbine as part of a topping power cycle. The pressure and temperature of the producer gas is sufficiently reduced after this to admit it into the fuel cell. High temperature exhaust gas exiting the cathode of the fuel cell enters a heat recovery steam generator, which is part of a bottoming cycle in the integrated plant. Thus, electricity is generated at three locations in the plant for an overall thermodynamic efficiency, reaching 60% or more.

References

1. Brown, R. C. 2003. *Biorenewable Resources: Engineering New Products from Agriculture*. Iowa State Press, Ames, IA.
2. Tillman, D. A. 1991. *The Combustion of Solids Fuels and Wastes*. Academic Press, San Diego, CA.
3. Kumar, J. V. and Pratt, B. C. 1996. Compositional analysis of some renewable biofuels. *American Laboratory*, 28, 8, 15–20.
4. Bain, R. L., Overend, R. P., and Craig, K. R. 1998. Biomass-fired power generation. *Fuel Processing Technology*, 54, 1–3, 1–16.
5. Jenkins, B. M., Baxter, L. L., Miles, T. R., and Miles, T. R. 1998. Combustion properties of biomass. *Fuel Processing Technology*, 54, 17–46.
6. Annamalai, K., Sami, M., and Wooldridge, M. 2001. Co-firing of coal and biomass fuel blends. *Progress in Energy and Combustion Science*, 27, 171–214.
7. Reed, T. ed. 1981. Biomass Gasification: Principles and Technology, Noyes Data Corp., Park Ridge, NJ.
8. Klass, D. L. 1998. *Biomass for Renewable Energy, Fuels, and Chemicals*, pp. 274–276. Academic Press, San Diego, CA.
9. Hall, D. O. and Overend, R. P. eds. 1987. Biomass Regenerable Energy, Wiley, New York.
10. Higman, C. and van der Burgt, M. 2003. *Gasification*. pp. 10–11, Elsevier, Amsterdam.
11. Milne, T. A., Abatzoglou, N., and Evans, R. J. 1998. Biomass Gasifier "Tars": Their Nature, Formation, and Conversion, National Renewable Energy Laboratory Report, NREL/TP-570-25357.
12. Bridgwater, A. V. 1995. The technical and economic feasibility of biomass gasification for power generation. *Fuel*, 74, 631–653.
13. Henrich, E. and Weirich, F. 2004. Pressurized entrained flow gasifiers for biomass. *Environmental Engineering Science*, 21, 1, 53–64.
14. Organ, A. J. 1992. *Thermodynamic and Gas Dynamics of the Stirling Cycle Machine*. Cambridge University Press, Cambridge, UK.
15. Singer, J. G. ed. 1981. Combustion: Fossil Power Systems, Combustion Engineering Corp., Windsor, CT.
16. Poullikkas, A. 2005. An overview of current and future sustainable gas turbine technologies. *Renewable and Sustainable Energy Reviews*, 9, 5, 409–443.
17. Larminie, J. and Dicks, A. 2003. *Fuel Cell Systems Explained. 2nd Ed.* Wiley, West Sussex, UK.
18. Fuel Cell Handbook, October 2000 (CD-ROM), *5th Ed.*, US Department of Energy National Energy Technology Laboratory, Morgantown, WV/Pittsburgh, PA.
19. Stevens, D. J. 2001. Hot Gas Conditioning: Recent Progress with Larger-Scale Biomass Gasification Systems, National Renewable Energy Laboratory Report NREL/SR-510-29952.
20. Williams, R. H. and Larson, E. D. 1993. Advanced gasification-based biomass power generation, In *Renewable Energy: Sources for Fuels and Electricity*, Johansson, T. B., Kelly, H., Reddy, A. K. N., and Williams, R. H., eds., pp. 729–785. Island Press, Washington, DC.
21. Jonsson, M. and Yan, J. 2005. Humidified gas turbines-A review of proposed and implemented cycles. *Energy*, 30, 7, 1013–1078.

22.3 Biofuels

Robert C. Brown

22.3.1 Introduction

Almost 25% of energy consumption in the U.S. goes to transportation. More than half of this amount comes from imported petroleum. Thus, development of transportation fuels from biorenewable resources is a priority if decreased dependence on foreign sources of energy is to be achieved. Table 22.23 lists properties of both traditional and bio-based transportation fuels.[1–5]

Traditional transportation fuels are classified as gasoline, diesel fuel, or jet fuel. Gasoline is intended for spark-ignition (Otto cycle) engines; thus, it is relatively volatile but resistant to autoignition during compression. Diesel fuel is intended for use in compression ignition (Diesel cycle) engines; thus, it is less volatile compared to gasoline and more susceptible to autoignition during compression. Jet fuel is designed for use in gas turbine (Brayton cycle) engines, which are not limited by autoignition characteristics but otherwise have very strict fuel specifications for reasons of safety and engine durability.

Gasoline is a mixture of hundreds of different hydrocarbons obtained from a large number of refinery process streams that contain between 4 and 12 carbon atoms with boiling points in the range of 25°C–225°C. Most of the mixture consists of alkanes with butanes and pentanes added to meet vapor pressure specifications. A few percent of aromatic compounds are added to increase octane number, the figure of merit used to indicate the tendency of a fuel to undergo premature detonation within the combustion cylinder of an internal combustion engine. The higher the octane number, the less likely a fuel will detonate until exposed to an ignition source (electrical spark). Premature denotation is responsible for the phenomenon known as engine knock, which reduces fuel economy and can damage an engine. Various systems of octane rating have been developed, including research octane and motor octane numbers. Federal regulation in the U.S. requires gasoline sold commercially to be rated using an average of the research and motor octane numbers. Gasoline rated as "regular" has a commercial octane number of about 87 while premium grade is 93.

Diesel fuel, like gasoline, is also a mixture of light distillate hydrocarbons but with lower volatility and higher viscosity. Because diesel fuel is intended to be ignited by compression rather than by a spark, its autoignition temperature is lower than for gasoline. The combustion behavior of diesel fuels are conveniently rated according to cetane number, an indication of how long it takes a fuel to ignite (ignition delay) after it has been injected under pressure into a diesel engine. A high cetane number indicates short ignition delay; for example, No. 2 diesel fuel has cetane number of 37–56 while gasoline has a cetane number less than 15.

Jet fuel is designated as either Jet A fuel or Jet B fuel. Jet A fuel is a kerosene type of fuel with relatively high flash point whereas Jet B fuel is a wide boiling range fuel, which more readily evaporates.

Bio-based transportation fuels, also known as biofuels, are currently dominated by ethanol and biodiesel. Ethanol, by virtue of its high octane number, is suitable for use in spark-ignition engines, while the high cetane numbers of biodiesel, which are methyl or ethyl esters formulated from vegetable or animal fats, make them suitable for use in compression-ignition engines. However, there are other candidate liquid biofuels including methanol, mixed alcohols, and Fischer–Tropsch (F–T) liquids, as well as gaseous biofuels including hydrogen, methane, ammonia, and dimethyl ether, which will also be discussed.

22.3.2 Ethanol

Ethanol can be produced by the fermentation of sugar or starch crops. A fuel ethanol market has been developed in Brazil based on sugarcane[6] while the U.S. has relied on cornstarch for commercial

TABLE 22.23 Comparison of Transportation Fuels

Fuel Type	Fossil Fuel-Derived		Biomass-Derived							
	Gasoline[1]	No. 2 Diesel Fuel[1]	Methanol[1]	Ethanol[1]	Methyl Ester[2] (from soybean oil)	Fischer–Tropsch A[2]	Hydrogen[3]	Methane[3]	Ammonia[4]	Dimethyl Ether[2]
Specific gravity [a]	0.72–0.78	0.85	0.796	0.794	0.886	0.770	0.071 (liq)	0.422 (liq)	0.682 (liq)	0.660
Kinematic viscosity at 20°C–25°C (mm^2/s)	0.8	2.5	0.75	1.51	3.9	2.08	105[b]	16.5[b]	14.7[c]	0.227
Boiling point range (°C)	30–225	210–235	65	78	339	164–352	−253	−162	−33.4[c]	−24.9
Flash point (°C)	−43	52	11	13	188	58.5		−184	—	—
Autoignition temperature (°C)	370	254	464	423	—	—	566–582	540	630	235
Octane no. (research)	91–100	—	109	109	—	—	>130	>120	110	—
Octane no. (motor)	82–92	—	89	90	—	—	—	—	—	—
Cetane no.	<15	37–56	<15	<15	55	74.6	—	—	—	>55
Heat of vaporization (kJ/kg)	380	375	1185	920	—	—	447	509	1371[c]	402[d]
Lower heating value (MJ/kg)	43.5	45	20.1	27	37	43.9	120	49.5	18.8	28.88

[a] Measured at 16°C except for liquefied gases, which are saturated liquids at their respective boiling points.
[b] Munson, B. et al. 1994. Table 1.6 *Fundamentals of Fluid Mechanics*, 2nd Ed., Wiley, NY.
[c] Perry, R. and Green, W. 1984. *Perry's Chemical Engineers' Handbook*, 6th Ed., McGraw-Hill, NY, chap. 3.
[d] Kajitani, S. et al. 1997. p. 35 of Engine Performance and Exhaust Characteristics of Direct-Injection Diesel Engine Operated with DME, Society of Automotive Engineers Inc.

production of fuel ethanol. Technologies are also being developed to convert lignocellulose into sugars or syngas, a mixture of carbon monoxide (CO) and hydrogen, either of which can be fermented into ethanol.

On a volumetric basis, ethanol has only 66% of the heating value of gasoline. Thus, the range of a vehicle operating on pure ethanol is theoretically reduced by a corresponding amount and, accordingly, meaningful comparisons of the cost of gasoline and ethanol should be made on the basis of energy delivered ($/GJ) instead of fuel volume ($/l). However, fuel economy depends on many complex interactions between a fuel and the combustion environment within an engine, which some argue improves the relative performance of ethanol.[7] For example, the higher octane number for ethanol compared to gasoline (109 vs. 91–101) allows engines to be designed to run at higher compression ratios, which improves both power and fuel economy. Estimates for efficiency improvements in engines optimized for ethanol instead of gasoline range from 15 to 20%, resulting in a driving range approaching 80% of that of gasoline.[8]

Internal combustion engines can be fueled on pure ethanol (known as neat alcohol or E100) or blends of ethanol and gasoline. Brazil employs 190 proof ethanol (95 vol% alcohol and 5% water), which eliminates the energy consuming step of producing anhydrous ethanol. In the United States two ethanol–gasoline blends are common, E85 contains 85% ethanol and 15% gasoline and E10 contains only 10% ethanol with the balance being gasoline. The advantage of E10 is that it can be used in vehicles with engines designed for gasoline; however, its use is accompanied by a loss in fuel economy (as measured in km/l or miles/gal) compared to gasoline, amounting to 2%–5%.[9]

A significant problem with ethanol–gasoline blends is water-induced phase separation. Water contaminating a storage tank or pipeline is readily absorbed by ethanol, resulting in a lower water-rich layer and an upper hydrocarbon-rich layer, which interferes with proper engine operation. Water contamination is a problem that has not been fully addressed by the refining, blending, and distribution industries; thus transportation of ethanol–gasoline blends in pipelines is not permitted in the United States and long-term storage is to be avoided.[10]

22.3.3 Ethanol from Starch Crops

Starch is a polymer that accumulates as granules in many kinds of plant cells where they serve as a storage carbohydrate. Mechanical grinding readily liberates starch granules. The hydrogen bonds between the basic units of maltose in this polymer are easily penetrated by water, making depolymerization, and solubilization relatively easy.

Hydrolysis, the process by which water splits a larger reactant molecule into two smaller product molecules, is readily accomplished for starch. Acid catalyzed hydrolysis in “starch cookers” at temperatures of 150°C–200°C proceeds to completion in seconds to minutes. In recent years, enzymatic hydrolysis has supplanted acid hydrolysis due to higher selectivity.

Starch is a glucose polymer with two main components: amylose, a liner polymer of glucose with alpha—1,4 linkages, and amylopectin, a branched chain including alpha—1,6 linkages at the branch points.[11] Thus, enzymatic saccharification of starch requires two enzymes. The enzyme amylase hydrolyzes starch to maltose in a process known as liquefaction. The enzyme maltase hydrolyzes maltose into glucose in a process known as saccharification. The consumption of either acid or enzymes for starch hydrolysis is less than 1:100 by weight, making the cost of hydrolysis only a small part of the cost of starch fermentation.

Cereal grains, such as corn, wheat, and barley, are the most widely used sources of starch for fermentation.[11] The cell walls of grains must be disrupted to expose starch polymers before they can be hydrolyzed to fermentable sugars (that is, monosaccharides and disaccharides). Grain starch consists of 10–20 wt% amylose and 80–90 wt% amylopectin, both of which yield glucose or maltose on hydrolysis. Although the amylose is water soluble, the amylopectin is insoluble, requiring a “cooking” operation to solubilize it in prior to hydrolysis.

Cereal grains also contain other components, such as protein, oil and fiber, which may be of sufficient value to recover along with the starch. For example, gluten, a mixture of plant proteins occurring in cereal grains, chiefly wheat and corn, is of value as an adhesive and animal feed. If these components are to be separately recovered, extensive pretreatment, known as wet milling, is required before the starch is hydrolyzed and fermented. Under some circumstances, separation of plant components is not economically justified; simpler dry grinding is employed to release starch polymers and the whole grain is fermented. Of the 12.9 gigaliters (3.4 billion gal) of fuel ethanol produced in the United States in 2004, about two-thirds was from dry grinding while the remaining one-third was from wet milling. The capital investment for dry milling is less than that for a comparably sized wet-milling plant. However, the higher value of its by-products, greater product flexibility, and simpler ethanol production can make a wet-milling plant a more profitable investment.

22.3.3.1 Dry Grinding of Corn

Dry grinding for ethanol production[12] uses a roller mill to grind grain into a meal, which exposes the starch. The meal is slurried with water to form a mash, which is cooked with enzymes to release sugars, followed by fermentation to ethanol. The fibrous residue remaining upon completion of fermentation is recovered from the base of the beer stripping column, mixed with yeast and other unfermented residues, and dried to a co-product known as distillers' dried grains and solubles (DDGS). This coproduct, containing about 25 wt% protein and residual oil, is a valuable feed for cattle. Profitability of a corn-to-ethanol plant is strongly tied to the successful marketing of DDGS A dry grind ethanol plant is illustrated schematically in Figure 22.22.[13]

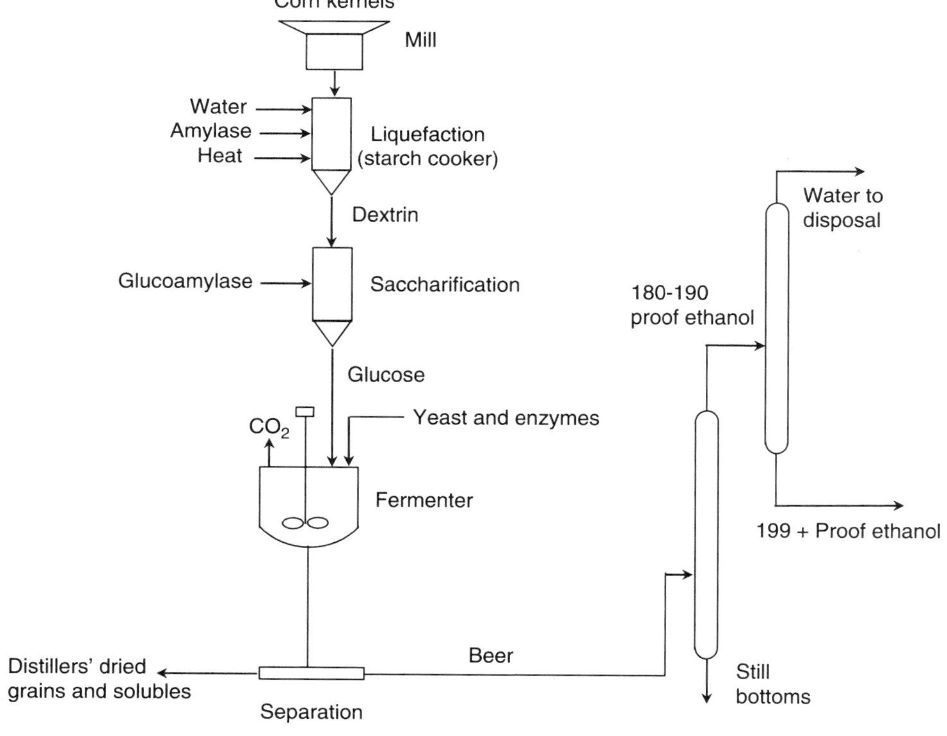

FIGURE 22.22 Dry grinding of corn. (From Brown, R. C., *Biorenewable Resources: Engineering New Products from Agriculture*, Blackwell Publishing, Ames, IA, 2003.)

A typical dry milling plant will produce about 9.5–9.8 L (2.5–2.6 gal) of ethanol per bushel of corn processed. Yields of co-products per bushel of corn are 7.7–8.2 kg (17–18 lb) of DDGS and 7.3–7.7 kg (16–17 lb) of carbon dioxide evolved from fermentation, the latter of which can be sold to the carbonated beverage industry. As a rule of thumb, the three products are produced in approximately equal weight per bushel, with each accounting for approximately one-third of the initial weight of the corn.

22.3.3.2 Wet Milling of Corn

Wet milling[14] has the advantage that it separates plant components into carbohydrate (starch), lipids (corn oil), a protein-rich material (gluten), and fiber (hulls). This gives a company access to higher value markets as well as provides flexibility in the use of starch as a food product or in the production of fuel ethanol.

The wet milling process is illustrated in Figure 22.23.[13] The corn is cleaned and then conveyed into steep tanks where it is soaked in a dilute solution of sulfur dioxide for 24–36 h, which swells and softens the corn kernels. Some of the protein and other compounds are dissolved in the resulting corn steep liquor, which represents an inexpensive source of nitrogen and vitamins.

After separating the corn from the steep liquor, the wet kernels are coarsely ground to release the hull and germ from the endosperm. Hydrocyclones or screens separate the germ from the rest of the components. After drying, oil is extracted from the germ by either solvents or a screw press, leaving a residual oil cake The hull and endosperm pass through rotating disc mills that grind the endosperm into fine fractions of starch and gluten while the hull yields coarser fiber particles, which can be screened out from the finer fractions. Centrifugal separators separate the lighter gluten from the starch.

The starch can be used directly as a food product or for industrial manufacturing processes, especially papermaking. The starch can also be converted to monosaccharides for the production of food or fuel, depending on relative market demand. Saccharification by amylase enzymes yields corn syrup, a glucose

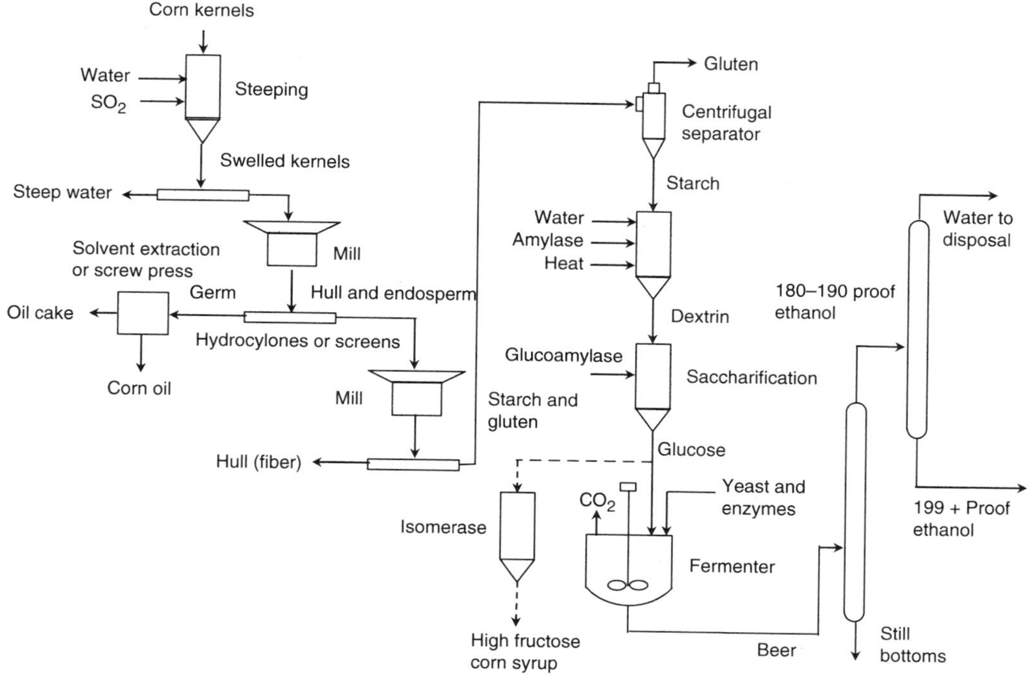

FIGURE 22.23 Wet milling of corn. (From Brown, R. C., *Biorenewable Resources: Engineering New Products from Agriculture*, Blackwell Publishing, Ames, IA, 2003.)

solution that can be directly fermented to fuel ethanol Alternatively, treated with isomerase enzymes, the glucose is partially converted to fructose to yield a liquid sweetener known as high fructose corn syrup (HFCS). In plants that can alternate between fuel ethanol and HFCS production, relatively more ethanol is produced in the winter while relatively more HFCS is produced in the summer.

The gluten product, known as corn gluten meal, contains 60% protein and is used primarily as poultry feed. The fiber from the hulls is combined with other by-products, such as the oil cake, steep water solubles, and excess yeast from stillage, dried and sold as corn gluten feed. Containing 21% or more of protein, it is primarily used as feed for dairy cattle.

A typical wet milling plant will produce 9.5–9.8 L (2.5–2.6 gal) of ethanol per bushel of corn processed. Yields of other coproducts per bushel of corn are 0.7 kg (1.7 lb) of corn oil, 1.4 kg (3 lb) of corn gluten meal (60% protein), 5.9 kg (13 lb) of corn gluten feed (21% protein), and 7.7 kg (17 lb) of carbon dioxide. Like dry milling, the three products of ethanol, feed, and carbon dioxide are produced in approximately equal weight per bushel, with each accounting for approximately one-third of the initial weight of the corn.

22.3.4 Ethanol from Lignocellulosic Feedstocks

Much of the carbohydrate in plant materials is structural polysaccharides, providing shape and strength to the plant. The hydrolysis of polysaccharides in cell walls is more difficult than the hydrolysis of storage polysaccharides like starch. This structural material, known as lignocellulose, is a composite of cellulose fibers embedded in a cross-linked lignin-hemicellulose matrix.[15] Depolymerization to basic plant components is difficult because lignocellulose is resistant to both chemical and biological attack.

A variety of physical, chemical, and enzymatic processes have been developed to fractionate lignocellulose into the major plant components of hemicellulose, cellulose, and lignin.[16] The hemicellulose fraction is readily hydrolyzed to pentoses (five carbon sugars) but pentoses are difficult to ferment. The cellulose exists as both amorphous and crystalline forms, which hydrolyze to hexoses (six carbon sugars). Crystalline cellulose is recalcitrant to hydrolysis. However, the resulting hexoses are readily fermented. Distillation can recover the desired products of fermentation. Lignin, which is not susceptible to biological transformation, can be chemically upgraded or, more frequently, simply burned as boiler fuel. The steps of pretreatment, hydrolysis, fermentation, and distillation in the production of bio-based products from lignocellulose are described in the following sections.

22.3.4.1 Pretreatment

Pretreatment is one of the most costly steps in conversion of lignocellulose to sugars, accounting for about 33% of the total processing costs.[17] Pretreatments often produce biological inhibitors, which impact the cost of fermenting the resulting sugars Accordingly, much attention is directed at developing low cost and effective pretreatments.

An important goal of all pretreatments is to increase the surface area of lignocellulosic material, making the polysaccharides more susceptible to hydrolysis. Thus, comminution, or size reduction, is an integral part of all pretreatments. Primary size reduction employs hammer mills to produce particles that can pass through 3 mm screen openings.

The mechanisms by which pretreatments improve the digestibility of lignocellulose are not well understood.[17] Pretreatment effectiveness has been correlated with removal of hemicellulose and lignin. Lignin solubilization is beneficial for subsequent hydrolysis, but may also produce derivatives that inhibit enzyme activity. Some pretreatments reduces crystallinity of cellulose, which improves reactivity, but this does not appear to be the key for many successfully pretreatments.

A large variety of pretreatment processes have been developed.[18] Biological pretreatments employ microorganisms that produce lignin-degrading enzymes (ligninase). Steam explosion involves saturation of the pores of plant materials with steam followed by rapid decompression; the explosive expansion of steam reduces the plant material to separated fibers, presumably increasing the accessibility of polysaccharides to subsequent hydrolysis. Ammonia fiber explosion (AFEX) is similar to steam explosion except that liquid ammonia is employed. It is very effective on agricultural residues but has not been successful in pretreating woody biomass.

22.3.4.2 Hydrolysis

Three basic methods for hydrolyzing structural polysaccharides in plant cell walls to ferment sugars that are available: concentrated acid hydrolysis, dilute acid hydrolysis, and enzymatic hydrolysis.[16,19] The two acid processes hydrolyze both hemicellulose and cellulose with very little pretreatment beyond comminution of the lignocellulosic material to particles of about 1 mm in size. The enzymatic process must be preceded by extensive pretreatment to separate the cellulose, hemicellulose, and lignin fractions.

Concentrated acid hydrolysis is based on the discovery over a century ago that carbohydrates in wood will dissolve in 72% sulfuric acid at room temperature, leaving behind the lignin fraction. For fermentation, the solution of oligiosaccharides is diluted to 4% H_2SO_4, and heated at the boiling point for four hours, or in an autoclave at 120°C for one hour to yield monosaccharides. Following neutralization with limestone, the sugar solution can be fermented. Concentrated acid hydrolysis is relatively simple and is attractive for its high sugar yields, which approach 100% of theoretical hexose yields.

Dilute acid hydrolysis (about 1% acid by weight) greatly reduces the amount of acid required to hydrolyze lignocellulose. The process is accelerated by operation at elevated temperatures: 100°C–160°C for hemicellulose and 180°C–220°C for cellulose. Unfortunately, the high temperatures cause oligiosaccharides released from the lignocellulose to decompose, greatly reducing yields of simple sugars to only 55%–60% of the theoretical yield. The decomposition products include a large number of microbial toxins, such as acetic acid and furfural, which inhibit fermentation of the sugars. The need for corrosion resistant equipment and low concentrations of sugars from some reactor systems also adversely impact the cost of sugars.

Enzymatic hydrolysis was developed to utilize both cellulose and hemicellulose better from lignocellulosic materials. Pretreatment solubilizes hemicellulose under milder conditions than those required for acid hydrolysis of cellulose. Subsequent enzymatic hydrolysis of the cellulose does not degrade pentoses released during prehydrolysis. Cellulose is a homopolysaccharide of glucose linked by β-1,4′-glycosidic bonds. Thus, enzymatic hydrolysis of cellulose proceeds in several steps to break glycosidic bonds by the action of a system of enzymes known as cellulase. The system of enzymes also usually contains hemicellulase to hydrolyze any hemicellulose not solubilized by prehydrolysis.

22.3.4.3 Fermentation

Simultaneous saccharification and fermentation (SSF) has been developed for fermenting sugars released from lignocellulose.[16,19] The SSF process combines hydrolysis (saccharification) and fermentation to overcome end product inhibition that occurs during hydrolysis of cellobiose. By combining hydrolysis and fermentation in the same reactor, glucose is rapidly removed before it can inhibit further hydrolysis. The SSF process is illustrated in Figure 22.24.[13] The biomass feedstock is milled and then prehydrolyzed to yield a mixture of pentoses, primarily xylose and arabinose, and fiber. The mixture is neutralized with limestone and mixed with cellulase and hemicellulase enzymes, which are either purchased commercially or produced on site, yeast, and nutrients. The cellulose and any remaining hemicellulose are solubilized to hexose (glucose) and pentoses (xylose and arabinose), which are immediately fermented to ethanol. The rate-limiting step is the hydrolysis of cellulose to glucose. The optimum temperature for the hydrolysis/fermentation reactor is a compromise between the optimum temperature for cellulase activity and the best temperature for the yeast. Lignin is separated from the mixture and used as boiler fuel.

The beer is distilled to ethanol in a process identical to that employed after sugar or starch fermentations. Energy consumption in the distillation process is partly responsible for criticism that ethanol production consumes more energy than it produces. Although there is basis for this criticism in older plants, modern plants pay close attention to energy consumption. Some plants are reported to use as little as 5.6 MJ of steam per liter of ethanol produced, with a total energy consumption of

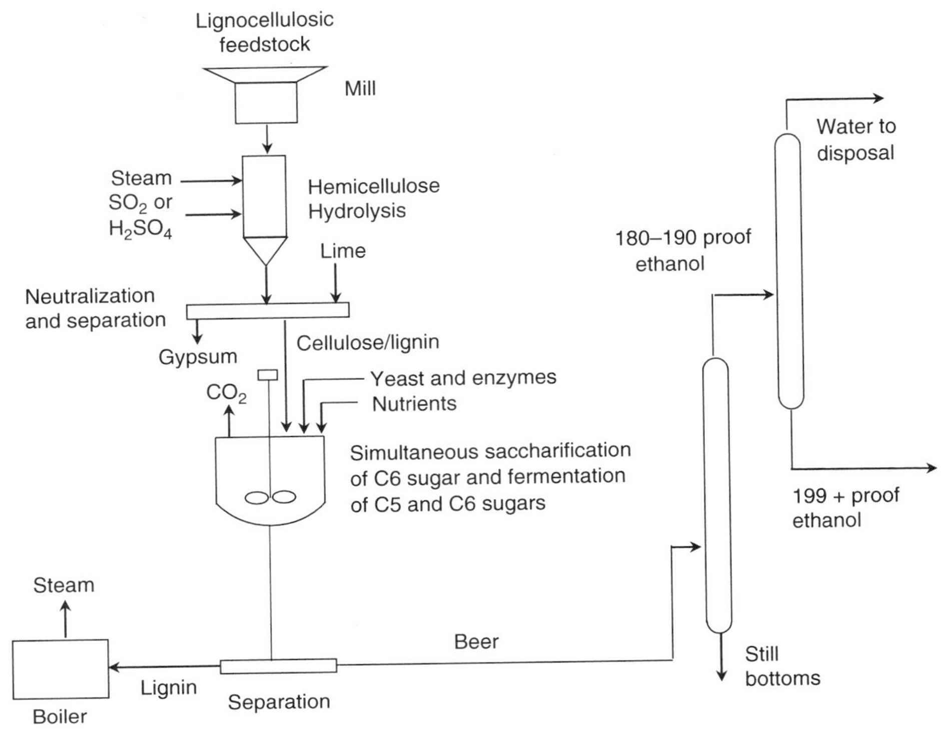

FIGURE 22.24 Enzymatic hydrolysis of lignocellulosic biomass. (From Brown, R. C., *Biorenewable Resources: Engineering New Products from Agriculture*, Blackwell Publishing, Ames, IA, 2003.)

11.1–12.5 MJ/L of product ethanol. A recent analysis of energy usage in corn-to-ethanol plants is found in Ref. 20.

22.3.5 Biodiesel

Vegetable oils, which are triglycerides of fatty acids, have been recognized long as potential fuels in diesel engines. Compared to petroleum-based diesel fuels, vegetable oils have higher viscosity and lower volatility, which results in fouling of engine valves and less favorable combustion performance, especially in direct-injection engines.[21] The solution to this problem is to convert the triglycerides into methyl esters or ethyl esters of the fatty acids, known as biodiesel, and the by-product 1,2,3-propanetriol (glycerol).

Table 22.23 illustrates that fuel properties of biodiesel are very similar to petroleum-based diesel. Only the specific gravity and viscosity of biodiesel are slightly higher than for diesel while the cetane numbers and heating values are comparable. Significantly higher flash points for biodiesel represent greater safety in storage and transportation. Biodiesel can be used in unmodified diesel engines with no excess wear or operational problems. Tests in light and heavy trucks showed few differences other than a requirement for more frequent oil changes because of the build-up of ester fuel in engine crankcases.[21]

Triglycerides, also known as fats and oils, are esters of glycerol and fatty acids, which are long-chain carboxylic acids containing even numbers of carbon atoms.[22] The acid fractions of triglycerides can vary in chain length and degree of saturation. Fats, which are solid or semi-solid at room temperature, have a high percentage of saturated acids, whereas oils, which are liquid at room temperature, have a high percentage of unsaturated acids. Plant-derived triglycerides are typically oils containing unsaturated fatty acids, including oleic, linoleic, and linolenic acids.

A wide variety of plant species produce triglycerides in commercially significant quantities, most of it occurring in seeds.[23] Average oil yields range from 150 L/ha for cottonseed to 814 L/ha for peanut oil although intensive cultivation might double these numbers. Soybeans are responsible for more than 50% of world production of oilseed, representing 48–82 million bbl/year. However, the Chinese tallow tree, cultivated in the southern U.S., has the potential for several fold higher productivity than soybeans and is particularly attractive for its ability to grow on saline soils that are not currently used for agriculture.

Extraction of seed oil is relatively straightforward. The seeds are crushed to release the oil from the seed. Mechanical pressing is used to extract oil from seeds with oil content exceeding 20%. Solvent extraction is required for seeds of lower oil content. The residual seed material, known as meal, is used in animal feed.

Triglycerides are also recovered as a coproduct of the pulping of pinewood by the kraft process.[24] The esters of both fatty and resin acids are saponified to sodium salts and recovered as soap foam on the surface of the black (pulping) liquor. These salts are acidified to form a mixture of 30% fatty acids, 35% resin acids, and 35% unsaponifiable esters known as tall oil.

Microorganisms, including yeasts, fungi, and algae are also potential sources of triglycerides.[25] Anaerobic yeasts and fungi accumulate triglycerides during the latter stages of growth when nutrients other than carbon begin to be exhausted. Algae, which grow over a wide range of temperatures in high-salinity water, can produce as much as 60% of their body weight as lipids when deprived of key nutrients such as silicon for diatoms or nitrogen for green algae. They employ relatively low substrate concentrations, in the order of 10–40 g/L. Unlike ethanol production, product recovery is relatively simple because of the sequestration of the oil in the algae. One suggestion is to build algae ponds in the desert Southwest United States where inexpensive flat land, water from alkaline aquifiers, and carbon dioxide from power plants could be combined to generate triglyceride-based fuel.

The higher viscosity and lower volatility of triglycerides compared to diesel fuel leads to coking of the injectors and rings of diesel engines. Chemical modification of triglycerides to methyl or ethyl esters yields excellent diesel-engine fuel. Biodiesel is the generic name given to these modified vegetable oils and animal fats. Suitable feedstocks include soybean, sunflower, cottonseed, corn, groundnut (peanut), safflower, rapeseed, waste cooking oils, and animal fats. Waste oils or tallow (white or yellow grease) can also be converted to biodiesel.

Transesterification describes the process by which triglycerides are reacted with methanol or ethanol to produce methyl esters and ethyl esters, respectively, along with the coproduct glycerol.[26] For example, one triglyceride molecule reacts with three methanol molecules to produce one molecule of 1,2,3-propanetriol (glycerol) and three ester molecules:

$$\underset{\text{Triglyceride}}{\begin{array}{l} R1{-}C({=}O){-}O{-}CH_2 \\ R2{-}C({=}O){-}O{-}CH \\ R3{-}C({=}O){-}O{-}CH_2 \end{array}} + \underset{\text{Methanol}}{3\,CH_3OH} \longrightarrow \underset{\text{Glycerol}}{\begin{array}{l} HO{-}CH_2 \\ HO{-}CH \\ HO{-}CH_2 \end{array}} + \underset{\text{Methyl ester}}{R1{-}C({=}O){-}O{-}CH_3} + R2{-}C({=}O){-}O{-}CH_3 + R3{-}C({=}O){-}O{-}CH_3 \tag{22.14}$$

Near quantitative yields of methyl (or ethyl) esters can be produced in one hour at room temperature using 6:1 molar ratios of alcohol and oil when catalyzed by 1% lye (NaOH or KOH).

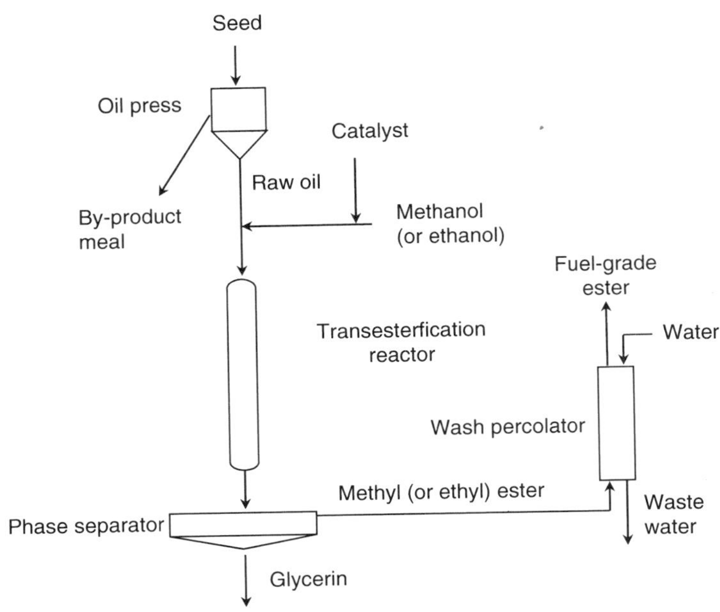

FIGURE 22.25 Conversion of triglycerides to methyl (or ethyl) esters and glycerol. (From Brown, R. C., *Biorenewable Resources: Engineering New Products from Agriculture*, Blackwell Publishing, Ames, IA, 2003.)

The lye also serves as a reactant in the conversion of esters into salts of fatty acids. These salts are familiarly known as soaps and the process is called saponification (soap-forming). Small amounts of soap are also produced by the reaction of lye with fatty acids. Upon completion, the glycerol and soap are removed in a phase separator. A flow sheet for a biodiesel production facility is given in Figure 22.25.[13]

22.3.6 Transportation Fuels from Biomass-Derived Syngas

The producer gas resulting from gasification can be used to manufacture a variety of liquid transportation fuels including methanol, ethanol, mixed alcohols, and F–T liquids. It can also be used to produce hydrogen (H_2), methane (CH_4), dimethyl ether (CH_3OCH_3), and ammonia (NH_3), which are gaseous compounds at ambient conditions but can be compressed or liquefied for use as transportation fuels. A relatively pure mixture of CO and hydrogen is usually preferred for synthesizing these compounds. Since raw producer gas can also contain various amounts of light hydrocarbons, tar, particulate matter, and trace contaminants, such as sulfur, chlorine, and ammonia, some downstream treatment of the gas stream may be required to produce the desired proportions of CO and H_2.

22.3.6.1 Methanol from Syngas

As shown in Table 22.23, the fuel properties of methanol are similar to those of ethanol: narrow boiling point range, high heat of vaporization, and high octane number. It has only 49% of the volumetric heating value of gasoline. As a transportation fuel, it has many of the same advantages and disadvantages as ethanol.

Methanol is formed by the exothermic reaction of one mole of CO with two moles of hydrogen:[27]

$$CO + 2H_2 \rightarrow CH_3OH \tag{22.15}$$

Low temperatures and high pressures thermodynamically favor the production of methanol. Current commercial operations use a fixed catalytic bed operated at 250°C and 60–100 atmospheres with gas recycle to remove the large amount of heat released by this exothermic reaction. More recently, liquid phase slurry reactors have been introduced to improve contact between syngas and catalyst as well as enhance the removal of heat from the reactor.

Biomass gasification does not necessarily yield the H_2/CO ratio of 2.0 required for methanol synthesis. Hydrogen enrichment can be achieved by passing syngas and steam over a catalytic bed, which promotes the water–gas shift reaction:

$$CO + H_2O \rightarrow CO_2 + H_2 \tag{22.16}$$

Low temperatures thermodynamically favor this exothermic reaction. To obtain satisfactory reaction rates, catalysts are employed in one or more fixed bed reactors operated in the temperature range of 250°C–400°C.

However, methanol is considerably more toxic than ethanol. Recent rulings banning the closely related and similarly toxic fuel additive methyl tertiary butyl ether (MTBE) from many states because of groundwater contamination makes methanol an unlikely replacement for gasoline.

22.3.6.2 Alcohols from Syngas

22.3.6.2.1 Catalytic

Efforts in Germany during World War II to develop alternative motor fuels discovered that iron-based catalysts could yield appreciable quantities of water soluble alcohols from syngas, especially ethanol:[28]

$$CO + 3H_2 \rightarrow CH_3CH_2OH \tag{22.17}$$

These early efforts yielded liquids containing as much as 45%–60% alcohols of which 60%–70% was ethanol. Working at pressures of around 50 bar and temperatures in the range of 220°C–370°C, researchers have developed catalysts with selectivity to alcohols of over 95%, but production of pure ethanol has been elusive.

Because the product typically contains a mixture methanol, ethanol, 1-propanol, and 2-propanol, some researchers have advocated the use of "mixed alcohols" as transportation fuels. One advantage is the ability to use lower H_2:CO ratios than is required for methanol or F–T synthesis:[29]

$$nCO + 2nH_2 \rightarrow C_nH_{2n+1}OH + (n-1)H_2O \tag{22.18}$$

with n typically ranging from 1 to 8. The process was commercialized in Germany between 1935 and 1945 but eventually abandoned because of the increased availability of inexpensive petroleum. An extensive review of mixed alcohol synthesis technology is found in Ref. 30.

An alternative approach to obtaining neat ethanol from syngas is to first synthesize methanol and subsequently react this product with additional syngas:[31]

$$CH_3OH + 2CO + H_2 \rightarrow CH_3CH_2OH + CO_2 \tag{22.19}$$

Direct carbonylation of methanol has the advantage of yielding ethanol without coproduct water, which would eliminate energy-intensive distillations. The cost-effectiveness of this approach to ethanol synthesis has not been proven.

22.3.6.2.2 Biological

Certain microorganisms, are known as unicarbontrophs, able to grow on one carbon compounds as the sole source of carbon and energy.[32] Acetogens can convert CO or mixtures of CO and H_2 to fatty acids and, in some cases, alcohols. *Clostridium Ijungdahli*, a gram-positive, motile, rod-shaped anaerobic bacterium isolated from chicken waste, has received particular attention for its ability to co-metabolize

CO and H_2 to form acetic acid (CH_3COOH) and ethanol (CH_3CH_2OH). The wild-type strain of *C. ljundahlii* produced an ethanol-to-acetate ratio of only 0.05 with maximum ethanol concentration of 0.1 g/L. This ratio is very sensitive to acidity; in decreasing pH to 4.0 and increased the ratio to 3.0. Other adjustments to the culture media and operating conditions nearly eliminated acetate production and increased ethanol concentration to 48 g/L after 25 days.

This gasification/fermentation route to bio-based fuels from lignocellulosic feedstocks has several advantages compared to hydrolytic/fermentation techniques.[33] Gasification allows very high conversion of feedstock to usable carbon compounds (approaching 100%) whereas hydrolysis only recovers about half the lignocellulose as fermentable sugars. Gasification yields a uniform product (a gaseous mixture of CO, CO_2, and H_2) regardless of the biomass feedstock employed whereas hydrolysis yields a product dependent on the content of cellulose, hemicellulose, and lignin in the feedstock. Finally, since the syngas is produced at high temperatures, gasification yields an inherently aseptic carbon supply for fermentation.

Biological processing of syngas has several advantages compared to chemical processing.[33] The H_2/CO ratio is not critical to biological processing of syngas, thus making unnecessary the water–gas shift reaction to increase the hydrogen content of biomass-derived syngas. Whereas catalytic syngas reactors require high temperatures and pressures, biocatalysts operate near ambient temperature and pressure. Also, biocatalysts are typically more specific than inorganic catalysts.

Syngas fermentation faces several challenges before commercial adoption.[33] Syngas bioreactors exhibit low volumetric productivities due, in part, to low cell densities. Cell recycle or immobilization of cells in the bioreactor are possible solutions to this problem. Mass transfer of syngas into the liquid phase is also relatively slow. In commercial-scale aerobic fermentations, mass transfer of oxygen is generally the rate limiting process. The problem will be exacerbated for syngas fermentations since the molar solubility of CO and H_2 are only 77% and 65% of that of oxygen, respectively. Dispersion of syngas into microbubbles of 50 μm diameter will be important to successful design of multiphase bioreactors.

22.3.6.3 Fischer–Tropsch Liquids

Production of hydrocarbons from syngas can be directly accomplished by F–T synthesis, which reacts and polymerizes syngas to light hydrocarbon gases, paraffinic waxes, and alcohols according to the generalized reaction:[30]

$$CO + 2H_2 \rightarrow -CH_2- + H_2O \tag{22.20}$$

Additional processing can produce diesel fuel and gasoline. Both methanol synthesis and F–T synthesis require careful control of the H_2/CO ratio to satisfy the stoichiometry of the synthesis reactions as well as to avoid deposition of carbon on the catalysts (coking). An optimal H_2/CO ratio of 2:1 is maintained through the water–gas shift reaction (Equation 22.16). Product distributions are functions of temperature, feed gas composition (H_2/CO), pressure, catalyst type, and catalyst composition. Depending on the types and quantities of F–T products desired, either low (200°C–240°C) or high temperature (300°C–350°C) synthesis is used with either an iron (Fe) or cobalt catalyst (Co).

The technology was extensively developed and commercialized in Germany during World War II when it was denied access to petroleum-rich regions of the world. Likewise South Africa, faced with a world oil embargo during their era of apartheid, employed F–T technology to sustain its national economy. A comprehensive bibliography of F–T literature can be found on the Web.[34]

22.3.6.4 Gaseous Transportation Fuels

The ideal transportation fuel is a stable liquid at ambient temperature and pressure that can be readily vaporized and burned within an engine. However, some gaseous compounds are also potential transportation fuels if their density can be substantially increased by compression. Among these gaseous transportation fuels are hydrogen, methane, ammonia, and dimethyl ether.

Hydrogen can be manufactured from syngas via the water–gas shift reaction (Equation 22.16). This moderately exothermic reaction is best performed at relatively low temperatures in one or more stages with the aid of catalysts. Although this might be one of the most cost-effective ways to produce hydrogen fuel, the physical characteristics of hydrogen present challenges in its use as transportation fuel In particular, its low density even under cryogenic or high pressure conditions limits on-board storage of this fuel. Its wide flammability range also presents unique safety problems in its use in transportation systems.[35]

Methane can be coaxed to be the main product of gasification in a process known as hydrogasification:[36]

$$C + 2H_2 \rightarrow CH_4 \tag{22.21}$$

$$CO + 3H_2 \rightarrow CH_4 + H_2O \tag{22.22}$$

These exothermic reactions require low temperatures, high pressures, and large quantities of hydrogen to favor complete conversion to methane. Thus, the process requires separate generation of hydrogen and catalysts to achieve reasonable reaction rates. It has been demonstrated at the commercial scale using coal as the carbonaceous fuel. Although more easily pressurized or liquefied than hydrogen, its density is still too low to be an attractive transportation fuel except in some urban mass transit applications.[37]

Anhydrous ammonia (NH_3) is a gas at ambient conditions but is readily liquefied at room temperature by storage at 10 bars pressure, achieving 87% of the density of gasoline.[4] In fact, it has nearly the same density, boiling point, and octane number as propane, which has been widely employed as a portable fuel source.

Ammonia is produced by the Haber process at 200 bar and 500°C[38]:

$$N_2 + 3H_2 \rightarrow 2NH_3 \tag{22.23}$$

As a widely employed agricultural fertilizer, the United States already has in place production, storage, and distribution infrastructure for its use.

Ammonia has been tested as fuel in spark ignition engines, diesel engines, and gas turbines. In tests dating back to the 1960s, near theoretical performance was achieved with ammonia if it was partially dissociated to achieve 1% hydrogen concentration at the engine intake.[39] Somewhat surprising for this nitrogen-rich fuel, nitrogen oxide emissions were lower than those obtained from octane fuel. Because of its lower heating value, an ammonia-fueled vehicle would require a fuel tank about 2.4 times larger than for a propane-fueled vehicle.

Dimethyl ether, like liquefied petroleum gas (LPG) is a non-toxic, flammable gas at ambient conditions that is easily stored as a liquid under modest pressures.[40] It is currently used as an aerosol propellant in the cosmetic industry, but has excellent potential as a fuel for heating, cooking, and power. It is particularly attractive as a substitute for petroleum-based diesel fuel since it has comparable cetane number but yields essentially zero particulate emissions and low NO_x emissions. It is produced either directly from syngas or indirectly through the dehydration of methanol by reactions at high pressure over catalysts. Of course, this syngas route also allows it to be produced from fossil fuels, and much of the current interest in this alternative fuel arises from the possibility of manufacturing it from inexpensive stranded natural gas.

22.3.6.5 Energy Return from Renewable Fuels

Some researchers have expressed concern that renewable fuels return less energy than the fossil energy used to produce them. This criticism has been particularly leveled against grain ethanol but more recently the energy return on any biomass-derived transportation fuel has come into question. The ratio of energy returned to fossil fuel invested is defined as:

$$R_E = \frac{E_{out}}{E_{f,in}} \tag{22.24}$$

where E_{out} = energy content of a unit of motor fuel, $E_{f,in}$ = fossil energy input to produce a unit of motor fuel.

Some authors[41] refer to this ratio as the "energy return on investment," but this name is more commonly associated with another kind of energy ratio subsequently described.

Notice that this definition is not equivalent to the classical energy efficiency for a thermodynamic process:

$$\eta = \frac{E_{out}}{E_{in}} \tag{22.25}$$

where E_{out} = energy content of a unit of motor fuel, E_{in} = energy input (both fossil and renewable) to produce a unit of motor fuel.

Neither should it be confused with energy return on investment (EROI), first introduced in the 1950s as a way to account for all the energy expended in the manufacture of an energy product, including the energy to extract, transport, process, and distribute the product, an accounting now incorporated into a life cycle assessment[42]:

$$\text{EROI} = \frac{E_{out}}{E_{M,\,in}} \tag{22.26}$$

where E_{out} = energy content of a unit of energy product, $E_{M,in}$ = energy consumed to manufacture a unit of energy product (excluding chemical enthalpy of feedstock).

The concept was originally formulated to compare the energy consumed in the manufacture of various kinds of durable and non-durable goods. For this purpose it provides a useful alternative to standard economic evaluations for decision making. However, when the concept was applied to energy products, such as electricity and motor fuels, the chemical enthalpy of the feedstock was not included in the EROI calculation. The chemical enthalpy of a fuel is usually the single largest energy input in its processing; thus, the advantage of fossil fuels compared to renewable fuels is often overstated in EROI comparisons because the production of energy products from fossil fuels often consume relatively smaller amounts of energy in their manufacture (this is particularly true for petroleum and natural gas).

The R_E is preferred when evaluating how effective an energy product is in displacing fossil fuels and reducing greenhouse gas emissions. An $R_E > 0.76–0.81$ (the range for the refining of gasoline) indicates at least some nominal advantage over petroleum-derived fuels while an $R_E > 1$ indicates that more "renewable" energy in the form of motor fuel was produced than fossil fuel was consumed in its production.

A large number of studies have investigated R_E in the production of ethanol from corn grain. Wang[43] summarized these results as a function of year of publication. As shown in Figure 22.26, R_E has been climbing over the years, but there is considerable scatter in the values. These values range from 0.44 to 2.1. Averaging over the values reported by 14 *different* study groups (to avoid replicating values of the same study groups reported in different publications) yields an energy ratio of 1.3. The differences appear largely to arise over disagreements on the amount of fertilizer applied to corn crops, the yield of corn crops, the ethanol yield from corn, and the amount of process heat required within ethanol plants.[44] In general, higher E_R values can be expected from large, modern ethanol plants. In comparison, the E_R for production of gasoline from petroleum is only 0.76–0.81.[41]

Evaluations of R_E for cellulosic ethanol manufacture are less common. Hammerschlag[41] reports that three studies found $r_E > 4.4$ while only one study, by Pimentel and Patzek,[45] report r_E to be as low as 0.69. This discrepancy appears to arise from Pimentel and Patzek's assumption that fossil fuels rather than

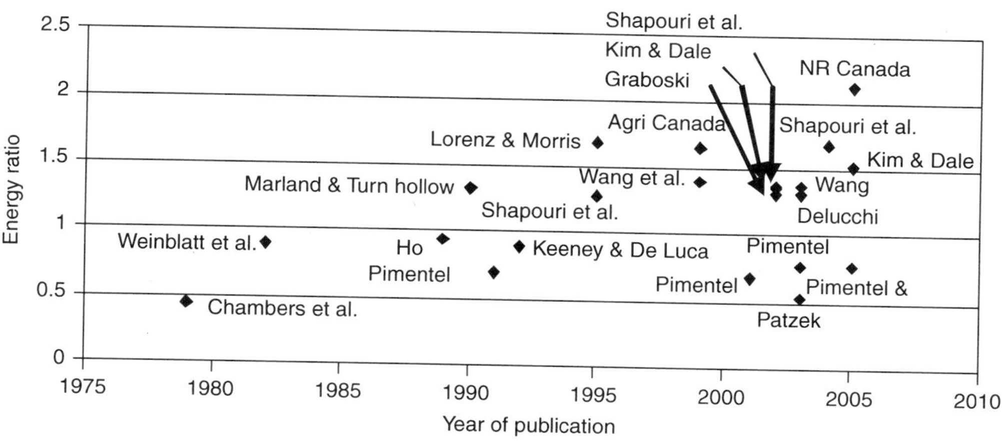

FIGURE 22.26 ER values reported by various investigators over a twenty-five year interval. (Adapted from Wang, W., *NGCA Renewable Fuels Forum*, National Press Club, Australia, August 23, 2005.)

lignin by-product will be used for process heat in the cellulose ethanol plant. With proper energy integration, R_E will likely exceed 4.4 in the production of ethanol from cellulose.

References

1. Borman, G. L. and Ragland, K. W. 1998. *Combustion Engineering*, pp. 25–60. McGraw Hill, New York.
2. National Renewable Energy Laboratory, Advanced Vehicles and Fuels Research, Petroleum-Based Fuels Property Database, http://www.engineeringtoolbox.com/fuels-properties-24_839qframed.html
3. National Renewable Energy Laboratory, Alternative Fuels, General Table of Fuel Properties, http://www.eere.energy.gov/afdc/altfuel/fuel_properties.html
4. MacKenzie, J. J. and Avery, W. H. 1996. Ammonia fuel: The key to hydrogen-based transportation. *Proceedings of the 31st Intersociety Energy Conversion Engineering Conference*, IECEC 96, Part 3 (of 4), Washington, DC, USA, IEEE, Piscataway, NJ, USA.
5. Teng, H. et al. 2001. Thermochemical characteristics of dimethyl ether—An alternative fuel for compression ignition engines. In *New Developments in Alternative Fuels for CI Engines SP-1608*, G. J. Thompson and B. T. Jett, eds., pp. 179–184. Society of Automotive Engineers, Warrendale, PA.
6. Rosillo-Calle, F. and Cortez, L. A. B. 1998. Towards ProAlcool II—A review of the Brazilian bioethanol programme. *Biomass and Bioenergy*, 14, 2, 115–124.
7. Bailey, B. K. 1996. Performance of ethanol as a transportation fuel. In *Handbook on Bioethanol: Production and Utilization*, C. E. Wyman, ed., Taylor & Francis, Washington, DC.
8. Lynd, L. R., Cushman, J. H., Nichols, R. J., and Wyman, C. E. 1991. Fuel ethanol from cellulosic biomass. *Science*, 251, 1318–1323.
9. Shadis, W. J. and McCallum, P. W. 1980. A Comparative Assessment of Current Gasohol Fuel Economy Data, Paper 800889, Society of Automotive Engineers, Detroit, MI, August.
10. Klass, D. L. 1998. *Biomass for Renewable Energy, Fuels, and Chemicals*, pp. 401–402. Academic Press, San Diego, CA.
11. Wayman, M. and Parekh, S. R. 1990. Cereal grains. *Biotechnology of Biomass Conversion: Fuels and Chemicals from Renewable Resources. Open University Press, Philadelphia, PA. Chap. 4.*
12. Watson, S. A. and Ramstad, P. E. eds. 1987. *Corn: Chemistry and Technology*, American Association of Cereal Chemists, St. Paul, Minnesota, MN.

13. Brown, R. C. 2003. *Biorenewable Resources: Engineering New Products from Agriculture*. Blackwell Publishing, Ames, IA.
14. Blanchard, P. H. 1992. *Technology of Corn Wet Milling and Associated Processes*. Elsevier, New York.
15. Rowell, R. M., Young, R. A., and Rowell, J. R. 1997. *Paper and Composites from Agro-Based Resources,*. CRC Press, Boca Raton, FL.
16. Schell, D., McMillan, J. D., Philippidis, G. P., Hinman, N. D., and Riley, C. 1992. Ethanol from lignocellulosic biomass, In *Advances in Solar Energy: Volume 7*, K. Boer, ed., pp. 373–448. American Solar Energy Society, Albuquerque, NM.
17. Lynd, L. R. 1996. Overview and evaluation of fuel ethanol from cellulosic biomass: Technology, economics, the environment, and policy. *Annual Review of Energy and Environment*, 21, 403–465.
18. Mosier, N., Wyman, C., Dale, B., Elander, R., Lee, Y. Y., Holtzapple, M., and Ladisch, M. 2005. Features of promising technologies for pretreatment of lignocellulosic biomass. *Bioresource Technology*, 96, 6, 673–686.
19. Wayman, M. and Parekh, S. R. 1990. Ethanol from wood and other cellulosics. *Biotechnology of Biomass Conversion: Fuels and Chemicals from Renewable Resources. Open University Press, Philadelphia, PA. Chap. 4.*
20. Shapouri, H., Duffield, J. A., and Wang, M. 2003. The energy balance of corn ethanol revisited. *Transactions of the ASAE*, 46, 4, 959–968.
21. Krawczyk, T. 1999. Specialty oils. *INFORM—International News on Fats, Oils and Related Materials*, 10, 6, 552.
22. Ouellette, R. J. 1998. *Organic Chemistry: A Brief Introduction. 2nd Ed.* Prentice Hall, Englewood Cliffs, NJ.
23. Lipinsky, E. S., McClure, T. A., Kresovich, S., Otis, J. L., and Wagner, C. K. 1984. *Fuels and Chemicals from Oilseeds*, E. B., Shultz, Jr. and R. P., Morgan, eds., AAAS Selected Symposium 91, Westview Press Boulder, CO., chap. 11.
24. Klass, D. L. 1998. *Biomass for Renewable Energy, Fuels, and Chemicals*, p. 356. Academic Press, San Diego, CA.
25. Klass, D. L. 1998. *Biomass for Renewable Energy, Fuels, and Chemicals*, pp. 341–344. Academic Press, San Diego, CA.
26. Ma, F. and Hanna, M. A. 1999. Hanna biodiesel production: A review. *Bioresource Technology*, 70, 1, 1–15.
27. Beenackers, A. A. C. M. and Van Swaaij, W. P. M. 1984. Methanol from wood: Process principles and technologies for producing methanol from biomass. *International Journal of Solar Energy*, 2, 349–367.
28. Klass, D. L. 1998. *Biomass for Renewable Energy, Fuels, and Chemicals*, pp. 427–429. Academic Press, San Diego, CA.
29. Forzatti, P., Tronconi, E., and Pasquon, I. 1991. Higher alcohol synthesis. *Catalysis Reviews-Science and Engineering*, 33, 1–2, 109–168.
30. Spath, P. L. and Dayton, D. C. 2003. Preliminary Screening—Technical and Economic Assessment of Synthesis Gas to Fuels and Chemicals with Emphasis on the Potential for Biomass-Derived Syngas, National Renewable Energy Laboratory Report NREL/TP-510-34929, December.
31. Chem. Eng. News. 1982. Technology: Catalyst Converts Methanol to Dry Ethanol, 60 (38), 41, September 20.
32. Zeikus, J. G., Kerby, R., and Krzycki, J. A. 1985. Single-carbon chemistry of acetogenic and methanogenic bacteria. *Science*, 227, 1167–1173.
33. Worden, R. M, Bredwell, M. D., and Grethlein, A. J. 1997. *Engineering Issues in Synthesis-Gas Fermentations, Fuels and Chemicals from Boimass*, pp. 320–336. ACS Symposium Series No. 666, American Chemical Society, Washington, DC.
34. Fischer–Tropsch Archive, 2005. www.fischer-tropsch.org
35. Romm, J. J. 2004. *The Hype About Hydrogen: Fact and Fiction in the Race to Save the Climate*. Island Press, Washington, DC.

36. Reed, T. ed. 1981. *Biomass Gasification: Principles and Technology*, Noyes Data Corp., Park Ridge, NJ.
37. Vieira de Carvalho, A. J. 1982. Natural gas and other alternative fuels for transportation purposes. *Energy*, 10, 2, 187–215.
38. Satterfield, C. N. 1991. *Heterogeneous Catalysis in Industrial Practice,. 2nd Ed.*, McGraw-Hill, Inc., New York.
39. Starkman, E. S. et al. 1966. *Ammonia as a Spark Ignition Engine Fuel: Theory and Application*, SAE Paper No. 660155, pp. 765–784.
40. Sorenson, S. C. 2001. Dimethyl ether in diesel engines: Progress and perspectives. *Journal of Engineering for Gas Turbines and Power*, 123, 3, 652–658.
41. Hammerschlag, R. 2006. Ethanol's energy return on investment: A survey of the literature 1990—present, Environmental Science and Technology.
42. Cottrell, W. F. 1955. *Energy and Society,*. McGraw-Hill, New York.
43. Wang, W. 2005. Energy and greenhouse gas emissions impacts of fuel ethanol,. NGCA Renewable Fuels Forum, National Press Club, Australia, August 23.
44. Shapouri, H., Duffield, J. A., and Wang, M. 2003. The energy balance of corn ethanol revisited. *Transactions of the American Society of Agricultural Engineers*, 46, 4, 959–968.
45. Pimentel, D. and Patzek, T. 2005. Ethanol production using corn, switchgrass, and wood; biodiesel production using soybean and sunflower. *Natural Resource Research*, 14, 1, 65–76.

23

Geothermal Power Generation

23.1 Introduction 23-2

23.2 Definition and Use of Geothermal Energy 23-2

23.3 Requirements for Commercial Geothermal Power Production 23-3
Definition of a Geothermal Resource (Heat, Permeability, Water) • Improving a Geothermal Resource through Human Intervention • Geothermal Energy as a "Renewable" Resource • Economic Access • Economic Access: Power Plant Cost

23.4 Exploration and Assessment of Geothermal Resources 23-15
Overview • Exploration and Discovery • Risk of Exploration

23.5 Management of the Geothermal Resource for Power Production 23-18
Goals of Resource Management • Resource Chemistry • Barriers to Resource Management • Resource Characterization

23.6 Geothermal Steam Supply (from Wellhead to Turbine) 23-25
Overview of Steam/Brine Separation for Steam Turbine Supply • Multipressure Steam Flash and Separation • Steam Pipeline Operation • Steam Washing Prior to Steam Turbine Admission • Turbine Washing to Remove Scale Buildup

23.7 Geothermal Power Production—Steam Turbine Technologies 23-32
Overview of Geothermal Power Generation Technologies • Steam Turbine Conversion • Condensing Steam Turbine Process • Design of Geothermal Turbines • Design of Heat-Rejection Systems • Geothermal Condenser Gas-Removal Systems

23.8 Geothermal Power Production—Binary Power Plant Technologies 23-38
Binary Power Plant Advantages • Types of Binary Systems • Binary Power Plants for Pressurized Geothermal Brine • Integrated Steam Turbine and Binary Power Plants

23.9 Environmental Impact 23-43
Geothermal Power Plant Emissions • Land Use
23.10 Additional Information on Geothermal Energy 23-46
References 23-46

Kevin Kitz
U.S. Geothermal Inc.

23.1 Introduction

Roman mythology holds that humans obtained fire from the gods on Mount Olympus to meet their needs for light and heat. In much of today's industrialized world, the use of electricity has edged out fire for these age-old needs. However, the convenient use of electricity is now faced with increasing cost and concerns about the availability and environmental consequences of the fuels that are needed to produce it. Can those Roman gods again come to mankind's aid?

Deep below Mount Olympus, Vulcan toils over his forge. The forge glows red as Vulcan hammers out weapons for the gods, including Jupiter's own thunderbolts. Vulcan's forge is stoked by geothermal energy that humans, too, have used for the production of electricity since 1904, starting in the very homeland of the Romans.

Geothermal power production in the U.S. approached 18,000 GWh in 2005,[1] or an average of more than 2000 MW over the year. It does so on an around-the-clock basis, providing baseload power to several western utilities in California, Nevada, Hawaii, Utah and starting in 2006, Alaska. The power generated has very low emissions and is immune to price fluctuations in the fossil fuel markets. Similar benefits are derived around the world from plants operating in 24 countries worldwide, including Italy, the Philippines, Indonesia, Kenya, New Zealand, and Iceland. In Iceland, geothermal power not only provides 16.6% of the electric power generated every year, but 54% of the total primary energy use in the country, including the 87% of households that are heated by geothermal water.[2] Worldwide direct use of geothermal energy is documented in 76 countries.[3]

This chapter examines geothermal power technologies and the issues involved in further utilization of geothermal energy for the production of electric power.

23.2 Definition and Use of Geothermal Energy

Geothermal energy is present on Earth from two sources:

1. As heat that flows upward and outward across the entire surface of the world from the very deep (mantle and core) radioactive decay of uranium, thorium, and potassium. In most regions of the world, this energy flux is too small to be commercially useful for any purpose.
2. As localized heat resulting from the movement of magma into the earth's crust. In some areas of the world, most frequently along the so-called "Ring of Fire", this localized heat, with high heat flux and high temperatures, can be found between the surface and 10,000 ft. below ground. Where such heat flux meets the requisite conditions, geothermal energy can be developed for multiple and varied purposes. Where temperatures are sufficiently high, geothermal energy may be used for electric power generation. This form of geothermal energy is the subject of this chapter.

Geothermal energy was first used for experimental power generation in Larderello, Italy, on July 4, 1904, by Prince Piero Ginori Conti. However, commercial development followed slowly thereafter. The Larderello site also saw the first commercial geothermal power plant (250 kW) in 1913, as well as the first large-power installation in 1938 (69 MW). It would be 20 years before the next large geothermal power installation would be built: halfway around the world in Wairakei, New Zealand, with the first unit—commissioned in 1958—that, under a steady development program, grew to 193 MW of installed capacity by 1963. In the U.S., the installation of the first unit (11 MW) of what was to become the largest geothermal power complex in the world, The Geysers, in Sonoma County, California, occurred in 1960.

Over the next quarter-century, a total of 31 turbine generator sets were installed at The Geysers, with a nameplate capacity of 1890 MW. Plant retirements and declining steam supply have since reduced generation at The Geysers to an annual average generation of 1020 MW from 1421 MW of installed capacity—still the largest geothermal field in the world.[1]

Since those early efforts, Lund et al. report that a total of 2564 MW of geothermal power generation capacity is currently installed in the U.S., generating approximately 2000 MW of annual average generation. In 2005, worldwide annual geothermal power was estimated at 56,875 GWh from 8932 MW of installed capacity.[3] These values put the U.S. and worldwide geothermal power plant capacity factors* at 78% and 73%, respectively.

Geothermal energy is also utilized in direct-heat uses for space heating, recreation and bathing, and industrial and agricultural uses. Geothermal energy in direct-use application is estimated to have an installed capacity of 12,100 MW thermal, with an annual average energy use of 48,511 GWh/year energy equivalent.[3] This excludes ground-coupled heat pumps (GCHPs, see below). In the U.S., the first geothermal heating district was installed in Boise, Idaho, in 1892, and is still in operation today.

GCHPs are also often referred to as *geothermal*. GCHP units are reported by Lund to have 15,721 MW of installed capacity and 24,111 GWh of annual energy, representing 56.5% and 33.2% of worldwide direct use, respectively.[3] It should be noted that although these numbers are reported as "geothermal," in many instances the ground temperature is primarily controlled not by the flow of heat from below the surface, but by the annual average ambient temperature of the location, which is a solar phenomenon. From the numbers above, it can be seen that GCHPs provide an additional worldwide energy production of almost 50% more than that provided by geothermal electric power generation.

This chapter will cover the technologies and issues in the utilization of geothermal energy for electric power generation, with a particular focus on the issues facing the development of new capacity, both technical and economic. Although the technologies of geothermal energy for power and for direct use are quite different, the issues covered in this chapter related to reservoir issues, and the economic factors affecting or controlling the development and maintenance of the reservoirs are often the same or similar.

23.3 Requirements for Commercial Geothermal Power Production

For new geothermal power to be installed, or an existing geothermal power plant to continue operating, commercial geothermal power production requires two major elements: a geothermal resource and economic access.

23.3.1 Definition of a Geothermal Resource (Heat, Permeability, Water)

A geothermal resource for power production comprises three major elements: heat, sufficient reservoir permeability, and water.

1. Heat is clearly the first element for commercial power generation. Most commercial geothermal resources produce fluids from reservoirs with a resource temperature of at least 320°F. However, there are examples of geothermal fields with lower temperature fluids. Nonetheless, resource temperature is a good first indicator of economic viability, see Table 23.1. As can be calculated from the table, the flow requirements for a 60 MW plant are 18 million pounds per hour of 300°F geothermal liquid, but only 4 million pounds per hour of geothermal liquid if the reservoir temperature is 450°F.

*Capacity factor is defined as the actual generation produced compared to the theoretical generation that would be produced in one year with the power plant running at full rated capacity.

TABLE 23.1 Power Generation Potential from a Range of Resource Temperatures

1000 lbm Geothermal Fluid of	Electrical Generation (kWh)	Power Plant Type[a] (see Later Sections)
Liquid at 300°F	3.3	Binary
Liquid at 350°F	5.6	Single-flash steam
Liquid at 400°F	10.4	Double-flash steam
Liquid at 450°F	14.5	Double-flash steam
Steam at 350°F	53.5	Dry steam

[a] The column labeled "power plant type" is the technology used as a basis for the power generation calculation. All plant types are used over a larger range of temperatures than that indicated in the table.

Source: From DiPippo, R. 2005. *Geothermal Power Plants—Principles, Applications and Case Studies*, p. 424.

2. *Permeability* describes the ability of the reservoir fluid (water or steam) to flow through the rock formation. Permeability allows deep-seated geothermal heat sources to create a geothermal resource through the convection (flow under the influence of heating and cooling) of hot water or steam. The convection of the geothermal fluids through the reservoir heats a large volume of rock, thereby storing a large quantity of energy over a period of tens of thousands of years. Geothermal reservoir permeability is dynamic, with the hotter fluids dissolving minerals and increasing permeability, and the cooling fluids depositing minerals and restricting permeability. Reservoir permeability is what allows the extraction of that stored heat through a relatively few number of wells in commercial geothermal developments.

   Matrix permeability is the ability of the fluid to flow through the bulk rock itself. Fresh water wells, natural gas wells, and oil wells are frequently observed to obtain a significant portion of their total flow from matrix permeability. In other words, the fluids enter the wellbore as uniform flow along an area of unfractured rock. Sufficient matrix permeability is rare in geothermal wells that will support commercially significant flow rates.

   Fracture permeability is low most geothermal wells which obtain the majority of total wellbore inflow through naturally occurring fractures in the rock. Pervasive fracturing allows a single well to obtain flow contributions from a large area at high flow rates, even where the matrix permeability is low. In most cases, the fractures extend over large areas and volumes as a result of tectonically active areas, but some fields have been discovered and developed that essentially comprise a single fault system (which may be comprised of multiple fractures in the rock). In such fields, only wellbores that cross the fault produce geothermal fluid at commercial rates.
3. Water is the motive fluid for geothermal power production. It may be brought to the surface as steam from one of the few (but typically large) geothermal steam fields, such as Larderello (Italy), The Geysers (California), and Kamojang (Indonesia). A water-steam combination may also be produced to the surface from high-temperature liquid-dominated or liquid-and-steam reservoirs. In low- or moderate-temperature resources, pressurized water may be pumped to the surface using downhole pumps.

The development of new geothermal power plants and the expansion and maintenance of existing geothermal fields depends on the three elements of heat, permeability, and water. The three occur simultaneously in relatively few places around the world, but there remain many thousands of megawatts of undeveloped worldwide and U.S. resources. Over the long term, the natural resource base for geothermal power could be supplemented with human intervention to create new systems or enhance existing systems.

23.3.2 Improving a Geothermal Resource through Human Intervention

The terms *hot dry rock* (HDR) or *hot fractured rock* (HFR) refer to a family of experimental technologies that are not yet commercially proven. The objective is to establish one or more of the missing elements of a commercial geothermal resource (specifically, permeability or water) where a heat source already exists. HDR experiments have been undertaken as research projects by the U.S. Department of Energy and similar agencies in Europe and Japan. The problems are daunting and the costs are high; U.S. DOE funding, for example, has diminished substantially from its maximum in the 1980s.

The concept behind HDR and HFR is to drill a well into the hot rock and then pump water into the well at very high pressure, causing the rock to fracture. The fractures provide a heat transfer surface and flow path allowing water to be pumped from the surface into an injection well, circulated through the man-made fractures, and ultimately recovered in a production well some distance away from the injection well. The theory is straightforward; unfortunately, the implementation to date has not been.

As of 2005, the first privately-funded HFR development had commenced at the Cooper Basin in southern Australia, and it will be watched closely for its success. The developer's plans call for an initial installation of a 3 MW power plant if the fracturing process is successful. [5]

If HDR and HFR technology is developed and implemented in significant quantities over the next 20 years, the geothermal reservoir management strategies and the energy-to-power conversion technologies will be much the same as that described in the rest of this chapter.

Between the naturally-occurring resource base and the potential man-made resource base of HDR/HFR is the enhanced geothermal system (EGS). EGS technologies seek to supplement a naturally occurring geothermal resource primarily by the addition of more liquid, or by stimulation of wells to tie into a larger, naturally occurring fracture network. The goal of EGS is to extend the life or capacity of existing fields, rather than the creation of an entirely new resource, as is being attempted at the Cooper Basin described above. There have been many studies, evaluations, and proposals to date, and the use of EGS at The Geysers project—including those plants owned and operated by Calpine* and NCPA†—has been a proven success.

At The Geysers, two pipelines bring 77,000 m^3/day (20 million gallons/day) of treated wastewater from adjoining cities to the mountains where The Geysers facilities are located. The water serves to replace that which has been depleted from The Geysers over its approximately 50 year operating history. Stark et al. report that tracer tests show that 40% of the injected water is recovered as steam within a year. The injection of the water from just one of the two pipelines (delivering about 55% of the total wastewater) is forecast to result in an annual average generation increase of 84 MW. With a parasitic pump load of 9 MW, the annual average benefit is roughly 75 MW.[6]

23.3.3 Geothermal Energy as a "Renewable" Resource

Throughout this chapter, declines in flow rates from geothermal wells and output from geothermal power plants are discussed. What does this mean in terms of whether geothermal energy is renewable?

> *In recent years many thoughtful papers have been published on the renewability and sustainability of geothermal energy... However, no universally accepted definitions of the words "renewabilty" and "sustainability" seem to exist and definitions used often have ambiguities...*[7]

From a thermodynamic point of view, it may seem that the only true renewable geothermal development would be one in which the extraction rate is the same as the natural heat influx rate into the system. However, reservoir simulations and field observations frequently reveal that the natural heat influx rate increases as production occurs, due to pressure changes that allow more hot liquid to flow into the system. Therefore, operating at a "nonrenewable" level increases the ultimate energy extraction from the resource. More importantly, a true "renewable" level of energy extraction would very often be subeconomic, and is therefore of little interest in the development of geothermal resources for society's benefit.

---

*Calpine Corporation, http://www.calpine.com.

†Northern California Power Agency, http://www.ncpa.com.

Another definition argues that, as long as the power or heat usage from the resource continues at a constant level for hundreds of years, it approximates a true renewable resource, although it may be termed a "sustainable resource." Again, this is an interesting theoretical discussion, but not one that actually is put into practice in the development of most geothermal resources, for both the reasons of economics and an inability to know what this actual level would be.

Sanyal (2004) argues that sustainable geothermal development occurs if the project maintains its output, including make-up well drilling, over the amortization period of the power plant.[7] This definition recognized the critical role that economics plays in actual natural resource development decisions. In short, projects are developed to meet economic requirements. This is true of any new power project, whether geothermal, another renewable source, or a fossil-fired resource. Section 23.3.4, below, will illustrate why the question of sustainable development for a 50-year period, let alone a 300-year period, will not play a role in the development decisions for a particular geothermal resource. The reason, in summary, is that even at a mere 50 years in the future, there are no meaningful economic consequences to development decisions made today when using a discounted cash flow analysis.

More valuable than the theoretical discussion of whether geothermal energy is renewable, sustainable, both, or neither, is to look at the history of geothermal development. Geothermal use at Larderello, Italy, is over 200 years old, starting with mineral extraction in the early 1800s and including almost 100 years of commercial power generation. During the last 100 years of power generation, turbines have been renewed or replaced, power output has grown and shrunk and grown again, and new methods of extracting more energy have been developed. In fact, geothermal power generation in 2003 was higher than any other year, with 5.3 billion kWh produced.[8]

Many further decades of geothermal power generation are expected from the Larderello region, as well as Wairakei, The Geysers, and other fields that are approaching 50 years of power generation. Geothermal power-generating facilities only very rarely cease operation, although few will reach their fiftieth anniversary at their maximum generation level.

23.3.4 Economic Access

Given that a geothermal resource exists—whether naturally occurring or developed with EGS or HDR technologies—the resource will be economically viable for power generation only if the four elements of economic access are met: wellhead energy cost, electricity transmission, a market for the electric power, and the power plant cost.

23.3.4.1 Economic Access: Wellhead Energy Cost

Wellhead energy cost is the total cost to bring useable geothermal energy to the surface and to return the cooled geothermal fluids to the reservoir. The thermodynamic condition (pressure, temperature, and flow rate) of the geothermal energy at the surface will influence the capital cost of the power plant. The wellhead energy cost is the result of the combined effects of the productivity of the well, the cost to drill the well, and the enthalpy (or available energy) of the fluid that is produced from the well, among other aspects. These factors vary widely from field to field, and even within a single field.

- Well productivity is the ability of the well to bring fluid to the surface at a temperature and pressure useful for power generation. Clearly, the more fluid that is produced by each well, the fewer wells are needed for a given power plant size, and therefore the lower the total cost of the geothermal well-field development. Less obvious is the fact that highly-productive wells can also have a major impact on the cost and design of the power plant itself.

  Highly-productive wells allow the power plant inlet pressure to be increased (and thereby increase the available energy) on plants that directly use the geothermal steam in a turbine. An increase in the power plant inlet pressure can allow the size of the plant to be increased, because last-stage turbine blades can only be built to a limited maximum size in geothermal service and in binary power plant turbines as well (see Section 23.5 and Section 23.6). Alternatively, for plants contractually limited to a particular megawatt capacity, e.g., 60 MW, higher inlet pressures reduce the amount of fluid that must be extracted from the reservoir and delivered to the power plant, thereby reducing the size of the heat rejection system and the injection capacity (two major components of geothermal power costs, as will be discussed in later sections).

In binary power plants (see Section 23.6) with pumped wells, highly productive wells not only reduce the direct drilling costs, but also reduce the number of pumps that must be installed (at an installed 2005 cost of roughly $500,000 each). Even more importantly, the pumps can be set at a shallower depth, thereby reducing the pump parasitic loads. The energy for the production pump parasitic loads (400–1000 hp per pump) is supplied by the power plant itself. Again, the productivity of the wells will either reduce the size of the plant that must be built to meet contractual and pump parasitic obligations, or can increase the amount of power available for sale from a particular power plant. The combined effect of fewer wells and more power sales dramatically increase the economic viability of a new or expansion geothermal binary power plant development. Well injectivity is a similar issue for the disposal of the residual geothermal liquids for both binary and steam power plants.

- The energy production rate (a combination of mass flow, temperature, and steam content) of production wells vary from field to field, often varying substantially within a given field itself, and almost always changes over time for a particular well. In steam power plants, high-enthalpy wells can result either from a high-temperature portion of the field, or as a result of what is termed "excess steam" production. Excess steam occurs when the enthalpy of the produced fluid is higher than the enthalpy of the reservoir fluid as a whole due to the greater mobility of the steam through the reservoir rock than liquid water. In pumped brine binary power plants, high-enthalpy wells are simply wells with higher flowing temperatures. For both steam and binary power plants, higher enthalpy wells result in lower power plant cost, as the benefits of the higher enthalpy are found in lower drilling costs (fewer wells), lower heat-rejection system costs (higher available energy per unit mass into the plant), and lower brine-injection costs (less produced fluid).
- The depth of geothermal resources is highly variable, with most geothermal resources produced between depths of 1,500–7,000 ft. below surface, although resources commonly exist outside of this range. One example of a shallow resource is the Salt Wells field in Nevada, in which the production and injection wells are between 450 and 750 ft. deep. One particular well at that field is reported to be only 470 ft. deep and to be capable of between 4 and 5 MW of power from 140°C (284°F) fluid. The drilling time for this well was only 12 days.[9] The cost of the well is on the order of hundreds of thousands of dollars. By contrast, make-up wells drilled at one field in the Philippines almost 25 years after the start of production had a true vertical depth of over 10,000 ft., and an even greater measured depth due to the lateral reach of the well. The cost of these wells between 2001 and 2003 was on the order of $3 million to $5 million dollars each. At the Puna geothermal project in Hawaii, drilling is difficult and the location remote, so the daily drilling cost is high. Wells drilled in 2005 cost approximately $6 million, and were drilled to a depth of about 6,000 ft.

There is no threshold value for the wellhead energy cost below which geothermal development is viable and above which it is not. Rather, there is a continuum in which the effects of transmission, market, and the power plant cost affect to varying degrees the ability to construct a new geothermal plant.

For existing power plants, the wellhead energy cost determines whether or not a make-up well will be economical, and ultimately the end of make-up well drilling. After make-up well drilling ends, geothermal power plants enter a period of slow decline in the output of the power plant over the coming years and decades. Although the plant is no longer able to achieve its initial rated capacity, it will continue to operate economically, literally for decades. There are very few examples of geothermal resources that, once in operation, shut down completely.

23.3.4.2 Economic Access: Electricity Transmission

With a suitable geothermal resource and a viable market for geothermal power, the link between the two is transmission. As with other renewable and nonrenewable energy sources, transmission can be a significant issue in the viability of development of a new geothermal resource, or in the expansion of an existing geothermal power facility.

Occasionally, geothermal resources are found in close proximity to existing transmission access. Examples include Steamboat Springs, Nevada and Raft River, Idaho. For these developments, the cost of transmission access is low, and transmission line construction and interconnection does not play a major role in the development of the resource.

However, geothermal resources often occur at substantial distances from existing transmission, and entail the need for considerable capital expenditures on the transmission system to deliver the power to the customer. If the field is large enough, the long transmission lines and expenses can be justified. Two separate subsea cables were built in the Philippines to transport hundreds of megawatts of electricity generated from geothermal resources from islands with small native loads to the main load center on the island of Luzon.

The cost of the transmission line has most often been the responsibility of the utility that was receiving the power (e.g., Pacific Gas and Electric for the power generated at The Geysers), whereas in more recent years, the development of the power line has been the responsibility of the geothermal developer in the U.S. The power price offered to geothermal and other renewable developers may fail to recognize this cost (e.g., the Idaho PURPA posted rate), yet the cost must still be borne by the developer, and as such, the cost of transmission construction can become a substantial burden to the project economics. A 107-mile-long dedicated transmission line was built and paid for by a consortium of geothermal power developers in the Imperial Valley, California, to deliver some 500 MW of power to a common customer, Southern California Edison. After the line was built, it was turned over to the local utility, the Imperial Irrigation District, for ongoing operation and maintenance.

When a transmission system is used to deliver power and the transmission lines are owned by an entity other than the ultimate purchaser of the power, the use of that transmission system incurs wheeling costs. Wheeling costs comprise a monthly reservation/use fee and a power loss that can be paid either with funds or with power delivered to the transmission entity. Typically, wheeling charges are not charged by the utility that actually is receiving the power under contract. "Pancaked charges" refers to the circumstances in which power crosses multiple transmission segments owned by different entities, incurring separate transmission charges through each segment. The losses assessed by each utility usually are not based on actual losses for the power being transmitted, but represent a system-wide average apportioned out to the users of the system on a pro-rata basis. In the U.S., wheeling protocols are established by the FERC (Federal Energy Regulatory Commission).

Common wheeling charges are 1%–5% for losses, plus $1,000–$3,000 per MW-month for the reservation costs for firm capacity. Therefore, wheeling through more than one, and often even only one, utility becomes prohibitive. As a result, the viability of development is strongly influenced by the willingness of the utility in whose control area the power plant is located to accept the power. Delivery to the closest utility generally eliminates wheeling fees and losses and improves the economic viability of the plant. On the other hand, having only a single party with whom to negotiate a contract puts the developer in a difficult position.

In deregulated markets, such as the U.S., transmission reservations are the mechanism by which a power plant ensures access to the market for the life of the contract. The ability to obtain that reservation is critical to development, or else the geothermal development may face greatly increased transmission costs for transmission upgrades or longer transmission lines. One of the advantages of geothermal power as opposed to wind power is the efficiency of transmission usage. Both resources must reserve transmission capacity for their peak delivery, e.g., 100 MW. Wind will typically deliver between 25 and 40 MW on an annual average basis, while geothermal will typically deliver about 90–95 MW. On the transmission system peak hours, the difference in utilization will be even greater, due to the intermittency of the wind resource and the typical incentives in a geothermal contract to be on-line during peak hours. A geothermal power plant can be expected to be on-line and at near-full capacity for over 95% of the peak hours.

Where transmission system capacity is limited, geothermal and other baseload resources will make much greater use of that capacity. Comparisons of new generation costs seldom account for the efficiency and hidden costs of transmission usage, which is unfortunate in the evaluation of the economics of geothermal power generation, because it is so efficient in its use of transmission capacity.

23.3.4.3 Economic Access: Viable Market

Power generation, regardless of the technology, is a highly specialized form of manufacturing. As with all manufactured products, there must be a place to sell the manufactured product. Therefore, for new geothermal power to come online, regardless of how low the wellhead energy cost and even if transmission is readily available and affordable, without a market, there is no development. Market consists of three elements: willing buyer, price vs. characteristics, and contract provisions. The first two are much the same for all power developments, whether renewable or otherwise. For geothermal power, there are unique contract provisions that must be present for realistic expansion of the role of geothermal power in the electrical demand of the United States.

23.3.4.3.1 Willing Buyer

The willing buyer must be just that: "willing" and a "buyer." In the U.S., this rare beast has been found in a number of geothermal development habitats over the last 40 years. More recently, the captive breeding program seems to be showing success. A much larger number of willing buyers have been observed, as many U.S. electric utilities have been stung first by wild swings in natural gas prices, second by 2004–2005 gas prices climbing from $4 or less to over $10 per million Btu, with long-term price forecasts for gas remaining high, and lastly by state-mandated renewable energy targets.

From 1960 through roughly the mid-1980s, geothermal development in the U.S. (and in many places around the world) was most commonly undertaken by two parties: the resource developer and the local utility. The resource developer drilled wells and built cross-country pipelines to deliver a flow of steam or mixed brine and steam to the power plant boundary. The resource developer also was responsible for injection of the residual liquids, both brine and condensed steam. The local electric utility—i.e., the willing buyer—used the steam to generate power. Under this U.S. development model, the utility met the mandate of its monopoly franchise to supply power. The utility also met its investor's objective of investing in new facilities, the power plant and transmission lines, for which it was able, under its monopoly, to earn a regulated rate of return on the invested capital dollars. However, an inherent inefficiency was created in this development model by having two owners who each had different economic drivers, separate operating groups that worked side by side, double administration costs, and other factors. One common and unfortunate outcome of the two-owner model occurred because the steam was sold on a $/MWh basis instead of on the energy content of the steam. The consequence was that the power plant owner built inefficient power plants because inefficiency is cheaper than efficiency, if one only looks at the power plant cost. Of course, the consequence was that the steam developer had to drill additional risky and expensive wells to supply additional steam for the inefficient power plant. Those few contracts that had steam sales based on energy content saw exactly the opposite effect in power plant design. The power plants tended to be very efficient to minimize steam use (fuel cost). The net result of two-owner operation was that the price of geothermal power was comparatively high and resource utilization was inefficient.

With the widespread occurrence of independent power producers (IPP) in the U.S. in the mid-1980s, a regulated utility could theoretically obtain a lower price through the integrated operation of geothermal field and power plant and could meet its mandate for power supply, but by buying power from the IPP, the utility could not meet the objectives of its shareholders for new investment. The IPP developer's objective of building a power plant put it in direct competition with the utility for the return on investment in new power plant capacity. Under the IPP model, each megawatt of capacity installed by an IPP directly equates to less investment by the utility in baseload power (potentially $700,000–$1,700,000 per MW). Consequently, because the utility faces a reduced rate base (the capital on which the utility is allowed to earn a return), the IPP faces an intrinsic unwillingness in the utility to procure the output of a new geothermal (or any other type of fuel) plant. Although geothermal development should, in theory, thrive under the IPP model, the reality is that utilities in the U.S. have shown little interest in bringing on new geothermal development unless forced to do so in the interest of the ratepayer by governmental authority.

From the mid-1980s through the mid-2000s, many geothermal power plants were built exclusively because of the Federal PURPA* regulations and the consequent requirement of state public utility commissions for the utility to buy the power at a published avoided cost. For example, California's Standard Offer Number Four resulted in the construction of several of the Salton Sea and Heber Units, among others, in the late 1980s and early 1990s. Idaho's Posted Rate 10 MW PURPA contracts resulted in the development of the Raft River Geothermal Power Plants in southern Idaho, with anticipated operation dates of 2007 through 2009. In California and Nevada, the passage of Renewable Portfolio Standards has also spurred utilities to contract for geothermal power to comply with state law. Both PURPA and the RPS have created utility buyers, but they have not necessarily been "willing." The state RPSs and the prospect of a federal RPS is now emerging as a major factor in the decision of utilities to sign contracts for IPP-developed geothermal (and other renewable) power plants, in spite of the fact that it is against the interests of their shareholders, though not their ratepayers.

23.3.4.3.2 Price of Delivered Power

The price vs. the characteristics for the electrical energy is the most obvious component of a viable market. The characteristics of geothermal power are very attractive and rare in the renewable power industry. It is price-stable, has low to negligible environmental impact, and has a steady, predictable, and reliable output.

The unit price paid for geothermal-generated electrical power must cover the expense of the power plant and the wellfield development, as well as the often long and expensive exploration phase. In addition, the price paid per MWh must satisfy annual cash flow requirements and the return on investment sought by the geothermal developer. As most contracts from the late 1990s onward are not in the public domain, it is hard to know what constitutes an adequate price for electricity generated from geothermal energy. A few that have emerged are described below.

In 2003, the power purchase prices awarded to three geothermal power plants after a round of renewables bidding in Nevada were inadvertently released. The contracts were for both the power output and environmental attributes ("green tags," carbon credits, etc.). Two were for expansions to existing power plants. One price was less than $45/MWh, and the other was more than $52/MWh. The third contract was for a greenfield power plant at under $50/MWh. All were for resources with temperatures greater than 320°F. Only the power plant that bid the highest price was ever built. The other plants proposed power purchase prices too low to actually accomplish development drilling, build the power plant, and close financing. The prices bid for these plants were offered before the 2004–2005 worldwide run-up in steel prices that had a tremendous negative impact on the cost of building a geothermal power plant with its miles of steel well casing, miles of steel surface piping, and enormous steel heat exchangers.

In January 2005, the Idaho Public Utility Commission approved a 20-year PURPA contract to supply power from the U.S. Geothermal, Inc.† Raft River Geothermal Power Plant for an initial price of $52.70, escalating at 2.3% per year.[10] This gives an equivalent levelized price of $60.99. Although Raft River is a greenfield development, the production and injection wells were drilled in the late 1970s, with those costs written off by previous owners.

*PURPA is the Public Utility Regulatory Policies Act.

†U.S. Geothermal, Inc., http://www.usgeothermal.com.

In May 2005, two 10 MW contracts were announced at a fixed price of $57.50/MWh with an annual escalation of 1.5%. These contracts were for expansions to existing power plants at two different fields in southern California (Heber and East Mesa). The price includes the value for the energy, as well as the environmental attributes. The contract also allowed for the sharing of U.S. federal production tax credits (PTCs) with the power purchasers.[11] The levelized price, excluding the PTCs, would be about $63. Speaking at the groundbreaking of an additional 8 MW expansion of the Heber facility, Robert Foster, the President of Southern California Edison, said, "I like... the geothermal plant... because we know that these plants produce reliable power 24×7. Unlike other forms of renewable energy, I know that this energy will always be there and we don't need to have shadow generation to support geothermal power. Even though we like to have dispatchable* power, we can forego dispatchability for the reliability of geothermal power."[11]

An international benchmark is the Sarulla, Indonesia, development, in which the state oil company drilled the wells for the power plant and the power plant developer was responsible for the power plant and pipeline system (excluding the cost of transmission and drilling.) The power price was bid at approximately $45/MWh in 2005 by two companies. Price escalations and adjustment factors are not available at the time of this writing.

An adequate selling price for the generated electrical power in a geothermal development is determined in part by resource temperature (higher temperature means lower power plant price), wellhead energy costs, power plant size (smaller requires higher price) and whether the new power plant is greenfield or an expansion (expansion can take advantage of existing infrastructure and personnel costs). Within the geothermal community, the general rule of thumb is that for geothermal resources above 300°F, a true grassroots geothermal power plant must sell power for $60–$65 per MWh for at least 20 years, for an all-in project (wells, pipelines, power plant, and transmission). Expansions to existing facilities can reduce the cost by 10%–15%,[12] or roughly a $5–$15 per MWh reduction, with most of the savings being in a lower drilling failure rate, use of the existing transmission line, and sharing of operating personnel and spare well capacity. Figure 23.1 provides representative cost elements in a geothermal power development.

23.3.4.3.3 Contract Provisions for Operation

Contract provisions for a viable geothermal power market are two fold: those governing once the plant is up and operating, and those that govern prior to the operation date.

Baseload. The nature of geothermal power plants lead to the general requirement that geothermal plant contracts allow them to operate *baseloaded*† at full available capacity. Three factors lead to this requirement: low variable costs (see Table 23.2), high plant availability, and high fuel/plant capacity factor.

Geothermal plants have low variable costs, which are those costs resulting from changes in plant output. Therefore, any dispatch order that reduces the output of the geothermal plant simply raises the cost of power during noncurtailed periods to recover the fixed costs. The geothermal power plant operates most effectively and at greatest economic efficiency at full baseload.

Geothermal power generation typically operates at a 95%–99% plant availability (hours capable of operation per year divided by hours per year). What this means is that as long as the plant has permission to operate, it will generate power, because it also has a high fuel availability. By contrast, wind generation also has high mechanical availability, but because the wind is not blowing all the time (no or low "fuel" availability), the plants do not generate, even though they are "available." Although there is little difference in wind and geothermal generation plant mechanical availability, the capacity factors are very different because of the nature of the fuel supply.

---

*A dispatchable plant can be ordered to vary its load from zero to full load to meet the needs of the electric power grid.

†The term baseload refers to the minimum amount of power continuously required by a utility's customers over a 24-h or annual basis. A baseloaded plant is one that remains on-line at all hours of the day at maximum or near maximum output to serve that demand.

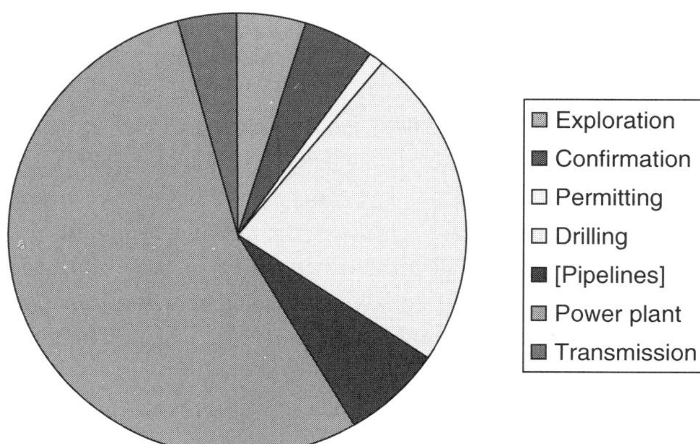

FIGURE 23.1 Typical cost breakdown of geothermal power projects. (From Hance, C. N. 2005. *Factors Affecting the Costs of Geothermal Power Development*. Geothermal Energy Association and the U.S. Department of Energy, Washington, DC).

Geothermal power plants typically operate at high capacity factor (MWh generated per year divided by the product of hours per year and plant capacity).* The high mechanical availability and the steady flow of energy from the production wells means that geothermal power plants as a class are unsurpassed by any other generation technology, whether renewable, fossil-fueled, nuclear, or hydro.

Seasonal Pricing. Another contractual provision in many geothermal power supply contracts is for the utility to have seasonal pricing. For example many California and Nevada geothermal contracts have prices that are higher in the summer and lower in the spring and fall. This provides the geothermal plant operator incentive for the power plant to be online in the peak summer hours, when replacement power is the most expensive for the utility. The reliable and predictable summer peak output is one of the advantages for a utility (and its ratepayers) to have available energy sources using biomass, geothermal power, and solar generation. Because summer peak power spot market costs and the cost for simple-cycle gas turbine generation can commonly rise to over $150/MWh in the summer, the reliable fixed-cost prices of nonintermittent resources reduce the cost of meeting summer peak loads.

23.3.4.3.4 Contract Provisions for New Geothermal Development

Unique contract provisions are advantageous for *development* of new grassroots geothermal power plants as opposed to the expansion of an existing geothermal power plant. The expansion of an existing geothermal power plant is based on the historical performance of the geothermal resource. Therefore, operators of existing plants can commit to definitive start-up schedules and subsequent liquidated damages (LD) for missing those schedules. This is not so with new geothermal plants. Unique contract provisions can assist in the development of a grassroots geothermal power plant by recognizing and mitigating the risks, uncertainties, and costs of such development.

One unique challenge to the development of a grassroots geothermal power plant can be a contractual "chicken and egg" situation in which two events must occur, neither of which can occur before the other has been satisfied. The first of these is that it is very difficult and expensive, if not almost impossible, to secure funding for exploration drilling without a power sales contract. Geothermal energy, in comparison to natural gas or oil, is neither transportable nor does it have an

*The fuel-capacity factor is useful in addition to the plant-capacity factor, because after make-up drilling is completed, the power plant mechanical capacity remains essentially constant, but the resource capacity declines. Consequently, the power plant capacity factor drops each year, even though the power plant may be converting essentially 90–99% of the available fuel supply. The output and performance of the plant remains highly predictable, however, on both an hourly and annual basis.

automatic market after its discovery. Therefore, before risking the large capital for an exploration program, investors (whether as shareholders or venture capital) want a utility commitment to buy the geothermal power if a commercial resource is discovered. On the other hand, it is often not possible to sign a power sales contract without a proven resource, because the power sales contract may contain a large LD clause for failure to deliver the contracted power by a contracted date, and may include the requirement to post a large bond.

One solution for those utilities that would like to add geothermal power to their mix of resources to consider contracts with geothermal developers that recognize the risk, uncertainty, and cost in the discovery of a new geothermal field. Such contracts would not contain penalties for failure to discover a resource, but will assure the developer a market should their exploration efforts be fruitful. Provisions of the contract to protect the utility can include the following:

1. A price acceptable to the utility
2. An expiration date two to five years beyond the date on which the contract is signed, by which time the developer must produce a resource discovery report or forfeit the contract
3. A requirement for a notice of intent from the developer to deliver first power to the utility after two to three years
4. The requirement for bonds or LDs for failure to deliver after the submission of the notice of intent.

The delay between the notice of intent and first power is of minor consequence for the developer, as the cycle time from resource confirmation to commercial operation will generally be two years or more, and the delay gives the utility the necessary time to time to plan for the delivery of the power. There are no statistics on success rates for grassroots exploration to power plant commercial operation under such a contract scenario, but it is likely that a utility that wants 50 MW of geothermal power should plan on signing resource discovery and power purchase contracts for between 75 and 100 MW from two or more exploration prospects.

23.3.5 Economic Access: Power Plant Cost

Power plant costs comprise two elements: the capital cost and the operating cost.

23.3.5.1 Capital Costs

Capital costs vary according to four major variables:

1. Resource temperature: Higher temperature resources have lower $/MW capital cost. Table 23.1 indicates the first benefit of higher temperature, namely lower flows, and hence, fewer wells and smaller components are needed. However, the reduction of the size of the heat rejection system (condenser and cooling system) required per megawatt of capacity is also of great value. Higher-temperature fuel sources require much less cooling than lower-temperature sources. This is illustrated by a comparison of fossil-fuel power plants and geothermal plants in which geothermal power plants may have a heat rejection system eight times as large per megawatt as that in a combustion-turbine combined cycle plant.
2. Power technology: Power plants using a binary process generally have a cost several hundred dollars per kilowatt higher than power plants using only steam turbine technology. However, this is not universally true, and binary power technology plants have won numerous open bids around the world over suppliers of steam turbine technology.
3. Wet or dry cooling: Steam flash power plants generally use wet cooling with the water supplied by the condensed geothermal steam. Binary power plants can use wet or dry cooling. In most climates, wet cooling gives an advantage in net plant output over dry cooling, and also causes less capacity degradation than dry cooling in the peak summer hours. Because these peak hours are generally the utility's most expensive hours, wet cooling offers utilities a benefit. However, there is often not water available (neither condensed geothermal steam nor meteoric waters) for a binary power plant cooling system, and therefore, dry cooling is used.

4. Plant size: Larger plants and larger machines are less expensive per kilowatt of installed capacity than smaller plants and smaller machines. This occurs both because of the general economy of scale of machine size, as well as a distribution of fixed costs (e.g., civil engineering, roads, site preparation, instrumentation, insulation, and paint) over a greater plant size.

23.3.5.2 Operating Costs

Geothermal plants have low variable costs, which are direct costs resulting from changes in plant output. Fixed operating costs are those costs that exist even if the power plant is not operating, e.g., personnel and interest on loans. If output is reduced, the staff costs are the same, the fuel costs are the same (i.e., the amortization of the well costs), and chemical use changes negligibly. Thus, there are no cost savings with a reduction in geothermal plant output other than the royalty paid to the holder of the geothermal rights. For all intents and purposes, a geothermal power plant has only fixed operating costs. Table 23.2 illustrates the difference in variable costs between a geothermal plant and a combined cycle plant using a combustion turbine. The table illustrates why baseloaded operation is optimal for geothermal power plant cost structure whereas a dispatchable operation fits the cost structure of fossil-fired power plants.

Not only does reducing the output from a geothermal plant not reduce operating costs, but throttling a geothermal turbine to reduce its power output actually has a negative effect on long-term operations, because it increases scale formation at the turbine inlet. Reducing the flow from geothermal wells on a regular basis, as would be required if the plant is not operated at baseload, can also induce damage in both the wellbore and the steel well casing in certain geothermal fields due to thermal or pressure cycling.

TABLE 23.2 Comparison of Fixed and Variable Costs for Geothermal and Fossil-Fired Plants

Cost Category	Geothermal Power Plant with Wells and Piping		Fossil-Fuel Power Plant	
	Variable Cost	Fixed Cost	Variable Cost	Fixed Cost
Fuel costs to increase plant output from 50% to 100%	Geothermal royalties (1%–5% of gross power sales)	Drill wells and build pipelines, mostly before start-up of plant	Essentially entire additional fuel cost	Not applicable
Cost/Savings to stop plant and restart later	Low cost savings other than royalties. Low additional costs other than ongoing fixed costs	May have to vent steam to air rather than shut-in wells, and expose wells to thermal cycling damage	Fuel savings when stopped. Fuel costs to keep warm or cold start plus start/stop related maintenance costs	Not applicable
Fixed operating personnel costs (number of MW that carry cost of each staff person)	Not applicable	10–13 people for 0–15 MW =1–1.5 MW per person	Not applicable	15–20 people for 250–550 MW = 13–36 MW per person
		13–18 people for 40–120 MW =2–8 MW per person		
Fixed capital costs	Not applicable	$1,400–$3,000 per kW of capacity	Not applicable	$700–$1300 per kW of capacity

23.4 Exploration and Assessment of Geothermal Resources

Geothermal resources capable of sustaining commercial electrical production require specialized methods of exploration and assessment. To finance the high initial capital cost of a geothermal power plant and wellfield, the equity and debt providers must be assured that a geothermal reservoir is capable of sustaining the expected amount of energy production for a period of 20–30 years. Likewise, the purchaser of power, typically a local electrical utility, must have adequate assurances to justify entering into a long-term power purchase agreement (PPA). The geothermal industry has developed a suite of exploration and reservoir assessment techniques which provide the assurances required by all parties involved in a particular project.

Even when an active geothermal system is found, many geothermal discoveries do not lead to commercial development for a host of reasons, as discussed previously. This does not necessarily mean that a particular geothermal resource is forever noncommercial. Raft River (Idaho), Salt Wells (Nevada), and other resources under development in 2005 were discovered, drilled, and assessed in the 1970s and 1980s and were deemed noncommercial. Today, these resources are under commercial development, through a combination of improvements in exploration techniques, changing electrical power and natural gas markets, and advances in geothermal power technology.

Although this section is written from the perspective of geothermal development for power production, the same considerations and approaches are used for geothermal resources developed for direct-use applications.

23.4.1 Overview

The discovery of a geothermal resource and the assessment that it will be capable of sustaining 20–30 years of commercial production is a complex and costly undertaking, and has been the subject of entire books and many journal articles. This section is meant to provide a brief overview of the subject, much like learning geography while looking out the window of the space shuttle at the earth passing by below. In that brief glance, the shape of some of the continents would be largely visible, and one might be able to discern some forest, desert, and cities.

The process of geothermal exploration generally begins with the observation of surface manifestation of geothermal heat; for example, hot springs, fumaroles, or surface deposits of silica (sinters). However, some geothermal resources have no surface manifestations, and are discovered by accident when drilling takes place for purposes unrelated to geothermal development. Several commercial geothermal fields in the western United States have been discovered through the drilling of irrigation wells or mineral exploration holes.

From this initial surface exploration, detailed geoscientific work is undertaken to characterize and assess the size of the geothermal resource underlying the visible manifestations. After a promising resource is identified by the geoscientific study, wells must be drilled to prove the existence of the essential elements of a commercially viable geothermal resource, as described in Section 23.3.1, i.e., heat, permeability, and fluid. The critical first step in the drilling campaign is to drill a discovery well that is capable of producing commercial quantities of hot fluid. With the successful completion and testing of one or more additional wells, the process of exploration ends and the resource is considered "proven."

Finally, the remaining wells for the development must be drilled. Once the geothermal resource is under production, pressure, temperature and chemistry of the reservoir are monitored to evaluate and optimize resource production, and to make plans for further development (Section 23.5).

23.4.2 Exploration and Discovery

Surface geoscientific investigations are the first step in the process of discovery. These investigations include geochemical analysis of hydrothermal manifestations (hot springs, fumaroles, mudpots, etc.) and surface geologic mapping, including the type and extent of rock units, hydrothermal mineralization and

TABLE 23.3 Geophysical Investigations for Geothermal Exploration

Exploration Tool	Objective	Indicative of
• Resistivity	Detection of a clay cap, transition depth to high temperature mineralization	Convecting geothermal system that has deposited minerals to seal-off the top of the system
• Seismic reflection	Faulting, structure	Flow paths (good drilling targets), structural block volume
• Microearthquake monitoring	Identify active faulting and fracturing	Fracture permeability, fracture density, injection and production drilling targets
• Magnetics	Loss of iron minerals	Convecting geothermal system
• Geodesy	Active surface deformation	Rapid natural subsidence overlying active rift zones
• Self potential	Map natural ground voltage	Shallow active hydrothermal systems
• Microgravity	Faulting, mineral deposits	Flow paths, convecting geothermal system
• Shallow Wells	Temperature gradient	High heat flux, possible max T at depth, lateral reservoir extents

surface expression of faulting, fracturing, and other structural features. Surface mapping techniques have been greatly enhanced by the use of remote sensing (satellite imagery). Together, these methods may indicate the ultimate resource temperature, provide clues as to the resource structure, and provide evidence of the vigor and extent of the resource. If a water sample can be obtained from a hot spring, the geochemistry, via various chemical "geothermometers," can provide an indication of the maximum temperature to which the water was heated on its journey to the surface. However, the surface geoscience does not provide a good definition of the depth to the commercial geothermal reservoir, nor can it define its lateral extent or ultimate productivity. Not all hot springs are indications of commercial geothermal power resources, even if the geothermometers indicate a high-temperature origin of the fluid; the geothermometers can be erroneous, or, even if high-temperature water is available, it may not be producible at commercially viable flow rates.

The second step is geophysical investigation to map the lateral extent, depth, and distribution of permeability of the active geothermal system. These investigations and their objective may include those listed in Table 23.3.

The cost of conducting and interpreting a typical suite of geophysical surveys can vary greatly, depending on the individual characteristics of a particular exploration project. The total cost is dependent on the methods chosen, overall size of the area to be explored, the maximum depth to be explored, surface topography and the remoteness of the site. For example, an effective geophysical program to explore and define a small geothermal reservoir in Nevada may cost as little as $150,000, whereas a program in the remote jungles of Indonesia may cost in excess of $1,500,000.

The third, and final, step is the riskiest and most expensive portion of the exploration program: the exploration drilling to determine if the resource is commercially viable. The exploration drilling program can involve both full-size development wells (commonly final casing size of 9 to 13 in. with reservoir hole diameter of 8 to 12 in.) and slim holes (final casing size of 6 in. or less and reservoir hole diameter of 5.5 in. or less). Slim hole wells have the advantage of costing only about two-thirds to three-fourths that of a full-size well, but have little commercial value, even if drilled into the heart of the resource. Depending on terrain, remoteness, road requirements, possible maximum resource temperature, depth, and other factors, the cost of the first exploration well will often exceed $2,000,000, and may reach $5,000,000 for deep or remote locations. Unexpected drilling problems can result in enormous cost overruns or the total loss of a wellbore and all the investment in it. Rare shallow resources may see exploration wells at less than $1 million. Follow-up wells, if drilled at the same time, will save on the order of several hundred thousand dollars as a result of shared mobilization and construction costs.

Typically, three to eight wells will be required to prove and delineate an undeveloped geothermal resource, depending on the efficacy of the exploration program used to target the wells, the size of the first power plant, and other factors that cannot be predicted prior to drilling. Sometimes a nonproductive well will eventually be used for injection, so the investment may not be altogether wasted if a particular well fails to encounter commercial production. There are instances when, even after a multiwell exploration program, the resource that is discovered cannot be economically developed at all.

After the first (or, preferably, two or more) well(s) have been drilled and found to flow with commercial temperatures and flow rates, a long-term flow test and an interference test will be conducted. The budget for the flow test, with cross-country pipelines, staff, instrumentation, chemical sampling, interpretation, logistics, travel, and so on, can cost \$200,000–\$500,000, although long multiwell tests can incur greater costs.

The final stage of the geothermal reservoir assessment process is the analysis of the well and reservoir flow testing data. This analysis is done using established reservoir engineering techniques in which the behavior of the geothermal reservoir is numerically simulated at the proposed production rates for the expected duration of the power plant life, typically 20–30 years. The numerical model is used to investigate different production and injection strategies to optimize the resource development strategy. Positive results of this analysis are an essential factor for obtaining the debt and equity financing required for completing the wellfield and initiating power plant construction.

23.4.3 Risk of Exploration

A modest exploration budget of \$2–\$10 million, staged over the phases above, may suffice for many geothermal resources, but much greater budgets are well documented. Western GeoPower Corp. is a company seeking to commercialize the South Meager Creek Geothermal Project in British Columbia. Because they are a public company, various aspects of their development work are available in the public domain and can be used as a case study.[13]

The company raised over U.S. \$24 million for corporate expenses and a three-well exploration program in the remote area where the resource was located; the program was executed in 2004–2005. As of the time of this writing, the three wells had been drilled, but not flow-tested. These exploration wells were targeted on the basis of previously-drilled wells and surface exploration over the previous 20 years since the resource was first identified. Temperatures in the three wells were announced to have ranged from 240°C to 260°C (460°F–500°F), which demonstrates that the thermal basis for the targeting was sound. Western GeoPower has not yet announced whether the other elements of a commercial geothermal resource (permeability and water), per Section 23.3.1, have been discovered.

Regardless of whether or not a geothermal power plant is eventually developed at South Meager Creek, the exploration program demonstrates the potentially high risks and high costs of grassroots geothermal resource exploration. Not only was there substantial cost, but it has taken two decades for the exploration project to get to its current state.

Both geothermal and wind resources are renewable energy sources that must be converted at the location at which they are found, but it is much more difficult to demonstrate a viable commercial geothermal resource than a commercial wind resource. The cost of an array of wind exploration towers, the data collection, and analysis is miniscule by comparison with the cost of geothermal resource exploration. Additionally, the time to prove a commercial wind resource is much shorter. These cash and time costs not only affect the ease and extent of exploration, but the power price that must be obtained to pay for the exploration. The resource discovery process is by far the greatest disadvantage and barrier to expanding geothermal power generation. The costs and risks are why geothermal exploration is most often undertaken outside of the U.S. with strong government support or with government funding. In the U.S., continued DOE support of exploration and discovery through cost-share programs will be necessary to bring significant quantities of new geothermal resources under development.

In spite of the barriers to exploration and discovery of commercial geothermal resources, they are discovered, developed, and electric power is delivered to the transmission grid. The remainder of this chapter is dedicated to these processes, from the resource (fuel) management issues to the power cycles.

23.5 Management of the Geothermal Resource for Power Production

Geothermal resource management initially seeks to ascertain the initial and long-term behavior of the geothermal reservoir so as to select the optimal power process for the particular geothermal resource, including the changing conditions that might be encountered over the first 20 years. Second, geothermal resource management seeks to design the geothermal resource development to meet short- and long-term economic goals that may include some or all of the following: minimizing invested capital cost in the wells and the power plant, maximizing the power plant installed capacity that the resource can support, and minimizing operating costs.

The complexities and issues of geothermal resource management have no equivalent in any other electric power generation technology, whether conventional or renewable.

- The fuel supply is located 2,000–10,000 ft. below the surface, and may be more than 10 cubic miles in volume.
- The response of the resource to development plays a huge role in the overall economic and generation performance of the resource, but the details of the response are largely unknowable until after the resource is under full production for a number of years.
- The "fuel supply" almost invariably changes with time, and as a result, the power plant either must be operated off-design or modified to meet the new conditions.
- Through much of the resource life, make-up wells will need to be sited and drilled to maintain adequate geothermal energy supply to the power plant, which is a combination of flow rate and pressure or temperature.

Although the management of natural gas and oil fields face many of the same issues, fossil-fuel power plants are seldom tied to a single and solitary gas or oil resource.

In this section, the issues involved in managing the geothermal resource, and how the characteristics of the fuel supply interact with the design and operation of the power plant, will be examined. Case studies illustrate successes, changes, and occasional failures in geothermal resource management.

Section 23.5 begins with the large goals, examines some of the subgoals, and ends with the beginning—the characterization of the geothermal resource for the selection of the power process.

23.5.1 Goals of Resource Management

Geothermal energy production is the science and technology of heat recovery. At commercially-significant extraction and power generation rates, the heat extraction rate from the production wells is almost always far greater than the natural heat addition rate that created the geothermal resource. The energy balance is closed by extraction of the heat-in-place in the volume of the reservoir, i.e., cooling of the geothermal resource.

$$\begin{aligned}&\textit{Heat produced out of the geothermal system per year}\\&-\textit{Natural heat inflow to geothermal system per year} = \textit{Cooling of geothermal system per year}.\end{aligned}$$

The geothermal resource volume is large, however, so the cooling is a long, slow process that allows the geothermal resource to be productive for many decades. (Some of the issues of renewability and sustainability in geothermal power are addressed in Section 23.3.4.) Therefore, the overall strategic goal of resource management is efficient heat mining (cooling) of the geothermal resource. Four of the most important resource management subgoals include (1) minimizing the capital cost of the development, (2) residual brine management (if applicable), (3) injection placement to enhance production, and (4) geothermal fluid monitoring and control, including chemistry, rate, pressure, and temperature.

23.5.1.1 Minimizing Development Capital Cost

To achieve heat recovery from the resource, geothermal fluids (steam or brine) must move through and be heated by the rock. Boiling is the most efficient mechanism, because of the high heat transfer rates of vaporization and the high heat capacity of steam. The native, in-place water is the primary mechanism for many years of the production. Geothermal reservoir rock porosities are most commonly in the range of 3%–10%, but may be as high as 20% in sedimentary systems.

As discussed in Section 23.3, matrix flow of fluid through the bulk rock occurs at only very low rates, so much of the heat of the reservoir volume must transfer by the much slower process of conduction through the rock to the fracture surfaces, where liquid is mobile and heating or vaporization can more readily occur. (This also explains part of the difficulty of creating the HDR/HFR geothermal resources of Section 23.3.2.) The native-state fluid mass is not sufficient to achieve the goal of heat mining the reservoir. This can only be achieved by flowing supplemental water through the resource. Therefore, the placement of the residual geothermal fluids through the injection wells can have a large effect on the overall capability of the field over the long term. From an ideal heat transfer perspective, the optimal development would be to have many production wells drilled into the hottest areas of the field, with many injection wells ringing the production area to sweep heat and provide pressure support to the production wells. Unfortunately, the need to manage and limit the capital cost of the development intrudes on this Utopian development, and forces geothermal resource developers to seek compromises to the ideal.

Figure 23.2 illustrates a deliverability curve for a hypothetical well producing a mixture of steam and brine. The curve segments are as follows:

- Segment A to B is an unstable flow regime (not all wells have this segment).
- Segment B to C is increasing flow due to greater differential pressure between wellbore and reservoir.
- Segment C to D is the same as B to C, except for an increasingly dominant frictional pressure loss in the wellbore or fractures feeding the wellbore that reduces flow rate at a faster rate than in the segment from B to C.

In the Utopian geothermal development, the well would be operated somewhere just to the right of point B, because point B has the highest available energy per unit mass, and will therefore allow the greatest cumulative power generation over time from the resource. However, there is not much flow at

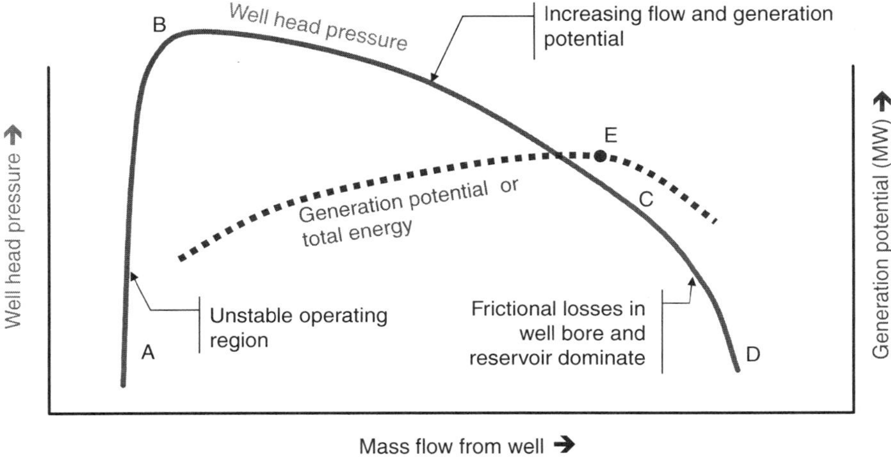

FIGURE 23.2 A hypothetical free-flowing geothermal production well deliverability curve.

point B, and thus a large number of wells and long lengths of connecting pipelines would be required. This is an unacceptable barrier to economic commercial development, in part because of the high capital, and in part because the additional generation occurs so far in the future that its present worth value is minimal, as discussed in Section 23.5.3. Therefore, a much more common optimization of the wells is to maximize the energy production from the wells (shown as point E on the dashed line in Figure 23.2); doing so minimizes the amount of money invested in production wells because of the high energy production per well. The disadvantages to this strategy stem from larger flows per megawatt-hour of generation that result in higher injection well costs (more spent fluid because of lower available energy per pound), higher power plant heat rejection costs (lower useful energy per unit mass), lower ultimate field capacity, and more make-up drilling sooner. The choice of the exact operating point on this curve is much more complex than a single number, but these two extremes (point B and point C) illustrate two of the important issues. A third issue is the provisions of the power sales contract. Power sales contracts that do not limit the output capacity of the field, and in which the field capacity can be fully developed in a short period of time, will tend to push the optimal operating point closer to point B because point B maximizes the ultimate field capacity. Those in which the power sales limit is much smaller than the capacity limit of the field will tend to push the operating point toward point C.

The knowledge of how a resource will respond to development can often be used to minimize the capital cost of the development. For example, at the Salak geothermal field in Indonesia, the reservoir modeling prior to production indicated that there was a region of the reservoir that would quickly evolve from producing a mixture of steam and brine to almost entirely steam production. Wells and a power plant were located in this area to minimize piping costs and to capture the benefit of lower brine handling and injection costs that resulted from the production of nearly dry steam.

23.5.1.2 Residual Brine Management

A large variety of operating practices has developed around the world over the use and disposal of brine, from primarily the two-phase (steam and brine) reservoirs. This is because in two-phase reservoirs, the reservoir pressure is primarily controlled by the steam pressure in the reservoir, whereas in primarily liquid reservoirs, the reservoir pressure is strongly influenced by the injection of the liquids back into the reservoir. A table of options and applications for use of the residual brine is presented in Table 23.4.

In geothermal reservoirs, both heat mining and pressure support are managed by injection well placement. When this is accomplished according to objective, the results can be seen and monitored in the production well characteristics. Injection well placement is best accomplished in a manner similar to Goldilocks and the three bears: not too close (rapid cooling of produced fluids), not too far (no benefit), but just right (reservoir management).

When separated brine is disposed of through injection, the brine is generally injected into either the bottom of the reservoir or outside the reservoir. The location of the brine disposal is a difficult task, and in the history of many geothermal fields, brine injection locations have had to be adjusted at least once during the life of the field, usually because the injection wells are too close and cold-water breakthrough is observed. If the injection is ideally placed, it provides both pressure support for the production wells, as well as "mining" of the stored heat of the reservoir rocks as the injectate flows from the high-pressure zone of the injection to the low-pressure zone of the production well. However, if the injection well is too close or if the injection well encounters a highly permeable crack linking the production and injection areas, cold injectate can enter a production area, and quickly reduce the enthalpy (temperature or steam content) of the fluid produced by that zone. Therefore, in many instances, the preferred option is to inject some portion of the total injectate volume in steam/brine systems effectively outside of the geothermal reservoir, sacrificing the benefits of in-field injection, but also avoiding the potentially severe consequences of enthalpy loss at the production wells. As a result of the need to move injection distant from production, it is common in geothermal fields to have injection pipelines several miles long that carry the brine to distant injection wells.

TABLE 23.4 Uses of Separated Brine in Worldwide Geothermal Power Operations

Option	Use of Separated Brine	Fields in Which it is Used[a]
1	Surface disposal	Wairakei, New Zealand (river disposal). Cerro Prieto, Mexico (evaporation ponds). Tiwi, Philippines (Ocean disposal in early years of operation). Svartsengi, Iceland (The Blue Lagoon swimming, bathing, and health spa)
2	District or sensible heating	Nesjavellir, Iceland (Indirect heat exchange with fresh water to supply district heating in Reykjavik)
3	Separated hot brine injected back in geothermal reservoir after a single flash due to silica saturation	Bacon–Manito, Philippines. Tiwi, Philippines (current operation). Salak, Indonesia
4	Separated hot brine used as source for binary power plant, and then reinjected	Mak-Ban, Philippines. Rotokawa, Kawerau, and Mokai, New Zealand. Brady Hot Spring, Nevada
5	Hot brine flashed multiple times for steam to power plant, and then brine reinjected	Salton Sea, California. Mt. Apo, Philippines. Hatchobaru, Japan
6	Minerals extraction, followed by reinjection	Salton Sea, California. (After a four-stage flash process, silica is precipitated to stabilize the brine for injection. The silica is disposed in landfill. A multiyear, multimillion dollar effort to recover zinc was abandoned in 2004.)

[a] Examples, not a complete list.

As with almost every situation in geothermal resource development not tied to conservation of energy or mass, there are exceptions to the long injection pipeline option. The particular characteristics of the deep vertical fracturing of the Steamboat Hills (Nevada) geothermal resource have allowed a large-capacity injection well to operate without negative consequences to nearby production wells. In most geothermal fields, such geographic proximity would have resulted in cold water breakthrough to the production well within a very short time, perhaps on the order of days or months.

23.5.1.3 Clean-Water Injection Placement to Enhance Steam Production

The dryout of a region of a reservoir can occur when the water present in the region is no longer sufficient to remove the heat that is stored in the rock. At that point, the adoption of a new clean-water injection strategy can improve a number of the negative consequences of that dryout: specifically, loss of production, superheat, and acid gas production.

As dryout occurs, the temperature of the rock and the heat stored therein remain high, but there is no water to capture or transfer the heat. Production declines, and as it does, the pressure drops. Steam passing through rock that has a temperature higher than the saturation temperature of the steam's pressure becomes superheated. Superheated steam, upon mixing at the surface with wells that are not superheated, can evaporate the water from the brine flowing from other wells. If the brine rates are low enough and the evaporation is high enough, minerals will precipitate in the pipelines or separation vessels.

Superheated steam is also capable of carrying hydrochloric acid gases produced from certain rock formations out of the reservoir. Because the gases are extremely hydrophilic, they can only pass through the reservoir and into the steel-lined wellbore if no water is present. The hydrochloric gases are harmless to the steel as long as the steam is superheated. They can be removed at the surface by one of several methods: mixing with large quantities of water, mixing with smaller quantities of brine (which is naturally buffered against changes in pH), or, in some instances, by caustic injection into the pipeline. However, if scrubbing of the hydrochloric acid is not accomplished before the onset of condensation, the rate of corrosion at the condensation point can be so high that the steel pipe wall can be thinned to failure in less than a year.

Injection into the reservoir can be used to solve all three problems. Adding water to the dried-out region through injection not only provides the means to increase production and heat-mining, it eliminates the superheat, and with the elimination of the superheat, also eliminates the acid gas production and pipeline scaling.

23.5.2 Resource Chemistry

The geothermal water and steam that are produced contain the chemical signatures of the rocks and processes (e.g., boiling) through which they have passed. Measuring the steam and brine chemistry is an important tool in the exploration process for new geothermal resources, in determining the brine processing requirements of the power cycle, and as an ongoing activity in developed geothermal resources for analyzing the response of the resource to production and for resource management planning. Examples include:

- In the exploration phase, geothermal water is collected from surface springs or fumaroles for clues as to the nature of the geothermal resource. For example, the relative concentrations of sodium, calcium, magnesium, and potassium in the water tell geochemists whether the fluid started out hot and then cooled (indicating a high-temperature resource elsewhere and indicating commercial geothermal resource potential) or whether the observed temperature of the fluid is close to the maximum temperature of the resource (not hot enough for power production).
- The extreme example of geothermal fluid chemistry is the Salton Sea (California) geothermal brine that contains more than 200,000 ppm total dissolved solids (TDS), or 20%, in its native state in the reservoir. The high TDS is a result of the high temperatures and the marine sediments that comprise the geothermal reservoir. The brine is produced and is flashed to steam at four pressures ($\sim$250, 125, 20, and 0 psig). In so doing, the brine becomes supersaturated with respect to silica and other minerals and salts, resulting in large chemical processing facilities necessary to prepare the brine for injection back into the geothermal reservoir. The silica treatment processes can either reduce the pH to prevent precipitation in the equipment or, alternatively, the silica can be removed from the brine by the controlled precipitation of the silica in tanks and vessels.
- Silica concentration is primarily a function of the reservoir temperature, but the concentration of silica determines how far the brine temperature can be lowered (how much energy can be extracted) before the onset of silica scaling. The power process is frequently designed to prevent silica from reaching saturation. In such cases, the operating point of the well is pushed to the left of point C in Figure 23.2, regardless of other considerations. The injection of acid or chemicals to inhibit the precipitation of silica can be used to lower the minimum allowable brine temperature and extract more energy from the brine. Another strategy to extract more energy from the brine is to use a binary power plant after the first steam flash. This strategy is discussed in more detail in Section 23.7.
- Tritium (an artifact of atmospheric nuclear weapons testing) detected in the geothermal steam or water indicates that young groundwater is entering the geothermal reservoir as a result of geothermal exploitation. Native-state geothermal water, which has typically been underground for tens of thousands of years, contains no detectable tritium. Tritium from groundwater may occur in the geothermal reservoir as a result of intentional injection programs to replenish water, such as at The Geysers, or the result of unwanted cold water influx from the surface. At Tiwi (Philippines), the cold water in question has quenched several square miles of productive

resource. The advancing surfacewater could be detected in the tritium concentrations long before the temperature indicated a cooling process was occurring at the well.

- Increasing salt concentrations and other mineral concentrations indicate the return of injected brine from flash steam plants. Again, this can occur long before a change in production temperature is measured. If the rate of cooling is negligible or small, these chemical signatures indicate successful heat mining through the injection program.

23.5.3 Barriers to Resource Management

In most human endeavors, the best results come by starting with the end in mind. This is true of geothermal resource development and resource management as well. By understanding the particular resource the wells can be placed to ideally collect and convert the geothermal energy stored in the reservoir. Unfortunately, in practice, the ideal placement of the production and injection wells may be either impractical or even impossible for three reasons: incomplete information, capital limitations and sunk costs, and economic impetus.

23.5.3.1 Incomplete Information

First, many aspects of the geothermal resource are simply not known in the early stages of development of a new resource. Such incomplete information includes whether cold water from injection will short-circuit to the production wells, how strong the natural influx to the system will be, or where that influx will actually occur. In fact, the exact area and volume of the resource is commonly not even known until ten years or more after the onset of production.

The Geysers facility illustrates this point. The unit 1 power plant went into operation in 1960, and the final power plant—the J. W. Aidlin Plant that began operation in 1989—brought the total installed capacity to 1,890 MW, although deratings and retirements commenced soon thereafter. However, reservoir information and analysis was not complete, even in the mid-1980s. Two additional power plants were ordered and never completed because there was insufficient steam for their operation. Therefore, the geothermal resource management strategy is dynamic and evolving from the beginning. As more information becomes available about the resource over time, the knowledge of the resource improves, and the decisions about its management more informed.

A surprising and interesting example of this evolving knowledge has occurred in recent years at Larderello, Italy. Almost 100 years after the start of power generation, it now appears that the historical development in the Larderello area is at the top of a much larger and deeper regional system, perhaps 400 km^2 in size and between 3,000 and 4,000 m deep (10,000–13,000 ft.). Plans have been announced to begin drilling and developing this deep hot resource in 2007 with 11 deep-production wells.[8]

23.5.3.2 Capital Limitations and Sunk Costs

There may be capital limitations and sunk costs that must be taken into account in well placement decisions. For example, suppose that it would be ideal to inject the cooled brine to the north of the power plant, but an unsuccessful exploration well will serve as an injection well to the east, saving over a million dollars. In one development in Indonesia, a well that was targeted as an injection well was found, after being drilled, to have a very large production capacity. However, pipelines were already under construction, the well was distant from the power plant, and production well capacity had already been essentially completed. What was needed for plant operation was injection capacity, so the well was used for injection. Time and capital constraints forced the decision, even though it was against the ideal resource management strategy. This example also illustrates the first point about incomplete information at the onset of the project.

23.5.3.3 Economic Impetus

The third driver for well placement decisions is the economic impetus. The economic test criterion used for investment decisions is discounted cash flow, in which decisions are based largely on achieving the highest net present value or the lowest net present cost in which all costs and earnings are brought back to the present at an assumed discount or interest rate. For public (government) development, that discount rate may vary from 5% to 15%; for development by private corporations, the rate for normal equipment and operational investments is typically 9%–12%, but for resource development issues is not uncommonly as high as 20%, due to the greater risk and uncertainty. The lower the discount rate, the more important future performance is to the total value of the project, as viewed by "today's" decision makers. The converse is also true, and in fact dominates spending decisions for resource management at the start of the development. Suppose that the publishers of this text elected to develop a new geothermal power plant named "CRC #1." They are sharp businesspeople, and are told by their resource development experts that there are several resource development options that can be pursued. Both strategies produce the same income with the same expenses for the first 15–20 years. However, the second strategy will produce significantly more value in the outer years because of better resource management. They perform a discounted cash flow calculation and develop Table 23.5. The conclusion that is reached is that, for an investment as risky as resource performance in 15–20 years, the corporation will use a discount rate of 15%–20%. The table shows that the resource managers have a budget to optimize the resource development of between $0.25 and $1.2 million (the highlighted area in Table 23.5) to create $10 million of value. In most cases, this will not be sufficient to implement an enhanced resource management plan. At the extreme, in looking at resource development strategies that would create the $10 million value in 50 years, the table shows that less than $10,000 of additional spending today can be justified at a 15% discount rate and only $1,000 at a 20% discount rate. In short, there is virtually no economic impetus for resource development decisions that yield a benefit after 50 years.

Even if not implemented at the onset of the resource development, some resource management strategies will be able to be implemented at a later date. However, other decisions are irreversible, such as inefficiency in the conversion of the resource to electrical power. Conversion inefficiency directly reduces the total generation from the resource (megawatt-hours), but may also ultimately lead to a reduction in the installed capacity of the field (megawatts). While this is clear from the beginning, if such additional generation does not occur for 30 years, Table 23.5 demonstrates that the generation is of inconsequential value at the time of initial development. A scenario under which this occurs is where the generation contract for the geothermal field is limited to a given output.

Suppose that a PPA is for 100 MW average over the year. With the high-efficiency power plant, the resource is forecast to provide 100 MW for 40 years before beginning its decline. With the lower-efficiency (and less expensive) power plant, the plant is forecast to be able to maintain 100 MW for 25 years. In terms of today's decision making, the value of the additional 15 years of full-capacity generation are of marginal value, and the economic calculations will almost certainly show that the power plant and resource strategy with the lower capital cost is the preferred investment strategy "today."

The same calculations also apply to the efficiency of nonrenewable natural resources, such as natural gas and petroleum. Straight discounted cash flow calculations provide little incentive for the

TABLE 23.5 Amount of Spending that can be Justified "Today" for a Savings or Earnings of $10 Million Dollars in the Future

Discount Rate Scenario	Economically Justified Spending "Today" at a Discount Rate of				
	0%	4%	11%	15%	20%
$10 MM value in 15 years	$10,000,000	$5,552,645	$2,090,043	$1,228,945	$ 649,055
$10 MM value in 20 years	$10,000,000	$4,563,869	$1,240,339	$611,003	$ 260,841
$10 MM value in 30 years	$10,000,000	$3,083,187	$436,828	$151,031	$42,127
$10 MM value in 50 years	$10,000,000	$1,407,126	$54,182	$9,228	$1,099

conservation of depletable natural resources for a time frame greater than 20 years. However, at the end of the 20 years, the impact of the inefficiency can be very large on the future value.

23.5.4 Resource Characterization

An early task of the geothermal resource management team is to characterize the resource based on the exploration and initial development wells. The characteristic of the fluids that will be produced from the well must be determined: steam only, brine and steam, or brine only. Subsequent tasks must address the design issues within each resource type.

Table 23.6 provides a list of some of the design issues and options that derive from this first basic resource characterization. The characteristics of the wells will determine the options for the power cycle, and from that the configuration of the surface facilities. As illustrated in Table 23.6, there are three types of resources that are used in the two dominant power plant types: steam power plants and binary power plants, as discussed in detail in subsequent sections.

23.6 Geothermal Steam Supply (from Wellhead to Turbine)

The first two columns of Table 23.6 discuss resources in which steam is produced from the wells, as opposed to the third column, in which only pressurized brine is produced. In almost every case, the steam that is produced from these wells is used in a steam turbine. In this section, the technology of handling geothermal steam from the wellhead to the geothermal steam turbine is discussed.

23.6.1 Overview of Steam/Brine Separation for Steam Turbine Supply

In the vast majority of geothermal fields in the world, a two-phase mixture of steam and brine is brought to the surface through the production wells. The brine must be separated from the steam before delivering the steam to the turbine. Separators of many different configurations are in use throughout the world, but the most common is a vertical vessel with a tangential entry to centrifugally separate the steam and brine. The level of purity of the steam varies between that which meets the geothermal standard of 0.1–0.5 ppm of chloride to over 10 ppm chloride.

Steam purity is typically measured by the chloride content of the steam condensate. However, the chloride itself is seldom intrinsically a problem to the alloys employed in a geothermal steam turbine. Rather, chloride is a convenient measure of the co-contaminants that are much harder to measure: mostly silica, but other minerals as well. Where these minerals are carried in the steam to the turbine, they may precipitate primarily on the first-stage stationary blades, reducing efficiency and eventually restricting plant capacity as well. The consequences and treatment of these mineral deposits are further discussed in Section 23.6.5.

With typical geothermal brine chloride levels, the separators need to achieve over 99.5% dryness to achieve the lower steam chloride concentration objective. The greatest challenge in doing so is to minimize the pressure drop through the vessel. The pressure drop has two negative consequences. The first is that the pressure drop results in an unrecoverable loss of available energy (reduced efficiency). Second, it results in a higher pressure at the wellhead, which reduces flow from the well (moving the operating point toward “B” in Figure 23.2), thereby requiring more wells (greater capital cost) to supply a given steam demand.

Where high-chloride (high-silica) steam is produced from the separators, steam washing must be employed to clean the steam before admission to the turbine, with a consequent parasitic steam loss as described below. While the commitment to, and investment in, steam purity at the discharge of the flash

TABLE 23.6 Geothermal Facilities Design Parameters Resulting from the Geothermal Resource Type

Wells Flow: Design Issue	Steam Only	Steam and Brine Two-Phase Flow	Brine only
		Representative Fields	
	Geysers, California Lardarello, Italy Kamojang, Indonesia	Cerro Prieto, Mexico Coso, California Wairakei, New Zealand Mak-Ban, Philippines Miravalles, Costa Rica	East Mesa, California Steamboat Springs, Nevada Raft River, Idaho
		Production Well Design	
Production of fluid to surface by	Free flow	Free flow	Pumped
Well design—flow	Minimize pressure drop (velocity) to maximize flow	Maximize flow, but must maintain high velocity to lift liquid out of the well	12-in. inside diameter (min) to deepest expected pump installation depth
Well design—pressure	Maximum reservoir pressure (Generally less than 600 psi, and falls with time)	Saturation pressure at maximum temperature. (Generally less than 400 psi, but can be 2000 psi.)	Often less than 200 psi at surface and less than 600 psi for pumps at the zero flow condition
Production pump parasitic load	Not applicable	Not applicable	Usually between 5 and 8% of gross generator output, but can be as high as 15%
		Surface Facility Design	
Production pipeline network	Steam only to power plant. Build with large cross-section for low pressure drop. Condensate knockout pots must be provided	Two-phase flow to separators requires high velocities and pressure drop to prevent slugging. After separator, two pipelines of single phase steam and separated brine	Single-phase liquid flow to binary power plant or separator. Size pipe for low pressure drop. Water hammer at start-up is an issue
Brine steam separator	Not applicable	Required. installed vessel cost greater than $600,000 per 10–20 MW	Only required if steam turbine to be used
Steam washing system to remove impurities and condensate	Ususlly required on a continuous basis. Installed cost greater than $500,000 per 10–90 MW	Same	Not required

(continued)

TABLE 23.6 (*Continued*)

Wells Flow: Design Issue	Steam Only	Steam and Brine Two-Phase Flow	Brine only
		Power Plant Design	
Power plant location	Center of initial and long-term production. Usually accomplished	Same goal. Not uncommonly, the center of production (make-up well drilling) moves away from the power plant site as new resource is discovered, resulting in large pipeline costs and available energy loss	Same goal. Usually accomplished, as these tend to be smaller and better-defined geothermal resources
Steam turbine	Single pressure entry	Size as per other geothermal steam turbines	Not commonly used
	Size from to 5–125 MW. Most commonly 25–75 MW	(1) Single inlet pressure, as for steam only resource (2) Multiple inlet pressures on same or multiple turbines (needs multiflash design of brine system)	Size is less than 40 MW
Binary power plant Size has historically been 1–8 MW per turbine/generator set. Since 2004, size has increased to 20 MW per T/G	Not applicable	(1) Can be used for recovery of energy from brine instead of multiple flashes (2) Can be used instead of steam turbine on small plants	Most commonly used instead of steam turbine
Geothermal combined cycle power plant	Has not been used in practice, but could be	(1) Steam turbine with binary plant used as steam turbine condenser (2) Brine energy recovered in separate binary plant	Not applicable
Effect of brine chemistry on total available energy	Not applicable	Determines lowest temperature to which brine can be flashed without scaling Temperature can be lowered by adding acid	Determines the lowest brine temperature at discharge of binary plant heat exchangers to avoid scaling. Acid addition lowers temperature
Condenser	Generally sub-atmospheric. Direct contact of cooling water and steam, unless H_2S abatement is required, then shell and tube often used	Same, unless a geothermal combined cycle unit is used	Same for steam-only plants Generally supra-atmospheric for binary plants
Condenser gas removal equipment	Large to handle 0.5 to 8% NCG in steam. Sized for NCGs and air infiltration	Same	Small. Sized only for minimal breakdown of hydrocarbon and air infiltration

(*continued*)

TABLE 23.6 (*Continued*)

Wells Flow: Design Issue	Steam Only	Steam and Brine Two-Phase Flow	Brine only
		Power Plant Design (continued)	
	Creates large parasitic steam or power load	Not applicable for geothermal-combined cycle units	
Cooling System	Water-cooled, usually uses condensed steam as make-up water for cooling tower	Same for steam turbine plants Most often air-cooled for binary and geothermal combined cycle plants	Usually air-cooled condenser, unless surface water or steam condensate is available. Air cooling creates a large parasitic load and summer capacity loss
Abatement of H_2S gas contained in geothermal steam	If required by regulators	Same	Not required for an all-binary system, as H_2S never leaves brine
		Surplus Fluid Design	
Brine injection pipeline	Not applicable	Often long and expensive. Lengths of 0.5 to 1 mile from production area, or longer	Same as steam/brine system
Brine injection wells	Not applicable	Required, generally 1 well per 10–40 MW. (Often negative wellhead pressure)	Required, generally 1 well per 3–10 MW (usually positive wellhead pressure)
Cooling water blowdown well	Usually required. One per power plant	Same. Should not mix with brine because of corrosion/precipitation if condensate is used in cooling tower Not required for geothermal combined cycle, as steam condensate is not oxygenated and can be mixed into brine	Not required for binary-only power plant
Fluid injection or disposal strategy	Inject cooling tower blowdown to regenerate steam (80% is evaporated in cooling tower). Supplement with fresh water injection. Water injection can minimize the production of hydrochloric acid gas with the steam	(1) Generally, all brine and condensate must be injected into or outside of the active reservoir for reservoir management (heat mining, steam flow and pressure support) and environmental compliance	Always reinjected to maintain a high reservoir liquid level because of the production well pumps

(*continued*)

TABLE 23.6 (*Continued*)

Wells Flow: Design Issue	Steam Only	Steam and Brine Two-Phase Flow	Brine only
		Surplus Fluid Design (continued)	
		(2) Rarely, some may be surface disposed to evaporation pond, river, or ocean to reduce capital and operating costs, but may be against best reservoir management practice	
Parasitic load of injection	Cooling tower blowdown pump load is very small, approaching zero	Generally pump load is small because of two-phase reservoir	Pump loads can consume 3%–7% of total gross generator output

separation vessel varies from field to field, it is, in fact, better in practice to "put a fence at the top of the cliff, than an ambulance at the bottom."[14] In other words, the best practice in geothermal design is to first ensure high-efficiency separation (the fence) and not rely on the steam-washing system (the ambulance) at the power plant to prevent scaling of the turbine. Steam washing is defined and described below in Section 23.6.5.

Figure 23.3 illustrates a modern high-efficiency, low-pressure-drop geothermal steam separator designed by Sinclair Knight Mertz Consultants.* The two-phase mixture of steam and brine from the wells on the right enters the vessel through a special lemniscate entry (providing lower entrance turbulence than a tangential entry). The brine is separated by the centrifugal motion of the steam flowing upward. The steam enters an exit pipe at the top, flowing downward to exit through the bottom of the vessel. The brine exits tangentially through the vessel side-wall below the inlet.

23.6.2 Multipressure Steam Flash and Separation

The number of steam pressures (flashes) that are taken from the brine varies from field to field. Each time the brine is flashed, additional energy is extracted from the brine by delivering a supply of steam to the power plant at a lower pressure, but the concentration of minerals in the residual brine is increased. Many fields use only one flash to avoid silica saturation of the brine. Multiple flashes can quickly drive the brine past the silica saturation point by the combined effect of cooling and concentration due to the extraction of steam. Where multiple-flash separation is used, it is usually limited to two flash pressures. At the CalEnergy† Salton Sea facilities, turbines are driven off of as many as three separate flashes, producing turbine inlet pressures of roughly 250, 100, and 10 psig (actual pressures vary among the various plants) with a final flash at atmospheric pressure that is vented, but a large silica precipitation facility is employed to desupersaturate the silica from the brine to prevent pipeline and injection well fouling.

Figure 23.4 illustrates a multiflash steam separation and cleaning process from the well to the turbine inlets. Brine from the first separator is flashed to the lower pressure across the level control valve of the higher pressure separator. Commonly, each flow of steam is washed with geothermal steam condensate from the power plant before entering the turbine. Residual brine is injected.

---

*Sinclair Knight Mertz Consulting, http://www.skmconsulting.com.

†CalEnergy Generation, http://www.calenergy.com.

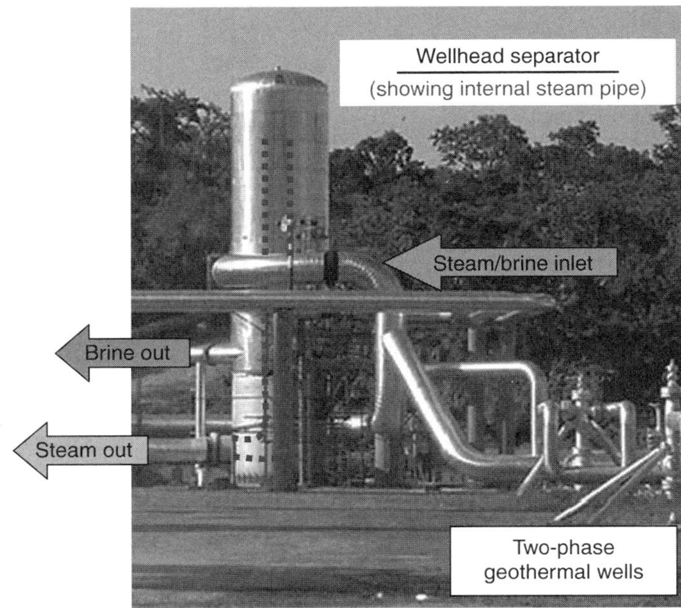

FIGURE 23.3 Annotated photo of a high-efficiency steam separator.

23.6.3 Steam Pipeline Operation

Steam pressures used in geothermal power generation are often less than 120 psig at the turbine inlet, and the steam pipelines are sized to keep steam velocities low. Thus, on large-steam turbine installations, there may be many miles of 30–42 in. pipeline snaking along roads and hillsides to bring the steam to the power plant.

These long pipelines introduce various operational problems. The pressure drop that occurs over such long distances represents an unrecoverable loss of available energy.* The long distances and the transport of saturated steam can also lead to considerable condensation in the pipelines, especially during windy, rainy, or snowy weather. Condensed steam must be removed from the pipeline either in knockout pots with steam traps, or in the steam scrubbing vessel near the turbine inlet to prevent water damage to the turbine.

23.6.4 Steam Washing Prior to Steam Turbine Admission

After the steam is brought to the edge of the power plant from the dry steam wells or the separators, the steam at many facilities is washed by injecting steam condensate from the power plant condenser into the steam line. Typical flow rates look to achieve a total moisture of 1%–2.5%, or about 10–25 thousand lbs/h (up to 50 gal/min) for a common 55 MW power plant using about 1 million lbs/h of steam. For each 100 pounds of condensate injected into the steam flow, approximately 22 pounds of steam is condensed. This condensed steam also helps to scrub the vapor, but the condensation is essentially a parasitic load. In a large facility, with poor steam quality, the wash water can represent as much as one megawatt of steam flow. The condensation and any injected water are removed in the power plant moisture removal vessel that immediately precedes the turbine.

*A wellhead separator pressure of 160 psia and a turbine inlet pressure of 110 psia are commonly seen. Assuming a constant condenser pressure of 101 mm Hga (4 in.), this pressure drop between wellhead and turbine represents up to a 7% loss of available energy.

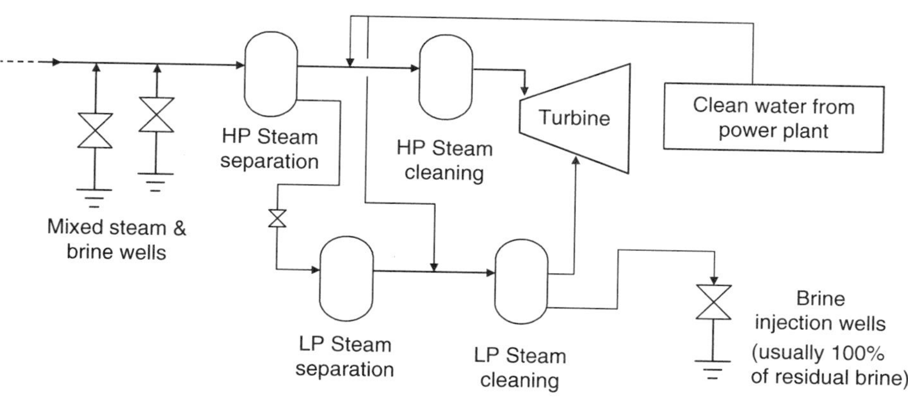

FIGURE 23.4 Geothermal double-flash process wih steam cleaning.

The moisture removal vessel can be either a centrifugal separator or a vane-type demister. Pad-type demisters are not used because they are prone to scaling. The Porta-Test Whirlyscrub® V Gas Scrubber* is used upstream of many U.S. and worldwide geothermal steam turbines to ensure low-moisture steam is delivered to the turbine.

When high-efficiency steam separation is successfully implemented for wells producing a two-phase mixture of steam and brine, steam washing is unnecessary on a continuous basis. Even if a high-efficiency separator is used, a moisture removal vessel is still required to protect the turbine from entrained water and/or solids that may originate in the pipeline between the separator and the moisture removal vessel.

23.6.5 Turbine Washing to Remove Scale Buildup

The consequence of insufficient steam purity and contaminant cleanup is that the turbine first-stage nozzle diaphragms will develop a layer of scale that restricts the steam passage. A schematic illustration of the turbine pressure and generation response to inlet nozzle scale build-up and removal through turbine washing is shown in Figure 23.5. As scaling occurs, the operator will allow the turbine inlet pressure to increase to maintain plant contracted output. As the scaling and pressure build up, the pressure limit of the turbine casing is eventually reached, and at that point, further scaling of the turbine results in reduction of the of turbine output due to decreased steam flow.

Turbine scaling occurs as the steam passes across the first-stage stationary blades. Entrained microdroplets of liquid with dissolved impurities evaporate as they cross the blades. The impurities are left behind, and scale buildup is the result.

Over the years, various experiments have been undertaken to remove the built-up scale with online turbine washing. Early efforts took the unit off-line and injected water at reduced rotating speeds. Later, low load injection of water into the turbine inlet was tried. Eventually, however, it became the widely accepted practice to inject water into the turbine at almost full-load to clean off scaling and restore design operating conditions. This practice has evolved in part by consideration of the actual operation of a low-pressure geothermal turbine. In normal operation, the saturated steam enters the turbine with condensation forming as the steam passes through the turbine to produce work. A typical geothermal steam turbine operating around 120 psia inlet pressure and 2 psia condenser pressure (4 in. Hg, absolute) has an exhaust moisture of 14–16% less interstage moisture drains. Therefore, adding 1%–2% moisture

*http://www.natcogroup.com/

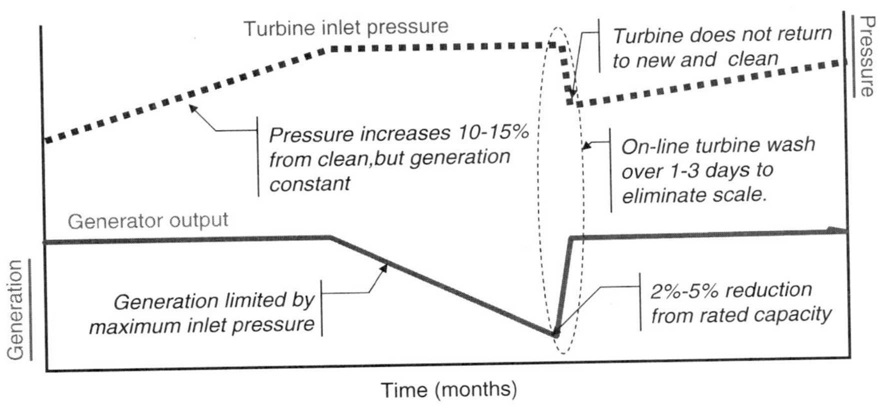

FIGURE 23.5 Response of turbine to scaling of the first stage nozzles and to an on-line turbine wash (schematic-not to scale).

to the steam entering the turbine for short periods of time does not represent a large deviation from normal operating conditions.

23.7 Geothermal Power Production—Steam Turbine Technologies

23.7.1 Overview of Geothermal Power Generation Technologies

Two conversion technologies dominate the geothermal power industry: Direct steam turbines in an open-loop Rankine cycle configuration, and binary power plants using a hydrocarbon (pentane or butane) in the boilers and turbines in a closed-loop organic Rankine cycle (ORC). Geothermal combined cycle (GCC) units marry the two technologies. Ammonia/water binary power plant technologies have been proposed as a potential third power cycle option, with one such plant having been built as of 2006, and additional plants planned for operation in the coming years. Table 23.7 lists the 2004 breakdown of worldwide installed capacity.

Steam turbines represent 95% of the installed capacity of geothermal power plants worldwide, including the steam turbines of the integrated steam turbine and binary power plants. Binary power plant technologies, discussed in Section 23.8, represent the remaining 5% of installed worldwide capacity.

23.7.2 Steam Turbine Conversion

Geothermal power plants using steam as the motive fluid are open-cycle systems in which geothermal energy in the form of steam is admitted to the power plant, the energy is extracted, and the residual brine and condensed steam are then discharged from the plant to surface disposal, or injection, as described above. Gases contained in the steam are vented to the atmosphere. Because there are no fuel handling systems, boilers, superheaters, ash systems, or other systems related to the burning of fuel, geothermal steam power plants are very simple. It is referred to as an "open" cycle because the condensed steam is not returned to a boiler in a closed-loop circuit. The steam from the geothermal resource is delivered to the turbine inlet and may originate from either dry steam wells or from the separated steam of wells producing a brine/steam mixture.

There are three main steam power processes in use around the world: steam turbines without condensers (atmospheric discharge), steam turbines with subatmospheric condensers, and steam turbines integrated with binary power plants.

TABLE 23.7 Breakdown of Geothermal Power Technologies Worldwide

Resource Type	Power Plant Type	Installed Capacity	Number of Units
Dry steam	Steam turbine	2460 MW	63
Flash steam	Steam turbine	5831 MW	210
Pressurized brine, including residual brine from a flash steam plant	Organic Rankine cycle (ORC)	274 MW	154
Steam/brine	Integrated steam turbine and ORC, or ORC only	306 MW	39
Pressurized brine	Binary-Kalina cycle (ammonia)	2 MW	1
	Total	8873 MW	467

Source: From DiPippo, R. 2005. *Geothermal Power Plants—Principles, Applications and Case Studies*, pp. 404–405.

In a few installations worldwide, the steam turbine may exhaust directly to atmosphere. It is an inefficient mode of operation since 50% of the available energy is lost when discharged at atmospheric pressure. From an economic perspective, the loss of 50% of the available energy does not justify long-term operation of atmospheric discharge turbines, because with only the addition of a condenser and cooling tower, the output of the facility can be roughly doubled with no additional resource development. However, atmospheric discharge turbines are sometimes used to prove reservoir capability because of rapid installation and low cost. In other places, they have been used when CO_2 content in the steam was over 5%. The use of atmospheric discharge turbines provides another example of the trade-offs that occur between long-term efficiency benefits and economic constraints as discussed in Section 23.5.3.

Geothermal combined-cycle steam/binary power plants also use a steam turbine that discharges at slightly above atmospheric pressure. However, instead of the steam venting to the atmosphere, the steam is condensed and energy recovered in the vaporizers and preheaters of the binary power plant. In this way, the energy of the exhaust steam is captured in the power conversion cycle of the binary power plant. More details on this process and advantages of this "combined cycle" technology are discussed in Section 23.8.4.

Most of the total installed worldwide geothermal generating capacity is from steam turbines in which the geothermal steam is admitted to the turbine at between 15 and 250 psia and exits the turbine at subatmospheric pressures of between 1.0 and 2.5 psia. The condenser is cooled by water from a cooling tower. The steam rate (pounds/kilowatt-hour) of the turbines is determined by the turbine efficiency (usually slightly better than 80%, including entrance and leaving losses), inlet pressure, and the condenser pressure. Figure 23.6 is a photo of the world's largest-capacity (110-MW net rated output) single shaft geothermal steam turbine, in the Fuji Electric* shop before its installation at the Wayang Windu (Indonesia) geothermal power project. The design inlet and condenser pressures are 150 and 1.7 psia, respectively, using a 27-in. long last stage blade.[15]

23.7.3 Condensing Steam Turbine Process

Figure 23.7 is a general illustrations of a dry steam power process. Steam is produced from the geothermal reservoir through multiple wells and delivered to the power plant through cross-country pipelines. The steam is always passed through a moisture removal vessel before entering the turbine to ensure that no particles or slugs of water reach the turbine. In many installations around the world, the steam is washed before entering this vessel to remove any impurities (see Section 23.6.4).

The basic process for a single-flash steam process is much the same, as shown in Figure 23.8. The key difference is that a two-phase mixture of brine and steam are produced from the geothermal reservoir to

*Fuji Electric Systems Co., Ltd., http://www.fesys.co.jp.

FIGURE 23.6 The Wayang Windu 110 MW geothermal turbine. (Photo courtesy Fuji Electric Systems Co., Ltd.)

the surface. A centrifugal separator removes the steam for delivery to the power plant (Section 23.6.1), and the brine is injected back into the reservoir.

Figure 23.9 is a typical process-flow diagram for a geothermal power plant using a steam turbine and a condenser. In this illustration, steam is supplied from dry steam wells or a flash separation process at only one pressure (i.e., there is no low pressure turbine inlet). The steam is also used directly in the gas removal ejectors. Steam exiting the turbine is condensed in a direct contact condenser that is supplied with cold water from a cooling tower. The condensed steam and the cooling water are returned to the cooling tower. Approximately 80% of the steam to the power plant is evaporated in the cooling tower, and the remaining 20% forms the cooling tower blowdown. In hot arid climates, the amount of water evaporated can be greater than 100% during the hottest hours of the day and greater than 90% over a 24 hour period. The steam contains noncondensable gases (NCG), primarily CO_2, but also H_2S and other trace gases. These gases must be removed from the condenser to keep a low turbine exhaust pressure.

In Figure 23.9 geothermal steam is used as the motive force for the first-stage ejector to compress the NCG. The ejector steam is condensed in the direct-contact ejector condenser (the intercondenser), with the condensed steam and cooling water being returned to the main condenser. The NCG from the

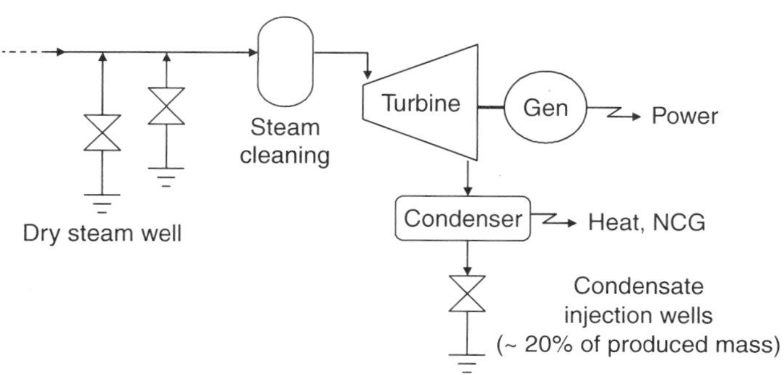

FIGURE 23.7 Geothermal dry steam power process (water cooled).

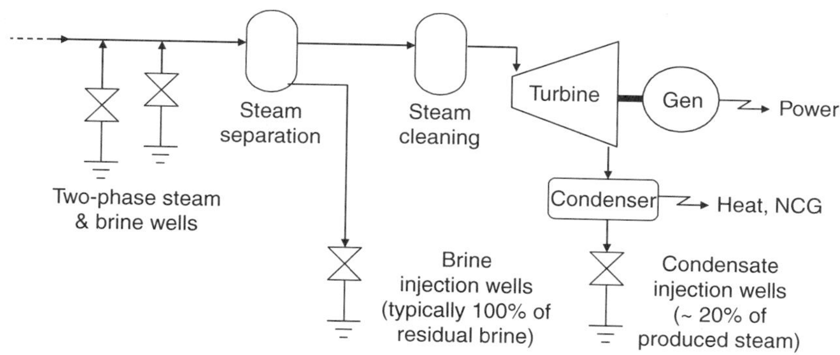

FIGURE 23.8 Geothermal single-flash steam power process (water cooled).

intercondenser is compressed in a second stage to atmospheric pressure either by the vacuum pump or by a second-stage ejector with an aftercondenser. The amount of steam used by the ejector and the size of the vacuum pump motor is related to the size of the power plant and the NCG content of the steam. The vacuum pump is the primary system because it is more efficient than an ejector. The second-stage ejector is shown as a backup system because it has lower capital costs than a standby stainless steel compressor. Stainless steel is required throughout the NCG system because of the corrosiveness of moist CO_2 gases. The NCGs are dispersed using the cooling tower fans. If H_2S abatement is required, it is installed between the discharge of the second stage of compression and the cooling tower. The gas removal system is discussed in greater detail in Section 23.7.6.

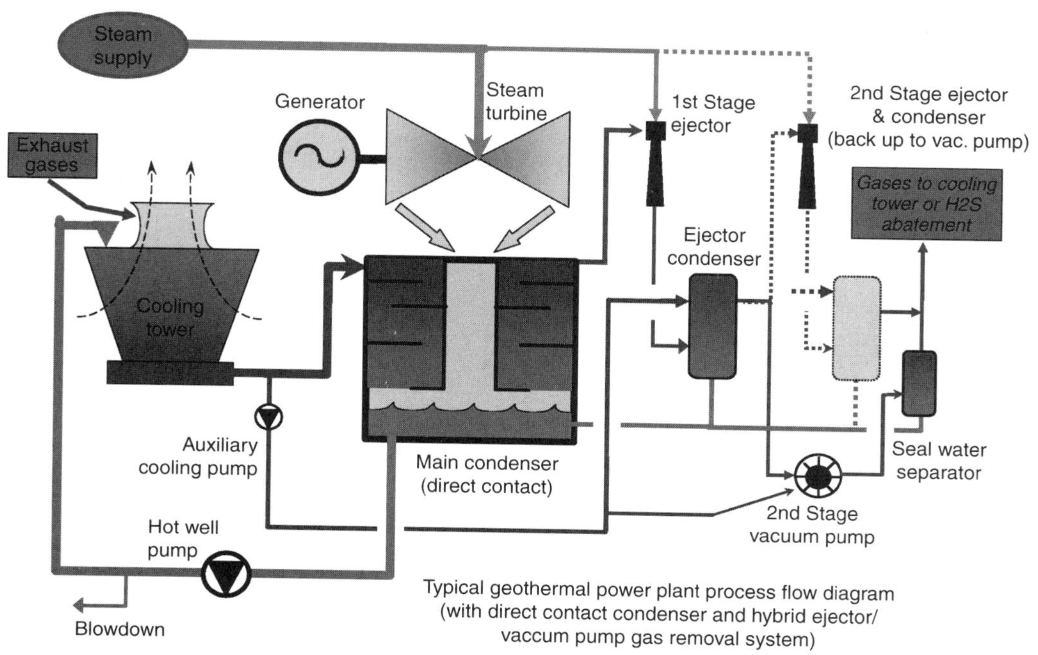

FIGURE 23.9 Geothermal steam power plant with direct contact condenser. (From Williamson, K. H., Gunderson, R. P., Hamblin, G. M., Gallup, D. L., and Kitz, K., *Proceedings of the IEEE*, 89 (12), 1783–1792, 2001.)

23.7.4 Design of Geothermal Turbines

Geothermal steam can be a virtual chemical soup that varies from field to field. In most cases, the steam is saturated as it enters the turbine, and has a moisture content approaching or even exceeding 15% as it leaves the turbine for subatmospheric discharge turbines. The combination of scaling tendency, potential corrosivity, and high moisture content heavily influences the design of geothermal turbines.

Geothermal turbines generally use the same alloys as low pressure turbines in fossil fuel power plants, but with small variations in the alloy composition of the blades and rotors to enhance the alloy's ability to withstand the corrosive environment of geothermal steam. Detailed design can be just as important as alloy selection. For example, by using long-radius transitions and avoiding notches in the design of geothermal turbine parts, especially in the blade attachment areas, the stresses are more evenly applied and the blades and rotors are much less susceptible to cracking and failure.

The last-stage turbine blades are exposed to high moisture content, which requires methods to minimize blade leading edge erosion. Both erosion-resistant alloy shields and blade-surface hardening are used to resist the impingement of water on the last-stage blades. In addition to moisture, the specter of corrosion once again strongly influences the design of geothermal turbines at the last stage.

Last-stage blade size is a key element in the maximum size geothermal turbine that can be built. Geothermal turbine last-stage blades are designed for lower combined stresses than equivalent blades in conventional power plants. Because a large portion of last-stage blades stresses is proportional to the steam flow, this results in lower steam flows (per square foot of exhaust area) than in conventional power plants. High moisture content—the result of using saturated steam at the turbine inlet—tends to limit blade size to about 660 mm (26 in.) for turbines operating at 3600 rpm (60 Hz power), and its equivalent of about 765 mm (30 in.) for turbines operating at 3000 rpm (50 Hz power). The combined effect of the limited blade length and the reduced flow rate per unit area limit the maximum size that can be achieved in a geothermal turbine/generator. Thus, it is common to see two to three 55–70 MW turbines installed in a single powerhouse, rather than a single 200 MW unit. As discussed above, the largest two-flow geothermal turbine/generator (two rows of last-stage blades with steam inlet in the middle) ever built was Wayang Windu in Indonesia, with a rated net output of 110 MW.

23.7.5 Design of Heat-Rejection Systems

A second area affecting geothermal competitiveness is the heat-rejection system comprising the condenser, cooling tower, and cooling water pumps. The heat-rejection system of geothermal power plants can be up to eight times as large as a combustion-turbine combined-cycle power plant per kilowatt of output because of the low available energy per mass of steam at the inlet to the geothermal turbine. The heat-rejection system of any geothermal power plant, whether for a steam turbine or binary plant, represents a very large capital cost burden for the project.

Where H_2S abatement is required, modern plants are most often built with shell-and-tube condensers. Thus, shell-and-tube is found in most steam condensers in U.S. geothermal power plants. Outside of the U.S., the majority of geothermal steam condensers are direct-contact. In this system, steam exiting the turbine encounters either water sprayed from nozzles or droplets from trays. Heat transfer is much more efficient. The higher the gas content of the steam, the larger the advantage that direct-contact condensers bring. The disadvantages are that the cooling water is exposed to the CO_2, H_2S, and ammonia, thereby providing substantial biological growth potential in the circulating water. Counteracting this is the high blowdown rate in a direct-contact geothermal cooling system.

Cooling tower make-up water comes from the condensed geothermal steam. In most circumstances, between 70% and 80% of the steam that enters the turbine is eventually evaporated in the cooling tower. The remaining 30%–20% steam is condensed and not evaporated, and is the source of the cooling system blowdown. Although the low cycles of concentration in the cooling water help control

TABLE 23.8 NCG Amount and Composition in Steam at Tiwi and Mak-Ban, Philippines

Field	Percent Gas in Steam (wt%)	Gas Composition (Mole %)								
		CO_2	H_2S	NH_3	CH_4	H_2	N_2	O_2	He	As, Hg
Tiwi	2.8%	97.9	1.24	0.04	0.05	0.31	0.45	0.04	0.001	Trace
Mak-Ban	0.5%	88.7	6.9	0.08	0.22	1.24	2.61	0.10	0.006	Trace

biological growth in the system, at the same time, the high blowdown rates result in it being relatively expensive to dose the water with biocides, since the biocides are quickly washed out of the system. Additionally, the presence of H_2S and NH_3 in the cooling water substantially reduces the efficacy of conventional biocide treatments. Cooling tower blowdown is most commonly injected back into the geothermal reservoir.

23.7.6 Geothermal Condenser Gas-Removal Systems

In addition to the challenges of mechanical design of the steam turbine and the size of the cooling and condenser system, the steam also contains large quantities of noncondensable gases (NCG). Average NCG compositions for two fields in the Philippines are shown in Table 23.8.

As a consequence of high NCG content in the geothermal steam, the gas-removal systems of geothermal power plants dwarf the gas-removal systems of conventional power plants. For example, the power plants at Tiwi, with a gas content of about 3%, can use either a two-stage steam jet gas ejector that consumes 110,000 lbs of steam per hour, or a hybrid system of first-stage ejector and second-stage liquid ring vacuum pump (consuming 33,000 lbs of steam per hour and 0.6 MW of electrical load). This is summarized in tabular form in Table 23.9. The two systems were designed for different gas loads, but both systems place a tremendous parasitic load on the plant. The two-stage ejector consumes an additional 11% steam over that used by the turbine, while the ejector hybrid consumes only 3.4% of the turbine steam, but also consumes 1.2% of the net electric output of the facility.

The higher the NCG content is above 1% in the steam, the greater the incentive to cut the parasitic steam usage with mechanical gas compression. This may take the form of ejector hybrids, as discussed above, or to eliminate the ejector altogether using an all-mechanical compressor. As in most aspects of geothermal design, a variety of solutions is employed.

Whatever the particular compressor system used, the additional cost of compressor systems is usually justified by the reduction of production wells required to supply the ejectors with steam, by an increase in the turbine steam supply at a geothermal field that is short of steam, or looking to maximize installed capacity. However, the electric parasitic load can have considerable negative implications on plant revenue, and for this reason, it is sometimes avoided on newer installations. The reason that the electric loads of mechanically driven compressors can have large impacts on revenue is found in the size limitation of the geothermal turbine, as discussed above. If the turbine size is limited (e.g., by last-stage blade size), then adding parasitic electric load reduces the power that can be sold, and hence reduces

TABLE 23.9 Parasitic Steam and Power for Condenser Gas Removal at Tiwi, Philippines, for one 55 MW Unit

Option	Design Gas Load % Steam	Parasitic Steam		Parasitic Power	
		Ejector Flow (lbs/h)	% of Turbine Flow	kW	% of Rated Capacity
Two-stage ejector	2.8%	110,000	11.1%	Not applicable	
Ejector hybrid	2.8%	33,000	3.4%	635	1.2%

revenue. This is, once again, an example of the need to sacrifice efficiency in the short term to obtain higher net present value in a discounted cash flow calculation.

23.8 Geothermal Power Production—Binary Power Plant Technologies

Section 23.7 presented the technologies and strategies of the geothermal power plant that directly uses geothermal steam in the turbine. In Section 23.8, the use of geothermal steam and brine as an energy source but not as the power plant working fluid (that which drives the turbine) is discussed. The power plants that use these two fluids are known as *binary power plants*. All of the closed-loop working fluid cycles used in geothermal energy production, with a few exceptions, are Rankine cycles with a hydrocarbon as the working fluid.

23.8.1 Binary Power Plant Advantages

If one were to consider an idealized geothermal system in which the thermodynamics of pure water and steam played the dominant role in decision-making, there may not be a large role for the binary geothermal power plant. However, one does not have to delve too far into the reality of geothermal development before the advantages of the binary geothermal power plant in some applications become apparent. Some of these reasons have already been discussed in the preceding pages. Two of the most important answers to the question "Why binary?" are provided below.

23.8.1.1 Silica Solubility

As noted in several earlier sections, avoiding the precipitation of silica is a requirement for successful long-term geothermal well and pipeline operations. The precipitation of silica is a function of three factors: time, concentration, and temperature. Time is easily extended by the addition of acid to the geothermal brine, but for inexplicable reasons, this straightforward and easy solution is widely ignored. That leaves concentration and temperature as the factors within the control of the power process designer.

A brine that starts out at 450°F in the reservoir, when introduced into a steam separator at a pressure of 135 psia, will flash off approximately 12% of the initial water mass as steam. If there is a second flash at 35 psia, the total water mass flashed to steam is 21%. In such a flash process, two factors are at work increasing the potential of forming silica scale: mineral concentration in the brine and lower brine temperature. The physical properties that inhibit scale precipitation are that amorphous silica solubility is greater than the quartz silica solubility, and the often-slow kinetics of amorphous silica precipitation.

In contrast, the geothermal heat is transferred to the working fluid in a binary power plant without concentrating the silica by flashing off steam from the brine. Therefore, the temperature of the brine can be lowered further before the amorphous silica precipitation temperature is reached. Because of the silica solubility limit, the binary power plant can extract more energy from the produced brine (lower temperature) than a flash plant is capable of, especially where pH modification of the brine (acid addition) is not used. This can be true either on wells that are pumped and in which pressurized brine is delivered to the power plant, or after the brine has been flashed once and the binary plant is then installed to make use of the heat in the residual brine.

23.8.1.2 Lower Parasitics vs. Cycle Inefficiency

In a binary plant, a heat exchanger must be used to transfer the heat from the geothermal fluid to the power plant working fluid. As such, there is an inherent loss of efficiency from the process, as well as the capital cost of the heat exchanger itself. There is also a cycle feed pump, a significant power consumer that has no counterpart at all in a geothermal steam turbine power plant. On the other hand, in a plant directly using geothermal steam, the process of removing NCGs from the condenser is also an inherent loss of efficiency and a significant capital cost that does not exist in a binary power plant. How the two options balance out is dependent on site-specific factors, but conceptually, there are offsetting advantages and disadvantages, making the thermodynamic and economic balance closer than it might appear at first glance.

The exception to the offsetting advantages of binary plants versus direct steam plants occurs when the use of binary also forces the use of air-cooled condensers instead of water-cooled condensers and evaporative cooling towers. The water-cooled direct contact condenser has lower capital cost, higher efficiency, and lower parasitic loads than an air-cooled condenser system in all but the coldest climates.

23.8.1.3 Compactness

Binary power plants operate with working pressures much higher than those used for low-temperature flash power plants. This translates to smaller pipes and eliminates flash separators and scrubbers on pressurized brine systems; additionally, the condensers have smaller volumes because of the much higher condenser pressures of the working fluids.

23.8.2 Types of Binary Systems

There are two types of binary power systems currently in use, both of which are based on hydrocarbons, and a third currently under development that is based on using an ammonia–water mixture as the working fluid. A fourth binary power system based on commercial refrigeration equipment is also under development, with a 200 kW demonstration project starting operation in Alaska at Chena Hot Springs in 2006. All of these power cycles are similar to that shown in Figure 23.10.

23.8.2.1 Subcritical Pentane Binary Power Plants

The geothermal fluids (brine or steam) are used to boil pentane as the power cycle working fluid in an ORC. The pentane passes through a turbine coupled to a generator and is then condensed. The cooled geothermal fluid is injected back into the geothermal reservoir.

The use of pressurized geothermal water from pumped wells in a subcritical pentane-based power plant is the most prevalent method to utilize geothermal resources from 280° to 320°F. This system is also used to extract additional energy and power from the residual brine from single-flash flash plants all around the world.

Where two-phase geothermal fluid is available instead of pressurized brine, the two-phase fluids can be used in one of two ways. In some instances, both the steam and the brine are used in the vaporizers and preheaters. On larger facilities, it is more common to see a mixture of steam turbines and ORC plants. This type of integrated facility is made up of a pressurized brine facility of the type described above and a

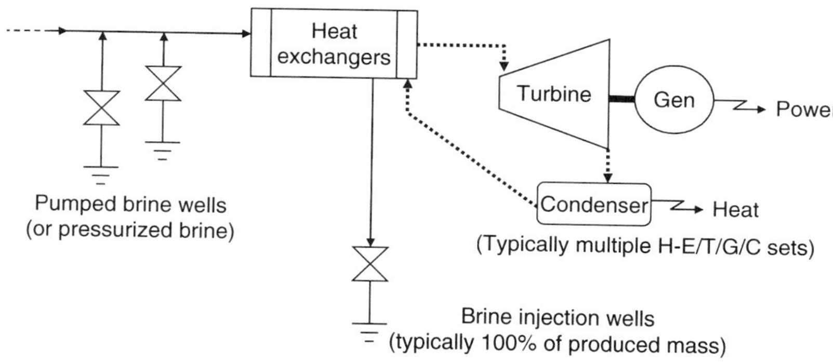

FIGURE 23.10 Pressurized brine binary power process (air or water cooled).

geothermal combined-cycle facility. In the geothermal combined-cycle facility, geothermal steam is first passed through a steam turbine and then condensed in a binary power plant.

23.8.2.2 Supercritical Hydrocarbon Binary Power Plants

Several hydrocarbon-based binary power plants have been built that use supercritical isobutane as the working fluid in an ORC. The pinch point is eliminated in the vaporizer in the supercritical regime. Although there is a theoretical thermodynamic benefit to a supercritical power cycle, the market has not embraced these plants because they tend to be much more expensive than the subcritical plants.

23.8.2.3 Ammonia–Water Binary Power Plants

Several different power cycles have been proposed based on ammonia-water systems. The potential advantages and disadvantages that all ammonia systems share over hydrocarbon-based binary systems include:

- Higher heat-capacity fluid means smaller heat exchangers and less parasitic load from the working fluid cycle feed pumps.
- Higher pressures result in smaller vapor equipment components.
- The boiling point changes with water/ammonia concentration, providing a boiling point glide instead of a single pinch point temperature, increasing thermodynamic efficiency.
- Condensation temperature changes with concentration, again providing a glide instead of a pinch point and a higher thermodynamic efficiency.
- The design of the heat exchangers is more complex to achieve the glides.
- There is a tradeoff of ammonia toxicity for hydrocarbon flammability.
- Ammonia has been used in refrigeration cycles for a century, providing a wealth of practical operating knowledge, even though few ammonia-based power plants exist.
- Ammonia vapor and steam have similar molecular weights (17 and 18 g/mol, respectively), which theoretically means that conventional steam turbines could be used, with little modification; if true, it would allow a rapid step up in turbine size to 30 MW or more and for the competitive bidding for supply of the turbine.
- Ammonia cycles are more complex, with more exchangers and cycle pumps.

In spite of the net benefit of the potential advantages, only one 2 MW geothermal plant has been installed to date, in Husavic, Iceland, using one of many Kalina cycle ammonia-water technologies. Plans to install additional Kalina cycle plants at Salt Wells, Nevada, and a 42 MW plant at Cove Fort, Utah, have been announced.

23.8.3 Binary Power Plants for Pressurized Geothermal Brine

Ormat Technologies* has developed a binary power plant approach, using subcritical pentane-based Rankine cycle technology, that commercially dominates the pressurized brine market. Because of that dominance, their technology will be used to illustrate the principles of the binary power plant design in this section. The technology has also been used in some instances in which the wells produce a mixture of steam and brine. These same basic power plants are now also being used to generate power from natural gas pipeline compressors exhaust gas waste heat.

Ormat started out designing and building small binary plants. One of their first installations was at the East Mesa field in California near the Mexican border. The ORMESA I plant began operation in 1987 and

*Ormat Technologies, http://www.ormat.com.

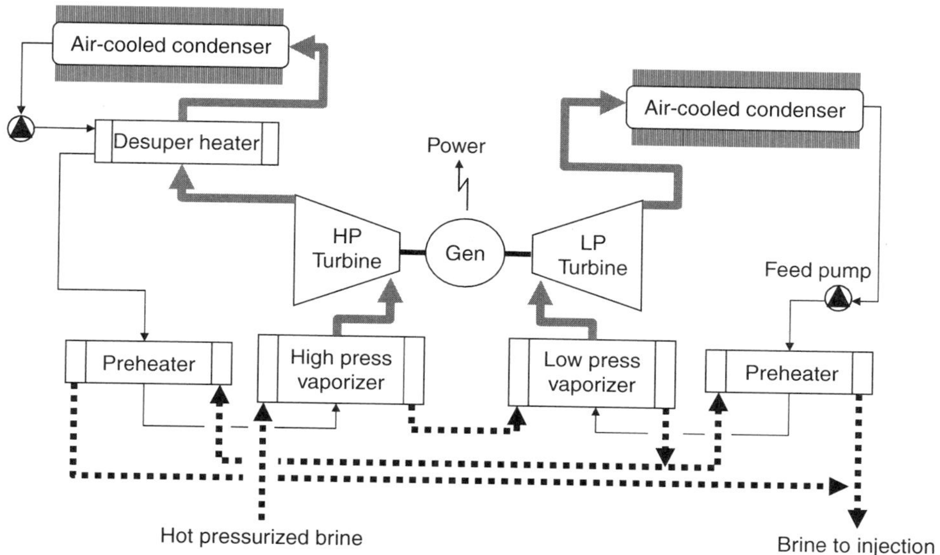

FIGURE 23.11 Ormat two-level subcritical pentane power cycle.

was comprised of 26 individual boiler/turbine/condenser units, with a total output of 24 MW. Since the East Mesa installation, Ormat has steadily increased the size of their units in the pursuit of both economy of scale and efficiency, and hence, lower installed cost ($/kW). Initially, Ormat technology was based on a modular design, in which the economy of scale of a single large plant was traded off in favor of the economy of scale of many smaller standardized, modular, factory-assembled skid-mounted units that required minimal field construction. Ormat now also offers a larger T/G unit comprised of a high- and low-pressure turbine coupled to opposite ends of a single generator, with the capability to deliver 20 MW to the grid.

The basic ORC process design used by Ormat is illustrated in Figure 23.11. The hot pressurized geothermal brine, whether from a pumped well or as residual brine from a flash plant, enters the high-pressure (HP) vaporizer to generate HP pentane vapor for the HP turbine. The geothermal fluid exits the HP turbine and enters the low-pressure (LP) vaporizer, which generates vapor for the LP turbine. From the LP vaporizer, the geothermal fluid splits between the preheaters of the LP and HP vaporizers. The pentane for the LP turbine is pumped out of the condenser to the preheaters and into the vaporizers. Pentane vapor passes through the turbine and into the condenser. The HP pentane cycle is similar, except that the pentane from the condenser is first pumped to a desuperheater located between the turbine discharge and the condenser.

Examples of where pressurized brine from the wells is used directly in the binary power plant include Soda Lake and Steamboat Springs in Nevada, as well as Heber and East Mesa in California.

Binary power plants have been installed to capture the waste energy of several geothermal power plant hot brine discharges. Examples include Brady Hot Springs (Nevada), Mak-Ban (Philippines), Los Azufres (Mexico), Miravalles (Costa Rica), and Wairakei (New Zealand). There also remain many locations where binary power plants could be installed on existing steam flash plants.

23.8.4 Integrated Steam Turbine and Binary Power Plants

In other locations, steam turbine(s) and binary power units have been combined into a single facility. Integrated facilities are installed at fields where the geothermal fluid is produced to the surface as a mix of steam and brine, rather than pumped and pressurized brine. The largest integrated facilities are those of

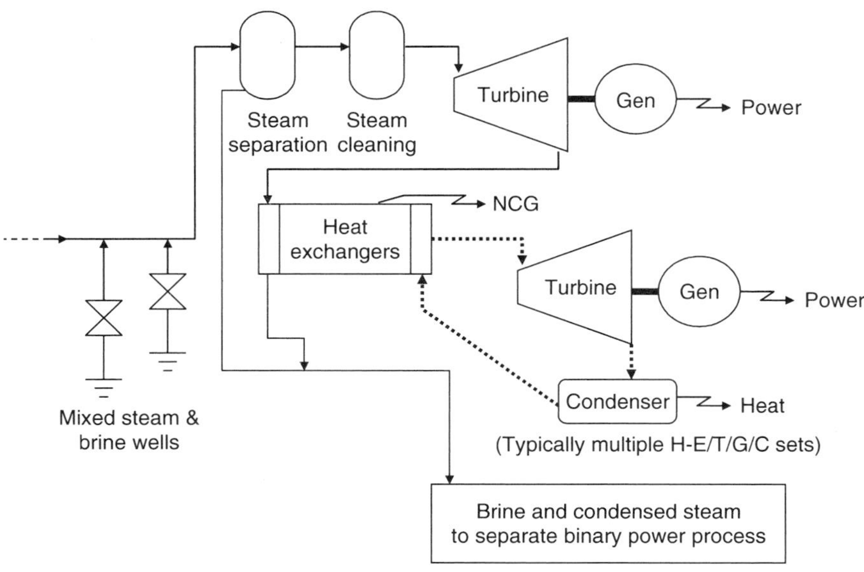

FIGURE 23.12 Integrated brine steam combined cycle binary power process (air cooled).

Ormat Technologies, and are based on the integration of the pressurized brine ORC and what is referred to as *GCC* plants. See Figure 23.12.

23.8.4.1 The Geothermal Combined Cycle Process

The brine/steam mix flows from the production well to a separator. Brine from the separator is directed to a pressurized brine binary power plant, as described above.

Steam from the separator is cleaned and delivered to a steam turbine. The steam turbine faces the same inlet steam quality issues as were discussed in Section 23.8. However, because the turbine discharges at near-atmospheric pressures, the previously discussed concerns about last-stage blade length and moisture-induced erosion are eliminated. With the elimination of these two issues, cycle optimization of the geothermal resource can be undertaken without a cap on the maximum turbine inlet pressure. The available mass and the optimized pressure of the steam result in varying steam turbine sizes and outputs at different geothermal fields.

At the larger GCC units (see Table 23.10), the steam that discharges from the steam turbine is divided among multiple binary units. Each binary unit condenses the steam in a process of heating and

TABLE 23.10 Examples of Ormat Geothermal Combined Cycle Projects

Year	Project Name & Location	Project Net Output	Description of the Air-Cooled Power Plant
1996	Upper Mahiao, Leyte, Philippines	125 MW	Four geothermal steam turbines that feed multiple binary units for 120 MW of net output. An additional 5 MW of power are generated by other binary units using the geothermal brine
2000	Mokai, New Zealand	60 MW	One geothermal steam turbine that feeds multiple binary units for 50 MW of net output. 10 MW of power are generated by binary units using the geothermal brine

vaporizing the binary unit working fluid. The condensed steam is mixed back into the brine to reduce silica scaling potential. In some instances, small steam turbines are coupled to one end of a generator, with the steam discharged used in a binary plant with the pentane turbine coupled to the other end of the generator. In most cases, the condensers for the binary power plant are air-cooled.

23.8.4.2 Silica Solubility Limits

In fields with a steam/brine mix, the GCC competes directly with multiflash steam technology. The GCC can produce greater output where flash plants have elected to use a single flash to control silica (SiO_2) scaling, resulting in a brine discharge temperature of 300°F or greater. The same temperature limitation of the single-flash technology does not apply to the GCC, so the GCC is able to capture more energy from a given brine flow and therefore generate more power. Although it may at first seem counter-intuitive, the hotter the resource and thus the more steam that is produced in the flash process, the more advantage the GCC will have.

In a flash process, hotter brine results in a greater silica concentration in the reservoir fluid. Hotter brine also results in more flash, and therefore more concentration of the silica. The silica solubility limit controls the separated brine temperature, and therefore the minimum allowable flash pressure and temperature. Thus, because of the silica concentration, higher geothermal resource temperatures also result in a higher brine temperature that must be discharged from the flash plant.

Although the integrated binary plant also uses the initial flash in the steam turbine of the GCC, the condensed steam is mixed back into the brine flow. As a result, the final brine temperature can be lower without risking silica scaling. Consequently, more energy is available to the integrated binary power plant than is available to the flash-only plant. The binary plant advantage would be reduced if more geothermal operators were willing to use acid to slightly lower the pH of the brine and retard the kinetics of the silica precipitation.

23.8.4.3 Compare and Contrast

The GCC technology has some additional advantages and disadvantages over a single- or multiple-flash plant when compared side-by-side for a particular geothermal resource. The large advantage that the GCC obtains in greater energy availability due to the silica solubility limits is discussed in the paragraph above. The larger units that Ormat has developed in recent years will also help to make them more competitive through an economy of scale. Another advantage is obtained by discharging the steam turbine at slightly above atmospheric pressure. In this way, the NCGs are vented without the cost or parasitic load of the gas removal equipment that is needed when the steam turbine exhaust is at sub-atmospheric pressure, as occurs with the steam flash process. Since the steam condensate is mixed back into the brine with the GCC, the cost of an oxygenated fluid disposal well may be eliminated.

The advantages that the steam-turbine process holds over the GCC are: (1) the ability to use large equipment (up to a single 110-MW turbine); (2) no need for large heat exchangers (vaporizers and preheaters) and the intrinsic thermodynamic and capital cost penalty they entail; (3) the steam flash process uses wet cooling from the condensed steam, whereas the GCC uses dry cooling, resulting in a higher final condenser temperature. Thus, the steam process regains some of the thermodynamic disadvantage incurred by the silica solubility limit and the gas removal system parasitic steam load by having a lower power plant heat-rejection temperature and, thereby, greater available energy.

23.9 Environmental Impact

The generation of electric power will have an environmental impact, regardless of the energy source used. A large unregulated strip mine and unabated emissions from an old coal plant would probably represent the extreme negative end of the spectrum in its environmental impact to land, water, and air, but would not be particularly relevant to discussions of environmental impacts of power generation in the U.S. Understanding the other end of the spectrum is much more complex. Solar and wind generation would seem to have zero emissions, but both of these require fossil-fired generation to back them up during periods when these generation technologies are unavailable or erratic. These "firming" requirements increase emissions from fossil-fired plants, and result in them not being the zero-emission technology that they might otherwise seem to be. What this shows is that even renewable technologies have negative environmental consequences. Society's need for new sources of electric power generation, which cannot be served by conservation and efficiency, will result in negative environmental impacts.

Like other renewable technologies, geothermal power generation must be evaluated on a site-specific basis to accurately determine its environmental impacts. Two areas of environmental impact relevant to the use of geothermal energy, emissions and land use, are discussed below.

23.9.1 Geothermal Power Plant Emissions

Air emissions and clean air are the most-discussed environmental benefits of renewable-energy electric power generation, especially with the concern of the effect of greenhouse gases on global warming. Like other renewable energy sources, the emissions from geothermal power are generally low compared to fossil-fired generation. However, certain geothermal resources with high gas content may approach that of a natural gas fired power plant with regards to CO_2 emissions, but this is a rare exception. The concentrated CO_2 discharged from such units does make CO_2 capture feasible. The actual emissions from geothermal power plants are dependent on the highly variable gas content of the geothermal fluids that are being produced and the nature of the geothermal power process. Table 23.11 gives representative emissions of geothermal power plants compared to fossil-fired generation sources.

TABLE 23.11 Emissions from Generation Sources

Plant/Fuel	Emissions Rate (lbs/MWh)				
	NO_x	SO_2	CO_2	Particulates	Notes
Conventional Power Plants					
Coal	4.3	10.4	2,190	2.2	
Coal—life cycle	7.4	14.8	—	20.3	—[a]
Natural gas	3.0	0.2	1,210	0.1	—[b]
Avg. of all U.S. plants	3.0	6.0	1,390	—	Ref. 18
Geothermal Power Plants					
Pumped geothermal brine	0	0	0	Negligible	—[c]
Geothermal flash steam plants	0	Negligible with abatement to 0.35	60 (avg) some plants <120 to >900	Negligible	—[d]
Dry steam resources	Trace	Negligible with abatement	90 (Geysers) some plants <120 to > 900	Negligible	

[a] Kagel reports that this includes the emissions from the mining and transportation of coal to the power plant.

[b] Kagel reports as an average of direct fired, combined cycle, and simple cycle plants.

[c] Cooling tower drift only, if water cooling is used. Air-cooled plants have zero particulate emissions.

[d] 150–900 lbs/MWh of CO_2 is added to Kagel's reported numbers for both flash and steam plants. It is based on 0.7%–5% CO_2 in the geothermal steam and 18–20 klbs per MWh steam usage in the power plant. These are realistic, though nonproject-specific, values for some high-gas geothermal resources worldwide.

Source: From Kagel, A. et al. 2005. *A Guide to Geothermal Energy and the Environment*, p. 39. Geothermal Energy Association, Washinton, DC.

Pumped binary power plants have no emissions, except for fugitive leaks of the working fluid. As discussed above, the gases are still present in the brine, but because the pressure is never lowered below the bubble pressure of the gases, they remain in solution and are injected back into the geothermal resource.

23.9.1.1 CO_2 Emissions

Flash-steam power plants have emissions on the order of magnitude described, but vary according to the particular resource characteristics and whether the power process is single-flash, double-flash, or coupled to a binary system. Most geothermal flash steam resources have NCG content of between 0.5% and 2% in the steam, which, when coupled with either double flash or a binary system, results in the reported values. However, for those systems that operate with only a single flash and have NCG content of up to 5% in the steam, the CO_2 emissions can reach the 900 pounds per MWh reported in the above table.

Like the flash steam plants, the emissions from a dry steam power plant are dependent on the particular resource being considered.

23.9.1.2 H_2S and SO_x Emissions

For geothermal power plants with some form of H_2S abatement, sulfur emissions are negligible to very low, as shown in the table. For those plants without H_2S abatement, the released H_2S eventually converts to SO_2, and can be on the same order of magnitude as U.S. coal plant emissions for geothermal resources with high-H_2S content in the NCGs.

Except for California and the western U.S., few geothermal plants worldwide operate H_2S abatement systems. Stretford™ and Lo-Cat™ are the most common H_2S abatement technologies for gas-phase H_2S emissions. The use of bioreactors is an emerging technology for abatement, one in which sulfur-loving bacteria are used to convert the H_2S to sulfate. Bioreactors have been independently developed and put into operation at the Salton Sea, California, and in Wairakei, New Zealand.

23.9.1.3 NO_x Emissions

The reason NO_x emissions are reported as "negligible" instead of zero for the dry steam plants is a consequence of the early H_2S abatement technology installed at The Geysers that incinerates the H_2S and then scrubs it out of the gas stream using the cooling water. The incineration also produces small amounts of nitrogen oxides. Since those early efforts at H_2S abatement, the geothermal industry has primarily used a cold catalytic conversion of H_2S to elemental sulfur, which has no NO_x emissions.

23.9.1.4 Particulate Emissions

Air-cooled plants have no particulate emissions, whereas water-cooled plants have only the evaporated minerals remaining from cooling tower drift, the quantity of which will be a function of the TDS in the cooling water. This quantity of particulate is negligible.

23.9.2 Land Use

Compared to other renewable energy sources, geothermal power developments have low land use and visual impact. The total geothermal development of wells and power plant may occur over many square miles, but the actual land occupied by the facilities is small. Some negative impacts may be attributed when a geothermal development occurs in a roadless area. There is a visual impact from the pipelines running from wells to power plant, and from the cooling tower steam plume on a cold day. When an air-cooled power cycle is used, geothermal power generation achieves its minimum visual impact. The fenced area of the power plant itself may be no larger than a couple of acres for up to 20 MW of net generation. The height of the air-cooled condenser, the highest point in the plant, is less than 25 ft. above ground level. Overall, the land-use impact of geothermal development is small. Table 23.12 provides the total land use of various renewable and coal power technologies.

TABLE 23.12 30-Year Land Use

Energy Source	Land Use (m^2/GWh)
Coal, including mining	3642
Solar thermal	3561
Central station photovoltaic	3237
Wind, including roads	1335
Geothermal, including roads and pipelines	404

Source: From Brophy, P., *Renewab. Energy*, 10, 374, 1997.

23.10 Additional Information on Geothermal Energy

This chapter has touched on only some of the many topics in the field of geothermal energy—a subject to which entire books have been devoted. Some additional geothermal energy information sources are provided in this section.

For a broad view on the subject of geothermal energy, including many direct-use applications such as aquaculture, district heating, industrial heat use, environmental issues and financing, a recent book, *Geothermal Energy—Utilization and Technology*, edited by Mary Dickson and Mario Fanelli, is an excellent resource. Each chapter has been written by an expert of geothermal subject matter. At the end of each chapter are questions, answers, and references regarding the particular subject matter of that chapter.

Another recent book, *Geothermal Power Plants—Principles, Applications and Case Studies*, authored by Ronald DiPippo, provides some coverage of geothermal exploration, drilling, and reservoir engineering, with a primary focus on geothermal power generation. The book covers the conversion technologies, thermodynamics, equipment, and operation, and includes several detailed case studies of particular geothermal power plant developments.

In addition, there are a number of excellent geothermal resources on the World Wide Web.

Geo-Heat Center http://geoheat.oit.edu/	Focus is direct use. Excellent online library. Software and databases
Geothermal Resources Council www.geothermal.org/	GRC bulletin and transaction articles. Must be a member of the GRC to download articles
Geothermal Energy Association www.geo-energy.org	The U.S. geothermal industry trade association several excellent broad research articles available for download including three papers on Geothermal Energy Costs, Environmental Impact, and Employment
International Geothermal Association http://iga.igg.cnr.it/index.php	Downloadable papers from the last two world geothermal conferences

References

1. Lund, J. W., Bloomquist, R., Gordon, B., Tonya, L., and Renner, J. 2005. The United States of America country update—2005. *Geothermal Resources Council Transactions*, 29, 817–830.
2. Ragnarsson, Á. 2005. Geothermal development in Iceland (2000–2004). In *Proceedings of the World Geothermal Congress*, 11 pp.
3. Lund, J. W. 2005. Worldwide utilization of geothermal energy—2005. *Geothermal Resources Council Transactions*, 29, 831–836.
4. DiPippo, R. 2005. *Geothermal Power Plants—Principles, Applications and Case Studies*, Elsevier Advanced Technology.
5. Wyborn, D., de Graaf, L., and Hann, S. 2005. Enhanced geothermal development in the Cooper Basin, South Australia. *Geothermal Resources Council Transactions*, 29, 151–155.

6. Stark, M. A., Box, W. T. Jr., Beall, J. J., Goyal, K. P., and Pingol, A. S. 2005. The Santa Rosa-Geysers recharge project, Geysers geothermal field, California. *Geothermal Resources Council Transactions*, 29, 144–150.
7. Sanyal, S. 2004. Defining the sustainability and renewabilty of geothermal power. *Geothermal Resource Council Bulletin*, July/August, 149–151.
8. Capetti, G. and Ceppatelli, L. 2005. Geothermal power generation in Italy, 2000–2004 update report. In *Proceedings of the World Geothermal Congress 2005*, 7 pp.
9. State of Nevada, Division of Minerals. 2005. Nevada geothermal update. http://minerals.state.nv.us/
10. Idaho Public Utilities Commission Order # 29692, Case No. IPC-E-05-1. January 24, 2005. http://www.puc.idaho.gov
11. Ormat Technologies News Release, May 6, 2005. http://www.ormat.com/index_news.htm (accessed December, 2005).
12. Hance, C. N. 2005. *Factors Affecting the Costs of Geothermal Power Development*. Geothermal Energy Association and the US Department of Energy, Washington, DC.
13. Western GeoPower. 2005. Corporation website. http://www.geopower.ca/newsreleases.htm (accessed December, 2005).
14. Phil Lory of Sinclair Knight Mertz Consulting, Personal communication.
15. Fuji Electric Systems 2005. *Wayang Windu Project Information Bulletin, GEC82-12,* Fuji Electric Systems, Tokyo, Japan.
16. Williamson, K. H., Gunderson, R. P., Hamblin, G. M.., Gallup, D. L., and Kitz, K. 2001. Geothermal power technology. *Proceedings of the IEEE*, 89, (12), 1783–1792.
17. Kagel, A., Bates, D., and Gawell, K. 2005. *A Guide to Geothermal Energy and the Environment*. Geothermal Energy Association, Washinton, DC.
18. Kagel's source reported as US EPA. Average power plant emissions from EPA 2000 emissions data. http://www.epa.gov/cleanenergy/egrid/highlights
19. Kagel, 51. Citing data from Brophy, P. 1997. Environmental advantages to utilization of geothermal energy. *Renewable Energy*, 10, 374.

24

Waste-to-Energy Combustion

24.1 Introduction 24-1

24.2 Waste Quantities and Characteristics 24-2
Waste Quantities • Waste Characteristics

24.3 Design of WTE Facilities.. 24-6
General Features • Fuel Handling • Combustion Principles • Furnaces • Boilers • Residue Handling and Disposal • Other Plant Facilities

24.4 Air Pollution Control Facilities 24-24
Particulate Control • Gaseous Emission Control • Organic Compound Control • Trace Metals

24.5 Performance 24-32

24.6 Costs.. 24-34

24.7 Status of Other Technologies.. 24-36
Modular Systems • Fluidized Beds • Pyrolysis and Gasification

24.8 Future Issues and Trends 24-38

References.. 24-38

Charles O. Velzy
Private Consultant

Leonard M. Grillo
Grillo Engineering Company

24.1 Introduction

One of the most serious issues facing urbanized areas today is development of cost-effective environmentally acceptable disposal of the community's solid waste. The solid waste generated in a community may be collected by private companies or governmental entities, or portions by both, but the assurance that the waste is ultimately disposed of in an environmentally safe manner is a governmental responsibility.

Solid waste management is a major issue in the United States, because of increasing concerns with environmental problems. One potential solution is to use municipal solid waste, which, for all practical purposes is a renewable commodity, for the generation of electricity. An analysis by Penner and Richards[54] showed that incineration of municipal waste, even after 30% of the waste was recycled, could provide as much electric power as eight large nuclear or coal generating stations. Their analysis further concluded that this could provide 1%–2% of the total electric energy needs in the United States at prices competitive with coal-fired base load power plants.

The basic technology for modern waste-to-energy combustion was developed in Europe during the 1960s and 1970s. This technology, which has been modified and improved since its development, has been widely implemented in the United States. However, despite the fact that incineration of solid waste can decrease its volume ninefold and ameliorate the final waste disposal into landfills, the full potential of utilizing solid waste for energy production is not being realized because of widespread fears regarding environmental pollution. In preparing this chapter, the realities of the situation have been taken into account and the discussion emphasizes the prevention of pollution as much or more than the production of power from waste. Waste-to-energy combustion in modern facilities with adequate environmental safeguards and careful monitoring has been shown to be a safe and cost-effective technology that is likely to increase in importance during the next decade.

Two conditions usually point to the use of combustion processes in treating municipal solid waste prior to ultimate disposal: the waste is collected in an urbanized area with little or no conveniently located land for siting of sanitary landfills (need for volume reduction); and markets exist for energy recovered from the combustion process, and possibly for reclaimed materials, with the energy attractively priced. Even some rural areas are currently considering waste to energy facilities.

Modern waste-to-energy (WTE) plants reflect significant advances that have been made in addressing the technical and practical difficulties of material handling, combustion control, and flue gas cleanup. In the early days of waste incineration, when air pollution regulations were undemanding or nonexistent, relatively simple, fixed-grate plants operating on a single- or two-shift basis were common. However, with increasingly stringent air pollution control regulations, more complex plants requiring continuous operation are now being built.

24.2 Waste Quantities and Characteristics

Municipal solid waste (MSW) as used herein refers to solid waste collected from residences and commercial, light industrial and institutional waste. It does not include heavy industrial waste, which is another problem and varies widely in quantity and characteristics, depending on the industry and specific industrial plant. Changes in packaging practices and improvements in the general standard of living have resulted in significant increases in the quantities of solid waste generated over the past 50 years. Additionally, increasing emphasis on and participation in the recycling of wastes by local communities has resulted in significant variations in quantity and characteristics of MSW at the local community level. All of these factors must be considered when planning a WTE facility.

24.2.1 Waste Quantities

In the United States approximately 120 million tons of MSW were generated in 1970, increasing to 220 million tons in 1998[25] MSW generation is projected to increase to almost 260 million tons by the year 2010.[27] At the local level the quantity of solid waste generated varies geographically, daily, and seasonally, according to the effectiveness of the local recycling initiatives, and differences in socioeconomic conditions.[60]

Over the past 40 years, numerous studies by EPA,[23] APWA,[2] and others[67] have indicated that urbanized areas in this country generate approximately 2.0 pounds per capita per day of MSW from residences and another 2.0 pounds per capita per day from commercial and institutional facilities, on a national average basis. Thus, a typical community of 100,000 inhabitants would generate about 200 tons per day of gross MSW discards averaged over a 1-year time period.

These projections are subject to adjustments related to specific community characteristics. Thus, communities in the south, with longer active growing seasons than those in the north, tend to produce and collect more yard waste. Recent requirements for on-site disposal and/or composting of yard waste is changing this variable. Rural communities tend to produce less waste per capita than highly urbanized areas because of their greater potential for on-site waste disposal. In the past, the communities in the north tended to produce more waste in the winter due to the prevalence of heating of homes with solid fuels, which produced large quantities of ashes for disposal. Variations in yard wastes and ashes produced from home heating with solid fuels are also examples of variations in MSW quantities related to seasonal effects. Seasonal variations in MSW generation have been noted to range $\pm$15% from the average, while daily variations in waste collections may range up to $\pm$50% from the average, depending largely on number of collections per week. Daily variations in waste quantities are more important in designing certain plant components, while geographic and seasonal effects are more important in establishing plant size.

Waste quantities are also affected by the effectiveness of local recycling initiatives and by socio-economic conditions. EPA studies have indicated an increase, nationally, in waste recycling from 6.6% in 1960–1970, to 16.2% in 1990, [24] and 28.2% in 1998.[26] The national recycling rate is projected to increase to about 34% by the year 2010,[27] at which time it is expected to level off. A community in New England[9] projected a 14.0% drop in MSW generation between 1989 and 1991 due to recycling activities in the community, and a 6.0% drop due to the recession during this period in this region of the country. A 15% drop in MSW collections was noted in a Long Island community during a recessionary period in 1972. The impact of recycling on MSW generation should be considered in plant sizing, while the impact of recessionary periods on plant economics may require specific consideration during project planning.

24.2.2 Waste Characteristics

It is important in approaching the design of a WTE facility that one consider the potential variations in both physical and chemical composition of MSW. Historically, one of the most troublesome areas in WTE plants has been materials handling systems. To successfully select materials handling system components and design an integrated process, one must have adequate information on the variability and extremes of the physical size and shape the solid waste facility must handle, the bulk density and angle of repose of the material, and the variation in noncombustible content. This information generally is not available from published surveys and reports, and can only be secured through inspection of the MSW in the field. Materials handling equipment for refuse feeding and residue handling must be large enough and oriented properly to pass the largest bulky items in the MSW, and large enough and rugged enough to handle the quantities of materials required to meet plant design capacity, or the plant will experience expensive periods of down time and might have to be derated.

In the design of the furnace/boiler portion of WTE facilities, the refuse characteristics of interest are the calorific value, moisture content, proportion of noncombustibles, and other components (such as heavy metals, chlorine, and sulfur) whose presence during combustion will result in the need for flue gas cleanup. The capacity of a WTE furnace boiler is roughly inversely proportional to the calorific or heating value of the waste. Table 24.1 illustrates the variation in waste characteristics that has been observed in studies defining the average solid waste composition in the United States over the past 40 years.

As indicated in Table 24.1, approximately 35%–40% of the combustible fraction of MSW is composed of cellulosic material such as paper and wood. This percentage has remained relatively constant over the past 20 years, even after taking into account the greater than fourfold increase in recycling percentage achieved over this period. The remainder of the combustible content is composed of various types of plastics, rubber, and leather. The heat released by burning cellulose is approximately 8000 Btu/lb (on a dry basis), while that released by plastics, rubber, and leather is significantly higher on a per pound, dry basis. Heat released by burning garbage (on a dry basis) is only slightly less than cellulose. However, the moisture content of garbage has been observed to range from 50 to 75%, by weight, while that of the cellulosic fraction of MSW usually ranges from 15 to 30%.

In recent years, it has been observed that the higher heating value (HHV) of the combustible portion of MSW (moisture and ash free) averages about 9400 Btu/lb. Considering the recent changes in MSW composition following recycling (increase in plastics while cellulosic material has remained relatively constant), this moisture and ash free HHV has probably increased to 9500 Btu/lb. Taking 9500 Btu/lb as the moisture and ash free heat content of MSW, Table 24.2 illustrates the variation in as-received heat

TABLE 24.1 Waste Characteristics (in percent)

	Oceanside N Y. 1966–1967	1970[a]		1998		2010[a]	
		Generated	After Recovery	Generated	After Recovery	Generated	After Recovery
Paper materials	32.72–53.33	36.6	33.2	38.2	31.1	36.9	27.1
Plastics	2.45–8.82	2.4	2.6	10.2	13.4	12.0	16.7
Rubber & leather	—	2.5	2.4	3.1	3.8	2.8	3.4
Textiles	2.24–3.97	1.7	1.7	3.9	4.7	4.5	5.6
Wood	1.22–6.58	3.1	3.3	5.4	7.1	5.3	7.1
Food (garbage)	7.23–16.70	10.6	11.3	10.0	13.6	11.3	16.2
Yard wastes[b]	0.26–33.33	19.2	20.5	12.6	9.6	11.5	7.0
Noncombustible	22.47–14.36	24.1	24.9	16.6	16.7	15.7	17.0
Total	—	100.0	100.0	100.0	100.0	100.0	100.0
Total weight (in mil. tons)	—	121.0	113.0	220.2	158.1	260.0	171.6
Recovery (recycling & composting)	—	6.6		28.2		34.0	

Note: See indicated references for further details.

[a] Details may not add to totals due to rounding

[b] Including grass, dirt and leaves

Source: From Marjorie A. Franklin. 2002. Chapter 5, Solid Waste Stream Characteristics in Handbook of Solid Waste Management, *2nd Ed.*, McGraw Hill; Marjorie A. Franklin. Sept. 2004 personal communication; Kaiser, E. R., Zeit, C. D., and McCaffery, J. B. 1968. Municipal Incinerator Refuse and Residue. Proceedings 1968 National Incinerator Conference, pp. 142–153, ASME, New York.

TABLE 24.2 Variation in Heat Content of MSW

Noncombustible %	15		25	
	Comb. %	Heat Cont. Btu/lb	Comb. %	Heat Cont. Btu/lb
Moisture %				
20	65	6125	55	5225
30	55	5225	45	4275
40	45	4275	35	3325
50	35	3225	25	2375

content that one could expect in MSW with moisture content ranging from 20 to 50% by weight and noncombustible content ranging from 25% by weight (earliest period) to approximately 15% by weight (currently).

Moisture content is a highly important and also a highly variable characteristic of waste materials. The moisture content of MSW is generally around 25%, but has been observed to vary from 15 to 70%. This variation may be due, for example, to seasonal variations in precipitation, the nature of the waste (e.g., grass clippings vs. paper) and the method of storage and collection (e.g., open vs. closed containers/trucks). Thus, after a heavy rain, the moisture content of the solid waste may be so high that it may be difficult to sustain combustion. The combustion of solid waste usually can proceed without supplementary fuel when the heat value is greater than 3500–4000 Btu/lb (8140–9300 kJ/kg). This type of variation in MSW composition must be considered in the design of WTE facilities.

The ultimate or elemental analysis of the combustible portion of the MSW refers to the chemical analysis of the waste for carbon, hydrogen, oxygen, sulfur, chlorine, and nitrogen. This information is used to estimate the heat content of waste, moisture, and ash free; to predict the composition of the flue gases; and, from the last three elements (sulfur, chlorine, and nitrogen), to assess the possible impact of waste combustion on air pollution. A typical analysis of solid waste is presented in Table 24.3.

TABLE 24.3 Analysis of Solid Waste

	Percent by Weight	
	West Europe	U.S.
Proximate analysis		
Combustible	42.1	50.3
Water	31.0	25.2
Ash and inert material	26.9	24.5
Total	100	100
Ultimate (elemental) analysis of combustibles		
Carbon	51.1	50.9
Hydrogen	7.1	6.8
Oxygen	40.1	40.3
Nitrogen	1.2	1.0
Sulfur	0.5	0.4
Chlorine	—	0.6
Total	100	100

Gross heat value, as fired = 3870 Btu/lb (9000 kJ/kg).

Source: From Domalski, E. S., Ledford A. E., Jr., Bruce, S. S., and Churney, K. L. The Chlorine Content of Municipal Solid Waste from Baltimore County, Maryland, and Brooklyn, NY, *Proceedings 1986 National Waste Processing Conference*, Denver, CO, June 1–4, pp. 435–448, ASME, New York; Seeker, W. R., Lanier, W. S., and Heap, M. P. 1987. *Municipal Waste Combustion Study: Combustion Control of Organic Emissions EPA Research Triangle Park, NC*; Suess, M. J., et al. 1985. *Solid Waste Management Selected Topics 1985 WHO Regional Office for Europe, Copenhagen, Denmark.*

24.3 Design of WTE Facilities

The primary function of a WTE facility is to reduce solid waste to an inert residue with minimum adverse impact on the environment. Thermal efficiency, in terms of maximizing the capture of energy liberated in the combustion process, is of secondary importance. WTE facilities are usually classified as mass-burn systems, or refuse-derived fuel (RDF) systems.

24.3.1 General Features

24.3.1.1 Types of Facilities

Mass-burn systems are large facilities (usually over 200 tons per day) that burn, as-received, unprocessed MSW which is extremely heterogeneous. Most mass-burn systems burn the waste in a single combustion chamber under conditions of excess air (i.e., more than is needed to complete combustion) (see Figure 24.1). The waste is burned on a sloping, moving grate, which helps agitate the MSW and mixes it with combustion air. Many different proprietary grate systems exist.

In refuse-derived fuel (RDF) systems, usually large facilities, the MSW is first processed (see Figure 24.2) by mechanical means to produce a more homogeneous material prior to introduction into a furnace/boiler. Several types of RDF can be made—coarse, fluff, powder, and densified. These differ in complexity and horsepower requirements of the waste processing facilities, size of particle produced, and whether or not the material is compacted under pressure into pellets, briquettes, or similar forms. The coarse type of RDF is the most common form produced at this time.

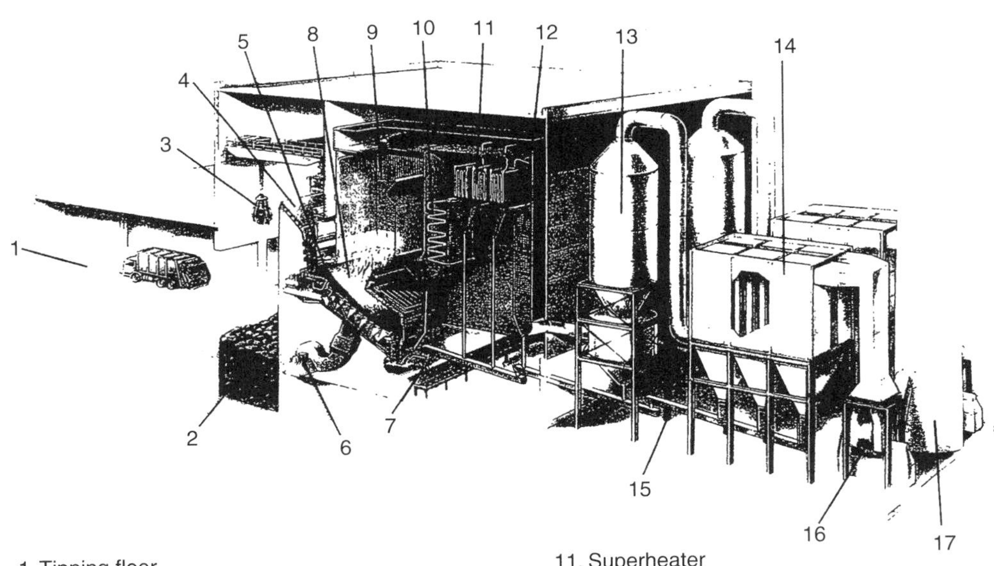

1. Tipping floor
2. Refuse holding pit
3. Feed crane
4. Feed chute
5. Martin stroker grate
6. Combustion air fan
7. Martin residue discharger and handling system
8. Combustion chamber
9. Radiant zone (furnace)
10. Convection zone
11. Superheater
12. Economizer
13. Dry gas scrubber
14. Baghouse or electrostatic precipitator
15. Fly ash handling system
16. Induced draft fan
17. Stack

FIGURE 24.1 A typical Covanta facility.

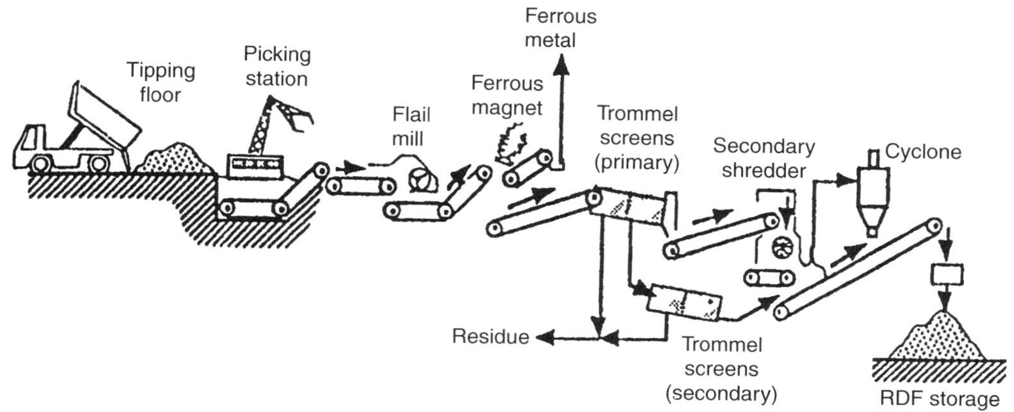

FIGURE 24.2 RDF processing system.

RDF can be burned in one of the two types of boilers. It can be used as the sole or primary fuel in dedicated boilers (see Figure 24.3) or it can be co-fired with conventional fossil fuels in existing industrial or utility boilers. One advantage of these systems is that RDF can be produced at one location for use at a nearby off-site boiler, allowing for flexibility in locating processing facilities. Also, some materials, such as steel and glass, can be recovered for recycling during the initial processing step.

Mass-burn and RDF systems together account for 86 of the 98 currently operating waste-to-energy facilities and 98% of the waste combustion capacity. Modular units, described briefly later, account for the other 12 units and 2% of the waste combustion capacity.[42]

24.3.1.2 Operation and Capacity

The capacity to be provided in a facility is a function of the area and population to be served; and the rate of refuse production for the population served. A small plant (100 tons per day) without energy recovery might not be operated on weekends. For capacities above 400 tons per day, or any plant with energy recovery, economic and/or equipment operating considerations usually dictate three-shift operation, seven days per week.

If collection records, preferably by weight, are available for the community, forecasts for determining required plant capacity can be made with reasonable accuracy. If records are not available, refuse quantities for establishing plant size may be approximated by assuming a refuse generation rate of 4.0 pounds per capita per day when there is little or no waste from industry. Of course, in planning for plant capacity, the impact of local recycling activities on both quantity and characteristics of MSW must be considered as discussed earlier.

Other factors must be taken into account in establishing the size or capacity of a facility. Should the facility serve only one community, or should it be regional and serve several communities? What are the possible benefits of economies of scale? What is the impact of the cost of hauling refuse to a central point on overall project economics? There is substantial evidence available at present to show that implementation becomes much more difficult as the number of separate political jurisdictions is increased. Imposition of regional plans on local jurisdictions to achieve economies of scale, where it cannot be conclusively demonstrated that such regional plans make sound economic sense based on the total cost of the solid waste management plan, including the cost of transporting the solid waste to the regional facility, is, at best, unwise. Economies of scale in these projects have tended to be illusory, while haul costs to gather sufficient waste together to achieve the economies of scale have tended to be ignored in developing total project economics.

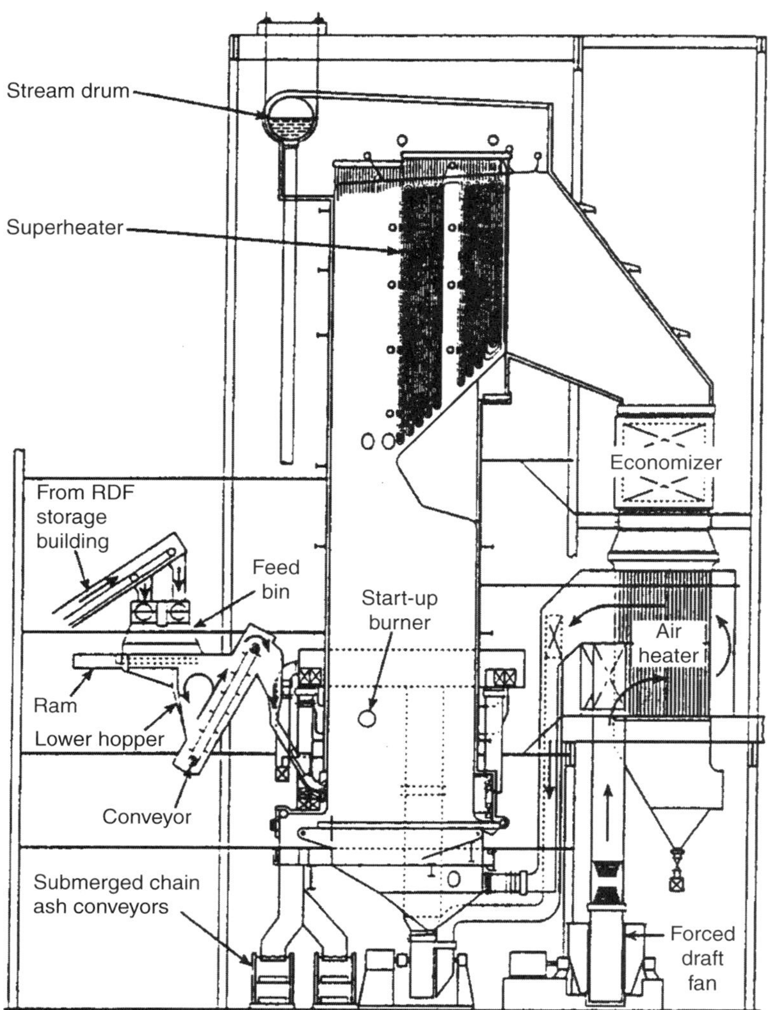

FIGURE 24.3 West Palm Beach, FL, RDF-fired boiler by the Babcock and Wilcox Company.

24.3.1.3 Siting

One of the key issues to face in implementing a WTE project is locating a site for the facility. Since MSW is usually delivered to these plants by truck, inevitably there will be substantial truck traffic in the vicinity of the plants. The equipment and processes used in these plants are industrial in nature. They are generally noisy at the source and tend to produce dust and odors. These facts indicate that it is desirable to site such plants in industrial or commercial areas.[46] It has been contended, as cited in a 1989 OTA report[68] "that sites are sometimes selected to avoid middle- and higher-income neighborhoods that have sufficient resources to fight such development."

Plants should be located near major highways to minimize the impact from increased truck traffic. As shown by operating WTE plants in Europe and the United States, it is possible to control all nuisance conditions by proper attention to the details of plant design. The local impact of truck deliveries to the plant can be minimized by providing sufficient length of access road so that refuse truck queuing does not take place on public highways. Odors and noise can be confined to the plant building. Odors and fugitive dust can be destroyed by collecting plant air and using it for combustion air supply. Noise should be attenuated at the source to maintain healthful working conditions. In all cases, there is no need to adversely impact the surrounding neighborhood. Proper attention to architectural treatment can result in a structure that blends into its surroundings; if sited in industrial or commercially zoned areas.

Since considerable vertical distance is frequently required in the passage of MSW through a mass-burn WTE plant, there is an advantage in a sloping or hillside site. Collection trucks can deliver MSW at the higher elevation, while residue trucks operate at the lower elevation, requiring minimum site regrading. This consideration does not generally apply to RDF plants.

24.3.2 Fuel Handling

24.3.2.1 Refuse Receipt, Processing, and Storage

Scales, preferably integrated into an automated record keeping system, should be provided to record the weight of MSW delivered to the WTE plant. Either the entire tipping area or individual tipping positions should be enclosed so as to control potential nuisance conditions in the vicinity of the plant, such as blowing papers, dust, and/or odors. The number of tipping positions provided should take into consideration the peak number of trucks expected per hour at the facility and should be located so the trucks have adequate time and room to maneuver to and from the dumping positions while minimizing queue time.

Collections usually are limited to one 8-hour daily shift 5 days per week (sometimes with partial weekend collection) while burning will usually be continuous, so ample storage must be provided. This usually requires 2–3 days of refuse storage at most WTE plants. Seasonal and cyclic variations should also be a consideration in establishing plant storage requirements.

Refuse storage in large mass-burn plants is normally in long, relatively narrow, and deep pits extending along the front of the furnaces. It will generally be necessary to rehandle the refuse dumped from the trucks. In some mass-burn plants and in RDF plants, floor dumping and storage of the MSW either on the dump floor or in shallow, relatively wide pits is common practice.

When computing the dimensions required for storage of as-received MSW, the required volume may be determined based on an MSW bulk density of from 400 to 600 lb/yd^3 (240–360 kg/m^3).[2,61] Other factors to consider in sizing and laying out refuse storage facilities in WTE plants is that refuse flows very poorly and can maintain an angle of repose greater than 90°. Thus, MSW is commonly stacked in storage facilities to maximize storage capacity.

Sizing a refuse storage pit requires the use of empirical data, judgment, site constraints, and knowledge of plant layout and operations. The pit should be at least long enough to provide sufficient truck tipping positions so that the trucks are not unduly delayed in discharging their waste (tipping) into the pit. It has been found in practice that it takes an average of 10 min for the truck to perform the tipping operation. This time may be shorter for packer trucks and longer for transfer vehicles. Each tipping position must provide at least 14 ft. of unobstructed width for this operation. Ideally, 20 ft. should be allowed for each tipping position to allow for convenient truck access and space for armored building support columns.

The pit should be capable of holding a minimum of 3 days' storage at the facility's maximum continuous rating. The desired volume should be based on a bulk density of about 500 lb/cy of waste. The dimensions of the pit will be dependent on site constraints and the facility design. The pit is usually at least as long as the width of all the boilers it feeds. The depth will be dependent on groundwater conditions, but should be 30–45 ft. deep if possible. The total volume of storage should equal the volume in the pit up to the tipping floor level plus the volume above the tipping floor assuming the waste is stacked at a 45° from the charging floor wall to the tipping floor. The remainder of the pit will be used for the grapple to move waste away from the tipping positions. After the waste has been stacked, the grapple can remove the material against the tipping floor wall and form a trench, since the waste will maintain a vertical face when stacked.

At most newer and more successful RDF plants, after receiving the MSW in a floor dump type of operation, the MSW is loaded onto conveyors that carry the material to flail mills or trommels with bag-breaking blades. These facilities break apart the bags containing the waste, allowing glass and some metals to be separated from the remaining MSW. The separated MSW, primarily the light combustible fraction, is then reduced in size. Removal of the glass prior to the size reduction process alleviates the problem, experienced in earlier plants, of contamination of combustible material with glass shards.

Processes to produce powdered fuel or RDF fuel pellets, although interesting, have not been developed to a state of commercial availability. A process to produce RDF by "hydro-pulping" after being attempted in two full-scale plants, was not commercially successful.

24.3.2.2 Refuse Feeding

Batch feeding of MSW, practiced in the past in mass-burn plants, is undesirable and is not practiced in modern plants. In the larger mass-burn plants, the solid waste is usually moved from the storage pit to a charging hopper by a traveling bridge crane and an orange-peel type of grapple. The grapple size is established by a duty cycle analysis, taking into account the quantity of material that must be moved from the pit to the furnaces, the distances over which the material must be moved, allowable crane speeds, and the need to rehandle (mixing and/or stacking) material in the pit. Grapples can range in size from 1.5 to 8 yd^3 (1–6 m^3) capacity and larger.

The crane used in this service should be capable of meeting the severest of duty requirements.[53] The load lifting capability is established by adding to the grapple weight, 1.5 times the volumetric capacity of the grapple times a density of MSW of 600–800 lb/yd^3 (360–480 kg/m^3).[40] In the past, the crane has been operated from an air-conditioned cab mounted on the bridge. However, crane operation is now centralized in a fixed control room, usually located at the charging floor elevation and either over the tipping positions opposite the charging hoppers or in the vicinity of the charging hoppers.

In modern mass-burn plants, the MSW is deposited from the crane grapple into a charging hopper. The charging hopper, which is built large enough to prevent spillage on the charging floor and with slopes steep enough to prevent bridging, is placed on top of a vertical feed chute that discharges the MSW into the furnace. The feed chute may be constructed of water cooled steel plates or steel plates lined with smooth refractory material. The chute is normally at least 4 ft. (1.2 m) wide, to pass large objects with a minimum of bridging, and the width of the furnace. It is kept full of refuse to prevent uncontrolled admission of air into the furnace. The refuse is fed from the bottom of the feed chute into the furnace by a portion of the mechanical grate, or by a ram. The ram generally provides better control of the rate of feed into the furnace than the older technique of using a portion of the mechanical grate for refuse feed.

In RDF plants, conveyors, live bottom bins, and pneumatic handling of the size-reduced MSW combustible material have been utilized. The fuel material is usually blown into these furnaces, where it is partially burned while in suspension, with combustion being completed on grates at the bottom of the furnace. These fuel feeding systems are generally more complex than the mass-burn systems.

24.3.3 Combustion Principles

Combustion is the rapid oxidation of combustible substances with release of heat. Oxygen is the sole supporter of combustion. Carbon and hydrogen are by far the most important of the combustible substances. These two elements occur either in a free or combined state in all fuels—solid, liquid, and gaseous. Sulfur is the only other element considered to be combustible. In combustion of MSW, sulfur is a minor constituent with regard to heating value. However, it is a concern in design of the air pollution control equipment. The only source of oxygen considered here will be the oxygen in the air around us.

Table 24.4 displays the elements and compounds that play a part in the combustion process. The elemental and molecular weights displayed are approximate values which are sufficient for combustion calculations. Nitrogen is listed as chemical nitrogen N_2, with a molecular weight of 28.0 and as atmospheric nitrogen, N_{2atm}, which is a calculated figure to account for trace constituents of dry air. Water occurs as a vapor in air and in the products of combustion and as a liquid or vapor constituent of MSW fuel.

TABLE 24.4 Elements and Compounds Encountered in Combustion

Substance	Molecular Symbol	Molecular Weight	Form	Density (lb/ft.^{3})
Carbon	C	12.0	Solid	—
Hydrogen	H_2	2.0	Gas	0.0053
Sulfur	S	32.1	Solid	—
Carbon monoxide	CO	28.0	Gas	0.0780
Oxygen	O_2	32.0	Gas	0.0846
Nitrogen	N_2	28.0	Gas	0.0744
Nitrogen atmos.	N_{2atm}	28.2	Gas	0.0748
Dry air		29.0	Gas	0.0766
Carbon dioxide	CO_2	44.0	Gas	0.1170
Water	H_2O	18.0	Gas/liquid	0.0476
Sulfur dioxide	SO_2	64.1	Gas	0.1733
Oxides of nitrogen	NO_x	—	Gas	—
Hydrogen chloride	HCl	36.5	Gas	0.1016

Source: From Hecklinger, R. S. 1996. *The Engineering Handbook*, CRC Press, Inc., Boca Raton, FL.

A U.S. standard atmosphere of dry air has been defined as a mechanical mixture of 20.947% O_2, 78.086% N_2, 0.934% Ar (argon), and 0.033% CO_2 by volume.[17] The percentages of argon and carbon dioxide in air can be combined with chemical nitrogen to develop the following compositions of dry air by volume and by weight that can be used for combustion calculations:

Constituent	% by Volume	% by Weight
Oxygen, O_2	20.95	23.14
Atmospheric nitrogen, N_{2atm}	79.05	76.86

Atmospheric air also contains some water vapor. The level of water vapor in air, or its humidity, is a function of atmospheric conditions. It is measured by wet and dry bulb thermometer readings and a psychrometric chart. If specific data are not known, the American Boiler Manufacturers Association recommends a standard of 0.013 pounds of water per pound of dry air, which corresponds to 60% relative humidity and a dry bulb temperature of 80°F.

Table 24.5 displays the chemical reactions of combustion. These reactions result in complete combustion; that is, the elements and compounds unite with all the oxygen with which they are capable of entering into combination. In actuality, combustion is a more complex process in which heat in the combustion chamber causes intermediate reactions leading up to complete combustion.

TABLE 24.5 Chemical Reactions of Combustion

Combustible	Reaction
Carbon	$C+O_2=CO_2$
Hydrogen	$2H_2+O_2=2H_2O$
Sulfur	$S+O_2=SO_2$
Carbon monoxide	$2CO+O_2=2CO_2$
Nitrogen	$N_2+O_2=2NO$
Nitrogen	$N_2+2O_2=2NO_2$
Nitrogen	$N_2+3O_2=2NO_3$
Chlorine	$4Cl+2H_2O=4HCl+O_2$

Source: From Hecklinger, R. S. 1996. *The Engineering Handbook*, CRC Press, Inc., Boca Raton, FL.

An example of intermediate steps to complete combustion would be when carbon reacts with oxygen to form carbon monoxide and, later in the combustion process, the carbon monoxide reacts with more oxygen to form carbon dioxide. The combined reaction produces precisely the same result as if an atom of carbon combined with a molecule of oxygen to form a molecule of carbon dioxide in the initial reaction. An effectively controlled combustion process results in well less than 0.01% of the carbon in the fuel leaving the combustion chamber as carbon monoxide; and the remaining 99.99% of the carbon in the fuel leaves the combustion process as carbon dioxide. It should also be noted with regard to Table 24.5 that some of the sulfur in a fuel may combust to SO_3 rather than SO_2 with a markedly higher release of heat. However, it is known that only a small portion of the sulfur will combust to SO_3 and some of the sulfur in fuel may be in the form of pyrites (FeS_2), which do not combust at all. Therefore, only the SO_2 reaction is given. Also, some nitrogen is converted to oxides of nitrogen (NO_x), and some chlorine is converted to hydrogen chloride in the presence of moisture in the flue gas. While these components do not factor into the combustion calculations, they are important for the purpose of establishing air pollution control requirements.

Factors directly affecting furnace design are the moisture and the combustible content of the solid waste to be burnt and the volatility of the material to be burnt. The means for temperature control and sizing of flues and other plant elements should be based on design parameters that result in large sizes. Combustion controls should provide satisfactory operation for loads below the maximum rated capacity of the units.

The combustible portion of MSW is composed largely of cellulose and similar materials originating from wood, mixed with appreciable amounts of plastics and rubber, as well as some fats, oils, and waxes. The heat released by burning dry cellulose is approximately 8000 Btu/lb, while that released by certain plastics, rubber, fats, oils, and so on, may be as high as 17,000 Btu/lb. If MSW consists of five parts cellulose and one part plastics, rubber, oil, and fat, the heat content of the dry combustible matter only is approximately 9500 Btu/lb.

The heat released in combustion of basic combustible substances is displayed in Table 24.6. The heating value of a substance can be expressed either as higher (or gross) heating value or as lower (or net) heating value. The higher heating value takes into account the fact that water vapor formed or evaporated in the process of combustion includes the latent heat of vaporization, which could be recovered if the products of combustion are reduced in temperature sufficiently to condense the water vapor to liquid water. The lower heating value is predicated on the assumption that the latent heat of vaporization will not be recovered from the products of combustion.

The heat released during combustion may be determined in a bomb calorimeter, a device with a metal container (bomb) immersed in a water jacket. A 1 g MSW sample is burned with a known quantity of oxygen, and the heat released is determined by measuring the increase in temperature of the water in the water jacket. Since the bomb calorimeter is cooled to near ambient conditions, the heat recovery measured includes the latent heat of vaporization as the products of combustion are cooled and condensed in the bomb. That is, the bomb calorimeter inherently measures higher heating value (HHV). It has been customary in the United States to express heating value as HHV. In Europe and elsewhere, heating value is frequently expressed as the lower heating value (LHV).

TABLE 24.6 Heat of Combustion

Combustible	Molecular Symbol	Heating Value (Btu per Pound)	
		Gross	Net
Carbon	C	14,100	14,100
Hydrogen	H_2	61,100	51,600
Sulfur	S	3980	3980
Carbon monoxide	CO	4350	4350

Source: From Hecklinger, R. S. 1996. *The Engineering Handbook*, CRC Press, Inc., Boca Raton, FL.

Heating value can be converted from HHV to LHV if weight decimal percentages of moisture and hydrogen (other than the hydrogen in moisture) in the fuel are known, using the following formula:

$$LHV_{Btu/lb} = HHV_{Btu/lb} - [\%H_2O + (9 \times \%H_2)] \times (1050 \quad Btu/lb) \tag{24.1}$$

$$LHV_{J/kg} = HHV_{J/kg} - [(9 \times \%H_2) + \%H_2O] \times (2240 \quad KJ/kg) \tag{24.2}$$

For example (using data from Table 24.9),

$$\begin{aligned} LHV_{Btu/lb} &= HHV_{Btu/lb} - [\%H_2O + (9 \times \%H_2)] \times 1050 \quad Btu/lb \\ LHV_{Btu/lb} &= 4940 - [0.30 + (9 \times 0.047)] \times 1050 \\ LHV_{Btu/lb} &= 4940 - [0.30 + 0.42] \times 1050 \\ LHV_{Btu/lb} &= 4940 - 756 = 4184 \quad Btu/lb \end{aligned}$$

Another method for determining the approximate higher heating value for MSW is to perform an ultimate analysis and then apply Dulong's formula:

$$HHV_{(Btu/lb)} = 14,544\ C + 62,028\left(H_2 - \frac{O_2}{8}\right) + 4050\ S \tag{24.3}$$

where C, H_2, O_2, and S represent the decimal proportionate parts by weight of carbon, hydrogen, oxygen, and sulfur in the fuel. The term $O_2/8$ is a correction used to account for hydrogen which is already combined with oxygen in the form of water. For example (using data from Table 24.9)

$$\begin{aligned} HHV_{(Btu/lb)} &= 14,544\ C + 62,028\left(H_2 - \tfrac{O_2}{8}\right) + 4050\ S \\ HHV &= 14,544 \times 0.257 + 62,028\left(0.047 - \tfrac{0.21}{8}\right) + 4050 \times 0.001 \\ HHV &= 3738 + 62,028(0.047 - 0.026) + 4.0 \\ HHV &= 3738 + 62,028 \times 0.021 + 4.0 \\ HHV &= 3738 + 1303 + 4 = 5045 \quad Btu/lb \end{aligned}$$

An alternate method of estimating the HHV is to multiply the approximate dry combustable HHV of 9,500 Btu/lb by the weight fraction of combustibles:

$$\begin{aligned} HHV &= 9500 \times (1 - \text{moisture} - \text{ash}) \\ HHV &= 9500 \times (1 - 0.30 - 0.18) \\ HHV &= 9500 \times 0.52 = 4940 \quad Btu/lb \end{aligned}$$

The American Society for Testing and Materials (ASTM) publishes methods for determining the ultimate analysis of solid fuels such as MSW.[3] The ultimate analysis of a fuel is developed through measures of carbon, hydrogen, sulfur, nitrogen, ash, and moisture content Oxygen is normally determined "by difference"; that is, once the percentages of the other components are measured, the remaining material is assumed to be oxygen. For solid fuels, such as MSW, it is frequently desirable to determine the proximate analysis of the fuel. The procedure for determining the proximate analysis is also prescribed by ASTM.[3] The qualities of the fuel measured in percentage by weight are moisture, volatile matter, fixed carbon, and ash. This provides an indication of combustion characteristics of a solid fuel. As a solid fuel is heated to combustion, first the moisture in the fuel evaporates, then some of the combustible constituents volatilize (gasify) and combust as a gas with oxygen, and the remaining combustible constituents remain as fixed carbon in a solid state and combust with oxygen to form carbon dioxide. The material remaining after combustion is complete is the ash. MSW, with a high percentage of volatiles and a low percentage of fixed carbon, burns with much flame.

Table 24.7 displays ignition temperatures for combustible substances in MSW. The ignition temperature is the temperature to which the combustible substance must be raised before it will unite in chemical combination with oxygen. Thus, the temperature must be reached and oxygen must be present for combustion to take place. Ignition temperatures are not fixed temperatures for a given substance. The actual ignition temperature is influenced by combustion chamber configuration, oxygen

TABLE 24.7 Ignition Temperatures

Combustible	Molecular Symbol	Ignition Temperature °F
Carbon (fixed)	C	650
Hydrogen	H_2	1080
Sulfur	S	470
Carbon monoxide	CO	1170

Source: From Stultz, S. C. and Kitto, J. B. eds. 1992. *Steam: Its Generation and Use, 40th Ed.*, The Babcock and Wilcox Co., Barberton, OH.

fuel ratio, and synergistic effect of multiple combustible substances. The ignition temperature of MSW is the ignition temperature of its fixed carbon component. The volatile components of MSW are gasified but not ignited before the ignition temperature is attained.

The oxygen, nitrogen, and air data displayed in Table 24.8 represent the weight of air theoretically required to completely combust one pound of a combustible substance. The weight of oxygen required is the ratio of molecular weight of oxygen to molecular weight of the combustion constituent as displayed in Table 24.5. The weights of nitrogen and air required are calculated from the percentage by weight constituents of dry air. In actuality, to achieve complete combustion, air in excess of the theoretical requirement is required for complete combustion to increase the likelihood that all of the combustible substances are joined with sufficient oxygen to complete combustion. The level of excess air required in the combustion of MSW depends on the configuration of the combustion chamber, the nature of the fuel firing equipment, and the effectiveness of mixing combustion air with the MSW. An excess air level of 80% is commonly associated with combustion of MSW in modern WTE facilities. Excess air is generally monitored using an oxygen analyzer at the economizer outlet. The type of analyzer used at waste-to-energy facilities generally reports percent wet oxygen. The dry oxygen can be estimated by assuming 15% moisture in the flue gas using the following equation:

$$\text{dry oxygen} = (\text{wet oxygen})/(1 - \text{percent moisture}/100), \text{ or dry oxygen} = (1.176)(\text{wet oxygen})$$

Excess air can be approximated by the following equation:

$$\text{Excess air} = 55.2 - 10.46 \times (\text{dry } O_2) + 1.4 \times (\text{dry } O_2)^2$$

where dry O_2 is the percentage dry oxygen in the flue gas.

Excess air serves to dilute and thereby reduce the temperature of the products of combustion. The reduction of temperature tends to reduce the heat energy available for useful work. Therefore, the actual excess air used in the combustion process is a balance between the desire to achieve complete combustion and the objective of maximizing the heat energy available for useful work.

TABLE 24.8 Theoretical Combustion Air

Combustible	Pounds Per Pound of Combustible					
	Required for Combustion			Products of Combustion		
	O_2	N_{atm}	Air	CO_2	H_2O	N_{atm}
Carbon	2.67	8.87	11.54	3.67		8.87
Hydrogen	8.00	26.57	34.57		9.00	26.57
Sulfur	1.00	3.32	4.32	2.00		3.32
Carbon monoxide	0.57	1.89	2.46	1.57		1.89

Source: From Hecklinger, R. S. 1996. *The Engineering Handbook*, CRC Press, Inc., Boca Raton, FL.

TABLE 24.9 Sample Calculation for Municipal Solid Waste (MSW)

Air Calculations (80% Excess Air)				
Ultimate Analysis				
Substance	Fraction % by Weight	Oxygen Required for Combustion, lb/lb of Element	Theoretical Oxygen, lb/lb of Element	Theoretical Dry Air, lb/lb of Element
Carbon	0.279	2.67	0.745[a]	3.218[b]
Hydrogen	0.037	8.00	0.296[a]	1.279[b]
Oxygen	0.209	—	—	—
Nitrogen	0.005	—	—	—
Sulfur	0.002	1.00	0.002[a]	0.009[b]
Ash	0.187			
Fuel moisture	0.281			
Total	1.000		1.043	4.505
Less oxygen in fuel			(0.209)	(0.903)[c]
Air				
Required at 100% theoretical air			0.834	3.603
180% of theoretical air (80% excess air)			1.501	6.485
Excess			0.667	2.882
HHV = 5100 Btu/lb				

Products of Combustion		
	lb/lb of Element	lb of Product
Carbon dioxide	3.67	1.024[d]
Moisture from hydrogen	9.00	0.333[d]
Oxygen		0.667[e]
Nitrogen		4.989[f]
Sulfur dioxide	2.00	0.004[d]
Moisture from fuel	1.00	0.281[d]
Moisture from air		0.084[g]
Total moisture		0.698[h]
Total		7.382

LHV = 5100 − [0.281 + (9 × 0.037)] × 1050 = 4455 BTU/lb

Temperature developed in combustion = 60 + 4455/[(0.698 × 0.55[i]) + (7.382 − 0.698) × 0.28[j]] = 2035°F

Check:

Total products of combustion = 180% of theoretical air + moisture from air + fraction percent by weight of C, H, O, N, S, and moisture

7.382 = 6.485 + 0.084 + 0.279 + 0.037 + 0.209 + 0.005 + 0.002 + 0.281 = 7.382

[a] Weight percent of element times oxygen required for combustion.
[b] Theoretical oxygen times 4.32.
[c] Amount of theoretical air due to oxygen in fuel.
[d] Weight percent of element times lb/lb of element.
[e] Excess oxygen.
[f] 180% of theoretical dry air times 0.7686 plus weight percent of nitrogen in fuel.
[g] Moisture in combustion air = 0.013 times 180% of theoretical air (0.013 lb moisture per lb of dry air at 80°F and 60% relative humidity).
[h] Total of moisture from combustion of hydrogen, moisture in fuel, and moisture from air.
[i] Heat capacity of water vapor.
[j] Heat capacity of dry flue gas.

It is frequently useful to know the temperature attained by combustion. The heat released during combustion heats the products of combustion to a calculable temperature. It must be understood that the calculation procedure presented here assumes complete combustion and that no heat is lost to the surrounding environment. Thus, it is a temperature that is useful to compare one combustion process with another. The heat available for heating the products of combustion is the lower heating value of the fuel. The increase in temperature is the lower heating value divided by the mean specific heat of the products of combustion. The mean specific heat is a function of the constituent products of combustion ($W_{P.C.}$) and the temperature. To approximate the theoretical temperature attainable, one can use a specific heat of 0.55 Btu/lb per °F for water vapor (W_{H_2O}) and 0.28 Btu/lb per °F for the other gaseous products of combustion ($W_{P.C.} - W_{H_2O}$). Thus, the formula approximating the temperature attained during combustion is

$$T_{comb} = T + \frac{LHV_{Btu/lb}}{0.55W_{H_2O} + 0.28(W_{P.C.} - W_{H_2O})} \tag{24.4}$$

For example (using data from Table 24.9)

$$T_{comb} = T_{ambient} + \frac{\mathrm{LHV}_{Btu/lb}}{0.55W_{H_2O} + 0.28(W_{P.C.} - W_{H_2O})}$$

If the ambient temperature is assumed to be 60°F, then

$$T_{comb} = 60 + \frac{4184}{0.55 \times 0.81 + 0.28(7.55 - 0.81)}$$

$$T_{comb} = 60 + \frac{4184}{0.45 + 1.89} = 60 + \frac{4184}{2.34}$$

$$T_{comb} = 1848 \cong 1850°F$$

Typical combustion calculations are provided in Table 24.9 for MSW to determine the products of the combustion process. Each of the combustible substances combines and completely combusts with oxygen as displayed in Table 24.5. The weight ratio of oxygen to the combustible substance is the ratio of molecular weights. Table 24.8 displays the weight or volume of oxygen theoretically required for complete combustion of one pound of the combustible substance. Sulfur dioxide from combustion of sulfur in fuel is combined with CO_2 in the sample calculation as a matter of convenience. If desired, a separate column can be prepared for sulfur dioxide in the products of combustion. Oxygen in the fuel combines with the combustible substances in the fuel, thereby reducing the quantity of air required to achieve complete combustion. The sample calculation uses the weight percentages of oxygen to reduce the theoretical air requirements and the nitrogen in the products of combustion. The decimal percentage of excess air is multiplied by the total theoretical air requirement to establish the weight of excess air and the total air requirement including excess air.

24.3.4 Furnaces

While the general principles of a modern waste combustor burning as-received MSW are common to all types, the specific solid waste combustion process is rather complex. The waste is heated by contact with hot combustion gases or preheated air, and by radiation from the furnace walls. Drying occurs in a temperature range of 122°F–302°F (50°C–150°C). At higher temperatures, volatile matter is formed by complicated thermal decomposition reactions. This volatile matter is generally combustible and, after ignition, produces flames. The remaining material is further degased and burns much more slowly. In an RDF furnace (see Figure 24.3), most of the volatile matter and some of the fixed carbon is burned in suspension while the remaining fixed carbon is combusted on a grate at the bottom of the furnace.

The complexity of the combustion of solid waste streams results from the nature of the decomposition and burning reactions and their association with heat transfer, air flow, and diffusion. In most waste combustors, combustion takes place while the solids are supported on and conveyed by a grate. Since the early 1960s, most MSW incinerators have incorporated one of a number of available proprietary grate systems that allow continuous feed of unscreened waste into and movement through furnaces with integral boiler facilities. The grate performs several functions: provides support for the refuse, admits underfire air though openings in the grate surface, transports the solid waste from feed mechanism to ash quench, agitates the bed, and serves to agitate and redistribute the burning mass.

The basic design factors which determine furnace capacity are grate area and furnace volume. Also, the available capacity and method of introducing both underfire and overfire air will influence, to a lesser extent, furnace capacity. Required grate area, in a conservative design, is normally determined by limiting the burning rate to between 60 and 70 lb/ft.^{2} hr (290–340 kg/m^2 hr) of grate area.[2] This is based on limiting the heat release rate loading on the grate to 250,000–300,000 Btu/ft.^{2} of grate per hr (2.8–3.4 GJ/m^2/hr).

Furnace volume required is established by the rate of heat release from the fuel. Thus, furnace volume is generally established by using heat release rates ranging from 12,500 to 20,000 Btu/ft.^{3}/hr (450–750 MJ/m^3/hr), with the lower heat release rate being more desirable from the standpoint of developing a conservative design. A conservative approach to design in this area is desirable because of probable periodic operation above design capacity to meet short-term higher than normal refuse collections and possible receipt of high heat-content waste.

Water wall units burning as-received MSW have been built as small as 75–100 tons per day (68–91 tons/day) capacity. However, the cost per ton of rated capacity of such units is relatively high. A more common unit size for both mass-burn and RDF furnaces is 250–1000 tons/day (225–900 tons/day), while water wall mass-fired units have been built as large as 750–1200 tons/day (675–1090 tons/day) capacity.[7]

The primary objective of a mechanical grate in a mass-burn furnace is to convey the refuse from the point of feed through the burning zone to the point of residue discharge with a proper depth of fuel and sufficient retention time to achieve complete combustion. The refuse bed should be agitated so as to enhance combustion. However, the agitation should not be so pronounced that particulate emissions are unreasonably increased. The rate of movement of the grate or its parts should be adjustable to meet varying conditions or needs in the furnace.

In the United States over the past 20 years, several types of mechanical grates have been used in continuous feed furnaces burning as-received MSW. These include reciprocating grates (see Figure 24.4), rocking grates (see Figure 24.5), roller grates (see Figure 24.6), and water wall rotary combustors for mass-burn units, and traveling grates for RDF units (see Figure 24.3). The reciprocating grates, rocking grates, and roller grates agitate and move the refuse material through the furnace by the movement of the grate elements and the incline of the grate bed. Additional agitation is obtained, particularly in the reciprocating grate, by drops in elevation between grate sections. The rotary combustor slowly rotates to tumble the refuse material, which is conveyed through the inside of the cylinder. The combustor is inclined from the horizontal so that gravity assists in moving the material through the unit. The traveling grate conveys the refuse through the furnace on the grate surface. Stirring is accomplished by building the grate in two or more sections, with a drop between sections to agitate the material.

Other grate systems have been developed in Europe for burning as-received MSW, some of which are currently being utilized in plants being constructed or in operation in the United States. The roller grate, or so-called Dusseldorf System (see Figure 24.6), uses a series of 5 or more rotating cylindrical grates, or drums, placed at a slope of about 30°.[57] The refuse is conveyed by the surface of the drums, which rotate in the direction of refuse flow, and is agitated as it tumbles from drum to drum. Underfire air is introduced through the surface of the drums. Both the Von Roll and the Martin grates use a reciprocating motion to push the refuse material through the furnace. However, in the Martin grate (see Figure 24.7), the grate surface slopes steeply down from the feed end of the furnace to the ash discharge end and the grate sections push the refuse uphill against the flow of waste, causing a gentle tumbling and agitation of the fuel bed.

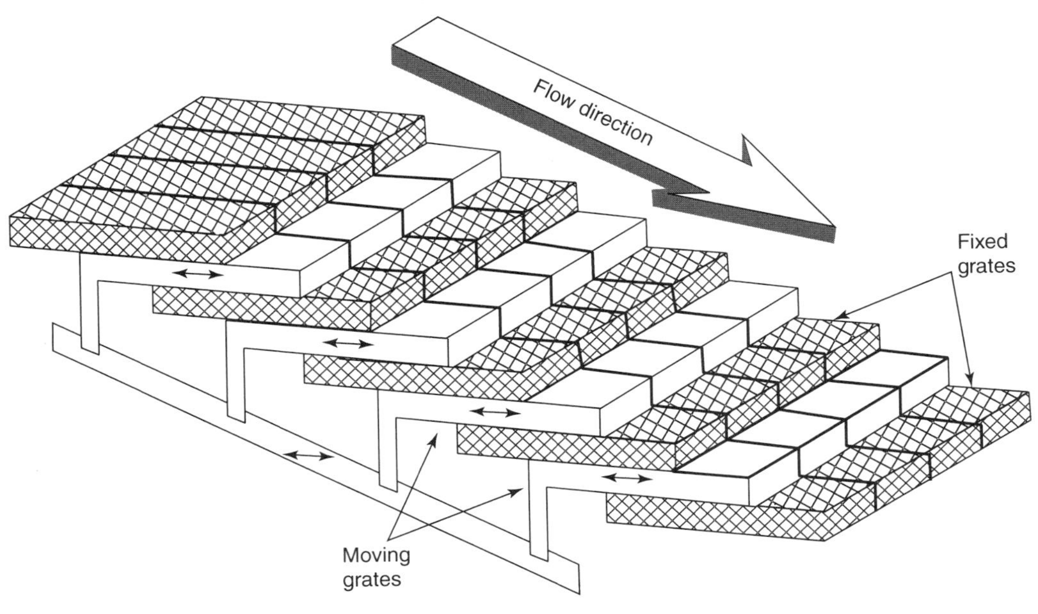

FIGURE 24.4 Reciprocating grates.

Another variable feature in the various grate designs is the percentage of open area to allow for passage of underfire air.[74] These air openings vary from approximately 2 to 30% of the grate surface area. The smaller air openings tend to limit the quantity of siftings dropping through the grates and create a pressure drop that assists in controlling the point of introduction of underfire air. RDF grates generally have a smaller

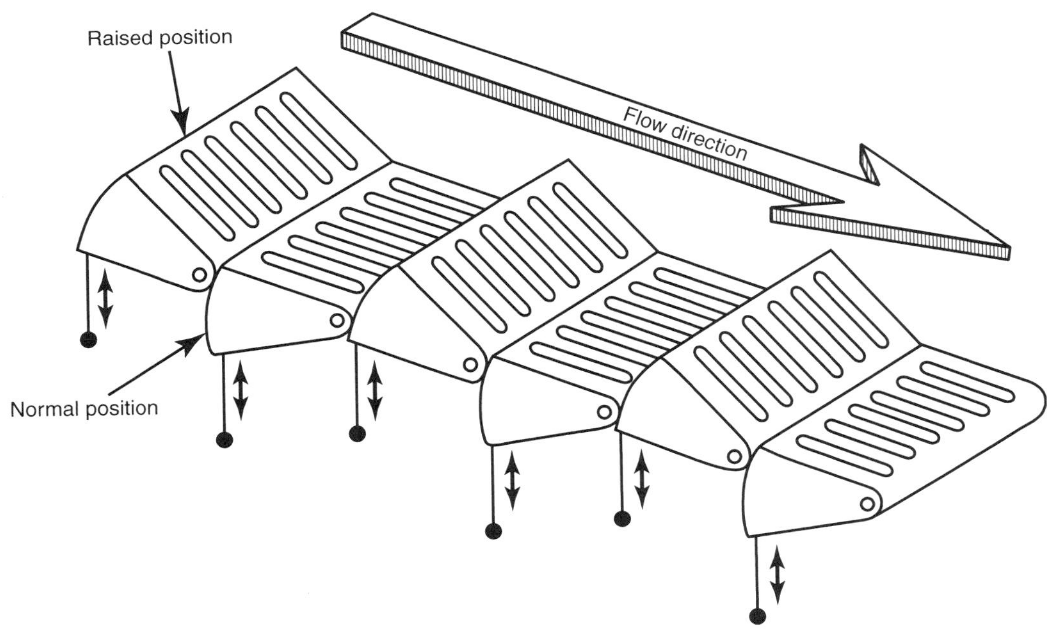

FIGURE 24.5 Rocking grates.

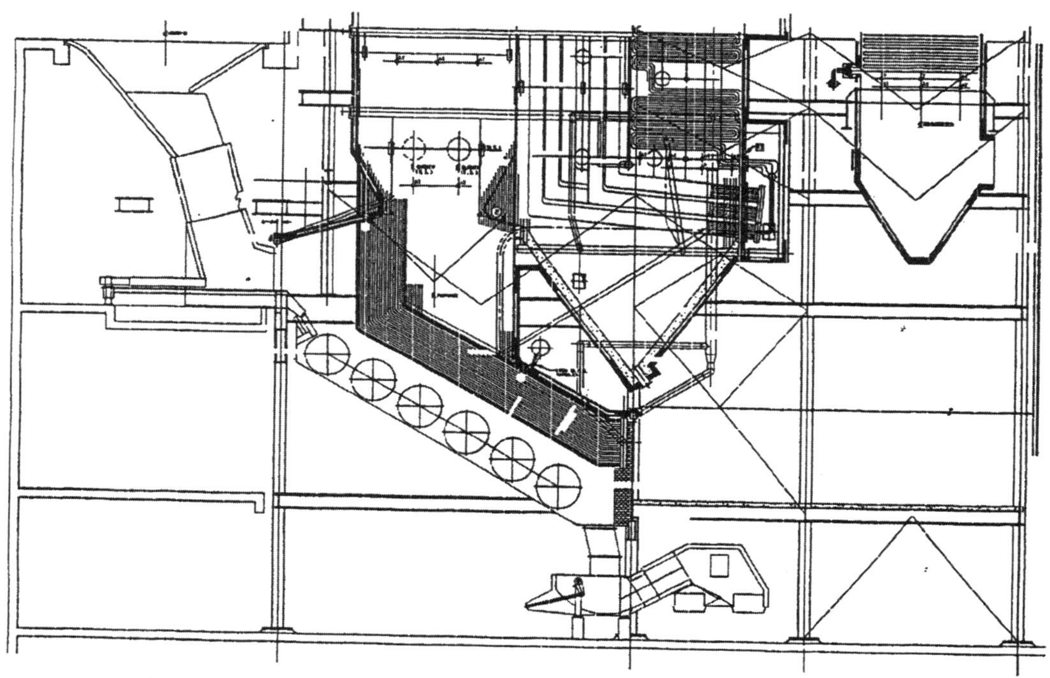

FIGURE 24.6 Roller grate system by Covanta Energy Group.

percentage of air openings. Larger air openings make control of underfire air more difficult but allow for continuous removal of fine material, which could interfere with the combustion process, from the fuel bed.

Furnace configuration is largely dictated by the type of grate used. In the continuous feed mechanical grate system, the furnace is rectangular in plan and the height is dependent upon the volume required by the limiting rate of heat release cited earlier. An optimum furnace configuration would provide sufficient volume for retention of gases in the high-temperature zone of maximum fuel volatilization long enough to ensure complete combustion, and would be arranged so that the entire volume is effectively utilized. Temperatures are usually high enough with present-day refuse for proper combustion. Turbulence should be provided by a properly designed overfire air system.

With present-day mass-fired water wall furnaces, the use of refractories in furnace construction has been minimized but not eliminated. Refractory materials may be used to line charging chutes, provide a transition enclosure between the top of the grates and the bottom of the water walls, a protective coating on the water wall tubes, and an insulating layer between the hot gases and the metal walls of flues downstream of the primary combustion chamber. Refractory brick used in a charging chute must be able to withstand high temperatures, flame impingement, thermal shock, slagging, spalling, and abrasion. The protective coating on the water wall tubes must be relatively dense castable material with a relatively high heat conductivity.[75] Insulating refractories used in flues downstream from the boilers, on the other hand, should have a low heat conductivity.

Refractories are generally classified according to their physical and chemical properties, such as resistance to chemical attack, hardness, strength, heat conductivity, porosity, and thermal expansion.[50] The material may be cast in brick in a variety of shapes and laid up with air-setting or thermal-setting mortar, or may be used in a moldable or plastic form. Material used in waste combustor construction includes "high duty" and "superduty" fireclay brick, phosphate-bonded alumina material, and silicon carbide, among others. In selecting the proper materials for application in this type of service, because the variety of materials is so great and the conditions of service so varied and severe,[18] advice of a recognized manufacturer should be sought.

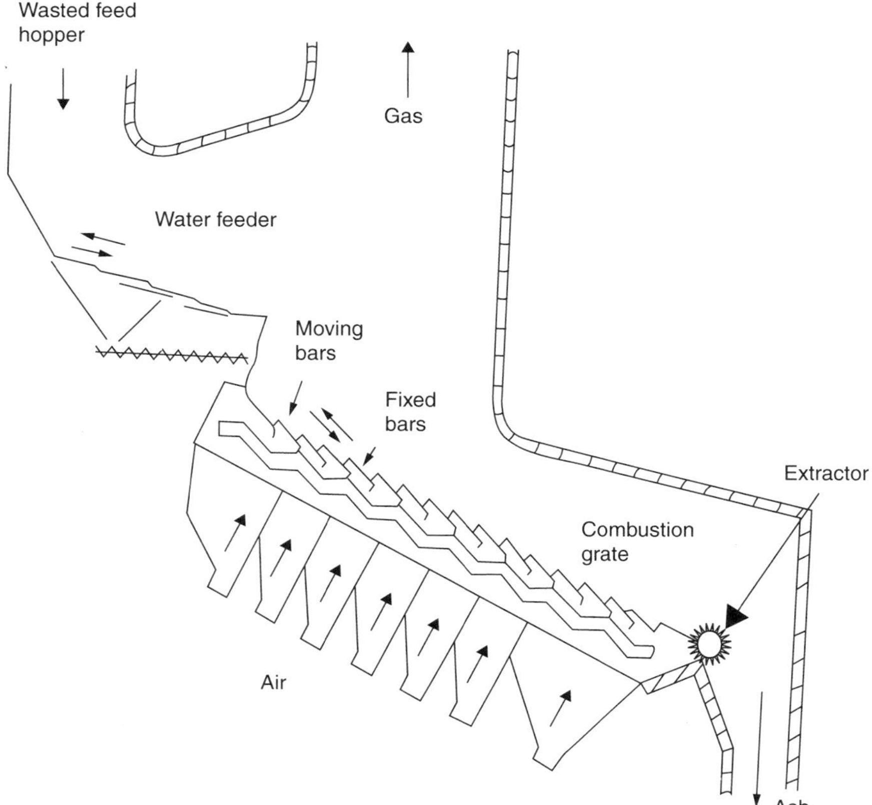

FIGURE 24.7 Martin system. (From Braun, H., Metzger, M., and Vogg, H. Zur Problematik der Quecksilber-Abscheidung aus Rauchgasen von Mullverbrennungsanlagen, Vol. 1, Teil 2.)

As indicated in the section on combustion calculations, the combustion process requires oxygen to complete the reactions involved in the burning process. The air that must be delivered in the furnace to supply the exact amount of oxygen required for completion of combustion is called the stoichiometric air requirement. Additional air supplied to the furnace is called excess air and is usually expressed as a percentage of the stoichiometric requirements.

The total air supply capacity in a waste combustor must be greater than the stoichiometric requirement for combustion because of imperfect mixing and to assist in controlling temperatures, particularly with dry, high heat-content refuse. The total combustion air requirements can range from 6 to 8 lb of air/lb of refuse for mass-fired water wall furnaces, and slightly less for RDF facilities.

In the modern mass-burn mechanical grate furnace chamber, at least two blower systems should be provided to supply combustion air to the furnace—one for underfire or undergrate air and the other for overfire air. Underfire air, admitted to the furnace from under the grates and through the fuel bed, is used to supply primary air to the combustion process and to cool the grates.

Overfire air may be introduced in two levels. Air introduced at the first level, called secondary air, immediately above the fuel bed, is used to promote turbulence and mixing, and to complete the combustion of volatile gases driven off the bed of burning solid waste. The second row of nozzles, which are generally located higher in the furnace wall, allow the introduction of air, called tertiary air, into the furnace to promote additional mixing of gases and for temperature control.

Blower capacities should be divided so that the underfire air blower is capable of furnishing half to two-thirds of the total calculated combustion air requirements, while the overfire blower should have a capacity of about half of the total calculated air requirements. Setting these capacities requires some judgment related to assessing how great a variation is anticipated in refuse heat contents during plant operation. Variable frequency drivers on the fan motors or dampers on fan inlets and air distribution ducts should be provided for control purposes.

Pressures on underfire air systems in mass-burn units for most U.S. types of grates will normally range from 2 to 5 in. of water. European grates frequently require a higher pressure. The pressure on the overfire air should be high enough that the air, when introduced into the furnace, produces adequate turbulence without impinging on the opposite wall. This is normally accomplished by the use of numerous relatively small (1½–3 in. in diameter) nozzles at pressures of 20 in. of water or higher.

Recirculated flue gas is sometimes used in part for underfire air and tempering air. Using recirculated flue gas as combustion air reduces the quantity of fresh air needed, thereby increasing thermal efficiency and minimizing thermal NO_x formation. It can also be used as tempering air to control the temperature of the flue gas entering the boiler.[14] The duct work for recirculated flue gas is highly susceptible to corrosion due to the presence of acid gases. For this reason, it is critical that the recirculated gas be taken from the system after the pollution control devices.

In an RDF-fired, spreader-stoker type of unit, the combustible material is generally introduced through several air-swept spouts in the front water wall, is partially burned in suspension, and then falls onto a grate on which combustion is completed as the partially burned material is conveyed to the residue discharge under the front water wall face of the furnace. Densified RDF can also be burned in such units. The RDF can furnish all the combustible input to the system, or it may be co-fired with a fossil fuel, generally coal.

Some combustion air in RDF-fired units is introduced with the fuel through the air-swept feed spouts to distribute the fuel on the grate. Additional air is introduced into the furnace higher in the water wall area to enhance turbulence and mixing in the unit and/or to control temperatures. This additional combustion air supply is similar to the tertiary air utilized in the mass-burn units.

24.3.5 Boilers

Substantial quantities of heat energy may be recovered during the thermal destruction of the combustible portions of MSW. Systems that have been successfully used to recover this energy include mass-fired refractory combustion chambers followed by a convection boiler section; a mass-fired water wall unit where the water wall furnace enclosure forms an integral part of the boiler system and an RDF semisuspension-fired spreader-stoker/boiler unit. Each system has apparent advantages and disadvantages.

24.3.5.1 Refractory Furnace with Waste Heat Boiler

In a refractory furnace waste heat boiler unit, energy extraction efficiencies are generally lower, assuming the same boiler outlet temperatures, than with the other systems. Approximately 50%–65% of the heat generated in the combustion process may be recovered with such systems. These units can produce approximately 2–3 lbs steam/lb of normal MSW (heat content = 4500 Btu/lb), versus 3 or more lb/lb MSW in mass-fired water wall units. This lower efficiency of steam generation is caused by larger heat losses due to higher excess air quantities needed with such units to control furnace temperatures so that furnace refractories are not damaged. However, the boilers in such units, if properly designed and operated, generally are less susceptible to boiler tube metal wastage problems than the other systems listed earlier.

24.3.5.2 Mass-Fired Water Wall Units

Mass-fired water wall units are the most widely utilized type of heat recovery unit in this field today. In this type of unit, the primary combustion chamber is fabricated from closely spaced steel tubes through which water circulates. This water wall lined, primary combustion chamber incorporated into the overall boiler system is followed by a convection type of boiler surface. It has been found desirable in these plants to coat a substantial height of the primary combustion chamber, subject to higher temperatures and flame impingement, with a thin coating of a silicon carbide type of refractory material or inconel and to limit average gas velocities to under 15 ft./s (4.5 m/s) in this portion of the furnace. Gas velocities entering the boiler convection bank should be less than 30 ft./s (9.0 m/s).[75] Efficiency of heat recovery in such units has been found to range generally from 65 to 75%, with steam production usually above 3 lb of steam/lb of normal as-received MSW. Water table studies have been used occasionally in some larger units to check on combinations of furnace configuration and location of overfire combustion of air nozzles.[29]

24.3.5.3 RDF-Fired Water Wall Units

As pointed out earlier, RDF may be burned in a semisuspension-fired spreader-stoker/boiler unit where the RDF is introduced through several air-swept spouts in the front water wall, partially burns in suspension, and then falls on a grate on which combustion is completed. In this type of unit, the water wall lined primary combustion enclosure (furnace) may be followed by a superheater (usually), a convection boiler heat transfer surface, and (sometimes) an economizer surface.

Efficiencies of RDF-fired boilers generally range from 65 to 80% of the heat input from the RDF. Steam production from RDF would normally be expected to be somewhat greater than 3 lb of steam/lb of RDF. However, when one takes into account the combustible material lost in the processing of as-received MSW to produce the RDF, steam production normally will fall to about 3 lb of steam/lb of as-received MSW.[37]

If the energy recovered from the combustion of as-received MSW or RDF is to be used to produce electricity, some superheating is desirable, if not necessary. Since boiler tube metal wastage in these plants is, at least partially, a function of tube metal temperature[11] and steam is a less efficient cooling medium than water, superheater surface is more prone to metal wastage problems than other areas of boiler tubing. Tube metal temperatures, above which metal wastage can be a significant operational problem, are generally thought to range from 650 to 850°F (345°C–455°C). These temperatures are lower than those for maximum efficiency of electrical generation by steam driven turbines. It is desirable to consider this in facility design to reduce plant downtime and minimize maintenance costs.

In the 1980s and early 1990s a so-called full-suspension combustion concept was attempted in which finely shredded combustible material from MSW was blown into the furnace through nozzles located one-half to two-thirds up the height of the water wall furnace enclosure. In this type of unit most of the RDF, usually composed of smaller sized particles than in the semisuspension-fired unit was supposed to burn in suspension. This concept was not successful due to problems related to the additional handling of the RDF and greater power required to achieve a finer shred. Also, some boilers seemed to experience a greater tendency for slag formation in the boiler. While the concept initially anticipated that the RDF would completely burn in suspension, experience indicated that this does not occur. Accordingly, dump grates became a necessity in such furnace boiler units to allow for completion of combustion prior to water quenching of the residue. This concept has been abandoned.

24.3.6 Residue Handling and Disposal

The residue from a well-designed, well-operated mass-fired incinerator burning as-received refuse will include the noncombustible material in the MSW and usually somewhat less than 3% of the combustibles. The nature of this material will vary from relatively fine, light ash, burned tin cans, and partly melted glass to large, bulky items such as lawn mowers and bicycles.

In most modern WTE plants bottom ash residue is discharged from the end of the furnace grate through a chute into a trough filled with water. Removal from the trough may be either by a ram discharger onto a conveyor or by a flight conveyor to an elevated storage hopper from which it is discharged to a truck. If a water-filled trough with a flight conveyor is used, normally two troughs are provided, arranged so that the residue can be discharged through either trough. The second trough serves as a standby. Fly ash, residue collected in the air pollution control equipment downstream of the furnace/boiler, is usually returned back to and mixed with the bottom ash.

A key feature in the design of ash discharge facilities is provision for sealing the discharge end of the furnace to prevent uncontrollable admission of air. This seal is usually provided by carrying the ash discharge chute at least 6 in. (15 cm) below the water surface in the receiving trough. In the design of the conveyor mechanism, the proportions should be large because the material frequently contains bulky metal items and wire, potential causes of jamming. Also, the residue material tends to be extremely abrasive. A grizzly screen is often used to remove oversized bulky materials from the residue prior to its being loaded onto trucks for delivery to the landfill.

Residue is usually taken to a landfill for final disposal. Many modern facilities dispose of their residue at monofills (landfills that accept WTE plant residue only). The volume of material remaining for ultimate disposal will range from 5 to 15% of that received at the plant. Many plants currently operating in the United States that weigh MSW received at the plant and residue discharged from the furnaces, indicate that the weight of MSW is only reduced by from 40 to 50%. However, as much a one-third of the residue weight in these plants may be attributed to incomplete drainage of the material prior to its discharge into the final transportation container. The ram-type ash discharger used in European and most large U.S. plants generally achieves much better dewatering of residues than older water-filled trough, ash drag residue handling systems. These systems can achieve 65%–75% weight reduction.

The main components of ash are inert materials of low solubility, such as silicates, clay, and sand. Aluminum, calcium, chlorine, iron, selenium, sodium, and zinc are major elements in all particles and, along with carbon, can comprise over 10% by weight of the ash.[22]

A broad range of trace metals and organic compounds may be found in fly and bottom ash. Data on ash composition are difficult to compare, however, because they reflect different types and sizes of facilities, unknown sample collection methodology and sample size at each facility, interlab variation in testing procedures (even using the same test), and the heterogeneous nature of MSW itself. In addition, the presence of a substance in ash does not mean that it will enter the environment. Its fate depends on its solubility, how the ash is managed, and whether the ash is subject to conditions that cause leaching.[48]

Metals tend to be distributed differently in fly and bottom ash. Most volatile and semivolatile metals, such as arsenic, mercury, lead, cadmium, and zinc, tend to be more concentrated or “enriched” in fly ash.[59,71] Less volatile metals, such as aluminum, chromium, iron, nickel, and tin, typically are concentrated in bottom ash.[58,59]

Organic chemicals also exhibit differing distributions. Dioxin/furan and polychlorinated biphenyls (PCBs) tend to be enriched in fly ash, while other chemicals such as polycyclic-aromatic hydrocarbons (PAHs) and phthalates tend to be concentrated in bottom ash.[71] Concentrations of dioxins/furans in fly ash exhibit a wide range, but they are significantly lower in ash from modern facilities than in ash from older incinerators.[34,72,80]

From a regulatory standpoint, a number of different testing procedures have been developed and utilized by regulatory agencies over the past several years in an attempt to predict the behavior of MSW residues deposited in landfills. Most of these methods were developed to predict leaching characteristics of residues deposited in landfills with raw or as-received MSW. Test results using these methods have been quite variable. However, as pointed out earlier, most modern WTE facilities dispose of their residue in monofills.[68] Tests of leachate from such monofills indicate metals concentrations below extraction procedure (EP) toxicity limits, and in most cases below U.S. drinking water standards.[68] Most test data show little or no leaching of organic chemicals.[35,68]

Following a court decision in the mid-1990s that ash residue from combustion of MSW is not exempt from the rules and regulations for hazardous waste, regulators have required testing of ash residues as they are discharged from the plant, i.e., separately if bottom ash and fly ash are discharged separately, and combined if they are combined prior to discharge from the plant. Bottom ash is alkaline and usually tests as nonhazardous, while fly ash is acidic and frequently tests as hazardous. When the fly ash is mixed with bottom ash prior to discharge from the plant, the alkaline bottom ash neutralizes the smaller quantities of acidic fly ash. The mixture tests as nonhazardous and can be disposed of in the normal monofill.

24.3.7 Other Plant Facilities

The balance of the plant equipment is similar to that used in fossil fuel-fired boiler facilities. However, there are differences. Thus, the combustion of MSW produces a highly corrosive environment for boiler tube materials. Metal chlorides are believed to be primarily responsible for boiler tube corrosion problems.[43] The most important factors in high temperature corrosion are metal temperature, gas temperature, temperature gradient between gas temperature and metal temperature, deposit characteristics, and temperature fluctuations.[1] For this reason, boiler tubes are generally fabricated using corrosion-resistant alloys. Boiler tube shields or weld overlay cladding of boiler tubes with inconel are also used in highly corrosive/erosive areas.[43]

Some waste-to-energy facilities incorporate an air heater to preheat combustion air. Finned tubes plug quickly due to the large quantity of flyash in the flue gas. These air heaters are always of the bare tube design.

Since thermal efficiency is not an overriding concern in waste-to-energy facilities, many plants have one, or at most two, feedwater heaters. Some have only a deaerator for feedwater heating, unlike conventional power plants which have several stages of feedwater heaters.

24.4 Air Pollution Control Facilities

Potential emissions from the combustion of MSW may be broadly classified into particulates, gaseous emissions, organic compounds, and trace metals. The concern is to reduce emissions so as to adequately protect public health. This is achieved using good combustion practice and equipment specially designed to remove the targeted pollutants.

During the 1980s and 1990s, the emission requirements for air pollution control equipment became more stringent as the USEPA promulgated new standards. The most recent standards were published in the December 19, 1995 Federal Register as 40 CFR Part 60 Subpart Cb[6,73]. These standards established emission limits for large (over 248 tons per day [225 Mg/d]) and small (under 248 tons per day [225 Mg/d]) combustor units. Table 24.10 and Table 24.11 contain a summary of those standards.

24.4.1 Particulate Control

Particulates have been a matter of concern and regulatory agency attention for some time. The initial concern was from the standpoint of reducing gross particulate emissions that were both an aesthetic and a potential public health problem. Current interest and concern is directed toward better control of submicron-size particles[76] and other pollutants.

Electrostatic precipitators were the most commonly used gas cleaning device for particulate emission control in municipal waste combustors in the 1970s and early 1980s. They were designed to achieve high collection efficiencies (99% or higher) and meet the air emissions standards at the time. As emission standards became more stringent for particulates, fabric filters became more prevalent. Many electrostatic precipitators were replaced with fabric filters due to the 1995 regulations.

Fabric filters can operate at high efficiency, even in the submicron particle size range. They became widely used in the late 1980s because of the increasing emphasis of regulatory agencies on acid gas control and lower particulate emission levels. Baghouses are more effective than electrostatic precipitators for acid gas scrubbing when preceded by a spray dryer. The original bags used in these facilities had a limited life at high temperatures. Experiments using different materials of construction have led to longer bag life.

The scrubber/fabric filter control systems have been shown to be capable of operating at a particulate emission level of 20 mg/Nm^3 (0.009 gr/dscf) and lower (see Table 24.12). The material selected for the

TABLE 24.10 Emission Limits for Large Combustor Units

Parameter	Limit	Conditions
Particulates	27 mg/DSCM	Corrected to 7% oxygen
Opacity	10%	6-min average
Cadmium	0.04 mg/DSCM	Corrected to 7% oxygen
Lead	0.49 mg/DSCM	Corrected to 7% oxygen
Mercury	Lesser of 0.08 mg/DSCM or 85% removal	Corrected to 7% oxygen
SO_2	Lesser of 31 ppmv or 75% removal	Corrected to 7% oxygen 24-hour geometric mean
HCl	Lesser of 31 ppmv or 95% removal	Corrected to 7% oxygen
Dioxin/furans	60 ng/DSCM–ESP 30 ng/DSCM—all others	Corrected to 7% oxygen
NO_x	220 ppmv—water wall	Corrected to 7% oxygen
	250 ppmv—rotary water wall	Corrected to 7% oxygen
	250 ppmv—RDF	Corrected to 7% oxygen
	240-fluidized bed	Corrected to 7% oxygen

filter bags can have an important effect on filtering efficiency and the emission level thus achieved. In general, test results to date for the scrubber/fabric indicate lower particulate emissions than those for electrostatic precipitators on WTE plants. However, in general, electrostatic precipitators have not been designed to meet emission levels as low as those specified for fabric filter installations. Electrostatic precipitators following spray drying absorbers in Europe have been tested at particulate emission levels of 1–8 mg/Nm^3 (0.00045–0.0036 gr/dscf). The reliability and overall economics of the various control processes must be considered when making a selection of equipment to meet these very low emission control requirements. Data are available[68] on emission levels for approximately 30 different specific elements, many of them heavy metals. Elements found to occur in stack emission from municipal waste combustors are lead, chromium, cadmium, arsenic, zinc, antimony, mercury, molybdenum, calcium, vanadium, aluminum, magnesium, barium, potassium, strontium, sodium, manganese, cobalt, copper, silver, iron, titanium, boron, phosphorus, tin, and others.

Since the condensation point for metals such as lead, cadmium, chromium, and zinc is above 570°F (300°C), the removal efficiency for such metals is highly dependent on the particulate removal efficiency. Some metal compounds, particularly chlorides such as $AsCl_3$ at 252°F (122°C) and $SnCl_4$ at 212°F (100°C), have condensation points below 300°C. For such compounds, particulate collection

TABLE 24.11 Emission Limits for Small Combustor Units

Parameter	Limit	Conditions
Particulates	70 mg/DSCM	Corrected to 7% oxygen
Opacity	10%	6-min average
Cadmium	0.10 mg/DSCM	Corrected to 7% oxygen
Lead	1.6 mg/DSCM	Corrected to 7% oxygen
Mercury	Lesser of 0.08 mg/DSCM or 85% removal	Corrected to 7% oxygen
SO_2	Lesser of 80 ppmv or 50% removal	Corrected to 7% oxygen 24-hour geometric mean
HCl	Lesser of 250 ppmv or 50% removal	Corrected to 7% oxygen
Dioxin/furans	125 ng/DSCM	Corrected to 7% oxygen
NO_x	No limit	Corrected to 7% oxygen

TABLE 24.12 Particulate Emissions from Municipal Waste Combustors

	Particulates (gr/dscf)[a] @ 12% CO_2
Plant G (1983); EP	0.0321
Plant T (1984); DS, BH	0.012
Plant M (1984); DS, EP	0.0104
Plant W (1985); DS, BH	0.004
Plant P (1985); EP	0.0163
Plant T (1986); EP	0.007
Plant M (1986); DS, BH	0.007

EP, electrostatic precipitator; DS, dry scrubber; BH, bag house.

[a] Grains per dry standard cubic foot.

Source: From Velzy, C. O. 1987. *U.S. Experience in Combustion of Municipal Solid Waste*, Atlanta, GA, February 20.

temperatures will be a factor in collection efficiency. High removal (over 99%) has been observed for most metals for highly efficient (over 99%) particulate removal systems operating at appropriate temperatures.

24.4.2 Gaseous Emission Control

Gaseous emissions such as SO_2, HCl, CO, NO_x and hydrocarbons have recently become a concern in municipal waste combustors and their emissions are now regulated. Acid gas emissions are controlled by scrubbing devices. Carbon monoxide, NO_x, and hydrocarbons are controlled by good combustion practice. Oxides of nitrogen in some cases also require control equipment in the form of selective noncatalytic reduction (SNCR) to reduce NO_x to acceptable levels.

Common gaseous emission factors, based on tests at a number of waste-to-energy plants, are shown in Table 24.13. High carbon monoxide and hydrocarbon emissions are caused by incomplete combustion and/or upsets in combustion conditions. High nitrogen oxide emissions are generally caused by high combustion temperatures. Hydrogen chloride (and hydrogen fluoride) and sulfur oxides, on the other hand, are directly a function of the chlorine (fluorine) and sulfur content in the fuel. The highest emissions, cited in Table 24.13, are from older, poorly controlled plants without significant pollution control equipment.

TABLE 24.13 Gaseous Emission Factors for Municipal Waste Combustors (lb/ton)

	New York Incinerators 1968–1969	Test Results U.S. Plants 1971–1978	Martin Plants 1984–1986	EPA Data Base Tests Through 1988	
				Mass-Burn	RDF
Carbon monoxide	—	3.7–9.3	0.2	0.06–16.2	1.0–5.2
Nitrogen oxides	—	0.5–2.2	5.0–6.0	0.5–4.5	2.5–3.2
Hydrocarbons	0.1–22.1	1.1	0.015–0.006	0.01–0.1	0.005–0.01
Hydrochloric acid	1.4–8.6	4.6–14.5	5.0–0.2	0.05–5.7	0.02–9.3
Sulfur oxides	1.3–8.0	0.8–2.2	1.0–2.0	0.05–4.8	0.05–2.3

Source: From Carrotti, A. A. and Smith, R. A. 1974. *Gaseous Emissions from Municipal Incinerators.*; Cooper Engineers, Inc. 1985. *Air Emissions Tests of a Deutsche Babcock Anlagen, Dry Scrubber System at the Munich North Refuse-Fired Power Plant*; Hahn, J. L., VonDemfange, H. P., Zurlinden, R. A., Stianche, K. F., Seelinger, R. W. (Ogden Martin Systems, Inc.), and Weiand, H., Schetter, G., Spichal, P., and Martin, J.E. (Martin GmbH), 1986. *National Waste Processing Conference*, Denver, CO, June 1–4, 1986. ASME, New York; Murdoch, J. D. and Gay, J. L. 1989. *Material Recovery with Incineration, Monmouth County, N.J.*, p. 329, Tulsa, OK, August 14–17; Velzy, C. O. 1985. *Standards and Control of Trace Emissions from Refuse-Fired Facilities*, Madison, WI, November 20–22.

Hydrogen chloride (and hydrogen fluoride) and sulfur oxides are best removed by acid gas scrubbing devices using chemical treatment. Initial efforts at acid gas control used wet collectors. However, this type of flue gas cleaning equipment is subject to problems such as corrosion, erosion, generation of acidic waste water, wet plumes, and, not least, high operating cost. Because of these problems, various semiwet and dry methods of cleaning flue gases have been developed and installed. These methods of gas treatment are based on the injection of slurried or powdered lime, limestone, or dolomite; adsorption; and absorption; followed by chemical conversion.[68] Since the reactivity of these lime materials is rather low, a multiple of the stoichiometric quantity is normally required to obtain a satisfactory cleaning effect. High removal efficiencies can be achieved for HCl, but reduction of SO_2 and SO_3 is more difficult to achieve and maintain. Slaked lime is highly reactive and stoichiometric ratios of 1.2–1.7 have been used for 97%–99% HCl removals and 60%–90% SO_2 reductions, depending on operating conditions and particulate collector (fabric filters having demonstrated higher removal efficiencies than electrostatic precipitators).

Lime injection into a scrubber/fabric filter system has resulted in removal efficiencies of 90%–99% for HCl and 70%–90% SO_2, provided that the flue gas temperature and the stoichiometric ratio for lime addition are suitable. This combination of processes has reduced HCl levels below 20 ppm and SO_2 to levels below 30 ppm for MSW waste-to-energy plants. This technology has also been extensively used in other applications for acid gas removal. The scrubber/electrostatic precipitor combination has been shown to provide about 90% HCl removal, but typically less SO_2 removal (about 50%). Since this removal efficiency does not meet the most recent regulations, many electrostatic precipitators have been replaced with baghouses. Lime injection into the furnace has also been tested with some success (about 50%–70% efficiency), but fails to meet the most recent regulations.

Some sampling to determine HF removal has been reported. In general, HF removal normally follows HCl removal (i.e., is usually over 90%–95%).

Carbon monoxide and hydrocarbon emissions are best controlled by maintaining proper combustion conditions. Nitrogen oxide emissions are controlled by ammonia injection or by use of combustion control techniques such as limitation of combustion temperatures or recirculation of flue gases. Note in the last column of Table 24.13 that attempts to limit hydrocarbon emissions by improving combustion conditions and raising furnace operating temperatures seem to have resulted in increasing the level of NO_x emissions.

Selective noncatalytic reduction (SNCR) appears to be the most practical method of reducing NO_x emissions for most municipal waste combustors. SNCR involves the use of ammonia to reduce NO_x to nitrogen and water. The SNCR reaction occurs at a temperature of 1600°F–2100°F. At lower temperatures, a catalyst is required to promote the reaction (selective catalytic reduction, or SCR). SCR is not used on municipal waste combustors. Tests conducted at a municipal waste combustor demonstrated that NO_x emission levels of 150 ppmv (45%–55% reduction) can be achieved with SNCR.[6]

Thermodynamic equilibrium considerations indicate that under excess air conditions and with temperatures of 1472°F (800°C) and higher, maintained in a completely mixed reactor for a suitable period of time, emissions of organic or hydrocarbon compounds should be at nondetectable levels. However, measurements at operating plants, particularly those constructed prior to the early to mid-1980s, indicated significant emissions of trace organic or hydrocarbon compounds, some of which are toxic. These tests indicated that the basic objective of combustion control, thorough mixing of combustion products with oxygen at a temperature that is sufficiently high to provide for the rapid destruction of all organic or hydrocarbon compounds, had not been achieved in these early WTE plants.

If the fuel, or the gas driven off of the fuel bed, is not adequately mixed with air, fuel-rich pockets will exist containing relatively high levels of hydrocarbons, which then can be carried out of the combustion system. Kinetic considerations indicate that such hydrocarbons can be destroyed rapidly in the presence of oxygen at elevated temperatures. Also, if too much combustion air is introduced into the combustion chamber, either in total or in a particular area of the chamber, temperatures will be reduced, combustion reactions can be quenched, and hydrocarbons carried out of the combustion system. Achieving the goal of proper combustion control, destruction of all hydrocarbon compounds to form carbon dioxide and water, will minimize emission of potentially toxic substances as well as other compounds that may be precursors and capable of forming toxic compounds downstream in cooler regions of the boiler.

Table 24.13 shows that the hydrocarbons can vary significantly, frequently over relatively short periods of time, based on measurements at older municipal waste combustors. The highest levels shown in this table occurred in one of the older plants and no doubt indicate very poor combustion conditions. Tests at modern WTE plants indicate consistently low levels of hydrocarbons, which are indicative of good combustion control. In modern, well-designed and -operated plants, photochemical oxidants and PAH are in concentrations too low to cause any known adverse health effects. Tests[55,66] for other substances that might be of concern, such as polychlorinated biphenyls (PCBs), generally have found levels discharged to the atmosphere to be so low as to have a negligible impact on the environment.

24.4.3 Organic Compound Control

Organic compounds for which emission data are available include polychlorinated dibenzodioxins (PCDDs), polychlorinated dibenzofurans (PCDFs), chlorobenzenes (CBs), chlorophenols (CPs), polycyclic aromatic hydrocarbons (PAHs), and PCBs. A number of other organic compounds, including aldehydes, chlorinated alkanes, and phthalic acid esters, have also been identified in specific testing programs. Since dioxin/furan emissions have generated the most interest over the past several years, there are more data for these compounds, in particular for the tetrahomologues, and especially the 2,3,7,8-substituted isomers. The other compounds have been less frequently reported in the literature. The reason for this emphasis is the toxicity of dioxin/furan to laboratory animals and the perceived risk to humans.

Upset conditions in energy-from-waste plants can lead to local air-deficient conditions resulting in the emission of organic compounds. PAHs are formed during fuel-rich combustion as a consequence of free radical reactions in the high temperature flame zone. In addition, it is found that in the presence of water cooled surfaces, such as found in oil-fired home-heating furnaces, a high fraction of the polycyclic compounds are oxygenated. Similar free radical reactions probably take place in fuel-rich zones of incinerator flames yielding PAH, oxygenated compounds such as phenols, and, in the presence of chlorine, some dioxin/furan. The argument for the synthesis of dioxin/furan at temperatures of 400°F–800°F is supported by the increase in the concentration of these pollutants across a heat recovery boiler downstream of the refractory lined combustion chamber of a waste combustor.

This free radical mechanism appears to be the dominant source of dioxin/furan in municipal waste combustors. These compounds may also be present as contaminants in a number of chemicals, most notably chlorinated phenols and polychlorinated biphenyls (PCBs). Their presence in MSW results from the use of these chemicals, discontinued in some cases, as fungicides and bactericides for the phenol derivatives, or the use of PCBs as heat exchanger and capacitor fluids. These compounds are expected to survive in a furnace combustion chamber only if large excesses in the local air flow cool the gases to below the decomposition or reaction temperature. Dioxin/furan can also be produced by condensation reactions involving the chlorinated phenols and biphenyls. The observed formation of dioxin/furan when fly ash from MSW incinerators is heated to temperatures of 480°F (250°C) suggest such catalyzed condensation reactions of chlorinated phenols. PCBs are precursors to furan, and pyrolysis of PCBs in laboratory reactors at elevated temperatures for a few seconds has yielded furan.

The available test data clearly show that dioxin/furan exit the boilers and, depending on the emission control devices employed, some fraction enters the atmosphere either as gases or adsorbed onto particulates. In addition, the solids remaining behind in the form of fly ash and bottom ash contain most of the same compounds, and these become another potential source of discharge to the environment.

Emission data for total dioxin/furan generally fall into three main categories:

- Low emissions, in the range of 20–130 ng/Nm^3

- Medium emissions, from 130 to 1000 ng/Nm^3
- High emissions, over 1000 ng/Nm^3

Average dioxin/furan emission from older plants ranges from about 500 to 1,000 ng/Nm^3. The lower emission levels tend to be associated with newer, well-operated mass-fired facilities such as water wall plants, and with modular, starved or controlled air types of incinerators (see Table 24.14). In most test programs, adequate operating data were not collected to correlate emissions with operations. Researchers in the field theorize that combustion conditions can play a role in minimizing emissions, and several studies[63,68,70] were conducted in Canada and the United States to define that role more exactly.

Emissions from MSW combustion contain small amounts of many different dioxin/furan isomers. While individual dioxin/furan isomers have widely differing toxicities, the 2,3,7,7-TCDD isomer, present as a small proportion of the total dioxin/furan, is of greatest known toxicological concern. Based on animal studies it has been generally concluded that other 2,3,7,8-substituted dioxin/furan isomers, in addition to the 2,3,7,8-TCDD, are also likely to be of toxicological concern. A method for expressing the relative overall toxicological impact of all dioxin/furan isomers, as so-called "2,3,7,8-TCDD toxic equivalents," was developed in the mid-1980s[4] and has been used by the EPA intermittently in its regulatory efforts since this time.

In this method, emissions are sampled, extracted, and analyzed for all constituent isomers of dioxin/furan. A system of toxicity weighting factors from the existing toxicological data (based almost entirely on animal studies) is applied to each constituent dioxin/furan isomer and the results are summed to arrive at the 2,3,7,8-TCDD toxic equivalent. An example of dioxin/furan test results expressed as 2,3,7,8-TCDD toxic equivalents, using three different systems of weighting factors, is shown in Table 24.15.

Emission control systems consisting of a scrubber/fabric filter have been evaluated for dioxin emissions.[36] Recently dioxin removal efficiencies exceeding 99% were obtained, which resulted in dioxin concentrations at the stack that approach the detection limit of the sampling and analytical equipment currently available. Emissions of furan followed a similar range of values as dioxin with the scrubber/high efficiency particulate removal combination reducing furan to very low or nondetectable levels. Additional reductions of over 50% can be achieved by activated carbon injection.[6]

Some limited data on emissions of CB, CP, PCB, and PAH are available. Most sampling programs for dioxin/furan have unfortunately neglected to analyze for these compounds. Maximum levels from Canadian studies[36] are as included in Table 24.16 along with some data from tests on U.S. plants.[35] The scrubber/fabric filter technology generally demonstrated removal rates of 80%–99% for these compounds in these studies.

TABLE 24.14 Summary of Average Total PCDD/PCDF Concentrations from MSW Combustion in Modern Plants (ng/Nm^3, dry, at 12% CO_2)

	Total PCDD + PCDF
Peekskill, NY, electrostatic precipitator only (1985)	100.25
Wurzburg, FRG, dry scrubber-baghouse (1985)	49.95
Tulsa, OK, electrostatic precipitator only (1986)	34.45
Marion Co., OR, dry scrubber-baghouse (1986)	1.55

Source: From Hahn, J. L., VonDemfange, H. P., Zurlinden, R. A., Stianche, K. F., Seelinger, R. W. (Ogden Martin Systems, Inc.), and Weiand, H., Schetter, G., Spichal, P., and Martin, J. E. 1986. (*Martin GmbH*) *1986 National Waste Processing Conference*, Denver, CO, June 1–4, ASME, New York; NYS Department of Environmental Conservation Bureau of Toxic Air Sampling, Division of Air Resources. 1986. *Preliminary Report on Westchester RESCO RRF*, January 8; Ogden Projects, Inc. 1986. *Environmental Test Report, Walter B. Hall Resource Recovery Facility*, October 20; Ogden Projects, Inc. 1986. *Environmental Test Report, Marion County Solid Waste-to-Energy Facility*, December 5.

TABLE 24.15 Toxic Equivalent emissions by U.S. EPA, Swedish, and California Methods[83]

Facility	Toxic Equivalents ng/Nm^3 at 12% CO_2		
	U.S. EPA	Swedish	California
Peekskill, NY	1.62	3.83	9.73
Tulsa, OK	0.7	1.74	4.75
Wurzburg, FRG	0.37	0.81	2.11
Marion Co., OR	0.11	0.16	0.29
From WHO Workshop; Naples, Italy			
Max. from avg. oper.	25.0	52.78	134.5
Achievable with no acid gas cleaning	0.9	2.2	5.94

Source: From Hahn, J. L., Sussman, D. B. 1986. *Dioxin Emissions from Modern, Mass-Fired, Stoker/Boilers with Advanced Air Pollution Control Equipment*. Fukuoka, Japan, September; NYS Department of Environmental Conservation Bureau of Toxic Air Sampling, Division of Air Resources. 1986. *Preliminary Report on Westchester RESCO RRF*, January 8; Ogden Projects, Inc. 1986. *Environmental Test Report, Walter B. Hall Resource Recovery Facility*, October 20.; Ogden Projects, Inc. 1986. *Environmental Test Report, Marion County Solid· Waste-to-Energy Facility*, December 5.; World Health Organization. 1986. Report on PCDD and PCDF Emissions from Incinerators fox Municipal Sewage, Sludge and Solid Waste—Evaluation of Human Exposure, from WHO Workshop, Naples, Italy, March.

Very few studies report on other organic products in the flue gas. Some data from tests on older plants have been reported for aldehydes and certain other volatile hydrocarbons.[5] Such data are not available for newer plants.

The conventional combustion gas measurements include CO, total hydrocarbons (THCs), CO_2, and H_2O. Both CO and THC have been of interest as potential surrogates for dioxin/furan emissions; however, no strong correlations have been found in previous studies. In fact, very few studies have attempted to determine CO and dioxin/furan emission data for several operating conditions on the same combustor to develop a correlation. On the other hand, several authors have attempted to correlate CO and dioxin/furan data obtained from several different facilities. From such comparisons, it appears that low CO levels (below 100 ppm) are associated with low dioxin/furan emissions.[41] High CO levels of several 100 ppm and even over 1000 ppm have been associated with high dioxin/furan emissions. During poor or upset combustion conditions, CO levels of several thousand ppm have been observed and THC levels have risen from a typical 1–5 ppm to 100 ppm and more. Since one of the measures available to combustor operators to optimize combustion control is to minimize CO production, one would assume from these general correlations that this would also tend to minimize dioxin/furan emissions, along with emissions of other trace chlorinated hydrocarbons. THC is not as useful as CO as an indicator of proper

TABLE 24.16 Organic Emissions (ng/Nm^3)

Chemical Emitted	U.S. Plants	Canadian Pilot Plant	
	Before Particulate Removal	Before Particulate Removal	After Scrubber/Fabric Filter
CB	10,000–500,000	17,000	3000
CP	22,000–80,000	30,000	8000
PCB	—	700	Nondetectable (ND)
PAH	ND to 5,600,000	30,000	130

Source: From Battelle Columbus Labs. 1982. *Characterization of Stack Emissions from Municipal Refuse-to-Energy Systems*. National Technical Information Service, PB87-110482, October.; Hay, D. J., Finkelstein, A., Klicius, R. (Environment Canada), and Marenlette, L. (Flakt Canada, Ltd.). The National Incinerator Testing and Evaluation Program: An Assessment of (A) Two-Stage Incineration (B) Pilot Scale Emission Control, Report EPS 3/UP/2, September 1986. Hay, D.J., Finkelstein, A., Klicius, R. (Environment Canada), and Marenlette, L. (Flakt Canada, Ltd.). 1986. The National Incinerator Testing and Evaluation Program: An Assessment of (A) Two-Stage Incineration (B) Pilot Scale Emission Control, Report EPS 3/UP/2, September.

TABLE 24.17 Lime Addition with Baghouse, Percent Removal of Organics

	Dry System				Wet/Dry System	
	110°C	125°C	140°C	200°C	140°C	140°C Recycle
CB	95	98	98	62	> 99	> 99
PCB	72	> 99	> 99	54	> 99	> 99
PAH	84	82	84	98	> 99	79
CP	97	99	99	56	99	96

Source: From Hay, D. J., Finkelstein, A., Klicius, R. (Environment Canada), and Marenlette, L. (Flakt Canada, Ltd.) 1986. The National Incinerator Testing and Evaluation Program: An Assessment of (A) Two-Stage Incineration (B) Pilot Scale Emission Control, Report EPS 3/UP/2, September.

combustion because of problems in sampling to consistently obtain a representative sample for analysis at the analytical instrument.

Table 24.17 shows operating results achieved using dry and semidry lime injection followed by a baghouse for removal of trace organic pollutants from waste combustor emissions.

24.4.4 Trace Metals

Trace metals are not destroyed during combustion, and the composition of wastes therefore provides, on a statistical basis, the measure of the total inorganic residue. The distribution of trace metals between bottom ash and ash carried over to the air pollution control device is dependent upon the design and operation of the combustor and the composition of the feed. The amount of ash carried up and with the flue gases discharged from a burning refuse bed increases with increasing underfire air rate and with bed agitation. Modular incinerators (described later in "Status of Other Technologies") with low underfire air flow rates tend to have lower particulate emissions than conventional mass-burn units and RDF units for this reason. In addition, the amount of ash carried from the combustion chamber will be influenced by the particle size of the inorganic content of the MSW.

The distribution of trace metals between the different components of refuse has a strong influence on the fate of the trace metals. For example, TiO_2 used as a pigment in paper products, has a particle size of about 0.2 μm and will tend to be carried off by the flue gases passing through the refuse bed, whereas TiO_2 present in glass will remain with the glass in the bottom ash. Up to 20% of the inorganic content of the waste will be entrained from burning refuse beds to form fly ash particles. The remainder will end up in the residue.

Volatile metals and their compounds, usually present in trace amounts in the feed, will vaporize from the refuse and condense in the cooler portions of a furnace either as an ultra fine aerosol (size less than 1 μm) or on the surface of the fly ash, preferentially on the surface of the finer ash particles. A large fraction of certain elements in the feed, such as mercury, will be volatilized.

Since mercury is a very volatile metal, it exists in vapor phase at temperatures as low as 68°F–122°F (20°C–50°C). Several studies have indicated that sufficient cooling of the flue gas (typically below 140°C, based on test results conducted to date) and a highly efficient particulate removal system to remove the particles on which the mercury has been adsorbed[8,10,81] are both required to achieve high mercury removal. High mercury removal has been obtained for the scrubber/fabric filter system, provided that the flue gas is adequately cooled (see Table 24.18).

Test results with carbon injection at two municipal waste combustors demonstrated that the EPA emission guidelines of 80 mg/dscm or 85% reduction in mercury emissions can be achieved with a spray dryer, fabric filter and carbon injection.[6]

Elements such as sodium (Na), lead (Pb), zinc (Zn), and cadmium (Cd), will be distributed between the volatiles and the residue in amounts that depend on the chemical and physical form in which the

TABLE 24.18 Lime Addition with Baghouse Mercury Concentrations (μg/km^3 @ 8% O_2)

Operation Dry System	Inlet	Outlet	%Removal
230°F (110°C)	440	40	90.9
260°F (125°C)	480	23	97.9
285°F (140°C)	320	20	93.8
390°F (>200°C)	450	610	0
Wet–dry system			
140°C	290	10	94.7
140°C ← recycle	350	19	94.7

Source: From Hay, D. J., Finkelstein, A., Klicius, R. (Environment Canada), and Marenlette, L. (Flakt Canada, Ltd.). 1986. *The National Incinerator Testing and Evaluation Program: An Assessment of (A) Two-Stage Incineration (B) Pilot Scale Emission Control*, Report EPS 3/UP/2, September.

elements are present. For example, sodium in glass will be retained in the residue but that in common salt will partially disassociate and be discharged with the emission gases.

Some of the data on metal emissions that are available from tests on resource recovery plants is shown in Table 24.19. Note that the emission of trace metals can be dramatically limited at WTE plants by the use of high-efficiency particulate control devices that are installed on modern facilities.

While sampling for metal emissions is fairly well established, in order to obtain enough sample to analyze at highly controlled sources, samples times are extremely long, sometimes over 8 hours using the U.S. EPA Method 5 sample train (relatively low sample rate). Several studies[30,31,44] of waste combustor emissions in the United States in the 1970s concluded that "municipal incinerators can be major sources of Cd, Zn, Sb and possibly Sn…." This conclusion is based on the relative concentration of these materials in the total suspended particulate catch. However, two of the three plants tested in these studies utilized inefficient air pollution control facilities. Thus, particulate emissions in these plants were relatively high when compared to the German and Japanese plant data in Table 24.19, which is similar to emission data from most modern mass-burn and RDF waste-to-energy plants in the United States.[68] Note also in Table 24.19 that, as the efficiency of particulate control improves, trace metal emissions generally decrease, and in most cases decrease significantly. Even though there is ample evidence from test data[19] to indicate that heavy metals tend to concentrate on the finer particulates, there is also evidence from test results to show that at high particulate removal efficiency (99%±), high trace metal removal (99%±) is achieved. Thus, the conclusion in these studies of waste combustors quoted earlier is not valid for WTE plants utilizing efficient air pollution control devices.

The important operating parameters for such equipment are flue gas temperature and composition, contact time, relative velocity of particles and gas stream, and possible activation of particles. See Table 24.18 and Table 24.20 for operating results achieved using dry and semidry lime injection followed by a baghouse for removal of heavy metal pollutants from waste combustor emissions.

24.5 Performance

Mass burning of MSW is the most highly developed and commercially proven combustion process presently available for reducing the volume of MSW prior to ultimate disposal of residuals on the land, and for extracting energy from the waste.[71] Hundreds of such plants, incorporating various grate systems and boiler concepts, which differ in details of design, construction, and quality of operation, have been built throughout the world since the mid-1960 s. Mass-burn systems are generally furnished with a guaranteed availability of 85%, while in practice availabilities of 90%–95% have been achieved.[68] Availability cannot approach 100% because standard maintenance practice requires periodic shutdowns.

TABLE 24.19 Trace Metal Emissions Test Results

	Japanese Plant Uncontrolled	Braintree Mass. Part. Rem. Eff. 74+%	Particulate Removal Efficiency 99+%				
			German Plants	Japanese Plants	Tulsa, OK	Dry Scrubber, Fab. Filter	
						Marion, Co., OR	Pilot Plant Canada
Arsenic (As) (lb/Ton × 10^{-3})	< 0.4	0.125	0.09	< 0.0016	—	—	0.00033–0.00064
Beryllium (Be) (lb/Ton × 10^{-3})	< 0.3	0.00027	0.002	< 0.0016[a]	0.000025	0.000021	—
Cadmium (Cd) (lb/Ton × 10^{-3})	0.7	1.30	0.25	0.11	—	—	ND-0.006
Chromium (Cr) (lb/Ton × 10^{-3})	16.0	0.34	0.185	0.026	—	—	ND-0.016
Lead (Pb) (lb/Ton × 10^{-3})	17.0	42.4	10.0	0.1	3.5	0.29	ND-0.08
Mercury (Hg) (lb/Ton × 10^{-3})							
on particulates	0.5	0.11	0.067	0.03	—	—	—
vapor phase	0.8	4.38[a]	—	0.90	3.5	2.9	0.16–9.83
Selenium (Se) (lb/Ton × 10^{-3})	< 0.3	—	—	< 0.0016	—	—	—
Particulates (lb/Ton)	25.7	1.3	0.5	0.19	0.13	0.16	< 0.01

[a] Total on particulate and vapor phase.

Source: From Clark, L., *Case History of a 240 Ton Day Resource Recovery Project: Part II*, pp. 235–248, ASME, New York.; Hahn, J. L. and Sofaer, D. S. 1988. *Variability of NO_x Emissions from Modern Mass-Fired Resource Recovery Facilities*, Dallas, TX, June.; Ogden Projects, Inc. 1986. *Environmental Test Report, Walter B. Hall Resource Recovery Facility*, October 20.; Ogden Projects, Inc. 1986. *Environmental Test Report, Marion County Solid Waste-to-Energy Facility*, December 5.; Velzy, C. O. 1985. *Standards and Control of Trace Emissions from Refuse-Fired Facilities*, Madison, WI, November 20–22.

TABLE 24.20 Lime Addition with Baghouse, Metal Concentrations ($\mu g/km^3$ @ 8% O_2)

Metal	Inlet	Outlet	Removal
Zinc	77,000–108,000	5–10	96→99.99
Cadmium	1000–3500	1.0–0.6	96→99.96
Lead	34,000–44,000	1–6	95→99.98
Chromium	1400–3100	0.2–1	>99.92
Nickel	700–2500	0.4–2	>99.81
Arsenic	80–150	0.02–0.07	>99.95
Antimony	800–2200	0.2–0.6	>99.92
Mercury	190–480	10–610	0→90

Source: From Carlsson, K. 1986. *Waste Management & Research*, 4; Hay, D. J., Finkelstein, A., Klicius, R. (Environment Canada), and Marenlette, L. (Flakt Canada, Ltd.). 1986. *The National Incinerator Testing and Evaluation Program: An Assessment of (A) Two-Stage Incineration (B) Pilot Scale Emission Control*, Report EPS 3/UP/2, September.

The newest mass-burn facilities seem capable of achieving high reliability, based on their performance in Europe, Japan, and the United States.

Refuse-derived fuel (RDF) facilities became popular during the 1970s. The early plants were generally designed with the intent to remove and recycle metals, glass, and other marketable materials, with the remaining fraction, RDF, to be burned in an existing boiler as a replacement fuel. Those types of facilities all failed and are no longer in operation. The main reasons for failure were that the recycled materials were highly contaminated with waste and were not marketable, and the boilers were not designed to handle the inconsistent RDF that was being fed to them.

The RDF approach quickly evolved to produce a fuel with a known specification to be burned in a dedicated boiler designed specifically to bum that fuel. The materials which were removed were sold, if possible, or landfilled. The primary difference in philosophy between the two types of RDF plants was that the early ones treated the RDF as the "waste" that contaminated the recovered materials, and the newer generation treated the recovered materials as the "waste" that contaminated the RDF. The newer generation RDF facilities which were designed in this manner have been successful.

Fluidized bed technology has been used outside the U.S. to combust solid waste for several years. One advantage of fluidized bed combustion is that the boiler is more efficient than those in mass-burn or spreader-stoker facilities. Also, fluidized bed combustion produces lower NO_x emissions than other incineration methods. Although lower, these NO_x emissions must still be controlled with additional air pollution control equipment, as with other combustion facilities. Fluidized bed combustion also has the advantage of being able to add limestone with the sand in the bed to assist in acid gas removal. However, a scrubber is still needed to reduce emissions to permitted levels.

The major disadvantage to a fluidized bed facility is that it does not have a long-term proven track record in the U.S. Also, the size of the units are small when compared to the size needed for typical U.S. waste-to-energy facilities.

24.6 Costs

It is extremely difficult to obtain accurate, consistent, and comparable WTE plant construction cost data from which to develop information which might be useful in predicting a planned new plant's construction cost during the study stage of a project. However, a 1988 study[56] has developed such data (appropriate for the time frame of mid-1987), which is confirmed in general by this author's personal experience. This study indicates that for the upper 90% confidence limit for the smallest facility, and the largest facility, the construction costs would range as indicated below:

1. A small modular combustion unit with a waste heat boiler and a capacity of less than 250 TPD—$68,000 and $40,000 per ton of daily MSW processing capability. (In most instances, such plants

don't incorporate the same degree of equipment redundancy, and/or the same quality of equipment as the larger plants.)

2. A small refractory wall furnace with waste heat boiler and dry scrubbers of between 200 and 500 TPD capacity—$90,000 and $70,000 per ton of daily MSW processing capability.
3. A small, field erected, water wall congeneration or electric generation facility with dry scrubbers of between 500 and 1,500 TPD capacity—$112,000 and $85,000 per ton of daily MSW processing capability.
4. A large, field erected, water wall congeneration or electric generation facility with a dry scrubber between 2,000 and 3,000 TPD capacity—$129,000 and $112,000 per ton of daily MSW processing capacity.

In this study,[56] the construction costs were said to include the vendor quote for construction plus contingency, utility interconnection expenses, and any identified allowances clearly associated with the construction price. All other costs, such as land acquisition, interest during construction, development costs, and management fees were not considered or included, where known, due to their highly project specific nature.

The following specific observations were made by the authors at the conclusion of this study.[56]

1. Capital construction price decreases with increasing size within size ranges and increases with a higher-value energy product.
2. Price is also affected by the construction, procurement, and air pollution control methods employed.
3. Refuse-derived fuel and mass burning water wall facilities are competitively priced with each other.

The effect of plant capacity on capital costs or mass-burn plants is shown in Figure 24.8. Capital costs for other types of plants are similar.

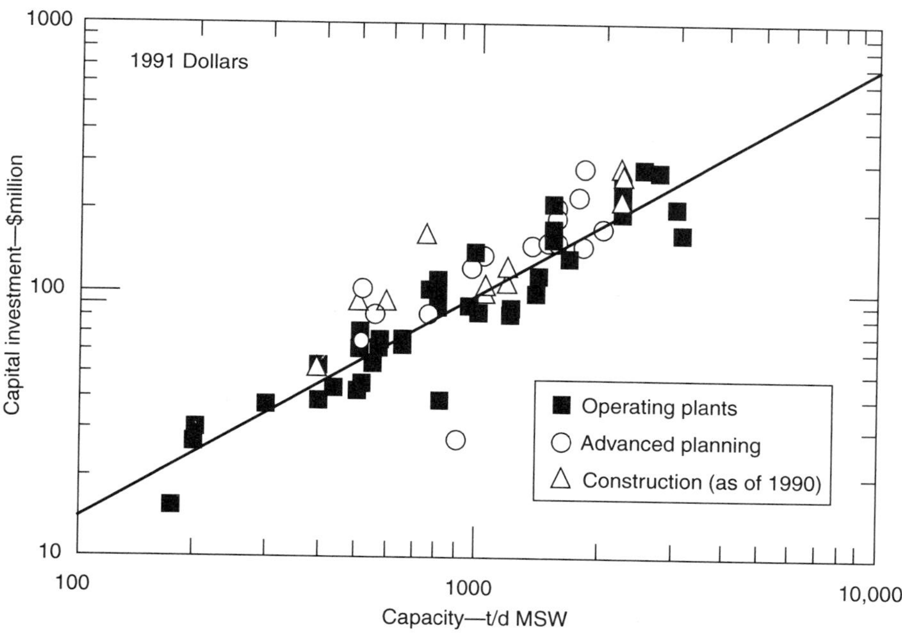

FIGURE 24.8 Effect of plant capacity on capital costs for mass-burn facilities (electricity only), excluding costs associated with collection (e.g., trucks).

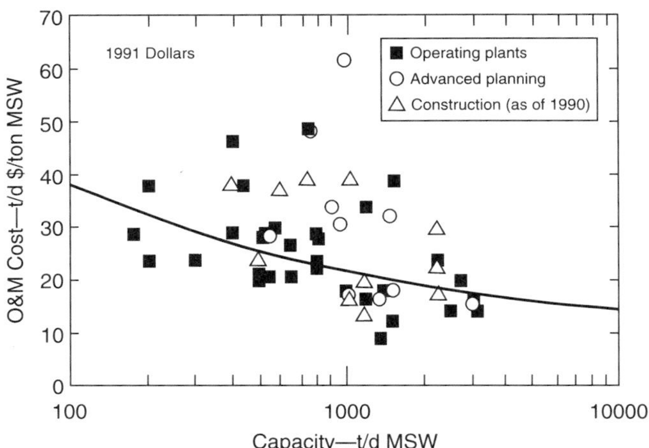

FIGURE 24.9 Effect of plant capacity on O&M costs for mass-burn facilities (electricity producing only), excluding costs associated with collection (e.g., trucks).

With respect to operating costs and/or tipping fees, information is even more difficult to obtain from which to develop costs for planning purposes. Costs cited in the literature from 1989 through 1994 range from \$40 to \$80 per ton of daily rated MSW processing capability.[20,45,48,78,79] Tipping fees on Long Island, which has generally high labor rates, high power costs, and very long hauls for residue disposal, have been noted to range up to \$110 per ton of daily rated MSW processing capability.[28] Plants in other parts of the country where plant operating cost elements are significantly lower have been found to have tipping fees closer to \$40 per ton.[9] Thus, tipping fees for a specific facility would have to be developed based on cost factors for that specific plant.

The effect of plant capacity on O&M costs for mass-burn plants is shown in Figure 24.9. Information is so limited on other types of plants, and the costs are so dependent on local conditions that we do not feel that curves developed for other types of plants would be useful.

24.7 Status of Other Technologies

Several other technologies have been used to a small extent to burn MSW and beneficially use the energy produced in the combustion process. Their use in the future depends on numerous factors, perhaps the most important of which is full-scale demonstration of successful, reliable operation, after which total operational costs are shown to be competitive with mass-burn and/or RDF combustion costs.

24.7.1 Modular Systems

Modular systems, generally utilized in smaller plants, are assemblies of factory-prefabricated major components joined together in the field to form a total operational system. They have been built in individual unit sizes up to just over 100 tons per day capacity, combined into plants of just over 400 tons per day capacity. Modular systems are similar to mass-burn systems in that they combust unprocessed MSW, but they feature two combustion chambers, and the MSW is charged into the system with a hydraulic ram and combustion takes place on a series of stationary hearths. MSW is pushed from one hearth to the next by hydraulic rams. Two types of modular systems have been built and operated: starved air and excess air.

The primary chamber of a starved air modular system is usually operated in a slightly oxygen-deficient ("starved air") environment. The volatile portion of the MSW is vaporized in this chamber and the resulting gases flow into the secondary chamber. The secondary chamber operates in an excess air condition, and combustion of the gases driven off the MSW is essentially completed in this chamber. An excess air modular system operates in a manner similar to a field erected boiler system, with excess air injected into the primary chamber.

One advantage of these units, as indicated in the section on costs, is low cost. Another advantage is that factory prefabrication of major system components can result in shortening of the field construction time. One disadvantage of the two-chamber modular system is that waste burn out in the residue is not always complete, which increases ash quantities and reduces the efficiency of energy recovery.[70] Energy recovery efficiency is also reduced due to generally higher "excess air" levels carried in these units. Also, combustion control is generally less effective in this type of unit, increasing the possibility of discharge of trace organic emissions. As pointed out earlier, these types of units generally utilize a lower quality of equipment and include less redundancy than larger mass-fired water wall and RDF WTE plants. Modular plants are responsible for about 2% of the total MSW burned at this time in the United States.

24.7.2 Fluidized Beds

Fluidized bed combustion (FBC) differs from mass-burn and RDF combustion in that the fuel is burned in "fluid suspension"—entrained along with particles of sand in an upward flow of turbulent air at a temperature controlled to 1500°F–1600°F (816°C–971°C). To date, it has been used primarily to burn sewage sludge, industrial waste, and coal and has been used to combust RDF in one facility in the U.S. Fluidized bed combustion of MSW is more commonly used in some European countries.

"Bubbling" FBC designs retain the material near the bottom of the furnace, while "circulating" designs allow bed material to move upward and then be returned near the bottom of the bed for further combustion. The reason for the interest in this combustion technique to burn RDF is the potential for these designs to provide more consistent combustion because of the extreme turbulence and the proximity of the RDF waste particles to the hot sand particles.[47] Such systems also require lower combustion temperatures than mass-burn and current RDF systems.

24.7.3 Pyrolysis and Gasification

Pyrolysis is the chemical decomposition of a substance by heat in the absence of oxygen. It generally occurs at relatively low temperatures (900°F–1100°F, compared with 1,800°F for mass-burn). The heterogeneous nature of MSW makes pyrolysis reactions complex and difficult to control. Besides producing larger quantities of solid residues that must be managed for ultimate disposal, pyrolysis produces liquid tar and/or gases that are potentially marketable energy forms. The quality of the fuel products depends on the material fed into the reactor (e.g., moisture, ash, cellulose, trace constituent content) and operating conditions (e.g., temperature and particle size).[68]

Gasification is similar to pyrolysis in that it is the chemical decomposition of the substance by heat in the absence of oxygen. However, gasification occurs at temperatures of approximately 2200°F (1200°C). The reaction produces a synthetic gas with a heat content of approximately 250 BTU/cf. The approximate composition by volume on a dry basis is 25%–42% carbon monoxide, 25%–42% hydrogen, 10%–25% carbon dioxide, and 3%–4% nitrogen and other constituents. The synthetic gas is then cooled rapidly to reduce formation of dioxins and cleaned.[62]

Several prototype pyrolysis facilities were built in the 1970s with grants from EPA. These facilities were unable to produce quality fuels in substantial quantities. No one in the United States has yet successfully developed and applied the pyrolysis or gasification technology to MSW combustion. However, the use of pyrolysis and gasification for MSW management still attracts attention in other countries. Additional reading for pyrolysis and gasification technologies is available.[84,85]

24.8 Future Issues and Trends

It has been demonstrated by actual experience that modern mass-burn and RDF-fired MSW WTE plants can be designed and operated with reasonable assurance of continuous service and without adversely affecting nearby neighborhood property values. Allegations that WTE plant sites are situated adjacent to neighborhoods of low-income, disadvantaged, or minority populations ignore the specific technical siting criteria outlined earlier (i.e., adjacent to major highways, low land cost, industrial type area, etc.) which are generally followed in siting such facilities. Such areas frequently are closer to low-income neighborhoods than to middle- and higher-income neighborhoods.[68]

In 1994 the Supreme Court found that a local community could not force the MSW from that community to be taken to a specific facility such as a WTE plant.[26] This court decision was a major blow to the WTE industry bringing most planning and construction of new facilities to a halt in the mid-1990s. Many communities, when considering solid waste disposal options, have opted for low cost landfill disposal because of concern over impacts of higher cost WTE alternatives on taxes. At some point in the not too distant future, as the current landfills are rapidly filled, WTE technology will have to be utilized to solve the solid waste disposal problem. Some signs of this new interest in WTE technology are already occurring.

Another issue facing WTE plants is the uncertain future of regulatory requirements, both from the standpoint of legislation and from that of the regulatory agencies. In the past, legislation has been passed by Congress calling for Best Available Control Technology (BACT), then Lowest Achievable Emission Requirements (LAER), and then, most recently, Maximum Achievable Control Technology (MACT). The impact of this legislation, each calling for significant reductions in allowable emissions (absent any indication of the existence of a significant public health problem or benefit), has been to require extensive retrofits of existing plants and addition of equipment to proposed new plants, all at substantial expense without proven benefit and, in many cases, without prior proof of operational viability. Most facilities have opted to upgrade the air pollution control equipment and continue to operate.

Several positive actions are occurring in the field. Thus, most project developers have recognized the desirability of implementing a proactive program early in the project planning process to involve the public, particularly those in the vicinity of the proposed facility.[60] Also, the potential for materials recycling, which had been overenthusiastically embraced a number of years ago (state recycling goals as high as 70%, with some local communities projecting that their entire quantity of MSW could be managed through recycling), is gradually being recognized.[46] Franklin Associates[25] projects an increase in the recycling rate of from 22% in 1993 to 30% by the year 2000. Much of this increase in recycling is to come from increases in recovery of paper materials and diversion of yard wastes to composting. The impact of these changes in waste composition on energy available at WTE plants will be minimal, with the reduction in moisture content due to diversion of yard wastes being a positive factor.

The need to generate electric energy and safely manage the MSW generated by modern civilization, particularly in the vicinity of major metropolitan areas, together with the proven performance of modem WTE plants, indicate that this technology will be utilized to dispose of a portion of this country's solid waste and provide electricity.

References

1. Albina, D., Millrath, K., and Themelis, N., "Effects of Feed Composition on Boiler Corrosion in Waste-to-Energy Plants." *Proceedings of the 12th Annual North American waste to Energy Conference.* ASME, New York, pp. 99–109, 2004.
2. Public Administration Service, *Municipal Refuse Disposal. 3rd Ed.,* American Public Works Association, Chicago, IL, 1970.
3. "Fuels: Coal and Coke," *Annual Book of ASTM Standards*, Section 5, Vol. 05.05, American Society for Testing and Materials (ASTM), Philadelphia, PA.
4. Barnes, D. "Rump Session." *Chemosphere*, 14 (6–7), 987–989, 1985.

5. Battelle Columbus Labs. "Characterization of Stack Emissions from Municipal Refuse-to-Energy Systems." National Technical Information Service, PB87-110482, October, 1981.
6. Bauer, J. P., and Wofford, J. "Compliance with the New Emissions Guidelines for Existing Municipal Waste Combustion Facilities." *Proceedings 1996 National Waste Processing Conference*, ASME, New York, pp. 9–18, 1996.
7. Beltz, P. R., Engdahl, R. B., Dartoy, J., "Evaluation of European Refuse-Fired Energy Systems Design Practices, Summary, Conclusions and Inventory," prepared for U.S. EPA by Battelle Laboratories, Columbus, OH, 1979.
8. Bergstrom, J. G. T., "Mercury Behavior in Flue Gases," *Waste Management and Research*, 4(1), 1986.
9. Bilmes, J. S., "Impact of the Recession and Recycling on Solid Waste Processing Facilities in New England," *Proceedings 1992 National Waste Processing Conference*, ASME, New York, pp. 351–360, May, 1992.
10. Braun, H., Metzger, M., Vogg, H., *Zur Problematik der Quecksilber-Abscheidung aus Rauchgasen von Mullverbrennungsanlagen*. Vol.1, Teil 2, 1985.
11. Bryers, R. W. Ed., *Ash Deposits and Corrosion due to Impurities in Combustion Gases*, Hemisphere Publishing Corp., New York, 1978.
12. Carlsson, K., "Heavy metals from energy-from-waste plants—comparison of gas cleaning systems," *Waste Management & Research*, 4(1), 1986.
13. Carrotti, A. A. and Smith, R. A., *Gaseous Emissions from Municipal Incinerators*. USEPA Publication No. SW-18C, 1974.
14. Clark, L., "Case History of a 240 Ton Day Resource Recovery Project: Part II," *Proceedings of the 1996 National Waste Processing Conference*, ASME, New York, pp. 235–248, 1996.
15. Cooper Engineers, Inc., *Air Emissions Tests of a Deutsche Babcock Anlagen*, Dry Scrubber System at the Munich North Refuse-Fired Power Plant, February, 1985.
16. Cooper Engineers, Inc., *Air Emissions and Performance Testing of a Dry Scrubber* (Quench Reactor). Dry Venturi and Fabric Filter System Operating on Flue Gas From Combustion of Municipal Solid Waste in Japan, West County Agency of Contra Costa County Waste Co-Disposal/Energy Recovery Project, May, 1995.
17. Lide, D. R., *CRC Handbook of Chemistry and Physics. 71st Ed.*, CRC Press, Boca Raton, FL, 1990.
18. Criss, G. H. and Olsen, R. A., "The chemistry of incinerator slags and their compatibility with fireclay and high alumina refractories," *Proceedings 1968 National Incinerator Conference*, New York, ASME, pp. 75–88, 1968.
19. Dannecker, W., "Schadstoffmessungen bei mullverbrennungsanlagen." *VGB Krazftwekstechnick*, 63 (3), 1983.
20. Davis, C. F., "Annual Snapshot of Six Large Scale RDF Projects," *Proceedings 1992 National Waste Processing Conference*, ASME, New York, pp. 75–88, May, 1992.
21. Domalski, E. S., Ledford A. E., Jr., Bruce, S. S., and Churney, K. L., "The Chlorine Content of Municipal Solid Waste from Baltimore County, Maryland, and Brooklyn, NY," *Proceedings 1986 National Waste Processing Conference*, ASME, New York, pp. 435–448, June, 1986.
22. Eighmy, T. T., Collins, M. R., DiPietro, J. V., and Guay, M. A., "Factors Affecting Inorganic Leaching Phenomena from Incineration Residues," paper presented at Conference on Municipal Solid Waste Technology. San Diego, CA, 1989.
23. Franklin Associates, Ltd., Characterization of Municipal Solid Waste in the United States, 1960–2000, Report prepared for U.S. EPA, NTIS No. PB87-17323, Prairie Village, KS, July 25, 1986.
24. Franklin Associates, Ltd., Characterization of Municipal Solid Waste in the United States, 1960 to 2000 (Update 1988), Report prepared for U.S. EPA, Office of Solid Waste and Emergency Response, Prairie Village, KS, March 30, 1988.
25. Franklin Associates, Ltd., Characterization of Municipal Solid Waste in the United States. (1994 Update), Report prepared for U.S. EPA Office of Solid Waste and Emergency Response, Prairie Village, KS, November, 1994.

26. Franklin, M. A., "Solid Waste Stream Characteristics," *Handbook of Solid Waste Management, 2nd Ed.*, Chapter 5, McGraw Hill, 2002.
27. Franklin, M. A., Personal communication, September, 2004.
28. Freeman, D., "Waste-to-Energy Fights the Fires of Opposition," *World Wastes*, 37(1), 24, 1994.
29. Fryling, G. R. Ed., *Combustion Engineering*, Combustion Engineering, Inc., New York, 1966.
30. Greenberg, P. R., Zoller, W. H., and Gordon, G. E., "Composition and size distributions of particles released in refuse incineration," *Environmental Science and Technology*, ACS, 12(5), 566–573, 1978.
31. Greenberg, R. R., Gordon, G. E., Zoller, W. H., Jacko, R. B., Neuendorf, D. W., and Yost, K. J., "Composition of Particles from the Nicosia Municipal Incinerator," *Environmental Science and Technology*, ACS, 12(11), 1329–1332, 1978.
32. Hahn, J. L., VonDemfange, H. P., Zurlinden, R. A., Stianche, K. F., and Seelinger, R. W. (Ogden Martin Systems, Inc.), and Weiand, H., Schetter, G., Spichal, P., and Martin, J. E. (Martin GmbH), *Proceedings of the 1986*. National Waste Processing Conference, ASME, New York, June, 1986.
33. Hahn, J. L., Sussman, D. B., "Dioxin Emissions from Modern, Mass-Fired, Stoker/Boilers with Advanced Air Pollution Control Equipment." Presented at Dioxin'86, Fukuoka, Japan, September 1986.
34. Hahn, J. L. and Sofaer, D. S., "Variability of NO_x Emissions from Modern Mass-Fired Resource Recovery Facilities." Paper No. 88-21.7 presented at 81st Annual Meeting of Air Pollution Control Association, Dallas, TX, June 1988.
35. Hasselriis, F., "The Environmental and Health Impact of Waste Combustion—the Rush to Judgment Versus Getting the Facts First," *Proceedings 1992 National Waste Processing Conference*, ASME, New York, pp. 361–369, May, 1992.
36. Hay, D. J., Finkelstein, A., Klicius, R. (Environment Canada), and Marenlette, L., (Flakt Canada, Ltd.), "The National Incinerator Testing and Evaluation Program: An Assessment of (A) Two-Stage Incineration (B) Pilot Scale Emission Control," Report EPS 3/UP/2, September, 1986.
37. Hecklinger, R. S., "The Relative Value of Energy Derived from Municipal Refuse." *Journal of Energy Resources Technology*, 101, 251–259, 1979.
38. Hecklinger, R. S., "Combustion." *The Engineering Handbook* CRC Press, Inc., Boca Raton, FL, 1996.
39. Junk, G. A. and Ford, C. S., "A Review of Organic Emissions from Selected Combustion Processes," *Chemosphere*, 9, 187–230, 1980.
40. Kaiser, E. R., Zeit, C. D., and McCaffery, J. B., "Municipal incinerator refuse and residue." *Proceedings 1968 National Incinerator Conference*, ASME, New York, pp. 142–153, 1968.
41. Kilgroe, J. D., Lanier, W. S., and Von Alten, T. R., "Development of Good Combustion Practice for Municipal Waste Combustors." *Proceedings 1992 National Waste Processing Conference*, ASME, New York, pp. 145–156, May, 1992.
42. Kiser, J. V. L. and Zannes, M., "The 2000 L.W.S.A.Directory of Waste-to-Energy Plants." Integrated Waste Services Association, 2002.
43. Lai, G., "Corrosion Mechanisms and Alloy Performance in Waste-to-Energy Boiler Combustion Environments." *Proceedings of the 12th Annual North American Waste to Energy Conference*, ASME, New York, pp. 91–98, 2004.
44. Law, S. L. and Gordon, G. E., "Sources of metals in municipal incinerator emissions." *Environmental Science and Technology*, ACS, 13(4), 432–438, 1979.
45. Leavitt, C., "Calculating the Costs of Waste Management," *World Wastes*, 37(4), 42, 1994.
46. Michaels, A., "Solid Waste Forum," *Public Works*, p. 72, March, 1995.
47. Minott, D. H., "Operating Principles and Environmental Performance of Fluid-Bed Energy Recovery Facilities," Paper No. 88-21.9 presented at 81st Annual Meeting of Air Pollution Control Association, June 1989.
48. Murdoch, J. D. and Gay, J. L., "Material recovery with incineration. Monmouth County, N. J.," *Proceedings, 27th Annual International Solid Waste Exposition*, SWANA, Silver Springs, MD (Pub. #GR-0028), p. 329, August, 1989.

49. NYS Department of Environmental Conservation, Bureau of Toxic Air Sampling, Division of Air Resources, "Preliminary Report on Westchester RESCO RRF," January 8, 1986.
50. Norton, F. H., "Refractories," *Marks Standard Handbook for Mechanical Engineers, 8th Ed.*, McGraw-Hill, New York, pp. 6.172–6.173, 1978.
51. Ogden Projects, Inc., Environmental Test Report, Walter B. Hall Resource Recovery Facility, October 20, 1986.
52. Ogden Projects, Inc., Environmental Test Report, Marion County Solid Waste-to-Energy Facility, December 5, 1986.
53. O'Malley, W. R., "Special factors involved in specifying incinerator cranes." *Proceedings 1968 National Incinerator Conference*, ASME, New York, pp. 114–123, 1968.
54. Penner, S. S. and Richards, M. B., Estimates of Growth Rates for Municipal-Waste Incineration and Environmental Controls Costs for Coal Utilization in the U.S., *Energy: The International Journal*, 14, 961, 1989.
55. Richard, J. J. and Junk, G. A., "Polychlorinated Biphenyls in Effluents from Combustion of Coal/Refuse," *Environmental Science and Technology*, ACS, 15(9), 1095–1100, 1981.
56. Rigo, H. G. and Conley, A. D., "Waste-to-Energy Facility Capital Costs," *Proceedings 1988 National Waste Processing Conference*, ASME, New York, pp. 23–28, May, 1988.
57. Rogus, C. A., "An Appraisal of Refuse Incineration in Western Europe." *Proceedings 1966 National Incinerator Conference*, ASME, New York, pp. 114–123, 1966.
58. Sawell, S. E., Bridle, T. R., and Constable, T. W., "Leachability of Organic and Inorganic Contaminants in Ashes from Lime-Based Air Pollution Control Devices on a Municipal Waste Incinerator." Paper No. 87-94A presented at 80th Annual Meeting of Air Pollution Control Association, New York, June, 1987.
59. Sawell, S. E. and Constable, T. W., "NITEP Phase IIB: Assessment of Contaminant Leachability from the Residues of a Mass Burning Incinerator," Vol. VI of National Incinerator Testing and Evaluation Program, The Combustion Characterization of Mass Burning Incinerator Technology, Environment Canada, Toronto, Canada, August, 1988.
60. Scarlett, T., "WTE Report Focuses on Socioeconomics," *World Wastes*, 37(5), 14, 1994.
61. Scherrer, R. and Oberlaender, B., "Refuse Pit Storage Requirements," *Proceedings 1990 National Waste Processing Conference*, ASME, New York, pp. 135–142, June, 1990.
62. Schilli, J., "The Fourth Dimension for Waste Management in the United States: Thermoselect Gasification Technology and the Hydrogen Energy Economy." *Proceedings of the 12th Annual North American Waste to Energy Conference*, ASME, New York, pp. 251–258, 2004.
63. Seeker, W. R., Lanier, W. S., and Heap, M. P., *Municipal Waste Combustion Study: Combustion Control of Organic Emissions*, EPA, Research Triangle Park, NC, January, 1987.
64. Stultz, S. C. and Kitto, J. B., Eds., Steam: Its Generation and Use, *40th Ed.*, The Babcock and Wilcox Co., Barberton, OH, 1992.
65. Suess, M. J. et al., Solid Waste Management Selected Topics. WHO Regional Office for Europe, Copenhagen, Denmark, 1985.
66. Timm, C. M., *Sampling Survey Related to Possible Emission of Polychlorinated Biphenyls from the Incineration of Domestic Refuse*, NTIS Publication No. PB-251–285, 1975.
67. U.S. Congress, Office of Technology Assessment, "Chapter 3, Generation and Composition of MSW," *Facing Amercia's Trash, What Next for Municipal Solid Waste*, OTA-A-424, U.S. Government Prinitng Office, Washington, DC, October, 1989.
68. U.S. Congress, Office to Technology Assessment, "Chapter 6, Incineration," Facing America's Trash, What Next for Municipal Solid Waste. OTA-A-424, U.S. Government Prinitng Office, Washington, DC, October, 1989.
69. U.S. Environmental Protection Agency, *Fourth Report to Congress, Resources Recovery and Waste Reduction*, Report No. SW-600, Washington, DC, 1977.
70. U.S. Environmental Protection Agency, *Municipal Waste Combustion Study, Report to Congress*, EPA/530-SW-87-021a, Washington, DC, 1987.

71. U.S. Environmental Protection Agency, *Characterization of Municipal Waste Combustor Ashes and Leachates from Municipal Solid Waste Landfills, Monofills and Codisposal Sites*, prepared by NUS Corporation for Office of Solid Waste and Emergency Response, EPA/530-SW-87-028A, Washington, DC, 1987.
72. U.S. Environmental Protection Agency, *Municipal Waste Combustion Multipollutant Study, Characterization Emission Test Report, Office of Air Quality Planning and Standards*, EMB Report No. 87-MIN-04, Research Triangle Park, NC, September, 1988.
73. U. S. Environmental Protection Agency, Standards for Performance for Stationary Sources and Emission Guidelines for Existing Sources—Municipal Waste Combustors, 40 CFR, Part 60, Subpart Cb, December 19, 1995.
74. Velzy, C. O., "The Enigma of Incinerator Design," *Incinerator and Solid Waste Technology*, ASME, New York, pp. 65–74, 1975.
75. Velzy, C. O., 30 Years of Refuse Fired Boiler Experience, presented at Engineering Foundation Conference, Franklin Pierce College, July 17,1978.
76. Velzy, C. O., "Trace Emissions in Resource Recovery—Problems, Issues and Possible Control Techniques." *Proceedings of the DOE-ANL Workshop; Energy from Municipal Waste: State-of-the-Art and Emerging Technologies*, Argonne National Laboratory Report ANL/CNSV-TM-137, February, 1984.
77. Velzy, C. O., "Standards and Control of Trace Emissions from Refuse-Fired Facilities." *Municipal Solid Waste as a Utility Fuel, EPRI Conference Proceedings*, November 20–22, 1985.
78. Velzy, C. O., "U.S. Experience in Combustion of Municipal Solid Waste," presented at APCA Specialty Conference on Regulatory Approaches for Control of Air Pollutants, February 20, 1987.
79. Velzy, C. O., "Considerations in Planning for an Energy-from-Waste Incinerator," presented at New Techniques and Practical Advice for Problems in Municipal Waste Management, Toronto, Ontario, Canada, April 1–2, 1987.
80. Visalli, J. R., "A Comparison of Dioxin, Furan and Combustion Gas Data from Test Programs at Three MSW Incinerators." *Journal of the Air Pollution Control Association*, 37 (12), 1451–1463, 1987.
81. Vogg, H., Braun, H., Metzger, M., and Schneider, J., "The Specific Role of Cadmium and Mercury in Municipal Solid Waste Incineration," *Waste Management and Research*, 4(1), 1986.
82. Vogg, H., Metzger, M. and Stieglitz, L., Recent Findings on the Formation and Decomposition of PCDD/PCDF in Solid Municipal Waste Incineration, ISWA/WHO, Specialized Seminar, Copenhagen, Denmark, January, 1987.
83. World Health Organization, Report on PCDD and PCDF Emissions from Incinerators for Municipal Sewage, Sludge and Solid Waste—Evaluation of Human Exposure, from WHO Workshop, Naples, Italy, March, 1986.
84. Malkow, T., "Novel and Innovative Pyrolysis and Gasification Technologies for Energy Efficient and Environmentally Sound MSW Disposal," *Waste Management*, 24, 53–79, 2004.
85. Bridgewater, A., Ed., Pyrolysis and Gasification of Biomass and Waste, CPL Press, Newbury, UK, 2003.

25

Ocean Energy Technology

Desikan Bharathan
National Renewable Energy Laboratory
Federica Zangrando
National Renewable Energy Laboratory

25.1 Ocean Thermal Energy Conversion 25-1
25.2 Tidal Power 25-2
25.3 Wave Power 25-2
25.4 Concluding Remarks 25-3
Defining Terms 25-3
References 25-3
For Further Information 25-4

The ocean contains a vast renewable energy potential in its waves and tides; in the temperature difference between cold, deep waters and warm surface waters; and in the salinity differences at river mouths (SERI 1990; Cavanagh, Clarke, and Price 1993; WEC 1993). Waves offer a power source for which numerous systems have been conceived. Tides are a result of the gravity of the sun, the moon, and the rotation of the Earth working together. The ocean also acts as a gigantic solar collector, capturing the energy of the sun in its surface water as heat. The temperature difference between warm surface waters and cold water from the ocean depths provides a potential source of energy. Other sources of ocean energy include ocean currents, salinity gradients, and ocean-grown biomass.

The following subsections briefly describe the status and potential of the various ocean energy technologies, with emphasis placed on those with a near-term applicability.

25.1 Ocean Thermal Energy Conversion

Ocean thermal energy conversion (OTEC) technology is based on the principle that energy can be extracted from two reservoirs at different temperatures (SERI 1989). A temperature difference of as little as 20°C can yield usable energy. Such temperature differences prevail between ocean waters at the surface and at depths up to 1000 m in tropical and subtropical latitudes between 24° north and south of the equator. Here, surface water temperatures range from 22 to 29°C, while temperatures at a depth of 1000 m range from 4 to 6°C. This constitutes a vast, renewable resource, estimated at 10^{13} W, for potential base load power generation.

Past research has been concentrated on two OTEC power cycles, namely, **closed cycle** and **open cycle**, for converting this thermal energy to electrical energy (Avery and Wu 1994). Both cycles have been demonstrated, but no commercial system is in operation. In a closed-cycle system, warm seawater is used to vaporize a working fluid such as ammonia flowing through a heat exchanger (evaporator). The vapor expands at moderate pressures and turns a turbine. The vapor is condensed in a condenser using cold seawater pumped from the ocean depths and the condensed fluid is pumped back to the evaporator to repeat the cycle. The working fluid remains in a closed system and is continuously circulated.

In an open-cycle system, the warm seawater is "flash" evaporated in a vacuum chamber to make steam at an absolute pressure of about 2.4 kPa. This steam expands through a low-pressure turbine coupled to a generator to produce electricity. The steam is then condensed by using cold seawater. If a surface condenser is used, condensed steam is separated as desalinated water.

Effluent from a closed-cycle or an open-cycle system can be further processed to make more desalinated water through evaporator/condenser systems in subsequent stages. For combined production of power and water, these systems can be competitive with conventional systems in several coastal markets.

25.2 Tidal Power

The energy from tides is derived from the kinetic energy of water moving from a higher to a lower elevation, similar to hydroelectric plants. High tide can provide the potential energy for seawater to flow into an enclosed basin or estuary and then be discharged at low tide (Ryan 1980). Electricity is produced from the gravity-driven inflow or outflow or both through a turbogenerator. The tidal resource is variable but predictable, and there are no technical barriers for deployment of this technology. Because costs are strongly driven by the civil works required to dam the reservoir, only a few sites around the world have the proper conditions of tides and landscape to use this technology. Recent developments in marine and offshore construction have helped reduce construction time and cost; however, the economics of tidal power production remains uncompetitive to conventional energy systems.

The highest tides in the world can reach above 17 m, as in the Bay of Fundy between Maine and Nova Scotia, where it is projected that up to 10,000 MW could be produced by tidal systems in this bay alone. A minimum tidal range (difference between mean high and low tides) of 5 m is required for plants using conventional hydroelectric equipment. More recently, low-head hydroelectric power equipment has proved adaptable to tidal power and new systems for 2-m ranges have been proposed.

A few tidal power stations are operating in France, the former U.S.S.R., China, and Canada. The largest and longest-operating plant is the 240-MW tidal power station on the Rance River estuary in northern France (Banal and Bichon 1981), which has operated with 95% availability since 1968. The 400-kW tidal plant in Kislaya Bay on the Barents Sea in the former U.S.S.R. has been operating since 1968; at this favorable site, only a 50-m-wide dam was needed to close the reservoir. The 20-MW Canadian plant at Annapolis on the Bay of Fundy has operated reliably since 1984. A number of small turbine generator plants of up to 4 MW are also installed on the China coastline.

25.3 Wave Power

Waves contain significant power that can be harnessed by shore-mounted or offshore systems. Offshore installations will have larger incident power on the device but require more costly installations. A myriad of wave-energy converter concepts have been devised, transforming wave energy into other forms of mechanical (rotary, oscillating, or relative motion), potential, or pneumatic energy, and ultimately into electricity; very few have been tested at sea.

The power per unit frontal length of the wave is proportional to wave height squared and to wave period, with their representative values on the order of 2 m and 10 s. The strong dependence on wave height makes the resource highly variable, even on a seasonal and a yearly average basis. The northeastern Pacific and Atlantic coasts have average yearly incident wave power of about 50 kW/m, while near the tip of South America the average power can reach 100 kW/m. Japan, on the other hand, receives an average of 15 kW/m. Waves during storms can reach 200 kW/m, but large waves are unsafe for operation. Because of their severity, they impose severe constraints and increase system costs. Overall, the amount of power that could be harvested from waves breaking against world coastlines is estimated to be on the order of the current global consumption of energy. However, total installed capacity amounts to less than 1 MW worldwide.

A commonly deployed device is the oscillating water column (OWC), which has so far been mounted on shore but is also proposed for floating plants. It consists of an air chamber in contact with the sea so that the water column in the chamber oscillates with the waves. This motion makes the air flow in and out of the chamber, turning a turbine. A Wells turbine uses symmetrical airfoil blades in a simple rotor to extract power from the airflow in both directions. Flywheels are often used to smooth out fluctuating delivered energy by the waves.

Two of the largest wave-energy power plants were built at Toftestallen, near Bergen, Norway. A Norwegian company, Norwave A.S., built a 350-kW tapered channel (Tapchan) device in 1984, which survived a severe storm in 1989 (the 500 kW multiresonant OWC plant built by Kvaerner Brug A.S. did not). The channel takes advantage of the rocky coastline to funnel waves through a 60-m-wide opening into a coastal reservoir of 5500 m^2, while maintaining civil engineering costs to a minimum. Wave height increases as the channel narrows over its 60-m length, and the rising waves spill over the 3-m-high channel walls, filling the reservoir. Continuous wave action maintains the reservoir level at a relatively constant elevation above sea level, providing potential energy for a low-head hydroelectric Kaplan turbogenerator. Estimates by Norwave to rebuild an identical plant at this site suggested capital costs of $3500/kW installed, and energy costs of 8¢/kWh, at a plant capacity factor of 25% (ASCE 1992). In recent efforts, the National Institute of Ocean Technology, India, has installed a 150-kW wave-energy conversion device in the southern tip of India.

25.4 Concluding Remarks

Among the many ocean energy prospects, OTEC, tides, and tapered channel wave-energy converters offer the most near-term potential and possess applicability for a large variety of sites. To realize their potential, additional research and development is required. Ocean resources can be most suitable to address humanity's future needs for water and hydrogen for fuel cells.

Defining Terms

Ocean thermal energy conversion (OTEC): A system that utilizes the temperature difference between the seawater at the surface and at depths.

Closed-cycle OTEC: Uses a working fluid in a closed cycle.

Open-cycle OTEC: Uses steam flashed from the warm seawater as the working fluid that is condensed and exhausted.

References

ASCE, 1992. In *Ocean Energy Recovery, the State of the Art*, R. J. Seymour, ed., American Society of Civil Engineers, New York.

Avery, W. H. and Wu, C. 1994. *Renewable Energy from the Ocean, a Guide to OTEC*. Oxford University Press, New York.

Banal, M. and Bichon, A. 1981. Tidal energy in France: the Rance Estuary tidal power station—some results after 15 years of operation, Paper K3, 2nd Symposium on Wave and Tidal Energy, Cambridge, England, September.

Cavanagh, J. E., Clarke, J. H., and Price, R. 1993. Ocean energy systems. In *Renewable Energy, Sources for Fuels and Electricity*, T. B. Johansson, H. Kelley, A. K. N. Reddy, and R. H. Williams, eds., Island Press, Washington, DC, chap. 12.

Ryan, P. R. 1979/80. Harnessing power from tides: state of the art. *Oceanus*, 22 (4), 64–67.

SERI, 1989. Ocean thermal energy conversion: an overview, Solar Energy Research Institute, SERI/SP-220-3024, Golden, CO.

SERI, 1990. The potential of renewable energy: an interlaboratory white paper, Solar Energy Research Institute, SERI/TP-260-3674, Golden, CO.

WEC, 1993. *Renewable Energy Resources—Opportunities and Constraints 1990–2020*. World Energy Council, London, England.

For Further Information

CEC, 1992. Energy technology status report, California Energy Commission, Sacramento, CA, 1992.

Funakoshi, H., Ohno, M., Takahashi, S., and Oikawa, K. 1993. Present situation of wave energy conversion systems. *Civil Eng. Jpn.*, 32, 108–134.

26

Fuel Cells

Xianguo Li
University of Waterloo

26.1 Introduction 26-1
26.2 Principle of Operation for Fuel Cells 26-2
26.3 Typical Fuel Cell Systems 26-3
26.4 Performance of Fuel Cells 26-4
Reversible Cell Potential • Energy Conversion Efficiency • Practical Fuel Cell Efficiency and Energy-Loss Mechanisms • Efficiency Loss in Operating Fuel Cells: Stoichiometry, Utilization, and Nernst Loss
26.5 Fuel Cell Electrode Processes 26-25
26.6 Cell Connection and Stack Design Considerations 26-27
26.7 Six Major Types of Fuel Cells 26-29
Alkaline Fuel Cells • Polymer-Electrolyte-Membrane Fuel Cells • Direct-Methanol Fuel Cells • Phosphoric-Acid Fuel Cells • Molten-Carbonate Fuel Cells • Solid-Oxide Fuel Cells
26.8 Summary 26-44
References 26-45

26.1 Introduction

A fuel cell is an electrochemical device that converts the chemical energy of a fuel and oxidant directly into electric energy. Such a direct one-step conversion avoids the inefficient multistep processes involved in heat engines via combustion, thus eliminating the emission of chemical pollutants. Besides being efficient and clean, fuel cells are also compatible with renewable energy sources and carriers for sustainable development and energy security. Fuel cells offer additional advantages for both mobile and stationary applications, including quiet operation without vibration and noise; they are therefore candidates for onsite applications. Their inherent modularity allows for simple construction and operation with possible applications for dispersed, distributed, and portable power generation. Their fast response to changing load conditions while maintaining high efficiency makes them ideally suited to load-following applications. Their potential high efficiency also represents less chemical, thermal, and carbon dioxide emissions for the same amount of energy conversion and power generation. Consequently, fuel cells are often regarded as one of the advanced energy technologies of the future.

At present, fuel cells are routinely used in space applications, and have been under intensive development for terrestrial use, such as for utilities and zero-emission vehicles. There exist a variety of fuel cells, which are classified based on their operating temperature and the type of ion migrating through the electrolyte. However, the choice of electrolyte defines the properties of a fuel cell. Therefore, fuel cells are often named by the nature of their electrolyte. There are presently six major fuel cell technologies at varying stages of development and commercialization: alkaline, phosphoric acid, polymer electrolyte membrane, molten carbonate, solid oxide, and direct methanol fuel cell (DMFC). This chapter provides a summary of fundamentals and the state-of-the-art technology for these types of fuel cells. Detailed information regarding their electrochemical reactions, operation principles, construction and design, and specific areas of application can be found elsewhere [1–7].

26.2 Principle of Operation for Fuel Cells

A fuel cell is composed of three active components: a fuel electrode (anode), an oxidant electrode (cathode), and an electrolyte sandwiched between them. Figure 26.1 illustrates the basic operational principle of fuel cells with a typical acid electrolyte, where it is seen that molecular hydrogen is delivered from a gas-flow stream to the anode (or fuel electrode), and it reacts electrochemically in the anode as:

$$\text{Anode reaction}: \ \mathrm{H_2} \Rightarrow 2\mathrm{H}^+ + 2e^- \tag{26.1}$$

The hydrogen (fuel) is oxidized at the anode-electrolyte interface into hydrogen ions or protons, H^+, and an electron, e^-. The protons migrate through the (acid) electrolyte while the electrons are forced to transfer through an external circuit, both arriving at the cathode (or oxidant electrode). At the cathode, the protons and electrons react with the oxygen supplied from an external gas-flow stream to form water:

$$\text{Cathode reaction}: \ \frac{1}{2}\mathrm{O_2} + 2\mathrm{H}^+ + 2\mathrm{e}^- \Rightarrow \mathrm{H_2O} \tag{26.2}$$

Thus, oxygen is reduced into water at the cathode by combining with H^+ and e^-. Now, both the electric current and mass transfer form a complete circuit. The electrons go through the external electric circuit and do work on the electric load, constituting the useful electric energy output from the fuel cell. At the same time, waste heat is also generated due to the electrochemical reactions occurring at the anode and the cathode, as well as due to protons migrating through the electrolyte and electrons transporting in the solid portion of the electrodes and the external circuit. As a result, the overall cell reaction can be

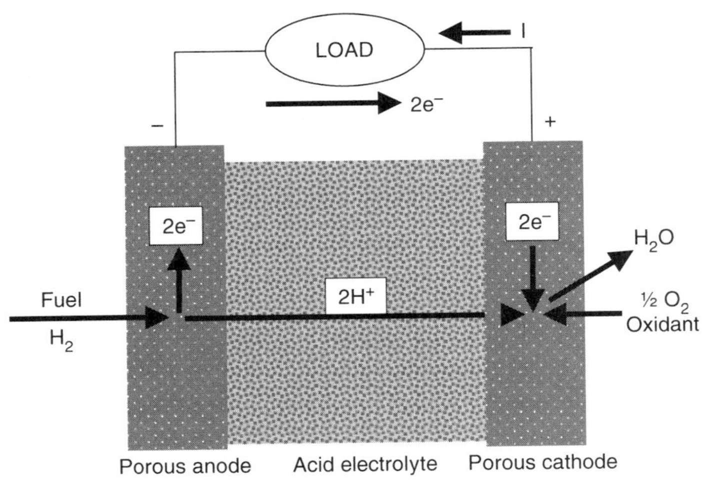

FIGURE 26.1 Schematic of a typical acid-electrolyte fuel cell.

obtained by summing the above two half-cell reactions to yield:

$$\text{Overall cell reaction}: \ H_2 + \frac{1}{2}O_2 \Rightarrow H_2O + W + \text{waste heat}, \quad (26.3)$$

where W is the useful electric energy output from the fuel cell. Although the half-cell reactions may be quite different in different types of fuel cells (described later), the overall cell reaction remains the same as the equation shown above.

Therefore, the by-products of the electrochemical reactions described above are water and waste heat. They should be removed continuously from the cell to maintain its continuous isothermal operation for electric power generation. This need for the continuous removal of water and heat results in the so-called *water and heat (thermal) management* that may become the two critical issues for the design and operation of some types of fuel cells. In general, they are difficult tasks.

26.3 Typical Fuel Cell Systems

In general, a fuel cell power system involves more than just a fuel cell itself because fuel cells require a steady supply of qualifying fuel and oxidant as reactants for continuous generation of electric power. The oxidant is usually pure oxygen for specialized applications such as space and some military applications, and is almost invariably air for terrestrial and commercial applications. Depending on the specific types of fuel cells, both fuel and oxidant streams must meet certain purity requirements for fuel cell operations. Therefore, a fuel cell power system is usually composed of a number of subsystems for fuel processing, oxidant conditioning, electrolyte management, cooling or thermal management, and reaction product removal. A schematic of a typical rudimentary fuel cell system is illustrated in Figure 26.2. Normally, a power-conditioning unit is required to convert the DC electric power into AC power because fuel cells generate DC power whereas most electric equipment operates on AC. The waste heat produced in the fuel cell power section is often integrated through a series of heat exchangers into the fuel cell system for better energy efficiency, and it is also possible for some types of fuel cells to use the waste heat as the heat source for either cogeneration or bottoming cycles for additional electric energy generation.

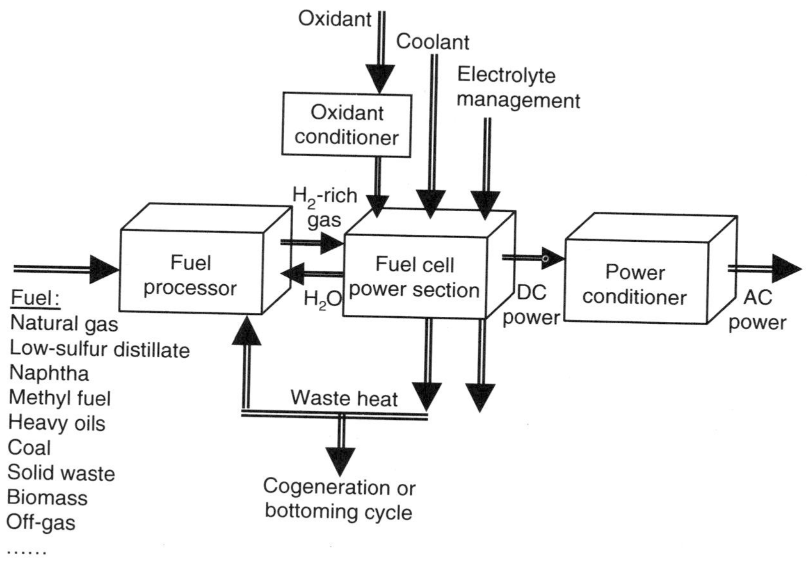

FIGURE 26.2 Schematic of a typical rudimentary fuel cell system.

The cogeneration of heat and hot water (and sometimes steam) along with electricity increases the overall energy efficiency of the fuel cell system to as much as 85% or more. Heat is critical to human survival, such as for space heating and household use. Both heat and steam are significantly important commodities in industrial processes and in many other practical applications.

The DC-to-AC inverter is a fairly mature technology due to the incorporation of semiconductor and integrated-circuit technology, and its conversion efficiency is very high, as much as 96% for MW-sized power plants. The fuel processor converts the primary and/or portable fuel (such as natural gas, low-sulfur distillate, naphtha, methyl fuel—mostly methanol, heavy oils, coal, solid waste, biomass, etc.) into H_2 and CO. These secondary fuels (H_2 and CO) are considerably more electrochemically active in the electrochemical cell stack than the primary fuels. Even though fuel processing technology is highly advanced and efficient, it typically accounts for a third of the power-plant size, weight, and cost for the hydrocarbon-fueled, fuel cell power plants. The electrochemical fuel cell stack accounts for approximately another third of the size, weight, and cost. The ancillary components and subsystems associated with air supply, thermal management, water recovery and treatment, cabinet ventilation, and system control and diagnostics (or often referred to as the *balance of the plant*) accounts for the remaining third. In fuel cell systems, the most important subsystem is the electrochemical fuel cell stack, and the fuel processor is the second major subsystem if hydrocarbon fuels are used as the primary fuel.

26.4 Performance of Fuel Cells

Although numerous studies aimed at developing fuel cells as practical power sources have been conducted, some confusion and misconceptions exist about the thermodynamic performance of fuel cells in comparison to heat engines [2,8–10]. In this section, the fundamental principles—the first and second laws of thermodynamics—will be used to derive the idealized best possible performance—specifically, the reversible cell potential and the reversible energy conversion efficiency of fuel cells. The effects of the operating conditions, such as temperature, pressure, and reactant concentrations, on the cell performance will be discussed. The maximum possible efficiency for fuel cells will be compared with the Carnot efficiency, the maximum possible efficiency for heat engines against which fuel cells are competing for commercial success. Then, the possibility of greater than 100% efficiency for fuel cells will be analyzed and ruled out based on the fundamental principles. Finally, various energy-loss mechanisms in a fuel cell will be described, including both reversible and irreversible losses; the amount and rate of heat generation in an operating fuel cell will be derived; various forms of efficiency for fuel cells will be defined; and further energy losses in operating fuel cells will be considered as the Nernst loss due to limited utilization (or stoichiometry) of the reactants supplied to the cell.

26.4.1 Reversible Cell Potential

In a fuel cell, the chemical energy of a fuel and an oxidant is converted directly into electrical energy. The maximum possible electrical energy output and the corresponding electrical potential difference between the cathode and anode are achieved when the fuel cell is operated under thermodynamically reversible conditions. This maximum possible cell potential is called the *reversible cell potential* and it is an important parameter for fuel cells; it is derived later in this section.

A thermodynamic system* model is shown in Figure 26.3 for the analysis of fuel cell performance. It is a control-volume system for the fuel cell to which fuel and oxidant streams enter and product or exhaust streams exit. The fuel cell is located inside a thermal bath to maintain the desired system temperature, T.

*A *thermodynamic system*, or simply *system*, is (in thermodynamics) a collection of matter under study or analysis, whereas the term *fuel cell system* in fuel cell literature usually denotes the fuel cell power plant that consists of fuel cell stack(s) and auxiliary equipment (also called *balance of the plant*)—see Section 26.3, for example. In this chapter, a fuel cell system may imply both meanings. However, the context will indicate which it is meant to be.

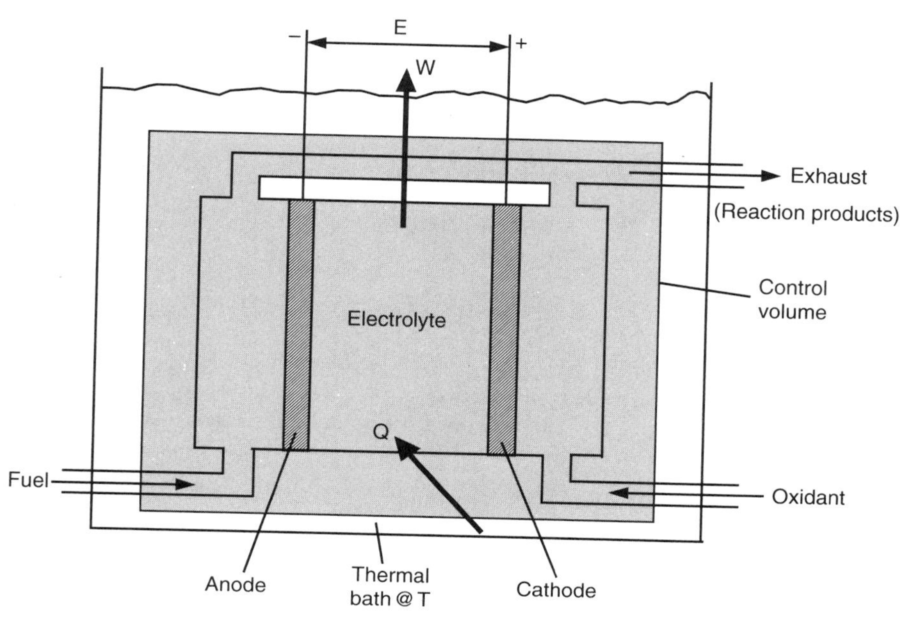

FIGURE 26.3 A thermodynamic model of fuel cell system.

The reactant streams (fuel and oxidant) and the exhaust stream are considered to have the same T and pressure, P. It is assumed that the fuel and oxidant inflows and the exhaust outflow are steady; the kinetic and gravitational potential energy changes are negligible. Furthermore, the overall electrochemical reactions occurring inside the fuel cell system boundary is described as follows:

$$\text{fuel (e.g., } H_2) + \text{oxidant (e.g., } O_2) \Rightarrow \dot{W} + \dot{Q} + \text{product,} \tag{26.4}$$

where $\dot{W}$ is the rate of work done by the system and $\dot{Q}$ is the rate of heat transferred into the system from the surrounding constant-temperature thermal bath that may or may not be in thermal equilibrium with the fuel cell system at T and P. For hydrogen–oxygen fuel cells, the reaction product is water. The first and second laws of thermodynamics, respectively, can be written for the present fuel cell system as

$$\frac{dE_{CV}}{dt} = [(\dot{N}h)_F + (\dot{N}h)_{Ox}]_{in} - [(\dot{N}h)_{Ex}]_{out} + \dot{Q} - \dot{W}, \tag{26.5}$$

$$\frac{dS_{CV}}{dt} = [(\dot{N}s)_F + (\dot{N}s)_{Ox}]_{in} - [(\dot{N}s)_{Ex}]_{out} + \frac{\dot{Q}}{T} + \dot{\wp}_s, \tag{26.6}$$

where $\dot{N}$ is the molar flow rate, h is the (absolute) enthalpy per unit mole, s is the specific entropy on a mole basis, and $\dot{\wp}_s$ is the rate of entropy generation due to irreversibilities. The subscripts "F," "Ox," and "Ex" stand for fuel, oxidant, and exhaust stream, respectively.

For a steady process, there are no temporal changes in the amount of energy, E_{CV}, and entropy, S_{CV}, within the control volume system; hence, $dE_{CV}/dt=0$ and $dS_{CV}/dt=0$. Therefore, Equation 26.5 and

Equation 26.6 can be simplified as follows:

$$\dot{N}_F(h_{in} - h_{out}) + \dot{Q} - \dot{W} = 0, \tag{26.7}$$

$$\dot{Q} = -T\wp_s - \dot{N}_F T(s_{in} - s_{out}), \tag{26.8}$$

where

$$h_{in} = \left(h_F + \frac{\dot{N}_{Ox}}{\dot{N}_F} h_{Ox}\right)_{in} \text{ and } h_{out} = \frac{\dot{N}_{Ex}}{\dot{N}_F} h_{Ex}. \tag{26.9}$$

h_{in} is the amount of enthalpy per mole of fuel carried into the system by the reactant inflow, and h_{out} is the amount of enthalpy per mole of fuel taken out of the system by the exhaust stream. Similarly,

$$s_{in} = \left(s_F + \frac{\dot{N}_{Ox}}{\dot{N}_F} s_{Ox}\right)_{in} \text{ and } s_{out} = \frac{\dot{N}_{Ex}}{\dot{N}_F} s_{Ex}, \tag{26.10}$$

are the amount of entropy per mole of fuel brought into and carried out of the system by the reactant inflow and the outgoing exhaust stream containing the reaction products, respectively.

Substitution of Equation 26.8 into Equation 26.7 yields

$$\dot{W} = \dot{N}_F(h_{in} - h_{out}) - \dot{N}_F T(s_{in} - s_{out}) - T\dot{\wp}_s. \tag{26.11}$$

Let

$$w = \frac{\dot{W}}{\dot{N}_F}; \quad q = \frac{\dot{Q}}{\dot{N}_F}; \text{ and } \wp_s = \frac{\dot{\wp}_s}{\dot{N}_F} \tag{26.12}$$

represent, respectively, the amount of work done, heat transferred, and entropy generated per unit mole of fuel. Equation 26.8 and Equation 26.11 then become

$$q = -T\wp_s - T(s_{in} - s_{out}) = T\Delta s - T\wp_s \tag{26.13}$$

$$w = (h_{in} - h_{out}) - T(s_{in} - s_{out}) - T\wp_s. \tag{26.14}$$

Because the enthalpy and entropy changes for the fuel cell reaction are defined as

$$\Delta h = h_{out} - h_{in} \text{ and } \Delta s = s_{out} - s_{in}, \tag{26.15}$$

Equation 26.14 can also be expressed as

$$w = -\Delta h + T\Delta s - T\wp_s = -[(h - Ts)_{out} - (h - Ts)_{in}] - T\wp_s. \tag{26.16}$$

From the definition of the Gibbs function (per mole of fuel), $g = h - Ts$, Equation 26.14 or Equation 26.16 can also be written as

$$w = -(g_{out} - g_{in}) - T\wp_s = -\Delta g - T\wp_s. \tag{26.17}$$

By the second law of thermodynamics, entropy can be generated but can never be destroyed. Hence, $\wp_s \geq 0$, and also absolute temperature, *T*, is greater than zero by the third law of thermodynamics. The maximum possible work (i.e., useful energy) output from the present system therefore occurs when $\wp_s = 0$ (under the thermodynamically reversible condition) because the change in the Gibbs function is usually negative for useful fuel cell reactions. Therefore, from Equation 26.17 it is clear that the maximum possible work output from the present fuel cell system is equal to the decrease in the Gibbs function, or

$$w_{\max} = -\Delta g \tag{26.18}$$

for all reversible processes, irrespective of the specific fuel cell involved. In fact, it should be pointed out that in the derivation of Equation 26.17 and Equation 26.18, no specifics about the control-volume system have been stipulated; hence, these equations are valid for any energy conversion systems.

For a fuel cell system, the electrical energy output is conventionally expressed in terms of the cell potential difference between the cathode and the anode. Because (electrical) potential is the (electrical) potential energy per unit (electrical) charge, its SI unit is J/C, which is more often called volts (V). Potential energy is defined as the work done when a charge is moved from one location to another in the electrical field, normally external circuits. For the internal circuit of fuel cells, such as that shown in Figure 26.3, *electromotive force* is the terminology often used, which is also defined as the work done by transferring one Coulomb positive charge from a low potential to a high potential. Hence, electromotive force also has the SI units of J/C, or V. Here, the term *cell potential*, instead of *electromotive force*, is adopted from this point forward; the notation *E* will be used to represent the cell potential. Because electrons are normally the particles that carry electrical charge, the work done by a fuel cell may be expressed as:

$$w\ (\text{J/mol fuel}) = E \times (\text{Coulombs of electron charge transferred/mol fuel}),$$

or

$$w = E(nN_0 e) = E(nF), \tag{26.19}$$

where n is the number of moles of electrons transferred per mole of fuel consumed, N_0 is Avogadro's number (6.023×10^{23} electrons/mol electron), and e is the electric charge per electron (1.6021×10^{-19} Coulomb/electron). $N_0 e$ is equal to 96,487 Coulomb/mol electron and is often referred to as the *Faraday constant*, *F* The cell potential becomes, from Equation 26.17,

$$E = \frac{w}{nF} = \frac{-\Delta g - T\wp_s}{nF}. \tag{26.20}$$

Hence, the maximum possible cell potential, or the reversible cell potential, E_r, becomes

$$E_r = -\frac{\Delta g}{nF}. \tag{26.21}$$

From the reversible cell potential given above, Equation 26.20 can also be rewritten as

$$E = E_r - \frac{T\wp_s}{nF} = E_r - \eta, \tag{26.22}$$

where

$$\eta = \frac{T \wp_s}{nF} \tag{26.23}$$

is the cell voltage loss due to irreversibilities (or entropy generation). Clearly, the actual cell potential can be calculated by subtracting the cell voltage loss from the reversible cell potential. Alternatively, the amount of entropy generated per mole of fuel consumed can be determined as

$$\wp_s = \frac{nF\eta}{T} = \frac{nF(E_r - E)}{T}. \tag{26.24}$$

Thus, the amount of entropy generated for the fuel cell reaction process, which represents the degree of irreversibility (the degree of deviation from the idealized reversible condition), can be measured after the cell potential, E, and the cell operating temperature, T, are known.

Note that the Gibbs function is a thermodynamic property that is determined by state variables such as temperature and pressure. Therefore, the change in the Gibbs function for the fuel cell reaction discussed here is

$$\Delta g = \Delta h - T\Delta s, \tag{26.25}$$

which is also a function of T and P. The specific effect of the operating conditions, such as temperature, pressure, and reactant concentrations, on the reversible cell potential will be presented in the next section. If the reaction occurs at the standard reference temperature and pressure (25°C and 1 atm), the resulting cell potential is usually called the *standard reversible cell potential*, E_r°:*

$$E_r^\circ(T_{ref}) = -\frac{\Delta g(T_{ref}, P_{ref})}{nF}. \tag{26.26}$$

If pure hydrogen and oxygen are used as reactants to form water as product, then E_r° (25°C) = 1.229 V for the product water in liquid form, and E_r° (25°C) = 1.185 V if the product water is in vapor form. The difference in E_r° corresponds to the energy required for the vaporization of water. It might be pointed out that any fuel containing hydrogen (including hydrogen itself, hydrocarbons, alcohols, and to a lesser extent, coal) has two values for Δg and Δh, one higher and one lower, depending on whether the product water is in the form of liquid or vapor. Therefore, care should be taken when referring to reversible cell potential and energy efficiency to be discussed later in Section 26.4.3.

The standard reversible cell potential, E_r°, can be determined for any other electrochemical reactions. Some of the potential fuel cell reactions and the resulting E_r° are shown in Table 26.1 along with other relevant parameters. From this table, it might be noted that E_r° should be approximately above 1 V for the reaction to be realistic for fuel cell applications. This is because if E_r° is much less than 1 V, and considering the cell voltage loss that inevitably occurs in practical fuel cells due to irreversibilities, the actual cell potential might become too small to be useful for practical applications. Therefore, the rule of thumb is, for any proposed fuel and oxidant, to calculate E_r° and determine if E_r° is on the order of 1 V or larger before proceeding to any further work on it.

26.4.1.1 The Effect of Operating Conditions on Reversible Cell Potential

The most important operating conditions that influence fuel cell performance are the operating temperature, pressure, and reactant concentrations. For many useful electrochemical reactions, the

*In the literature, the superscript «o» sometimes denotes the value at the standard reference condition of 25°C and 1 atm; and sometimes it also refers to parameters evaluated at 1 atm. The latter meaning has been adopted in the present chapter.

TABLE 26.1 Standard Enthalpy and Gibbs Function of Reaction for Candidate Fuels and Oxidants, and Corresponding Standard Reversible Cell Potential as well as Other Relevant Parameters (at 25°C and 1 atm)

Fuel	Reaction	n N	$-\Delta h$ (J/mole)	$-\Delta g$ (J/mole)	E_r^o (V)	η^a (%)
Hydrogen	$H_2+0.5O_2 \rightarrow H_2O\ (\lambda)$	2	286.0	237.3	1.229	82.97
	$H_2+Cl_2 \rightarrow 2HCl(aq)$	2	335.5	262.5	1.359	78.33
	$H_2+Br_2 \rightarrow 2HBr(aq)$	2	242.0	205.7	1.066	85.01
Methane	$CH_4+2O_2 \rightarrow CO_2+2H_2O(\lambda)$	8	890.8	818.4	1.060	91.87
Propane	$C_3H_8+5O_2 \rightarrow 3CO_2+4H_2O(\lambda)$	20	2221.1	2109.3	1.093	94.96
Decane	$C_{10}H_{22}+15.5O_2 \rightarrow 10CO_2+11H_2O(\lambda)$	66	6832.9	6590.5	1.102	96.45
Carbon	$CO+0.5O_2 \rightarrow CO_2$	2	283.1	257.2	1.333	90.86
Carbon monoxide	$C(s)+0.5O_2 \rightarrow CO$	2	110.1	137.3	0.712	124.18[b]
Carbon	$C(s)+O_2 \rightarrow CO_2$	4	393.7	394.6	1.020	100.22[b]
Methanol	$CH_3OH(\lambda)+1.5O_2 \rightarrow CO_2+2H_2O(\lambda)$	6	726.6	702.5	1.214	96.68
Formaldehyde	$CH_2O(g)+O_2 \rightarrow CO_2+H_2O(\lambda)$	4	561.3	522.0	1.350	93.00
Formic Acid	$HCOOH+0.5O_2 \rightarrow CO_2+H_2O(\lambda)$	2	270.3	285.5	1.480	105.62[b]
Ammonia	$NH_3+0.75O_2 \rightarrow 1.5H_2O(\lambda)+0.5N_2$	3	382.8	338.2	1.170	88.36
Hydrazine	$N_2H_4+O_2 \rightarrow 2H_2O(\lambda)+N_2$	4	622.4	602.4	1.560	96.77
Zinc	$Zn+0.5O_2 \rightarrow ZnO$	2	348.1	318.3	1.650	91.43
Sodium	$Na+0.25H_2O+0.25O_2 \rightarrow NaOH\ (aq)$	1	326.8	300.7	3.120	92.00

[a]Energy conversion efficiency.

[b]There is a conceptual problem with these efficiency data; see Section 26.4.3.5 for explanation.

Source: From Appleby, A. J., in Fuel Cell Systems, eds. L. J. M. J. Blomen and M. N. Mugerwa, Plenum Press, New York, 1993.

entropy change is negative and is almost constant with the change of temperature to a good approximation, provided the temperature change $T - T_{\text{ref}}$ is not too large. Then, the effect of temperature on the reversible cell potential may be written as [1]

$$E_r(T,P) = E_r(T_{\text{ref}},P) + \left(\frac{\Delta s(T_{\text{ref}},P)}{nF}\right)(T - T_{\text{ref}}). \tag{26.27}$$

It must be emphasized that the expression given in Equation 26.27 is an approximation. Strictly speaking, the reversible cell potential at any temperature and pressure should be determined from Equation 26.2 by first calculating the property changes for the particular fuel cell reaction involved. Such a procedure has been followed for the hydrogen and oxygen reaction to form gaseous water, and the results are presented in Figure 26.4. Clearly, the reversible cell potential indeed decreases almost linearly as temperature is increased over a large temperature range. However, it is noted that the reversible cell potential is larger for the product water in the liquid state at low temperatures, but it decreases much faster than the gaseous water as the product when temperature is increased. At temperatures slightly above about 373 K, the reversible cell potential for liquid water as product actually becomes smaller. This may seem odd, but it is because at such high temperatures, pressurization is necessary to keep the water product in liquid form as the reactants (hydrogen and oxygen) are fed at 1 atm. Also notice that the critical temperature for water is about 647 K, beyond which a distinct liquid state does not exist for water, hence the shorter curve shown for the liquid water case in Figure 26.4.

As previously discussed, the entropy change for most fuel cell reactions is negative; consequently, the reversible cell potential decreases as temperature is increased as shown in Figure 26.4. However, for some reactions, such as

$$C(s) + \frac{1}{2}O_2(g) \rightarrow CO(g),$$

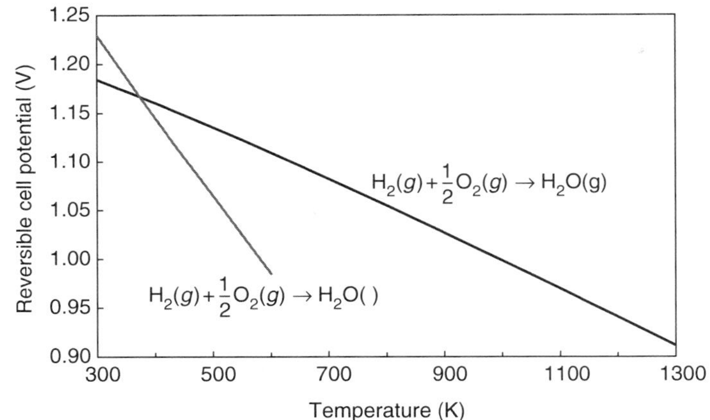

FIGURE 26.4 Effect of temperature on the reversible cell potential of a hydrogen–oxygen fuel cell for the reaction of $H_2 + \frac{1}{2}O_2 \rightarrow H_2O$ at a pressure of 1 atm.

the entropy change is positive. For example, $\Delta s = 89$ J/(mol fuel · K) at the standard reference temperature and pressure. As a result, the reversible cell potential for this type of reaction will increase with temperature, which is clearly seen in Figure 26.5 for the reversible cell potential as a function of temperature for a number of important fuel cell reactions.

The effect of pressure on reversible cell potential can be expressed as [1]

$$E_r(T, P) = E_r(T, P_{ref}) - \frac{\Delta N \Re T}{nF} ln\left(\frac{P}{P_{ref}}\right), \tag{26.28}$$

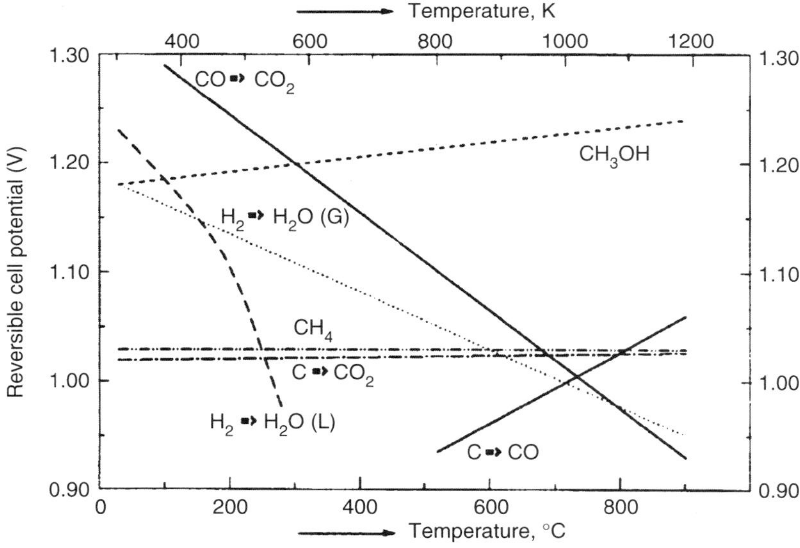

FIGURE 26.5 Standard reversible cell potential, *Er*, as a function of temperature for the most important fuel cell reactions at the pressure of 1 atm. (From Barendrecht, E., Electrochemistry of fuel cells, In *Fuel cell systems*, L. J. M. J. Blomen and M. N. Mugerwa, eds., Plenum Press, New York, pp. 73–79, 1993.)

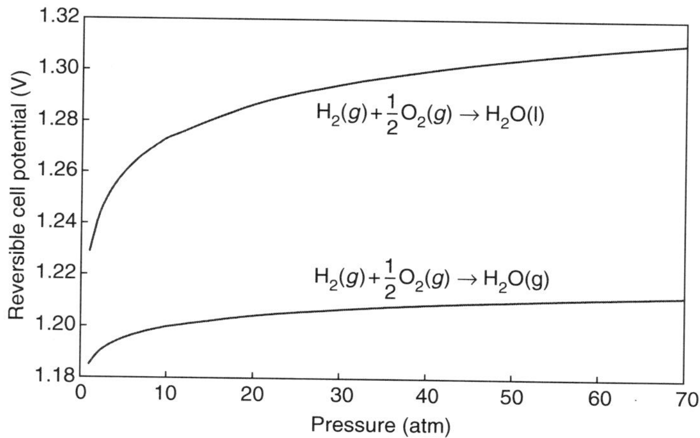

FIGURE 26.6 Standard reversible cell potential, E_r, as a function of pressure for the fuel cell reaction of $H_2(g) + \frac{1}{2}O_2 \rightarrow H_2O$ at the temperature of 25°C.

where ΔN represents the total number of mole changes for all the gaseous species involved in the fuel cell reaction. This equation indicates that the pressure dependence of the reversible cell potential is a logarithmic function, and therefore the dependence becomes weaker as pressure P is increased. Figure 26.6 shows the effect of pressure on the reversible cell potential for the hydrogen–oxygen fuel cell for both liquid and vapor water as the reaction product. It is seen that the reversible cell potential rapidly increases with pressure for low-pressure values; the increase gradually slows for higher pressure values. Also, the pressure effect is larger for liquid water as the product because of the larger coefficient arising from the larger change in the number of moles between the product and the reactant, ΔN.

It might be also emphasized that the pressure effect is small on the reversible cell potential at low temperatures, as shown in the above example. However, this effect increases significantly at high temperatures because the pressure effect coefficient, $-\Delta N \Re T/(nF)$, is directly proportional to temperature.

It might be pointed out that for high-temperature fuel cells, the dependence of the actual cell potential E on the pressure closely follows the results given in Equation 26.28, whereas a significant deviation occurs for the low-temperature fuel cells. The difference arises from the fact that at high temperatures, the reaction kinetics are fast and pressurization primarily increases the reactant concentration, hence better performance. At low temperatures, the reaction kinetics are slow and higher reactant concentration does not yield a proportional increase in the cell potential due to the significant cell potential loss associated with the slow kinetics.

When the fuel, oxidant, and exhaust streams contain chemically inert gas species, the reversible cell potential will be lowered due to the dilution effect of the inert species; this can be derived as [1]

$$E_r(T,P_i) = E_r(T,P) - \frac{\Re T}{nF} \ln K, \tag{26.29}$$

where P_i is the partial pressure of the reactant in the fuel and oxidant streams, and

$$K = \prod_{i=1}^{N_g} \left(\frac{P_i}{P}\right)^{(\nu_i''-\nu_i')/\nu_F'} \tag{26.30}$$

is defined similar to the equilibrium constant for partial pressure (but they are not the same); N_g is the total number of gas species in the reacting system, excluding the solid and liquid species; ν_i' and ν_i'' are the number of moles of species i in the reactant and product mixtures, respectively. Equation 26.29 is the general form of the Nernst equation, representing the effect of the reactant and product concentrations on the reversible cell potential. When the reactant streams contain inert diluents for a given operating temperature and pressure, the diluents will cause a voltage loss for the reversible cell potential, and the amount of loss is generally called the Nernst loss, and its magnitude is equal to the second term on the right-hand side of Equation 26.29.

26.4.2 Energy Conversion Efficiency

26.4.2.1 The Definition of Energy Conversion Efficiency

The efficiency for any energy conversion process or system is often defined as*(26.31)

$$\eta = \frac{\text{Useful energy obtained}}{\text{Energy available for conversion that is an expense}}.$$

Based on this definition, it is well known that 100% energy conversion efficiency is possible by the first law of thermodynamics, but is not possible by the second law of thermodynamics for many energy conversion systems that produce power output, such as steam turbines, gas turbines, and internal combustion engines, which involve irreversible losses of energy. These thermal energy conversion systems are often referred to as heat engines.

26.4.2.2 Reversible Energy Conversion Efficiency for Fuel Cells

For the present fuel cell system described in Figure 26.3, the energy balance equation, Equation 26.5, can be written, on a per unit mole of fuel basis, as:

$$h_{in} - h_{out} + q - w = 0 \text{ or } h_{in} - h_{out} = -q + w, \tag{26.32}$$

which indicates that the enthalpy change, $-\Delta h = h_{in} - h_{out}$, provides the energy available for conversion into the useful energy exhibited as work here, and it is the expense to be paid for the useful work output. At the same time, waste heat, q, is also generated, which represents a degradation of energy. The amount of waste heat generated can be determined from the second-law expression, Equation 26.6 or Equation 26.13, as

$$q = T\Delta s - T\wp_s \tag{26.33}$$

and the useful energy output as work is, from Equation 26.17, or combining Equation 26.32 and Equation 26.33

$$w = -\Delta g - T\wp_s \tag{26.34}$$

Therefore, the energy conversion efficiency for the fuel cell system described in Figure 26.3 becomes, according to Equation 26.31,

$$\eta = \frac{w}{-\Delta h} = \frac{\Delta g + T\wp_s}{\Delta h}. \tag{26.35}$$

Note that both Δh and Δg are negative for power generation systems, including fuel cells, as is clearly shown in Table 26.1. By the second law, the entropy generation per unit mole of fuel is

$$\wp_s \geq 0 \tag{26.36}$$

*Note that in literature η is commonly used as efficiency in thermodynamics; whereas it is also conventionally used as overpotential, or voltage loss, for fuel cell analysis, as in electrochemistry. In the present chapter, η is used for both in order to be consistent with the respective convention, and its meaning would become clear from the context.

and the equality holds for all reversible processes, whereas entropy is always generated for irreversible processes. Therefore, the maximum possible efficiency allowed by the second law is, when the process is reversible (i.e., $\wp_s = 0$),

$$\eta_r = \frac{w_{max}}{-\Delta h} = \frac{\Delta g(T, P)}{\Delta h(T, P)}. \tag{26.37}$$

Because both the enthalpy and Gibbs function changes depend on the system temperature and pressure, the same holds for the energy conversion efficiency. It should be pointed out that in the above derivation, no assumption specifically related to a fuel cell has been made; the only assumption made is that the energy conversion system for power production is reversible for all processes involved. Thus, Equation 26.37 is valid for any power production system, be it an electrochemical converter like a fuel cell or a conventional thermal-energy converter like a heat engine, as long as the process is reversible. Therefore, it may be called the *reversible efficiency* because it is the maximum possible efficiency that is allowed by the second law of thermodynamics. In what follows, it will be demonstrated that the maximum possible efficiency for conventional heat engines, the well-known Carnot efficiency, is really the reversible efficiency applied specifically to the conventional thermal power cycles, thus is equivalent to Equation 26.37.

26.4.2.3 Carnot Efficiency: The Reversible Energy Conversion Efficiency for Heat Engines

Consider a heat engine operating between two temperature thermal energy reservoirs (TER), one at a high temperature T_H and the other at a low temperature T_L, as shown in Figure 26.7. The heat engine obtains energy from the high-temperature TER in the form of heat with the quantity q_H; a portion of this heat is converted to work output w and the remainder is rejected to the low-temperature TER in the amount of q_L as waste heat. Applying the first and second laws to the heat engine,

$$\text{First law}: w = q_H - q_L, \tag{26.38}$$

$$\text{Second law}: \wp_{s,HE} = \frac{q_L}{T_L} - \frac{q_H}{T_H}, \tag{26.39}$$

where $\wp_{s,HE}$ represents the amount of entropy production during the energy conversion process by means of the heat engine. From Equation 26.39, the amount of heat rejection can be determined as

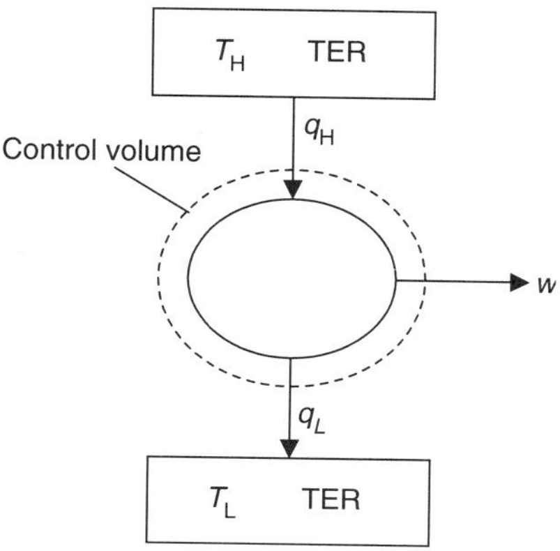

FIGURE 26.7 Thermodynamic system model of heat engines operating between two temperature thermal-energy reservoirs.

$$q_L = \frac{T_L}{T_H} q_H + T_L \wp_{s,HE}. \qquad (26.40)$$

The efficiency for the heat engine is, by the definition of Equation 26.31,

$$\eta = \frac{w}{q_H} \qquad (26.41)$$

Substituting Equation 26.38 and Equation 26.40 into Equation 26.41 yields

$$\eta = 1 - \frac{T_L}{T_H} - \frac{T_L}{q_H} \wp_{s,HE}. \qquad (26.42)$$

As pointed out earlier, the second law of thermodynamics dictates that the entropy generation within the heat engine can never be negative; at most, it can vanish under the thermodynamically reversible condition. Therefore, the maximum possible efficiency for the heat engine is achieved if the process is reversible ($\wp_{s,HE}=0$),

$$\eta_{r,HE} = 1 - \frac{T_L}{T_H}. \qquad (26.43)$$

This is the familiar Carnot efficiency, giving the upper bound for the efficiency of all heat engines. Because $T_L < T_H$, and $T_L \neq 0$ by the third law of thermodynamics, and because the high-temperature T_H is finite, 100% efficiency is not possible by the second law for any energy conversion system that produces power output using heat engines, such as steam and gas turbines, internal combustion engines, etc. because of the second law requirement that the entropy generation term must never be negative. In contrast, 100% efficiency is possible by the first law that merely states the principle of energy conservation.

26.4.2.4 The Equivalency of Carnot and Fuel Cell Reversible Efficiency

As shown above, both the Carnot efficiency and the reversible energy conversion efficiency for fuel cell, Equation 26.37, are the maximum possible efficiency allowed by the second law; the former is applied specifically to heat engines, whereas the latter is derived for fuel cells. Therefore, they must be related somehow as they both are the maximum possible efficiency dictated by the second law. In this subsection, it is demonstrated that they are actually equivalent, just expressed in a different form under a suitable condition for the comparison.

Suppose for a heat engine the high temperature TER is maintained at T_H by the combustion of a fuel with an oxidant, both reactants are originally at the temperature of T_L, as shown schematically in Figure 26.8. It is assumed that both the fuel and oxidant are the same as used in Figure 26.3 for the derivation of fuel cell performance, the combustion process is carried out at the same system pressure P in a controlled manner such that the combustion products leave the TER at the pressure P and temperature T_L. Neglecting the changes in the kinetic and gravitational potential energy, the first and second laws become for the high temperature TER:

$$\text{First law}: q_H = h_R - h_P = -\Delta h(T_L, P), \qquad (26.44)$$

$$\text{Second law}: \wp_{s,TER} = (s_P - s_R) + \frac{q_H}{T_H} = \Delta s(T_L, P) + \frac{q_H}{T_H}, \qquad (26.45)$$

where the subscript "R" and "P" represent the reactant and product mixture, respectively.

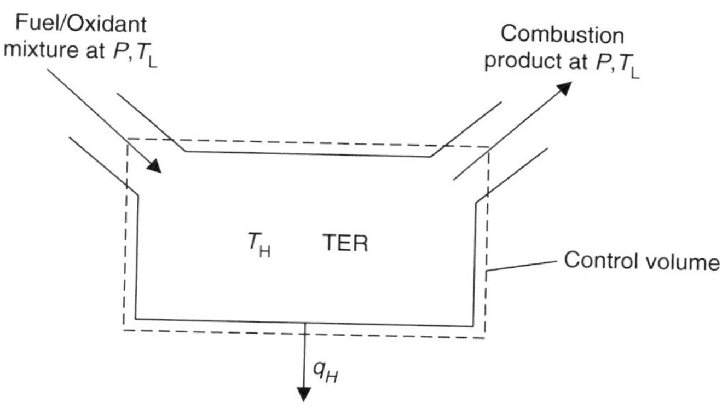

FIGURE 26.8 Thermodynamic system model of high temperature thermal energy reservoir (TER) maintained by combustion process of a fuel/oxidant mixture.

After rearranging, Equation 26.45 gives the temperature T_H resulting from the combustion process

$$T_H = \frac{q_H}{\wp_s, \text{TER} - \Delta s(T_L, P)} \tag{26.46}$$

Substitution of Equation 26.44 and Equation 26.46 into Equation 26.42 leads to

$$\eta = \frac{\Delta g(T_L, P)}{\Delta h(T_L, P)} + \frac{T_L}{\Delta h(T_L, P)}(\wp_{s,\text{HE}} + \wp_{s,\text{TER}}) \tag{26.47}$$

where $\Delta g(T_L, P) = \Delta h(T_L, P) - T_L \Delta s(T_L, P)$ is the change in the Gibbs function between the reaction product and reactant. If all the processes within the heat engine and high temperature TER are reversible ($\wp_{s,\text{HE}} = 0$ and $\wp_{s,\text{TER}} = 0$), then Equation 26.47 reduces to

$$\eta_r = \frac{\Delta g(T_L, P)}{\Delta h(T_L, P)} \tag{26.48}$$

which is exactly the same as Equation 26.37—the efficiency expression derived for fuel cells. Note that for the combustion process to be reversible (i.e., $\wp_{s,\text{HE}} = 0$), there should theoretically be no product dissociations and no incomplete combustion products or by-products (such as pollutants) formed, and the perfect combustion products should consist of stable chemical species only, as would be obtained from an ideal and complete stoichiometric reaction. Therefore, it may be stated that any reversible heat engine operating under the maximum temperature limit allowed by a "perfect" combustion of a fuel/oxidant mixture has the same efficiency as that of a reversible isothermal fuel cell using the same fuel and oxidant and operating at the same temperature as that of the low temperature TER. Or simply stated, the maximum possible efficiency is the same for both fuel cells and heat engines.

It should be emphasized that the combustion process is inherently irreversible, and other irreversibilities occur in the heat engine as well, so that the actual efficiency for heat engine is much lower than the maximum allowed by the second law. Similarly, fuel cells can never achieve, although it is quite possible to achieve very closely, the maximum possible efficiency allowed by the second law. Therefore, the actual energy efficiency for fuel cells is typically much higher than the heat engine. The various mechanisms of irreversible losses in fuel cells will be described later in this chapter.

26.4.2.5 The Possibility of Over 100% Fuel Cell Efficiency: Is It Real or Hype?

It is well known that no heat engine could have efficiency of 100% or more, including the ideal Carnot efficiency, as discussed earlier. However, it has been reported that the ideal fuel cell efficiency, η_r, according to Equation 26.37, could, in principle, be over 100% for some special fuel cell reactions (e.g., [2] and [8]), even though it is unachievable in practice. This has also sometimes been used as evidence that fuel cells could have higher energy efficiencies than the competing heat engines. Is this realistic even under the thermodynamically reversible condition? The answer is no. With the following analysis, it can be shown that this is really due to a conceptual error in stretching the application of Equation 26.37 beyond its validity range.

Consider the thermodynamic model system used for fuel cell analysis, as shown in Figure 26.3. The amount of heat transfer from the surrounding thermal bath to the fuel cell system is given in Equation 26.33 for practical fuel cells. Under the thermodynamically reversible condition, the amount of heat transfer becomes

$$q = T\Delta s = \Delta h - \Delta g. \tag{26.49}$$

For most of fuel cell systems, Δs is negative (i.e., $\Delta s < 0$, just like Δh and Δg), indicating that heat is actually transferred from the fuel cell to the ambient environment, or heat is lost from the fuel cell system, rather than the other way around. Hence, the reversible efficiency, according to Equation 26.37,

$$\eta_r = \frac{\Delta g}{\Delta h} = \frac{\Delta h - T\Delta s}{\Delta h} = 1 - \frac{T\Delta s}{\Delta h} < 1, \tag{26.50}$$

is less than 100%, as it should be by the common perception of the parameter called *efficiency*.

However, for some special reactions, such as

$$C(s) + \frac{1}{2}O_2(g) \rightarrow CO(g), \tag{26.51}$$

the entropy change, Δs, is positive. Physically, it indicates that the fuel cell absorbs heat from the ambient and converts it completely into electrical energy along with the chemical energy of the reactants. This is equivalent to stating that the less useful form of energy—heat—is converted completely into the more useful form of energy—electric energy—without the generation of entropy (i.e., reversible condition) during the conversion process when Equation 26.37 is used for the efficiency calculation; such a process is clearly a violation of the second law. Therefore, the reversible efficiency for this particular fuel cell reaction becomes larger than 100%, that is, a physically impossible result, when Equation 26.37 is utilized for the efficiency calculation for this type of fuel cell reaction. According to Equation 26.37, the reversible fuel cell efficiency for the reaction shown in Equation 26.51 would be equal to $\eta_r = 124\%$ at the standard temperature and pressure, 163% at 500°C and 1 atm, and 197% at 1000°C and 1 atm.

The root of the problem from the straightforward application of Equation 26.37 leading to the physically impossible result of over 100% energy efficiency is as follows: At atmospheric temperature for fuel cell operations, the energy from the thermal bath (or the atmosphere) as heat may be free. But at elevated temperatures, external means must be employed to keep the thermal bath at temperatures above the ambient atmospheric temperature, which constitutes an expense. Therefore, the heat from the thermal bath to the fuel cell system is no longer a free-energy input, rather it is part of the energy input that has to be paid for. By definition, Equation 26.31, the efficiency definition for fuel cells, must be modified accordingly, such that the ideal reversible efficiency will be no longer over 100% for fuel cells. Thus, the reversible fuel cell efficiency shown in Equation 26.37 is only valid for fuel cell reactions where the entropy change between the product and reactant is negative (hence, heat is lost from the fuel cell), and it cannot be applied for reactions with positive entropy change, such as that given in Equation 26.51.

26.4.3 Practical Fuel Cell Efficiency and Energy-Loss Mechanisms

From the preceding analysis, it is clear that energy loss in fuel cells occurs under both reversible and irreversible conditions. Each type of energy loss mechanism and its associated expression for energy conversion efficiency in fuel cells are described in the following section.

26.4.3.1 Reversible Energy Loss and Reversible Energy Efficiency

The energy loss in fuel cells under reversible conditions is equal to the heat transferred (or lost) to the environment, as given in Equation 26.49,

$$q = T\Delta s = \Delta h - \Delta g, \tag{26.49}$$

because of the negative entropy change for the fuel cell reaction. The associated energy conversion efficiency, which has been called the *reversible energy conversion efficiency*, has been derived and given in Equation 26.37 or Equation 26.50. Combining Equation 26.49 with Equation 26.50 yields

$$\eta_r = \frac{\Delta g}{\Delta h} = \frac{\Delta g}{\Delta g + T\Delta s}. \tag{26.52}$$

Dividing the numerator and the denominator by the factor (nF), and utilizing Equation 26.21, Equation 26.52 becomes [1]

$$\eta_r = \frac{E_r}{E_r - T\left(\frac{\partial E_r}{\partial T}\right)_P}. \tag{26.53}$$

Therefore, when the entropy change is negative, as described earlier, the reversible efficiency, η_r, is less than 100% and the reversible cell potential decreases with temperature; and according to Equation 26.53, the reversible efficiency, η_r, also decreases with temperature. For example, for H_2 and O_2 reaction forming gaseous water at 1 atm pressure,

$$\left(\frac{\partial E_r}{\partial T}\right)_P = -0.2302 \times 10^{-3}\, V/K$$

at 25°C, and the reversible efficiency is about 95% at 25°C, and it becomes 88% at 600 K and 78% at 1000 K. Figure 26.9 illustrates the reversible efficiency as a function of temperature for the hydrogen and oxygen reaction with gaseous water as the reaction product. It is seen that the reversible efficiency decreases almost linearly. For most fuel cell reactions,

$$\left(\frac{\partial E_r}{\partial T}\right)_P = -(0.1 \sim 1.0) \times 10^{-3}\, V/K$$

at 25°C and 1 atm, hence the reversible efficiency is typically around 90%.

However, for the reaction of carbon and oxygen to form carbon monoxide, as shown in Equation 26.51, the entropy change is positive, and the reversible cell potential increases with temperature, as presented previously; hence, the reversible efficiency will also increase with temperature, according to Equation 26.53. But as discussed previously, the efficiency expression, Equation 26.52 or Equation 26.53, is not valid for such reactions.

From the reversible energy efficiency (Equation 26.37) and dividing the numerator and denominator by the factor (nF) gives, after utilizing the reversible cell potential (Equation 26.21):

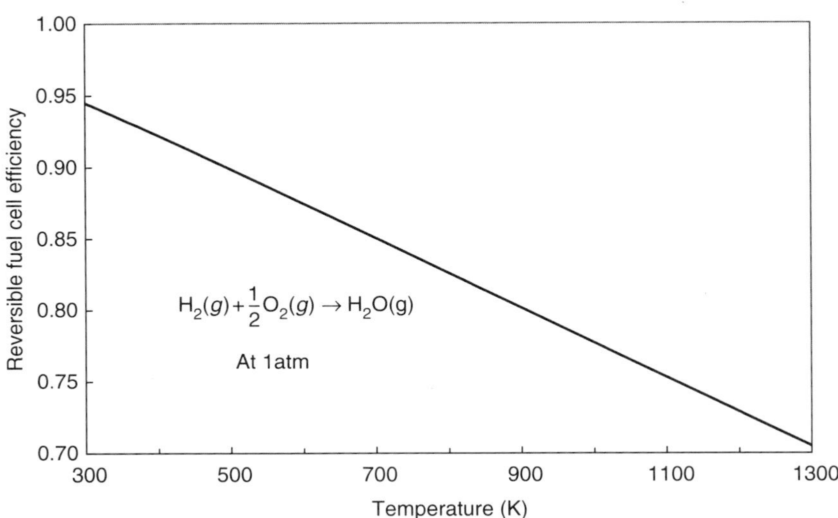

FIGURE 26.9 The reversible fuel cell efficiency (based on LHV) as a function of temperature for the reaction of $H_2 + \frac{1}{2}O_2 \rightarrow H_2O(g)$ occurring at 1 atm pressure.

$$\eta_r = \frac{E_r}{(-\Delta h/nF)} = \frac{E_r}{E_{tn}}, \tag{26.54}$$

where

$$E_{tn} = -\frac{\Delta h}{nF} \tag{26.55}$$

is called *thermoneutral voltage* (or potential)—a voltage a fuel cell would have if all the chemical energy of the fuel and oxidant is converted to electric energy (i.e., 100% energy conversion into electricity). For example, for the reaction

$$H_2(g) + \frac{1}{2}O_2(g) \rightarrow H_2O(\ell),$$

$E_{tn} = 1.48$ V and the corresponding reversible efficiency is $\eta_r = 83\%$ at 25°C and 1 atm, whereas at the same temperature and pressure, for the reaction

$$H_2(g) + \frac{1}{2}O_2(g) \rightarrow H_2O(g),$$

$E_{tn} = 1.25$ V and the corresponding reversible efficiency is $\eta_r = 95\%$.

From the above discussion, it is noted that for the hydrogen and oxygen reaction, the reversible cell efficiency can differ by as much as 14%, depending on whether the product water is liquid or vapor, or whether the higher or lower heating value is used for the efficiency calculation under identical operating conditions. For most of the hydrocarbon fuels that contain hydrogen (including hydrogen itself, hydrocarbons, alcohols, and to a lesser extent coal), there exist two values for the change in the enthalpy and Gibbs function, e.g.,

$$\text{For natural gas (methane, } CH_4\text{)}: \quad \frac{\text{LHV}}{\text{HHV}} = 0.90;$$

$$\text{For coals of typical hydrogen and water content}: \quad \frac{\text{LHV}}{\text{HHV}} = 0.95 \sim 0.98.$$

Therefore, different efficiency values result, depending on which heating value ($-\Delta h$) is used for the efficiency calculation. Typically in fuel cell analysis, the HHV is used unless stated otherwise, and this convention will be used throughout this chapter unless explicitly stated otherwise.

It should be emphasized that from the preceding analysis it is known that for most fuel cell reactions, the reversible efficiency, η_r, decreases as the fuel cell operating temperature is increased. This effect is important in considering high-temperature fuel cells, namely the molten-carbonate fuel cells and the solid-oxide fuel cells. For example, Figure 26.9 indicates that the reversible cell efficiency is reduced to less than 70% (based on LHV) for the hydrogen and oxygen reaction at the typical operating temperature of 1000°C for solid-oxide fuel cells, as opposed to around 95% at 25°C as discussed above. This significant reduction in the reversible cell efficiency seems to work against high-temperature fuel cells. However, the irreversible losses, to be described below, decrease drastically as temperature is increased, so that the practical fuel cell performance (such as efficiency and power output under practical operating condition) increases. Therefore, further analysis should be carried out for efficiency under practical operating conditions rather than the idealized reversible condition; this is the focus of the following discussion.

26.4.3.2 Mechanism of Irreversible Energy Losses

For fuel cells, the reversible cell potential and the corresponding reversible efficiency are obtained under the thermodynamically reversible condition, implying that there is no rigorous occurrence of continuous reaction or electrical-current output. For practical applications, a useful amount of work (electrical energy) is obtained only when a reasonably large current, I, is drawn from the cells because the electrical energy output is through the electrical power output that is defined as

$$\text{Power} = EI \quad \text{or} \quad \text{Power Density} = EJ. \tag{26.56}$$

However, both the cell potential and efficiency decrease from their corresponding (equilibrium) reversible values because of irreversible losses when current is increased. These irreversible losses are often called *polarization, overpotential*, or *overvoltage** in the literature, and they originate primarily from three sources: activation, ohmic, and concentration polarization. The actual cell potential as a function of current is the result of these polarizations; therefore, a plot of the cell potential vs. current output is conventionally called a *polarization curve*. It should be noted that the magnitude of electrical current output depends largely on the active cell area; therefore, a better measure is the current density, J, (A/cm^2) instead of current, I, itself; and the unit A/cm^2 is often used rather than A/m^2 as the unit for the current density because square meter is too large to be used for fuel cell analysis.

A typical polarization curve is illustrated in Figure 26.10 for the cell potential as a function of current density. The ideal cell-potential-current relation is independent of the current drawn from the cell, and the cell potential remains equal to the reversible cell potential. The difference between the thermoneutral voltage and the reversible cell potential represents the energy loss under the reversible condition (the reversible loss). However, the actual cell potential is smaller than the reversible cell potential and decreases as the current drawn is increased due to the three mechanisms of irreversible losses: activation, ohmic, and concentration polarization. The activation polarization, η_{act}, arises from the slow rate of electrochemical reactions, and a portion of the energy is lost (or spent) on driving up the rate of electrochemical reactions to meet the rate required by the current demand. The ohmic polarization,

*The terms *polarization, overpotential*, and *overvoltage* have been loosely used in literature to denote cell potential (or voltage) loss. However, they do have some subtle differences [1] that are neglected here so that these words might be used interchangeably.

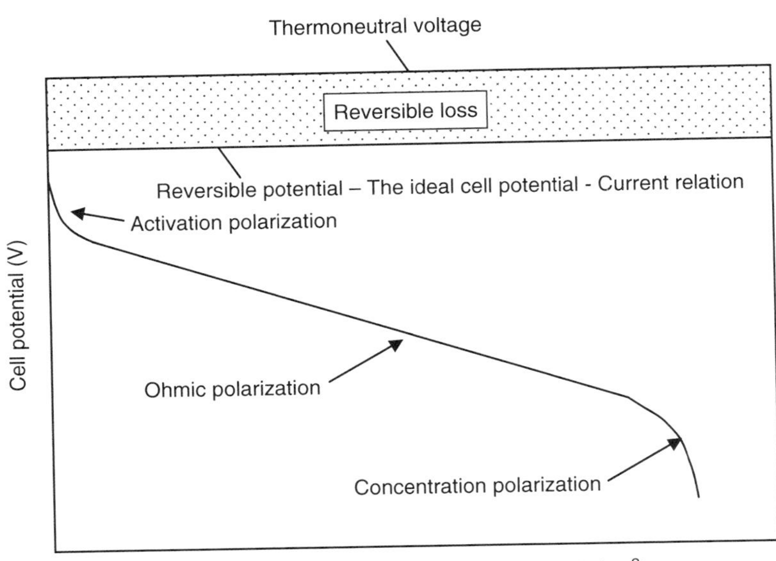

FIGURE 26.10 Schematic of a typical polarization curve. The cell potential for a fuel cell decreases as the current drawn from the cell is increased due to activation, ohmic and concentration polarizations.

η_{ohm}, arises due to electrical resistance in the cell, including ionic resistance to the flow of ions in the electrolyte and electronic resistance to the flow of electrons in the rest of the cell components. Normally, the ohmic polarization is linearly dependent on the cell current. Concentration polarization, η_{conc}, is caused by the slow rate of mass transfer resulting in the depletion of reactants in the vicinity of active reaction sites and the overaccumulation of reaction products that block the reactants from reaching the reaction sites. It usually becomes significant, or even prohibitive, at high current density, when the slow rate of mass transfer is unable to meet the high demand required by the high current output. As shown in Figure 26.10, concentration polarization is often the cause of the cell potential decreasing rapidly to zero. The current (density) corresponding to the zero cell potential is often called the *limiting current* (density), and evidently it is controlled by the concentration activation. From Figure 26.10, it is also clear that activation polarization is dominant at small current densities, whereas concentration polarization is predominant at high current densities. The linear drop in the cell potential due to resistance loss occurs at intermediate current densities, and practical fuel cell operation is almost always located within the ohmic polarization region. It should be emphasized that these three loss mechanisms actually occur simultaneously in an operating cell, despite their different influences at the different current density conditions.

Figure 26.10 also indicates that even at zero current output from the fuel cell, the actual cell potential is smaller than the idealized reversible cell potential. This small difference in cell potential is directly related to the chemical potential difference between the cathode and anode. Consequently, even at zero external load current, there are electrons delivered to the cathode, where oxygen ions are formed, and the electrons migrate through the electrolyte to the anode where they deionize to release electrons. The released electron migrates back to the cathode to continue the process or "exchange." The ionization/deionization reactions proceeding at a slow rate yield an extremely small current, often called *exchange current*, I_0, or *exchange current density*, J_0, and the cell potential is reduced below the reversible cell potential. Therefore, exchange current arises from the electrons migrating through the electrolyte rather than through the external load, and about $0.1\sim0.2$ V of cell potential loss results from the exchange process. Consequently, the efficiency of a real fuel cell is approximately $(8\sim16)\%$ lower than the reversible cell efficiency, η_r, even at close to zero current output.

The exchange current density, J_0, is very small; it is at least about 10^{-2} A/cm^2 for H_2 oxidation at the anode, and about 10^{-5} times lower for O_2 reduction at the cathode. In comparison, the O_2 reduction process at the cathode is so slow that competing anodic reactions play a significant role, such as oxidation of the electrocatalyst, corrosion of the electrode materials, and oxidation of organic impurities in the electrode structure. All these anodic reactions result in the corrosion of electrodes, thereby limiting the cell life unless appropriate countermeasures are taken.

It should be pointed out that the cell potential loss resulting from the exchange current diminishes when the current drawn through the external load is increased beyond a certain critical value. As the external current is increased, the cell potential decreases as shown in Figure 26.10. Consequently, the driving force for the exchange current is reduced, leading to a smaller exchange current; this is the only form of energy loss that decreases when the external current is increased.

26.4.3.3 Amount and Rate of Waste Heat Generation

From the above discussion, it becomes clear that the actual cell potential E is lower than the reversible cell potential E_r, and the difference is due to the potential losses arising from the above irreversible loss mechanisms. Therefore,

$$E = E_r - (\eta_{act} + \eta_{ohm} + \eta_{conc}). \tag{26.57}$$

The irreversible energy loss as heat (or waste heat generation) per mole fuel consumed can be easily obtained, because the entropy generation is, according to Equation 26.24 and Equation 26.57:

$$\wp_s = \frac{nF(E_r - E)}{T} = \frac{nF(\eta_{act} + \eta_{ohm} + \eta_{conc})}{T}. \tag{26.58}$$

Then, Equation 26.13 for the total heat loss from the fuel cell becomes

$$q = T\Delta s - T\wp_s = \underbrace{T\Delta s}_{\text{Reversible Loss}} \underbrace{-nF(\eta_{act} + \eta_{ohm} + \eta_{conc})}_{\text{Irreversible Losses}} \tag{26.59}$$

Because the entropy change is negative ($\Delta s < 0$) for most fuel cell reactions, the heat transfer is negative as well, implying that energy is lost as heat from the fuel cell shown in Figure 26.3 for both reversible and irreversible losses.

Because $T\Delta s = \Delta h - \Delta g$ by the definition of the Gibbs function change for fuel cell reactions, Equation 26.59 can be written as

$$\frac{q}{nF} = \frac{T\Delta s}{nF} - (\eta_{act} + \eta_{ohm} + \eta_{conc}) = \frac{\Delta h - \Delta g}{nF} - (\eta_{act} + \eta_{ohm} + \eta_{conc}). \tag{26.60}$$

Considering the definition for the thermoneutral voltage and the reversible cell potential, the above expression becomes

$$\frac{q}{nF} = -E_{tn} + E_r - (\eta_{act} + \eta_{ohm} + \eta_{conc}). \tag{26.61}$$

Combining with Equation 26.57, Equation 26.61 reduces to

$$\frac{q}{nF} = -(E_{tn} - E). \tag{26.62}$$

Hence, the equivalent cell potential loss due to the energy loss in the fuel cell as heat is equal to the difference between the thermoneutral voltage and the actual cell potential.

The rate of heat loss per mole of fuel consumed in the fuel cell, Equation 26.59, can be expressed as an equivalent power loss:

$$P_{\text{Heat Loss}} = I\left(\frac{q}{nF}\right) = -I(E_{\text{tn}} - E) = I\left[\frac{T\Delta s}{nF} - (\eta_{\text{act}} + \eta_{\text{ohm}} + \eta_{\text{conc}})\right]. \tag{26.63}$$

This expression is important in determining the cooling requirements of fuel cell stacks.

26.4.3.4 Various Forms of Irreversible Energy Conversion Efficiency

After the above description of the irreversible energy losses, several forms of energy efficiency that would be useful in the analysis of fuel cell performance may now be introduced.

(1) Voltage Efficiency, η_E:

The voltage efficiency is defined as:

$$\eta_E = \frac{E}{E_r}. \tag{26.64}$$

Because the actual cell potential, E, is compared with the maximum possible cell potential, E_r, allowed by the second law, the voltage efficiency is really a specific form of the exergy efficiency, representing the degree of departure of the cell operation from the idealized thermodynamically reversible condition. As shown in Equation 26.57, $E < E_r$, hence $\eta_E < 1$.

(2) Current Efficiency η_I:

The current efficiency is a measure of how much current is produced from a given amount of fuel consumed in the fuel cell reaction; it is defined as

$$\eta_I = \frac{I}{nF\left(\frac{dN_F}{dt}\right)}. \tag{26.65}$$

where $\frac{dN_F}{dt}$ represents the rate of fuel consumption in the fuel cell (mol/s). The current efficiency would be less than 100% if part of the reactants participate in non-current-productive side reactions, called parasitic reactions, such as reactant crossover through the electrolyte region, incomplete conversion of reactants to the desired products, reaction with the cell components, or even reactant leakage from the cell compartment due to a sealing problem, etc. For example, in direct-methanol fuel cells, about 20% of the liquid methanol can cross over to the cathode side through the proton-conducting polymer membrane, implying the current efficiency is only about 80% for such cells. However, for most practical fuel cells, especially at operating conditions where the current output is sufficiently larger than zero (without the effect of exchange current discussed previously), the current efficiency is about 100%. This is because for practical fuel cells, all the parasitic reactions are undesirable and would have been removed by appropriate design.

(3) Overall Free Energy Conversion Efficiency, η_{FC}:

The overall free-energy conversion efficiency is defined as the product of the reversible efficiency, voltage, and current efficiency:

$$\eta_{FC} = \eta_r \times \eta_E \times \eta_I. \tag{26.66}$$

If the current efficiency is 100%, as is often the case for well-designed practical fuel cells, substituting the definitions for the various efficiencies into the above equation leads to

$$\eta_{FC} = \frac{E}{E_{tn}}. \tag{26.67}$$

Therefore, the overall free-energy conversion efficiency is really the overall efficiency for energy conversion process occurring within the fuel cell. Because the thermoneutral voltage is a fixed value for a given fuel and oxidant under a given operating condition of temperature and pressure, the overall energy conversion efficiency for fuel cells is proportional to the actual cell potential,

$$\eta_{FC} \sim E. \tag{26.68}$$

This is an important result. After the actual cell potential is determined, the energy conversion efficiency of the fuel cell is known as well. This is the primary reason that in the fuel cell literature, the cell polarization curve is almost always given without specifically showing the cell energy efficiency as a function of the current. Further, Equation 26.68 implies that the fuel cell efficiency will depend on the current output in the same way as the cell potential, i.e., decrease as the current output is increased.

(4) Fuel Cell System Efficiency, η_s

Because a fuel cell system is composed of one or multiple fuel cell stacks and auxiliary equipment that would also have its own energy efficiency of η_{aux}, the total fuel cell system efficiency is

$$\eta_s = \eta_{FC} \times \eta_{aux}. \tag{26.69}$$

26.4.4 Efficiency Loss in Operating Fuel Cells: Stoichiometry, Utilization, and Nernst Loss

In an operating fuel cell, reactant composition changes between the inlet and outlet of the fuel cell along the flow path over the electrode surface because reactants are consumed to yield current output and reaction products are formed along the way. The change in reactant composition results in additional loss of cell potential beyond those losses described in the preceding section. This potential loss arises from the fact that the cell potential E adjusts to the lowest electrode potential given by the Nernst equation, Equation 26.29, for the various reactant compositions at the exit of the anode and cathode chambers. This is because electrodes are usually made of good electronic conductors and they are consequently isopotential surfaces. The cell potential E may not exceed the minimum local value set by the Nernst equation. This additional cell potential loss is often called the *Nernst loss*, which is equal to the difference between the inlet and exit Nernst potentials determined from the inlet and exit reactant compositions. According to Equation 26.29, this additional cell potential loss due to the consumption of reactants in the cell is, when the reactant streams are arranged in a concurrent flow,

$$\eta_N = \frac{\Re T}{nF} lnK_{out} - \frac{\Re T}{nF} lnK_{in} = \frac{\Re T}{nF} ln\frac{K_{out}}{K_{in}}. \tag{26.70}$$

In the case of a fuel cell where both fuel and oxidant flow is in the same direction (concurrent), the minimum Nernst potential occurs at the flow outlet. When the reactant flows are in counter-flow, cross flow, or more complex arrangements, it becomes difficult to determine the location of the minimum potential due to the reactant consumption. Appropriate flow channel design for the anode and cathode side can minimize the Nernst loss.

Equation 26.70 also implies that the Nernst loss will be extremely large, approaching infinity if all the reactants are consumed in the in-cell electrochemical reaction leading to zero reactant concentration at the cell outlet. To reduce the Nernst loss to an acceptable level for practical fuel cell operations, reactants are almost always supplied in greater than the stoichiometric amounts required for the desired current production. The actual amount of reactants supplied to a fuel cell is often expressed in terms of a parameter called *stoichiometry*, S_t:

$$S_t = \frac{\text{Molar flow fate of reactants supplied to a fuel cell}}{\text{Molar flow rate of reactants consumed in the fuel cell}} = \frac{\dot{N}_{in}}{\dot{N}_{consumed}}. \tag{26.71}$$

For example, for proton-exchange-membrane fuel cells, typical operation uses $S_t \approx (1.1 \sim 1.2)$ for H_2 and St$\approx$2 for O_2 (pure or in air). Therefore, stoichiometry really represents the actual flow rate for the reactant delivered to the fuel cell. Because there are normally at least two types of reactants for a fuel cell, one as fuel and another as oxidant, stoichiometry can be defined for either reactant.

Alternatively, reactant flow rate can be expressed in terms of a parameter called *utilization*, U_t:

$$U_t = \frac{\text{Molar flow rate of reactants consumed in a fuel cell}}{\text{Molar flow rate of reactants supplied into the fuel cell}} = \frac{\dot{N}_{\text{consumed}}}{\dot{N}_{\text{in}}}. \tag{26.72}$$

Clearly, stoichiometry and utilization are inversely proportional to each other.

For properly designed practical fuel cells, no reactant crossover or leakage out of the cell may occur in general; therefore, the rate of reactant consumed within the cell is equal to the difference between the molar flow rate into and exiting the cell. For example, the stoichiometry for the fuel may be expressed as

$$S_{t,F} = \frac{\dot{N}_{F,\text{in}}}{\dot{N}_{F,\text{consumed}}} = \frac{\dot{N}_{F,\text{in}}}{\dot{N}_{F,\text{in}} - \dot{N}_{F,\text{out}}} = \frac{1}{U_{t,F}}. \tag{26.73}$$

Effect of reactant utilization on the reversible cell potential is illustrated in Figure 26.11. It is seen that the reactant composition at the cell outlet decreases, hence the reversible cell potential decreases as well when the utilization factor is increased. The decrease is rapid when utilization goes beyond about 90%. In practical fuel cell operation, 100% utilization (or unity stoichiometry) will result in reactant concentrations vanishing at the cell exit, then the Nernst loss becomes dominant and the cell potential is reduced to zero; this is certainly an undesirable situation that must be avoided. Therefore, typical operation requires that the utilization be about (80$\sim$90)% for fuel, and 50% for oxidant to balance the Nernst loss with the parasitic losses associated with the reactant supply.

As shown in Equation 26.70, the additional Nernst loss due to the reactant depletion in the cell is directly proportional to the cell operating temperature. Figure 26.12 shows the reversible cell potential at

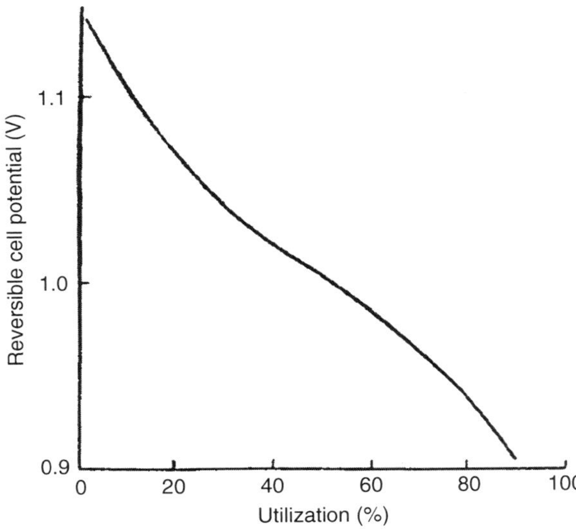

FIGURE 26.11 Reversible cell potential as a function of reactant utilization (both fuel and oxidant utilizations are set equal) for a molten carbonate fuel cell operating at 650̂C and 1 atm. Reactant compositions at the cell inlet: 80% H_2/20% CO_2 mixture saturated with H_2O(g) at 25̂C for the fuel gas; and 60% CO_2/30% O_2/10% inert gas mixture for the oxidant gas [4].

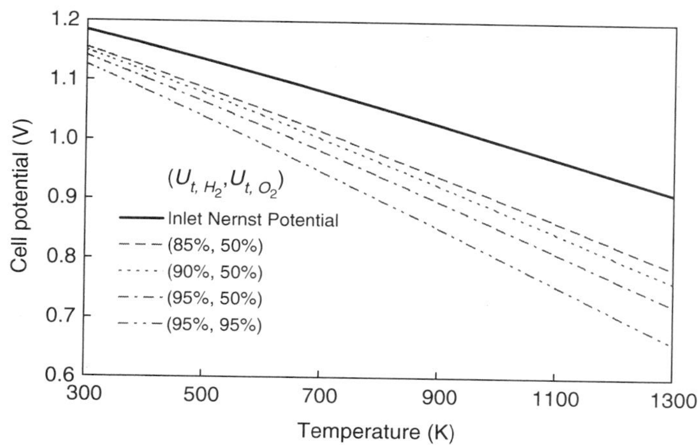

FIGURE 26.12 Inlet and outlet Nernst potential as a function of temperature and utilization for the reaction of $H_2(g) + \frac{1}{2}O_2(g) \rightarrow H_2O(g)$ at 1 atm.

the cell inlet and outlet for the hydrogen and oxygen reaction forming gaseous water product at 1 atm as a function of temperature. The outlet Nernst potentials (i.e., the reversible cell potential at the cell outlet) are determined for oxygen utilization of 50%, and hydrogen utilizations of 85%, 90%, and 95%, respectively, as well as for the utilization of 95% for both hydrogen and oxygen. It is clearly seen that the outlet Nernst potential decreases when either utilization or temperature is increased.

If pure hydrogen is used as fuel, the anode compartment can be designed as a dead-end chamber for hydrogen supply. Similarly, if pure oxygen is used as oxidant, a dead-end cathode compartment can be employed. However, inert impurities in the reactant gas will accumulate at the anode and cathode compartments, and they must be removed either periodically or continuously to maintain good fuel cell performance. Periodic purging or continuous bleeding can be implemented for this purpose, but this results in a small loss of fuel, and hence less than 100% utilization.

From the above discussion, it is evident that 100% utilization for reactants is practically an unwise design. Because in-cell fuel utilization will never be 100% in practice, the determination of in-cell energy conversion efficiency and the cell potential must take utilization factor into consideration. If the fuel exiting the fuel cell is discarded (not recirculated back to the cell or not utilized for other useful purpose such as providing heat for fuel preprocessing), then the overall energy conversion efficiency must be equal to the overall fuel cell efficiency given in Equation 26.66 multiplied by the utilization to take into account the fact that not all the fuel is being used for electric energy production.

26.5 Fuel Cell Electrode Processes

The thermodynamic process described in the previous section for a fuel cell is a gross underestimate of what happens in reality. It has been identified that there are many physical (i.e., transport of mass, momentum, and energy) and chemical processes involved in the overall electrochemical reactions in the porous fuel cell electrodes that influence the performance of fuel cells. The transport processes involving the mass transfer of reactants and products play a prominent role in the performance of porous electrodes in fuel cells, and those involving heat transfer and thermal management are important in fuel cell systems. Some of the important physical and chemical processes occurring in porous fuel cell electrodes during electrochemical reactions for liquid electrolyte fuel cells are [4]:

1. First, the reactant stream consists of a multicomponent gas mixture. For example, the fuel stream typically contains hydrogen and water vapor, as well as carbon dioxide and even some carbon monoxide, whereas the oxidant stream usually contains oxygen, nitrogen, water vapor, carbon

dioxide, etc. The molecular reactant (such as H_2 or O_2) is transferred to the porous electrode surface from the reactant stream through the mechanism of convection; it is then transported through the porous electrode, primarily by diffusion, to reach the gas/electrolyte interface.
2. The reactant dissolves into the liquid electrolyte at the two-phase interface.
3. The dissolved reactant then diffuses through the liquid electrolyte to arrive at the electrode surface.
4. Some pre-electrochemical homogeneous or heterogeneous chemical reactions may occur, such as electrode corrosion reaction, or impurities in the reactant stream may react with the electrolyte.
5. Electroactive species (which could be reactant itself as well as impurities in the reactant stream) are adsorbed onto the solid electrode surface.
6. The adsorbed species may migrate on the solid electrode surface, principally by the mechanism of diffusion.
7. Electrochemical reactions then occur on the electrode surface wetted by the electrolyte—the so-called three-phase boundary, giving rise to electrically charged species (or ions and electrons).
8. The electrically charged species and other neutral reaction product such as water, still adsorbed on the electrode surface, may migrate along the surface due to diffusion in what has been referred to as *post-electrochemical surface migration*.
9. The adsorbed reaction products become desorbed.
10. Some post-electrochemical homogeneous or heterogeneous chemical reactions may occur.
11. The electrochemical reaction products (neutral species, ions, and electrons) are transported away from the electrode surface, mainly by diffusion but also influenced, for the ions, by the electric field set up between the anode and cathode. The electron motion is dominated by the electric field effect.
12. The neutral reaction products diffuse through the electrolyte to reach the reactant gas–electrolyte interface.
13. Finally, the products will be transported out of the electrode and the cell in the gas form.

Normally, any of the above thirteen processes can influence the performance of a fuel cell, exhibiting the complex nature of a fuel cell operation. For well-designed practical fuel cells, the above electrode processes might be grouped into the following three major steps:

1. Delivery of molecular reactant to the electrode surface: This step involves a number of physical and chemical processes preceding the electrochemical reaction, and it generally includes the transport of molecular reactant from the gas-phase supply outside the electrode structure to the liquid electrolyte surface, typically through the porous electrode structure, then the dissolution of molecular reactant into the liquid electrolyte, followed by the diffusion of the dissolved reactant through the electrolyte to the electrode surface, and finally the adsorption of the reactant on the electrode surface.

   It might be pointed out that the delivery of the electrons and ions to the electrode surface is as important as the delivery of molecular reactant because they are all needed for the electrochemical reactions to occur at the electrode surface.

2. Reduction or oxidation of the adsorbed molecular reactant at the electrode surface in the presence of electrolyte: This is the step for electrochemical reactions that produces electric current. The reactions only occur at the electrode surface that is covered by the electrolyte, hence, often referred to as the *three-phase boundary*, or active reaction sites. Significant increase in the reactive sites is essential for good fuel cell performance.
3. Removal of the reaction products from the electrode surface for the regeneration of the reaction sites and for the continuous production of electric current. This is especially important for low-temperature acid-electrolyte fuel cells, because product water is formed at the cathode and is in the liquid form. Liquid water accumulation in the porous electrode structure may block the transport of molecular oxygen to the reaction sites, severely hindering the process described in step 1 above, thereby degrading the cell performance due to oxygen starvation at the reaction sites. Such a

phenomenon is often referred to as the *water flooding* of electrodes, and is a critical issue for PEM fuel cells.

26.6 Cell Connection and Stack Design Considerations

The potential of a working cell is typically around 0.7 ~ 0.8 V, and it is normally too small for practical applications because of the limited power available from a single cell. Therefore, many individual cells are connected (or stacked) together to form a fuel cell stack. Although many stacking configurations are possible, the overpotential associated with the transport processes discussed in this chapter imposes limitations and technical difficulties, making cell stacking one of the significant technical challenges in the drive for the fuel cell commercialization. We will briefly discuss the stacking options below and the associated transport-related issues.

The electrical connection among the individual cells may be arranged in parallel or in series, as shown in Figure 26.13. The parallel connection still provides a low voltage output from the stack, but a very high current output because the stack current is the sum of the current produced in each cell. Such an extremely large current flow will cause an excessively large ohmic voltage loss in the stack components and at the surface contacts among the components. Thus, parallel connection is typically avoided unless for small current or power applications.

Series connection can have two typical arrangements: unipolar and bipolar, as shown in Figure 26.14. The unipolar design has one fuel stream supplying fuel to two anodes for the two adjacent cells, and one oxidant stream delivering oxidant to two cathodes for the two adjacent cells. This arrangement of one reactant stream serving for two adjacent electrodes simplifies the reactant flow channel design. However,

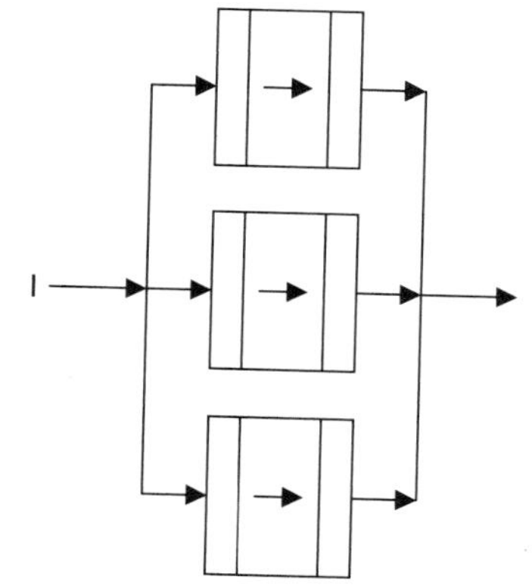

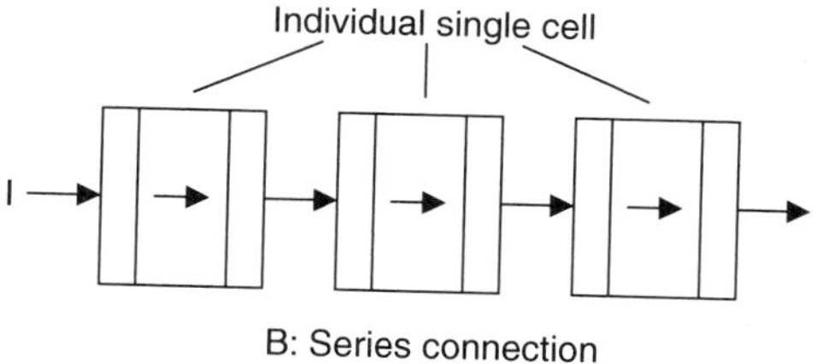

FIGURE 26.13 Cell connection (stacking) configurations. (a) Parallel connection; (b) series connection. The arrow in the diagram represents the direction of the current flow.

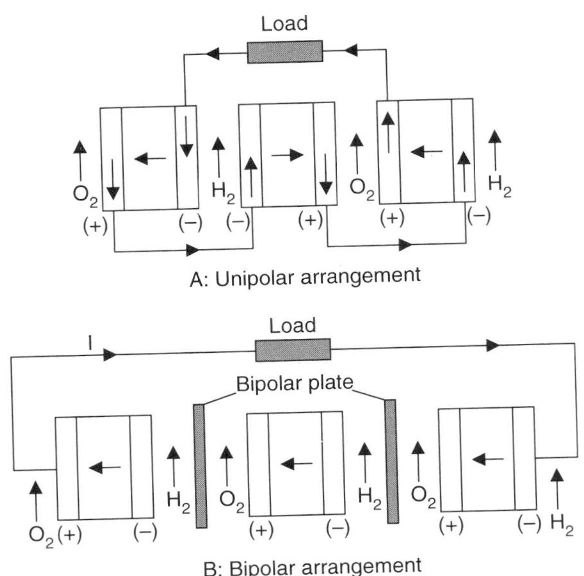

FIGURE 26.14 Series cell connection (stacking) configurations. (a) Unipolar arrangement with the edge collection of the current generated in each cell; (b) bipolar arrangement with the end-plate collection of the current.

it forces the electrical current generated in each cell to be collected at the edge of the electrodes. Because the electrode is very thin (<1 mm), whereas the other electrode dimensions (in the direction of the current flow) are at least on the order of centimeters or larger, the ohmic resistance tends to be very large. Thus, edge collection of current, although used in early fuel cell stack designs, is generally avoided in recent fuel cell stack development, primarily due to the excessively large ohmic voltage losses.

The bipolar arrangement has the current flow normal to the electrode surface, instead of along the electrode surface as in the unipolar arrangement, thus the current flow path is very short, whereas the cross-sectional area available for the current flow is very large. The ohmic voltage loss for this case is very small in comparison, and bipolar design is favored in recent fuel cell stack technology. However, this end-plate collection of current results in the complex design for the reactant flow channels and complex organization for the reactant stream, and the bipolar plate has to fulfill several functions simultaneously to obtain a good overall stack performance. Bipolar plates serve as current collectors, and are responsible for reactant delivery to the electrode surface, cell reaction product removal (such as water), and for the integrity of the cell/stack. These functions for the bipolar plates may be contradictory to each other, and optimal design for the bipolar plates represents one of the significant technical challenges for practical fuel cells.

A typical configuration of a bipolar plate is shown in Figure 26.15. The reactant flow follows the flow channels made on the bipolar plate, thus distributing the reactant over the electrode surface. On the other hand, the land between the adjacent flow channels serves as the passage for the current flow from one cell to the next. Therefore, a wide flow channel is beneficial for the reactant distribution over the electrode surface and for the reaction product removal, whereas a wide land is beneficial for the electron flow and for the mechanical integrity of the cells and the entire stack. Normally, the same cell unit as shown in Figure 26.15c is repeated to form a stack. The plate at the end of the stack only has flow channels on one side of its surfaces.

From the cell repeating unit in a stack, it is clear that the reactant concentration decreases along the flow direction following the flow channel design, and into the electrode due to convectional and diffusional mass transfer. Thus, the concentration field is three dimensional, and so is the electrical field due to the flow channels made on the bipolar plate surfaces, the nonuniform rate of electrochemical

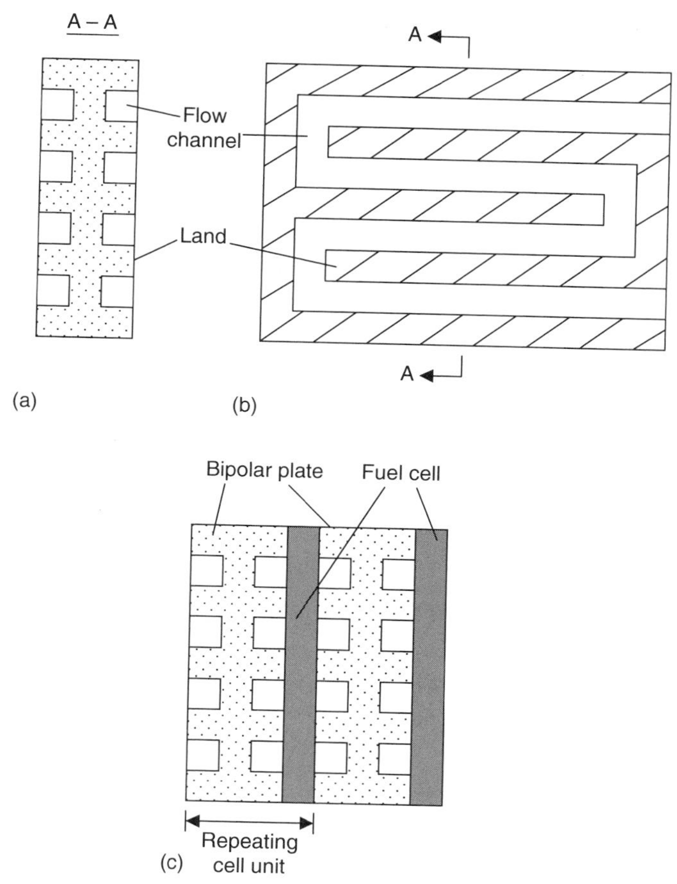

FIGURE 26.15 Typical configuration of a bipolar plate. (a) Cross-sectional view; (b) face view; (c) repeating cell unit in a stack.

reactions in the catalyst layers, and the three-dimensional distribution of the reactant concentrations and the temperature. Therefore, an accurate prediction of the cell performance requires a three-dimensional analysis based on solving the conservation equations governing the transport phenomena of the reactant flow, species concentration, temperature, and electric fields, incorporating the in-cell electrochemical reaction processes. This poses significant challenges to fuel cell designers.

26.7 Six Major Types of Fuel Cells

Fuel cell technology has been developed and improved dramatically in the past few years, and this has once again captured both the public's and industry's attention concerning the prospect of fuel cells as practical power sources for terrestrial applications. At present, fuel cell technology is being routinely used in many specific areas, notably in space explorations, where fuel cell operates on pure hydrogen and oxygen with over 70% efficiency with drinkable water as the only by-product. There are now approximately over 200 fuel cell units for terrestrial applications operating in 15 countries. Impressive technical progress has been achieved in terms of higher power density and better performance as well as reduced capital and maintenance and operation cost; this is driving the development of competitively priced fuel cell-based power generation systems with advanced features for terrestrial use, such as utility power plants and zero-emission vehicles. In view of decreasing fossil-fuel reserves and increasing energy demands worldwide, fuel cells will probably become one of the major energy technologies with fierce international competition in the twenty-first century.

The major terrestrial commercial applications of fuel cells are electric power generation in the utility industry, and as a zero-emission powertrain in the transportation sector. For these practical applications, the efficiencies of fuel cells range some where between 40% and 65% based on the lower heating value (LHV) of hydrogen. Typically, the cell electric potential is only about 1 V across a single cell, and it decreases due to various loss mechanisms under operational conditions. Thus, multiple cells are required to be connected together in electrical series to achieve a useful voltage for practical purposes, and these connected cells are often referred to as a *fuel cell stack*. A fuel cell system consists of one or multiple fuel cell stacks connected in series and/or parallel, and the necessary auxiliaries whose composition depends on the type of fuel cells and the kind of primary fuels used. The major accessories include thermal management (or cooling) subsystem, fuel supply, storage and processing subsystem, and oxidant (typically air) supply and conditioning subsystem.

In this section, a summary of the state-of-the-art technology for the six major types of fuel cells is presented, including

- Alkaline fuel cells (AFCs)
- Polymer-electrolyte-membrane fuel cells (PEMFCs)
- Direct-methanol fuel cells (DMFCs)
- Phosphoric-acid fuel cells (PAFCs)
- Molten-carbonate fuel cells (MCFCs)
- Solid-oxide fuel cells (SOFCs)

Five of these are classified based on their electrolytes used, including the alkaline, phosphoric-acid, polymer-electrolyte-membrane, molten-carbonate, and solid-oxide fuel cell. The direct-methanol fuel cell is classified based on the fuel used for electricity generation. Table 26.2 provides a summary of the operational characteristics and application of the six major types of fuel cell.

26.7.1 Alkaline Fuel Cells

Alkaline fuel cells (AFCs) give the best performance among all the fuel cell types under the same or similar operating conditions when running on pure hydrogen and oxygen. Hence, they are among the first fuel cells to have been studied and taken into development for practical applications, and they are the first type of fuel cells to have reached successful routine applications, mainly in space programs such as space shuttle missions in the United States and similar space exploration endeavors in China and Europe, where pure hydrogen and oxygen are used as reactants. Because of their success in space programs, AFCs are also the type of fuel cells on which probably the largest number of fuel cell development programs have begun in the world in an effort to bring them to terrestrial applications, particularly in Europe. However, almost all of the AFC development programs have come to an end. At present, the few activities related to AFC R&D are in Europe.

The AFCs have the highest energy conversion efficiency among all types of fuel cells under the same operating conditions if pure hydrogen and pure oxygen are used as the reactants. That was one of the important reasons that AFCs were selected for the US space shuttle missions. The AFCs used in the shuttle missions are operated at about 200°C for better performance (i.e., high energy conversion efficiency of over 70% and high power density that is critical for space applications), and the alkaline electrolyte is potassium hydroxide (KOH) solution immobilized in an asbestos matrix. As a result, the AFCs operate at high pressure to prevent the boiling and depletion of the liquid electrolyte. Consequently, these severe operating conditions of high temperature and high pressure dictate extremely strict requirement for cell-component materials that must withstand the extreme corrosive oxidizing and reducing environments of the cathode and anode. To meet these requirements, precious metals such as platinum, gold, and silver are used for the

TABLE 26.2 Operational Characteristics and Technological Status of Various Fuel Cells

Type of Fuel Cells	Operating Temperature (°C)	Power Density (mW/cm^2) (Present) Projected	Projected Rated Power Level (kW)	Fuel Efficiency (chemical to electrical)	Lifetime Projected (hours)	Capital Cost Projected (U.S.$/kW)	Areas of Application
AFC	60–90	(100–200) > 300	10–100	40–60	> 10,000	> 200	Space, mobile
PAFC	160–220	(200) 250	100–5000	55	> 40,000	> 3,000	Dispersed & distributed power
PEMFC	50–80	(350) > 600	0.01–1000	45–60	> 40,000	> 200	Portable, mobile, space, stationary
MCFC	600–700	(100) > 200	1000–100,000	60–65	> 40,000	1,000	Distributed power generation
SOFC	800–1000	(240) 300	100–100,000	55–65	> 40,000	1,500	Baseload power generation
DMFC	90	(230) ?	0.001–100	34	> 10,000	> 200	Portable, mobile

construction of the electrodes, although these precious metals are not necessary for the electrochemical reactions leading to electric power generation. Each shuttle flight contains a 36-kW AFC power system. Its purchase price is about US$28.5 million, and it costs NASA an additional $12–19 million annually for operation and maintenance. Although the manufacturer claims about 2400 h of lifetime, NASA's experience indicates that the real lifetime is only about 1200 h. With sufficient technology development, 10,000 h are expected as the life potential (or upper limit) for the AFC system. This belief is based on the nature of the AFC systems and the data accumulated on both stacks and single cells.

The typical working temperature of AFC power systems aimed at commercial and terrestrial applications ranges from 20 to 90°C, and the electrolyte is a KOH solution (30–35%). There are four different cell types investigated:

1. Cell with a free liquid electrolyte between two porous electrodes
2. ELOFLUX cell with liquid KOH in the pore systems
3. Matrix cell where the electrolyte is fixed in the electrode matrix
4. The falling film cell

Many technical challenges have been encountered in the development of AFCs. The most important ones are:

- Preparation method of the electrodes: the electrodes consist of porous material that is covered with a layer of catalyst. In general, it is very difficult to distribute the catalyst at the surface and to produce a defined pore system for the transport of the reactants.
- Costs of the electrodes, stacks, and fuel cell systems: the preparation of electrodes with noble metal catalysts is very expensive. In general, the electrodes are manufactured in small-scale production with high overhead costs.
- Lifetime of the electrode/degradation: the electrolyte is very corrosive and the catalyst materials are sensitive to high polarization. Using nickel and silver as catalysts, to reduce the costs of the fuel cell, leads to a high degradation of these catalysts.
- Diaphragm made of asbestos: the diaphragm of low-temperature AFCs is made of asbestos. But this material is hazardous, and in some countries its use is banned. Therefore, new diaphragms should be developed, but it is difficult to find a material with a similar performance in alkaline electrolyte.
- Carbon-dioxide-contaminated fuel and oxidant streams (carbonating of electrolyte and electrodes): the electrolyte intolerance of carbon dioxide is the most important disadvantage of air-breathing AFCs with reformate gases from primary fossil fuels.

Other problems associated with the AFCs power systems are the concerns for the safety and reliability of AFC power systems. For example, the liquid KOH electrolyte contained in an asbestos matrix can only withstand a 5-psi limit of pressure differential between the anode and cathode reactant gases. This dictates the need for sophisticated pinpoint pressure control during the operation, including transient, startup, and shut-down processes. It is also a safety issue because of the likelihood of the reactants mixing in the AFC system with the possibility of a serious fire. In terms of general safety considerations, the use of the corrosive potassium hydroxide electrolyte in the AFCs represents the need for hazardous material handling, and the handling of asbestos matrix poses a potential hazard to one's health. With flowing reactant gases, the potential for the gradual loss of the liquid electrolyte, drying of the electrolyte matrix, reactant crossover of the matrix and ensuing life-limiting reactant mixing (or actual AFC stack failure due to fire) is very real in the AFC system.

The major technical challenge is that alkaline electrolytes, like potassium or sodium hydroxide, do not reject carbon dioxide, even the 300 ~ 350 ppm of carbon dioxide in atmospheric air is not tolerated (carbon dioxide concentration in both cathode and anode gases must be less than 10 ~ 100 ppm by volume), whereas terrestrial applications almost invariably require the use of atmospheric air as oxidant due to technical and economic considerations. For municipal electric applications, hydrocarbon fuels, especially natural gas, are expected to be the primary fuel, and their reformation into hydrogen-rich gases invariably contain a significant amount of carbon dioxide, e.g., steam-reforming of the natural gas results in the reformate gas consisting approximately of 80% hydrogen, 20% carbon dioxide, and a trace amount of other components such as carbon monoxide. Carbonaceous products of aging and corrosion shorten AFC life; they degrade the alkaline electrolyte. Whether originating as impurities in the gaseous reactants or from some fuel cell materials, oxides of carbon will chemically react with the alkaline electrolyte and produce irreversible decay that will decrease performance and shorten life. Consequently, AFCs are currently restricted to specialized applications where pure hydrogen and oxygen are utilized. The revival of this technology will depend almost completely on the successful curing of "CO_2 syndrome"—the efficient and economic scrubbing of carbon dioxide, even though claims have been made of successful resolution of carbon dioxide poisoning in the AFCs. This is especially true for utility applications, although some optimistic estimate indicates that the AFC stack costs are similar to all other low-temperature fuel cell systems, and the production costs for the AFC systems seem to be the lowest. A price of about US$400–500/kW has been quoted using today's technologies and today's knowledge in large-scale production. However, small-scale commercial production cost is estimated to be 5–10 times higher.

26.7.2 Polymer-Electrolyte-Membrane Fuel Cells

26.7.2.1 Introduction

The polymer-electrolyte-membrane fuel cells (PEMFC) are also called solid-polymer (electrolyte) fuel cell or proton exchange membrance fuel cell. It is perhaps the most elegant of all fuel cell systems in terms of design and mode of operation. It was the first type of fuel cell that was put into practical application in the Gemini space missions from 1962 to 1966. It consists of a solid polymeric membrane acting as the electrolyte. The solid membrane is an excellent proton conductor, sandwiched between two platinum-catalyzed porous carbon electrodes. It has fast-start capability and yields the highest output power density among all types of the fuel cells. Because of the solid membrane as the electrolyte, there is no corrosive fluid spillage hazard, and there is lower sensitivity to orientation. It has no volatile electrolyte and has minimal corrosion concerns. It has truly zero pollutant emissions with potable liquid product water when hydrogen is used as fuel. As a result, the PEMFC is particularly suited for vehicular power applications, although it is also being considered for stationary power applications to a lesser degree.

The proton conducting polymer membrane belongs to a class of materials called *ionomers* or *polyelectrolytes*, which contain functional groups that will dissociate in the presence of water. The dissociation produces ions fixed to the polymer and simple counterions that can freely exchange with ions of the same sign from the solution. The current available polyelectrolytes have cations as the counterions. In the case of hydrogen, the cations are protons. Therefore, the membrane must be fully hydrated to have adequate ionic conductivity. As a result, the fuel cell must be operated under conditions where the product water does not evaporate faster than it is produced, and the reactant gases, both hydrogen and oxygen, need to be humidified. Therefore, water and thermal management in the membrane become critical for efficient cell performance, are fairly complex, and require dynamic control to match the varying operating conditions of the fuel cell. Because of the limitation imposed by the membrane and problems with water balance, the operating temperature of PEMFCs is usually less than 120°C, typically 80°C. This rather low operating temperature requires the use of noble metals as catalysts in both the anode and cathode side with generally higher catalyst loadings than those used in PAFCs.

Currently the polymer electrolyte used is made of perfluorinated sulfonic acid membrane that is essentially acid in solid polymeric form. Therefore, PEMFCs are essentially acid-electrolyte fuel cells, with operational principle essentially the same as those of PAFCs. As a result, most PEMFC design, material selection, component fabrication, etc. are similar to those of PAFCs. The only difference is the humidification of reactant gases dictated by the membrane performance. Reactant humidification is often achieved by a number of techniques, e.g., by passing gas stream through a water column, by using in-stack humidification section of cell and membrane arrangement, and by spraying water into the reactant streams. In the early stage of the PEMFC development, the membranes were based on polystyrene, but since 1968 a Teflon-based product named Nafion by DuPont, is used. This offers high stability, high oxygen solubility, and high mechanical strength.

26.7.2.2 Basic Operating Principle

The polymer-electrolyte-membrane (PEM) is essentially an acid electrolyte. The PEM fuel cell requires hydrogen gas as the fuel, and oxygen (typically air) as the oxidant. The half-cell reactions are

$$\text{Anode}: H_2 \rightarrow 2H^+ + 2e^- \tag{26.74}$$

$$\text{Cathode}: \frac{1}{2}O_2 + 2H^+ + 2e^- \rightarrow H_2O, \tag{26.75}$$

and the overall cell reaction is

$$H_2 + \frac{1}{2}O_2 \rightarrow H_2O + \text{Heat Generated} + \text{Electric Energy}. \tag{26.76}$$

The current PEMFCs use perfluorinated sulfonic acid membrane (almost exclusively Nafion from DuPont) as the proton-conducting electrolyte, carbon paper or cloth as the anode and cathode backing layers, and platinum or its alloys, either pure or supported on carbon black, as the catalyst. The bipolar plate with the reactant gas flow fields is often made of graphite plate. The stoichiometry is around 1.1 to 1.2 for the fuel, and 2 for the oxidant (oxygen). The PEMFCs usually operate at about 80°C and 1–8 atm pressure. The pressures, in general, are maintained equal on either side of the membrane. Operation at high pressure is necessary to attain high power densities, particularly when air is chosen as the cathodic reactant.

To prevent the membrane dryout leading to local hot-spot (and crack) formation, performance degradation, and lifetime reduction, both fuel and oxidant streams are fully humidified, and the operating temperature is limited by the saturation temperature of water corresponding to the operating pressure. The liquid water formed at the cathode does not dissolve in the electrolyte membrane, and is usually removed from the cell by the excessive oxidant gas stream. The accumulation of liquid water in the cathode backing layer blocks the oxygen transfer to the catalytic sites, thus resulting in the phenomenon called *water-flooding* that causes performance reduction. Local hot and cold spots will cause the evaporation and condensation of water. Thus, an integrated approach to thermal and water management is critical to PEMFCs' operation and performance, and a proper design must be implemented.

26.7.2.3 Acceptable Contamination Levels

As an acid-electrolyte fuel cell operating at low temperature, the PEMFC is primarily vulnerable to carbon monoxide poisoning. Even a trace amount of CO drastically reduces the performance levels, although CO poisoning effect is reversible and does not cause permanent damages to the PEMFC system. Furthermore, the performance reduction due to CO poisoning takes a long time (on the order of two hours) to reach steady state. This transient effect may have profound implication for transportation applications. Therefore, the PEMFC requires the use of a fuel virtually free of CO (must be less than a few ppm). Also, high-quality water that is free of metal ions should be used for the cell cooling and reactant humidification to avoid the contamination of the membrane electrolyte. This requirement has a severe implication on the materials that can be used for cell components. On the other hand, carbon dioxide does not affect PEMFC operation and performance except through the effect of reactant dilution (the Nernst loss).

26.7.2.4 Major Technological Problems

For practical applications, PEMFC performance in terms of energy efficiency, power density (both size and weight) and capital cost must be further improved. This can be accomplished by systematic research in:

i. New oxygen-reduction electrocatalysts: this includes the reduction of precious-metal platinum and its alloys loading from 4 mg/cm^2 to 0.4 mg/cm^2 or lower without affecting the long-term performance and the lifetime, and the development of CO-tolerant catalysts.
ii. New types of polymer electrolyte with higher oxygen solubility, thermal stability, long life and low cost. A self-humidified membrane or a polymer without the need of humidification will be ideal for PEMFC operation and performance enhancement with significant simplification of system complexities and reduction of the cost.
iii. Profound changes in oxygen (air) diffusion electrode structure to minimize all transport-related losses. The minimization of all transport losses is the most promising direction for PEMFC performance improvement.
iv. Optimal thermal and water management throughout the individual cells and the whole stack to avoid local hot and dry spot formation and to avoid water-flooding of the electrode.

In addition to the above issues, the development of low-cost, light-weight materials for construction of reactant gas flow fields and bipolar plates is one of the major barriers to PEMFCs' large-scale commercialization. The successful solution of this problem will further increase the output power density. Additional issues include optimal design of flow fields with the operating conditions, and an appropriate selection of materials and fabrication techniques. It has been reported that over 20% improvement in the performance of PEMFC stacks can be obtained just by appropriate design of flow channels alone. The current leading technologies for bipolar plate design include: injection-molded carbon-polymer composites, injection molded and carbonized amorphous carbon, assembled three-piece metallics, and stamped unitized metallics.

26.7.2.5 Technological Status

PEMFCs have achieved a high power density of over 1 kW/kg and 0.7 kW/L, perhaps the highest among all types of the fuel cells currently under development. It is also projected that the power density may be further improved, up to 2–3 kW/L, with unitized metallic (stainless steel) bipolar plates. The capital cost has been estimated to vary from the most optimistic of $1,500/kW to the most pessimistic of $50,000/kW at the current technology, and is projected to reach approximately $200–$300/kW, assuming a 10–20-fold reduction in the membrane and catalyst cost and also considering mass production. A target of $30/kW has been set for PEMFC power unit running on pure hydrogen for transportation applications. It is expected that PEMFC technology is about five to ten years from commercialization, and pre-commercial demonstration for buses and passenger vehicles are underway with increasing intensity; the first demonstration for residential combined heat and power application just began at the end of 1999. However, application of PEMFCs in powering portable and mobile electronics such as laptops has already begun.

26.7.2.6 Applications

PEMFCs have a high power density, a variable power output, and a short startup time due to low operating temperature; the solid polymer electrolyte is virtually corrosion-free, and can withstand a large pressure differential (as high as 750 psi reported by NASA) between the anode and cathode reactant gas streams. Hence, PEMFCs are suitable for use in the transportation sector. Currently, they are considered the best choice for zero-emission vehicles as far as present-day-available fuel cell technologies are concerned. Their high power density and small size make them primary candidates for light-duty vehicles, though they are also used for heavy-duty vehicles. For high-profile automobile application, pure hydrogen and air are used as reactants at the present. However, conventional gasoline and diesel engines are extremely cheap, estimated to cost about $30–$50/kW. Therefore, the cost of PEMFC systems must be lowered at least by two to three orders of magnitude to be competitive with the conventional heat engines in the transportation arena.

For electricity generation from the hydrocarbon fuels, a reformer with carbon monoxide and sulfur cleaning is necessary. It is estimated that the cost of the reforming system is about the same as the fuel cell stack itself, which is also the same as the cost of other ancillary systems. Apart from the high cost, the optimal chemical to electric conversion efficiency is around 40%–45%, and the low operating temperature makes the utilization of the waste heat difficult, if at all possible, for the reforming of hydrocarbon fuels, cogeneration of heat, and combined cycles. On the other hand, conventional thermal power plants with combined gas and steam turbines have energy efficiencies approaching 60% with a very low capital cost of US$1,000/kW. Therefore, the best possible application of the PEMFC systems for the utility industry is the use of PEMFCs in the size of tens to hundreds of kW range for remote regions, with the possibility for residential combined heat and power application.

In addition, NASA is conducting a feasibility study of using the PEMFC power systems for its space programs (mainly space shuttle missions) in place of its current three 12-kW AFC power modules. As discussed in the AFCs section, NASA is motivated by the extremely high cost, low lifetime, and maintenance difficulty associated with its current AFC systems. Currently, NASA's feasibility study is in its second phase by using parabolic flight tests in airplanes to simulate low-gravity environment. If all goes well, NASA will conduct real-time tests in shuttles in a couple of years.

26.7.3 Direct-Methanol Fuel Cells

26.7.3.1 Introduction

All the fuel cells reviewed in this chapter for commercial applications require the use of gaseous hydrogen directly or liquid/solid hydrocarbon fuels, e.g., methanol, reformed to hydrogen as the fuel. Pure oxygen or oxygen in air is used as the oxidant. Hence, these fuel cells are often referred to as hydrogen–oxygen or hydrogen-air types of fuel cells. The use of gaseous hydrogen as a fuel presents a number of practical problems, such as storage system weight and volume, as well as handling and safety issues, especially for consumer and transportation applications. Although liquid hydrogen has the highest energy density, the liquefaction of hydrogen requires roughly one-third of the specific energy, and the required thermal insulation increases the volume of the reservoir significantly. The use of metal hydrides decreases the specific energy density and the weight of the reservoir becomes excessive. The size and weight of a power system are extremely important for transportation applications, as they directly affect the fuel economy and vehicle capacity, although they are less critical for stationary applications. The low volumetric energy density of hydrogen also limits the distance between vehicle refueling.

Methanol as a fuel offers ease of handling and storage, and potential infrastructure capability for distribution. Methanol also has a higher theoretical energy density than hydrogen (5 kWh/L compared with 2.6 kWh/L for liquid hydrogen). Easy refueling is another advantage for methanol. However, in the conventional hydrogen-air or hydrogen–oxygen fuel cells, a reformer is needed that adds complexity and cost, as well as production of undesirable pollutants such as carbon monoxide. The addition of a reformer also increases response time.

Therefore, direct oxidation of methanol is an attractive alternative in view of its simplicity from a system point of view. The direct-methanol fuel cells utilizing proton-exchange membrane (PEM) have the capability of efficient heat removal and thermal control through the circulating liquid, and elimination of humidification required to avoid membrane dryout. These two characteristics have to be accounted for in the direct and indirect hydrogen systems that impact their volume and weight and, consequently, the output power density.

26.7.3.2 Basic Operating Principle

The direct-methanol fuel cell allows the direct use of an aqueous, low-concentration (3%), liquid methanol solution as the fuel. Air is the oxidant. The methanol and water react directly in the anode chamber of the fuel cell to produce carbon dioxide and protons that permeate the PEM and react with the oxygen at the cathode. The half-cell reactions are:

$$\text{Anode}: CH_3OH + H_2O \rightarrow CO_2 + 6H^+ + 6e^- \tag{26.77}$$

$$\text{Cathode}: 6H^+ + 6e^- + \frac{3}{2}O_2 \rightarrow 3H_2O, \tag{26.78}$$

and the net cell reaction is

$$CH_3OH + \frac{3}{2}O_2 \rightarrow CO_2 + 2H_2O. \tag{26.79}$$

Because a PEM (typically Nafion 117) is used as the electrolyte, the cell operating temperature must be less than the water boiling temperature to prevent dryout of the membrane. Typically, the operating temperature is around 90°C, and the operating pressure ranges from one to several atmospheres.

26.7.3.3 Acceptable Contamination Levels

The system is extremely sensitive to carbon monoxide (CO) and hydrogen sulfide (H_2S). Carbon monoxide may exist as one of the reaction intermediaries, and can poison the catalyst. There are arguments concerning whether CO is present in the anode during the reaction. Sulfur may be present if methanol is made from petroleum oils, and needs to be removed.

26.7.3.4 Major Technological Problems

The PEM used in the DMFCs is Nafion 117, which is the same as employed in the PEMFCs. Although it works well in both types of cells, it is expensive with only one supplier. Because the electrolyte in DMFCs is essentially acid, expensive precious metals (typically platinum or its alloys) are used as the catalyst. However, the most serious problem is the so-called *methanol crossover*. This phenomenon is caused by the electro-osmotic effect. When the protons migrate through the electrolyte membrane, a number of water molecules are dragged along with each proton, and because methanol is dissolved in liquid water on the anode side, methanol is dragged through the membrane electrolyte to reach the cathode side together with the protons and water. Fortunately, the methanol at the cathode is oxidized into carbon dioxide and water at the cathode catalyst sites, producing no safety hazards. But the methanol oxidation in the cathode does not produce useful electric energy. The development of a new membrane with low methanol crossover is a key to the success of DMFCs.

Such a low-methanol-crossover membrane has a number of advantages. First, it reduces the methanol crossover, enhancing fuel utilization and energy efficiency. Second, it reduces the amount of water produced at the cathode, leading to a lower activation and concentration polarization, thus allowing higher cell voltage at the same operating current. Third, it allows higher methanol concentration in the fuel stream, resulting in better performance.

26.7.3.5 Technological Status

DMFCs are the least developed among all the fuel cell technologies. Although methanol itself has simpler storage requirements than hydrogen and is simpler to make and transport, its electrochemical activity is much slower than that of hydrogen, i.e., its oxidation rate is about four orders of magnitude lower than that of hydrogen. Also, the conversion takes place at low temperature (about 80–90°C), and the contaminant problem is a serious issue.

The state-of-the-art performance is an energy conversion efficiency of 34% (from methanol to electricity) at 90°C using 20 psig air, at a cell voltage of 0.5 V (corresponding to a voltage efficiency of 42%) together with the methanol crossover accounting for 20% of the current produced (equivalent to a fuel efficiency of 80%). This 20% methanol crossover occurs when the fuel stream used is an aqueous solution containing only 3% methanol. It has been projected that with the better membrane under development and improvement of the membrane electrode assembly, a cell voltage of 0.6 V can be achieved with only 5% methanol crossover. This is equivalent to 50% voltage efficiency and 95% fuel efficiency, resulting in an overall stack efficiency of 47% (from methanol to electricity). The DMFC system efficiency will be lower due to the necessary auxiliary systems.

The DMFC power system is under-developed, and until now no one could demonstrate any feasibility for commercialization. It remains at a scale of small demonstration in the sub-kW range. As such, no system cost estimate is available or has ever been carried out. However, the current DMFCs basically use the same cell components, materials, construction, and fabrication techniques as the PEMFCs; therefore, it is expected that the system and component costs will be similar to those of the PEMFCs. It is said that one company in the world has recently been formed to explore the potential of the DMFC systems, and to develop DMFC for transportation applications. However, the DMFC system is at least ten years away from any realistic practical applications, judging from the progress of other types of fuel cells in the past.

26.7.3.6 Applications

DMFCs offer the potential for high power density and cold-start capabilities, a convenience for on-board fuel storage and compatibility with existing refueling infrastructure. Therefore, DMFCs are the most attractive for applications where storage or generation of hydrogen causes significant effort and has a negative impact on the volume and weight of the system. As a result, DMFCs have a great potential for transportation applications ranging from automobiles, trains and ships, etc. For utility applications, small DMFCs units have potential for use in residential and office buildings, hotels and hospitals, etc. for the combined electricity and heat supply (cogeneration). Because methanol can be made from agricultural products, the use of methanol is also compatible with renewable energy sources to allow for sustainable development.

26.7.4 Phosphoric-Acid Fuel Cells

The phosphoric-acid fuel cell (PAFC) is the most advanced type of fuel cell and is considered to be "technically mature" and ready for commercialization after nearly 30 years of R&D and over half a billion dollars expenditure. Therefore, the PAFC has been referred to as the first-generation fuel cell technology. Unlike the alkaline-fuel cell systems that were primarily developed for space applications, the PAFC was targeted initially for terrestrial applications with the carbon-dioxide-containing air as the oxidant gas and hydrocarbon-reformed gas as the fuel for electrochemical reactions and electric power generation.

The basic components of a phosphoric-acid fuel cell are the electrodes consisting of finely dispersed platinum catalyst or carbon paper, SiC matrix holding the phosphoric acid, and a bipolar graphite plate with flow channels for fuel and oxidant. The operating temperature ranges between 160°C–220°C and it can use either hydrogen or hydrogen produced from hydrocarbons (typically natural gas) or alcohols as the anodic reactant. In the case of hydrogen produced from a reformer with air as the anodic reactant, a temperature of 200°C and a pressure as high as 8 atm are required for satisfactory performance. PAFCs are advantageous from a thermal management point of view. The rejection of waste heat and product water is very efficient in this system and the waste heat at about 200°C can be used efficiently for the endothermic steam-reforming reaction. The waste heat can also be used for space heating and hot-water supply.

However, the PAFC cannot tolerate the presence of carbon monoxide and H_2S, which are commonly present in the reformed fuels. These contaminants poison the catalyst and decrease its electrochemical catalytic activity. A major challenge for using natural-gas-reformed fuel, therefore, lies in the removal of carbon monoxide to a level of less than 200–300 ppm. Carbon monoxide tolerance is better at the operating temperature of above 180°C. However, removal of sulfur is still essential. Further, the PAFC has a lower performance, primarily due to the slow oxygen reaction rate at the cathode. Therefore, PAFC is typically operated at higher temperature (near 200°C) for better electrochemical reactivity and for smaller internal resistance, which is mainly due to the phosphoric-acid electrolyte. As a result, PAFC exhibits the problems of both high- and low-temperature fuel cells, but possibly none of the advantages of either options.

The PAFC system is the most advanced fuel cell system for terrestrial applications. Its major use is in on-site integrated energy systems to provide electrical power in apartments, shopping centers, office buildings, hotels, and hospitals, etc. These fuel cells are commercially available in the range from 24 V, 250-W portable units to 200-kW on-site generators. PAFC systems of 0.5–1.0 MW are being developed for use in stationary power plants of 1–11 MW capacity. The power density of PAFC system is about 200 mW/cm^2, and the power density for 36-kW brassboard PAFC fuel cell stack has been reported to be 0.12 kW/kg and 0.16 kW/L. The most advanced PAFC system costs about US$3,000/kW (the best technology possible for the PAFCs), whereas the conventional thermal power generation system costs only about US$1,000/kW. Thus, it is believed that the PAFC is not commercially viable at present, even though the US DOE and DOD have been subsidizing half of the cost ($1,500/kW) to gain operational and maintenance experience for practical fuel cell systems. Although Japan seems determined to push ahead, interest in the PAFC systems is waning in the United States and Europe.

26.7.5 Molten-Carbonate Fuel Cells

26.7.5.1 Introduction

The molten-carbonate fuel cell (MCFC) is often referred to as the second-generation fuel cell because its commercialization is normally expected after the PAFC. It is believed that the development and technical maturity of the MCFC is about 5–7 years behind the PAFC. At present, the MCFC has reached the early demonstration stage of pre-commercial stacks, marking the transition from fundamental and applied R&D towards product development. MCFCs are being targeted to operate on coal-derived fuel gases or natural gas. This contrasts with the phosphoric-acid fuel cells discussed earlier that prefer natural gas as primary fuel.

The MCFC operates at higher temperature than all the fuel cells described thus far. The operating temperature of the MCFC is generally around 600–700°C, typically 650°C. Such high temperatures produce high-grade waste heat that is suitable for fuel processing, cogeneration, or combined cycle operation, leading to higher electric efficiency. It also yields the possibility of utilizing carbonaceous fuels (especially natural gas) directly, through internal reforming to produce the fuel (hydrogen) ultimately used by the fuel cell electrochemical reactions. This results in simpler MCFC systems (i.e., without external reforming or fuel processing subsystem), less parasitic load, and less cooling power requirements, hence higher overall system efficiency. The high operating temperature reduces voltage losses due to reduced activation, ohmic, and mass-transfer polarization. The activation polarization is reduced to such an extent that it does not require expensive catalysts as low-temperature fuel cells do, such as PAFCs and PEMFCs. It also offers great flexibility in the use of available fuels, say, through in situ reforming of fuels. It has been estimated that the MCFC can achieve an energy conversion efficiency of 52–60% (from chemical energy to electrical energy) with internal reforming and natural gas as the primary fuel. Some studies have indicate that the MCFC efficiency of methane to electricity conversion is the highest attainable by any fuel cell or other single pass/simple cycle generation scheme.

26.7.5.2 Basic Operating Principle

A MCFC consists of two porous gas-diffusion electrodes (anode and cathode), and a carbonate electrolyte in liquid form. The electrochemical reaction occurring at the anode and the cathode is

$$\text{Anode}: \quad H_2 + CO_3^{2-} \rightarrow H_2O + CO_2 + 2e^- \tag{26.80}$$

$$\text{Cathode}: \quad \frac{1}{2}O_2 + CO_2 + 2e^- \rightarrow CO_3^{2-}, \tag{26.81}$$

and the net cell reaction is

$$H_2 + \frac{1}{2}O_2 \rightarrow H_2O \tag{26.82}$$

Besides the hydrogen oxidation reaction at the anode, other fuel gases such as carbon monoxide, methane, and higher hydrocarbons, etc., are also oxidized by conversion to hydrogen. Although direct electrochemical oxidation of carbon monoxide is possible, it occurs very slowly compared to that of hydrogen. Therefore, the oxidation of carbon monoxide is mainly via the water–gas shift reaction

$$CO + H_2O \rightleftharpoons CO_2 + H_2, \tag{26.83}$$

that, at the operation temperature of the MCFC, equilibrates very rapidly at catalysts such as nickel. Therefore, carbon monoxide becomes a fuel, instead of a contaminant as in the previously-described low-temperature fuel cells. Direct electrochemical reaction of methane appears to be negligible. Therefore, methane and other hydrocarbons must be steam-reformed, which can be done either in a separate reformer (external reforming) or in the MCFC itself (so-called *internal reforming*).

As a result, water and carbon dioxide are important components of the feed gases to the MCFCs. Water, produced by the main anode reaction, helps to shift the equilibrium reactions to produce more hydrogen for the anodic electrochemical reaction. Water must also be present in the feed gas, especially in low-Btu (i.e., high CO content) fuel mixtures, to avoid carbon deposition in the fuel gas flow channels supplying the cell, or even inside the cell itself. Carbon dioxide, from the fuel exhaust gas, is usually recycled to the cathode as it is required for the reduction of oxygen.

The MCFCs use a molten alkali carbonate mixture as the electrolyte, which is immobilized in a porous lithium aluminate matrix. The conducting species is carbonate ions. Lithiated nickel oxide is the material of the current choice for the cathode, and nickel, cobalt and copper are currently used as anode materials, often in the form of powdered alloys and composites with oxides. As a porous metal structure, it is subject to sintering and creeping under the compressive force necessary for stack operation. Additives such as chromium or aluminum form dispersed oxides and thereby increase the long-term stability of the anode with respect to sintering and creeping. MCFCs normally have about 75–80% fuel (hydrogen) utilization.

26.7.5.3 Acceptable Contamination Levels

MCFCs do not suffer from carbon monoxide poisoning; in fact, they can utilize carbon monoxide in the anode gas as the fuel. However, they are extremely sensitive to the presence of sulfur (<1 ppm) in the reformed fuel (as hydrogen sulfide, H_2S) and oxidant gas stream (SO_2 in the recycled anode exhaust). The presence of HCl, HF, HBr, etc. causes corrosion, whereas trace metals can spoil the electrodes. The presence of particulates of coal/fine ash in the reformed fuel can clog the gas passages.

26.7.5.4 Major Technological Problems

The main research efforts for the MCFCs are focused on increasing the lifetime and endurance, and reducing the long-term performance decay. The main determining factors for the MCFC are electrolyte loss, cathode dissolution, electrode creepage and sintering, separator-plate corrosion, and catalyst poisoning for internal reforming.

Electrolyte loss results in increased ohmic resistance and activation polarization, and it is the most important and continuously active factor in causing the long-term performance degradation. It is primarily a result of electrolyte consumption by the corrosion/dissolution processes of cell components, electric-potential-driven electrolyte migration and electrolyte vaporization. Electrolyte evaporation (usually Li_2CO_3 and/or K_2CO_3) occurs either directly as carbonate or indirectly as hydroxide.

The cathode consists of NiO, which slowly dissolves in the electrolyte during operation. It is then transported towards the anode and precipitates in the electrolyte matrix as Ni. These processes lead to a gradual degradation of cathode performance and the shorting of the electrolyte matrix. The time at which shorting occurs depends not only, via NiO solubility, on the CO_2 partial pressure and the cell temperature, but also the matrix structure, i.e., on the porosity, pore size, and in particular, thickness of the matrix. Experience indicates that this cell shorting mechanism tends to limit stack life to about 25,000 h under the atmospheric reference gas conditions, and much shorter for real operating conditions.

Electrode, especially anode, creepage, and sintering (i.e., a coarsening and compression of electrode particles), result in increased ohmic resistance and electrode polarization. NiO cathodes have quite satisfactory sinter and creepage resistance. Creep resistance of electrodes has important effect on maintaining low contact resistance of the cells and stacks. The corrosion of the separator plate depends on many factors, such as the substrate, possible protective layers, composition of the electrolyte, local potential and gas composition, and the oxidizing and reducing atmospheres at the cathode and anode, respectively. Poisoning of the reforming catalyst occurs for direct internal reforming MCFCs. It is caused by the evaporation of electrolyte from the cell components and condensation on the catalyst which is the coldest spot in the cell, and by liquid creep within the cell.

26.7.5.5 Technological Status

MCFC technology is in the first demonstration phase, and under the product development with full-scale systems at the 250-kW to 2-MW range. The short-term goal is to reach a lifetime of 25,000 h, and the ultimate target is 40,000 h. It is estimated that the capital cost is about US$1000–1600/kW for the MCFC power systems. The cost breakdown is, at full scale production levels, about 1/3 for the stack, and 2/3 for the balance of the plant. It is also generally accepted that the cost of raw materials will constitute about 80% of total stack costs. Although substantial development efforts supported by fundamental research are still needed, the available knowledge and number of alternatives will probably make it possible to produce pre-commercial units in the earlier part of the coming decade at a capital cost of US$2000–4000/kW. Pre-competitive commercial units may be expected some years later by which time further cost reduction to full competitiveness will be guided by extensive operating experience and increased volume production.

26.7.5.6 Applications

The MCFC is being developed for their potential as baseload utility generators. However, their best application is in distributed power generation and cogeneration (i.e., for capacities less than 20 MW in size), and in this size range MCFCs are 50%–100% more efficient than turbines—the conventional power generator. Other applications have been foreseen, such as pipeline compressor stations, commercial buildings, and industrial sites in the near term and repowering applications in the longer term. Due to its high operation temperature, it only has very limited potential for transportation applications. This is because of its relatively low power density and long start-up times. However, it may be suitable as a powertrain for large surface ships and trains.

26.7.6 Solid-Oxide Fuel Cells

26.7.6.1 Introduction

Solid-oxide fuel cells (SOFC) have emerged as a serious alternative high-temperature fuel cell, and they have been often referred to as the third-generation fuel cell technology because their commercialization is expected after the PAFCs (the first generation) and MCFCs (the second generation).

SOFC is an all-solid-state power system, including the electrolyte, and it is operated at high temperature of around 1000°C for adequate ionic and electronic conductivity of various cell components. The all-solid-state cell composition makes the SOFC system simpler in concept, design and construction; two-phase (gas–solid) contact for the reaction zone reduces corrosion and eliminates all the problems associated with the liquid electrolyte management. The high-temperature operation results in fast electrochemical kinetics (i.e., low activation polarization) and no need for noble-metal catalysts. The fuel may be gaseous hydrogen, H_2/CO mixture, or hydrocarbons because the high-temperature operation makes the internal in situ reforming of hydrocarbons with water vapor possible. It is specially noticed that CO is no longer a contaminant, rather it becomes a fuel in SOFCs. Even with external reforming, the SOFC fuel feedstock stream does not require the extensive steam reforming with shift conversion as it does for the low-temperature fuel cell systems. More importantly, the SOFC provides high-quality waste heat that can be utilized for cogeneration applications or combined cycle operation for additional electric power generation. The SOFC operating condition is also compatible with the coal gasification process, which makes the SOFC systems highly efficient when using coal as the primary fuel. It has been estimated that the chemical to electrical energy conversion efficiency is 50–60%, even though some estimates go as high as 70%–80%. Also, nitrogen oxides are not produced, and the amount of carbon dioxide released per kWh is around 50% less than for power sources based on combustion because of the high efficiency.

26.7.6.2 Basic Operating Principle

As mentioned earlier, both hydrogen and carbon monoxide can be oxidized in the SOFCs directly. Therefore, if hydrogen or hydrogen-rich gas mixture is used as fuel, and oxygen (or air) is used as oxidant, the half cell reaction becomes

$$\text{Anode}: H_2 + O^{2-} \rightarrow H_2O + 2e^- \tag{26.84}$$

$$\text{Cathode}: 2e^- + \frac{1}{2}O_2 \rightarrow O^{2-}, \tag{26.85}$$

and the overall cell reaction becomes

$$H_2 + \frac{1}{2}O_2 \rightarrow H_2O. \tag{26.86}$$

However, if carbon monoxide is provided to the anode instead of hydrogen, the anode reaction becomes

$$\text{Anode}: CO + O^{2-} \rightarrow CO_2 + 2e^-, \tag{26.87}$$

with the cathode reaction remaining the same, the cell reaction becomes

$$CO + \frac{1}{2}O_2 \rightarrow CO_2. \tag{26.88}$$

If the fuel stream contains both hydrogen and carbon monoxide as is the case for hydrocarbon reformed gas mixture, especially from the gasification of coal, the oxidation of hydrogen and carbon monoxide occurs simultaneously at the anode, and the combined anode reaction becomes

$$\text{Anode}: aH_2 + bCO + (a+b)O^{2-} \rightarrow aH_2O + bCO_2 + 2(a+b)e^-. \tag{26.89}$$

Consequently, the corresponding cathode and overall cell reaction become

$$\text{Cathode}: \frac{1}{2}(a+b)O_2 + 2(a+b)e^- \rightarrow (a+b)O^{2-} \tag{26.90}$$

$$\text{Cell}: \frac{1}{2}(a+b)O_2 + aH_2 + bCO \rightarrow aH_2O + bCO_2. \tag{26.91}$$

The solid electrolyte in SOFCs is usually yttria-stabilized zirconia (YSZ), thus a high-operating temperature of around 1000°C is required to ensure adequate ionic conductivity and low ohmic resistance. This is especially important because the cell open-circuit voltage is low, compared with low-temperature fuel cells, typically around 0.9–1 V under the typical working conditions of the SOFCs. The high-temperature operation of the SOFCs makes the activation polarization very small, resulting in the design operation in the range dominated by the ohmic polarization. The conventional material for the anode is nickel–YSZ–cermet and cathode is usually made of lanthanum strontium manganite. Metallic current collector plates of a high-temperature, corrosion-resistant, chromium-based alloy are typically used.

26.7.6.3 Acceptable Contamination Levels

Because of high temperature, the SOFCs can better tolerate impurities in the incoming fuel stream. They can operate equally well on dry or humidified hydrogen or carbon monoxide fuel or on mixtures of them. But hydrogen sulfide (H_2S), hydrogen chloride (HCl) and ammonia (NH_3) are impurities typically found in coal gasified products, and each of these substances is potentially harmful to the performance of SOFCs. The main poisoning factor for SOFCs is H_2S. Though the sulfur tolerance level is approximately two orders of magnitude greater than other fuel cells, the level is below 80 ppm. However, studies have shown that the effect of hydrogen sulfide (H_2S) is reversible, meaning that the cell performance will recover if hydrogen sulfide is removed from the fuel stream or clean fuel is provided after the contaminant poison has occurred.

26.7.6.4 Major Technological Problems

The high-temperature operation of the SOFCs places stringent requirements on materials used for cell construction, and appropriate materials for cell components are very scarce. Therefore, the key technical challenges are the development of suitable materials and the fabrication techniques. Of the material requirements, the most important consideration is the matching of the thermal expansion coefficients of electrode materials with that of the electrolyte to prevent cracking or delamination of SOFC components either during high temperature operation or heating/cooling cycles. One of the remedies for the thermal expansion mismatch is to increase the mechanical toughness of the cell materials by developing either new materials or doping the existing materials with SrO and CaO.

The electrode voltage losses are reduced when the electrode material possesses both ionic and electronic conductivities (so-called *mixed conduction*), for which the electrochemical reactions occur throughout the entire surface of the electrode rather than only at the three-phase interface of, e.g., the cathode, the air (gas phase), and the electrolyte. Therefore, it is important for performance enhancement to develop mixed-conduction materials for both the cathode and anode which have good thermal expansion match with the electrolyte used and good electrical conductivity to reduce the ohmic polarization which dominates the SOFC voltage losses.

Another focus of the current development is the intermediate temperature SOFCs operating at around 800°C a lower for better matching with the bottoming turbine cycles and lessening requirements for the cell component materials. Again appropriate materials with adequate electrical conductivity are the key areas of the development effort, and thermal expansion matching among the cell components is still necessary.

26.7.6.5 Technological Status

There are three major configurations for SOFCs: the tubular, flat plate, and monolithic. Even though the SOFC technology is in the developmental stage, the tubular design has gone through development since the late 1950s, and is now being demonstrated at user sites in a complete operating fuel cell power unit of nominal 25-kW (40-kW maximum) capacity. The flat plate and the monolithic designs are at a much earlier development status typified by subscale, single-cell, and short-stack development (up to 40 cells). The present estimated capital cost is US$1,500/kW, but is expected to be reduced with improvements in technology. Therefore, the SOFCs may become very competitive with the existing technology for electric power generation. However, it is believed that the SOFC technology is at least five to ten years away from commercialization.

26.7.6.6 Applications

SOFCs are very attractive in electrical utility and industrial applications. The high operating temperature allows them to use hydrogen and carbon monoxide from natural-gas steam reformers and coal gasification plants, a major advantage as far as fuel selection is concerned. SOFCs are being developed for large ($>$ 10 MW, especially 100–300 MW) baseload stationary power plants with coal as the primary fuel. This is one of the most lucrative markets for this type of fuel cells.

A promising field for SOFCs is the decentralized power supply in the MW range, where the SOFC gains interest due to its capability to convert natural gas without external reforming. In the range of one to some tenths of a MW, the predicted benefits in electrical efficiency of SOFC-based power plants over conventional methods of electricity generation from natural gas can only be achieved by an internal reforming SOFC. Therefore, internal reforming is a major target of present worldwide SOFC development. $SOFC_s$ are also being developed as auxiliary power for large transport trucks.

26.8 Summary

This chapter is focused on fuel cells, including the basic principle of operation, system composition, and balance of plants, the performance and the design considerations, as well as the state-of-the-art technology. The performance of fuel cell is analyzed in terms of the cell potential and energy conversion efficiency under the idealized reversible and practical irreversible conditions, and misconceptions regarding fuel cell energy efficiency are clarified. The effect of operating conditions, namely, temperature, pressure, and reactant concentration, on the reversible cell potential is also given. It is shown that both fuel cells and heat engines have the same maximum theoretical efficiency, which is equivalent to the Carnot efficiency, when operating on the same fuel and oxidant. However, fuel cells have less irreversibilities in practice, resulting in higher practical efficiencies. Further, possibilities of over 100% fuel cell efficiency is ruled out from the fundamental principles. Both reversible and irreversible energy-loss mechanisms are described for fuel cells, expression for waste heat generation in a fuel cell is derived, and various forms of fuel cell efficiency are defined. Finally, the Nernst potential loss arising from the reactant consumption in practical cells is considered, and issues related to reactant utilization are outlined. Then important physical and chemical processes occurring in fuel cell electrodes are provided, that relate to the transport phenomena and electrochemical reactions for current generation. These processes affect how the cells are connected together to form fuel cell stack of different sizes for desired power output.

Finally, the characteristics, technological status and preferred area of applications are summarized for each of the six major types of fuel cells. AFCs have the best performance when operating on pure hydrogen and oxygen, its intolerance of carbon dioxide hinders its role for terrestrial applications. Significant progress is being made for polymer-electrolyte-membrane fuel cell, although it is still too expensive to be competitive in the marketplace. However, the PEM fuel cell is believed to be the most promising candidate for transportation applications because of its high power density, fast startup, high efficiency, and easy and safe handling. But until its cost is lowered by at least orders of magnitude, it will not be economically acceptable. Due to the difficulty of on-board fuel (hydrogen) storage and the lack of infrastructure for fuel (hydrogen) distribution, DMFCs are believed by some to be the most appropriate technology for vehicular application. PEMFCs are expected to be five to ten years away from commercialization, whereas the DMFCs are at the early stage in their technological development. DMFCs also have considerable potential for portable applications. Phosphoric acid fuel cell is the most commercially developed fuel cell, operating at intermediate temperatures. PAFCs are being used for combined heat and power applications with high energy efficiency. The high-temperature fuel cells like MCFCs and SOFCs may be most appropriate for cogeneration and combined cycle systems (with gas or steam turbine as the bottoming cycle). The MCFCs have the highest energy efficiency attainable from methane to electricity conversion in the size range of 250 kW to 20 MW; whereas the SOFCs are best suited for baseload utility applications operating on coal-derived gases. It is estimated that the MCFC technology is about five to ten years away from commercialization, and the SOFCs are probably years afterwards.

References

1. Li, X. 2006. *Principles of Fuel Cells,* Taylor & Francis, New York.
2. Appleby, A. J. and Foulkes, F. R. 1989. *Fuel Cell Handbook,* Van Nostrand Reinhold, New York.
3. Blomen, L. J. M. J. and Mugerwa, M. N. eds. 1993. *Fuel Cell Systems*, Plenum Press, New York.
4. Hirschenhofer, J. H., Stauffer, D. B., and Engleman, R. R. 1994. *Fuel Cells: A Handbook,* US Department of Energy, Washington, DC, Rev. 3.
5. Kordesch, K. and Simader, G. 1996. *Fuel Cells and their Applications,* VCH, New York.
6. Appleby, A. J. 1996. Fuel cell technology: Status and future prospects, *Energy,* 21, 521–653.
7. Vielstich, W., Gasteiger, H., and Lamm, A. eds. 2003. *Handbook of Fuel Cells—Fundamentals, Technology, Applications*, Wiley, New York.
8. Appleby, A. J. 1993. Characteristics of fuel cell systems, In *Fuel Cell Systems*, L. J. M. J. Blomen and M. N. Mugerwa, eds., Plenum Press, New York.
9. Li, X. 1997. Fuel cells and their thermodynamic performance, In *Proceedings of the Energy Week Conference & Exhibition, Book V, Vol. 4,* pp. 346–356. ASME and API, Houston, TX.
10. Li, X. 2002. Fuel cells—the environmentally friendly energy converter and power generator, *Int. J. Global Energy Issues*, 17, 68–91.
11. Barendrecht, E. 1993. Electrochemistry of fuel cells, In *Fuel cell systems*, L. J. M. J. Blomen and M. N. Mugerwa, eds., Plenum Press, New York, pp. 73–79.

27

Direct Energy Conversion

27.1 Thermionic Energy Conversion 27-1
Introduction • Principles of Thermionic Energy Conversion • Types of Thermionic Converters • Converter Output Characteristics • Thermodynamic Analysis • Design Transition to Space Reactors—Concluding Remarks

Defining Terms 27-6

References 27-7

For Further Information 27-7

27.2 Thermoelectric Power Conversion 27-7
Introduction • Thermoelectric Effects • Thermoelectric Applications • Additional Considerations

Nomenclature 27-13

Defining Terms 27-14

References 27-14

For Further Information 27-14

27.3 Magnetohydrodynamic Power Generation 27-15
Introduction • Electrical Conductivity Considerations • Generator Configurations and Loading • Components • Systems • Heat Sources, Applications, and Environmental Considerations • Technology Status • Future Prospects

For Further Information 27-19

Mysore L. Ramalingam
UES, Inc.

Jean-Pierre Fleurial
Jet Propulsion Laboratory/California Institute of Technology

William D. Jackson
HMJ Corporation

27.1 Thermionic Energy Conversion

Mysore L. Ramalingam

27.1.1 Introduction

Thermionic energy conversion (TEC) is the process of converting heat directly to useful electrical work by the phenomenon of thermionic electron emission. This fundamental concept can be applied to a cylindrical version of the planar converter, considered the building block for space nuclear power systems (SNPS) at any power level. Space nuclear reactors based on TEC can produce power in the range of 5 kWe–5 MWe, a spectrum that serves the needs of current users such as National Aeronautics and Space Administration (NASA), United States Air Force (USAF), United States Department of Energy (USDOE), and Ballistic Missile Defense Organization (BMDO). Electrical power in this range is currently being considered for commercial telecommunication satellites, navigation, propulsion, and planetary exploration missions.

The history of thermionic emission dates back to the mid-1700s when Charles Dufay observed that electricity is conducted in the space near a red-hot body. Although Thomas Edison requested a patent in the late 1800s, indicating that he had observed thermionic electron emission while perfecting his electric light system, it was not until the 1960s that the phenomenon of TEC was adequately described theoretically and experimentally (Hatsopoulos and Gryftopoulos 1973). These pioneering activities have led to the development of thermionic SNPS that could potentially be augmented by Brayton and Stirling cycle generators to produce additional power from waste heat in NASA manned lunar and martian exploration missions (Ramalingam and Young 1993).

27.1.2 Principles of Thermionic Energy Conversion

Figure 27.1 represents a schematic of the essential components and processes in an elementary thermionic converter (TC). Electrons "boil-off" from the emitter material surface, a refractory metal such as tungsten, when heated to high temperatures (2000 K) by a **heat source**. The electrons then traverse the small interelectrode gap, to a colder (1000 K) collector surface where they condense, producing an output voltage that drives the current through the electrical load and back to the emitter. The flow of electrons through the electrical load is sustained by the temperature difference and the difference in **surface work functions** of the electrodes.

Surface Work Function. In a simple form, the energy required to separate an electron from a metal surface atom and take it to infinity outside the surface is termed the electron work function or the work function of the metal surface. The force experienced by an electron as it crosses an interface between a metal and a rarefied vapor can be represented by the **electron motive**, Ψ, which is defined as a scalar quantity whose negative gradient at any point is a measure of the force exerted on the electron at that point (Langmuir and Kingdon 1925). At absolute zero the kinetic energy of the **free electrons** would occupy quantum energy levels from zero to some maximum value called the Fermi level. Each energy level contains a limited number of free electrons, similar to the electrons contained in each electron orbit surrounding the nucleus of an atom. Fermi energy, μ, corresponds to the highest energy of all free electrons at absolute zero. At temperatures other than absolute zero some of the free electrons begin to experience energies greater than that at the Fermi level. Thus, the electron work function Φ, would be defined as

$$\Phi = \Psi_T - \mu \tag{27.1}$$

where Ψ_T represents the electron motive or energy at some temperature, T, above absolute zero.

Interelectrode Motive Distribution. Figure 27.2 provides a schematic representation of the electron motive distribution in the interelectrode space of a thermionic converter. Under ideal conditions of particle transport, the motive varies linearly from Ψ_{EM}, the motive just outside the emitter, to Ψ_{CO},

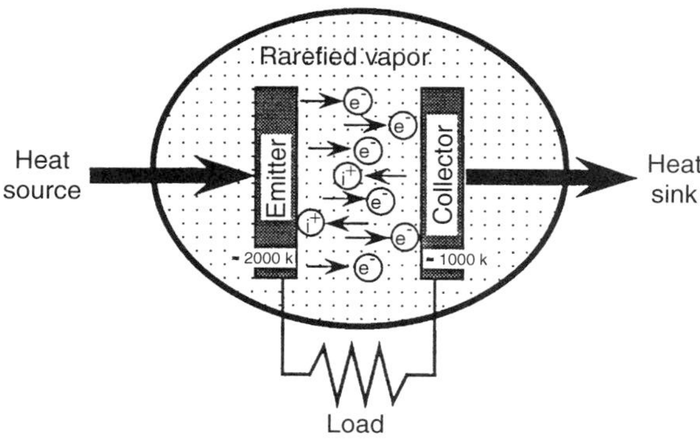

FIGURE 27.1 Schematic of an Elementary TEC.

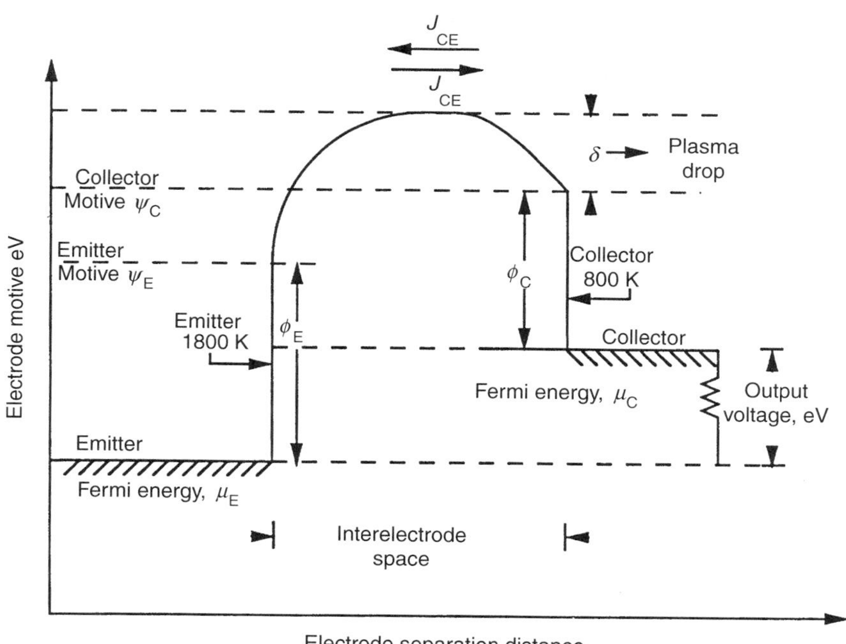

FIGURE 27.2 Electron motive distribution in the interelectrode gap.

the motive outside the collector surface. The magnitudes of the Fermi energies of the emitter and collector relative to $\Psi_{\rm EM}$ and $\Psi_{\rm CO}$ are clearly indicated. The internal voltage drop of the converter is defined as;

$$\Delta V = \frac{\Psi_{\rm EM} - \Psi_{\rm CO}}{e} \tag{27.2}$$

In a conventional thermionic converter, the emitter and collector are not at the same temperature, but to a good approximation, the output voltage, neglecting **lead losses** and **particle interaction losses**, can be represented by the relationship.

$$V = \frac{\mu_{\rm CO} - \mu_{\rm EM}}{e} \tag{27.3}$$

Since a real thermionic converter has an ionizing medium to improve its performance, a similar motive distribution can be defined for the ions. It is sufficient to state that the ion interelectrode motive has a slope equal and opposite to the corresponding electron interelectrode motive. The ions are, therefore, decelerated when the electrons are accelerated and vice versa.

Electron Saturation Current. In the absence of a strong influence from an external electrical source, the electron current ejected from a hot metal at the emitter surface into the vacuum ionizing medium is termed the *electron saturation current*. As this quantity depends on the number of free electrons $N(\varepsilon_x)$, Fermi–Dirac statistics provide the means to compute the number of free electrons, $N(\varepsilon_x)\,{\rm d}\varepsilon_x$, incident on a unit area within the metal in unit time with energies corresponding to the motion normal to the area, between ε_x and $\varepsilon_x + {\rm d}\varepsilon_x$. For energies greater than the Fermi energy, the functional dependence of $N(\varepsilon_x)$ on ε_x is given by (Fowler 1955)

$$N(\varepsilon_x) \approx [4\pi m_e kT/h^3][\exp\{-\varepsilon_x - \mu/kT\}] \tag{27.4}$$

where m_e is the mass of the electron $= 9.108 \times 10^{-28}$ g and h is Planck's constant $= 4.140 \times 10^{-15}$ eV s.

The electron saturation current density, J_{sat}, for a uniform surface, is found by integrating $N(\varepsilon_x)$ in the range of ε_x from Ψ_T to infinity for all $\Psi_T - \mu > kT$, which is the case for almost all materials and practical temperatures. The result of the integration yields

$$J_{sat} = AT^2 \exp\left[-\frac{(\Psi_T - \mu)}{kT}\right] \tag{27.5}$$

or

$$J_{sat} = AT^2 \exp\left[-\frac{\Phi}{kT}\right] \tag{27.6}$$

where A is the Richardson constant ≈ 120 A/cm^2 K^2.

Equation 27.6, which is the most fundamental and important relationship for the design of a thermionic converter, is called the Richardson–Dushmann Equation (Richardson 1912). On similar lines, the ion saturation current density for a converter with an ionizing medium is given by the relationship (Taylor and Langmuir 1933):

$$\tilde{J}_{iSat} = \frac{ep_g}{(2\pi m_g kT_g)^{0.5}(1 + 2\exp\{[V_i - \Phi/kT]\})} \tag{27.7}$$

where p_g, T_g, m_g, and V_i are the pressure, temperature, mass, and first ionization energy, respectively, of the ionizing medium.

27.1.3 Types of Thermionic Converters

Thermionic converters can be broadly classified as vacuum thermionic converters and vapor thermionic converters, depending on the presence of an ionizing medium in the interelectrode gap. In vacuum thermionic converters the interelectrode space is evacuated so that the space is free of particles other than electrons and the two electrodes are placed very close together, thereby neutralizing the negative space charge buildup on the electrode surface and reducing the total number of electrons in transit. Due to machining limitations, vacuum converters have been all but completely replaced by vapor-filled thermionic converters. In vapor-filled thermionic converters, the interelectrode space is filled with a rarefied ionizing medium at a vapor pressure generally on the order of 1–10 Torr. The vapor generally used is cesium as it is the most easily ionized of all stable gases and this can be provided through an external two-phase reservoir or an internal graphite reservoir (Young et al. 1993). The vapor neutralizes the negative space charge effect by producing positive ions at the electrode surfaces and gets adsorbed on the surfaces, thereby altering the work function characteristics.

27.1.4 Converter Output Characteristics

Figure 27.3 represents the output current–voltage characteristics for various modes of operation of the vacuum and vapor-filled thermionic converters. Characteristics obtained by not considering particle interactions in the interelectrode gap are generally considered ideal output characteristics. The figure essentially displays three types of converter output current–voltage characteristics, an ideal characteristic, an ignited mode characteristic, and an unignited mode characteristic. For an ideal converter in the interelectrode space the net output current density consists of the electron current density, J_{EMCO} flowing from emitter to collector diminished by the electron current density J_{COEM} flowing from collector to emitter and the ion-current density $\tilde{J}_{iEMCO}$ flowing from emitter to collector. Thus,

$$J_{net} = J_{EMCO} - J_{COEM} - \tilde{J}_{iEMCO} \tag{27.8}$$

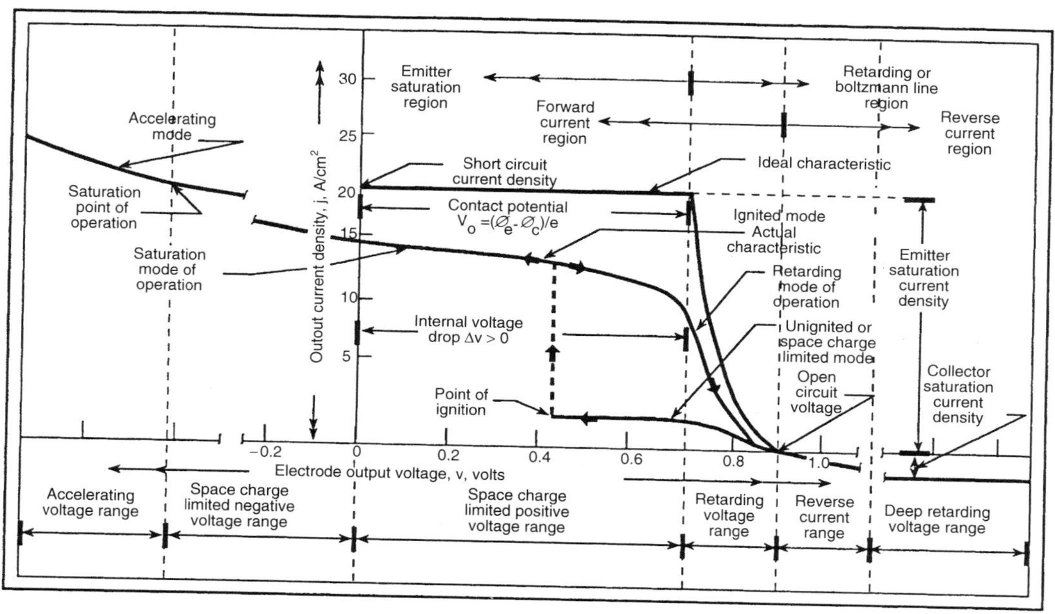

FIGURE 27.3 Thermionic diode output current density characteristics and nomenclature.

By expressing the individual terms as functions of ϕ, T, and V,

$$J_{\text{net}} = AT_{\text{EM}}^2 \exp\left[-\frac{\Phi_{\text{EM}}}{kT_{\text{EM}}}\right] - AT_{\text{CO}}^2 \exp\left[-\frac{(\Phi_{\text{EM}} - eV)}{kT_{\text{CO}}}\right] - \tilde{J}_{\text{EMS}} \exp[-(\Psi_{\text{EM}} - \Psi_{\text{CO}})kT_{\text{EM}}] \tag{27.9}$$

for $eV < \Phi_{\text{EM}} - \Phi_{\text{CO}}$

Similar relationships can be generated for various types of thermionic converters.

27.1.5 Thermodynamic Analysis

In thermodynamic terms a thermionic converter is a heat engine that receives heat at high temperature, rejects heat at a lower temperature, and produces useful electrical work while operating in a cycle analogous to a simple vapor cycle engine. Based on the application of the first law of thermodynamics to the control volumes around the emitter (Houston 1959; Angrist 1976),

$$\text{Energy in} = \text{Energy out} \tag{27.10}$$

i.e.,

$$q_{\text{CB}} + q_{\text{JH}} + q_{\text{HS}} = q_{\text{EC}} + q_{\text{WB}} + q_{\text{CD}} + q_{\text{RA}} \tag{27.11}$$

where, by using the terminology in Figure 27.2, each of the terms in Equation 27.11 can be elaborated as follows:

(a) Energy supplied by back emission of the collector:

$$q_{\text{CB}} = J_{\text{COEM}}[\Phi_{\text{CO}} + \delta + V + (2kT_{\text{CO}}/e)] \tag{27.12}$$

(b) Energy supplied by joule heating of lead wires and plasma:

$$q_{\mathrm{JH}} = 0.5[J_{\mathrm{EMCO}} - J_{\mathrm{COEM}}]^2(R_{\mathrm{LW}} + R_{\mathrm{PL}}) \tag{27.13}$$

(c) Energy dissipated by electron cooling:

$$q_{\mathrm{EC}} = J_{\mathrm{EMCO}}[\Phi_{\mathrm{CO}} + \delta + V - \Phi_{\mathrm{EM}} + (2kT_{\mathrm{EM}})/e] \tag{27.14}$$

(d) Energy dissipated due to phase change by electron evaporation:

$$q_{\mathrm{WB}} = J_{\mathrm{EM}}\Phi_{\mathrm{EM}} \tag{27.15}$$

(e) Energy dissipated by conduction through the lead wires and plasma:

$$q_{\mathrm{CD}} = \Delta T\left[\frac{K_{\mathrm{LW}}A_{\mathrm{LW}}}{A_{\mathrm{e}}L_{\mathrm{LW}}} + \frac{K_{\mathrm{PL}}A_{\mathrm{PL}}}{A_{\mathrm{e}}L_{\mathrm{IG}}}\right] \tag{27.16}$$

Here, K represents thermal conductivity, LW = lead wires, PL = plasma, and IG = interelectrode gap.

(f) Energy dissipated by radiation from emitter to collector:

$$q_{\mathrm{RA}} = 5.67 \times 10^{-12}(T_{\mathrm{EM}}^2 - T_{\mathrm{CO}}^4)(\varepsilon_{\mathrm{EM}}^{-1} + \varepsilon_{\mathrm{CO}}^{-1} - 1)^{-1} \tag{27.17}$$

Substitution for the various terms in Equation 27.10 yields q_{Hs}, the energy supplied to the emitter from the heat source.

The thermal efficiency of the thermionic converter is now expressed as

$$\eta_{\mathrm{TH}} = [V(J_{\mathrm{EMCO}} - J_{\mathrm{COEM}})/q_{\mathrm{HS}}] \tag{27.18}$$

27.1.6 Design Transition to Space Reactors—Concluding Remarks

All the fundamentals discussed so far for a planar thermionic converter can be applied to a cylindrical version which then becomes the building block for space power systems at any power level. In a thermionic reactor, heat from the nuclear fission process produces the temperatures needed for thermionic emission to occur. The design of a thermionic SNPS is a user-defined compromise between the required output power and the need to operate reliably for a specified lifetime. Based on the type of contact the emitter has with the nuclear fuel, the power systems can be categorized as "in-core" or "out-of-core" power systems. At this stage it suffices to state that the emitter design for in-core systems is extremely complex because of its direct contact with the hot nuclear fuel.

Defining Terms

Electron motive: A scalar quantity whose negative gradient at any point is a measure of the force exerted on an electron at that point.

Free electrons: Electrons available to be extracted from the emitter for thermionic emission.

Heat source: Electron bombardment heating of the emitter.

Lead losses: Voltage drop as a result of the built-in resistance of the leads and joints.

Particle interaction losses: Voltage drop in the interelectrode gap as a result of particle collisions and other interactions.

Surface work function: A measure of the electron-emitting capacity of the surface.

Thermionic energy conversion: Energy conversion from heat energy to useful electrical energy by thermionic electron emission.

References

Angrist, S. W. 1976. *Direct Energy Conversion. 3rd Ed.* Allyn and Bacon, Boston.

Fowler, R. H. 1955. *Statistical Mechanics. 2nd Ed.* Cambridge University Press, New York.

Hatsopoulos, G. N. and Gyftopoulos, E. P. 1973. *Thermionic Energy Conversion., Vol. 1*, MIT Press, Cambridge, MA.

Houston, J. M. 1959. Theoretical efficiency of the thermionic energy converter. *J. Appl. Phys.*, 30, 481–487.

Langmuir, I. and Kingdon, K. H. 1925. Thermionic effects caused by vapors of alkali metals. *Proc. R. Soc. Lond. A*, 107, 61–79.

Ramalingam, M. L. and Young, T. J. 1993. The power of thermionic energy conversion. *Mech. Eng.*, 115 (9), 78–83.

Richardson, O. W. 1912. Some applications of the electron theory of matter. *Philos. Mag.*, 23, 594–627.

Taylor, J. B. and Langmuir, I. 1933. The evaporation of atoms, ions and electrons from cesium films on tungsten. *Phys. Rev.*, 44, 423–458.

Young, T. J., Thayer, K. L., and Ramalingam, M. L. 1993. Performance simulation of an advanced cylindrical thermionic fuel element with a graphite reservoir, presented at 28th AIAA Thermophysics Conference, Orlando, FL.

For Further Information

Cayless, M. A. 1961. Thermionic generation of electricity. *Br. J. Appl. Phys.*, 12 433–442.

Hatsopoulos, G. N. and Gryftopoulos, E. P. 1979. *Thermionic Energy Conversion, Vol. 2*, MIT Press, Cambridge, MA.

Hernquist, K. G., Kanefsky, M., and Norman, F. H. 1959. Thermionic energy converter. *RCA Rev.*, 19 244–258.

Ramalingam, M. L. 1993. The Advanced Single Cell Thermionic Converter Program, WL-TR-93-2112, USAF Technical Report, Dayton, OH.

Rasor, N. S. 1960. Figure of merit for thermionic energy conversion. *J. Appl. Phys.*, 31 163–167.

27.2 Thermoelectric Power Conversion

Jean-Pierre Fleurial

27.2.1 Introduction

The advances in materials science and solid-state physics during the 1940s and 1950s resulted in intensive studies of thermoelectric effects and related applications in the late 1950s and through the mid-1960s (Rowe and Bhandari 1983). The development of semiconductors with good thermoelectric properties made possible the fabrication of thermoelectric generators and refrigerators. Being solid-state devices, thermoelectric systems offer some unique advantages, such as high reliability, long life, small-size and no-vibrations refrigerators, and can be used in a wide temperature range, from 200 to 1300 K. However, because of their limited conversion efficiencies, these devices have remained confined to specialized applications. As the following sections will emphasize, the performance of those devices is closely associated with the magnitude of the **dimensionless figure of merit**, ZT, of the thermoelectric semiconductor.

ZT represents the relative magnitude of electrical and thermal cross-effect transport in materials. State-of-the-art thermoelectric materials, known since the early 1960s, have been extensively developed. Although significant improvements of the thermoelectric properties of these materials have been achieved, a maximum *ZT* value close to 1 is the current approximate limit over the whole 100–1500 K temperature range (Figure 27.4). To expand the use of thermoelectric devices to a wide range of applications will require improving *ZT* by a factor of two to three. There is no theoretical limitation on the value of *ZT*, and new promising approaches are now focusing on the investigation of totally different materials and the development of novel thin film heterostructure.

27.2.2 Thermoelectric Effects

Thermoelectric devices are based on two transport phenomena: the Seebeck effect for power generation and the Peltier effect for electronic refrigeration. If a steady temperature gradient is applied along a conducting sample, the initially uniform charge carrier distribution is disturbed as the free carriers located at the high-temperature end diffuse to the low-temperature end. This results in the generation of a back emf which opposes any further diffusion current. The open-circuit voltage when no current flows is the Seebeck voltage. When the junctions of a circuit formed from two dissimilar conductors (n- and p-type semiconductors) connected electrically in series but thermally in parallel are maintained at different temperatures T_1 and T_2, the open-circuit voltage V developed is given by $V = S_{pn}(T_1 - T_2)$, where S_{pn} is the Seebeck coefficient expressed in $\mu V\ K^{-1}$.

The complementary Peltier effect arises when an electrical current I passes through the junction. A temperature gradient is then established across the junctions and the corresponding rate of reversible heat absorption $\dot{Q}$ is given by $\dot{Q} = \Pi_{pn} I$, where Π_{pn} is the Peltier coefficient expressed in $W\ A^{-1}$ or V. There is actually a third, less-important phenomenon, the Thomson effect, which is produced when an electrical current passes along a single conducting sample over which a temperature gradient is maintained. The rate of reversible heat absorption is given by $\dot{Q} = \beta I(T_1 - T_2)$, where β is the Thomson coefficient expressed in $V\ K^{-1}$. The three coefficients are related by the Kelvin relationships:

$$S_{pn} = \frac{\Pi_{pn}}{T} \quad \text{and} \quad \frac{dS_{pn}}{dT} = \frac{\beta_p - \beta_n}{T} \tag{27.19}$$

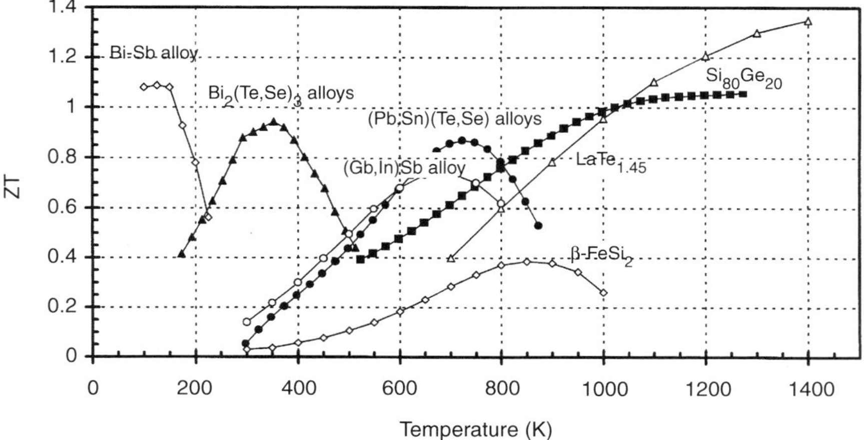

FIGURE 27.4 Typical temperature variations of *ZT* of state-of-the-art n-type thermoelectric alloys.

27.2.3 Thermoelectric Applications

The schematic of a thermoelectric device, or module, on Figure 27.5, illustrates the three different modes of operation: power generation, cooling, and heating. The *thermoelectric module* is a standardized device consisting of several p- and n-type legs connected electrically in series and thermally in parallel, and bonded to a ceramic plate on each side (typically alumina). The modules are fabricated in a great variety of sizes, shapes, and number of thermoelectric couples and can operate in a wide range of currents, voltages, powers, and efficiencies. Complex, large-scale thermoelectric systems can be easily designed and built by assembling various numbers of these modules connected in series or in parallel depending on the type of applications.

Power Generation. When a temperature gradient is applied across the thermoelectric device, the heat absorbed at the hot junction (Figure 27.5, hot side $T_h - T_1$ and cold side, $T_c - T_2$) will generate a current through the circuit and deliver electrical power to the load resistance R_L (Harman and Honig 1967). The conversion efficiency η of a thermoelectric generator is determined by the ratio of the electrical energy, supplied to the load resistance, to the thermal energy, absorbed at the hot junction, and is given by

$$\eta = \frac{R_L I^2}{S_{pn} I T_h + K(T_h - T_c) - (1/2)RI^2} \quad (27.20)$$

where K is the thermal conductance in parallel and R is the electrical series resistance of one p–n thermoelectric couple. The electrical power P_L generated can be written as

$$P_L = \frac{S_{pn}(T_h - T_c)^2 R_L}{(R_L + R)^2} \quad (27.21)$$

The thermoelectric generator can be designed to operate at maximum power output, by matching the load and couple resistances, $R_L = R$. The corresponding conversion efficiency is

$$\eta_P = \frac{T_h - T_c}{(3/2)T_h + (1/2)T_c + (1/4)Z_{pn}^{-1}} \quad (27.22)$$

where Z_{pn} is the figure of merit of the p–n couple given by

$$Z_{pn} = \frac{S_{pn}^2}{RK} \quad (27.23)$$

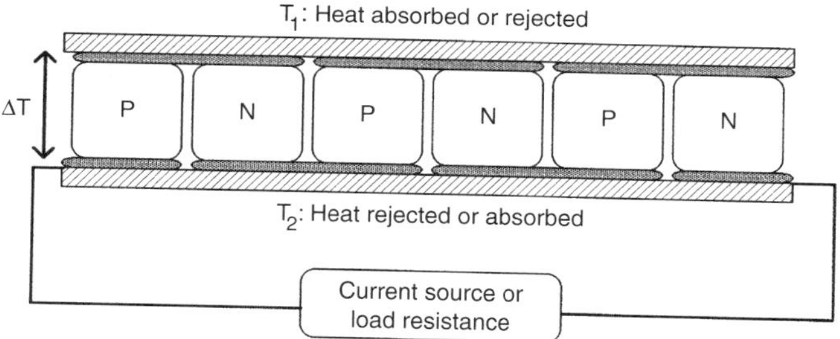

FIGURE 27.5 Schematic of a thermoelectric module.

The figure of merit can be optimized by adjusting the device geometry and minimizing the RK product. This results in Z_{pn} becoming independent of the dimensions of the **thermoelectric legs.** Moreover, if the p- and n-type legs have similar transport properties, the figure of merit, $Z_{pn}=Z$, can be directly related to the Seebeck coefficient S, electrical conductivity σ or resistivity ρ, and thermal conductivity λ of the thermoelectric material:

$$Z = \frac{S^2}{\rho\lambda} = \frac{S^2\sigma}{\lambda} \tag{27.24}$$

The maximum performance η_{max} of the generator is obtained by optimizing the load-to-couple-resistance ratio, leading to the maximum energy conversion efficiency expressed as

$$\eta_{max} = \frac{T_h - T_c}{T_h}\,\frac{\sqrt{1+Z_{pn}T_{av}}-1}{\sqrt{1+Z_{pn}T_{av}}+(T_c/T_h)} \tag{27.25}$$

It must be noted that the maximum efficiency is thus the product of the Carnot efficiency, less than unity, and of a material-related efficiency, increasing with increasing Z_{pn} values as illustrated in Figure 27.6.

Refrigeration. When a current source is used to deliver electrical power to a thermoelectric device, heat can be pumped from T_1 to T_2 and the device thus operates as a refrigerator (Figure 27.5, hot side $T_h=T_2$ and cold side, $T_c=T_1$). As in the case of a thermoelectric generator the operation of a thermoelectric cooler depends solely upon the properties of the p–n thermocouple materials expressed in terms of the figure of merit Z_{pn} and the two temperatures T_c and T_h (Goldsmid 1986). The conversion efficiency or coefficient of performance, COP, of a thermoelectric refrigerator is determined by the ratio of the cooling power pumped at the cold junction to the electrical power input from the current source and is given by

$$\text{COP} = \frac{S_{pn}T_cI-(1/2)RI^2-K(T_h-T_c)}{S_{pn}(T_h-T_c)I+RI^2} \tag{27.26}$$

There are three different modes of operation which are of interest to thermoelectric coolers. A thermoelectric cooler be designed to operate at maximum cooling power, $Q_{c\ max}$, by optimizing the value

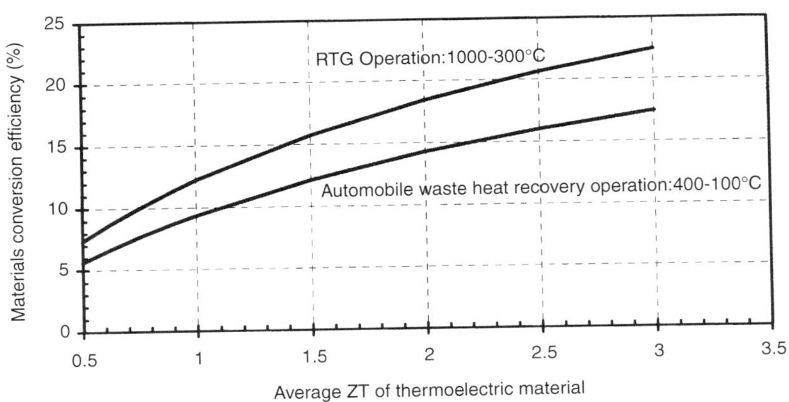

FIGURE 27.6 Maximum conversion efficiency η_{max} as a function of the average material figure of merit ZT, calculated using Equation 27.25 for two systems operating in different temperature ranges: the radioisotope generator (RTG) used for deep space missions and an automobile waste heat recovery generator.

of the current:

$$I_{Q_{\mathrm{c\,max}}} = \frac{S_{\mathrm{pn}}T_{\mathrm{c}}}{R} \quad \text{and} \quad Q_{\mathrm{c\,max}} = \frac{1}{2}\frac{(ST_{\mathrm{c}})^2}{R} - K(T_{\mathrm{h}} - T_{\mathrm{c}}) \tag{27.27}$$

Similarly, the conditions required for operating at maximum efficiency, $\mathrm{COP}_{\mathrm{max}}$, across a constant temperature gradient, are determined by differentiating Equation 27.26 with respect to I, with the solution:

$$I_{\mathrm{COP}_{\mathrm{max}}} = \frac{K(T_{\mathrm{h}} - T_{\mathrm{c}})_{\mathrm{c}}}{S_{\mathrm{pn}}T_{\mathrm{av}}}\left(1 + \sqrt{1 + Z_{\mathrm{pn}}T_{\mathrm{av}}}\right) \tag{27.28}$$

$$\mathrm{COP}_{\mathrm{max}} = \frac{T_{\mathrm{c}}}{T_{\mathrm{h}} - T_{\mathrm{c}}} - \frac{\sqrt{1 + Z_{\mathrm{pn}}T_{\mathrm{av}}} - (T_{\mathrm{h}}/T_{\mathrm{c}})}{\sqrt{1 + Z_{\mathrm{pn}}T_{\mathrm{av}}} + 1} \tag{27.29}$$

By reversing the input current to the device, the thermoelectric refrigerator can become a heat pump, with T_1 being the hot junction temperature. The expression of the maximum conversion efficiency of the heat pump is very similar to Equation 27.29 because of the following relationship:

$$(\mathrm{COP}_{\mathrm{max}})^{\text{heat pump}} = 1 + (\mathrm{COP}_{\mathrm{max}})^{\text{refrigerator}} \tag{27.30}$$

The maximum COP expression in Equation 27.29 is similar to the one derived for the conversion efficiency η of a thermoelectric generator in Equation 27.25. However, there is a major difference between the $\mathrm{COP}_{\mathrm{max}}$ and η_{max} parameters. Clearly, η_{max} increases with increasing ΔT values but is limited by the Carnot efficiency (Equation 27.22) which is less than 1, while $\mathrm{COP}_{\mathrm{max}}$ in Equation 27.20 increases with decreasing ΔT values and can reach values much larger than 1. Figure 27.7 represents the variations of the $\mathrm{COP}_{\mathrm{max}}$ of a thermoelectric cooling device optimized for working voltage and geometry as a function of average ZT values and temperature differences (hot junction temperature at 300 K). The average ZT value for current state-of-the-art commercially available materials (Bi_2Te_3-based alloys) is about 0.8. For example, it can be seen that a $\mathrm{COP}_{\mathrm{max}}$ of 4 is obtained for a $(T_{\mathrm{h}} - T_{\mathrm{c}})$ difference of 10 K, meaning that to pump 8 W of thermal power only 2 W of electrical power needs to be provided to the thermoelectric cooling device. This also means that 10 W of thermal power will be rejected at the hot side of the cooler.

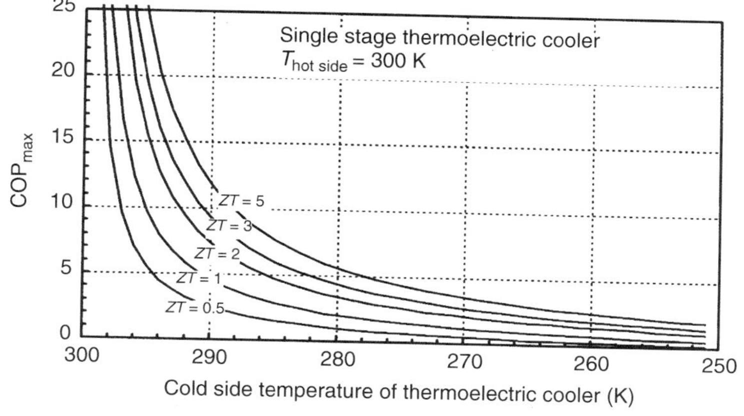

FIGURE 27.7 Maximum material coefficient of performance $\mathrm{COP}_{\mathrm{max}}$ of a single-stage thermoelectric cooler calculated using Equation 27.29 as a function of the cold-side temperature (hot-side temperature of 300 K). Curves corresponding to various values of the average material figure of merit are displayed.

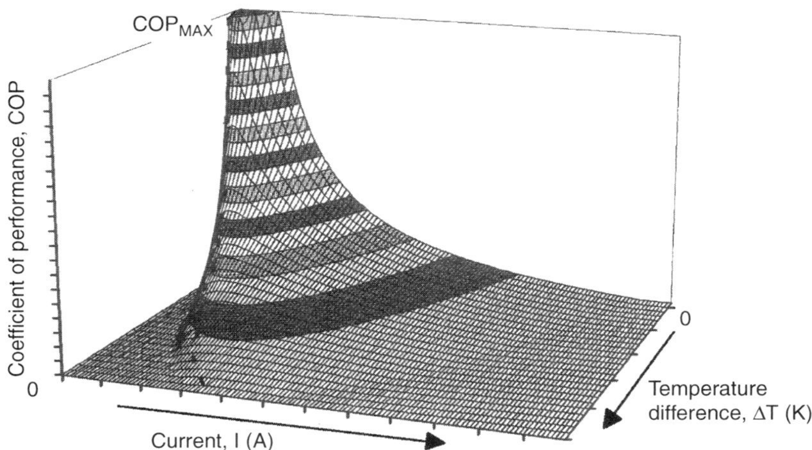

FIGURE 27.8 Three-dimensional plot of the variations of the COP of a thermoelectric cooler as a function of the operating current and the temperature difference.

The operation of a thermoelectric refrigerator at maximum cooling power necessitates a substantially higher input current than the operation at maximum efficiency. This is illustrated by calculating the variations of the maximum COP and cooling power with the input current and temperature difference which have been plotted in Figure 27.8 and Figure 27.9. The calculation was based on the properties of a thermoelectric cooler using state-of-the-art Bi_2Te_3-based alloys, and the arbitrary units are the same for both graphs. It can be seen that $I_{COP_{max}}$ increases while $I_{Q_{c\ max}}$ decreases with increasing ΔT. Also, it is possible to operate at the same cooling power with two different current values.

Finally, the third problem of interest for thermoelectric coolers is to determine the maximum temperature difference, ΔT_{max}, that can be achieved across the device. As shown on Figure 27.9, by operating at maximum cooling power and extrapolating Equation 27.27 to $Q_{c\ max}=0$, ΔT_{max} is given by

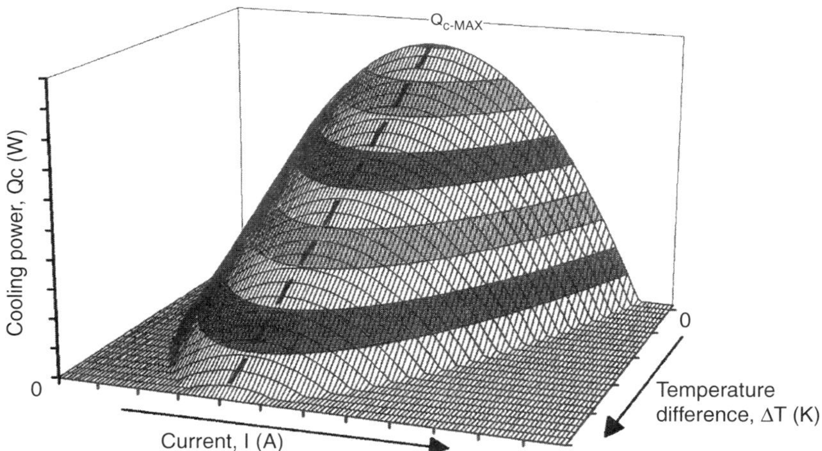

FIGURE 27.9 Three-dimensional plot of the variations of the cooling power of a thermoelectric cooler as a function of the operating current and the temperature difference.

$$\Delta T_{max} = \frac{1}{2} Z_{pn} T_c^2 \quad \text{and} \quad T_{c\ min} = \frac{\sqrt{1 + 2Z_{pn}T_h} - 1}{Z_{pn}} \tag{27.31}$$

where $T_{c\ min}$ corresponds to the lowest cold-side temperature achievable. If the cooler operates at a ΔT close to ΔT_{max} or higher, it becomes necessary to consider a cascade arrangement with several **stages**. The COP of an n-stage thermoelectric cooler is optimized if the COP of each stage, COP_i, is the same, which requires $\Delta T_i/T_{i-1}$ to be the same for each stage. The overall maximum COP is then expressed as

$$COP_{max} = \frac{1}{\left(\prod_{i=1}^{n} (1 + (1/COP_i)) - 1\right)} \tag{27.32}$$

27.2.4 Additional Considerations

When considering the operation of an actual thermoelectric device, several other important parameters must be considered. The thermal and electrical contact resistances can substantially degrade the device performance, in particular for short lengths of the thermoelectric legs. For example, the conversion efficiency of a radioisotope generator system is about 20% lower than the value calculated in Figure 27.6 for the thermoelectric materials only. The electrical contact resistance arises from the connection (see Figure 27.5) of all the legs in series. Typical values obtained for actual generators and coolers are 10–25 μΩ cm^2. The thermal contact resistance is generated by the heat-transfer characteristics of the ceramic plates and contact layers used to build the thermoelectric module. The heat exchangers and corresponding heat losses should also be taken into account.

In addition, the transport properties of the thermoelectric materials vary with temperature, as illustrated in Figure 27.4. When a thermoelectric device is operating across a wide temperature range, these variations should be factored in the calculation of its performance.

Nomenclature

COP coefficient of performance
COP_{max} maximum coefficient of performance
COP_i coefficient of performance of the *i*th stage of a multistage thermoelectric cooler
I current intensity
$I_{COP_{max}}$ current intensity required to operate a thermoelectric cooler at maximum conversion efficiency
$I_{Q_c\ max}$ current intensity required to operate a thermoelectric cooler at maximum cooling power
K thermal conductance
Q rate of reversible heat absorption
R electrical resistance
R_L load resistance
P_L electrical power delivered to the load resistance
S Seebeck coefficient
S_{pn} Seebeck coefficient of a p–n couple of thermoelements
T_1 temperature
T_2 temperature
T_{av} average temperature across the thermoelectric device
T_c cold-side temperature of a thermoelectric device
T_{cmin} minimum cold-side temperature which can be achieved by a thermoelectric cooler
T_h hot-side temperature of a thermoelectric device
V voltage; open-circuit voltage

Z thermoelectric figure of merit
Z_{pn} thermoelectric figure of merit of a p–n couple of thermoelements
ZT dimensionless thermoelectric figure of merit
β Thomson coefficient
β_p Thomson coefficient for the p-type thermoelement
β_n Thomson coefficient for the n-type thermoelement
ΔT temperature difference across a thermoelectric device
ΔT_{max} maximum temperature difference which can be achieved across a thermoelectric cooler
η thermoelectric conversion efficiency
η_{max} maximum thermoelectric conversion efficiency
λ thermal conductivity
Π_{pn} Peltier coefficient
ρ electrical resistivity

Defining Terms

Coefficient of performance: Electrical to thermal energy conversion efficiency of a thermoelectric refrigerator, determined by the ratio of the cooling power pumped at the cold junction to the electrical power input from the current source.

Dimensionless figure of merit: The performance of a thermoelectric device depends solely upon the properties of the thermoelectric material, expressed in terms of the dimensionless figure of merit *ZT*, and the hot-side and cold-side temperatures. *ZT* is calculated as the square of the Seebeck coefficient times the absolute temperature divided by the product of the electrical resistivity to the thermal conductivity. The best *ZT* values are obtained in heavily doped semiconductors, such as Bi_2Te_3 alloys, PbTe alloys, and Si–Ge alloys.

Stage: Multistage thermoelectric coolers are used to achieve larger temperature differences than possible with a single-stage cooler composed of only one module.

Thermoelectric leg: Single thermoelectric element made of n-type or p-type thermoelectric material used in fabricating a thermoelectric couple, the building block of thermoelectric modules. The geometry of the leg (cross-section-to-length ratio) must be optimized to maximize the performance of the device.

Thermoelectric module: Standardized device consisting of several p- and n-type legs connected electrically in series and thermally in parallel, and bonded to a ceramic plate on each. The modules are fabricated in a great variety of sizes, shapes, and number of thermoelectric couples.

References

Goldsmid, H. J. 1986. *Electronic Refrigeration*. Pion Ltd, London.

Hannan, T. C. and Honig, J. M. 1967. *Thermoelectric and Thermomagnetic Effects and Applications*. McGraw-Hill, New York.

Rowe, D. M. and Bhandari, C. M. 1983. *Modern Thermoelectrics*. Reston Publishing, Reston, VA.

For Further Information

The *Proceedings of the Annual International Conference on Thermoelectrics* are published annually by the International Thermoelectric Society (ITS). These proceedings provide the latest information on thermoelectric materials research and development as well as thermoelectric devices and systems. The ITS also publishes a semiannual newsletter. For ITS membership or questions related to thermoelectrics, you may contact the current ITS secretary: Dr. Jean-Pierre Fleurial, Jet Propulsion Laboratory, MS 277-212, Pasadena, CA 91109. Phone (818)-354-4144. Fax (818) 393-6951. E-mail jean-pierre.fleurial@jpl.nasa.gov

Also, the *CRC Handbook of Thermoelectrics*, edited by D.M. Rowe was published by CRC Press Inc., Boca Raton, FL, became available in 1996. This handbook covers all current activities in thermoelectrics.

27.3 Magnetohydrodynamic Power Generation

William D. Jackson

27.3.1 Introduction

The discipline known as magnetohydrodynamics (MHD) deals with the interactions between electrically conducting fluids and electromagnetic fields. First investigated experimentally by Michael Faraday in 1832 during his efforts to establish the principles of electromagnetic induction, application to energy conversion yields a heat engine which has its output in electrical form and, therefore, qualifies as a direct converter. This is generally referred to as an MHD generator, but could be better described as an electromagnetic turbine as it operates on a thermodynamic cycle similar to that of a gas turbine.

The operating principle is elegantly simple, as shown in Figure 27.10. A pressurized, electrically conducting fluid flows through a transverse magnetic field in a channel or duct. Electrodes located on the channel walls parallel to the magnetic field and connected through an external circuit enable the motionally induced "Faraday electromotive force" to drive an electric current through the circuit and thus deliver power to a load connected into it. Taking the fluid velocity as **u** and the magnetic flux density as **B**, the intensity of the motionally induced field is **u**×**B**. The current density, **J**, in the channel for a

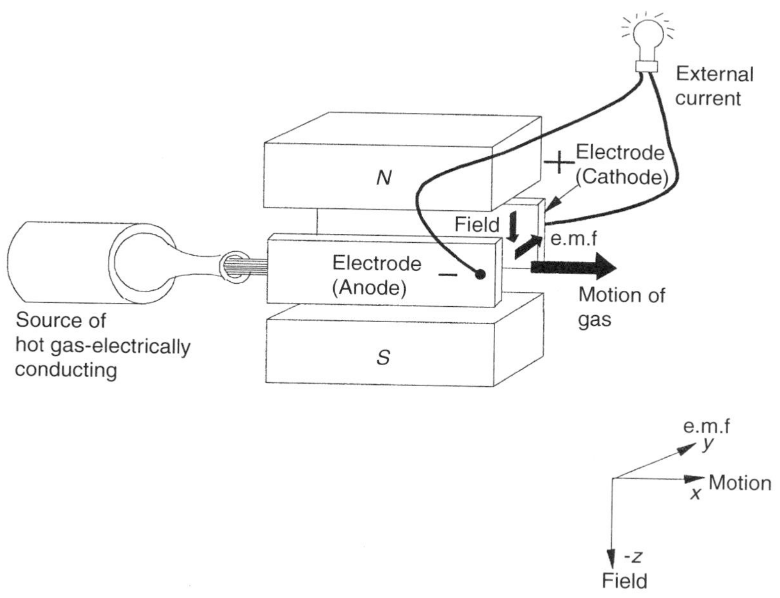

FIGURE 27.10 Principle of electromagnetic turbine or MHD generator.

scalar conductivity σ is then given by Ohm's law for a moving conductor as

$$\mathbf{J} = \sigma[\mathbf{E} + \mathbf{u} \times \mathbf{B}]$$

By taking the coordinates of Figure 27.10 and assuming that the quantities are constant, the power density flow from the MHD generator is, using $\mathbf{E} \cdot \mathbf{J}$

$$w_e = \sigma u_x^2 B^2 k(1 - k)$$

where $k = Ez/u \times B$ is the "loading factor" relating loaded electric field to open circuit induction and is used in the same manner as the regulation of an electrical machine is applied to its terminal voltage.

It is instructive at this point to determine the power density of an MHD generator using values representative of the most commonly considered type of MHD generator. Combustion gas with $\sigma = 10$ S/m, a flow velocity of 800 m/s and an applied field of 5 T for maximum power transfer ($k = 0.5$) yields w_e as 40 MW/m^3. This value is sufficiently attractive to qualify MHD generators for bulk power applications. An intensive, worldwide development effort to utilize this and other MHD generator properties has been conducted since the late 1950s. However, this has not yet led to any significant application of MHD generators. To understand how this has come about and what still needs to be accomplished to make MHD attractive will now be summarized.

27.3.2 Electrical Conductivity Considerations

Two approaches have been followed to obtain adequate ionization and, therefore, conductivity in the working fluid. What may be termed the mainline approach to achieving electrical conductivity is to add a readily ionizable material to "seed" the working fluid. Alkali metals with ionization potentials around 4 V are obvious candidates, although a lower value would be highly desirable. A potassium salt with an ionization potential of 4.09 eV has been widely used because of low cost but cesium with a 3.89-eV value is the preferred seed material when the running time is short or the working fluid is recycled. There are two methods of ionization:

1. Thermal ionization in which recombination ensures a common temperature for electrons, ions, and neutrals; the mass action law (Saha equation) is followed; and the heat of ionization in electron volts is the ionization potential; and
2. Extrathermal or nonequilibrium ionization where electrons and heavy particles are at different temperatures and the concept of entwined fluids (electron, ion, and neutral gases) is involved.

The former is applicable to diatomic combustion gases while the latter occurs in monatomic (noble) gases but is also observed in hydrogen. Only a small amount of seed material is required and is typically around 1% of the total mass flow for maximum conductivity.

The existence of mutually perpendicular **E** and **B** fields in an MHD generator is of major significance in that the electrons are now subjected to the Hall effect. This causes electrons and, therefore, electric currents to flow at an angle with respect to the **E** field in the collision-dominated environment of the MHD generator. The presence of a significant Hall effect requires that the electrical boundary conditions on the channel be carefully selected and also introduces the possibility of working fluid instabilities driven by force fluctuations of electrical origin. A further source of fluctuations and consequent loss of conductivity occurs when nonequilibrium ionization is employed due to current concentration by Joule heating. This latter effect is controlled by operating only in a regime where, throughout the channel, complete ionization of the seed material is achieved.

27.3.3 Generator Configurations and Loading

The basic consequence of the Hall effect is to set up **E** fields in both transverse and axial directions in the generator channel and these are generally referred to as the Faraday and Hall fields, respectively.

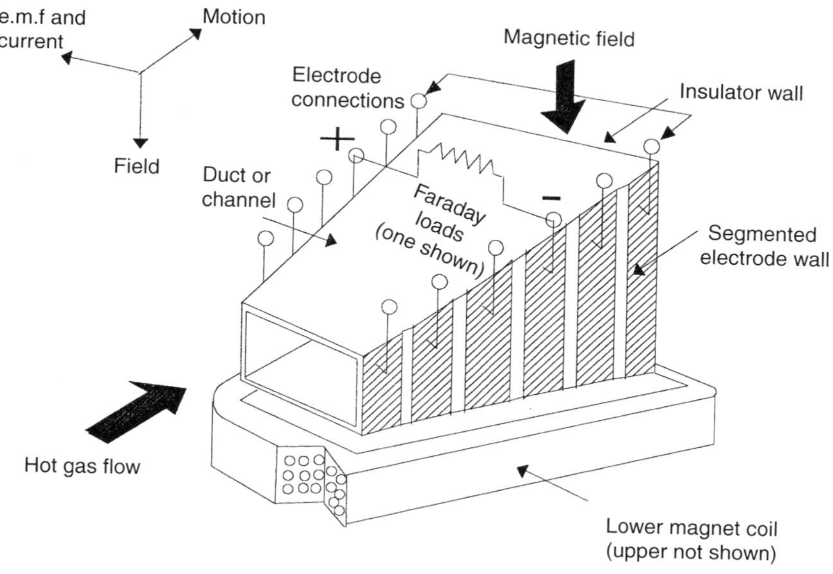

FIGURE 27.11 Basic Faraday linear MHD generator.

The direction of the Faraday field depends on the magnetic field; the Hall field depends on the Hall parameter and is always directed toward the upstream end of the channel. These considerations, in turn, lead to the MHD generator having the characteristics of a gyrator—a two-terminal pair power-producing device in which one terminal pair (Faraday) is a voltage source and the other (Hall) is a current source dependent in this case on the Hall parameter. Electric power can be extracted from either the Faraday or Hall terminals, or both.

This has resulted in several electrical boundary conditions being utilized with the axial flow channel as shown in Figure 27.11. These are most readily understood by treating each anode-cathode pair as a generating cell. The segmented Faraday configuration is then simply a parallel operation of cells which leads to the apparently inconvenient requirement of separate loading of individual cells: the Hall connection utilizes a single load by series connection but depends on the Hall parameter for its performance; and the diagonal connection connects the cells in series-parallel and so avoids Hall parameter dependence while retaining the single load feature. In all three linear configurations, the channel walls are electrically segmented to support the Hall field, and experience has shown that this must be sufficiently finely graded so that no more than 50 V is supported by the interelectrode gaps to avoid electrical breakdown.

The MHD generator is a linear version of the homopolar machine originally demonstrated by Faraday and is, as a practical matter, confined to DC generation. Accordingly, some form of DC–AC conversion using power electronics is required for the vast majority of applications. The single load feature loses significance in this situation as the "power conditioning" can readily be arranged to combine (consolidate) the individual cell outputs before conversion to the required AC system conditions. Indeed, to maximize power extraction while limiting interelectrode voltages and controlling electrode currents (to ensure adequate lifetime), the power conditioning is arranged to extract power from both Faraday and Hall terminal pairs.

An alternative geometry is to set up a radial flow (usually but not necessarily outward) with the disk configuration of Figure 27.11d. The result is a Hall generator, which is generally favored for nonequilibrium ionization as it avoids the inevitable nonuniformities associated with electrode segmentation with their proclivity for promoting ionization instabilities. A measure of Faraday performance is achievable through the introduction of swirl, and additional ring electrodes enable power conditioning to control (and optimize) the radial electric field.

27.3.4 Components

An MHD generator per se requires several components to make up a complete powertrain. In addition to the power conditioning needed for DC–AC conversion these include a magnet, seed injector, combustor with fuel and oxidizer supply or an input heat exchanger, nozzle, compressor, diffuser, exhaust gas-cleaning system (for once-through systems), and controls. The need to accommodate a channel between the poles of a magnet qualifies the MHD generator as a large-air-gap machine.

27.3.5 Systems

Power systems incorporating MHD generators are either of the once-through (open-cycle) or working fluid recycle (closed-cycle) type, and the complete MHD system described in the previous section can either be stand-alone or incorporated into a more complex configuration such as a binary cycle. In the latter case, the high-temperature operation of the MHD unit makes it a topping cycle candidate and it is in this mode that MHD has received most system consideration. An MHD generator operated directly on ionized combustion gas is usually associated with an open cycle while nonequilibrium ionization with seeded noble gases and LMMHD are invariably considered for closed-cycle operation. An exception is nonequilibrium ionization in cesium-seeded hydrogen which has received some attention for open-cycle operation.

27.3.6 Heat Sources, Applications, and Environmental Considerations

A heat source capable of providing from 1000 K for LMMHD to over 3000 K for open-cycle systems for power production is a candidate. Rocket motor fuels, all fossil fuels, high-temperature nuclear reactors cooled with hydrogen and biomass are suitable for open cycles, while closed cycles can be driven through a heat exchanger by any of these combustion sources. A high-temperature nuclear reactor, probably helium cooled, is also a feasible source for MHD and in the early stages of development of the process received much attention. With the abandonment of efforts to develop commercial reactors to operate at temperatures above 1200 K, attention has focused on high-energy fuels for pulse power (few seconds) operation and coal for utility power generation.

While the debate over the role of fossil energy in the long-term electricity generation scenario continues, it is established that coal is a major resource which must be utilized at maximum efficiency to limit CO_2 production and must be combusted in a manner which reduces SO_2 and NO_x effluents to acceptable levels. The use of MHD generators significantly advances all of these objectives. Briefly, it was first observed that the "electromagnetic turbine" has the major advantage that it cannot only provide the highest efficiency of any known converter from the Carnot viewpoint but that its operation is not adversely affected by coal slag and ash. Indeed, slag provides an excellent renewable coating for the channel walls and increases lifetime.

System calculations have shown that, when coupled as a topping cycle to the best available steam plant technology, a thermal efficiency with coal and full environmental control is 52.5%. When coupled into a ternary cycle with either a gas turbine or fuel cells and a steam turbine, efficiencies upward of 60% are possible.

27.3.7 Technology Status

A pulse-type MHD generator was successfully built and operated by Avco (now Textron Defense Industries) in 1963 and a complete natural gas-fired pilot plant with a nominal 20-MW MHD generator was commissioned in the U.S.S.R. on the northern outskirts of Moscow in 1971. In the decade of the 1980s, development was focused on coal firing as a result of the oil crises of the 1970s and in the U.S.

progressed to the point where the technology base was developed for a demonstration plant with a 15-MW MHD generator.

27.3.8 Future Prospects

The two particular attributes of the MHD generator are its rapid start-up to multimegawatt power levels for pulse power applications and its ability to provide a very high overall thermal efficiency for power plants using coal while meeting the most stringent environmental standards. The first has already been utilized in crustal exploration, and the second must surely be utilized when coal is the fuel of choice for electric power production. In the meantime, MHD has been established as a viable technology on which further development work will be conducted for advanced applications such as the conversion system for thermonuclear fusion reactors.

For Further Information

The following proceedings series contain a full and complete record of MHD generator and power system development:

Proceedings of the Symposia for the Engineering Aspects of Magnetohydrodynamics, held annually in the U.S. since 1960 (except for 1971 and 1980).

Proceedings of 11 International Conferences on Magnetohydrodynamic Electrical Power Generation, held in 1962, 1964, 1966, 1968, 1971, 1975, 1980, 1983, 1986, 1989, and 1992. The 12th conference will be held in Yokahama, Japan in October 1996.

A1

The International System of Units, Fundamental Constants, and Conversion Factors

Nitin Goel
Intel Technology India Pvt. Ltd.

The International system of units (SI) is based on seven base units. Other derived units can be related to these base units through governing equations. The base units with the recommended symbols are listed in Table A1.1. Derived units of interest in solar engineering are given in Table A1.2.

Standard prefixes can be used in the SI system to designate multiples of the basic units and thereby conserve space. The standard prefixes are listed in Table A1.3.

Table A1.4 lists some physical constants that are frequently used in solar engineering, together with their values in the SI system of units.

Conversion factors between the SI and English systems for commonly used quantities are given in Table A1.5.

TABLE A1.1 The Seven SI Base Units

Quantity	Name of Unit	Symbol
Length	Meter	m
Mass	Kilogram	kg
Time	Second	s
Electric current	Ampere	A
Thermodynamic temperature	Kelvin	K
Luminous intensity	Candela	cd
Amount of a substance	Mole	mol

TABLE A1.2 SI Derived Units

Quantity	Name of Unit	Symbol
Acceleration	Meters per second squared	m/s^2
Area	Square meters	m^2
Density	Kilogram per cubic meter	kg/m^3
Dynamic viscosity	Newton-second per square meter	$N\ s/m^2$
Force	Newton ($=1$ kg m/s^2)	N
Frequency	Hertz	Hz
Kinematic viscosity	Square meter per second	m^2/s
Plane angle	Radian	rad
Potential difference	Volt	V
Power	Watt($=1$ J/s)	W
Pressure	Pascal ($=1$ N/m^2)	Pa
Radiant intensity	Watts per steradian	W/sr
Solid angle	Steradian	sr
Specific heat	Joules per kilogram–Kelvin	J/kg K
Thermal conductivity	Watts per meter–Kelvin	W/m K
Velocity	Meters per second	m/s
Volume	Cubic meter	m^3
Work, energy, heat	Joule ($=1$ N/m)	J

TABLE A1.3 English Prefixes

Multiplier	Symbol	Prefix	Multiplier	Multiplier Symbol
10^{12}	T	Tera	10^3	M (thousand)
10^9	G	Giga	10^6	MM (million)
10^6	m	Mega		
10^3	k	Kilo		
10^2	h	Hecto		
10^1	da	Deka		
10^{-1}	d	Deci		
10^{-2}	c	Centi		
10^{-3}	m	Milli		
10^{-6}	μ	Micro		
10^{-9}	n	Nano		
10^{-12}	p	Pico		
10^{-15}	f	Femto		
10^{-18}	a	Atto		

TABLE A1.4 Physical Constants in SI Units

Quantity	Symbol	Value
Avogadro constant	N	6.022169×10^{26} kmol^{-1}
Boltzmann constant	k	1.380622×10^{-23} J/K
First radiation constant	$C_1=2\pi hC^2$	3.741844×10^{-16} W m^2
Gas constant	R	8.31434×10^{3} J/kmol K
Planck constant	h	6.626196×10^{-34} J s
Second radiation constant	$C_2=hc/k$	1.438833×10^{-2} m K
Speed of light in a vacuum	C	2.997925×10^{8} m/s
Stefan–Boltzmann constant	σ	5.66961×10^{-8} W/m^2 K^4

TABLE A1.5 Conversion Factors

Physical Quantity	Symbol	Conversion Factor
Area	A	1 ft.^{2} = 0.0929 m^2
		1 acre = 43,560 ft.^{2} = 4047 m^2
		1 hectare = 10,000 m^2
		1 square mile = 640 acres
Density	ρ	1 lb$_m$/ft.^{3} = 16.018 kg/m^3
Heat, energy, or work	Q or W	1 Btu = 1055.1 J
		1 kWh = 3.6 MJ
		1 Therm = 105.506 MJ
		1 cal = 4.186 J
		1 ft. lb$_f$ = 1.3558 J
Force	F	1 lb$_f$ = 4.448 N
Heat flow rate, refrigeration	q	1 Btu/h = 0.2931 W
		1 ton (refrigeration) = 3.517 kW
		1 Btu/s = 1055.1 W
Heat flux	q/A	1 Btu/h ft.^{2} = 3.1525 W/m^2
Heat-transfer coefficient	h	1 Btu/h ft.^{2} F = 5.678 W/m^2 K
Length	L	1 ft. = 0.3048 m
		1 in. = 2.54 cm
		1 mi = 1.6093 km
Mass	m	1 lb$_m$ = 0.4536 kg
		1 ton = 2240 lbm
		1 tonne (metric) = 1000 kg
Mass flow rate	$\dot{m}$	1 lb$_m$/h = 0.000126 kg/s
Power	$\dot{W}$	1 hp = 745.7 W
		1 kW = 3415 Btu/h
		1 ft. lb$_f$/s = 1.3558 W
		1 Btu/h = 0.293 W
Pressure	p	1 lb$_f$/in.^{2} (psi) = 6894.8 Pa (N/m^2)
		1 in. Hg = 3,386 Pa
		1 atm = 101,325 Pa (N/m^2) = 14.696 psi
Radiation	I	1 langley = 41,860 J/m^2
		1 langley/min = 697.4 W/m^2
Specific heat capacity	c	1 Btu/lb$_m$ °F = 4187 J/kg K
Internal energy or enthalpy	e or h	1 Btu/lb$_m$ = 2326.0 J/kg
		1 cal/g = 4184 J/kg
Temperature	T	T(°R) = (9/5)T(K)
		T(°F) = [T(°C)](9/5) + 32
		T(°F) = [T(K) − 273.15](9/5) + 32
Thermal conductivity	k	1 Btu/h ft. °F = 1.731 W/m K
Thermal resistance	R_{th}	1 h °F/Btu = 1.8958 K/W
Velocity	V	1 ft./s = 0.3048 m/s
		1 mi/h = 0.44703 m/s
Viscosity, dynamic	μ	1 lb$_m$/ft. s = 1.488 N s/m^2
		1 cP = 0.00100 N s/m^2
Viscosity, kinematic	ν	1 ft.^{2}/s = 0.09029 m^2/s
		1 ft.^{2}/h = 2.581 × 10^{-5} m^2/s
Volume	V	1 ft.^{3} = 0.02832 m^3 = 28.32 L
		1 barrel = 42 gal (U.S.)
		1 gal (U.S. liq.) = 3.785 L
		1 gal (U.K.) = 4.546 L
Volumetric flow rate	$\dot{Q}$	1 ft.^{3}/min (cfia) = 0.000472 m^3/s
		1 gal/min (GPM) = 0.0631 l/s

A2

Solar Radiation Data

Nitin Goel
Intel Technology India Pvt. Ltd.

The altitude and azimuth of the sun are given by

$$\sin \alpha = \sin L \sin \delta_s + \cos \phi \cos \delta_s \cos h_s \quad (1)$$

and

$$\sin a_s = -\cos \delta_s \sin h_s / \cos \alpha \quad (2)$$

where α = altitude of the sun (angular elevation above the horizon)
L = latitude of the observer
δ_s = declination of the sun
h_s = hour angle of sun (angular distance from the meridian of the observer)
a_s = azimuth of the sun (measured eastward from north)

From Eqs. (1) and (2) it can be seen that the altitude and azimuth of the sun are functions of the latitude of the observer, the time of day (hour angle), and the date (declination).

Figure A2.1(*b*–*g*) provides a series of charts, one for each 5° of latitude (except 5°, 15°, 75°, and 85°) giving the altitude and azimuth of the sun as a function of the true solar time and the declination of the sun in a form originally suggested by Hand. Linear interpolation for intermediate latitudes will give results within the accuracy to which the charts can be read.

On these charts, a point corresponding to the projected position of the sun is determined from the heavy lines corresponding to declination and solar time.

To find the solar altitude and azimuth:

1. Select the chart or charts appropriate to the latitude.
2. Find the solar declination δ corresponding to the date.
3. Determine the *true solar time* as follows:
   (a) To the *local standard time* (zone time) add 4′ for each degree of longitude the station is east of the standard meridian or subtract 4′ for each degree west of the standard meridian to get the *local mean solar time.*
   (b) To the *local mean solar time* add algebraically the equation of time; the sum is the required *true solar time.*
4. Read the required altitude and azimuth at the point determined by the declination and the true solar time. Interpolate linearly between two charts for intermediate latitudes.

It should be emphasized that the solar altitude determined from these charts is the true geometric position of the center of the sun. At low solar elevations terrestrial refraction may considerably alter the apparent position of sun. Under average atmospheric refraction the sun will appear on the horizon when it actually is about 34′ below the horizon; the effect of refraction decreases rapidly with increasing solar elevation. Since sunset or sunrise is defined as the time when the upper limb of the sun appears on the horizon, and the semidiameter of the sun is 16′, sunset or sunrise occurs under average atmospheric refraction when the sun is 50′ below the horizon. In polar regions especially, unusual atmospheric refraction can make considerable variation in the time of sunset or sunrise.

The 90°N chart is included for interpolation purposes; the azimuths lose their directional significance at the pole.

Altitude and azimuth in southern latitudes. To compute solar altitude and azimuth for southern latitudes, change the sign of the solar declination and proceed as above. The resulting azimuths will indicate angular distance from *south* (measured eastward) rather than from north.

(a)

FIGURE A2.1 Description of method for calculating true solar time, together with accompanying meteorological charts, for computing solar-altitude and azimuth angles, (a) Description of method; (b) chart, 25°N latitude; (c) chart, 30°N latitude; (d) chart, 35°N latitude; (e) chart, 40°N latitude; (f) chart, 45°N latitude; (g) chart, 50°N latitude. Description and charts reproduced from the "Smithsonian Meteorological Tables" with permission from the Smithsonian Institute, Washington, D.C.

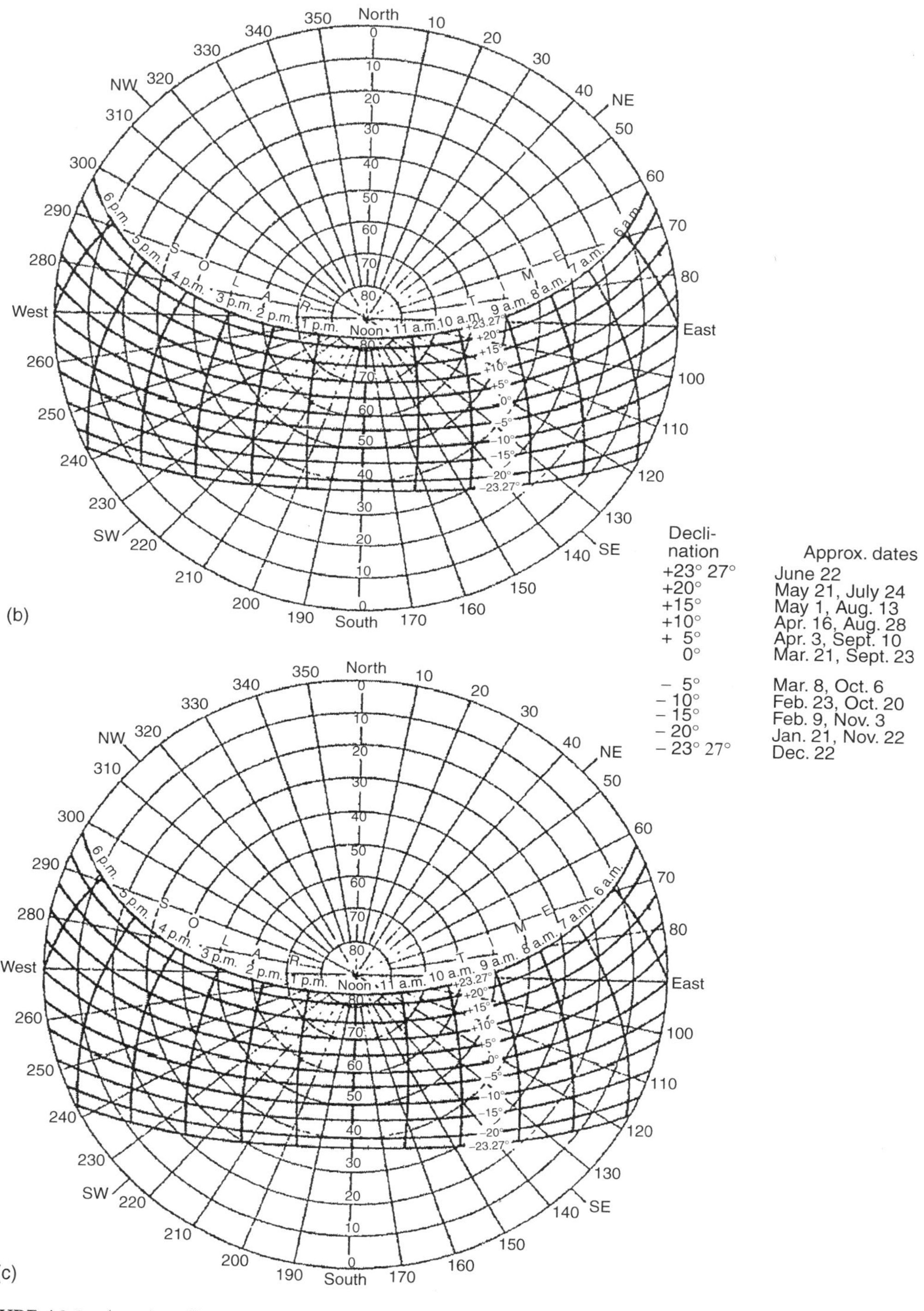

FIGURE A2.1 *(continued)*

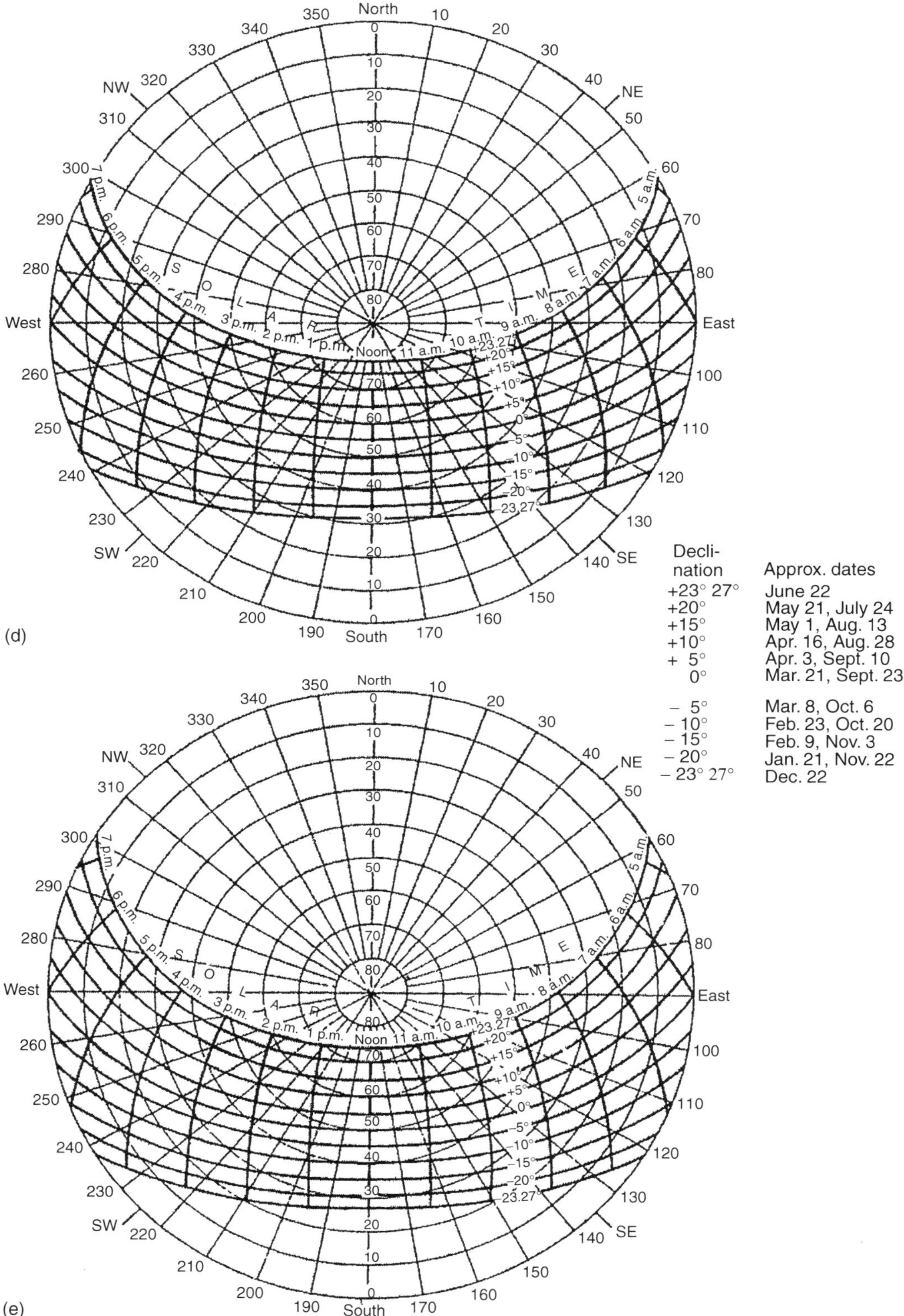

FIGURE A2.1 *(continued)*

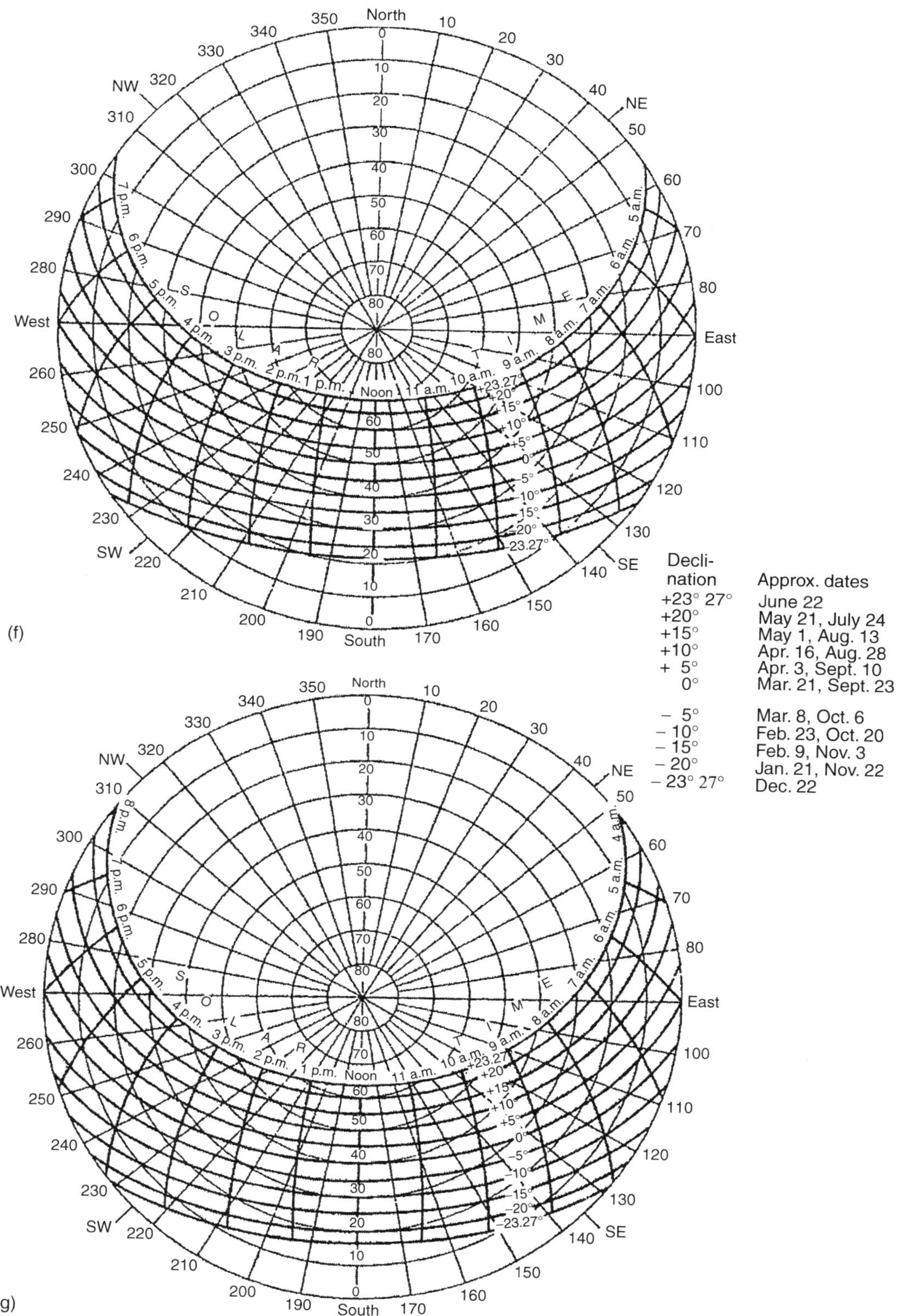

FIGURE A2.1 (*continued*)

TABLE A2.1 Solar Irradiance for Different Air Masses

Wavelength	Air Mass; $\alpha=0.66$; $\beta=0.085$[a]				
	0	1	4	7	10
0.290	482.0	0.0	0.0	0.0	0.0
0.295	584.0	0.0	0.0	0.0	0.0
0.300	514.0	4.1	0.0	0.0	0.0
0.305	603.0	11.4	0.0	0.0	0.0
0.310	689.0	30.5	0.0	0.0	0.0
0.315	764.0	79.4	0.1	0.0	0.0
0.320	830.0	202.6	2.9	0.0	0.0
0.325	975.0	269.5	5.7	0.1	0.0
0.330	1059.0	331.6	10.2	0.3	0.0
0.335	1081.0	383.4	17.1	0.8	0.0
0.340	1074.0	431.3	24.9	1.8	0.1
0.345	1069.0	449.2	33.3	2.5	0.2
0.350	1093.0	480.5	40.8	3.5	0.3
0.355	1083.0	498.0	48.4	4.7	0.5
0.360	1068.0	513.7	57.2	6.4	0.7
0.365	1132.0	561.3	68.4	8.3	1.0
0.370	1181.0	603.5	80.5	10.7	1.4
0.375	1157.0	609.4	89.0	13.0	1.9
0.380	1120.0	608.0	97.2	15.6	2.5
0.385	1098.0	609.8	104.5	17.9	3.1
0.390	1098.0	623.9	114.5	21.0	3.9
0.395	1189.0	691.2	135.8	26.7	5.2
0.400	1429.0	849.9	178.8	37.6	7.9
0.405	1644.0	992.8	218.7	48.2	10.6
0.410	1751.0	1073.7	247.5	57.1	13.2
0.415	1774.0	1104.5	266.5	64.3	15.5
0.420	1747.0	1104.3	278.9	70.4	17.8
0.425	1693.0	1086.5	287.2	78.9	20.1
0.430	1639.0	1067.9	295.4	81.7	22.6
0.435	1663.0	1100.1	318.4	92.2	26.7
0.440	1810.0	1215.5	368.2	111.5	33.8
0.445	1922.0	1310.4	415.3	131.6	41.7
0.450	2006.0	1388.4	460.3	152.6	50.6
0.455	2057.0	1434.8	486.9	165.2	56.1
0.460	2066.0	1452.2	504.4	175.2	60.8
0.465	2048.0	1450.7	515.7	183.3	65.1
0.470	2033.0	1451.2	527.9	192.0	69.8
0.475	2044.0	1470.3	547.3	203.7	75.8
0.480	2074.0	1503.4	572.6	218.1	83.1
0.485	1976.0	1443.3	562.4	219.2	85.4
0.490	1950.0	1435.2	572.2	228.2	91.0
0.495	1960.0	1453.6	592.9	241.9	98.7
0.500	1942.0	1451.2	605.6	252.7	105.5
0.505	1920.0	1440.1	607.6	256.4	108.2
0.510	1882.0	1416.8	604.4	257.8	110.0
0.515	1833.0	1384.9	597.3	257.6	111.1
0.520	1833.0	1390.0	606.1	264.3	115.2
0.525	1852.0	1409.5	621.3	273.9	120.7
0.530	1842.0	1406.9	626.9	279.4	124.5
0.535	1818.0	1393.6	627.7	282.8	127.4
0.540	1783.0	1371.7	624.5	284.4	129.5
0.545	1754.0	1354.2	623.2	286.8	132.0
0.550	1725.0	1336.6	621.7	289.2	134.5
0.555	1720.0	1335.7	625.5	293.0	137.3

(continued)

TABLE A2.1 ***(Continued)***

Wavelength	Air Mass; $\alpha=0.66$; $\beta=0.085$[a]				
	0	1	4	7	10
0.560	1695.0	1319.2	622.0	293.3	138.3
0.565	1705.0	1330.0	631.3	299.6	142.2
0.570	1712.0	1338.4	639.5	305.6	146.0
0.575	1719.0	1346.9	647.8	311.6	149.6
0.580	1715.0	1346.7	652.0	315.7	152.8
0.585	1712.0	1347.3	656.6	320.0	156.0
0.590	1700.0	1340.7	657.7	322.6	158.3
0.595	1682.0	1329.4	656.4	324.1	160.0
0.600	1660.0	1319.6	655.8	325.9	162.0
0.605	1647.0	1311.0	661.3	333.6	168.2
0.610	1635.0	1307.9	669.6	342.8	175.5
0.620	1602.0	1294.2	682.4	359.9	189.7
0.630	1570.0	1280.9	695.6	377.8	205.2
0.640	1544.0	1272.1	711.4	397.9	222.5
0.650	1511.0	1257.1	723.9	416.9	240.1
0.660	1486.0	1244.2	730.2	428.6	251.6
0.670	1456.0	1226.8	733.8	438.9	262.5
0.680	1427.0	1209.9	737.4	449.5	273.9
0.690	1402.0	1196.2	742.9	461.3	286.5
0.698	1374.6	1010.3	546.1	311.8	181.6
0.700	1369.0	1175.3	743.7	470.6	297.7
0.710	1344.0	1157.4	739.2	472.1	301.5
0.720	1314.0	1135.1	731.7	471.6	304.0
0.728	1295.5	1003.1	582.3	351.7	212.5
0.730	1290.0	1117.8	727.1	479.0	307.7
0.740	1260.0	1095.1	718.9	471.9	309.8
0.750	1235.0	1076.6	713.2	472.4	313.0
0.762	1205.5	794.0	357.1	163.6	69.1
0.770	1185.0	1039.2	700.8	472.7	318.8
0.780	1159.0	1019.4	693.6	472.0	321.1
0.790	1134.0	1000.3	686.7	471.4	323.6
0.800	1109.0	981.2	679.4	470.5	325.8
0.806	1095.1	874.4	547.7	355.9	234.4
0.825	1048.0	931.6	654.3	459.6	322.8
0.830	1036.0	921.8	649.3	457.3	322.1
0.835	1024.5	912.4	644.4	455.2	321.5
0.846	998.1	476.2	181.0	85.9	44.2
0.860	968.0	506.4	212.0	107.4	58.3
0.870	947.0	453.8	174.7	84.0	43.8
0.875	436.5	449.2	173.4	83.6	43.7
0.887	912.5	448.6	178.3	87.7	46.7
0.900	891.0	448.9	183.7	92.3	50.0
0.907	882.8	455.2	190.9	97.6	53.7
0.915	874.5	461.5	198.5	103.2	57.5
0.925	863.5	279.0	73.6	28.0	12.1
0.930	858.0	221.8	46.9	15.4	6.0
0.940	847.0	313.4	95.0	39.6	18.5
0.950	837.0	296.5	86.3	35.0	16.0
0.955	828.5	321.1	102.3	44.1	21.2
0.965	811.5	344.4	120.4	55.1	27.8
0.975	794.0	576.9	346.0	224.6	150.1
0.985	776.0	544.6	316.1	201.2	132.4
1.018	719.2	617.5	391.0	247.5	156.7
1.082	620.0	512.9	290.4	164.4	93.1

(continued)

TABLE A2.1 *(Continued)*

Wavelength	Air Mass; $\alpha=0.66$; $\beta=0.085$[a]				
	0	1	4	7	10
1.094	602.0	464.1	303.1	210.8	149.9
1.098	596.0	503.7	304.1	183.6	110.9
1.101	591.8	504.8	362.7	267.3	198.8
1.128	560.5	135.1	27.7	9.1	3.6
1.131	557.0	152.2	35.3	12.6	5.3
1.137	550.1	143.1	31.7	11.0	4.5
1.144	542.0	191.2	57.4	24.2	11.6
1.147	538.5	174.5	48.2	19.3	8.8
1.178	507.0	399.3	195.1	95.4	46.6
1.189	496.0	402.2	214.5	114.4	61.0
1.193	492.0	424.0	310.8	233.3	176.6
1.222	464.3	391.8	235.3	141.3	84.9
1.236	451.2	390.8	254.1	165.2	107.4
1.264	426.5	329.2	209.7	140.0	94.3
1.276	416.7	342.6	238.6	172.6	126.3
1.288	406.8	347.3	216.1	134.4	83.7
1.314	386.1	298.3	137.6	63.5	29.3
1.335	369.7	190.6	85.0	46.7	27.7
1.384	343.7	5.7	0.1	0.0	0.0
1.432	321.0	44.6	5.4	1.3	0.4
1.457	308.6	85.4	20.6	7.7	3.3
1.472	301.4	77.4	17.4	6.2	2.6
1.542	270.4	239.3	165.9	115.0	79.7
1.572	257.3	222.6	168.1	130.4	102.1
1.599	245.4	216.0	166.7	131.5	104.5
1.608	241.5	208.5	157.4	122.1	95.7
1.626	233.6	206.7	160.7	127.5	101.9
1.644	225.6	197.9	152.4	120.1	95.5
1.650	223.0	195.7	150.9	119.1	94.7
1.676	212.1	181.9	114.8	72.4	45.7
1.732	187.9	161.5	102.5	65.1	41.3
1.782	166.6	136.7	75.6	41.8	23.1
1.862	138.2	4.0	0.1	0.0	0.0
1.955	112.9	42.7	14.5	6.8	3.6
2.008	102.0	69.4	35.8	17.7	6.4
2.014	101.2	74.7	45.5	28.8	17.8
2.057	95.6	69.5	41.3	25.3	14.8
2.124	87.4	70.0	35.9	18.4	9.5
2.156	83.8	66.0	32.3	15.8	7.7
2.201	78.9	66.1	49.1	38.0	29.7
2.266	72.4	61.6	46.8	36.8	29.3
2.320	67.6	57.2	43.2	33.8	26.8
2.338	66.3	54.7	39.9	30.4	23.4
2.356	65.1	52.0	36.3	26.5	19.6
2.388	62.8	36.0	18.7	11.7	7.8
2.415	61.0	32.5	15.8	9.4	6.0
2.453	58.3	29.6	13.7	7.9	5.0
2.494	55.4	20.3	6.8	3.2	1.7
2.537	52.4	4.6	0.4	0.1	0.0
2.900	35.0	2.9	0.2	0.0	0.0
2.941	33.4	6.0	1.0	0.3	0.1
2.954	32.8	5.7	0.9	0.3	0.1
2.973	32.1	8.7	2.2	0.9	0.4
3.005	30.8	7.8	1.8	0.7	0.3

(continued)

TABLE A2.1 ***(Continued)***

Wavelength	Air Mass; $\alpha=0.66$; $\beta=0.085$[a]				
	0	1	4	7	10
3.045	28.8	4.7	0.7	0.2	0.1
3.056	28.2	4.9	0.8	0.2	0.1
3.097	26.2	3.2	0.4	0.1	0.0
3.132	24.9	6.8	1.7	0.7	0.3
3.156	24.1	18.7	12.6	8.9	6.3
3.204	22.5	2.1	0.2	0.0	0.0
3.214	22.1	3.4	0.5	0.1	0.0
3.245	21.1	3.9	0.7	0.2	0.1
3.260	20.6	3.7	0.6	0.2	0.1
3.285	19.7	14.2	8.5	5.1	2.8
3.317	18.8	12.9	6.9	3.5	1.3
3.344	18.1	4.2	0.9	0.3	0.1
3.403	16.5	12.3	7.8	5.1	3.2
3.450	15.6	12.5	8.9	6.7	5.0
3.507	14.5	12.5	9.9	8.1	6.7
3.538	14.2	11.8	8.8	6.9	5.5
3.573	13.8	10.9	5.4	2.6	1.3
3.633	13.1	10.8	8.3	6.7	5.5
3.673	12.6	9.1	6.1	4.6	3.5
3.696	12.3	10.4	8.2	6.7	5.6
3.712	12.2	10.9	9.0	7.6	6.5
3.765	11.5	9.5	7.2	5.9	4.8
3.812	11.0	8.9	6.7	5.4	4.4
3.888	10.4	8.1	5.6	4.0	2.9
3.923	10.1	8.0	5.6	4.2	3.1
3.948	9.9	7.8	5.5	4.0	3.0
4.045	9.1	6.7	4.1	2.6	1.5
Total Wm^3	1353	889.2	448.7	255.2	153.8

W/m^2 μm; H_2O 20 mm; O_3 3.4 mm.

[a] The parameters α and β are measures of turbidity of the atmosphere. They are used in the atmospheric transmittance equation $\bar{\tau}_{atm} = e^{-(C_1+C_2)m}$; C_1 includes Rayleigh and ozone attenuation; $C_2 \equiv \beta/\lambda^{\alpha}$.

Source: From Thekaekara, M. P. 1974. *The Energy Crisis and Energy from the Sun*. Institute for Environmental Sciences.

TABLE A2.2 Monthly Averaged, Daily Extraterrestrial Insolation on a Horizontal Surface (Units: Wh/m^2)

Latitude (deg)	January	February	March	April	May	June	July	August	September	October	November	December
20	7415	8397	9552	10,422	10,801	10,868	10,794	10,499	9791	8686	7598	7076
25	6656	7769	9153	10,312	10,936	11,119	10,988	10,484	9494	8129	6871	6284
30	5861	7087	8686	10,127	11,001	11,303	11,114	10,395	9125	7513	6103	5463
35	5039	6359	8153	9869	10,995	11,422	11,172	10,233	8687	6845	5304	4621
40	4200	5591	7559	9540	10,922	11,478	11,165	10,002	8184	6129	4483	3771
45	3355	4791	6909	9145	10,786	11,477	11,099	9705	7620	5373	3648	2925
50	2519	3967	6207	8686	10,594	11,430	10,981	9347	6998	4583	2815	2100
55	1711	3132	5460	8171	10,358	11,352	10,825	8935	6325	3770	1999	1320
60	963	2299	4673	7608	10,097	11,276	10,657	8480	5605	2942	1227	623
65	334	1491	3855	7008	9852	11,279	10,531	8001	4846	2116	544	97

TABLE A2.3a Worldwide Global Horizontal Average Solar Radiation (Units: MJ/sq.m-day)

Position	Lat	Long	January	February	March	April	May	June	July	August	September	October	November	December
Argentina														
Buenos Aires	34.58 S	58.48 W	24.86	21.75	18.56	11.75	8.71	7.15	7.82	8.75	14.49	16.66	24.90	21.93
Australia														
Adelaide	34.93 S	138.52 E	20.99	17.50	20.15	18.27	17.98	—	18.81	19.64	20.11	20.88	20.57	20.72
Brisbane	27.43 S	153.08 E	25.36	22.22	13.25	16.61	12.23	11.52	9.70	15.10	17.61	19.89	—	—
Canberra	35.30 S	148.18 E	28.20	24.68	20.56	14.89	10.29	6.62	—	12.33	16.88	24.06	26.00	25.77
Darwin	12.47 S	130.83 E	26.92	23.40	18.13	13.62	9.30	7.89	9.41	11.15	14.85	18.87	23.43	22.34
Hobart	42.88 S	147.32 E	—	—	—	10.09	7.26	6.04	5.72	9.21	13.54	18.12	—	—
Laverton	37.85 S	114.08 E	22.96	20.42	15.59	13.40	7.48	6.10	6.54	10.43	13.24	18.76	—	—
Sydney	33.87 S	151.20 E	21.09	21.75	17.63	13.63	9.78	8.79	7.62	12.84	16.93	22.10	—	—
Austria														
Wien	48.20 N	16.57 E	3.54	7.10	8.05	14.72	16.79	20.87	19.89	17.27	12.55	8.45	3.51	2.82
Innsbruck	47.27 N	11.38 E	5.57	9.28	10.15	15.96	14.57	17.65	18.35	17.26	12.98	9.08	4.28	3.50
Barbados														
Husbands	13.15 N	59.62 W	19.11	20.23	—	21.80	19.84	20.86	21.55	22.14	—	—	18.30	16.56
Belgium														
Ostende	51.23 N	2.92 E	2.82	5.75	9.93	15.18	16.74	16.93	18.21	18.29	11.71	6.15	2.69	1.97
Melle	50.98 N	3.83 E	2.40	4.66	8.41	13.55	14.23	13.28	15.71	15.61	10.63	5.82	2.40	1.59
Brunei														
Brunei	4.98 N	114.93 E	19.46	20.12	22.71	20.54	19.74	18.31	19.38	20.08	20.83	17.51	17.39	18.12
Bulgaria														
Chirpan	42.20 N	25.33 E	6.72	6.79	8.54	13.27	17.25	17.39	19.85	14.61	12.53	8.52	5.08	5.09
Sofia	42.65 N	23.38 E	4.05	6.23	7.93	9.36	12.98	19.73	19.40	17.70	14.71	6.44	—	3.14
Canada[a]														
Montreal	45.47 N	73.75 E	4.74	8.33	11.84	10.55	15.05	22.44	21.08	18.67	14.83	9.18	4.04	4.01
Ottawa	45.32 N	75.67 E	5.34	9.59	13.33	13.98	20.18	20.34	19.46	17.88	13.84	7.38	4.64	5.04
Toronto	43.67 N	79.38 E	4.79	8.15	11.96	14.00	18.16	24.35	23.38	—	15.89	9.40	4.72	3.79
Vancouver	49.18 N	123.17 E	3.73	4.81	12.14	16.41	20.65	24.04	22.87	19.08	12.77	7.39	4.29	1.53
Chile														
Pascua	27.17 S	109.43 W	19.64	16.65	—	11.12	9.52	8.81	10.90	12.29	17.19	20.51	21.20	22.44
Santiago	33.45 S	70.70 W	18.61	16.33	13.44	8.32	5.07	3.66	3.35	5.65	8.15	13.62	20.14	23.88
China														
Beijing	39.93 N	116.28 W	7.73	10.59	13.87	17.93	20.18	18.65	15.64	16.61	15.52	11.29	7.25	6.89
Guangzhou	23.13 N	113.32 E	11.01	6.32	4.04	7.89	10.53	12.48	16.14	16.02	15.03	15.79	11.55	9.10

(continued)

TABLE A2.3a *(Continued)*

Position	Lat	Long	January	February	March	April	May	June	July	August	September	October	November	December
Harbin	45.75 N	126.77 E	5.15	9.54	17.55	20.51	20.33	17.85	19.18	16.09	13.38	14.50	10.50	6.98
Kunming	25.02 N	102.68 E	9.92	11.26	14.38	18.00	18.53	17.37	11.95	18.47	15.94	12.45	11.96	13.62
Lanzhou	36.05 N	103.88 E	7.30	12.47	10.62	18.91	17.40	20.40	20.23	17.37	13.23	10.21	8.22	6.43
Shanghai	31.17 N	121.43 E	7.44	10.31	11.78	14.36	14.23	16.79	14.63	11.85	15.96	12.03	7.73	8.70
Columbia														
Bogota	4.70 N	74.13 W	17.89	—	19.37	16.58	14.86	—	15.42	18.20	17.05	14.58	14.20	16.66
Cuba														
Havana	23.17 N	82.35 W	—	14.70	18.94	20.95	22.63	18.83	21.40	20.19	16.84	16.98	13.19	13.81
Czech														
Kucharovice	48.88 N	16.08 E	3.03	5.85	9.88	14.06	20.84	19.24	21.18	19.41	13.61	6.11	3.47	2.12
Churanov	49.07 N	13.62 E	2.89	5.82	9.24	13.18	21.32	15.68	20.51	19.49	12.84	5.68	3.36	2.99
Hradec Kralov	50.25 N	15.85 E	3.51	5.94	10.58	15.95	20.42	18.43	17.17	17.92	11.86	6.27	2.45	1.89
Denmark														
Copenhagen	55.67 N	12.30 E	1.83	3.32	7.09	11.12	21.39	24.93	—	13.92	10.10	5.20	2.81	1.23
Egypt														
Cairo	30.08 N	31.28 E	10.06	12.96	18.49	23.04	21.91	26.07	25.16	23.09	21.01	—	11.74	9.85
Mersa Matruh	31.33 N	27.22 E	8.38	11.92	18.47	24.27	24.17	—	26.67	26.27	21.92	18.28	11.71	8.76
Ethiopia														
Addis Ababa	8.98 N	38.80 E	—	11.39	—	12.01	—	—	—	6.33	9.35	11.71	11.69	11.50
Fiji														
Nandi	17.75 S	177.45 E	20.82	20.65	20.25	18.81	15.68	14.18	15.08	16.71	19.37	20.11	21.78	25.09
Suva	48.05 S	178.57 E	20.37	17.74	16.22	13.82	10.81	12.48	11.40	—	—	18.49	19.96	20.99
Finland														
Helsinki	60.32 N	24.97 E	1.13	2.94	5.59	11.52	17.60	16.81	20.66	15.44	8.44	3.31	0.97	0.63
France														
Agen	44.18 N	0.60 E	4.83	7.40	10.69	17.12	19.25	20.42	21.63	20.64	15.56	8.41	5.09	5.01
Nice	43.65 N	7.20 E	6.83		11.37	17.79	20.74	24.10	24.85	24.86	15.04	10.99	7.08	6.73
Paris	48.97 N	2.45 E	2.62	5.08	7.21	12.90	14.84	13.04	15.54	16.30	10.17	5.61	3.14	2.20
Germany														
Bonn	50.70 N	7.15 E	2.94	5.82	8.01	14.27	15.67	14.41	18.57	17.80	11.70	6.15	3.42	1.90
Nuremberg	53.33 N	13.20 E	3.23	6.92	9.08	15.69	15.71	18.21	21.14	17.98	12.43	8.15	2.79	2.51
Bremen	53.05 N	8.80 E	2.36	4.93	8.53	14.52	14.94	14.52	19.40	15.02	10.48	6.27	2.80	1.66
Hamburg	53.63 N	10.00 E	1.97	3.96	7.59	12.32	14.11	12.69	19.00	14.11	10.29	6.45	2.33	1.43
Stuttgart	48.83 N	9.20 E	3.59	7.18	9.22	15.81	17.72	17.44	22.21	19.87	12.36	7.81	3.19	2.54
Ghana														
Bole	9.03 N	2.48 W	18.29	19.76	19.71	19.15	16.61	—	—	13.68	16.29	17.27	17.33	15.93

Accra	5.60 N	0.17 W	14.82	16.26	18.27	16.73	18.15	13.96	13.86	13.49	15.32	19.14	18.16	14.23
Great Britain														
Belfast	54.65 N	6.22 W	2.00	3.60	6.85	12.00	15.41	15.09	15.46	13.56	11.49	4.63	2.34	1.24
Jersey	49.22 N	2.20 W	2.76	5.65	9.51	14.98	18.51	17.83	18.14	18.62	12.98	6.16	3.26	2.83
London	51.52 N	0.12 W	2.24	3.87	7.40	12.01	12.38	13.24	16.59	16.23	12.59	5.67	2.87	1.97
Greece														
Athens	37.97 N	23.72 E	9.11	10.94	15.70	20.91	23.85	25.48	24.21	23.08	19.03	13.29	5.98	6.64
Sikiwna	37.98 N	22.73 E	7.60	8.16	11.99	21.06	22.62	24.32	23.56	21.73	17.30	11.75	9.45	6.35
Guadeloupe														
Le Raizet	16.27 N	61.52 W	14.88	18.10	20.55	19.69	20.26	20.65	20.65	20.24	18.47	17.79	13.49	14.38
Guyana														
Cayenne	4.83 N	52.37 W	14.46	14.67	16.28	17.57	—	14.92	17.42	18.24	20.52	—	22.69	17.04
Hong Kong														
King's Park	22.32 N	114.17 W	12.34	7.39	6.94	9.50	11.38	13.60	16.70	17.06	15.91	16.52	14.19	10.00
Hungary														
Budapest	47.43 N	19.18 E	2.61	7.46	11.14	14.46	20.69	19.47	21.46	19.72	12.88	7.96	2.95	2.47
Iceland														
Reykjavik	64.13 N	21.90 W	0.52	2.02	6.25	11.77	13.07	14.58	16.83	11.35	9.70	3.18	1.00	0.65
India														
Bombay	19.12 N	72.85 E	18.44	21.00	22.72	24.52	24.86	19.75	15.84	16.00	18.19	20.38	19.18	17.81
Calcutta	22.53 N	88.33 E	15.69	18.34	20.09	22.34	22.37	17.55	17.07	16.55	16.52	16.90	16.35	15.00
Madras	13.00 N	80.18 E	19.09	22.71	25.14	24.88	23.89	—	18.22	19.68	19.S1	16.41	14.76	15.79
Nagpur	21. 10 N	79.05 E	18.08	21.01	22.25	24.08	24.79	19.84	15.58	15.47	17.66	20.10	18.98	17.33
New Delhi	28.58 N	77.20 E	14.62	18.25	20.15	23.40	23.80	19.16	20.20	19.89	20.08	19.74	16.95	14.22
Ireland														
Dublin	53.43 N	6.25 W	2.51	4.75	7.48	11.06	17.46	19.11	15.64	13.89	9.65	5.77	2.93	—
Israel														
Jerusalem	31.78 N	35.22 E	10.79	13.01	18.08	23.79	29.10	31.54	31.83	28.79	25.19	20.26	12.61	10.71
Italy														
Milan	45.43 N	9.28 E	—	6.48	10.09	13.17	17.55	16.32	18.60	16.86	11.64	5.40	3.52	2.41
Rome	41.80 N	12.55 E	—	9.75	13.38	15.82	15.82	18.89	22.27	21.53	16.08	8.27	6.41	4.49
Japan														
Fukuoka	33.58 N	130.38 E	8.11	8.72	10.95	13.97	14.36	12.81	13.84	16.75	13.92	11.86	10.05	7.30
Tateno	36.05 N	140.13 E	9.06	12.17	11.00	15.78	16.52	15.26	—	—	—	9.60	8.55	8.26
Yonago	35.43 N	133.35 E	6.25	7.16	10.87	17.30	16.72	15.44	17.06	19.93	12.41	10.82	7.50	5.51
Kenya														
Mombasa	4.03 S	39.62 E	22.30	22.17	22.74	18.49	18.31	17.41	—	18.12	21.03	22.97	21.87	21.25
Nairobi	1.32 S	36.92 E	—	24.10	21.20	18.65	14.83	15.00	13.44	14.12	19.14	19.38	16.90	18.27

(continued)

TABLE A2.3a *(Continued)*

Position	Lat	Long	January	February	March	April	May	June	July	August	September	October	November	December
Lithuania														
Kaunas	54.88 N	23.88 E	1.89	4.43	7.40	12.97	18.88	18.74	21.41	15.79	10.40	5.64	1.80	1.10
Madagascar														
Antanarivo	18.80 S	47.48 E	15.94	13.18	13.07	11.53	9.25	8.21	9.32	—	—	16.43	15.19	15.62
Malaysia														
Kualalumpur	3.12 N	101.55 E	15.36	17.67	18.48	16.87	15.67	16.24	15.32	15.89	14.62	14.13	13.54	11.53
Piang	5.30 N	100.27 E	19.47	21.35	23.24	20.52	18.63	19.32	17.17	16.96	15.93	16.01	18.35	17.37
Martinique														
Le Lamentin	14.60 N	61.00 W	17.76	20.07	22.53	21.95	22.42	21.23	20.86	21.84	20.23	19.87	14.08	16.25
Mexico														
Chihuahua	28.63 N	106.08 W	14.80	—	—	—	26.94	26.28	24.01	24.22	20.25	19.55	10.57	15.79
Orizabita	20.58 N	99.20 E	19.49	23.07	27.44	27.35	26.04	25.05	—	27.53	21.06	17.85	15.48	12.93
Mongolia														
Ulan Bator	47.93 N	106.98 E	6.28	9.22	14.34	18.18	20.50	19.34	16.34	16.65	14.08	11.36	7.19	5.35
Uliasutai	47.75 N	96.85 E	6.43	10.71	14.83	20.32	23.86	20.46	21.66	17.81	15.97	10.92	7.32	5.08
Morocco														
Casablanca	33.57 N	7.67 E	11.46	12.70	15.93	21.25	24.45	25.27	25.53	23.60	19.97	14.68	11.61	9.03
Mozambique														
Maputo	25.97 S	32.60 E	26.35	23.16	19.33	20.54	16.33	14.17	—	—	—	22.55	25.48	26.19
Netherlands														
Maastricht	50.92 N	5.78 E	3.20	5.43	8.48	14.82	14.97	14.32	18.40	17.51	11.65	6.51	3.01	1.72
New Caledonia														
Koumac	20.57 S	164.28 E	24.89	21.15	16.96	18.98	15.67	14.55	15.75	17.62	22.48	15.83	27.53	26.91
New Zealand														
Wilmington	41.28 S	174.77 E	22.59	19.67	14.91	9.52	6.97	4.37	5.74	7.14	12.50	16.34	19.07	24.07
Christchurch	43.48 S	172.55 E	23.46	19.68	13.98	8.96	6.47	4.74	5.38	6.94	13.18	17.45	18.91	24.35
Nigeria														
Benin City	6.32 N	5.60 E	14.89	17.29	19.15	17.21	16.97	15.04	10.24	12.54	14.37	15.99	17.43	15.75
Norway														
Bergen	60.40 N	5.32 E	0.46	1.33	3.18	8.36	19.24	16.70	16.28	10.19	6.53	3.19	1.36	0.35
Oman														
Seeb	23.58 N	58.28 E	12.90	14.86	21.22	22.22	25.30	24.02	23.46	21.66	20.07	18.45	15.49	13.12
Salalah	17.03 N	54.08 E	16.52	16.92	18.49	20.65	21.46	16.92	8.52	11.41	17.14	18.62	16.42	—
Pakistan														
Karachi	24.90 N	67.13 E	13.84	—	—	19.69	20.31	16.62	—	—	—	—	12.94	11.07
Multan	30.20 N	71.43 E	12.29	15.86	18.33	22.35	22.57	21.65	20.31	20.44	20.57	15.91	12.68	10.00

Islamabad	33.62 N	73.10 E	10.38	12.42	16.98	22.65	—	25.49	20.64	18.91	14.20	15.30	10.64	8.30
Peru														
Puno	15.83 S	70.02 W	14.98	12.92	16.08	20.03	17.45	17.42	15.74	15.32	16.11	16.18	14.24	13.90
Poland														
Warszawa	52.28 N	20.97 E	1.73	3.83	7.81	10.53	19.22	17.11	20.18	15.00	10.65	4.95	2.39	1.68
Kolobrzeg	54.18 N	15.58 E	2.50	3.25	8.86	15.21	20.79	20.50	17.19	16.46	7.95	5.75	1.78	1.18
Portugal														
Evora	38.57 N	7.90 W	9.92	12.43	17.81	18.69	23.57	29.23	28.75	23.77	20.17	—	6.81	4.57
Lisbon	38.72 N	9.15 W	9.24	11.60	17.52	18.49	24.64	29.02	28.14	22.20	19.76	13.56	7.18	4.83
Romania														
Bucuresti	44.50 N	26.13 E	7.05	10.22	12.04	16.53	18.97	22.16	23.19	—	17.17	9.55	4.82	—
Constania	44.22 N	28.63 E	5.62	9.28	14.31	20.59	23.23	25.80	27.98	24.22	16.91	11.89	6.19	5.10
Galati	45.50 N	28.02 E	6.09	9.33	14.31	17.75	21.77	22.74	25.55	19.70	14.05	11.26	6.32	5.38
Russia														
Alexandovsko	60.38 N	77.87 E	1.34	4.17	9.16	17.05	21.83	21.34	20.26	13.05	10.16	4.68	1.71	0.68
Moscow	55.75 N	37.57 E	1.45	3.96	8.09	11.69	18.86	18.12	17.51	14.17	10.92	4.03	2.28	1.29
St. Petersburg	59.97 N	30.30 E	1.03	3.11	4.88	12.24	20.59	21.55	20.43	13.27	7.83	2.93	1.16	0.59
Verkhoyansk	67.55 N	133.38 E	0.21	2.25	7.61	15.96	19.64	—	—	14.12	7.59	3.51	0.54	—
St. Pierre & Miquelon														
St. Pierre	46.77 N	56.17 W	4.43	6.61	12.50	17.57	18.55	17.84	19.95	16.46	12.76	8.15	3.69	3.33
Singapore														
Singapore	1.37 N	103.98 E	19.08	20.94	20.75	18.20	14.89	15.22	13.92	16.66	16.51	15.82	13.81	12.67
South Korea														
Seoul	37.57 N	126.97 E	6.24	9.40	10.34	13.98	16.35	17.49	10.65	12.94	11.87	10.35	6.47	5.14
South Africa														
Cape Town	33.98 S	18.60 E	27.47	25.57	—	15.81	11.44	9.08	8.35	13.76	17.30	22.16	26.37	27.68
Port Elizabeth	33.98 S	25.60 E	27.22	22.06	19.01	15.29	11.79	11.13	10.73	13.97	18.52	23.09	23.15	27.26
Pretoria	25.73 S	28.18 E	26.06	22.43	20.52	16.09	15.67	13.67	15.19	18.65	21.62	21.75	24.82	23.43
Spain														
Madrid	40.45 N	3.72 W	7.73	10.53	15.35	21.74	22.81	22.05	26.27	22.90	18.89	10.21	8.69	5.56
Sudan														
Wad Madani	14.40 N	33.48 E	21.92	24.01	23.43	25.17	23.92	23.51	22.40	22.85	21.75	20.47	20.19	19.21
Elfasher	13.62 N	25.33 E	21.56	21.84	24.54	25.29	24.31	24.15	22.87	21.19	22.58	23.85	—	—
Shambat	15.67 N	32.53 E	23.90	27.38	—	27.45	23.21	26.15	23.55	25.46	24.05	23.51	23.82	22.53
Sweden														
Karlstad	59.37 N	13.47 E	1.26	3.13	5.02	14.01	19.90	16.70	20.92	14.14	10.52	3.98	1.47	0.94
Lund	55.72 N	13.22 E	1.97	3.47	6.66	12.48	17.83	13.38	18.74	14.99	10.39	5.45	1.82	1.21
Stockholm	59.35 N	18.07 E	1.32	2.69	4.75	13.21	15.58	14.79	20.52	14.48	10.50	4.04	1.19	0.83

(continued)

TABLE A2.3a *(Continued)*

Position	Lat	Long	January	February	March	April	May	June	July	August	September	October	November	December
Switzerland														
Geneva	46.25 N	6.13 E	2.56	7.21	9.46	17.07	20.98	19.78	22.38	20.50	13.62	8.44	3.31	2.87
Zurich	47.48 N	8.53 E	2.31	7.02	7.54	15.04	16.33	16.73	20.28	18.32	12.52	7.18	2.64	2.29
Thailand														
Bangkok	13.73 N	100.57 E	16.67	19.34	23.00	22.48	20.59	17.71	18.02	16.04	16.23	16.81	18.60	16.43
Trinidad & Tobago														
Crown Point	11.15 N	60.83 W	13.05	15.61	15.17	16.96	17.61	15.37	13.16	13.08	12.24	8.76	—	—
Tunisia														
Sidi Bouzid	36.87 N	10.35 E	7.88	10.38	13.20	17.98	25.12	26.68	27.43	24.33	18.87	12.11	9.37	6.72
Tunis	36.83 N	10.23 E	7.64	9.88	14.79	31.61	25.31	26.03	26.60	20.37	19.58	12.91	9.35	7.16
Ukraine														
Kiev	50.40 N	30.45 E	2.17	4.87	11.15	12.30	20.49	—	18.99	18.55	9.72	9.84	3.72	2.52
Uzbekistan														
Tashkent	41.27 N	69.27 E	7.27	10.81	15.93	23.60	25.21	29.53	28.50	26.68	20.76	13.25	8.61	4.59
Venezuela														
Caracas	10.50 N	66.88 W	14.25	13.56	16.30	15.56	15.69	15.56	16.28	17.11	17.04	15.14	14.74	13.50
St. Antonio	7.85 N	72.45 W	11.78	10.54	10.65	12.07	12.65	21.20	14.68	15.86	16.62	15.32	12.28	11.28
St. Fernando	7.90 N	67.42 W	14.92	16.82	16.89	—	—	14.09	13.78	14.42	14.86	15.27	14.25	13.11
Vietnam														
Hanoi	21.03 N	105.85 E	5.99	7.48	8.73	13.58	19.10	21.26	19.85	19.78	20.67	14.78	12.44	13.21
Yogoslavia														
Beograd	44.78 N	20.53 E	4.92	6.27	10.64	14.74	20.95	22.80	22.09	20.27	15.57	11.24	6.77	4.99
Kopaonik	43.28 N	20.80 E	7.03	10.93	14.75	12.78	13.54	20.43	22.48	—	20.14	11.61	6.26	4.64
Portoroz	45.52 N	13.57 E	5.11	7.84	13.75	17.30	23.66	22.31	25.14	21.34	13.40	8.98	6.04	3.92
Zambia														
Lusaka	15.42 S	28.32 W	16.10	18.02	20.24	19.84	17.11	16.37	19.45	20.72	21.68	23.83	23.85	20.52
Zimbabwe														
Bulawayo	20.15 S	28.62 N	20.03	22.11	21.03	18.09	17.15	15.36	16.46	19.49	21.55	23.44	25.08	23.46
Harare	17.83 S	31.02 N	19.38	19.00	19.22	17.67	18.35	16.10	14.55	17.87	21.47	23.98	19.92	21.88

Note: Data for 872 locations is available from these sources in 68 countries.

[a] Source for Canadian Data: Environment Canada: Internet address: http://www.ec.gc.ca./envhome.html

Source: From Voeikov Main Geophysical Observatory, Russia: Internee address: http://wrdc-mgo.nrel.gov/html/get_data-ap.html

TABLE A2.3b Average Daily Solar Radiation on a Horizontal Surface in U.S.A. (Units: MJ/sq. m-day)

Position	January	February	March	April	May	June	July	August	September	October	November	December	Average
Alabama													
Birmingham	9.20	11.92	15.67	19.65	21.58	22.37	21.24	20.21	17.15	14.42	10.22	8.40	16.01
Montgomery	9.54	12.49	16.24	20.33	22.37	23.17	21.80	20.56	17.72	14.99	10.90	8.97	16.58
Alaska													
Fairbanks	0.62	2.77	8.31	14.66	17.98	19.65	16.92	12.36	7.02	3.20	1.01	0.23	8.74
Anchorage	1.02	3.41	8.18	13.06	15.90	17.72	16.69	12.72	8.06	3.97	1.48	0.56	8.63
Nome	0.51	2.95	8.29	15.22	18.97	19.65	16.69	11.81	7.72	3.63	0.99	0.09	8.86
St. Paul Island	1.82	4.32	8.52	12.72	14.08	14.42	12.83	10.33	7.84	4.54	2.16	1.25	7.95
Yakutat	1.36	3.63	7.72	12.61	14.76	15.79	14.99	12.15	7.95	3.97	1.82	0.86	8.18
Arizona													
Phoenix	11.58	15.33	19.87	25.44	28.85	30.09	27.37	25.44	21.92	17.60	12.95	10.56	20.56
Tucson	12.38	15.90	20.21	25.44	28.39	29.30	25.44	24.08	21.58	17.94	13.63	11.24	20.44
Arkansas													
Little Rock	9.09	11.81	15.56	19.19	21.80	23.51	23.17	21.35	17.26	14.08	9.77	8.06	16.24
Fort Smith	9.31	12.15	15.67	19.31	21.69	23.39	23.85	24.46	17.26	13.97	9.88	8.29	16.35
California													
Bakersfield	8.29	11.92	16.69	22.15	26.57	28.96	28.73	26.01	21.35	15.90	10.33	7.61	18.74
Fresno	7.61	11.58	16.81	22.49	27.14	29.07	28.96	25.89	21.12	15.56	9.65	6.70	18.62
Long Beach	9.99	12.95	17.03	21.60	23.17	24.19	26.12	24.08	19.31	14.99	11.24	9.31	17.83
Sacramento	6.93	10.68	15.56	21.24	25.89	28.28	28.62	25.32	20.56	14.54	8.63	6.25	17.72
San Diego	11.02	13.97	17.72	21.92	22.49	23.28	24.98	23.51	19.53	15.79	12.26	10.22	18.06
San Francisco	7.72	10.68	15.22	20.44	24.08	25.78	26.46	23.39	19.31	13.97	8.97	7.04	16.92
Los Angeles	10.11	13.06	17.26	21.80	23.05	23.74	25.67	23.51	18.97	14.99	11.36	9.31	17.72
Santa Maria	10.22	13.29	17.49	22.26	25.10	26.57	26.91	24.42	20.10	15.67	11.47	9.54	18.62
Colorado													
Boulder	7.84	10.45	15.64	17.94	17.94	20.47	20.28	17.12	16.07	12.09	8.66	7.10	14.31
Colorado Springs	9.09	12.15	16.13	20.33	22.26	24.98	23.96	21.69	18.51	14.42	9.99	8.18	16.81
Connecticut													
Hartford	6.70	9.65	13.17	16.69	19.53	21.24	21.12	18.51	14.76	10.68	6.59	5.45	13.74
Delaware													
Wilmington	7.27	10.22	13.97	17.60	20.33	22.49	21.80	19.65	15.79	11.81	7.84	6.25	14.65
Florida													
Daytona Beach	11.24	13.85	17.94	22.15	23.17	22.03	21.69	20.44	17.72	14.99	12.15	10.33	17.38

(continued)

TABLE A2.3b ***(Continued)***

Position	January	February	March	April	May	June	July	August	September	October	November	December	Average
Jacksonville	10.45	13.17	17.03	21.12	22.03	21.58	21.01	19.42	16.69	14.20	11.47	9.65	16.47
Tallahassee	10.33	13.29	16.92	21.24	22.49	22.03	20.90	19.65	17.72	15.56	11.92	9.77	16.81
Miami	12.72	15.22	18.51	21.58	21.46	20.10	21.10	20.10	17.60	15.67	13.17	11.81	17.38
Key West	13.17	16.01	19.65	22.71	22.83	22.03	22.03	21.01	18.74	16.47	13.85	15.79	18.40
Tampa	11.58	14.42	18.17	22.26	23.05	21.92	20.90	19.65	17.60	16.01	12.83	11.02	17.49
Georgia													
Athens	9.43	12.38	16.01	20.21	22.03	22.83	21.80	20.21	17.26	14.42	10.45	8.40	16.29
Atlanta	9.31	12.26	16.13	20.33	22.37	23.17	22.15	20.56	17.49	14.54	10.56	8.52	16.43
Columbus	9.77	12.72	16.47	20.67	22.37	22.83	21.58	20.33	17.60	14.99	11.02	9.09	16.62
Macon	9.54	12.61	16.35	20.56	22.37	22.83	21.58	20.21	17.26	14.88	10.90	8.86	16.50
Savanna	9.99	12.72	16.81	21.01	22.37	22.60	21.80	19.76	16.92	14.65	11.13	9.20	16.58
Hawaii													
Honolulu	14.08	16.92	19.42	21.24	22.83	23.51	23.74	23.28	21.35	18.06	14.88	13.40	19.42
Idaho													
Boise	5.79	8.97	13.63	18.97	23.51	26.01	27.37	23.62	18.40	12.26	6.70	5.11	15.90
Illinois													
Chicago	6.47	9.31	12.49	16.47	20.44	22.60	22.03	19.31	15.10	10.79	6.47	5.22	13.85
Rockford	6.70	9.77	12.72	16.58	20.33	22.49	22.15	19.42	15.22	10.79	6.59	5.34	14.08
Springfield	7.50	10.33	13.40	17.83	21.46	23.51	23.05	20.56	16.58	12.26	7.72	6.13	15.10
Indiana													
Indianapolis	7.04	9.99	13.17	17.49	21.24	23.28	22.60	20.33	16.35	11.92	7.38	5.79	14.76
Iowa													
Mason City	6.70	9.77	13.29	16.92	20.78	22.83	22.71	19.76	15.33	10.90	6.59	5.45	14.31
Waterloo	6.81	9.77	13.06	16.92	20.56	22.83	22.60	19.76	15.33	10.90	6.70	5.45	14.20
Kansas													
Dodge City	9.65	12.83	16.69	21.01	23.28	25.78	25.67	22.60	18.40	14.42	10.11	8.40	17.49
Goodland	8.97	11.92	16.13	20.44	22.71	25.78	25.55	22.60	18.28	14.08	9.65	7.84	17.03
Kentucky													
Lexington	7.27	9.88	13.51	17.60	20.56	22.26	21.46	19.65	16.01	12.38	7.95	6.25	14.54
Louisville	7.27	10.22	13.63	17.83	20.90	22.71	22.03	20.10	16.35	12.38	7.95	6.25	14.76
Louisiana													
New Orleans	9.77	12.83	16.01	19.87	21.80	22.03	20.67	19.65	17.60	15.56	11.24	9.31	16.35
Lake Charles	9.77	12.83	16.13	19.31	21.58	22.71	21.58	20.33	18.06	15.56	11.47	9.31	16.58
Maine													
Portland	6.70	9.99	13.78	16.92	19.99	21.92	21.69	19.31	15.22	10.56	6.47	5.45	13.97
Maryland													

Baltimore	7.38	10.33	13.97	17.60	20.21	22.15	21.69	19.19	15.79	11.92	8.06	6.36	14.54
Massachusetts													
Boston	6.70	9.65	13.40	16.92	20.21	22.03	21.80	19.31	15.33	10.79	6.81	5.45	14.08
Michigan													
Detroit	5.91	8.86	12.38	16.47	20.33	22.37	21.92	18.97	14.76	10.11	6.13	4.66	13.63
Lansing	5.91	8.86	12.49	16.58	20.21	22.26	21.92	18.85	14.54	9.77	5.91	4.66	13.51
Minnesota													
Duluth	5.68	9.31	13.74	17.38	20.10	21.46	21.80	18.28	13.29	8.86	5.34	4.43	13.29
Minneapolis	6.36	9.77	13.51	16.92	20.56	22.49	22.83	19.42	14.65	9.99	6.13	4.88	13.97
Rochester	6.36	9.65	13.17	16.58	20.10	22.15	22.15	19.08	14.54	10.11	6.25	5.11	13.74
Mississippi													
Jackson	9.43	12.38	16.13	19.87	22.15	23.05	22.15	19.08	14.54	10.11	6.25	5.11	13.74
Missouri													
Columbia	8.06	10.90	14.31	18.62	21.58	23.62	23.85	21.12	16.69	12.72	8.29	6.70	15.56
Kansas City	7.95	10.68	14.08	18.28	21.24	23.28	23.62	20.78	16.58	12.72	8.40	6.70	15.44
Springfield	8.52	11.02	14.65	18.62	21.24	23.05	23.62	21.24	16.81	13.17	8.86	7.27	15.67
St. Louis	7.84	10.56	13.97	18.06	21.12	23.05	22.94	20.44	16.58	12.49	8.18	6.59	15.22
Montana													
Helena	5.22	8.29	12.61	17.15	20.67	23.28	25.21	21.24	15.79	10.45	6.02	4.43	14.20
Lewistown	5.22	8.40	12.72	17.15	20.33	23.05	24.53	20.78	15.10	10.22	5.91	4.32	13.97
Nebraska													
Omaha	7.50	10.33	13.97	18.06	21.24	2.40	23.51	20.56	16.01	11.81	7.61	6.13	15.10
Lincoln	7.33	10.10	13.65	16.22	19.26	21.21	22.15	18.87	15.44	11.54	7.76	6.20	14.16
Nevada													
Elko	7.61	10.56	14.42	18.85	22.71	25.67	26.69	23.62	19.31	13.63	8.29	6.70	16.58
Las Vegas	10.79	14.42	19.42	24.87	28.16	30.09	28.28	25.89	22.15	17.03	12.15	9.88	20.33
Reno	8.29	11.58	16.24	21.24	25.10	27.48	28.16	24.98	20.56	14.88	9.31	7.38	17.94
New Hampshire													
Concord	6.81	10.11	13.97	16.92	20.21	21.80	21.80	19.08	14.99	10.45	6.47	5.45	14.08
New Jersey													
Atlantic City	7.38	10.22	13.97	17.49	20.21	21.92	21.24	19.19	15.79	11.92	8.06	6.36	14.54
Newark	6.93	9.77	13.51	17.26	19.76	21.35	21.01	18.85	15.33	11.36	7.27	5.68	13.97
New Mexico													
Albuquerque	11.47	14.99	19.31	24.53	27.60	29.07	27.03	24.76	21.12	17.03	12.49	10.33	19.99
New York													
Albany	6.36	9.43	12.95	16.69	19.53	21.46	21.58	18.51	14.65	10.11	6.13	5.00	13.51
Buffalo	5.68	8.40	12.15	16.35	19.76	22.03	21.69	18.62	14.08	9.54	5.68	4.54	13.29
New York City	6.93	9.88	13.85	17.72	20.44	22.03	21.69	19.42	15.56	11.47	7.27	5.79	14.31

(continued)

TABLE A2.3b *(Continued)*

Position	January	February	March	April	May	June	July	August	September	October	November	December	Average
Rochester	5.68	8.52	12.26	16.58	19.87	21.92	21.69	18.51	14.20	9.54	5.68	4.54	13.29
North Carolina													
Charlotte	8.97	11.81	15.67	19.76	21.58	22.60	21.92	19.99	16.92	13.97	9.99	8.06	16.01
Wilmington	9.31	12.15	16.24	20.44	21.92	22.60	21.58	19.53	16.69	14.08	10.56	8.52	16.13
North Dakota													
Fargo	5.79	9.09	13.17	16.92	20.56	22.37	23.17	19.87	14.31	9.54	5.68	4.54	13.74
Bismarck	6.12	9.75	13.88	17.43	21.45	23.01	24.06	20.12	15.21	10.61	6.28	4.84	14.39
Ohio													
Cleveland	5.79	8.63	12.04	16.58	20.10	22.15	21.92	18.97	14.76	10.22	6.02	4.66	13.51
Columbus	6.47	9.09	12.49	16.58	19.76	21.58	21.12	18.97	15.44	11.24	6.81	5.34	13.74
Dayton	6.81	9.43	12.83	17.03	20.33	22.37	22.37	19.65	15.90	11.47	7.04	5.45	14.20
Youngstown	5.79	8.40	11.92	15.90	19.19	21.24	20.78	18.06	14.31	10.11	6.02	4.77	13.06
Oklahoma													
Oklahoma City	9.88	1.25	16.47	20.33	22.26	24.42	24.98	22.49	18.17	14.54	10.45	8.74	17.15
Oregon													
Eugene	4.54	7.04	11.24	15.79	19.99	22.37	24.19	21.01	15.90	9.65	5.11	3.75	13.40
Medford	5.34	8.52	13.17	18.62	23.39	26.23	27.82	23.96	18.62	11.92	6.02	4.43	15.67
Portland	4.20	6.70	10.68	15.10	18.97	21.24	22.60	19.53	14.88	9.20	4.88	3.52	12.61
Pacific Islands													
Guam	16.35	17.38	19.65	20.78	20.56	19.76	18.28	17.49	17.49	16.58	15.79	15.10	17.94
Pennsylvania													
Philadelphia	7.04	9.88	13.63	17.26	19.99	22.03	21.46	19.42	15.67	11.58	7.72	6.02	14.31
Pittsburgh	6.25	8.97	12.61	16.47	19.65	21.80	21.35	18.85	15.10	10.90	6.59	5.00	13.63
Rhode Island													
Providence	6.70	9.65	13.40	16.92	19.99	21.58	21.24	18.85	15.22	11.02	6.93	5.56	13.97
South Carolina													
Charleston	9.77	12.72	16.81	21.12	22.37	22.37	21.92	19.65	16.92	14.54	11.02	9.09	16.58
Greenville	9.20	12.04	15.90	19.99	21.58	22.60	21.58	19.87	16.81	14.08	10.22	8.18	16.01
South Dakota													
Pierre	6.47	9.54	13.85	17.94	21.46	24.08	24.42	21.46	16.35	11.24	7.04	5.45	14.99
Rapid City	6.70	9.88	14.20	18.28	21.46	24.19	24.42	21.80	16.92	11.81	7.50	5.79	15.33
Tennessee													
Memphis	8.86	11.58	15.22	19.42	22.03	23.85	23.39	21.46	17.38	14.20	9.65	7.84	16.24
Nashville	8.29	11.13	14.65	19.31	21.69	23.51	22.49	20.56	16.81	13.51	8.97	7.15	15.67
Texas													
Austin	10.68	13.63	17.03	19.53	21.24	23.74	24.42	22.83	18.85	15.67	11.92	9.99	17.49

Brownsville	10.33	13.17	16.47	19.08	20.78	22.83	23.28	21.58	18.62	16.13	12.38	9.88	17.03
El Paso	12.38	16.24	20.90	25.44	28.05	28.85	26.46	24.30	21.12	17.72	13.63	11.47	20.56
Houston	9.54	12.26	15.22	18.06	20.21	21.69	21.35	20.21	17.49	15.10	11.02	8.97	15.90
San Antonio	10.88	13.53	16.26	17.35	21.10	23.87	24.92	22.81	19.22	15.52	11.50	9.98	17.24
Utah													
Salt Lake City	6.93	10.45	14.76	19.42	23.39	26.46	26.35	23.39	18.85	13.29	8.06	6.02	16.47
Vermont													
Burlington	5.79	9.20	13.06	16.47	19.87	21.69	21.80	18.74	14.42	9.43	5.56	4.43	13.40
Virginia													
Norfolk	8.06	10.90	14.65	18.51	20.78	22.15	21.12	19.42	16.13	12.49	9.09	7.27	15.10
Richmond	8.06	10.90	14.76	18.62	20.90	22.49	21.58	19.53	16.24	12.61	8.97	7.15	15.22
Washington													
Olympia	3.63	6.02	9.99	14.20	18.06	20.10	21.12	18.17	13.63	7.95	4.32	3.07	11.70
Seattle	3.52	5.91	10.11	14.65	19.08	20.78	21.80	18.51	13.51	7.95	4.20	2.84	11.92
Yakima	4.88	7.95	12.83	17.83	22.49	24.87	25.89	22.26	16.92	10.68	5.56	4.09	17.76
West Virginia													
Charleston	7.04	9.65	13.40	17.15	20.21	21.69	20.90	18.97	15.56	11.81	7.72	6.02	14.20
Elkins	6.93	9.43	12.83	16.35	19.08	20.56	19.99	18.06	14.88	11.13	7.27	5.79	13.51
Wisconsin													
Green Bay	6.25	9.31	13.17	16.81	20.56	22.49	22.03	18.85	14.20	9.65	5.79	4.88	13.74
Madison	6.59	9.88	13.29	16.92	20.67	22.83	22.37	19.42	14.76	3.41	6.25	5.22	14.08
Milwaukee	6.47	9.31	12.72	16.69	20.78	22.94	22.60	19.42	14.88	10.22	6.25	5.11	13.97
Wyoming													
Rock Springs	7.61	10.90	15.10	19.42	23.17	26.01	25.78	22.94	18.62	13.40	8.40	6.70	16.58
Sendan	6.47	9.77	13.97	17.94	20.90	23.85	24.64	21.69	16.47	11.24	7.15	5.56	14.99

Source: From National Renewable Energy Laboratory, U.S.A.; Internet Address: http://rredc.nrel.gov/solar.

TABLE A2.4 Reflectivity Values for Characteristic Surfaces (Integrated Over Solar Spectrum and Angle of Incidence)

Surface	Average Reflectivity
Snow (freshly fallen or with ice film)	0.75
Water surfaces (relatively large incidence angles)	0.07
Soils (clay, loam, etc.)	0.14
Earth roads	0.04
Coniferous forest (winter)	0.07
Forests in autumn, ripe field crops, plants	0.26
Weathered blacktop	0.10
Weathered concrete	0.22
Dead leaves	0.30
Dry grass	0.20
Green grass	0.26
Bituminous and gravel roof	0.13
Crushed rock surface	0.20
Building surfaces, dark (red brick, dark paints, etc.)	0.27
Building surfaces, light (light brick, light paints, etc.)	0.60

Source: From Hunn, B. D. and Calafell, D. O. 1977. *Solar Energy*, Vol. 19, p. 87; see also List, R. J. 1949. *Smithsonian Meteorological Tables*, 6th Ed., pp. 442–443. Smithsonian Institution Press.

A3

Properties of Gases, Vapors, Liquids and Solids

Nitin Goel

Intel Technology India Pvt. Ltd.

TABLE A3.1 Properties of Dry Air at Atmospheric Pressures between 250 and 1000 K

T^a (K)	ρ (kg/m^3)	c_p (kJ/kg K)	μ (kg/m s$\times10^5$)	ν (m^2/s$\times10^6$)	k (W/m K)	α (m^2/s$\times10^4$)	dPr
250	1.4128	1.0053	1.488	9.49	0.02227	0.13161	0.722
300	1.1774	1.0057	1.983	15.68	0.02624	0.22160	0.708
350	0.9980	1.0090	2.075	20.76	0.03003	0.2983	0.697
400	0.8826	1.0140	2.286	25.90	0.03365	0.3760	0.689
450	0.7833	1.0207	2.484	28.86	0.03707	0.4222	0.683
500	0.7048	1.0295	2.671	37.90	0.04038	0.5564	0.680
550	0.6423	1.0392	2.848	44.34	0.04360	0.6532	0.680
600	0.5879	1.0551	3.018	51.34	0.04659	0.7512	0.680
650	0.5430	1.0635	3.177	58.51	0.04953	0.8578	0.682
700	0.5030	1.0752	3.332	66.25	0.05230	0.9672	0.684
750	0.4709	1.0856	3.481	73.91	0.05509	1.0774	0.686
800	0.4405	1.0978	3.625	82.29	0.05779	1.1951	0.689
850	0.4149	1.1095	3.765	90.75	0.06028	1.3097	0.692
900	0.3925	1.1212	3.899	99.3	0.06279	1.4271	0.696
950	0.3716	1.1321	4.023	108.2	0.06525	1.5510	0.699
1000	0.3524	1.1417	4.152	117.8	0.06752	1.6779	0.702

[a] Symbols: K = absolute temperature, degrees Kelvin; $\nu=\mu/\rho$; ρ = density; c_p specific heat capacity; $\alpha=c_p\rho/k$; μ = viscosity; k = thermal conductivity; Pr = Prandtl number, dimensionless. The values of μ, k, c_p, and Pr are not strongly pressure-dependent and may be used over a fairly wide range of pressures.

Source: From Natl. Bureau Standards (U.S.) Circ. 564, 1955.

TABLE A3.2 Properties of Water (Saturated Liquid) between 273 and 533 K

T			c_p (kJ/kg °C)	ρ (kg/m^3)	μ (kg/m s)	k (W/m °C)	Pr	$(g\beta\rho^2 c_p/\mu k)(m^{-3}{}^\circ C^{-1})$
K	°F	°C						
273	32	0	4.225	999.8	1.79×10^{-3}	0.566	13.25	
277.4	40	4.44	4.208	999.8	1.55	0.575	11.35	1.91×10^{9}
283	50	10	4.195	999.2	1.31	0.585	9.40	6.34×10^{9}
288.6	60	15.56	4.186	998.6	1.12	0.595	7.88	1.08×10^{10}
294.1	70	21.11	4.179	997.4	9.8×10^{-4}	0.604	6.78	1.46×10^{10}
299.7	80	26.67	4.179	995.8	8.6	0.614	5.85	1.91×10^{10}
302.2	90	32.22	4.174	994.9	7.65	0.623	5.12	2.48×10^{10}
310.8	100	37.78	4.174	993.0	6.82	0.630	4.53	3.3×10^{10}
316.3	110	43.33	4.174	990.6	6.16	0.637	4.04	4.19×10^{10}
322.9	120	48.89	4.174	988.8	5.62	0.644	3.64	4.89×10^{10}
327.4	130	54.44	4.179	985.7	5.13	0.649	3.30	5.66×10^{10}
333.0	140	60	4.179	983.3	4.71	0.654	3.01	6.48×10^{10}
338.6	150	65.55	4.183	980.3	4.3	0.659	2.73	7.62×10^{10}
342.1	160	71.11	4.186	977.3	4.01	0.665	2.53	8.84×10^{10}
349.7	170	76.67	4.191	973.7	3.72	0.668	2.33	9.85×10^{10}
355.2	180	82.22	4.195	970.2	3.47	0.673	2.16	1.09×10^{11}
360.8	190	87.78	4.199	966.7	3.27	0.675	2.03	
366.3	200	93.33	4.204	963.2	3.06	0.678	1.90	
377.4	220	104.4	4.216	955.1	2.67	0.684	1.66	
388.6	240	115.6	4.229	946.7	2.44	0.685	1.51	
399.7	260	126.7	4.250	937.2	2.19	0.685	1.36	
410.8	280	137.8	4.271	928.1	1.98	0.685	1.24	
421.9	300	148.9	4.296	918.0	1.86	0.684	1.17	
449.7	350	176.7	4.371	890.4	1.57	0.677	1.02	
477.4	400	204.4	4.467	859.4	1.36	0.665	1.00	
505.2	450	232.2	4.585	825.7	1.20	0.646	0.85	
533.0	500	260	4.731	785.2	1.07	0.616	0.83	

Source: Adapted from Brown, A. I. and S. M. Marco. 1958. Introduction to Heat Transfer, 3d Ed., McGraw-Hill Book Company, New York.

TABLE A3.3 Emittances and Absorptances of Materials

Substance	Short-Wave Absorptance	Long-Wave Emittance	*a ε*
Class I substances: Absorptance to emittance ratios less than 0.5			
Magnesium carbonate, $MgCO_3$	0.025–0.04	0.79	0.03–0.05
White plaster	0.07	0.91	0.08
Snow, fine particles, fresh	0.13	0.82	0.16
White paint, 0.017 in. on aluminum	0.20	0.91	0.22
Whitewash on galvanized iron	0.22	0.90	0.24
White paper	0.25–0.28	0.95	0.26–0.29
White enamel on iron	0.25–0.45	0.9	0.28–0.5
Ice, with sparse snow cover	0.31	0.96–0.97	0.32
Snow, ice granules	0.33	0.89	0.37
Aluminum oil base paint	0.45	0.90	0.50
White powdered sand	0.45	0.84	0.54
Class II substances: Absorptance to emittance ratios between 0.5 and 0.9			
Asbestos felt	0.25	0.50	0.50
Green oil base paint	0.5	0.9	0.56
Bricks, red	0.55	0.92	0.60
Asbestos cement board, white	0.59	0.96	0.61
Marble, polished	0.5–0.6	0.9	0.61
Wood, planed oak	–	0.9	–
Rough concrete	0.60	0.97	0.62
Concrete	0.60	0.88	0.68
Grass, green, after rain	0.67	0.98	0.68
Grass, high and dry	0.67–0.69	0.9	0.76
Vegetable fields and shrubs, wilted	0.70	0.9	0.78
Oak leaves	0.71–0.78	0.91–0.95	0.78–0.82
Frozen soil	–	0.93–0.94	–
Desert surface	0.75	0.9	0.83
Common vegetable fields and shrubs	0.72–0.76	0.9	0.82
Ground, dry plowed	0.75–0.80	0.9	0.83–0.89
Oak woodland	0.82	0.9	0.91
Pine forest	0.86	0.9	0.96
Earth surface as a whole (land and sea, no clouds)	0.83	10^{10}	–
Class III substances: Absorptance to emittance ratios between 0.8 and 1.0			
Grey paint	0.75	0.95	0.79
Red oil base paint	0.74	0.90	0.82
Asbestos, slate	0.81	0.96	0.84
Asbestos, paper		0.93–0.96	
Linoleum, red–brown	0.84	0.92	0.91
Dry sand	0.82	0.90	0.91
Green roll roofing	0.88	0.91–0.97	0.93
Slate, dark grey	0.89	–	
Old grey rubber		0.86	–
Hard black rubber	–	0.90–0.95	
Asphalt pavement	0.93	–	–
Black cupric oxide on copper	0.91	0.96	0.95
Bare moist ground	0.9	0.95	0.95
Wet sand	0.91	0.95	0.96
Water	0.94	0.95–0.96	0.98
Black tar paper	0.93	0.93	1.0
Black gloss paint	0.90	0.90	1.0
Small hole in large box, furnace, or enclosure	0.99	0.99	1.0
"Hohlraum," theoretically perfect black body	1.0	1.0	1.0
Class IV substances: Absorptance to emittance ratios greater than 1.0			
Black silk velvet	0.99	0.97	1.02

(continued)

TABLE A3.3 ***(Continued)***

Substance	Short-Wave Absorptance	Long-Wave Emittance	$a\ \varepsilon$
Alfalfa, dark green	0.97	0.95	1.02
Lampblack	0.98	0.95	1.03
Black paint, 0.017 in. on aluminum	0.94–0.98	0.88	1.07–1.11
Granite	0.55	0.44	1.25
Graphite	0.78	0.41	1.90
High ratios, but absorptances less than 0.80			
Dull brass, copper, lead	0.2–0.4	0.4–0.65	1.63–2.0
Galvanized sheet iron, oxidized	0.8	0.28	2.86
Galvanized iron, clean, new	0.65	0.13	5.0
Aluminum foil	0.15	0.05	3.00
Magnesium	0.3	0.07	4.3
Chromium	0.49	0.08	6.13
Polished zinc	0.46	0.02	23.0
Deposited silver (optical reflector) untarnished	0.07	0.01	
Class V substances: Selective surfaces[a]			
Plated metals:[b]			
Black sulfide on metal	0.92	0.10	9.2
Black cupric oxide on sheet aluminum	0.08–0.93	0.09–0.21	
Copper ($5\times10{\sim}^{5}$ cm thick) on nickel or silver-plated metal			
Cobalt oxide on platinum			
Cobalt oxide on polished nickel	0.93–0.94	0.24–0.40	3.9
Black nickel oxide on aluminum	0.85–0.93	0.06–0.1	14.5–15.5
Black chrome	0.87	0.09	9.8
Particulate coatings:			
Lampblack on metal			
Black iron oxide, 47 μm grain size, on aluminum			
Geometrically enhanced surfaces:[c]			
Optimally corrugated greys	0.89	0.77	1.2
Optimally corrugated selectives	0.95	0.16	5.9
Stainless-steel wire mesh	0.63–0.86	0.23–0.28	2.7–3.0
Copper, treated with NaClO, and NaOH	0.87	0.13	6.69

[a] Selective surfaces absorb most of the solar radiation between 0.3 and 1.9 μm, and emit very little in the 5–15 μm range–the infrared.

[b] For a discussion of plated selective surfaces, see Daniels, *Direct Use of the Sun's Energy*, especially chapter 12.

[c] For a discussion of how surface selectivity can be enhanced through surface geometry, see K. G. T. Hollands, July 1963. Directional selectivity emittance and absorptance properties of vee corrugated specular surfaces, *J. Sol. Energy Sci. Eng, vol. 3.*

Source: From Anderson, B. 1977. *Solar Energy*, McGraw-Hill Book Company. With permission.

TABLE A3.4 Thermal Properties of Metals and Alloys

Material	k, Btu/(hr)(ft.)(°F)				c, Btu/(lb_m)(°F)	ρ, $lb_m/ft.^3$	α, $ft.^2/hr$
	32°F	212°F	572°F	932°F	32°F	32°F	32°F
Metals							
Aluminum	117	119	133	155	0.208	169	3.33
Bismuth	4.9	3.9	…	…	0.029	612	0.28
Copper, pure	224	218	212	207	0.091	558	4.42
Gold	169	170	…	…	0.030	1,203	4.68
Iron, pure	35.8	36.6	…	…	0.104	491	0.70
Lead	20.1	19	18	…	0.030	705	0.95
Magnesium	91	92	…	…	0.232	109	3.60
Mercury	4.8	…	…	…	0.033	849	0.17
Nickel	34.5	34	32	…	0.103	555	0.60
Silver	242	238	…	…	0.056	655	6.6
Tin	36	34	…	…	0.054	456	1.46
Zinc	65	64	59	…	0.091	446	1.60
Alloys							
Admiralty metal	65	64					
Brass, 70% Cu, 30% Zn	56	60	66	…	0.092	532	1.14
Bronze, 75% Cu, 25% Sn	15	…	…	…	0.082	540	0.34
Cast iron							
Plain	33	31.8	27.7	24.8	0.11	474	0.63
Alloy	30	28.3	27		0.10	455	0.66
Constantan, 60% Cu, 40% Ni	12.4	12.8			0.10	557	0.22
18-8 Stainless steel,							
Type 304	8.0	9.4	10.9	12.4	0.11	488	0.15
Type 347	8.0	9.3	11.0	12.8	0.11	488	0.15
Steel, mild, 1% C	26.5	26	25	22	0.11	490	0.49

Source: From Kreith, F. 1997. *Principles of Heat Transfer*, PWS Publishing Co., Boston.

TABLE A3.5 Thermal Properties of Some Insulating and Building Materials

Material	Average, Temperature, °F	k, Btu/(hr)(ft.)(°F)	c, Btu/(lb_m)(°F)	ρ, $lb_m/ft.^3$	a, $ft.^2/hr$
		Insulating Materials			
Asbestos	32	0.087	0.25	36	−0.01
	392	0.12		36	−0.01
Cork	86	0.025	0.04	10	−0.006
Cotton, fabric	200	0.046			
Diatomaceous earth, powdered	100	0.030	0.21	14	−0.01
	300	0.036	...		
	600	0.046	...		
Molded pipe covering	400	0.051	...	26	
	1600	0.088	...		
		Glass Wool			
Fine	20	0.022	...		
	100	0.031	...	1.5	
	200	0.043	...		
Packed	20	0.016	...		
	100	0.022	...	6.0	
	200	0.029	...		
Hair felt	100	0.027	...	8.2	
Kaolin insulating brick	932	0.15	...	27	
	2102	0.26	...		
Kaolin insulating firebrick	392	0.05	...	19	
	1400	0.11	...		
85% magnesia	32	0.032	...	17	
	200	0.037	...	17	
Rock wool	20	0.017	...	8	
	200	0.030	...		
Rubber	32	0.087	0.48	75	0.0024
		Building Materials			
Brick					
Fire-clay	392	0.58	0.20	144	0.02
	1832	0.95			
Masonry	70	0.38	0.20	106	0.018
Zirconia	392	0.84	...	304	
	1832	1.13	...		
Chrome brick	392	0.82	...	246	
	1832	0.96	...		
		Concrete			
Stone	−70	0.54	0.20	144	0.019
10% Moisture	−70	0.70	...	140	−0.025
Glass, window	−70	−0.45	0.2	170	0.013
Limestone, dry	70	0.40	0.22	105	0.017
		Sand			
Dry	68	0.20	...	95	
10% H_2O	68	0.60		100	
		Soil			
Dry	70	−0.20	0.44	...	−0.01
Wet	70	−1.5	...		−0.03
		Wood			
Oak ⊥ to grain	70	0.12	0.57	51	0.0041
∥ to grain	70	0.20	0.57	51	0.0069
Pine ⊥ to grain	70	0.06	0.67	31	0.0029
∥ to grain	70	0.14	0.67	31	0.0067
Ice	32	1.28	0.46	57	0.048

Source: From Kreith, R. 1997. *Principles of Heat Transfer*, PWS Publishing Co.

TABLE A3.6 Saturated Steam and Water–SI Units

Temperature (K)	Pressure (MN/m^2)	Specific Volume (m^3/kg)		Specific Energy Internal (kJ/kg)		Specific Enthalpy (kJ/kg)			Specific Entropy (kJ/kg.K)	
		v_f	v_g	u_f	u_g	h_f	h_{fg}	h_g	s_f	s_g
273.15	0.0006109	0.0010002	206.278	−0.03	2375.3	−0.02	2501.4	2501.3	−0.0001	9.1565
273.16	0.0006113	0.0010002	206.136	0	2375.3	+0.01	2501.3	2501.4	0	9.1562
278.15	0.0008721	0.0010001	147.120	+20.97	2382.3	20.98	2489.6	2510.6	+0.0761	9.0257
280.13	0.0010000	0.0010002	129.208	29.30	2385.0	29.30	2484.9	2514.2	0.1059	8.975
283.15	0.0012276	0.0010004	106.379	42.00	2389.2	42.01	2477.7	2519.8	0.1510	8.9008
286.18	0.0015000	0.0010007	87.980	54.71	2393.3	54.71	2470.6	2525.3	0.1957	8.8279
288.15	0.0017051	0.0010009	77.926	62.99	2396.1	62.99	2465.9	2528.9	0.2245	8.7814
290.65	0.0020000	0.0010013	67.004	73.48	2399.5	73.48	2460.0	2533.5	0.2607	8.7237
293.15	0.002339	0.0010018	57.791	83.95	2402.9	83.96	2454.1	2538.1	0.2966	8.6672
297.23	0.0030000	0.0010027	45.665	101.04	2408.5	101.05	2444.5	2545.5	0.3545	8.5776
298.15	0.003169	0.0010029	43.360	104.88	2409.8	104.89	2442.3	2547.2	0.3674	8.5580
302.11	0.004000	0.0010040	34.800	121.45	2415.2	121.46	2432.9	2554.4	0.4226	8.4746
303.15	0.004246	0.0010043	32.894	125.78	2416.6	125.79	2430.5	2556.3	0.4369	8.4533
306.03	0.005000	0.0010053	28.192	137.81	2420.5	137.82	2423.7	2561.5	0.4764	8.3951
308.15	0.005628	0.0010060	25.216	146.67	2423.4	146.68	2418.6	2565.3	0.5053	8.3531
309.31	0.006000	0.0010064	23.739	151.53	2425.0	151.53	2415.9	2567.4	0.5210	8.3304
312.15	0.007000	0.0010074	20.530	163.39	2428.8	163.40	2409.1	2572.5	0.5592	8.2758
313.15	0.007384	0.0010078	19.523	167.56	2430.1	167.57	2406.7	2574.3	0.5725	8.2570
314.66	0.008000	0.0010084	18.103	173.87	2432.2	173.88	2403.1	2577.0	0.5926	8.2287
316.91	0.009000	0.0010094	16.203	183.27	2435.2	183.29	2397.7	2581.0	0.6224	8.1872
318.15	0.009593	0.0010099	15.258	188.44	2436.8	188.45	2394.8	2583.2	0.6387	8.1648
318.96	0.010000	0.0010102	14.674	191.82	2437.9	191.83	2392.8	2584.7	0.6493	8.1502
323.15	0.012349	0.0010121	12.032	209.32	2443.5	209.33	2382.7	2592.1	0.7038	8.0763
327.12	0.015000	0.0010141	10.022	225.92	2448.7	225.94	2373.1	2599.1	0.7549	8.0085
328.15	0.015758	0.0010146	9.568	230.21	2450.1	230.23	2370.7	2600.9	0.7679	7.9913
333.15	0.019940	0.0010172	7.671	251.11	2456.6	251.13	2358.5	2609.6	0.8312	7.9096
333.21	0.020000	0.0010172	7.649	251.38	2456.7	251.40	2358.3	2609.7	0.8320	7.9085
338.15	0.025030	0.0010199	6.197	272.02	2463.1	272.06	2346.2	2618.3	0.8935	7.8310
342.25	0.030000	0.0010223	5.229	289.20	2468.4	289.23	2336.1	2625.3	0.9439	7.7686
343.15	0.031190	0.0010228	5.042	292.95	2469.6	292.98	2333.8	2626.8	0.9549	7.7553
348.15	0.038580	0.0010259	4.131	313.90	2475.9	313.93	2221.4	2635.3	1.0155	7.6824
349.02	0.040000	0.0010265	3.993	317.53	2477.0	317.58	2319.2	2636.8	1.0259	7.6700
353.15	0.047390	0.0010291	3.407	334.86	2482.2	334.91	2308.8	2643.7	1.0753	7.6122
354.48	0.050000	0.0010300	3.240	340.44	2483.9	340.49	2305.4	2645.9	1.0910	7.5939

358.15	0.057830	0.0010325	2.828	355.84	2488.4	355.90	2296.0	2651.9	1.1343	7.5445
359.09	0.060000	0.0010331	2.732	359.79	2489.6	359.86	2293.6	2653.5	1.1453	7.5320
363.10	0.070000	0.0010360	2.365	376.63	2494.5	376.70	2283.3	2660.0	1.1919	7.4797
363.15	0.070140	0.0010360	2.361	376.85	2494.5	376.92	2283.2	2660.1	1.1925	7.4791
366.65	0.080000	0.0010386	2.087	391.58	2498.8	391.66	2274.1	2665.8	1.2329	7.4346
368.15	0.084550	0.0010397	1.9819	397.88	2500.6	397.96	2270.2	2668.1	1.2500	7.4159

Subscripts: *f* refers to a property of liquid in equilibrium with vapor; *g* refers to a property of vapor in equilibrium with liquid; *fg* refers to a change by evaporation. Table from Bolz, R. E. and G. L. Tuve, eds. 1973. *CRC Handbook of Tables for Applied Engineering Science*, 2nd Ed., Chemical Rubber Co., Cleveland, Ohio.

TABLE A3.7 Superheated Steam–SI Units

Pressure (MN/m^2) (Saturation Temperature)[a]		Temperature									
		50°C 323.15 K	100°C 373.15 K	150°C 423.15 K	200°C 473.15 K	300°C 573.15 K	400°C 673.15 K	500°C 773.15 K	700°C 973.15 K	1000°C 1273.15 K	1300°C 1573.15 K
0.001	*v*	149.093	172.187	195.272	218.352	264.508	310.661	356.814	449.117	587.571	726.025
(6.98°C)	*u*	2445.4	2516.4	2588.4	2661.6	2812.2	2969.0	3132.4	3479.6	4053.0	4683.7
(280.13 K)	*h*	2594.5	2688.6	2783.6	2880.0	3076.8	3279.7	3489.2	3928.7	4640.6	5409.7
	s	9.2423	9.5129	9.7520	9.9671	10.3443	10.6705	10.9605	11.4655	12.1019	12.6438
0.002	*v*	74.524	86.081	97.628	109.170	132.251	155.329	178.405	224.558	293.785	363.012
(17.50°C)	*u*	2445.2	2516.3	2588.3	2661.6	2812.2	2969.0	3132.4	3479.6	4053.0	4683.7
(290.65 K)	*h*	2594.3	2688.4	2793.6	2879.9	3076.7	3279.7	3489.2	3928.7	4640.6	5409.7
	s	8.9219	9.1928	9.4320	9.6471	10.0243	10.3506	10.6406	11.1456	11.7820	12.3239
0.004	*v*	37.240	43.028	48.806	54.580	66.122	77.662	89.201	112.278	146.892	181.506
(28.%°C)	*u*	2444.9	2516.1	2588.2	2661.5	2812.2	2969.0	3132.3	3479.6	4053.0	4683.7
(302.11 K)	*h*	2593.9	2688.2	2783.4	2879.8	3076.7	3279.6	3489.2	3928.7	4640.6	5409.7
	s	8.6009	8.8724	9.1118	9.3271	9.7044	10.0307	10.3207	10.8257	11.4621	12.0040
0.006	*v*	24.812	28.676	32.532	36.383	44.079	51.774	59.467	74.852	97.928	121.004
(36.16°C)	*u*	2444.6	2515.9	2588.1	2661.4	2812.2	2969.0	31323	3479.6	4053.0	4683.7
(309.31 K)	*h*	2593.4	2688.0	2783.3	2879.7	3076.6	3279.6	3489.1	3928.7	4640.6	5409.7
	s	8.4128	8.6847	8.9244	9.1398	9.5172	9.8435	10.1336	10.6386	11.2750	11.8168
0.008	*v*	18.598	21.501	24.395	27.284	33.058	38.829	44.599	56.138	73.446	90.753
(41.51°C)	*u*	2444.2	2515.7	2588.0	2661.4	2812.1	2969.0	3132.3	3479.6	4053.0	4683.7
(314.66 K)	*h*	2593.0	2687.7	2783.1	2879.6	3076.6	3279.6	3489.1	3928.7	4640.6	5409.7
	s	8.2790	8.5514	8.7914	9.0069	9.3844	9.7107	10.0008	10.5058	11.1422	11.6841
0.010	*v*	14.869	17.196	19.512	21.825	26.445	31.063	35.679	44.911	58.757	72.602
(45.81°C)	*u*	2443.9	2515.5	2587.9	2661.3	2812.1	2968.9	3132.3	3479.6	4053.0	4683.7
(318.96 K)	*h*	2592.6	2687.5	2783.0	2879.5	3076.5	3279.6	3489.1	3928.7	4640.6	5409.7
	s	8.1749	8.4479	8.6882	8.9038	9.2813	9.6077	9.8978	10.4028	11.0393	11.5811
0.020	*v*	7.412	8.585	9.748	10.907	13.219	15.529	17.838	22.455	29.378	36.301
(60.06°C)	*u*	2442.2	2514.6	2587.3	2660.9	2811.9	2968.8	3132.2	3479.5	4053.0	4683.7
(333.21 K)	*h*	2590.4	2686.2	2782.3	2879.1	3076.3	3279.4	3489.0	3928.6	4640.6	5409.7
	s	7.8498	8.1255	8.3669	8.5831	8.9611	9.2876	9.5778	10.0829	10.7193	11.2612
0.040	*v*	3.683	4.279	4.866	5.448	6.606	7.763	8.918	11.227	14.689	18.151
(75.87°C)	*u*	2438.8	2512.6	2586.2	2660.2	2811.5	2968.6	3132.1	3479.4	4052.9	4683.6
(349.02 K)	*h*	2586.1	2683.8	2780.8	2878.1	3075.8	3279.1	3488.8	3928.5	4640.5	5409.6
	s	7.5192	7.8003	8.0444	8.2617	8.6406	8.9674	9.2577	9.7629	10.3994	10.9412
0.060	*v*	2.440	2.844	3.238	3.628	4.402	5.174	5.944	7.484	9.792	12.100
(85.94°C)	*u*	2435.3	2510.6	2585.1	2659.5	2811.2	2968.4	3131.9	3479.4	4052.9	4683.6
(359.09 K)	*h*	2581.7	2681.3	2779.4	2877.2	3075.3	3278.8	3488.6	3928.4	4640.4	5409.6
	s	7.3212	7.6079	7.8546	8.0731	8.4528	8.7799	9.0704	9.5757	10.2122	10.7541
0.080	*v*	1.8183	2.127	2.425	2.718	3.300	3.879	4.458	5.613	7.344	9.075
(93.50°C)	*u*	2431.7	2508.7	2583.9	2658.8	2810.8	2968.1	3131.7	3479.3	4052.8	4683.5
(366.65 K)	*h*	2577.2	2678.8	2777.9	2876.2	3074.8	3278.5	3488.3	3928.3	4640.4	5409.5
	s	7.1775	7.4698	7.7191	7.9388	8.3194	8.6468	8.9374	9.4428	10.0794	10.6213
0.100	*v*	1.4450	1.6958	1.9364	2.172	2.639	3.103	3.565	4.490	5.875	7.260

(99.63°C)	*u*	2428.2	2506.7	2582.8	2658.1	2810.4	2967.9	3131.6	3479.2	4052.8	4683.5
(372.78 K)	*h*	2572.7	2676.2	2776.4	2875.3	3074.3	3278.2	3488.1	3928.2	4640.3	5409.5
	s	7.0633	7.3614	7.6134	7.8343	8.2158	8.5435	8.8342	9.3398	9.9764	10.5183
0.200	*v*	0.6969	0.8340	0.9596	1.0803	1.3162	1.5493	1.7814	2.244	2.937	3.630
(120.23°C)	*u*	2409.5	2496.3	2576.9	2654.4	2808.6	2966.7	3130.8	3478.8	4052.5	4683.2
(393.38 K)	*h*	2548.9	2663.1	2768.8	2870.5	3071.8	3276.6	3487.1	3927.6	4640.0	5409.3
	s	6.6844	7.0135	7.2795	7.5066	7.8926	8.2218	8.5133	9.0194	9.6563	10.1982
0.300	*v*	0.4455	0.5461	0.6339	0.7163	0.8753	1.0315	1.1867	1.4957	1.9581	2.4201
(133.55°C)	*u*	2389.1	2485.4	2570.8	2650.7	2806.7	2965.6	3130.0	3478.4	4052.3	4683.0
(406.70 K)	*h*	2522.7	2649.2	2761.0	2865.6	3069.3	3275.0	3486.0	3927.1	4639.7	5409.0
	s	6.4319	6.7965	7.0778	7.3115	7.7022	8.0330	8.3251	8.8319	9.4690	10.0110
0.400	*v*	0.3177	0.4017	0.4708	0.5342	0.6548	0.7726	0.8893	1.1215	1.4685	1.8151
(143.63°C)	*u*	2366.3	2473.8	2564.5	2646.8	2804.8	2964.4	3129.2	3477.9	4052.0	4682.8
(416.78 K)	*h*	2493.4	2634.5	2752.8	2860.5	3066.8	3273.4	3484.9	3926.5	4639.4	5408.8
	s	6.2248	6.6319	6.9299	7.1706	7.5662	7.8985	8.1913	8.6987	9.3360	9.8780
0.500	*v*		0.3146	0.3729	0.4249	0.5226	0.6173	0.7109	0.8969	1.1747	1.4521
(151.86°C)	*u*		2461.5	2557.9	2642.9	2802.9	2963.2	328.4	3477.5	4051.8	4682.5
(425.01 K)	*h*		2618.7	2744.4	2855.4	3064.2	3271.9	3483.9	3925.9	4639.1	5408.6
	s		6.4945	6.8111	7.0592	7.4599	7.7938	8.0873	8.5952	9.2328	9.7749

[a] Symbols: v = specific volume, m^3/kg; u = specific internal energy, U/kg; h = specific enthalpy, kJ/kg; s = specific entropy, kJ/K kg.

Source: From Bolz, R. E. and G. L. Tuve, Eds. 1973. *CRC Handbook of Tables for Applied Engineering Science*, 2nd Ed., Chemical Rubber Co., Cleveland, Ohio.

A4

Ultimate Analysis of Biomass Fuels

Nitin Goel
Intel Technology India Pvt. Ltd.

Material	C[a] (%)	H_2[a] (%)	O_2[a] (%)	N_2[a] (%)	S[a] (%)	A[a] (%)	HHV (kJ/kg)[b]
			Agricultural Wastes				
Bagasse (sugarcane refuse)	47.3	6.1	35.3	0.0	0.0	11.3	21,255
Feedlot manure	42.7	5.5	31.3	2.4	0.3	17.8	17,160
Rice hulls	38.5	5.7	39.8	0.5	0.0	15.5	15,370
Rice straw	39.2	5.1	35.8	0.6	0.1	19.2	15,210
			Municipal Solid Waste				
General	33.9	4.6	22.4	0.7	0.4	38.0	13,130
Brown paper	44.9	6.1	47.8	0.0	0.1	1.1	17,920
Cardboard	45.5	6.1	44.5	0.2	0.1	3.6	18,235
Corrugated boxes	43.8	5.7	45.1	0.1	0.2	5.1	16,430
Food fats	76.7	12.1	11.2	0.0	0.0	0.0	38,835
Garbage	45.0	6.4	28.8	3.3	0.5	16.0	19,730
Glass bottles (labels)	0.5	0.1	0.4	0.0	0.0	99.0	195
Magazine paper	33.2	5.0	38.9	0.1	0.1	22.7	12,650
Metal cans (labels, etc.)	4.5	0.6	4.3	0.1	0.0	90.5	1,725
Newspapers	49.1	6.1	43.0	0.1	0.2	1.5	19,720
Oils, paints	66.9	9.6	5.2	2.0	0.0	16.3	31,165
Paper food cartons	44.7	6.1	41.9	0.2	0.2	6.9	17,975
Plastics							
General	60.0	7.2	22.6	0.0	0.0	10.2	33,415
Polyethylene	85.6	14.4	0.0	0.0	0.0	0.0	46,395
Vinyl chloride	47.1	5.9	18.6	(Chlorine=28.4%)			20,535
Rags	55.0	6.6	31.2	4.6	0.1	2.5	13,955
Rubber	77.7	10.3	0.0	0.0	2.0	10.0	26,350
			Sewage				
Raw sewage	45.5	6.8	25.8	3.3	2.5	16.1	16,465
Sewage sludge	14.2	2.1	10.5	1.1	0.7	71.4	4,745
			Wood and Wood Products				
Hardwoods							
Beech	51.6	6.3	41.5	0.0	0.0	0.6	20,370
Hickory	49.7	6.5	43.1	0.0	0.0	0.7	20,165

(continued)

Material	C^a (%)	$H_2{}^a$ (%)	$O_2{}^a$ (%)	$N_2{}^a$ (%)	S^a (%)	A^a (%)	HHV (kJ/kg)[b]
Maple	50.6	6.0	41.7	0.3	0.0	1.4	19,955
Poplar	51.6	6.3	41.5	0.0	0.0	0.6	20,745
Oak	49.5	6.6	43.4	0.3	0.0	0.2	20,185
Softwoods							
Douglas fir	52.3	6.3	40.5	0.1	0.0	0.8	21,045
Pine	52.6	6.1	40.9	0.2	0.0	0.2	21,280
Redwood	53.5	5.9	40.3	0.1	0.0	0.2	21,025
Western hemlock	50.4	5.8	41.4	0.1	0.1	2.2	20,045
Wood products							
Charcoal (made at 400°C)	76.5	3.9	15.4	0.8	0.0	3.4	28,560
Charcoal (made at 500°C)	81.7	3.2	11.5	0.2	0.0	3.4	31,630
Douglas fir bark	56.2	5.9	36.7	0.0	0.0	1.2	22,095
Pine bark	52.3	5.8	38.8	0.2	0.0	2.9	20,420
Dry sawdust pellets	47.2	5.5	46.3	0.0	0.0	1.0	20,500
Ripe leaves	40.5	6.0	45.1	0.2	0.1	8.1	16,400
Plant wastes							
Brush	42.5	5.9	41.2	2.0	0.1	8.3	18,370
Evergreen trimmings	49.5	6.6	41.2	1.7	0.2	0.8	6,425
Garden plants	48.0	6.8	41.3	1.2	0.3	2.4	8,835
Grass	48.4	6.8	41.6	1.2	0.3	1.7	18,520

[a] All percentages on moisture-free basis.
[b] 1 kJ/kg = 0.43 Btu/lbm.

Index

1,2,3-propanetriol. *See* glycerol
1954 Atomic Energy Act, **16**–2
1990 Clean Air Act Amendments, **2**–12
2,3,7,8-TCDD toxic equivalents, **24**–29
40 CFR Part 60, **24**–24

A

Abnormal combustion, **10**–13
Absolute efficiency, **11**–17
Absorbed glass mat lead-acid batteries, **15**–9
Absorbers
  efficiency of, **19**–7
  flat-plate solar collectors, **18**–4
  loss from, **19**–60
  materials for, properties and efficiencies of, **19**–68
  metal, **19**–65
  volumetric, **19**–62, **19**–67
Absorptances, **A3**–4
Absorption charge mode, **20**–14
Absorption chillers, **18**–124
Absorption coefficient, thin-film technologies, **20**–30
Absorption-cooling refrigeration cycles, **18**–120
ABWR, **16**–8
AC-to-DC rectifiers, **15**–4
Acceptance angle, **19**–23
Acetogens, **22**–10, **22**–61
  two-stage digestion, **22**–20
Acid gas production, geothermal reservoirs, **23**–21
Acid gas scrubbing, **24**–27
  particulate control with, **24**–24
Acid rain, **2**–14
Acid-electrolyte fuel cells, **26**–2, **26**–34
Acoustic flow meters, **11**–17
Active pitch control, **21**–20
Active solar systems, **18**–23
  simulation programs for, **18**–39
Active space heating, system design for, **18**–39
Actuator disk model, **21**–10
Additives, use of to increase octane ratings, **10**–19
Adiabatic efficiency, **9**–3
Adiabatic saturators, **9**–9
Adjustable blade propeller turbines, **11**–4
Advanced boiling-water reactor. *See* ABWR
Advanced CANDU Reactor, **16**–10
Advanced gas-cooled reactors. *See* AGRs
Advanced heavy-water reactor. *See* AHWR
Advanced Turbine System program, **13**–21
Advanco/Vanguard dish-Stirling system, **19**–89
Aerobic composting process, comparison with anaerobic process, **22**–22
Aeroderivative gas turbines, **9**–6
Aerodynamics
  unsteady, **21**–13
  wind turbines, **21**–4
    modeling of, **21**–9, **21**–15
AFEX, **22**–56
Africa, technical potential of CSP in, **19**–4
Agricultural animal waste
  degradation of with high-solids anaerobic digestion, **22**–20
  use of as feedstock for anaerobic digestion process, **22**–2
Agricultural residues, degradation of with high-solids anaerobic digestion, **22**–20
Agricultural sector, geothermal energy use for, **23**–3
Agricultural wastes, ultimate analysis of, **A4**–2
Agriculture residue recovery, **3**–2
AGRs, **16**–9
AHWR, **16**–10
Air
  dry, properties of, **A3**–2
  use of as a thermal storage fluid, **19**–73
Air conditioning
  absorption, **18**–124
  desiccant systems, **18**–126
Air density, **6**–1
Air heat transfer, **18**–100
Air heaters, in WTE facilities, **24**–24
Air pollution
  control facilities, **24**–24
  control requirements for WTE combustion components, **24**–11
Air return ratio, **19**–70
Air solar collectors, **18**–2
  absorbers used in, **18**–5
  advantages and disadvantages of, **18**–3

thermal storage in, **18**–33
Air starvation, **3**–7
Air-cooled generators, **8**–14
Air-cooled volumetric receivers, advantages of, **19**–66
Airfoils, **21**–4
stall, power limitation using, **21**–19
Akzo Nobel, development of flexible amorphous solar cells by, **20**–41
Alcohols, from syngas, **22**–61
Algae, as sources of triglycerides, **22**–59
Alkali, effect of on the direct combustion of biomass, **22**–40
Alkaline fuel cells, **26**–2, **26**–30, **26**–44
All-or-nothing auxiliary heaters, **18**–27
Alloys, thermal properties of, **A3**–6
Alpha-configuration, **12**–4
Altitude angle, **20**–13
solar, **5**–4
American Solargenix, collector design, **19**–49
Ammonia, **22**–51
anhydrous, **22**–63
conversion of nitrogen to, **22**–8
Ammonia fiber explosion. *See* AFEX
Ammonia-water binary power plants, **23**–40
Ammonia-water systems, **18**–125
double effect, **18**–126
Amorphous cellulose, **22**–56
Amorphous silicon (*a*-Si), **20**–11, **20**–31
deposition techniques for, **20**–35
properties of, **20**–35
solar cells, **20**–34
configurations of, **20**–36
flexible, **20**–41
stability and recombination issues in, **20**–37
Anaerobic attached growth reactors, **22**–19
Anaerobic bacteria, **22**–9
nutrient requirements for, **22**–10
physical and chemical factors that affect, **22**–11
synthesis of ethanol from syngas using, **3**–16
Anaerobic composting process
biological characteristics of humus produced from, **22**–30
chemical characteristics of humus produced from, **22**–29
comparison with aerobic process, **22**–22
Anaerobic contact process, **22**–19
Anaerobic digestion, **3**–5, **3**–13, **22**–2
biochemical reactions in, **22**–13
commercial-scale technologies, **22**–28
fundamentals of, **22**–9
high-solids process, **22**–10
in-vessel process, **22**–4
utilization of byproducts from, **22**–24
indicators of unbalance operation of, **22**–16
modes of operation for, **22**–19
monitoring, **22**–16
plants, **22**–30
reactor types used for, **22**–16
stages of, **22**–12
Analytical approach, solar system design using, **18**–39
AndaSol plant, **19**–49
Anderson-Shulz-Flory mechanism, **3**–15
Anhydrous ammonia, **22**–63
Anhydrous ethanol, **22**–53
Animal manures, methane yields from, limitation of by polymeric lignin, **22**–12
Animals fats, modification of to biodiesel, **22**–59
Annual lighting energy saved, **18**–118
determination of, **18**–104
Annual normal incident radiation, **19**–3
Annual thermal efficiency, parabolic trough collectors, **19**–49
Annular flow, **19**–40
Anthracite, **2**–2, **13**–3
deposits of, **2**–6
Antiknock
index, **10**–18
quality, **10**–19
Antireflection glazings, **18**–51
AP-reactors, **16**–9
Aperture controls, **18**–102, **18**–117
API gravity scale, **2**–24
Apparent window transmittance, **19**–59
Arabinose, **22**–57
Arco Solar, CIS thin film technology innovations at, **20**–47
ARDISS project, **19**–42
Area-velocity measurements, **11**–17
Ash
components of, **24**–23
content of coal, **2**–4
discharge facilities, design of, **24**–23
fusibility, **2**–6
in organic substrates, **22**–4
testing of residues in, **24**–23
Ash-fouling, **22**–40
ASHRAE, Standard 93-77, **18**–7, **18**–10
Asia, ABWR development in, **16**–9
Asphalt, **2**–20
ASTM, coal specifications, **2**–4
Atmosphere of dry air, **24**–11
Atmospheric attenuation, **19**–55
Atmospheric discharge turbines, **23**–33
Atmospheric fluidized bed combustion, **13**–14
Atmospheric optical depth, **5**–4
Atomic Energy Commission, **16**–2
Atomic layer deposition, **20**–46
use of on CIGS solar cells, **20**–48
ATS heliostat, **19**–57
Attemperators, **8**–7
Attenuation losses, **19**–55
Australia
geothermal development in, **23**–5
integration of LF systems with coal-fired technology in, **19**–17
Autoignition, **10**–5, **22**–51
Auxiliary heat requirements, **18**–66
Auxiliary heaters, **18**–27
parallel, **18**–43
Axial flow propeller turbines, **11**–4
Azimuth angle, **20**–13
difference, **18**–104
solar, **5**–4
Azimuth-elevation tracking, **19**–87
Azimuthal spacing of heliostats, **19**–55

B

Backup system, passive solar heating, **18**–62
BACT, **24**–38
Balance of system components, **20**–17
Balance of the plant, **26**–4
Balanced digestion, **22**–16
Bandgap
  amorphous silicon, **20**–35
  limits imposed on solar cells due to, **20**–2
Bang-bang motion, **12**–5
Baseload utility generators, use of MCFCs in, **26**–41
Baseloading, **23**–11, **23**–14
Bat collisions, **21**–27
Batch digestion studies, **22**–5
Batch reactor, **22**–17
Bath heating, **18**–54
Bathing, geothermal energy use for, **23**–3
Batteries. *See also* specific types of batteries
  mobile, **15**–3
  secondary, **15**–8
Battery electric vehicles, **15**–3
Beam normal radiation, **5**–4
Beam radiation, **18**–9, **19**–9
Bearings, **12**–8
Belgium, uprating of nuclear plants in, **16**–6
BEM model, **21**–12
Best Available Control Technology. *See* BACT
Beta-configuration, **12**–5
Billboard external receivers, **19**–62
Binary geothermal plants, **7** – 4
Binary power plants, **23**–7, **23**–32, **23**–38
  ammonia-water, **23**–40
  capital costs, **23**–13
  compactness of, **23**–39
  integration of steam turbines in, **23**–41
  pressurized geothermal brine, **23**–40
  sub-critical pentane, **23**–39
  supercritical hydrocarbon, **23**–40
Binary thermodynamic cycles, **8**–5
Biochemical methane potential assay. *See* BMP assay
Biochemical transformations, **22**–13
Biodegradability, **22**–3
  methods to estimate, **22**–5
Biodegradability of the organic fraction of MSW. *See* BOF/MSW
Biodegradable volatile solids. *See* BVS
Biodiesel, **3**–8, **15**–20, **22**–51, **22**–58
  fuel properties of, **22**–58
Bioenergy recovery system, **22**–3
Bioenergy, gas turbine engines for, **22**–47
Biofuel plantations, **3**–3
Biofuels, **1**–5, **22**–51
Biogases, **22**–8
  characteristics of, **22**–24
  utilization of, **22**–26
Biological pretreatments, **22**–56
Biological processes, loading rates for, **22**–7
Biomass, **3**–1
  combustion applications, **3**–5
  conversion of with gas turbines, **22**–47
  conversion technologies, **3**–4
  crops, primary productivity and solar efficiency of, **15**–19
  direct combustion of, materials suitable for, **22**–40
  electric power generation from, **3**–8
  energy demand for, **1**–1
  energy potential of, **3**–3
  facility supply considerations, **3**–3
  fuels
    direct combustion of, **22**–37
    ultimate analysis of, **A4**–2
  gasification, **22**–61
    use of fluidized beds for, **22**–44
  gasification of, **3**–9
  recycling of, **22**–19
  resources, forestry, **3**–2
  social impacts of, **3**–3
  solids, thermochemical energy storage in, **15**–18
  thermal gasification of, **22**–41
  total sustainable potential of, **1**–5
  use of as feedstock for anaerobic digestion process, **22**–2
  use of in combined cycle power systems, **22**–48
Biomass-derived syngas. *See also* syngas
  energy return on, **22**–63
  transportation fuels from, **22**–60
Biomass/coal, cofiring operation, **3**–9
Biorefinery facilities, **3**–3
Biosolids. *See also* humus material
  digested, characteristics of, **22**–27
Bipolar fuel stack arrangement, **26**–28
Bird collisions, **21**–27
Bituminous coal, **2**–2, **13**–3
  deposits of, **2**–6
  tests for suitability of, **2**–6
Black liquor, **3**–8
Black lung disease, **2**–14
Blade element momentum model. *See* BEM model
Blade pitch control, **21**–19
Blade torsion, modeling for, **21**–17
Blade velocity, **21**–5
Blade-bending moment predictions, **21**–15
Blading, steam turbines, **8**–8
Blended feeds, use of in cofiring, **3**–9
Blocking, **19**–54
Blow-by loss, **12**–8
Blowers, in combustion furnaces, **24**–20
Blowouts, **2**–14
BMP assay, **22**–6
Boeing Corporation, CIS thin film technology innovations at, **20**–47
Boeing process, **20**–49
BOF/MSW, **22**–4
  anaerobic transformation of, **22**–13
  biological characteristics of humus produced from, **22**–30
  chemical characteristics of humus produced from, **22**–29
  codigestion with WWTPS, **22**–21
  degradation of with high-solids anaerobic digestion, **22**–20
  estimating gas production from, **22**–7
  uses for biogas produced from, **22**–27
Boiler feed pumps, **8**–13

Boiler tubes, corrosion of in WTE facilities, **24**–24
Boilers, **3**–8, **22**–39, **24**–21
steam, **8**–5
Boiling-water reactors (BWR), **16**–4
ABWR, **16**–8
ESBWR, **16**–9
Bomb calorimeters, **24**–11
Boost pressures, **10**–21
Bottom ash, **24**–22
metals in, **24**–23
Bottom dead center, **10**–3
Bottoming cycle systems, **22**–48
use of waste heat in, **26**–3
Bottoming cycles, **8**–5
Bowed steam turbine blades, **8**–8
Box turbines, **21**–3
Brady Hot Springs, pressurized brine power plant at, **23**–41
Brake mean effective pressure, **10**–11
Brake work, **10**–11
Brake-specific fuel consumption, **10**–11
Brayton cycle, **9**–1, **9**–4, **13**–24, **22**–47
engines, **22**–51
use of in GFR nuclear reactors, **16**–15
Brayton cycle engines, **3**–8
Brine chemistry, **23**–22
Brine management, residual, **23**–20
Brine, pressurized, use of in binary power plants, **23**–41
Brown coal, **13**–3
Bubbling fluidized bed combustion, **24**–37
Bubbling fluidized beds, **3**–11, **13**–14
Bubbly flow, **19**–39
Buffer storage, **18**–24
Building materials, thermal properties of, **A3**–7
Buildings
use of DMFC units in, **26**–38
use of photovoltaic panels on, **20**–2
Bulb turbines, **11**–5
Bulk density, **3**–4
Buried contact PV cells, **20**–11, **20**–24
BVS, **22**–4
fraction of in organic wastes, **22**–6
organic loading rate, **22**–7

C

C/N ratio, **22**–7
Cadmium, **15**–9
environmental concerns regarding, **20**–52
in combustion flue gas, **24**–31
Cadmium sulfide, **20**–31
deposition of on CIGS solar cells, **20**–48
Cadmium telluride. *See* CdTe
Caking, **2**–4
California, Standard Offer Number Four, effect of on geothermal power plants, **23**–10
Calorific value
biomass gases, **3**–10, **3**–12
coal, **2**–4
Canada
CANDU reactor, **16**–5
next generation heavy-water reactor development in, **16**–10
South Meager Creek Geothermal Project, **23**–17
CANDU reactor, **16**–4
next generation, **16**–10
Capacitors, **15**–7
Capacity factors, power generation plants, **23**–11
Capital costs
development, minimization of, **23**–19
geothermal power plants, **23**–13
Capital limitations, geothermal resource management and, **23**–23
Car-washing facilities, use of solar process heat in, **18**–55
Carbon content of coal, **2**–4
Carbon dioxide, emissions, **13**–6, **13**–9
Carbon fiber, use of for wind turbines, **21**–23
Carbon injection, use of to reduce mercury emissions, **24**–31
Carbon monoxide, **3**–7
control of in flue gas emissions, **24**–27
measurement of in combustion gas, **24**–30
Carbon to nitrogen ratio. *See* C/N ratio
Carbon, monoxide, **10**–15
Carbon-oxygen reaction, **22**–42
Carburetors, **10**–12
Carnot cycle, **8**–3, **12**–1
Carnot cycle efficiency, **19**–7, **22**–47, **26**–13, **26**–44
equivalency of with fuel cell reversible efficiency, **26**–14
Cascaded humid air turbines, **13**–23
Catalyst poisoning, **26**–40
Catalytic converters, **10**–17
Cathode dissolution, **26**–40
Cavitation, **11**–12
Cavity receivers, **19**–62
convective heat loss calculation for, **19**–61
saturated steam, **19**–75
water-steam, **19**–64
CBs, **24**–28
CCGT power plants, use of solar energy in, **19**–82
CdTe, **20**–29, **20**–31
absorber layer, **20**–44
deposition techniques for, **20**–46
environmental concerns, **20**–52
solar cells, **20**–41
device structure, **20**–43
efficiency of, **20**–43
flexible, **20**–46
problems of electrical back contact and stability in, **20**–45
use of in PV cells, **20**–3
Cell potential, **26**–7
Cellulose, **22**–56
pyrolysis of, **22**–37
Cellulosic ethanol, **15**–20
Cellulosic material, percentage of in municipal solid waste, **24**–3
Center-feed solar field layouts, **19**–33
Centering ports, **12**–8
Central Appalachian Basin, high-sulfur coal from, **2**–13
Central feed solar field layout, **19**–46
Central receiver systems. *See* CRS
Central station generation options, **13**–24

Centrifugal compressors, use of in turbochargers, **10**–22
Centrifugal forces, **8**–8
Centrifugal type supercharger, **10**–21
Ceramic monolith volumetric absorbers, **19**–67
Cereal grains, use of to produce ethanol, **22**–54
CESA-1 plant, **19**–62, **19**–64, **19**–66, **19**–81, **19**–84
Cetane index, **10**–19
CFD models, **21**–13, **21**–14
CGS, solar cells, **20**–46
Chalcopyrites, use of for thin-film solar cells, **20**–46
Char, **3**–10
Charge controllers, **20**–14, **22**–38
Charge pressure, **12**–2
Chemical energy, **3**–1
Chemical reaction latent heat storage, **15**–18
Chemical vapor deposition methods. *See* CVD methods
Chemostat studies, **22**–6
Chernobyl, **16**–14
China
  electrical capacity additions in, **1**–4
  increase in energy use in, **1**–1
  wind energy installations in, **21**–1
Chinese tallow tree, **22**–59
Chlorobenzenes. *See* CBs
Chlorophenols. *See* CPs
CHP. *See also* cogeneration
  biomass applications for, **3**–7
CIEMAT, **19**–67, **19**–81
CIGS, **20**–29, **20**–31
  absorbers, deposition and growth of, **20**–49
  alternative growth processes, **20**–50
  deposition
    methods for, **20**–48
    nonvacuum methods for, **20**–50
  environmental concerns regarding, **20**–52
  solar cells, **20**–46
    configurations of, **20**–48
    flexible, **20**–51
    properties of, **20**–47
    sodium incorporation into, **20**–49
Circulating fluid beds, **13**–14
Circulating fluidized beds, **3**–11
Circumsolar irradiance, **19**–11
CIS, **20**–29
  solar cells, **20**–46
    properties of, **20**–47
  use of in PV cells, **20**–3
CIS Solar Technologies, CIGS solar cells developed by, **20**–51
CISCuT, CIGS solar cells developed by, **20**–51
Clean coal technologies, **13**–4
  Europe, **13**–8
  Japan, **13**–7
  United States, **13**–6
Clean-water injection, enhancing steam production with, **23**–21
Clear-sky illuminance, **18**–109
Clearance volume, **10**–8
Clearness index, monthly, **18**–16
Clearness number, **5**–4
Closed cycle gas turbines, **9**–7
Closed cycle, OTEC power, **25**–1
Closed fuel cycle, **16**–20
Closed heaters, **8**–11
Closed-loop organic Rankine cycle configuration, **23**–32
Closed-loop solar thermal systems
  definition of, **18**–25
  description of, **18**–26
  design methods for, **18**–38
  earth contact cooling, **18**–99
  multipass (*see also* multipass systems)
    system design, **18**–43
Closed-loop volumetric air technologies, **19**–17, **19**–74
Clostridium ijungdahli, **22**–61
Cloud cover, prediction of, **20**–13
CN, **10**–19
Co-evaporation process, **20**–49
Co-firing, **22**–40
  disadvantages of, **22**–41
CO_2 emissions, **1**–5
  avoidance of with CSP plants, **19**–4
  dry steam power plants, **23**–45
  flash-steam power plants, **23**–45
  geothermal power plants, **23**–44
Coal, **1**–3, **2**–1, **2**–16, **13**–3
  analysis and properties of, **2**–2
  characteristics of by rank, **2**–3
  cleaning of, **13**–5
  content of, **2**–2, **2**–4
  energy demand for, **1**–1
  gasification, **13**–2, **13**–16
  reactivity of, **2**–6
  reserves, **2**–6
  sulfur emissions from, **7**–5
  terminology, **2**–7, **2**–15
  tests for suitability of, **2**–6
  transportation of, **2**–8
  U.S. demonstrated reserve base, **2**–9
  use of in steam-fired power plants, **8**–1
  world recoverable reserves of, **2**–11
Coal-fired Rankine plants, use of solar multiples with, **19**–3
Coalification, **2**–1, **2**–14
Cocurrent moving bed gasifiers, **3**–11
COE, wind, **21**–26
Coefficients of performance. *See* COP
Coextruded tubing, **13**–6
Cofiring, **3**–9
Cogeneration, **3**–7. *See also* CHP
  use of waste heat, **26**–3
Coke, **2**–6
  petroleum, **13**–15
Coking, **2**–4
Cold pistons, **12**–4
Cold streams, **18**–52
Collector acceptance angle, **19**–23
Collector field technology, **19**–52
Collector geometrical end losses, **19**–26
Collector time constant, **18**–10
Collector-heat exchanger
  correction factor, variation of, **18**–32
  performance, **18**–30
COLON SOLAR prototype, **19**–57, **19**–74
Combined cycle, **13**–2

gas turbine steam engines, **9**–8
power plants, **13**–24, **14**–1, **22**–48
combined heat and power from, **14**–7
single-shaft, **14**–4
systems, integration of power towers into, **19**–75
thermodynamics, **14**–2
Combined heat and power. *See* CHP
Combustion, **13**–4
abnormal, **10**–13
air requirements for, **24**–20
ash, components of, **24**–23
biomass applications, **3**–6
biomass, efficiency of, **3**–7
chamber shape, **10**–17
chemical reactions of, **24**–11
coal, **2**–14
control, **13**–11, **24**–11, **24**–28
efficiency, **9**–3
elements and compounds encountered in, **24**–11
emission limits for, **24**–25
extraction of work or electricity from biomass using, **15**–19
fibrous plant material, **22**–37
fluidized bed, **13**–13
furnaces, **24**–16
gas, measurements, **24**–30
gaseous emission factors for MSW, **24**–26
heat of, **24**–12
in compression ignition engines, **10**–14
normal, **10**–13
premixed homogeneous, **10**–3
pressurized fluidized bed, **9**–2, **13**–15
principles of, **24**–10
products of, **10**–15
fuel cells, **26**–15
selective noncatalytic, **13**–12
systems, pressurized fluidized bed, **9**–2
temperature attained during, **24**–16
turbine-based power plants, **13**–20
Combustion turbine/combined cycle plants, **15**–14
Combustors, **22**–38
gas turbine, **9**–10
Comfort conditions, **18**–95
Commercial solar systems, **18**–24
Community district heating, **3**–7. *See also* district heating
Compact steam generators, **19**–29
Complete-mix reactors, **22**–18
Complex organic substrate, pathway of in anaerobic digestion, **22**–12
Composting, aerobic *vs.* anaerobic, **22**–22
Compound parabolic concentrators. *See* CPCs
Compressed, natural gas, **2**–24
Compressed-air energy storage systems, **15**–14
Compression ignition engines, **3**–8, **10**–1, **10**–4
Compression ratio, **10**–7
Compressors
in geothermal plants, **23**–37
use of in turbochargers, **10**–22
Computer simulations, solar system design using, **18**–39
Concentrated acid hydrolysis, **22**–57
Concentrating solar collectors, **18**–3, **18**–24
beam quality, **19**–9
Concentrating solar power. *See* CSP
Concentrating solar thermal power. *See* CSP
Concentration polarization, **26**–20
Concentration ratio, **18**–3
Concentrators, in dish-Stirling systems, **19**–87
Condensate pumps, **8**–13
Condenser-coupled absorption system, **18**–126
Condensers, steam turbines, **8**–12
Condensing steam turbines, **23**–33
Conductive absorber loss, **19**–60
Conductors, sizing of in PV systems, **20**–19
Cone optics, **19**–11
Confinement, **17**–2
CONSOLAR, **19**–50, **19**–85
Constant pressure
heat addition, **10**–8
line, **9**–4
method of turbocharging, **10**–22
Constant-volume heat addition, **10**–7
Constrained exogenous parameters, **18**–37
Content of coal, **2**–2
Conti, Prince Piero Ginori, **23**–2
Continuous feed mechanical grate system, **24**–19
Continuous flow reactor, **22**–17
Continuous stirred-tank reactors. *See* CSTRs
Control of emissions, **10**–16
Control systems for passive solar heating systems, **18**–62
Controlled circulation boilers, **8**–6
Controls
design of for solar thermal systems, **18**–29
wind turbine, **21**–22
Convection/ventilation, **18**–95
stack-effect/solar chimney concepts for, **18**–96
Convective absorber loss, **19**–60
Convective heat loss, calculation of for solar towers, **19**–61
Convective loss heat transfer coefficient, **19**–60
Conventional thermal power plants, energy efficiencies of, **26**–36
Conversion, **13**–5
efficiency, **27**–10
fusion energy, **17**–2
ocean thermal energy, **25**–1
thermionic energy, **27**–1
uranium, **16**–22
Conversion factors, **A1**–4
Converters, **12**–6
thermionic, types of, **27**–4
Cooking, biomass applications for, **3**–7
Cool Water plant, **13**–16
Coolers in Stirling engines, **12**–3
Cooling systems
earth contact, **18**–99
liquid-desiccant, **18**–128
passive, **18**–59
Cooling towers, **23**–37
Cooling, thermoelectric, **27**–9
Coolth, **18**–59
Cooper Basin, HFR development at, **23**–5
COP, **27**–10
Copper indium diselenide. *See* CIS

Copper sulfide, **20**–29, **20**–31
Copper-gallium-diselenide. *See* CGS
Copper-indium gallium diselenide. *See* CIGS
Corn
  dry grinding of, **22**–54
  ethanol production from, **3**–14
  wet milling of, **22**–55
Coso geothermal field, **7**–6
Cost of energy. *See* COE
Cost Reduction Study for Solar Thermal Power Plants (World Bank), **19**–4, **19**–91
Costa Rica, pressurized brine power plant in, **23**–41
Costs
  heliostats, reduction of, **19**–57
  solar thermal systems, **18**–47
  wind turbines, **21**–24
Counterflow moving bed gasifiers, **3**–10
CPCs, **18**–3, **18**–13, **18**–24, **19**–13
  use of for industrial process heat, **18**–51
CPs, control of in emissions from MSW combustion, **24**–28
Creep, thermal, **12**–9
Crop residues, use of as feedstock for anaerobic digestion process, **22**–2
Cross draft gasifiers, **3**–11
Cross-flow scavenging, **10**–4
Cross-head, **12**–6
CRS, **19**–15, **19**–73
  control of, **19**–58
  operation of with heat storage systems, **19**–17
  solar thermal power plants, **19**–50
  technology description, **19**–51
  use of air as working fluid in, **19**–80
Crucible growth, silicon, **20**–9
Crude oil, **2**–24, **10**–18
Crystalline cellulose, **22**–56
Crystalline silicon, **20**–24
  disadvantages of, **20**–29
  ingot production methods, **20**–8
  solar cells, **20**–34
    manufacture of, **20**–8
CSP
  market opportunities for, **19**–91
  plants, **19**–2
    operational strategy for, **19**–5
  systems
    characteristics of, **19**–16
    efficiency of, **19**–6
    energy payback time of, **19**–4
    integration of with conventional power plants, **19**–3
    low-cost, **19**–4
    use of reflective parabolic concentrators in, **19**–13
CSTRs, **22**–18
CT, **13**–20
Cu(InGa)Se_2, solar cells, **20**–46
Current efficiency, **26**–22
Current-voltage characteristics, **27**–4
Cut-off ratio, **10**–9
CVD methods, **20**–11, **20**–46
  amorphous silicon, **20**–35
CVs, control of in emissions from MSW combustion, **24**–28
Cycle feed pumps, **23**–38
Cycle inefficiency, **23**–38
Cycle life, **15**–4
Cycles, **13**–7
  selection of, **13**–6
Cyclic fatigue loads, **21**–23
Cyclone boilers, **3**–9
Cylinder volume, **10**–8
Cylinders, use of in steam turbines, **8**–10
Cylindrical external receivers, **19**–62
Czochralski method of silicon ingot production, **20**–8

D

D-T reaction, **17**–1
Daily solar thermal storage, **18**–24
Daily utilizability, **18**–18
  Phibar method for determining, **18**–19
Dairy manure, use of in inoculation, **22**–3
Dangling bonds passivation, **20**–34, **20**–37
Darrieus turbines, **21**–3
  structural dynamics, **21**–18
Daylighting, **18**–59
  controls, **18**–117
  economics, **18**–118
  system controls, **18**–117
  system design, **18**–102
    methods for, **18**–104
DC electric power, conversion of to AC, **26**–3
DC-to-AC converters, **15**–4
DC-to-AC inverters, **26**–4
DDGS, **22**–54
Deaerator, **8**–12
Decentralized power supply, use of SOFCs for, **26**–44
Declination angle, **20**–13
Declination, solar, **5**–3
Decommissioning, **16**–29
Degraded sunshape, **19**–11
Degrees of freedom, **21**–17
Delivered power, price of, **23**–10
DELSOL3, **19**–56
Demonstrated reserve base, **2**–8
  U.S. coal, **2**–9
Demonstrated resources, **2**–8
Deposition techniques
  amorphous silicon, **20**–35
  CdTe, **20**–45
  CIGS, **20**–48
Depth of discharge, **15**–4
Desiccant cooling, **18**–120
  ventilation cycle, **18**–130
Desiccant dehumidification systems, **18**–126
Desiccants, types of, **18**–127
Design methods, levels of, **18**–38, **18**–65
Design point, definition of, **19**–31
Desulfurization of flue gas, **13**–2
Detailed design methods, **18**–39, **18**–65
Deterministic loads, **21**–16
Deuterium and tritium reaction. *See* D-T reaction
Diesel
  engines, **3**–8, **10**–1, **10**–4, **22**–51

fuel, **10**–18, **10**–19
Diffuse radiation, **5**–4, **18**–9, **18**–59, **19**–9
use of in daylighting system design, **18**–104
Digestate material, **22**–24
Digester gas, **22**–24
Digester operations, stabilizing, **22**–16
Dilatometric methods for testing of coal, **2**–6
Dilute acid hydrolysis, **22**–57
Dimensionless, figure of merit, **27**–7
Dimethyl ether, **22**–51, **22**–63
Dimming controls for electric lighting, **18**–103
Dioxin/furan emissions. *See also* dioxins; furans
measurement of, **24**–30
Dioxins
control of in emissions from MSW combustion, **24**–28
in combustion ash, **24**–23
measurement of, **24**–30
Direct combustion
biomass, disadvantages of process, **22**–40
fibrous plant material, **22**–37
Direct coupling, **18**–28, **18**–54
Direct drive generators, **21**–21
Direct electric storage, **15**–7
Direct gain passive solar systems, **18**–62, **18**–74
Direct heat, geothermal energy use for, **23**–3
Direct injection, **10**–5
Direct irradiance, **19**–11
Direct radiation, **5**–4
Direct solar radiation, incidence angle of, **19**–26
Direct steam generation. *See* DSG
Direct thermal storage, **15**–15
Direct-methanol fuel cells. *See* DMFC
Direct-return solar field layouts, **19**–32
Directly illuminated parabolic dish receivers, **19**–87
Directly irradiated receivers, **19**–62
Discounted cash flow analysis, **18**–38
Discovered resources, **2**–19
Dish-Stirling systems, **19**–85
description of, **19**–86
developments of, **19**–89
Dish/engine systems, **19**–15, **19**–17
Displacement volume, **10**–10
Displacers, **12**–4
DISS project, **19**–42
Distillers' dried grains and solubles. *See* DDGS
Distributed grid technologies, **15**–3
Distribution, systems for passive solar heating systems, **18**–62
District heating, **3**–7
District-heating systems, **15**–3
DMFC, **26**–2, **26**–44
acceptable contamination levels for, **26**–37
basic operating principle, **26**–36
technological problems of, **26**–37
DOE
Dish Engine Critical Components, **19**–89
Solar Two project, **15**–15
Domestic hot water, use of closed-lop solar systems for, **18**–25
Domestic solar thermal systems, **18**–26
water heating, system design, **18**–42
Doping control, **20**–42
Double reheat, **13**–15
Double tank systems, **18**–28
Double-acting pistons, **12**–4
Double-effect absorption systems, **18**–126
Double-flash process, **23**–31
Double-glazed antireflection collectors, efficiency curves, **18**–52
Double-loop heat exchangers, heat collection decrease caused by, **18**–31
Double-multiple stream-tube model, **21**–12
Downdraft gasification, **3**–10
Downdraft gasifiers, **22**–43
Draft tube, **11**–1
design of, **11**–15
Drag devices, translating, **21**–7
Drag force, **21**–5
Drain cooler approach, **8**–12
Drain-back systems, **18**–28
Drain-down systems, **18**–29
DRANCO process, **22**–31
Dresden Nuclear Power Plant, **16**–2
Drilling costs, geothermal, **23**–7
Drives
swash plate, **12**–6
wobble-plate, **12**–6
Drum-type boilers, **8**–6
Dry air
properties of, **A3**–2
standard atmosphere of, **24**–11
Dry anaerobic composting process. *See* DRANCO process
Dry condenser, **8**–12
Dry cooling, effect of on capital costs, **23**–13
Dry grinding, starch, **22**–54
Dry milling, **3**–14
Dry natural gas, **2**–23. *See also* natural gas
Dry organic substrates, **22**–4. *See also* organic substrates
Dry steam power plants, CO_2 emissions from, **23**–45
Dry steam power process, **23**–33
Drying, **15**–15
DSG, **19**–19, **19**–30, **19**–38, **19**–49
processes, **19**–41
solar field design, **19**–45
Dual-medium storage systems, thermal energy storage in, **19**–38
Duct firing, **14**–6
Duseldorf System, **24**–17
Duty-cycle analysis, use of in fuel handling design for WTE facilities, **24**–10
Dye-dilution method, **11**–17
Dye-sensitized solar cells, **20**–31, **20**–51

E

E10, **22**–53
E100, **22**–53
E85, **22**–53
Earth contact cooling, **18**–99
Earth-sun relationships, **5**–2
East Mesa, expansion to, **23**–11
Economic access, definition of, **23**–6
Economic and simplified boiling-water reactor. *See* ESBWR

Economizer, **8**–6
ECR, **20**–36
EDF-CNRS, CIGS solar cells developed by, **20**–51
Edge-defined film-fed growth process. *See* EFG process
EDLC, **15**–7
Efficiency, **9**–2, **15**–4
  absolute, **11**–17
  biomass, **3**–7
  relative, **11**–17
  thermal, **8**–1
  volumetric, **10**–11
EFG process, **20**–9
EGS, **23**–5
Electric double layer capacitor. *See* EDLC
Electric generators, **8**–14
Electric lighting, photosensor control of, **18**–117
Electric power generation
  environmental impact of, **23**–43
  from geothermal energy, **23**–2
Electric power grid, integration of wind turbines into, **21**–21
Electric storage, direct, **15**–7
Electric vehicles, **15**–3
Electric-lighting dimming controls, **18**–103
Electrical back contact, problems of in CdTe solar cells, **20**–45
Electrical capacity, **1**–4
Electrical energy, **15**–2
Electrical power, generation of, wind turbine subsystems, **21**–20
Electricity, **1**–3
  consumption, contribution of wind power to, **21**–1
  emissions from generation sources, **23**–44
  from renewable energy technologies, **15**–2
  generated from geothermal energy, price for, **23**–10
  generation
    parabolic trough collectors, **19**–33
    with solar thermal energy, **19**–45
  production, with concentration solar thermal technologies, **19**–3
  transmission, geothermal power, **23**–8
Electrochemical capacitors, **15**–7
Electrochemical energy storage, **15**–8
Electrochemical fuel cell stack, **26**–4
Electrode creepage, **26**–40
Electrode processes, fuel cell, **26**–25
Electrodeposition, **20**–46
Electrolytes
  loss of in MCFCs, **26**–40
  use of in fuel cells, **26**–2
Electrolytic hydrogen, **15**–13
Electromagnetic turbines, **27**–18
Electromotive force, **26**–7
Electron cyclotron resonance reactor. *See* ECR
Electron drift, **20**–4
Electron motive, **27**–1
Electron-hole pair, **20**–5
Electrons
  free, **27**–2
  saturation current of, **27**–3
Electrostatic precipitators, **13**–11
  particulate control with, **24**–24
Elemental mercury, **13**–13
ELOFLUX cells, **26**–32
Emissions, **1**–3
  carbon dioxide, **13**–6
  control of, **10**–16
  from pulverized coal plants, **13**–9
  limits for, **24**–24
  NO_x, **9**–11
  thermionic electron, **27**–2
  unburned hydrocarbon, **10**–14
Emittances, **A3**–4
Empirical correlation approach, solar system design using, **18**–39
Endogenous parameters, **18**–37
Eneas, Aubrey, **19**–3
Energy
  balance, geothermal, **23**–18
  conservation
    distinction from passive solar systems, **18**–59
    role of, **1**–7
  conversion, **1**–10
    efficiency
      definition of, **26**–12
      irreversible, forms of, **26**–22
    process, **15**–2
  crops, **15**–18
    biodegradability of, **22**–6
    high yield, **3**–2
    use of as feedstock for anaerobic digestion process, **22**–2
  density, **15**–5
  dumping, **18**–26
  forecast of future mix for, **1**–9
  global consumption of, **1**–1
  loss
    irreversible, mechanism of, **26**–19
    mechanisms, **26**–17
  production rate, **23**–7
  recovery, **22**–23
  solar, **5**–1, **12**–9
  storage systems, **15**–2 (*see also* storage; specific types of storage systems)
    characterization of, **15**–4
    electrochemical, **15**–8
    thermochemical, **15**–18
  use by sectors, **1**–3
  value, computation of, **22**–23
  wind, potential of, **6**–6
Energy Information Administration (EIA), **2**–7, **2**–16
Energy Return on Investment. *See* EROI
Engines
  compression ignition, **10**–1, **10**–4
  diesel, **10**–1, **10**–4
  efficiency, **9**–3
  four-stroke, **10**–2
  gas turbine, **22**–47
  internal combustion, **10**–1
  knock, **10**–13, **10**–15
  prime movers, **3**–8
  spark ignition, **10**–1
  Stirling, **12**–1

free-piston converters, **12**–6
kinematic, **12**–6
power control in, **12**–4
two-stroke, **10**–2
English prefixes for units, **A1**–2
Enhanced geothermal system. *See* EGS
Enrichment facilities, uranium, **16**–22
Enthalpy
effectiveness, **18**–131
fuel cells, **26**–6
Entrained
flow, **3**–11
flow reactor, **22**–44
Entropy fuel cells, **26**–6
EnviroDish, **19**–90
Environmental concerns
combined cycle power plants, **14**–8
geothermal constraints, **7**–4
MHD applications, **27**–18
wind turbines, **21**–26
WTE combustion, **24**–2
Enzymatic
hydrolysis, **22**–57
saccharification, **22**–53, **22**–55
EPR, **16**–9
Equalization charge, **20**–14
Equation of time, **5**–4
Equivalence ratio, **10**–12
Ericsson, John, **19**–3
EROI, **22**–64
ESBWR, **16**–9
ET-100 PTCs, **19**–45
Ethanol, **3**–8, **3**–13, **15**–20, **22**–51
cellulosic, sawdust as a source of, **15**–20
dry grinding for production of, **22**–54
from starch crops, **22**–53
polymeric composition of biomass feedstocks, **3**–5
Ethanol-gasoline blends, water-induced phase separation in, **22**–53
Euler equations, use of for analysis of turbines, **11**–14
EUROCIS consortium, CIS thin film technology innovations at, **20**–47
Eurodish concentrator, **19**–87
Europe
generation-III light water reactor development in, **16**–9
management of MSW in, **22**–29
supercritical plant design in, **13**–8
technical potential of CSP in, **19**–4
wind energy resources, **6**–5
European pressurized-water reactor. *See* EPR
European Union, industrial process heat demand in, **18**–50
EUROTROUGH project, **19**–42
collector design, **19**–49
Evacuated tubular collectors, **18**–12
use of for industrial process heat, **18**–51
Evaporative coolers, **18**–97
Excess air WTE systems, **24**–36
Excess steam production, **23**–7
Exchange current density, **26**–20
Excitation, **8**–15
Exogenous parameters, **18**–37
Expansion process, **9**–3
Exploration
drilling programs, **23**–16
geothermal resources, risk of, **23**–17
measuring resource chemistry during, **23**–22
External combustion
gas turbines, **9**–2
steam engines, **9**–2
External receivers, **19**–62
tubular, **19**–62
External reforming, **26**–40
Extraterrestrial solar radiation, **5**–2
Extrathermal ionization, **27**–16

F

F-chart design method, **18**–39, **18**–42
F-class machines, **13**–21
F-T liquids, **22**–51, **22**–62
Fabric filters, particulate control with, **24**–24
Faceted heliostats, **19**–58
Falling film cells, **26**–32
Faraday constant, **26**–7
Faraday field, **27**–16
Fast-breeder reactors, **16**–11
Fast-neutron reactors, **16**–11
Fatigue failure, **21**–16
Fatty acids, **22**–58
use of in phase-change drywall, **15**–17
Federal Energy Regulatory Commission (FERC) (U.S.), wheeling protocols, **23**–8
Feedstock, seasonal switching, **3**–3
Feedwater booster pumps, **8**–13
Feedwater heaters, in WTE facilities, **24**–24
Fermentation
lignocellulosic, **22**–57
syngas, **22**–62
Fibrous plant material, **22**–37
Finished petroleum productions, **2**–20
Finite-element codes, **21**–17
Finite-element-based equations of motion, **21**–17
Finland, uprating of nuclear plants in, **16**–6
Firming requirements, **23**–44
First-generation fuel cell technology, **26**–38. *See also* phosphoric acid fuel cells
First-level methods of design, **18**–65. *See also* simple design methods
Fischer-Tropsch liquids, **3**–8, **3**–15
Fission, **16**–1
nuclear, **17**–1
Fixed bed gasifiers, **3**–10
Fixed carbon content, **2**–2, **2**–4, **2**–15
Fixed costs, **23**–14
Fixed-blade propeller turbines, **11**–4
Fixed-pitch stall control, **21**–20
Fixed-wake model for wind turbine vorticity, **21**–14
Flame kernel, **10**–13
Flame propagation, **10**–13
Flaming combustion, **22**–38

Flash steam, geothermal, 7–5
Flash-steam systems, **23**–29
  CO_2 emissions from, **23**–45
  use of parabolic trough collectors with, **19**–30
Flat plate solar collectors, **18**–3, **18**–24
  absorbers, **18**–4
  determining individual hourly utilizability fraction for, **18**–17, **18**–19
  performance improvements, **18**–11
  use of for industrial process heat, **18**–51
  use of for solar process heat, **18**–55
Flexcell, development of flexible amorphous solar cells by, **20**–41
Flexible solar cell technologies, **20**–30, **20**–41, **20**–46, **20**–51. *See also* specific cell types
Float zone method of silicon ingot production, **20**–8
Flooded cell batteries, **15**–9
Flow battery technology, **15**–6, **15**–11
Flow reactors, entrained, **22**–44
Flow testing, geothermal resources, **23**–17
Flue gas
  cleaning equipment for, **24**–27
  desulfurization, **13**–2
  recirculated, use of as combustion air, **24**–21
Fluidized bed combustion, **22**–39, **24**–34, **24**–37
Fluidized bed gasifiers, **22**–44
Fluidized beds, **3**–8, **3**–10
  circulating, **13**–14
  combustion boilers, **13**–13
  pressurized, **9**–2, **13**–15
Flux-line trackers, **19**–19
Fly ash, **24**–23
  composition of from co-firing, **22**–41
  formation of dioxin/furan in, **24**–28
  metals in, **24**–23
Flywheel energy-storage systems, **15**–14
Foam volumetric absorbers, **19**–67
Food waste
  degradation of with high-solids anaerobic digestion, **22**–20
  use of as feedstock for anaerobic digestion process, **22**–2
Forced convection, **19**–60
Forestry, residues, degradation of with high-solids anaerobic digestion, **22**–20
Fossil fuels. *See also* specific types of fuels
  central station generation options, **13**–24
  energy demand for, **1**–1
  energy return from biomass-derived fuel, **22**–63
  savings of due to CSP plants, **19**–4
Fossil-fired plants, cost comparison with geothermal power plants, **23**–14
Four-stroke engines, **10**–2
Fracture permeability, **23**–4
Frameless PV modules, **20**–32
France
  LMFBR in, **16**–5
  Themis, **19**–62
Francia, Giovanni, **19**–3
Free ammonia toxicity, anaerobic digestion, **22**–16
Free electrons, **27**–2
Free energy conversion efficiency, **26**–22
Free liquid electrolyte fuel cells, **26**–32
Free radical reactions in MSW combustion, **24**–28
Free swelling index, **2**–6
Free-piston engines, **12**–6, **19**–88
Free-piston Stirling converters, **12**–6
Free-wake model for wind turbine vorticity, **21**–14
Fresnel reflectors, **18**–3, **19**–15
Front electrical contact
  CdTe solar cells, **20**–43
  CIGS solar cells, **20**–48
Fuel cell stack
  design considerations, **26**–27
  electrochemical, **26**–4
Fuel cell systems, **26**–3
  energy efficiency of, **26**–4
  thermodynamic model of, **26**–5
Fuel cells, **3**–8, **22**–47
  efficiency loss in, **26**–23
  efficiency of, **26**–15
    ideal, **26**–16
    practical, **26**–17
  electrode processes in, **26**–25
  integrated coal gasification, **13**–25
  operational characteristics and technological status of, **26**–31
  principle of operation for, **26**–2
  reversible energy conversion efficiency for, **26**–12
  types of, **26**–29
Fuel electrodes, **26**–2
Fuel gases, heating values of, **3**–10
Fuel handling, WTE facilities, **24**–9
Fuel processor, **26**–4
Fuel-air mixture, emissions control and, **10**–16
Fuel-air ratio, **10**–12
Fuel-capacity factor, **23**–12
Fuel/air ratio, importance of for gasifier performance, **22**–44
Fuels, **3**–8, **10**–17, **15**–6
  classification of, **2**–19
  comparison chart, **22**–52
  diesel, **10**–19
  dry prepared, **3**–7
  fusion, **17**–1
  loading, **3**–2
  metering control, **10**–12
  nuclear, sources of and world reserves, **4**–1
  properties of, **15**–6
  ultimate analysis of in WTE, **24**–13
Fuji, development of flexible amorphous solar cells by, **20**–41
Full computational fluid dynamics models. *See* CFD models
Full-span pitch controls, **21**–20
Fully-adjustable turbines, **11**–4
Fungi, as sources of triglycerides, **22**–59
Furans
  control of in emissions from MSW combustion, **24**–28
  in combustion ash, **24**–23
  measurement of, **24**–30
Furling, power limitation using, **21**–19
Furnace boilers, refuse characteristics for design of, **24**–3
Furnaces
  combustion, **24**–17
  configuration of, **24**–19

refractory wall, costs of, **24**–35
Fusion
energy conversion, **17**–2
fuel, **17**–1
magnetic, **17**–2

G

Gallium arsenide, **20**–29
use of in PV cells, **20**–3
Gamma-configuration, **12**–5
Gas bearings, **12**–8
Gas centrifugation, **16**–22
Gas turbine engines, **3**–8
Gas turbines, **9**–1
classes of for power generation, **22**–47
combustors, **9**–10
efficiency of, **9**–3
exhaust gas temperature, **14**–2
H series, **13**–22
materials used for, **9**–10
solar air preheating system for, **19**–84
steam-injected, **9**–8
Gas-cooled fast-reactor system. *See* GFR
Gas-cooled reactors, **16**–5
GFR, **16**–15
Gas-heating value, **22**–45
Gas-phase reactions, **22**–42
Gas-removal systems, **23**–37
Gas-turbine-modular helium reactor. *See* GT-MHR
Gaseous biofuels, **22**–51
Gaseous diffusion, **16**–22
Gaseous emission control, **24**–26
Gaseous transportation fuels, **22**–62
Gases, medium calorific value, **3**–10, **3**–12
Gasification of coal, **13**–2, **13**–16
Gasification technologies, **15**–20, **22**–41. *See also* thermal gasification
lignocellulosic feedstocks, **22**–62
WTE facilities, **24**–37
Gasifiers, **3**–10, **22**–42
commercialization of, **3**–12
fuel/air ratio in, **22**–44
producer gas composition from, **22**–43
thermodynamic efficiency of, **22**–45
types of, **3**–11
Gasoline, **2**–20, **10**–18, **22**–51
spark ignition engines, **3**–8
GAST project, **19**–62, **19**–80
GCC, **23**–32, **23**–42
comparisons of with other geothermal plant technologies, **23**–43
GCHPs, **7**-4, **23**–3
Generalized solar load ratio, **18**–91
Generalized yearly correlations, **18**–38
Generation IV International Forum, **16**–14
Generation-III nuclear power reactors, **16**–8
Generation-IV nuclear technologies, **16**–11
Generators
auxilliaries, **8**–15
electric, **8**–14
efficiency of, **21**–21
magnetohydrodynamic, **27**–16
steam, **8**–5 (*see also* boilers)
Geometric concentration ratio, **19**–6, **19**–23
Geometrical losses, **19**–24
Geophysical investigation, costs of, **23**–16
Geopressured-geothermal systems, **7**–2
Geoscientific studies, **23**–15
Geothermal, **7** – 1
applications, **7**–4
brine, pressurized, **23**–40
condenser gas-removal systems, **23**–37
heat pump, **7**-4
liquids, **23**–3
power generation capacity, **7**–2
reservoirs, **23**–4
residual brine management, **23**–20
steam supply, **23**–25
turbines, design of, **23**–36
wells, element chemistry of, **7** – 7
Geothermal combined cycle. *See* GCC
Geothermal energy, **15**–15
distributed grid technologies, **15**–3
operating conditions for, **7**-6
potential, **7** – 2
price for electricity generated from, **23**–10
renewability of, **23**–5
sources of, **23**–2
sustainability of, **23**–6
transmission of, **23**–8
types of systems, **7**-2
wellhead energy cost, **23**–6
worldwide generation of, **23**–3
Geothermal power
contract provisions for, **23**–11
efficiency of transmission usage compared to wind power, **23**–8
operations, uses of separated brine in, **23**–21
plants
capital costs of, **23**–13
contract provisions for, **23**–12
cost comparisons with fossil-fired plants, **23**–14
design parameters for, **23**–26
emissions from, **23**–44
expansions to, **23**–11
land use, **23**–45
production
binary power plant technologies, **23**–38
commercial requirements for, **23**–3
steam turbine technologies, **23**–32
worldwide, **23**–2
viable market for, **23**–9
Geothermal resources
characterization of, **23**–25
chemistry of, **23**–22
definition of, **23**–3
depth of, **23**–7
discovery process for, **23**–17
exploration and discovery of, **23**–15

improving through human intervention, **23**–4
management of
barriers to, **23**–23
complexities and issues, **23**–18
goals of, **23**–18
risk of exploration, **23**–17
Germany
dish-Stirling systems in, **19**–89
uprating of nuclear plants in, **16**–6
use of solar process heat in, **18**–55
GFDI, **20**–18
GFR, **16**–15
GFRP composites, use of for wind turbines, **21**–23
Gibbs free energy, **26**–6
conversion of solar heat to, **19**–2
Gieseler plastometer test, **2**–6
Glass transmittances, **18**–108
Glazings, **18**–51
Global efficiency, **19**–27
Global energy consumption, **1**–1
Global inventory of mercury, **13**–13
Global Market Initiative, **19**–91
Global Solar, CIS thin film technology innovations at, **20**–47
Glow discharge, **20**–11
Glow plugs, **10**–6
Glow-discharge CVD, **20**–36
Glowing combustion, **22**–38
Glucose, **22**–57
Glycerol, **22**–58
GM-100, **19**–57
Governors, control of turbine speed with, **11**–11
Grain starch, **22**–53
Graphite, **2**–2
Grapples, sizing of for WTE facilities, **24**–10
Grassroots geothermal development, contract provisions for, **23**–12
Grate-fired systems, **22**–39
Grates, **24**–17
Gray-King assay test, **2**–6
Greece, use of solar process heat in, **18**–55
Green energy, **3**–13
Greenfield development, geothermal power pricing and, **23**–10
Greenhouse effect, **18**–4
Greenhouse gas emissions
coal and, **2**–14
geothermal energy and, **7**–4
Grid. *See* electric power grid
Grid-connected solar systems, **19**–50
Grid-tied inverters, **20**–16
Grizzly screen, **24**–23
Gross calorific value, **2**–2, **2**–15
Ground fault detection and interruption device. *See* GFDI
Ground-coupled heat pumps. *See* GCHPs
Ground-reflected radiation, **18**–9
incident illuminance, **18**–105, **18**–107
GT-MHR, **16**–11
GUDE project, **19**–42

H

H series gas turbine, **13**–22
H turbines, **21**–3
H_2S abatement, **23**–36, **23**–45
Haber process, **22**–63
Half-life, **17**–1
Hall effect, **27**–16
Harmonic excitation, **8**–8
HAWTs, **21**–2
configuration of, **21**–3
dynamics codes for, **21**–17
momentum models for, **21**–10
performance of, **21**–12
small, structural dynamics of, **21**–16
yaw subsystem of, **21**–22
HDR, **23**–4
Head, **11**–1, **11**–5
Heat
as a geothermal resource, **23**–3
engines, **27**–5
flow, **7** – 1
pumps, ground-coupled, **7**–4
source, **27**–2, **27**–18
Heat demand, **18**–50
distribution of by temperature range, **18**–51
industrial processes, **18**–54
Heat engines
efficiency of, **26**–15
reversible energy conversion efficiency for, **26**–13
Heat exchangers, **8**–11
double-loop, heat collection decrease caused by, **18**–31
in binary power plants, **23**–38
in Stirling engines, **12**–3
optimization of, **18**–32
use of in closed-loop solar thermal systems, **18**–28
use of in thermosyphon systems, **18**–62
Heat loss coefficient, **19**–24
Heat recovery, **14**–2
assessment, **18**–52
steam generators, **9**–8, **14**–2
Heat transfer, analysis of, **18**–64
earth contact heating/cooling, **18**–100
Heat-rejection systems, design of, **23**–36
Heat-transfer fluid technology. *See* HTF technology
Heaters
closed, **8**–11
open, **8**–11
reflux, **12**–3
Heating
earth contact, **18**–99
thermoelectric, **27**–9
Heating and drying
in direct combustion, **22**–37
in thermal gasification, **22**–41
Heating systems, passive, **18**–59
Heating values, **15**–6
Heavy metal pollutants, removal of from MSW combustor emissions, **24**–32
Heavy-water reactors, **16**–10

Heber, expansion to, **23**–11
HELIOS, **19**–55
Heliostat field control systems. *See* HFCS
Heliostat fields, **19**–50
  tracking control, **19**–52
Heliostats, **18**–3
  cost reduction of, **19**–57
  development of, **19**–56
  drive characteristics, **19**–58
  glass-metal, **19**–75
  spacing of, **19**–55
Helium
  use of as a coolant in generation-IV nuclear reactors, **16**–15
  use of as a working gas in Stirling engines, **12**–2
Hemicellulose, **22**–56
Heterogeneous process of combustion, **10**–14
Hexose, **22**–57
HFCS, **19**–58
HFR, **23**–4
HHV, **3**–5, **15**–6, **24**–3
  conversion of to LHV, **24**–13
  efficiency of, **9**–3
  measurement of, **24**–11
High-cycle fatigue database, **21**–23
High-efficiency flat plate solar collectors, use of for industrial process heat, **18**–51
High-efficiency steam separation, **23**–31
High-enthalpy wells, **23**–7
High-level radioactive wastes, managing spent fuel from, **16**–27
High-pressure-ratio cycle, **9**–4
High-solids anaerobic digestion, **22**–10, **22**–20
High-sulfur coal, reserves of, **2**–6
High-temperature gas-cooled reactors. *See* HTGR
Higher heating value. *See* HHV
Hill diagram, **11**–7
HITREC-SOLAR project, **19**–67
  volumetric receivers used in, **19**–70
Hole drift, **20**–4
Honeycomb material, use of to improve collector performance, **18**–11
Horizontal illuminance, **18**–107
Horizontal sky illuminance, **18**–113
Horizontal windmills, **21**–3. *See also* HAWTs
Horizontal-axis machine, **21**–2
Horizontal-axis wind turbines. *See* HAWTs
Hot dry rock. *See* HDR
Hot dry rock geothermal systems, **7**–2
Hot fractured rock. *See* HFR
Hot pistons, **12**–4
Hot streams, **18**–52
Hot-wire CVD. *See* HWCVD
Hottel-Whillier-Bliss equation, **18**–6
Hour angle, **20**–13
HTF technology, **19**–29, **19**–33, **19**–45, **19**–49. *See also* thermal oils
HTGR, **16**–10
Humid air turbine, **9**–9
Humid air turbines, **13**–22
Humidity effectiveness, **18**–129
Humus material, **22**–24. *See also* biosolids
  characteristics of, **22**–27
  production of from BOF/MSW
    biological characteristics of, **22**–30
    chemical characteristics of, **22**–29
  uses for, **22**–30
  utilization of, **22**–28
HWCVD, **20**–36
Hybrid cooling systems, **18**–132
Hybrid models for wind turbines, **21**–13, **21**–15
Hybrid PV systems, **20**–15
Hybrid solar cells, **20**–40
Hybrid systems, **18**–59, **19**–75
Hybrid-electric vehicles, **1**–5, **15**–3
  use of Stirling engines in, **12**–9
Hydration-dehydration, **15**–17
Hydraulic efficiency, **11**–9
Hydraulic turbines, description of, **11**–1
Hydrocarbon fuels, electricity generation from, **26**–35
Hydrocarbon-based binary power plants, **23**–40
Hydrocarbons, **2**–24
  control of in flue gas emissions, **24**–27
  unburned, **10**–15
    emissions of, **10**–13
Hydrogasification, **22**–63
Hydrogen, **1**–5, **22**–51
  electrolytic, **15**–13
  manufacture of from syngas, **22**–63
  storage technologies, **15**–13
  use of as a working gas in Stirling engines, **12**–2
  use of for generator cooling, **8**–14
  use of in fuel cells, **26**–25
Hydrogen chloride, removal of from flue gas, **24**–27
Hydrogen fluoride, removal of from flue gas, **24**–27
Hydrogen-air fuel cells, **26**–36
Hydrogen-oxygen fuel cells, **26**–36
Hydrogenation reaction, **22**–42
Hydrolysis
  lignocelluose, **22**–57
  starch, **22**–53
Hydrolytics, **22**–10
Hydropower installations, components of, **11**–1
Hydrothermal, resources, **7**–2, **7**–61

I

Iceland, geothermal power production in, **23**–2
Idaho, Posted Rate 10 MW PURPA contracts, effect of on geothermal power plants, **23**–10
Ideal diesel cycle, **10**–8
Ideal fuel cell efficiency, **26**–16
Identified accessible base, **7**–3
IEA-SSPS-CRS project, **19**–62
IGCC, **3**–8, **13**–17, **13**–26, **14**–1
  Buggenum, **13**–19
  Polk County, **13**–18
IGCC power, use of biomass in, **22**–48
IGHAT plants, **13**–23
Ignition
  delay period, **10**–5, **10**–14

quality, **10**–19
surface, **10**–13
Ignition temperature, combustible substances in MSW, **24**–13
III-V materials, multijunction solar cells based on, **20**–29. *See also* specific materials
ILGAR, deposition on CIGS solar cells using, **20**–48
Illuminance, **18**–118
measurement of, **18**–102
workplane, **18**–104
Impulse turbines, **11**–2
performance characteristics, **11**–8
In situ leaching, **4**–2
In-core power systems, **27**–6
In-vessel anaerobic digestion process, **22**–4. *See also* anaerobic digestion
commercial-scale technologies, **22**–28
utilization of byproducts from, **22**–24
Incentives, photovoltaics, **20**–23
Incidence angle
direct solar radiation, **19**–26
modifier, **18**–7, **18**–9, **19**–27
Incident direct sky illuminance, **18**–105
Incident ground-reflected illuminance, **18**–105
Incident radiation, annual, **19**–3
Incineration and injection process, geothermal, **7**—5
Independent power producers (IPPs), **23**–9
Index tests, **11**–18
India
electrical capacity additions in, **1**–4
fast-breeder reactors in, **16**–11
increase in energy use in, **1**–1
wind energy installations in, **21**–1
Indicated mean effective pressure, **10**–10
Indicated resources, **2**–8
Indicated work, **10**–10
Indicator diagram, **10**–10
Indirect coupling, **18**–28
Indirect gain passive solar systems, **18**–62
Indirect injection, **10**–5
Indirect parabolic dish receivers, **19**–87
Indirectly heated gasification, **22**–42, **22**–45
Indirectly heated receivers, **19**–62
INDITEP project, **19**–43
Indium phosphide, **20**–29
Individual hourly utilizability, **18**–15
Indonesia
geothermal power production in, **23**–2, **23**–33
power price for geothermal energy, **23**–11
Induction generators, **21**–20
Industrial process heat applications, **19**–29
integration of solar thermal systems in, **18**–51
parabolic trough collector designs for, **19**–20
system design for, **18**–42
use of solar thermal systems for, **18**–26, **18**–49
Industrial production processes, integration of solar heat into, **18**–52
Industrial sector
biomass applications in, **3**–6
geothermal energy use for, **23**–3
potential for solar thermal systems in, **18**–50
Industrial solar systems, **18**–24
Inferred resources, **2**–8
Injection process, **19**–40
Injection wells, **23**–20
Inoculation, **22**–3
Insolation, **5**–2
daily extraterrestrial on a horizontal surface, **A2**–10
effect of day-to-day changes in, **18**–14
Insulation, thermal properties of materials for, **A3**–7
Intake pressurization, **10**–20
Integrated gasification combined cycle technology. *See* IGCC
Integrated gasification humid air turbine plants. *See* IGHAT plants
Integrated solar combined cycle system. *See* ISCCS
Intercooled recuperated gas turbines, **9**–7
Intercooled steam-injected gas turbine. *See* ISTIG
Intercoolers, **9**–7
Intercooling, **10**–22, **13**–24
Interelectrode motive distribution, **27**–2
Intermediate-level radioactive wastes, **16**–27
Intermittent flow, **19**–39
Internal combustion coal-fired gas turbine, **9**–2
Internal combustion engines, **3**–8, **9**–2, **10**–1
actual cycles of, **10**–10
use of ethanol in, **22**–53
Internal reforming, **26**–40
International reactor innovative & secure. *See* IRIS
International system of units. *See* SI
International Thermonuclear Experimental Reactor. *See* ITER
Inverters, **20**–16
bypass switch at, **20**–18
Ion layer gas reaction. *See* ILGAR
Ionic mercury, **13**–13
Ionization, thermal, **27**–16
Ionomers, use of for PEMFC, **26**–33
IRIS, **16**–9
Iron making, **2**–6
Irradiance
circumsolar, **19**–11
fluctuation in, **19**–3
Irreversible energy conversion efficiency, **26**–44
forms of, **26**–22
Irreversible energy loss, mechanism of, **26**–19
ISCCS, **19**–49, **19**–84
ISET, CIGS solar cells developed by, **20**–51
Isolated gain passive solar systems, **18**–64
ISTIG, **22**–48
Italy
geothermal power production in, **23**–2
wind energy installations in, **21**–1
ITER, **17**–2

J

J.W. Aidlin Plant, incomplete resource information at, **23**–23
Japan
fast-breeder reactors in, **16**–11
generation-III light water reactor development in, **16**–9

LMFBR in, **16**–5
ultrasupercritical plant design in, **13**–7
wind energy installations in, **21**–1
Jet fuel, **2**–20
Junction activation treatment, **20**–44

K

Kamojang, geothermal steam fields in, **23**–4
Kaplan propeller turbines, **11**–4
Kenya, geothermal power production in, **23**–2
Kerosene, **2**–20
Kinematic engines, **19**–88
Kinematic Stirling engines, **12**–6
Klimep, **10**–19
Knitted-wire volumetric absorbers, **19**–67
Knock, **10**–13
in CI engines, **10**–15
Knock-limited indicated mean effective pressure. *See* klimep
Kockums 4-95 Stirling engine-based PCU, **19**–89
KOH electrolyte, **26**–32
Kompogas system, **22**–33
Koppers-Totzek gasifier, **3**–11
Korea. *See* South Korea
Korean next generation reactor, **16**–9

L

LAER, **24**–38
Land use, geothermal power plants, **23**–45
Landfills
biodegradation in, **22**–4
disposal of WTE residue at, **24**–23
leaching characteristics of MSW residues in, **24**–23
Laptops, use of PEMFCs to power, **26**–35
Larderello site
geothermal power plant at, **23**–2
geothermal steam fields at, **23**–4
incomplete resource information at, **23**–23
Large eddy simulation method, **11**–14
Last-stage turbine blades, **23**–36
Latent heat storage of, **15**–6
Latent heat, storage of, **15**–17
Law of detailed balance, **20**–4
LCR, **18**–67
method of design, **18**–73
reference systems for, **18**–74
tables for representative cities, **18**–77
Leachate, characteristics of in landfills and monofills, **24**–23
Lead losses, **27**–3
Lead, in combustion flue gas, **24**–31
Lead-acid batteries, **15**–8
flooded, charging in V arrays, **20**–14
Lead-cooled fast reactor. *See* LFR
Legislation, emissions, **24**–38
Lenses, **19**–13
LFR, **16**–18
LHV, **3**–5, **13**–2, **18**–6, **24**–12
conversion to from HHV, **24**–13
efficiency of, **9**–3
Lift devices, translating, **21**–7
Lift line model, **21**–13
Lift surface model, **21**–13
Light dependent degradation, **20**–37. *See also* SWE
Light loss factor, **18**–108, **18**–114
Light trapping, **20**–39
Light-water graphite-moderated reactors, **16**–6
Light-water reactors, **16**–8
Lighting
controls, **18**–117
terms and units, **18**–102
Lignin, **22**–4, **22**–56
measurement of in organic wastes, **22**–5
pyrolysis of, **22**–37
solubilization of, **22**–12
Lignite, **2**–2, **13**–3
deposits of, **2**–6
Lignocellulosic crops
ethanol from, **22**–56
fermentation of, **22**–57
gasification of, **22**–62
hydrolysis of, **22**–57
pretreatment of, **22**–56
Lignocellulosics, **3**–6
liquid fuels and bioproducts from, **3**–14
Limb-darkened distribution, **19**–10
Lime
removal of heavy metal pollutants using, **24**–32
use of for removal of trace organic pollutants, **24**–31
use of to clean flue gas, **24**–27
Limestone scrubbers, **13**–9
Limiting current, **26**–20
Line-axis concentrating collectors, use of for industrial process heat, **18**–51
Linear current boosters, **20**–15, **20**–19
Linear Fresnel reflector systems, **19**–15. *See also* Fresnel reflectors
integration of combined-cycle plants, **19**–17
Linear seal, **12**–6
Linoleic acids, **22**–58
Linolemic acids, **22**–58
Liquefied natural gas. *See* LNG
Liquefied refinery gases, 2–20. *See also* LNG
Liquid desiccants, **18**–127
physical properties of, **18**–130
Liquid electrolyte fuels, **26**–25
Liquid KOH electrolyte, **26**–32
Liquid sodium, use of as a thermal storage fluid, **19**–73
Liquid-desiccant cooling systems, **18**–128
Liquid-filled solar collectors, **18**–2, **18**–33
absorbers used in, **18**–5
advantages and disadvantages of, **18**–3
testing of, **18**–8
Liquid-metal-cooled fast-breeder reactors. *See* LMFBR
Liquids
sensible heat storage in, **15**–15
storage of thermal energy in, **18**–33
Lithium bromide-water systems, **18**–124
double effect, **18**–126
Lithium ion batteries, **15**–9
Lithium polymer batteries, **15**–9

LMFBR, **16**–9
LMMHD, **27**–18. *See also* magnetohydrodynamics
LNG, **2**–21, **2**–24
Lo-Cat H_2S abatement, **23**–45
Load analysis
  rule of thumb estimates, **18**–67
  wind turbines, **21**–16
Load collector ratio. *See* LCR
Load distribution, predictions of, **21**–14
Local controllers, types of for parabolic trough concentrators, **19**–18
Logging residues, **3**–2
Long-term batch digestion, **22**–5
Longwall mining, **2**–14
Los Azufres, pressurized brine power plant at, **23**–41
Low activation advanced materials, **17**–1
Low specific speed device, **11**–2
Low-level radioactive wastes, **16**–27
Low-pressure-ratio cycle, **9**–4
Low-solids anaerobic digestion, **22**–19
Low-sulfur coal, **2**–12
  reserves of, **2**–6
Lower heating value. *See* LHV
Lower parasitics, **23**–38
Lowest Achievable Emission Requirements. *See* LAER
LPG, **22**–63
LS-1 collectors, characteristics of, **19**–48
LS-2 collectors, characteristics of, **19**–48
LS-3 collectors, **19**–49
  characteristics of, **19**–48
  thermal loss from, **19**–24
Lubricants, **2**–20
Lumen method of sidelighting, **18**–105. *See also* sidelighting
Lumen method of skylighting, **18**–113
Lumen, definition of, **18**–102
Luminous fluxes, **18**–108
Lumped mass equations of motion, **21**–17
LUZ International, **19**–3, **19**–45

M

Macerals, **2**–1, **2**–15
MACT, **24**–38
Magmatic systems, **7**–2
Magnetic, fusion, **17**–2
Magnetohydrodynamics, **27**–15
  generator configuration and loading, **27**–16
Magnox reactors, **16**–5
Mak-Ban, pressurized brine power plant at, **23**–41
Make-up wells, **23**–7
Manure
  degradation of with high-solids anaerobic digestion, **22**–20
  use of in inoculation, **22**–3
Martin grates, **24**–17, **24**–20
Mass burn furnaces, sizes of, **24**–17
Mass-burn plants, **24**-6
  effect of plant capacity on capital costs for, **24**–35
  effect of plant capacity on O&M costs for, **24**–36
  performance of, **24**–32
  refuse storage at, **24**–9
Mass-fired waterwall units, **24**–21
Materials handling systems, selection of for WTE facilities, **24**–3
Materials, low activation advanced, **17**–1
Matrix fuel cells, **26**–32
Maximum Achievable Control Technology. *See* MACT
Maximum power point of the cell, **20**–7
Maximum power tracker, **20**–16
Maximum turbine inlet temperature, **9**–9
Measured resources, **2**–7
Mechanical availability, power generation plants, **23**–11
Mechanical energy, **15**–2
  storage, **15**–13
Mechanical work, conversion of solar heat to, **19**–2
Medium calorific value gases, **3**–10
Medium temperature collectors, development of for industrial process heat, **18**–51
Memory effect, **15**–8
  nickel-cadmium batteries, **15**–10
Mercury
  control of, **13**–13
  removal of from flue gas, **24**–31
Mesophilic digestion, **22**–20
Metal absorbers, **19**–65
  volumetric, **19**–67
Metal chalcogenides. *See also* CdTe
  use of on solar cells, **20**–41
Metal emissions, **24**–32
Metal organic chemical vapor deposition. *See* MOCVD
Metallurgical coals, **2**–6
Metals
  in combustion ash, **24**–23
  thermal properties of, **A3**–6
METAROZ project, **19**–81
Methane, **22**–51, **22**–63
  production of from biological processes, **22**–7
Methanogens
  morphological characteristics of, **22**–10
  nutrient requirements for, **22**–10
  two-stage digestion, **22**–20
Methanol, **26**–36
  direct carbonylation of, **22**–61
  from syngas, **22**–60
Methyl tertiary butyl ether. *See* MTBE
Mexico, pressurized brine power plant in, **23**–41
Micromorph solar cells, **20**–35
Microorganisms
  action of, **3**–5
  as sources of triglycerides, **22**–59
Microturbines, **9**–1
Mid-level design methods, **18**–39, **18**–65
Miller cycle engine, **10**–1
Milling process, uranium, **16**–22
Minable coal, **2**–8
Mine drainage, **2**–14
Mine tailings, radioactive, **16**–26
Mineral concentrations, geothermal reservoirs, **23**–23
Mineral matter, **2**–4
Mining

coal, environmental aspects of, **2**–14
uranium, **4**–2, **16**–21
Miravalles, pressurized brine power plant at, **23**–41
Mixed conduction, **26**–43
Mobile electronics, use of PEMFCs to power, **26**–35
MOCVD, **20**–46
deposition on CIGS solar cells using, **20**–48
Modal equations of motion, **21**–17
Modal formulations, wind turbines, **21**–17
Model testing, hydraulic turbines, **11**–12
Modeling, wind turbine aerodynamics, **21**–9
Modular WTE units, **24**–7, **24**–36
construction costs for, **24**–34
particulate emissions from, **24**–31
Moisture content
biomass, effect of on direct combustion, **22**–40
solid waste, **24**–5
Moisture content of coal, **2**–4
Molten carbonate fuel cells, **26**–2, **26**–39, **26**–44
acceptable contamination levels for, **26**–40
basic operating principle, **26**–39
energy conversion efficiency of, **26**–39
technological problems of, **26**–40
use of in baseload utility generators, **26**–41
Molten salt technology, **19**–17, **19**–74
use of in the Solar Tres project, **19**–80
Molten salt thermal storage, **19**–49
Molten salt tubular receivers, **19**–64
Molten salts, use of in dual-medium thermal storage systems, **19**–38
Molten-salt reactor. *See* MSR
Molybdenum, use of as a back contact on CIGS cells, **20**–48
Momentum model, **21**–10
limitations of, **21**–14
Monofills, leaching characteristics of MSW residues in, **24**–23
Monolithic integration, **20**–30
Monolithic modules, **20**–40
Monolithic solid storage, **15**–17
Monthly clearness index, **18**–16. *See also* clearness index
Mouchot, Augustin, **19**–3
MOX fuel, **16**–23
MSR, **16**–19
MTBE, **22**–61
Multiblade water pumpers, **21**–2
Multibody dynamic codes, **21**–17
Multicrystaline silicon PV cells, manufacture of, **20**–9
Multieffect absorption systems, **18**–126
Multiflash steam separation, **23**–29, **23**–43
Multijunction solar cells, **20**–2
Multijunction-thin-film fabrication, **20**–11
Multipass systems
design methods for, **18**–38
design of, **18**–43
use of closed-lop solar systems for, **18**–25
Multiple stream-tube model, **21**–12
Multiple-junction solar cells
architecture of, **20**–40
development of, **20**–39
Multishaft CC arrangements, **14**–6
Municipal solid waste
analysis of, **24**–5
classification of combustion emissions from, **24**–24
combustible portion of, **24**–11
combustion
calculations, **24**–16
dioxin/furan emission from, **24**–29
process, **24**–16
trace metals in, **24**–31
combustors, gaseous emission factors for, **24**–26
composition of, **22**–3
generation of in US, **24**–2
heat content of, **24** – 5
higher heating value of, **24**–3 (*See also* HHV)
ignition temperature of combustible substances in, **24**–14
incineration of, potential energy from, **24**–1
incinerators, grates in, **24**–17
methane yields from, limitation of by polymeric lignin, **22**–12
refractory volatile solids in, **22**–5
residue handling and disposal, **24**–22
seasonal variations in, **24**–3
storage of at WTE facilities, **24**–9
ultimate analysis of, **A4**–2
use of as feedstock for anaerobic digestion process, **22**–2
use of combustion process to treat, **24**–2
Municipal wastewater. *See also* wastewater
anaerobic digestion of, **22**–19

N

N-type doping control, **20**–42
Naphtha, **2**–20, **10**–19
Natural circulation boilers, **8**–6
Natural convection/ventilation, **18**–95, **19**–60
Natural gas, **2**–16, **2**–24
combined cycle power plants, **14**–1, **14**–4
electricity production using, **19**–46
energy demand for, **1**–1
liquids, **2**–20
production measurement, **2**–23
reserves and resources, **2**–21
Natural gas combined cycle power plants, use of solar multiples with, **19**–3
Navier-Stokes equations, use of with CFD models, **21**–14
NCG, **23**–45
in geothermal systems, **23**–37
removal of from binary power plants, **23**–38
Neat alcohol, **22**–53
Nernst loss, **26**–12, **26**–23, **26**–44
Net building load coefficient. *See* NLC
Net energy savings, passive systems, **18**–60
Net primary productivity, **15**–19
Net skylight transmittance, **18**–113
Netherlands
IGCC plants in, **13**–19
wind energy installations in, **21**–1
Nevada, geothermal power plants in, **23**–41
New Mexico Wind Energy Center, **21**–24
New Source Performance Standards, **22**–40

New source performance standards, **3–7**
New Zealand
  geothermal power production in, **23–2**
  pressurized brine power plant in, **23–41**
Next-generation nuclear technologies, **16–8**
  flexible fuel requirements of, **16–10**
Nickel metal hydride batteries, **15–10**
Nickel-cadmium batteries, **15–9**
Nitrate-based molten salts, use of as a thermal storage fluid, **19–73**
Nitrogen, conversion of to ammonia, **22–8**
NLC, **18–66**
No-storage solar systems, **18–24**
Nocturnal cooling systems, **18–97**
Noise
  control of near WTE facilities, **24–8**
  wind turbines, **21–26**
Non-condensable gases. *See* NCG
Nonconcentrating collectors, **18–3**
Noncondensible gas, **7-5**
Nonequilibrium ionization, **27–16**
Nonimaging concentrators, **18–13**
Nontracking solar collectors, **18–24**
  types of, **18–3**
Nonvented Trombe-wall systems, **18–75**
Normal combustion process, **10–13**
Normal incident radiation, annual, **19–3**
North fields, **19–53**
NO_x
  control of in flue gas emissions, **24–27**
  emissions, **9–11**, **10–15**
    control of, **13–11**
    from pulverized coal plants, **13–88**
  emissions of from dry steam plants, **23–45**
NREL, thin film technology innovations at, **20–47**, **20–51**
Nuclear fission, **17–1**. *See also* fission
Nuclear fuels
  processing of, **4–2**
  sources and reserves of, **4–1**
Nuclear power, **2–16**, **4–1**, **16–1**
  economics of, **16–29**
  fuel cycle, **16–20**
  generation-III reactors, **16–8**
  generation-IV reactors, **16–11**
  growth of, **16–6**
  passive safety systems, **16–8**
  plant-life extension, **16–6**
  reactor technologies (*see also* reactors)
    development of, **16–2**
    under construction, **16–7**
    worldwide units, **16–3**
  steam-fired plants, **8–2**
  thermal reactors, advanced, **16–12**
Nuclear waste, **16–26**
Nuisance conditions, control of near WTE facilities, **24–8**
Numerical simulation
  analysis of turbines using, **11–14**
  solar system modeling using, **18–34**
Nutrient deficiency, anaerobic digestion, **22–16**

O

Ocean thermal energy conversion (OTEC), **25–1**
Octane
  number, **10–20**
  rating, **10–18**
  research number, **10–18**
Odors, control of near WTE facilities, **24–8**
Off-peak electricity use, **19–48**
Ohmic polarization, **26–20**
Oil, **1–5**, **2–16**
  energy demand for, **1–1**
  world refining capacity, **2–21**
Oilseed, **22–59**
Oleic acids, **22–58**
Once-through boilers, **8–6**
Once-through cycle, **16–20**
Once-through process, **19–40**
  control of, **19–43**
Once-through superheated water-steam receivers, **19–63**
One-day repetitive method, solar system design using, **18–39**
Open cycle, OTEC power, **25–1**
Open fuel cycle, **16–20**
Open heaters, **8–11**
Open pit mining, uranium, **4–2**
Open ponds, **18–97**
Open-loop Rankine cycle configuration, **23–32**
Open-loop solar thermal systems
  definition of, **18–25**
  design methods for, **18–38**
  single-pass, system design, **18–44**
Open-loop volumetric air technologies, **19–17**, **19–74**
Open-pit technology, uranium mining, **16–22**
Operating costs, geothermal power plants, **23–14**
Optical concentration, **19–9**
Optical efficiency, **19–53**
  mapping of, **19–56**
Optical losses, **19–24**
Optical tower height, **19–52**
Optimum pressure ratio, **9–5**
Orbital combustion process, **10–4**
ORC, **23–32**
Organic chemicals, in combustion ash, **24–23**
Organic compound control, MSW combustion, **24–28**
Organic loading rate, **22–7**
Organic overloading, anaerobic digestion, **22–16**
Organic Rankine cycle. *See* ORC
Organic solar cells, **20–31**, **20–51**
Organic substrates
  biodegradable volatile solids of, **22–4**
  complex, pathway of in anaerobic digestion, **22–12**
Organic wastes
  biodegradability of, **22–6**
  use of as feedstock for anaerobic digestion process, **22–2**
Orimulsion, **13–3**
Ormat Technologies, binary power plant technology developed by, **23–40**
ORMESA I plant, **23–40**
Oscillating water column, **25–3**
Otto cycle

engine, **22**–51
ideal, **10**–7
supercharged *vs.* naturally aspirated, **10**–20
Otto spark ignition engine, **3**–8. *See also* Otto cycle
Out-of-core power systems, **27**–6
Overall free energy conversion efficiency, **26**–22
Overfire air, **24**–20
Overpotential, **26**–19
Overvoltage, **26**–19
Oxidants, in fuel cells, **26**–3
Oxidation converters, **10**–17
Oxides of nitrogen, **10**–15. *See also* NO_x
Oxygenators, **10**–19

P

P-i-n configuration, **20**–36
instability of, **20**–39
P-n junction, **20**–3, **20**–36
illuminated, **20**–5
P-type doping control, **20**–42
Packed-bed dehumidifier, **18**–129
PAHs
control of in emissions from MSW combustion, **24**–28
in combustion ash, **24**–23
Pancaked charges, **23**–8
Paper and paperboard products
percentage of in municipal solid waste, **24**–3
use of as feedstock for anaerobic digestion process, **22**–2
Parabaloids, **18**–3
Parabolic concentrators, sizing of, **19**–13
Parabolic dish receivers, **19**–87
Parabolic trough collectors, **18**–3, **19**–15, **19**–17
annual thermal efficiency of, **19**–49
connections used in, **19**–22
efficiencies and energy balance in, **19**–27
electricity generation with, **19**–33
geometrical losses in, **19**–24
industrial applications for, **19**–29
net output thermal power, **19**–28
operational principles and components of, **19**–18
optical losses in, **19**–24
orientation of, **19**–21
performance parameters and losses in, **19**–23
receiver pipes for, **19**–19
solar fields, sizing and layout of, **19**–31
solar reflectors used in, **19**–20
state-of-the-art solar power plants using, **19**–45
thermal loss in, **19**–24
thermal storage systems for, **19**–35
use of for industrial process heat, **18**–52
Parabolic trough power plants, **19**–18
elements of, **19**–33
Paraffins, use of in phase-change drywall, **15**–17
Parallel fuel cell connections, **26**–27
Parallel-powered combined cycle power plants, **14**–1
Parallel-sided steam turbine blades, **8**–8
Partial depletion effect, **18**–26
Partial oxidation, **15**–20
reactors, **3**–10
Partial-span control, **21**–20
Particle interaction losses, **27**–3
Particle receivers, **19**–62
Particulate control, **24**–24
Particulate removal efficiency, **24**–32
Particulates, **10**–16
emissions of from geothermal power plants, **23**–45
emissions of from pulverized coal plants, **13**–9
Passive heating systems
design approaches, **18**–65
reference systems for, **18**–74
types of, **18**–62
Passive pitch control, **21**–20
Passive safety systems, **16**–8
Passive solar systems, **18**–23
costs, **18**–60
distinction from energy conservation, **18**–59
LCR method of design for, **18**–73
Passive solar water heaters, **18**–62
Passive space cooling, design of, **18**–94
Payback period, wind energy, **21**–24
PCBs
control of in emissions from MSW combustion, **24**–28
in combustion ash, **24**–23
PCDDs, control of in emissions from MSW combustion, **24**–28
PCDFs, control of in emissions from MSW combustion, **24**–28
PCU systems, **19**–85
Peak electricity use, **19**–48
Peak optical efficiency, **19**–24, **19**–27
Peak power limitations, wind turbines, **21**–18
Peak sun hours, **20**–13
Pebble-bed modular reactors, **16**–11
Pebble-bed storage systems, **18**–33
PECVD, **20**–36
Peltier effect, **27**–8
Pelton-type impulse turbines, **11**–2
PEM fuel cell, **26**–33, **26**–44
acceptable contamination levels for, **26**–34
basic operating principle, **26**–34
technological issues of, **26**–35
PEMFC, **26**–38
Pennington cycle, **18**–127
Pentane, use of as power cycle working fluid, **23**–39
Pentoses, **22**–57
Per capita energy use, **1**–8. *See also* energy
Permanent magnet generators, **21**–21
Permeability, **23**–4
Persia, early windmills in, **21**–2
Petrochemical feedstock, **2**–20
Petroleum, **2**–16
coke, **13**–15
sulfur emissions from, **7**–5
Phase change materials, storage of thermal energy in, **18**–33
Phase-change materials, latent heat storage in, **15**–17
Phibar f-chart method, **18**–43
Phibar method, **18**–19
PHOEBUS, **19**–67, **19**–80
Phosphoric acid fuel cells, **26**–2, **26**–33, **26**–38
costs of, **26**–39

Photon capture, **20**–5
Photosensors, **18**–117
Photosynthesis, **3**–1
efficiency of, **3**–1
Photovoltaic conversion, **20**–1
Photovoltaic effect, **20**–6
Photovoltaic panels. *See also* PV cells
costs of, **20**–2
Photovoltaics, **15**–2. *See also* PV systems
conversion efficiencies for, **20**–8
distributed grid technologies, **15**–3
latest developments in, **20**–24
space applications of, materials for, **20**–29
thin-film technology for, **20**–28 (*see also* thin films; specific technologies)
Phthalates, in combustion ash, **24**–23
Physical constants, **A1**–3
Physical vapor deposition, use of on CIGS solar cells, **20**–48
PIC, **3**–7
Pig manure, use of in inoculation, **22**–3
Pinch analysis, **18**–52
Pinch point, **14**–3
Pipe heat transfer, **18**–100
Pipelines
scaling, **23**–22
steam, operation of, **23**–30
Piping losses, **8**–4
Piston/displacer configurations of Stirling engines, **12**–4
Pistons, **12**–4
seals, **12**–8
Pit turbines, **11**–5
Pitch control, **21**–19
Pitch-regulated wind turbines, **21**–19
Pitch-to-feather control, **21**–19
Pitch-to-stall control, **21**–19
Pitting loss, **11**–12
Planar reflectors, use of to improve collector performance, **18**–12
Planar thermionic converters, **27**–6
Plant capacity factors, **23**–11
Plant matter
fibrous, **22**–37
thermochemical energy storage in, **15**–18
Plant polymers, biomass from, **3**–13
Plant-life extension, **16**–6
Plasma, **17**–4
Plasma-enhanced chemical vapor deposition. *See* PECVD
Plug-flow reactors, **22**–19
Plug-in hybrid vehicles, **15**–3
Plutonium, **16**–23
use of in GT-MHR reactors, **16**–11
Polar tracking, **19**–87
Polarization, **26**–19
Pollution
concerns regarding WTE combustion, **24**–2
control requirements for WTE combustion components, **24**–11
Polychlorinated biphenyls. *See* PCBs
Polychlorinated dibenzodioxins. *See* PCDDs
Polychlorinated dibenzofurans. *See* PCDFs
Polycrystalline ingot production, **20**–10
Polycyclic aromatic hydrocarbons. *See* PAHs
Polyelectrolytes, use of for PEMFC, **26**–33
Polymer electrolyte membrane fuel cells, **26**–2
Polymer sealing rings, **12**–8
Polymeric composition, biomass, **3**–5
Polysulfide bromide batteries, **15**–12
Polytropic efficiency, **9**–3
Porous fuel cell electrodes, **26**–25
Porta-Test Whirlyscrub V Gas Scrubber, **23**–31
Portable electronics, use of PEMFCs to power, **26**–35
Positive displacement type supercharger, **10**–20
Post-electrochemical surface migration, **26**–26
Postcombustion, **13**–5
options for control of nitrogen oxide emissions, **13**–11
Powder River Basin, low-sulfur coal from, **2**–13
Power
coefficient, **21**–5
conditioning, **27**–17
control, spark ignition engines, **10**–10
conversion systems, **19**–33
cycles
binary thermodynamic, **8**–5
bottoming, **8**–5
Brayton, **9**–4
Carnot, **8**–3
high-pressure-ratio, **9**–4
low-pressure-ratio, **9**–4
OTEC, **25**–1
Rankine, analysis of, **8**–2
regenerative, **8**–4
simple, **9**–4
topping, **8**–5
delivered, price of, **23**–10
density, **10**–4, **15**–6
generation
fuels for, **13**–2
thermoelectric, **27**–7
generation boilers, **3**–8
generation systems, gas turbine combined cycle, **9**–1
plants
combined cycle, **14**–1
combustion turbine-based, **13**–20
Cool Water, **13**–16
gasification, **13**–16
geothermal, capital costs, **23**–14
pulverized-coal, **13**–5
purchase agreement, **23**–15
quality, **15**–3
tidal, **25**–2
towers, **19**–15, **19**–50, **19**–58
calculation of convective heat loss for, **19**–61
experimental, **19**–74
integration of into combined cycle plants, **19**–75
operating temperature and flux ranges of, **19**–63
technology for, **19**–17
turbines, **11**–4
wave, **25**–2
Precombustion, **13**–4
chambers, **10**–5
Precombustors, **3**–12
Precursor materials, selinization of, **20**–49

Prefixes for units, English, **A1–2**
Premixed homogeneous combustion, **10–3**
Prescribed-wake model for wind turbine vorticity, **21–14**
Pressure
  boost, **10–21**
  effect of on reversible cell potential, **26–9**
Pressure drop
  in solar collectors, **18–11**
  wellhead, **23–25**, **23–30**
Pressure ratio, optimum, **9–5**
Pressure-time technique, **11–17**
Pressurized fluidized bed combustion, **13–15**
Pressurized fluidized bed combustion system, **9–2**
Pressurized geothermal brine, **23–40**
Pressurized heavy-water reactors, **16–4**
Pressurized-water reactors, **16–3**
Primary energy consumption, **1–2**
Primary energy demand, **1–9**
Primary productivity, biomass crops, **15–19**
Prime movers, **3–8**
PRISM reactor, **16–11**
Process baths, **18–54**
Process heat
  demand, **18–55**
  production, solar thermal plants for, **18–55**
PRODISS project, **19–42**
Producer gas, **22–60**
  improving heating value of, **22–45**
Production functions, **18–36**
Products of combustion, **10–15**
Products of incomplete combustion. *See* PIC
Proved reserves, **2–19**
Proximate analysis
  biomass, **3–5**
  coal, **2–2**, **2–4**, **2–15**
  solid fuels, **24–13**
PS10, **19–50**, **19–74**, **19–77**
Public Utility Regulatory Policy Act (PURPA) (U.S.), effect of on geothermal power plants, **23–10**
Pulp and paper industry, use of advanced power generation cycles in, **3–8**
Pulse method of turbocharging, **10–22**
Pulse-type MHD generator, **27–18**
Pulverized coal boilers, **3–9**, **22–39**
Pulverized-coal plants, **13–5**
Pumped brine binary power plants, energy production rate in, **23–7**
Pumped hydro, **15–13**
Pumped-storage schemes, turbine analysis using, **11–14**
Pumps, **8–13**
  losses due to in Rankine cycle, **8–4**
PUREX process, **16–25**
Push rods, **12–6**
PV arrays, orientation of, **20–13**
PV cells. *See also* solar cells; specific cell types
  efficiency of, **20–2**
  p-n junction, **20–3**
  properties of, **20–7**
PV modules
  fabrication of, **20–12**
  frameless, **20–32**
  monolithic, **20–40**
PV pumping systems, **20–19**
PV systems
  components of, **20–15**
  configurations of, **20–14**
  efficiencies of, **20–25**
  installed costs of, **20–28**
  reliability of, **20–24**
  sample costs, **20–23**
  standalone, **20–20**
  utility-interactive, **20–18**
Pyranometers, **5–6**
Pyrheliometers, **5–6**
Pyrolysis, **3–9**, **15–20**
  in direct combustion, **22–37**
  in thermal gasification, **22–41**
Pyrolytic WTE facilities, **24–37**
Pyroprocessing process, **16–25**

Q

Quicklime, use of in spray driers, **13–11**

R

Radial staggered heliostat field layout, **19–55**
Radiant energy, **5–1**
Radiation
  beam normal, **5–4**
  diffuse, **5–4**
  direct, **5–4**
  distribution, effect of on collector performance, **18–15**
  extraterrestrial solar, **5–2** (*see also* solar radiation)
  statistics, **18–16**
Radiative cooling systems, **18–97**
Radioactive waste, **4–3**
Radioactive wastes
  managing, **16–27**
  types of, **16–26**
Raft River Geothermal Power Plants, **23–10**
Ram-type ash dischargers, **24–23**
Random loads, **21–16**
Rankine cycle
  analysis of, **8–2**
  coal-fired, **13–24**
  engines, **3–8**
Rankine steam power cycle, **22–46**, **23–32**
Raw producer gas, **22–60**
Ray tracing analysis, **19–11**
Rayleigh distribution, approximation of wind speed density using, **6–2**
RCU, **18–114**
RDF furnaces, sizes of, **24–17**
RDF plants, **24–6**
  performance of, **24–34**
  refuse feeding systems in, **24–10**
  refuse storage at, **24–9**
RDF-fired waterwall units, **24–22**
Reactant utilization, **26–24**

Reaction turbines, **11**–3
    performance characteristics, **11**–9
Reactivity, coal, **2**–6
Reactors, **16**–3. *See also* nuclear power
    ABWR, **16**–8
    AHWR, **16**–10
    boiling-water, **16**–4
    EPR, **16**–9
    fast-breeder, **16**–11
    fast-neutron, **16**–11
    gas-cooled, **16**–5
    generation-III, **16**–8
    generation-IV, **16**–11
    GT-MHR, **16**–11
    heavy-water, **16**–10
    HTGR, **16**–10
    IRIS, **16**–9
    LFR, **16**–18
    light-water, **16**–8
    light-water graphite-moderated, **16**–6
    LMFBR, **16**–5
    MSR, **16**–19
    pressurized heavy-water, **16**–4
    pressurized-water, **16**–3
    PRISM, **16**–11
    SFR, **16**–18
    slurry, **3**–16
    space, **27**–6
    SWR, **16**–10, **16**–17
    thermal, **10**–17
        advanced, **16**–12
    tokamak, **17**–2
    VHTR, **16**–16
Real-time operating system. *See* RTOS
Receiver and power control system. *See* RPSCS
Receiver pipes, for parabolic trough collector systems, **19**–19
Receiver spillage, **19**–53, **19**–55
Reciprocating grates, **24**–17
Recirculation process, **19**–40
    control of, **19**–43
Recombination effects, **20**–37
Recoverable coal, **2**–8
Recoverable reserves, **2**–19
Recoverable wood residues, **3**–2
Recreation, geothermal energy use for, **23**–3
Recuperative cycle, **13**–24
Recuperators, **9**–6
Recycle flow reactors, **22**–19
Recycling, rate of in US, **24**–3
Reduction converters, **10**–17
REFCOM system, **22**–21
Reflection, Snell's law of, **19**–11
Reflective parabolic concentrators, sizing of, **19**–13
Reflective solar concentrators, **19**–50
Reflective towers, **19**–85
Reflectivity values, **A2**–22
Reflectors. *See also* specific types of reflectors
    shapes of, **18**–13
    use of in parabolic trough collectors, **19**–20
    use of to improve collector performance, **18**–12
Reflux heaters in Stirling engines, **12**–3
Reforming, **15**–20
REFOS development program, **19**–72
Refractory furnaces with waste heat boilers, **24**–21
Refractory material, classification of, **24**–19
Refractory volatile solids. *See* RVS
Refractory wall furnaces, costs of, **24**–35
Refrigerant-absorbent pairs, **18**–124
Refrigeration, thermoelectric, **27**–10
Refuse
    feeding, **24**–10
    storage of at WTE facilities, **24**–9
Refuse-derived fuel systems. *See* RDF plants
Regenerative cycle, **8**–4, **13**–24
Regenerators, **9**–6
    in Stirling engines, **12**–4
Reheat burners, **9**–7
Reheat, double, **13**–15
Reheaters, **8**–7
Relative efficiency, **11**–17
Relative-area design method, **18**–42
Renewable energy, **1**–3. *See also* specific renewable energy sources
    present status and potential of, **1**–5
Renewable fuels, energy return from, **22**–63
Renewable Portfolio Standards (U.S.), effect of on geothermal power plants, **23**–10
Repowered plants, **13**–24
Reprocessing, worldwide capacity, **16**–25
Research octane number, **10**–18
Reserve base, **2**–8
Reserves
    coal, **2**–8, **2**–19
    nuclear, **4**–2
Reservoirs, geothermal, **23**–4. *See also* geothermal, reservoirs
Residential sector, biomass use in, **3**–7
Residential solar systems, **18**–24
Residual brine management, **23**–20
Residues used for biomass energy, **3**–1
Resolution, solar system modeling using, **18**–34
Resonance problems, wind turbines, **21**–17
Resource developers, geothermal power, **23**–9
Resource temperature, geothermal, capital costs related to, **23**–13
Resources, **2**–7
    geothermal, **7**–1
    non-renewable, **2**–16
    wind energy, **6**–2
Reverse-return solar field layouts, **19**–32
Reversible cell potential, **26**–4
    effect of operating conditions on, **26**–8
    effect of pressure on, **26**–9
    effect of temperature on, **26**–9
Reversible energy
    conversion efficiency, **26**–12, **26**–17, **26**–44
        equivalency of with Carnot cycle efficiency, **26**–14
    loss, **26**–17
Reynolds averaged Navier-Stokes models
    use of for analysis of turbines, **11**–14
    use of with CFD models, **21**–14
RF glow discharge cell fabrication method, **20**–11

RF sputtering, deposition on CIGS solar cells using, **20**–48
Richardson-Dushmann equation, **27**–4
Rim angles, **19**–14, **19**–23
Ring of Fire, **23**–2
Ringbom configuration, **12**–5
Rinia configuration, **12**–4
Rock storage systems, **18**–33, **18**–97
Roll-bond collectors, **18**–4
Roll-to-roll manufacturing
  a-Si cells, **20**–11
  CIGS solar cells, **20**–51
  flexible solar modules, **20**–30
Roller grates, **24**–17
Room coefficient of utilization. *See* RCU
Room surface direct depreciation. *See* RSDD
Room-and-pillar mining, **2**–14
Roots blower, **10**–21
Rotary combustor grates, **24**–17
Rotational energy, conversion of into electricity, **21**–20
Rotors, steam turbine, **8**–9
RPSCS, **19**–58
RSDD, **18**–108, **18**–114
RTOS, **19**–58
Rule-of-thumb engineering estimates, **18**–38, **18**–65
  passive system design, **18**–66
  skylight placement, **18**–116
  south-facing window overhang, **18**–94
  thermal mass storage, **18**–67
Runner, **11**–2
Russia
  fast-breeder reactors in, **16**–11
  generation-III light water reactor development in, **16**–10
  GT-MHR, **16**–11
  LMFBR in, **16**–5
RVS, **22**–4

S

S-809 airfoil, **21**–7
  profile and performance characteristics of, **21**–8
Sac volume, **10**–15
Saccharification of starch, **22**–53, **22**–55, **22**–57
Safety controller for wind turbines, **21**–22
Salinity-gradient ponds, **15**–16
Salt concentrations, geothermal reservoirs, **23**–23
Salt velocity method, **11**–17
Salton Sea
  geothermal brine in, **23**–22
  multiflash steam separation at, **23**–29
Salton Sea geothermal system, **7**–6
Sanlúcar heliostats, **19**–57
Sanyo, development of flexible amorphous solar cells by, **20**–41
Saturated steam, **19**–17, **A3**–8
  use of as a thermal storage fluid, **19**–73
Sawdust, as a source of cellulosic ethanol, **15**–20
Scaling, **23**–31
Scaling formulae, **11**–5
Scavenging, **10**–2
  cross-flow, **10**–4
Schottky junction, **20**–3
Screen printing deposition, **20**–46
Scrubbers, wet, **13**–9
Scrubbing, **13**–9
  particulate control with, **24**–24
SDD, **18**–114
Sealed gel cell batteries, **15**–9
Seals, **12**–6
Seasonal pricing, geothermal power plants, **23**–12
Seasonal solar thermal storage, **18**–24
Second-generation fuel cells, **26**–39. *See also* molten carbonate fuel cells
Secondary batteries, **15**–8
Seebeck effect, **27**–8
Seed oil, extraction of, **22**–59
SEGS plants, **19**–3, **19**–45
  basic characteristics of, **19**–47
  profitability of, **19**–49
  thermal loss in collectors at, **19**–24
Selective catalytic reduction, **13**–11
Selective non-catalytic reduction. *See* SNCR
Selective noncatalytic combustion, **13**–12
Selective surfaces, **18**–4
Self-discharge time, **15**–4
Selinization of precursor materials, **20**–49
Semi-Kaplan turbines, **11**–4
Semiconductor materials, bandgaps of, **20**–30
Sensible heat storage, **15**–6, **15**–15
Separator-plate corrosion, **26**–40
Series fuel cell connections, **26**–27
SFR, **16**–18
Shading control, **18**–94, **18**–117, **19**–54
Shadow-band trackers, **19**–19
Shadowing, **19**–54
Shallow geothermal resources, **23**–7
Shed vorticity, **21**–13
Shell Middle Distillate Synthesis process, **3**–16
Shell Solar
  CIGS solar modules developed by, **20**–50
  thin-film technology innovations at, **20**–47
Shell-and-tube heat exchangers, **8**–12
Shippingport Atomic Power Station, **16**–2
Showa Shell, CIGS solar modules developed by, **20**–50
Shuman, Franc, **19**–3
SI, **A1**–2
Side reflectors, use of to improve collector performance, **18**–12
Sidelighting, **18**–104
  architectural features, **18**–103
  lumen method of, **18**–105
Siemens configuration, **12**–4, **13**–20
Siemens, CIS thin film technology innovations at, **20**–47
Silica
  concentration, **23**–22
  solubility of in binary power plants, **23**–38, **23**–43
Silicon
  amorphous, **20**–11 (*see also* amorphous silicon (*a*-Si))
  thin-film, **20**–34
  use of in PV cells, **20**–3
Similitude, **11**–5
Simple cycle, **9**–21

Simple design methods, **18**–38
Simple harmonic motion, **12**–5
Simultaneous saccharification and fermentation. *See* SSF
Single flash steam process, **23**–33
Single-glazed antireflection collectors, efficiency curves, **18**–52
Single-medium storage systems, thermal energy storage in, **19**–36
Single-pass solar thermal systems, **18**–26
  system design, **18**–44
Sintering, **26**–40
Site assessment, WTE facilities, **24**–8, **24**–38
Sky diffuse factor, **5**–5
Skylight, **18**–59
  net transmittance, **18**–113
  rules of thumb for placement of, **18**–116
Skylight direct depreciation. *See* SDD
Skylighting, lumen method of, **18**–113
Slag, **2**–6
Slaked lime, use of to clean flue gas, **24**–27
Slats, use of to improve collector performance, **18**–12
Slim hole wells, **23**–16
SLR method
  passive solar system design, **18**–91
  reference system correlation parameters, **18**–92
Slurry reactors, **3**–16
Small-scale gasifiers, **3**–13
SMES, **15**–7
SNCR, **24**–27
Snell's law of reflection, **19**–11
Sodium nitrate, sensible heat storage in, **15**–15
Sodium, in combustion flue gas, **24**–31
Sodium-cooled fast reactor. *See* SFR
Sodium-nickel-chloride batteries. *See* zebra batteries
Sodium-sulfur batteries, **15**–10
Soil
  heat transfer, **18**–99
  temperature and properties, **18**–101
SOLAIR project, **19**–67, **19**–70, **19**–80
Solar air heating systems, **18**–24
Solar altitude angle, **18**–104
Solar beam radiation, **18**–59
  use of in daylighting system design, **18**–104
Solar capacity, prediction of, **19**–4
Solar cells, **20**–2. *See also* PV cells; specific cell types
  cadmium telluride, **20**–42 (*see also* CdTe)
  configurations of, **20**–36
  Cu(InGa)Se_2, **20**–46
  efficiencies of on rigid substrates, **20**–33
  hybrid, **20**–40
  limits imposed on due to bandgap, **20**–2
  manufacture of, **20**–8
  micromorph, **20**–35
  multiple-junction, **20**–39
  tandem, **20**–39
  thin film efficiencies, **20**–32
Solar chimneys, design concepts, **18**–96
Solar collectors
  characteristics of, **18**–10
  classification of, **18**–2
  designs for, **19**–49
  in passive solar heating systems, **18**–62
  long-term performance of, **18**–14
  multieffect absorption systems, **18**–126
  performance parameters, corrections to, **18**–30, **18**–32
  piping losses, **18**–32
  shading losses, **18**–32
  types of, **18**–3
Solar concentration, **19**–6
Solar concentrators. *See also* concentrating solar collectors
  beam quality, **19**–9
  in dish-Stirling systems, **19**–87
  maximum concentration ratio for, **19**–14
  stretched membrane, use of in dish-Stirling systems, **19**–89
Solar constant, **5**–2
Solar control, **18**–94
Solar cooling, **18**–120
Solar dairies, **18**–55
Solar declination, **5**–3
Solar desiccant dehumidification, **18**–126
Solar efficiencies, **15**–19
Solar electric generating system plants. *See* SEGS plants
Solar energy, **5**–1, **12**–9
  exergetic value of, **19**–2
  plant matter as a storage medium for, **15**–18
Solar fields, sizing and layout of, **19**–31
Solar fraction, **18**–35
Solar heat, integration of into industrial production processes, **18**–52
Solar hot-water heating systems, **18**–60
Solar hybrid liquid-desiccant systems. *See also* liquid desiccants
  costs of, **18**–132
Solar hybrid power plants, **19**–82. *See also* hybrid systems
Solar insolation. *See* insolation
Solar irradiance for different air masses, **A2**–6
Solar One, **19**–17
  external cylindrical tubular receiver used in, **19**–65
  use of external tubular receivers in, **19**–62
  use of once-through superheated water-steam receiver in, **19**–63
Solar ponds, **15**–16
Solar power plants
  initiatives for, **19**–49
  state-of-the-art, **19**–45
Solar process heat. *See also* industrial process heat applications
  existing plants, **18**–55
  potential for, **18**–49
  system concepts, **18**–56
Solar radiation
  data, **5**–6, **A2**–2
  horizontal surfaces, **5**–5
  measurements of, **5**–6
  tilted surfaces, **5**–5
  United States, horizontal average, **A2**–17
  wind and, **6**–1
  worldwide, horizontal average, **A2**–11
Solar receivers, **19**–2, **19**–50, **19**–58
  convection losses from, **19**–61
  design of, **19**–11
  efficiency of, **19**–6, **19**–59

in dish-Stirling systems, **19**–87
PHOEBUS-type, effect of ARR in, **19**–70
selection of, **19**–62
tower, operating temperature and flux ranges of, **19**–63
volumetric atmospheric, **19**–80
Solar reflectors, use of in parabolic trough collectors, **19**–20
Solar savings fraction, **18**–67
Solar systems
costs, **18**–47
design methods for, **18**–38
design of, **18**–2
design recommendations for, **18**–46
modeling of, **18**–33
sizing methodology, **18**–36
Solar thermal
energy, **15**–2, **15**–15
distributed grid technologies, **15**–3
generating electricity with, **19**–45
market opportunities, **19**–91
power plants
central-receiver, **19**–50
factors limiting development of, **19**–3
flow diagram for, **19**–2
process heat production, **18**–55
technologies for, **19**–15
systems
classification of, **18**–22
integration of in industrial processes, **18**–51
modeling of, **18**–5
storage, **18**–33
Solar tower power plants. *See* STPP
Solar transients, **19**–3
Solar Tres, **19**–50, **19**–80
technical specifications and design performance of, **19**–81
Solar Two, **19**–17
molten salt tubular receivers, **19**–64, **19**–76
use of external tubular receivers in, **19**–63
Solar-load ratio method. *See* SLR method
Solar-supplemented systems, **18**–23
economic analysis, **18**–38
sizing methodology for, **18**–36
SOLGAS, **19**–50, **19**–74
SOLGATE, **19**–50, **19**–84
SOLHYCO, **19**–85
Solid desiccants, **18**–127
Solid fuels
combustion of, **22**–37
ultimate analysis of, **24**–13
Solid oxide fuel cells, **26**–2, **26**–41, **26**–44
acceptable contamination levels for, **26**–43
basic operating principle, **26**–42
technological issues of, **26**–43
Solid polymer (electrolyte) fuel cells, **26**–33
Solid waste. *See* municipal solid waste
Solid-fuel combustors, **22**–39
Solid-gas reactions, in thermal gasification, **22**–41
Solid-liquid phase changes, **15**–18
Solid-solid phase changes, **15**–17
Solids
sensible heat storage in, **15**–17
storage of thermal energy in, **18**–33
Solution mining, uranium, **16**–22
Soot, **10**–16
Source circuit, **20**–17
South Africa
dish-Stirling systems in, **19**–89
HTGR development in, **16**–11
use of F-T liquids in, **22**–62
South fields, **19**–53
South Korea, generation-III light water reactor development in, **16**–9
South Meager Creek Geothermal Project, **23**–17
Southern Europe, technical potential of CSP in, **19**–4
Soybeans, **22**–59
Space heating, **15**–15
active, system design for, **18**–39
geothermal energy use for, **23**–3
solar air systems for, **18**–24
use of closed-lop solar systems for, **18**–25
Space programs
use of alkaline fuel cells in, **26**–30
use of PAFCs in, **26**–38
Spain
AndaSol plant, **19**–49
cavity receiver projects in, **19**–62
CRS plants in, **19**–17
heliostat developments in, **19**–57
uprating of nuclear plants in, **16**–6
wind energy installations in, **21**–1
Spark ignition engines, **3**–8, **10**–1, **22**–51
four-stroke, **10**–2
power control in, **10**–10
two-stroke, **10**–3
Spark timing, **10**–12
Specific energy, **15**–5
Specific power, **15**–6
Specific speed, **11**–6
Speed, wind, distribution of, **6**–2
Spent-fuel
storage of, **16**–26
transportation, **16**–26
waste management policies for, **16**–28
Spray attemperators, **8**–7
Spray drying, **13**–9
Spray pyrolysis deposition, **20**–46
Sputtering, **20**–11, **20**–35, **20**–45
SSF, **22**–57
Stability, problems of in CdTe solar cells, **20**–45
Stack effect, **18**–95
Stack emissions, MSW combustors, elements found in, **24**–25
Staebler-Wronsky effect. *See* SWE
Stagnation temperature, **18**–10, **19**–7
Stall, **21**–7
control, **21**–19
Stall-regulated wind turbines, **21**–19
Stand-alone inverters, **20**–16
Stand-alone solar thermal systems, **18**–22
Standard atmosphere of dry air, **24**–11
Standard candle, **18**–102
Starch, ethanol from, **22**–53
Starved air WTE systems, **24**–36

Stationary flat-plate collectors, **18**–3. *See also* flat plate solar collectors
Steady-state procedure, **18**–7
Steam
  boilers, **8**–5
    components, **8**–6
  chemistry, **23**–22
  coals, **2**–6
  cooling, **9**–10
  cycle, **3**–8
  engines, **3**–8
  explosion, **22**–56
  generators, **8**–5, **19**–33 (*see also* boilers)
    compact, **19**–29
    heat recovery, **14**–2
  injection, **13**–24
  power plants, **22**–47
    components of, **8**–1
    energy production rate in, **23**–7
  power processes, **23**–32
  production, clean-water injection to enhance, **23**–21
  purity, measurement of, **23**–25
  separation, high-efficiency, **23**–31
  supercritical pressures, **13**–6
  turbine power plants
    capital costs of, **23**–13
    steam/brine separation for, **23**–25
  turbines, **8**–7
    arrangement of, **8**–9
    condensers in, **8**–12
    condensing, **23**–33
    integration of in binary power plants, **23**–41
    materials for, **8**–10
    technologies for, **23**–32
  washing, **23**–25, **23**–30
Steam-bending forces, **8**–8
Steam-injected gas turbine cycle. *See* STIG cycle
Steam-injected gas turbines, **9**–8
Steinmuller heliostat ASM-150, **19**–57
STEM project, **19**–42
STEOR, **19**–75
Stepwise steady-state simulations, solar system modeling using, **18**–35
STIG cycle, **22**–48
Still gas, **2**–21
Stirling cycle, **12**–1, **22**–46
  mechanical implementation of, **12**–4
Stirling engines, **3**–8, **12**–1
  heat exchange in, **12**–3
  hybrid-electric applications of, **12**–9
  kinematic, **12**–6
  power control in, **12**–4
  use of in solar applications, **19**–87
Stoichiometry, **26**–23
Stoker boilers, **3**–9
Storage
  capacities, **15**–3
  comparison of media for, **19**–79
  direct thermal, **15**–15
  heat, industrial processes, **18**–54
  in passive solar heating systems, **18**–62
  mechanical energy, **15**–13
  sensible heat, **15**–15
  solar thermal, **18**–24
  technologies, characterization of, **15**–4
  thermal energy, **15**–3
    rule of thumb estimates for, **18**–67
  thermochemical energy, **15**–18
STPP
  open-air volumetric receivers in, **19**–82
  tower projects, **19**–73
Stratification coefficient, **18**–33
Stratified flow, **19**–40
Stream-tubes, **21**–10
Stretched membrane concentrators, use of in dish-Stirling systems, **19**–89
Stretched-membrane heliostats, **19**–57
Stretford H_2S abatement, **23**–45
Stretford process, **7**–5
String ribbon technology, **20**–9
Sub-bituminous coals, **13**–3
Subcritical cycles, **13**–6
Subcritical pentane binary power plants, **23**–39
Subsidence, **2**–14
Substrate biodegradability, **22**–4
Substrate configurations, amorphous silicon solar cells, **20**–37
Sugar
  bioproductions from, **3**–14
  cane, ethanol production from, **3**–13
  industry, use of IGCC technology in, **3**–8
Sulfate inhabitation, **22**–12
Sulfide, response of on methanogens, **22**–12
Sulfur dioxides, emissions of from pulverized coal plants, **13**–9
Sulfur oxides, removal of from flue gas, **24**–27
Sun
  determining the position of, **20**–13
  illuminance, **18**–105, **18**–113
  tracking, **19**–9
Sun-window geometry, **18**–104
Sunk costs, geothermal resource management and, **23**–23
Sunlight, **18**–59
Sunshape, **19**–10
  degraded, **19**–11
Sunspace systems, **18**–76
Super off-peak electricity use, **19**–48
Supercapacitors, **15**–7
Supercharging, **10**–20
Supercritical cycles, **13**–7
Supercritical hydrocarbon binary power plants, **23**–40
Supercritical steam
  power plants, **8**–2
  pressures, **13**–6
Supercritical units, **13**–5
Supercritical-water-cooled reactor. *See* SWR
Superheated steam, **23**–21, **A3**–10
  use of as a thermal storage fluid, **19**–73
Superheated water-steam receivers, **19**–63
Superheaters, **8**–6
Superstrate configurations
  amorphous silicon solar cells, **20**–37
  CdTe solar cells, **20**–43

Supervisory controller for wind turbines, **21**–22
Surface geoscientific investigations, **23**–15
Surface ignition, **10**–13
Surface work functions, **27**–2
Surface-type condensers, **8**–12
Suspension burners, **22**–39
Swamp coolers, **18**–97
Swash plate drives, **12**–6
SWE, **20**–37, **20**–39
Sweden, uprating of nuclear plants in, **16**–6
Swelling, **2**–4
Swirl chambers, **10**–5
Swirl ratio, **10**–5
Switzerland, uprating of nuclear plants in, **16**–6
SWR, **16**–10, **16**–17
Synchronous generators, **21**–20
Syngas, **3**–15, **13**–20, **15**–20, **22**–53
  alcohols from, **22**–61
  biological processing of, **22**–62
  biomass-derived, transportation fuels from, **22**–60
  methanol from, **22**–60
  synthesis of ethanol from, **3**–16

T

T-S coordinates, **9**–3
Tail vanes, **21**–22
  power limitation using, **21**–19
Tandem solar cells, development of, **20**–39
Tar, **22**–42
Tax incentives, geothermal power plants, **23**–11
TCO material
  choice of for amorphous silicon solar cells, **20**–38
  use of with CdTe solar cells, **20**–43
TDOM technique, **22**–6
Technology Program Solar Air Receiver project. *See* TSA project
Temper embrittlement, **13**–6
Temperature
  dependence of solar energy system efficiency on, **19**–6
  effect of on anaerobic bacteria, **22**–11
  effect of on reversible cell potential, **26**–9
Temperature-entropy coordinates. *See* T-S coordinates
Terminal temperature difference, **8**–12
Test procedures, solar collectors, **18**–7
TFTR, **17**–2
THCs, measurement of in combustion gas, **24**–30
The Geysers, **23**–2
  geothermal steam fields at, **7**–5, **23**–4
  H_2S abatement technology installed at, **23**–45
  incomplete resource information at, **23**–23
  tritium content at, **23**–22
  use of EGS at, **23**–5
The Philippines
  geothermal power production in, **23**–2
  pressurized brine power plant in, **23**–41
Themis, **19**–62
  molten salt tubular receivers, **19**–64
Thermal creep, **12**–9
Thermal efficiency, **8**–1
Thermal energy, **15**–3
  storage of, **15**–15, **19**–4
    dual-medium storage systems, **19**–38
    single-medium storage systems, **19**–36
Thermal fluids, **19**–73
Thermal gasification, **3**–8
  biomass, **22**–41
  fibrous plant material, **22**–37
Thermal ionization, **27**–16
Thermal loss coefficient, **19**–24
Thermal management, **26**–3
Thermal oils, **19**–19
  storage tanks using, **19**–36
Thermal power plants, energy efficiencies of, **26**–36
Thermal properties of metals and alloys, **A3**–6
Thermal reactors, **10**–17
  advanced, **16**–12
Thermal storage, **18**–33. *See also* solar thermal
  beds, **15**–17
  molten salt, **19**–49
  systems for parabolic trough collectors, **19**–35
  wall passive solar systems, **18**–62
Thermionic converters
  thermodynamic analysis of, **27**–5
  types of, **27**–4
Thermionic energy conversion, **27**–1
  principles of, **27**–2
Thermo-fluid systems. *See* thermosyphon systems
Thermochemical energy storage, **15**–18
Thermocline storage systems, **15**–16
Thermodynamic method of determining turbine efficiency, **11**–17
Thermodynamic performance, comparison of fuel cells and heat engines, **26**–4
Thermodynamic systems, **26**–4
Thermoelectirc, power conversion, **27**–7
Thermoelectric
  effects, **27**–8
  legs, **27**–10
  module, **27**–9
Thermoneutral voltage, **26**–18
Thermophilic bacteria, **22**–11
Thermophilic digestion, **22**–20
Thermosyphon systems, **18**–29, **18**–60
  design considerations, **18**–61
Thin films, **20**–8
  efficiencies of, **20**–32
  production of by dendritic web growth, **20**–11
Thin silicon, **20**–24
Thin-film PV technology, **20**–28
  advantages of, **20**–30
  amorphous silicon, companies active in the development of, **20**–42
  environmental concerns regarding, **20**–52
  production cost of, **20**–31
  silicon, **20**–34
Third-generation fuel cells, **26**–41. *See also* solid oxide fuel cells
Thomson effect, **27**–8
Thorium, **4**–2
  resources, **16**–21

use of in heavy-water reactors, **16**–10
Three Mile Island, **16**–14
Three-phase boundary, **26**–26
Three-stage co-evaporation process, **20**–49
Three-stone stove, **3**–7
Three-way catalytic converters, **10**–17
Tidal power, **25**–2
Tipping fees, **24**–36
Tiwi, tritium in reservoir at, **23**–22
Tokamak, **17**–2
Tokamak Fusion Test Reactor. *See* TFTR
Top dead center, **10**–2
Toplighting, **18**–104
architectural features, **18**–103
Topping cycles, **8**–5
systems, **22**–48
Topping-up auxiliary heaters, **18**–27
Total hydrocarbons. *See* THCs
Total solar radiation, **20**–13
Toxic overloading, anaerobic digestion, **22**–16
Trace metals, in MSW combustion, **24**–31, **24**–33
Tracking axis, **19**–18
Tracking mirrors, **19**–52. *See also* heliostats
Tracking mounts, for PV arrays, **20**–13
Tracking solar collectors, **18**–24
types of, **18**–3
Trailing vorticity, **21**–13
Transesterification, **22**–59
Translation velocity, **21**–5
Transmission
geothermal energy, **23**–8
reservations, **23**–8
Transparent conducting oxides. *See* TCO material
Transportation, **1**–5
energy storage for, **15**–1
fuels
biobased, **22**–51
biomass-derived syngas, **22**–60
comparison of, **22**–52
gaseous, **22**–62
Transportation sector
use of DMFC for, **26**–38
use of PEMFCs for, **26**–35
Trap density, **20**–37
Traveling grates, RDF systems, **24**–17
Trickle-type flat plate solar collectors, **18**–4
Triglycerides, **22**–58
Triple-glazed antireflection collectors, efficiency curves, **18**–52
Tritium, in geothermal steam, **23**–22
TRNSYS, **18**–39
Trombe-wall-type passive solar systems, **18**–64, **18**–75
True digestible organic matter technique. *See* TDOM technique
TSA project, **19**–67, **19**–81
receiver efficiencies, **19**–69
Tube-and-sheet collectors, **18**–4
Tubing, coextruded, **13**–6
Tubular collectors. *See also* specific types of collectors
evacuated, **18**–12
Tubular receivers, **19**–62
heat transfer principles in, **19**–64
molten salt, **19**–64
Turbines. *See also* specific turbine types
analysis of using numerical simulation, **11**–14
bulb, **11**–5
efficiency of, **9**–3
power available, **11**–5
electromagnetic, **27**–18
field tests for, **11**–17
fully-adjustable, **11**–4
gas, **9**–1
aeroderivative, **9**–6
efficiency of, **9**–3
humid air, **9**–10, **13**–22
hydraulic, **11**–1
evaluation with respect to tailwater elevations, **11**–12
impulse, **11**–2
Pelton-type, **11**–2
performance characteristics, **11**–8
materials used for, **9**–10
maximum inlet temperature, **9**–9
pit, **11**–5
power, **11**–4
propeller, **11**–4
reaction, **11**–3, **11**–9
scaling, **23**–31
speed regulation of, **11**–10
steam, **8**–7, **14**–5
blade arrangements, **8**–10
condensers in, **8**–12
materials for, **8**–10
rotors in, **8**–10
test loop for, **11**–13
transient response of, **10**–22
use of in coal combustion plants, **13**–20
Turbo-lag, **10**–22
Turbocharging, **10**–20
methods of, **10**–22
Two-stage anaerobic composting, **22**–22
Two-stage digestion, **22**–20
Two-stroke engines, **10**–2
Two-tank sensible heat storage, **15**–16

U

Ultimate analysis
biomass, **3**–5
coal, **2**–4, **2**–15
solid fuels, **24**–13
Ultracapacitors, **15**–7
Ultrasupercritical units, **13**–5
Unburned hydrocarbon, **10**–15
emissions of, **10**–14
Underfire air, **24**–20
Underground mining, uranium, **4**–2, **16**–22
Unfired boiler systems, **19**–29
Unicarbontrophs, **22**–61
Uninterruptible power supplies, **15**–3
Unipolar fuel stack arrangement, **26**–28
Unit size, **15**–4
United Kingdom, wind energy installations in, **21**–1

United Solar, development of flexible amorphous solar cells by, **20**–41
United States
  central station options for fossil fuels, **13**–24
  coal reserves in, **2**–6
  CRS plants in, **19**–17
  demand for wind energy in, **21**–2
  dish-Stirling systems in, **19**–89
  electric energy needs in, **24**–2
  electric power generation in, fuels for, **13**–2
  electrical capacity of geothermal fields in, **7**–2
  fast-breeder reactors in, **16**–11
  geothermal power production in, **23**–2
  heat flow in, **7**–1
  HTGR development in, **16**–11
  IGCC plants in, **13**–18
  light-water nuclear reactor development in, **16**–8
  production tax credits, **23**–11
  standard atmosphere of dry air, **24**–11
  supercritical plant design in, **13**–6
  uprating of nuclear plants in, **16**–6
  wind energy installations in, **21**–1
  wind energy resources, **6**–4
Unsteady aerodynamics, **21**–13
Updraft gasification, **3**–10
Updraft gasifiers, **22**–42
UPS inverters, **20**–17
Uranium, **4**–1, **16**–1
  conversion and enrichment, **16**–22
  depleted, **4**–2
  enrichment facilities, **16**–24
  fuel fabrication and use, **16**–23
  mining and milling, **16**–21
  reprocessing, **16**–23
  reserves, **4**–2
  resources, **16**–21
  spent-fuel storage, **16**–26
  use of in heavy-water reactors, **16**–10
UREX+ process, **16**–25
Utilities, as buyers of geothermal power, **23**–9
Utility shaping, **15**–3
Utility-interactive PV systems, **20**–18, **20**–22
Utility-scale steam power plants, **22**–47
Utilizability
  annual time scales, **18**–21
  factor, **18**–16
    Phibar method of determining, **18**–19
  monthly time scales, **18**–19
Utilization, **26**–24
  coefficient, **18**–114

V

Vacuum evaporation, **20**–49
Vacuum receiver pipes, for parabolic trough collector systems, **19**–19
Vacuum thermionic converters, **27**–4
Valorga system, **22**–31
  schematic of, **22**–33
Valves, **8**–11
Vanadium redox flow batteries, **15**–12
Vapor thermionic converters, **27**–4
Vapor-compression refrigeration cycle, **18**–121
  solar cooling using, **18**–120
Variable costs, **23**–14
Variable exogenous parameters, **18**–37
Variable-angle swash plate drive, **12**–6
Variable-speed turbines, structural dynamics of, **21**–17
Variable-speed wind turbines, **21**–21
VAWTs, **21**–2
  configuration of, **21**–3
  momentum models for, **21**–12
  performance of, **21**–12
  structural dynamics of, **21**–18
Vegetable oils, **22**–58
  modification of to biodiesel, **22**–59
Vehicular power, use of PEMFC for, **26**–33
Vented Trombe-wall systems, **18**–75
Ventilation, generators, **8**–14
Vertical illuminance, **18**–106
Vertical windmills, **21**–3. *See also* VAWTs
Vertical-axis wind turbines. *See* VAWTs
Very-high-temperature reactor. *See* VHTR
VHTR, **16**–16
Viscosity, diesel, **10**–19
Volatile gases, due to pyrolysis, **22**–38
Volatile matter content, **2**–2, **2**–15
Volatile solids, in organic substrates, **22**–4
Volatility, **10**–18
Voltage
  depression, **15**–8
  efficiency, **26**–22
Volume energy storage capacity, comparison of media, **19**–79
Volume reduction, computation of, **22**–24
Volumetric absorbers, **19**–62
Volumetric atmospheric receivers, **19**–80
Volumetric effect, **19**–62
Volumetric efficiency, **10**–11
Volumetric receivers, **19**–62
  air-cooled, advantages of, **19**–66
  heat transfer principles in, **19**–64
  metal absorbers in, **19**–65
Von Roll grates, **24**–17
Vortex-based models, **21**–13
  limitations of, **21**–14
VP-1, **19**–19
VVER reactors, **16**–10

W

Wairakei, pressurized brine power plant at, **23**–41
Wankel rotary engine, **10**–1
Waste
  characteristics of, importance of in WTE facility design, **24**–3
  quantities of in United States, **24**–2
Waste heat
  boilers, refractory furnaces with, **24**–21
  from MCFCs, **26**–39
  generation, amount and rate of, **26**–21

use of by fuel cells, **26**–3
Waste management, policies for spent fuel, **16**–28
Waste-to-energy cogeneration systems. *See* WTE combustion
Wastewater treatment plant sludge. *See* WWTPS
Wastewater, anaerobic digestion of, **22**–19
Water, **A3**–8
as motive fluid for geothermal power production, **23**–4
contamination, ethanol-gasoline blends, **22**–53
properties of, **A3**–3
sensible heat storage in, **15**–15
sprays, **18**–97
storage systems, **18**–33
Water and heat management, **26**–3
Water-dominated geothermal systems, **7**–4. *See also* hydrothermal
Water-gas shift reaction, **22**–63
Water-steam cavity receiver, **19**–64
Waterwall
cogeneration facility, costs of, **24**–35
furnace units, **24**–17
mass-fired, **24**–21
RDF-fired, **24**–22
rotary combustion, mass burn systems, **24**–17
systems, **18**–76
Wave power, **25**–2
Waxes, **2**–20
Wayang Windu geothermal power project, **23**–33
WDD, **18**–108
Websites
geothermal energy, **23**–5, **23**–10, **23**–29, **23**–33, **23**–46
Industrial Solar Technologies, **19**–20
NREL, PV, **20**–32
SOLEL Solar Systems, **19**–20
Solitem, **19**–20
Valorga system, **22**–32
wind mapping, **6**–3
Weidman receiver, **19**–75
Well productivity, **23**–6
Well-mixed reactors, **22**–18. *See also* CSTRs
Wellhead
energy cost of for geothermal energy, **23**–6
pressure drop, **23**–25, **23**–30
Wells
geothermal, optimization of, **23**–20
placement decisions, economic impetus for, **23**–23
Western Europe, management of MSW in, **22**–29
Western GeoPower Corp., **23**–17
Wet cooling, effect of on capital costs, **23**–13
Wet milling, **3**–14, **22**–55
Wet scrubbers, lime/limestone, **13**–9
Wheeling charges, **23**–8
Whole-tree burners, **22**–39
Wicket gates, **11**–2
control of turbine speed with, **11**–11
Willing buyers, **23**–9
Wind
characterization of, **6**–6
cost of energy, **21**–26
energy potential, **6**–6
energy resource, **6**–2
farms, **21**–24
origins of, **6**–1
power, **6**–1, **15**–2
efficiency of transmission usage compared to geothermal power, **23**–8
power density of, **21**–18
shear, **6**–2
speed distribution, **6**–2
Wind energy
cost to generate, **21**–1
payback period, **21**–24
worldwide installed capacity, **21**–1
Wind turbines, **21**–2
aerodynamics of, **21**–4
analysis, **21**–9
basic configurations of, **21**–4
CFD modeling of, **21**–14
controls subsystem, **21**–22
costs of, **21**–24
components, **21**–25
environmental concerns, **21**–26
hybrid models for, **21**–15
installations of, **21**–23
loads on, **21**–16
materials used for, **21**–23
performance of, **21**–12
power predictions for, **21**–15
resonance problems of, **21**–17
structural dynamic considerations, **21**–16
subsystems of, **21**–20
variable speed, **21**–21
visual impact of, **21**–26
vortex-based models for, **21**–13
Windmills, **21**–2, **21**–4
Window controls, **18**–109
Window dirt depreciation. *See* WDD
Wire-mesh volumetric absorbers, **19**–67, **19**–69
Wobble-plate drive, **12**–6
Wood, **2**–16
percentage of in municipal solid waste, **24**–3
recoverable residues of, **3**–2
ultimate analysis of, **A4**–2
use of for heating, **3**–7
Working fluids, **18**–2, **27**–16
Working gases, **12**–2
Workplane illuminance, **18**–104, **18**–108, **18**–115
World Bank
Cost Reduction Study for Solar Thermal Power Plants, **19**–4, **19**–91
Global Environment Facility, **19**–49
World Energy Assessment, **19**–91
World energy consumption, **2**–17
World total energy demand, **1**–2
Worm-gear drive systems, **19**–58
WTE combustion, potential contribution of the US electrical energy needs, **24**–2
WTE facilities
boiler tube corrosion in, **24**–24
construction costs for, **24**–34
design of, **24**–6
importance of MSW characteristics, **24**–3

operation and capacity, **24**–7
fuel handling, **24**–9
regulatory requirements affecting, **24**–38
residue handling and disposal in, **24**–22
siting, **24**–8, **24**–38
types of, **24**–6
Wurth Solar, CIS thin film technology innovations at, **20**–47
WWTPS, **22**–20
codigestion of with BOF/MSW, **22**–21
use of as feedstock for anaerobic digestion process, **22**–2

X

Xylose, **22**–57

Y

Yard trimmings, use of as feedstock for anaerobic digestion process, **22**–2
Yaw subsystem, **21**–22
Yawed flow, **21**–13
Yearly normalized energy surface. *See* YNES
Yearly production function, **18**–37
Yeasts, as sources of triglycerides, **22**–59
YNES, **19**–56
YSZ electrolytes in SOFCs, **26**–42

Z

Zebra batteries, **15**–10
Zero cell potential, **26**–20
Zero sulfur fuels, **3**–15
Zinc bromide batteries, **15**–12
Zinc oxide, use of on CIGS solar cells, **20**–48
Zinc phosphide, **20**–29
Zinc, in combustion flue gas, **24**–31
ZSW, thin-film technology innovations at, **20**–47